Handbook of
GENE LEVEL
DIAGNOSTICS
in Clinical Practice

Victor A. Bernstam
BioSol Ltd.
Ann Arbor, Michigan

CRC Press
Boca Raton   Ann Arbor   London   Tokyo

Library of Congress Cataloging-in-Publication Data

Bernstam, Victor A.
    Handbook of gene level diagnostics in clinical practice / Victor A. Bernstam.
      p. cm.
    Includes bibliographical references and index.
    ISBN 0-8493-6824-3
    1. Genetic disorders—Diagnosis. 2. Human chromosome abnormalities—Diagnosis. 3. Human
cytogenetics. 4. Prenatal diagnosis.   I. Title.
    [DNLM: 1. Genetics, Medical—methods. 2. Genetic Techniques. 3. Hereditary Diseases—
diagnosis. 4. Hereditary Diseases—genetics. 5. Prenatal Diagnosis—methods.   QZ 50 B531h]
RB155.6.B47   1992
616′.042—dc20
DNLM/DLC
for Library of Congress   92-19271
                                                          CIP

International Standard Book Number 0-8493-6824-3

Library of Congress Card Number 92-19271
Printed in the United States   2 3 4 5 6 7 8 9 0
Printed on acid-free paper

Foreword

Dr. Victor A. Bernstam has written the timely, comprehensive *CRC Handbook of Gene Level Diagnostics in Clinical Practice*. It is a monumental piece of work which presents the state-of-the-art molecular genetic assays with specific applications in the evaluation and diagnosis of a variety of diseases.

The College of American Pathologists Resource Committee on Molecular Pathology has identified seven areas of clinical laboratory testing involving DNA/RNA nucleotide assays:

1. Molecular oncology
2. Molecular genetic diseases
3. Molecular HLA/histocompatibility
4. Parentage testing by DNA polymorphisms
5. Forensic identity testing by DNA polymorphisms
6. *In situ* hybridization assays
7. Molecular infectious diseases

This handbook discusses the various assay procedures and their specific applications to clinical conditions. It will be useful to both clinical laboratory scientists and physicians who order the laboratory tests and need information on the usefulness of the molecular biological tests as well as their interpretation.

The molecular nucleotide assays in clinical laboratories will grow at an explosive and logarithmic rate. There are many assays utilizing synthetic nucleotides with extremely sensitive nonisotopic detection methods suitable to automation. Many current chemiluminescent methods are more sensitive than certain radioisotopic assays.

The amplification assays such as polymerase chain reaction (PCR) have extremely high sensitivity and specificity, and can detect single copies of the target gene.

The widespread use of the newer nucleotide molecular assays in clinical laboratories awaits the commercial development of:

1. Sensitive nonisotopic chemiluminescent assays utilizing synthetic nucleotides
2. Rapid batch assays with automation
3. Simplified automated amplification assays (in development) including PCR, self-sustaining sequence replication (3SR), ligase amplification reaction (LAR), etc.

One would predict that future nucleotide assays may have certain advantages over immuno-chemical assays in that extensive new information on gene sequence and structure will become available from the "Human Genome Project". Subsequent synthesis of specific nucleotide probes for assays will be developed more rapidly, whereas the production of many specific monoclonal antibodies to detect the corresponding gene products would be more expensive and time consuming.

Dr. Bernstam is to be congratulated in writing this book. He has spent countless hours of effort to bring forward important information which future scientists in the clinical arena will need.

Robert M. Nakamura, M.D.
Chairman, Department of Pathology
Scripps Clinic and Research Foundation
Adjunct Professor of Pathology
University of California
School of Medicine
La Jolla, California

To
Elmer, Luda, Inna, Luba and Shura

Preface

It all started with my early preoccupation with philosophy and fascination with the possibility of learning Nature's ways by trying to understand the "philosophy of the cell". Some of the secrets can be gleaned from the logic of the cell's interaction with other cells, its response to adverse environment, injury and its repair as well as from its involvement in disease.

The explosive accumulation of new facts, concepts and molecular regulatory phenomenology of cellular behavior makes my dream of seeing molecular biology applied to patient management a realistic possibility in the near future.

This book focuses primarily on the recent advances in cellular molecular biology and molecular genetics which enhance our potential to recognize diseases, to monitor the efficacy of therapeutic efforts, and even to anticipate the onset of some conditions.

In a field as dynamic as molecular biology the information condensed in this book should be regarded only as a panoramic, "freeze-frame" shot of representative areas in gene level diagnostics at the time of this writing. I hope it proves useful to pathologists and primary care physicians in various disciplines, laboratory scientists and biotechnology specialists, students and residents. My task will be fulfilled if the readers find this book objectively directing their attention to the unparalleled potential of molecular diagnostics available to the practicing physician.

Gene level diagnostics are here to stay and I firmly believe that patient-related benefits will continue to grow exponentially. I would like to think that this book may help advance the rising standards of medical care.

Victor A. Bernstam
Ann Arbor, Michigan
June 1992

Acknowledgments

I am grateful for the advice and encouragement of Drs. A. Gotz, K. McClatchey, J. Plafker and L. Simson. It is a pleasure to thank Drs. R. V. Lloyd and F. Whitehouse for reviewing parts of the manuscript. The friendly cooperation, efficiency and professionalism of Jeff Holtmeier, Marsha Baker, Cathy Walker, Suzanne Lassandro, Monique Power, Debbie Berman, Susan McColl and many others at CRC Press, Inc. made this project a timely reality and a memorable and pleasant experience.

I would not be able to write this book without the faith and support of my entire family. Elmer Bernstam, in particular, has contributed most generously his time and talents. His resourcefulness, uncompromising judgement and humor made him an invaluable partner at all stages in this work and I am enormously grateful for his commitment and help.

The Author

Victor A. Bernstam, President of BioSol, Ltd. (Ann Arbor, MI) received his M.D. from the Leningrad (now St. Petersburg) Medical Institute, USSR in 1966, and his Ph.D. in cellular molecular biology from the USSR Academy of Sciences Komarov Botanical Institute in Leningrad in 1973. Since 1960 he has conducted research on brain biochemistry, molecular mechanisms of hormonal induction of gene expression, heat shock effects on protein biosynthesis, membrane phenomena in experimental and clinical atherosclerosis, and laboratory diagnostics. Dr. Bernstam has held positions as a Research Scientist at the Komarov Botanical Institute and the Institute of Cytology of the USSR Academy of Sciences in Leningrad, as well as in the Cellular Chemistry Laboratory at the University of Michigan School of Public Health in Ann Arbor. His medical experience includes work as an Emergency Physician in Leningrad, and later following American Pathology Board Certification as a Pathologist concentrating on clinical special chemistry and molecular pathology. Dr. Bernstam served as a Medical Director of Chelsea Medical Laboratory, MDS Laboratories, and later was one of the organizers and, subsequently, Medical Director of Warde Medical Laboratory (Ann Arbor, MI) where he established and headed a Molecular Diagnostics Group. He has published and lectured on GABA metabolism, heat shock research and apolipoprotein testing for coronary risk assessment. He has developed and patented a novel ligand-removing, clarifying reagent, Liposol®. He is a member of the College of American Pathologists, American Association for Clinical Chemistry, New York Academy of Sciences, and the honorary society Sigma Xi. He continues research on laboratory diagnostics.

Table of Contents

Part 1
General Aspects of Gene Level Evaluation:
Survey of Techniques and Applications

Introduction

The avalanche of new diagnostic possibilities at the gene level presents a challenge to the medical community in selecting reliable molecular technologies that can benefit patient management. Clearly, a "wholesale" transfer of molecular biological tools to clinical diagnostic laboratories may not offer better diagnostic modalities than the existing methods. Immunological approaches are frequently fully adequate for clinical purposes, however, in a number of conditions DNA probes offer a substantially higher level of patient care, as in the diagnosis of infectious and genetic diseases.

The advent of gene level diagnostics introduced new definitions of **infection** and the need for reappraisal of the previously held beliefs and views on etiology of some diseases, the role of viral factors, and the efficacy of treatments. Ultrasensitive diagnostic methods such as the polymerase chain reaction (PCR) allow the identification of single virally infected cells among 10^6 cells, and if in the case of human immunodeficiency virus (HIV) the etiological role of the virus appears to be established, the situation is less certain with HPV (see Part 6). Again, in infectious disease diagnosis PCR has established new criteria of **diagnostic sensitivity** and **performance standards** for other diagnostic methods. It may eventually replace some of the existing methods, stimulate the refinement of others, and serve to complement the existing ones in specific cases (Peter, 1991).

In **genetic disorders**, the introduction of new molecular probes offers not only diagnostic but also predictive possibilities, raising new hopes, along with concerns, for the appropriate application of this powerful technology. This necessitates the development of practical guidelines for the acceptability of a given molecular biological test as a clinical diagnostic tool. Efforts to develop appropriate **standards** have been reported (Lebo et al., 1990a). The DNA Committee of the Pacific Southwest Regional Genetics Network addressed four general issues: how to disseminate information about DNA testing to consumers, third party payers, etc.; the need for quality assurance and licensure standards for DNA testing laboratories and personnel; the establishment of adequate reimbursement mechanisms on a national scale; and the practical aspects of the molecular technology transfer, its evaluation, and the assessment of its potential impact and role in clinical genetics (Lebo et al., 1990a).

With respect to the appropriateness of a new genetic test for regular clinical care, for disease genes that are clones the Committee suggested that "a disease test should be moved from the research test list to the clinical test list when at least 70% of matings are expected to be informative" (Lebo et al., 1990a, p. 586). When a disease is diagnosed by linked polymorphisms this requirement is a more rigorous one, necessitating a definite answer with a confidence level of at least 95%, since "the result of a DNA test is usually the only clinical criterion used to choose a reproductive option" (Lebo et al., 1990a, p. 586).

The objective of this book is to provide a reasonably wide overview of the current status of possibilities in using gene level evaluation of a disease. The emphasis is placed on the **demonstration of various methodological choices and possibilities** that gene probes may offer to enhance patient management. The details of molecular aspects of pathogenesis of various conditions are discussed only to illustrate the scope and depth of diagnostically relevant information. No attempt is made to offer comprehensive coverage of the molecular pathology of all the conditions discussed. It is hoped that the reader may benefit from this review by better appreciating the advantages, limitations, and drawbacks of using gene level diagnostics in resolving individual clinical problems. The sophistication and the progressively increasing user-friendliness of gene level techniques will hopefully be understood as an ongoing process that cannot be captured with a satisfactory degree of completeness at any given moment. The overview presented here will serve its purpose if it enhances the vantage point of those interested in the current developments in this exciting field of medicine and helps to understand, foresee, and apply gene level techniques in clinical practice.

Before discussing specific methodological approaches to gene level diagnosis, selected aspects of studies on gene structure and function will be mentioned briefly that may have particularly novel impacts on molecular diagnostics in clinical practice. While the results of some areas of research, such as that of the Human Genome Project appear at this time somewhat removed from practical application in molecular diagnostics, the results of studies on the structural and functional characteristics of repetitive DNA, tumor suppressor and metastasis suppressor genes, and mitochondrial genes already have immediate impact on disease evaluation at the gene level. Regulation of gene expression through amplification, the degree of methylation, and the phenomenon of genome imprinting may soon become a part of the vocabulary of disease evaluation in routine clinical practice.

Although **serum tumor markers** continue to be a useful tool for management of cancer patients (e.g., Bates and Longo, 1987; Chia et al., 1988; Longo, 1987), their use in cancer detection by screening asymptomatic individuals is very limited (Pluygers et al., 1986; Makuch and Muenz, 1987; Neville, 1986). The majority of **antigenic determinants used as tumor markers are not organ specific** and **cannot be used for definitive diagnosis**, but only in monitoring the therapy of cancer patients (Bates and Longo, 1987; Chia et al., 1988; Schwartz, 1990). Furthermore, the **low specificity** of antigenic tumor markers is due to marked variations in their levels associated with a variety of physiologic and toxicologic conditions as well as nonmalignant diseases (Touitou and Bogdan, 1988), including those leading to immunologic disturbances (Nakao and Kozma, 1988). Antigenic tumor markers find wider use in diagnostic tissue analysis by immunohistochemistry complementing morphologic characteristics independently evaluated by a pathologist using light and electron microscopy (Triche, 1987).

Elucidation of the **genes encoding the commonly used antigenic tumor markers** provides for a better understanding of their relevance to specific disease conditions. One example is the family of 10 genes constituting the prototype tumor marker, the **carcinoembryonic antigen (CEA)** (Inazawa and Abe, 1990; Thomas et al., 1990; Willcocks and Craig, 1990). CEA has been shown to function as an accessory adhesion molecule and to be associated with a tendency of colonic tumor cells to form liver metastases; however, its role in the normal adult colon still remains unsolved (Thomas et al., 1990).

The recognized role of certain oncogenes (e.g., *ras* and *myc*) in different stages of carcinogenesis draws attention to the possibility of using their evaluation in the context of diagnostic and/or prognostic patient management (e.g., Field and Spandidos, 1990).

1.1.1 HUMAN GENOME PROJECT

The international research project that started with discussions held in 1985 on the appropriateness of a concerted effort toward eventually deciphering the nucleotide sequence of the entire human genome was eventually organized and concrete results have begun to appear (Watson and Cook-Deegan, 1991).

The Human Genome Project (HGP) has several objectives that unite research programs of NIH, the DOH, the Howard Hughes Medical Institute as well as parallel efforts in the private sector in the U.S. and similar efforts in other countries. In defining these objectives for the years 1991 to 1995, the National Center for Human Genome Research emphasized several interrelated

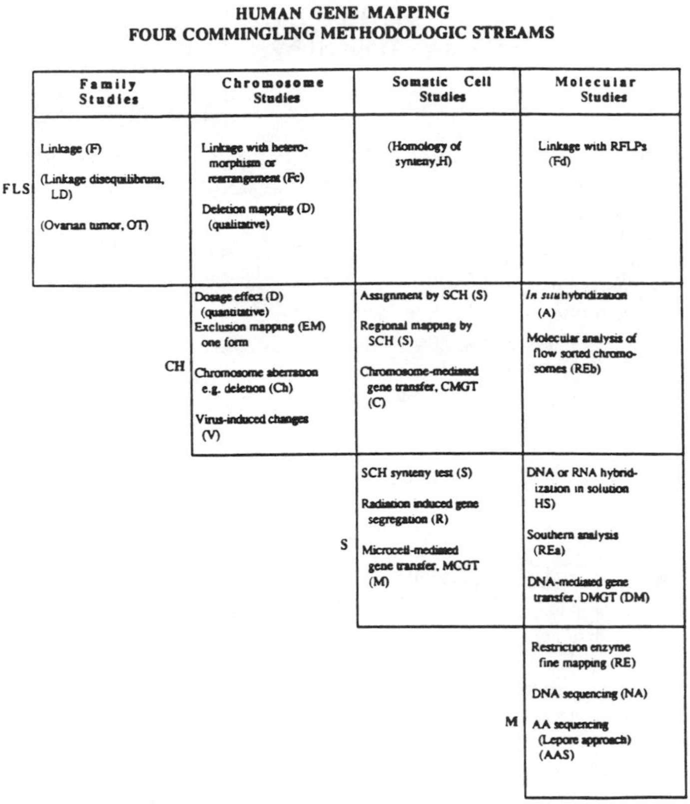

HUMAN GENE MAPPING
FOUR COMMINGLING METHODOLOGIC STREAMS

	Family Studies	Chromosome Studies	Somatic Cell Studies	Molecular Studies
FLS	Linkage (F) (Linkage disequilibrium, LD) (Ovarian tumor, OT)	Linkage with hetero-morphism or rearrangement (Fc) Deletion mapping (D) (qualitative)	(Homology of synteny,H)	Linkage with RFLPs (Fd)
CH		Dosage effect (D) (quantitative) Exclusion mapping (EM) one form Chromosome aberration e.g. deletion (Ch) Virus-induced changes (V)	Assignment by SCH (S) Regional mapping by SCH (S) Chromosome-mediated gene transfer, CMGT (C)	*In situ* hybridization (A) Molecular analysis of flow sorted chromo-somes (REb)
S			SCH synteny test (S) Radiation induced gene segregation (R) Microcell-mediated gene transfer, MCGT (M)	DNA or RNA hybrid-ization in solution HS) Southern analysis (REa) DNA-mediated gene transfer, DMGT (DM)
M				Restriction enzyme fine mapping (RE) DNA sequencing (NA) AA sequencing (Lepore approach) (AAS)

FIGURE 1.1. Interrelationship of methodological approaches to human gene mapping. (From McKusick, V. A. (1991). FASEB J. 5:12–20. With permission.)

goals (Understanding Our Genetic Inheritance. The U.S. Human Genome Project: the First Five years FY 1991–1995 National Technical Information Service [1990]):

1. the development of **high-resolution genetic maps** of the human genome, eventually at the resolution of 1 centimorgan (cM)
2. the establishment of overlapping physical (**contig**) maps of all human chromosomes and of the DNA of selected study organisms at the resolution of reference DNA landmarks spaced at 100-kb intervals. This would allow an easy access to cloning specific DNA regions by various investigators without the need to store reference DNA material physically
3. the determination of the **complete sequence of human DNA** and that of selected model organisms, emphasizing the early characterization of regions containing genes of interest
4. the development of appropriate adaptation of achievements of **informatics** in collecting, storing, distributing, as well as in analyzing and retrieving the data generated by physically discrete research teams
5. the creation and utilization of **appropriate technologies** to achieve these goals

By now almost 2000 genes of the estimated 50,000 to 100,000 genes in the human have been mapped to specific chromosomal locations (McKusick, 1991). **Human gene mapping** proceeds along the four major avenues: family studies, chromosome studies, somatic cell studies, and molecular studies (McKusick, 1991) (Figure 1.1). Guidelines for human linkage maps elaborating the mode of presentation and standardization of these have been developed (Keats et al.,

TABLE 1.1
Diseases Mapped by RFLPs[a]

Disease	Chromosomal location
Charcot-Marie-Tooth disease	1q,17,Xq13
Usher syndrome	1q
van der Woude lip-pit syndrome	1q
Aniridia	2p, 11p
Waardenburg syndrome	2q
von Hippel-Lindau syndrome	3p
Huntington disease	4p
Facioscapulohumeral muscular dystrophy	4q
Spinal muscular atrophy, several types	5q
Adenomatous polyposis of colon	5q
Hemochromatosis	6p
Juvenile myoclonic epilepsy	6p
Spinocerebellar ataxia (one form)	6p
Craniosynostosis	7p
Greig craniopolysyndactyly syndrome	7p
Cystic fibrosis	7q
Langer-Giedion syndrome	8q
Friedreich ataxia	9q
Torsion dystonia	9q
Tuberous sclerosis	9q, 11q
Nail-patella syndrome	9q
Multiple endocrine neoplasia, type II	10q
Wilms tumor-l	11p
Multiple endocrine neoplasia, type I	11q
Ataxia-telangiectasia	11q
Retinoblastoma	13q
Wilson disease	13q
Marfan syndrome	15q
Polycystic kidney disease	16p
Batten disease	16p
Cataract, Marner type	16q
Neurofibromatosis	17q
Myotonic dystrophy	19q
Malignant hyperthermia	19q
Alzheimer disease (one form)	21q
Acoustic neuroma, bilateral	22q
Duchenne muscular dystrophy	Xp
Retinitis pigmentosa (two forms)	Xp
Wiskott-Aldrich syndrome	Xp
Alpon syndrome	Xq

[a] List of mapped Mendelian disorders for which the biochemical basis was
    not previously known (mapping as a first step toward basic understanding).

From McKusick, V. A. (1991). FASEB J. 5:12–20. With permission.

1991). So far the most productive approach to human gene mapping has been **somatic cell hybridization** followed by *in situ* **hybridization (ISH)**. In fact, demonstration of a specific chromosomal location of a cloned gene in a panel of DNAs from somatic cell hybrids together with ISH is a prerequisite of complete characterization of a gene. Identification of the **location of a gene** by **restriction fragment length polymorphism (RFLP)** with respect to other established DNA markers complements its definition. A number of diseases in which the product of the mutant gene had not yet been identified have been mapped by RFLPs (Table 1.1). The number of human genetic diseases so far remains unknown, and it is argued that the frequently mentioned number of 3,000–5,000 is actually unfounded (Billings, 1991).

Some of the methodologies helpful in gene mapping, such as the use of **somatic cell hybrids** (e.g., Delattre et al., 1991; Golubic et al., 1991), also combined with PCR analysis (Cox and Lehrach, 1991) and **chromosome assignment** of specific genes to chromosomes **by PCR** (e.g., Dionne et al., 1990), **radiation-induced gene segregation (radiation mapping)** (e.g., Cox et al., 1989; Goodfellow et al., 1990), **yeast artificial chromosomes (YAC)** (e.g., Burke et al., 1987), **microcell-mediated gene transfer** (e.g., Fournier and Ruddle, 1977; Satoh et al., 1991), and **chromosome and DNA-mediated gene transfers** (for a review, see Bentley et al., 1988), will hardly be employed in **clinical gene level diagnosis** in the very near future. Other strategies, such as physical **mapping by repetitive sequence fingerprinting** (Stallings et al., 1990), develop approaches helpful in identity studies. The possibility of genetic mapping by single sperm typing allows establishment of linkage relationships among DNA polymorphisms not resorting to pedigree analysis (Arnheim et al., 1991).

Other techniques used in developing the "applied anatomy of the human genome" (McKusick, 1991) include **RFLP** and **sequence tagged sites (STS)** (Olson et al., 1989); **allele-specific oligonucleotides (ASO)** screening procedures, especially using **PCR** (Saiki et al., 1985, 1986); **restriction enzyme allele recognition** (Mullis and Faloona, 1987) and **cleavage mismatch detection** (Myers et al., 1985a,b); **multiplex PCR deletion detection** (Chamberlain et al., 1988); **competitive oligonucleotide priming (COP)** (Ballabio et al., 1990; Gibbs et al., 1989a) and analysis of the **genomic distribution of α satellite DNA sequences** (Choo et al., 1991). These and many others may be used in clinical diagnostics, and some are already used in gene level diagnosis (see below). Specific genes can be isolated in their active or inactive state by the technique of **nucleoprotein hybridization** (Vincenz et al., 1991). Combination of two-dimensional gel electrophoresis of proteins followed with microsequencing of these, cloning of the cDNAs, comparison and assignment of partial sequences to the DNA sequences eventually results in the assignment of these to the chromosome locations (Celis et al., 1991).

Immunological detection of nucleic acids and/or their hybrids is particularly appealing to routine diagnostic laboratories (e.g., Stollar and Rashtchian, 1987; Traincard et al., 1989). Simplification of nucleic acid **handling procedures** adaptable to clinical and forensic diagnostic laboratories has been achieved at various stages of gene level testing. An example of such a simplification is the recently introduced and already commercially available technique of isolation of mRNA from eukaryotic cells and RNA-DNA hybridization using **magnetic beads** coupled with streptavidin (Hornes and Korsnes, 1990).

Again, **quantification of DNA hybridization** has been made possible in an adaptation of the **slide immunoenzymatic assay (SIA-DNA)** that can be used in routine clinical laboratories and even in field studies (Conway de Macario et al., 1990), or in DNA probe methods utilizing **sulfonated cytosines** with the detection of these by enzyme immunoassay formats suitable to automation (Bains, 1991). Using a cloned probe p17h8 (D17Z1), which detects highly repetitive, primate-specific **α satellite DNA**, minute amounts of human genomic DNA can be quantitated in forensic specimen within less than 4 h with a sensitivity at subnanogram levels (Waye et al., 1989).

An increasing number of protocols are being developed for **nonradioactive PCR-based detection systems** applicable to clinical diagnostics (e.g., Emanuel, 1991; Mantero et al., 1991; Tabibzadeh et al., 1991).

1.1.2 SEQUENCE-TAGGED SITES

An important aspect of the HGP that has immediate relevance not only to human genome mapping but also to the earliest clinical diagnostic utilization of the forthcoming information is the proposed referencing of defined DNA. The **sequence-tagged sites** (STS) proposal (Olson et al., 1989) aimed to allow merging of the genetic and physical maps of the human genome gathered by diverse methods in different laboratories into a single consensus physical map. The essence of the STS concept is the **use of short tracts of single copy DNA** in a given region of

DNA that can be easily recognized and recovered at any time by PCR. An STS will identify its physical location on the genome map by a 200- to 500-bp sequence unique to a given DNA fragment. It can be easily recognized by PCR primers complementary to opposite strands and opposite ends of the sequence tract of interest.

One particular advantage of this referencing strategy is that the **identification code** of a particular segment of DNA can be stored in a **database**, and anyone interested in the particular stretch of DNA can use a set of primers specifically defined by the appropriate STS. The STS proposal envisions eventual spacing of 100 kb between neighboring STSs, requiring a total of 30,000 STSs. At this resolution, one-step recovery would be possible for most regions of the genome using the YAC system for cloning. STSs can also serve as **reference points** for testing mapped disease genes. The importance of STS has generated novel methods to establish readily applicable strategies for creating human STS expression map of the genome. An example of one such strategy is the design of PCR primers for 3′ untranslated regions of human cDNAs allowing rapid assignment of genes to specific chromosomes and chromosome regions (Wilcox et al., 1991).

The practical utility and, consequently, commercial potential of sequence data has led to controversial efforts seeking patent coverage of defined sequences of coding regions of the genome — the so-called **expressed sequence tags (EST).** Heated discussions followed application for patents at NIH in the U.S. (Anderson, 1992) and later the U.K. (Aldhous, 1992) although the French Research Minister voiced his opposition to cDNA patenting (Curien, 1991) supported by the concerns expressed by the American Society of Human Genetics (1991).

1.1.3 REPETITIVE DNA

Although a discussion of **repetitive DNA** is beyond the scope of this book, some of its features impacting the molecular diagnostics at this time should be summarized. The origin, structure, and function of repetitive DNA continues to be the subject of intensive research (for reviews see, e.g., Hardman, 1986; Jeffreys et al., 1985a–c; 1987, 1990; Moyzis et al., 1987; Vogt, 1990). Repetitive DNA is composed of **tandemly repeated sequences**, which show an extraordinarily high degree of diversity (see also Section 1.2.5). The possible role of repetitive sequences in regulating the expression of genes is suggested by the variable degree of methylation in somatic cells and the germ line (Hardman, 1986). The so-called **"middle repetitive" DNA** comprises up to 40% of the DNA in eukaryotic genomes and consists of numerous families of lower copy number repetitive sequences (Britten and Kohne, 1968). Repetitive DNA is distributed in the human genome essentially in two ways: large tandem block structures (up to several thousand nucleotides) are composed of head-to-tail arrangements of various sequence units with a highly variable copy number (type I). Type II repeated sequences are composed of small variable tandem sequence repeats in which the same basic sequence, or **core unit**, can be clustered or interspersed in the genome. Three main classes of repetitive DNA are recognized based primarily on the reassociation kinetics of these sequences in hybridization experiments (Britten and Kohne, 1968): the **highly repetitive DNA** with more than 10^5 copies per haploid genome, the **middle repetitive sequences** with 102 to 105 copies per haploid genome, and the **low repetitive sequence** class with all sequence families represented in 2 to 100 copies per haploid genome. A substantially different picture emerges from the analysis of the genome by restriction mapping and sequence determinations compared to that gained by reassociation kinetics only (Vogt, 1990). Repetitive DNA can be easily resolved from the rest of the DNA in CsCl density gradients, forming additional (usually four) satellite peaks characterized by its unique buoyant density.

In situ hybridization of various fractions of satellite DNA to human metaphase chromosomes indicates that it is primarily located in the **constitutive heterochromatin** of the centromeric regions of chromosomes. Some chromosomes (2, 3, 4, 6, 8, 11, and 18) show no evidence of satellite DNA, and chromosomes 10, 12, 16, 19, and X contain only small amounts of satellite

DNA (Vogt, 1990). Using restriction analysis, the fine structure of repetitive elements can be resolved. Some satellite sequences have been found to be characteristic of a given chromosome such as those defined for chromosome 16 (Moyzis et al., 1987), and the Y chromosome (reviewed in Vogt, 1990). A family of repetitive sequences related to the α-satellite family of the African green monkey and therefore termed, the **alphoid DNA sequence** family has a mean core unit of 171 bp and is predominantly located in the centromeric regions of chromosomes (Arn and Jabs, 1990). A notable alphoid DNA *Sau*3A subfamily is important for its high degree of RFLP among humans (Vogt, 1990). Other sequence families (724, D22Z3, 71-7B) have so far not been utilized in molecular diagnostic applications.

Repetitive sequence families of type II are composed of simple sequence motifs such as $(\mathbf{T})_n/ (\mathbf{A})_n$, $(\mathbf{G})_n/(\mathbf{C})_n$, $(\mathbf{TA})_n$, $(\mathbf{TG})_n$, $(\mathbf{GC})_n$, $(\mathbf{TC})_n$, $(\mathbf{GGA})_n$, $(\mathbf{GCA})_n$, and minisatellite sequence families such as $(\mathbf{TTTTA})_n$ and $(\mathbf{TTAGGG})_n$ (Vogt, 1990). Although the exact origin and role of various families of repetitive sequences such as *Alu* and **L1** belonging to the **short interspersed repeat (SINE) family** in human DNA remain a subject of intense studies and debate, certain similarities in their structure with those of retroviral elements and transposons have been noted (e.g., Hardman, 1986). Among the approximately 500,000 *Alu* family members there are some 500 that constitute a human-specific subfamily (Batzer and Deininger, 1991). These elements can now be relatively easily identified by PCR amplification using *Alu*-specific primers corresponding to the consensus sequences of the *Alu* family (e.g., Bernard et al., 1991; Lengauer et al., 1990). The highly conserved nature of tandemly repeated DNA sequences apparently carries specific, so far not fully disclosed, functions in the genome. Definite tissue-specific changes in repetitive DNA have been observed during the developmental process (Yokota et al., 1989). Aside from being the **area of more frequent mutational events** such as insertions, deletions, amplifications, and oncogene translocations (e.g., Krowczynska et al., 1990), one of the proposed roles of repetitive DNA is in influencing the conformation of chromosome loops with a definite locus-specific organization of their folding structure in the nucleus (Vogt, 1990). These "super-structures" are thought to attract specific nuclear proteins in a nonrandom fashion expressed in the so-called "chromatin folding code" (Vogt, 1990). These higher order chromatin structures are believed to influence various chromatin functions defined in a "gene expression code", a "replication code", and the "recombination code" (Vogt, 1990).

Consistent with the association of repeated sequences with mutation sites, analysis of DNA sequences in the vicinity of deletion breakpoints revealed that the junction regions show nonrandom sequence elements both at the nucleotide and dinucleotide sequence levels (Krawczak and Cooper, 1991). **Direct repeats** of between 2 and 8 bp are in the immediate vicinity of almost all deletions causing human genetic disease. **Inverted repeats** were not as frequent in the junction regions, whereas **symmetrical sequence motifs** were overrepresented at sites of single base deletions (Krawczak and Cooper, 1991). A model of the slipped mispairing mechanism accounting for this phenomenon is proposed, although predicted deletion of one whole repeat copy is very rare. It is believed that the relative importance of various mechanisms resulting in gene deletions is influenced by local primary and secondary DNA structure, making precise prediction of a gene deletion extremely difficult (Krawczak and Cooper, 1991).

1.1.4 RECOMBINATIONAL HOT SPOTS

DNA sequences capable of assuming the left-handed Z conformation in areas enriched in GC and CA/GT appear to be the hotspots for spontaneous deletions (Freund et al., 1989). In fact, this phenomenon seems to occur in fragile X syndrome (Kraemer et al., 1991; Oberlé et al., 1991).

Studies with viral insertion and transfection experiments have demonstrated that rearrangements commonly occur following **integration of DNA sequences** (reviewed in Murnane, 1990). The resulting **instability of the genome** is associated with loss or amplification of the integrated sequences and neighboring DNA. It appears that although integration frequently occurs in regions containing repetitive cellular DNA, random integration events also happen and lead to a general enhancement in the rate of recombination around the integration site. Amplification and the formation of episomal intermediates often accompany the destabilization of the area of integration. It appears that a combination of sequences within the integrated and the adjacent DNA is needed for an unstable region to arise. One of the means of detecting an area of instability can be an RFLP near the integration site, providing a diagnostic tool. What still remains undetermined is whether the recombination hot spots arising from integration of sequences are associated with the same degree of genomic instability as that resulting from amplification of endogenous genes (Murnane, 1990).

Aside from mutations due to deletions other mechanisms operating at the chromosomal translocation level are recognized that result in the acquisition or loss of a function in affected cells, leading to a pathological condition. Most examples have so far been defined in hematological malignancies. Transcription deregulation of the juxtaposed cellular gene leads to inappropriate levels of a gene product compared with those in normal cells at an equivalent stage of differentiation (Cleary, 1991). Several genes activated by chromosomal translocation in lymphoid malignancies code for **helix-loop-helix** (**HLH**) **proteins** displaying DNA binding and dimerization with heterologous members of this family of proteins. This may represent an essential aspect of an enhancer binding regulatory system. *myc* protooncogene produces the prototypical HLH protein. Other genes coding for **oncogenic regulatory proteins** found in leukemia and other conditions include the *Hox* **11 homeobox gene**, **Ttg/Rhom-1** and **Rhom-2** (these proteins contain cysteine-rich, so-called **LIM domains**). Other types of translocations occur not in immunoglobulin or T cell receptor genes — the prototype being the *bcr-abl* fusion — but rather in introns, and **chimeric proteins** with unique new functions absent in the wild type are formed as a result. These represent gain or loss of function mutations mediated by chromosomal mechanisms (Cleary, 1991).

The immediate practical consequences of the elucidation of specific molecular events of chromosomal translocations resulting in the formation of chimeric proteins are seen in **acute promyelocytic leukemia (APL)**. **Translocation t(15;17)** leads to the formation of a truncated product, **retinoic acid receptor α (RAR α)**. These leukemias proved to respond dramatically to retinoid treatments. The abnormal fusion protein acts by a dominant negative mechanism interfering with promyelocytic differentiation. Chimeric RAR α exhibits altered trans-activation properties that, compared with the wild type, are promoter- and cell-specific. Chimeric proteins may also act as competitive inhibitors of their normal cellular counterparts (Cleary, 1991). The identification of the site of fusion using molecular or immunological techniques now has important implications for the management of patients with APL.

Another aspect of changes in gene expression relevant to disease evaluation involves **oncogenic conversion by altered transcription factors**. Among these are c-*jun*, c-*fos*, c-*rel*, and c-*myc*, the so-called "immediate early" genes activated early in mitogenic stimulation of cells and apparently leading to a cascade of cellular functions associated with initiation and promotion of growth. Other protooncogenes, such as *c-erbA*, may be involved in promoting differentiation — an opposing influence. Yet other genes, such as *AP*-1 and *NF-k*B, appear to have a wide range of targets, their products acting as pleiotropic transcription factors or as general second messengers. The inhibiting activity of v-*erb*A acting as a negative oncogene on transcription counteracts the normally expressed c-*erb*A activity that appears to suppress transformation. In this sense, c-*erb*A function resembles that of tumor suppressor genes, the inactivation of which is associated with tumorigenesis (Cleary, 1991). An excellent collection of reviews of the mechanisms of hematological malignancies and molecular techniques for the diagnosis and management of patients has been presented (Cossman, 1990).

1.1.5 TUMOR SUPPRESSOR GENES

Over the past few years it has become clear that a number of genes display **recessive characteristics** in tumorigenesis. Several conditions have been described in which such

recessive oncogenes, antioncogenes, or **tumor suppressor genes** appear to play the key role, defined or inferred in **retinoblastoma**, **Wilms' tumor, osteosarcoma,** as well as in **colon, renal, lung,** and **breast cancers**, discussed in more detail in appropriate sections.

Basic research on tumor suppressor genes (**TSGs**) in some aspects extends and refines our understanding of the biology of oncogenes and its comprehensive coverage is outside the scope of this book. Numerous detailed reviews of this field have been appearing (e.g., Benedict et al., 1990a,b; Bishop, 1991; Carbone and Levine, 1990; Den Otter et al., 1990; Freeman et al., 1990; Green, 1989; Hansen and Cavenee, 1988; Hunter, 1991; Israel, 1989; Klein, 1987, 1990; Knudson, 1989; Lane and Benchimol, 1990; S. W. Lee et al., 1991; Levine et al., 1991; Marshall, 1991; Sager, 1989; Stanbridge, 1990; Stanbridge and Nowell, 1990; Weinberg, 1989).

The refinement of our understanding of the best characterized TSGs so far, *Rb* and *p53*, and of their biochemical functions will certainly contribute to more efficient diagnosis, monitoring, and eventually treatment of malignancies. In spite of the diversity of histiotypes, in which both oncogenic and tumor suppressor activities have been demonstrated, the emerging evidence supports the view that a number of common characteristics can be recognized. The general features of tumor suppressor activities observed in experimental systems and in human tumors are briefly summarized.

The dominant nature of **activated oncogenes** appears to be universally counterbalanced by, or intimately interconnected with, the function of the **recessive genes** (Harris, 1990; Hunter, 1991; Stanbridge, 1990). These appear to encode proteins controlling fundamental cellular functions of **signal reception, transduction,** and **propagation,** as well as **progression through the cell cycle** and the **initiation of DNA replication** and **transcription**. The loss of function of these genes at various levels (alteration or loss of the gene, defects in transcription, as well as production of dysfunctional final products) leads to carcinogenesis (Sager, 1989; Scrable et al., 1990).

Some of the common features shared by diverse types of oncogenic proteins include their **ability to interact with other oncoproteins** in bringing about **immortalization** of cells; their interaction with specific **target genes**; and the presence of **similar amino acid sequences** among diverse nuclear oncoproteins (Green, 1989). The phenomenon of **suppression of malignancy** was first discovered in fusion experiments, which demonstrated that loss of genetic material may lead to malignancy (Harris, 1990). Subsequent research with individual chromosomes and revertants suggested that imposition of **terminal differentiation** leads to suppression of malignancy.

The distinction between oncogenes, recessive genes, TSGs, and antioncogenes is viewed as artificial, reflecting operational differences that had led to the recognition of specific cellular functions, rather than intrinsically different structural and functional entities (Harris, 1990). It appears that **pleiotropic alterations** of a gene structure may result in the expression of either stimulating or suppressing activities in the process of differentiation. The operational distinction between oncogenes and TSGs may be further blurred by the possibility that products of some mutant genes act as **allosteric inhibitors** of their normal counterparts (Wyke and Poole, 1990).

Numerous examples are known in which a cooperation between dominantly acting oncogenes and inactivating TSGs can be discerned in the **multistage process of carcinogenesis** (Marshall, 1991). The multiplicity of genetic alterations apparently interacting at some point in carcinogenesis is more evident among TSGs rather than oncogenes (e.g., inactivation of both the *RB* and *p53* loci in lung cancer and osteosarcomas; *p53* and *DCC* in colon cancer, etc.) (Marshall, 1991). In a single tumor, multiple genetic changes can be demonstrated both in the activation of protooncogenes and in alterations of TSGs (Hunter, 1991; Marshall, 1991).

Loss of function of TSGs may be brought about by focal deletions, larger "interstitial" deletions, by the loss of the entire chromosome, or chromosomal translocations with mutations of TSGs occurring more likely on the paternal than the maternal chromosome (reviewed in Bishop, 1991).

p53 appears to play a role in the cell cycle control (Deppert et al., 1990; Steinmeyer et al., 1991). Some TSG products are localized to the nucleus (p53, RB1, WT1), whereas others resemble cell adhesion molecules such as DCC. Loss or mutations of the *p53* gene have been linked to the inherited susceptibility to cancer development, **Li-Fraumeni syndrome** (Srivastava et al., 1990), and associated with **lymphoid malignancies** (**Burkitt lymphoma** and **chronic lymphocytic leukemia**) (Gaidano et al., 1991).

Ectopic TSGs introduced into transformed or cancerous cells, with their counterparts either defective or absent, lead to apparent **restoration of normalcy**, at least in part (Bishop, 1991; Hicks et al., 1991; Hunter, 1991; Johnson et al., 1991; Levine et al., 1991; Marshall, 1991; Mercer et al., 1991; Reihaus et al., 1990). Although the exact mechanism of action of any TSG has not been defined yet, some of these at least may trigger a **cascade type of action**, and it appears that in spite of apparently diverse points of initial action, many **TSGs interfere with the process of transformation** in a convergent fashion, eventually targeting similar effector mechanisms (Bassin and Benade, 1990; Wyke and Poole, 1990). Among these general pathways are the *ras*-dependent transformation interrupted by **Krev-1** and **RAP2** as well as the **18-kb DNA fragment** isolated from preneoplastic rat 208F cells transformed with the human EJ-Ha-*ras* gene and pSV2*neo*. Although this 18-kb sequence has not yet been assigned to a specific location, the induction of **revertant phenotype** in cells transfected by this DNA fragment was enhanced 10,000-fold (Bassin and Benade, 1990).

Some of the tumor suppressor activities observed during induction of the "**flat" phenotype in revertants** by interferon, in the course of **retransformation** produced by 5-azadeoxycytidine-2 (5-Azad C), suggest a role for **DNA methylation** in the expression of transformation-related genes (Bassin and Benade, 1990). Again, analysis of the contribution of **chromatin topology** and DNA methylation to the transcriptional specificity in proviral activation suggests that these mechanisms are similar to the regulatory activities operating in development and differentiation (Wyke and Poole, 1990). It is thought that the genes, which may be affected in such trans-acting functions, behave as activators or suppressors of transformation pathways by producing either local or general changes in gene expression (Wyke and Poole, 1990).

The notion of **multistage carcinogenesis** has been widely accepted, and experimental and clinical evidence supports this concept at least in a few well-known cases. The "**two-hit" Knudson model** of the development of retinoblastoma (Knudson, 1971, 1986) envisioned the alteration of both alleles of the RB gene as necessary for the disease to occur. While this concept was fundamental in establishing the existence of putative recessive genes, or TSGs, a **multihit** (three-, four-, or five-hit) **process** leading to TSG alteration has been recently proposed (Den Otter et al., 1990).

According to that view, malignization may indeed start with two mutations in antioncogenes or TSGs; however, four mutations are usually required for the development of tumors in adults. Support for this hypothesis is provided by calculations based on several assumptions: the number of somatic cells produced in a human lifetime (2^{54}), an average mutation frequency per gene per generation of cell division (10^{-6}), and the incidence of cancer during a human lifetime (1.25 to 2.5% of the population develops cancer due to endogenous specific cancer mutations) (Den Otter et al., 1990). The 2-hit hypothesis predicts about 10^9 times more *de novo* arising tumor cells than are actually observed in reality. In contrast, ignoring the role of activated oncogenes, the requirement for four recessive mutations in TSGs implies that these occur at two different loci (Den Otter et al., 1990).

The other corollary of the multihit hypothesis is that an inherited defective antioncogene apparently is responsible for a wide range of tumor types (Den Otter et al., 1990). The authors admit the uncertainties in the assumptions they make, but allowing for three or five mutations needed, they emphasize the point that the required number of mutations is certainly not a two-hit phenomenon (Den Otter et al., 1990).

Stipulations of any model of a process as complex as carcinogenesis unavoidably lead to oversimplifications. Even the other well-studied example of the involvement of TSGs, Wilms' tumor, suggests a more complex picture of genetic lesions than the two-hit inactivation of *RB*1 alleles in retinoblastoma (Stanbridge, 1990). I believe that any model postulating a specific number of triggering events, such as mutations in recessive or dominantly acting genes, must address the interactive complexity of the consequences of such mutations. Such interactions may lead not only to an **avalanche effect**, itself mimicking apparent mutational events elsewhere, but can actually result also in functional aberrations of the initially uninvolved genes. The consequent functional inactivation of another gene, directly or indirectly resulting from the initial mutation, may present itself as an independent event. This can be interpreted as another "initial" mutation distorting the actual number of mutations appearing to be required for the initiation of tumorigenesis.

Simplifications implying only **straightforward progression of a dysfunction** from a gene to the effector target, albeit an understandable and necessary step in the unraveling of the functional complexity of cellular processes, do not give full credit to potential **pleiotropic** and **reciprocal interactions** between the genes and their products involved. It is interesting to note, in this regard, the reported phenomenon of **co-translational modification** of wild-type *p53* by the mutant *p53* which induces the wild-type protein to assume a mutant conformation which lacks the growth suppressor function of wild-type *p53* (Milner and Medcalf, 1991). Further sophistication of testable models of carcinogenesis will have to incorporate the possibility of inaccurate estimates of the pleiotropic aspects of oncogene and TSG action. Quantitative models of tumor suppression or activation will certainly have to account also for the demonstrated phenomenon of suppression of malignancy in a dose-dependent manner demonstrated in experimental systems (Stanbridge, 1990).

A warning against oversimplification and equating tumor suppressor activities with antioncogenes has been voiced (Stanbridge, 1990). TSGs, although appearing to act in a dominant pattern similar to oncogenes, do not directly interfere with the expression of oncogenes, and in this sense do not behave as antioncogenes (Stanbridge, 1990). Examples of **negative regulation** of some oncogenes by TSGs can be found, however. A case in point is HeLa + fibroblast hybrids, in which the expression of a 75-kDa membrane-associated glycophosphoprotein (p75) displays almost an absolute correlation with tumorigenic expression (Stanbridge, 1990). Nontumorigenic hybrids do not express this antigen, whereas HeLa cells and all the tumorigenic segregants reexpress the antigen. It is believed that *p75* is a critical oncogene identical to **intestinal alkaline phosphatase**, which is negatively regulated by a TSG located on **chromosome 11**. It appears that certain oncogenes can be negatively regulated in trans by TSG products acting at transcriptional and posttranscriptional levels (Stanbridge, 1990).

One of the promising approaches to the **identification and isolation** of TSGs is the method of **differential** or **subtractive hybridization**, which is based on the comparison of gene expression in two related cell populations (Sambrook et al., 1989). Although molecular **cloning of TSGs** at this time appears to be more difficult than the cloning of cellular oncogenes (Stanbridge, 1990), direct identification of candidate TSGs is being attempted.

An example of the use of subtractive hybridization is the isolation of TSGs from Syrian hamster embryo (SHE) cells undergoing carcinogen-induced neoplastic transformation (Cizdziel et al., 1991). A series of genes encoding proteins highly homologous to the chondrocyte-specific collagens type II and type IX were found to be down-regulated during carcinogen-induced immortalization of chondrocyte-like cell lines selected from the mixed embryo cell population. Along with their progression toward tumorigenicity, the cells lost their ability to express the chondrocyte differentiation markers, concomitant with the loss of their ability to suppress the tumorigenicity of the BP6T sarcoma cell line (Cizdziel et al., 1991).

In an elegant study with human breast tissue using subtractive hybridization, the genes uniquely or preferentially expressed in one of a pair of closely related cell populations were selected and recovered as cDNAs from normal breast cells after subtraction with mRNA from breast carcinoma cells (S. W. Lee et al., 1991). Among the isolated genes were those encoding the **gap-junction protein connexin 26**, two different **keratins, glutathione-S-transferase**, and an unknown gene expressed in normal mammary epithelial cell strains, but absent from tumor-derived cell lines. This appeared to be a Ca^{2+}-**binding protein of the *S100* gene family**, tentatively related to calcyclin (S. W. Lee et al., 1991).

An important distinction to be considered in the analysis of tumor suppressor activities is emphasized by S. W. Lee and co-workers (1991). While traditionally the loss of TSG expression has been related only to alterations of the gene itself, it should be recognized that the **regulatory events** serving to maintain that gene integrity can also determine its expression. The authors distinguish **two classes of TSG**: class I includes those that lost function by mutation, and class II are the genes whose expression is lost due to a mutation occurring elsewhere (S. W. Lee et al., 1991). As alluded to above, these workers emphasize that "the regulation of class II gene expression may be determined by a class I suppressor gene and its mutation may inhibit expression of several class II genes coordinately" (S. W. Lee et al., 1991, p. 2828). According to this model, it is expected that the **regulatory gene must be a *positively activating regulator***, which is operational in normal cells and non-functional in tumor cells. In cell hybrids, this regulatory gene is active and its presence is manifested by allowing the expression of class II genes that maintain tumor suppression.

According to these criteria, *RB, WT1*, and *p53* are class I TSGs, which are oncogenic when their function is lost, most likely affecting a whole subset of downstream class II genes (S. W. Lee et al., 1991). In fact, *p53* has been shown to exert its control over the cell cycle progression through a selective down-regulation of **proliferating cell nuclear antigen** mRNA and protein expression (Mercer et al., 1991). With respect to **clinical utility**, class II genes appear to be of particular interest, because they are not lost and their function may be restored or up regulated. Some of the genes differentially expressed in normal cells and absent in tumors may serve as useful diagnostic markers (S. W. Lee et al., 1991). Recent studies in mice made homozygous for the **null allele of the *p53* gene** suggest that the *p53* gene may not be essential for normal embryonic development, although its absence predisposes the animal to the development of neoplasias, but an oncogenic mutant form of *p53* is not obligatory for many types of tumors to emerge (Donehower et al., 1992).

Although not absolute, the frequent association of *p53* mutations with neoplasia has been proposed to be utilized for specific identification of malignancies by immunohistochemical detection of the *p53* products in diagnostic cytopathological evaluations (Hall et al., 1991).

1.1.6 GENOMIC IMPRINTING

Genomic imprinting (GI) is a common occurrence in mammals, including humans (Hall, 1990a,b). GI encompasses the disease phenomenology stemming from chromosomal abnor-malities such as translocations, inversions, and duplications only if these occur in chromosomes transmitted from either parent (the mother or the father) consistently manifested for a given genetic disorder. In other words, GI leads to modification of phenotype determined by a particular allele depending on the origin of the gamete (Sapienza, 1991). In certain cases what formerly had been interpreted as expression of autosomal recessive inheritance has now been traced to specific molecular events such as submicroscopic deletions, and the expression of the chromosomal defect that is imprintable can be expected only when it is transmitted from a parent of one sex but not the opposite sex (Hall, 1990a,b). One variant of imprinting is **uniparental disomy**; a striking example of it is the male-to-male transmission of hemophilia A, in which the male child inherits both the X and the Y chromosome from the father (references in Hall, 1990a,b).

In **maternal imprinting**, regardless of the sex of the offspring the affected gene is not phenotypically expressed. The male line of the family may carry the gene without expression and the offspring of the male line who inherits the genes will express it and manifest the phenotype. On the contrary, similar to the mother, her nonmanifesting daughter's children will not have the phenotypic expression (Hall, 1990a,b). Thus, it is the transmitting ("skipped"), nonmanifesting individuals who are the clue to the type of imprinting, maternal or paternal: in maternal imprinting, a male is nonmanifesting who transmits the genes to manifesting offspring, and vice versa, in paternal imprinting, females transmit the trait, being nonmanifesting carriers themselves.

Thus, **imprinting amounts to functionally turned-off segments of chromosomes** and appears to be a dominant function as reflected in the proposed nomenclature (Hall, 1990a, 1992). Examples of genomic imprinting include Beckwith-Wiedemann syndrome, narcolepsy (references in Hall, 1990a), embryonal rhabdomyosarcoma (Scrable et al., 1989), Wilms' tumor, and Prader-Willi and Angelman syndromes (Sapienza, 1991; Smeets et al., 1992). The adverse consequences of uniparental disomy have been demonstrated (Hall, 1990a, 1992), some of which may arise through balanced translocations (Smeets et al., 1992).

Genomic imprinting is viewed in the context of a larger phenomenon of dominance modification of phenotype (Sapienza, 1991). Analysis of experimental and clinical evidence in molecular oncology points to the critical role of **modifier genes** located on alleles at tumor suppressor loci (Sapienza, 1991). With respect to the possible molecular mechanisms accounting for GI as related to tumor phenotypes, the inactivation of an allele may not necessarily mean a true mutation event (Sapienza, 1991). The imprinting of the paternal tumor suppressor allele has been demonstrated in a number of tumors: Wilms' tumor, osteosarcoma, bilateral retinoblastoma, and embryonal rhabdomyosarcoma. As mentioned above, inactivation of an allele by GI reflects the activity of the modifier genes that generate the imprint (Sapienza, 1991). Furthermore, it appears that individuals vary in the way they imprint their genomes; in addition to some cells in an individual having no imprint, GI may appear, or be modified if already present, after fertilization (Sapienza, 1991). Variability of the expressivity of disease phenotypes such as observed in Huntington's disease, myotonic dystrophy, neurofibromatosis, and other conditions may be related to some of the above characteristics of GI, such as cellular mosaicism, in the expression of the affected allele (Sapienza, 1991). Moreover, different classes of modifiers of variegation of expressivity of disease phenotypes can be recognized, including those sensitive to the position of the alleles, to its dose, and to the cumulative effect of modifiers. Depending on whether the maternal or paternal disomy of chromosome 11 is inherited in the **Beckwith-Weidemann syndrome**, the phenotypic characteristics of this condition vary significantly (Henry et al., 1991). Nuclear transfer experiments reveal the potential for reversal of GI and genomic activation in differentiated somatic cells (DiBerardino, 1989). The importance of recognizing the phenomenon of GI for patient management is exemplified by the demonstration of erroneous prenatal evaluations based solely on chromosome counts. In some conditions, this may mask the uniparental disomy underlying disease phenotypes when both chromosomes of a pair come from either parent (Hall, 1992).

1.1.7 DNA METHYLATION

As a general rule, DNA in terminally differentiated somatic cells has a fixed methylation pattern reproducible after replication. On the other hand, the methylation pattern of genes expressing tissue-specific and housekeeping activities varies during development (Frank et al., 1991). **DNA methylation** has been implicated in serving as cell memory mechanisms for maintaining the high fidelity of the proper state of differentiation (Riggs, 1989). In transfection experiments, the process of specific gene demethylation was shown to be rapid, cell-type specific, and to involve a variety of **CpG island** sequences (Frank et al., 1991). CpG islands are known to occur in all cell types and mark the 5'-end of genes, so that methylation-free CpG islands occur in transcriptionally active DNA. m5C projects a methyl group in to the major groove of B-form DNA, thereby preventing the binding of transcription factors (Strobl, 1990). Conversely, activation of specific transcription sites during tissue differentiation or hormonal stimulation is accompanied by demethylation within the 5′ or 3′ regions of the activated gene. So far, no tissue- or cell-specific differences in the DNA methyltransferase gene activities have been discovered. Transcriptionally inactive satellite DNA has been shown to contain several times more **methyl-5-cytosine** (**m5C**) concentrated in CpG islands than bulk DNA and to be enriched in methylated CpT, TpC, and CpC dinucleotides (Strobl, 1990).

The **correlation between the extent of methylation and transcriptional inactivation** is not absolute and, for example, in the chicken lysozyme gene no such correlation could be observed in different tissues (Wolfl et al., 1991). This new evidence comes from experiments in which all methylated cytosines have been identified by sequencing rather than through the conventional technique of restricting the CpG sites by **methylation-sensitive endonucleases** (Saluz and Jost, 1989). As an example, the lysozyme gene was found to be hypomethylated in erythrocytes that are in a terminal differentiated state, in which chromatin is nontranscribable (Wolfl et al., 1991). On the other hand, following the level of methylation of the rat γ-crystallin gene at different developmental stages, a clear correlation between the extent of demethylation of the promoter and 5′ gene regions and the expression of the gene has been observed (Peek et al., 1991). Likewise, a positive correlation between site-specific methylation and prolactin gene activity has been found in human lymphoid cell lines (Gellersten and Kempf, 1990).

DNA methylation may exert its transcriptional inactivation influence indirectly through **methyl-CpG binding proteins** (e.g., Wang et al., 1986; Boyes and Bird, 1991). Even partial methylation of chromatin affecting the coding or noncoding DNA strand is sufficient to block expression of the hemimethylated chromatin (Deobagkar et al., 1990). The methylation/demethylation switch may also function at the protein level in repeatedly activatable cellular systems, for example, the G proteins as shown by their regulatory role in visual signal transduction (Perez-Sala et al., 1991).

As mentioned above, DNA methylation is one of the mechanisms of GI as dramatically shown by the dynamics of **parental-specific methylation** of an imprinted transgene during gametogenesis and embryogenesis (Chaillet et al., 1991). The **dynamic nature of methylation** apparently is involved in the inheritance of epigenetic defects such as produced by DNA damage leading to heritable abnormalities in gene expression potentially associated with carcinogenesis and aging (Holliday, 1987). **Sperm-specific methylation patterns** have also been demonstrated in the c-Ha-*ras*-1, insulin, and *RB* genes representing imprinting of the parental chromosomes (Ghazi et al., 1990). The dynamic pattern of methylation of the testis-specific genes has also been traced in the mouse male germ cell line, ranging from the completely unmethylated state of some genes to an undermethylated state in pachytene spermatocytes and round spermatids to **remethylation** at later stages of spermatogenesis (Ariel et al., 1991). As expected, a variable state of methylation has been observed in human testicular cancer detectable in hypoxanthine phosphorybosyltransferase and the phosphoglycerate kinase gene loci by *Hpa*II/*Msp*I analysis (Peltomaki, 1991). Interestingly, teratocarcinomas showed generalized **hypermethylation** and complete demethylation, suggesting that methylation plays a role in the development of testicular tumors of germ cell origin (Peltomaki, 1991), and in female somatic and embryonal carcinomas (Bartlett et al., 1991).

Evidence has been presented that **hypomethylation** also occurs in some forms of autoimmune diseases (systemic lupus erythematosis and rheumatoid arthritis) (Richardson et al., 1990). In fact, demethylation of regulatory genes is thought to be one of the possible mechanisms involved in cellular oncogene activation in autoimmune diseases (Sibbit, 1991).

Methylation has been shown to regulate the transforming potential of the Ha-*ras* gene (Borrello et al., 1987). A more direct approach to understanding the role of methylation in carcinogenesis has been taken in analyzing the activity of the human DNA methyltransferase gene which codes for the enzyme catalyzing DNA methylation (El-Deiry et al., 1991).

Interestingly, the expression of this gene is low in normal human cells, increased up to 50-fold in virally transformed cells, and markedly elevated (several 100-fold) in human cancer cells. Moreover, an increase in the DNA methyltransferase gene expression precedes the development of colonic neoplasia. The level of this gene transcript in premalignant lesions is also markedly increased but less so than in cancers (El-Deiry et al., 1991). It thus appears that determination of the degree of methylation of specific genes may be useful in disease evaluation for patient management.

A discussion of current theories of carcinogenesis is beyond the scope of this book. Numerous reviews analyzing various aspects of cancer research, the products of which progressively find their way into cancer diagnostics and patient management, appear regularly (e.g., Beresford, 1988; Bowden, 1990; Cooper and Stratton, 1991; Crist and Kun, 1991; Duesberg, 1987; Flood et al., 1988; Freeman et al., 1990; Harris, 1986; Hollingsworth and Lee, 1991; Israel, 1990; Loeb, 1989; Mulvihill, 1989; Nowell, 1991; Pierotti and Della Porta, 1989; Pitot and Dragan, 1991; Preston-Martin et al., 1990; Renan, 1990; Stanbridge and Nowell, 1990; C. J. Thiele, 1990; Weinstein, 1988; Woll, 1991). Major emphasis has been placed on deciphering the molecular pathology of the metastatic process (e.g., Fidler, 1990; Killion and Fidler, 1989; Mareel et al., 1991; Vleminckx et al., 1991). The successes of cancer research can be translated into improved cancer diagnosis and patient management (e.g., Fletcher et al., 1991; Trent, 1989; Wyke, 1990). Several excellent books dedicated to the molecular biological approach to diagnosis have appeared and are highly recommended (e.g., Cossman, 1990; Fenoglio-Preiser and Willman, 1991; Groopman and Skipper, 1991; Lindsten and Pettersson, 1991; Lynch and Tautu, 1991; Orlic, 1989; Peschle, 1987; Rowland et al., 1989; Sorg, 1990).

1.1.8 MITOCHONDRIAL DNA

Evaluation of **mitochondrial DNA (mtDNA)** appears to be acquiring a progressively larger role in the diagnosis of human disease. The best known examples are a base change in mtDNA associated with **Leber's hereditary optic neuropathy** (Singh et al., 1989; Wallace et al., 1988) and large deletions seen in patients with various encephalomyelopathies such as **Kearns-Sayre syndrome** (e.g., Holt et al., 1988a,b; Ozawa et al., 1988; Zeviani et al., 1990). A whole range of diseases traceable to alterations in mtDNA has been uncovered (e.g., Rowland et al., 1989).

The mitochondrial genome is located in the cytoplasm and is almost exclusively maternally transmitted, with the occasional paternal transmission of phenotypic abnormalities explainable on the basis of the assumption that certain enzyme subunits may be encoded by nuclear DNA and thereby become affected (Egger and Wilson, 1983). The molecular characterization of human mtDNA has been accomplished relatively recently (Anderson et al., 1981). The circular mtDNA molecule is transcribed and the posttranscriptional processing is performed with the help of proteins and RNA molecules encoded by the nuclear DNA (Lestienne and Norby, 1990; Wallace, 1991).

Determining the molecular details of mtDNA mutations affecting mitochondrial gene expression is an area of intensive and promising research. Major insights have been gained already, not only in the molecular pathology of neuromuscular disorders, ophthalmoplegia and encephalopathies (e.g., Grossman, 1990; Hammans et al., 1991; Hess et al., 1991; Holt et al., 1990; Lauber et al., 1991; Nakase et al., 1990; Norby et al., 1991; Sahashi et al., 1990; Schapira et al., 1990a,b; Wallace, 1991), and cardiomyopathies (e.g., M. Tanaka et al., 1990), but also in various aspects of endurance training (e.g., Dionne et al., 1991), aging and degenerative diseases (e.g., Linnane et al., 1989, 1990), and liver diseases (Berrez et al., 1991). The complexity of unraveling details of mitochondrially related disorders is due to the impaired nuclear control over mtDNA replication or transcription superimposed on intrinsically mitochondrial derangements (e.g., Zeviani et al., 1990).

It appears that **mitochondrial tRNA** genes are the hot spots for point mutations causing neuromuscular diseases (Lauber et al., 1991). Precise identification of molecular defects is important in mtDNA-related disorders. For example, in the Kearns-Sayre syndrome the deletions causing the condition may vary in length and position in the mitochondrial genome between affected individuals, but be similar in different tissues of the same person, indicating that they arise during or before embryogenesis (L. Nelson et al., 1989).

mtDNA defects can result from pharmacological therapies, such as in zidovudine-induced myopathy in (AIDS) patients (Arnaudo et al., 1991) and in mitochondrially inherited susceptibility to aminoglycoside antibiotic-induced ototoxicity (Hu et al., 1991). The importance of better understanding the **mtDNA-nuclear DNA regulatory interactions** is essential not only for the precise diagnosis of mtDNA-related conditions, but is also emphasized by the recently discovered mediation of the effects of certain drugs via this interaction. Thus, the selective inhibition of the expression of mitochondrial genes by interferon appears to be mediated by an interferon-responsive nuclear gene that encodes a product that in turn regulates mitochondrial gene expression (Shan et al., 1990).

Analysis of mtDNA for clinical diagnostic purposes can be performed following isolation by simplified procedures (Wiesner et al., 1991). The isolated mtDNA can be subjected to restriction digestion, Southern analysis, or evaluated by PCR-based assays (e.g., Lestienne et al., 1990; Norby et al., 1991). The use of **confocal scanning laser microscopy** for evaluation of mtDNA has also been proposed (Ruiters et al., 1991).

A Survey of Newer Gene Probing Techniques

1.2.1 INTRODUCTION

The entire spectrum of diagnostic challenges, ranging from **identification of pathogens** and the recognition of **distinction between norm and pathology** to tracing **genealogy** of individuals and their **identity characteristics** as well as definitive identification of the source of the biological material in forensic practice, has come to rely on the unique and fundamental nature of information contained in the genome.

The genomic chemistry of human pathogens is sufficiently less complex than the human DNA to qualify for the definition of being a relatively simpler analytical target. Establishing the distinction between normal and neoplastic tissue or that between two individuals presents a formidable challenge. In fact, since most of the **genomic DNA sequences** are shared among humans, the importance of useful genetic markers in meeting these challenges cannot be overestimated. Until relatively recently almost all **polymorphic markers** were limited to gene products identified by electrophoretic or serologic techniques. Molecular biological technology now allows the polymorphisms to be defined at the level of the genome and used for diagnostic purposes.

In this chapter, a brief discussion will be given of **DNA variability** (**polymorphism**) and the newer approaches to analyze and use this unique molecular genetic information in linkage studies, parentage testing, and forensic practice. A large number of publications referred to throughout this discussion are available and deal with this subject at various levels of detail and sophistication; the reader should consult these for a more detailed coverage.

1.2.2 DETECTION OF KNOWN MUTATIONS

Theoretically speaking, gene level diagnosis can be at least tentatively established by existing molecular biological techniques for over 500 genetic disorders for which chromosomal assignments have been determined (Cooper and Schmidtke, 1989; McKusick, 1991) and the database of established DNA alterations associated with human disease is being constantly updated. Furthermore, since **over 5000 genes have been defined** at least in some detail, direct evaluation of these for clinical diagnosis is also possible. The search for disease susceptibility genes proceeds along different routes, including the epidemiologic approach (Khoury et al., 1990). The pace of introduction of gene level diagnosis in real medical practice, however, lags behind for a number of reasons. Nevertheless, the enormous speed of accumulation of practically usable information on human genes associated with diseases makes the introduction of gene level diagnosis in routine clinical practice technologically a realistic possibility in the near future.

Some of the methods already in use or being developed for the characterization of the human genome at the level of physical and genetic mapping, and certainly the determination of specific nucleotide sequences, will undoubtedly find their way into the practice of molecular diagnostics.

For simplicity of presentation all the methods briefly described here are grouped into (1) those primarily utilized for the detection of known alterations of DNA, (2) methods for scanning DNA when the target alteration or mutation has not been defined, (3) techniques employed in linkage analysis, and (4) approaches currently used predominantly in gene-mapping studies, but which may eventually find application in clinical practice. These distinctions are not absolute, and a given method may be used for diagnostic purposes either in its present format, or modified, to suit other applications. Several reviews of these techniques have appeared (e.g., Barrell, 1991; Billings et al., 1991; McKusick, 1991; Rose, 1991; Rossiter and Caskey, 1990, 1991).

New developments in the methodology of gene analysis have been increasing almost exponentially in number and versatility of their analytic potential. This discussion gives only a brief survey of representative technologies. For more detailed coverage numerous reviews and monographs should be consulted (e.g., Donnis-Keller, 1991; Keller and Manak, 1989; Kriegler, 1990; Lebo et al., 1990a; Peter, 1991; Rose, 1991; D. Thiele, 1990; White, 1991; Wright and Wynford-Thomas, 1990).

Technology for the **detection of variations in DNA structure** is undergoing a continuing evolution toward progressively higher degrees of precision, simplification, automation, and informative content. With the introduction of the **polymerase chain reaction (PCR)**, this and potentially other gene amplification methods have revolutionized the very concept of the molecular diagnostic field. Because numerous excellent publications fully dedicated to the methodological aspects of gene-mapping strategies are available (e.g., Donnis-Keller, 1991; Kirby, 1990), this chapter will briefly emphasize the more recent developments, selected only to illustrate a panorama of approaches, some of which will soon undoubtedly take their place in the clinical applications of gene technology.

1.2.2.1 SOUTHERN AND NORTHERN BLOTTING

DNA sequence alterations associated with diseases can be produced by either focal **single-base modifications**, or through **deletions, duplications, insertions, translocation, methylation**, and so on. One established technique that is capable of identifying such DNA alterations is **Southern blotting** (Southern, 1975), which reveals relatively major sequence differences detectable by **restriction endonuclease digestion**. The classic example of this approach is the detection of the β sickle globin allele responsible for sickle cell anemia (Geever et al., 1981). This approach is the basis for the **restriction fragment length polymorphism (RFLP)** analysis as well. If transcription is not affected, the detection of an altered gene can be done by identifying and studying the mRNA produced using **Northern analysis**.

Recent modifications of **nucleic acid extractions** yielding high molecular weight DNA and RNA transcripts up to 10 kb in length (Raha et al., 1990), the use of **glass powder** suspension for the extraction of RNA or DNA, simplifying the procedures for PCR amplifications (Yamada et al., 1990), as well as **automated methods for DNA extraction** from clinical specimens (Taylor et al., 1990) are all bringing recombinant DNA technology closer to the clinical diagnostic laboratory. In fact, attempts are being made to develop an automated chemical system similar to Southern analysis based on **solution hybridization** and **solid-phase capture chemistry** for specific gene detection (Mayrand et al., 1990). An even further simplified procedure, one that does not require extraction of genomic DNA (Schwartz et al., 1990), allows the detection of specific mutations by **PCR** *directly from dried blood spots* on Guthrie cards.

1.2.2.2 ALLELE-SPECIFIC OLIGONUCLEOTIDE PROBES

In order to detect changes in DNA sequences at a single-base level, several other techniques have been introduced. **Allele-specific oligonucleotide (ASO) probes** form a duplex with the nucleic acid to be tested, which is extremely unstable, even if only a single nucleotide mismatch occurs (Conner et al., 1983). The absence of a duplex in the expected position signals the presence of a mutation. The efficiency of this approach was initially demonstrated by the detection of sickle cell β-globin allele under appropriate hybridization conditions by 19-base-long oligonucleotides Hβ19A′ and Hβ19S. The ASO technique for sickle cell trait determination, although similar to analysis by restriction digestion with *Mst*II, proved to be better in distinguishing the β^c from the β^s allele, which could not be recognized by *Mst*II. This ASO technique can generally be used for detecting point mutations that do not create a recognition site for restriction enzyme digestion, provided an appropriate oligonucleotide probe is constructed.

A modified technique, the **reverse ASO**, offers the benefit of easier standardization of testing by manufacturing uniform lots of immobilized oligonucleotides to which the unknown material is added (Saiki et al., 1989). Besides, in this format many alleles can be screened at the same time. Further enhancement of the selectivity of the ASO probe, and thus the specificity of the assay, can be achieved by **oligonucleotide competition** and an **increase in the stringency** of the hybridization and washing conditions (Nozari et al., 1986; Wu et al., 1989).

1.2.2.3 GENOMIC SUBTRACTION

This technique, which isolates the DNA that is absent from deletion mutants, subtracts the DNA present in both the wild-type and the deletion mutant genomes from a mixture of the wild-type and mutant DNA (Straus and Ausubel, 1990). The mutant DNA is biotinylated, allowed to reassociate with the wild-type DNA, and the hybrids are then "subtracted" from the mixture by capturing them on avidin via the biotin attached to the mutant DNA. The DNA not present in the deletion mutant is thereby left behind in the solution, and after several cycles of removal of the DNA corresponding to the mutant DNA, the remaining unbound DNA, representing that DNA that is absent in the deletion mutant, is amplified by PCR.

The potential clinical applications of this approach include the detection of pathogen DNA in infected tissue; it can also be used for the isolation of various disease genes as well (Straus and Ausubel, 1990). An appealing procedure using a **subtraction hybridization** with photoactivatable biotin, streptavidin binding, and organic extraction for solution hybridization has been proposed for the analysis of mammalian genomic DNA (Barr and Emanuel, 1990). This technique can also be applied to **deletion cloning**.

1.2.2.4 ALLELE-SPECIFIC PCR

As an extension of the earlier developed ASO hybridization assays, **allele-specific PCR (AS-PCR)** allows for direct determination of the genotype by the presence or absence of an amplified fragment specific for the allele of interest (Wu et al., 1989). Using nonradioactive primers for the normal and the sickle cell β globin alleles in genomic DNA, the sickle cell mutation could be detected without hybridization, ligation, or restriction enzyme cleavage. Two 14-mer allele-specific primers, H β 14S and H β 14A, complementary to the 5′ end of the sickle cell and normal β globin genes were synthesized. The two primers differed from each other by a single nucleotide at the 3′ end — a 3′-T in the sickle cell primer and a 3′-A in the normal one. One additional primer, BGP2, complementary to the opposite strand 3′ of the allele-specific primers, was used as the second primer for PCR. The authors propose a dual labeling of the primers — one labeled with a fluorescent label and the other with biotin. The AS-PCR product then can be captured on a streptavidin-agarose column or in titration plate wells, while the fluorescent label may be used for quantitating the amount of the amplified fragment (Wu et al., 1989). Another widely used technique is based on the introduction of artificially created restriction sites by **site directed mutagenesis** in the PCR amplification step, as utilized, for example, for the diagnosis of **phenylketonuria** (Eiken et al., 1991).

Direct haplotyping is possible using AS-PCR without the need for pedigree studies when allelic sequences at the polymorphic priming site are known (Ruano and Kidd, 1989). This was demonstrated at two polymorphic β-globin loci. One locus was identified by the presence or absence of a 630-bp band following AS-PCR with a specific set of primers. The other locus was identified following *Taq*I digestion of the amplified product. Other modifications of this procedure subsequently were reported employing **single-molecule dilution** (**SMD**) (Ruano et al., 1990) and the so-called "**booster**" **PCR** (Ruano et al., 1989).

A more recent elaboration of this approach offering a faster procedure utilizes pairs of allele-specific PCR primers to amplify differentially each haplotype (Sarkar and Sommer, 1991). In this procedure, termed **double PCR amplification of specific alleles** (**PASA**), the detection of mutant alleles is based on the finding that primers mismatched within two bases of their 3′ ends are capable of unequivocally distinguishing two different alleles by the presence or absence of the amplification products in agarose gel electrophoresis (Sarkar et al., 1990a,b; Sommer et al., 1989). If the desired allele is present, the allele-specific primer gives rise to an abundance of the amplified product, whereas if a different allele is present there is no amplification of the segment tested for (Sarkar and Sommer, 1991).

The PASA approach has been tested in the detection of mutations responsible for hemophilia B (Bottema et al., 1990), and phenylketonuria (Sommer et al., 1989). The introduction of 5% formamide into the amplification mixture markedly enhanced the specificity of PCR (Sarkar et al., 1990b). A significant increase in the sensitivity of detection of mutations can be achieved using the technique of **whole-genome PCR**, whereby total human genomic DNA is cleaved into fragments only several hundred base pairs long and ligated to the so-called "**catch-linkers**". These serve as primers for PCR amplification that produces DNA fragments for specific enrichment and selection of the sequences of interest (Kinzler and Vogelstein, 1989).

A practically important assay has been recently introduced using AS-PCR from **Guthrie cards** for the common mutation at position 985 of the **medium-chain acyl-CoA dehydrogenase** (**MCAD**) **gene**, designated K329E, that is linked to **sudden infant death syndrome** and a condition resembling **Reye's syndrome** (Matsubara et al., 1991). Using this assay for presymptomatic diagnosis on newborns, appropriate dietary management may be instituted to prevent these fatal conditions, and widespread screening for the **K329E mutations** is advocated in populations with a relativley high prevalence of it. In the general population the K329E mutation frequency is 1 in 68, making it as common as **phenylketonuria** for which mass screening programs have been adopted in many countries (Matsubara et al., 1991).

1.2.2.5 COMPETITIVE OLIGONUCLEOTIDE PRIMING-PCR

By analogy, a recently developed technique uses a **mixture of primers**, which compete for the binding site on the template sequence so that the highest degree of complementarity would exclude the mismatched primer. This technique, called **competitive oligonucleotide priming-PCR** (**COP-PCR**), is used for specific allele recognition where either competing radiolabeled (Ballabio et al., 1990; Gibbs et al., 1989a) or fluorescently tagged (Chehab and Kan, 1989) primers identify the tested mutation.

1.2.2.6 OLIGONUCLEOTIDE LIGATION ASSAY

Another established technique requiring the knowledge of the target sequence is the **oligonucleotide ligation assay** (**OLA**) (Landegren et al., 1988). Two synthetic oligonucleotides are joined head to tail over the target sequence by the enzyme **DNA ligase**. If a mismatch occurs due to alteration (mutation) of nucleotides in the target sequence at the junction region the ligation fails to occur.

A dramatic enhancement of the specificity of the detection of point mutations using a dot-blot format can be achieved using ligation by **T4 DNA ligase** of the adjacently hybridizing oligonucleotide probes (Alves and Carr, 1988). The described technique which calls for immersion of gels with fractionated target DNA into ligase solution, markedly simplifies the detection of point mutations as an alternative to amplification of target DNA.

1.2.2.7 LIGATION AMPLIFICATION REACTION

This method of **allele-specific detection** utilizes the ligation of oligonucleotide pairs that are complementary to adjacent sites on the template DNA (Wu and Wallace, 1989a,b). The bacteriophage T4 DNA ligase is capable of joining adjacent oligonucleotide pairs annealed to a complementary DNA sequence in a very specific fashion (Wu and Wallace, 1989a). The presence of an altered base in the template DNA at the point of juncture of the two oligonucleotides markedly reduces the efficiency of ligation by this enzyme (see also Landegren et al., 1988; Alves and Carr, 1988). The presence of 200 mM NaCl and/or 2 to 5 mM spermidine further suppresses the ligation at the junctional mismatch site.

Based on this property of the T4 DNA ligase, which in addition to being a method for the detection of point mutations is also a novel amplification procedure, the **ligation amplification reaction (LAR)** has been developed (Wu and Wallace, 1989a,b). When a single pair of oligonucleotides is used the amplification increases in a linear fashion. An **exponential increase** in the LAR product is obtained when two pairs of oligonucleotides, one complementary to the upper strand and one to the lower strand of a target sequence, are present (Wu and Wallace, 1989a,b). The products of LAR serve as templates for subsequent rounds of ligation feeding into the exponential growth of the product yield much like that in PCR.

1.2.2.8 LIGASE DETECTION REACTION AND LIGASE CHAIN REACTION

Another addition to amplification methods for DNA analysis is the use of **thermostable DNA ligase**, which both amplifies DNA and discriminates a **single-base substitution** (Barany, 1991). Thermostability of the enzyme allows for the reaction to linearly increase the product. The reaction is composed of two parts—one is the **ligase detection reaction (LDR)**, the product of which is then further amplified in a **ligase chain reaction (LCR)**, in which both strands of genomic DNA are used as targets for oligonucleotide hybridization. Similar to PCR, the ligation products from one round of ligation become templates for another round and the amount of product exponentially increases when repeated thermal cycling is maintained. The diagnostic utility of this assay lies in the fact that the **ligation/amplification cycle** is interrupted by the occurrence of even a single-base mismatch as was demonstrated in comparison of normal β^A and sickle β^S-globin genotypes from 10-μL blood samples (Barany, 1991) (Figure 1.2).

1.2.2.9 TRANSCRIPTION-BASED AMPLIFICATION SYSTEM

Transcription-based amplification system (TAS) is primary designed to produce RNA copies of a target DNA or RNA sequence (Kwoh et al., 1989). First, short nucleotide sequences, the **polymerase-binding sequences (PBSs)**, are recognized and positioned on the 3′ side of the target sequence by a DNA-dependent RNA polymerase. To achieve this, an oligonucleotide primer containing two domains is used. Depending on the target sequences to be amplified either a reverse transcriptase or a DNA polymerase is employed in a **primer-extension reaction**. A second oligonucleotide primer following a primer-extension reaction complements the product of the first primer-extension reaction to form a double-stranded PBS-containing cDNA copy of the target sequence (Figure 1.3). A number of RNA polymerases (**T7**, **T3**, or **SP6**) can be used to recognize the PBS domains.

One of the advantages of the TAS is in the efficiency of the transcription step, which can produce from 10 to 1000 RNA copies per DNA template. A large-scale amplification of target nucleic acid sequences can be accomplished by a relatively few cycles of amplification, because the RNA produced during the first TAS cycle serves as target for subsequent TAS cycles, similar to the situation in PCR.

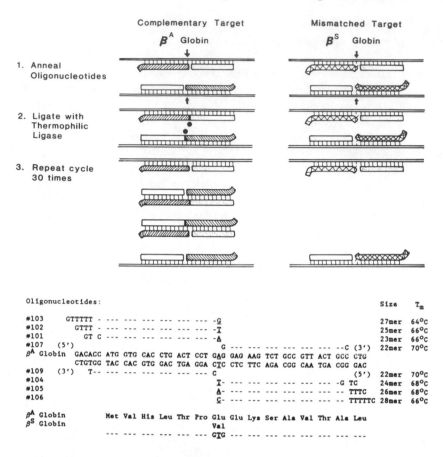

FIGURE 1.2. Diagram and example of DNA amplification/detection using LCR. *Upper*: DNA is heat denatured, and four complementary oligonucleotides are hybridized to their target at a temperature near their melting temperature (65°C; t_m). Thermostable ligase will covalently attach only adjacent oligonucleotides that are perfectly complemetary to the target (*left*). Products from one round of ligations become targets for the next round, and thus products increase exponentially. Oligonucleotides containing a single-base mismatch at the junction do not ligate efficiently and, therefore, do not amplify the product (*right*). *Lower*: Nucleotide sequence and corresponding translated sequence of the oligonucleotides used in detecting β^A- and β^S-globin genes. Oligonucleotides 101 and 104 detect the β^A target, whereas oligonucleotides 102 and 105 detect the β^S target when ligated to labeled oligonucleotides 107 and 109, respectively. Oligonucleotides 103 and 106 were designed to assay the efficiency of ligation of G-T or G-A and C-A or C-T mismatches when using β^A- or β^S-globin gene targets, respectively. Oligonucleotides have calculated t_m values of 66–70°C, just at or slightly above ligation temperature. The diagnostic oligonucleotides (101-106) contained slightly different length tails to facilitate discrimination of various products when separated on polyacrylamide denaturing gel. (From Barany, F. (1991). Proc. Natl. Acad. Sci. U.S.A. *88*:189–193. With permission.)

In addition to the fewer cycles of amplification required to achieve a large-scale increase in copy number (10^6 copies produced in 4 TAS cycles), the product of the TAS is in part **single-stranded RNA**, which does not have to be denatured prior to detection by hybridization. An adaptation of the TAS protocol to the detection of human immunodeficiency virus type 1 (HIV-1) RNA with the identification of the reaction products by a bead-based sandwich hybridization has been described (Kwoh et al., 1989).

1.2.2.10 SELF-SUSTAINED SEQUENCE REPLICATION
While the TAS system requires temperature shifts for optimal reaction rates at the DNA synthesis and RNA transcription steps, a modification of this system using three different enzymes is an **isothermal process**. The **self-sustained sequence replication (3SR) system** in essence reproduces the strategy employed by retroviral replication (Guatelli et al., 1990). The

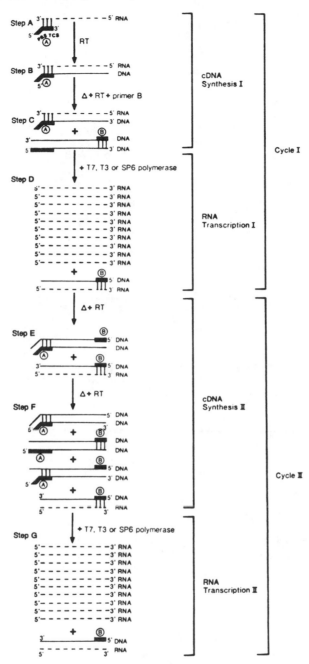

FIGURE 1.3. Transcription-based amplification system (TAS). A two-cycle scheme for amplifying an RNA target sequence (- - -) using sequential cDNA synthesis and RNA transcription is displayed as seven steps. A target nucleic acid molecule RNA (or denatured DNA) is hybridized to a primer oligonucleotide (primer A) that contains a polymerase-binding site (PBS) (for T7, T3, or SP6 polymerase) and a target complementary sequence (TCS) (step A). Reverse transcriptase (RT) elongates primer A to yield a newly synthesized DNA strand complementary to the target RNA (step B). The RNA·DNA heteroduplex is denatured by heat (Δ) and oligonucleotide B is annealed to the newly synthesized DNA strand containing the PBS. Reverse transcriptase is added to produce a double-stranded cDNA and a new DNA·RNA heteroduplex (step C). Incubation of the double-stranded cDNA with T7 (T3 or SP6) RNA polymerase results in the synthesis of multiple RNA transcripts from the PBS-containing double-stranded DNA template (step D). Some of this RNA is immediately converted to RNA·DNA heteroduplex by RT (still in the reaction mixture from step C) using oligonucleotide B as a primer. Further amplification of target sequences can be obtained by a second cycle of cDNA synthesis (steps E and F) and RNA transcription (step G). (From Kwoh, D. Y. et al. (1989). Proc. Natl. Acad Sci. U.S.A. *86*:1173–1177. By permission of the publisher.)

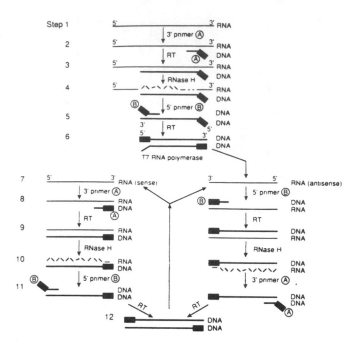

FIGURE 1.4. Self-sustained sequence replication (3SR). The 3SR reaction depends on a continuous cycle of reverse transcription and transcription reactions to replicate an RNA target by means of cDNA intermediates. Oligonucleotides A and B prime DNA synthesis and encode the promoter sequence for the T7 RNA polymerase (black boxes). Steps 1 through 6 depict the synthesis of a double-stranded cDNA, which is a transcription template for T7 RNA polymerase. Complete cDNA synthesis is dependent on the digestion of the RNA in the intermediate RNA·DNA hybrid (step 4) by RNase H. Transcription-competent cDNAs yield antisense RNA copies of the original target (step 7, right). These transcripts are converted to cDNAs containing double-stranded promoters on both ends in an inverted repeat orientation (steps 7 through 12). These cDNAs can yield either sense or antisense RNAs, which can reenter the cycle. Thin lines, RNA; thick lines, DNA; RT, reverse transcription. (From Guatelli, J. C. et al. (1990). Proc. Natl. Acad. Sci. U.S.A. *87*:1874–1878. With permission .)

three enzymes carrying out the cDNA synthesis and RNA transcription simultaneously are **avian myeloblastosis virus (AMV) reverse transcriptase**, *Escherichia coli* **RNase H** that dissociates RNA-DNA duplexes, and **T7 RNA polymerase** (Figure 1.4). Each priming nucleotide contains in its sequence a region for T7 RNA polymerase binding, an area for the preferred transcriptional initiation, and a sequence complementary to the target sequence.

The main application of 3SR is best suited for the specific detection and amplification of RNA sequences. In contrast to both PCR and TAS, the *in vitro* 3SR reaction operates at 37°C and, consequently, denaturation of double-stranded DNA cannot be achieved and, therefore, it cannot serve as the target template. On the other hand, the 3SR system would be ideal for the evaluation of transcriptional activity of specific genes. A predenaturation step can be included for amplification of DNA templates by 3SR with the same efficiency.

The efficiency of 3SR appears to be higher than that of PCR, because a 10^5-fold amplification has been observed within 15 min of 3SR, and, theoretically, can be achieved by PCR within 85 min under ideal reaction conditions.

Another technique for the amplification of broad classes of cDNA, using T7 RNA polymerase, has been described in application to the analysis of neural gene expression (Van Gelder et al., 1990).

1.2.2.11 QB AMPLIFICATION

QB amplification is one of the first systems introduced for **amplification of the probe** rather than the target. The probe contains RNA bacteriophage replication sequences and an enzyme "replicase", the RNA-dependent RNA polymerase from the **bacteriophage QB** (Chu et al., 1986). QB probes are constructed to contain two components — a portion of the QB phage RNA (MDV RNA) joined to a sequence complementary to target RNA or target DNA. Following hybridization of QB probes to the target, the formed hybrids are purified from the unhybridized probe and QB replicase is added.

QB amplification achieves a billion-fold increase of the probe reporter in 30 min without temperature cycling (Lizardi et al., 1988). *Specificity of QB amplification cannot be determined from the amplified product*, however, and confirmation of positive results must be performed. One suggested combination for clinical use would be screening by QB amplification followed by confirmation with PCR (Gillespie, 1990).

1.2.3 DETECTION OF UNDEFINED DNA ALTERATIONS: SCANNING FOR NEW MUTATIONS

1.2.3.1 RNASE PROTECTION ASSAY

While ASO and OLA require the knowledge of expected alteration in the targeted nucleic acid sequence, other methods have been developed in which unknown single-base alterations in heteroduplexes due to point mutations have been identified by cleaving DNA with the **single strand-specific S1 nuclease** (Shenk et al., 1975). Single-base mismatches in **RNase protection assays** (RPA) have been recognized by cleaving RNA at the point of mismatch by **RNase A** in RNA-DNA heteroduplexes (Myers et al., 1985a,b), or in RNA-RNA duplexes (Winter et al., 1985; Gibbs and Caskey, 1987). Only about one third of all possible single-base mutations can be detected by RPAs (Gibbs and Caskey, 1987). The efficiency of mismatch detection can be almost doubled by reversing the procedure and labeling the mutant RNA instead of the wild-type RNA or DNA (Gibbs and Caskey, 1987).

1.2.3.2 CHEMICAL CLEAVAGE METHODS

Chemicals such as **carbodiimide** were used to cleave single-base pair mismatches in DNA-DNA heteroduplexes (Novack et al., 1986). The **chemical mismatch cleavage** method utilizing **hydroxylamine** and **osmium tetroxide** enjoyed wider application (Cotton et al., 1988). **Heteroduplex DNA** containing mismatches is first incubated with osmium tetroxide to detect T and C mismatches, or with hydroxylamine to detect C mismatches, and then exposed to **piperidine** to cleave the DNA at the modified mismatched base. This methodology is suitable for screening all possible mutational changes and has been used for detecting **point mutations** in hemophilia B (Montandon et al., 1989), in the *p53* gene in lung cancer (Curiel et al., 1990) and in human ornithine transcarbamylase gene (Grompe et al., 1989). One of the criticisms directed at the chemical cleavage mismatch method has been that it may fail to detect T-T mismatches in certain sequences (Bhattacharyya and Lilley, 1989). No such failure, however, has been detected in the screening for the *p53* gene point mutations (Curiel et al., 1990).

1.2.3.3 NUCLEIC ACID ALTERATIONS DETECTED BY ELECTROPHORETIC MOBILITY

1.2.3.3.1 Denaturing Gradient Gel Electrophoresis

The resolution of DNA in gel electrophoresis relative to a number of variables such as gel concentration, electric field, and other experimental conditions has been thoroughly analyzed (e.g., Epplen, 1991; Rickwood and Hames, 1982; Slater et al., 1989; Slater and Noolandi, 1989). Various modifications of gel electrophoresis (Lane et al., 1990), **particularly denaturing gradient gel electrophoresis (DGGE)**, have been used to characterize DNA molecules. DGGE is capable of resolving DNA molecules differing by even a single base. The method is based on **differential melting** of heteroduplexes compared with that of homoduplexes. **Heteroduplexes** are formed when one mutant and one wild-type DNA (or RNA) strand associated in a duplex, whereas **homoduplexes** are formed by wild-type strands. Generally, the conformation of a DNA molecule is determined by the relative abundance of specific nucleotides (Fischer and Lerman, 1983; Myers et al., 1985a,b, 1987).

The **strength of bonding** of GC pairs is greater than that of AT, and under the same denaturing conditions (e.g., urea, formamide, or temperature) duplexes enriched in GC will retain their conformation longer than AT-rich regions. This **dissociation of a double helix** under denaturing conditions, such as in a gradient of denaturants in electrophoretic gels, will result in an **abrupt reduction of migration**. Mismatches or substitutions found in heteroduplexes reduce the strength of complementary strand bonding, which can be compared under denaturing electrophoretic conditions to that of the wild-type DNA (Fischer and Lerman, 1983; Myers et al., 1985a,b, 1987). Using reversible cross-linkers DGGE has been made more convenient for the subsequent Southern transfer of genomic DNA fragments to be screened with specific probes (Borrensen et al., 1988). This powerful technique has been used, for example, to identify polymorphisms useful in genetic linkage analysis in application to the proximal region of chromosome 21 (Burmeister et al., 1991).

DNA strands separate first in the low-melting domains, thus altering migration of the entire molecule, which may contain a mutation in the high-melting domains. This may lead to loss of detection of DNA alteration in the high-melting domains due to interference from the low-melting domains. To induce a more gradual and complete melting of the target sequences they are modified by the addition of GC sequences known as **GC-clamps**, which can be easily added by PCR using primers carrying a GC-rich sequence at the 5′ end (Sheffield et al., 1989). In this modification, variations in the melting profile of the entire sequence can be evaluated. An example of the application of DGGE is the rapid detection of latent carriers of a subtype of acute intermittent porphyria that cannot be diagnosed by porphobilinogen deaminase assays (Bourgeois et al., 1992).

Other denaturing agents, such as temperature, can be used to reveal variations in nucleic acid duplexes reflected in their electrophoretic mobility that led to the development of **temperature gradient gel electrophoresis** (Riesner et al., 1988; Rosenbaum and Riesner, 1987; Wartell et al., 1990). The utility of these techniques for clinical diagnostic applications at this time appears to be limited.

1.2.3.3.2 Combination of Amplification and Single-Strand Conformation Polymorphism Analysis

The electrophoretic mobility of single-stranded nucleic acids in nondenaturing polyacrylamide gels depends not only on size but also on sequence characteristics, being sensitive to even single-base substitutions (Orita et al., 1989a). In the initial version of this technique, sample DNA was cut by restriction enzymes, denatured, separated by polyacrylamide gel electrophoresis under nondenaturing conditions followed by conventional Southern blotting and hybridization to a probe. It is argued that under nondenaturing conditions the single-stranded (ss) DNA conformation is stabilized by interstrand interactions that influence DNA mobility in the gel. Therefore, the conformation reflected in altered mobility is determined, in turn, by changes in the sequence, justifying the term **single-strand conformation polymorphism** (SSCP).

A combination of SSCP with PCR amplification of a target sequence (**PCR-SSCP**) allows for rapid and sensitive detection of most sequence changes, including single-base substitutions, without the need for restriction enzyme digestion, blotting, and hybridization to probes (Orita et al., 1989b). The PCR-SSCP technique requires that the target sequences be known in order to design primers, and can be used for the detection of oncogene activation, prenatal evaluation for the presence or absence of particular alleles, or in linkage analysis for a known sequence. Applicability of this method has been shown in the analysis of the *Alu* repeats and the *ras* gene. In some cases, however, spurious results were observed when primers for the *ras* gene were located in exon sequences and in the analysis of *Alu* repeats in the β-tubulin locus. Another drawback of PCR-SSCP is that the **effect of sequence change on electrophoretic mobility cannot be predicted**, and some sequence alterations may not appreciably affect the mobility. In certain instances, however, PCR-SSCP allows for rapid and simple screening of mutations.

Further development of this technique was designed to isolate a single sequence variant from a mixture of two or more alleles (Suzuki et al., 1991). In practice the products of the PCR amplification frequently contain very low levels of nucleic acids carrying the mutations of interest among fragments of normal and mutated alleles present in various ratios. The direct sequencing approach in such cases is not feasible. Thus, mutated alleles first must be isolated. Furthermore, the task becomes even more complicated when one allele carries a double mutation. A combination of PCR-SSCP with separation of the strands using the Pharmacia Phastgel system allows a simplified, nonradioactive method for the detection of single-base mutations as demonstrated in the identification of a rare variant of Tay-Sachs disease (Ainsworth et al., 1991).

A multistage procedure designed to specifically identify a mutated sequence in a mixture of amplified fragments involves the use of PCR-SSCP for amplification and separation of each sequence variant by gel electrophoresis, elution of separated ss DNA fragments and asymmetric amplification of each sequence species, followed by direct sequencing. The efficiency of this approach was demonstrated by the detection (in a surgically excised specimen of lung cancer tissue) of a point mutation in exon 1 of the Ki-*ras*2 gene that could not be identified by the conventional method of direct sequencing of the genomic DNA (Suzuki et al., 1991). In addition, it allows for the successful identification of the sequence of a minor constituent present in a mixture at a concentration below 3%. This compares favorably with the lower limit of sequence determination when the direct method is used, which requires the minor constituent to be present at a concentration of 25% (Suzuki et al., 1991). Successful application of PCR-SSCP to donor-recipient bone-marrow matching relied on amplification of the second exon of an HLA class II gene followed by electrophoresis in ion-denaturing polyacrylamide gel (Summers et al., 1992).

1.2.3.4 A COLOR COMPLEMENTATION ASSAY

While PCR-SSCP analysis can be used in clinical practice, the **color complementation assay** may have a greater appeal for routine applications. This rapid, relatively simple method using a nonradioactive PCR-based color complementation assay is capable of detecting infectious agents and various mutations in human DNA (Chehab and Kan, 1989).

Using **fluorescent oligonucleotide primers** the simultaneous amplification of two or more DNA segments of specific genomic DNA can be performed. As a result of this reaction several (up to five) products of different colors are generated. The fluorescence of each dye corresponding to its amplified DNA locus is evaluated on a fluorimeter. Although the relative fluorescence values obtained on the fluorimeter are not quantitative, the main advantage of this system is that the identification of the PCR-amplified products is enormously simplified. For example, when the diagnosis can be established by the presence or absence of an allele such as in gene deletions, chromosome translocations, and in infectious diseases, only two colors are needed — one for the test and one as an internal control for the PCR reaction. This color complementation assay was successfully demonstrated in a variety of conditions: the α globin deletion in hydrops fetalis, the 14;18 translocation in follicular lymphoma, detection of cytomegalovirus (CMV), β-thalassemia due to a 4-bp deletion or a single-base pair substitution and the 3-bp deletion in cystic fibrosis (Chehab and Kan, 1989).

1.2.3.5 EXON-SCANNING TECHNIQUE

The challenge of screening a large number of long stretches of DNA or RNA for **unknown sequence alterations** as needed in the linkage analysis and isolation of "candidate genes" cannot be met with conventional techniques. Southern and Northern blotting may fail to detect small genetic lesions, whereas ASO probes, which are helpful in identifying known polymorphisms, cannot be used to screen large genes or mRNAs for unknown sequence alterations (Kaufman et al., 1990). Again, direct DNA sequencing is not a method of choice in screening for genetic alterations in disorders of unknown etiology. An exon-scanning technique uses RNA probes derived from cDNA templates (**cRNAs**) to identify lesions in suspect genes. The cRNA probe forms a heteroduplex with the exons of the target gene, but not with the introns, which form a ssDNA loop that may constitute most of the gene. The sites of cRNA probes not hybridized to the exon DNA are cleaved by RNase A.

The basic difference from the RNase protection assay (RPA) (Myers et al., 1985a,b) is in using cRNA probes derived from cDNA templates, rather than colinear RNA transcribed from cloned genomic DNA. In the original RPA, the analysis of large genes would require many contiguous RNA probes from cloned genomic DNAs. These must cover the entire length of a suspect gene, in which introns and repetitive sequences in probes derived from genomic DNA would interfere with the analysis. In the **exon-scanning system**, the entire coding region composed of exons scattered over large lengths of DNA can be screened for point mutations encountered only in the exons (Kaufman et al., 1990).

Some of the deficiencies of this technique, tested by its ability to detect hemoglobinopathies and the mutation in the ornithine aminotransferase gene, included erroneous cleavage of the cRNA probe at perfectly matched base pairs, at intron/exon junctions, and RNase nibbling at the ends of cRNA fragments. Following identification of the suspect exon by this technique, PCR amplification and sequencing can be added to further characterize the implicated gene regions (Kaufman et al., 1990) (Figure 1.5). Among the advantages of the exon-scanning technique is the small amount of DNA needed (that can be obtained from lymphocytes or amniotic fluid sample), its ability to use both sense and antisense cRNA probes to evaluate both DNA strands, and the fact that many DNA samples can be analyzed on a single gel. Since it screens only exons, the polymorphisms occurring in introns may not complicate the analysis. Besides, this technique can essentially map the implicated exon with a sequence alteration for further characterization by PCR amplification and sequencing.

The greatest use of this approach may be in the evaluation of candidate genes in heterogeneous disorders, for the characterization of expressed genes identified by linkage analysis, as well as in clinical and prenatal diagnosis (Kaufman et al., 1990). Using transfection systems, genomic DNA segments of 20 kb or more can be screened in a single transfection by selecting from cloned genomic DNA those RNA sequences, exons, that are flanked by functional 5' and 3' splice sites (Buckler et al., 1991). By inserting random segments of chromosomal DNA into an intron present in a mammalian expression vector, transfecting these, and then screening cytoplasmic mRNA for the splice sites by PCR, the exons acquired from the genomic fragment are identified and then amplified. The method can be useful in isolating a wide spectrum of genes.

1.2.3.6 MULTIPLEX PCR

Several target DNA sequences can be amplified by PCR simultaneously in a system termed **multiplex PCR** (Chamberlain et al., 1988). First examples of its application included analysis of nine regions of the dystrophin gene for deletions (Chamberlain et al., 1988, 1990) and an eight-fragment multiplex PCR amplification of the human hypoxanthine-guanine phosphoribosyltransferase (*HPRT*) gene in Lesch-Nyhan syndrome (Gibbs et al., 1990).

1.2.3.7 AMPLIFICATION AND SEQUENCING

Amplification of the targeted DNA sequences by PCR followed with direct sequencing of the amplified products has been widely used (e.g., Engelke et al., 1988; Stoflet et al., 1988). Among

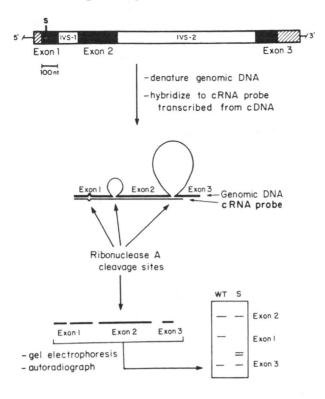

FIGURE 1.5. Flow diagram of exon scanning. A labeled cRNA probe transcribed from a wild-type cDNA is hybridized to denatured genomic DNA. In the resulting cRNA·DNA hybrid, the exon sequences of the DNA base pair with the cRNA probe. The intron sequences, which may constitute most of the gene, loop out as single-stranded DNA. The RNA·DNA hybrid is then treated with RNase A, which digests all unhybridized probe and cleaves the cRNA at exon/intron junctions, generally producing exon-sized cRNA fragments. RNase A also cleaves at mismatched base pairs. The cRNA fragments are fractionated by size by electrophoresis under denaturation conditions and visualized by autoradiography. The example shows a β-globin gene from an *HbS/HbS* homozygote. The loss of exon 1 and the appearance of two new bands localizes the *HbS* mutation relative to the ends of exon 1. (From Kaufman, D. L. et al. (1990). Genomics *8*:656–663. With permission.)

the many reported applications are the characterization of c-Ki-*ras* oncogene alleles (McMahon et al., 1987), β-thalassemia mutations (C. Wong et al., 1987), mitochondrial DNA (Wrischnik et al., 1987), the HLA-DQA locus (Gyllensten and Erlich, 1988), mutations leading to the Lesch-Nyhan syndrome (Gibbs et al., 1989b) and α_1-antitrypsin deficiency (Newton et al., 1988), and the analysis of DNA polymorphisms (Yandell and Dryja, 1989) to name only a few.

The technique of detecting point mutations by PCR followed by sequencing of the amplified product has been markedly improved (Haliassos et al., 1989). In the original format, PCR coupled with direct sequencing, although being very sensitive, failed to detect a mutation if it was present in less than 5% of the cells tested. Using a modified primer during the PCR an **artificial restriction site** can be introduced into the amplified product. This creates a known, easily detected RFLP indicative of a given mutation. A 1000-fold increase in the sensitivity of the method is thus achieved that can be used for screening purposes, or for monitoring the effects of therapy in oncology. This modification allowed the detection of 20% more mutations at codon 12 in the Ki-*ras* oncogene DNA from colorectal cancers that could not be identified by other techniques (Haliassos et al., 1989).

Significantly more definitive information on the polymorphisms under study is obtained when the amplified DNA segments are directly sequenced. Immobilization of the PCR products via biotinylated ASO primers and steptavidin-coated magnetic beads drastically improves this methodology (Kaneoka et al., 1991). As shown by the analysis of HLA-DRB genes, the solid-phase DNA sequencing following PCR amplification allows for rapid search of polymorphisms. There is no need for DNA cloning of genomic or PCR-amplified DNA prior to sequencing as required in earlier protocols.

Improvements are being continuously reported in simplifying PCR-sequencing procedures (e.g., Parkinson and Cheng, 1989; Ruano and Kidd, 1991) universally directed toward **automation** (e.g., Creasey et al., 1991; Davis et al., 1991; Mardis and Roe, 1989; C. Martin et al., 1991; Toneguzzo et al., 1989; Tracy and Mulcahy, 1991; Wilson et al., 1990). DNA sequences can be amplified and sequenced from a single sperm for the determination of genetic disease characteristics (Arnheim et al., 1990). Even the unsurpassed resolution of the **tunneling microscope** is contemplated for use in DNA sequencing (Lindsay and Philipp, 1991).

1.2.3.8 QUANTITATIVE AMPLIFICATIONS

The majority of strategies for amplification either of the target or the probe are qualitative methods primarily designed to establish the presence or absence of a nucleic acid characteristic in a condition under study. A significant enhancement of the informative content of an amplification scheme would result from quantification of the initial amount of the template DNA or RNA. To achieve this goal some of the proposed approaches use coamplification, restriction digestion of the amplified product, and size analysis on agarose gels as well as **competitive PCR** (Becker-Andre and Hahlbrock, 1990; Gilliland et al., 1990). Other methods employ hybridization techniques followed by isotopic or immunoenzymatic estimates of the generated product (Coutlee et al., 1990; Syvanen et al., 1988).

A novel technique, termed **DIANA** — for **d**etection of **i**mmobilized, **a**mplified **n**ucleic **a**cids (Lundeberg et al., 1990; Wahlberg et al., 1990) — uses a qualitative colorimetric assay of *in vitro*-amplified DNA following capture of the PCR products on **streptavidin-coated magnetic beads**. By introducing competitive titration with an *in vitro*-cloned sequence of the target into which the *lac* operator sequence is inserted, quantification of specific DNA can be performed in clinical samples (Lundeberg et al., 1991) (Figure 1.6). This assay allowed the detection of *Plasmodium falciparum* DNA in the range of 20 to 50,000 parasites per sample, making the diagnosis of malaria infection possible not only in the acute but also in the chronic phase.

1.2.3.9 AUTOMATED AMPLIFICATION ASSAYS

Wide acceptance of PCR amplification for diagnosis will be markedly enhanced by automating the multistep procedure of temperature cycling, addition of reagents, shortening the amplification times, increasing the specificity of PCR assays, and allowing for a large number of samples to be processed simultaneously.

Although **thermocyclers** were introduced in 1987, continuing efforts are being made to achieve a maximally automated system for a wide range of specific uses. A new Perkin-Elmer Cetus system (Norwalk, CT), the GeneAmp PCR System 9600, has been developed ensuring a better temperature control, eliminating the need for a mineral oil overlay to control evaporation, allowing for compatibility with multichannel liquid-handling devices, and minimizing contamination among many other features (Haff et al., 1991).

Since **temperature control** of the PCR is critical for the specificity and speed of the reaction, a system capable of very rapid temperature changes achieved by hot air instead of water circulation is particularly attractive (Wittwer et al., 1989, 1990; Wittwer and Garling, 1991). In addition, the typical reaction time can be reduced to 15 min or less (Wittwer et al., 1990).

Detection of specific DNA sequences by PCR coupled with the discrimination of allelic sequence variants by a colorimetric **oligonucleotide ligation assay** (**OLA**) has been automated (Nickerson et al., 1990). A variety of applications of this non-isotopic system can be visualized and the system has been demonstrated in the detection of sickle cell hemoglobin, cystic fibrosis

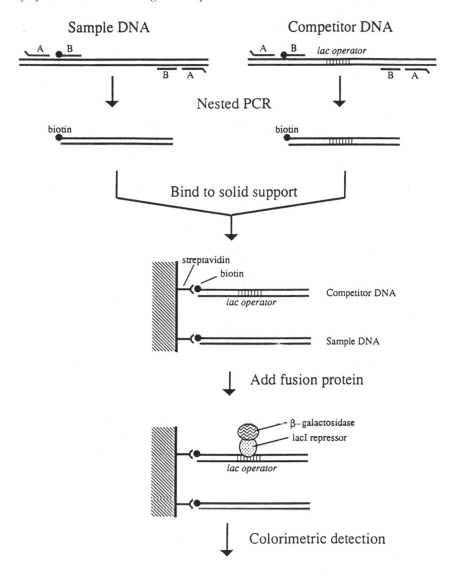

FIGURE 1.6. Diagram showing the principle of the quantitative DIANA, in which a solid support is used to capture the biotinylated DNA fragment obtained after a nested PCR, with outer primers A and inner primers B. A LacI repressor-β-galactosidase fusion protein is used for the subsequent colorimetric detection of a *lac* operator sequence, which is introduced into a cloned version of the target sequence. A known amount of this cloned DNA is coamplified with the target DNA. The detection is based on the specific interaction of the LacI β-galactosidase fusion protein with the *lac* operator sequence and the subsequent hydrolysis of the substrate *o*-nitrophenyl-β-D-galactoside (ONPG) to give the colorimetric response. (From Lundeberg, J. et al. (1991). *BioTechniques 10*:68–75. With permission .)

(ΔF508 mutation) as well as in the genetic linkage mapping of gene segments of the human T cell receptor β chain locus. The **automated PCR/OLA system** eliminates the need to measure DNA fragment sizes, the use of radioisotopes, as well as the requirement for high-quality DNA, particularly when polymorphic human sequence-tagged sites (STSs) become available for automated typing, e.g., in forensic practice. The entire enzyme-linked immunosorbent assay (ELISA)-based assay can be performed in **microtiter plates**, and no centrifugation or electrophoresis is needed. **Digoxigenin-tagged oligonucleotides** serve as reporter molecules in the ELISA assay of the duplexes captured via their biotinylated oligonucleotides (Nickerson et al., 1990).

1.2.4 SELECTED CHROMOSOME ANALYSIS TECHNIQUES APPLICABLE TO GENE LEVEL DIAGNOSIS IN CLINICAL PRACTICE

Specific assignment of defined DNA sequences to chromosomal locations represents a direct form of genome mapping that links genetic, REA, yeast artificial chromosome (YAC)-based methods, and large cosmid strategies (Billings et al., 1991). At the present stage of technology, chromosomal mapping is capable of directly placing genetic sites within or outside of bands (5 Mb), or at distances as close as 100 kb apart using *in situ* hybridization (ISH) techniques (Billings et al., 1991). The extensive efforts related to the pursuit of the objectives of the Human Genome Project (HGP) fully exploit the already existing technology of fine physical assignment of defined DNA segments to chromosomes. In addition, the continuing developments in all aspects of genome mapping strategies will undoubtedly introduce new as well as modify currently prevailing approaches.

At present, essentially two convergent strategies are used: the **top down approach** identifies and makes assignments of progressively smaller fragments of DNA from a given chromosome (Billings et al., 1991). **Bottom up strategy** identifies and aligns, in a continuum of sequences, overlapping sets of DNA clones (contigs). Techniques known as **chromosome walking** and **jumping**, and **probe walking** (Washio et al., 1989), are used in these strategies. In reality, a combination of these convergent types of techniques — bottom up and top down — are employed for specific gene identification and mapping.

Which particular **mapping techniques** will find their way into clinical gene level diagnostics is impossible to predict with certainty. At present, the practical utility of ISH and its variants is fully recognized and a number of diagnostic protocols already exist (see below). Some of the newer techniques include **pulsed field gel electrophoresis** (**PFGE**), capable of resolving large fragments of DNA (up to 10 Mb) (Cantor et al., 1988; Fan et al., 1989; Gardiner, 1991), **controlled partial digestion** of DNA with select restriction endonucleases, such as *Not*I and *Mlu*I, which recognize **CpG islands** rarely scattered throughout the human genome (Hanish and McClelland, 1990; Ito and Sakaki, 1988; Melmer and Buchwald, 1990; Sved and Bird, 1990), the **use of YAC vectors** containing large fragments of DNA up to 1 Mb in size (Burke et al., 1987); **cell hybridization** and **microcell-mediated gene transfer** (for a review, see Bentley et al., 1988), and, of course, PCR (for a review, see Rose, 1991).

1.2.4.1 ANALYSIS OF DNA FROM SPECIFIC CHROMOSOMAL REGIONS

Genetic linkage analysis is capable of determining the physical location of a mutant allele. The isolation of the genes, however, may not be possible using the current cloning strategies due to limitations on the size of the restriction fragments that can be cloned in appropriate vectors. A novel approach of **microcloning region-specific chromosomal DNA** has been developed (Kao and Yu, 1991; Saunders et al., 1989). Single bands are dissected from polytene chromosomes and digested with *Sau*3A, followed by ligation of oligonucleotide adapters to these fragments. The latter provide convenient primers for PCR in this "microamplification" approach that can be more efficiently used for establishing **chromosomal walks**.

An alternative approach uses the PCR amplification of restriction fragments obtained from **chromosome dissection** at a specific region (Johnson, 1990). This rather complicated micromanipulation procedure is combined with sequence-independent amplification of DNA. Restriction digestion of chromosomes can be followed by PCR amplification of the relatively short DNA sequences from a given chromosome fragment to which a complementary **oligonucleotide linker** is attached at both ends (Johnson, 1990). The linker serves as an identical primer site on each strand of the restriction fragments, allowing for efficient PCR amplification.

Although this procedure was developed for *Drosophila* polytene chromosomes, the general approach may be useful for amplification of any regions of chromosomes, eliminating the need

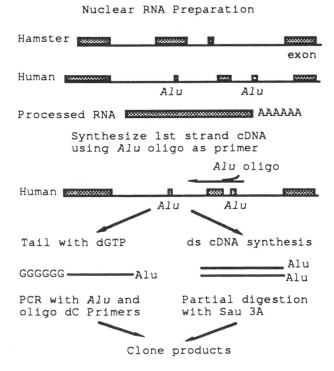

FIGURE 1.7. Isolation of human expressed sequences. (From Corbo, L. et al. In: Biotechnology and Human Genetic Predisposition to Disease. Wiley-Liss, New York, pp. 187–197. Copyright © 1990. Reprinted by permission of Wiley-Liss, a division of John Wiley & Sons, Inc.)

to clone the restricted chromosome fragments. The collection of DNA molecules representing unique and repetitive sequences obtained in this manner is influenced to a lesser extent by the topography of restriction sites, and offers an easier approach to the identification of genes of interest than conventional techniques such as chromosome walking (Johnson, 1990).

A new strategy for **isolating expressed sequences** from defined human chromosomal regions has been devised using **somatic cell hybrids** (Corbo et al., 1990). In hybrid cell lines, the complexity of the genomic region under study is reduced. In addition, this approach takes advantage of the fact that short repetitive stretches of *Alu* **sequences** are frequently reflected in **heterologous nuclear RNA (hnRNA)**. The latter fact allows for the construction of human-specific *Alu* oligonucleotide primers to produce cDNA. A diagram summarizing this approach is given in Figure 1.7. This approach allowed the isolation of transcribed sequences from the distal region of the X chromosome long arm. Hybridization confirmed that all the sequences isolated from the hybrid cell lines were indeed human (Corbo et al., 1990).

Direct amplification of a microdissected **Giemsa-banded chromosomal segment** has been described in application to a cloned human brain sodium channel (α subunit) gene sequence from chromosome 2q22-q23 (Han et al., 1991). This technique, termed **chromosomal micro-dissection PCR (CM-PCR)**, provides a relatively simple, precise, and direct approach to gene mapping. Several advantages of CM-PCR over traditional techniques of gene mapping include the rapidity of the assay, the ability to work with rehydrated archival tissue, and the fact that CM-PCR produces a level of resolution less ambiguous than that obtained by ISH with radioactive probes.

Although CM-PCR is hardly applicable to routine gene level diagnosis in clinical practice, the demonstrated possibility of studying a specific chromosomal region that can be isolated from an individual offers a potential for diagnostic evaluations. The CM-PCR technique can offer a direct approach to the analysis of chromosomal aberrations (deletions, duplications, transloca-tions, inversions) encountered in various human malignancies. Also, CM-PCR can be used in linkage analysis by simultaneously amplifying loci of interest and identifying their allelic states (Han et al., 1991).

The diagnostic utility of the continuously accumulating basic information about the human genome is constantly demonstrated. For example, on the basis of sequence data from the **GenBank**, PCR primers have been designed for the 22 autosomes and the X chromosome (Theune et al., 1991). Sets of **specific chromosome primers** can be used in multiplexing PCR amplifications for a rapid and detailed analysis of human chromosomes in somatic cell hybrids and radiation-induced hybrids. A similar list of PCR primers for each chromosome that has very little overlap with those described by Theune and co-workers (1991) has been given by other researchers (Abbott and Povey, 1991). The two listings of PCR primers provide a basis for the comprehensive analysis of human chromosomes. This information, containing **STSs** for all the human chromosomes, allows for the detection of most chromosomal aberrations applicable to molecular cytogenetic diagnosis.

An alternative approach has been described that is based on two complete sets of small inserts, complete digest DNA libraries for each of the 24 human chromosomal types constructed at the National Laboratory Gene Library Project (Van Dilla and Deaven, 1990). The available information is used for genome mapping studies as well as in the search for RFLP markers linked to various genetic diseases. In addition, these libraries are used as a source of unique sequence probes for "**chromosome painting**" by **fluorescent ISH** (**FISH**). The practical utility of these probes is very high for the detection of chromosome aberrations in metaphase and interphase nuclei, so that they can be used as highly specific cytochemical stains for individual chromo-somes (Trask, 1991). In fact, this technology has reached the commercial stage. Chromosome paints developed by Drs. J. Gray and D. Pinkel at Lawrence Livermore National Laboratory have been converted into a reagent by Imagenetics (Naperville, IL) and is now sold by Life Technologies (Bethesda, MD) (Gebhart, 1992). It is expected that by the end of 1992 the paints for all human chromosomes will be available.

1.2.4.2 AUTOMATED KARYOTYPING

Attempts to standardize and automate the analysis of human chromosomes have been going on for more than 25 years and a summary of the advances made over the past 15 years by one group of researchers has been published (Mayall et al., 1990; van Vliet et al., 1990). The **Athena semi automated karyotyping system** analyzes metaphase spreads. It provides automated segmentation of metaphase images into individual chromosomes and automated measurements of each banded chromosome, resulting in a specific description of the karyotype according to the standard Paris convention (van Vliet et al., 1990). The actual processing time for automatic segmentation and classification is on the order of 1 to 2 min.

Another approach to the identification of individual chromosomes is attempted using **multiple dye-DNA interactions** followed by high resolution of the total and peak intensities of the dyes (Arndt-Jovin and Jovin, 1990). This serves to produce patterns translated into karyotype images without the use of banding.

An interesting and promising system of karyotype analysis is representation of chromosomes in **stylized images resembling bar codes** (Martin et al., 1990). This approach can be used in conjunction with automated karyotyping. A two-step approach is used for diagnostic purposes: first, all chromosomes are recognized by a minimum of characteristic parameters; then the full banding pattern with maximum resolution is used to detect chromosome aberrations. This approach needs further refinement but appears to offer a workable diagnostic system for wide use in clinical practice.

1.2.4.3 *IN SITU* HYBRIDIZATION IN CHROMOSOME ANALYSIS

Newer techniques using ISH for the chromosomal and/or gene level evaluation of diseases allow visualization of the target site relative to high-resolution banding. In cancer cytogenetics, in particular, this approach may be integrated into the clinical management of patients not only with hematological malignancies but with solid tumors as well (Atkin, 1989; Dal Chin and Sandberg, 1989; Pathak and Goodacre, 1986; Sandberg et al., 1988; Sandberg and Turc-Carel, 1987; Trent, 1989).

Chromosomal aberrations such as deletions, translocations, the formation of isochromosomes, and other structural rearrangements have been traditionally identified by banding techniques. **High-resolution banding** techniques are continually being improved to the level appropriate for use in routine clinical laboratories (Latos-Bielenska and Hameister, 1988).

In some clinical situations, the presence of chromosomal abnormalities such as a balanced chromosomal translocation can be confirmed by **quantitative hybridization** dosage studies, which demonstrated in one case that the mother had a double dose of DNA in the suspect deleted area of chromosome 5 (Smart et al., 1989).

ISH and Southern blotting complemented high-resolution banding techniques in establishing the exact molecular cause of gonadal dysgenesis (Cantrell et al., 1989), or aberrations in sexual development of patients with Y; autosome translocations (Andersson et al., 1988). DNA probe analysis has been successful in the majority of cases of sex reversal even when no detectable genetic defects could be demonstrated by conventional cytogenetic methods (de la Chapelle, 1988).

Assessment of efficiency of a routine banding procedure performed on amniotic fluid cell culture revealed that **banding resolution** of at least 400 bands could be achieved on all preparations (Kao et al., 1990). Occasional cells showed a banding resolution between a 550- and an 850-band level. Retrospective analysis of consistent chromosomal abnormalities in 9069 human neoplasms revealed that the breakpoints of the primary abnormalities were concentrated within only 71 of the 329 chromosomal bands (Mitelman and Heim, 1988). Furthermore, 27 of the 41 well-recognized oncogene sites have been consistently involved in neoplasia-associated rearrangements. Others, however, find no statistical association between **common fragile sites** and nonrandom chromosome breakpoints in cancer cells, at the 400-band resolution level (Sutherland and Simmers, 1988).

A significant advance in cytogenetics relevant to practical molecular diagnostics is the development of a new technique termed **interphase cytogenetics** (Deville et al., 1988), which allows to recognitize specific chromosomal regions in interphase tumor cells by ISH. The advantage of using ISH on interphase chromosomes is that interphase chromatin is not as compact as metaphase chromatin. This not only allows a better resolution by ISH, but also expands the range of diagnostic possibilities of ISH in clinical laboratories. It removes the need to use cell culture techniques and synchronization of cells, which are cumbersome and time-consuming procedures. Besides, they potentially can introduce aberrations at the DNA or chromosome level absent in the original material.

The basic methodologies of ISH have been extensively covered (Hofler, 1987; Singer et al., 1986; Uhl, 1989). The progressive introduction of **nonradioactive** and **multiple labels**, the use of **shorter oligonucleotide probes**, both **sense** and **antisense RNA probes**, the extension of hybridization strategies previously employed in solid support or soluble hybridization procedures to tissue sections and chromosomes, the **combination of microdissection** of target chromosomal regions **with amplification**, promising efforts at **automation** of ISH - all these newer developments enormously enhance the practical utility of ISH for gene level diagnosis.

A survey of the more recent methodological advances in ISH will be given here, some of which can be applied to patient management. The impact of ISH and its variants on diagnostic pathology is hard to overestimate. Emerging combinations of ISH with flow cytometry, electron microscopy (EM), and confocal microscopy (e.g., Lichter et al., 1990) may eventually evolve into diagnostic modalities of practical utility as well. In the meantime, construction of specific, "custom-designed" probes tagged by easily and rapidly quantifiable nonradioactive labels such as digoxigenin allow the application of **ISH on frozen sections** (Maggiano et al., 1991), bringing this technique even closer to diagnostic laboratories. The use of *in situ* **hybridization (ISH)** for the detection of infectious agents will be addressed in the discussion of specific diseases (see below).

Methods for the detection of nucleic acids in **formalin-fixed, paraffin-embedded tissues** by amplification systems such as PCR are particularly attractive for diagnostic laboratories involved in gene level probing. An example of applying PCR to amplification of specific DNA sequences is its use on routine cytogenetic chromosome spreads fixed in methanol/glacial acetic acid (Jonveaux, 1991). Although depurination of DNA occurs as a result of this fixation PCR amplification can be performed on preparations stored for many years. Whereas ISH on routine formalin-fixed sections has been used for a variety of determinations (e.g., Hankin and Lloyd, 1989), PCR amplification from fixed tissue adds a new dimension to tissue analysis for infectious and genetic diseases.

Successful amplification of target sequences has been described from **formalin-** or **B-5-fixed bone marrow specimens** (Crisan et al., 1990), from **boiled** clinical specimens using fast multiplex PCR amplification of HPV sequences (Vandenvelde et al., 1990), and in **formalin-fixed, paraffin-embedded** liver tissue for the analysis of gene expression and the detection of RNA viruses (Weizsacker et al., 1991). A detailed, systematic study of the effects of various fixatives on the efficiency of PCR amplification has been presented (Greer et al., 1991). The best suited tissue fixatives for optimizing subsequent PCR analysis were found to be acetone or 10% buffered neutral formalin. Zamboni's, Clarke's, paraformaldehyde, formalin-alcohol-acetic acid, and methacarn were found to compromise amplification efficiency. The worst results were observed with tissues processed with Carnoy's, Zenker's, or Bouin's fixatives (Greer et al., 1991).

One of the challenging tasks facing gene level diagnosis (as well as any type of diagnosis, for that matter) is the establishment of **distinction between the benign and malignant process**. Although all of diagnostic pathology must deal with this issue, some attempts have been made to develop an objective criterion defining a population of cells characterized by proliferative activity usually associated with "deviant" cellular behavior. An **assay for DNA polymerase α** activity has been described in application to paraformaldehyde-postfixed frozen sections (Tsutsumi et al., 1990). This immunocytochemical assay appears to be informative not only in fully developed neoplasms but also in their precursor lesions. An ISH evaluation of the expression of the **histone 3 (*H3*) gene** that is absent in resting cells has been proposed for "the unambiguous identification of cycling cells in any tissues, normal or diseased" (Chou et al., 1990).

1.2.4.3.1 Nonradioactive ISH

Although radioactive labels allow ISH to reach very high levels of sensitivity capable of identifying DNA sequences a few hundred base pairs in length, the disadvantages associated with the handling of radioactive materials, the need for autoradiography and lengthy exposure times, limited shelf life of radioisotopes, and frequently extensive background make alternative reporter molecules much more attractive for diagnostic laboratories. ISH is frequently combined with conventional immunohistochemistry, particularly when gene expression is evaluated at the mRNA and protein levels simultaneously (e.g., Bakkus et al., 1989; Close et al., 1990; Hankin and Lloyd, 1989; Jimenez-Garcia et al., 1989; Kawata et al., 1990; Warford, 1988).

Several major classes of nonradioactive probes have been developed over the years, allowing for the direct and indirect (multistep) visualization of nucleic acid probes (for reviews, see

Hopman et al., 1986a,b, 1987, 1988a,b; Singer et al., 1986). Chemical modification of nucleic acids with **acetylaminofluorene** (AAF) (Landegent et al., 1984; Tchen et al., 1984) allows labeling of both DNA and RNA, double or single stranded, with the detection of AAF by poly- or monoclonal antibodies (Berger, 1986; Hopman et al., 1988a; Simmons et al., 1989; Terenghi et al., 1987). **Mercuriation** of nucleic acids is performed after ISH is completed using haptens with SH groups that bind the mercury (Hopman et al., 1987).

Various reporter molecules can be introduced as haptens, for example, **trinitrophenyl**, **biotinyl**, and **fluorescent** groups. Biotinylation of nucleic acid probes for ISH has received wide acceptance, however, in some cases the biotin label has been found to reduce hybridization sensitivity. **Photochemical biotinylation** has also been described (Forster et al., 1985). Biotinylated nucleic acid probes are useful in identifying the highly repetitive, middle repetitive, and large single-copy sequences (15–40 kb long) (Hopman et al., 1988a).

Optimization of conditions for ISH after varying pretreatment, hybridization, and labeling protocols using biotinylated probes followed by image analysis of ISH have been described (Larsson and Hougaard, 1990). Biotinylation of oligonucleotides and various hybridization procedures has been described in great detail (e.g., Barr and Emanuel, 1990; Brigati et al., 1983; Cook et al., 1988; Elias et al., 1989; Eweida et al., 1989; Folsom et al., 1989; Guitteny et al., 1988; Kincaid and Nightingale, 1988; Oelmuller et al., 1990; Reisfeld et al., 1987; Riley et al., 1986; Shroyer and Nakane, 1982; Soh and Pestka, 1990; Zwadyk et al., 1986). Biotinylated **alphoid repeat probes** have been used in ISH in the clinical diagnostic laboratory to detect the centromeres of chromosomes 13, 21, 14, and 22 (Kolvraa et al., 1991).

Introduction of the **immuno-gold-silver detection** of biotinylated probes (Cremers et al., 1987; Jackson et al., 1989) applicable to formalin-fixed, paraffin-embedded tissue dramatically increases the ISH sensitivity. A **streptavidin-gold detection system**, however, failed to reveal the same degree of sensitivity, apparently due to **steric hindrance** of the bulky streptavidin-gold complex (Jackson et al., 1989).

Numerous other nonradioactive probes for ISH have been described such as **dimerized thymidine** probes (Nakane et al., 1987), DNA **modification at the cytosine** residues by bisulfite-catalyzed **transamination** (Sverdlov et al., 1974; Viscidi et al., 1986), **5-bromouridinylated oligonucleotides** for analysis of DNA and RNA both for ISH and membrane hybridization assays (Ortigao et al., 1989; Jirikowski et al., 1990), **acridinium ester** labeled DNA oligonucleotide probes (Septak, 1989) and oligonucleotides covalently linked to **porphyrins** (Doan et al., 1987a), and **oligo-[α]-thymidylate** covalently linked to an azidoproflavine derivative (Doan et al., 1987b). A system based on **sulfonylation** of DNA directly in tissue sections followed by immunocytochemical detection of modified DNA shows a very high specificity for DNA without cross-reactivity with RNA and superior staining qualities compared to Feulgen staining of DNA (Fritz et al., 1989).

Fluorochromes have long been used to label nucleic acid and oligonucleotide probes for ISH (e.g., Bauman et al., 1980; Bertin et al., 1990; Evans et al., 1991; van Prooijen-Knegt et al., 1982), and now **multicolor fluorescence** has become commonplace for the simultaneous detection of different nucleic acid sequences. **Double hybridization** (Cremer et al., 1986; Haase et al., 1985; Hopman et al., 1986b, 1988b; Mullink et al., 1989) and **triple hybridization** procedures (Bresser and Evinger-Hodges, 1987) have been described. A triple hybridization protocol uses three differently haptenized (AAF, mercurated, and biotinylated) chromosome-specific repetitive probes combined with three fluorochromes: blue-amino methyl coumarin acetic acid (AMCA), green-fluorescein isothiocyanate (FITC), and red-TRITC (Nederlof et al., 1989). Three-color hybridization was applied to study chromosomal abnormalities, such as are encountered in human solid tumors, using interphase nuclei of tumor cells.

Multiplicity of specific probes using different reporter molecules allows the simultaneous identification of a number of parameters and may present an attractive option for gene-level diagnosis. At present, by combining double-haptenized probes and three fluorochromes *up to seven specific targets* can be evaluated simultaneously (Nederlof et al., 1990). Since the choice of targets can be made to represent different aspects of, for example, chromosomal structure, a substantially more objective assessment of chromosomal abnormalities becomes possible. Combined with the **digitalizing image evaluation** that can be automated, this approach may lead to a standardized, rapid tool for molecular cytogenetic evaluation of patient specimens in clinical practice. **Bioluminescent ISH** systems based on luciferin derivatives display a very high sensitivity and are amenable to quantitation (for a review, see Geiger et al., 1989).

An important new **chemoluminescent** labeling system using derivatives of **1,2-dioxetane** in combination with various enzymatic triggering combinations allowing multiplex reactions to be monitored simultaneously has been introduced (Beck and Koster, 1990; Nelson and Kacian, 1990; Pollard-Knight et al., 1990). Extraordinary sensitivity, diversity, low cost, and the ability of this chemoluminescent system to be coupled with other labeling methods makes dioxetane detection a likely candidate for widespread use in clinical diagnostic assays.

Another labeling technique of nucleic acid probes for ISH is based on the incorporation of the nucleotide analog **digoxigenin-11-d-UT**, a derivative of a plant steroid found in digitalis. The hybrid signal is detected by antibodies raised against digoxigenin-11-d-UT which do not show any cross-reactivity with animal antigens and produce fewer nonspecific signals than biotinylated labels (Herrington et al., 1989a,b; Farquharson et al., 1990; Maggiano et al., 1991). An exhaustive analysis of DNA labeling and detection systems based on digoxigenin has been presented (Holtke et al., 1990; Kessler et al., 1990; Muhlegger et al., 1990; Seibl et al., 1990).

A hybridization signal detection system based on **digoxigenin-antidigoxigenin** interaction avoids two major disadvantages of the biotin-streptavidin system: (1) the ubiquitous presence of biotin (vitamin H), which in eukaryotic as well as prokaryotic cells contributes to the occurrence of nonspecific signals, and (2) the nonspecific binding of streptavidin to immobilizing supports used in hybridization such as nitrocellulose or nylon (Zischler et al., 1989a,b). Digoxigenin reporter molecules allow for reduced background noise in filter hybridization and ISH.

Analysis of human chromosomes using digoxigenated **simple repeats (CAC)$_5$** as probes revealed a pattern of banding resembling that of R bands (Zischler et al., 1991). Likewise, digoxigenin-labeled probes have been used for ISH of the **M13 minisatellite tandem repeat sequences** that show an R band-like pattern on human metaphase chromosomes (Christmann et al., 1991). Digoxigenin labeling of oligonucleotides has been used also for **fingerprinting analysis** (Schafer et al., 1988; Zischler et al., 1989a,b). The PCR has been used to produce single-stranded or double-stranded cDNA probes for ISH (Scully et al., 1990), as well as for generating vector-free digoxigenin-dUTP-labeled (Liesack et al., 1990), and biotin-11-dUTP-labeled probes (Weier et al., 1990).

A **double-labeling ISH** approach using various combinations of radioactive and nonradioactive reporters at the same time has also been described in the analysis of mRNAs, chromosomal DNA sequences, and proteins (e.g., Singer et al., 1987).

Along with the construction of modified synthetic oligonucleotides with various reporter systems (Huynh-Dinh, 1990), ingenious new **nonradioactive detection methods for nucleic acids** keep appearing. One such novel system uses the principle of **sandwich hybridization** with the primary probe without a label and a second probe, complementary to a portion of the primary probe tagged with biotin or any other nonradioactive reporter molecule for "**universal probe system**" (Nakagami et al., 1991). Another novel use of ISH with various non-radioactive probes has been described for DNA sequence mapping by transmission and scanning electron microscopy (Narayanswami and Hamkalo, 1991; Thiry et al., 1991), and confocal microscopy (Albertson et al., 1991) (see also Sections 2.4.3.8 and 2.4.3.9). Efforts directed at the development of new, versatile labeling and detection systems applicable to a variety of tasks in basic and clinical studies with nucleic acids never stop (for a review of developments in analytical methodology for immunoassays and DNA hybridizations, see Diamandis, 1990). An excellent review addresses the analysis of genes and chromosomes by nonisotopic ISH (Lichter et al., 1991).

1.2.4.3.2 Banding and ISH

Until recently, whenever ISH signal had to be assigned to a specific chromosomal location, a sequential protocol of ISH and then banding, with the two images subsequently superimposed, has been employed (Cherif et al., 1989; Lawrence et al., 1988a,b, 1989; Takahashi et al., 1989; Viegas-Pequignot et al., 1989a,b). A significant improvement allowing for the **simultaneous visualization of R-banded chromosomes and hybridization signal** generated by probes as small as 1 kb has been described in 23 unique DNA segments mapped by ISH to the long arm of chromosome 11 (Fan et al., 1990). The procedure uses a standard fluorescent microscope without laser excitation, complex computer analysis of the data, and image amplification or overlay of separate photographic images.

Larger cosmid probes combined with **competitive suppression of hybridization to repetitive sequences** could be successfully used. A number of the probes have been designed for marked chromosomal sites (11q13 and 11q23 regions) associated with various disease or growth control genes (hematologic neoplasias, Ewing's sarcoma, and ataxia-telangiectasia). This procedure combines biotin-labeled probes and staining of chromosomes with both **propidium iodide** and **4',6-diamidino-2-phenylindole (DAPI)**, producing **R** and **Q banding** patterns, respectively. It allows unambiguous chromosome identification of various markers useful in the practical analysis of chromosomal alterations.

A similar procedure combines nonisotopic ISH with identification of R bands by incorporation of **bromodeoxyuridine** into cells synchronized with thymidine (Takahashi et al., 1990). Precise localization of the human type II collagen gene (COL2A1) was achieved by the detection of fluorescent signals on R-banded prometaphase chromosomes stained with propidium iodide. A combination of Giemsa (GTG) banding followed by nonradioactive ISH on the same metaphase spreads of cells from solid tumors has been described using eight mid-repetitive DNA probes specific for the pericentromeric regions of chromosomes 1, 6, 7, 10, 16, 17, and 18 and the X chromosome (Smit et al., 1990).

Chromosomal *in situ* suppression hybridization (CISS) (see Section 1.2.4.3.7) can be combined with classical Giemsa (GTG) banding within one protocol allowing for the high-resolution mapping of specific sequences as demonstrated by identifying translocated Y chromosomal sequences in the chromosome complement of an XX male (Klever et al., 1991). A potentially clinically useful technique of **replicational banding** can identify genes that are actively expressed; it offers a particular advantage in prenatal diagnosis of developmental abnormalities (Cheung et al., 1990).

1.2.4.3.3 Repetitive DNA-ISH

The advantages of precise chromosomal characterization and analysis of chromosomal alterations in interphase nuclei can be utilized in clinical practice for markedly simplified cytogenetic studies. Chromosome-specific probes containing only **repetitive (satellite) DNA** hybridize predominantly to the **centromeric** regions of the chromosome (see Section 1.1).

Hybridization probes constructed from **chromosome-specific libraries** that specifically stain individual chromosomes have also been developed (Cremer et al., 1988; Pinkel et al., 1988). In these protocols, total genomic DNA is used as a competitor to exclude the dispersed repetitive sequences from participation in ISH. By using a DNA clone representing a subpopulation of the repetitive human alphoid DNA family of the centromeric region of chromosome 18, and varying the stringency of ISH when either only this chromosome or the entire chromosome complement could be identified over interphase nuclei, various chromosomal aberrations can be identified, as demonstrated by the diagnosis of trisomy 18 with the probe L1.84 (Cremer et al., 1986).

Likewise, the pericentric region of human chromosome 17 has been analyzed by a cDNA clone containing the entire alphoid repeat (Meyne and Moyzis, 1989). Using a monomer of the higher order repeat, the chromosomal specificity of ISH was improved. **Shorter probes**, such as a 42-nucleotide oligomer of a divergent region of monomer 1, further improved the chromosome specificity, suggesting a methodology for precise identification of individual chromosomes using alphoid repeats. Biotinylated DNA probes have been shown to allow **high-resolution mapping of satellite DNA** (Manuelidis et al., 1990). A number of practical applications of ISH with repetitive DNA probes can be visualized. An example is the screening for the effects of aneuploidy-inducing agents **griseofulvin** and **benomyl** (Raimondi et al., 1989).

1.2.4.3.4 Single-Copy ISH

A pioneering approach of **chromosome-specific "painting"**, also in interphase nuclei, using biotinylated DNA library probes pinpoints fine chromosomal aberrations associated with various pathological conditions (Cremer et al., 1988). In one such study, DNA probes complementary to single-copy and repetitive sequences have been used for the characterization of chromosomes 1, 4, 7, 18, and 22 (Cremer et al., 1988).

A variety of single-copy genes can be detected by ISH and a combination of ISH with routine histology and DNA analysis has been demonstrated in tissues as well as cytospins prepared from cerebrospinal fluid, pleural fluid, bone marrow, peripheral blood, and cell lines (Seibel et al., 1988).

Simultaneous hybridization with different probes combined with **digital image analysis** allows the visualization of complete chromosomes, deletions, and translocated segments of chromosomes in very complex karyotypes (Cremer et al., 1988). Detection of very weak fluorescent signals generated by ISH of single-copy DNA sequences using biotinylated probes is possible due to enhancement of the detection sensitivity by **intensified fluorescence digital imaging microscopy** (Viegas-Pequignot et al., 1989a).

An example of the application of ISH for the analysis of specific genes is the identification of the gene for the human type 5, tartrate-resistant, iron-containing acid phosphatase isoenzyme (B. S. Allen et al., 1989). Elevated levels of this isozyme are seen in leukocytes and splenocytes in hairy cell leukemia, the spleens of patients with Hodgkin's disease, Gaucher disease, and the sera of patients with active bone turnover.

Specific **erythropoietin (*Epo*) gene expression** has been evaluated by a 1.2-kb biotinylated *Epo* probe detected with a streptavidin-gold reagent (Vogt et al., 1989). This technique allows the estimation of *Epo* gene expression not only in pathologically altered kidneys sections, but also at the level of macrophages that serve as regulator cells at the sites of erythropoiesis. Many more examples of the use of ISH for the evaluation of single-copy DNA sequences will be referred to below in the course of discussion of specific gene-level diagnoses. A thorough coverage of the use of oligonucleotide cDNA probes for ISH has been given (Uhl, 1989).

1.2.4.3.5 Single-Copy RNA ISH

In situ hybridization has been used extensively for the detection not only of the general class of mRNA populations by detecting poly(A) sequences (Pringle et al., 1989), but also for the identification of a large number of specific messages. Both radioactive and nonradioactive probes have been used (for a review of earlier studies, see Hopman et al., 1988a; Singer et al., 1986). Combining **microfluorometry** and ISH, the simultaneous evaluation of the specific gene expression (mRNA level) and the DNA content in individual cells becomes possible without distorting the morphological picture (Tang et al., 1989). Using **riboprobes** to study the expression of colony-stimulating factor, granulocyte-macrophage and interleukin-3 mRNA, a sensitivity comparable to that of Northern blots can be achieved (Williamson et al., 1989).

In some cases, ISH demonstrates sensitivity even higher than that of Northern blotting. Chemical transformation of oral epithelium leads to a marked increase in the **transforming growth factor** α mRNA expression that is detectable by ISH but not by Northern blotting (Chang et al., 1989). mRNA-ISH has been used for two- and three-dimensional **visualization of specific gene expression** in a variety of cell types (Lawrence et al. 1988a,b, 1989; Multhaupt et al., 1989; Terenghi et al., 1987).

Detailed protocols of **single-copy ISH** adapted to different types of tissues as well as numerous examples of its application have been published (Altar et al., 1989; Bruns, 1989; Gerfen, 1989; Giaid et al., 1989; Higgins and Mah, 1989; McCabe and Pfaff, 1989; Siegel, 1989; Simmons et al., 1989; Sparkes, 1990; Watts and Swanson, 1989; Wilson and Higgins, 1990).

1.2.4.3.6 Quantitative ISH

Although the demonstration of specific gene expression in individual cells can provide valuable information for diagnosis, determining the **magnitude of expression** significantly enhances the diagnostic utility of ISH. A detailed methodological study of ISH, performed by Lawrence and Singer (1985), allowed them to offer a rapid quantitative approach to ISH. Using synthetic oligonucleotide probes constructed for different regions of an mRNA allows for markedly enhanced sensitivity of quantitative ISH (Taneja and Singer, 1987). Comprehensive discussions of various aspects of **quantitative ISH** using radiolabeled probes have been presented (Gerfen, 1989; Uhl, 1989) as well as graphical and statistical approaches to ISH data analysis (McCabe et al., 1989).

Recently, a quantitative ISH procedure using biotinylated probes has been developed and demonstrated in the measurement by computerized image analysis of the relative levels of cellular mRNA for proopiomelanocortin in the anterior lobe of the pituitary (Larsson et al., 1991). Although this procedure does not allow the determination of absolute values for mRNA levels, the relative quantitation can be informative in evaluating specific gene expression.

1.2.4.3.7 Competitive ISH: "Chromosome Painting"

"Chromosome painting" uses large pools of cloned genomic sequences from a single human chromosome as probe, and by performing a preannealing step in the presence of an excess of sonicated total human DNA, complete staining of a given chromosome can be achieved in metaphase and interphase nuclei (Cremer et al., 1988; Kievits et al., 1990; Landegent et al., 1987; Lichter et al., 1988, 1990a, 1991; Nederlof et al., 1989; Pinkel et al., 1988). In general, the **competitive** *in situ* **hybridization (CISH)** approach is used to enhance the specificity in detecting target sequences. When applied to tissues, CISH can be helpful in reducing the contribution of signal from repetitive DNA through the addition of competitor DNA, by analogy with the technique used in Southern hybridizations (Landegent et al., 1987). A marked increase in the ratio of specific-to-nonspecific hybridization signal can be achieved under optimal preannealing conditions (Lichter et al., 1988).

A variant of competitive ISH, called **chromosomal** *in situ* **suppression (CISS)-hybridization**, using biotinylated phage DNA library inserts from sorted human chromosomes has been used to detect aberrations of chromosomes in irradiated peripheral lymphocytes (Cremer et al., 1990), laser-microdissected chromosomes (Lengauer et al., 1991) and hybrid cells (Boyle et al., 1990) (see also Section 1.2.4.3.2).

1.2.4.3.8 Confocal Microscopy and ISH

A study of the **spatial organization** of chromosomes within the nucleus has been markedly advanced by the technique of optical sectioning of intact cells and stepwise movement of the microscope focus through successive planes of the DAPI-stained nucleus (Agard and Sedat, 1983). The introduction of confocal microscopy eliminated the need to conduct complex reconstruction and analysis of cells, formerly used to eliminate convolutional distortions produced by the older techniques (Lichter et al., 1988). Combining FISH with confocal scanning laser microscopy, a **three-dimensional topography** of specific nucleic acid sequences and their transcripts can be analyzed in single live cells (Puppels et al., 1990; van Dekken et al., 1990a,b).

1.2.4.3.9 EM-ISH

Like confocal microscopy-ISH, the EM-ISH combination at this time is primarily utilized in basic research. Even single-base mutations can be detected in large DNA fragments by immuno electron microscopy (Ganguly et al., 1989). The simultaneous determination of at least two sequences can be achieved by a modification of ISH at the EM level (Narayanswami and Hamkalo, 1990). In addition to the specific assignment of genes, the location of a sequence is expected to provide insights into its properties and function by **high-resolution sequence mapping**. The basis for the detection is immunogold tagging of nonradioactively labeled probes (biotin-, digoxigenin-, or dinitrophenyl-coupled deoxyribonucleotides incorporated by nick translation, terminal transferase addition, gap-filling reactions, or primer extension).

A comparison of the specificity and signal intensity achieved with biotinylated dsDNA and single-stranded, strand-specific RNA probes used in a combination of light microscopy and EM with ISH assays on the same portion of the poliovirus genome revealed the **RNA probes to be more sensitive and produce lower background than DNA probes** (Troxler et al., 1990). The protocols developed can be used without modifications for light microscopy ISH on semithick sections or ultrathin sections for EM-ISH. Immunogold detection affords the highest sensitivity of the assays (Troxler et al., 1990). DNA hybridization even using biotinylated probes can be combined with EM to study fine aspects of nucleic acids or nucleoproteins (e.g., Brantley and Beer, 1989; Hutchison, 1984; Oakes and Lake, 1990; Oakes et al., 1990; Sogo and Thoma, 1989). Recently, **scanning tunneling microscopy** has been applied to the analysis of surface structures of nucleic acids at atomic level resolution (e.g., Arscott and Bloomfield, 1990; Denley, 1990).

1.2.4.3.10 Flow Cytometry-ISH (FC-FISH)

Although fluorescent labels as such, or secondary detection systems based on fluorescence, have been widely used in ISH for a long time, a combination of this modality with flow cytometry (FCM) that is also capable of resolving fluorescent properties of particles has not been equally popular so far. Only a few studies on the detection of repetitive DNA in isolated nuclei (Trask et al., 1985, 1988), or suspended chromosomes (Dudin et al., 1987), and RNA populations (Bauman et al., 1980, 1981; Bauman and Bentvelzen, 1988) have been performed. Using rare-cutting restriction endonucleases such as *Not*I, very large DNA fragments can be obtained, sorted by FCM, and subjected to appropriate analysis (Minoshima et al., 1990).

The main obstacle was the **loss of morphologic integrity** of ISH-processed cells that interfered with informative FCM analysis. This drawback was markedly reduced when cellular morphology could be well retained in cells of erythroid lineage following new procedures developed for FC-FISH (Bayer and Bauman, 1990).

The wide use of FCM for clinical diagnostic purposes may offer a suitable background for introducing its combination with ISH as a method of gene level testing in clinical practice.

1.2.4.3.11 Automation of ISH

Automation of the multistep, labor-intensive ISH procedures will undoubtedly facilitate wider acceptance of this modality into diagnostic laboratory practice. So far efforts to automate ISH have been reported either in the processing of glass slides (Unger et al., 1988) or in the development of automated evaluation of the final results of ISH (Stolz et al., 1989).

Automation of the **slide processing** aspect of ISH has extended logically from the developments in automation of conventional immunohistochemistry. The only automated system described to date is the adaption of capillary gap technology and the modification of the Fisher Code-On Immunology System (Unger et al., 1988). Understandably, a higher reproducibility and consistency of the results using biotinylated nick-translated HPV types 6 and 11 DNA probes were obtained in the automated system compared to manually processed slides.

An automated, quantitative technique for the **determination of grain densities** of ISH slides can be performed using the high-resolution image analysis system IPS KONTRON (Stolz et al., 1989). Again, an adaptation of the existing technology for high-resolution image analysis employed for laboratory evaluation of blood cell counts, cervical smears, radiological images, chromosome analysis, and quantitation in histopathology has been applied to the assessment of ISH results. The advantage of this system over similar attempts to quantitate ISH lies in the more specific analysis of individual cells, which also adjusts the results to the background labeling level. It thus becomes possible to automatically quantitate cell populations displaying differing levels of a given gene expression that may have diagnostic significance.

When an X centromere probe that recognizes the alphoid satellite sequences is used for ISH, the automated detection of fragile X chromosomes is possible (Piper et al., 1990). This procedure accomplishes a fully automatic metaphase identification, digitization at high resolution, segmentation, and analysis, offering an **almost fully automated diagnostic ISH system**. Automated morphological detection combined with digitized fluorescence microscopy has been shown to accomplish automated assessment of DNA ploidy, yielding results similar to those obtained by FCM (Lockett et al., 1990).

1.2.5 DNA POLYMORPHISMS

1.2.5.1 SITE POLYMORPHISMS

Molecular analysis of chromosomal DNA using restriction endonuclease digestion revealed that DNA polymorphisms occur frequently, and, in the case of the human β-globin gene cluster, can provide useful codominant markers (Jeffreys, 1979). The RFLPs detected are due to base substitution or microdeletion or insertion that leads to the presence or absence of the recognition site for the enzyme used. The overall variability of DNA is low, and the variable regions are not uniformly distributed (Jeffreys et al., 1987). Nevertheless, the **informative potential** of DNA polymorphisms by far exceeds that of **protein polymorphisms**, since the coding sequences for proteins occupy only a small portion of the genome, whereas DNA polymorphisms in the coding and noncoding regions are distributed throughout the entire genome. Furthermore, certain noncoding regions display very high variability, in part resulting from silent mutations (Jeffreys et al., 1987).

Judicious choice of restriction enzymes, such as *Taq*I, that cleave at sequences containing **highly mutable sites** (e.g., at CpG islands; Cooper and Krawczak, 1990), combined with the use of long probes that cover multiple restriction sites of a given locus, can disclose RFLPs of virtually unlimited variety. Conventional methodology subsequent to enzyme digestion calls for separation of the DNA fragments produced. A number of newer technical developments allow for markedly increased resolving power, distinguishing fragments within a single base difference. This complements and enhances the resolvable fine polymorphic variations in genomic DNA.

A marked improvement in resolution of DNA fragments can be achieved with **high-performance capillary electrophoresis (HPCE)**, which combines high resolving power and speed (Gordon et al., 1988; Karger, 1989; Karger et al., 1989). The high resolving power is suitable for sequence determination because it allows discrimination at the single-nucleotide level (Cohen et al., 1990; Drossman et al., 1990; Heiger et al., 1990; Swerdlow and Gesteland, 1990). The high electric field that can be applied to the capillary system increases both the speed and resolution of the separation. In addition, the method can be automated, making it potentially useful for large-scale sequencing tasks such as required for the Human Genome Project (Cohen et al., 1990; Swerdlow et al., 1990).

Thus, relative to slab-gel electrophoresis, which is currently widely used for separating fragments of DNA produced by restriction digestion for the analysis of RFLPs, VNTRs (see Section 1.2.5.2) and amplification products, HPCE offers the advantages of speed, resolution, the possibility of automation, and a subattomole level of detection (Cohen et al., 1990). While slab-gel electrophoretic systems allow resolution of fragments above 400 bp in length, HPCE is capable of separating fragments ranging in size from under 100 bp up to 12,000 bp (Heiger et al., 1990). Resolution of restriction fragments in this range of sizes is important in clinical, diagnostic, and forensic applications of RFLPs, PCR product analysis, and in genome mapping. Practical diagnostic application of HPCE has been reported in the analysis of PCR-amplified HIV-1 *gag* gene and HLA DQα (Mayer et al., 1991).

RFLPs have been used extensively for studies on association (linkage) between disease loci and polymorphic markers. The construction of detailed linkage maps of human chromosomes has been advancing at an ever-increasing speed following the pioneering observations demonstrating this possibility (Botstein et al., 1980; White et al., 1985).

1.2.5.2 HYPERVARIABLE LOCI

The first report of a highly variable locus, in which DNA polymorphism was due to DNA rearrangements rather than to base substitutions or modifications, identified an arbitrary selected region showing a significant frequency of DNA sequence variation (Wyman and White, 1980). In that study, recombinant repetitive DNA probes cloned in the λ phage vector Charon 4A were used to screen a library of 15 to 20-kb segments of human genome in order to detect clones that were free of repetitive human DNA. With the **probe/restriction enzyme pairs** used in that initial study at least 16 distinct allelic combinations per chromosome have been defined and their usefulness in linkage analysis was demonstrated (Wyman and White, 1980).

Regions displaying a high level of DNA polymorphisms, known as **hypervariable regions** (**HVRs**), have been subsequently identified in various parts of the genome. Some of the first HVRs were discovered in a region of the human insulin gene (Bell et al., 1982), near the c-Ha-*ras*-1 oncogene (Capon et al., 1983), in and around the α-globin gene cluster (Higgs et al., 1981; Goodbourn et al., 1983; Jarman et al., 1986; Proudfoot et al., 1982), 3′ of the apolipoprotein B gene (Huang and Breslow, 1987), and 5′ to the joining segments of the heavy-chain immunoglobulin gene, (Silva et al., 1987).

In contrast to **site polymorphisms**, where changes in the DNA sequence produced by base modifications, insertions, or deletions eliminate the site of restriction enzyme cut, the different length of HVRs is due to the variable number of tandemly repeated sequences (Jeffreys et al., 1987). These repeats, called **minisatellites**, can be identified in an unknown mixture of DNA fragments by hybridization of a short **unit sequence** to the DNA of interest. Hypervariability of minisatellites apparently results from either **unequal recombination** between misaligned minisatellites, or **slippage** in the process of replication that produces a higher or lower number of these tandemly repeated units (Jeffreys et al., 1987). HVRs have also become known as **variable number of tandem repeats (VNTRs)**.

The usefulness of site polymorphisms for linkage analysis is limited because only two alleles can exist for each RFLP and they become noninformative in homozygotes for a particular allele (Thein and Wainscoat, 1986). In contrast, VNTRs show extremely high variability in length, and the frequency of heterozygotes may be on the order of 80 to 100% (Jeffreys et al., 1987).

One of the **minisatellite probes (33.15)** consisted of 29 repeats of a 16-bp core sequence, while a related probe **(33.6)** produced a different pattern of hybridization due to differences in their **core units** (Jeffreys et al., 1985a–c). The resulting patterns are **individual-specific DNA "fingerprints"** and each band (fragment) represents a different locus. Although chromosomal assignments of these fragments are not known, the characteristic and reproducible pattern can be used for an individual identification. Furthermore, since the majority of the bands are not genetically linked to each other, and are inherited in a Mendelian fashion with heterogeneity of almost 100%, these patterns can be used for linkage analysis of genetic disorders. A specific pattern can also be used for generating probes from the fragments (bands) displaying an informative association (linkage) with a given genetic trait. Polymerase chain reaction amplification directly from agarose gels makes this approach a highly practical and fast route to molecular genetic probing.

1.2.5.3 RFLP PROBES

The conventional methodology for restriction enzyme digestion of chromosomal DNA, agarose gel electrophoresis, transfer of the separated fragments by Southern blotting to membranes, and the hybridization conditions of various stringency have been well described in numerous publications (e.g., Ausubel et al., 1991; Bishop and Rawlings, 1987; Davis et al., 1986; Hames and Higgins, 1985; Maniatis et al., 1989; Perbal, 1988).

1.2.5.4 SINGLE LOCUS PROBES

High stringency conditions, which allow the formation of hybrids with only the highest degree of complementarity of the probe to the target sequence, are used in detecting restriction fragments from a single genetic locus. Naturally, longer probes will require a higher degree of complementarity to form a stable hybrid than shorter probes annealed under the same stringency conditions. Higher stringency hybridization is used to reveal RFLPs that show two alleles (fragments) from both chromosomes, if a person is heterozygous for the probe.

Single-locus probes have been produced in large numbers by cloning DNA of interest and have been used in the early stages of human genome mapping (e.g., Donnis-Keller et al., 1987). Likewise, oligonucleotide probes corresponding to the **consensus sequences** of the tandem repeats of various hypervariable loci have been constructed. Such VNTR probes, each derived from a single locus, have identified a large number of polymorphisms spanning the entire human genome. The use of these probes in gene mapping has been described in detail (Donnis-Keller, 1991; Nakamura et al., 1987).

It is estimated that for the RFLP probes to be efficient in genome mapping they should be separated from each other by no more than 0.2 morgans (M). Since the human genome is estimated to be 33 M in length, only as few as 150 evenly distributed RFLPs were initially thought to be sufficient to cover the entire genome (Botstein et al., 1980). To compensate for the uneven distribution of RFLPs, and lack of their informativeness in many families, alternative estimates raised the requirement for RFLP marker loci for the task of mapping the human genome by an order of magnitude (Lange and Boehnke, 1982).

A **combination of RFLP analysis with PCR amplification of target sequences** is an approach receiving much attention. Numerous variants have been explored (e.g., Gibbs et al., 1989a,b; Newton et al., 1989; Weber and May, 1989; C. Wong et al., 1987; Wu et al., 1989). The simplicity of this approach for analysis of the inheritance of alleles within a pedigree was again demonstrated in the case of the interphotoreceptor retinoid-binding protein (*IRBP*) gene (Saperstein and Nickerson, 1991). In contrast to Southern blotting, this combination of **PCR coupled to restriction digestion (PCRD)**, does not require probing of the products following digestion, reducing the labor required by a factor of 10, which makes linkage analysis acceptable for a small clinical laboratory.

Automation of the analysis of DNA polymorphisms offers a desirable modality not only for large-scale mapping efforts such as are involved in the HGP, but can offer a better controlled, faster, and reproducible tool for individual identification in forensic practice or parentage testing.

In addition to VNTRs, useful for high-resolution genetic linkage studies (Jeffreys et al., 1987) and yielding DNA **fingerprinting from single cells** when PCR technology is applied, allelic variations in $(dC-dA)_n$ **repeats** can also be easily detected with PCR. The wide distribution of this class of repeats, estimated to number up to 100,000 in the human genome, justified the efforts to develop an automated system for genotyping individuals with respect to $(dC\text{-}dA)_2$ polymorphisms (Diehl et al., 1990).

1.2.5.5 MULTIPLE LOCI PROBES

Hybridization patterns generated under less stringent conditions allow the probes to form less complementary hybrids and the RFLPs produced reveal fragments derived from multiple chromosome loci.

As mentioned above, this approach was introduced by Jeffreys and colleagues (1985a,b, 1986) when two related HVR probes derived from the **"core" consensus repeat sequences** were used to generate RFLP patterns. Scores of bands (alleles) have been produced in such patterns of *Hin*fI-digested chromosomal DNA derived from as many as 30 to 50 independent loci (Jeffreys et al., 1985b). While **reduced stringency** hybridization and wash conditions were used by the Jeffreys group to identify multiple-locus individual specific patterns, a **pAC365 probe** detects loci on 16 human chromosomes, producing a multiple-locus RFLP pattern under conditions of **high stringency**.

Since **each band represents a different locus**, the DNA "fingerprints" produced by multiple locus probes can trace a number of loci simultaneously. This imparts an extremely high degree of uniqueness to a given pattern, since the high degree of variability (polymorphism) at any locus (band) identified is very high and the number of bands is large. The probability that all fragments (bands) detected in one individual are present in the RFLP pattern from another individual using probes 33.6 and 33.15 has been calculated to be less than 5×10^{-10}. Thus, with the **exceptions of identical twins**, who can share a large number of alleles, these DNA fingerprints are **totally individual-specific** even within one family (Jeffreys et al., 1985b). The search for additional multilocus probes continues (e.g., Ip et al., 1989) (see also Section 1.2.5.11.2).

1.2.5.6 ANALYSIS OF DNA POLYMORPHISMS USING PCR

The number of emerging PCR-based techniques for analysis of genomic characteristics is increasing almost exponentially, and only selected representative approaches will be mentioned here. Analysis of DNA sequences is now possible at the **single-cell level** (Arnheim et al., 1990) and, given the strategem for inferring haplotypes from PCR-amplified samples (Clark, 1990), the convenience of PCR-based evaluations in clinical diagnostic practice is readily apparent, particularly when noninvasive **sampling from buccal mucosa** has been shown to be sufficient for direct genetic analysis (Lench et al., 1988). Prerequisites for such an analysis are specific amplification and sequencing that have been achieved using both cloned dsDNA and genomic DNA (e.g., Roux and Dhanarajan, 1990).

A general strategy for rapid screening of genomic DNA for sequence variation from many individuals has been devised as a two-step procedure using thermostable *Taq* DNA polymerase (Levedakou et al., 1989). First, the genomic sequences flanking the coding region of a target gene starting from the cDNA sequence are amplified. Then sequencing primers are synthesized based on the cDNA sequence of the gene. The same *Taq* DNA polymerase is then used for sequence analysis of amplified linear dsDNA under conditions similar to those used for the amplification reaction. This feature, when combined with fluorescently labeled primers, allows for easy automation of the entire cycle, offering an attractive tool for large-scale screening of DNA variations applicable to clinical diagnosis.

PCR has been suggested to produce stocks of single or multiple probes in clinical diagnostic laboratories for more efficient and rapid Southern analysis in various RFLP studies (Tarleton and Schwartz, 1991). By combining **reverse transcription (RT)** with PCR, a practical, rapid, and sensitive approach for **isozyme genotyping** is obtained (Mocharla et al., 1990). Interestingly, direct PCR amplification can be performed on lyophilized tissues and cells, markedly simplifying the process of genotyping in diagnostic laboratories (Wisotzkey et al., 1990). Furthermore, a simplified **ligase-free subcloning** procedure has been developed that allows incorporation of the PCR-amplified products directly into the vector when cloning is needed (Shuldiner et al., 1991).

Multiple polymorphic loci can be amplified simultaneously in a procedure using **amplified sequence polymorphisms (ASPs)** (Skolnick and Wallace, 1988). Another modification of PCR technique can be used to **identify species** and **genera**. This approach is based on the sharing of sequence motifs of **tRNA genes**, which occur in multiple copies dispersed throughout the genome in most species (Welsh and McClelland, 1991). It was expected that **consensus tRNA primers** could generate fingerprints that might be relatively stable in evolution since the organization of tRNA genes has changed little in evolution. This version of **PCR-generated fingerprints** was termed **tDNA-PCR** to indicate that DNA coding tRNAs is amplified to give fingerprints useful in identification at the species or genus level. Reflecting the relatedness at the DNA level with respect to a particular, highly conserved gene, this approach offers a very simple, universal way to compare reliably genomes of organisms at the species/genus level (Welsh and McClelland, 1991; Welsh et al., 1991). Consequently, the method can be of use in the epidemiology of human diseases. Yet another version of PCR fingerprinting has been used for selection of HLA-matched unrelated bone marrow donors (Clay et al., 1991).

An important application of **PCR amplification for detecting polymorphisms at VNTR loci** allows to use these highly specific polymorphic markers in a highly simplified format (Decorte et al., 1990). Alleles up to 2000 bp can be detected on ethidium bromide-stained agarose gels. The method can be used for typing **allelic variations at any hypervariable regions**. This approach has the additional advantage over Southern blot analysis traditionally used for this purpose in that it can distinguish small differences (11 to 70 bp) between large DNA fragments. The method is suitable for a variety of **medical** and **forensic** applications as well as in **paternity determinations**, with the results available within 2 days (Decorte et al., 1990).

Other polymorphic loci such as the **HLA-DQ-α locus** have been studied by PCR amplification with subsequent probing of amplified DNA for the allelic diversity with nonradioactively labeled oligonucleotides (Helmuth et al., 1990). The probes labeled with horseradish peroxidase were hybridized in a **dot-blot** format to immobilized amplified DNA (Bugawan et al., 1988) or, alternatively, the probes were immobilized on a filter and hybridized to biotin-labeled PCR-amplified DNA in a **reverse dot-blot** technique (Saiki et al., 1989). This approach allows for rapid and relatively simple typing of a large number of samples.

In contrast to identifying alleles by their electrophoretic mobility, the definition achieved by hybridization with allele-specific probes eliminates the necessity of precise calibration of fragment (allele) sizes. And, although potentially "blank" alleles — those not amplifiable with the primers used or not detectable with the probes used — may cause a problem, this was not the case in a survey of over 2000 individuals and over 200 cell lines (Helmuth et al., 1990). The wide range of frequencies of DQ-α alleles in different populations suggests that these polymorphisms can be used as markers in population studies (Helmuth et al., 1990).

Compared to the VNTR system defined by RFLPs, an **allele-specific typing system**, such as for the DQ-α locus, in addition to identifying discrete traits transmitted in a Mendelian fashion with the permanence and distinctness maintained over millions of years, allows for better discrimination of alleles not limited by the resolution of the bands in a gel electrophoresis system (Helmuth et al., 1990).

Parenthetically, the conventional **submarine gel electrophoresis** used to separate PCR amplification products suffers from **altered migration of bands** depending on the concentration of the same sample of DNA. Furthermore, the resolution of shorter fragments such as 242-bp DQ-α in the presence of fragments longer than 1500 bp requires that agarose gels of different concentrations be used (Allen RC et al., 1989). **Rehydratable polyacrylamide gels** and **silver staining** markedly improve the resolution to within 1.6 4 bp in the range of 100 and 500 bp (R. C. Allen et al., 1989).

The **discriminatory power** of the DQ-α markers is less than that attainable with some VNTR systems (Balazs et al., 1989). It is, however, counterbalanced by the practical utility of this PCR-based typing, which can rapidly generate information starting from minute amounts of material. The convenience of the reverse blot format, where all the probes come fixed to a typing strip, has been translated into a commercial product. This nonradioactive system has been widely used in forensic casework (Helmuth et al., 1990).

1.2.5.7 INTERSPERSED REPETITIVE SEQUENCES-PCR (IRS-PCR)

Identification of specific human DNA may be complicated by the presence of nonhuman DNA in studies of somatic cell hybrids or in forensic practice. The presence of repetitive elements in the human genome distinguishes it from that of nonmammalian species (Britten and Kohne, 1968; Davidson et al., 1973, 1975; Jelinek and Schmid, 1982; Weiner et al., 1986). Combined biochemical, cytological, and recombinant DNA approaches together with computational analysis indicate that the distribution of **interspersed repetitive DNA sequences** in the human genome can be best described by models assuming a **random distribution** (Moyzis et al., 1989).

An average distance between repetitive domains is 3 kb, however, however, some sequences such as the *Alu* family display varying density. *Alu* elements represent the major family of **short interspersed repeats** (**SINEs**) in mammalian genomes, and, characteristically, there are approximately 10^6 copies of a 300-bp sequence scattered throughout the human genome approximately every 3 to 4 kb (Britten et al., 1988). *Alu* sequences are frequently interspersed in the human genome (Britten et al., 1988; Jurka and Smith, 1988; Labuda and Striker, 1989; Zuckerkandl et al., 1989) and **flank anonymous DNA segments** harboring yet undisclosed polymorphisms. ISH studies on human metaphase chromosomes indicate that *Alu* sequences are at least 50-fold less abundant in centromeric heterochromatic regions. Interestingly, as determined by ISH, the members of the other family of repetitive sequences, **L1**, are **in an inverse relationship** to *Alu* repeats with respect to their cytologically identifiable locations on chromosomes (Korenberg and Rykowski, 1988). While *Alu* sequences are underrepresented in centromeric regions, other repetitive sequences tandemly arranged are found there (Moyzis et al., 1987). Abundance measurements for other major repetitive sequences, L1 and GT/AC (Rich et al., 1984) families, indicate that these should be located, on average, every 30 to 60 kb throughout the human genome (Moyzis et al., 1989).

Application of **interspersed repetitive sequences PCR (IRS-PCR)** directed toward both *Alu* and L1 sequences allows for a **chromosome-specific pattern of amplification products** to be easily analyzed on agarose gels run with ethidium bromide (Ledbetter et al., 1990b). This "**PCR-karyotype**" approach can substantially simplify the characterization of somatic cell hybrids. The chromosome-specific banding pattern of PCR-karyotype products in agarose electrophoresis offers a convenient alternative to conventional cytogenetic procedures. Furthermore, **monochromosomal and subchromosomal reference panels** have been prepared for a dot-blot hybridization protocol helpful in detailed characterization of chromosomal material. In certain comparative studies, the abundance of *Alu*-primed amplification products may be a limiting factor, and predigestion of template DNAs with restriction enzymes prior to *Alu*-PCR ("**restricted *Alu*-PCR**") has been proposed (Guzzetta et al., 1991). This allows, for example, the isolation of specific markers from a given chromosomal region.

The method of **amplified sequence polymorphisms (ASPs)** can be applied, when sequence information is available, to perform detailed scrutiny of differences in the genome patterns being compared. An alternative approach, when prior knowledge of the DNA sequence of each polymorphic locus to be amplified is not required, helps to identify polymorphisms flanked by *Alu* sequences (Sinnett et al., 1990). It appears that approximately 100,000 *Alu*-flanked relatively short (1 kb) DNA segments occur in the genome that can be amplified by PCR.

Amplification of these "**alumorphs**" using *Alu*-specific primers allows the detection, and analysis if combined with restriction enzyme digestion, of multiple (10 to 15) polymorphic loci in a single experiment (Sinnett et al., 1990). Granted that **alumorphs are less informative than the RFLP or VNTR** markers, they are **more uniformly scattered** throughout the genome than minisatellites or VNTR markers, which tend to cluster in the proterminal regions of human autosomes (Armour et al., 1989; Royle et al., 1988). This feature makes alumorphs attractive markers for gene mapping, as well as helpful adjuncts in forensic investigations and population studies (Sinnett et al., 1990).

A PCR protocol has been developed based on *Alu* repeats to identify **human-specific sequences** in somatic cell hybrids using primers directed at the human *Alu* repeat sequences (D. L. Nelson et al., 1989, 1990). To distinguish human DNA from rodent DNA, advantage is taken of the 31-bp insert found in the second monomer of the 300-bp human *Alu* repeat, which is present only in primates. Amplification of this insert permits specific identification of human DNA in a mixture of heterologous material.

By analogy with the amplification of *Alu* repeats, the other major class of IRSs, the **L1 element**, has also been used for PCR-based analysis (Ledbetter et al., 1990a,b). This **long interspersed repetitive sequence (LINE)** is present at 10^4 to 10^5 copies per genome in mammals (Singer and Skowronski, 1985). The human L1 elements have also been known as *Kpn*I sequences, and the number of complete 6-kb elements is less than 10^4 due to truncation (Singer and Skowronski, 1985). Analysis of L1 sequences suggests that most L1 sequences were derived from a small number of structural genes apparently evolved by pseudogene formation (retroposition), transposition, gene conversion, and RNA recombination (Scott et al., 1987).

Thus, primers constructed to a variety of IRS types can be used for the identification and amplification of defined chromosomal regions by IRS-PCR as well as for cytogenetic studies on human chromosomes by FISH (Lichter et al., 1990a). They can also be used to serve as **STSs** to identify reference points in the physical mapping of the genome (Olson et al., 1989). **Telomeric repetitive sequences** featuring the **(TTAGGG)$_n$ motif** have been identified and cloned (Moyzis et al., 1988). A new class of DNA length polymorphisms associated with *Alu* sequences (*Alu* **sequence-related polymorphisms**) has been described using PCR amplification of a **(TTA)$_n$ repeat motif** in the 3-hydroxy-3-methylglutaryl coenzyme A (HMG-CoA) reductase gene (Zuliani and Hobbs, 1990).

1.2.5.8 AMPLIFICATION USING ARBITRARY PRIMERS

An interesting and promising procedure to study DNA polymorphisms has been developed using **amplification with arbitrary primers** (J. G. K. Williams et al., 1990). This procedure for amplification of genomic DNA, distinct from conventional PCR, *does not require specific sequence information* because it uses primers of arbitrary nucleotide sequences. These primers detect polymorphisms that are inherited in a Mendelian fashion, and can be used to construct genetic maps for DNA fingerprinting and other applications. The polymorphisms generated in this manner are termed **random amplified polymorphic DNA (RAPD)** markers. Nucleotide sequencing, hybridizations, and specific probes are not necessary when RAPD markers are used, offering a significant advantage over other, more traditional methods. A universal set of primers can be used for genomic analysis and the information can be easily transferred among various collaborators. Most significantly, the entire process can be readily automated and, according to the authors, genetic maps using RAPD markers can be generated more efficiently and with greater marker density than by RFLP or targeted PCR-based techniques (J. G. K. Williams et al., 1990).

An essentially similar approach using arbitrary primers for PCR amplification has been described by others to generate fingerprints (Welsh and McClelland, 1990). This **arbitrary primed PCR (AP-PCR)** requires no prior sequence information. Two cycles of low-stringency amplification are followed by PCR at higher stringency, generating polymorphic fingerprints. This method was demonstrated on a number of *Staphylococcus* and *Streptococcus* strains and three varieties of mice and proved to be species and subspecies specific. Occasional amplified products contained prominent derivatives of plasmids, reducing somewhat the uniqueness of the patterns produced. Subsequent elaboration of the primer selection proved AP-PCR to be able to distinguish between strains of almost any organism as demonstrated by strain identification and genetic mapping in the mouse (Welsh et al., 1991). It appears that the **method can be applied virtually to any species, including man** (Welsh and McClelland, 1990).

1.2.5.9 MINISATELLITE VARIABLE REPEATS

Amplification of hypervariable loci by PCR significantly increases the sensitivity of single-locus minisatellite probe analysis (Boerwinkle et al., 1989; Horn et al., 1989; Jeffreys et al., 1988a,b). Variant sequences have been identified in the tandemly repeated units arising through a number of possible mechanisms, including base substitution (Z. Wong et al., 1986, 1987). These **minisatellite variable repeats (MVRs)** may have different patterns of variation in minisatellite alleles, and so far very little is known about the extent of their allelic variation (Jeffreys et al., 1990, 1991).

A further refinement of the analysis of minisatellites can be achieved by **mapping variant repeat units within amplified alleles** (Jeffreys et al., 1990, 1991). This approach dramatically enhances the number of different alleles that can be distinguished in a population. For example, in the human hypervariable locus D1S8 this PCR-based technique allows the distinction of over 10^{70} allelic states. Two rounds of PCR amplification of genomic DNA are performed: first, two parental alleles are isolated by PCR, and then much shorter mutant alleles produced by internal deletions within a variable repeat unit are amplified. The amplified products are then end-labeled by cleavage with *Eco*RI followed by fill-in labeling. A combination of two restriction digest patterns is compared: one resulting from restriction with *Hin*fI which cleaves once in all repeat units, and the other produced by *Hae*III, which cleaves at the A→G transition.

Partial digestion with *Hin*fI followed by separation of the fragments by electrophoresis, allows the number of minisatellite repeat units to be determined. Comparison of the partial digest patterns produced by *Hae*III and *Hin*fI enables each repeat unit to be evaluated for the presence of a mutation reflected in the presence of a restriction site for *Hae*III (Jeffreys et al., 1990, 1991).

Jeffreys and colleagues (1990, 1991) applied this technique to the smallest MS 32 allele. The allelic patterns produced are highly reproducible. Importantly, for the first time, an objective criterion (the presence of a restriction site) can be applied to comparing different band patterns as an alternative to the side-by-side comparison usually performed for Southern blot analysis in identity determinations. Importantly, the **MVR map** can be encoded in a **binary code**, in which restriction fragments cut by *Hae*III are designated 1 and the noncleaved repeats are designated 0, or in a two-letter code such as *a* for repeat cut, and *t* for repeat not cut. This allows the entire ladder of bands in an electrophoretic gel to be expressed in a highly objective manner, and the coded information can be stored and manipulated as if it were a DNA sequence (Jeffreys et al., 1990, 1991). Several minisatellites have been identified suitable for MVR-PCR and even for multiplex MVR-PCR where length variation does not interfere in the analysis of sequence variation (Jeffreys et al., 1991).

This MVR mapping can also be used for estimating somatic and germ-line mutation rates as they occur within a single individual. Until now, such analysis could be accomplished only in clonal cell populations such as tumors (Jeffreys et al., 1990, 1991). Among other applications, MVR mapping may revolutionize direct determination of the effects of environmental agents on human mutation rates. A number of fundamental questions relating to human evolution can be addressed with this powerful technique as well (Jeffreys et al., 1990, 1991; Dover, 1990).

1.2.5.10 DNA FINGERPRINTING ANALYSIS OF SOMATIC DNA CHANGES IN TUMORS

Changes in the genomic DNA in mammalian tumors have been traditionally implicated by cytogenetic studies. These have been supplemented by the detection of tumor-associated chromosomal aberrations such as **loss of heterozygosity** using RFLPs (see below). The majority of earlier RFLP associations have been established using single-copy DNA probes with known chromosomal location. When informative, such an analysis yields specific assignment of a disease to a given chromosomal location.

Alternatively, application of the DNA fingerprinting technique utilizing probes for various repetitive sequences in the genomic DNA allows assignment of genotypic traits to reproducible fingerprinting patterns. This approach, until recently, was used predominantly in **paternity testing, forensic applications**, and some **cell line analysis** (Masters et al., 1988). A variety of probes have been used to detect repetitive sequences. Among those, wild-type **M13 phage DNA** proved to be helpful as a universal marker for DNA fingerprinting in humans, animals, plants, and microorganisms.

Human minisatellite probes have also proved useful in the analysis of **genomic abnormalities** in tumor cells. It turned out that differences between tumors and unaffected tissues were reflected in the **intensity of hybridization bands** and in the emergence of **novel bands** detectable in tumor material, these differences appearing to be tumor-specific rather than tissue-specific (Thein et al., 1987; White et al., 1990; Uchida et al., 1990). Wild-type M13 phage DNA proved not only widely useful for DNA fingerprinting (Chen et al., 1990; Georges et al., 1988; Ryskov et al., 1988; Vassart et al., 1987), but also helpful in the detection of genomic alterations of human (Hayward et al., 1990; Lagoda et al., 1989; Pakkala et al., 1988) and animal (Muller et al., 1990) tumor cells.

Using an **oligonucleotide probe (GTG)$_5$**, DNA fingerprints have been generated from intracranial tumor tissues removed surgically that demonstrated tumor-specific chromosomal aberrations not identifiable in corresponding unaffected tissues (Nurnberg et al., 1991). Further characterization was achieved by additional hybridization with the **(GT)$_8$** and **(GATA)$_4$** probes. Thus, it was established that in many gliomas, the amplification unit contained two simple repetitive DNA fingerprint loci, **(CAC/GTG)$_n$** and **(CA/GT)$_n$**, in addition to the EGFR gene (Nurnberg et al., 1991). This approach provides unique data for the **definitive identification of certain tumors** that can be used in clinical diagnostic laboratory. Restriction analysis can also be helpful in **establishing histogenesis** in some poorly differentiated carcinomas, as shown in a series of midline carcinomas of uncertain histology (Motzer et al., 1991).

An obvious advantage of the characterization of tumors by DNA fingerprinting compared to classical cytogenetic methods is that such an analysis does not require metaphase chromosomes. Although this method does not allow the identification of a specific genomic alteration responsible for the tumor development, a reproducible and specific fingerprinting pattern that can be associated with a given type of neoplasia may provide a relatively simple and rapid diagnostic tool for clinical practice. Considering that **DNA amplification is rare in normal human cells** in contrast to the situation in tumors, in which amplification of oncogenes and of genes associated with drug resistance phenotypes occurs (Wright et al., 1990a), **analysis of DNA amplification combined with DNA fingerprinting** may find its way into clinical practice in the evaluation of specific tumors. Furthermore, DNA fingerprinting also offers a relatively simple and sensitive tool for **monitoring the course of tumor progression** in addition to **assessing tumor clonality** (Hayward et al., 1990).

Understandably, the utility of this approach wholly depends on the **reproducibility of the restriction enzyme digestion** of the tumor tissue. The fine differences between the digestion patterns of normal and tumor tissues generated by a given enzyme are interpreted as specific and reproducible diagnostic criteria. This implies, in turn, that the restriction enzymes cleave DNA in tumor and unaffected cells with the same specificity and efficiency. This premise, however, may not always hold (Parkes et al., 1990). Comparison has been made of the digestion efficiency of nine restriction enzymes (*Hin*dIII, *Eco*RI, *Eco*RV, *Bgl*I, *Bgl*II, *Kpn*I, *Xba*I, *Bam*HI, and *Pst*I) on DNA extracted from transformed and nontransformed NIH 3T3 cells in culture and DNA extracted from nude mice tumors. DNA from cultured cells was digested by all enzymes to completion. Three enzymes, *Hin*dIII, *Kpn*I, and *Xba*I, consistently **failed to completely digest** the tumor DNA, and, importantly, ethidium bromide staining failed to reveal this. It appears that anomalous restriction patterns obtained with tumor DNA were not due to alterations of the tumor DNA itself, but rather were caused by the presence of a factor(s) in preparations of tumor DNA preventing the cleavage of particular sites in substrate DNA (Parkes et al., 1990). These observations should always be considered when comparing digests of normal and tumor DNAs for any purpose and appropriate controls should be included.

1.2.5.11 DNA POLYMORPHISMS IN IDENTITY TESTING

1.2.5.11.1 Forensic Applications

As discussed in Section 1.2.5.9, hybridization patterns produced by the minisatellite probes 33.6 and 33.15 display a very high degree of polymorphism. Since the introduction of HVR probes by Jeffreys and colleagues (Jeffreys et al., 1985a,b; Wong et al., 1986), they have been widely applied to **identity determinations** (Gill et al., 1985, 1987; Jeffreys et al., 1985c). Various problems arising in application of DNA technology to forensics have been the subject of a symposium (Ballantyne, 1989) and numerous press reports. Although **socio-legal issues** related to the use of recombinant DNA techniques in identity determinations required in forensic practice will certainly outlive the current technical problems (Gill, 1990), a brief summary of the technical issues is in order.

Analysis of nucleic acid material in forensic practice may have to be performed on specimens presented as stains (blood, semen, body fluids), tissue fragments (skin squames, hair), vaginal swabs, aspirates, and even bone (Hagelberg et al., 1991). Frequently, the amount of specimen is not adequate for sufficient DNA or RNA to be extracted for "conventional" processing by gel electrophoresis. Prior to performing restriction digestions an aliquot of the specimen is evaluated to determine the state of the DNA (degree of its degradation), and if the DNA is predominantly represented by fragments less than 2 kb in size it is considered too far degraded for forensic RFLP analysis (Baird, 1989). Again, to ensure that the DNA recovered from the presented specimen is not contaminated by bacterial DNA, it is probed by *Alu* repeat sequences characteristic of mammalian origin (see Section 1.2.5.7).

DNA is then digested, usually by *Pst*I, and tested with DNA probes specific for the Y and X chromosomes. Following separation of the DNA digests on 0.9% agarose and transfer to a nylon membrane, the Southern blot is hybridized with two nonoverlapping radioactively labeled probes — **pAC255**, the D2S44 locus with fragment sizes from 6.5 to 17.0 kb, and then the **pAC256** probe, which recognizes the D17S19 locus with fragment sizes from 2.0 to 6.0 kb (Baird, 1989). Following autoradiography with these two probes, the blot is stripped of them and hybridizations are repeated with other highly polymorphic DNA probes: **pAC225, pAC254, pAC061**, and **pAC404**. Finally, the membrane is challenged with a **bacterial ribosomal gene probe** labeled with the vector to reveal the possible bacterial contamination of the specimen. Size markers are run in the same gels at several locations to provide reference landmarks for comparison of molecular weights and specific band positions. This is the procedure reportedly employed at Lifecodes Corporation (Valhalla, NY) (Baird, 1989).

Subsequent analysis of fragments (alleles) revealed by the polymorphic probes is interpreted within the framework of the known **frequency distributions of specific DNA fragments** for a given population (American blacks, Caucasians, Hispanics). Following that, the frequency of the evolving pattern is determined using the Hardy-Weinberg equation, and the patterns generated from various sources are used for comparison. The importance of establishing extensive databases for various ethnic and racial groups ensuring reliable analysis of DNA polymorphisms in forensic applications has been repeatedly debated (Austad, 1992; Chakraborty and Kidd, 1991; Lewontin and Hartl, 1991; Risch and Devlin, 1992; Roberts, 1991; Wills, 1992).

Another major U.S. laboratory providing DNA support to forensic investigations is Cellmark Diagnostics (Germantown, MD). This facility reportedly used four highly polymorphic **locus-specific minisatellite probes MS1**, **MS31**, **MS43**, and **g3** (Cotton et al., 1989). The probability of two unrelated individuals possessing the same genotype with these probes is calculated to be 3.4×10^{-12}. The virtual uniqueness of these patterns obviates the need for sizing alleles or determining specific loci (Jeffreys et al., 1990). At the present time, forensic laboratories predominantly use Southern analysis with the minisatellite and VNTR single-locus probes, and population databases for these probes are being developed (Balazs et al., 1989).

A recently conducted assessment of the use of DNA technology in forensic practice (Office of Technology Assement [OTA], 1990) concluded that methodological errors, including poor sample handling, incomplete DNA digestion, incomplete Southern transfer of DNA to a membrane, poor probe labeling, and inadequate autoradiography, **may fail to produce expected results, but will not lead to an error in identification**. Validation studies at the Federal Bureau of Investigation (FBI) also indicated that adverse exposure of the specimen and chemical and bacterial degradation also *lead to no result, rather than to an error in identification* (Caskey and Hammond, 1989).

Furthermore, the OTA finds that, when properly performed, DNA technologies per se are reliable with acceptable reproducibility of results within one laboratory and among different laboratories (OTA, 1990). The OTA highlights the following considerations for choosing forensic single-locus probes: a DNA probe for forensic use should (1) contain DNA sequences that detect only **one chromosomal locus** under a reasonable range of hybridization conditions, (2) produce **well-characterized patterns** (such as defined size range, number of bands, and relative band intensities) so that unexpected patterns can be recognized, and (3) detect a **single polymorphic fragment per allele** so that each person's test yields either one or two fragments, depending on whether an individual is homozygous or heterozygous, respectively (OTA, 1990, p. 62).

The OTA goes on to emphasize that probes that may detect alleles of varying number or intensity on Southern blots should be avoided in order to minimize the problem of identifying the true bands that comprise an RFLP pattern. Weak intensity bands and a dirty background produced by certain probes lead to uncertainty about the patterns generated and are to be avoided. Furthermore, a series of probes from different chromosomes, or reasonably distant sites on the same chromosome, should be selected to ensure that the alleles revealed are independently segregating. The probes selected should be sufficiently widely used and available to other scientists to confirm their alleged properties and, finally, the probe should be characterized with respect to its chromosomal assignment filed with the Human Gene Mapping Workshop.

Interestingly, **two-dimensional electrophoresis** of restricted DNA fragments probed by minisatellite core sequence fragments has been able to resolve up to 625 separate spots per probe per human individual (Uitterlinden et al., 1989). This technique offers a powerful tool for generating individualized fingerprints helpful not only in identity determinations, but also in the studies of disease associations, mutations, and other processes related to cancer and aging. It remains to be seen whether the technical aspects of two-dimensional DNA electrophoresis can be standardized to the degree acceptable in identity determinations.

Recently, a procedure for **in-gel hybridization** with nonradioactive $(CAC)_5$ oligonucleotide probes has been described, markedly simplifying the process of fingerprinting (Zischler et al., 1991). Amplification of simple repeat DNA sequences by PCR followed by hybridization with

$(CAC)_5/(GTG)_5$, $(CT)_8$, and $(GACA)_4$ **probes** can be used for the analysis of forensic stains using a specific probe combination depending on the amount of the recovered material and the informativeness of the results obtained with a given probe (Roewer et al., 1991).

An important aspect in forensic casework is the ability to declare that two or more DNA patterns match. **Mixing experiments**, although capable of revealing minor discrepancies in the patterns being compared, are not routinely performed by forensic laboratories due to the frequently limited amount of the available DNA and difficulty in quantitating it when preparing an informative mix; however, many scientists believe that **mixing tests are the "gold standard" for DNA typing of forensic specimens** (OTA, 1990). As an alternative to mixing, identity is inferred by comparing the **positions of bands** and, thereby, **their size** in two separate lanes. Due to possible minor displacements in the positions of bands in different lanes of electrophoresis gels, resulting in "**band shifts**", the best estimate of an allele size is obtained by comparing it to a set of **internal lane controls** — polymorphic markers. Scientists differ as to the interpretation of matches based on the analysis of bands. The important issue is how close the two alleles must be in size (as identical measurements are not achievable) to be declared a match based on the "molecular rule" reference. It is generally agreed that fragment lengths should be within a specified, and always observed, **range of deviation** observed empirically under a given set of reproducible experimental conditions when known forensic samples are repeatedly tested (OTA, 1990).

Multilocus probes, initially used in an **immigration case** in the United Kingdom in 1985 (Jeffreys et al., 1985a), are viewed by some as requiring more sample than single-locus probes, and, in some cases, the large number of bands produced may complicate forensic analysis. The **current consensus is that multilocus probe analysis is *not* the most appropriate technique for criminal casework** (OTA, 1990).

Newer methods such as PCR amplification substantially expand the range of application of DNA technology to forensic identifications. Even degraded specimens can be used for PCR of samples containing picogram quantities of DNA, which is inadequate for Southern analysis. A degree of restraint in the use of PCR for criminal casework is founded on the technical aspects of this technology, which may interfere with identity determinations (see Section 1.2.7). The major concern when using PCR is **contamination** produced by the reaction itself and proper controls and precautions must be observed. The **misincorporation** of a nucleotide into the amplified product, occurring once or twice per 20,000 to 1 million bases, happens randomly, and is not considered a problem because the entire population of the molecules generated is examined rather than only a single molecule, thereby reducing the relative contribution of a misincorporation. The minuscule amount of starting material, as may frequently be the case in forensic practice, can potentially present a problem if misincorporation occurs early in the amplification reaction. **Preferential amplification** of one allele over the other by PCR due to the condition of a forensic sample may result in spurious identification of the person as homozygous for a trait. Currently, the only genetic system generally employed in forensic casework using PCR is the HLA DQ-α-1 system, although it is not as discriminating as RFLP analysis.

DNA technology is developing at a very rapid rate, to say the least, and the limitations of the present-day methods of identity testing may be obsolete in the near future. The introduction of nonradioactive labels along with the improvements in DNA assay methodology may substantially widen the scope of laboratories entering the field. The **maintenance of proficiency** in forensic DNA testing may become more accessible, particularly if and when semi- and fully automated DNA typing systems become available.

An important advance in the standardization of DNA genotyping methodology has been the introduction of **minisatellite variable repeats (MVRs)** by Jeffreys and colleagues (1990) discussed above (see Section 1.2.5.9). Not only is the even higher degree of uniqueness of the fingerprints produced by this method available for more secure identity testing, but the possibility of objective interpretation of the restriction patterns in this system offers an unparalleled and reproducible "bar coding" of an individual. Since no subjective steps such as size measurement of fragments is required, the reading of patterns and conversion of these into a digitalized format offers an exchangeable, reproducible, and unique identification system particularly attractive for forensic practice. The **digitalized transformant** of a given pattern can be stored and objectively compared to any set of similarly presented patterns for computerized criminal casework as if these were DNA sequences themselves with their inherent individual uniqueness. In certain cases, the identification of the source of material as belonging to humans or animals of different species may be important. A protocol termed **FINS (Forensically Informative Nucleotide Sequencing)** allows the identification of the animal origin of DNA under study using PCR, sequencing of the amplified product, and computerized phylogenetic analysis of the sequence patterns (Bartlett and Davidson, 1992).

Statistical considerations in the application of single locus and multilocus probes for identity testing and disease associations have been widely discussed and the interested reader should consult original publications for detailed coverage (Cohen, 1990; Evett et al., 1989; Lynch, 1990; Martinez and Goldin, 1990; Morris et al., 1989; Risch, 1990 a–c; Yassouridis and Epplen, 1991). **Computerized assessment of hypervariable DNA profiles** has been developed and the initial results suggest that band matching should be performed only after analysis of the errors arising in the process of electrophoretic separation of fragments of various sizes (Gill et al., 1990).

1.2.5.11.2 Paternity Testing

The role of DNA probe technology in parentage testing has firmly established this modality as an adjunct to testing genetic polymorphisms expressed in the products of gene expression, predominantly **proteins**. Some authors maintain that the range of genetic markers composed of red cell antigens, red cell enzymes, serum protein and HLA polymorphisms is fully adequate for efficient paternity testing service (Dunn et al., 1989). Nevertheless, the low mutation rate, estimated to be 0.0038 per DNA fingerprint band per gamete using the Jeffreys probes 33.6 and 33.15, allows for unambiguous paternity testing even when HLA typing cannot distinguish between tentative fathers (Helminen et al., 1988). Numerous, detailed discussions of the underlying principles, procedures, and the statistical apparatus employed in routine parentage testing using DNA probing have been presented (e.g., Allen et al., 1990; Cawood, 1989; Devlin et al., 1990; Dykes, 1987; Gazit and Gazit, 1990; Kirby, 1990; Melvin et al., 1988; Smouse and Chakraborty, 1986; Thein and Wainscoat, 1986; Watkins, 1988). Guidelines for standardization of parentage testing laboratories have been issued by the American Association of Blood Banks (1989).

As a rule, in parentage testing, the high informative power of HVRs is used. Currently, a large number of probes are available, combinations of which allow over 99% of men falsely accused in paternity suits to be excluded (for a review see Allen et al., 1990). Some of the widely used probes in paternity testing are those for the D14S1 locus, the flanking region of HRAS-1 (Baird et al., 1986), the five probes for HVR loci (D2S44, D14S1, D14S13, D17S79, and DXYS14) (Balazs et al., 1989), and pa3′HVR for the D16S85 locus (R. W. Allen et al., 1989). A number of different restriction enzymes are capable of revealing chromosomal fragments informative with these probes in parentage testing (Allen et al., 1990), so long as the enzyme cuts outside an HVR displaying the polymorphism tested and nonisotopic probes are available (Dykes, 1988).

In certain disputed paternity cases, technical considerations related to the resolving power of agarose electrophoresis used in separating the restricted fragments in single-locus probe testing come into play. **Variability in the estimates of the length of allele fragments** (usually within

0.6 to 1.0%) must be entered into the statistical evaluation of the observed frequency of a restriction fragment in the population (see also Gill et al., 1990). Using the measurement of fragment lengths of alleles at the D14S1 locus as an example, an approach to calculate the probability of paternity incorporating the allele measurement error has been developed (Gjertson et al., 1988). An averaging approach that combines close alleles into the so-called "**frequency bins**" is used (Allen et al., 1990). The technical aspects of allele size analysis in paternity testing are well discussed elsewhere (Allen et al., 1990). Advantages of using DNA probes have been clearly demonstrated in a case of disputed paternity in which 7 probes had been used to produce the **cumulative paternity index** of 1.4×10^6, which was 316 times higher than that obtainable from the 23 standard blood group and HLA markers (Yokoi et al., 1990).

A more informative pattern is generated by probing Southern blots under reduced stringency conditions to reveal fragments derived from multiple chromosomal loci - **multilocus probing** (see Section 1.2.5.5). Save for the case of identical twins, such multilocus probes provide virtually unique individual identification (Jeffreys et al., 1985a–c). Technically, comparison of patterns of the child, the mother, and the alleged father are made by aligning the generated band ladders and excluding bands not shared by the mother and the child. Again, a variety of probes can be used, some being the same probes employed for single-locus testing, such as the pa3′ HVR probe, as well as the M13 phage. The latter is informative due to 15-bp repeats that appear to be highly represented in the genomes of a variety of species, including human (Devor et al., 1988; Vassart et al., 1987; see above).

The statistics of calculating the probability of paternity using multilocus probes are different from those applied to statistical treatment of data obtained with single-locus probes (Allen et al., 1990). A general formula applicable to the usual trio cases for calculating the probability of paternity using DNA fingerprinting with the Jeffreys probes has been described (Honma and Ishiyama, 1989). **In cases when either the putative father and/or mother are absent, single-locus probing involving relatives becomes much less informative, whereas DNA finger-printing with multilocus probes is still effective.** This is due to the large number of highly polymorphic loci inherited via Mendelian fashion that are revealed in the patterns simultaneously (Evett et al., 1989; Jeffreys et al., 1987; Morris et al., 1989).

Although HLA system and red blood cell (RBC) antigen typing is useful in cases of disputed paternity when the alleged father is diseased prior to testing (Smith et al., 1989), the combined probability of paternity equals 99.998% (Odelberg et al., 1988). RFLP analysis using *Hin*fI and *Sau*3A single digests and the minisatellite DNA probes 15.1.11.4 and 6.3 gave in one case nearly conclusive evidence that the putative father was the biologic father of the child (Odelberg et al., 1988). Improvement of single-locus DNA probe informativity in a disputed case of paternity, in which one of two brothers appeared to be the father, was achieved by the use of a second restriction enzyme that excluded one of the brothers (Chiafari and Wenk, 1990).

Minisatellite probes have been successfully used on amniocytes or chorionic villi for **prenatal determination of familial relationships** (Kovacs et al., 1990). Reliable and reproducible protocols have been developed for the routine DNA fingerprinting also applicable to paternity testing using the Jeffreys probes 33.15 and 33.6 (Smith et al., 1990). Again, analysis of mutant bands in offspring examined in over 100 families confirmed that the mutant bands did not significantly affect the use of minisatellite probes in paternity testing. A sufficient number of additional informative markers is available that can establish the relationship.

A general formula describing the probability relationships for the determination of parentage or other blood relationship in cases in which critical family members are unavailable has been proposed (Honma and Ishiyama, 1990). This formula presents a generalized relationship, based on Bayes' theorem, for calculating the probability of parentage, grandparentage, brotherhood, and other relationships with absent key family members.

An emerging approach to determining familial relationships based on the maternal inheritance of the **mitochondrial genome** uses PCR amplification and sequencing of the D loop

(Orrego and King, 1990). So far, however, not enough population data are available to make this approach widely acceptable yet.

The **issue of mutations** in establishing paternity disputes is far from being completely resolved according to some authors (Kobilinsky and Levine, 1988, 1989).

1.2.6 FLOW CYTOMETRY AND IMAGE ANALYSIS

Flow cytometry (FCM) has been used extensively not only for the relatively crude estimates of DNA content in abnormal (benign and malignant) cells (e.g., Grogan and Collins 1990; Keren, 1989), but also for the more sophisticated experimental applications in molecular genetics (e.g., Crissman and Steinkamp, 1990; Jett et al., 1990; van Dilla et al., 1990). One of the prerequisites of FCM is the presentation of the sample for analysis as a composite of discrete cells or subcellular components. In fact, this disadvantage in the analysis of solid tumors is substantially eliminated in the application of FCM to hematological malignancies (Keren, 1989) or clinical cytology (Melamed and Staiano-Coico, 1990; Rijken et al., 1991). Nevertheless, **in equivocal cytologic cases, the additional benefit of FCM, or image analysis, appears to be questionable, leaving cytologic examination as the definitive technique** (Rijken et al., 1991). It seems that until assessment of the specific gene expression becomes available on a routine basis for FCM and image cytometry (ICM) techniques, these will remain essentially of limited diagnostic value in clinical practice.

Techniques have been devised to disintegrate even formalin-fixed, paraffin-embedded tissue, making informative FCM measurements possible (for a review see, e.g., Hedley, 1989). Although still lacking the necessary standardization and support from extensive comparative clinical studies, this approach has been widely exploited to develop correlations between DNA ploidy and the biological behavior of tumors (see appropriate sections in this book; Grogan and Collins, 1990).

Comparison of FCM with ICM for DNA quantitation in solid tumors revealed that ICM was able to identify some aneuploid populations where FCM failed (Bauer et al., 1990). On the other hand, selected cases could be detected by FCM, but not by ICM. Thus, while FCM may fail because of cell loss during processing, ICM shows a somewhat higher sensitivity due to visual identification of target cells by the operator (Bauer et al., 1990). Flow cytometry has been shown to offer simultaneous assessment of nuclear (e.g., DNA content), cytoplasmic, and cell surface parameters (e.g., cytoplasmic antigens or oncogenes) (e.g., Dent et al., 1989; Stokke et al., 1991; Schmid et al., 1991; van Dilla et al., 1990). The clinical utility of such determinations is evident in the evaluation of the DNA index in breast cancer samples, which offers a distinct advantage over the biochemical assays on homogenates (Remvikos et al., 1991).

A more refined evaluation of aneuploidy can be achieved by **fluorescent ISH (FISH)**; however, the automation available in FCM offers a definite advantage for clinical applications. One of the reasons for the frequently observed discrepancies in the assessment of DNA ploidy of tumors is the **wide scatter of FCM readings** on repeated aliquots of the same tissue sample due to the cellular heterogeneity of tumors (Eastmond and Pinkel, 1990). A detailed study performed on primary breast carcinomas has shown that, on average, at least four samples are needed for reliable determination of DNA ploidy of primary tumors by FCM (Beerman et al., 1991). Flow cytometry of nucleic acids may, theoretically, offer a powerful tool for clinical diagnosis, provided more specific tumor markers can be determined in representative cell populations. A recent effort to use the relative abundance of single-stranded vs. double-stranded RNA in solid nonhematopoietic neoplasms indicated that double-stranded RNA content may be a useful parameter that complements DNA ploidy in the evaluation of solid tumors (El-Naggar et al., 1991).

1.2.7 QUALITY CONTROL IN GENE LEVEL DIAGNOSIS

The range of quality control (QC) issues in clinical application of gene level diagnosis is as broad and varied as the methodologies employed and the diseases evaluated. Comprehensive coverage of this topic is beyond the scope of this book and only selected aspects of QC in various diagnostic approaches will be addressed.

The rapidly expanding field of gene level diagnostics offers an increasing number of clinically useful tools whose wide acceptance will be progressively facilitated by the user-friendly methodologies the industry is developing. In spite of the traditional "black box" approach that the clinical testing technology has had as its goal achieved through the introduction of maximum automation and computerization of the interpretation of the results, molecular biological diagnostics will require a different level of facilities, equipment, and most importantly, training of technologists and physicians.

1.2.7.1 FACILITIES

Organization of a molecular diagnostics laboratory, either within an existing clinical laboratory structure, or as a separate, free-standing facility, requires that a higher level of precautions against **pollution of the environment**, **cross-contamination of specimens** (e.g., by amplified products), and **protection of personnel** be taken into consideration. One must be concerned with, and ensure prevention of, a possible escape of potentially hazardous material into the laboratory space or the environment, as well as try to minimize the interference of the ever-present **nucleases** with DNA and, especially, RNA assays. **Cross contamination** of the specimens evaluated by PCR is a special problem discussed below (see Section 1.2.7.7).

When organizing such a laboratory, particular attention should be given to the features of construction, interior, benchwork placement, storage, and testing spaces. Among other things, this requires that special attention be given to creating the maximally affordable **dust-free environment** with air conditioning designed in such a way that relatively **low air flows** would still be adequate for appropriate air exchange. Work surfaces should be able to withstand strong cleansing solutions such as **bleach** and sterilization by **ultraviolet light** without creating cracks or other defects of the bench surface. Where possible, separate rooms should be dedicated to certain procedures in handling specimens and/or nucleic acid preparations, as in PCR procedures.

Although progressively moving away from radioactive labels, a large number of procedures may still be expected to use **radioactive reporter molecules**, and appropriate regulations observed elsewhere in clinical laboratories dealing with radioactive material also apply here. It is desirable to have a **liquid scintillation counter**, not only for certain testing protocols but also for monitoring the quality of reagents and supplies. Other essential equipment includes adequate **refrigeration** and **deep freezer** (−70°C) space, preparative **high-speed centrifuges**, **table top centrifuges**, a **vertical luminary flow hood** with dedicated exhaust conduit, UV-equipped table top **hoods**, a **laser densitometer**, a **dark room**, a **water-purification system**, and a **glass washing** facility. A list of other laboratory equipment for a molecular biology laboratory can be found in a number of manuals (e.g., Ausubel et al., 1989–1991; Perbal, 1988; Sambrook et al., 1989).

In our experience, the time invested in the search for the most appropriate equipment suitable for a given laboratory space and function pays off in a substantial reduction of required capitalization, and in savings of space without sacrifice of quality and versatility of the equipment acquired. Farkas (1990) shares experience in establishing a diagnostic molecular pathology laboratory in a large acute care hospital.

Transfer of technology from basic research and biotechnology into clinical diagnostic laboratories will be further accelerated with the wide introduction of nucleic acid probes, automation, and **chemo-** and **bioluminescence**-based detection systems. An example of the research-oriented equipment which has a great potential in the clinical diagnostics field is the **Phast System** (Pharmacia) for **electrophoresis** and **isoelectric focusing (IEF).** A call can be made that the wide introduction of the Phast System into the clinical diagnostic laboratory is long overdue. With respect to standardization of protocols for protein electrophoresis and isoelectric focusing, experience suggests that this programmable system using precast mini gels offers an attractive modality to achieve operator-independent standardization and reproducibility of assays. Adaptation of IEF protocols for routine protein electrophoresis of clinical specimens in many cases allows one to avoid separate immunoelectrophoretic identification of immuno-globulins. IEF of hemoglobins by the Phast System is, in this author's opinion, far superior to the currently accepted methodologies in routine clinical laboratories (Bernstam, 1992).

1.2.7.2 QUALITY CONTROL IN FLOW CYTOMETRY AND IMAGE CYTOMETRY

Detailed discussion of QC considerations in FCM and ICM at the stages of **sample procurement** and **preparation**, **instrument standardization**, as well as **data analysis** and **interpretations** can be found in several dedicated publications (e.g., Coon et al., 1987; Grogan and Collins, 1990; Horan et al., 1990; Keren, 1989; McCoy et al., 1990, Vindelov and Cristensen, 1990). One of the important considerations for informative gene level determinations by FCM and ICM in clinical practice is the **quality of nuclear and/or cytoplasmic nucleic acids** recovered from patient samples. Comparison of DNA content estimates performed by FCM on **fine needle aspirates (FNA)** of solid tumors with those derived from whole tumor samples demonstrated acceptability of retrieval of the material by FNA only if multiple (no less than three) aspirates are assayed (Sloan et al., 1990).

Demonstration of the **feasibility of extracting DNA from paraffin-embedded tissue for FCM** has stimulated widespread use of this approach in diagnostic pathology (reviewed in Coon et al., 1987; Merkel and McGuire, 1990; Raber and Barlogie, 1990). This modality offers a number of advantages, such as the ability to perform DNA FCM following histopathologic evaluation of the specimen, selection of the best areas for DNA measurements, accumulation of similar or related material for retrospective studies, and so on (Coon et al., 1987).

DNA FCM on paraffin-embedded tissue also has a number of limitations, some of which may seriously hamper interlaboratory comparisons as well as meaningful assessments of the DNA content and comparison with either previous or later evaluations on similar material. Analysis of the reproducibility of DNA FCM on formalin-fixed paraffin-embedded tissue suggested that ploidy assessment performed even on two aliquots from the same paraffin block material may vary significantly (Price and Herman, 1990). The CVs for the G_0G_1 peaks of such samples were relatively high (mean, 8.58 to 9.7). Consequently, each laboratory is advised to determine its own **mean channel differences** between replicate aliquots to develop criteria for the interpretation of DNA aneuploidy in clinical tumor specimens (Price and Herman, 1990).

Variability in the DNA FCM of solid tumors (e.g., Kallioniemi, 1988) also accounts for the **interlaboratory discrepancies** that have been addressed in a few studies (Coon et al., 1988; Homburger et al., 1989; Joensuu and Kallioniemi, 1989; van Thiel et al., 1989). In a study of DNA FCM on paraffin-embedded human breast carcinomas without intratumor heterogeneity, 12% of the laboratories reported different DNA ploidy and **DNA index (DI)** values (Kallioniemi et al., 1990). These variations were not due to lack of consensus on interpretation of data or the criteria chosen for aneuploidy, but stemmed from analytical variations, differences in the detection rates of near-diploid and tetraploid DIs. Nevertheless, the use of DIs offers a better means of evaluating malignancy with lower intra- and interobserver variation (83.9 and 82.2%, respectively) than histopathologic grading (65 and 57% reproducibility, respectively) (Bocking et al., 1989). Clearly, coordinated efforts similar to those conducted by Coon and colleagues (1988) are needed for better QC of DNA measurements by FCM.

Meaningful **interpretations of the DNA content in isolated nuclei** recovered from paraffin-embedded tissue must take into account (1) the substantial decrease in the **propidium iodide** fluorescence (the most commonly used stain for this purpose) produced by formalin fixation (Becker and Mikel, 1990; Coon et al., 1987) as well as (2) the dependence of the DNA signal on the **state of chromatin dispersion** (Becker and Mikel, 1990). The decrease in the fluorescence signal from previously formalin-fixed cells is apparently due to the reduced accessibility of DNA to the dye, which is produced by **cross-linking of the chromatin proteins**. Importantly, the degree of decrease in FCM fluorescence varies in specimens with different ploidy whereas ICM evaluation does not suffer from this artifact (Becker and Mikel, 1990).

Other artifactual phenomena include **ion-dependent alterations in the chromatin structure** produced by changes in tonicity. The DNA signal shows opposite shifts produced by changes in tonicity in the FCM and ICM assays. Detailed analysis of the effects of various fixatives on DNA measurements in colonic adenocarcinoma, squamous carcinoma of the lung, mammary adenocarcinoma, and plasma cell infiltrates in the spleen revealed that **some fixatives** (e.g., **Zenker's** and **B-5**) **significantly affected DNA FCM** (Herbert et al., 1989).

The evaluation of DNA ploidy by FCM usually makes use of the relative DNA content in the form of a **DI** calculated either with respect to the fluorescence of chick or trout red blood cells as **calibrators**, or as the ratio of fluorescence intensity between abnormal and normal nuclei. The calculated DI reflects the degree of aneuploidy (Grogan and Collins, 1990). Alternatively, fluorescein-conjugated **calibration beads** and blank particles can be used to derive a standardized electronic signal (Givan et al., 1988). A detailed discussion has been presented dealing with the various potential sources of biological and instrumental variability in FCM, and the use of standards and controls aimed at minimizing the adverse effects of this variability (Horan et al., 1990).

Another source of error in the assessment of DNA content by FCM is the marked variability of the **staining characteristics of the nuclei** isolated from deparaffinized, rehydrated, and disaggregated tissue for **Feulgen staining** (Schimmelpenning et al., 1990). **Disaggregation of tissue** significantly contributes to variations in the Feulgen stainability that can reach 300%. The process of disaggregation itself, rather than the duration or method of formalin fixation, and the length of storage appear to be the major contributors to this variability. These findings emphasize the need to use appropriate **internal standard** cells, and external standard cells appear to be unacceptable (Schimmelpenning et al., 1990). The low levels of fluorescent signals seemed to result from the **enzymatic digestion step** of the disaggregation. Since no such variations were noted in stained tissue sections not subjected to disaggregation, it is recommended that when DNA FCM is performed on paraffin-embedded material, a tissue section cut from a paraffin block of the series should be run together with the disaggregated preparations in one batch as a staining control (Schimmelpenning et al., 1990).

The Feulgen staining does not appear to vary as much in **cytology preparations** not requiring disaggregation, and a method has been proposed to destain slides stained by the **Papanicolaou method** followed by Feulgen staining for DNA quantification by ICM (Gurley et al., 1990). However, when Feulgen staining is used for the nuclear DNA quantification by ICM the nature of fixative agents is critical. A comparison of nine fixatives proved the **Riguad fixative** to be the best for automatic cytophotometry of Feulgen-stained nuclei (Giroud and Montmasson, 1989). Analysis of the staining characteristics of various commercial dyes used in Feulgen staining has also been addressed (Schulte and Wittekind, 1989).

DNA FCM is not an acceptable technique at this time for the assessment of chromosomal abnormalities detectable by cytogenetic methods (Shackney et al., 1990). Flow cytometry failed to detect solid tumors displaying numerical chromosomal abnormalities in up to 65% of patients. However, since FCM offers substantial savings in time and labor over those required for cytogenetic studies, efforts are being directed toward major improvements in the analysis of chromosomal aberrations to make this modality clinically useful. A combination of

DNA FCM and ICM is suggested, particularly in the presence of aneuploid populations demonstrated by FCM to have wide CVs for the G_0/G_1 peaks, which should be confirmed by ICM (McFadden et al., 1990).

1.2.7.3 QUALITY CONTROL IN ISH

The range of diagnostic situations in which electrophoresis and various hybridization methodologies are or can be used is virtually limitless and, accordingly, appropriate QC procedures must be adapted to each specific situation. In a broader context, attention should be directed to some general requirements mandatory for producing informative gene-level evaluation. In some form or other, hybridization of nucleic acids in solution or on solid supports (membranes, cells, tissue, or microplates), and separation of nucleic acids or their fragments (e.g., such as restriction digests) are a part of the majority of molecular diagnostic procedures.

The importance of **maintaining optimal procedural parameters** called for from assay to assay cannot be overemphasized. For example, minor deviations in the **salt concentration** or **temperature** may alter the stringency of hybridization either way, thus giving information, if at all, on the wrong set of genetic determinants than those being tested for. Minor mistakes in the **gel preparation**, or conditions of the **electrophoretic runs**, will fail to reveal the sought for presence of informative alleles and so on. Strict adherence to appropriate protocols is absolutely mandatory unless deviations from the established procedures are fully researched and proven to enhance the method.

In view of the increasing role of ISH in clinical diagnosis, some of the QC concerns will be mentioned below. Meticulous attention to technical details in performing ISH is a prerequisite for making this modality an important and powerful tool for clinical diagnosis (Stoler and Ratliff, 1990). The **specificity of identification** of a given nucleic acid target in a tissue rests largely on the controls used. Suggested general test procedures aimed at ensuring the specificity of ISH under varying conditions of pretreatment, hybridization, and probe labeling conditions have been discussed (e.g., Bresser and Evinger-Hodges, 1987; Larsson and Hougaard, 1990; McCabe and Pfaff, 1989). Among those are (1) **preincubation** of some tissue with "cold" probe prior to addition of labeled probes to have a "competition" or "blocking" control, (2) hybridization to a **heterologous probe** and other tissue not expected to have the same target nucleic acids, (3) inclusion of a "positive control" with overabundance of the target, (4) nuclease (**RNase** and **DNase**) digestion, and (5) hybridization with **the vector** without the probe. Additional tests include those for "positive chemography" (signals generated not by the probe but other chemicals used in the procedure), "negative chemography" (loss of signal due to the chemicals used), and selection, when possible, of the "antisense" strands when using cRNA probes to compare these to a control probe ("sense" strand).

A partial list of procedures helpful in optimizing ISH deals with a number of frequently encountered problems (McCabe and Pfaff, 1989). Examples include variations in **tissue fixation** that may lead to **excessive cross-linking**, thereby blocking accessibility of the nucleic acid target for the probe; on the other hand, **proteolytic treatment**, if excessive, although opening up the target may lead to **topological distortions** affecting the ISH results (McCabe and Pfaff, 1989). An interesting interference in ISH performed on human peripheral and central nervous system tissues has been reported to occur from **lipofuscin**, which causes false-positive signals with a variety of DNA probes (Steiner et al., 1989). Appropriate nuclease treatment and controls without any probes helped to resolve the problem. Another interference in ISH is the **non-specific binding of DNA to eosinophils**, which can be effectively eliminated by pretreatment of slides with **carbol chromotrope** (Patterson et al., 1989). This problem may be encountered, for example, in the diagnostic identification of viral nucleic acids in peripheral blood, as in HIV or CMV diagnosis.

Strong **background staining** may be encountered in ISH procedures using **biotin-labeled probes**, particularly when applied to liver sections (Grody et al., 1987). In hepatocytes, in contrast to many other types, biotin is found in a protein-bound form and is not eliminated during processing and fixation. It is, therefore, necessary to use biotin-blocking procedures, while **nonspecific signals from glycogen** can be removed by **amylase** pretreatment of the liver sections (Grody et al., 1987). Increased specificity of any hybridization, including ISH, is also achieved through the use of **multiple and competitive probes**; this is thoroughly discussed in numerous publications (e.g., Altar et al., 1989; Gillespie, 1990; Kajimura et al., 1990; Miller et al., 1989).

Temperature control is of paramount importance in hybridization. When single-stranded RNA or synthetic oligonucleotides are used as probes attention should be given to the possible nonenzymatic hydrolysis of ssRNA in aqueous solutions at neutral pH and elevated temperatures (Tenhunen, 1989). The **duplex stability** between oligonucleotides and nucleic acids can be predicted based on the behavior of synthetic oligonucleotides with defined sequences at elevated temperatures (e.g., Albretsen et al., 1988; Kajimura et al., 1990; McGraw et al., 1990).

1.2.7.4 QUALITY CONTROL IN RESTRICTION ENDONUCLEASE ANALYSIS

The widespread use of DNA and RNA restriction endonuclease analysis (REA) from the study of microorganisms to tumor tissues to identity testing, raises certain procedural concerns that deserve particular mention here. One of the assumptions implicit in comparing tumors and normal tissues is that restriction digestion of the extracted DNA is equally efficient in both instances. As referred to above (see Section 1.2.5.10), this assumption has proved to be invalid at least in one instance, when DNA extracted from nude mouse tumors (but not normal tissues) was shown to be incompletely digested by three of the nine commonly used restriction enzymes (*Hind*III, *Kpn*I, and *Xba*I) (Parkes et al., 1990). This deficiency was not detectable by ethidium bromide staining of the gels. The presence of specific **restriction enzyme inhibitors** in the tumors was thought to account for this phenomenon because these aberrant digestion patterns were not observed in restriction digests of DNA isolated from cell lines derived from these tumors. The presence of tumor DNA even interfered with digestion of λ DNA by *Hind*III, but not *Bam*HI.

Another **potential artifact in restriction analysis** of tumors or other tissues, particularly retrieved from immunocompromised individuals, may be spurious DNAs in blot hybridizations arising from **bacterial contamination** (Howell and Kaplan, 1987). This can be an especially significant interference in fingerprint studies using vector sequences for probing repetitive DNA polymorphisms, because bacterial contamination can produce false-positive signals with plasmid probes (e.g., Westneat et al., 1988). One should be aware of yet another possible artifact arising from the **failure of some restriction enzymes** (e.g., *Bam*HI and *Hind*III) **to cut tissues exposed to prolonged (24 h) formalin fixation** (Warford et al., 1988).

A detailed set of **standards and controls for quality assurance of DNA fingerprinting** as used for identity testing has been developed by Technical Working Group on DNA Analysis Methods (TWGDAM) and is available (Mudd et al., 1989). This document emphasizes the required level of personnel training, documentation, materials and equipment, fundamental aspects of the analytical procedures, controls, and standards, as well as interpretation. A list of recommendations on the criteria of DNA testing in paternity testing and forensics has been published (Executive Committee of the International Society for Forensic Hemogenetics, 1989).

Yet another group of specific standards has been proposed by the American Association of Blood Banks (1989), including those specifying DNA loci, that must be met to obtain reportable results for parentage testing using DNA polymorphism. Validation of DNA loci by family studies is required. The chromosomal location of the polymorphic loci should be traceable to the Yale Gene Library, or to the International Human Gene Mapping Workshop. The use of appropriate restriction enzymes and probes as well as the conditions of hybridization and sizes of alleles should be documented in the literature. Other requirements include the knowledge of the type of polymorphism being detected, assurances as to the completeness of endonuclease digestion of DNA, adequacy of size markers, and the presence of appropriate human DNA controls. Selection of optimal restriction enzyme/probe combinations can be assessed by computer programs that predict the ability of a particular restriction enzyme to produce the most efficient digestion of a particular DNA molecule (e.g., Sands et al., 1990).

Southern blot analysis in its various modifications may still be used for clinical molecular genetic diagnosis for some time to come, despite the revolutionary development in the amplification of diagnostic targets. In this context, some aspects of standardization of molecular hybridization procedures are of interest for the QC-conscious diagnostician.

1.2.7.5 QUALITY CONTROL IN ISOLATION OF NUCLEIC ACIDS

In a clinical molecular diagnostic laboratory, the isolation of nucleic acids from a large number of tissue samples may present a logistic challenge particularly when attempting to ensure the high quality of the nucleic acid extracted. In addition to the protocols suggested in the reference manuals mentioned above, one can see a plethora of methods, modifications, or suggestions on improving the existing protocols appearing almost daily. A small representative sampling of such reports, helpful in clinical molecular diagnostic laboratory, is given below.

Isolation of nucleic acids from blood samples continues to be the subject of numerous studies. Of all the anticoagulants commonly used in blood preservation, **acid citrate dextrose (ACD)** solution appears to be superior to **ethylenediaminetetraacetic acid (EDTA)** and **heparin** in not affecting the yield of DNA (Gustafson et al., 1987). Undegraded DNA can be isolated in high yield from frozen blood kept in ACD for 5 days at 23°C. Up to three cycles of freezing and thawing have negligible effects of the yield of DNA. No changes in restriction pattern obtained with three enzymes (*Eco*RI, *Hin*dIII, and *Xba*I) could be seen following storage of DNA in solution at temperatures up to 37°C for 6 months (Madisen et al., 1987). Progressive DNA degradation follows exposure to 45°C for 6 to 7 weeks. Although DNA can be extracted from blood samples stored at –70°C for at least 2 months, or at 23°C for over a week, and used for genetic studies, the yield of high molecular weight DNA is reduced (Madisen et al., 1987).

For a number of purposes, DNA and RNA may have to be isolated from a large number of small samples virtually at the same time. Procedures specifically adapted to yield undegraded nucleic acids, preferably without the use of conventional phenol extraction, have been published (e.g., Grimberg et al., 1989; Johns and Paulus-Thomas, 1989; Miller et al., 1988; Pearse and Wu, 1988).

Isolation of high molecular weight DNA and RNA from either cultured cells or small biopsy samples may be a common task in a diagnostic molecular laboratory. Simplifications and adaptations of conventional procedures claiming to give reproducibly high yields of **undegraded nucleic acids** suitable for REA abound (e.g., Adell and Ogbonna, 1990; Boom et al., 1990; Emmett and Petrack, 1988; Iversen et al., 1987; Meese and Blin, 1987; Reymond, 1987; Rupp and Locker, 1988). In gene level diagnosis of infectious agents it may be required to isolate DNA from **gram-positive** or **acid-fast bacteria**. The traditional techniques may fail and require prolonged procedures because of the cell wall thickness in these organisms. **Mechanical lysis procedures** require special equipment which, in addition, can disrupt the DNA; culturing in the presence of special additives such as glycin, ampicillin or cycloserine to obtain protoplasts is more a research technique than a diagnostic approach; enzymatic disgestion with **lysozyme, lysostaphin, pronase** or **proteinase K** is lengthy and expensive, and the most efficient approach is thermal shock technique (–196°C/+100°C) followed by 3% SDS lysis that works on most gram-positive bacteria but fails in *Mycobacteria* and *Corynebacteria* (Bollet et al., 1991). Instead, an ingenious procedure using microwave oven yields 20 to 200 μg of high molecular weight DNA from 200 μg of wet weight bacteria (Bollet et al., 1991). The only variable in this procedure is the length of heating exposure that yields DNA suitable for REA or PCR. Convenient modifications of DNA analysis for RFLPs allow the isolation of DNA from cells,

digestion, separation of the restricted fragments, and subsequent blotting for Southern analysis
— **all in the agarose gel** (e.g., Mage et al., 1988; Pollman and Zuccarelli, 1989; Yanamandra
and Lee, 1989). This approach may have a particular appeal to clinical diagnostic laboratories.
A very useful modification of a technique for the isolation of nucleic acids from tissues
undergoing regular processing treatment and paraffin embedding in histological laboratories
has been developed (Sato et al., 1991). In fact isolation of nucleic acids in a condition suitable
for various molecular biological assays can be accomplished from formalin-fixed, paraffin-
embedded tissue (e.g., Heller et al., 1991; Stanta and Schneider, 1991).

1.2.7.6 NUCLEIC ACID MEASUREMENTS

A frequent source of error is the determination of nucleic acid concentrations, either in
solution in the presence of interfering substances, or on gels and blots. A host of **spectroscopic**
and **dye-binding, radioactive** (e.g., Kuroda et al., 1988), **chemoluminescent digoxigenin-
based probes** (e.g., Lanzillo, 1990, 1991; Nevinny-Stickel et al., 1991), and **bioluminescent
techniques** (e.g., Hauber et al., 1989) aimed at developing reproducible linear methods for DNA
and RNA quantitation at picogram (Kuroda et al., 1988) and even attomole (Lanzillo, 1991)
levels continue to be developed. Errors in estimates of DNA concentration based on **ethidium
bromide fluorescence** in one- or two-dimensional gels may be very large, and the following
conditions should be observed: the reproducibility of the electrophoretic runs must be ensured,
the films must be calibrated by an internal standard and appropriate background controls must
be selected to determine the baseline for integration (Ribeiro et al., 1989). Meticulous attention
to the photographic process, limitations of the film used, and scanning using high-resolution
densitometry is a must to obtain reliable quantitation of DNA from agarose gels.

An alternative to the photographic system requiring detection of ethidium bromide-stained
DNA fragments in agarose gels is DNA-DNA hybridization with sulfonated probes or the **direct
sulfonation procedure** (Chapman and Brown, 1989; Fritz et al., 1989; Lebacq et al., 1988).
Although being highly sensitive (10 pg DNA per band) and nonradioactive, this approach should
be used only when samples are relatively free of contaminating nucleic acids. In order to
minimize variations in the DNA concentration measurements in multiple small-volume samples
(particularly prone to error due to viscosity of DNA), an adaptation of a fluorescent mircotiter
plate reader has been reported that gives a linear response over the range of 250 to 2000 ng (Riley
et al., 1989).

Optimization of the **transfer of nucleic acids** from gels onto various membranes for
Southern or Northern analysis is critical for standardization of these basic diagnostic techniques.
The increased sensitivity of **nylon membranes**, compared to that of **nitrocellulose membranes**, has been demonstrated, and procedural modifications allowing for greater uniformity
of blots and reproducibility of transfers have been described (e.g., Khandjian, 1987; Twomey
and Krawetz, 1990). Nonradioactive biotinylated DNA probes have been used to improve the
detection of unique sequences in Southern blots, particularly in clinical diagnostic laboratories
(Gregersen et al., 1987). The sensitivity of hybridization methods is drastically influenced by
the **mode of retention** and **capture of nucleic acid fragments**. A quantitative hybridization has
been developed using retention of nucleic acids by **dA-tailed capture probes** (Hunsaker et al.,
1989; Morrissey et al., 1989; Morrissey and Collins, 1989). Multiple modifications of procedures for quantitation of nucleic acids are appearing continuously (e.g., Barron, 1989; Lion et
al., 1989).

1.2.7.7 QUALITY CONTROL IN AMPLIFICATION SYSTEMS

The best studied amplification system that has already been tested in a wide range of
diagnostic applications is PCR, although its use in the clinical diagnostic laboratories is still
restricted due to patent regulations at the time of this writing. Elimination of some potentially

troublesome aspects of PCR assays, and the development of more standardized and reproducible protocols for clinical laboratories, are additional considerations delaying wide introduction of PCR in diagnostic laboratories.

Although the basic QC parameters of PCR have been addressed repeatedly (e.g., Kitchin et al., 1990; Lo et al., 1988), systematic studies concerned with QC of PCR in specific diagnostic applications are relatively few (e.g., Linz et al., 1990; Mullis and Faloona, 1987; Syvanen et al., 1988). Significant differences are observed in the **amplification efficiency** of *Taq* polymerase from various commercial sources (Linz et al., 1990). At least 50 times more template DNA was required with one of the tested enzymes compared to others in order to generate a product band of similar intensity (Linz et al., 1990). The wrong choice of the **primer concentration** may completely block the amplification, and higher primer concentrations are not always optimal (Linz et al., 1990). The **effects of temperature** and **primer length** on the specificity and efficiency of PCR have been described (e.g., Wu et al., 1991; Yap and McGee, 1991).

The **conformation of the template DNA** (supercoiled DNA supports lower amplification rates than linearized), the optimum range of **elongation temperature** and its effect on the size of the amplified product, the effect of **hybridization temperature** on the efficiency of amplification with difference primer sets, and the **number of cycles** (20 to 30) were all parameters that had to be optimized for reproducible amplification of HIV2 *pol* sequences (Linz et al., 1990).

Clearly, the **primer sequences** are of critical importance, not only in ensuring the specificity of PCR amplification, but also for the very possibility of the reaction to proceed, sometimes even with a single mismatch at the 3′ end (Linz et al., 1990). **Mismatches** at the 5′ end were generally without major negative consequences. It is recommended that parallel reactions be used with degenerate primers or mixtures of 3′ degenerate primers to prevent failure of amplification due to 3′ mismatches.

The absolute requirement of *Taq* polymerases for **magnesium ion** appears to display two maxima at low (below 5 mmol/l) and high (10 to 12 mmol/l) concentrations (Linz et al., 1990). Because maximal efficiency of the reaction can be achieved at the lower concentration, and enzyme batches that require the higher Mg^{2+} concentration usually display lower activity, it is believed that the high Mg^{2+} requirement may represent the presence of **contaminating chelating compounds** in the enzyme preparation (Linz et al., 1990).

One of the sources of variation in the specificity of PCR is the **fidelity of the DNA polymerase** used in the PCR. The error rates in the amplified products generated by PCR using different DNA polymerases have been compared between T4, modified T7, Klenow fragment of polymerase I, and Thermus aquaticus (*Taq*) DNA polymerase (Keohavong and Thilly, 1989). The error rate induced in the 104-bp fragment of exon 3 of the human hypoxanthine-guanine phosphoribosyltransferase gene ranged from 1.3×10^{-4} for the Klenow fragment and 2.1×10^{-4} for *Taq* polymerase down to approximately 3.4×10^{-5} for modified T7 and 3×10^{-6} for T4 DNA polymerase. These values agree with those reported previously (Saiki et al., 1988a,b; Tindall and Kunkel, 1988). Calculations reveal that at low concentrations of starting material (templates, targets), the **probability of an error in a PCR amplification is around 1%** (Krawczak et al., 1989).

Significant **variability in the efficiency of PCR** from paraffin-embedded tissues is associated with the particular **fixatives** used and the length of fixation time (Ben-Ezra et al., 1991). Some fixatives (**ethanol** and **Omnifix**) did not affect amplification of the β-actin gene, whereas all specimens fixed in **Bouin's** or **B-5** were negative. Fixation in **formalin** and **Zenker's** solution led to variability in the PCR product. Interestingly, with almost all fixatives tested the PCR signal for RNA was less affected by fixation than that for DNA. This could be due to the greater abundance of β-actin mRNA than DNA molecules per cell. The other possible reason is that a shorter polynucleotide is preferentially amplified over the longer one (156-bp RNA product vs. 250-bp for DNA) (Ben-Ezra et al., 1991). This study confirmed the previously reported detrimental effects of cross-linking fixatives (Jackson et al., 1990) and formalin fixation (Jackson et al., 1990; Rogers et al., 1990) on the efficiency of PCR.

An important source of false-positive results in PCR amplification is the well-recognized **extraneous contamination** of the reaction mixture by positive control material as well as by the very products of amplification in the so-called "**carry over**". A number of measures to reduce the posssibility of these types of contamination have been widely publicized (Kwok and Higuchi, 1989). These include **physical isolation** of PCR preparatory stages from the products of amplification (samples are prepared in rooms or biosafety hoods separate from those where the amplification is performed); **autoclaving** water and buffer solutions that do not deteriorate during autoclaving; use of **disposable labware** (pipette tips and microcentifuge tubes); allocation into **single doses** of all the **reagents already premixed** that are consumed in a single run; use of **gloves at all stages** of the preparation and performance of the reaction, and use of **positive displacement pipettes**; and addition of **DNA negative controls**, which cannot be overemphasized. Treatment of reusable equipment (gel apparatus, etc.) with 1 *M* HCl serves to depurinate any residual DNA (Kwok and Higuchi, 1989).

However, simple sterilization or denaturation is not adequate to prevent the amplified PCR products ("amplicons") from contaminating subsequent reactions, because a single strand, no matter how short, of nucleic acid can be readily amplified and can withstand sterilizing temperatures (Cimino et al., 1991). The carryover problem presents a particular concern in clinical diagnostic laboratories when similar amplifications are routinely repeated.

A new "post-PCR sterilization" approach has been proposed based on the blockage of *Taq* DNA polymerase in the amplification reaction by a **photochemically modified base** in the amplified polynucleotide strand (Cimino et al., 1991). This method calls for the introduction of **isopsoralen derivatives** that form cyclobutane adducts with pyrimidine bases photoactivated after amplification. For optimum sterilization, the protocol must be specifically adapted to each amplicon system. It must consider the tolerable level of carryover and the expected magnitude of amplification matched with a practical level of isopsoralen concentration. A description of this procedure in the context of a retroviral target testing in clinical samples has been given (Isaacs et al., 1991). With respect to **inactivation by ultraviolet light**, its efficiency is higher with larger DNA segments (>700 bp) compared to the shorted ones (<250 bp), and it is lower in eliminating dried DNA (Sarkar and Sommer, 1990a,b).

Since **positive controls**, by definition, contain a large number of target sequences, the possibility of contamination from positive controls may be high. Therefore in some applications positive controls may be omitted, and Southern analysis, or **sequencing**, is used as quality control in addition to the identification of the product on sizing gels. In this context, males may be excluded from even performing PCR assays for **Y-specific sequences** (Lo et al., 1989).

False-negative results obviously may occur when primers fail to anneal, and, depending on which primer (wild type or mutant) failed, the assessment of zygosity by **allele-specific primers** might misidentify a heterozygote as a homozygote mutant or wild type (Wright and Wynford-Thomas, 1990).

Nonspecific amplifications may occur with primers annealing also to nontarget sequences. Monitoring of this problem is best accomplished by using **multiple primers** to various sequences within the DNA area of interest and strict monitoring of the expected band size. **Smearing patterns** and **wrong band positions** on the gels may suggest nonspecific priming. Of course, direct sequencing or Southern analysis may be helpful in identifying the problem. Another source of error in PCR amplification has been reported to be the formation of chimeric cDNA clones due to the reverse transcriptase activity of *Taq* polymerase (Brakenhoff et al., 1991).

An intrinsic source of contamination of the PCR process is the presence of DNA in *Taq* polymerase preparations that cannot be traced to either *E. coli* or *T. aquaticus*, nor can it be successfully eliminated by DNase, restriction enzyme digestion, or $CsCl_2$ density gradient centrifugation (Rand and Houck, 1990). This contamination results in the synthesis of a 165-bp fragment in the absence of any added template. The contamination is thought to occur during the purification process, or enters with the reagents added to the enzyme. This problem can be resolved by running a **no-template control** and choosing other primers (Rand and Houck, 1990).

An effort to combat the 10 to 100% false-positive rate in one study, when testing for HSV and human herpesvirus-6 (HHV-6), included **wipe tests** similar to those used to monitor radioactive contaminations (Cone et al., 1990). Each suspect surface was wiped, and the rinsing solution was amplified with primer sets derived from HHV-6. The laboratory locations found to harbor DNA contributing to false-positive results included **freezer door handles**, **freezer shelf**, and room **door handles**. Replacement of contaminated objects and cleaning with 1 M HCl proved effective, and subsequent retesting proved to be negative, suggesting that the wipe test may be helpful in monitoring contamination in the PCR process.

Another approach to avoid false positivity in PCR is by using **anticontamination primers**. In the case of HPV PCR, primers from the sites flanking the HPV cloning site were used to avoid detection of cloned HPV plasmids (van den Brule et al., 1989). The validity of this approach was confirmed in a subsequent study using a different set of primers to other HPV sequences (Beyer-Finkler et al., 1990).

In the course of any PCR amplification of two or more homologous alleles using the same primers, the generation of **heteroduplexes** may be expected, and, in specific cases, this artifact may be useful for the identification of deletion/insertions, if these are sufficiently large (Nagamine et al., 1989).

Other amplification processes for the detection of nucleic acid sequences are relatively far from clinical application at the time of this writing (see Section 1.2.2). Among alternatives to PCR, the **Q beta replicase system** has received the widest publicity (Kramer and Lizardi, 1989; 1990; Lizardi et al., 1988; Lomeli et al., 1989). This method relies on the fidelity of hybridization of a given probe inserted into an RNA, which is then efficiently amplified by a dedicated enzyme. Although the process is very rapid and produces enough RNA to be detected without radioactive labels, in practice the sensitivity of the assay is limited by the persistent presence of nonhybridized probe, reducing the limit of detection to about 10,000 molecules of target (Kramer and Lizardi, 1989, 1990). Improvement in the methodology promises a substantial increase in sensitivity that can make this technology a highly appealing one for a number of objectives.

Prenatal Diagnosis

1.3.1 INTRODUCTION

A variety of **inherited diseases** are being diagnosed prenatally or early in childhood using restriction analysis (REA), DNA and oligonucleotide probes and, lately, the polymerase chain reaction (PCR). The revolutionary changes introduced by molecular cytogenetics into the development of both the databases and specific probes for clinical diagnosis cannot be overestimated. The majority of the diagnostic achievements in this area remain, however, in the realm of research or specialized laboratories. Simplification of the procedural aspects of molecular diagnostics such as construction of specific nucleic acid probes with nonradioactive labels, introduction of PCR combined with automation, and the development of new detection systems (colorimetric and fluorescent) makes gene probe diagnostics a potentially viable alternative to conventional methods (Connor, 1989; Ostrer, 1989).

While some inherited single gene disorders can be diagnosed using a simple approach such as PCR with fluorimetric detection of the products (e.g., in sickle cell disease), the majority of cases may require a combination of different methods. For example, early prenatal diagnosis of 21-hydroxylase deficiency using a specific 21-OH DNA probe should be combined with genotyping of the fetus by HLA-DNA probes on chorionic villus samples (Raux-Demay et al., 1989). The diagnosis of familial amyloid polyneuropathy relies on REA of white blood cell DNA followed by hybridization of a cloned prealbumin DNA to a Southern blot of the digests (Harada et al., 1989a,b).

Similarly, REA combined with specific oligonucleotide 20-mer probes has been applied to the diagnosis of neuronopathic Gaucher's disease (Tsuji et al., 1987). Prenatal diagnosis of α-antitrypsin deficiency can be accomplished either by restriction fragment length polymorphism (RFLP) analysis or by hybridization using M- and Z-specific oligonucleotides (Hejtmancik et al., 1986). Despite the fact that the results of both methods were in agreement for all samples tested, RFLP analysis was considered to be more reliable under routine laboratory conditions (Hejtmancik et al., 1986). However, in families in which no siblings are available for comparison the use of oligonucleotide probes may be preferred.

Application of molecular biological tools has spread to, and will undoubtedly intensify in, the **prenatal determination of paternity, fetal sexing, preimplantation diagnosis, and *in vitro* fertilization**. In many cases, a combination of traditional methods of prenatal diagnosis (ultrasonography, amniotic fluid cells, and chorionic villus sampling) with the new molecular biological techniques is being used (Chervenak et al., 1986; Orkin, 1987; Pearson, 1987). Analysis of fetal DNA from the circulating nucleated erythrocytes in maternal blood has been described (see below; also see Bianchi et al., 1990). Reports summarizing the experience some centers have accumulated so far in prenatal testing at the gene level emphasize the advantages of molecular diagnostics over enzymatic and other biochemical techniques (Northrup et al., 1990; Upadhyaya et al., 1990).

It can be expected that, given the development of adequate molecular tools, the prenatal evaluation of **predisposition** to various diseases will also find widespread use. The identification of a **carrier status** or **fetal infection** is performed on a routine basis. A wide range of new ethical, moral, and legal issues is raised by the availability of molecular diagnosis that reduces the level of uncertainty in predicting unfavorable developments later in life. These pose dilemmas to society of a magnitude never encountered before (e.g., Annas and Elias, 1990; Dunstan, 1988; Wertz and Fletcher, 1989a,b) and frequently reflecting religious restrictions (e.g., Brown, 1990).

It has become progressively more evident that the availability of prenatal diagnosis per se does not necessarily ensure the universal acceptance of this modality, even by populations at risk, as exemplified by the parental attitude to neonatal and prenatal screening for cystic fibrosis (CF) (Al-Jader et al., 1990). Efforts to confront these issues are made and guidelines relevant to the general and more specific aspects of molecular genetic evaluation are being developed (Bird, 1989; Kazazian, 1985, 1990; Michael and Buckle, 1990; Wertz and Fletcher, 1989a,b; Wexler, 1989).

This discussion will highlight only selected areas of prenatal gene level testing. Various approaches and their advantages and limitations in the testing for DMD/BMD, Huntington's disease, cystic fibrosis, and hemoglobinopathies are referred to in more detail in respective sections. It has been emphasized that prenatally ascertained cases of sex chromosome anomalies deserve to be followed up which may provide, among other insights, unique information on genetic imprinting that could not be gained in earlier studies (Evans et al., 1991). It is argued that the finding of low maternal AFP associated with an increased incidence of 45,X makes it imperative that analysis of sex chromosome anomalies be performed (Evans et al., 1991).

1.3.2 FETAL SEXING

The **definitive identification of sex** at various stages of human development can be of interest for a variety of reasons. Obviously, in **sex-linked inherited conditions** the determination of the sex of the child may offer an option to discontinue a pregnancy if the trait is going to be expressed. Moreover, in the practice of *in vitro* fertilization, molecular diagnostics allow for sexing at the eight-cell stage, as will be discussed below.

The first probes specific for repeat sequences of the heterochromatic region of the long arm of chromosome Y have been used for dot-blot (Gill, 1987; Lau et al., 1984) and *in situ* hybridization (ISH) in sex diagnosis (Lau, 1985; Lau and Schonberg, 1984). When (1) this heterochromatic region of the Y chromosome is absent, as in some normal, fertile males, (2) translocations occur in apparently normal females, or (3) the child is a Turner mosaics, the probes directed to the heterochromatic region of the Yq region may give an incorrect diagnosis of sex (Witt and Erickson, 1989).

A large number of DNA probes for unique and repetitive sequences are available for the identification of Y chromosome DNA. A probe (**pS4**) generated from restriction digestion of the chromosome by *Hae*III has been constructed and shown to detect repeated sequences on Yq12 with a sensitivity of 5 ng (McDonough et al., 1986). Combining pS4 with the probe developed to a unique sequence in the euchromatic region of the Y chromosome, **4B-2** (Burk et al., 1985), analysis of defective gametogenesis associated with translocated Y chromosome has been performed in a pedigree with recurrent abortion (Tho et al., 1987). Genomic DNA isolated from the proband, her amniocytes, and her daughters was used in this study. Specific probes for the Y chromosome (p75/79, Y84, Y190) and the X chromosome (pSV2X5) allow fine analysis of Turner's syndrome patients by combining ISH and DNA hybridization with conventional G banding techniques (Cooper et al., 1991).

Restriction analysis performed on chorionic villi may be affected by **maternal tissue contamination** (Niazi et al., 1981). With a combination of two probes, one λKC8, derived from a unique sequence localized on the short arm of the X chromosome, and another probe, L1.28, also located on the short arm of the X chromosome, the absence of maternal contamination of chorionic villi could be demonstrated (Elles et al., 1983). The electrophoretic patterns of restriction fragments characteristic of various allelic distributions were used to establish the presence or absence of the maternal alleles in the chorionic villi DNA.

A commercially available probe, **pHY2.1** (Amersham, Arlington Heights, IL), recognizes 2000 copies on the Y chromosome and 100 copies elsewhere in the gene (West et al., 1987, 1988). This probe and a similar probe (pY431a) allowed for sex determination using fluorescent hybridization on uncultured amniotic fluid cells (Guyot et al., 1988), and even on degraded DNA from old blood stains (Gill, 1987). This probe can also be used for ISH evaluation of Y/autosome translocations in metaphase spreads or interphase nuclei (Ellis et al., 1990). Sex chromosome-specific probes such as biotinylated **pY3.4** combined with indirect immunofluorescence for its detection are helpful in monitoring the presence of host cells in peripheral blood of bone marrow transplant recipients for the development of graft-versus-host disease or relapse after transplantation (Przepiorka et al., 1991).

1.3.2.1 FETAL SEXING BY POLYMERASE CHAIN REACTION-BASED ASSAYS

The next generation of tests for fetal sex determination is based on PCR amplification of selected sequences of either X or Y chromosomes. A rapid fetal sexing PCR procedure that can be used on any specimen, even collected as dried blood samples (Guthrie cards), is based on the 5.5-kb *Eco*RI sequences of the **alphoid satellite family** located in the pericentromeric regions of the Y chromosome and is highly characteristic for these **Y alphoid repeats** (Witt and Erickson, 1989). The PCR-based assay is an extension of the earlier developed sex diagnosis procedure for blotted DNA using the probe Y97 specific for the Y chromosomal alphoid repeats (Stalvey and Erickson, 1987). Two sets of oligonucleotide primers were used: Y1, Y2 flanking the 170-bp fragment of the alphoid repeats of the Y chromosome; and X1, X2 flanking the 130-bp fragment of alphoid repeats of the X chromosome. Specificity of the amplification was such that no 170-bp product (the Y chromosome specific) could be detected in female DNA from dried spots. The sensitivity of this assay was 5 to 10 µl of spotted blood (Witt and Erickson, 1989).

Another PCR-based assay for fetal sexing identifies a single and discrete 530-bp fragment using two oligonucleotide primers specific for the **ZFY gene** cloned from a region of the human Y chromosome (Sasi et al., 1991). This assay can also detect *ZFY* DNA sequences in XX males as well as in female patients with Turner syndrome, or in the detection of male cells in the recipient of bone marrow transplantation when the donor is a female (Sasi et al., 1991). While these procedures are definitely attractive for a number of clinical and forensic uses when sex determination is sought, their application to maternal blood, although possible, has not yet been reported.

Lo and coworkers (1989) described the first method for sex determination of the fetus by testing the **maternal blood**. Blood was collected from 19 pregnant women at 9–11 weeks and at 32–41 weeks. The primers were Y1.1 and Y1.2 flanking a 149-bp fragment of a Y-specific repeat sequence (Kogan et al., 1987). The first PCR product was then reamplified with another set of primers, Y1.3 and Y1.4, which flank a 102-bp fragment internal to Y1.1 and Y1.2. Extreme precautions were taken to minimize the chance of contamination: DNA was extracted under **class II containment conditions**, the reagents were predigested, no DNA was extracted from men during these experiments and even all blood samples were taken and handled by women. Despite these rigorous measures, amplification beyond 20 cycles in the second nest consistently yielded false-positive results. Adjusting the second amplification to only 15 to 20 cycles was sufficient to have positive results in a mother who carried a male fetus, whereas a mother with a female fetus tested negative. In the case of a male fetus at risk of a sex-linked inherited condition this procedure would have to be followed by chorionic villus biopsy or amniocentesis to evaluate an abnormal sex-linked gene (Lo et al., 1989).

Similar **background contamination** was also experienced during the second amplification of the nested PCR for CMV detection (Porter-Jordan et al., 1990). Fragmented DNA was thought to be the source of that contamination. Likewise, when the second nest was limited to only 20 cycles, this background was eliminated. One of the sources of erroneous results appears to be the extension of the fragmented DNA in a reaction known as "jumping PCR". This leads to a mosaic of connected sequences spanning the entire genome, which is then amplified to yield false-positive results (Porter-Jordan and Garrett, 1990). Neither predigestion of reagents with *Eco*RI nor autoclaving could eliminate this interference. Because the shorter fragments of DNA can still be present, they can be extended in the second nest of the reaction.

Using a similar approach, other investigators encountered problems with **reproducibility** (Holzgreve et al., 1990). Choosing a different probe for PCR amplification, **27A**, derived from a unique Y-specific sequence, the detection limit of male cells after Southern hybridization was 10 pg male DNA, corresponding to one to two male cells. The disappointing result was the finding of weak false positivity in 6 out of 10 control females. This happened in spite of isolating DNA under class II containment conditions and predigestion of all respects. The only effective measure, although impractical, appears to be **ultrafiltration** of all reagents in addition to **class II containment** and **predigestion** with restriction enzymes. Referring to the narrow "window" (15 to 20 cycles), reportedly optimal in avoiding false-positive results, the authors express their concern that the **variability of the amplification efficiency** combined with the **variability in fetal cell concentration in maternal circulation** may compromise the acceptable reproducibility of the assay (Holzgreve et al., 1990).

Further refinement of the PCR-based fetal sex determination using maternal blood assay came from Lo and coworkers (1990a). Targeting a single-copy sequence that is part of a gene expressed in testicular tissue, a set of new Y primers (Y1.5, Y1.6, Y1.7, Y1.8) was used. The primers Y1.5 and Y1.6 bracket a 239-bp Y-specific sequence, whereas the other pair of primers, Y1.7 and Y1.8, flank a 198-bp sequence internal to Y1.5 and Y1.6. The first amplification ran for 40 cycles followed with the inner nest of 25 cycles (Lo et al., 1990a). In contrast to the previously reported nested PCR that used Y1.2/Y1.3 followed by Y1.3/Y1.4, which consistently yielded false-positive results when amplification was prolonged, the new primers produced false-positive results due to what appeared to be a sampling error, or inadequacy of anticontamination regimen in some cases. The use of multiple specimens from each patient may be needed along with further refinement before the procedure can be used routinely (Lo et al., 1990a).

Discounting the probability of the "jumping PCR" as responsible for the false-positive results, Lo and coworkers (1990b) believe that **persisting male cells in maternal circulation from previous pregnancy may account for the "internal" contamination (Hook, 1990)**. Other causes include autosomal cross-reactivity of Y primers (Nakagome et al., 1990) and the "vanishing" male twin of a female sibling (Mueller et al., 1990). In practice, however, contamination may be the main problem (Lo et al., 1990b).

No coamplification of maternal DNA sequences could be observed when careful dissection of trophoblast tissue from maternal dicidua was carried out in the prenatal diagnosis of β-thalassemia in 300 couples at risk (Rosatelli et al., 1990). The amplified product was probed with phosphorus- or horseradish peroxidase-labeled allele-specific oligonucleotide (ASOs) in a dot-blot hybridization format. An important noninvasive protocol has been realized by isolating fetal nucleated erythrocytes from maternal circulation by FCM which could then be used in PCR-based identification of Y-specific DNA and ISH with chromosome-specific probes to establish cases of trisomy 21 and trisomy 18 (Price et al., 1991).

1.3.2.2 SEX DETERMINATION GENES

Sex identification of a **fetus** or **embryo**, in particular, would be unequivocal if the very gene(s) responsible for the sex determination could be the targets of such an evaluation. So far, all the reported sex evaluation studies relied on the identification of repeated or unique sequences of the Y or X chromosomes, rather than the genes now considered to be the **candidate sex-determinant genes**. Although the gene responsible for sex determination has not yet been firmly defined, studies appear to be coming closer to its identification (Berta et al., 1990; Gubbay et al., 1990; Sinclair et al., 1990).

Sex determination is essentially equated to testis determination in mammals, including humans (McLaren, 1988). The gene, termed **TDF**, for **testis-determining factor**, has been mapped to a short segment of the Y chromosome adjacent to the pseudoautosomal boundary within an area located 35 kb distal to the boundary (Sinclair et al., 1990). It is likely that more than one gene is required for male or female sex determination. However, attempts to identify and clone *TDF* are underway using all the information on the detailed mapping of the Y chromosomes generated by deletion analysis, meiotic maps of the pseudoautosomal region (an area shared by the X and Y chromosomes), as well as a long-range restriction map linking the first two maps (Sinclair et al., 1990).

Previously, one of the candidate genes for *TDF*, named **ZFY**, was isolated from a 140-kb region, but later established to be identical to *TDF*. Sinclair and coworkers (1990) have found an **open reading frame (ORF)** that is part of a conserved, Y-specific gene, named *SRY*, for **sex-determining region Y**. It encodes a testis-specific transcript (Sinclair et al., 1990).

Genetic studies on some XY females with **gonadal dysgenesis** as well as in normal males identified mutations within the *SRY* gene responsible for sex reversal in the affected individuals (Berta et al., 1990). This provided "compelling evidence that *SRY* is required for male sex determination" (Berta et al., 1990). Amplification of *SRY* sequences from DNA of XY females or normal males by PCR revealed that in the majority of the XY females, except those with *de novo* mutations, the *SRY* gene appeared normal. This observation suggested that, although *SRY* is required for male sex determination, this is not the "ultimate" *TDF* gene yet (Berta et al., 1990).

Essentially similar findings have been made in mice (Koopman et al., 1990). Evidence supporting the key role of *SRY* in sex determination acting as *TDF* is coming from studies on *de novo* mutations in this gene (Jager et al., 1990). At the same time, exceptions to this evidence also are found, suggesting that, having discounted **ZFY**, **H-Y** antigen, or **Bkm** as the earlier proposed candidates for TDF, "we must look even further before the secret of sex is solved" (Ferguson-Smith et al., 1990). In the meantime, the protein encoded by *SRY* has been found to contain a **high mobility group** (HMG) box responsible for specific DNA-binding activity of this protein (Harley et al., 1992). Mutations in the HMG box have been found in XY females suggesting that specific DNA-binding activity encoded by *SRY* is essential for sex determination (Harley et al., 1992).

1.3.2.3 PREIMPLANTATION ANALYSIS AND SEXING OF HUMAN EMBRYOS

Identification of sex of human embryos can be accomplished by cytogenetic analysis of metaphase chromosomes as well as by ISH (Penketh et al., 1989) after *in vitro* fertilization. Biotinylated Y-specific sequences available commercially (Amersham) and used as probes for ISH allow undelayed processing of female embryos for transfer to the mother's uterus when X-linked disorders are to be avoided. The entire procedure can be accomplished within 48 h. In the male interphase nuclei, this probe (**pHY 2.1**) has an efficiency of 66% in detecting the Y body, which is comparable to that of sex determination by fluorescent staining of metaphase chromosomes (Penketh et al., 1989).

Since transfer of the biopsied embryos to the mother's uterus should be performed as fast as possible, a **PCR-based sexing technique** has been developed (Handyside et al., 1989). Human embryos at the 6 to 10-cell cleavage stage, 3 days following *in vitro* fertilization, were manipulated and a single cell was removed through a hole in the zona pellucida. Oligonucleotide primers selected from the published 3.4 repeat sequences on the Y chromosome were selected. To prevent false-positive results, all the buffers were ultrafiltrated and predigested with *Eco*RI to remove potential DNA contamination. A cell was judged male by the presence on electrophoresis gels of a 149-bp fragment of the Y specific repeat, and female if the band was absent or if it was very faint. The entire procedure took 5 h (Handyside et al., 1989). The apparent success of this approach is demonstrated by its application to two known cases of couples at risk of transmitting adrenoleukodystrophy and X-linked mental retardation (Handyside et al., 1990). Following testing, both women eventually were carrying normal female twins.

The **micromanipulation of preimplantation embryos** has been successfully performed to allow nondestructive embryo biopsy on which ISH with probes specific for chromosome X and chromosome 3 was performed (Grifo et al., 1990). It appears that the fast pace of accumulated experience in these micromanipulation techniques combined with ISH and/or PCR approaches summarized above make the preimplantation genetic diagnosis suitable for clinical application when medical indications justify the risks for the woman and her fetus associated with these diagnostic procedures (Buster and Carson, 1989).

1.3.3 PRENATAL DETERMINATION OF PATERNITY

The parentage of a fetus may be requested in pregnancies resulting from a **sexual assault**, or with a history of **multiple sexual partners**. Until recently, polymorphic protein markers such as HLA antigens evaluated on **amniotic cells** or **chorionic villus** samples, red cell antigens, and serum enzymes have been used for this purpose (Callaway et al., 1986; Pollack et al., 1981; Tuch et al., 1985). Material obtained at abortions has also been used to establish paternity by these methods (Reisner et al., 1988). One of the major drawbacks of this approach lies in the poor expression of some of the protein markers in fetal tissues (Pollack et al., 1981).

In contrast, genotypic evaluation of fetal material provides substantially better evidence of highly specific polymorphisms helpful in human identity testing. So far the most widely used approach to genotyping has been **molecular genetic fingerprinting** (see above). This approach relies primarily on the restriction patterns generated with probes for short, tandemly repeated sequences found in the hypervariable regions (Kirby, 1990). As discussed above, the resulting patterns are highly specific for individuals and offer a much greater accuracy than that achieved on the basis of protein polymorphisms (Helminen et al., 1988; Jeffreys et al., 1985 a–c).

Although widely used in paternity determinations on postnatal or aborted material, the fingerprinting method has also been shown to be a simple, reliable, and accurate method for prenatal paternity testing when performed on material routinely obtained at **amniocentesis**, or on **chorionic villus samples** (Kovacs et al., 1990). Undoubtedly, application of PCR to this task will soon follow.

Yet another use of REA in prenatal diagnosis is the **analysis of zygosity** in a twin tubal gestation. In one such study, RFLPs of the twins proved them to be dizygotic, raising the possibility that many twins formerly diagnosed as monozygotic may have been, in fact, dizygotic (Neuman et al., 1990). Although of largely academic significance (only 89 such cases have been reported so far), this specific application illustrates yet another use of the molecular genetic approach that allows accurate description of biological phenomena of potential significance to clinical practice.

Application of molecular analysis to the study of **human evolution** produces intriguing evidence concerning maternal and paternal lineages of humans (Cann et al., 1987; Casanova et al., 1985, Ngo et al., 1986; Oakey and Tyler-Smith, 1990; Tyler-Smith et al., 1988; Tyler-Smith and Brown, 1987). Using **alphoid satellite DNA** (DYZ3), the **Y chromosome 190-bp repeat** (DYZ5), and **poxY1** probes, Oakey and Tyler-Smith (1990) distinguished 33 haplotypes, each of which defined a paternal lineage. The distribution of variants in different lineages appears to be nonrandom. Discussion of current issues in evolutionary science is beyond the scope of this book; it appears, however, that regardless of whether the study of sex and autosomal chromosomes confirms the hypothetical ancestry of modern humans traceable to one or two individuals, this aspect of evolutionary science will eventually contribute to better understanding of specific genetic processes.

1.3.4 PRENATAL GENE LEVEL DIAGNOSIS OF HEMOGLOBINOPATHIES: RECENT DEVELOPMENTS

Already in 1988 there were over 30 centers involved in prenatal diagnosis of hematologic diseases in the world (Alter, 1988). In addition to amniocytes usually retrieved at gestation week 16, fetal blood sampling (gestation weeks 18 to 20), and chorionic villus sampling (gestation weeks 8 to 12), the more recent studies use cordocentesis. Significantly receptive attitude to and utilization of prenatal genetic evaluation for hemoglobinopathies has been reported in unselected patients (Loader et al., 1991; Rowley et al., 1991a,b).

1.3.4.1 SICKLE CELL ANEMIA

Identification of **β-globin point mutations** directly (Chang and Kan, 1982; Conner et al., 1983; Goossens et al., 1983; Old et al., 1982; Orkin, 1984; Orkin et al., 1982; Williamson et al., 1981) and indirectly, by linkage analysis (Kan and Dozy, 1978; Orkin and Kazazian, 1984), has long been performed during the first trimester on the material obtained by amniocentesis or chorionic villus biopsy. These earlier methods, based on REA of the β-globin gene (for reviews, see Alter, 1987; Kazazian, 1985; Orkin, 1984), are much more labor intensive than the newer assays based on PCR.

One such PCR assay identifies **Hb β^A, β^S** as well as **Hb β^C** alleles (Embury et al., 1987). Following amplification, the genomic DNA sequences are hybridized with a 40-mer radiolabeled oligonucleotide probe to the β^A noncoding strand. The annealed hybrids are then sequentially digested by two restriction enzymes, *Dde*I and *Hin*fI. The **GTG sickle mutation** of the normal **GAG codon** introduces a single base pair mismatch that inhibits cleavage by the restriction enzyme. This can be detected by electrophoresis of the restriction products followed by autoradiography. A modification of this approach was used for the identification of β^C allele, in which the *Dde*I cut was prevented to various extents as revealed when different hybridization probes were used. Comparison of the cleavage products generated by the two enzymes digesting two different kinds of duplexes led to the diagnosis of hemoglobin C disease (Embury et al., 1987).

A simplified PCR for the sickle cell mutation uses amplification of a 294-bp segment of the β-globin gene followed by digestion of the amplified products by *Oxa*NI. This enzyme that has a recognition site that is eliminated by the A→T mutation at codon 6 in the sickle gene (Chehab et al., 1987). An attractive feature of this procedure is the identification of an abnormal band by ethidium bromide or silver staining without radioactive reagents, allowing the diagnosis to be established within 3 to 4 h. Likewise, a deletion of the α-globin gene responsible for α-thalassemia can be identified (Chehab et al., 1987). A combination of PCR amplification of β-globin target sequences followed by *Cvn*I digestion offers the same specificity and sensitivity as ASO hybridization, and, in addition, it is much faster (6–8 h vs. 1–2 weeks) (Crisan et al., 1989).

1.3.4.2 THALASSEMIA SYNDROMES

1.3.4.2.1 Prenatal Diagnosis of β-Thalassemias

In contrast to sickle cell anemia, which represents a consequence of a single mutation, β-thalassemia is a collection of different mutations with similar phenotypic characteristics (Orkin, 1984). Thus, in the Mediterranean variant of β-thalassemia, some 20 mutations have been characterized. Of these, 12 mutations are so rare that only a few individuals from different countries have been identified, whereas 8 mutations are very common (Cao et al., 1989).

Over 90 mutations have been identified to date (Kazazian, 1990; Ristaldi et al., 1989). **Direct identification** of such mutations by REA would require that the mutation under study affect the restriction site. Unfortunately, only a few mutations can be recognized in this way. The most common ones (those at intervening sequence −1, position 110, and in codon 39) cannot be identified by REA only (Orkin, 1984). By combining REA with hybridization using short synthetic oligonucleotides labeled to a high specific activity, prenatal identification of the mutations can be performed under appropriate stringency conditions (Loi et al., 1986; Orkin et al., 1983; Pirastu et al., 1983; Thein et al., 1985). The sensitivity of the method is proportional to the specific radioactivity of the synthetic oligonucleotide probe (Orkin, 1984).

Strict adherence to appropriate hybridization and washing conditions is mandatory, as is the prerequisite knowledge of the mutations to be tested, since the probes designed for assays in Mediterranean populations will fail in analysis of Asians due to clustering of specific mutations in different areas of the genome in particular ethnic groups (Orkin, 1984).

Indirect method of detecting β-thalassemia mutations relies, similar to other indirect DNA assays, on linked polymorphisms. There is no single polymorphism pattern, however, that can be widely applied to analysis of families at risk for β-thalassemia (Ko et al., 1990; Orkin, 1984). One of the alternatives until recently has been the search for appropriate polymorphisms within a family and a large number of such studies have been performed on a referral basis (Alter, 1988; Boehm et al., 1983; Kazazian, 1985).

One strategy for prenatal detection of β-thalassemia mutations in prospective parents utilizes **amplification of DNA** followed by **dot-blot analysis** with a set of ASO probes complementary to the most common mutations in a given population (Ristaldi et al., 1989). The high sensitivity of this technique (0.05 μg of DNA) allows for a rapid (12 to 24 h) identification of mutations even in a crude amniotic cell lysate, thereby obviating the need for culturing the amniotic cells. Moreover, due to marked amplification of the target DNA sequences to be tested, nonradioactively labeled oligonucleotide probes can be used. A similar procedure combining PCR amplification, restriction digestion, and agar gel electrophoresis, instead of dot-blot hybridization, can also be used for the identification of β-thalassemia mutations (Di Marzo et al., 1988).

As in the case of fetal sexing, the possibility of **noninvasive** (with respect to the fetus) **prenatal testing** for a hemoglobinopathy has been successfully demonstrated (Camaschella et al., 1990). Maternal blood, or more precisely, the **nucleated fetal cells in maternal circulation**, have been used in three pregnancies at risk for β-thalassemia/Hb Lepore disease to show either the presence or absence of the Lepore-specific DNA fragments in the fetus by PCR amplification. These findings were subsequently confirmed by traditional approaches. As a precaution against possible contamination by targeted sequences, the authors prepared DNA in Milan and performed PCR manipulations in Turin. All manipulations were performed under class II containment and in a positive case DNA was obtained in several split isolation procedures (Camaschella et al., 1990). One limitation of this procedure at this time is that **only paternal alleles would be detected** while a maternal allele inherited by the fetus would not be identified. Again, only positive results provide the definitive diagnosis, whereas negative results may stem from technical failures, low number, or even absence, of fetal cells in the maternal circulation, or failure of the primers and the reaction conditions to provide the required specificity and sensitivity of the assay (Camaschella et al., 1990). At least in an experimental system, PCR-based analysis has successfully identified the normal β-globin gene, allowing exclusion of β-thalassemia at preimplantation stage by sampling of the embryo (Lindeman et al., 1990).

Certain types of β-globin mutations in various parts of the world cluster into haplotypes that provide a basis for prenatal diagnosis of thalassemia in a clinical setting (Kazazian, 1990; Nagel and Ranney, 1990). An example of a study enabling the direct detection of mutant β-globin genes is the analysis of the **spectrum of mutations** causing β-thalassemia in Thailand, which are somewhat different from those found among the Chinese (Thein et al., 1990). A combination of synthetic oligonucleotide probe hybridization, amplification of the genomic DNA by PCR, followed by direct sequencing as well as cloning and sequencing of the β-globin genes has been successful in identifying the β-globin mutation in 97% of the 116 β-thalassemia genes evaluated (Thein et al., 1990).

1.3.4.2.2 Prenatal Diagnosis of α-Thalassemias

α-Thalassemia, a heterogeneous group of inherited microcytic anemias, results from decreased α-globin chain synthesis (Higgs et al., 1989; Weatherall, 1986). This is a common genetic disease in Southeast Asia resulting in **hydrops fetalis** when all four α-globin genes are either deleted or inactive. Intrauterine fetal demise occurs in the third trimester, with fetal red cells carrying nonfunctional hemoglobin composed of γ-globin tetramers (**Hb Bart's**). Based on the detailed structural information of the α-globin gene cluster deletions, a prenatal identification of the predominant haplotype found in Southeast Asians (alpha°-thal-1) has been developed (Lebo et al., 1990b).

Since the earliest possible diagnosis is beneficial to the mother, the new approach using **PCR** and **dual restriction analysis** offers a significant advantage over the red blood cell (RBC) analysis, globin chain electrophoresis, and single restriction enzyme analysis formerly used for the diagnosis of α-thalassemia. Prior to testing the chorionic villi or amniocytes, hemoglobin electrophoresis of the parents is performed to detect hemoglobin H, which would demonstrate the presence of a single active α-globin gene in one of the parents. α-Globin gene is analyzed by two restriction enzymes in carriers with microcytosis in order to distinguish the arrangement of two or three α-globin genes, and to establish whether prenatal testing is indicated (Lebo et al., 1990b).

Combinations of the restriction enzymes and the expected restriction patterns in some of the most important α-globin gene mutations have been published (Lebo et al., 1990b). Different allele-specific α-globin primers for the PCR amplification step of the evaluation allow the distinction of the α^+-thal-2 ($-\alpha/-\alpha$) deletion from the heterozygous α°-thal-1 ($-,-/\alpha\alpha$) variant, thereby indicating that the latter person is not at risk for a homozygous hydrops fetalis. This procedure stems from the earlier proposal by Chehab and coworkers (1987) to use PCR amplification in testing for the presence of any α globin genes. The PCR assays for α thalassemia, however, are not capable of identifying all fetal genotypes. They offer, nevertheless, a fast and reliable approach to distinguishing the most prevalent mutations (Lebo et al., 1990b).

1.3.5 PRENATAL GENE LEVEL DIAGNOSIS OF COAGULATION DISORDERS

1.3.5.1 HEMOPHILIA A

Hemophilia A, a sex-linked bleeding disorder largely related to defects in **factor VIII (FVIII)** cofactor activity, has served as a paradigm of inherited bleeding disorders. The severity of FVIII activity deficits varies among patients in any given hemophilia pedigree and, unlike single mutation disorders, it is caused by a variety of mutations at the FVIII locus. The FVIII gene is located at the telomere of the long arm of X chromosome at Xq28. A number of different genotypes have been identified by RFLP analysis. The recognition of female carriers and prenatal diagnosis relies on RFLPs in and near the FVIII locus (Antonarakis et al., 1985; Gitschier et al., 1985).

These **restriction site polymorphisms** can be analyzed with Southern blots, whereas **sequence polymorphisms** can be identified by ASO hybridization to genomic Southern blots under high stringency conditions. Both types of polymorphisms can be also analyzed by PCR (Kogan et al., 1987; Kogan and Gitschier, 1990). Many of the FVIII gene polymorphisms are in linkage disequilibrium with one another, complicating the indirect analysis of the carrier status (Kogan and Gitschier, 1990). The polymorphism evaluation and fetal sexing by PCR in hemophilia A cases provide useful information in about 70% of hemophilia pedigrees (Kogan and Gitschier, 1990). Interpretation of the results requires that an affected male (or unambiguous female carrier) be available to identify the mutant allele in the fetus (Lawn, 1991). Naturally, the mother must be heterozygous for the polymorphism used for diagnosis. In assessing the polymorphism data, attention should be given to the possibility of misleading results due to **cross-over** between indirect, linked probes and the FVIII gene locus when only extragenic probes are used (Peake and Bloom, 1986).

Using intragenic *Hind*III polymorphism of exons 79 and 20 of the FVIII gene, a prenatal diagnosis of hemophilia A can be established in the first trimester of pregnancy (Wehnert et al., 1990). The *Hind*III RFLP is in strong linkage disequilibrium with the *Bcl*II RFLP used earlier (Kogan et al., 1987), so that both RFLPs give essentially similar information in pedigree studies and in prenatal diagnosis. The latter approach generates a constant 480-bp fragment that can be used as an internal control of restriction digestion (Wehnert et al., 1990).

1.3.5.2 HEMOPHILIA B

Hemophilia B, an X-linked recessive bleeding disorder clinically resembling hemophilia A, is produced by decreased **FIX** activity in plasma. A recent review of the **molecular genetics of hemophilia B** emphasizes the variety of genetic lesions causing the disease (Diuguid and Furie, 1991; Furie and Furie, 1992). It appears that up to one third of cases of hemophilia B are due to **new mutations**, which must be precisely identified in a given pedigree for the prenatal diagnosis to be reliable. A number of polymorphisms have been recognized, predominantly in the Caucasian population, that allow the status of 89% of the potential carrier population to be determined with 99.9% certainty (Diuguid and Furie, 1991).

Although we are not yet aware of a PCR-based procedure for hemophilia B diagnosis, it is within the capability of this technique, when applied to other disorders, to pinpoint the precise mutation in a given family. Based on this information, specific oligonucleotide probes can be constructed to screen for it in potential carriers or fetuses. It remains to be seen whether this technique can be used on maternal blood as in the case of thalassemia or fetal sex determination.

The molecular genetic aspects of various coagulation factor deficiencies are being widely studied (Roberts and Lozier, 1991; Furie and Furie, 1992), however, their use in clinical practice for prenatal diagnosis appears to be limited at present.

1.3.5.3 VON WILLEBRAND DISEASE

von Willebrand factor (vWF) is a multimeric glycoprotein synthesized and secreted into plasma by endothelial cells and accumulated in platelet α granules (Lynch, 1991). Recently, a highly informative (heterozygosity rate 75%) VNTR sequence has been identified in intron 40 of the vWF gene (Bignell et al., 1990). PCR amplification of leukocyte DNA from the cord blood was able to show that the infant was homozygous for the vWF VNTR marker, thus providing unequivocal information that the mutant maternal allele had been inherited (Bignell et al., 1990). Gene level analysis for von Willebrand disease (vWD) has been accomplished by amplification of the vWF gene segment containing TCTA repeats to successfully establish prenatal diagnosis of a severe case of vWD (references in Lynch, 1991).

1.3.6 PRENATAL GENE LEVEL DIAGNOSIS OF PLATELET DISORDERS

Neonatal alloimmune thrombocytopenia is a serious bleeding disorder with a risk of intracranial bleeding *in utero* or during delivery. The alloantigen most frequently responsible for the condition is **HPA-la (Zw^a** or **P1A)** (Kaplan, 1988). When the mother is **HPA-1(a–)** and

HLA DRw52a positive, the fetus is potentially at risk (Kuijpers et al., 1990). The alloantigens can be tested on platelets, but when these are not available the amplification of a sequence of noncoding DNA directly upstream of the HPA-1a site can be used for analysis. This amplified product is a 170-bp fragment, the identification of which has been sought in a fetus and 14 healthy individuals (Kuijpers et al., 1990). In this clinical case fetal blood was retrieved at the 28th week of gestation. The amplified products were then digested with *Nci*I to yield an RFLP that demonstrated three fragments in heterozygous **HPA-1 (a+, b+)** (106-, 64- and 170-bp fragments), only a 106-bp fragment in **HPA-1 (a+, b–)**, and 64- and 106-bp fragments in **HPA-1 (a–, b+)**. The fetus proved to be heterozygous. This method demonstrates the feasibility of typing for the platelet HPA-1 polymorphism based on the amplification of DNA from lymphocytes, chorionic villi, or other fetal tissues when platelets are not available (Kuijpers et al., 1990).

1.3.7 PRENATAL DIAGNOSIS OF SELECTED INFECTIOUS DISEASES

1.3.7.1 CYTOMEGALOVIRUS INFECTION

Congenital cytomegalovirus (CMV) infection is among the leading congenital viral infections, affecting up to 1% of live births (Stagno et al., 1982). Infection of the fetus was previously thought to occur only during the acute phase of a primary maternal infection similar to congenital rubella and toxoplasmosis. In fact, congenital CMV infection has been observed as a result of reactivation of the latent virus during recurrent maternal infections even in mothers with substantial humoral immunity. The primary maternal CMV infection leads to **demonstrable fetal involvement** in only 50% of cases. The role of gestational age at the time of infection in determining the severity of symptoms in the offspring is still not resolved (Stagno et al., 1982). The significance of **intrauterine diagnosis** of CMV transmission acquires additional importance because there are many genetic variants of CMV circulating in the general population. Some of these demonstrate an oncogenic potential. The efforts aimed at developing a subunit vaccine can benefit from more efficient prenatal diagnosis of CMV.

Estimates of **vertical transmission** of CMV indicate that about 2 to 4% of pregnant seronegative women pass a primary infection to their fetuses, and 10 to 20% of seroimmune women infect fetuses through a recurrent CMV infection (Doerr, 1987). The experience in prenatal diagnosis of congenital CMV infection accumulated over the past 20 years emphasizes the usefulness of **amniotic fluid analysis** as an adjunct procedure (Grose and Weiner, 1990). Although predominantly immunological in nature, the detection methods undoubtedly will soon be complemented by gene probes for CMV, as demonstrated by DNA ISH (Sorbello et al., 1988). For a discussion of the DNA probe diagnostics of CMV see Section 6.3.1.4.

The speed, definitiveness, and progressive simplification of PCR-based methods will make prenatal diagnosis of fetal CMV infection an important tool in fighting this common cause of postnatal morbidity. The need for prenatal diagnosis of CMV is severalfold: (1) it will contribute to the understanding of the natural history of congenital CMV; (2) asymptomatic infected neonates will be detected, and the earliest recognition of late sequelae will be improved; (3) better understanding of the processes underlying the development of severe late sequelae will be possible; and (4) application of fetal therapy will be enhanced (Grose and Weiner, 1990).

1.3.7.2 HUMAN IMMUNODEFICIENCY VIRUS (HIV)-1

The significance of prenatal diagnosis of HIV-1 infection of the fetus may be emphasized by the documented predominant route of transmission of the virus **transplacentally** (Pahwa, 1988). In which period of pregnancy (i.e., intrauterine, intrapartum, or postpartum) the infection of the fetus/newborn occurs has not been definitively established (Pizzo, 1990). Isolation of the virus from a 20-week-old fetus (Jovaisas et al., 1985), as well as pathological features of the placenta and the fetus, suggest that infection may occur early in pregnancy, perhaps during the first or second trimester (Marion et al., 1989). Defining the precise time and route of vertical transmission of HIV-1 to the fetus or neonate (Davis, 1988) can help to develop preventive measures. **Intrapartum transmission** is a possibility, although **postnatal infection via breast milk** has already been documented (Oxtoby, 1988).

The conventional methods of viral diagnosis such as culturing HIV-1, detection of the p24 antigens, and other immunologically identifiable markers are largely unreliable for a variety of reasons, including the differences in the clinical course of HIV-1 infection in the pediatric population (Andiman, 1989; Blanche et al., 1989; Chan et al., 1990; Falloon et al., 1989; Husson et al., 1990; Pahwa, 1988; Pizzo, 1990; Schochetman et al., 1989).

PCR amplification of the HIV-1 sequences **in the fetus** appears to hold the greatest promise due to its inherent sensitivity and specificity. HIV-1 DNA in the **newborns** and **children** has been identified by PCR in uncultured peripheral blood mononuclear cells (PBMC) in children of HIV-infected mothers (Laure et al., 1988). Positive identification of the virus was made in 6 of 14 symptom-free newborn infants. Of these, only one infant had **HIV-1 antigenemia**. The viral DNA was also detected in 5 of 10 children (2 to 5 years old) of infected mothers who had become seronegative 12 to 15 months after birth. No false-positive or false-negative results were observed.

Efforts to maximize the specificity of PCR detection of HIV-1 infection involved the use of **multiple primers** in a **multistage PCR**. A two-stage PCR using four outer primer pairs complementary to different parts of the HIV-1 genome was used simultaneously in the first reaction, and in the second PCR nest another set of four inner primer pairs was used in four separate reactions (P. Williams et al., 1990). Furthermore, the positive amplified products of PCR were then tested in more detail in yet one more PCR using two sets of primers that spanned the hypervariable region of the *env* gene. Products of amplification of this reaction were then run on polyacrylamide gels to create a **pattern of amplified DNA length variation**. These patterns were found to be **characteristic of each child** and persisted for at least 3 to 7 months. The authors believe that by comparing the patterns of the mother during pregnancy with those of the infant, the distinctive HIV-1 DNA length-variant patterns can be helpful in identifying the timing of vertical transmission (P. Williams et al., 1990).

We are not aware of any reports so far on the **prenatal identification of HIV-1 in the fetus** by PCR or by other techniques with a continuous monitoring through pregnancy, birth, and postnatally. Several problems inherent in such approaches can be recognized. The exquisite sensitivity of PCR is frequently its major drawback due to amplification of contaminating nucleic acid material (see Section 1.2.7.7). In the case of fetuses of infected mothers, the very process of **invasive retrieval** of a sample for amplification may introduce maternal tissue, and thus lead to false-positive results. One possible solution to this dilemma would be a concomitant amplification of strictly maternal genes unlikely to be present in fetal tissues at a given gestational stage. Another alternative may be the concomitant amplification of the integration sites of the virus into the host genome, if these were shown to be different in the mother and the fetus (P. Williams et al., 1990).

Other **false-positive results** may stem from amplification of **non-HIV-1 sequences**; however, this interference can be largely diminished (P. Williams et al., 1990). On the other hand, **false-negative** results may result either from inadequate sensitivity of the assay simply due to the **vanishingly low viral load** in the fetus or from **sampling errors**. The importance of all these concerns combined with **problems of interpretation** of the results of very sensitive techniques such as PCR has been emphasized (Best and Banatvala, 1990; Husson et al., 1990).

Interestingly, in the case of **HTLV-1**, analysis of cord blood from 20 seropositive women by PCR revealed no evidence of HTLV-1 infection, suggesting, among other things, that no significant contamination of the cord blood sample by maternal blood occurs that may interfere with PCR amplification (Ville et al., 1991).

1.3.7.3 VARICELLA ZOSTER VIRUS INFECTION

In utero diagnosis of congenital varicella zoster virus (VZV) infection can be accomplished by retrieving chorionic villi which are then probed by PCR using VZV-specific primers (e.g., ORF-63) (Isada et al., 1991). Both culture assays and Southern analysis of neonatal brain tissue for VZV proved to be negative. Although the presence of the virus was detected, this did not correlate with the manifestation of the infection or its extent (Isada et al., 1991). The cause of the discrepancy may lie in the fact that the VZV latency site is in sensory nerve ganglia and the detection of viral DNA in placental tissue does not correlate with fetal disease (Isada et al., 1991).

1.3.8 PRENATAL MOLECULAR GENETIC DIAGNOSIS OF CHROMOSOME INSTABILITY SYNDROMES

This group of genetic disorders of different etiology and manifestations is united by a common feature — **cellular hypersensitivity to mutagenic chemicals**. This property is used for prenatal diagnosis (Auerbach, 1987). The principle diseases in this group include **Fanconi's anemia** (FA), **ataxia-telangiectasia** (A-T), **Bloom syndrome** (BS), **xeroderma pigmentosum** (XP), and **Cockayne's syndrome** (CS). These disorders are associated with a variety of developmental defects, from growth retardation to immunodeficiency and premature aging. The common approach to the diagnosis of these disorders has been the demonstration of mutagen hypersensitivity in cultured amniotic cells or cultured trophoblast cells from chorionic villi. Failure to repair DNA damage has also been used for diagnostic purposes (see Section 1.5.7).

1.3.9 OTHER EXAMPLES OF DISORDERS FOR WHICH THE DISEASE LOCUS HAS BEEN IDENTIFIED

1.3.9.1 α-ANTITRYPSIN DEFICIENCY

α-Antitrypsin (AAT) deficiency leads to early onset **emphysema** and **fatal cirrhosis** of the liver (Eriksson, 1965; Erickson et al., 1986; Larsson, 1978; Sharp et al., 1969). AAT deficiency occurs at a frequency of 1 in 7600 North American whites (Cox et al., 1976), and in about 1 in 2000 to 3000 Northern Europeans (Fagerhol and Cox, 1981). The genetics of AAT deficiency have been well studied. Prenatal diagnosis of AAT deficiency is indicated in cases when the parents have already had a child with severe neonatal hepatitism or in cases with a family history of severe emphysema.

The majority of individuals with normal levels of the **proteinase inhibitor (PI)** are homozygous for the *M* **allele**, whereas those with reduced levels of AAT usually have the **Z allele**. The Z allele is produced as a result of a G→A transition in the AAT gene and can be detected by oligonucleotide hybridization (Kidd et al., 1983). When the Z allele is present, a variant AAT is synthesized, and the most common form of the disease results from the **PIZZ phenotype** accompanied by deficient release of the enzyme from the liver (Carrell et al., 1982). People with an **SZ phenotype** are at an elevated risk for emphysema, whereas several rare alleles can result in a profound AAT deficiency leading to **panacinar emphysema** in 80 to 90% of the PIZZ individuals (Carrell and Owen, 1980; Janus et al., 1985).

A large role in the eventual development of emphysema in these individuals is given to environmental factors such as **smoking** (Janus et al., 1985; Larsson, 1978). The PIZZ phenotype is manifested very early in life, so that 14 to 17% of PIZZ infants have neonatal hepatitis with a poor prognosis in about one third of cases (Moroz et al., 1976; Sveger, 1976; Sveger and Thelin, 1981). Moreover, a PIZZ fetus has a 40% chance of developing severe cirrhosis in a pedigree in which the previous sibling had a severe liver disease (Cox and Mansfield, 1987).

Prenatal diagnosis of AAT deficiency has long been performed by **PI typing of fetal blood** samples (Jeppsson et al., 1981) retrieved at fetoscopy, that procedure itself carrying a risk of fetal loss of up to 5% (Rocker and Laurence, 1981). An oligonucleotide hybridization technique has been developed to detect the point mutation in the *M* and *Z* alleles (Kidd et al., 1983, 1984). Subsequently, this technique was extended to the evaluation of the *S* allele as well (Nukiwa et al., 1986). Cox and coworkers (1985; Cox and Mansfield, 1985) described several RFLPs in and around the *AAT* gene with some RFLPs being in strong linkage disequilibrium with the *Z* allele, particularly evident when using *Ava*II digestion (Hejtmancik et al., 1986). Comparison of the results of an RFLP analysis with those obtained by **oligonucleotide hybridization**, suggests that the latter approach is preferable in families in which no siblings are available for RFLP analysis. In informative kindreds, the **linkage analysis** appears as accurate and reliable as oligonucleotide hybridization and is technically easier (Hejtmancik et al., 1986).

Chorionic villus biopsy is the preferred source of DNA for prenatal diagnosis of the AAT deficiency (Schwartz et al., 1989). It appears that the traditional RFLP analysis using *Ava*II and the genomic 3′ 6.5-kb probe is particularly useful since the majority of families are informative. It is applicable to various types of AAT deficiency, and is an acceptable method for a routine diagnostic laboratory (Schwartz et al., 1989). As in all linkage studies, family analysis and samples from appropriate members of the pedigree are required, usually delaying the results. The **direct oligonucleotide hybridization** for the most common point mutations, particularly in the *Z* allele, is a much more convenient procedure for the laboratories capable of performing Southern analysis (Meisen et al., 1988).

The speed and sensitivity of prenatal identification of the AAT mutations has been markedly improved by PCR amplification (Abbott et al., 1988; Bruun Petersen et al., 1988; Newton et al., 1988; Schwartz et al., 1989). Less than 1 µg of fetal tissue is required and the result can be available within 72 h (Schwartz et al., 1989). In addition to the *Ava*II polymorphism, two other enzymes are used (*Mae*III and *Bst*EII) to increase the frequency of informative families. By amplifying the sequence containing the *Bst*EII polymorphic site followed by REA of the PCR amplification products, a rapid and relatively simple RFLP analysis is performed. In informative families with an affected child, the entire procedure may require less than 8 h for the definitive diagnosis to be made.

1.3.9.2 ADULT POLYCYSTIC KIDNEY DISEASE

Adult polycystic kidney disease (APKD) is an autosomal dominant **monogenic** disorder. It has a prevalence of about 1:1000, with a highly variable age of onset and high expressivity in adult life (Zerres et al., 1984). While detection of renal cysts by ultrasonography in adults may reveal asymptomatic heterozygotes, this modality is unreliable for fetal diagnosis (Pretorius et al., 1987). The *APKD* gene has been assigned to **chromosome 16 (16p13.3-p13.1)** and a number of RFLPs in and near the **PKD1 locus** have been identified and used for prenatal diagnosis (Breuning et al., 1987; Reeders et al., 1985–1988; Watson et al., 1987). Another locus yet to be mapped (**PKD2**) has also been identified (Kimberling et al., 1988; Romeo et al., 1988). Prenatal diagnosis of APKD by RFLP analysis has been performed on DNA obtained from chorionic villus biopsy as early as 9 weeks (Reeders et al., 1986) and 10 weeks of gestation (Novelli et al., 1989). Prenatal diagnosis of APKD by DNA probes has been supported by the echographic observation of the earliest manifestations of APKD at 29 weeks of pregnancy and was later confirmed by autopsy findings (Ceccherini et al., 1989).

1.3.9.3 TAY-SACHS DISEASE

Until recently the only approach to prenatal diagnosis of Tay-Sachs disease (TSD) was the measurement of **hexosaminidase A activity** in cultured amniocytes or chorionic villus samples (Kaback et al., 1977). The identification of **HEXA mutations**, accounting for the majority of TSD alleles in Ashkenazi Jews, made a direct DNA-based diagnosis possible (Myerowitz, 1988; Myerowitz and Costigan, 1988; Navon and Proia, 1989; Paw et al., 1989; for a review, see Mahuran, 1991). Prenatal diagnosis would be indicated particularly in couples who either already had an affected child, or have a positive family history. In a reported pregnancy at risk, DNA from peripheral blood lymphocytes was restricted with *Hae*III and a 4-bp insertion mutation in exon 11 was found in both parents (Triggs-Raine et al., 1990a). At 10 weeks gestation, chorionic villi were obtained and fetal DNA was restricted with *Hae*III. Comparison with the parental pattern revealed that the fetus was heterozygous for the 4-bp insertion allele, the pregnancy was allowed to proceed, and the infant proved to be healthy.

The higher specificity and predictive value of DNA-based tests for the identification of Tay-Sachs carriers compared to enzyme-based diagnostics emphisizes the value of DNA tests (Triggs-Raine et al., 1990b). A combination of PCR amplification of mutant alleles followed by cleavage of a *Dde*I restriction site generated by the mutation can be used to identify carriers of the disease (Arpaia et al., 1988). Recently, other frequently encountered mutations have been defined in Ashkenazi Jews leading to defects in **sphingomyelin metabolism** and resulting in Niemann-Pick disease (Levran et al., 1991), and in the **glucocerebrosidase gene** responsible for Gaucher's disease (Firon et al., 1990).

1.3.9.4 ORNITHINE TRANSCARBAMYLASE DEFICIENCY

Ornithine transcarbamylase (OTC) is the second enzyme of the urea cycle that is involved in **detoxification** of the products of **nitrogen metabolism**, primarily in the liver, to yield an excretable product, **urea** (Fenton, 1989). OTC deficiency is an X-linked disorder with a severe, often fatal **hyperammonia intoxication** in hemizygous affected males. The neurologic prognosis of affected boys and some heterozygous girls is poor due to ensuing coma and death. Prenatal diagnosis of this disorder using conventional biochemical techniques is difficult since the enzyme is not expressed in amniocytes and the metabolic products of OTC deficiency are not detectable in the amniotic fluid (Fenton, 1989).

Four RFLPs have been identified (Fox et al., 1986a; Rozen et al., 1985) that make prenatal diagnosis possible in about 80% of obligate gene carriers, given that a linkage can be established (Fox et al., 1986a). Probes generated with the restriction enzymes *Msp*I, *Bam*HI, *Taq*I (Fox et al., 1986b), as well as *Pst*I (Liechti et al., 1990), allow the prediction of affected fetuses. A detailed description of molecular genetic diagnosis of OTC deficiency has been given (Fenton, 1989).

Two RFLPs produced by *Msp*I and *Bam*HI are informative in 74% of known female carriers, whereas by using *Taq*I site alterations the mutation can be detected directly in 17% of affected males or heterozygous females, suggesting that all families should also be screened for *Taq*I alterations (Spence et al., 1989). New mutations may account for the failure of the known RFLPs to provide diagnostic information. In parents of affected individuals in whom DNA analysis fails to identify the mutation, the biochemical testing appears to provide the most information. In such cases, DNA analysis of antecedents in an affected family may prove useful. Although measurement of urinary orotidine excretion following administration of allopurinol can be used to identify women who are heterozygous for OTC deficiency (Hauser et al., 1990), the characterization of the *OTC* gene and some of the mutations associated with the disease allows not only the screening of female carriers, but also the performance of prenatal diagnosis of OTC deficiency (Matsuda et al., 1989). The technique of **chemical mismatch cleavage** in combination with PCR has been proposed for prenatal diagnosis of OTC deficiency (Grompe et al., 1991).

1.3.9.5 HEREDITARY AMYLOIDOSIS (HA)

Hereditary amyloidosis (HA) is an autosomal dominant disorder (**familial amyloidotic polyneuropathy type II**) manifested in a late onset carpal tunnel syndrome, progressive peripheral neuropathy, vitreous opacities, gastrointestinal dysfunction, and eventual death usually following cardiomyopathy. This is one of a number of familial amyloidotic polyneuropathies related to mutations of the plasma protein **transthyretin (TTR)** (Herbert,

1989). **DNA restriction analysis** has been used for the identification of presymptomatic carriers and prenatal determination of heterozygosity. **PCR amplification** of an exon 3 region containing a fragment of the *TTR* gene can be performed on DNA extracted from amniocytes of a fetus at risk of carrying the serine-84 prealbumin (*TTR*) gene (Nichols et al., 1989). Allele-specific oligonucleotides and restriction analysis of the PCR amplified products have been used to determine prenatally the carrier status of a fetus for the HA gene for the first time (Nichols et al., 1989).

1.3.9.6 SEVERE COMBINED IMMUNODEFICIENCY AND DI GEORGE SYNDROME

Severe combined immunodeficiency (SCID) comprises a number of sex-linked and autosomal genetic defects in cellular and humoral immunity (Gelfand and Dosch, 1983). The underlying immunologic dysfunction, resulting in severe and persistent infections in infancy causing early death, includes a decreased number of T cells, unresponsiveness to mitogens, as well as hypogammaglobulinemia. In over 80% of cases, the genetic cause is unknown, and prenatal diagnosis of SCID had been relying on lymphocyte functional studies in fetal blood samples (Blakemore et al., 1987; Durandy and Griscelli, 1983).

Progress in prenatal diagnosis of SCID came from studies on the inactivation pattern of the X chromosome in the T cells of a woman, which appears to be nonrandom in cases of the X-linked form of SCID (Conley et al., 1988). Furthermore, genetic linkage studies identified the X-linked SCID locus on the proximal long arm of the **X chromosome (Xq13.1-21.1)** (Puck et al., 1989). So far, no specific genetic alteration responsible for the X-linked SCID has been identified, but polymorphic DNA markers have proved their usefulness in prenatal diagnosis (Puck et al., 1990).

In **Di George syndrome** characteristically presenting with developmental abnormalities and cellular immunodeficiency related to hypoplastic thymus, at least three loci consistently deleted have been defined — D22S75, D22S66, and D22S259 in **chromosome 22q11** allowing for the use of RFLP and dosage analysis with appropriate DNA probes (Driscoll et al., 1991). The use of FISH for the deleted region is expected to provide substantially better prenatal evaluation of pregnancies at risk for this syndrome.

1.3.10 CONCLUDING REMARKS

The above-cited examples of prenatal diagnosis using gene level evaluation of fetuses at risk for inherited or congenital conditions are by no means exhaustive. Not only can more diseases be identified antenatally then discussed here, albeit with different degrees of certainty, but an ever-increasing number of these are becoming amenable to gene level diagnosis, with progressively increasing definitiveness in many cases.

The increased reliability, speed, and informativeness of prenatal diagnosis of a number of conditions previously identified antenatally only by RFLP analysis (Ostrer, 1989) will soon be achieved by amplification of the genome areas of interest combined with ASO analysis and direct sequencing of the PCR-amplified products. Further examples of such an enhanced specificity and certainty of prenatal identification of gene alterations can be seen in the detection of **phenylketonurea** in the fetus, in which RFLP analysis had been inconclusive (Huang et al., 1990), and in the prenatal detection of heterozygotes in the diagnosis of **cystic fibrosis** (Lemna et al., 1990; also see Chapter 2.5).

Chapter 1.4

Aging

1.4.1 INTRODUCTION

Evaluation of pathological conditions at the gene level is inevitably affected by the ongoing **physiological alterations** in cells and tissues involved in the disease process. Although so far relatively little is known of the impact of some of these physiological processes on molecular genetic analysis of disease conditions, a brief discussion of some common physiological processes potentially affecting interpretation of molecular pathological findings seems to be in order. Included in this discussion are **DNA repair**, **aging**, **drug resistance**, **environmental effects** on the genome, and **metastasis**.

Advances in the understanding of the biological processes in the cell, tissues, and organisms at the molecular level have raised the practical possibility of assessing some of the changes occurring in the normal process of aging, and, hopefully, distinguishing them from aberrations associated with disease. So far, **no unifying theory of aging has been developed**, and the plausibility of attaining a single theory of aging is far from certain. In the meantime, the study of molecular processes peculiar to the aging cell and organism is conducted on cultured cells, animal models, and on tissues of older humans (for a review see, "Molecular Biology of Aging, 1990"). It seems likely that in the foreseeable future our understanding of the molecular processes associated with normal aging of organisms will be understood in sufficient detail to allow these to be incorporated in the analysis of and in the differentiation of pathological process from normal physiologic changes of aging. Here only some of the representative lines of research on molecular mechanisms of aging at the genome level will be briefly discussed.

1.4.2 GENETIC INSTABILITY

The main theories on the evolution of aging consider at least *three* alternatives (Kirkwood, 1989). First, it had been postulated that aging is directly beneficial at the species or group level to prevent overcrowding, among other things, and specific genes may have evolved to **control life span**. Support for this view has not been strong. Second, aging may be due to the **late expression of mutations** accumulated in the germ line, and selection may exert very little pressure on these potentially deleterious genetic alterations because individuals may not live long enough for these accumulated mutations to manifest themselves in evolution. Experimental support for that view is also lacking.

The third category of genes invoked in the mechanism of aging consists of those that offer some **short-term benefit in a trade-off for the potential for long-term survival** (Williams, 1957, as cited in Kirkwood, 1989). According to Kirkwood (1989), the latter hypothesis, particularly in application to the mechanisms of cellular maintenance and repair, describes the most plausible mechanisms to explain the aging process. An extension of the "disposable soma" theory (Kirkwood, 1989) is the prediction that, first, aging is due to an accumulation of **unrepaired somatic defects** and, second, the rate of aging can be modulated by altering the level of important maintenance and repair processes. Citing the observation that senescence is rare in natural populations, the **homeostasis of maintenance and repair functions** is thought to be the point of application of the damage and defective repair, and it is in this context that **genetic instability is viewed as a major cause of aging** (Kirkwood, 1989).

Experimental verification of the assumptions of the **somatic mutation theory** fails, however, to provide strong support for it (Kirkwood, 1989). Although the "classical" somatic mutation theory may not offer a satisfactory explanation of aging, the development of some of its postulates envisions **mutation interactions, general error theory**, the **role of transposable elements**, and **epimutations** involving demethylation (loss of 5-methylcytosine, in particular) as having a role in the aging process (Kirkwood, 1989). The importance of efficient DNA repair contributing to longevity lends support of the notion that genetic instability is involved in aging.

Several categories of DNA-related processes have been considered in the context of genetic instability and aging (Slagboom, 1990; Vijg and Papaconstantinou, 1990). The search for candidate genes determining aging and longevity in humans begins with the study of lower organisms. Some of these include genes controlling **protein synthesis, free radical scavenging, mitochondrial functions** (see below), as well as those encoding **reparatory systems**, particularly DNA repair. DNA damage can be expressed in various **epigenetic effects** such as the **poly-ADP-ribosylation** of chromosomal proteins resulting in modulation of gene expression. One of the events in this chain can be the induction of protooncogene activity such as c-*fos*, which can lead to poly-ADP-ribosylation. Additional mechanisms of genetic instability appear to involve **DNA recombination** (Bollag et al., 1989), **transposon-related inactivation** of genes (Murray, 1990), and **DNA amplification** (Gaubatz and Flores, 1990).

1.4.3 PROGRAMMED CELL DEATH

Although the role of programmed cell death in the process of aging is not unequivocally established as one of cardinal causes of senescence, the study of the cell death phenomenology and the control mechanisms governing this process under various conditions promises to at least offer insights into some manageable aspects of aging and oncogenesis (for a review, see Lockshin and Zakeri, 1990; Williams, 1991).

Current research efforts are aimed at establishing reproducible biosynthetic patterns occurring in cells undergoing involution. Although intermediate forms of cell death between the extremes of **necrosis** (resulting from **acute trauma** to the cell) and **apoptosis** (physiologic cell death) certainly exist, the phenomenon of "programmed cell death" is defined as a "subset for which a sequence of steps can be recognized" (Lockshin and Zakeri, 1990, p. B136).

Analysis of protein synthesis in dying cells suggests that an **active genetic death program** may be responsible for cell death, justifying the search for genes controlling cell death. Thus, **transforming growth factor** (TGF)-β_1 can inhibit cell proliferation and lead to increased apoptosis of cultured uterine epithelial cells, indicating the role of growth regulatory genes in cell death (Rotello et al., 1991). Active protein synthesis, however, does not appear to be an invariable prerequisite for the execution of apoptosis in some systems, as shown by inhibitor studies of apoptosis induced by mild hyperthermia in human and murine cultured tumor cells (Takano et al., 1991).

One of the popular model systems for the molecular genetic studies of cell death is the nematode *Caenorhabditis elegans*. The *ced* (cell death) **mutation** in this organism prevents almost all of the programmed cell deaths, and reconstitution experiments confirm that the **death program is an active process** (reviewed in Lockshin and Zakeri, 1990). Protein synthetic patterns suggest that specific enzymes may be brought into function to execute an orderly shutdown of the cell life.

Not all tissues seem to require active protein synthesis for dying, and lymphocytes and hepatocytes undergo apoptosis in what appears to be a predetermined process. The activity of

lysosomal enzymes and the **calcium-activated endonuclease (CaN)**, in particular, have been linked to cell death in a variety of tissues (Lockshin and Zakeri, 1990). A host of other gene products become expressed in the dying cell (*c-myc, c-fos,* **heat shock protein 70, testosterone-repressed prostate message 2 [TRPM-2], transglutaminases**, etc.) (see also Fargnoli et al., 1990). Studies on the mechanisms of **neuronal death** related to their interaction with the environment indicate that programmed cell death is operational in neurons in response to numerous influences, which include changed anatomic, hormonal, growth factor, and energy environment (Lockshin and Zakeri, 1990).

Association of the products of the **major histocompatibility complex (MHC)** with nonimmune systems has been observed to affect mixed function oxidases, reproductive senescence, DNA repair, and free radical scavenging enzymes (Walford, 1990). Interaction with the α_3 domain of MHC class I in mature T lymphocytes until a certain stage in their differentiation process may lead to deletion of programmed cell death (Sambhara and Miller, 1991). A related active area of study focuses on the role of interaction of MHC antigens in tumorigenesis (Fentestein and Garrido, 1986).

1.4.4 ROLE OF CARBOHYDRATE METABOLISM IN AGING

Decrease in diverse physiologic functions of the cell has been linked to the age-related **slowing** and **impairment of carbohydrate metabolism** (for a review, see Tollefsbol, 1987). The impaired induction of enzymes of carbohydrate metabolism is considered to be the culprit of the general slowing of carbohydrate metabolic pathways underlying the overall diminution in the production of anabolic metabolites and energy compounds. The intracellular impairments traceable to the reduction in carbohydrate metabolism are believed to accompany various pathologies associated with the aging process, such as **glucose intolerance** and **diabetes** (Tollefsbol, 1987). It is believed that the age-related decrease in gene expression contributes to cellular senescence rather than vice versa, which accounts for the decreased enzyme induction in aging (Tollefsbol, 1987).

1.4.5 THE OXIDATIVE STRESS HYPOTHESIS AND THE FREE RADICAL THEORY OF AGING

Another theory, or hypothesis, visualizes oxidative damage as the central cause of multifactorial deleterious consequences of **peroxidation** of cellular membrane lipids, inactivation of enzymes, and DNA damage (for a review, see Sohal and Allen, 1990). In the latest version of this hypothesis, the most fundamental concept is that aging and development are a continuum, and *aging constitutes the terminal stage of development*. As such, like other stages of ontogeny, aging is characterized by changes in gene expression, and, consequently, by an alteration in the state of differentiation. According to the **oxidative stress hypothesis**, the level of oxidative stress (which is the balance between prooxidants and antioxidants) increases during cellular differentiation and aging so that *aging is not a genetically programmed phenomenon*, but results from the consequences of oxidative stress on genetic programs (Sohal and Allen, 1990).

The experimental evidence pertinent to the oxidative stress hypothesis is interpreted to indicate that although life span is not necessarily shortened directly as a result of oxidative stress, the organisms must make adaptive adjustments "in the balance between anti-oxidant defenses, metabolic potential, and their rate of oxidant generation" (Sohal and Allen, 1990, p. 503). The age-associated changes in gene expression are found in changes in the **rate of transcription** and **processing of mRNA** and, most notably, in the **rate of translation of mRNA**. At this point, all the data related to the effects of oxidative stress on aging as mediated through its effects on gene expression are **not conclusive** and numerous alternative interpretations of the data are possible.

The role of **direct oxidative DNA damage** under the influence of **environmental agents** (e.g., mutagens and carcinogens) has also been considered (Demple, 1990).

The **free radical theory** of aging is yet another actively pursued line of research on aging claiming to provide practically verifiable means of prolonging the maximum life span due to prevention of, and/or elimination of, free radical accumulation in the organism (Harman, 1991). This theory includes cancer and atherosclerosis, as well as other common degenerative diseases, among the growing number of **free radical diseases** which can be counteracted, to a degree, by the administration of antioxidants (Harman, 1991).

1.4.6 ROLE OF DNA METHYLATION IN AGING

As noted above (see Section 1.1), most organisms appear to exhibit a loss of **5-methyldeoxycytosine (5mC)** during aging, and the extent of this loss inversely correlates with the life span potential of a particular species (Wilson et al., 1987). Using restriction endonucleases *Hpa*II and *Msp*I, which preferentially cut sequences containing **CpG dinucleotides** predominantly methylated in high eukaryotes, a pattern of age-related changes in DNA methylation can be established. **Hypomethylation** has been consistently observed in total DNA and some specific genes that are transcribed by **RNA polymerase II** (Selker, 1990). On the other hand, DNA **methylation was found to increase with age** in several regions of the rRNA genes at their 5′ ends; this is associated with preferential inactivation of a large block of rRNA genes on chromosome 16 in aging mice (Swisshelm et al., 1990). **Age-related patterns of methylation** have also been observed in the c-*myc* gene and were found to be more related to biological aging rather than to chronological aging (Ono et al., 1990).

1.4.7 MITOCHONDRIAL DNA IN AGING

Somatic mitochondrial DNA (mtDNA) mutations have been observed in tissues of aging organisms and implicated in the reduced functional capacity of tissues characteristic of advanced age (Ames, 1989a,b; DiMauro et al., 1989; Linnane et al., 1989). Although large numbers of deletions in mtDNA are characteristic of deteriorated function of skeletal muscle in **Kearns-Sayre syndrome (KSS)** (Shoubridge et al., 1990), the threshold level for such mutations of mtDNA before deleterious effects are produced is not yet known. PCR amplification of mtDNA from adult mitochondria from the heart and brain indicated that even normal heart muscle and brain of adults contain low levels of a specific mtDNA deletions not observed in fetal heart or brain (Cortopassi and Arnheim, 1990). This observation lends support to the idea that **accumulation of specific mtDNA deletions** may be associated with aging.

At the **cellular level**, a deregulation of the normal biosynthetic process in the mitochondria brought about in lymphocytes by a **t(14;18) translocation** and resulting in juxtaposition of the *bcl*-2 gene with the immunoglobulin heavy chain locus, leads to **prolonged cell survival** (Hockenbery et al., 1990). It appears that **overexpression** of the *bcl*-2 gene **blocks the apoptotic death** of a pro-B lymphocyte cell line. This observation demonstrates a unique property of an oncogene localized to mitochondria that may interfere with programmed cell death without promoting cell division (Hockenbery et al., 1990). Furthermore, induction of *bcl*-2 expression in human B lymphocytes by **Epstein-Barr virus (EBV) latent membrane protein 1 (LMP 1)**, which specifically up regulates the cellular oncogene *bcl*-2 in mediating its effect on apoptosis, protects these cells from programmed cell death (Henderson et al., 1991).

Numerous other mechanisms are being studied that lead to **cellular immortalization**, including the effects of DNA tumor viruses SV40 (simian virus 40), HPV (human papilloma virus) 16 and 18, and type 5 adenovirus and tumor suppressor genes *RB*, and *p53* and their interaction with E1A, E1B, and *ras* (Shay et al., 1991).

1.4.8 CONCLUDING REMARKS

The illustrative examples of the types of studies on aging given above clearly suggest the state of relative infancy in the molecular understanding of aging. Nevertheless, a definite role for selective gene expression in the senescence of cells and organisms appears established at the nuclear and mitochondrial DNA level. The cause-effect relationships between environmental stresses, metabolic changes, and various levels of gene expression are far from being understood at this time. It is hoped that consistent phenomenology of changes at the gene level accompanying the normal aging process can be of help in defining precise causes of pathological conditions at the gene level in clinical practice. Molecular genetic analysis of inherited conditions displaying symptoms of premature aging will undoubtedly lead to better understanding of the mechanisms of normal aging. In **Werner's syndrome** (**WS**), a rare autosomal recessive disease, although the primary genetic defect has not been defined yet, linkage studies of the WS mutation point to a group of markers on chromosome 8 (Goto et al., 1992). An important biological phenomenon that appears to be directly involved in aging and in a variety of pathological processes is DNA repair of intrinsic and environmentally induced DNA damage (discussed in Chapter 1.5).

DNA Repair

1.5.1 INTRODUCTION

A disease process is a condition of **altered homeostasis**, either transient or permanent, affecting a number of physiological functions at different levels ranging from molecular to organismal. The damage underlying a disease process, be it an infection, a heritable mutation, or a physical-chemical injury, activates mechanisms aimed at restoring homeostasis. The cellular and molecular mechanisms immediately involved in the repair of the primary focus of injury have been studied extensively. In human pathology, the study of such responses has gained the most dramatic success in unraveling the **reparatory processes at the DNA level**. There is little doubt that with better understanding of the DNA repair of specific genes, the details of coordinated efforts of subcellular and cellular repair systems will soon follow, giving a coherent picture of a given pathological condition.

The molecular biological studies of DNA repair have been centered traditionally on the **effects of injuring agents** that can be reproducibly inflicted at a given damage dose, followed by quantitation of the derangements and the ensuing reparative activities at the molecular level. The model systems that have been studied in greatest detail use **UV** and **X irradiation**, **chemical mutagens**, and **chemotherapeutic agents**. The literature on the subject is immense (e.g., Echols and Goodman, 1990; Friedberg, 1985; Guttenplan, 1990; Kaina et al., 1991; Pacifici and Davies, 1991; Swisshelm et al., 1990). A number of periodicals are dedicated largely to DNA damage and repair studies (e.g., *Mutation Research* and *Radiation and Environmental Biophysics*).

Significant advances have been made in developing molecular technology to dissect fine structural changes in the DNA at a **specific gene level**. Several reviews on the subject have dealt with the developments in DNA repair research, which bring this field of study much closer to understanding not only rare human conditions such as **Xeroderma pigmentosum, Cockayne's syndrome,** and **Bloom syndrome**, but also the role of DNA repair in **aging** and **malignancy** (e.g., Ames, 1989a, b; Bohr, 1990; Bohr et al., 1989; Bootsma et al., 1988; Cleaver, 1984; Collins and Sedgwick, 1990; Duker and Gallagher, 1986; Frankenberg-Schwager, 1990; Hanawalt, 1990; Hoeijmakers et al., 1990; Schwaiger and Hirsch-Kauffmann, 1986; Topal, 1988; van Ankeren and Wheeler, 1985; Walker, 1985).

Although the assessment of DNA repair in pathological conditions at this time remains essentially a basic research endeavor, the emerging findings will make this area of molecular pathology in the near future increasingly important for evaluating disease processes in a clinical setting. Only the salient features of the current understanding of DNA repair mechanisms in mammalian cells and their relevance to human disease will be summarized here. Some of the methodological approaches to the evaluation of DNA damage and repair will also be discussed.

1.5.2 DNA DAMAGE

Damage of genomic DNA occurs as a result of application of physical (e.g., UV, X-ray irradiation, heat), chemical (e.g., known antigens, chemotherapeutics), or biological agents (e.g., viruses). All of these can damage DNA indirectly, usually through the **secondary changes** produced in the intracellular milieu, which could be specific or nonspecific in nature. Depending on the nature and the dose of an agent, a specific response of a cell, including that originating at the genome level, can be elicited by a largely **nonspecific external stimulus** such as heat shock (Alexandrov, 1977; Bernstam, 1978). Transduction of the external stimuli originating at the cell periphery through a network of intracellular mediators results in the specific response of various biosynthetic apparatuses of the cell (Bernstam, 1978).

Within the protein biosynthetic system of the cell, the initiation of translation and DNA replication appears to be relatively more sensitive to injury than other steps in protein biosynthesis (Bernstam, 1978). On restoration of normal conditions, the cell appears to restore first the most sensitive functions. Following the pattern of recovery of altered functions in the cell can disclose the inherent "philosophy" of the cell in maintaining homeostasis and provides insights into the ways cells respond to injury.

Investigation of the effects of specific agents on DNA structure and function and the recovery from the inflicted damage produced an impressive body of evidence at the molecular level. The concept of **organ specificity** of chemical damage accommodates the multiplicity of stages, factors, and target genes participating in the process of carcinogenesis (Dong and Jeffrey, 1990). As far as endogenously accumulated intracellular products involved in DNA damage are concerned, there are four significant processes: **oxidation**, **methylation**, **deamination**, and **depurination** (Ames, 1983; Saul and Ames, 1986). The existence of specific DNA repair glycosylases for oxidated, methylated, and deaminated adducts as well as the existence of a repair system for apurinic sites generated by spontaneous depurination has been demonstrated (Lindahl, 1982). According to the measurements of oxidative products accumulated and excreted by the body, the major type of DNA damage appears to be related to oxidation (Ames, 1989a,b; Ames et al., 1985; Ames and Saul, 1988).

1.5.2.1 ENVIRONMENTAL EFFECTS ON THE HUMAN GENOME

DNA technology has been applied successfully to the detection of pathogens in the environment as well as to the evaluation of their effects on the genome (for a review see, e.g., Sayler and Layton, 1990). **Nucleic acid probes** have helped in the assessment of the role of plasmids in determining the resistance of certain pathogens to specific environmental factors such as heavy metals, organic compounds, and specific interactions with other members of the biosphere.

The study of **environmental mutagenic effects** is expanding progressively and the relevant literature is vast. It is impossible in this discussion to give due credit to the entire body of accumulated evidence even by covering only the most prominent areas of research. Instead, only several pertinent examples of studies conducted at the gene level will be cited here to emphasize yet another potential for gene level diagnosis in clinical practice. For a review of the field see a compendium of papers collected in five volumes (Mendelson and Albertini, 1990).

Of particular interest to practicing clinicians is the issue of the **human genome-environment interaction**, since it has long been recognized that specific relationships could be recognized between **genetic susceptibility** and **risk factors** for disease, on the one hand, and the environment on the other (Ottman, 1990). **Epidemiologic methods** have been shown to apply to the investigation of these relationships and plausible models were proposed (Ottman, 1990). One of the models describes the interaction between the recessive gene for phenylketonuria (**PKU**), blood levels of phenylalanine, and mental retardation. Clearly, the determination of the genetic basis for the metabolic disorder has a direct practical benefit, as shown in **Xeroderma pigmentosum**, where avoidance of sun exposure may reduce the genetically determined predisposition to skin cancer.

Other models include the enhancing effect of **barbiturates** on the manifestations and consequences of **porphyria variegata**; **hemolytic episodes** provoked by **fava bean** consumption in glucose 6-phosphate dehydrogenase (**G6PD**)-deficient individuals, and the predisposi-

Model	Illustration	Example

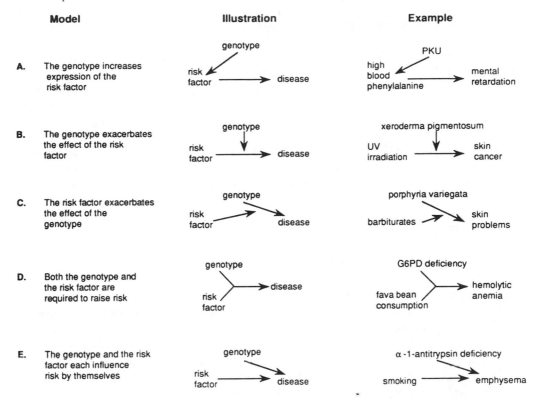

A. The genotype increases expression of the risk factor

B. The genotype exacerbates the effect of the risk factor

C. The risk factor exacerbates the effect of the genotype

D. Both the genotype and the risk factor are required to raise risk

E. The genotype and the risk factor each influence risk by themselves

FIGURE 1.8. Five hypothetical relationships between genetic susceptibility to disease and risk factors for disease identified in epidemiologic studies. The genetic susceptibility may be either polygenic or due to a dominant, recessive, or X-linked major locus. The risk factor may be only one of many factors associated with disease risk, and may itself have either genetic or nongenetic origins. (From Ottman, R. Genet. Epidemiol. 7:177–185. Copyright ©1990. Reprinted by permission of Wiley-Liss, a division of John Wiley & Sons, Inc.)

tion of α-1-antitrypsin-deficient persons, in whom **smoking** significantly increases the risk of developing **emphysema** (Ottman, 1990). These models may form a basis for directed epidemiologic analysis of gene-environment interactions. A large model study focuses on the possible induction of inherited defects in humans by exposure to environmental mutagens, such as in the long-term administration of **antischistosomal drugs** manifesting genotoxicity and carcinogenicity (Kramers et al., 1991) (Figure 1.8).

Another direction of epidemiologic studies addressing effects of the environment is in unraveling its complex influences on many disease-related risk factors associated with **familial environment** linked to cultural inheritance, and genetic predisposition (Rao and Wette, 1990). Direct assessment of effects of the environment on the human genome can be performed by demonstrating the formation of chromatid and isochromatid breaks and "minutes" by chromosome analysis in lymphocytes of persons exposed to **polychlorinated biphenyl (PCB)**. **Structural chromosome aberrations** were found in 55% of people living within a 2-km radius of a factory manufacturing capacitors in a small town in Slovenia, Yugoslavia (Tretjak et al., 1990).

Similar chromosomal aberrations have been previously reported for the effects of **dioxins** (Pocchiari et al., 1979) and **Agent Orange** (Kaye et al., 1985, as cited in Tretjak et al., 1990). Structural chromosomal aberrations (**dicentric chromosomes** and **acentric fragments**) were significantly increased in the peripheral blood lymphocytes of nurses handling **cytostatic drugs** without a safety cover compared to those using protection and nurses not handling cytostatics (Oestreicher et al., 1990). The frequency of chromatid breaks and sister chromatid exchanges was not significantly different between these groups.

Occupational exposure to genotoxic agents is likely to be more widely monitored at the chromosomal and molecular level. In the **nuclear industry**, chromosomal aberration analysis has long been used to monitor exposure to **external radiation** (Martin et al., 1991). Since sister chromatid exchanges (**SCE**) and chromosome aberrations arise from different DNA lesions, the measurement of both characteristics provides complementary information. In a study of blood lymphocyte cultures from workers occupationally exposed to uranium, an increase in dicentric chromosomes was noted and an interaction between exposure to uranium and smoking has been detected in a group of workers (Martin et al., 1991). An increase in SCEs attributable to smoking, however, was the same in the worker and control groups.

It appears that at the present time surprisingly minor effort is given to the assessment of environmental effects on the human genome in clinical practice. The vast arsenal of tools at the disposal of molecular geneticists can certainly be brought to disclose the short- and long-range effects of environmental factors on the realization of inherent predispositions to disease and the causation of "sporadic" cases traceable to genome alteration by the environment. It is hoped that wide application of gene level diagnosis in clinical practice will raise the level of awareness of the importance of environmental effects on the genome as well.

1.5.2.2 RADIATION DAMAGE

In molecular terms, ultraviolet (**UV**)-induced damage eliminated via the **nucleotide excision repair** process and recovery from **chemical adducts** are the best characterized reparatory processes. Excision repair has been dissected into consecutive steps, beginning with (1) *preincision recognition* of the damaged area of DNA, (2) *incision of the damaged DNA strand* in the vicinity of the site of damage, followed by (3) *excision of the defective site* with concomitant or sequential *degradation of the excised product*, (4) *repair replication* filling in the excised area with normal nucleotides, and finally (5) *ligation of the repaired segment* at its 3′ end to the undegraded stretch of DNA (Bohr, 1990; Bohr et al., 1989) (Figures 1.9–1.11).

As for the types of lesions that can be produced by ionizing radiation, UV irradiation (254 nm) has been shown to produce most commonly covalent linkage of two neighboring bases — **pyrimidine dimers**. A less common photoproduct that is assuming a progressively greater role in our understanding of fine derangements of DNA structure and repair is a 6-4 linkage between adjacent pyrimidines (**6-4 photoproduct**) (Bohr, 1990).

Ultraviolet-induced 6-4 photoproducts and their repair in XP-A cells can be evaluated quantitatively by **laser imaging microspectrofluorimetry** or by **autoradiography** with the help of a monoclonal antibody raised against 6-4 photoproducts (Mori et al., 1990). At very low levels of irradiation normal cells repair more than 80% of the initial damage within 4 h, whereas there is virtually no repair in XP-A cells even within 8 h. Some other common types of DNA damage in eukaryotic cells include **DNA-protein cross-links** (DPCs), **base damage**, **single-** and **double-strand breaks**, and **bulky lesions** (Frankenberg-Schwager, 1990).

Usually more than one type of damage occurs as a result of action of radiation or carcinogens. Understanding the direct DNA alteration induced by chemotherapeutic agents and the cellular efforts to repair the damage can contribute to the elucidation of the phenomenon of multidrug resistance developing in oncology patients (see below). The nuclear structure, accessibility of chromatin to endonuclease attack, as well as the size of the available pool of DNA precursors affect DNA excision repair in a cell cycle-related manner (Kaufman and Wilson, 1990) (Figure 1.12).

The **accessibility of specific regions of DNA** during various cell functions is influenced by cytoskeletal and nuclear protein structures, as demonstrated by the altered sensitivity of chromatin DNA in malignant cells to DNase I hydrolysis. It is believed that at least in some types of cancer the DNA sequences of regulatory significance for cell physiology, which are exposed in normal cells, become sequestered in a less exposed configuration that is reflected in the loss of the nuclear rim of exposed DNA (Krystosek and Puck, 1990).

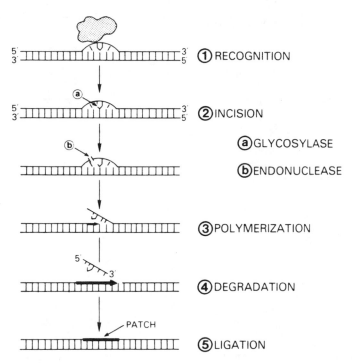

FIGURE 1.9. Steps in nucleotide excision repair. (a and b) Activities of pyrimidine dimer-specific endonucleases, T4 endonuclease V. (From Bohr et al. (1989). Lab. Invest. *61*:143–161. © by The U.S. and Canadian Academy of Pathology, Inc. With permission.)

1.5.2.3 ALKYLATING AGENTS

This is a large group of chemicals that includes **carcinogens** such as N-methyl-N'-nitro-N-nitrosoguanidine (**MNNG**) and ethylnitrosourea (**ENU**) as well as many widely used chemotherapeutic drugs (e.g., **cyclophosphamide, melphalan, busalfan,** and **chlorambucil**). Alkylating agents are efficient mutagens that produce a number of lesions in DNA, such as alkylation at the N^7 position of guanine, the O^6 position of guanine, and the N^3 position of adenine (Boiteux et al., 1984; Yarosh, 1985). In rodents, an age-related increase in the steady state levels of 7-methylguanine is observed following exposure to alkylating agents, indicating the lower efficiency of reparative systems in older animals (Tan et al., 1990).

In one type of lesions, for example, an alkyl group on the O^6 position of guanine is removed by a protein acceptor molecule, O^6-**alkylguanine-DNA alkyltransferase (AGT)**, that transfers the alkyl group onto this protein (Maher et al., 1990). Analysis of cell lines showing varying efficiency in eliminating the consequences of alkylating agent toxicity indicated that the action of AGT and excision repair are responsible for protection against cell killing and accumulation of mutations (Maher et al., 1990).

Monofunctional alkylating agents, such as O^6-**methylguanine (m^6-Gua)** produce methylation of bases in DNA, and a specific repair enzyme, **m^6-Gua-DNA methyltransferase**, affords protection to most human cell lines. Alkylating agents produce **mismatch base pairing**, which is corrected by proteins binding specifically to DNA mismatches. Some of the proteins appear to be inducible in mononuclear cells exposed to alkylating agents, and they apparently may play a role in the development of drug resistance during cancer therapy (Karran and Stephenson, 1990; Laval, 1990; Maher et al., 1990). A rapid **genotoxicity test** measures DNA strand breaks by unwinding DNA strands in alkaline solutions and evaluating the dissociated strands by **hydroxyapatite elution chromatography**; its potential clinical usefulness has been demonstrated in mouse lymphoma cells treated *in vitro* with some 78 compounds (Garberg et al., 1988).

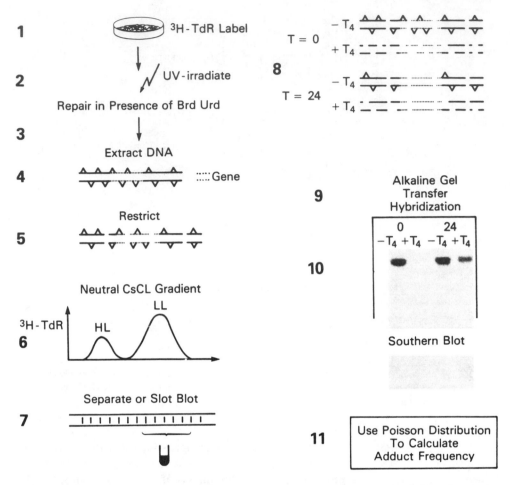

FIGURE 1.10. Flow diagram of steps in evaluating DNA repair by measuring frequency of pyrimidine dimers using restriction analysis: 1, cells prelabeled; 2, cells irradiated by ultraviolet; 3, repair; 4, DNA extraction; 5, restriction; 6, separation of replicated from parental DNA; 7, gradient fractionation; 8, reaction with pyrimidine dimer-specific endonuclease (T4 endonuclease V); 9, alkaline gel electrophoresis; 10, Southern transfer and hybridization; 11, calculations. (From Bohr et al. (1989). Lab Invest. *61*:143–161. © by the U.S. and Canadian Academy of Pathology, Inc. With permission.)

1.5.2.4 HEAVY METAL COMPOUNDS

Cis (II) platinum diamminedichloride (**cisplatin**) and **carboplatinum** are widely used antitumor agents that interfere with DNA-related functions through the formation of **intra-** and **interstrand cross-links**. Details of the cisplatin action have not been finalized yet, and much effort is being directed toward the study of the resistance that develops in many cases during cisplatin therapy (Section 1.6.4).

1.5.2.5 ANTIBIOTICS

A variety of antitumor antibiotics targeted to DNA have been introduced into clinical practice. The specific effects on DNA include **single-** and **double-strand breaks** (e.g., bleomycin, daunomycin, adriamycin), **intercalation** (mitomycin C), **inhibition** of **transcription** and **replication** (actinomycin D), and **interaction** with **topoisomerases** leading to strand breaks (epipodophyllotoxins, etoposide [VP16] and teniposide [VM26]) (Bohr, 1990).

1.5.2.6 FRAMESHIFT MUTAGENESIS

Acridines represent a model class of compounds known to form DNA-acridine complexes by **reversible binding** (Ashby et al., 1989). Through insertion or deletion of a base, a **shift out**

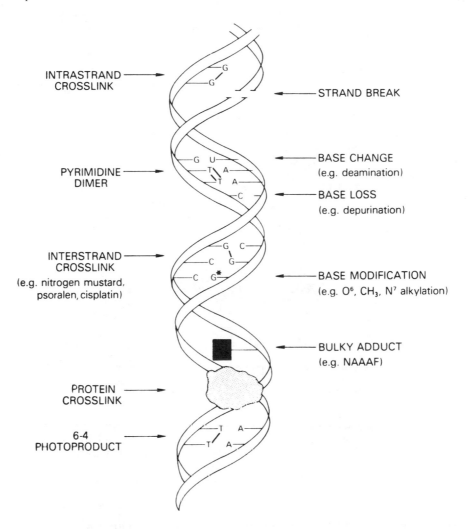

FIGURE 1.11. Common types of damage in DNA. (From Bohr et al. (1989). Lab Invest. *61*:143–161. © by The U.S. and Canadian Academy of Pathology, Inc. With permission.)

of frame occurs in the particular genetic message due to a distortion of an uninterrupted meaningful sequence of triplets. Most antitumor **intercalating agents** appear to exert their cytotoxicity through the stabilization of the cleavable complexes between mammalian **topoisomerase II** and DNA, preventing the passage of other double strands through the cleavage site during replication (Nelson et al., 1984). A variety of specific cellular repair mechanisms are involved in counteracting these effects of acridines, and they appear to be different from the enzymes repairing the damage produced by radiation and chemical agents (Ferguson and Denny, 1990).

1.5.3 DNA REPAIR GENES

The enzymatic factors involved in eliminating the consequences of various types of damage have been defined and the corresponding genes are known as **DNA repair genes**. The complex interactions between oxygen availability, linear energy transfer of the radiation applied, cell cycle stage, overall metabolic balance, hyperthermia, and the DNA reparatory systems of the cell determine the outcome of radiation-induced damage (Collins and Sedgwick, 1990; Frankenberg-Schwager, 1990).

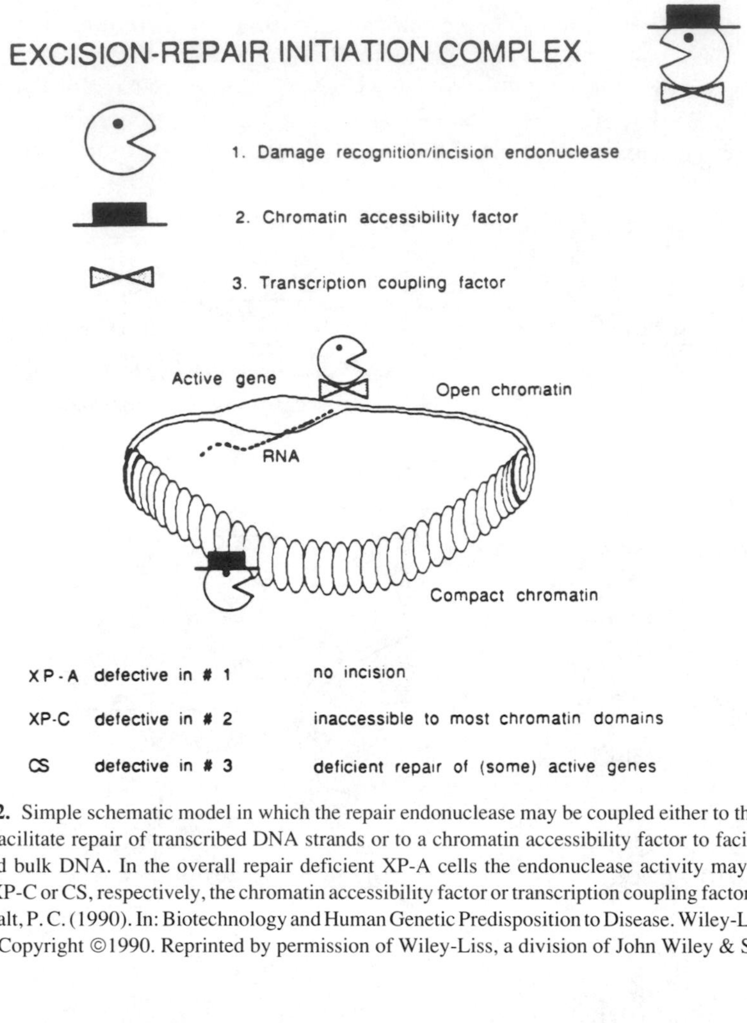

FIGURE 1.12. Simple schematic model in which the repair endonuclease may be coupled either to the transcription apparatus to facilitate repair of transcribed DNA strands or to a chromatin accessibility factor to facilitate access to nontranscribed bulk DNA. In the overall repair deficient XP-A cells the endonuclease activity may be missing or defective. In XP-C or CS, respectively, the chromatin accessibility factor or transcription coupling factor may be faulty. (From Hanawalt, P. C. (1990). In: Biotechnology and Human Genetic Predisposition to Disease. Wiley-Liss, New York, pp. 135–145. Copyright ©1990. Reprinted by permission of Wiley-Liss, a division of John Wiley & Sons, Inc. With permission.)

DNA repair genes have been implicated from the study of **complementation groups** of various **Xeroderma pigmentosum (XP)** cell strains. Following fusion of cells these complementation groups are able to restore the deficient DNA reparatory capacity as reflected in the efficiency of recovery of cells from UV damage (Bohr, 1990). More than 13 genes appear to be involved in the early steps of the excision of UV-induced damage in mononuclear cells (Bohr, 1990). Significant progress has been made in cloning and characterizing DNA repair genes such as *ERCC*-**1**, the *den* **V** gene, **DNA ligase I** gene (defective in Bloom syndrome), and the human repair gene coding for O^6-**methylguanine**-DNA **methyltransferase** (Collins and Sedgwick, 1990).

1.5.4 PREFERENTIAL DNA REPAIR

An important concept with solid experimental support describes heterogeneity of DNA repair, which is reflected in the preferential repair of active genes (Bohr et al., 1989). Most of these data have been obtained in analyzing gene-specific repair after UV irradiation of CHO cells (Bohr et al., 1985). In normal CHO cells, repair of the whole genome may take up to 24 h, whereas recovery of the essential gene for dihydrofolate reductase (**DHFR**) occurs much faster. Repair is seen primarily in the transcription of DNA strands and certain genomic regions rather than being determined by the accessibility of the chromatin structure (Mellon et al., 1986, 1987). Furthermore, the level of methylation of selected areas in the genome may be coordinated with the preferential DNA repair activity (Ho et al., 1989). Preferential repair of the **actively**

transcribed genes has been shown in the c-*abl* and c-*mos* protooncogenes (the former being much more actively transcribed), and in the **methallothionine gene** (Bohr, 1990).

Again, since repair in the methallothionine gene is much more efficient when the gene is activated, an interaction between transcription and DNA repair is apparent. Preferential repair activity is also found in the areas of a gene involved in more frequent structural rearrangements, as is the case with the c-*myc* gene (Bohr, 1990). A marked difference in repair efficiency has been observed in the 5′-flanking region of the c-*myc* gene, where many translocation breakpoints are present. Such a correlation may eventually have practical application, because in the c-*myc* gene preferential repair in certain regions of the gene varies with the inherent resistance of the organism. Thus, cells from the more resistant mice repair the damage much more efficiently than do those from sensitive animals.

Preferential repair of actively transcribed genes has been demonstrated in UV-irradiated fibroblasts from the XP complementation group C (Kantor et al., 1990). The removal of pyrimidine dimers is increased from specific genomic restricting fragments, indicating the preferential repair of certain genomic regions in cells where most of the DNA remained unrepaired (Kantor et al., 1990).

It would be appropriate to note here that the preferential repair of actively transcribed genes and of the areas in the genome with the higher rate of structural rearrangements is similar to the tendency of cells in general to repair first the damage to the most sensitive cellular function (Alexandrov, 1977; Bernstam, 1978). As mentioned above, in the recovery of protein biosynthesis from a heat shock it is the initiation of translation that is damaged and recovered first, followed by the resumption of transcription (Bernstam, 1978). It appears that regulatory loops linking recovery of translation to preferential repair of the actively transcribed genes may also be operational in human disease.

1.5.5 DNA REPAIR AND DRUG RESISTANCE

The phenomenon of multidrug resistance accounts for a large proportion of chemotherapeutic failures. Molecular events underlying this inherent or acquired property of cells, including tumor cells, involve various aspects of cellular metabolism from membrane permeability to the conformational rearrangements of chromatin. In this context, the study of the formation of **cisplatin adducts** in specific regions of the genome and of **interstrand cross-links** acquires practical significance (Bohr, 1990). Different types of chemotherapeutic agents display selective preference in interacting with coding or noncoding areas of the genome as well as with active vs. nonactive genes. This gives hope that gene-specific pharmacokinetics of designed chemotherapeutic agents may allow a precisely targeted interference with selected gene functions (Bohr, 1990).

1.5.5.1 TOPOISOMERASES

Topoisomerases regulate the **superhelical configuration** of cellular DNA. The enzymes are able to relax the supercoils in the chromatin by passing DNA strands through one another: single-stranded DNA (ssDNA) by topoisomerase I (**TI**), and double-stranded DNA (dsDNA) in the case of topoisomerase II (**TII**) (Wang, 1985). DNA topoisomerases are involved in **DNA replication** and **transcription** in mammalian cells. The involvement of these enzymes in DNA repair is likely, although details of their role in determining the resistance of cells to radiation damage are only beginning to emerge.

TII induces dsDNA cleavage and antitumor antibiotics such as streptonigrin promote this activity (Yamashita et al., 1990). This enzyme alters the superhelical density of DNA through **a concerted breakage of both DNA strands**, the formation of an **intermediate bridging complex** of the enzyme and the **double-strand breaks (DSB)**, and, eventually, **resealing** of the DNA strand breaks (Wang, 1985). In the presence of antitumor drugs, which inhibit TII, such intermediate cleavable complexes accumulate in the genome.

There is a direct correlation between the failure of cells to repair DSBs induced by radiation and their sensitivity to drugs that stabilize the cleavable complexes (Evans et al., 1989; Jeggo et al., 1989). Such cleavable complexes and associated DSBs are normally repaired by TII in the absence of inhibitors, and, indeed, cytotoxic DSBs are induced by topoisomerase inhibitors (Caldecott et al., 1990). The cleavable complexes become stabilized and the repair systems of the cell are aimed at eliminating the potentially cytotoxic DSBs.

TI, capable of producing **single-strand breaks** (**SSB**), is specifically inhibited by the antitumor agent camptothecin, which traps TI-DNA covalent complexes in which one enzyme molecule is covalently bound to the 3′ terminus of an SSB (Hsiang et al., 1985; Thomsen et al., 1987). All the DNA strand breaks produced by camptothecin are protein linked, and camptothecin-induced SSBs and DNA-protein complexes represent drug-stabilized TI complexes (Covey et al., 1989). It is believed that the mechanism of cell death produced by camptothecin is due to the collision between moving replicating forks and the camptothecin-induced and stabilized cleavable complexes between DNA and TI (Hsiang et al., 1989 a,b).

It is not yet clear how these enzymes are involved in DNA repair of radiation damage, but a combination of both TI and TII inhibitors produced a marked reduction of gene-specific repair in the hamster DHFR gene after UV damage (Bohr, 1990). In another experimental system, the effects of camptothecin and VP16 (the inhibitors of TI and TII, respectively) were different, so that radiation repair was blocked by TI but not TII (Musk and Steel, 1990).

Clearly, much more needs to be learned about the role of topoisomerases in DNA repair. Nevertheless, it is already evident that in the near future, the evaluation of cellular resistance to specific chemical or physical agents in clinical patient management can be performed at the gene-specific level.

In the meantime, quantitative relationships between the cytotoxicity due to DNA strand breaks and DNA immunoreactivity to anti-DNA monoclonal antibodies can be exploited to predict the sensitivity of individual patients to alkylating agents at the genomic level (Frankfurt et al., 1990).

1.5.6 ASSESSMENT OF DNA DAMAGE AND REPAIR

The whole armament of molecular biological tools has been used in DNA repair research and summaries of earlier methodological approaches have been presented (Friedberg, 1985; Friedberg and Hanawalt, 1988). There are several major groups of methods. The detection of DNA lesions and evaluation of the cellular reparatory efforts is accomplished by chemical, physical, enzymatic, and immunological techniques. Antibodies have been produced to quantitate the DNA adducts of various carcinogens (**benzo[a]pyrene, aflatoxin B_1, ethylnitrosourea**) as well as **thymidine dimers** and **thymine glycols** produced by radiation (Leadon, 1988). DNA lesions can be converted to a break in the DNA and then identified by **sequencing** (Brash, 1988). As mentioned above, DNA damage and its repair can be assessed at the individual gene level (Bohr and Okumoto, 1988; Vos, 1988).

A variety of DNA sequence probes developed to characterize mutations and their repair are well described in Friedberg and Hanawalt (1988). Fine aspects of DNA repair are studied in numerous experimental laboratory systems, but those of potential interest for molecular diagnostic pathology should be applicable to material isolated from the patient. A number of techniques can be used on patient DNA samples, including **alkaline elution** or **alkaline sucrose gradient** analysis of DNA (e.g., Doerjer et al., 1988; Kawamura and Preisler, 1989; Olive et al., 1988). This may give information on the efficiency of DNA ligation in the process of DNA repair, for example, in suspected Bloom syndrome (Lasko et al., 1990). Quantitation of DNA DSBs produced by various agents can be measured now by **pulsed-field gel electrophoresis** and its modifications (Blocher, 1990; Denko et al., 1989; Gardiner, 1991; Iliakis et al., 1990; Stamato and Denko, 1990; Stewart et al., 1988).

Since DNA damage and repair have traditionally been assessed on an averaged tissue or in cell population specimens, the fine details of the process may have been hidden by the heterogeneity of responses of constituent cellular populations. In order to analyze molecular details of DNA damage and repair, methods have been devised for **molecular dosimetry of DNA damage** at the level of individual cells or tissue sections employing, for example, quantitative immunofluorescence microscopy (Baan et al., 1990).

The **PCR restriction site mutation method** has been introduced for the study of mutations in mammalian cells (Parry et al., 1990). First, nonmutant DNA is cut at restriction sites, then the mutant DNA sequences are amplified. A number of technical aspects are to be painstakingly elaborated before this approach can be transferred to clinical applications. Some of the potential problems are the variable efficiency of nonmutant DNA cleavage, fidelity of PCR amplification, removal of restricted sequences in order not to contaminate the mutant DNA being amplified, and so on (Bridges et al., 1990).

One ingenious method **eliminates unmutated sequences** using biotinylated oligonucleo-tides that bind to nonmutated DNA sequences, and the formed duplexes are then separated by capturing these on streptavidin-coated magnetic beads (Bridges et al., 1990). Adaptation of a variety of specific probe methods, chromatographic evaluations of adducts and dimers, measurement of structural alterations in DNA due to cross-linking and base modifications, as well as enzyme probes detecting damaged DNA can be developed for use on patient material.

A number of **human repair genes** [*ERCC*-1, *ERCC*-2, and X-ray repair cross-complement-ing 1 (*XRCC*-1)] have been identified by **complementing studies**, in which the **candidate genes** were able to correct repair deficiency of cultured cells and were assigned to **human chromo-some 19** (Thompson et al., 1989). An elaborate technique using a **cisplatin cross-linked plasmid shuttle vector** has been devised for use as a specific probe to assess the DNA repair capacity of XP cells (Chu et al., 1990; Chu and Berg, 1987). Shuttle systems that introduce vector DNA into mammalian cells are used widely to evaluate the mutagenic activity as well as the reparative capacity of target cells (Calos, 1988; Ganesan and Spivak, 1988).

A number of *in vivo* and *in vitro* assay systems have been devised for the study of **purified repair proteins** introduced in repair-deficient cells or subcellular systems (Hoeijmakers et al., 1990; Keeney and Linn, 1990). Analysis of DNA repair efficiency and replicative synthesis in peripheral blood lymphocytes following their UV irradiation *in vitro* can be used to evaluate the effects of the *in vivo* exposure of an individual to antineoplastic drugs (Celotti et al., 1990).

1.5.7 DNA REPAIR IN HUMAN DISORDERS

Bohr (1990; Bohr et al., 1989) distinguishes **DNA repair-deficient** disorders, conditions suspect for DNA repair deficiency, conditions with a possible DNA repair defect and diseases with demonstrated **hypersensitivity** to DNA damage, in which DNA repair may be important (Table 1.2). Only the salient features of the major conditions will be mentioned here since detailed discussions have appeared (Arlett, 1986; Bohr et al., 1989; Timme and Moses, 1988).

1.5.7.1 XERODERMA PIGMENTOSUM

Xeroderma pigmentosum (XP) is a rare autosomal recessive disorder manifested by exces-sive solar damage to the skin due to deficient excision of **UV-induced DNA photoproducts** (Cleaver, 1968; Setlow et al., 1969). Increased frequency of skin cancers (basal cell carcinoma and malignant melanoma) and internal malignancies affecting the lungs and brain make XP a paradigm of a pathological condition revealing that carcinogenesis involves DNA damage, and that DNA repair plays a major role in maintaining cell homeostasis. Eight **complementation groups** (A through H) and a variant (XP-V) have been identified, the four most common being A, C, D, and variant (Cleaver, 1968, 1990; Cleaver and Kraemer, 1989; Friedberg, 1985). Some of the XP complementation groups are rare and are represented by a single individual (Cleaver,

TABLE 1.2
DNA Damage-Sensitive Human Hereditary Disorders with Established or Potential DNA Repair Deficiency

Disorder	Risk of cancer	Clinical features	Characteristics
DNA repair deficiency			
Xeroderma pigmentosum	++++	Photosensitivity	Hypersensitive to UV[a]
		Neurological abnormalities	Defective excision repair
		Mental impairment	Nine complementation groups
			Increased chromosome breakage and sister chromatid exchanges after UV
Cockayne's syndrome	None	Dwarfism	Hypersensitivity to UV
		Neurological abnormalities	Lack of preferential DNA repair
		Mental retardation	Impaired post-UV DNA and RNA synthesis
		Photosensitivity	
Bloom syndrome	++++	Telangiectasia	Ligase I deficiency
		Photosensitivity	Spontaneous chromosome breaks, aberrations and sister chromatid exchanges. Altered glycosylase
		Immunodeficiency	
Probable defect			
Fanconi's anemia	++++	Skeletal abnormalities	Spontaneous chromosome damage
		Bone marrow hypofunction	Hypersensitivity to cross-linking agents
		Mental deficiency	
		Leukemia	Possible deficiency in cross-link repair
Ataxia telangiectasia	+++	Telangiectasia	Hypersensitivity to X-irradiation and some carcinogens
		Cerebellar ataxia	
		Immunodeficiency	Spontaneous chromosome rearrangements
		Neurologic abnormalities	X-ray-resistant DNA synthesis
			Possible deficiency in repair of strand breaks
Suspected DNA repair defect			
Tricothiodystrophy (PIBID)	None	Telangiectasia	Possible defective excision repair
		Sulfur deficient hair	
		Neurological abnormalities	
		Mental impairment	
Dyskeratosis congenita	++	Hyperpigmentation	Possible defective cross-link repair
		Telangiectasis	
		Leukoplakia	
Nevoid basal cell carcinoma	+++	Skeletal abnormalities	Increased rate of spontaneous chromosome breaks and sister chromatid exchanges
		Mental impairment	
			Possible deficiency in UV repair
Gardner's syndrome	++++	Polyps in colon	Increased rate of sister chromatid exchanges after UV
		Osteomas	
		Dental deformities	
No known repair defect			
Dysplastic nevus syndrome	++++	Dysplastic nevi	Hypersensitive to UV
			Increased sister chromatid exchanges
			Chromosomal rearrangements after UV
Retinoblastoma	++++	Hypersensitive to ionizing irradiation	Hypersensitivity to ionizing irradiation
		Ocular malignancy	Spontaneous chromosomal deletion
		Dysmorphic features	

[a] UV, Ultraviolet irradiation.

From Bohr, V. A. et al. (1989). Lab. Invest. *61*:143–161. © by The U.S. and Canadian Academy of Pathology, Inc. With permission.

1990). In the nine complementation groups the biochemical defect is in the removal of pyrimidine dimers, whereas XP-V is deficient in **postreplication repair (PRR)** (Boyer et al., 1990). PRR eliminates daughter-strand gaps and bypasses lesions in the affected DNA template (D'Ambrosio and Setlow, 1978; Park and Cleaver, 1979).

The deficiency in the classical forms of XP appears to affect an early step in the repair process, before the removal and degradation of the excised segment of DNA (Cleaver, 1969); however, fine details of this process are not fully understood (Cleaver, 1990; Kraemer et al., 1987; Kraemer and Slor, 1985; Lambert and Lambert, 1985). The failure of endonucleases in XP-A cells to excise the damaged areas of DNA early in the repair process does not appear to reside in aberrations of the enzymes themselves (Tsongalis et al., 1990). Some of the repair genes for XP-A are located on **chromosome 9** as shown by alleviation of the DNA repair defect by transfer of human chromosome 9 in hybrid cell experiments (Ishizaki et al., 1990; Kaur and Athwal, 1989). Although there is no preferential band assignment of chromosomal rearrangements encountered in all XP patients, a constant tendency toward chromosome instability in XP mutation carriers is observed (Casati et al., 1991).

A further refinement of the details of DNA repair mechanisms came from studies in which extracts from human cell lines such as XP cells could be added to cell-free systems (Hansson et al., 1990). It appears that XP cells fail to accomplish the **initial incision**, but are capable of carrying out later steps of repair. The first *XP* genes of the mouse (Tanaka et al., 1989) as well as a human gene (K. Tanaka et al., 1990) have been cloned and others will soon follow (Arrand et al., 1989; Cleaver, 1990). The broad spectrum of clinical symptoms and cellular phenotypes in repair-deficient cells can be explained by a combination of single gene defects involving DNA repair inherited in a simple Mendelian fashion (Cleaver, 1990).

This view contrasts with the opposing notions suggesting that multiple gene loci operate simultaneously in determining XP and that the disorder may be, in fact, composed of several different clinical abnormalities (Lambert and Lambert, 1985; Lehmann and Norris, 1989). In this context, it is of interest that preferential gene-specific repair of the actively transcribed *DHFR* gene does occur in XP-A group cell lines after UV irradiation (Bohr, 1990). An immunological defect has been suggested to play a crucial role in determining dermatological symptoms of XP patients (Lehmann and Bridges, 1990).

The **molecular diagnosis of XP** is based on the hypersensitivity of cultured cells to UV radiation due to **defective repair of UV-induced pyrimidine dimers** in DNA. Prenatal diagnosis relies on the **analysis of DNA repair** of cultured amniotic fluid or trophoblast cells following UV irradiation. The **unscheduled DNA synthesis** is evaluated by autoradiography of $[H]^3$ thymidine incorporation into nonreplicating cells.

A more complex situation arises when the proband has a **variant form** of XP. Then the PRR defect is detected by measuring the **conversion** of low to high molecular weight DNA after irradiation of cells using fractionation of DNA on **alkaline sucrose gradients** following pulse labeling of DNA. It is a relatively common research procedure, but somewhat complicated to perform in a clinical laboratory setting; however, it provides clinically relevant information for diagnosing XP (Kraemer and Slor, 1985).

1.5.7.2 BLOOM SYNDROME

Bloom Syndrome (BS) is manifested in low birth weight, stunted growth, normally proportioned dwarfed body, photosensitivity, butterfly telangiectasia of the face, and high susceptibility to infection resulting from moderate to severe immunodeficiency (Friedberg, 1985). Chromosomal abnormalities such as **sister chromatid exchanges (SCE)** are 10 to 15 times more frequent in BS than in normal cells; **mitotic crossing over** and **nonhomologous translocations** are common and BS patients have a very high risk (25%) for developing hematopoietic cancer (German, 1983).

Fibroblasts from BS patients display a **decreased UV resistance**, pointing to a defect in DNA repair that has been traced to **DNA ligase I** and **glycosylase deficiency** (Chan et al., 1987; Seal et al., 1988; Willis and Lindahl, 1987). The pattern of preferential repair of the *DHFR* gene in BS is similar to that in normal cells (Bohr, 1990).

Prenatal diagnosis utilizes the characteristic property of BS cells to form **SCE** in cultured amniocytes (Auerbach, 1987).

1.5.7.3 COCKAYNE'S SYNDROME

This disease is transmitted in an autosomal recessive way and manifested in growth and developmental arrest leading to dwarfism, photosensitivity, deafness, intracranial calcifications, mental deficiency, sunken eyes, disproportionately long arms and legs, and a general appearance of **premature aging** (Friedberg, 1985). Progressive neurologic degeneration leads to death at an early age; however, there does not appear to be an increase in the incidence of malignancies.

Similar to XP, but to a lesser extent, Cockayne's syndrome (CS) fibroblasts and lymphocytes show **hypersensitivity to UV**. The characteristic molecular defect is *not* in the post irradiation repair synthesis of DNA, but in the **recovery of DNA and RNA synthesis in irradiated cells** (Lehmann et al., 1985; Mayne and Lehmann, 1982). Although repair capacity at the level of the genome does not appear to be altered in CS, the **preferential repair of active genes is defective** (Bohr, 1990; Bohr et al., 1989). This deficiency is manifested in the repair of a number of genes — multidrug resistance, DNA polymerase B, TII, and DHFR.

The **diagnostic procedure in prenatal diagnosis** is simple and evaluates the level of [^{3}H] uridine incorporation into RNA in amniotic fluid cells measured by scintillation counting or autoradiography.

1.5.7.4 FANCONI'S ANEMIA

Fanconi's anemia (FA) is a **pancytopenic** disorder, apparently transmitted in an autosomal recessive mode, and is accompanied by congenital and developmental abnormalities and a threefold increased risk of malignancy and death at an early age from hemorrhage, infection, or leukemia (Auerbach, 1987; Friedberg, 1985).

Chromosomal instability in FA with an increased frequency of spontaneous chromatid aberrations such as gaps, breaks, and chromosomal translocations (but not SCEs), and abnormal susceptibility to the action of DNA cross-linking agents have long been recognized (Sasaki and Tonomura, 1973). The **deficiency in repair of cross-linking damage** has not been firmly established so far (Bohr et al., 1989); however, disturbances in oxygen metabolism have been reported in cultured cells from FA patients (Gille et al., 1987). An increased incidence of acute nonlymphocytic leukemia (ANLL), liver tumors, and skin carcinomas is also reported in FA (German, 1983).

Variability of the phenotypic expression of FA precludes accurate clinical diagnosis and parents of affected children prefer **prenatal diagnosis** in subsequent pregnancies (Auerbach, 1987). Although the underlying molecular events are not fully defined, the prenatal diagnosis has been based on the hypersensitivity of FA cells to the blastogenic effect of **DNA cross-linking agents** such as **diepoxybutane (DEB), mitomycin C, cyclophosphamide, nitrogen mustard, psoralens followed by long wavelength UV irradiation (PUVA),** and **cis-platinum** (Auerbach et al., 1981, 1985). DEB-induced chromosomal breakage can be detected in cultured amniotic fluid cells and trophoblast cells from fetuses with FA when compared to control cells, thereby providing a reliable method of prenatal diagnosis of FA.

1.5.7.5 ATAXIA-TELANGIECTASIA

Ataxia-telangiectasia (A-T) occurs at a frequency of 1 in 40,000 live births, affects the nervous and immune systems and the skin, and is transmitted in an autosomal recessive mode of inheritance (Friedberg, 1985). Clinically, the dominating manifestation is a progressive ataxia due to **degeneration of Purkinje cells**, severe muscular discoordination and progressive mental retardation. Oculocutaneous telangiectasias, premature graying, and skin atrophy are accompanied by an immune dysfunction leading to increased susceptibility to infections.

As in the majority of other chromosomal breakage syndromes (BS, FA, XP, and Werner's syndrome), A-T patients have a markedly elevated incidence of malignancy (~10%) (Spector et al., 1982). A distinguishing chromosomal abnormality is a stable, **clonal translocation** involving **chromosomes 7** and **14** (McKinnon, 1987). A-T lymphocytes and cultured fibroblasts are exceedingly sensitive to the lethal effects of ionizing radiation and various other DNA-damaging agents causing DNA strand breaks.

Interestingly, despite the high hypersensitivity of A-T cells to X irradiation, DNA synthesis in these cells is much more resistant to X irradiation than is DNA synthesis in normal diploid cells (Painter, 1981). A-T cells display **radioresistant replication initiation** and **chain elongation** (Friedberg, 1985). The exact deficiency in the DNA repair system in A-T has not been identified so far.

Based on the observed lack of normal inhibition of DNA synthesis in irradiated A-T cells, **genetic complementation studies** have also been devised to diagnose A-T prenatally (Auerbach, 1987; Jaspers et al., 1988-1990; Jaspers and Bootsma, 1982; Murnane and Painter, 1982; Painter and Young, 1980).

A similar inherited disorder, **Nijmegen breakage syndrome**, also belongs to the so-called **chromosome instability syndromes**. The affected individuals suffer from immunodeficiency and have a 60- to 150-fold elevated risk of developing malignancies (Gatti and Swift, 1985; Lehmann et al., 1989; Swift et al., 1990).

Radioresistant DNA synthesis can be demonstrated in pregnancies with both disorders using autoradiographic detection of incorporated tritiated thymidine in cultured chorionic villus cells (Jaspers et al., 1990). This offers a better approach for prenatal diagnosis of these two syndromes than conventional cytogenetic procedures.

1.5.7.6 BASAL CELL CARCINOMA AND OTHER CONDITIONS

The multiple basal cell carcinoma (BCC) lesions in nevoid BCC syndrome (NBCCS) differ from the ordinary BCC lesions in their location and histology, with up to 75% of the affected persons developing malignant changes in the lesions (Bohr et al., 1989). Although there is a 25% reduction in maximal DNA repair as reflected in unscheduled DNA synthesis, the exact lesion has not been identified. Individuals who develop BCC in sun-exposed areas have been shown to accumulate a higher number of **UV-induced pyrimidine dimers** the removal of which is delayed (Alcalay et al., 1990).

Other dermatological conditions such as **dyskeratosis congenita, dysplastic nevus syndrome, trichothiodystrophy,** and **mer-phenotype,** although consistent with a deficiency in DNA repair, have not yet been defined in specific molecular terms (Bohr et al., 1989).

Evidence has been accumulating to implicate defective DNA repair systems as responsible for various clinical manifestations in a number of **primary neuronal degenerations** of children (Robbins, 1989). These include **Friedreich's ataxia** (Brumback, 1983; Brumback et al., 1986; Chamberlain and Lewis, 1982), **familial dysautonomia** (Scudiero et al., 1981), and **Usher's syndrome** (Nove et al., 1987) in children, as well as **Huntington's disease** (Nove et al., 1987; Scudiero et al., 1981), **Alzheimer's disease** (Robbins, 1987; Robbins et al., 1985; Scudiero et al., 1986), and **Parkinson's disease** (Robbins et al., 1985) in adults.

1.5.8 DNA REPAIR AND CANCER

The role of DNA repair in carcinogenesis is more than intuitively relevant. Given the abundant evidence on the specific nature of structural DNA damage produced by external or internal physical and chemical agents, and the involvement of diverse, and often precisely targeted, repair systems in the restoration of normal cellular functions, the probability of malignization as a result of defective DNA repair is certainly very high. Furthermore, location of damage sites in functional or structural association with **oncogenes** offers one of the plausible mechanisms of tumorigenesis.

In fact, the **methylating agent *N*-methyl-*N*-nitrosourea (MNU)** has a surprisingly specific effect on H-*ras* protooncogene in animal systems, producing the **G to A base change** (Barbacid, 1987). Possible explanations for mutations occurring at particular positions in the genome include presentation of particular base positions as easy targets to mutagens; furthermore, some DNA sequences are unaccessible to reparative activity of the cell and that may allow the lesions to persist (Topal, 1988).

Support for the existence of both mechanisms comes from chemical carcinogenesis studies. Preferential repair of UV-induced damage by the c-*abl* protooncogene is more active compared to that of c-*mos*; although it does not link specific gene repair to tumorigenesis, it does point to the importance of the differential reparatory capacity of individual protooncogenes (Madhani et al., 1986). The importance of gaining detailed understanding of the **repairability of protooncogenes** so far studied in experimental tumorigenesis cannot be overemphasized (Hanawalt, 1990).

The deficiency of repair of O^6-methyl-G produced by alkylating agents correlates with the high incidence of tumors in such cells (Pegg, 1984). The level of methyltransferase activities involved in repair of this type of damage, and of other reparatory systems coordinated within the cells to counteract the damage, varies in different human tissues (Topal, 1988). **Lowered DNA reparatory capacity** appears to play an important role in the development of some of the most commonly occurring human malignancies as suggested, for example, by the decreased repair of strand breaks in patients with secondary leukemia (Bohr and Koeber, 1985) and in mouse lymphoma cells (Evans et al., 1987).

A large, ongoing epidemiological study assesses the role of chemotherapeutic agents in causing secondary malignancies. Results of observations made on over 130,000 survivors of Hodgkin's disease, ovarian cancer and testicular cancer have been summarized (Kaldor et al., 1987). Because the nature and dosages of the insulting agents (chemotherapeutics) as well as other parameters of the secondary malignancies developed are known, this approach may provide some, although indirect, information on the role of DNA repair in tumorigenesis. One of the few strong cases in which DNA repair deficiency is known to be associated with malignancies is XP (Cleaver, 1990).

1.5.9 DNA REPAIR AND AGING

Analysis of the role of DNA repair in aging has appeared (Iversen, 1988; Tice and Setlow, 1985; Warner et al., 1987). Several theories linking DNA damage and repair with aging have been proposed (Hayflick, 1985; Kirkwood, 1989; Macieira-Coelho, 1984; Strehler, 1986; Vijg and Knook, 1987). The importance of DNA repair as a factor in the aging process is also implicated in studies on **oxidative damage** of DNA (Ames, 1989a,b). The efficiency of **nucleotide excision** declines with age (Vijg and Knook, 1987). **Accumulation of mutation defects** in DNA has been used in the majority of molecular studies on aging as a reflection of the reparative activity of cells and tissues at different ages (Tice and Setlow, 1985). The specific types of DNA damage that received particular attention have been **SSBs** and **DSBs** traditionally estimated on averaged cell populations (Mayer et al., 1989).

Applying a new **single-cell gel assay** methodology (Singh et al., 1988), capable of evaluating SSBs in individual cells following X irradiation of human lymphocytes, a distinctly **lower reparative capacity** of the cells from the older persons could be demonstrated (Singh et al., 1990). An **age-related decline** in DNA repair competence was detected in the 4 to 16% subpopulation of lymphocytes. Cells derived from a **progeria (Hutchinson-Gilford syn-**

drome) patient characterized by premature aging display a reduced capacity to repair UV excision damage (Wang et al., 1990). The role of DNA damage and repair in aging has been reviewed (e.g., Mullaart et al., 1990).

1.5.10 CONCLUDING REMARKS

Although experimental data on DNA repair in aging at this time offer little practical guidance with respect to diagnosis, monitoring, or prognosis of diseases in humans, the application of molecular biological techniques, such as PCR and its modifications, to uncovering the molecular contribution of the normal physiological process of aging to the disease condition under study may provide a valuable complementary approach to patient management. Comparative evaluation of DNA damage and repair in diseased tissues or cells isolated from patients with certain disorders may soon provide a basis for routine assessment of the response of an individual to a disease and/or therapy. An example of such an application of DNA repair data can be seen in the carrier detection in XP by measuring chromosomal aberrations in skin fibroblasts exposed to X irradiation (Parshad et al., 1990).

Chapter 1.6

Drug Resistance

1.6.1 INTRODUCTION

The development of optimal treatment protocols for bacterial infections requires knowledge of the susceptibility to a chosen drug(s). Testing for this parameter has become routine in clinical microbiological laboratories and is based predominantly on the phenotypic response of bacteria to a panel of antimicrobials. **Drug resistance in bacteria** varies with the organism, and effective laboratory assays for the presence and expression of the genes encoding the resistance factors are clearly of practical significance (Jacoby and Archer, 1991).

Newer testing methods are being introduced, such as an assay for erythromycin resistance using polymerase chain reaction (PCR) amplification of the sequences of *erm* genes encoding rRNA methylases that appear to be universal for the detection of the *erm* genes in gram-positive cocci (Arthur et al., 1990). In various strains of *Escherichia coli*, biotin-labeled DNA probes for the **dihydrofolate reductase (*DHRF*) gene** have been used to assess the plasmid-encoded **trimethoprim resistance** (Towner et al., 1991).

1.6.2 MULTIDRUG RESISTANCE GENES

In the case of **tumors**, failure of chemotherapy may become apparent in a clinical situation mostly on a trial-and-error basis. The advances of molecular biology allow a substantially more efficient and specific identification of potential failures of a chosen type of chemotherapy. An important concept underlying this approach is that of the *mdr* (**multidrug resistance**) **genes**, which control intrinsic or acquired resistance of tumor cells to a wide range of natural products and hydrophobic drugs such as **vincristine, actinomycin D**, and **adriamycin** (Arias, 1990; Benard et al., 1990; Deuchars and Ling, 1989). Obviously, the role of genetic factors is not limited only to tumor tissues. In fact, genetically determined **polymorphic drug metabolism** also accounts for the variable degree of β-blockade produced by the antiarrhythmic agent **propafenone**, a sodium channel blocker (Lee et al., 1990).

The human **P-glycoprotein gene** family is composed of two genes, *MDR*-1 and *MDR*-3, also called *MDR*-2, which are very similar in structure (Lincke et al., 1991). Interestingly, in different cell lines the extent of hypomethylation of the **CpG-rich sequences** at the 5′ ends of both genes roughly correlates with their transcriptional activity (Lincke et al., 1991). Of practical importance is the finding of specific sequences upstream and downstream of the initiation site, the deletion of which completely inhibits transcription of the *MDR*-1 gene (Cornwell, 1990). Moreover, *MDR*-1 transcription can be specifically inhibited by an oligonucleotide corresponding to sequence +46 to +58. These findings suggest the possibility of modulating the expression of the *MDR*-1 gene for clinical purposes by affecting the efficiency of the *MDR*-1 proximal promoter (Cornwell, 1990).

Only the *MDR*-1 gene has been demonstrated unequivocally to be responsible for the production of a 170,000-Da cell surface glycoprotein, called **P-glycoprotein**. This **multidrug transporter** is responsible for the energy-dependent **drug efflux** of various hydrophobic chemotherapeutic agents from tumor cells. It acts as a "hydrophobic vacuum cleaner" (Gottesman and Pastan, 1990), removing cytotoxic material through the cell membrane into the extracellular medium. Confirmation of the role of *MDR*-1 in determining the drug resistance has been provided in cell lines (e.g., Deuchars and Ling, 1989) as well as in transgenic mice, which developed resistance to daunomycin-induced leukopenia when human *MDR*-1 cDNA was introduced into the animals (Gottesman and Pastan, 1990).

Both *MDR*-1 and *MDR*-2 genes are located on **chromosome 7q21-1** and encode mRNAs of 4.5 and 4.1 kb, respectively. Transfection of only full-length *MDR*-1 cDNA into sensitive cells is capable of conferring the full MDR phenotype, whereas *MDR* 2 cDNA alone fails to do so. The MDR family of transmembrane transporters is related to the MHC class II region, linking the presentation of metabolized drugs at the cell surface to the functioning of the immune system (Deverson et al., 1990; Trowsdale et al., 1990).

P-Glycoprotein gene is **amplified** and **over expressed** in multidrug-resistant cell lines and the level of P-glycoprotein mRNA correlates with the **degree of resistance** (Capranico et al., 1989). The degree of overexpression of the gene, however, is not necessarily associated with a proportionate increase of the gene copy numbers. Over 500 human cancers have been tested for the expression of the *MDR*-1 gene and the *MDR*-1 mRNA levels were consistent with the extent of their resistance to daunomycin and/or with the eventual relapses following chemotherapy (Goldstein et al., 1989). This regularity was observed in **cancers of the kidney**, **liver**, **colon**, **adrenal**, **glands**, and **pancreas** as well as **neuroblastoma**, **lymphomas**, **leukemias**, and **ovarian cancer** (Goldstein et al., 1989).

The expression and activity of P-glycoprotein have been detected in human hematopoietic stem cells in the bone marrow as well, which may have implications for understanding and treatment of hematological malignancies (Chaudhary and Roninson, 1991). In fact, expression of the *MDR*-1 gene is observed in most **acute myeloid leukemia (AML)** patients who die early in the disease, indicating that *MDR*-1 gene expression is an unfavorable prognostic factor and multidrug resistance plays a role in AML (Pirker et al., 1991). Recently, transgenic mice normally not expressing the *MDR*-1 gene in their bone marrow have been induced to express this gene, allowing the testing of various cytotoxic drugs in these animals (Pasfan et al., 1991).

In **gastric adenocarcinoma**, the expression of *MDR*-1 was heterogeneous in individual cells and its expression was also noted in the intestinal metaplastic and dysplastic cells (Vollrath et al., 1991). In the case of adriamycin, the expression of **topoisomerase II mRNA** correlated better with resistance to the drug than did the expression of *MDR*-1 or **GST-π** (Kim et al., 1991). Comparison of the dose of **doxorubicin** causing 50% inhibition of growth (ID_{50}) with the mRNA levels transcribed from the *mdr*-1 and *gst-π* **(glutathione-S-transferase) genes** in human **breast tumors** showed that the tumors with high *mdr*-1 expression also had high ID_{50} values (Keith et al., 1990). A role of the *MDR*-1 gene expression has also been demonstrated in ovarian cancers with acquired resistance to doxorubicin treatment (Bourhis et al., 1989).

MDR-1 **transcripts** present in total RNA isolated from 92 invasive cervical cancers and from normal cervices **can be quantitated by slot-blot hybridization** using total RNA from the KB 8-5 cell line with 30 arbitrary units of *MDR*-1 mRNA as a probe (Riou et al., 1990). While a higher proportion of normal cervices (68%) than carcinomas (43%) displayed a characteristic 4.5-kb *MDR*-1 transcript, the majority of treated tumors (88%) exhibited the transcripts compared to only 35% of untreated cancers detectable by this approach. *The finding of overexpression of the MDR-1 gene correlates with the cellular resistance to cytotoxic drugs* and corresponds to that found in human **gastric** and **colorectal cell lines** (Chao et al., 1990; Park et al., 1990), **metastatic undifferentiated sarcoma of the liver** (Tawa et al., 1990), and **human ovarian cancer** cells (Parker et al., 1991).

Anticancer drug resistance is determined not only by the action of the "efflux pump"; the **enhanced DNA repair** (Parker et al., 1991) resembles the more efficient DNA/chromosome break repair in radioresistant tumor cell lines (Schwartz and Vaughan, 1989). The expression of the c-*myc* gene has been implicated in the acquisition of drug resistance in transfected NIH 3T3 cells (Niimi et al., 1991).

Flow karyotyping has been used to determine the DNA content of a six-gene amplicon in the multidrug-resistant Chinese hamster cell line CH^{r}B3 (Jongsma et al., 1990). This technique estimates deletions by flow cytometry (FCM) analysis of propidium iodide-stained chromosome suspensions. In one of the revertants of the precursor MDR cell line, the amplified DNA is cytogenetically visible as a large, **homogeneously staining region (HSR)** on chromosome 7q+. Interestingly, the **size of the HSR correlates with the degree of relative drug resistance** of the cells (Jongsma et al., 1990). In one of the revertants only DNA of the 7q+ HSR was lost, while the rest of the genome was unchanged. The revertant that lost its drug resistance was found to have also lost about 107 mbp, together with 11 to 24 copies of the P-glycoprotein gene (Jongsma et al., 1990). In fact, most of the human cells lines expressing *mdr* genes do not show amplification of the *MDR*-1 gene, suggesting the *overexpression of the gene occurs by mechanisms other than amplification* (Brown,1991). Moreover, cytotoxic drugs appear to target amplified genes differentially, as shown by the complex kinetics of amplified *DHFR* gene loss in methotrexate-resistant cells (Wani and Snapka, 1990).

In the meantime, the characterization of the *MDR*-1 gene product allows for the specific chemical circumvention or reversal of P-glycoprotein activity by verapamil and related drugs (e.g., Merry et al., 1989; Yusa and Tsuruo, 1989). **Verapamil** enhances the retention of antitumor agents by cells by competing for the binding sites on the P-glycoprotein transporter (Yusa and Tsuruo, 1989).

1.6.3 TOPOISOMERASES IN DRUG RESISTANCE

It appears that the relationship between the *MDR*-1 gene expression and drug resistance is rather complex. Thus, although alteration of topoisomerases can be seen as one of the molecular mechanisms involved in cellular differentiation, and P-glycoprotein has been related to the state of differentiation and, thereby, drug resistance, these two phenomena can be clearly dissociated under certain conditions (Gieseler et al., 1990). For example, in human neuroblastoma cell lines an increase in the *MDR*-1 gene expression associated with the retinoic acid-induced differentiation was not accompanied by a concomitant P-glycoprotein-mediated decrease in cytotoxic drug accumulation (Bates et al., 1989). This observation also suggests that *increased expression of the MDR-1 gene may not necessarily be associated with increased cytotoxic drug resistance*. Defective topoisomerases I and II (TI and TII) account for some aspects of the resistance of tumor cells to a group of chemotherapeutic drugs targeted to these enzymes (Tan et al., 1989). This resistance can be traced to the **rearrangements** in the topoisomerase gene and **hypermethylation** that resulted in defective transcription of the gene.

The alteration of topoisomerases is also viewed as an additional mechanism of transcriptional control of entire gene programs by **regulating the conformation of DNA** by twisting and unwinding the double strands (Gieseler et al., 1990). A variable pattern of an *Xmn*I RFLP of the TII locus on **chromosome 17** can be seen in individuals without malignant disease and exhibits an autosomal mode of inheritance (Zwelling et al., 1990). Genetic changes other than those affecting chromosome 17 account for the resistance of TII to the inhibiting actions of the specific TII-DNA intercalator, *m*-AMSA (Zwelling et al., 1990)

1.6.4 CISPLATIN RESISTANCE

Cisplatin is widely used alone or in combinational chemotherapy of cancers. Its most significant limitation, however, is the development of resistance, the mechanism of which is complex (Scanlon et al., 1988, 1989a). Cytotoxic effects of cisplatin on cells result from the formation of **cisplatin-DNA adducts**. The ability of the cell to repair this damage depends on the availability of repair enzymes and their substrates (deoxynucleotides). Cisplatin affects both **folate** and **methionine metabolism,** thus interfering with the supply of thymidine for repair of the damage. Cells relying more on endogenous folate metabolism are thus less affected by cisplatin toxicity (Scanlon et al., 1989a).

No transport carrier system specifically responsible for cisplatin influx and efflux from cells has been identified. The **detoxification role of glutathione (GSH)** in cisplatin resistance, although highly suggestive, remains to be defined (Howie et al., 1989; Lee et al., 1989; Lewis et al., 1989; Meijer et al., 1990); so is the contribution of **metallothionein**. Attention has also been focused on the role of DNA polymerase β in the **repair of cisplatin-induced DNA damage**.

In any case, the **repair enzymes** counteracting the effect of cisplatin-DNA complexes require deoxynucleotides. The dTMP synthase cycle supplying thymidine thus becomes rate limiting in DNA synthesis. Attempts are being directed at studying whether changes occur in the genes encoding various enzymes in this cycle in the course of development of cisplatin resistance (Scanlon et al., 1989a,b). Furthermore, *DNA repair enzymes are potentially involved in determining cisplatin resistance,* but so far no evidence of gene amplification for these enzymes in human cells has been obtained; DNA polymerase β mRNA, however, increases severalfold in cisplatin-resistant cell lines (Scanlon et al., 1989a,b).

Amplification of oncogenes c-*fos* and c-H-*ras* has been observed in patients who developed resistance to cisplatin treatment. In ovarian and breast carcinoma cell lines, the c-*myc* mRNA expression is selectively increased three- to fivefold in cisplatin-resistant cells. These findings alone are not sufficient to imply oncogenes are responsible for cisplatin resistance (Scanlon et al., 1989a).

1.6.5 DRUG RESISTANCE RELATED TO METABOLIC DEFECTS

The role of genetic factors in the individual response to drugs has been studied with respect to the inherent ability to oxidize a test drug. For this purpose, **debrisoquine**, **sparteine**, or **dextromethorphan** are used, and the ratio between the parent drug and its metabolites is determined (Heim and Meyer, 1990). Individuals classified as **poor metabolizers** of over 25 unrelated drugs are at risk for adverse drug reactions, the phenotype being inherited as an autosomal-recessive trait in 5 to 10% of European and North American populations (Heim and Meyer, 1990).

Routine debrisoquine phenotyping is recommended particularly for psychiatric patients (Balant et al., 1989). A link between the debrisoquine phenotype and susceptibility to lung cancer has also been suggested (Ayesh et al., 1984; Caporaso et al., 1989). The poor metabolizer phenotype is caused by the absence of a **cytochrome isozyme P450IID6** due to aberrant splicing of its pre-mRNA. Three mutant alleles of the P450IID6 gene locus (**CYP2D**) associated with the poor metabolizer phenotype have been identified by RFLPs (Gough et al., 1990; Skoda et al., 1988).

The evaluation of oxidation phenotype, although helpful in the selection of optimum doses of drugs and in the avoidance of adverse drug reactions and drug interactions, is not recommended for routine application prior to drug therapy because the clinical importance of sparteine-debrisoquine polymorphism is still not fully resolved (Lennard, 1990). In this context it is of interest to note that individual variations in genetic polymorphism evaluated by a metabolic ratio of debrisoquine/4-hydroxydebrisoquine are *not* influenced by environmental factors, such blood contituents as cholesterol and glucose, or some common therapeutic medications (Vincent-Viry et al., 1991).

The efficacy of chemotherapy is determined not only by the **permeability**, **retention**, and **efflux mechanisms** in which MDR plays such an important role. Cytotoxicity is primarily dependent on the interaction of the drugs with DNA and the ability of the cell to repair the damage produced (Epstein, 1990). Although acknowledging a certain degree of frustration in the search for breakthrough chemotherapeutic approaches in clinical oncology over the past few years, Epstein (1990) emphasizes the significant achievements reached in our understanding of the mechanisms of cytotoxicity and the development of specific *in vitro* and *in vivo* testing methods capable of better prediction and modulation of the effects of chemotherapy.

1.6.6 CONCLUDING REMARKS: CLINICALLY RELEVANT EVALUATION OF DRUG RESISTANCE AT THE GENE LEVEL

In vitro assays of synergistic or antagonistic effects of various cytotoxic drug combinations sometimes fail to reproduce or predict their clinical efficacy (Katz et al., 1990). **Chemotherapeutic drugs can be tested in transgenic mice that express the *MDR*-1 gene** in their bone marrow and faithfully reproduce the resistance to such drugs as cisplatin, methotrexate, and 5-fluorouracil as well as the reversal of drug resistance by verapamil, quinidine, and guanine (Mickisch et al., 1991). It is hoped that with the development of relatively simple and direct methods for the quantitation of DNA-drug interactions, better estimates of the clinical efficacy of various drug combinations can be achieved (Bellamy et al., 1990). Recently, a monoclonal antibody assay for the improved *in vitro* and *in vivo* detection of DNA modifications induced by cisplatin and carboplatin has been described (Tilby et al., 1991). Sequence analysis of the ***DHFR* gene** has been used to detect mutations possibly accounting for **methotrexate resistance** in human tumors (Dicker et al., 1989).

Clear evidence for the amplification of the **dTMP synthase gene** has also been presented by the PCR assay. An increased expression of the genes encoding dihydrofolate reductase, thymidylate synthase, and thymidine kinase has been demonstrated in ovarian and colon carcinoma patients (Kashani-Sabet et al., 1988; Scanlon et al., 1988; Scanlon and Kashani-Sabet, 1989). The utility of a PCR-based assay for the analysis of cisplatin resistance is founded on its ability to identify specifically potentially responsible genes, as well as on the high sensitivity of the assay, requiring as little as 2 ng of RNA (in the case of the DNA polymerase β gene assay) (Scanlon et al., 1989a).

Clinical utility of an assay of *MDR*-1 gene expression requires that the assay be both sensitive and quantitative. A PCR-based evaluation of various degrees of expression of the *MDR*-1 mRNA has been performed in colon carcinoma cell lines that express *mdr*-1/Pgp at levels comparable to those found in clinical samples (Murphy et al., 1990). Using coamplification of the *MDR*-1 gene with a control gene, β$_2$-microglobulin, to achieve normalization of the product, even small variations in the *MDR*-1 gene expression can be quantitated.

This approach, although labor intensive, offers a sensitive method for quantitating *MDR*-1 gene expression in small samples of RNA, which can be of practical use in clinical oncology (Murphy et al., 1990). Whereas RFLP analysis allows only 25% of poor metabolizers to be predicted using a number of restriction endonucleases, PCR assays complementing restriction analysis identify over 95% of individuals (Heim and Meyer, 1990), with only 1 to 2% of the poor metabolizers unidentified when two PCRs are used (Wolf et al., 1990).

A very sensitive PCR procedure for quantitation of *MDR*-1 mRNA has also been developed by other workers, revealing low levels of gene expression in the many tumors tested (Roninson et al., 1990). What remains to be determined, however, is the minimal level of *MDR*-1 gene expression capable of conferring resistance to chemotherapeutics relevant to the clinically observed resistance. Yet another PCR assay for measuring the levels of *MDR*-1 mRNA in clinical specimens demonstrated low levels of *MDR*-1 expression that are undetectable by conventional assays in most solid tumors and leukemias tested before chemotherapy (Noonan et al., 1990).

The *MDR*-1 gene, fortunately, is not activated in all the cancers tested so far. The evaluation of P-glycoprotein clearly has practical significance for the appropriate modification of chemotherapeutic protocols, as demonstrated in lymphoma, leukemia, neuroblastoma, and childhood soft-tissue sarcomas (Benard et al., 1990).

The clinical significance of the low level of expression of *MDR*-1 mRNA (one molecule per cell or lower) detectable only by PCR is still unclear. Even at these low levels, however, the detection of *MDR*-1 gene expression correlates with the proportion of acute nonlymphocytic leukemia patients who fail to respond to chemotherapy. Interestingly, PCR assay confirmed the absence of *MDR*-1 mRNA in Ewing's sarcomas, which are usually susceptible to chemotherapy. This finding was in sharp contrast to the low level of *MDR* gene expression in the other, more refractory types of sarcoma (Noonan et al., 1990). A very important finding opens an entirely new avenue for exploring the phenomenon of multidrug resistance which appears to be linked to the well-recognized cellular oncogene c-Ha-*ras*-1 and the *p53* tumor suppressor gene (Chin et al., 1992). The promoter of the human *MDR*-1 gene is a target for the products of both c-Ha-*ras*-1 and *p53* which are associated with tumor progression significantly, while a mutant *p53* stimulates the *MDR*-1 promoter, the wild-type *p53* specifically represses its expression (Chin et al., 1992). Thus, it appears that evaluation of the *MDR*-1 gene expression offers a useful tool for patient management.

Chapter 1.7

Metastasis

1.7.1 PHENOMENOLOGY OF METASTASIS

Although the study of molecular events involved in the metastatic process is relatively young, some emerging facts deserve attention as potential candidates for laboratory assays in the comprehensive management of oncological patients.

The process of metastasis comprises a set of diverse events predominantly considered in the context of the **spread of tumor cells to distant sites from the primary focus** as well as their **ability to establish secondary tumor growths** independent of the primary tumor. Analysis of the phenomenology of cancer metastasis is carried out at different levels: from tumor-host interactions, through factors involved in de-adhesion, migration, and colonization of distant sites, to the molecular genetic factors traceable to specific and general behavioral patterns of various metastasizing tumors (Fidler, 1987; Zetter, 1990).

Biological heterogeneity of primary and metastatic tumors extends to a variety of phenotypic characteristics such as receptor status, antigenic profiles, sensitivity to chemotherapeutic agents, and many others (Fidler, 1987). **Heterogeneity of cell populations** comprising the primary and secondary tumors is viewed as an integral aspect of the so far poorly defined, although amply documented, phenomenon of the specific interaction of tumor cells with the host tissue in an organ-specific manner (for reviews, see Fidler, 1987; Miller and Heppner, 1990; Zetter, 1990). The specificity of **intercellular interactions** in the process of metastasis is truly amazing, particularly considering that targeted metastatic spread occurs only when the transplants are placed in specific organ locations, supporting the so-called **"seed and soil" hypothesis** of Paget (Miller and Heppner, 1990). Selective organ- and intraorgan-specific homing of metastasizing cells of various tumors cannot be explained solely on the basis of blood vessel and lymphatic topography (Zetter, 1990).

1.7.2 CELLULAR AND TISSUE ELEMENTS IN METASTASIS

The experimental evidence points to a number of conditions likely to contribute to the site-specific tumor progression and metastasis. Among those are the action of inhibitors of cellular growth and specific growth factors, and the relative abundance of mitogenic factors in specific organ sites. An example of the latter is the production by stromal cells in the bone of a factor stimulating the growth of human prostatic carcinoma cells (Chackal-Roy et al., 1989).

An essential component of tumor growth is **neovascularization**, which develops in neoplasias larger than only a few millimeters. This process is supported by the **angiogenic factors** produced by the tumor cells as well as by activated lymphocytes and macrophages (Miller and Heppner, 1990). Permeation of a tumor mass by vessels promotes not only its own growth but also serves to provide mechanical routes for dissemination of tumor cells throughout the body. It is increasingly clear, however, that the **passive transport of tumor cells** to distant sites alone is not sufficient for the development of metastatic lesions.

Interaction of tumor cells with **extracellular matrix (ECM)** is an active process in which the coordinated action of various degradative enzymes (e.g., collagenase type IV, plasminogen activator, glycosidases, and cathepsin B) facilitates the **invasion, intravasation,** and later **extravasation** by tumor cells (Fidler, 1987; Liotta et al., 1991; Miller and Heppner, 1990). In addition to coordinated proteolytic activities, metastasizing tumor cells display **motility** and **tumor cell colony formation** eventually resulting in the establishment of secondary growths.

While tumorigenicity and metastatic potential may have common features, it is the elucidation of specific markers that are closely linked with the metastatic spread of the tumor that presents particular interest in patient management. A coverage of the complex regulatory interactions emerging from basic studies on tumor metastasis is beyond the scope of this discussion (for reviews, see e.g., Liotta et al., 1991; Ruoslahti, 1991; Zetter, 1990).

1.7.2.1 CELL ADHESION IN METASTASIS

It appears that the gene level evaluation of certain key elements controlling the metastatic process may provide diagnostic tools of practical value. As already mentioned, one of the critical determinants of site-specific metastasis is the process of **cell adhesion**. Experimental evidence suggests that the *metastatic cells preferentially adhere to endothelial cells from their target organs* as well as to other ECM elements (**fibronectin, laminin,** and **collagen types IV** and **V**), displaying fine characteristics of organ specificity (Zetter, 1990).

Since adhesion of metastatic tumor cells is mediated via a limited number of **specific peptide sequences,** it is possible, although not yet demonstrated, that genetic determinants in tumor cells responsible for such specific interactions can be defined. In that case, screening for such determinants in tumor cells may reveal a propensity to metastasis early in this process, and eventually serve to develop specific antimetastatic tools. In fact, expression of the genes for elastase and type IV collagenase appears to correlate significantly with the potential for pulmonary metastasis (Liotta and Stetler-Stevenson, 1989; Yusa et al., 1989). A receptor for elastin on the surface of tumor cells has been identified and its expression appears to correlate with organ-specific tumor metastasis in the lung (Zetter, 1990).

1.7.2.1.1 Integrins

The attachment of tumor cells to the **basement membrane** surface is mediated by **integrin** and nonintegrin-type receptors on the tumor cell surface (for reviews, see Dedhar, 1990; Mareel et al., 1990; Ruoslahti, 1991).

Integrins are a family of related **integral membrane glycoproteins** acting as the **primary mediators of cell-extracellular matrix adhesion,** and also as one of the many types of **molecular determinants in cell-cell adhesion** (Dedhar, 1990; Ruoslahti, 1991). The name *integrins* was coined to emphasize the role of these proteins in establishing the biological continuity between the intracellular cytoskeleton and the extracellular matrix (Ruoslahti, 1991). The importance of this class of molecules in diverse cellular functions is apparent not only from the standpoint of fundamental biological processes of growth and development, but also from the unraveling of their role in platelet aggregation, immune functions, tissue repair, and tumor invasion (Ruoslahti, 1991).

The **bridging** of the intracellular cytoskeleton with the extracellular matrix through integrins is cell type-specific and a varied assembly of its two basic subunits, α and β, spans the cellular membrane to form the extracellular ligand-binding site composed of sequences from both subunits. **Extracellular matrix ligands** for integrins include fibronectin, fibrin and fibrinogen, laminin, collagens, entactin, tenascin, thrombospondin, von Willebrand factor, and vitronectin. **Intercellular adhesion proteins** such as **ICAM-1, ICAM-2,** and **VCAM-1** also act as ligands for integrins (Ruoslahti, 1991).

The **integrin cDNAs** have been cloned and the amino acid sequences of many subunits have been derived. The recognition site for many of the integrins is the **tripeptide Arq-Gly-Asp (RGD)**. Interestingly, the highly active RGD peptides interfere with specific adhesion functions such as platelet aggregation via the **gpIIb/IIIa integrin** (Ruoslahti, 1991). Such proteins have been termed "**disintegrins**". Analysis of the role of integrins in human disease is only beginning on a systematic basis, and it is apparent that major aspects of tumor and metastasis phenomenology will be better understood when molecular events involving integrins are clarified.

In the meantime, correlations between the level of expression of specific integrins and phenotypic behavior of various tumors are being established and the role of cytokines and growth factors in modulating their expression is being studied (Dedhar, 1990). Evaluation of integrin protein regulation in human disease at the gene level so far has not been demonstrated in a clinical setting.

1.7.3 PROTEOLYTIC ACTIVITY OF TUMORS

The coordinated, highly localized, and multicomponent proteolytic activity of invading tumor cells is the intrinsic characteristic of metastatic tumors and is currently a focus of intense investigations (reviewed in Liotta et al., 1991). Of particular interest is the regulation of expression of the **matrix metalloproteinase gene family** which includes three categories of enzymes: interstitial collagenases, type IV collagenases (gelatinases), and stromelysins.

The observed positive correlation between type IV collagenase activity and tumor cell invasion acquires practical significance: an increase in this enzyme activity reflects the genetic induction of a metastatic phenotype (Liotta et al., 1991). The **72-kDa type IV collagenase mRNA** species, in particular, appears to correlate with the invasive potential of tumor cells. Its overexpression can be interpreted as a marker of many invasive and metastatic human cancers. The opposite leverage in the invasive phenotype can be seen in the expression of proteinase inhibitors. A group of **tissue inhibitors of metalloproteinases (TIMPs)** may function as **metastasis suppressors** and the degree of their expression can also characterize the metastatic potential of a given tumor (Liotta et al., 1991).

1.7.4 ROLE OF ONCOGENES, GROWTH FACTORS, AND TUMOR SUPPRESSOR GENES IN METASTASIS

Since the realization of the existence of separate, although at times overlapping, genetic controls determining tumorigenicity and metastatic potential, it has been found that oncogenes (such as *ras* and *myc*), capable of conferring a malignant phenotype, do not necessarily lead to the appearance of a metastatic phenotype (for reviews, see e.g., Cross and Dexter, 1991; Hart and Easty, 1991; Liotta et al., 1991; Wright et al., 1990b). It appears that invasion and metastasis require the activation of genes distinct from those determining uncontrolled growth of tumors (Liotta et al., 1991). Some of the oncogenes and growth factors have been shown to induce the expression of genes encoding proteinases associated with metastatic spread.

Among those are *c-fos*, which apparently is capable of suppressing metastasis indirectly through the histocompatibility system; the epidermal growth factor receptor (EGFR) and platelet-derived growth factor (PDGF), the enhanced expression of which is correlated with invasiveness; and a number of oncogenes other than *ras*, although probably acting via a common *ras*-dependent pathway (*mos, raf, src, fes, fms*) (Sobel, 1990). Numerous specific correlations of increased aggressiveness of individual types of tumor with the expression of oncogenes and specific RFLPs are discussed in this book in appropriate sections.

1.7.4.1 METASTASIS SUPPRESSOR GENES

Similar to the role of tumor suppressor genes, the inactivation of which promotes tumorigenesis, somatic cell hybridization and DNA transfection experiments as well as the isolation of cDNA clones by subtractive hybridization point to the existence of **metastasis suppressor genes** that inhibit invasion and metastasis (Fidler and Radinsky, 1990; Liotta et al., 1991; Sobel, 1990). While TIMPs (Section 1.7.3) may be viewed as tumor suppressor genes, since their up regulation enhances inhibition of metalloproteinases, evidence is accumulating on the existence of **recessive metastasis suppressor genes** as well. Particular interest in this regard has been generated by the differential expression of the *nm23* gene in several experimental metastatic models.

This gene was first identified due to the reduced level of expression of *nm23* mRNA in seven cell lines derived from a single K-1735 murine melanoma (Steeg et al., 1988). In the cell lines with high metastatic potential, the level of expression of the *nm23* gene is reduced by an order of magnitude compared to its level in cells with low metastatic propensity. A similar reciprocal relationship (low *nm23* mRNA level in highly metastatic cells) has also been observed in infiltrating ductal breast carcinomas from patients with lymph node metastases (Bevilacqua et al., 1989).

Furthermore, in a clinical correlation study, the loss of *nm23* mRNA has been associated with poor survival in breast cancer patients, reproducing the relationships also observed in mouse mammary tumors and chemical carcinogen-induced mammary tumors in rats (Liotta et al., 1991; Steeg and Liotta, 1990). Importantly, transfection of melanoma cells with *nm23* leads to a reduced incidence of primary tumor formation, marked reduction in tumor metastatic potential independent of tumor cell growth, and altered responses to the **cytokine transforming growth factor β-1**, suggesting that ***nm23* is capable of suppressing several aspects of carcinogenesis, including metastasis** (Leone et al., 1991).

Analogy of the *nm23* gene with the *Drosophila awd* gene suggests a role for *nm23* in signal transduction of cell-cell communication, the disruption of which by an aberrant *nm23* gene may result in the acquisition of an invasive phenotype by tumor cells (Liotta et al., 1991). This similarity extends to a class of enzymes, **nucleoside diphosphate (NDP) kinases**, participating in the formation of enzyme-bound high-energy phosphate intermediates (Liotta and Steeg, 1990; Steeg and Liotta, 1990).

At least two cellular functions known to be associated with cancer and development, **microtubule assembly/disassembly** and **signal transduction through G proteins**, can be affected by defective NDP kinases (Liotta et al., 1991). It is believed that members of the NDP kinase family, one type of which may be produced by *nm23*, could act as positive or negative regulators of oncogenic, developmental, or metastatic processes (Liotta et al., 1991). Indirect support for this view also comes from the identification of oncogenic potential of mutated G proteins participating in signal transduction (Landis et al., 1990).

1.7.5 CONCLUDING REMARKS

Our understanding of the intricate mechanisms underlying the multistage phenomenon of metastasis is still far from complete. The concept of **invasive phenotypes** visualizes a balance of I⁺ and I⁻ phenotypes as determining the M⁺ and M⁻ metastasis phenotypes on the basis of the expression of the sum total of multiple cellular and tissue characteristics (Mareel et al., 1990). The complexity of the phenomenon is difficult to overestimate. Nevertheless, in addition to significant correlations observed between the expression of certain known oncogenes, the existence of a set of specific metastasis-controlling genes appears to emerge. Among the likely candidates to offer practical tools for better patient management one can expect to find genes not only expressing dominant characteristics with respect to metastasis, but also metastasis suppressor genes. At the time of this writing, *nm23*, seems to be the best studied example of a recessively controlled metastasis suppressor gene, the expression of which correlates with well-defined metastatic characteristics of human tumors. It is not unreasonable to expect that evaluation of metastasis-related genes will soon become an integral part of clinical management of oncologic patients.

References

Abbott, C. M. and Povey, S. (1991). Development of human chromosome-specific PCR primers for characterization of somatic cell hybrids. Genomics 9:73–77.

Abbott, C. M. et al. (1988). Prenatal diagnosis of α-1-antitrypsin deficiency using polymerase chain reaction. Lancet ii:763–764.

Adell, K. and Ogbonna, G. (1990). Rapid purification of human DNA from whole blood for potential application in clinical chemistry laboratories. Clin. Chem. 36:261–264.

Agard, D. A. and Sedat, J. W. (1983). Three-dimensional architecture of a polytene nucleus. Nature (London) 302:676–681.

Ainsworth, P. J. et al. (1991). Diagnostic single strand conformational polymorphism (SSCP): a simplified non-radioactive method as applied to Tay-Sachs B1 variant. Nucleic Acids Res. 19:405.

Al-Jader, L. N. et al. (1990). Attitudes of parents of cystic fibrosis children towards neonatal screening and antenatal diagnosis. Clin. Genet. 38:460–465.

Albertson, D. G. et al. (1991). Mapping nonisotopically labeled DNA probes to human chromosome bands by confocal microscopy. Genomics 10:143–150.

Albretsen, C. et al. (1988). Optimal conditions for hybridization with oligonucleotides: a study with myc-oncogene DNA probes. Anal. Biochem. 170:193–202.

Alcalay, J. et al. (1990). Excision repair of pyrimidine dimers induced by simulated solar radiation in the skin of patients with basal cell carcinoma. J. Invest. Dermatol. 95:506–509.

Alexandrov, V. Ya. (1977). Cells, molecules and temperature. Conformational Flexibility of Macromolecules and Ecological Adaptation. Springer-Verlag, Berlin.

Aldhous, P. (1992). MRC follows NIH on patents. Nature 356:98.

Allen, B. S. et al. (1989). Localization of the human type 5, tartrate-resistant acid phosphatase gene by in situ hybridization. Genomics 4:597–600.

Allen, R. C. et al. (1989). Polymerase chain reaction amplification products separated on rehydratable polyacrylamide gels and stained with silver. BioTechniques 7:736–744.

Allen, R. W. et al. (1989). Characteristics of a DNA probe (pa3′HVR) when used for paternity testing. Transfusion 29:477–485.

Allen, R. W. et al. (1990). The application of restriction fragment length polymorphism mapping to parentage testing. Transfusion 30:552–646.

Altar, C. A. et al. (1989). In situ mRNA hybridization: standard procedures and novel approaches. In: Methods in Neurosciences, Vol. 1. Conn, P. M. (ed.) Academic Press, San Diego, CA, pp. 238–281.

Alter, B. P. (1987). Prenatal diagnosis of hematologic diseases, 1986 update. Acta Haematol. 78:137–141.

Alter, B. P. (1988). Prenatal diagnosis: general introduction, methodology, and review. Hemoglobin 12:763–772.

Alves, A. M. and Carr, F. J. (1988). Dot blot detection of point mutations with adjacently hybridizing synthetic oligonucleotide probes. Nucleic Acids Res. 16:8723.

American Association of Blood Banks (1989). Standards for parentage testing laboratories. American Association of Blood Banks, Arlington, VA.

American Society of Human Genetics (1991). The Human Genome Project and patents (letter). Science 254:1710–1712.

Ames, B. N. (1983). Dietary carcinogens and anticarcinogens, oxygen radicals and degenerative diseases. Science 221:1256–1264.

Ames, B. N. (1989a). Endogenous oxidative DNA damage, aging and cancer. Free Radical Res. Commun. 7:121–128.

Ames, B. N. (1989b). Endogenous DNA damage as related to cancer and aging. Mutat. Res. 214:41–46.

Ames, B. N. and Saul, R. L. (1988). Cancer, aging, and oxidative DNA damage. In: Theories of Carcinogenesis. Iversen, O. H. (ed.). Hemisphere, Washington, D. C., pp. 203–220.

Ames, B. N. et al. (1985). Oxidative DNA damage as related to cancer and aging: the assay of thymine glycol, thymidine glycol, and hydroxymethyluracil in human and rat urine. In: Molecular Biology of Aging: Gene Stability and Gene Expression. Sohal, R. S. et al. (eds.). Raven Press, New York, pp. 137–144.

Anderson, C. (1992). Patents, round two. Nature 355:665.

Anderson, S. et al. (1981). Sequence and organization of the human mitochondrial genome. Nature (London) 290:457–465.

Andersson, M. et al. (1988). Y;autosome translocations and mosaicism in the etiology of 45, X maleness: assignment of fertility factor to distal Yq11. Hum. Genet. 79:2–7.

Andiman, W. A. (1989). Virologic and serologic aspects of human immunodeficiency virus infection in infants and children. Semin. Perinatol. *13*:16–26.

Annas, G. J. and Elias, S. (1990). Legal and ethical implications of fetal diagnosis and gene therapy. Am. J. Med. Genet. *35*:215–218.

Antonarakis, S. E. et al. (1985). Hemophilia A: detection of molecular defects and of carriers by DNA analysis. N. Engl. J. Med. *313*:842–848.

Arias, I. M. (1990). Multidrug resistance genes, p-glycoprotein and the liver. Hepatology *12*:159–165.

Ariel, M. et al. (1991). Methylation patterns of testis-specific genes. Proc Natl. Acad. Sci. U.S.A. *88*:2317–2321.

Arlett, C. F. (1986). Human DNA repair defects. J. Inherited Metab. Dis *9* (Suppl. 1):69–84.

Arscott, P. G. and Bloomfield, V. A. (1990). Scanning tunnelling microscopy of nucleic acids and polynucleotides. Ultramicroscopy *33*:127–131.

Armour, J. A. L. et al. (1989). Sequences flanking the repeat arrays of human minisatellites: association with tandem and dispersed repeat elements. Nucleic Acids Res. *17*:4925–4935.

Arn, P.H. and Jabs, E. W. (1990). Characterization of human centromeric regions using restriction enzyme banding, alphoid DNA and structural alterations. Mol. Biol. Med. *7*:371–377.

Arnaudo, E. et al. (1991). Depletion of muscle mitochondrial DNA in AIDS patients with zidovudine-induced myopathy. Lancet *337*:508–510.

Arndt-Jovin, D. J. and Jovin, T. M. (1990). Multivariate chromosome analysis and complete karyotyping using dual labeling and fluroscence digital imaging microscopy. Cytometry *11*:80–93.

Arnheim, N. et al. (1990). PCR analysis of DNA sequences in single cells: single sperm gene mapping and genetic disease diagnosis. Genomics *8*:415–419.

Arnheim, N. et al. (1991). Genetic mapping by single sperm typing. Animal Genet. *22*:105–115.

Arpaia, E. et al. (1988). Identification of an altered splice site in Ashkenazi Tay-Sachs disease. Nature (London) *333*:85–86.

Arrand, J. E. et al. (1989). Molecular cloning and characterization of a mammalian excision repair gene that partially restores UV resistance to xeroderma pigmentosum complementation group D cells. Proc. Natl. Acad. Sci. U.S.A. *86*:6997–7001.

Arthur, M. et al. (1990). Detection of erythromycin resistance by the polymerase chain reaction using primers in conserved regions of *erm* rRNA methylase genes. Antimicrob. Agents Chemother. *34*:2024–2026.

Ashby, J. et al. (1989). Classification according to chemical structure, mutagenicity to *Salmonella* and level of carcinogenicity of a further 42 chemicals tested for carcinogenicity by the US National Toxicology Program. Mutat. Res. *223*:73–103.

Atkin, N. B. (1989). Solid tumor cytogenetics. Progress since 1979. Cancer Genet. Cytogenet. *40*:3–12.

Auerbach, A. D. (1987). Prenatal diagnosis of mutagen — hypersensitivity syndromes. Curr. Probl. Dermatol. *16*:197–209.

Auerbach, A. D. et al. (1981). Prenatal and postnatal diagnosis and carrier detection of Fanconi anemia by a cytogenetic method. Pediatrics *67*:128–135.

Auerbach, A. D. et al. (1985). Fanconi anemia: prenatal diagnosis in 30 fetuses at risk. Pediatrics *76*:794–800.

Austad, S. N. (1992). Forensic DNA typing (letter). Science *255*:1050.

Ausubel, F. M. et al. (1989–1991). Current Protocols in Molecular Biology. Wiley Interscience, New York.

Ayesh, R. et al. (1984). Metabolic oxidation phenotypes as markers for susceptibility to lung cancer. Nature (London) *312*:169–170.

Baan, R. A. et al. (1990). *In situ* detection of DNA damage in single cells or tissue sections by quantitative immunofluorescence microscopy. In: Mutation and the Environment, Part A. Wiley-Liss, New York, pp. 101–112.

Bains, W. (1991). Simplified format for DNA probe-based tests. Clin. Chem. *37*:248–253.

Baird, M. L. (1989). Quality control and quality assurance. In: Banbury Report 32: DNA Technology and Forensic Science. Ballantyne, J. et al. (eds.). Cold Spring Harbor Laboratory Press, Cold Spring Harbor, NY, pp. 175–190.

Baird, M. et al. (1986). Allele frequency distribution of two highly polymorphic DNA sequences in three ethnic groups and its application to the determination of paternity. Am. J. Hum. Genet. *39*:489–501.

Bakkus, M. H. C. et al. (1989). Detection of oncogene expression by fluorescent *in situ* hybridization in combination with immunofluorescent staining of cell surface markers. Oncogene *4*:1255–1262.

Balant, L. P. et al. (1989). Relevance of genetic polymorphism in drug metabolism in the development of new drugs. Eur. J. Clin. Pharmacol. *36*:551–554.

Balazs, I. et al. (1989). Human population genetic studies of five hypervariable DNA loci. Am. J. Hum. Genet. *44*:182–190.

Ballabio, A et al. (1990). PCR test for cystic fibrosis deletion [letter]. Nature (London) *343*:220.

Ballantyne, J. et al. (eds.) (1989). DNA Technology and Forensic Science. Cold Spring Harbor Laboratory Press, Cold Spring Harbor, NY, p. 368.

Barany, F. (1991). Genetic disease detection and DNA amplification using cloned thermostable ligase. Proc. Natl. Acad. Sci. U.S.A. *88*:189–193.

Barbacid, M. (1987). *ras* genes. Annu. Rev. Biochem. *56*:779–827.

Barr, F. G. and Emanuel, B. S. (1990). Application of a subtraction hybridization technique involving photoactivatable biotin and organic extraction to solution hybridization analysis of genomic DNA. Anal. Biochem. *186*:369–373.

Barrell, B. (1991). DNA sequencing: present limitations and prospects for the future. FASEB J. *5*:40–45.

Barron, C. (1989). An assay for quantitative nucleic acid hybridization on membrane filters. Anal. Biochem. *182*:280–282.

Bartlett, M. H. et al. (1991). DNA methylation of two X chromosome genes in female somatic and embryonal carcinoma cells. Somatic Cell. Mol. Genet. *17*:35–47.

Bartlett, S. E. and Davidson, W. S. (1992). FINS (Forensically informative nucleotide sequencing): a procedure for identifying the animal origin of biological specimens. BioTechniques *12*:408–411.

Bassin, R. H. and Benade, L. E. (1990). Defining the mechanisms of transformation through analysis of revertant cells. In: Tumor Suppressor Genes. Klein, G. (ed.). Marcel Dekker, New York, pp. 15–47.

Bates, S. E. and Longo, D. L. (1987). Use of serum tumor markers in cancer diagnosis and management. Semin. Oncol. *14*:102–138.

Bates, S. E. et al. (1989). Expression of a drug resistance gene in human neuroblastoma cell lines: modulation by retinoic acid-induced differentiation. Mol. Cell. Biol. *9*:4337–4344.

Batzer, M. A. and Deininger, P. L. (1991). A human-specific subfamily of *Alu* sequences. Genomics *9*:481–487.

Bauer, T. W. et al. (1990). A prospective comparison of DNA quantitation by image and flow cytometry. Am. J. Clin. Pathol. *93*:322–326.

Bauman, J. G. J. and Bentvelzen, P. (1988). Flow cytometric detection of ribosomal RNA in suspended cells by fluorescent *in situ* hybridization. Cytometry *9*:517–524.

Bauman, J. G. J. et al. (1980). A new method for fluorescence microscopical localization of specific DNA sequences by *in situ* hybridization of fluorochrome labelled RNA. Exp. Cell Res. *138*:485–490.

Bauman, J. G. J. et al. (1981). Cytochemical hybridization with fluorochome-labelled RNA. II. Applications. J. Histochem. Cytochem. *29*:238–246.

Bayer, J. A. and Bauman, J. G. J. (1990). Flow cytometric detection of beta-globin mRNA in murine hemopoietic tissues using fluorescent *in situ* hybridization. Cytometry *11*:132–143.

Beck, S. and Koster, H. (1990). Applications of dioxetane chemiluminescent probes to molecular biology. Anal. Chem. *62*:2258–2270.

Becker, R. L., Jr. and Mikel, U. V. (1990). Interrelation of formalin fixation, chromatin compactness and DNA values as measured by flow and image cytometry. Anal. Quant. Cytol. Histol. *12*:333–341.

Becker-Andre, M. and Hahlbrock, K. (1990). Absolute mRNA quantification using the polymerase chain reaction (PCR). A novel approach by a PCR aided transcript titration assay (PATTY). Nucleic Acids Res. *17*:9437–9446.

Beerman, H. et al. (1991). Flow cytometric analysis of DNA stemline heterogeneity in primary and metastatic breast cancer. Cytometry *12*:147–154.

Bell, G. I. et al. (1982). The highly polymorphic region near the human insulin gene is composed of simple tandemly repeating sequences. Nature (London) *295*:31–35.

Bellamy, W. T. et al. (1990). The clinical relevance of multidrug resistance. Cancer Invest. *8*:547–562.

Ben-Ezra, J. et al. (1991). Effect of fixation on the duplication of nucleic acids from paraffin-embedded material by the polymerase chain reaction. J. Histochem. Cytochem. *39*:351–354.

Benard, J. et al. (1990). Clinical significance of multiple drug resistance in human cancers. Anticancer Res. *10*:1297–1302.

Benbrook, D. et al. (1988). A new retinoic acid receptor identified from a hepatocellular carcinoma. Nature (London) *333*:669–672.

Benedict, W. F. et al. (1990a). Role of the retinoblastoma gene in the initiation and progression of human cancer. J. Clin. Invest. *85*:988–993.

Benedict, W. F. et al. (1990b). The retinoblastoma gene: its role in human malignancies. Cancer Invest. *8*:535–540.

Bentley, K. et al. (1988). A review of genomic physical mapping. Cancer Surv. *7*:267–294.

Beresford, W. A. (1988). A stromal role in epithelial metaplasia and cancer. Cancer J. *2*:145–147.

Berger, C. N. (1986). *In situ* hybridization of immunoglobulin-specific RNA in single cells of the B lymphocyte lineage with radiolabelled DNA probes. EMBO J. *5*:85–93.

Bernard, L. E. et al. (1991). Isolation of DNA fragments from a human chromosomal subregion by *Alu* PCR differential hybridization. Genomics *9*:241–246.

Bernstam, V. A. (1978). Heat effects on protein biosynthesis. Annu. Rev. Plant Physiol. *29*:25–46.

Bernstam, V. A. (1992). Unpublished observations.

Berrez, J. M. et al. (1991). Molecular analysis of a liver mitochondrial ornithine transcarbamylase deficiency. J. Inherited Metab. Dis. *14*:29–36.

Berta, P. et al. (1990). Genetic evidence equating SRY and the testis-determining factor. Nature (London) *348*:448–450.

Bertin, B. et al. (1990). Flow cytometric detection of yeast by *in situ* hybridization with a fluorescent ribosomal RNA probe. J. Microbiol. Methods *12*:1–12.

Best, J. M. and Banatvala, J. E. (1990). Congenital virus infections. More viruses are now known to infect the fetus. Br. Med. J. *300*:1151–1152.

Bevilacqua, G. et al. (1989). Association of low nm23 RNA levels in human primary infiltrating ductal breast carcinomas with lymph node involvement and other histopathological indicators of high metastatic potential. Cancer Res. *49*:5185–5190.

Beyer-Finkler, E. et al. (1990). Anti-contamination primers to improve specificity of polymerase chain reaction in human papillomavirus screening. Lancet *335*:1289–1290.

Bhattacharyya, A. and Lilley, D. M. J. (1989). Single base mismatches in DNA. Long- and short-range structure probed by analysis of axis trajectory and local chemical reactivity. J. Mol. Biol. *209*:583–597.

Bianchi, D. W. et al. (1990). Isolation of fetal DNA from nucleated erythrocytes in maternal blood. Proc. Natl. Acad. Sci. U.S.A. *87*:3279–3283.

Bignell, P. et al. (1990). Rapid neonatal diagnosis of von Willebrand's disease by use of the polymerase chain reaction. Lancet *336*:638–639.

Billings, P. (1991). How many genetic diseases? (letter). Lancet *338*:1603–1604.

Billings, P. R. et al. (1991). New techniques for physical mapping of the human genome. FASEB J. *5*:28–34.

Bird, S. J. (1989). Genetic testing for neurological diseases. A rose with thorns. Neurol. Clin. *7*:859–870.

Bishop, J. M. (1991). Molecular themes in oncogenesis. Cell (Cambridge, Mass) *64*:235–248.

Bishop, M. J. and Rawlings. C. J. (eds.) (1987). Nucleic Acid and Protein Sequence Analysis. A Practical Approach. IRL Press, Oxford.

Blakemore, K. et al. (1987). The value of fetal blood sampling in the prenatal diagnosis of severe combined immuno-deficiency syndrome. Am. J. Hum. Genet. *41* (Suppl.):267 A.

Blanche, S. et al., and the HIV Infection in Newborns French Collaborative Study Group (1989). A prospective study of infants born to women seropositive for human immunodeficiency virus type 1. N. Engl. J. Med. *320*:1643–1648.

Blocher, D. (1990). In CHEF electrophoresis a linear induction of dsb corresponds to a nonlinear fraction of extracted DNA with dose. Int. J. Radiat. Biol. *57*:7–12.

Bocking, A. et al. (1989). Representativity and reproducibility of DNA malignancy grading in different carcinomas. Anal. Quant. Cytol. Histol. *11*:81–86.

Boehm, C. D. et al. (1983). Prenatal diagnosis using DNA polymioephisus: report on 95 pregnancies at risk for sickle cell disease or β-thalassemia. N. Engl. J. Med. *308*:1054–1058.

Boerwinkle, E. et al. (1989). Rapid typing of tandemly repeated hypervariable loci by the polymerase chain reaction: application to the apolipoprotein B 3′ hypervariable region. Proc. Natl. Acad. Sci. U.S.A. *86*:212–216.

Bohr, V. A. (1990). DNA repair at the level of the gene: molecular and clinical considerations. J. Cancer Res. Clin. Oncol. *116*:384–391.

Bohr, V. A. and Koeber, L. (1985). DNA repair in lymphocytes from patients with secondary leukemia as measured by strand rejoining. Mutat. Res. *146*:219–225.

Bohr, V. A. and Okumoto, D. S. (1988). Analysis of pyrimidine dimers in defined genes. In: DNA Repair. A Laboratory Manual of Research Procedures, Vol. 3. Friedberg, E. C. and Hanawalt, P. C. (eds.). Marcel Dekker, New York, pp. 347–366.

Bohr, V. A. et al. (1985). DNA repair in an active gene: removal of pyrimidine dimers from the DHFR gene of CHO cells is much more efficient than the genome overall. Cell (Cambridge Mass) *40*:359–369.

Bohr, V. A. et al. (1989). Biology of disease. DNA repair and its pathogenetic implications. Lab. Invest. *61*:143–161.

Boiteux, S. et al. (1984). 3-Methyladenine residues in DNA induce *sfi*A in *Escherichia coli*. EMBO J. *3*:2569–2573.

Bollag, R. J. et al. (1989). Homologous recombination in mammalian cells. Annu. Rev. Genet. *23*:199–225.

Bollet, C. (1991). A simple method for the isolation of chromosomal DNA from Gram positive or acid-fast bacteria. Nucleic Acids Res. *19*:1955.

Boom, R. et al. (1990). Rapid and simple method for purification of nucleic acids. J. Clin. Microbiol. *28*:495–503.

Bootsma, D. et al. (1988). DNA repair in human cells: from genetic complementation to isolation of genes. Cancer Surv. *7*:303–315.

Borrello, M. G. et al. (1987). DNA methylation affecting the transforming activity of the human Ha-*ras* oncogene. Cancer Res. *47*:75–79.

Borrensen, A. L. et al. (1988). Detection of base mutations in genomic DNA using denaturing gradient gel electrophoresis (DGGE) followed by transfer and hybridization with gene specific probes. Mutat. Res. *202*:77–83.

Botstein, D. et al. (1980). Construction of a genetic linkage map in man using restriction fragment length polymorphisms. Am. J. Hum. Genet. *32*:314–331.

Bottema, C. D. K. et al. (1990). A past mutation at isoleucine-397 is now a common cause of moderate/mild hemophilia B. Br. J. Haematol. *75*:212–216.

Bourgeois, F. et al. (1992). Denaturing gradient gel electrophoresis for rapid detection of latent carriers of a subtype of acute intermittent porphyria with normal erythrocyte porphobilinogen deaminase activity. Clin. Chem. *38*:93–95.

Bourhis, J. et al. (1989). Expression of a human multidrug resistance gene in ovarian carcinomas. Cancer Res. *49*:5062–5065.

Bowden, G. T. (1990). Oncogene activation during multi-stage carcinogenesis. In: Mutation and the Environment, Part D. Wiley-Liss, New York, pp. 1–12.

Boyer, J. C. et al. (1990). Defective postreplication repair in xeroderma pigmentosum variant fibroblasts. Cancer Res. *50*:2593–2598.

Boyes, J. and Bird, A. (1991). DNA methylation inhibits trascription indirectly via a methyl-CpG binding protein. Cell (Cambridge, Mass) *64*:1123–1134.

Boyle, A. L. et al. (1990). Rapid analysis of mouse-hamster hybrid cell lines by *in situ* hybridization. Genomics *7*:127–130.

Brakenhoff, R. H. et al. (1991). Chimeric cDNA clones: novel PCR artifact. Nucleic Acids Res. *19*:1949.

Brantley, J. D. and Beer, M. (1989). Gene-specific labeling of chromatin for election microscopy. Gene Anal. Techn. *6*:75–78.

Brash, D. E. (1988). Quantitating DNA lesions at the DNA sequence level. In: DNA Repair. A Laboratory Manual of Research Procedures, Vol. 3. Friedberg, E. C. and Hanawalt, P. C. (eds.). Marcel Dekker, New York, pp. 327–346.

Bresser, J. and Evinger-Hodges, M. J. (1987). Comparison and optimization of *in situ* hybridization procedures yielding rapid, sensitive mRNA detections. Gene Anal. Tech. *4*:89–104.

Breuning, M. H. et al. (1987). Improved early diagnosis of adult polycystic kidney disease with flanking DNA markers. Lancet *ii*:1359–1361.

Bridges, B. A. et al. (1990). Possible methodologies for the detection and study of DNA sequence changes following mutagen exposure: magnetic enrichment in mutant DNA. Mutagenesis *5*:523–524.

Brigati, D. J. et al. (1983). Detection of viral genomes in cultured cells and paraffin-embedded tissue sections using biotin-labeled hybridization probes. Virology *126*:32–50.

Britten, R. J. and Kohne, D. E. (1968). Repeated sequences in DNA. Science *161*:529–540.

Britten, R. J. et al. (1988). Sources and evolution of human *Alu* repeated sequences. Proc. Natl. Acad. Sci. U.S.A. *85*:4770–4774.

Brown, J. (1990). Prenatal screening in Jewish law. J. Med. Ethics *16*:75–80.

Brown, R. (1991). Gene amplification and drug resistance. J. Pathol. *163*:287–292.

Brumback, R. A. (1983). Cellular hypersensitivity to ionizing radiation in Friedreich's ataxia. J. Neurol. Neurosurg. Psychiatry *46*:878–879.

Brumback, R. A. et al. (1986). The heart in Friedreich's ataxia. Arch. Neurol. *46*:189–192.

Bruns, G. A. P. (1989). Assigning genes to chromosomes: family studies, somatic cell hybridization, chromosome sorting, *in situ* hybridization, translocations. In: Molecular Genetics in Diseases of Brain, Nerve, and Muscle. Rowland, L. P. et al. (eds.). Oxford University Press, New York, pp. 156–171.

Bruun Petersen, K. et al. (1988). Detection of α-1-antitrypsin genotypes by analysis of amplified DNA sequences. Nucleic Acids Res. *16*:352.

Buckler, A. J. et al. (1991). Exon amplification: a strategy to isolate mammalian genes based on RNA splicing. Proc. Natl. Acad. Sci. U.S.A. *88*:4005–4009.

Bugawan, T. L. et al. (1988). The use of non-radioactive oligonucleotide probes to analyze enzymatically amplified DNA for prenatal diagnosis and forensic typing. BioTechnology *6*:943–947.

Burk, R. D. et al. (1985). Characterization and evolution of a single copy sequence from the human Y chromosome. Mol. Cell Biol. *5*:576–581.

Burke, D. T. et al. (1987). Cloning of large segments of exogenous DNA into yeast by means of artificial chromosome vectors. Science *236*:806–812.

Burmeister, M. et al. (1991). Identification of polymorphisms by genomic denaturing gradient gel electrophoresis: application to the proximal region of human chromosome 21. Nucleic Acids Res. *19*:1475–1481.

Buster, J. E. and Carson, S. A. (1989). Genetic diagnosis of the preimplantation embryo. Am. J. Med. Genet. *34*:211–216.

Caldecott, K. et al. (1990). DNA double-strand break repair pathways and cellular tolerance to inhibitors of topoisomerase II. Cancer Res. *50*:5778–5783.

Callaway, C. et al. (1986). HLA typing with cultured amniotic and chorionic villus cells for early prenatal diagnosis or parentage testing without one parent's availability. Hum. Immunol. *16*:200–204.

Calos, M. P. (1988). Use of *lacI* shuttle systems for analysis of mutation in human cells. In: DNA Repair. A Laboratory Manual of Research Procedures, Vol. 3. Friedberg, E. C. and Hanawalt, P. C. (eds.). Marcel Dekker, New York, pp. 277–294.

Camaschella, C. et al. (1990). Prenatal diagnosis of fetal hemoglobin Lepore-Boston disease on maternal peripheral blood. Blood *75*:2102–2106.

Cann, R. L. et al. (1987). Mitochondrial DNA and human evolution. Nature (London) *325*:31–36.

Cantor, C. R. et al. (1988). Pulsed-field gel electrophoresis of very large DNA molecules. Annu. Rev. Biophys. Biophys. Chem. *17*:287–304.

Cantrell, M. A. et al. (1989). Molecular analysis of 46, XY females and regional assignment of a new Y-chromosome-specific probe. Hum. Genet. *83*:88–92.

Cao, A. et al. (1989). β-Thalassemia mutations in Mediterranean populations. Br. J. Haematol. *71*:309–312.

Capon, D. J. et al. (1983). Complete nucleotide sequences of the T24 human bladder carcinoma oncogene and its normal homologs. Nature (London) *302*:33–37.

Caporaso, N. et al. (1989). Lung cancer risk, occupational exposure and the debrisoquine metabolic phenotype. Cancer Res. *49*:3675–3679.

Capranico, G. et al. (1989). P-glycoprotein, gene amplification and expression in multidrug-resistant murine P388 and B16 cell lines. Br. J. Cancer *59*:682–685.

Carbone, M. and Levine, A. S. (1990). Oncogenes, antioncogenes, and the regulation of cell growth. TEM May/June:248–253.

Carrell, R. W. et al. (1982). Structure and variation of human α-1-antitrypsin. Nature (London) *298*:329–334.

Carrell, R. W. and Owen, M. C. (1980). α-1-antitrypsin: structure, variation and disease. Essays Med. Biochem. *4*:83.

Casanova, M. et al. (1985). A human Y-linked DNA polymorphism and its potential for estimating genetic and evolutionary distance. Science *230*:1403–1406.

Casati, A. et al. (1991). Chromosome rearrangements in normal fibroblasts from xeroderma pigmentosum homozygotes and heterozygotes. Cancer Genet. Cytogenet. *51*:89–101

Caskey, C. T. and Hammond, H. (1989). DNA-based identification: disease and criminals. In: DNA Technology and Forensic Science. Ballantyne, J. et al. (eds.). Cold Spring Harbor Laboratory Press, Cold Spring Harbor, NY, pp. 127–142.

Cawood, A. H. (1989). DNA fingerprinting. Clin. Chem. *35*:1832–1837.

Ceccherini, I. et al. (1989). Autosomal dominant polycystic kidney disease: prenatal diagnosis by DNA analysis and sonography at 14 weeks. Prenatal. Diagn. *9*:751–758.

Celis, J. E. et al. (1991). Human cellular protein patterns and their link to genome DNA sequence data: usefulness of two-dimensional gel electrophoresis and microsequencing. FASEB J. *5*:2200–2208.

Cellotti, L. et al. (1990). Effects of *in vivo* exposure to antineoplastic drugs on DNA repair and replication in human lymphocytes. Mutat. Res. *245*:217–222.

Chackal-Roy, M. et al. (1989). Stimulation of human prostatic carcinoma cell growth by factors present in human bone marrow. J. Clin. Invest. *84*:43–50.

Chaillet, J. R. et al. (1991). Parental-specific methylation of an imprinted transgene is established during gametogenesis and progressively changes during embryogenesis. Cell (Cambridge, Mass) *66*:77–83.

Chakraborty, R. and Kidd, K. K. (1991). The utility of DNA typing in forensic work. Science *254*:1735–1739.

Chamberlain, J. S. et al. (1988). Deletion screening of the Duchenne muscular dystrophy locus via multiplex DNA amplification. Nucleic Acids Res. *16*:11141–11156.

Chamberlain, J. S. et al. (1990). Multiplex PCR for the diagnosis of Duchenne muscular dystrophy. In: PCR Protocols: A Guide to Methods and Applications. Innis, M., et al. (eds.). Academic Press, Orlando, FL, pp. 273–281.

Chamberlain, S. and Lewis, P. D. (1982). Studies of cellular hypersensitivity to ionizing radiation in Friedreich's ataxia. J. Neurol. Neurosurg. Psychiatry *45*:1136–1138.

Chan, J. Y. H. et al. (1987). Altered DNA ligase I activity in Bloom's syndrome cells. Nature (London) *325*:357–359.

Chan, M. M. et al. (1990). β2-microglobulin and neopterin: predictive markers for human immunodeficiency virus type 1 infection in children? J. Clin. Microbiol. *28*:2215–2219.

Chang, J. C. and Kan, Y. W. (1982). A sensitive new prenatal test for sickle-cell anemia. N. Engl. J. Med. *307*:30–32.

Chang, L. C. et al. (1989). Detection of transforming growth factor α messenger RNA in normal and chemically transformed hamster oral epithelium by *in situ* hybridization. Cancer Res. *49*:6700–6707.

Chao, C. C. K. et al. (1990). Increased *mdr* gene expression and decreased drug accumulation in a human colonic cancer cell line resistant to hydrophobic drugs. Biochem. Biophys. Res. Commun. *172*:842–849.

Chapman, R. W. and Brown, B. L. (1989). Two methods to detect DNA fragments produced by restriction enzymes. Anal. Biochem. *177*:199–202.

Chaudhary, P. M. and Roninson, I. B. (1991). Expression and activity of P-glycoprotein, a multidrug efflux pump, in human hematopoietic stem cells. Cell (Cambridge, Mass) *66*:85–94.

Chehab, F. F. and Kan, Y. W. (1989). Detection of specific DNA sequences by fluorescence amplification: a color complementation assay. Proc. Natl. Acad. Sci. U.S.A. *86*:9178–9182.

Chehab, F. F. et al. (1987). Detection of sickle cell anemia and thalassemias. Nature (London) *329*:293–294.

Chen, P. et al. (1990). Conditions for generating fingerprints using M13 phage DNA. Nucleic Acids Res. *18*:1065.

Cherif, D. et al. (1989). Detection of single-copy genes by nonisotopic *in situ* hybridization on human chromosomes. Hum. Genet. *81*:358–362.

Chervenak, F. A. et al. (1986). Advances in the diagnosis of fetal defects. N. Engl. J. Med. *315*:305–307.

Cheung, S. W. et al. (1990). Correlation between phenotypic expression of de novo marker chromosomes and genomic organization using replication banding. Prenatal. Diagn. *10*:717–724.

Chia, D. et al. (1988). The use of multiple tumor markers to monitor cancer patients. In: Altered Glycosylation in Tumor Cells. Alan R. Liss, New York, pp. 295–299.

Chiafari, F. A. and Wenk, R. E. (1990). Parentage analysis by endonuclease shattering of hypervariable DNA. Transfusion *30*:648–650.

Chin, K. V. et al. (1992). Modulation of activity of the promoter of the human *MDR*1 gene by *ras* and *p53*. Science *255*:459–462.

Choo, K. H. et al. (1991). A survey of the genomic distribution of α satellite DNA on all the human chromosomes, and derivation of a new consensus sequence. Nucleic Acids Res. *19*:1179–1182.

Chou, M. Y. et al. (1990). A rapid method to determine proliferation patterns of normal and malignant tissues by H3 mRNA *in situ* hybridization. Am. J. Pathol. *136*:729–733.

Christmann, A. et al. (1991). Non-radioactive *in situ* hybridization pattern of the M13 minisatellite sequences on human metaphase chromosomes. Hum. Genet. *86*:487–490.

Chu, B. C. F. et al. (1986). Synthesis of an amplifiable reporter RNA for bioassays. Nucleic Acids Res. *14*:5591–5603.

Chu, G. and Berg, P. (1987). DNA cross-linked by cisplatin: a new probe for the DNA repair defect in xeroderma pigmentosum. Mol. Biol. Med. *4*:277–290.

Chu, G. et al. (1990). How cells recognize damaged DNA: clues from xeroderma pigmentosum and yeast. In: Mutation and the Environment, Part A. Wiley-Liss, New York, pp. 275–282.

Cimino, G. D. et al. (1991). Post-PCR sterilization: a method to control carryover contamination for the polymerase chain reaction. Nucleic Acids Res. *19*:99–107.

Cizdziel, P. E. et al. (1991). Loss of a tumor suppressor gene function is correlated with down regulation of chondrocyte specific collagen expression in Syrian hamster embryo cells. Mol. Carcinogenesis *4*:14–24.

Clark, A. G. (1990). Inference of haplotypes from PCR-amplified samples of diploid population. Mol. Biol. Evol. *7*:111–122.

Clay, T. M. et al. (1991). PCR fingerprinting for selection of HLA matched unrelated marrow donors. Lancet *337*:1049–1052.

Cleary, M. L. (1991). Oncogenic conversion of transcription factors by chromosomal translocations. Cell *66*:619–622.

Cleaver, J. E. (1968). Defective repair replication of DNA in xeroderma pigmentosum. Nature (London) *218*:652–656.

Cleaver, J. E. (1969). Xeroderma pigmentosum: a human disease in which an initial stage of DNA repair is defective. Proc. Natl. Acad. Sci. U.S.A. *63*:428–435.

Cleaver, J. E. (1984). Defective DNA repair and cancer-prone disorders of man. In: Genes and Cancer. Alan R. Liss, New York, pp. 117–135.

Cleaver, J. E. (1990). Do we know the cause of xeroderma pigmentosum? Carcinogenesis *11*:875–882.

Cleaver, J. E. and Kraemer, K. H. (1989). Xeroderma pigmentosum. In: The Metabolic Basis of Inherited Disease, Vol. II. Scriver, C. R. et al. (eds.). McGraw-Hill, New York, pp. 2949–2971.

Close, P. M. et al. (1990). Zonal distribution of immunoglobulin-synthesizing cells within the germinal centre: an *in situ* hybridization and immunohistochemical study. J. Pathol. *162*:209–216.

Cohen, A. S. et al. (1990). Separation and analysis of DNA sequence reaction products by capillary gel electrophoresis. J. Chromatogr. *516*:49–60.

Cohen, J. E. (1990). DNA fingerprinting for forensic identification: potential effects on data interpretation of subpopulation heterogeneity and band number variability. Am. J. Hum. Genet. *46*:358–368.

Collins, A. and Sedgwick, S. (1990). Recent advances in DNA repair. A report of the 1989 meeting of the British Photobiology Society in association with the DNA Repair Network. Mutat. Res. *236*:139–143.

Cone, R. W. et al. (1990). Polymerase chain reaction decontamination: the wipe test. Lancet *336*:686–687.

Conley, M. E. et al. (1988). Nonrandom X chromosome inactivation in B cells from carriers of X chromosome-linked severe combined immunodeficiency. Proc. Natl. Acad. Sci. U.S.A. *85*:3090–3094.

Conner, B. J. et al. (1983). Detection of sickle cell β^s-globin allele by hybridization with synthetic oligonucleotides. Proc. Natl. Acad. Sci. U.S.A. *80*:278–282.

Connor, J. M. (1989). Genetic aspects of prenatal diagnosis. J. Inherized Metab. Dis. *12* (Suppl. 1):89–96.

Conway de Macario, E. et al. (1990). Adaptation of the slide immunoenzymatic assay for quantification of DNA hybridization: SIA-DNA. BioTechniques *8*:210–217.

Cook, A. F. et al. (1988). Synthesis and hybridization of a series of biotinylated oligonucleotides. Nucleic Acids Res. *16*:4077–4095.

Cooke, H. (1976). Repeated sequence specific to human males. Nature (London) *262*:182–186.

Coon, J. S. et al. (1987). Biology of disease: advances in flow cytometry for diagnostic pathology. Lab. Invest. *57*:453–479.

Coon, J. S. et al. (1988). Interinstitutional variability in DNA flow cytometric analysis of tumors: the National Cancer Institute's Flow Cytometry Network experience. Cancer *61*:126–130.

Cooper, C. et al. (1991). An investigation of siuy and dicentric chromosomes found in three Turner's syndrome patients using DNA analysis and *in situ* hybridisation with X and Y chromosome specific probes. J. Med. Genet. *28*:6–9.

Cooper, C. S. and Stratton, M. R. (1991). Soft tissue tumors: the genetic basis of development. Carcinogenesis *12*:155–165.

Cooper, D. N. and Krawczak, M. (1990). The mutational spectrum of single base-pair substitutions causing human genetic disease: patterns and predictions. Hum. Genet. *85*:55–74.

Cooper, D. N. and Schmidtke, J. (1989). Diagnosis of genetic disease using recombinant DNA. Hum. Genet. *83*:307–334.

Corbo, L. et al. (1990). Direct isolation of expressed sequences from defined chromosomal regions. In: Biotechnology and Human Genetic Predisposition to Disease. Wiley-Liss, New York, pp. 187–197.

Cornwell, M. M. (1990). The human multidrug resistance gene: sequences upstream and downstream of the initiation site influence transcription. Cell Growth Differ. *1*:607–615.

Cortopassi, G. A. and Arnheim, N. (1990). Detection of a specific mitochondrial DNA deletion in tissues of older humans. Nucleic Acids Res. *18*:6927–6933.

Cossman, J. (ed.) (1990). Molecular Genetics in Cancer Diagnosis. Elsevier, New York.

Cotton, R. G. H. et al. (1988). Reactivity of cytosine and thymine in single-base-pair mismatches with hydroxylamine and osmium tetroxide and its application to the study of mutations. Proc. Natl. Acad. Sci. U.S.A. *85*:4397–4401.

Cotton, R. W. et al. (1989). Current case experience with single-locus hypervariable probes. In: Banbury Report 32: DNA Technology and Forensic Science. Ballantyne, J. et al. (eds.). Cold Spring Harbor Laboratory Press, Cold Spring Harbor, NY, pp. 191–206.

Coutlee, F. et al. (1990). Enzyme immunoassay for detection of hybrids between PCR-amplified HIV-1 DNA and a RNA probe: PCR-EIA. AIDS Res. Hum. Retroviruses *6*:775–784.

Covey, J. M. et al. (1989). Protein-linked DNA strand breaks induced in mammalian cells by camptothecin, an inhibitor of topoisomerase I. Cancer Res. *49*:5016–5022.

Cox, R. D. and Lehrach, H. (1991). Genome mapping: PCR based meiotic and somatic cell hybrid analysis. BioEssays *13*:193–198.

Cox, D. R. et al. (1989). Segregation of the Huntington disease region of human chromosome 4 in a somatic cell hybrid. Genomics *4*:397–407.

Cox, D. W. et al. (1976). Protease inhibitors in patients with chronic obstructive pulmonary disease: the α-1-antitrypsin heterozygote controversy. Am. Rev. Respir. Dis. *113*:601–606.

Cox, D. W. et al. (1985). DNA restriction fragments associated with α-1-antitrypsin indicate a single origin for deficiency allele PIZ. Nature (London) *316*:79–81.

Cox, D. W. and Mansfield, T. (1985). Prenatal diagnosis for α-1-antitrypsin deficiency. Lancet *i*:230.

Cox, D. W. and Mansfield, T. (1987). Prenatal diagnosis of α-1-antitrypsin deficiency and estimates of fetal risk for disease. J. Med. Genet. *24*:52–59.

Creasey, A. et al. (1991). Application of a novel chemoluminescence-based DNA detection method to single-vector and multiplex DNA sequencing. BioTechniques *11*:102–109.

Cremer, T. et al. (1986). Detection of chromosome aberrations in the human interphase nuclus by visualization of specific target DNAs with radioactive and non-radioactive *in situ* hybridization techniques: diagnosis of trisomy 18 with probe L1. 84. Hum. Genet. *74*:346–352.

Cremer, T. et al. (1988). Detection of chromosome aberrations in metaphase and interphase tumor cells by *in situ* hybridization using chromosome-specific library probes. Hum. Genet. *80*:235–246.

Cremer, T. et al. (1990). Rapid metaphase and interphase detection of radiation-induced chromosome aberrations in human lymphocytes by chromosomal suppression *in situ* hybridization. Cytometry *11*:110–118.

Cremers, A. F. M. et al. (1987). Non-radioactive *in situ* hybridization. A comparison of several immunocytochemical detection systems using reflection-contrast microscopy and electron microscopy. Histochemistry *86*:609–615.

Crisan, D. et al. (1989). Prenatal diagnosis of hemoglobinopathies using polymerase chain reaction and allele-specific nucleotide probes. Clin. Chem. *35*:1854.

Crisan, D. et al. (1990). Polymerase chain reaction: amplification of DNA from fixed tissue. Clin. Biochem. *23*:489–495.

Crissman, H. A. and Steinkamp, J. A. (1990). Cytochemical techniques for multivariate analysis of DNA and other cellular constituents. In: Flow Cytometry and Sorting, 2nd ed. Wiley-Liss, New York, pp. 227–247.

Crist, W. M. and Kun, L. E. (1991). Common solid tumors of childhood. N. Engl. J. Med. *324*:461–471.

Cross, M. and Dexter, T. M. (1991). Growth factors in development, transformation, and tumorigenesis. Cell (Cambridge, Mass) *64*:271–280.

Curiel, D. T. et al. (1990). A chemical mismatch clearage method useful for the detection of point mutations in the *p53* gene in lung cancer. Am. J. Respir. Cell Mol. Biol. *3*:405–411.

Curien, H. (1991). The Human Genome Project and patents (letter). Science *254*:1710.

D'Ambrosio, S. M. and Setlow, R. B. (1978). Defective and enhanced postreplication repair in classical and variant xeroderma pigmentosum cells treated with *N*-acetoxy-3-acetylaminofluorene. Cancer Res. *38*:1147–1153.

Dal Chin, P. and Sandberg, A. A. (1989). Chromosome changes in soft tissue tumors: benign and malignant. Cancer Invest. *7*:63–76.

Davidson, E. H. et al. (1973). General interspersion of repetitive with nonrepetitive sequence elements in the DNA of *Xenopus*. J. Mol. Biol. *77*:1–23.

Davidson, E. H. et al. (1975). Comparative aspects of DNA organization in metazoa. Chromosoma *51*:253–259.

Davis, L. G. et al. (1986). Basic Methods in Molecular Biology. Elsevier, New York.

Davis, L. M. et al. (1991). Rapid DNA sequencing based upon single molecule detection. GATA *8*:1–7.

Davis, M. K. (1988). Vertical transmission of HIV [letter] JAMA *260*:30–31.

de la Chapelle, A. (1988). The complicated issue of human sex determination. Am. J. Hum. Genet. *43*:1–3.

Decorte, R. et al. (1990). Rapid detection of hypervariable regions by the polymerase chain reaction technique. DNA Cell Biol. *9*:461–469.

Dedhar, S. (1990). Integrins and tumor invasion. Bio Assays *12*:583–590.

Delattre, O. et al. (1991). Mapping of human chromosome with a panel of somatic cell hybrids. Genomics *9*:721–727.

Demple, B. (1990). Oxidative DNA damage: repair and inducible cellular responses. In: Mutation and the Environment, Part 1. Wiley-Liss, New York, pp. 155–167.

Den Otter, W. et al. (1990). Oncogenesis by mutations in anti-oncogenes: a view. Anticancer Res *10*:475–488.

Denko, N. et al. (1989). An asymmetric field inversion gel electrophoresis system for the separation of large DNAs. Anal. Biochem. *178*:172–176.

Denley, D. R. (1990). Practical applications of scanning tunneling microscopy. Ultramicroscopy *33*:83–92.

Dent, G. A. et al. (1989). Simultaneous paired analysis by flow cytometry of surface markers, cytoplasmic antigens, or oncogene expression with DNA content. Cytometry *10*:192–198.

Deobagkar, D. D. et al. (1990). Hemimethylation of DNA prevents chromatin expression. Proc. Natl. Acad. Sci. U.S.A. *87*:1691–1695.

Deppert, W. et al. (1990). Cell cycle control of p53 in normal (3T3) and chemically transformed (Meth A) mouse cells. II. Requirement for cell cycle progression. Oncogene *5*:1701–1706.

Deuchars, K. L. and Ling, V. (1989). P-glycoprotein and multidrug resistance in cancer chemotherapy. Semin. Oncol. *16*:156–165.

Deverson, E. V. et al. (1990). MHC class II region encoding proteins related to the multidrug resistance family of transmembrane transporters. Nature (London) *348*:738–741.

Deville, P. et al. (1988). Detection of chromosome aneuploidy in interphase nuclei from human primary breast tumors using chromosome specific repetitive DNA probes. Cancer Res. *48*:5825–5830.

Devlin, B. et al. (1990). No excess of homozygosity at loci used for DNA fingerprinting. Science *249*:1416–1420.

Devor, E. J. et al. (1988). A rapid method for confirming cell line identity: DNA "fingerprinting" with a minisatellite probe from M13 bacteriophage. BioTechniques *6*:200–202.

Di Marzo, R. et al. (1988). A rapid DNA method for first trimester prenatal diagnosis. Br. J. Haematol. *70*:504–505.

Diamandis, E. P. (1990). Analytical methodology for immunoassays and DNA hybridization assays — current status and selected systems — critical review. Clin. Chem. Acta *194*:19–50.

DiBerardino, M. A. (1989). Genomic activation in differentiated somatic cells. In: Developmental Biology, Vol. 6. DiBerardino, M. A. and Etkin, L. D. (eds.). Plenum Press, New York, pp. 175–198.

Dicker, A. P. et al. (1989). Sequence analysis of a human gene responsible for drug resistance: a rapid method for manual and automated direct sequencing of products generated by the polymerase chain reaction. BioTechniques *7*:830–837.

Diehl, S. R. et al. (1990). Automated genotyping of human DNA polymorphisms. Am. J. Hum. Genet. *47*:177A.

DiMauro, S. et al. (1989). Mitochondrial diseases. In: Molecular Genetics in Diseases of Brain, Nerve, and Muscle. Rowland, L. P. et al. (eds.). Oxford University Press, New York and Oxford, pp. 285–298.

Dionne, C. A. et al. (1990). Chromosome assignments by polymerase chain reaction techniques: assignment of the oncogene FGF-5 to chromosome 4. BioTechniques *8*:190–194.

Dionne, F. T. et al. (1991). Mitochondrial DNA sequence polymorphism, VO_{2max}, and response to endurance training. Med. Sci. Sports Exercise *23*:177–185.

Diuguid, D. L. and Furie, B. (1991). Molecular genetics of hemophilia B. In: Hematology. Basic Principles and Practice. Hoffman, R. et al. (eds.). Churchill Livingstone, London, pp. 1316–1324.

Doan, T. L. et al. (1987a). Sequence targeted chemical modifications of nucleic acids by complementary oligonucleotides covalently linked to porphyrins. Nucleic Acids Res. *15*:8643–8659.

Doan, T. L. et al. (1987b). Sequence-specific recognition, photo-crosslinking and clearage of the DNA double helix by an oligo-[α]-thymidylate covalently linked to an azidoproflavine derivative. Nucleic Acids Res. *15*:7749–7760.

Doerjer, G. et al. (1988). Biomonitoring of DNA damage by alkaline filter elution. Int. Arch. Occup. Environ. Health *60*:169–174.

Doerr, H. W. (1987). Cytomegalovirus infection in pregnancy. J. Virol. Methods *17*:127–132.

Dong, Z. and Jeffrey, A. M. (1990). Mechanisms of organ specificity in chemical carcinogenesis. Cancer Invest. *8*:523–533.

Donehower, L. A. et al. (1992). Mice deficient for p53 are developmentally normal but susceptible to spontaneous tumors. Nature *356*:215–221.

Donnis-Keller, H. (1991). Human Gene Mapping Techniques. Stockton Press, New York.

Donnis-Keller, H. et al. (1987). A genetic linkage map of the human genome. Cell (Cambridge, Mass) *51*:319–337.

Dover, G. A. (1990). Mapping "frozen accidents". Nature (London) *344*:812–813.

Driscoll, D. A. et al. (1991). Antenatal diagnosis of DiGeorge syndrome. Lancet *338*:1390.

Drossman, H. et al. (1990). High-speed separations of DNA sequencing reactions by capillary electrophoresis. Anal. Chem. *62*:900–903.

Dudin, G. et al. (1987). A method for nucleic acid hybridization to isolated chromosomes in suspension. Hum. Genet. *76*:290–292.

Duesberg, P. H. (1987). Latent cellular oncogenes: the paradox dissolves. J. Cell Sci. (Suppl. 7):169–187.

Duker, N. J. and Gallagher, P. E. (1986). Recent advances in molecular pathology. Detection of DNA damage in human cells and tissues using sequencing techniques. Exp. Mol. Pathol. *44*:117–131.

Dunn, D. S. et al. (1989). The usefulness of various polymorphisms in paternity testing. Experience with three southern African populations. S. Afr. Med. J. *76*:303–307.

Dunstan, G. R. (1988). Screening for fetal and genetic abnormality: social and ethical issues. J. Med. Genet. *25*:290–293.

Durandy, A. and Griscelli, C. (1983). Prenatal diagnosis of severe combined immunodeficiency and X-linked agammaglobulinemia. Birth Defects *19*:125–127.

Dykes, D. D. (1987). Parentage testing using restriction fragment length polymorphisms (RFLPs). In: Clinical Applications of Genetic Engineering. Lasky, L. C. and Edwards-Moulds, J. M. (eds.). American Association of Blood Banks, Arlington, VA, pp. 59–85.

Dykes, D. D. (1988). The use of biotinylated DNA probes in parentage testing: non-isotopic labelling and non-toxic extraction. Electrophoresis *9*:359–368.

Eastmond, D. A. and Pinkel, D. (1990). Detection of aneuploidy and ineuploidy-inducing agents in human lymphocytes using fluorescence *in situ* hybridization with chromosome-specific DNA probes. Mutat. Res. *234*:303–318.

Echols, H. and Goodman, M. F. (1990). Mutation induced by DNA damage: a many protein affair. Mutat. Res. *236*:301–311.

Eiken, H. G. et al. (1991). Application of natural and amplification created restriction sites for the diagnosis of PKU mutations. Nucleic Acids Res. *19*:1427–1430.

Egger, J. and Wilson, J. (1983). Mitochondrial inheritance in a mitochondrially mediated disease. N. Engl. J. Med. *309*:142–146.

El-Deiry, W. S. et al. (1991). High expression of the DNA methyltransferase gene characterizes human neoplastic cells and progression stages of colon cancer. Proc. Natl. Acad. Sci. U.S.A. *88*:3470–3474.

El-Naggar, A. K. et al. (1991). Single- and double-stranded RNA measurements by flow cytometry in solid neoplasms. Cytometry *12*:330–335.

Elias, J. M. et al. (1989). Sensitivity and detection efficiency of the peroxidase antiperoxidase (PAP), avidin-biotin peroxidase complex (ABC), and peroxidase-labelled avidin-biotin (LAB) methods. Am. J. Clin. Pathol. *92*:62–67.

Elles, R. G. et al. (1983). Absence of maternal contamination of chorionic villi used for fetal gene analysis. N. Engl. J. Med. *308*:1433–1435.

Ellis, P. M. et al. (1990). Relevance to prenatal diagnosis of the identification of a human Y/autosome translocation by Y-chromosome-specific *in situ* hybridization. Mol. Reprod. Dev. *25*:37–41.

Emanuel, J. R. (1991). Simple and efficient system for synthesis of non-radioactive nucleic acid probes using PCR. Nucleic Acids Res. *19*:2790.

Embury, S. H. et al. (1987). Rapid prenatal diagnosis of sickle cell anemia by a new method of DNA analysis. N. Engl. J. Med. *316*:656–661.

Emmett, M. and Petrack, B. (1988). Rapid isolation of total RNA from mammalian tissues. Anal. Biochem. *174*:658–661.

Engelke, D. R. et al. (1988). Direct sequencing of enzymatically amplified human genomic DNA. Proc. Natl. Acad. Sci. U.S.A. *85*:544–548.

Epplen, J. T. (ed.) (1991). Paper symposium: DNA fingerprinting. Electrophoresis *12*:101–232.

Epstein, R. J. (1990). Drug-induced DNA damage and tumor chemosensitivity. J. Clin. Oncol. *8*:2062–2084.

Erickson, S. et al. (1986). Risk of liver cirrhosis and primary liver cancer in α-1-antitrypsin deficiency. N. Engl. J. Med. *314*:736–739.

Eriksson, S. (1965). Studies in α-1-antitrypsin deficiency. Acta Med. Scand. *177* (Suppl. 432):5–85.

Evans, J. A. et al. (1991). Sex chromosome anomalies: prenatal diagnosis and the need for continued prospective studies. In: Birth Defects: Original Articles Series, Vol. 26. March of Dimes Birth Defects Foundation, pp. 273–281.

Evans, J. W. et al. (1991). The use of fluorescence *in situ* hybridization combined with premature chromosome condensation for the identification of chromosome damage. Br. J. Cancer *63*:517–521.

Evans, H. H. et al. (1987). Deficiency in DNA repair in mouse lymphoma strain L5178 Y-S. Proc. Natl. Acad. Sci. U.S.A. *84*:7562–7566.

Evans, H. H. et al. (1989). Relationship between topoisomerases II and radiosensitivity in mouse L5178 lymphoma strains. Mutat. Res. *217*:53–63.

Evett, I. W. et al. (1989). Paternity calculations from DNA multilocus profiles. J. Forensic Sci. *29*:249–254.

Eweida, M. et al. (1989). Highly sensitive and specific non-radioactive biotinylated probes for dot-blot, Southern and colony hybridizations. J. Virol. Methods *26*:35–44.

Executive Committee of the International Society for Forensic Hemogenetics. (1989). DNA Recommendation. Vox Sang. *57*:276–277.

Fagerhol, M. K. and Cox, D. W. (1981). The PI polymorphism: genetic, biochemical and clinical aspects of human α-1-antitrypsin. Adv. Hum. Genet. *11*:1–62.

Falloon, J. et al. (1989). Human immunodeficiency virus infection in children. J. Pediatr. *114*:1–30.

Fan, J. B. et al. (1989). Construction of a *Not*I restriction map of the fission yeast *Schizosaccharomyces pombe* genome. Nucleic Acids Res. *12*:2801–2817.

Fan, Y. S. et al. (1990). Mapping small DNA sequences by fluorescence *in situ* hybridization directly on banded metaphase chromosomes. Proc. Natl. Acad. Sci. U.S.A. *87*:6223–6227.

Fargnoli, J. et al. (1990). *In vivo* and *in vitro* studies on the heat shock response and aging. In: Molecular Biology of Aging. Alan R. Liss, New York, pp. 379–390.

Farkas, D. H. (1990). The establishment of a clinical molecular biology laboratory. Clin. BioTechnol. *2*:87–96.

Farquharson, M. et al. (1990). Detection of messenger RNA using a digoxigenin end labelled oligodeoxynucleotide probe. J. Clin. Pathol. *43*:424–428.

Fenoglio-Preiser, C. M. and Willman, C. L. (eds.) (1991). Molecular Diagnostics in Pathology. Williams & Wilkins, Baltimore.

Fentestein, H. and Garrido, F. (1986). MHC Antigens and Malignancy. Nature (London) *322*:502–503.

Fenton, W. A. (1989). Ornithine transcarbamylase: enzyme deficiency and biogenesis of mitochondrial proteins. In: Molecular Genetics in Diseases of Brain, Nerve, and Muscle. Rowland, L. P. et al. (eds.). Oxford University Press, New York, pp. 273–284.

Ferguson, L. R. and Denny, W. A. (1990). Frameshift mutagenesis by acridines and other reversibly-binding DNA ligands. Mutagenesis *5*:529–540.

Ferguson-Smith, M. A. et al. (1990). The secret of sex. Lancet *336*:809–810.

Fidler, I. J. (1987). Review: biologic heterogeneity of cancer metastases. Breast Cancer Res. Treat. *9*:17–26.

Fidler, I. J. (1990). Critical factors in the biology of human cancer metastasis: twenty-eighth GHA Clowes Memorial Award Lecture. Cancer Res. *50*:6130–6138.

Fidler, I. J. and Radinsky, R. (1990). Genetic control of cancer metastasis. J. Natl. Cancer Inst. *82*:164–168.

Field, J. K. and Spandidos, D. A. (1990). The role of *ras* and *myc* oncogenes in human solid tumors and their relevance in diagnosis and prognosis (review). Anticancer Res. *10*:1–22.

Firon, N. et al. (1990). Genotype assignment in Gaucher disease by selective amplification of the active glucocerebrosidase gene. Am. J. Hum. Genet. *46*:527–532.

Fischer, S. G. and Lerman, L. S. (1983). DNA fragments differing by single base pair substitutions are separated in denaturing gradient gels: correspondence with melting theory. Proc. Natl. Acad. Sci. U.S.A. *80*:1579–1583.

Fletcher, J. A. et al. (1991). Diagnostic relevance of clonal cytogenetic aberrations in malignant soft-tissue tumors. N. Engl. J. Med. *324*:436–443.

Flood, P. M. et al. (1988). The role of contrasuppression in tumor regression. Immunol. Rev. *7*:12–22.

Folsom, V. et al. (1989). Detection of DNA targets with biotinylated and fluorescinated RNA probes. Effects of the extent of derivatization on detection sensitivity. Anal. Biochem. *182*:309–314.

Forster, A. C. et al. (1985). Non-radioactive hybridization probes prepared by the chemical labelling of DNA and RNA with a novel reagent, photobiotin. Nucleic Acids Res. *13*:745–761.

Fournier, R. E. K. and Ruddle, F. H. (1977). Microcell-mediated transfer y murine chromosomes into mouse, Chinese hamster, and human somatic cells. Proc. Natl. Acad. Sci. U.S.A. *74*:319–323.

Fox, J. E. et al. (1986a). Identification and application of additional restriction fragment length polymorphisms at the human ornithine transcarbamoylase locus. Am. J. Hum. Genet. *38*:841–847.

Fox, J. et al. (1986b). Prenatal diagnosis of ornithine transcarbamoylase deficiency with use of DNA polymorphisms. N. Engl. J. Med. *315*:1205–1208.

Frank, D. et al. (1991). Demethylation of CpG islands in embryonic cells. Nature (London) *351*:239–241.

Frankenberg-Schwager, M. (1990). Induction, repair and biological significance of radiation-induced DNA lesions in eukaryotic cells. Radiat. Environ. Biophys. *29*:273–292.

Frankfurt, O. S. et al. (1990). Flow cytometric analysis of DNA damage and repair in the cells resistant to alkylating agents. Cancer Res. *50*:4453–4457.

Freeman, C. S. et al. (1990). An overview of tumor biology. Cancer Invest. *8*:71–90.

Freund, A.-M. et al. (1989). Z-DNA-forming sequences are spontaneous deletion hot spots. Proc. Natl. Acad. Sci. U.S.A. *86*:7465–7469.

Friedberg, E. C. (1985). DNA Repair. W. H. Freeman, San Francisco.

Friedberg, E. C. and Hanawalt, P. C. (eds.) (1988). DNA Repair: A Laboratory Manual of Research Procedures, Vol. 3. Marcel-Dekker, New York.

Fritz, P. et al. (1989). Immunocytochemical detection of sulphonylated DNA in tissue sections. An alternative to Feulgen staining of DNA. Histochemistry *93*:175–181.

Furie, B. and Furie, B. C. (1992). Molecular and cellular biology of blood coagulation. N. Engl. J. Med. *326*:800–806.

Gaidano, G. et al. (1991). p53 mutations in human lymphoid malignancies: association with Burkitt lymphoma and chronic lymphocytic leukemia. Proc. Natl. Acad. Sci. U.S.A. *88*:5413–5417.

Ganesan, A. K. and Spivak, G. (1988). Use of shuttle vectors for analysis of DNA damage processing in mammalian cells. In: DNA Repair. A Laboratory Manual of Research Procedures, Vol. 3. Friedberg, E. C. and Hanawalt, P. C. (eds.). Marcel Dekker, New York, pp. 295–310.

Ganguly, A. et al. (1989). Detection and location of single-base mutations in large DNA fragments by immunomicroscopy. Genomics *4*:530–538.

Garberg, P. et al. (1988). Evaluation of a genotoxicity test measuring DNA-strand breaks in mouse lymphoma cells by alkaline unwinding and hydroxyapatite elution. Mutat. Res. *203*:155–176.

Gardiner, K. (1991). Pulsed field gel electrophoresis. Anal. Chem. *63*:658–665.

Gatti, R. A. and Swift, M. (eds.) (1985). Ataxia-telangiectasia — Genetics, immunology and neuropathology of a degenerative disease of childhood. Kroc Found. Ser. *19*.

Gatti, R. A. et al. (1988). Localization of an ataxia-telangiectasia gene to chromosome 11q22-23. Nature (London) *336*:577–580.

Gaubatz, J. W. and Flores, S. W. (1990). Tissue-specific and age-related variations in repetitive sequences of mouse extrachromosomal circular DNAs. Mutat. Res. *237*:29–36.

Gazit, E. and Gazit, E. (1990). DNA fingerprinting. Isr. J. Med. Sci. *26*:158–162.

Gebhart, F. (1992). Lawrence Livermore Lab transfers chromosome paints technology. Genetic Engineering News, February, p. 25.

Geever, R. F. et al. (1981). Direct identification of sickle cell anemia by blot hybridization. Proc. Natl. Acad. Sci. U.S.A. *78*:5081–5085.

Geiger, R. et al. (1989). New, bioluminescence-enhanced detection systems for use in enzyme activity tests, enzyme immunoassays, protein blotting and nucleic acid hybridization. Mol. Cell. Probes *3*:309–328.

Gelfand, E. W. and Dosch, H. M. (1983). Diagnosis and classification of severe combined immunodeficiency disease. Birth Defects *19*:65–72.

Gellersten, B. and Kempf, R. (1990). Human prolactin gene expression: positive correlation between site-specific methylation and gene activity in a set of human lymphoid cell lines. Mol. Endocrinol. *5*:1874–1886.

Georges, M. et al. (1988). DNA fingerprinting in domestic animals using four different minisatellite probes. Cytogenet. Cell Genet. *47*:127–131.

Gerfen, C. R. (1989). Quantitation of *in situ* hybridization histochemistry for anlaysis of brain function. In: Methods in Neurosciences, Vol. 1. Conn, P. M. (ed.). Academic Press, San Oiego, pp. 79–97.

German, J. (1983). Patterns of neoplasia associated with the chromosome-breakage syndromes. In: Chromosome Mutation and Neoplasia. German, J. (ed.). Alan R. Liss, New York, pp. 97–134.

Ghazi, H. et al. (1990). Changes in the allelic methylation patterns of c-H-*ras*1, insulin and retinoblastoma genes in human development. Development (Cambridge, UK) (Suppl.):115–123.

Giaid, A. et al. (1989). *In situ* detection of peptide messenger RNA using complementary RNA probes. In: Methods in Neurosciences, Vol. 1. Conn, P. M. (ed.). Academic Press, San Diego, pp. 150–163.

Gibbs, R. A. and Caskey, C. T. (1987). Identification and localization of mutations at the Lesch-Nyhan locus by ribonuclease A cleavage. Science *236*:303–305.

Gibbs, R. A. et al. (1989a). Detection of single DNA base differences by competitive oligonucleotide priming. Nucleic Acids Res. *17*:2437–2448.

Gibbs, R. A. et al. (1989b). Identification of mutations leading to the Lesch-Nyhan syndrome by automated direct DNA sequencing of *in vitro* amplified cDNA. Proc. Natl. Acad. Sci. U.S.A. *86*:1919–1923.

Gibbs, R. A. et al. (1990). Multiplex DNA deletion detection and exon sequencing of the hypoxanthine phosphoribosyltransferase gene in Lesch-Nyhan families. Genomics *7*:235–244.

Gieseler, F. et al. (1990). Alteration of topoisomerase II action is a possible molecular mechanism of HL-60 cell differentiation. Environ. Health Perspect. *88*:183–185.

Gill, P. (1987). A new method for sex determination of the donor of forensic samples using a recombinant DNA probe. Electrophoresis *8*:35–38.

Gill, P. (1990). Case study. (A book review). Nature (London) 344:394.

Gill, P. et al. (1985). Forensic application of DNA "fingerprints". Nature (London) *318*:577–579.

Gill, P. et al. (1987). An evaluation of DNA fingerprinting for forensic purposes. Electrophoresis *8*:38–44.

Gill, P. et al. (1990). The analysis of hypervariable DNA profiles: problems associated with the objective determination of the probability of a match. Hum. Genet. *85*:75–79.

Gille, J. J. P. et al. (1987). Antioxidant status of Fanconi anemia fibroblasts. Hum. Genet. *77*:28–31.

Gillespie, D. (1990). The magic and challenge of DNA probes as diagnostic reagents. Vet. Microbiol. *24*:217–233.

Gilliland, G. et al. (1990). Competitive PCR for quantitation of mRNA. In: PCR Protocols: A Guide to Methods and Applications. Innis, M. A., et al. (eds.). Academic Press, San Diego, pp. 60–69.

Giroud, F. and Montmasson, M. P. (1989). Re-evaluation of optimal Feulgen reaction for automated cytology. Influence of fixatives. Anal. Quant. Cytol. Histol. *11*:87–95.

Gitschier, J. et al. (1985). Antenatal diagnosis and carrier detection of hemophilia A using factor VIII gene probe. Lancet *i*:1093–1094.

Givan, A. L. et al. (1988). A correction required for calulation of DNA ratios in flow cytometric analysis of ploidy. Cytometry *9*:271–274.

Gjertson, D. W. et al. (1988). Calculation of probability of paternity using DNA sequences. Am. J. Hum. Genet. *43*:860–869.

Goldstein, L. et al. (1989). Expression of multidrug resistance gene in human cancers. J. Natl. Cancer Inst. *81*:116–124.

Golubic, M. et al. (1991). Isolation of 37 single-copy DNA probes from human chromosome 6 and physical mapping of 11 probes by *in situ* hybridization. Genomics *9*:338–343.

Goodbourn, S. E. Y. et al. (1983). Molecular basis of length polymorphism in the human zeta-globin gene complex. Proc. Natl. Acad. Sci. U.S.A. *80*:5022–5026.

Goodfellow, P. J. et al. (1990). Generation of a panel of somatic cell hybrids containing unselected fragments of human chromosome 10 by x-ray irradiation and cell fusion: application to isolating the MEN2A region in hybrid cells. Somatic Cell. Mol. Genet. *16*:163–171.

Goossens, M. et al. (1983). Prenatal diagnosis of sickle cell anemia in the first trimester of pregnancy. N. Engl. J. Med. *309*:831–833.

Gordon, M. J. et al. (1988). Capillary electrophoresis. Science *242*:224–228.

Goto, M. et al. (1992). Genetic linkage of Werner's syndrome to five markers on chromosome 8. Nature *355*:735–738.

Gottesman, M. M. and Pastan, I. (1990). Mechanism and function of the multidrug transporter. Proc. Am. Assn. Cancer Res. *31*:517–518.

Gough, A. C. et al. (1990). Identification of the primary gene defect at the cytochrome P450 CYP2D locus. Nature (London) *347*:773–776.

Green, M. R. (1989). When the products of oncogenes and anti-oncogenes meet. Cell (Cambridge, Mass) *56*:1–3.

Greer, C. E. et al. (1991). PCR amplification from paraffin-embedded tissues. Effects of fixatives and fixation time. Am. J. Clin. Pathol. *95*:117–124.

Gregersen, N. et al. (1987). Improved methods for the detection of sequences in Southern blots of mammalian DNA by non-radioactive biotinylated DNA hybridization probes. Clin. Chem. Acta *169*:267–280.

Grifo, J. A. et al. (1990). Preembryo biopsy and analysis of blastomeres by *in situ* hybridization. Am. J. Obstet. Gynecol. *163*:2013–3019.

Griggin, S. (1988). *In situ* hybridization: visualizing brain messenger RNA. In: The Molecular Biology of Neurological Disease. Rosenberg, R. N. and Harding, A. E. (eds.). Butterworths, London, pp. 35–43.

Grimberg, J. et al. (1989). A simple and efficient non-organic procedure for the isolation of genomic DNA from blood. Nucleic Acids Res. *17*:8390.

Grody, W. W. et al. (1987). *In situ* viral DNA hybridization in diagnostic surgical pathology. Hum. Pathol. *18*:535–543.

Grogan, W. M. and Collins J. M. (1990). Guide to Flow Cytometry Methods. Marcel Dekker, New York.

Grompe, M. et al. (1989). Scanning detection of mutations in human ornithine transcarbamoylase by chemical mismatch cleavage. Proc. Natl. Acad. Sci. U.S.A. *86*:5888–5892.

Grompe, M. et al. (1991). Improved molecular diagnostics for ornithine transcarbamoylase deficiency. Am. J. Hum. Genet. *48*:212–222.

Groopman, J. D. and Skipper, P. L. (eds.) (1991). Molecular Dosimetry and Human Cancer: Analytical, Epidemiological, and Social Considerations. CRC Press, Boca Raton, FL.

Grose, C. and Weiner, C. P. (1990). Prenatal diagnosis of congenital cytomegalovirus infection: two decades later. Am. J. Obstet. Gynecol. *163*:447–450.

Grossman, L. I. (1990). Invited editorial: mitochondrial DNA in sickness and health. Am. J. Hum. Genet. *46*:415–417.

Guatelli, J. C. et al. (1990). Isothermal, *in vitro* amplification of nucleic acids by a multienzyme reaction modeled after retroviral replication. Proc. Natl. Acad. Sci. U.S.A. *87*:1874–1878.

Gubbay, J. et al. (1990). A gene mapping to the sex-determining region of the mouse Y chromosome is a member of a novel family of embryonically expressed genes. Nature (London) *346*:245–250.

Guitteny, A. F. et al. (1988). Histological detection of messenger RNAs with biotinylated synthetic oligonucleotide. J. Histochem. Cytochem. *36*:536–571.

Gurley, A. M. et al. (1990). Comparison of the Papanicolaou and Feulgen staining methods for DNA quantification by image analysis. Cytometry *11*:468–474.

Gustafson, S. et al. (1987). Parameters affecting the yield of DNA from human blood. Anal. Biochem. *165*:294–299.

Guttenplan, J. B. (1990). Mutagenesis by *N*-nitroso compounds: relationships to DNA adducts, DNA repair, and mutational efficiencies. Mutat. Res. *233*:177–187.

Guyot, B. et al. (1988). Prenatal diagnosis with biotinylated chromosome specific probes. Prenatal Diagn. *8*:485–493.

Guzzetta, V. et al. (1991). Isolation of region-specific and polymorphic markers from chromosome 17 by restricted *Alu* polymerase chain reaction. Genomics *9*:31–36.

Gyllensten, U. B. and Erlich, H. A. (1988). Generation of single-stranded DNA by the polymerase chain reaction and its application to direct sequencing of the HLA-DQA locus. Proc. Natl. Acad. Sci. U.S.A. *85*:7652–7656.

Haase, A. T. et al. (1985). Detection of two viral genomes in single cell by double-label hybridization *in situ* and color microradioautography. Science *227*:189–191.

Haff, L. et al. (1991). A high-performance system for automation of the polymerase chain reaction. BioTechniques *10*:102–112.

Hagelberg, E. et al. (1991). Identification of the skeletal remains of a murder victim by DNA analysis. Nature (London) *352*:427–429.

Haliassos, A. et al. (1989). Detection of minority point mutations by modified PCR technique: a new approach for a sensitive diagnosis of tumor-progression markers. Nucleic Acids Res. *17*:8093–8099.

Hall, J. G. (1990a). How imprinting is relevant to human disease. Development (Cambridge, UK) (Suppl.) 141–148.

Hall, J. G. (1990b). Genomic imprinting: review and relevance to human diseases. Am. J. Hum. Genet. *46*:857–873.

Hall, J. G. (1992). Genomic imprinting and its clinical implications. N. Engl. J. Med. *326*:827–829.

Hall, P. et al. (1991). p53 immunostaining as a marker of malignant disease in diagnostic cytopathology. Lancet *338*:513.

Hames, B. D. and Higgins, S. J. (1985). Nucleic Acid Hybridization. A Practical Approach. IRL Press, Oxford.

Hammans, S. R. et al. (1991). Mitochondrial encephalopathies: molecular genetic diagnosis from blood smaples. Lancet *337*:1311–1313.

Han, J. et al. (1991). Direct amplification of a single dissected chromosomal segment by polymerase chain reaction: a human brain sodium channel gene is on chromosome 2q22-q23. Proc. Natl. Acad. Sci. U.S.A. *88*:335–339.

Hanawalt, P. C. (1990). Role of gene expression in the fine structure of DNA damage processing. In: Biotechnology and Human Genetic Predisposition to Disease. Wiley-Liss, New York, pp. 135–145.

Handyside, A. H. et al. (1989). Biopsy of human preimplantation embryos and sexing by DNA amplification. Lancet *i*:347–349.

Handyside, A. H. et al. (1990). Pregnancies from biopsied human preimplantation embryos sexed by Y-specific DNA amplification. Nature (London) *344*:768–770.

Hanish, J. and McClelland, M. (1990). Methylase-limited partial *Not*I cleavage for physical mapping of genomic DNA. Nucleic Acids Res. *18*:3287–3291.

Hankin, R. C. and Lloyd, R. V. (1989). Detection of messenger RNA in routinely processed tissue sections with biotinylated oligonucleotide probes. Am. J. Clin. Pathol. *92*:166–171.

Hansen, M. F. and Cavenee, W. K. (1988). Tumor suppressors: recessive mutations that lead to cancer. Cell (Cambridge, Mass) *53*:172–173.

Hansson, J. et al. (1990). Complementation of the xeroderma pigmentosum DNA repair synthesis defect with *Escherichia coli* Uvr ABC proteins in a cell-free system. Nucleic Acids Res. *18*:35–40.

Harada, T. et al. (1989a). Genetic and clinical studies of Japanese patients with familial amyloid polyneuropathy. Eur. Neurol. *29*:48–52.

Harada, T. et al. (1989b). Clinical features and diagnosis by recombinant DNA techniques of familial amyloid polyneuropathy in Japan. Res. Comm. Chem. Pathol. Pharmacol. *65*:237–244.

Hardman, N. (1986). Structure and function of repetitive DNA in eukaryotes. Biochem. J. *234*:1–11.

Harley, V. R. et al. (1992). DNA binding activity of recombinant SRY from normal males and XY females. Science *255*:453–456.

Harman, D. (1991). The aging process: major risk factor for disease and death. Proc. Natl. Acad. Sci. U.S.A. *88*:5360–5363.

Harris, H. (1986). The genetic analysis of malignancy. J. Cell Sci. Suppl. *4*:431–444.

Harris, H. (1990). Tumor suppressor genes: studies with hybrid mouse cells. In: Tumor Suppressor Genes, Klein, G. (ed.). Marcel Dekker, New York, pp. 1–13.

Hart, I. R. and Easty, D. (1991). Identification of genes controlling metastatic behavior. Br. J. Cancer *63*:9–12.

Hauber, R. et al. (1989). New, sensitive, radioactive-free bioluminescence-enhanced detection system in protein blotting and nucleic acid hybridization. J. Biolumin. Chemilumin. *4*:367–372.

Hauser, E. R. et al. (1990). Allopurinol-induced orotidinuria. A test for mutations at the ornithine carbamoyltransferase locus in women. N. Engl. J. Med. *322*:1641–1645.

Hayflick, G. (1985). Theories of biological aging. Exp. Gerontol. *20*:145–159.

Hayward, N. et al. (1990). Detection of somatic mutations in tumors of diverse types by DNA fingerprinting with M13 phage DNA. Int. J. Cancer *45*:687–690.

Hedley, D. W. (1989). Flow cytometry using paraffin-embedded tissue: five years on. Cytometry *10*:229–241.

Heiger, D. N. et al. (1990). Separation of DNA restriction fragments by high performance capillary electrophoresis with low and zero crosslinked polyacrylamide using continuous and pulsed electric fields. J. Chromatogr. *516*:33–48.

Heim, M. and Meyer, U. A. (1990). Genotyping of poor metabolizers of debrisoquine by allele-specific PCR amplification. Lancet *336*:529–532.

Hejtmancik, J. F. et al. (1986). Prenatal diagnosis of α-1-antitrypsin deficiency by restriction fragment length polymorphisms, and comparison with oligonucleotide probe analysis. Lancet *ii*:767–769.

Heller, M. J. et al. (1991). An efficient method for the extraction of DNA from formalin-fixed, paraffin-embedded tissue by sonication. BioTechniques *11*:372–377.

Helminen, P. et al. (1988). Application of DNA fingerprints to paternity determinations. Lancet *ii*:574–576.

Helmuth, R. et al. (1990). HLA-DQ α allele and genotype frequencies in various human populations, determined by using enzymatic amplification and oligonucleotide probes. Am. J. Hum. Genet. *47*:515–523.

Henderson, S. et al. (1991). Induction of *bcl-2* expression by Epstein-Barr virus latent membrane protein 1 protects infected B cells from programmed cell death. Cell (Cambridge, Mass) *65*:1107–1115.

Henry, I. et al. (1991). Uniparental paternal disomy in a genetic cancer-predisposing syndrome. Nature *351*:665–667.

Herbert, D. J. et al. (1989). Effects of several commonly used fixatives on DNA and total nuclear protein analysis by flow cytometry. Am. J. Clin. Pathol. *91*:535–541.

Herbert, J. (1989). Familial amyloidotic polyneuropathies. In: Molecular Genetics in Diseases of Brain, Nerve, and Muscle. Rowland, L. P. et al. (eds.). Oxford University Press, New York, pp. 299–325.

Herrington, C. S. et al. (1989a). Interphase cytogenetics using biotin and digoxygenin-labelled probes. I. Relative sensitivity of both reporter molecules for HPV 16 detection in Caski cells. J. Clin. Pathol. *42*:592–600.

Herrington, C. S. et al. (1989b). Interphase cytogenetics using biotin and digoxigenin labelled probes. II. Simultaneous detection of two nucleic acid species in individual nuclei. J. Clin. Pathol. *42*:601–606.

Hess, J. F. et al. (1991). Impairment of mitochondrial transcription termination by a point mutation associated with the MELAS subgroup of mitochondrial encephalomyopathies. Nature (London) *351*:236–239.

Hicks, G. G. et al. (1991). Mutant p53 tumor suppressor alleles release *ras*-induced cell cycle growth arrest. Mol. Cell. Biol. *11*:1344–1352.

Higgins, G. A. and Mah, V. H. (1989). *In situ* hybridization approaches to human neurological disease. In: Methods in Neurosciences, Vol. 1. Conn, P. M. (ed.). Academic Press, San Diego, pp. 183–196.

Higgs, D. R. et al. (1981). Highly variable region of DNA flank the human α globin genes. Nucleic Acids Res. *9*:4213–4224.

Higgs, D. R. et al. (1989). A review of the molecular genetics of the human α-globin gene cluster. Blood *73*:1081–1104.

Ho, L. et al. (1989). Demethylation enhances removal of UV damage from the overall genome and from specific DNA sequences in CHO cells. Mol. Cell. Biol. *9*:1594–1603.

Hockenbery, D. et al. (1990). BcL-2 is an inner mitochondrial membrane protein that blocks programmed cell death. Nature (London) *348*:334–336.

Hoeijmakers, J. H. J. et al. (1990). Use of *in vivo* and *in vitro* assays for the characterization of mammalian excision repair and isolation of repair proteins. Mutat. Res. *236*:223–238.

Hofler, H. (1987). What's new in "*in situ* hybridization". Pathol. Res. Pract. *182*:421–430.

Holliday, R. (1987). The inheritance of epigenetic defects. Science *238*:163–170.

Hollingsworth, R. E. and Lee, W. H. (1991). Tumor suppressor genes: new prospects for cancer research. J. Natl. Cancer. Inst. *83*:91–96.

Holt, I. J. et al. (1988a). Deletions of mitochondrial DNA in patients with mitochondrial myopathies. Nature (London) *331*:717–719.

Holt, I. J. et al. (1988b). Mitochondrial DNA polymorphism in mitochondrial myopathy. Hum. Genet. *79*:53–57.

Holt, I. J. et al. (1990). A new mitochondrial disease associated with mitochondrial DNA heteroplasmy. Am. J. Hum. Genet. *46*:428–433.

Holtke, H. J. et al. (1990). Non-radioactive labelling and detection of nucleic acids. II. Optimization of the digoxigenin system. Biol. Chem. Hoppe-Seyler *371*:929–938.

Holzgreve, W. et al. (1990). Detection of fetal DNA in maternal blood by PCR. Lancet *335*:1220–1221.

Homburger, H. A. et al. (1989). Assessment of interlaboratory variability in analytical cytology. Arch. Pathol. Lab. Med. *113*:667–672.

Honma, M. and Ishiyama, I. (1989). Probability of paternity using DNA fingerprinting. Hum. Hered. *39*:165–169.

Honma, M. and Ishiyama, I. (1990). Application of DNA fingerprinting to parentage and extended family relationship testing. Hum. Hered. *40*:356–362.

Hook, E. B. (1990). Prenatal genetic diagnosis from maternal peripheral blood. Lancet *ii*:746.

Hopman, A. H. N. et al. (1986a). A non-radioactive *in situ* hybridization method based on mercurated nucleic acid probes and sulfhydryl-hapten ligands. Nucleic Acids Res. *14*:6471–6488.

Hopman, A. H. N. et al. (1986b). Bi-color detection of two target DNAs by non-radioactive *in situ* hybridization. Histochemistry *85*:1–4.

Hopman, A. H. N. et al. (1987). Mercurated nucleic acid probes, a new principle for non-radioactive *in situ* hybridization. Exp. Cell Res. *169*:357–368.

Hopman, A. H. N. et al. (1988a). Non-radioactive *in situ* hybridization. In: Molecular Neuroanatomy. Van Leeuwen et al. (eds.). Elsevier, Amsterdam and New York, pp. 43–68.

Hopman, A. H. N. et al. (1988b). *In situ* hybridization as a tool to study numerical chromosome aberrations in solid bladder tumors. Histochemistry *89*:307–316.

Horan, P. K. et al. (1990). Standards and controls in flow cytometry. In: Flow Cytometry and Sorting, 2nd Ed. Wiley-Liss, New York, pp. 397–414.

Horn, G. T. et al. (1989). Amplification of a highly polymorphic VNTR segment by the polymerase chain reaction. Nucleic Acids Res. *17*:2140.

Hornes, E. and Korsnes. L. (1990). Magnetic DNA hybridization properties of oligonucleotide probes attached to superparamagnetic beads and their use in the isolation of poly(A) mRNA from eukaryotic cells. GATA *7*:145–150.

Howell, M. D. and Kaplan, N. O. (1987). Spurious DNA blot hybridization resulting from bacterialcontamination of primary tissue preparations. Anal. Biochem. *161*:311–315.

Howie, A. F. et al. (1989). Expression of gluthathione-*S*-transferase B_1, B_2, Mu and Pi in breast cancers and their relationship to estrogen receptor status. Br. J. Cancer *60*:834–837.

Hsiang, Y. H. et al. (1985). Camptothecin induces protein-linked DNA breaks via mammalian DNA topoisomerase I. J. Biol. Chem. *260*:14873–14878.

Hsiang, Y. H. et al. (1989a). Arrest of replication forks by drug-stabilized topoisomerase I-DNA cleavable complexes as a mechanism of cell killing by camptothecin. Cancer Res. *49*:5077–5082.

Hsiang, Y. H. et al. (1989b). DNA toposomerase I-mediated DNA cleavage and cytotoxicity of camptothecin analogues. Cancer Res. *49*:4385–4389.

Hu, D. N. et al. (1991). Genetic aspects of antibiotic induced deafness: mitochondrial inheritance. J. Med. Genet. *28*:79–83.

Huang, L. S. and Breslow, J. L. (1987). A unique AT-rich hypervariable minisatellite 3′ to the apo B gene defines a high information restriction fragment length polymorphism. J. Biol. Chem. *262*:8952–8955.

Huang, S. Z. et al. (1990). Prenatal detection of an ARG → TER mutation at codon 111 of the PAH gene using DNA amplification. Prenatal Diagn. *10*:289–293.

Hunsaker, W. R. et al. (1989). Nucleic acid hybridization assays employing dA-tailed capture probes. II. Advanced multiple capture methods. Anal. Biochem. *181*:360–370.

Hunter, T. (1991). Cooperation between oncogenes. Cell (Cambridge, Mass) *64*:249–270.

Husson, R. N. et al. (1990). Diagnosis of human immunodeficiency virus infection in infants and children. Pediatrics *86*:1–10.

Hutchison, N. J. (1984). Hybridisation histochemistry: *in situ* hybridisation at the electron microscope level. In: Immunolabelling for Electron Microscopy, Polak and Varndell (eds.). Elsevier, Amsterdam, pp. 341–351.

Huynh-Dinh, T. (1990). Modified synthetic oligonucleotides as tool for molecular biology. Bull. Inst. Pasteur *88*:139–157.

Iliakis, G. et al. (1990). Induction and repair of DNA double-strand breaks in radiation-resistant cells obtained by transformation of primary rat embryo cells with the oncogenes H-*ras* and v-*myc*. Cancer Res. *50*:6575–6579.

Inazawa, J. and Abe, T. (1990). Carcinoembryonic antigen (CEA) family genes are located in human chromosome 19 at band q13. 2. Acta Histochem. Cytochem. *23*:411.

Ip, N. Y. et al. (1989). Discovery of a novel multilocus DNA polymorphism [DNF24]. Nucleic Acids Res. *17*:4427.

Isaacs, S. T. et al. (1991). Post-PCR sterilization: development and application to an HIV-1 diagnostic assay. Nucleic Acids Res. *19*:109–116.

Isada, N. B. et al. (1991). In utero diagnosis of congenital varicella zoster virus infection by chorionic villus sampling and polymerase chain reaction. Am. J. Obstet. Gynecol. *165*:1727–1730.

Ishizaki, K. et al. (1990). Human chromosome 9 can complement UV sensitivity of xeroderma pigmentosum group A cells. Mutat. Res. *235*:209–215.

Israel, M. A. (1989). Pediatric oncology: model tumors of unparalleled import. J. Natl. Cancer Inst. *81*:404–408.

Israel, M. A. (1990). Molecular approaches to oncologic pathology. Cancer Cells *2*:84–86.

Ito, T. and Sakaki, Y. (1988). A novel procedure for selective cloning of *Not*I linking fragments from mammalian genomes. Nucleic Acids Res. *16*:9177–9184.

Iversen, O. H. (ed.) (1988). Theories of Carcinogenesis. Hemisphere, Washington, D. C.

Iversen, P. L. et al. (1987). Rapid isolation of both RNA and DNA from cultured cells or whole tissues with a benchtop ultracentrifuge. BioTechniques *5*:521–523.

Jackson, D. P. et al. (1990). Tissue extraction of DNA and RNA and analysis by the polymerase chain reaction. J. Clin. Pathol. *43*:499–504.

Jackson, P. et al. (1989). *In situ* hybridization technique using an immuno-gold-silver staining system. Histochemical J. *21*:425–428.

Jacoby, G. A. and Archer, G. L. (1991). New mechanisms of bacterial resistance to antimicrobiol agents. N. Engl. J. Med. *324*:601–612.

Jager, R. J. et al. (1990). A human XY female with a frame shift mutation in the candidate testis-determining gene SRY. Nature (London) *348*:452–454.

Janus, E. D. et al. (1985). Smoking, lung function, and α-1-antitrypsin deficiency. Lancet *i*:152–154.

Jarman, A. P. et al. (1986). Molecular characterization of a hypervariable region downstream of the human α-globin gene cluster. EMBO J. *5*:1857–1863.

Jaspers, N. G. J. and Bootsma, D. (1982). Genetic heterogeneity in ataxia-telangiectasia studied by cell fusion. Proc. Natl. Acad. Sci. U.S.A. *79*:2641–2644.

Jaspers, N. G. J. et al. (1988). Patients with an inherited syndrome characterized by immunodeficiency, microcephaly and chromosomal instability: genetic relationship to ataxia-telangiectasia. Am. J. Hum. Genet. *42*:66–73.

Jaspers, N. G. J. et al. (1989). Genetic complementation analysis of ataxia-telangiectasi and Nijmegen breakage syndrome — a survey of 50 patients. Cytogenet. Cell Genet. *49*:259–263.

Jaspers, N. G. J. et al. (1990). First-trimester prenatal diagnosis of the Nijmegen breakage syndrome and ataxia-telangiectasia using an assay of radioresistant DNA synthesis. Prenatal Diagn. *10*:667–674.

Jeffreys, A. J. (1979). DNA sequence variants in the G-γ, A-γ, δ-, and β-globin genes of man. Cell (Cambridge, Mass) *18*:1–10.

Jeffreys, A. J. et al. (1985a). Positive identification of an immigration test case using DNA fingerprinting. Nature (London) *317*:818–819.

Jeffreys, A. J. et al. (1985b). Hypervariable "minisatellite" regions in human DNA. Nature (London) *314*:67–73.

Jeffreys, A. J. et al. (1985c). Individual specific "fingerprints" of human DNA. Nature (London) *316*:76–79.

Jeffreys, A. J. et al. (1986). DNA "fingerprints" and segregation analysis of multiple markers in human pedigrees. Am. J. Hum. Genet. *39*:11–24.

Jeffreys, A. J. et al. (1987). Highly variable minisatellites and DNA fingerprints. Biochem. Soc. Symp. *53*:165–180.

Jeffreys, A. J. et al. (1988a). Spontaneous mutation rates to new length alleles at tandem-repetitive hypervariable loci in human DNA. Nature (London) *332*:278–281.

Jeffreys, A. J. et al. (1988b). Amplification of human minisatellites by the polymerase chain reaction: towards DNA fingerprinting of single cells. Nucleic Acids Res. *16*:10953–10971.

Jeffreys, A. J. et al. (1990). Repeat unit sequence variation in minisatellites: a novel source of DNA polymorphism for studying variation and mutation by single molecule analysis. Cell (Cambridge, Mass) *60*:473–485.

Jeffreys, A. J. et al. (1991). Minisatellite repeat coding as a digital approach to DNA typing. Nature (London) *354*:204-209.

Jeggo, P. A. et al. (1989). Sensitivity of Chinese hamster ovary mutants defective in DNA double-strand break repair to topoisomerase II inhibitors. Cancer Res. *49*:7057–7063.

Jelinek, W. R. and Schmid, C. W. (1982). Repetitive sequences in eukaryotic DNA and their expression. Annu. Rev. Biochem. *51*:813–844.

Jeppsson, J. O. et al. (1981). Prenatal diagnosis of α-1-antitrypsin deficiency by analysis of fetal blood obtained at fetoscopy. Pediatr. Res. *15*:254–256.

Jett, J. H. et al. (1990). Ultrasensitive molecular-level flow cytometry. In: Flow Cytometry and Sorting, 2nd Ed. Wiley-Liss, New York, pp. 381–396.

Jimenez-Garcia, L. F. et al. (1989). Nucleologenesis: use of non-isotopic *in situ* hybridization and immunocytochemistry to compare the localization of rDNA and nucleolar proteins during mitosis. Biol. Cell *65*:239–246.

Jirikowski, G. F. et al. (1990). *In situ* hybridization of semithin Epon sections with BrdU labelled oligonucleotide probes. Histochemistry *94*:187–190.

Joensuu, H. and Kallioniemi, O. P. (1989). Different opinions on the classification of DNA histograms produced from paraffin-embedded tissue. Cytometry *10*:711–717.

Johns, M. B., Jr. and Paulus-Thomas, J. E. (1989). Purification of human genomic DNA from whole blood using sodium perchlorate in place of phenol. Anal. Biochem. *180*:276–278.

Johnson, D. H. (1990). Molecular cloning of DNA from specific chromosome regions by microdissection and sequence-independent amplification of DNA. Genomics *6*:243–251.

Johnson, P. et al. (1991). Expression of wild-type p53 is not compatible with continued growth of p53-negative tumor cells. Mol. Cell. Biol. *11*:1–11.

Jongsma, A. P. M. et al. (1990). Determination of the DNA content of a six-gene amplicon in the multidrug-resistant Chinese hamster cell line Ch-R-B3 by flow karyotyping. Cancer Res. *50*:2803–2807.

Jonveaux, P. (1991). PCR amplification of specific DNA sequences from routinely fixed chromosome spreads. Nucleic Acids Res. *191*:1946.

Jovaisas, E. et al. (1985). LAV/HTLV-III in 20 week fetus [letter]. Lancet *ii*:1129.

Jurka, J. and Smith, T. (1988). A fundamental division in the *Alu* family of repeated sequences. Proc. Natl. Acad. Sci. U.S.A. *85*:4775–4778.

Kaback, M. M. et al. (1977). Tay-Sachs disease heterozygotic screening and prenatal diagnosis, US experience and world perspective. Prog. Clin. Biol. Res. *18*:13–36.

Kaina, B. et al. (1991). Identification of human genes involved in repair and tolerance of DNA damage. Radiat. Environ. Biophys. *30*:1–19.

Kajimura, Y. et al. (1990). Application of long synthetic oligonucleotides for gene analysis: effect of probe length and stringency conditions on hybridization specificity. GATA *7*:71–79.

Kaldor, J. M. et al. (1987). Second malignancies following testicular cancer, ovarian cancer and Hodgkin's disease: an international collaborative study among cancer registries of the long-term effects of therapy. Int. J. Cancer *39*:571–585.

Kallioniemi, O. P. (1988). Comparison of fresh and paraffin-embedded tissue as starting material for DNA flow cytometry and evaluation of intratumor heterogeneity. Cytometry *9*:164–169.

Kallioniemi, O. P. et al. (1990). Interlaboratory comparison of DNA flow cytometric results from paraffin-embedded breast carcinomas. Breast Cancer Res. Treat. *17*:59–61.

Kan, Y. W. and Dozy, A. M. (1978). Antenatal diagnosis of sickle cell anemia by DNA analysis of amniotic fluid cells. Lancet *ii*:910–912.

Kaneoka, H. et al. (1991). Solid phase direct DNA sequencing of allele-specific polymerase chain reaction-amplified HLA-DR genes. BioTechniques *10*:30–34.

Kantor, G. J. et al. (1990). Selective repair of specific chromatin domains in UV-irradiated cells from xeroderma pigmentosum complementation group C. Mutat. Res. *235*:171–180.

Kao, Ft. and Yu, J. W. (1991). Chromosome microdissection and cloning in human genome and genetic disease analysis. Proc. Natl. Acad. Sci. U.S.A. *88*:1844–1848.

Kao, Y. S. et al. (1990). Banding resolution of amniotic cell chromosome preparations for prenatal diagnosis. Am. J. Clin. Pathol. *93*:765–770.

Kaplan, C. (1988). Prenatal treatment in neonatal alloimmune thrombocytopenia. Curr. Stud. Hematol. Blood Transfus. *55*:142–147.

Karger, B. L. (1989). High performance capillary electrophoresis. Nature (London) *339*:641–642.

Karger, B. L. et al. (1989). High performance capillary electrophoresis in the biological sciences. J. Chromatogr. *492*:585–614.

Karran, P. and Stephenson, C. (1990). Mismatch binding proteins and tolerance to alkylating agents in human cells. Mutat. Res. *236*:269–275.

Kashani-Sabet, M. et al. (1988). Detection of drug resistance in human tumors by *in vitro* enzymatic amplification. Cancer Res. *48*:5775–5778.

Katz, E. J. et al. (1990). Effect of topoisomerase modulators on cisplatin cytotoxicity in human ovarian carcinoma cells. Eur. J. Cancer *26*:724–727.

Kaufman, D. L. et al. (1990). Detection of point mutations associated with genetic diseases by an exon scanning technique. Genomics *8*:656–663.

Kaufmann, W. K. and Wilson, S. J. (1990). DNA repair endonuclease activity during synchronous growth of diploid human fibroblasts. Mutat. Res. *236*:107–117.

Kaur, G. P. and Athwal, R. S. (1989). Complementation of a DNA repair defect in xeroderma pigmentosum cells by transfer of human chromosome 9. Proc. Natl. Acad. Sci. U.S.A. *86*:8872–8876.

Kawamura, M. and Preisler, H. D. (1989). A new approach to the detection of DNA damage. Leuk. Res. *13*:391–398.

Kawata, M. et al. (1990). *In situ* hybridization histochemistry with oligodeoxynucleotide probe as a tool for the detection of neuropeptide mRNAs. Acta Histochem. Cytochem. *23*:307–325.

Kaye, C. I. et al. (1985). Evaluation of chromosomal damage in males exposed to agent orange and their families. J. Craniofac. Genet. Dev. Biol. (Suppl. I):259–265.

Kazazian, H. H., Jr. (1985). Gene probes: application to prenatal and postnatal diagnosis of genetic disease. Clin. Chem. *31*:1509–1513.

Kazazian, H. H., Jr. (1990). The thalassemia syndromes: molecular basis and prenatal diagnosis in 1990. Semin. Hematol. *27*:209–228.

Keats, B. J. B. et al. (1991). Guidelines for human linkage maps. An international system for human linkage maps (ISLM, 1990). Ann. Hum. Genet. *55*:1–6.

Keeney, S. and Linn, S. (1990). A critical review of permeabilized cell systems for studying mammalian DNA repair. Mutat. Res. *236*:239–252.

Keith, W. N. et al. (1990). Expression of mdr 1 and gst-TT in human breast tumors: comparison to *in vitro* chemosensitivity. Br. J. Cancer *61*:712–716.

Keller, G. H. and Manak, M. M. (1989). DNA Probes. Stockton Press, New York.

Keohavong, P. and Thilly, W. G. (1989). Fidelity of DNA amplification. Proc. Natl. Acad. Sci. U.S.A. *86*:9253–9257.

Keren, D. F. (ed.). (1989). Flow Cytometry in Clinical Diagnosis. ASCP Press, Chicago.

Kessler, C. et al. (1990). Non-radioactive labelling and detection of nucleic acids. I. A novel DNA labelling and detection system based on digoxigenin: anti-digoxigenin ELISA principle (digoxigenin system). Biol. Chem. Hoppe-Seyler *371*:917–927.

Khandjian, E. W. (1987). Optimized hybridization of DNA blotted and fixed to nitrocellulose and nylon membranes. BioTechnology *5*:165–167.

Khoury, M. J. et al. (1990). Epidemiologic approaches to the use of DNA markers in the search for disease susceptibility genes. Epidemiol. Rev. *12*:41–55.

Kidd, V. J. et al. (1983). α-1-Antitrypsin deficiency detection by direct analysis of the mutation in the gene. Nature (London) *304*:230–234.

Kidd, V. J. et al. (1984). Prenatal diagnosis of α-1-antitrypsin deficiency by direct analysis of the mutation site in the gene. N. Engl. J. Med. *310*:639–642.

Kievits, T. et al. (1990). Direct nonradioactive *in situ* hybridization of somatic cell hybrid DNA to human lymphocyte chromosomes. Cytometry *11*:105–109.

Killion, J. J. and Fidler, I. J. (1989). The biology of human metastasis. Semin. Oncol. *16*:106–115.

Kim, R. et al. (1991). Expression of MDR 1, GST-π and topoisomerase II as an indicator of clinical response to adriamycin. Anticancer Res. *11*:429–432.

Kimberling, W. J. et al. (1988). Linkage heterogeneity of autosomal dominant polycystic kidney disease. N. Engl. J. Med. *319*:913–918.

Kincaid, R. L. and Nightingale, M. S. (1988). A rapid non-radioactive procedure for plaque hybridization using biotinylated probes prepared by random primed labelling. BioTechniques *6*:42–49.

Kinzler, K. W. and Vogelstein, B. (1989). Whole genome PCR: application to the identification of sequences bound by gene regulatory proteins. Nucleic Acids Res. *17*:3645–3653.

Kirby, L. T. (1990). DNA Fingerprinting. An Introduction. Stockton Press, New York, p. 365.

Kirkwood, T. B. L. (1989). DNA mutation and aging. Mutat. Res. *219*:1–7.

Kitchin, P. A. et al. (1990). Avoidance of false positives. Nature (London) *344*:201.

Klein, G. (1987). The approaching era of the tumor suppressor genes. Science *238*:1539–1545.

Klein, G. (ed.). (1990). Tumor Suppressor Genes. Marcel Dekker, New York.

Klever, M. et al. (1991). Chromosomal *in situ* suppression hybridization after Giemsa banding. Hum. Genet. *86*:484–486.

Knudson, A. G. (1971). Mutation and cancer: statistical study of retinoblastoma. Proc. Natl. Acad. Sci. U.S.A. *68*:820–823.

Knudson, A. G. (1986). Genetics of human cancer. Annu. Rev. Genet. *20*:231–251.

Knudson, A. G., Jr. (1989). Hereditary cancers: clues to mechanisms of carcinogenesis. Br. J. Cancer *59*:661–666.

Ko, T.-M. et al. (1990). DNA polymorphism and globin chain analysis in the prenatal diagnosis of β-thalassemia major in Taiwan. Prenatal Diagn. *10*:237–244.

Kobilinsky, L. and Levine, L. (1988). Recent application of DNA analysis to issues of paternity. J. Forensic Sci. *33*:1107–1108.

Kobilinsky, L. and Levine, L. (1989). Discussion of "Recent application of DNA analysis to issues of paternity" [Letter to the Editor]. J. Forensic Sci. *35*:5–6.

Kogan, S. C. and Gitschier, J. (1990). Genetic prediction of hemophilia A. In: PCR Protocols. A Guide to Methods and Applications. Innis, M. A. et al. (eds.). Academic Press, San Diego, pp. 288–299.

Kogan, S. C. et al. (1987). An improved method for prenatal diagnosis of genetic diseases by analysis of amplified DNA sequences. Application to hemophilia A. N. Engl. J. Med. *317*:985–990.

Koopman, P. et al. (1990). Expression of a candidate sex-determining gene during mouse testis differentiation. Nature (London) *348*:450–452.

Korenberg, J. R. and Rykowski, M. C. (1988). Human genome organization: *Alu*, lines, and the molecular structure of metaphase chromosome bands. Cell (Cambridge, Mass) *53*:391–400.

Kovacs, B. W. et al. (1990). Prenatal determinations of paternity by molecular genetic "fingerprinting". Obstet. Gynecol. *75*:474–479.

Kowraa, S. et al. (1991). Application of fluorescence *in situ* hybridization techniques in clinical genetics: use of two alphoid repeat probes detecting the centromeres of chromosomes 13 and 21 or chromosomes 14 and 22, respectively. Clin. Genet. *39*:278–286.

Kraemer, K. H. and Slor, H. (1985). Xeroderma pigmentosum. Clin. Dermatol. *2*:33–69.

Kraemer, K. H. et al. (1987). Xeroderma pigmentosum. Cutaneous ocular and neurologic abnormalities in 830 published cases. Arch. Dermatol. *123*:241–250.

Kramer, F. R. and Lizardi, P. M. (1989). Replicatable RNA reporters. Nature (London) *339*:401–402.

Kramer, F. R. and Lizardi, P. M. (1990). Amplifiable hybridization probes. Ann. Biol. Clin. *48*:409–411.

Kramers, P. G. N. et al. (1991). Review of the genotoxicity and carcinogenicity of antischistosomal drugs; is there a case for a study of mutation epidemiology? Report of a task group on mutagenic antischistosomals. Mutat. Res. *257*:49–89.

Krawczak, M. and Cooper, D. N. (1991). Gene deletions causing human genetic disease: mechanisms of mutagenesis and the role of the local DNA sequence environment. Hum. Genet. *86*:425–441.

Krawczak, M. et al. (1989). Polymerase chain reaction: replication errors and reliability of gene diagnosis. Nucleic Acids Res. *17*:2197–2201.

Kremer, E. J. et al. (1991). Mapping of DNA instability at the Fragile X to a trinucleotide repeat sequence p(CCG)n. Science *252*:1711–1714.

Kriegler, M. (1990). Gene Transfer and Expression. A Laboratory Manual. Stockton Press, New York.

Krowczynska, A. M. et al. (1990). The human minisatellite consensus at breakpoints of oncogene translocations. Nucleic Acids Res. *18*:1121–1127.

Krystosek, A. and Puck, T. T. (1990). The spatial distribution of exposed nuclear DNA in normal, cancer, and reverse-transformed cells. Proc. Natl. Acad. Sci. U.S.A. *87*:6560–6564.

Kuijpers, R. W. A. M. et al. (1990). Typing of fetal platelet alloantigens when platelets are not available. Lancet *336*:1319.

Kuroda, S.-i. et al. (1988). A supersensitive dot-hybridization method: rapid and quantitative detection of host-derived DNA in recombinant products at the one picogram level. Biochem. Biophys. Res. Commun. *152*:9–14.

Kwoh, D. Y. et al. (1989). Transcription-based amplification system and detection of amplified human immunodeficiency virus type 1 with a bead-based sandwich hybridization format. Proc. Natl. Acad. Sci. U.S.A. *86*:1173–1177.

Kwok, S. and Higuchi, R. (1989). Avoiding false positives with PCR. Nature (London) *339*:237–238.

Labuda, D. and Striker, G. (1989). Sequence conservation in *Alu* evolution. Nucleic Acids Res. *17*:2477–2491.

Lagoda, P. J. L. et al. (1989). Increased detectability of somatic changes in the DNA from human tumors after probing with "synthetic" and "genome-derived" hypervariable multilocus probes. Hum. Genet. *84*:35–40.

Lambert, W. C. and Lambert, M. W. (1985). Co-recessive inheritance: a model for DNA repair, genetic disease and carcinogenesis. Mutat. Res. *145*:227–234.

Landegent, J. E. et al. (1984). α-Acetylaminofluorene-modified probes for the indirect hybridocytochemical detection of specific nucleic acid sequences. Exp. Cell Res. *153*:61–72.

Landegent, J. E. et al. (1987). Use of whole cosmid cloned genomic sequences for chromosomal localization by non-radioactive *in situ* hybridization. Hum. Genet. *77*:366–370.

Landegren, U. et al. (1988). A ligase-mediated gene detection technique. Science *241*:1077–1080.

Landis, C. A. et al. (1990). Oncogenic activation of G proteins by GTPase inhibiting mutations. Proc. Am. Assn. Cancer Res. *31*:505–506.

Lane, D. P. and Benchimol, S. (1990). p53: oncogene or anti-oncogene? Genes Dev. *4*:1–8.

Lane, T. et al. (1990). Quantized viral DNA packaging revealed by rotating gel electrophoresis. Virology *174*:472–478.

Lange, K. and Boehnke, M. (1982). How many polymorphic genes will it take to span the human genome? Am. J. Hum. Genet. *34*:842–845.

Lanzillo, J. J. (1990). Preparation of digoxigenin-labeled probes by the polymerase chain reaction. BioTechniques *8*:621–622

Lanzillo, J. J. (1991). Chemoluminescent nucleic acid detection with digoxigenin-labeled probes: a model system with probes for angiotensin converting enzyme which detect less than one attomole of target DNA. Anal. Biochem. *194*:45–53.

Larsson, C. (1978). Natural history and life expectancy in severe α-1-antitrypsin deficiency, Pi Z. Acta Med. Scand. *204*:345–351.

Larsson, L. I. and Hougaard, D. M. (1990). Optimization of non-radioactive *in situ* hybridization: image analysis of varying pretreatment, hybridization and probe labeling conditions. Histochemistry *93*:347–354.

Larsson, L. I. et al. (1991). Quantitative non-radioactive *in situ* hybridization. Model studies and studies on pituitary propiomelanocortin cells after adrenalectomy. Histochemistry *95*:209–215.

Lasko, D. D. et al. (1990). Eukaryotic DNA ligases. Mutat. Res. *236*:277–287.

Latos-Bielenska, A. and Hameister, H. (1988). Higher resolution banding techniques in the clinical routine. Clin. Genet. *33*:325–330.

Lau, Y. F. (1985). Detection of Y-specific repeat sequences in normal and variant human chromosomes using *in situ* hybridization with biotinylated probes. Cytogenet. Cell Genet. *39*:184–187.

Lau, Y. F. and Schonberg, S. (1984). A male-specific DNA probe detects heterochromatin sequences in a familial Yq-chromosome. Am. J. Hum. Genet. *36*:1394–1396.

Lau, Y. F. et al. (1984). A rapid screening test for antenatal sex determination. Lancet *i*:14–16.

Lauber, J. et al. (1991). Mutations in mitochondrial tRNA genes: a frequent cause of neuromuscular diseases. Nucleic Acids. Res. *19*:1393–1397.

Laure, F. et al. (1988). Detection of HIV1 DNA in infants and children by means of the polymerase chain reaction. Lancet *ii*:538–541.

Laval, F. (1990). Induction of proteins involved in the repair of alkylated bases in mammalian cells by DNA-damaging agents. Mutat. Res. *233*:211–218.

Lawn, R. M. (1991). Molecular genetics of factor VIII deficiency. In: Hematology. Basic Principles and Practice. Hoffman, R. et al. (eds.). Churchill Livingstone, New York, pp. 1284–1289.

Lawrence, J. B. and Singer, R. H. (1985). Quantitative analysis of *in situ* hybridization methods for the detection of actin gene expression. Nucleic Acids Res. *13*:1777–1799.

Lawrence, J. B. et al. (1988a). Intracellular distribution of histone mRNAs in human fibroblasts studied by *in situ* hybridization. Proc. Natl. Acad. Sci. U.S.A. *85*:463–467.

Lawrence, J. B. et al. (1988b). Sensitive high resolution chromatin and chromosome mapping *in situ*: presence and orientation of two closely integrated copies of EBV in a lymphoma line. Cell (Cambridge, Mass) *52*:51–61.

Lawrence, J. B. et al. (1989). Highly localized tracks of specific transcripts within interphase nuclei visualized by *in situ* hybridization. Cell (Cambridge, Mass) *57*:493–502.

Leadon, S. A. (1988). Immunological probes for lesions and repair patches in DNA. In: DNA Repair. A Laboratory Manual of Research Procedures, Vol. 3. Friedberg, E. C. and Hanawalt, P. C. (eds.). Marcell Dekker, New York, pp. 311–326.

Lebacq, P. et al. (1988). A new sensitive non-isotopic method using sulfonated probes to detect program quantities of specific DNA sequences on blot hybridization. J. Biochem Biophys. Methods *15*:255–266.

Lebo, R. V. et al. (1990a). Defined DNA diagnostic tests appropriate for standard clinical care. Am. J. Hum. Genet. *47*:583–590.

Lebo, R. V. et al. (1990b). Prenatal diagnosis of α-thalassemia by polymerase chain reaction and dual restriction enzyme analysis. Hum. Genet. *85*:293–299.

Ledbetter, S. A. et al. (1990a). Rapid isolation of DNA probes within specific chromosome regions by interspersed repetitive sequence polymerase chain reaction. Genomics *6*:475–481.

Ledbetter, S. A. et al. (1990b). "PCR-karyotype" of human chromosomes in somatic cell hybrids. Genomics *8*:614–622.

Lee, E. Y.-H. P. et al. (1991). Diverse mutations lead to inactivation of the retinoblastoma gene. In: Molecular Biology of the Retina: Basic and Clinically Relevant Studies. pp. 224–240.

Lee, F. Y. F. et al. (1989). Changes in cellular glutathione content during adriamycin treatment in human ovarian cancer — a possible indicator of chemosensitivity. Br. J. Cancer *60*:291–298.

Lee, J. T. et al. (1990). The role of genetically determined polymorphic drug metabolism in the beat-blockade produced by propafenone. N. Engl. J. Med. *322*:1764–1768.

Lee, S. W. et al. (1991). Positive selection of candidate tumor-suppressor genes by subtractive hybridization. Proc. Natl. Acad. Sci. U.S.A. *88*:2825–2829.

Lehmann, A. R. and Bridges, B. A. (1990). Sunlight-induced cancer: some new apsects and implications of the xeroderma pigmentosum model. Br. J. Dermatol. *122* (Suppl.) *35*:115–119.

Lehmann, A. R. and Norris, P. G. (1989). DNA repair and cancer: speculations based on studies with xeroderma pigmentosum, Cockayne's syndrome, and trichothiodystrophy. Carcinogenesis *10*:1353–1356.

Lehmann, A. R. et al. (1985). Prenatal diagnosis of Cockayne's syndrome. Lancet *i*:486–488.

Lehmann, A. R. et al. (1989). Meeting report; 4th international workshop on ataxia-telangiectasia. Cancer Res. *49*:6162–6163.

Lemna, W. K. et al. (1990). Mutation analysis for heterozygote detection and the prenatal diagnosis of cystic fibrosis. N. Engl. J. Med. *322*:291–296.

Lench, N. et al. (1988). Simple non-invasive method to obtain DNA for gene analysis. Lancet *i*:1356–1358.

Lengauer, C. et al. (1990). Painting of human chromosomes with probes generated from hybrid cell lines by PCR with *Alu* and L1 primers. Hum. Genet. *86*:1–6.

Lengauer, C. et al. (1991). Painting of defined chromosomal regions by *in situ* suppression hybridization of libraries from laser-microdissected chromosomes. Cytogenet. Cell Genet. *56*:27–30.

Lennard, M. S. (1990). Genetic polymorphism of sparteine/debrisoquine oxidation: a reappraisal. Pharmacol. Toxicol. *67*:273–283.

Leone, A. et al. (1991). Reduced tumor incidence, metastatic potential, and cytokine responsiveness of *nm*23-transfected melanoma cells. Cell (Cambridge, Mass) *65*:25–35.

Lestienne, P. and Norby, S. (1990). The structure, function and pathology of the human mitochondrial genome. Proc. 5th Eur. Congr. Biotechnol. 5th *1*:85–94.

Lestienne, P. et al. (1990). Polymerase chain reaction (PCR) as a tool for studying mitochondrial DNA mutations in human disease. Proc. Eur. Congr. Biotechnol. 5th *1*:406–409.

Levedakou, E. N. et al. (1989). A strategy to study gene polymorphism by direct sequence analysis of cosmid clones and amplified genomic DNA. BioTechniques *7*:438–442.

Levine, A. J. et al. (1991). The *p53* tumor suppressor gene. Nature (London) *351*:453–456.

Levran, O. et al. (1991). Niemann-Pick disease: a frequent missense mutation in the acid sphingomyelinase gene of Ashkenazi Jewish type A and B patients. Proc. Natl. Acad. Sci. U.S.A. *88*:3748–3752.

Lewis, A. D. et al. (1989). Glutathione-*S*-transferase isoenzymes in human tumors and tumor derived cell lines. Br. J. Cancer *60*:327 331.

Lewontin, R. C. and Hartl, D. L. (1991). Population genetics in forensic DNA typing. Science *254*:1745–1750.

Lichter, P. et al. (1988). Delineation of individual human chromosomes in metaphase and interphase cells by *in situ* suppression hybridization using recombinant DNA libraries. Hum. Genet. *80*:224–234.

Lichter, P. et al. (1990a). Fluorescence *in situ* hybridization with *Alu* and L1 polymerase chain reaction probes for rapid characterization of human chromosomes in hybrid cell lines. Proc. Natl. Acad. Sci. U.S.A. *87*:6634–6638.

Lichter, P. et al. (1990b). High resolution mapping of human chromosome 11 by *in situ* hybridization with cosmid clones. Science. *247*:64–69.

Lichter, P. et al. (1991). Analysis of genes and chromosomes by nonisotopic *in situ* hybridization. GATA *8*:24–35.

Liechti, S. et al. (1990). Prenatal exclusion of ornithine transcarbamylase (OTC) by using RFLP analysis. J. Inherited Metab. Dis. *13*:888–889.

Liesack, W. et al. (1990). Rapid generation of vector-free digoxigenin-d UTP labelled probes for non-radioactive hybridization using the polymerase chain reaction (PCR) method. Syst. Appl. Microbiol. *13*:255–256.

Lincke, C. R. et al. (1991). Structure of the human MDR3 gene and physical mapping of the human MDR locus. J. Biol. Chem. *266*:5303–5310.

Lindahl, T. (1982). DNA repair enzymes. Annu. Rev. Biochem. *51*:61–87.

Lindeman, R. et al. (1990). Exclusion of β-thalassemia by biopsy and DNA amplification in mouse pre-embryos. Prenatal Diagn. *10*:295–301.

Lindsay, S. M. and Philipp, M. (1991). Can scanning tunneling microscope sequence DNA? GATA *8*:8–13.

Lindsten, J. and Pettersson, U. (eds.) (1991). Etiology of Human Disease at the DNA level. Raven Press, New York.

Linnane, A. W. et al. (1989). Mitochondrial DNA mutations as an important contributor to ageing and degenerative diseases. Lancet *i*:642–645.

Linnane, A. W. et al. (1990). Mitochondrial gene mutation: the ageing process and degenerative diseases. Biochem. Int. *22*:1067–1076.

Linz, U. et al. (1990). Systematic studies on parameters influencing the performance of the polymerase chain reaction. J. Clin. Chem. Clin. Biochem. *28*:5–13.

Lion, T. et al. (1989). Rapid dot blot quantitation of viral DNA and amplified genes in less than 1000 human cells. DNA *8*:361–367.

Liotta, L. and Steeg, P. (1990). Clues to the function of nm23 and *awd* proteins in development, signal transduction, and tumor metastasis provided by studies of *Dictyostelium discoideum*. J. Natl. Cancer Inst. *82*:1170–1172.

Liotta, L. A. and Stetler-Stevenson, W. (1989). Metalloproteinases and malignant conversion: does correlation imply causality? J. Natl. Cancer Inst. *81*:556–557.

Liotta, L. A. et al. (1991). Cancer metastasis and angiogenesis: an imbalance of positive and negative regulation. Cell (Cambridge, Mass) *64*:327–336.

Lizardi, P. M. et al. (1988). Exponential amplification of recombinant-RNA hybridization probes. Biotechnology *6*:1197–1202.

Lo, Y.-M. D. et al. (1988). False-positive results and the polymerase chain reaction. Lancet *ii*:679.

Lo, Y.-M. D. et al. (1989). Prenatal sex determination by DNA amplification from maternal peripheral blood. Lancet *ii*:1363–1365.

Lo, Y.-M. D. et al. (1990a). Detection of single-copy fetal DNA sequence from maternal blood. Lancet *335*:1463–1464.

Lo, Y.-M. D. et al. (1990b). Prenatal genetic diagnosis from maternal peripheral blood. A reply. Lancet *ii*:746.

Loader, S. et al. (1991). Prenatal screening for hemoglobinopathies. II. Evaluation of counseling. Am. J. Hum. Genet. *48*:447–451.

Lockett, S. J. et al. (1990). Automatic measurement of DNA ploidy using fluorescence microscopy. In: Optical Microscopy for Biology. Wiley-Liss, New York, pp. 603–613.

Lockshin, R. A. and Zakeri, Z. F. (1990). Programmed cell death: new thoughts and relevance to aging. J. Gerontol. Biol. Sci. *45*:B135–140.

Loeb, L. A. (1989). Endogenous cancinogenesis: molecular oncology into the twenty-first century — presidential address. Cancer Res. *49*:5489–5496.

Loi, A. et al. (1986). Prenatal diagnosis of the most common Mediterranean β-thalassemia mutants by the oligonucleotide technique. Lancet *i*:274.

Lomeli, H. et al. (1989). Quantitative assays based on the use of replicatable hybridization probes. Clin. Chem *35*:1826–1831.

Longo, D. L. (1987). Tumor markers: current status of the quest — introductory overview. Semin. Oncol. *14*:85–88.

Lundeberg, J. et al. (1990). Rapid colorimetric detection of *in vitro* amplified DNA sequences. DNA Cell Biol. *9*:289–292.

Lundeberg, J. et al. (1991). Rapid colorimetric quantification of PCR-amplified DNA. BioTechniques *10*:68–75.

Lynch, D. C. (1991). Molecular genetics of von Willebrand disease. In: Hematology. Basic Principles and Practice. Hoffman, R. et al. (eds.). Churchill Livingstone, New York, pp. 1359–1361.

Lynch, H. T. and Tautu, P. (eds.) Recent Progress in the Genetic Epidemiology of Cancer. Springer-Verlag, Berlin.

Lynch, M. (1990). The similarity index and DNA fingerprinting. Mol. Biol. Evol. *7*:478–484.

Macieira-Coelho, A. (1984). A genomic reorganization during cellular senescence. Mech. Ageing Dev. *27*:257–262.

Madhani, H. D. et al. (1986). Differential DNA repair in transcriptionally active and inactive proto-oncogenes:c-*abl* and c-*mos*. Cell (Cambridge, Mass) *45*:417–423.

Madisen, L. et al. (1987). DNA banking: the effects of storage of blood and isolated DNA on the intergrity of DNA. Am. J. Med. Genet. *27*:379–390.

Mage, R. G. et al. (1988). Genetic analyses of restriction fragment length polymorphisms using high molecular weight DNAs from sperm or lymphocytes embedded in agarose. Gene Anal. Tech. *5*:94–96.

Maggiano, N. et al. (1991). Detection of mRNA and hnRNA using a digoxigenin labelled cDNA probe by *in situ* hybridization on frozen tissue sections. Histochemical J. *23*:69–74.

Maher, V. M. et al. (1990). Alkylation damage, DNA repair and mutagenesis in human cells. Mutat. Res. *233*:235–245.

Mahuran, D. J. (1991). The biochemistry of HEXA and HEXB gene mutations causing G_{m2} gangliosides. Biochim. Biophys. Acta *1096*:87–94.

Makuch, R. W. and Muenz, L. R. (1987). Evaluating the adequacy of tumor markers to discriminate among distinct populations. Semin. Oncol. *14*:89–101.

Maniatis, T. et al. (1989). Molecular Cloning. A Laboratory Manual. Cold Spring Harbor Laboratory, Cold Spring Harbor, New York.

Mantero, G. et al. (1991). DNA enzyme immunoassay: general methods for detecting products of polymerase chain reaction. Clin. Chem. *37*:422–429.

Manuelidis, L. et al. (1982). High-resolution mapping of satellite DNA using biotin-labeled DNA probes. J. Cell Biol. *95*:619–625.

Mardis, E. R. and Roe, B. A. (1989). Automated methods for single-stranded DNA isolation and dideoxynucleotide DNA sequencing reactions on a robotic workstation. BioTechniques *7*:840–850.

Mareel, M. M. et al. (1990). The invasive phenotypes. Cancer Metastasis Rev. *9*:45–62.

Mareel, M. M. et al. (1991). Mechanisms of Invasion and Metastasis. CRC Press, Boca Raton, FL.

Marion, R. W. et al. (1989). Human T-cell lymphotropic virus type III (HTLV-III) embryopathy: a new dysmorphic syndrome associated with intrauterine HTLV-III infection. Am. J. Dis. Child. *140*:638–640.

Marshall, C. J. (1991). Tumor suppressor genes. Cell (Cambridge, Mass) *64*:313–326.

Martin, A. O. et al. (1990). Stylized chromosome images. Cytometry *11*:40–50.

Martin, C. et al. (1991). Improved chemoluminescent DNA sequencing. BioTechniques *11*:110–113.

Martin, F. et al. (1991). A cytogenetic study of men occupationally exposed to uranium. Br. J. Ind. Med. *48*:98–102.

Martinez, M. and Goldin, L. R. (1990). Power of the linkage test for a heterogeneous disorder due to two independent inherited causes: a simulation study. Genet. Epidemiol. *7*:219–230.

Masters, J. R. W. et al. (1988). Bladder cancer cell line cross-contamination: identification using a locus-specific minisatellite probe. Br. J. Cancer *57*:284–286.

Matsubara, Y. et al. (1991). Prevalence of K329E mutation in medium-chain acyl-CoA dehydrogenase gene determined from Guthrie cards. Lancet *338*:552–553.

Matsuda, I. et al. (1989). Structure of the ornithine transcarbamylase (OTC) gene and DNA diagnosis of OTC deficiency. Clin. Chim. Acta *185*:283–290.

Mayall, B. H. et al. (1990). Experience with the Athena semiautomated karyotyping system. Cytometry *11*:59–72.

Mayer, A. et al. (1991). Separation and detection of DNA polynucleotides using capillary electrophoresis. Arch. Pathol. Lab. Med. *115*:1228–1234.

Mayer, P. J. et al. (1989). Age-dependent decline in rejoining of X-ray induced DNA double-strand breaks in normal human lymphocytes. Mutat. Res. *219*:95–100.

Mayne, L. V. and Lehman, A. R. (1982). Failure of RNA synthesis to recover after UV irradiation: an early defect in cells from individuals with Cockayne's syndrome and xeroderma pigmentosum. Cancer Res. *42*:1473–1478.

Mayrand, P. E. et al. (1990). Automation of specific human gene detection. Clin. Chem. *36*:2063–2071.

McCabe, J. T. and Pfaff, D. W. (1989). *In situ* hybridization: a methodological guide. In: Methods in Neurosciences, Vol. 1. Conn, P. M. (ed.). Academic Press, San Diego, pp. 98–126.

McCabe, J. T. et al. (1989). Graphical and statistical approaches to data analysis for *in situ* hybridization. In: Methods in Enzymology, Vol. 168. Conn, P. M. (ed.). Academic Press, San Diego, pp. 822–848.

McCoy, J. P. et al. (1990). Quality control in flow cytometry for diagnostic pathology. I. Cell surface phenotyping and general laboratory procedures. Am. J. Clin. Pathol. *93* (Suppl. 1):S27–37.

McDonough, P. G. et al. (1986). Use of two different deoxyribonucleic acid probes to detect Y chromosome deoxyribonucleic acid in subjects with normal and altered Y chromosomes. Am. J. Obstet. Gynecol. *154*:737.

McFadden, P. W. et al. (1990). Image analysis confirmation of DNA aneuploidy in flow cytometric DNA distributions having a wide coefficient of variation of the G_0/G_1 peak. Am. J. Clin. Pathol. *93*:637–642.

McGraw, R. A. et al. (1990). Sequence-dependent oligonucleotide-target duplex stabilities: rules from empirical studies with a set of twenty-mers. BioTechniques *8*:674–678.

McKinnon, P. J. (1987). Ataxia-telangiectasia: an inherited disorder of ionizing-radiation sensitivity in man. Hum. Genet. *75*:197–208.

McKusick, V. A. (1991). Current trends in mapping human genes. FASEB J. *5*:12–20.

McLaren, A. (1988). Sex determination in mammals. Trends Genet. *4*:153–157.

McMahon, G. et al. (1987). Characterization of c Ki-*ras* oncogene alleles by direct sequencing of enzymatically amplified DNA from carcinogen-induced tumors. Proc. Natl. Acad. Sci. U.S.A. *84*:4974–4978.

Meese, E. and Blin, N. (1987). Simultaneous isolation of high molecular weight RNA and DNA from limited amounts of tissues and cells. Gene Anal. Tech. *4*:45–49.

Meijer, C. et al. (1990). The role of glutathione in resistance to cisplatin in a human small cell lung cancer cell line. Br. J. Cancer *62*:72–77.

Meisen, C. et al. (1988). Prenatal diagnosis of α-1-antitrypsin deficiency using oligonucleotide probe analysis. Hum. Genet. *79*:190–192.

Melamed, M. R. and Staiano-Coico, L. (1990). Flow cytometry in clinical cytology. In: Flow Cytometry and Sorting, 2nd Ed. Wiley-Liss, New York, pp. 755–772.

Mellon, I. et al. (1986). Preferential DNA repair of an active gene in human cells. Proc. Natl. Acad. Sci. U.S.A. *83*:8878–8888.

Mellon, I. M. et al. (1987). Selective removal of transcription-blocking DNA damage from the transcribed strand of the mammalian DHFR gene. Cell (Cambridge, Mass) *51*:241–246.

Melmer, G. and Buchwald, M. (1990). Laboratory methods: use of short oligonucleotides to screen cosmid libraries for clones containing G/C-rich sequences. DNA Cell. Biol. *9*:377–385.

Meltzer, S. J. et al. (1990). Sequencing products of the polymerase chain reaction directly, without purification. BioTechniques *8*:142–148.

Melvin, J. R., Jr. et al. (1988). Paternity testing. In: Forensic Science Handbook, Vol. II. Saferstein, R. (ed.). Prentice-Hall, Englewood Cliffs, NJ, pp. 273–346.

Mendelson, M. L. and Albertini, R. J. (1990). Mutation and the Environment. Wiley-Liss, New York.

Mercer, W. E. et al. (1991). Growth suppression induced by wild-type p53 protein is accompanied by selective down-regulation of proliferating-cell nuclear antigen expression. Proc. Natl. Acad. Sci. U.S.A. *88*:1958–1962.

Merkel, D. E. and McGuire, W. L. (1990). Ploidy, proliferative activity and prognosis. DNA flow cytometry of solid tumors. Cancer *65*:1194–1205.

Merry, S. et al. (1989). Inherent adriamycin resistance in a murine tumor line: circumvention with verapamil and norverapamil. Br. J. Cancer *59*:895–897.

Meyne, J. and Moyzis, R. K. (1989). Human chromosome-specific repetitive DNA probes: targeting *in situ* hybridization to chromosome 17 with a 42-base-pair alphoid DNA oligomer. Genomics *4*:472–478.

Michael, M. and Buckle, S. (1990). Screening for genetic disorders: therapeutic abortion and IVF. J. Med. Ethics *16*:43–47.

Mickisch, G. H. et al. (1991). Transgenic mice that express the human multidrug-resistance gene in bone marrow enable a rapid identification of agents that reverse drug resistance. Proc. Natl. Acad. Sci. U.S.A. *88*:547–551.

Miller, F. R. and Heppner, G. H. (1990). Cellular interactions in metastasis. Cancer Metastasis Rev. *9*:21–34.

Miller, M. A. et al. (1989). Quantification of mRNA in discrete cell groups of brain by *in situ* hybridization histochemistry. In: Gene Probes. Conn, P. M. (ed.). Academic Press, San Diego, pp. 164–182.

Miller, S. A. et al. (1988). A simple salting out procedure for extracting DNA from human nucleated cells. Nucleic Acids Res. *16*:1215.

Milner, J. and Medcalf, E. A. (1991). Cotranslation of activated mutant p53 with wild type drives the wild-type p53 protein into the mutant conformation. Cell *65*:765–774.

Minoshima, S. et al. (1990). Isolation of giant DNA fragments from flow sorted human chromosomes. Cytometry *11*:539–546.

Mitelman, F. and Heim, S. (1988). Consistent involvement of only 71 of the 329 chromosomal bands of the human genome in primary neoplasia-associated rearrangements. Cancer Res. *48*:7115–7119.

Mocharla, H. et al. (1990). Coupled reverse transcription-polymerase chain reaction (RT-PCR) as a sensitive and rapid method for isozyme genotyping. Gene *93*:271–275.

Molecular Biology of Aging. (1990). UCLA Symp. Conf. on Molecular Biology of Aging Sante Fe, NM, 1989, Wiley-Liss, New York.

Montandon, A. J. et al. (1989). Direct detection of point mutations by mismatch analysis: application to hemophilia B. Nucleic Acids Res. *17*:3347–3358.

Mori, T. et al. (1990). *In situ* (6-4) photoproduct determined by laser cytometry and autoradiography. Mutat. Res. *236*:99–105.

Moroz, S. P. et al. (1976). Liver disease associated with α-1-antitrypsin deficiency in childhood. J. Pediatr. *88*:19–25.

Morris, J. W. et al. (1989). Biostatistical evaluation of evidence from continuous allele frequency distribution deoxyribonucleic acid (DNA) probes in reference to disputed paternity and identity. J. Forensic Sci. *34*:1311–1317.

Morrissey, D. V. and Collins, M. L. (1989). Nucleic acid hybridization assays employing dA-tailed capture probes. Single capture methods. Mol. Cell. Probes *3*:189–207.

Morrissey, D. V. et al. (1989). Nucleic acid hybridization assays employing dA-tailed capture probes. I. Multiple capture methods. Anal. Biochem. *181*:345–359.

Motzer, R. J. et al. (1991). Genetic analysis as an aid in diagnosis for patients with midline carcinomas of uncertain histologies. J. Natl. Cancer Inst. *83*:341–346.

Moyzis, R. K. et al. (1987). Human chromosome-specific repetitive DNA sequences: novel markers for genetic analysis. Chromosoma *95*:375–386.

Moyzis, R. K. et al. (1988). A highly conserved repetitive DNA sequence, (TTAGGG)n present at the telomeres of human chromosomes. Proc. Natl. Acad. Sci. U.S.A. *85*:6622–6626.

Moyzis, R. K. et al. (1989). The distribution of interspersed repetitive DNA sequences in the human genome. Genomics *4*:273–289.

Mudd, J. L. et al. (1989). Guidelines for a quality assurance program for DNA restriction fragment length polymorphism analysis. Crime Lab. Dig. *16*:41–59.

Mueller, U. W. et al. (1990). Isolation of fetal trophoblast cells from peripheral blood of pregnant women. Lancet *336*:197–200.

Muhlegger, K. et al. (1990). Non-radioactive labelling and detection of nucleic acids. IV. Synthesis and properties of digoxigenin-modified 2′-deoxyuridine-5′-triphosphates and a photoactivatable analog of digoxigenin (Photodigoxigenin). Biol. Chem. Hoppe-Seyler *371*:953–965.

Mullaart, E. et al. (1990). DNA damage metabolism and aging. Mutat. Res. *237*:189–210.

Muller, O. et al. (1990). Detection of genomic alterations in carcinogen-induced mouse liver tumors by DNA fingerprinting analysis. Mol. Carcinog. *3*:330–334.

Mullink, H. et al. (1989). Two color DNA *in situ* hybridization for the detection of two viral genomes using non-radioactive probes. Histochemistry *91*:195–198.

Mullis, K. B. and Faloona, F. A. (1987). Specific synthesis of DNA *in vitro* via a polymerase-catalyzed chain reaction. Methods Enzymol. *155*:335–350.

Multhaupt, H. et al. (1989). Cellular localization of induced human interferon-β mRNA by non-radioactive *in situ* hybridization. Histochemistry *91*:315–319.

Mulvihill, J. J. (1989). Prospects for cancer control and prevention through genetics. Clin. Genet. *36*:313–319.

Murnane, J. P. and Painter, R. B. (1982). Complementation of the defects in DNA synthesis in irradiated and unirradiated ataxia-telangiectasia cells. Proc. Natl. Acad. Sci. U.S.A. *79*:1960–1963.

Murphy, L. D. et al. (1990). Use of the polymerase chain reaction of the quantiation of *mdr1* gene expression. Biochemistry *29*:10351–10356.

Murray, V. (1990). Are transposons a cause of aging? Mutat. Res. *237*:59–63.

Musk, S. R. R. and Steel, G. G. (1990). The inhibition of cellular recovery in human tumour cells by inhibitors of topoisomerase. Br. J. Cancer *62*:364–367.

Myerowitz, R. (1988). Splice junction mutation in some Ashkenazi Jews with Tay-Sachs disease: evidence against a single defect within this ethnic group. Proc. Natl. Acad. Sci. U.S.A. *85*:3955–3959.

Myerowitz, R. and Costigan, F. C. (1988). The major defect in Ashkenazi Jews with Tay-Sachs disease is an insertion in the gene for the α chain of β-hexosaminidase. J. Biol. Chem. *263*:18567–18569.

Myers, R. M. et al. (1985a). Detection of single base substitutions in total genomic DNA. Nature (London) *313*:495–498.

Myers, R. M. et al. (1985b). Detection of single base substitutions by ribonuclease cleavage at mismatches in RNA:DNA duplexes. Science *230*:1242–1248.

Myers, R. M. et al. (1987). Detection and localization of single base changes by denaturing gradient gel electrophoresis. Methods Enzymol. *155*:501–527.

Nagamine, C. M. et al. (1989). A PCR artifact: generation of heteroduplexes. Am. J. Hum. Genet. *45*:337–339.

Nagel, R. L. and Ranney, H. M. (1990). Genetic epidemiology of structural mutations of the β-globin gene. Semin. Hematol. *27*:342–359.

Nakagami, S. et al. (1991). Non-radioactive detection of nucleic acid by the universal probe system. Anal. Biochem. *192*:11–16.

Nakagome, Y. et al. (1990). Prenatal sex determination. Lancet *335*:291.

Nakamura, Y. et al. (1987). Variable number of tandem repeat (VNTR) markers for human gene mapping. Science *235*:1616–1622.

Nakane, P. K. et al. (1987). *In situ* localization of mRNA using thymine-thymine dimerized cDNA. Acta Histochem. Cytochem. *20*:220–243.

Nakao, Y. and Kozma, L. (1988). IgG heavy-chain markers in endocrine and metabolic diseases. In: Immunogenetics of Endocrine Disorders, Alan R. Liss, New York, pp. 461–494.

Nakase, H. et al. (1990). Transcription and translation of deleted mitochondrial genomes in Kearns-Sayre syndrome: implications for pathogenesis. Am. J. Hum. Genet. *46*:418–427.

Narayanswami, S. and Hamkalo, B. A. (1990). High resolution mapping of *Xenopus laevis* 5S and ribosomal RNA genes by EM *in situ* hybridization. Cytometry *11*:144–152.

Narayanswami, S. and Hamkalo, B. A. (1991). DNA sequence mapping using electron microscopy. GATA *8*:14–23.

National Technical Information Service. (1990). Understanding our genetic inheritance. The U.S. Human Genome Project: the first five years FY 1991–1995. National Technical Information Service, U.S. Dept. of Commerce, DOE/ER-0452 P.

Navon, R. and Proia, R. L. (1989). The mutations in Ashkenazi Jews with adult G_{m2} gangliosidosis, the adult form of Tay-Sachs disease. Science *243*:1471–1474.

Nederlof, P. M. et al. (1989). Three-color fluorescence *in situ* hybridization for the simultaneous detection of multiple nucleic acid sequences. Cytometry *10*:20–27.

Nederlof, P. M. et al. (1990). Multiple fluorescence *in situ* hybridization. Cytometry *11*:126–131.

Nelson, D. L. et al. (1989). *Alu* polymerase chain reaction: a method for rapid isolation of human-specific sequences from complex DNA sources. Proc. Natl. Acad. Sci. U.S.A. *86*:6686–6690.

Nelson, D. L. et al. (1990). Specific amplification of human DNA. In: Biotechnology and Human Predisposition to Disease. Wiley-Liss, New York, pp. 177–186.

Nelson, E. M. et al. (1984). Mechanism of antitumor drug action. Poisoning of mammalian DNA topoisomerase II on DNA by 4′-(9-acridinylamino)methane-sulfon-*m*-aniside. Proc. Natl. Acad. Sci. U.S.A. *81*:1361–1364.

Nelson, I. et al. (1989). Mapping of heteroplasmic mitochondrial DNA deletions in Kearns-Sayre syndrome. Nucleic Acids. Res. *17*:8117–8124.

Nelson, N. C. and Kacian, D. L. (1990). Chemoluminescent DNA probes: a comparison of the acridinium ester and dioxetane detection systems and their use in clinical diagnostic assays. Clin. Chem. Acta *194*:73–90.

Neuman, W. L. et al. (1990). DNA analysis of unilaterial twin ectopic gestation. Obstet. Gynecol. *75*:479–483.

Neville, A. M. (1986). Tumor markers and their clinical value. Tumor Biol. *7*:83–90.

Nevinny-Stickel, C. et al. (1991). Nonradioactive HLA class II typing using polymerase chain reaction and digoxigenin-11-2′,3′-dideoxyuridinetriphosphate-labeled oligonucleotide probes. Hum. Immunol. *31*:7–13.

Newton, C. R. et al. (1988). Diagnosis of α-1-antitrypsin deficiency by enzymatic amplification of human genomic DNA and direct sequencing of polymerase chain reaction products. Nucleic Acids Res. *16*:8233–8243.

Newton, C. R. et al. (1989). Analysis of any point mutation in DNA. The amplification refractory mutation system (ARMS). Nucleic Acids Res. *17*:2503–2516.

Ngo, K. Y. et al. (1986). A DNA probe detecting multiple haplotypes of the human Y chromosome. Am. J. Hum. Genet. *38*:407–418.

Niazi, M. et al. (1981). Trophoblast sampling in early pregnancy: culture of rapidly dividing cells from immature placental villi. Br. J. Obstet. Gynaecol. *88*:1081–1085.

Nichols, W. C. et al. (1989). Prenatal detection of a gene for hereditary amyloidosis. Am. J. Med. Genet. *34*:520–524.

Nickerson, D. A. et al. (1990). Automated DNA diagnostics using an ELISA-based oligonucleotide ligation assay. Proc. Natl. Acad. Sci. U.S.A. *87*:8923–8927.

Niimi, S. et al. (1991). Resistance to anticancer drugs in NIH 3T3 cells transfected with c-*myc* and/or c-H-*ras* genes. Br. J. Cancer *63*:237–241.

Noonan, K. E. et al. (1990). Quantitative analysis of MDR1 (multidrug resistance) gene expression in human tumors by polymerase chain reaction. Proc. Natl. Acad. Sci. U.S.A. *87*:7160–7164.

Norby, S. et al. (1991). Mutation detection in Leber's heriditary optic neuropathy by PCR with allele-specific priming. Biochem. Biophys. Res. Commun. *175*:631–636.

Northrup, H. et al. (1990). Prenatal diagnosis of citrullinaemia: review of a 10 year experience including recent use of DNA analysis. Prenatal Diagn. *10*:771–779.

Novack, D. F. et al. (1986). Detection of single base-pair mismatches in DNA by chemical modification followed by electrophoresis in 15% polyacrylamide gel. Proc. Natl. Acad. Sci. U.S.A. *83*:586–590.

Nove, J. et al. (1987). Radiation sensitivity of fibroblast strains from patients with Usher's syndrome, Duchenne muscular dystrophy, and Huntington's disease. Mutat. Res. *184*:29–38.

Novelli, G. et al. (1989). Prenatal diagnosis of adult polycystic kidney disease with DNA markers on chromosome 16 and the genetic heterogeneity problem. Prenatal Diagn. *9*:759–767.

Nowell, P. C. (1991). How many human cancer genes? J. Natl. Cancer Inst. *83*:1061–1064.

Nozari, G. et al. (1986). Discrimination among the transcripts of the allelic human β-globin genes, β-^{A}, β-^{S}, and β-^{C}, using oligodeoxynucleotide hybridization probes. Gene *43*:23–28.

Nukiwa, T. et al. (1986). Evaluation of "at risk" α-1-antitrypsin genotype SZ with synthetic oligonucleotide gene probe. J. Clin. Invest. *77*:528–537.

Nurnberg, P. et al. (1991). Coamplification of simple repetitive DNA fingerprint fragments and the EGFR gene in human gliomas. Genes Chromosomes Cancer *3*:79–88.

O'Connell, P. et al. (1987). A primary genetic linkage map for human chromosome 12. Genomics *1*:93–102.

Oakes, M. I. and Lake, J. A. (1990). DNA-hybridization electron microscopy, localization of five regions of 16S rRNA on the surface of 30S ribosomal subunits. J. Mol. Biol. *211*:897–906.

Oakes, M.I. et al. (1990). DNA-hybridization electron microscopy. Tertiary structure of 16S rRNA. J. Mol. Biol. *211*:907–918.

Oakey, R. and Tyler-Smith, C. (1990). Y chromosome DNA haplotyping suggests that most European and Asian men are descended from one of two males. Genomics *7*:325–330.

Oberlé, I. et al. (1991). Instability of a 550-base pair DNA segment and abnormal methylation in fragile X syndrome. Science *252*:1097–1102.

Odelberg, S. J. et al. (1988). Establishing paternity using minisatellite DNA probes when the putative father is unavailable for testing. J. Forensic Sci. *33*:921–928.

Oelmuller, U. et al. (1990). Isolation of prokaryotic RNA and detection of specific mRNA with biotinylated probes. J. Microbiol. Methods *11*:73–84.

Oestreicher, U. et al. (1990). Chromosome and SCE analysis in peripheral lymphocytes of persons occupationally exposed to cytostatic drugs handled with and without use of safety covers. Mutat. Res. *242*:271–277.

Office of Technology Assessment (1990). Genetic Witness. Forensic uses of DNA tests. Congress of the United States Office of Technology Assessment, 196 pp.

Old, J. M. et al. (1982). First trimester fetal diagnosis for hemoglobinopathies: three cases. Lancet *ii*:1413–1416.

Olive, P. L. et al. (1988). Comparison between the DNA precipitation and alkaline unwinding assays for detecting DNA strand breaks and cross-links. Cancer Res. *48*:6444–6449.

Olson, M. et al. (1989). A common language for physical mapping of the human genome. Science *245*:1434–1435.

Ono, T. et al. (1990). Comparison of age-associated changes of c-*myc* gene methylation in liver between man and mouse. Mutat. Res. *237*:239–246.

Orita, M. et al. (1989a). Detection of polymorphisms of human DNA by gel electrophoresis as single-strand conformation polymorphisms. Proc. Natl. Acad. Sci. U.S.A. *86*:2766–2779.

Orita, M. et al. (1989b). Rapid and sensitive detection of point mutations and DNA polymorphisms using the polymerase chain reaction. Genomics *5*:874–879.

Orkin, S. H. (1984). Prenatal diagnosis of hemoglobin disorder by DNA analysis. Blood *63*:249–253.

Orkin, S. H. (1987). Progress through amplification. N. Engl. J. Med. *317*:1023–1025.

Orkin, S. H. and Kazazian, H. H., Jr. (1984). The mutation and polymorphism of the human β-globin gene and its surrounding DNA. Annu. Rev. Genet. *18*:131–171.

Orkin, S. H. et al. (1982). Improved detection of the sickle mutation by DNA analysis: application to prenatal diagnosis. N. Engl. J. Med. *307*:32–36.

Orkin, S. H. et al. (1983). Direct detection of the common Mediterranean β-thalassemia gene with synthetic DNA probes: an alternate approach for prenatal diagnosis. J. Clin. Invest. *71*:775–779.

Orlic, D. (ed.) (1989). Molecular and cellular controls of hematopoiesis. Ann. N.Y. Acad. Sci. Vol. 554.

Orrego, C. and King, M. C. (1990). Determination of familial relationships. In: PCR protocols: A Guide to Methods and Applications. Innis, M. A. et al. (eds.). Academic Press, San Diego, pp. 416–426.

Ortigao, R. et al. (1989). 5-Bromouridinylated oligonucleotide for hybridization analysis of DNA and RNA on membranes and *in situ*. Nucleosides Nucleotides *8*:805–813.

Osawa, T. et al. (1988). Maternal inheritance of deleted mitochondrial DNA in a family with mitochondrial myopathy. Biochem. Biophys. Res. Commun. *154*:1240–1247.

Ostrer, H. (1989). Prenatal diagnosis of genetic disorders by DNA analysis. Pediatr. Ann. *18*:701–713.

Ottman, R. (1990). An epidemiologic approach to gene-environment interaction. Genet. Epidemiol *7*:177–185.

Oxtoby, M. J. (1988). Human immunodeficiency virus and other viruses in human milk: placing the issues in broader perspectives. Pediatr. Infect. Dis. J. *7*:825–835.

Pacifici, R. E. and Davies, K. J. A. (1991). Protein, lipid and DNA repair systems in oxidative stress: the free-radical theory of aging revisited. Gerontology *37*:166–180.

Pahwa, S. (1988). Human immunodeficiency virus infection in children: nature of immunodeficiency, clinical spectrum and management. Pediatr. Infect. Dis. J. *7*:S61–S71.

Painter, R. B. (1981). Radioresistant DNA synthesis: an intrinsic feature of ataxia telangiectasia. Mutat. Res. *84*:183–190.

Painter, R. B. and Young, B. R. (1980). Radiosensitivity in ataxia-telangiectasia; a new explanation. Proc. Natl. Acad. Sci. U.S.A. *77*:7315–7317.

Pakkala, S. et al. (1988). Difference in DNA-fingerprints between remission and relapse in childhood acute lymphoblastic leukemia. Leuk. Res. *12*:757–762.

Park, J. G. et al. (1990). Chemosensitivity patterns and expression of human multidrug resistance-associated MDR1 gene by human gastric and colorectal carcinoma cell lines. J. Natl. Cancer Inst. *82*:193–198.

Park, S. D. and Cleaver, J. E. (1979). Postreplication repair: question of its definition and possible alteration in xeroderma pigmentosum cell strains. Proc. Natl. Acad. Sci. U.S.A. *76*:3927–3931.

Parker, R. J. et al. (1991). Acquired cisplatin resistance in human ovarian cancer cells is associated with enhanced repair of cisplatin — DNA lesions and reduced drug accumulation. J. Clin. Invest. *87*:772–777.

Parkes, J. L. et al. (1990). Resistance of tumor-derived DNA to restriction enzyme digestion. Cancer Invest. *8*:169–172.

Parkinson, C. and Cheng, S. Y. (1989). A convenient method to increase the number of readable bases in DNA sequencing. BioTechniques *7*:828–829.

Parry, J. M. et al. (1990). Restriction site mutation analysis, a proposed methodology for the detection and study of DNA base changes following mutagen exposure. Mutagenesis *5*:209–212.

Parshad, R. et al. (1990). Carrier detection in xeroderma pigmentosum. J. Clin. Invest. *85*:135–138.

Pastan, I. et al. (1991). Molecular manipulations of the multidrug transporter: a new role for transgenic mice. FASEB J. *5*:2523–2528.

Pathak, S. and Goodacre, A. (1986). Specific chromosome anomalies and predisposition to human breast, renal cell and colorectal carcinoma. Cancer Genet. Cytogenet. *19*:29–36.

Patterson, S. et al. (1989). DNA probes bind non-specifically to eosinophils during *in situ* hybridization: carbol chromotrope blocks binding to eosinophils but does not inhibit hybridization to specific nucleotide sequences. J. Virol. Methods *23*:105–109.

Paw, B. H. et al. (1989). Molecular basis of adult-onset and chronic G_{m2} gangliosidosis in patients of Ashkenazi Jewish origin: substitution of serine for glycine at position 269 of the α-subunit of β-hexosaminidase. Proc. Natl. Acad. Sci. U.S.A. *86*:2413–2417.

Peake, I. R. and Bloom, A. L. (1986). Recombination between genes and closely linked polymorphisms. Lancet *i*:1335.

Pearse, M. J. and Wu, L. (1988). Preparation of both DNA and RNA for hybridization analysis from limiting quantities of lymphoid cells. Immunol. Lett. *18*:219–224.

Pearson, P. L. (1987). Recombinant DNA, chromosomes and prenatal diagnosis. Curr. Probl. Derm. *16*:30–44.

Peek, R. et al. (1991). DNA methylation as a regulatory mechanism in rat γ-crystallin gene expression. Nucleic Acids Res. *19*:77–83.

Pegg, A. E. (1984). Methylation of the O^6-position of guanine in DNA is the most likely initiating event in carcinogenesis by methylating agents. Cancer Invest. *2*:223–231.

Peltomaki, P. (1991). DNA methylation changes in human testicular cancer. Biochim. Biophys. Acta *1096*:187–196.

Penketh, R. J. A. et al. (1989). Rapid sexing of human embryos by non-radioactive *in situ* hybridization: potential for preimplantation diagnosis of X-linked disorders. Prenatal Diagn. *9*:489–500.

Perbal, B. (1988). A Practical Guide to Molecular Cloning, 2nd Ed. Interscience, New York.

Perez-Sala, D. et al. (1991). Methylation and demethylation reactions of guanine nucleotide-binding proteins of retinal rod outer segments. Proc. Natl. Acad. Sci. U.S.A. *88*:3043–3046.

Peschle, C. (ed.) (1987). Normal and neoplastic blood cells: from genes to therapy. Ann. N.Y. Acad. Sci. 511.

Peter, J. B. (1991). The polymerase chain reaction: amplifying our options. Rev. Infect. Dis. *13*:166–171.

Pierotti, M. A. and Della Porta, G. (1989). DNA probes in the characterization of human solid tumors. In: Molecular Probes: Technology and Medical Applications. Albertini, A. et al. (eds.). Raven Press, New York, pp. 215–225.

Pinkel, D. et al. (1988). Fluorescence *in situ* hybridization with human chromosome-specific libraries: detection of trisomy 21 and translocations of chromosome 4. Proc. Natl. Acad. Sci. U.S.A. *85*:9138–9142.

Piper, J. et al. (1990). Automatic detection of fragile X chromosomes using an X centromere probe. Cytometry *11*:73–79.

Pirastu, M. et al. (1983). Prenatal diagnosis of β-thalassemia. Detection of a single nucleotide mutation on DNA. N. Engl. J. Med. *309*:284–287.

Pirker, R. et al. (1991). MDR 1 gene expression and treatment outcome in acute myeloid leukemia. J. Natl. Cancer Inst. *83*:708–712.

Pitot, H. C. and Dragan, Y. P. (1991). Facts and theories concerning the mechanisms of carcinogenesis. FASEB J. *5*:2280–2286.

Pizzo, P. A. (1990). Pediatric AIDS: problems within problems. J. Infect. Dis. *161*:316–325.

Pluygers, E. P. et al. (1986). Tumor markers for cancer detection I. Cancer Detect. Prev. *9*:495–504.

Pocchiari, F. et al. (1979). Human health effects from accidental release of TCDD at Seveso, Italy. Ann. N.Y. Acad. Sci. *320*:311–320.

Pollack, M. et al. (1980). Prenatal identification of paternity HLA typing helpful after rapes. JAMA *244*:1954–1956.

Pollack, M. S. et al. (1981). Technical and theoretical considerations in the HLA typing of amniotic fluid cells for prenatal diagnosis and paternity testing. Prenatal Diagn. *1*:183–195.

Pollard-Knight, D. et al. (1990). Non-radioactive nucleic acid detection by enhanced chemoluminescence using probes directly labelled with horseradish peroxidase. Anal. Biochem. *185*:84–89.

Pollman, M. J. and Zuccarelli, A. J. (1989). Rapid isolation of high-molecular-weight DNA from agarose gels. Anal. Biochem. *189*:12–17.

Porter-Jordan, K. and Garrett, C. T. (1990). Source of contamination in polymerase chain reaction assay. Lancet *335*:1220.

Porter-Jordan, K. et al. (1990). Nested polymerase chain reaction assay for the detection of cytomegalovirus overcomes false positives due to contamination with fragmented DNA. J. Med. Virol. *30*:85–91.

Preston-Martin, S. et al. (1990). Increased cell division as a cause of human cancer. Cancer Res. *50*:7415–7421.

Pretorius, D. H. et al. (1987). Diagnosis of autosomal dominant polycystic kidney disease *in utero* and in the young infant. J. Ultrasound Med. *6*:249–255.

Price, J. O. et al. (1991). Prenatal diagnosis with fetal cells isolated from maternal blood by multiparameter flow cytometry. Am. J. Obstet. Gynecol. *165*:1731–1737.

Price, J. and Herman, C. J. (1990). Reproducibility of FCM DNA content from replicate paraffin block samples. Cytometry *11*:845–847.

Pringle, J. H. et al. (1989). *In situ* hybridization demonstration of poly-adenylated RNA sequences in formalin-fixed paraffin sections using a biotinylated oligonucleotide poly d(T) probe. J. Pathol. *158*:279–286.

Proudfoot, N. J. et al. (1982). The structure of the human zeta-globin gene and a closely linked nearly identical pseudogene. Cell (Cambridge, Mass) *31*:553–563.

Przepiorka, D. et al. (1991). Use of a probe to repeat sequence of the Y chromosome for detection of host cells in peripheral blood of bone marrow transplant recipients. Am. J. Clin. Pathol. *95*:201–206.

Puck, J. M. et al. (1989). X-linked severe combined immunodeficiency: localization within the region Xq13. 1-q21. 1 by linkage and deletion analysis. Am. J. Hum. Genet. *44*:724–730.

Puck, J. M. et al. (1990). Prenatal diagnosis for X-linked severe combined immunodeficiency by analysis of maternal X-chromosome inactivation and linkage analysis. N. Engl. J. Med. *322*:1063–1066.

Puppels, G. J. et al. (1990). Studying single living cells and chromosomes by confocal Raman microspectroscopy. Nature *347*:301–303.

Raber, M. N. and Barlogie, B. (1990). DNA flow cytometry of human solid tumors. In: Flow Cytometry and Sorting, 2nd Ed. Wiley-Liss, New York, pp. 745–754.

Raha, S. et al. (1990). Simultaneous isolation of total cellular RNA and DNA from tissue culture cells using phenol and lithium chloride. GATA *7*:173–177.

Raimondi, E. et al. (1989). Aneuploidy assays on interphase nuclei by means of *in situ* hybridization with DNA probes. Mutagenesis *4*:165–169.

Rand, K. H. and Houck, H. (1990). *Taq* polymerase contains bacterial DNA of unknown origin. Mol. Cell Probes *4*:445–450.

Rao, D. C. and Wette, R. (1990). Environmental index in genetic epidemiology: an investigation of its role, adequacy, and limitations. Am. J. Hum. Genet. *46*:168–178.

Raux-Demay, M. et al. (1989). Early prenatal diagnosis of 21-hydroxylase deficiency using amniotic fluid 17-hydroxyprogesterone determination and DNA probes. Prenatal Diagn. *9*:457–466.

Reeders, S. T. et al. (1985). A highly polymorphic DNA marker linked to adult polycystic kidney disease on chromosome 16. Nature (London) *317*:542–544.

Reeders, S. T. et al. (1986). Prenatal diagnosis of autosomal dominant polycystic kidney disease with a DNA probe. Lancet *ii*:6–7.

Reeders, S. T. et al. (1987). A study of linkage heterogeneity in adult polycystic kidney disease. Hum. Genet. *76*:348–351.

Reeders, S. T. et al. (1988). Regional localization of the autosomal dominant polycystic kidney disease locus. Genomics *3*:150–155.

Reihaus, E. et al. (1990). Regulation of the level of the oncoprotein p53 in non-transformed and transformed cells. Oncogene *5*:137–145.

Reisfeld, A. et al. (1987). Non-radioactive hybridization probes prepared by the reaction of biotin hydrazide with DNA. Biochem. Biophys. Res. Commun. *142*:519–526.

Reisner, E. G. et al. (1988). Tests of genetic markers on aborted fetal material. J. Forensic Sci. *33*:1262–1266.

Remvikos, Y. et al. (1991). Progesterone receptor detection and quantitation in breast tumors by bivariate immunofluorescence DNA flow cytometry. Cytometry *12*:157–166.

Renan, M. J. (1990). Cancer genes: current status, future prospects, and applications in radiotherapy/oncology. Radiother. Oncol. *19*:197–218.

Reymond, C. D. (1987). A rapid method for the preparation of multiple samples of eukaryotic DNA. Nucleic Acids Res. *15*:8118.

Ribeiro, E. A. et al. (1989). Quantitative fluorescence of DNA-intercalated ethidium bromide on agarose gels. Anal. Biochem. *181*:197–208.

Rich, A. et al. (1984). The chemistry and biology of left-handed Z-DNA. Annu. Rev. Biochem. *53*:791–846.

Richardson, B. et al. (1990). Evidence for impaired T-cell DNA methylation in systemic lupus erythematosus and rheumatoid arthritis. Arthritis Rheum. *33*:1665–1673.

Rickwood, D. and Hames, B. D. (eds.) (1982). Gel Electrophoresis of Nucleic Acids. A Practical Approach. IRL Press, Oxford.

Riesner, D. et al. (1989). Temperature-gradient gel electrophoresis of nucleic acids; analysis of conformational transitions, sequence variations, and protein-nucleic acid interactions. Electrophoresis *10*:377–389.

Riggs, A. D. (1989). DNA methylation and cell memory. Cell Biophys. *15*:1–13.

Rijken, A. et al. (1991). Diagnostic value of DNA analysis in effusions by flow cytometry and image analysis. Am. J. Clin. Pathol. *95*:6–12.

Riley, J. et al. (1989). Rapid determination of DNA concentration in multiple samples. Nucleic Acids Res. *17*:8383.

Riley, L. K. et al. (1986). A method for biotinylating oligonucleotide probes for use in molecular hybridizations. DNA *5*:333–337.

Riou, G. F. et al. (1990). Expression of multidrug-resistance (MDR1) gene in normal epithelia and in invasive carcinomas of the uterine cervix. J. Natl. Cancer Inst. *82*:1493–1496.

Risch, N. (1990a). Linkage strategies for genetically complex traits. I. Multilocus models. Am. J. Hum. Genet. *46*:222–228.

Risch, N. (1990b). Linkage strategies for genetically complex traits. II. The power of affected relative pairs. Am. J. Hum. Genet. *46*:229–241.

Risch, N. (1990c). Linkage strategies for genetically complex traits. III. The effect of marker polymorphism on analysis of affected relative pairs. Am. J. Hum. Genet. *46*:242–253.

Risch, N. J. and Devlin, B. (1992). On the probability of matching DNA fingerprints. Science *255*:717–720.

Ristaldi, M. S. et al. (1989). Prenatal diagnosis of β-thalassemia in Mediterranean populations by dot blot analysis with DNA amplification and allele specific oligonucleotide probes. Prenatal Diagn. *9*:629–638.

Robbins, J. H. (1987). Incorrect priority claim for the DNA — damage hypothesis. Arch. Neurol. *44*:579–583.

Robbins, J. H. (1989). A childhood neurodegeneration due to defective DNA repair: a novel concept of disease based on studies of xeroderma pigmentosum. J. Child. Neurol. *4*:143–146.

Robbins, J. H. et al. (1985). Parkinson's disease and Alzheimer's disease: hypersensitivity to X rays in cultured cell lines. J. Neurol. Neurosurg. Psychiatry. *48*:916–923.

Roberts, H. R. and Lozier, J. N. (1991). Other clotting factor deficiencies. In: Hematology. Basic Principles and Practice. Hoffman, R. et al. (eds.). Churchill Livingstone, New York, pp. 1332–1342.

Roberts, L. (1991). Fight erupts over DNA fingerprinting. Sciene *254*:1721–1723.

Rocker, I. and Laurence, K. M. (1981). An assessment of fetoscopy. In: Fetoscopy. Rocker, I. and Laurence, K. M. (eds.). Elsevier, Amsterdam, pp. 301–309.

Roewer, L. et al. (1991). Hybridization and polymerase chain reaction amplification of simple repeated DNA sequences for the analysis of forensic stains. Electrophoresis *12*:181–186.

Rogers, B. B. et al. (1990). Analysis of DNA in fresh and fixed tissue by the polymerase chain reaction. Am. J. Pathol. *136*:541–548.

Romeo, G. et al. (1988). A second genetic locus for autosomal dominant polycystic kidney disease. Lancet *ii*:8–10.

Roninson, I. B. et al. (1990). Molecular biology and diagnosis of multidrug resistance. Proc. Am. Assn. Cancer Res. *31*:518–519.

Rosatelli, M. C. et al. (1990). Reliability of prenatal diagnosis of genetic diseases by analysis of amplified trophoblast DNA. J. Med. Genet. *27*:249–251.

Rose, E. A. (1991). Application of the polymerase chain reaction to genome analysis. FASEB J. *5*:46–54.

Rosenbaum, V. and Riesner, D. (1987). Temperature-gradient gel electrophoresis. Thermodynamic analysis of nucleic acids and proteins in purified form and in cellular extracts. Biophys. Chem. *26*:235–246.

Rossiter, B. J. F. and Caskey, C. T. (1990). Molecular scanning methods of mutation detection. J. Biol. Chem. *265*:12753–12756.

Rossiter, B. J. F. and Caskey, C. T. (1991). Molecular studies of human genetic disease. FASEB J. *5*:21–27.

Rotello, R. J. et al. (1991). Coordinated regulation of apoptosis and cell proliferation by transforming growth factor β 1 in cultured uterine epithelial cells. Proc. Natl. Acad. Sci. U.S.A. *88*:3412–3415.

Roux, K. H. and Dhanarajan, P. (1990). A strategy for single site PCR amplification of dsDNA: priming digested cloned or genomic DNA from an anchor-modified restriction site and a short internal sequence. BioTechniques *8*:48–57.

Rowland, L. P. et al. (eds.) (1989). Molecular Genetics in Diseases of Brain, Nerve and Muscle. Oxford University Press, New York and Oxford.

Rowley, P. T. et al. (1991a). Prenatal screening for hemoglobinopathies. I. A prospective regional trial. Am. J. Hum. Genet. *48*:439–446.

Rowley, P. T. et al. (1991b). Prenatal screening for hemoglobinopathies. IIII. Applicability of the health belief model. Am. J. Hum. Genet. *48*:452–459.

Royle, N. J. et al. (1988). Clustering of hypervariable minisatellites in the proterminal regions of human autosomes. Genomics *3*:352–360.

Rozen, R. et al. (1985). Gene deletion and restriction fragment length polymorphisms at the human ornithine transcarbamoylase locus. Nature (London) *313*:815–817.

Ruano, G. and Kidd, K. K. (1989). Direct haplotyping of chromosomal segments from multiple heterozygotes via allele-specific PCR amplification. Mutations by allele specific primers. Nucleic Acids Res. *17*:8392.

Ruano, G. and Kidd, K. K. (1991). Coupled amplification and sequencing of genomic DNA. Proc. Natl. Acad. Sci. U.S.A. *88*:2815–2819.

Ruano, G. et al. (1989). Biphasic amplification of very dilute DNA samples via "booster" PCR. Nucleic Acids Res. *17*:5407.

Ruano, G. et al. (1990). Haplotype of multiple polymorphisms resolved by enzymatic amplification of single DNA molecules. Proc. Natl. Acad. Sci. U.S.A. *87*:6296–6300.

Ruiters, M. H. J. et al. (1991). Confocal scanning laser microscopy of mitochondria: a possible tool in the diagnosis of mitochondrial disorders. J. Inherited Metab. Dis. *14*:45–48.

Ruoslahti, E. (1991). Integrins. J. Clin. Invest. *87*:1–5.

Rupp, G. M. and Locker, J. (1988). Purification and analysis of RNA from paraffin-embedded tissues. BioTechniques *6*:56–60.

Ryskov, A. P. et al. (1988). M13 phage DNA as a universal marker for DNA fingerprinting of animals, plants and microorganisms. FEBS Lett. *233*:388–392.

Sager, R. (1989). Tumor suppressor genes: the puzzle and the promise. Science *246*:1406–1412.

Sahashi, K. et al. (1990). Cytoplasmic body and mitochondrial DNA deletion. J. Neurol. Sci. *99*:291–300.

Saiki, R. K. et al. (1985). Enzymatic amplification of beta-globin genomic sequences and restriction site analysis for diagnosis of sickle cell anemia. Science *230*:1350–1354.

Saiki, R. K. et al. (1986). Analysis of enzymatically amplified beta globin and HLA-DQ α DNA with allele-specific oligonucleotide probes. Nature (London) *324*:163–166.

Saiki, R. K. et al. (1988a). Primer-directed emzymatic amplification of DNA with a thermostable DNA polymerase. Science *239*:487–491.

Saiki, R. K. et al. (1988b). Diagnosis of sickle cell anemia and β-thalassemia with enzymatically amplified DNA and non-radioactive allele-specific oligonucleotide probes. N. Engl. J. Med. *319*:537–541.

Saiki, R. K. et al. (1989). Genetic analysis of amplified DNA with immobilized sequence specific oligonucleotide probes. Proc. Natl. Acad. Sci. U.S.A. *86*:6230–6234.

Saluz, H. and Jost, J. P. (1989). A simple high-resolution procedure to study DNA methylation and *in vivo* DNA-protein interactions in a single-copy gene level in higher eukaryotes. Proc. Natl. Acad. Sci. U.S.A. *86*:2602–2606.

Sambhara, S.R. and Miller, R. G. (1991). Programmed cell death of T cells signaled by the T cell receptor and the α 3 domain of class I MHC. Science *252*: 1424–1427.

Sambrook, J. et al. (1989). Molecular Cloning: A Laboratory Manual. 2nd Ed. Cold Spring Harbor Laboratory Press, Cold Spring Harbor, NY.

Sandberg, A. A. and Turc-Carel, C. (1987). The cytogenetics of solid tumors. Relation to diagnosis, classification and pathology. Cancer *59*:387–395.

Sandberg, A. A. et al. (1988). Chromosomes in solid tumors and beyond. Cancer Res. *48*:1049–1059.

Sands, T. W. et al. (1990). A computer program to assist in the choice of restriction endonuclease for use in DNA analyses. Int. J. Biomed. Comput. *25*:39–52.

Saperstein, D. A. and Nickerson, J. M. (1991). Restriction fragment length polymorphism analysis using PCR coupled to restriction digests. BioTechniques *10*:488–489.

Sapienza, C. (1991). Genome imprinting and carcinogenesis. Biochim. Biophys. Acta *1072*:51–61.

Sarkar, G. and Sommer, S. (1990a). More light on PCR contamination [letter]. Nature (London) *347*:340–341.

Sarkar, G. and Sommer, S. (1990b). Shedding light on PCR contamination [letter]. Nature (London) *343*:27.

Sarkar, G. and Sommer, S. S. (1991). Haplotyping by double PCR amplification of specific alleles. BioTechniques *10*:436–440.

Sarkar, G. et al. (1990a). Characterization of polymerase chain reaction amplification of specific alleles. Anal. Biochem. *186*:64–68.

Sarkar, G. et al. (1990b). Formamide can dramatically improve the specificity of PCR. Nucleic Acids Res. *18*:7465.

Sasaki, M. S. and Tonomura, A. (1973). A high susceptibility of Franconi's anemia to chromosome breakage by DNA cross-linking agents. Cancer Res. *33*:1829–1836.

Sasi, R. et al. (1991). Prenatal sexing and detection of ZFY gene sequences in sex chromosome disorders by polymerase chain reaction. J. Clin. Lab. Analys. *5*:193–196.

Sato, Y. et al. (1991). The AMeX method: a multipurpose tissue-processing and paraffin embedding. III. Extraction and purification of RNA and application to slot-blot hybridization analysis. J. Pathol. *163*:81–85.

Satoh, H. et al. (1991). Introduction of new genetic markers on human chromosomes. Exp. Cell Res. *193*:5–11.

Saul, R. L. and Ames, B. N. (1986). Background levels of DNA damage in the population. In: Mechanisms of DNA Damage and Repair: Implications for Carcinogenesis and Risk Assessment. Simic, M. G. et al. (eds.). Plenum, New York, pp. 529–536.

Saunders, R. D. C. et al. (1989). PCR amplification of DNA microdissection from a single polytene chromosome band: a comparison with conventional microcloning. Nucleic Acids Res. *17*:9027–9037.

Sayler, G. S. and Layton, A. C. (1990). Environmental application of nucleic acid hybridization. Annu. Rev. Microbiol. *44*:625–648.

Scanlon, K. J. and Kashani-Sabet, M. (1989). Utility of the polymerase chain reaction in detection of gene expression in drug-resistant human tumors. J. Clin. Lab. Anal. *3*:323–329.

Scanlon, K. J. et al. (1988). Mechanisms for cisplatin-FVra synergism and cisplatin resistance in human ovarian carcinoma cells both *in vitro* and *in vivo*. In: Advances in Experimental Medicine and Biology, Vol. 244. Rustum, Y. and McGuire, J. (eds.). Plenum Press, New York, pp. 127–135.

Scanlon, K. J. et al. (1989a). Molecular basis of cisplatin resistance in human carcinomas: model systems and patients. Anticancer Res. *9*:1301–1312.

Scanlon, K. J. et al. (1989b). Differential gene expression in human cancer cells resistant to cisplatin. Cancer Invest. *7*:563–569.

Schafer, R. et al. (1988). DNA fingerprinting using non-radioactive oligonucleotide probes specific for simple repeats. Nucleic Acids Res. *16*:9344.

Schapira, A. H. V. et al. (1990a). Mitochondrial DNA analysis in Parkinson's disease. Movement Disorders *5*:294–297.

Schapira, A. H. V. et al. (1990b). Mitochondrial myopathy with a defect of mitochondrial-protein transport. N. Engl. J. Med. *323*:37–42.

Schimmelpenning, H. et al. (1990). Variations in Feulgen stainability of epithelial parenchymal cells extracted from paraffin-embedded salivary gland specimens. Cytometry *11*:475–480.

Schmid, I. et al. (1991). A gentle fixation and permeabilization method for combined cell surface and intracellular staining with improved precision in DNA quantification. Cytometry *12*:279–285.

Schochetman, G. et al. (1989). Serodiagnosis of infection with the AIDS virus and other human retroviruses. Annu. Rev. Microbiol. *49*:629–659.

Schulte, E. and Wittekind, D. (1989). Standardization of the Feulgen-Schiff technique. Staining characteristics of pure fuchsin dyes; a cytophotometric investigation. Histochemistry *91*:321–331.

Schwaiger, H. and Hirsch-Kauffmann, M. (1986). DNA repair in human cells: in Cockayne syndrome cells rejoining of DNA strands is impaired. Eur. J. Cell Biol. *41*:352–355.

Schwartz, E. I. et al. (1990). Polymerase chain reaction amplification from dried blood spots on Guthrie cards. Lancet *336*:639–640.

Schwartz, J. L. and Vaughan, A. T. M. (1989). Association among DNA/chromosome break rejoining rates, chromatin structure alterations, and radiation sensitivity in human tumor cell lines. Cancer Res. *49*:5054–5057.

Schwartz, M. K. (1990). Tumor markers: What is their role? Cancer Invest. *8*:439–440.

Schwartz, M. et al. (1989). Prenatal diagnosis of α-1-antitrypsin deficiency using polymerase chain reaction (PCR). Comparison of conventional RFLP methods with PCR used in combination with allele specific oligonucleotides or RFLP analysis. Clin. Genet. *36*:419–426.

Scott, A. F. et al. (1987). Origin of the human L1 elements: proposed progenitor genes deduced from a consensus DNA sequence. Genomics *1*:113–125.

Scrable, H. et al. (1989). A model for embryonal rhabdomyosarcoma tumorigenesis that involves genome imprinting. Proc. Natl. Acad. Sci. U.S.A. *86*:7480–7484.

Scrable, H. J. et al. (1990). Genetic and epigenetic losses of heterozygosity in cancer predisposition and progression. Adv. Cancer Res. *54*:25–62.

Scudiero, D. A. et al. (1981). Hypersensitivity to *N*-methyl-*N'*-nitro-*N*-nitrosoguanidine in fibroblasts from patients with Huntington disease, familial dysautonomia, and other primary neuronal degenerations. Proc. Natl. Acad. Sci. U.S.A. *78*:6451–6455.

Scudiero, D. A. et al. (1986). Alzheimer disease fibroblasts are hypersensitive to the lethal effects of a DNA-damaging chemical. Mutat. Res. *159*:125–131.

Scully, S. P. et al. (1990). The use of polymerase chain reaction generated nucleotide sequences as probes for hybridization. Mol. Cell. Probes *4*:485–495.

Seal, G. et al. (1988). Immunological lesions in human uracil DNA glycosylase: association with Bloom syndrome. Proc. Natl. Acad. Sci. U.S.A. *85*:2339–2343.

Seibel, N. L. et al. (1988). Detection of genes of interest in tissues and cells by *in situ* hybridization. J. Virol. Methods *21*:171–177.

Seibl, R. et al. (1990). Non-radioactive labelling and detection of nucleic acids. III. Applications of the digoxigenin system. Biol. Chem. Hoppe-Seyler *371*:939–951.

Selker, E. U. (1990). DNA methylation and chromatin structure: a view from below. Trends Biochem. *15*:103–107.

Septak, M. (1989). Acridinium ester-labelled DNA oligonucleotide probes. J. Biolumin. Chemilumin. *4*:351–356.

Setlow, R. B. et al. (1969). Evidence that xeroderma pigmentosum cells do not perform the first step in the repair of ultraviolet damage to their DNA. Proc. Natl. Acad. Sci. U.S.A. *64*:1035–1041.

Shackney, S. E. et al. (1990). Discrepancies between flow cytometric and cytogenetic studies in the detection of aneuploidy in human solid tumors. Cytometry *11*:94–104.

Shan, B. et al. (1990). Interferon selectively inhibits the expression of mitochondrial genes: a novel pathway for interferon-mediated responses. EMBO J. *9*:4307–4314.

Sharp, H. L. et al. (1969). Cirrhosis associated with α-1-antitrypsin deficiency: a previously unrecognized inherited disorder. J. Lab. Clin. Med. *73*:934.

Shay, J. W. et al. (1991). Defining the molecular mechanisms of human cell immortalization. Biochim. Biophys. Acta *1072*:1–7.

Sheffield, V. C. et al. (1989). Attachment of a 40 base pair G+C rich sequence (GC-clamp) to genomic DNA fragments by the polymerase chain reaction results in improved detection of single-base changes. Proc. Natl. Acad. Sci. U.S.A. *86*:232–236.

Shenk, T. E. et al. (1975). Biochemical method for mapping mutational alterations in DNA with SI nuclease: the location of deletions and temperature-sensitive mutations in simian virus 40. Proc. Natl. Acad. Sci. U.S.A. *72*:989–993.

Shoubridge, E. A. et al. (1990). Deletion mutants are functionally dominant over wild-type mitochondrial genomes in skeletal muscle fiber segments in mitochondrial disease. Cell (Cambridge, Mass) *62*:43–49.

Shroyer, K. P. and Nakane, P. K. (1982). Use of DNP-labelled cDNA for *in situ* hybridization. J. Cell Biol. *97*:337A.

Shuldiner, A. R. et al. (1991). Ligase-free subcloning: a versatile method to subclone polymerase chain reaction (PCR) products in a single day. Anal. Biochem. *194*:9–15.

Sibbit, W. L. Jr. (1991). Oncogenes, growth factors, and autoimmune diseases (review). Anticancer Res. *11*:97–114.

Siegel, R. E. (1989). Localization of neuronal mRNAs by hybridization histochemistry. In: Methods in Neurosciences, Vol. 1. Conn, P. M. (ed.). Academic Press, San Diego, pp. 136–149.

Silva, A. J. et al. (1987). Characterization of a highly polymorphic region 5′ to J_H in the human immunoglobulin heavy chain. Nucleic Acids Res. *15*:3845–3857.

Simmons, D. M. et al. (1989). A complete protocol for *in situ* hybridization of messenger RNAs in brain and other tissues with radiolabelled single-stranded RNA probes. J. Histotechnol. *12*:169–181.

Sinclair, A. H. et al. (1990). A gene from the human sex-determining region encodes a protein with homology to a conserved DNA-binding motif. Nature (London) *346*:240–244.

Singer, M. F. and Skowronski, J. (1985). Making sense out of LINES: long interspersed repeat sequences in mammalian genomes. Trends Biochem. Sci. *10*:119–122.

Singer, R. H. et al. (1986). Optimization of *in situ* hybridization using isotopic and non-isotopic detection methods. BioTechniques *4*:230–249.

Singer, R. H. et al. (1987). Double labelling *in situ* hybridization using non-isotopic and isotopic detection. Acta Histochem. Cytochem. *20*:589–599.

Singh, G. et al. (1989). A mitochondrial DNA mutation as a cause of Leber's hereditary optic neuropathy. N. Engl. J. Med. *320*:1300–1305.

Singh, N. P. et al. (1988). A simple technique for quantitation of low levels of DNA damage in individual cells. Exp. Cell Res. *175*:184–191.

Singh, N. P. et al. (1990). DNA damage and repair with age in individual human lymphocytes. Mutat. Res. *237*:123–130.

Sinnett, D. et al. (1990). Alumorphs — human DNA polymorphisms detected by polymerase chain reaction using *Alu*-specific primers. Genomics *7*:331–334.

Skoda, R. C. et al. (1988). Two mutant alleles of the human cytochrome P450 *db1* gene (P450 11 D1) associated with genetically deficient metabolism of debrisoquine and other drugs. Proc. Natl. Acad. Sci. U.S.A. *85*:5240–5243.

Skolnick, M. H. and Wallace, T. B. (1988). Simultaneous analysis of multiple polymorphic loci using amplified sequence polymorphisms (ASPs). Genomics *2*:273–279.

Slagboom, P. E. (1990). Meeting Report. The aging genome: determinant or target? Report of the EURAGE Meeting on "Genome instability and Aging", Nerja (Spain), 6–8 October, 1989. Mutat. Res. *237*:183–187.

Slater, G. W. and Noolandi, J. (1989). The biased reptation model of DNA gel electrophoresis: mobility vs molecular size and gel concentration. Biopolymers *28*:1781–1791.

Slater, G. W. et al. (1989). DNA gel electrophoresis: effect of field intensity and agarose concentration on band inversion. Biopolymers *28*:1793–1799.

Sloan, D. A. et al. (1990). The accuracy of fine needle aspiration biopsy for flow cytometric determination of tumor DNA content. J. Surg. Res. *49*:1–5.

Smart, R. D. et al. (1989). Confirmation of a balanced chromosomal translocation using molecular techniques. Prenatal Diagn. *9*:505–513.

Smeets, D. F. C. M. et al. (1992). Prader-Willi syndrome and Angelman syndrome in cousins from a family with a translocation between chromosomes 6 and 15. N. Engl. J. Med. *326*:807–811.

Smit, V. T. H. B. M. et al. (1990). Combined GTG-banding and non-radioactive *in situ* hybridization improves characterization of complex karyotypes. (With 2 color plates.) Cytogenet. Cell Genet. *54*:20–23.

Smith, J. C. et al. (1990). Highly polymorphic minisatellite DNA probes. Further evaluation for individual identification and paternity testing. J. Forensic Sci. Soc. *30*:3–18.

Smith, R. A. et al. (1989). Genetic marker analysis in cases of disputed paternity when the alleged father is deceased. Ann. Clin. Lab. Sci. *19*:332–336.

Smouse, P. E. and Chakraborty, R. (1986). The use of restriction fragment length polymorphisms in paternity analysis. Am. J. Hum. Genet. *38*:918–939.

Sobel, M. E. (1990). Metastasis suppressor genes. J. Natl. Cancer Inst. *82*:267–276.

Sogo, J. M. and Thoma, F. (1989). Electron microscopy of chromatin. Methods Enzymol. *170*:142–165.

Soh, J. and Pestka, S. (1990). Hybrid selection of mRNA with biotinylated DNA. GATA 7:80–86.

Sohal, R. S. and Allen, R. G. (1990). Oxidative stress as a causal factor in differentiation and aging: a unifying hypothesis. Exp. Gerontol. *25*:499–522.

Sommer, S. S. et al. (1989). A novel method for detecting point mutations or polymorphisms and its application to population screening for carriers of phenylketonuria. Mayo Clin. Proc. *64*:1361–1372.

Sorbello, A. F. et al. (1988). Rapid detection of cytometalovirus by fluorescent monoclonal antibody staining and *in situ* DNA hybridization in a dram vial cell culture system. J. Clin. Microbiol. *26*:1111–1114.

Sorg, C. (ed.) (1990). Molecular biology of B cell developments. S. Karger, Basel.

Southern, E. M. (1975). Detection of specific sequences among DNA fragments separated by gel electrophoresis. J. Mol. Biol. *98*:503–527.

Sparkes, R. S. (1990). Gene-mapping techniques. In: Molecular Neurobiological Techniques. Boulton, A. A. et al. (eds.). Humana Press, Clifton, NJ, pp. 153–162.

Spector, B. D. et al. (1982). Epidemiology of cancer in ataxia-telangiectasia. In: Ataxia-Telangiectasia — A Cellular and Molecular Link between Cancer, Neuropathology and Immune Deficiency. Bridges, B. A. and Harnden, D. G. (eds.). John Wiley & Sons, New York, pp. 103–138.

Spence, J. E. et al. (1989). Prenatal diagnosis and heterozygote detection by DNA analysis in ornithine transcarbamylase deficiency. J. Pediatr. *114*:582–588.

Srivastava, S. et al. (1990). Germ-line transmission of a mutated p53 gene in a cancer-prone family with Li-Fraumeni syndrome. Nature *348*:747–749.

Stagno, S. et al. (1982). Congenital cytomegalovirus infection. The relative importance of primary and recurrent maternal infection. N. Engl. J. Med. *306*:945–949.

Stallings, R. L. et al. (1990). Physical mapping of human chromosomes by repetitive sequence fingerprinting. Proc. Natl. Acad. Sci. U.S.A. *87*:6218–6222.

Stalvey, J. R. D. and Erickson, R. P. (1987). An improved method for detecting Y chromosomal DNA. Hum. Genet. *76*:240–243.

Stamato, T. D. and Denko, N. (1990). Asymmetric field inversion gel electrophoresis: a new method for detecting DNA double strand breaks in mammalian cells. Radiat. Res. *121*:196–205.

Stanbridge, E. J. (1990). Human tumor suppressor genes. Annu. Rev. Genet. *24*:615–657.

Stanbridge, E. J. and Nowell, P. C. (1990). Origins of human cancer revisited. Cell (Cambridge, Mass) *63*:867–874.

Stanta, G. and Schneider, C. (1991). RNA extracted from paraffin-embedded human tissues is amenable to analysis by PCR amplification. BioTechniques *11*:304–308.

Steeg, P. S. and Liotta, L. A. (1990). Reduced nm23 expression in tumor metastasis. Proc. Am. Assn. Cancer Res. *31*:504–505.

Steeg, P. S. et al. (1988). Evidence for a novel gene associated with low tumor metastatic potential. J. Natl. Cancer Inst. *80*:200–204.

Steel, C. M. (1990). Tracing disease genes through family studies. J. Pathol. *162*:7–13.

Steiner, I. et al. (1989). Effects of lipofuscin on *in situ* hybridization in human neuronal tissue. J. Virol. Methods *24*:1–10.

Steinmeyer, K. et al. (1990). Cell cycle control by p53 in normal (3T3) and chemically transformed (Meth A) mouse cells. I. Regulation of p53 expression. Oncogene *5*:1691–1699.

Stewart, G. et al. (1988). Transverse alternating field electrophoresis (TAFE). BioTechniques *6*:68–73.

Stoflet, E. S. et al. (1988). Genomic amplification with transcript sequencing. Science *239*:491–494.

Stokke, T. et al. (1991). Simultaneous assessment of chromatin structure, DNA content, and antigen expression by dual wavelength excitation flow cytometry. Cytometry *12*:172–178.

Stoler, M. H. and Ratliff, N. B. (1990). Potential and problems of the *in situ* molecular detection of viral genomes. Am. J. Clin. Pathol. *93*:714–716.

Stollar, B. D. and Rashtchian, A. (1987). Immunochemical approaches to gene probe assays. Anal. Biochem. *161*:387–394.

Stolz, W. et al. (1989). An automatic analysis method for *in situ* hybridization using high-resolution image analysis. Arch. Dermatol. Res. *281*:336–341.

Straus, D. and Ausubel, F. M. (1990). Genomic subtraction for cloning DNA corresponding to deletion mutations. Proc. Natl. Acad. Sci. U.S.A. *87*:1889–1893.

Strehler, B. L. (1986). Genetic instability as the primary cause of human aging. Exp. Gerontol. *21*:283–319.

Strobl, J. S. (1990). A role for DNA methylation in vertebrate gene expression? Mol. Endocrinol. *4*:181–183.

Summers, C. et al. (1992). Donor-recipient bone-marrow matching by single strand conformation polymorphism analysis. Lancet *339*:621.

Sutherland, G. R. and Simmers, R. N. (1988). No statistical association between common fragile sites and non-random chromosome breakpoints in cancer cells. Cancer Genet. Cytogenet. *31*:9–15.

Suzuki, Y. et al. (1991). Allele-specific polymerase chain reaction: a method for amplification and sequence determination of a single component among a mixture of sequence variants. Anal. Biochem. *192*:82–84.

Sved, J. and Bird, A. (1990). The expected equilibrium of the CpG dinucleotide in vertebrate genomes under a mutation model. Proc. Natl. Acad. Sci. U.S.A. *87*:4692–4696.

Sveger, T. (1976). Liver disease in α-1-antitrypsin deficiency detected by screening of 200,000 infants. N. Engl. J. Med. *294*:1316–1321.

Sveger, T. and Thelin, T. (1981). Four-year-old children with α-1-antitrypsin deficiency. Acta Pediatr. Scand. *70*:171–177.

Sverdlov, E. D. et al. (1974). Modification of cytidine residues with a bisulfite-*o*-methyl hydroxylamine mixture. Biochem. Biophys. Acta *340*:153–165.

Swerdlow, H. and Gesteland R. (1990). Capillary gel electrophoresis for rapid, high resolution DNA sequencing. Nucleic Acids Res. *18*:1415–1419.

Swerdlow, H. et al. (1990). Capillary gel electrophoresis for DNA sequencing. Laser-induced fluorescence detection with the sheath flow cuvette. J. Chromatogr. *516*:61–67.

Swift, M. et al. (1990). Effective testing of gene-disease associations. Am. J. Hum. Genet. *47*:266–274.

Swisshelm, K. et al. (1990). Age-related increase in methylation of ribosomal genes and inactivation of chromosome-specific rRNA gene clusters in mouse. Mutat. Res. *237*:131–146.

Syvanen, A. C. et al. (1988). Quantification of polymerase chain reaction products by affinity-based hybrid collection. Nucleic Acids Res. *16*:11327–11338.

Tabibzadeh, S. et al. (1991). Generation of nonradioactive bromodeoxyuridine labeled DNA probes by polymerase chain reaction. Nucleic Acids Res. *19*:2783.

Takahashi, E. et al. (1989). Human type II collagen gene (COL 2A1) assigned to chromosome 12q13. 1-q13. 2 by *in situ* hybridization with biotinylated DNA probe. Jpn. J. Hum. Genet. *34*:307–311.

Takahashi, E. et al. (1990). R-banding and nonisotopic *in situ* hybridization: precise localization of the human type II collagen gene (COL 2A1). Hum. Genet. *86*:14–16.

Takano, Y. S. et al. (1991). Apoptosis induced by mild hyperthermia in human and murine tumor cell lines: a study using electron microscopy and DNA gel electrophoresis. J. Pathol. *163*:329–336.

Tan, B. H. et al. (1990). Steady-state levels of 7-methylguanine increase in nuclear DNA of postmitotic mouse tissues during aging. Mutat. Res. *237*:229–238

Tan, K. B. et al. (1989). Nonproductive rearrangement of DNA topoisomerase I and II genes: correlation with resistance to topoisomerase inhibitors. J. Natl. Cancer Inst. *81*:1732–1735.

Tanaka, K. et al. (1989). Molecular cloning of a mouse DNA repair gene that complements the defect of group-A xeroderma pigmentosum. Proc. Natl. Acad. Sci. U.S.A. *86*:5512–5516.

Tanaka, K. et al. (1990). Analysis of a human DNA excision repair gene involved in group A xeroderma pigmentosum and containing a zinc-finger domain. Nature (London) *348*:73–76.

Tanaka, M. et al. (1990). Mitochondrial mutation in fatal infantile cardiomyopathy. Lancet *ii*:1452.

Taneja, K. and Singer, R. H. (1987). Use of oligodeoxynucleotide probes for quantitative *in situ* hybridization to action mRNA. Anal. Biochem. *166*:389–398.

Tang, F. J. et al. (1989). Simultaneous *in situ* detection of β-interferon mRNA and DNA in the same cell. J. Histochem. Cytochem. *37*:697–701.

Tarleton, J. and Schwartz, C. E. (1991). Using the polymerase chain reaction to maintain DNA probe inventories in clinical and diagnostic laboratories. Clin. Genet. *39*:121–124.

Tawa, A. et al. (1990). Increased expression of the multidrug-resistance gene in undifferentiated sarcoma. Cancer *66*:1980–1983.

Taylor, S. A. et al. (1990). Comparative study of automated versus manual extraction of DNA from clinical specimens. Am. J. Clin. Pathol. *93*:749–753.

Tchen, P. et al. (1984). Chemically modified nucleic acids as immunodetectable probes in hybridization experiments. Proc. Natl. Acad. Sci. U.S.A. *81*:3466–3470.

Tenhunen, J. (1989). Hydrolysis of single-stranded RNA in aqueous solutions — effect on quantitative hybridizations. Mol. Cell. Probes *3*:391–396.

Terenghi, G. et al. (1987). Localization of neuropeptide Y mRNA in neurons of human cerebral cortex by means of *in situ* hybridization with a complementing RNA probe. Proc. Natl. Acad. Sci. U.S.A. *84*:7315–7318.

Thein, S. L. and Wainscoat, J. S. (1986). The clinical applications of DNA polymorphisms. Dis. Markers *4*:203–217.

Thein, S. L. et al. (1985). Feasibility of prenatal diagnosis of β-thalassemia with synthetic DNA probes in two Mediterranean populations. Lancet *ii*:345–347.

Thein, S. L. et al. (1987). Detection of somatic changes in human cancer DNA by DNA fingerprint analysis. Br. J. Cancer *55*:353–356.

Thein, S. L. et al. (1990). The molecular basis of β-thalassemia in Thailand: application to prenatal diagnosis. Am. J. Hum. Genet. *47*:369–375.

Theune, S. et al. (1991). PCR primers for human chromosomes: reagents for the rapid analysis of somatic cell hybrids. Genomics *9*:511–516.

Thiele, C. J. (1990). Pediatric peripheral neuroectodermal tumors, oncogenes, and differentiation. Cancer Invest. *8*:629–639.

Thiele, D. (1990). The technique of polymerase chain reaction — a new diagnostic tool in microbiology and other scientific fields (review). Zentralbl. Bakt. *273*:431–454.

Thiry, M. et al. (1991). Localization of nucleolar chromatin by immunocytochemistry and *in situ* hybridization at the electron microscopic level. Electron Microsc. Rev. *4*:85–110.

Tho, S. P. T. et al. (1987). Use of single (4B-2) and repetitive copy (pS4) deoxyribonucleic acid (DNA) probes to characterize translocated Y DNA in a pedigree with recurrent abortion. Fertil. Steril. *48*:428–432.

Thomas, P. et al. (1990). The structure, metabolism and function of the carcinoembryonic antigen gene family. Biochim. Biophys. Acta *1032*:177–189.

Thompson, L. H. et al. (1989). Complementation of repair genes mutations on the hemizygous chromosome 9 in CHO: a third repair gene on human chromosome 19. Genomics *5*:670–679.

Thomsen, B. et al. (1987). Sequence specificity of DNA topoisomerase I in the presence and absence of camptothecin. EMBO J. *6*:1817–1823.

Tice, R. R. and Setlow, R. B. (1985). DNA repair and replication in aging organisms and cells. In: Handbook of the Biology of Aging, 2nd Ed. Finch, C. E. and Schneider, E. L. (eds.). Van Nostrand Reinhold, New York, pp. 173–224.

Tilby, M. J. et al. (1991). Sensitive detection of DNA modifications induced by cisplatin and carboplatin *in vitro* and *in vivo* using a monoclonal antibody. Cancer Res. *51*:123–129.

Timme, T. L. and Moses, R. E. (1988). Diseases with DNA damage-processing defects. Am. J. Med. Sci. *295*:40–48.

Tindall, K. R. and Kunkel, T. A. (1988). Fidelity of DNA synthesis by the *Thermus aquatius* DNA polymerase. Biochemistry *27*:6008–6013.

Tollefsbol, T. O. (1987). Gene expression of carbohydrate metabolism in cellular senescence and aging. Mol. Biol. Med. *4*:251–263.

Toneguzzo, F. et al. (1989). A system for on-line detection and resolution of radiolabeled DNA molecules and its application to automated DNA sequence analysis. BioTechniques *7*:866–877.

Topal, M. D. (1988). DNA repair, oncogenes and carcinogenesis. Carcinogenesis *9*.691–696.

Touitou, Y. and Bogdan, A. (1988). Tumor markers in non-malignant diseases. Eur. J. Cancer Clin. Oncol. *24*:1083–1091.

Towner, K. J. et al. (1991). Detection of novel trimethprim resistance determinants in the United Kingdom using biotin-labelled DNA probes. Epidemiol. Infect. *106*:63–70.

Tracy, T. E. and Mulcahy, L. S. (1991). A simple method for direct automated sequencing of PCR fragments. BioTechniques *11*:68–75.

Traincard, F. et al. (1989). Calibration of target amounts of DNA in hybridization experiments using monoclonal anti-nucleoside antibodies. Mol. Cell. Probes *3*:27–38.

Trask, B. et al. (1985). Detection of DNA sequences in nuclei in suspension by *in situ* hybridization and dual beam flow cytometry. Science *230*:1401–1403.

Trask, B. et al. (1988). Fluorescence *in situ* hybridization to interphase cell nuclei in suspension allows flow cytometric analysis of chromosome content and microscopic analysis of nuclear organization. Hum. Genet. *78*:251–259.

Trask, B. (1991). Fluorescence *in situ* hybridization: application in cytogenetics and gene mapping. Trends Genet. *7*:149–154.

Trent, J. M. (1989). Clinical correlations of chromosome change in human solid tumors: the tip of the iceberg? J. Natl. Cancer Inst. *81*:1852–1853.

Tretjak, Z. et al. (1990). Structural chromosomal aberrations and industrial waste. Lancet *i*:1288.

Triche, T. J. (1987). Morphologic tumor markers. Semin. Oncol. *14*:139–172.

Triggs-Raine, B. L. et al. (1990a). Prenatal exclusion of Tay-Sachs disease by DNA analysis. Lancet *335*:1164.

Triggs-Raine, B. L. et al. (1990b). Screening for carriers of Tay-Sachs disease among Ashkenazi Jews. A comparison of DNA-based and enzyme-based tests. N. Engl. J. Med. *323*:6–12.

Trowsdale, J. et al. (1990). Sequence encoded in the class II region of the MHC related to the "ABC" superfamily of transporters. Nature (London) *348*:741–744.

Troxler, M. et al. (1990). *In situ* hybridization for light and electron microscopy: a comparison of methods for the localization of viral RNA using biotinylated DNA and RNA probes. J. Virol. Methods *30*:1–14.

Tsongalis, G. J. et al. (1990). Electroporation of normal human DNA endonucleases into xeroderma pigmentosum cells corrects their DNA repair defect. Carcinogenesis *11*:499–503.

Tsuji, S. et al. (1987). A mutation in the human glucocerebrosidase gene in neuronopathic Gaucher's disease. N. Engl. J. Med. *316*:570–575.

Tsutsumi, Y. et al. (1990). DNA polymerase alpha. An immunohistochemical marker for proliferating cells in normal and neoplastic human tissues. Am. J. Clin. Pathol. *93*:643–650.

Tuch, B. E. et al. (1985). Typing of human fetal organs for the histocompatibility antigens A, B and DR. Pathology *17*:57–61.

Twomey, T. A. and Krawetz, S. A. (1990). Parameters affecting hybridization of nucleic acids blotted onto nylon or nitrocellulose membranes. BioTechniques *8*:478–551.

Tyler-Smith, C. and Brown, W. R. A. (1987). Structure of the major block of alphoid satellite DNA on the human Y chromosome. J. Mol. Biol. *195*:457–470.

Tyler-Smith, C. et al. (1988). Structure of a hypervariable tandemly repeated DNA sequence on the short arm of the human Y chromosome. J. Mol Biol. *203*:837–848.

Uchida, T. et al. (1990). DNA minisatellites demonstrate somatic DNA changes in most human bladder cancers. Cytogenet. Cell Genet. *53*:61–63.

Uhl, G. R. (1988). An approach to *in situ* hybridization using oligonucleotide cDNA probes. In: Molecular Neuro-anatomy. van Leeuwen et al. (eds.) Elsevier, New York, pp. 25–41.

Uhl, G. R. (1989). *In situ* hybridization: quantitation using radiolabelled hybridization probes. Methods Enzymol. *168*:741–752.

Uitterlinden, A. G. et al. (1989). Two-dimensional DNA fingerprinting of human individuals. Proc. Natl. Acad. Sci. U.S.A. *86*:2742–2746.

Unger, E. R. et al. (1988). Automation of *in situ* hybridization: application of the capillary action robotic workstation. J. Histotechnol. *11*:253–258.

Upadhyaya, M. et al. (1990). Chorionic villus sampling for prenatal diagnosis in Wales using DNA probes — 5 years' experience. Prenatal Diagn. *10*:593–603.

van Ankeren, S. C. and Wheeler, K. T. (1985). Relationship between DNA repair and cell recovery: importance of competing biochemical and metabolic processes. Radiat. Environ. Biophys. *24*:271–280.

van Dekken, H. et al. (1990a). Spatial topography of a pericentromeric region (1Q12) in hemopoietic cells studied by *in situ* hybridization and confocal microscopy. Cytometry *11*:570–578.

van Dekken, H. et al. (1990b). Three-dimensional reconstruction of pericentromeric (1Q12) DNA and ribosomal RNA sequences in HL60 cells after double-target *in situ* hybridization and confocal microscopy. Cytometry *11*:579–585.

van den Brule, A. J. C. et al. (1989). Use of anticontamination primers in the polymerase chain reaction for the detection of human papillomavirus genotypes in cervical scrapes and biopsies. J. Med. Virol. *29*:20–27.

van Dilla, M. A. and Deaven, L. L. (1990). Construction of gene libraries for each human chromosome. Cytometry *11*:208–218.

van Dilla, M. A. et al. (1990). Applications of flow cytometry and sorting to molecular genetics. In: Flow Cytometry and Sorting, 2nd Ed. Wiley-Liss, New York, pp. 563–603.

Van Gelder, R. N. et al. (1990). Amplified RNA synthesized from limited quantities of heterogeneous cDNA. Proc. Natl. Acad. Sci. U.S.A. *87*:1663–1667.

Van Prooijen-Knegt, A. C. et al. (1982). *In situ* hybridization of DNA sequences in human metaphase chromosomes visualized by an indirect fluorescent immunocytochemical procedure. Exp. Cell Res. *141*:397–407.

van Thiel, Th. P. H. et al. (1989). Reproducibility of low cytometric assessment of follicular tumors of the thyroid. J. Clin. Pathol. *42*:260–263.

van Vliet, L. J. et al. (1990). The Athena simi-automated karyotyping system. Cytometry *11*:51–58.

Vandenvelde, C. et al. (1990). Fast multiplex polymerase chain reaction on boiled clinical samples for rapid viral diagnosis. J. Virol. Methods *30*:215–228.

Vassart, G. et al. (1987). A sequence in M13 phage detects hypervariable minisatellites in human and animal DNA. Science *235*:683–684.

Viegas-Pequignot, E. et al. (1989a). Mapping of single copy DNA sequences on human chromosomes by *in situ* hybridization with biotinylated probes: enhancement of detection sensitivity by intensified-fluorescence digital imaging microscopy. Proc. Natl. Acad. Sci. U.S.A. *86*:582–586.

Viegas-Pequignot, E. et al. (1989b). Assignment of human desmin gene to band 2q35 by nonradioactive *in situ* hybridization. Hum. Genet. *83*:33–36.

Vijg, J. and Knook, D. L. (1987). DNA repair in relation to the aging process. J. Am. Geriatr. Soc. *35*:532–541.

Vijg, J. and Papaconstantinou, J. (1990). Aging and longevity genes: strategies for identifying DNA sequences controlling life span. J. Gerontol. *45*:B179–182.

Ville, Y. et al. (1991). Congenital hydrocephalus due to intrauterine HTLV-1 infection. Lancet *337*:248.

Vincent-Viry, M. et al. (1991). Relation between debrisoquine oxidation phenotype and morphological, biological, and pathological variables in a large population. Clin. Chem. *37*:327–332.

Vincenz, C. et al. (1991). Nucleoprotein hybridization: a method for isolating active and inactive genes as chromatin. Nucleic Acids Res. *19*:1325–1336.

Vindelov, L. L. and Christensen, I. J. (1990). A review of techniques and results obtained in one laboratory by an integrated system of methods designed for routine clinical flow cytometric DNA analysis. Cytometry *11*:753–770.

Viscidi, R. P. et al. (1986). Novel chemical method for the preparation of nucleic acids for non-isotopic hybridization. J. Clin. Microbiol. *23*:311–317.

Vleminckx, K. et al. (1991). Genetic manipulation of E-cadherin expression by epithelial tumor cells reveals an invasion suppressor role. Cell (Cambridge, Mass) *66*:107–119.

Vogt, Ch. et al. (1989). A role for the macrophage in normal hemopoiesis. III. *In vitro* and *in vivo* erythropoietin gene expression in macrophages detected by *in situ* hybridization. Exp. Hematol. *17*:391–397.

Vogt, P. (1990). Potential genetic functions of tandem repeated DNA sequence blocks in the human genome are based on a highly conserved "chromatin folding code". Hum. Genet. *84*:301–336.

Vollrath, V. et al. (1991). Multidrug resistance gene and P-glycoprotein expression in gastric adenocarcinoma and precursor lesions. Virchows Arch. B: Cell Pathol. *60*:133–138.

Vos, J. M. H. (1988). Analysis of psoralen monoadducts and interstrand crosslinks in defined genomic sequences. In: DNA Repair. A Laboratory Manual of Research Procedures, Vol. 3. Friedberg, E. C. and Hanawalt, P. C. (eds.). Marcel Dekker, New York, 367–398.

Wahlberg, J. et al. (1990). General colorimetric method for DNA diagnostics allowing direct solid phase genomic sequencing of the positive samples. Proc. Natl. Acad. Sci. U.S.A. *87*:6569–6573.

Walford, R. L. (1990). The major histocompatibility complex and aging in mammals. In: Molecular Biology of Aging. Alan R. Liss, New York, pp. 31–41.

Walker, G. C. (1985). Inducible DNA repair systems. Annu. Rev. Biochem. *54*:425–457.

Wallace, D. C. (1991). Mitochondrial genes and neuromuscular disease. In: Genes, Brain, and Behavior. McHugh, P. R. and McKusick, V. A. (eds.). Raven Press, New York, pp. 101–120.

Wallace, D. C. et al. (1988). Mitrochondrial DNA mutation associated with Leber's hereditary optic neuropathy. Science *242*:1427–1430.

Wang, J. (1985). DNA topoisomerases. Annu. Rev. Biochem. *54*:665–697.

Wang, R. Y. H. et al. (1986). Methylated DNA-binding protein from human placenta recognizes specific methylated sites on several prokaryotic DNAs. Nucleic Acids Res. *14*:9843–9860.

Wang, S. et al. (1990). Reduced DNA-repair capacity in cells orginating from a progeria patient. Mutat. Res. *237*:253–257.

Wani, M. A. and Snapka, R. M. (1990). Drug-induced loss of unstably amplified genes. Cancer Invest. *8*:587–593.

Warford, A. (1988). *In situ* hybridization: a new tool in pathology. Med. Lab. Sci. *45*:381–394.

Warford, A. et al. (1988). Southern blot analysis of DNA extracted from formol-saline fixed and paraffin wax embedded tissue. J. Pathol. *154*:313–320.

Warner, H. et al. (eds.) (1987). Modern Biological Theories of Aging, Vol. 31. Raven, New York.

Wartell, R. M. et al. (1990). Detecting base pair substitutions in DNA fragments by temperature-gradient gel electrophoresis. Nucleic Acids Res. *18*:2699–2705.

Washio, K. et al. (1989). Probe walking: development of novel probes for DNA fingerprinting. Hum. Genet. *83*:223–226.

Watkins, P. C. (1988). Restriction fragment length polymorphism (RFLP): applications in human chromosome mapping and genetic disease research. BioTechniques *6*:310–320.

Watson, J. D. and Cook-Deegan, R. M. (1991). Origins of the human genome project. FASEB J. *5*:8–11.

Watson, M. L. et al. (1987). Studies of genetic linkage between adult polycystic kidney disease and three markers on chromosome 16. J. Med. Genet. *24*:457–461.

Watts, A. G. and Swanson, L. W. (1989). Combination of *in situ* hybridization with immunohistochemistry and retrograde tract-tracing. In: Gene Probes. Conn, P. M. (ed.). Academic Press, San Diego, pp. 127–136.

Waye, J. S. et al. (1989). A simple and sensitive method for quantifying human genomic DNA in forensic specimen extracts. BioTechniques *7*:852–855.

Weatherall, D. J. (1986). The thalassemias: molecular pathogenesis. In: Hemoglobin: Molecular, Genetic and Clinical Aspects. Bunn, H. F. and Forget, B. G. (eds.). Saunders, Philadelphia, pp. 223–321.

Weber, J. L. and May, P. E. (1989). Abundant class of human DNA polymorphisms which can be typed using the polymerase chain reaction. Am. J. Hum. Genet. *44*:388–396.

Wehnert, M. et al. (1990). Prenatal diagnosis of hemophilia A by the polymerase chain reaction using the intragenic *Hin*dIII polymorphism. Prenatal Diagn. *10*:529–532.

Weier, H. U. G. et al. (1990). Synthesis of Y chromosome-specific labelled DNA probes by *in vitro* DNA amplification. J. Histochem. Cytochem. *38*:421–426.

Weinberg, R. A. (1989). Oncogenes, antioncogenes, and the molecular basis of multistep carcinogenesis. Cancer Res. *49*:3713–3721.

Weiner, A. M. et al. (1986). Nonviral retroposons: genes, pseudogenes, and transposable elements generated by the reverse flow of genetic information. Annu. Rev. Biochem. *55*:631–661.

Weinstein, I. B. (1988). The origins of human cancer: molecular mechanisms of carcinogenesis and their implications for cancer prevention and treatment — twenty seventh GHA Clowes Memorial Award Lecture. Cancer Res. *48*:4135–4143.

Weizsacker, F. V. et al. (1991). A simple and rapid method for the detection of RNA in formalin-fixed, paraffin-embedded tissues by PCR amplification. Biochem. Biophys. Res. Commun. *174*:176–180.

Welsh, J. and McClelland, M. (1990). Fingerprinting genomes using PCR with arbitrary primers. Nucleic Acids Res. *18*:7213–7218.

Welsh, J. and McClelland, M. (1991). Genomic fingerprints produced by PCR with consensus tRNA gene primers. Nucleic Acids Res. *19*:861–866.

Welsh, J. et al. (1991). Polymorphisms generated by arbitrarily primed PCR in the mouse: application to strain identification and genetic mapping. Nucleic Acids Res. *19*:303–306.

Wertz, D. C. and Fletcher, J. C. (1989a). Ethical issues in prenatal diagnosis. Pediatr. Ann. *18*:739–749.

Wertz, D. C. and Fletcher, J. C. (1989b). Ethical problems in prenatal diagnosis: a cross-cultural survey of medical geneticists in 78 nations. Prenatal Diagn. *9*:145–157.

West, J. D. et al. (1987). Sexing the human pre-embryo by DNA-DNA *in situ* hybridization. Lancet *ii*:1345–1347.

West, J. D. et al. (1988). Sexing whole human pre-embryo by *in situ* hybridization with a Y chromosome specific DNA probe. Hum. Reprod. *3*:1010–1019.

Westneat, D. F. et al. (1988). Improved hybridization conditions for DNA "fingerprints" probed with M13. Nucleic Acids Res. *16*:4161.

Wexler, N. S. (1989). The oracle of DNA. In: Molecular Genetics in Diseases of Brain, Nerve, and Muscle. Rowland, L. P. et al. (eds.). Oxford University Press, New York and Oxford, pp. 429–442.

White, J. J. et al. (1990). DNA alterations in prostatic adenocarcinoma and benign prostatic hyperplasia: detection by DNA fingerprint analyses. Mutat. Res. *237*:37–43.

White, R. L. (1991). Seeking tumor genes with DNA technology. Cancer *67*:2436–2438.

White, R. et al. (1985). Construction of linkage maps with DNA markers for human chromosomes. Nature (London) *313*:101–105.

Wiesner, R. J. et al. (1991). Purification of mitochondrial DNA from total cellular DNA of small tissue samples. Gene *98*:277–281.

Wilcox, A. S. et al. (1991). Use of 3′ untranslated sequences of human cDNAs for rapid chromosome assignment and conversion to STSs: implications for an expression map of the genome. Nucleic Acids Res. *19*:1837–1843.

Willcocks, T. C. and Craig, I. W. (1990). Characterization of the genomic organization of human carcinoembryonic antigen (CEA): comparison with other family members and sequence analysis of 5′ controlling region. Genomics *8*:492–500.

Williams, G. T. (1991). Programmed cell death: apoptosis and oncogenesis. Cell (Cambridge, Mass) *65*:1097–1098.

Williams, J. G. K. et al. (1990). DNA polymorphisms amplified by arbitrary primers are useful as genetic markers. Nucleic Acids Res. *18*:6531–6535.

Williams, P. et al. (1990). The polymerase chain reaction in the diagnosis of vertically transmitted HIV infection. AIDS *4*:393–398.

Williamson, D. J. et al. (1989). Detection of colony-stimulating factor mRNA in single T cells by *in situ* hybridization. J. Cell. Physiol. *139*:245–252.

Williamson, R. et al. (1981). Direct gene analysis of chorionic villi: a possible technique for first-trimester antenatal diagnosis of hemoglobinopathies. Lancet *ii*:1125–1127.

Willis, A. E. and Lindahl, T. (1987). DNA ligase I deficiency in Bloom's syndrome. Nature (London) *325*:355–357.

Wills, C. (1992). Forensic DNA typing (letter). Science *255*:1050.

Wilson, M. C. and Higgins, G. A. (1990). *In situ* hybridization. In: Molecular Neurobiological Techniques. Boulton, A. A. et al. (eds.). Humana Press, Clifton, NJ, pp. 239–284.

Wilson, R. K. et al. (1990). Optimization of asymmetric polymerase chain reaction for rapid fluorescent DNA sequencing. BioTechniques *8*:184–189.

Wilson, V. L. et al. (1987). Genomic 5-methyldeoxycytidine decreases with age. J. Biol. Chem. *262*:9948–9951.

Winter, E. et al. (1985). A method to detect and characterize point mutations in transcribed genes: amplification and overexpression of the mutant c-Ki-*ras* allele in human tumor cells. Proc. Natl. Acad. Sci. U.S.A. *82*:7575–7579.

Wisotzkey, J. D. et al. (1990). PCR amplification of 16S rDNA from lyophilized cell cultures facilitates studies in molecular systematics. Curr. Microbiol. *21*:325–327.

Witt, M. and Erickson R. P. (1989). A rapid method for detection of Y chromosomal DNA from dried blood specimens by the polymerase chain reaction. Hum. Genet. *82*:271–274.

Wittwer, C. T. and Garling, D. J. (1991). Rapid cycle DNA amplification: time and temperature optimization. BioTechniques *10*:76–83.

Wittwer, C. T. et al. (1989). Automated polymerase chain reaction in capillary tubes with hot air. Nucleic Acids Res. *17*:4353–4357.

Wittwer, C. T. et al. (1990). Minimizing the time required for DNA amplification by efficient heat transfer to small samples. Anal. Biochem. *186*:328–331.

Wolf, C. R. et al. (1990). Detection of debrisoquine hydroxylation phenotypes. Lancet *336*:1452–1453.

Wolfl, S. et al. (1991). Lack of correlation between DNA methylation and trascriptional inactivation: the chicken lysozyme gene. Proc. Natl. Acad. Sci. U.S.A. *88*:271–275.

Woll, P. J. (1991). Neuropeptide growth factors and cancer. Br. J. Cancer *63*:469–475.

Wong, C. et al. (1987). Characterization of β-thalassemia mutations using direct genomic sequencing of amplified single copy DNA. Nature (London) *330*:384–386.

Wong, Z. et al. (1986). Cloning a selected fragment from a human DNA "fingerprint": isolation of an extremely polymorphic minisatellite. Nucleic Acids Res. *14*:4605–4616.

Wong, Z. et al. (1987). Characterization of a panel of highly variable minisatellites cloned from human DNA. Ann. Hum. Genet. *51*:269–288.

Wright, J. A. et al. (1990a). DNA amplification is rare in human human cells. Proc. Natl. Acad. Sci. U.S.A. *87*:1791–1795.

Wright, J. A. et al. (1990b). Genetic regulation of metastatic progression. Anticancer Res. *10*:1247–1256.

Wright, P. A. and Wynford-Thomas, D. (1990). The polymerase chain reaction: miracle or mirage? A critical review of its uses and limitations in diagnosis and research. J. Pathol. *162*:99–117.

Wrischnik, L. A. et al. (1987). Length mutations in human mitochondrial DNA: direct sequencing of enzymatically amplified DNA. Nucleic Acids Res. *15*:529–542.

Wu, D. Y. and Wallace, R. B. (1989a). Specificity of the nick-closing activity of bacteriophage T4 DNA ligase. Gene *76*:245–254.

Wu, D. Y. and Wallace, R. B. (1989b). The ligation amplification reaction (LAR)-amplification of specific DNA sequences using sequential rounds of template-dependent ligation. Genomics *4*:560–569.

Wu, D. Y. et al. (1989). Allele-specific enzymatic amplification of beta-globin genomic DNA for diagnosis of sickle cell anemia. Proc. Natl. Acad. Sci. U.S.A. *86*:2757–2760.

Wu, D. Y. et al. (1991). The effect of temperature and oligonucleotide primer length on the specificity and efficiency of amplification by the polymerase chain reaction. DNA Cell Biol. *10*:233–238.

Wyke, J. A. (1990). Matching basic research to the management of cancer: the view from the other side of the fence. Br. J. Cancer *62*:341–347.

Wyke, J. and Poole, C. J. (1990). Regulated expression of retroviruses and their oncogenes: models for tumor suppression. In: Tumor Suppressor Genes. Klein, G. (ed.). Marcel Dekker, New York, pp. 49–73.

Wyman, A. and White, R. (1980). A highly polymorphic locus in human DNA. Proc. Natl. Acad. Sci. U.S.A. *77*:6754–6758.

Yamada, O. et al. (1990). A new method for extracting DNA or RNA for polymerase chain reaction. J. Virol. Methods *27*:203–210.

Yamashita, Y. et al. (1990). Induction of mammalian DNA topoisomerase II dependent DNA cleavage by antitumor antibiotic streptonigrin. Cancer Res. *50*:5841–5844.

Yanamandra, G. and Lee, M. L. (1989). Isolation and characterization of human DNA in agarose block. Gene Anal. Tech. *6*:71–74.

Yandell, D. W. and Dryja, T. P. (1989). Detection of DNA sequence polymorphisms by enzymatic amplification and direct genomic sequencing. Am. J. Hum. Genet. *45*:547–555.

Yap, E. P. H. and McGee, J. O. D. (1991). Short PCR product yields improved by lower denaturatiion temperatures. Nucleic Acids Res. *19*:1713.

Yarosh, D. B. (1985). The role of O^6-methylguanine-DNA methyltransferase in cell survival, mutagenesis and carcinogenesis. Mutat. Res. *145*:1–16.

Yassouridis, A. and Epplen, J. T. (1991). On paternity determination from multilocus DNA profiles. Electrophoresis *12*:221–226.

Yokoi, T. et al. (1990). An unexcluded paternity case investigated with hypervariable DNA loci. Transfusion *30*:819–823.

Yokota, H. et al. (1989). A tissue-specific change in repetitive DNA in rats. Proc. Natl. Acad. Sci. U.S.A. *86*:9233–9237.

Yusa, K. and Tsuruo, T. (1989). Reversal mechanism of multidrug resistance by verapamil: direct binding of verapamil to P-glycoprotein on specific sites and transport of verapamil outward across the plasma membrane of K562/ADM cells. Cancer Res. *49*:5002–5006.

Yusa, T. et al. (1989). Tumor cell interactions with elastin: implications for pulmonary metastasis. Am. Rev. Respir. Dis. *140*:1458–1462.

Zerres, K. et al. (1984). Cystic kidneys. Genetics, pathologic anatomy, clinical picture, and prenatal diagnosis. Hum. Genet. *68*:104–135.

Zetter, B. R. (1990). The cellular basis of site-specific tumor metastasis. N. Engl. J. Med. *322*:605–612.

Zeviani, M. et al. (1990). Nucleus-driven multiple large-scale deletions of the human mitochondrial genome: a new autosomal dominant disease. Am. J. Hum. Genet. *47*:904–914.

Zischler, H. et al. (1989a). Digoxigenated oligonucleotide probes specific for simple repeats in DNA fingerprinting and hybridization *in situ*. Hum. Genet. *82*:227–233.

Zischler, H. et al. (1989b). Non-radioactive oligonucleotide fingerprinting in the gel. Nucleic Acids Res. *17*:4411.

Zischler, H. et al. (1991). Oligonucleotide fingerprinting with $(CAC)_5$: nonradioactive in-gel hybridization and isolation of individual hypervariable loci. Electrophoresis *12*:141–146.

Zuckerkandl, E. et al. (1989). Maintenance of function without selection: *Alu* sequences as "cheap genes". J. Mol. Evol. *29*:504–512.

Zuliani, G. and Hobbs, H. H. (1990). A high frequency of length polymorphisms in repeated sequences adjacent to *Alu* sequences. Am. J. Hum. Genet. *46*:963–969.

Zwadyk, P., Jr. et al. (1986). Commercial detection methods for biotinylated gene probes: comparison with ^{32}P-labeled DNA probes. Curr. Microbiol. *14*:95–100.

Zwelling, L. A. et al. (1990). A restriction fragment length polymorphism for human topoisomerase II: possible relationship to drug-resistance. Cancer Commun. *2*:357–361.

Part 2
Gene Level Evaluation of Selected Disorders of the Heart, Lung, Gastrointestinal Tract, and Tumors of the Head and Neck

Chapter 2.1

Atherosclerosis and Coronary Heart Disease

2.1.1 INTRODUCTION

Epidemiological evidence provides solid support for the key role of lipids and lipoprotein disorders in the development of atherosclerosis and associated vascular abnormalities, coronary heart disease (CHD), and stroke. A number of factors have been recognized that contribute to the **multifactorial nature** of atherosclerosis: male gender, hypertension, obesity, family history of CHD, and so on. Analysis of **genetic factors** in atherosclerosis has been presented (e.g., Lusis and Sparkes, 1989; Sing and Moll, 1990).

Although the association between elevated lipid levels and the risk of CHD disease has been widely accepted, the specifics of such an association in women, the role of sex hormones in modulating the effects of elevated lipid levels (for a review, see Kalin and Zumoff, 1990), the importance of normalization of cholesterol level in children, the elderly, and in various diseases still remain poorly defined. On the other hand, stabilization and even regression of advanced atherosclerotic lesions as a result of lipid-lowering treatments have been reported in animal models (Wissler and Vesselinovitch, 1990) and humans (Blankenhorn, 1990). Again, although the benefits of lowering the cholesterol level have been widely advocated, the relative merits of various therapeutic regimens are not always clear and better understanding of the various mechanisms underlying hyperlipidemias by physicians and eventually the public is needed.

Advances in the molecular biology of hyperlipidemias have clarified some underlying mechanisms of lipid disorders and introduced a number of novel diagnostic modalities for their assessment. This progress can be eventually translated into a more specific therapy and prognosis. The diagnostic armament of practicing physicians in the near future will extend beyond evaluation of lipids and lipoproteins, including Lp(a) and apolipoproteins A, B, and E, to include in selected cases the analysis of *apo*A, B, and E gene polymorphisms, and of lipoprotein receptor gene defects, as well.

This chapter will summarize only some of the salient features of the molecular biology of lipoprotein disorders, and the currently available gene level diagnostic approaches for patient management. A number of comprehensive reviews on lipoprotein structure and function are available (e.g., Breslow, 1985, 1987–1989; Brown and Goldstein, 1986; Chan, 1989; Chan and Dresel, 1990; W. H. Li et al., 1988; Lusis, 1988; Mahley et al., 1991; Scott, 1990).

2.1.2 SYNOPSIS OF MOLECULAR BIOLOGY OF DYSLIPOPROTEINEMIAS

The fine balance of **lipid metabolism** in the body is controlled by a large number of proteins that, for simplicity, can be classified into three major categories: (1) **apolipoproteins**, the integral proteins determining the structure and specific interaction of lipid particles with other molecules and tissues: examples include **apo A-I, apo A-II, apo A-IV, apo C-I, apo C-II, apo C-III, apo D, apo E,** and **apo(a)**; (2) **lipid-modifying proteins** and **enzymes**, affecting

163

primarily the lipid moiety of lipid particles: **lipoprotein lipase (LPL), hepatic triglyceride lipase (HTGL), lecithin-cholesterol acyltransferase (LCAT), and cholesteryl ester transfer protein (CETP)**; and (3) proteins and enzymes of cellular lipid metabolism involved in **metabolic, catabolic,** and **transport functions** (the lipoprotein receptors): **low-density lipoprotein (LDL) receptor, chylomicron remnant receptor, scavenger receptor, lipoxygenases, phospholipases, fatty acid-binding protein (FABP), cellular retinol-binding protein (CRBP), and hormone-sensitive lipase (HSL)** (Chan and Dresel, 1990).

Chromosomal assignment of the corresponding genes are given in Table 2.1.

An interplay of genetic and environmental factors, with accompanying mutational events (e.g., Penn, 1990; Wakabayashi, 1990), determines the level of lipids in an individual, the risk of developing heart disease and stroke, as well as the degree of manifestation of dyslipoproteinemias. The **genetic component** is important and, in many situations, is predominantly responsible for the observed lipid disorders (Lusis, 1988). The so-called "candidate gene" approach (Lusis, 1988) has characterized the majority of current studies aimed at elucidating the molecular underpinnings of dyslipoproteinemias. Restriction analysis, DNA cloning, amplification, and sequencing, as well as genetic linkage data have provided a wealth of molecular evidence enabling specific diagnosis of a number of lipid disorders to be made at the gene level.

A summary of genetic defects in exogenous and endogenous lipid transport and in reverse cholesterol transport (Breslow, 1989) will be given here, emphasizing the clinical relevance of their diagnosis at the molecular level.

2.1.3 GENETIC DEFECTS IN EXOGENOUS LIPID TRANSPORT

Exogenous lipid transport includes the breakdown of dietary fats in the intestine and the formation of triglyceride-rich cholesterol-containing particles called **chylomicrons**. The proteins constitute about 7% of the particle and include the **intestinal form** of apo B (**B-48**), A-I, and **A-IV** located at the surface of chylomicrons. Following their formation, chylomicrons undergo several changes by acquiring **apo C-II** and **apo E** and gradually cleaving the lipids thus converting the particles into **chylomicron remnants** that are removed from circulation and catabolized in the liver. Genetic defects have been recognized in chylomicron secretion, processing, and clearance of the remnants (Breslow, 1989).

Chylomicron secretion is defective in two very rare conditions, **abetalipoproteinemia** and **homozygous hypobetalipoproteinemia**, occurring at a frequency of less than 1:1 million. Clinical manifestations in both conditions are related to fat malabsorption and consequent fat-soluble vitamin A, D, E, and K deficiencies, leading to ataxic neuropathy, retinitis pigmentosa, and acanthocytosis of the red blood cells (RBCs). The absence of apo B precludes the formation of chylomicrons and very low-density lipoprotein (VLDL) and LDL particles. Genetic linkage studies identified defects in the *apo* B gene in **homozygous abetalipoproteinemia**. In the other condition, abetalipoproteinemia, no defect of the *apo* B gene has been found and the genetic basis of the disease is not understood (Breslow, 1989). It appears that a posttranslational defect or a mutation in other gene(s) encoding factors involved in lipoprotein assembly or secretion are at fault (Pessah et al., 1991). A form of **familial hypobetalipoproteinemia** has also been described in which a dominantly transmitted mutation in a gene other than the *apo*B gene is responsible for the low plasma cholesterol and apo B-carrying lipoproteins (Fazio et al., 1991).

Processing of chylomicrons is defective in LPL deficiency and apo C-II deficiency — two rare conditions again occurring at a frequency of less than 1:1 million (Breslow, 1989). LPL gene polymorphisms appear to be associated with high-density lipoprotein (HDL) and total cholesterol levels (Heizmann et al., 1991). A number of specific mutations in the LPL gene result in the synthesis of enzymatically inactive proteins (e.g., Dichek et al., 1991; Henderson et al., 1990). Interestingly, the accumulation of large chylomicrons is not associated with increased risk of atherosclerosis, but rather leads to pancreatitis and eruptive xanthomas.

TABLE 2.1
Genes for Proteins Involved in Lipid Transport and Metabolism

Gene	Chromosomal location
I. Apolipoprotein genes	
apo A-I	11 (11q13-qter)
apo A-II	1 (1p21-qter)
apo A-IV	11 (11q13-qter)
apo B	2 (2p23-p24)
apo C-I	19 (19q)
apo C-II	19 (19q)
apo C-III	11 (11q13-qter)
apo D	3 (3q26.2-qter)
apo E	19 (19q)
apo(a)	6 (6q27)
II. Lipoprotein-modifying genes	
CETP	16 (16q12-16q21)
LCAT	16 (16q22)
LPL	8 (8p22)
HTGL	15 (15q21)
III. Genes related to cellular lipid metabolism	
LDL receptor	19 (19p)
LDL receptor-related protein	12 (12q13-q14)
HMG-CoA reductase	5 (5q13.1-q14)
HMG-CoA synthase	5 (5q14-p12)
Liver FABP	2 (2p12-q11)
Intestinal FABP	4 (4q28-q31)
CRBP	3
CRBP-II	3
HSL	19 (19cent-q13.3)

Note: FABP, fatty acid-binding protein; CRBP, cellular retinol binding protein; HSL, hormone-sensitive lipase.

From Chan and Dresel, 1990.

2.1.3.1 APOLIPOPROTEIN E

Apolipoprotein E is a 299-amino acid polypeptide with the receptor-binding region located between amino acids 140 and 160 (Innerarity et al., 1983, 1988; Lalazar et al., 1988; Weisgraber et al., 1983). In chylomicrons, apo E is on the surface of the lipid particle, acting as a ligand for the receptors on liver cells where receptor-mediated clearance of chylomicron remnants occurs. Apo E is also part of VLDL and HDL particles, where it mediates binding to the LDL receptor present on hepatic and extrahepatic cells. Defects in **chylomicron remnant clearance** are much more common and comprise a number of phenotypic groups determined by variation in the *apo* E gene. Mutations of the *apo* E gene affect receptor recognition and underlie **type III hyperlipoproteinemia**. Alteration of the *apo* E gene is not the only way this gene may affect lipid levels, and the predominant expression of certain allele combinations may have profound effects on lipid metabolism.

Three common alleles of the *apo* E gene have been recognized, ε4, ε3, and ε2, that can produce six phenotypes (for a review, see Breslow, 1988). The defective ε2 allele has a frequency of 8%, the most common allele is ε3 (77% frequency), and ε4 occurs at a frequency of 15%. The product of the ε2 allele fails to bind to receptors and, consequently, defective chylomicron removal from the circulation occurs in people homozygous or heterozygous for ε2. The most common phenotype is **E3/3** (60% of the population). Approximately 1% of the population is homozygous for the ε2 allele (E2/2) and still another 14% are heterozygous for the ε2 allele, having the E3/2 and E4/2 phenotypes.

The three isoforms of the *apo* E gene product, recognizable by isoelectric focusing, differ by cysteine-arginine interchanges at residues 112 and 158 in the protein. Apo E2 contains cysteine at both positions, which is thought to contribute to the poor recognition of the specific receptor. Notably, although this phenotype has just 1 to 2% of normal receptor-binding activity, less than 5% of people with the E2/2 phenotype are hyperlipidemic and disease expression is related to age, diet, hormonal status, and apparently the expression of other, as yet unidentified hyperlipidemia genes (Breslow, 1988).

Defective chylomicron remnant clearance occurs in at least three conditions, not all of which are associated with increased atherosclerosis and elevated risk of CHD. Although type III hyperlipidemia, associated with in the E2/2 phenotype, occurs in 1 in 5000 individuals that characteristically have elevated fasting levels of cholesterol and triglycerides, these patients may have either low, normal, or elevated LDL cholesterol levels. These people have an increased incidence of CHD or peripheral vascular disease and xanthomas.

Another variant of defective chylomicron clearance associated with the E2/2 phenotype does not present with fasting hyperlipidemia, the so-called **normolipidemic E2/2** individuals. They constitute up to 1% of the population, and although their chylomicron level is elevated their LDL level is decreased; they neither develop xanthomas nor have an elevated risk of CHD. One of the possible mechanisms accounting for this phenomenon appears to involve decreased delivery of cholesterol to the liver, up regulation of LDL receptors, and, therefore, enhanced clearance of LDL from plasma.

And, finally, the last group of people with defective chylomicron remnant clearance includes **E2 heterozygotes** (e.g., E4/2, E3/2). These individuals have a milder elevation of chylomicron remnants and intermediate-density lipoprotein (IDL) than that found in E2 homozygotes. Again, the beneficial effect of counterbalancing this increase by a decrease in LDL apparently accounts for the lack of increased risk of CHD.

The apo E phenotype affects the LDL cholesterol level in a number of ways, in which an interplay of E4 and E3 gene expression accounts for a wide spectrum of biochemical and clinical phenotypes. So far, in addition to the most common ε2, ε3, and ε4 alleles other alleles have been described (for a total of 11), the majority of which have amino acid substitutions replacing a positively charged amino acid with a neutral amino acid in the apo E receptor-binding domain.

The **E7** and **E5** mutations have been identified in Japanese, not Caucasians, and appear more frequently in individuals with hyperlipidemia and/or artherosclerosis. Apo E5 has a single base substitution that imparts two additional positive charges to the molecule (Tajima et al., 1988). There appears to exist a specific association between apo E isoforms and VLDL metabolism in hypertriglyceridemia, but the apo E phenotype effect on lipoprotein metabolism is independent of LDL receptor status as observed in patients with **familial hypercholesterolemia** (Dallongeville et al., 1991).

2.1.3.2 APOLIPOPROTEIN E POLYMORPHISM: POPULATION STUDIES

A study in the Finnish population revealed a higher frequency of the allele ε4 and a lower frequency of ε2 than that observed in other populations (Ehnholm et al., 1986). Total cholesterol, LDL cholesterol, and apo B levels were markedly higher in the E4/4 and, to a lesser degree, in the E4/3 phenotypes than in the most frequent E3/3 phenotype. The levels of these lipids and apolipoproteins tended to be lower in E2 heterozygotes.

In earlier population studies, the ε3 frequency was found to be highest among Japanese (Eto et al., 1989) and the ε2 frequency was highest in New Zealand (Wardell et al., 1982). In the Netherlands the *apo* E gene frequencies and phenotype distribution were essentially similar to those of the previously studied populations in Europe and the United States, with the exception of the ε4 allele, which was significantly lower than that in the Finnish population (Klasen et al., 1987). A strikingly high frequency of the ε4 allele has been observed among blacks in the United States and Nigeria (Kamboh et al., 1989), being almost twice as high as that reported in other populations (Zannis, 1986).

Analysis of *apo* E gene frequencies and their effects on plasma levels of apo E, apo B, and total cholesterol in 536 blood bank donors in Germany showed the effects of the ε4 allele to be the lowering of apo E levels on average by 0.19 mg/dl and the raising of apo B levels by 4.92 mg/dl and that of total cholesterol by 7.09 mg/dl (Boerwinkle and Utermann, 1988). The average effects of the ε2 allele were to raise apo E levels by 0.95 mg/dl, lower apo B levels by 9.46 mg/dl, and lower total cholesterol levels by 14.2 mg/dl. Analysis of apo E polymorphisms and their effect on the levels of lipids, lipoproteins, and apolipoproteins indicated that the apo E polymorphism can be considered as a two-locus, two-allele system (Hanis et al., 1991b). The practical value of determining apo E polymorphisms can also be seen in the observed influence of this polymorphism on the serum cholesterol response to dietary intervention (Manttari et al., 1991).

2.1.3.3 *apo* E GENE EXPRESSION, ATHEROSCLEROSIS, AND CORONARY HEART DISEASE

The well-documented effects of apo E polymorphism on the level of plasma lipid and lipoprotein levels in normal and hyperlipidemic individuals suggested a **role of apo E polymorphism in the predisposition to atherosclerosis** (for a review of earlier findings, see Davignon et al., 1988). Observations made in the studies on survivors of myocardial infarcts (MI) suggested that type III hyperlipidemia with the **E2/2 phenotype was overrepresented in the group of patients with MI**, even though the frequency of the E2/2 phenotype was the same in MI survivors and controls (Utermann et al., 1984). The **ε2 allele** was thought to protect against atherosclerosis in the absence of hyperlipoproteinemia. In contrast, the higher frequency of the ε2 allele in MI survivors in a Japanese study (Kameda et al., 1984).

The **ε4 allele may confer an increased risk of MI at an earlier age** (Cumming and Robertson, 1984). The presence of the ε4 allele appears to predispose patients to an earlier occurrence of MI; however, a host of other factors may modulate the effects of apo E polymorphism (Davignon et al., 1988). Equally inconclusive evidence has been obtained in patients undergoing **coronary angiography** for documented or suspected CHD, although the apo E phenotype distribution appears to be essentially similar in patients with and without CHD (Brenninkmeijer et al., 1984; Menzel et al., 1983; Reardon et al., 1985). Inadequacy in patient group selection, variations in treatment protocols, the presence of other risk factors (such as significant hyperlipidemia), and differences in age and sex distribution are believed to contribute to the observed conflicting results (Davignon et al., 1988).

A study of Finnish patients with angiographically confirmed CHD demonstrated the frequency of the ε4 allele was 1.4-fold higher than that in the normal Finnish population and twice as high as in other Caucasian populations (Kuusi et al., 1989). Among Japanese patients with CHD the frequencies of ε2 and ε4 alleles were significantly higher than in controls (Eto et al., 1989).

One of the proposed reasons for the apo E polymorphism-related differences in plasma and LDL cholesterol concentrations is believed to be due to **differential responses to dietary lipids** (Tikkanen et al., 1990). In a dietary intervention study, greater reductions in plasma cholesterol levels occurred in E2/2 individuals compared to those in other genotypes. These subjects also responded to the return to normal diet by a greater increase in plasma cholesterol than others, suggesting that the **effect of the apo E genotype on plasma cholesterol may be modulated by dietary lipids**. There is also a significant interaction between the apo E polymorphism, the presence of the ε4 allele, and weight gain with an increase in triglyceride levels; no such association is observed in individuals without the ε4 allele (Gueguen et al., 1989).

At the tissue level, in an animal model, cholesterol feeding was found to markedly increase *apo* E gene expression in atherosclerotically changed aortas (Crespo et al., 1990). A similar increase in *apo* E expression was also detected in human atheromata compared to that in unaffected areas of the aorta in the same individual. It is difficult to draw any conclusions about the role of *apo* E in the development of atherosclerotic plaques from these isolated observations.

Further improvements in the methods of detection of the *apo* E gene and its products will allow better appreciation of its interaction with other factors leading to the formation of atherosclerotic lesions.

2.1.3.4 *apo* E GENOTYPING METHODS

The three major apo E isoforms, E2, E3, and E4, and structural variants of apo E, can be evaluated by **isoelectric focusing (IEF)** of VLDL proteins (Gregg et al., 1983; Havekes et al., 1986; Kamboh et al., 1990; Rall et al., 1983; Stalenhoef et al., 1981; Stuyt et al., 1984; Utermann et al., 1984; Wardell et al., 1987; Weisgraber et al., 1981; Yamamura et al., 1984a,b, 1985; Zannis, 1986; Zannis and Breslow, 1981), IEF using **thrombin inhibitors** (Steinmetz, 1991), or by **two-dimensional gel electrophoresis** (Zannis and Breslow, 1981). These methods are quite laborious and have certain limitations.

Separation of isoforms is based on charge differences related to the presence of **cysteine** and **arginine** at specific sites in the apo E molecule. Thus, apo E2 (Cys-112, Cys-158), E3 (Cys-112, Arg-158), and E4 (Arg-112, Arg-158) and several other variants can be readily identified by IEF, whereas mutant proteins having neutral amino acid substitutions or neutralizing double mutants are not resolved. The site of mutation cannot be identified by these methods either.

2.1.3.4.1 Polymerase Chain Reaction in *apo* E Assays

Following cloning and sequencing of the *apo* E gene, the analysis of *apo* E alleles at the DNA level allowed the demonstration of structural alterations affecting apo E isoform receptor-binding capacities (Breslow, 1985, 1988; McLean et al., 1983). Further improvements in the routine detection of *apo* E alleles have come from the application of allele-specific oligonucleotides (ASOs) in hybridization and polymerase chain reaction (PCR) amplification. Mutations affecting the gene area encoding the **receptor-binding region** of apo E, both known or unknown, can be recognized by these molecular techniques (Smeets et al., 1988). Only minute quantities of DNA are needed for these methods. The discrimination of ε3-ε4 and ε2-ε3 alleles by the ASO hybridization approach is based on the **difference in thermal stability** between the hybrids formed by fully matched 19- or 20-mer oligonucleotides and duplexes with a minor G·T mismatch or duplexes with a more destabilizing mismatch (Smeets et al., 1988). In the **ASO hybridization** approach, specific sets of conditions optimizing the desired stringency of hybridization with every individual oligonucleotide probe must be maintained. On the other hand, **PCR amplification** produces much stronger signals that are less affected by nonhomologous hybridization. Combining PCR with **Southern blotting** allowed the identification of apo E isoforms and could be modified to recognize any mutation in the gene of interest (Smeets et al., 1988).

A similar and improved procedure using PCR amplification and ASOs (Weisgraber et al., 1988) generates only a single amplified *apo* E-specific product, rather than a number of different-sized bands, most of which may be unrelated to *apo* E (Smeets et al., 1988). The high concentration of unrelated amplification products precludes the use of **dot-blot** analysis, necessitating more laborious Southern blotting (Smeets et al., 1988). In contrast, when only a single *apo* E-related amplification product is generated, a **slot-blot** procedure can be used directly.

The benefits of combining DNA amplification with ASOs for the determination of *apo* E genotype are numerous. It is less labor intensive than IEF, assays can be performed on a much smaller sample, a large number of samples can be assayed, and unequivocal genotype assignments can be made (Weisgraber et al., 1988).

apo E genotyping and **automated sequence analysis of PCR products** and ASO probes have been used by other researchers as well (Emi et al., 1988; Houlston et al., 1989; Hixson and Vernier, 1990; Kontula et al., 1990). PCR amplification of the *apo* E gene region harboring the two variable nucleotides giving rise to **Arg→Cys interchange** in apo E isoforms has been combined with restriction digestion (Kontula et al., 1990). The **nucleotide substitutions** at those two loci give rise to RFLPs that can be detected by *Hha*I (Hixson and Vernier, 1990). The close proximity of the two variable loci (separated from each other by 138 bp) allows for convenient amplification of the region as a single PCR product.

A nonradioactive modification of PCR-based *apo* E genotyping using biotin label has also been developed (Syvanen et al., 1989). The subsequent modification of the method by the same authors allows for the identification of the variable nucleotide in the biotinylated and immobilized DNA by a **one-step primer extension reaction**, in which a single labeled nucleotide triphosphate, complementary to the nucleotide at the variable site, is incorporated.

A technique described as **restriction isotyping (restriction enzyme isoform genotyping)** avoids the use of costly and labor-intensive hybridization and sequence procedures and offers a simpler and faster method for typing the common apo E isoforms (Hixson and Vernier, 1990). Following PCR amplification of the *apo* E gene region spanning amino acid positions 112 and 158, where most of the substitutions producing isoforms E2, E3, and E4 occur, the amplification product is digested with *Hha*I. The resulting digestion products are then separated on polyacrylamide gels and specific patterns are produced corresponding to Arg→Cys interchanges. Thus, each genotype can be easily distinguished by a unique pattern of fragment sizes in all homozygous and heterozygous patients. **Visualization** of these patterns in stained gels does not required radioactive labels as in the case of ASO probes (Hixson and Vernier, 1990).

Although this approach cannot detect rare, unknown mutations, its simplicity and high efficiency make it a convenient method for laboratories performing evaluation of apo E gene effects. Automated sequencing of PCR amplification products is suitable for the identification of any mutations in the target DNA sequences (Emi et al., 1988). *Hha*I digestion, however, can misdiagnose a rare variant, as shown by its failure to recognize E1, which could be identified by a different restriction enzyme, *Taq*I (Kontula et al., 1990).

A novel PCR technique first amplifies the DNA fragment of interest with an oligonucleotide primer biotinylated at its 5′ end; the amplified product is then immobilized on an avidin matrix and double strands are separated (Syvanen et al., 1990). Then a one-step primer extension is performed to identify the variable nucleotide using a single nucleoside triphosphate (radioactively or nonradioactively labeled). **This procedure has been successfully applied to identify all possible combinations of the three *apo* E alleles** (Syvanen et al., 1990). Several other PCR-based protocols have been described for the analysis of *apo* E genotypes that do not use radioactive labels and can be performed in diagnostic clinical laboratories (e.g., Main et al., 1991; Wenham et al., 1991a,b).

2.1.4 GENETIC DEFECTS IN ENDOGENOUS LIPID TRANSPORT

Endogenous lipid transport starts with the formation of VLDL in the liver, followed by hydrolysis of its triglycerides by LPL, yielding IDL. These are then either taken up by the liver or further processed by HTGL to LDL which are cleared from the plasma by the LDL receptor (Breslow 1988, 1989). The various steps in this sequence are determined primarily by the apolipoproteins associated with the lipid particles. VLDL contains the liver form of apo B (apo B 100), apo C and apo E associated with triglycerides and cholesterol. IDL particles are recognized by **hepatic LDL receptors** that bind apo E on the IDL surface. Seventy percent of LDL particles enriched in cholesterol are cleared from plasma by LDL receptors on hepatocytes and extrahepatic cells (Breslow, 1989).

2.1.4.1 STRUCTURE, FUNCTION, AND MUTATIONS OF APOLIPOPROTEIN B

Apo B is essential for the assembly of VLDL particles and constitutes the protein component (ca. 25% by weight) of LDL particles. It is situated at the outer shell of a lipoprotein particle together with phospholipids and unesterified cholesterol, which surround the inner core composed of apolar neutral lipids — triglycerides and cholesterol esters (Breslow, 1985, 1988; Innerarity et al., 1990). Human apo B is a glycoprotein existing in two forms, **hepatic (B-100)** and **intestinal (B-48)**. B-100 is a simple polypeptide of apparent M_r 550,000 Da; it is encoded by a single-copy gene and is produced primarily in the liver. B-48 is approximately one half that molecular mass and is synthesized primarily in the small intestine (Higuchi et al., 1987).

It appears that a certain length of apo B-100 is necessary for the lipoprotein particle to enter the circulation, and a normal or nearly full-length apo B-100 is necessary to ensure normal VLDL secretion (Young et al., 1990). In contrast to other apolipoproteins such as A-I, A-II, and E, which transfer freely between lipoprotein particles, **apo B-100 remains associated with the same particle**. Apo B-100 appears to interact with lipids in a manner resembling that of intramembrane proteins via its hydrophobic β-sheet structure and segments of hydrophobic amino acids reminiscent of membrane-spanning regions of receptors (reviewed in Innerarity et al., 1990).

Modification of LDL particles through oxidation and acetylation and the association of apo B-containing lipoproteins with acidic glycosaminoglycans or specific proteoglycans contributes to the atherogenicity of these particles. There are **binding sites** in the apo B of LDL with a characteristically high affinity for proteoglycans and heparin (Camejo et al., 1988; Hirose et al., 1987; Weisgraber and Rall, 1987). Some of the heparin-binding sites are located in the region corresponding to the receptor-binding domain. The latter has been roughly identified in studies using monoclonal antibodies to epitopes involved in the binding of apo B-100 to the LDL receptor (see Breslow, 1988; Innerarity et al., 1990). The precise location of the receptor-binding region of apo B-100 remains to be defined. Two regions of apo B-100 containing clusters of positively charged amino acids between residues 3147–3157 and 3359–3367 are considered to be the LDL receptor-building region.

The **sequencing** of apo B-100 DNA allowed the determination of variation in both the quantity and quality of apo B genes. Five pairs of allelic variants designated Ag(c/g), Ag(x/y), Ag(t/z), Ag(a1/d), and Ag(h/i) have been defined immunologically that result from variations in the apo B protein (Berg et al., 1963, 1983, 1989a,b). One of these determinants, **Ag(x)**, has been linked to elevated cholesterol and triglyceride levels. Four of the antigen group (Ag) polymorphisms are associated with single amino acid substitutions. Restriction fragment length polymorphisms of apo B-100 for *Xba*I have been associated with myocardial infarct (Hegele et al., 1986).

Ma and coworkers (1987) have identified a different immunochemical polymorphism of Ag(c/g), which in a Finnish population is associated with elevated cholesterol levels (Aalto-Setala et al., 1988). A comparison between Ag and DNA polymorphisms has been performed to identify major apo B haplotypes in human populations that may be used to search for significant correlations between certain *apo* B alleles and lipid levels, atherosclerosis, and CHD (Dunning et al., 1988; Ma et al., 1988). A **Bsp 1286 I** RFLP identifies the Ag(c/g) locus of apo B-100 (Ma et al., 1989).

Informative RFLPs of the *apo* B gene have been found in **tandem repeat sequences in hypervariable minisatellites** in the *apo* B gene (Huang and Breslow, 1987). The high frequency of heterozygosity for this RFLP may be useful in disease association and in linkage analysis studies revealing the contribution of this locus to atherosclerosis susceptibility. Several more polymorphisms of the *apo* B gene have been identified (*Ava*II and *Bal*I), and the molecular basis and frequencies of the two previously reported RFLPs (*Hinc*II and *Pvu*II) were characterized (Huang et al., 1990b).

A number of studies have attempted to establish firm associations between apo B-100 polymorphisms and elevated lipid levels and CHD. While some of these demonstrated a positive correlation (Aalto-Setala et al., 1988, 1989; Friedl et al., 1991; Gavish et al., 1989a–c; Hegele et al., 1986; Law et al., 1986; Leren et al., 1988; Talmud et al., 1987), other researchers failed to observe an association in European (Darnfors et al., 1989; Ferns et al., 1988; Myklebost et al.,

1990; Wiklund et al., 1989) and in Asian (Aburatani et al., 1988; Mendis et al., 1991) populations. An extended study of RFLPs of the *apo* B gene has been performed on samples from individuals from the United Kingdom, Finland, Italy, Spain, and Africa (Houlston et al., 1990).

Importantly, variation in the *apo* B gene has been found to express itself differently in different populations. Among the two North European populations, a significant association was observed between LDL fractional catabolic rate and the apo B *Eco*RI and *Xba*I RFLP genotypes. In contrast, in the African population sample, a significant association was found between the *Xba*I RFLP and LDL apo B synthesis (rather than catabolism). Variations at both the apo A-I/C-III/C-IV and apo B loci were associated with lipoprotein and apolipoprotein levels in a sample of 209 children from Sezze in central Italy (Xu et al., 1990).

It appears that the **weight of the accumulated evidence favors an association between the *apo* B gene polymorphism and CHD**. Specific RFLPs established a significant association in 290 white men between apo B and serum levels of cholesterol, obesity, and CHD, but not with triglyceride and HDL level, smoking, or alcohol consumption (Rajput-Williams et al., 1988). Coronary heart disease was associated with minimum haplotypes involving *Xba*I and *Msp*I RFLPs. In addition to the earlier studies that pointed to such an association, later observations made on relatively large patient samples support this notion. In 205 patients with documented peripheral arterial disease, some of which had coexisting CHD, the frequency of the *R2* allele of the polymorphism detected with *Eco*RI, and the frequency of the X1 allele of the *Xba*I polymorphism, were significantly higher in the patient group (Monsalve et al., 1988). Since the frequency of these alleles in different patient groups suffering from peripheral disease alone or in combination with CDH was the same, the data obtained were interpreted as supporting the notion that variation at the *apo* B locus was one of the factors determining the predisposition to arterial disease, but it does not necessarily determine the site where the disease may affect the arterial system (Monsalve et al., 1988).

When 111 male Caucasians with premature CHD were compared with 122 elderly Caucasian males without CHD, the rare allele (*R1*) of the *Eco*RI RFLP in exon 29 that leads to an amino acid change (Glu → Lys-4154) was detected more frequently in the CHD group than in the control (Genest et al., 1990). The rare allele (*M2*) was also more frequent in the CHD group. Again, as reported earlier (Hegele et al., 1986), no statistically significant differences between the CHD and control groups were noted in their lipid, lipoprotein cholesterol, or apo AI and B levels. Although a correlation between the polymorphisms and CHD risk was established, the single-site polymorphisms observed in that study could hardly provide more information about the CHD risk of the patients than conventional lipid parameters (Genest et al., 1990).

A high-resolution method using **PCR amplification of hypervariable alleles of a minisatellite region** close to the *apo* B gene was used in 110 Austrian patients with severe CHD who had suffered a myocardial infarction (MI) and in 117 matched controls with no history of CHD (Friedl et al., 1990). Alleles containing 38, 44, 46, or 48 hypervariable elements displayed an association with CHD, elevated cholesterol, and apo B among patients and with elevated total triglycerides in the controls.

Thus, although the *apo* B gene polymorphisms do appear to be associated with CHD and in some cases with elevated levels of atherogenic lipids, their clinical utility at this time is not clear, since so far, this approach appears to offer essentially comparable information with that gained in serum lipid and apolipoprotein studies.

2.1.4.2 HETEROZYGOUS HYPOBETALIPOPROTEINEMIA AND ABETALIPOPROTEINEMIA

The most important genetic disorders resulting in clinically important dyslipidemias have been recognized to affect **VLDL secretion** or **formation**, and **LDL clearance**. Defective VLDL secretion occurs in a comparatively frequent condition called **heterozygous hypobeta-lipoproteinemia** at a frequency of 1 in 1000 in the general population. The specific mutations in the *apo* B gene result in premature stop codons in the defective allele, leading to interruption of apo B synthesis and reduction of its level in plasma (Talmud et al., 1989; Young et al., 1988). Consequently, affected individuals have a lowered LDL cholesterol level (30 to 50% normal) and, in effect, have a **lower risk of atherosclerosis**. Selective deficiencies of either hepatic (apo B-100) or intestinal (apo B-48) forms of apo B have also been described (Breslow, 1987).

Abetalipoproteinemia, characterized by low or undetectable concentrations of apo B and of the apo B-containing lipoproteins, is associated with severe fat malabsorption, acanthocytosis, progressive spinocerebellar degeneration and retinopathy. Although family studies indicate the autosomal recessive inheritance of a rare gene, the **molecular defect causing this condition has not yet been defined**. RFLPs for *Pvu*II, *Hinc*II, *Xba*I, *Eco*RI, and *Msp*I used in **linkage analysis** on two families with two children suffering from abetalipoproteinemia demonstrated that this condition was not caused by a defect in the *apo* B gene (Talmud et al., 1988). Apparently a defect in another gene involved in the synthesis or secretion of apo B-containing lipoproteins is responsible for this disorder. Similar conclusions were drawn from another study in eight affected families (Huang et al., 1990a). The search for the candidate gene continues, and one of the potentially promising approaches appears to be the analysis of the regions adjacent to the disease gene in children of consanguineous marriages (Huang et al., 1990a).

In a related condition, **Anderson's disease**, the affected individuals suffer from fat malabsorption, reduced plasma levels of cholesterol, triglycerides, and apo B, A-I, and C (Breslow, 1988). Amplification of genomic DNA by PCR followed by REA of the amplified products derived from **regions flanking the apo B gene 3′ hypervariable locus** has been performed in two affected families (Pessah et al., 1991). The results were incompatible with involvement of the *apo* B, gene suggesting a **posttranslational defect** or **a mutation in another gene** encoding a protein apparently involved in the processing, assembly, or secretion of apo B-containing lipoprotein particles.

2.1.4.3. FAMILIAL DEFECTIVE APOLIPOPROTEIN B-100

Another category of conditions with defective endogenous lipid transport includes two major disorders, **familial defective apo B-100 (FDB)** and **familial hypercholesterolemia (FH)**. Both disorders are characterized by defective LDL clearance. The first condition occurs at a frequency of 1 in 500 to 1 in 1000 in the general population and 1 in 50 to 1 in 100 in persons with upper decile LDL cholesterol. A mutation in the receptor-binding domain of apo B, changing an arginine codon at amino acid 3500 to glutamine, results in a protein with severely reduced binding activity (2 to 4% normal) (Innerarity et al., 1988, 1990; Soria et al., 1989).

The prevalence of **FDB** has been screened in 1100 patients in San Francisco, Montreal, Dallas, and Austria using competitive binding of plasma LDL to the apo B and E (LDL) receptors on cultured human fibroblasts, the binding of monoclonal antibody MB47 to LDL in a competitive radioimmunoassay (RIA) of whole plasma, as well as by the detection of a single base change at codon 3500 using PCR amplification of genomic DNA followed by ASO hybridization (Innerarity et al., 1990). Eleven probands who were heterozygous for FDB were discovered. Family studies extended this number to 42 but **no homozygotes for FDB have been found so far**. At a frequency of 1 in 500 in the general population, similar to that of FH, **FDB is one of the most common single-gene mutations responsible for a clinical abnormality** (Innerarity et al., 1990).

Unlike the situation with FH, where no single mutation is responsible for the observed phenotype, **in FDB the mutation at codon 3500 is by far the most common one**. Although no definitive impression can be formed about the effect of this mutation in FDB on the predisposition to atherosclerosis, 7 of 10 FDB heterozygotes have been found to have CHD (Tybjaerg-Hansen et al., 1990). These patients, however, were hypercholesterolemic, thus a direct causal relationship between FDB and this mutation cannot be established.

2.1.4.4 POLYMERASE CHAIN REACTION IN THE ANALYSIS OF *apo* B DNA POLYMORPHISM

A general method to type rapidly and accurately the 3'-flanking region of the *apo* B gene composed of a variable number of tandemly repeated short A+T-rich DNA sequences (VNTRs) has been described (Boerwinkle et al., 1989b, 1990; Boerwinkle and Chan, 1989). The amplification products are directly visualized in agarose gels. Some 12 alleles could be readily identified in a sample of 125 unrelated individuals. The sensitivity of this technique allows the discrimination of alleles varying by only 32 bp. The described technique can be directly applied to the study and rapid typing of other VNTRs in the human genome. Primers have also been described for the identification by PCR of an insertion/deletion polymorphism in the signal peptide region of apo B (Boerwinkle and Chan, 1989).

Typing different regions of the *apo* B gene-related DNA sequences by a combination of PCR amplification and hybridization with ASOs has been performed to analyze the association between the *apo* B gene polymorphisms and the apo B Ag immunological epitopes (Boerwinkle et al., 1990). The described PCR approach is substantially faster and less expensive than laborious Southern analyses, which are poorly suitable for large-scale population studies.

Other modifications of PCR-based identification of characteristic Apo B-100 mutations in FDB carriers include the use of sequence-modifying primers creating a recognizable RFLP in the amplified product, thus facilitating the detection of the target mutation (Motti et al., 1991). An even simpler protocol uses the **amplification refractory mutation system** (ARMS) with allele-specific primers exactly matching the target sequence (Wenham et al., 1991c). The assay is simple and fast, and the results are read from ethidium bromide-stained gel, making it appealing for use in clinical diagnostic laboratories.

2.1.4.5 FAMILIAL HYPERCHOLESTEROLEMIA

Defects in the LDL receptor account for the elevation of circulating LDL particles in another condition, familial hypercholesterolemia (**FH**), which occurs at a frequency of 1 in 500 in the general population, and 1 in 50 individuals with upper decile LDL cholesterol, similar to that of FDB (Brown and Goldstein, 1986).

Patients with FH develop xanthomas and an increased incidence of heart attacks at younger ages, which are thought to be related to the elevated levels of LDL in this condition. The disease is inherited as a dominant trait. The molecular basis of FH has been extensively studied, in particular by Brown and Goldstein and their coworkers (earlier work is reviewed in Brown and Goldstein, 1986; Goldstein et al., 1985). **FH heterozygotes** with one normal gene and one defective gene have long been shown to have one half of functional LDL receptors on the surface of their cells. **FH homozygotes** have virtually no LDL receptors (Brown and Goldstein, 1974). While FH heterozygotes occur at a frequency of 1 in 500, FH homozygotes are found at a frequency of 1 in 1 million (Slack, 1979).

The LDL receptor is synthesized in the rough endoplasmic reticulum as an integral membrane protein that later acquires sugars (for a review of studies on the LDL receptor, see Russell et al., 1989). The **LDL receptor precursor** (120,000 Da) is transported to the Golgi apparatus where maturation of the carbohydrate chains occurs to increase the molecular weight of the protein to 160,000. This **mature receptor** then migrates to specialized regions of the plasma membrane called **coated pits**. The receptor with its ligand enter the cell by **receptor-mediated endocytosis**, and the receptor dissociates from the ligand in the endosome and **recycles** back to the cell surface, where it is capable of a new round of LDL internalization. The cholesterol released from the LDL particles is used for the synthesis of cellular membranes, and it influences the expression of several genes involved in cholesterol metabolism, including the LDL receptor itself (Sudhof et al., 1987). Six domains have been discovered in the LDL receptor protein, which is 860 amino acids long: a signal sequence at the amino terminus, a ligand-binding domain, an epidermal growth factor precursor homology domain, a clustered O-linked sugar domain, a transmembrane domain, and a cytoplasmic domain at the carboxy terminus of the protein.

Over the years, the Brown and Goldstein laboratory and others have identified and provided molecular structural data to describe at least **four major classes of mutations** responsible for the FH phenotype (Esser et al., 1988; Esser and Russell, 1988; Goldstein and Brown, 1984; Horsthemke et al., 1986; Langlois et al., 1988; Lehrman et al., 1985). Class 1 mutations preclude the synthesis of the LDL receptor. Class 2 mutations encode LDL receptors deficient in their transport from ER to the cell surface. Class 3 mutations affect the LDL receptor proteins in their ability to bind LDL particles. Class 4 mutations produce LDL receptors that fail to cluster in coated pits on the cell surface and fail to internalize bound LDL.

The gene for the LDL receptor has been cloned (Russell et al., 1983). A number of mutations in the LDL receptor gene have been identified and their structural basis defined. An interesting feature of the LDL receptor mRNA is the presence of a very unusual 3′-untranslated region of 2 to 5 kb. It is long and has three copies of the middle-repetitive *Alu* family of DNA sequences, which is unusual for exons. It has been speculated that the presence of *Alu* sequences in the human LDL receptor mRNA may constitute the "hot spots" for genetic recombination accounting for significant variability of mutations associated with FH (Yamamoto et al., 1984).

The LDL receptor gene has been assigned by *in situ* hybridization (ISH) to bands p13.1–13.3 of **chromosome 19**, in agreement with the earlier studies using monoclonal antibodies on somatic cell hybrids (Lindgren et al., 1985). In individuals with the heterozygous and homozygous forms of FH up to 16 mutations are recognized, which include large deletions and insertions as well as single base changes resulting in nonsense or missense codons (Russell et al., 1989). These mutations are distributed all over the 45.5 kb of the LDL receptor gene and affect its various functions described above in the four biochemical classes of mutations.

Although variability of mutations of the LDL receptor gene resulting in the FH phenotype has been observed, some predominance of one type of mutation does occur. Such a prevailing deletion mutation has been detected in the French-Canadians, and consists of the removal of the promoter and exon 1 of the gene (Hobbs et al., 1987). It is likely that this particular mutation, the so-called **French-Canadian deletion**, accounting for up to 60% of the mutant genes responsible for FH in this ethnic group, is due to a **founder gene effect** given the history of relative reproductive isolation of the French-Canadians (Hobbs et al., 1987). Other common LDL receptor mutations were identified as **missense mutations** found in 14% of affected individuals (Leitersdorf et al., 1990).

Mutations of the LDL receptor gene among **Japanese kindreds with homozygous FH** were found to affect all four biochemical classes of mutations (synthesis, processing, ligand-binding activity, and internalization of the LDL receptor) (Funahashi et al., 1988). Four different rearrangements in the LDL receptor gene accounted for 17% of the mutations in the **Dutch population** sample of FH individuals (Top et al., 1990).

A **truncated receptor** failing to migrate to the cell surface results from a single nucleotide substitution that creates a premature termination codon at amino acid 660, eliminating 180 residues from the normal mature protein (Lehrman et al., 1987). This mutation is thought to account for the extraordinarily high incidence of FH in Lebanon. RFLPs in the LDL receptor gene defined by *Pvu*II and *Stu*I in South African Afrikaners also suggest the "founder gene effect" (Brink et al., 1987). *Pvu*II RFLPs have also been used in 37 Italian families to follow the inheritance of FH in a total of 79 patients (Daga et al., 1988). This RFLP allowed an unequivocal diagnosis to be made in 32.5% of the cases. Mutations in the LDL receptor locus account not only for FH (Hobbs et al., 1990), but certain DNA haplotypes in the human LDL receptor gene appear to be associated with normal serum cholesterol level; in particular, those containing the restriction site for *Pvu*II are associated with a reduction in plasma LDL cholesterol level (Schuster et al., 1990).

It appears that a large number of the LDL receptor gene mutations will be discovered that are associated with FH. Therefore, in a genetically heterogeneous population it may be very difficult to achieve a diagnosis of FH at the DNA level (Breslow, 1989). In fact, the potential **usefulness of RFLPs for diagnosis of FH** has been addressed in a study of 88 unrelated patients with FH in Germany (Schuster et al., 1989). The *Nco*I, *Apa*II, *Pvu*II, and *Stu*I RFLPs showed no difference in the relative allele frequency between the group of FH patients and controls for any of the polymorphisms. In that study, defective LDL receptor genes occurred on six different chromosomes as determined by the four RFLPs, suggesting that at least six different genetic defects could cause FH in the patients tested. Using linkage analysis, however, appropriate diagnosis can be established in over 90% of informative families using these four RFLPs.

It has been suggested that the **phenotypic expression of FH** in homozygotes appears to be dominated by the consequences of the genotypic variation at the LDL receptor locus (Thompson et al., 1989). In the case of heterozygous FH, however, not only the underlying mutation determines the phenotypic expression but also other influences such as gender, diet, and other forms of genetic polymorphism, including the apo E phenotype. Some evidence for the role of apo E polymorphisms in the phenotypic presentation of FH has been provided (Gylling et al., 1989). Female FH patients with the apo E3/3 phenotype were also found to respond better to simvastin treatment than male FH patients with the same apo E3/3 phenotype, as judged by LDL cholesterol level (de Knijff et al., 1990).

PCR assays can be used for the detection of specific mutations in the LDL receptor gene accounting for the FH phenotype. One such protocol is described for a duplex PCR amplification using two pairs of primers to identify the FH-Helsinki deletion mutation, which gives rise to a phenotype in which internalization of LDL particles by the cells is defective (Keinanen et al., 1990).

An unusual FH kindred has been described in which some relatives with a mutant LDL receptor gene had normal plasma cholesterol concentrations (Hobbs et al., 1989). Pedigree analysis suggested the existence of a **suppressor gene** that ameliorated or suppressed the hypercholesterolemic effect of the LDL receptor mutation.

Breslow (1989) emphasizes that even when taken together the combined heterozygous hypobetalipoproteinemia, FDB, and FH have a cumulative frequency of about 1 in 250 individuals, and cannot explain the **genetic variation in LDL cholesterol levels** in the general population. Other mutations involving the *apo* B gene have been proposed to exist that affect LDL cholesterol in plasma (Gavish et al., 1989a,b), the more so that a population-based sample suggested that significant *apo* B gene mutations exist in one in five persons (Gavish et al., 1989b). Consequently, variations at the *apo* B locus may account for a significant portion of genetic variations in the LDL cholesterol level. The actual alterations of the *apo* B gene accounting for this observation have not yet been identified.

2.1.4.6 LP(a)

Abnormalities in LDL metabolism may also be related to the existence of a related particle, **Lp(a)**, in which a large glycoprotein apo(a) is **covalently bound** to the apo B-100 moiety of LDL (Gaubatz et al., 1982; Morrisett et al., 1987; Trieu et al., 1991; Trieu and McConathy, 1990; Utermann, 1989; Utermann and Weber, 1983). Since the discovery of Lp(a) by K. Berg in 1963, the structure and biology of this complex lipoprotein particle have been extensively studied, in particular because of the observed **association of Lp(a) with atherosclerosis and CHD** (for a review, see Morrisett et al., 1987; Loscalzo, 1990; Utermann, 1989). Epidemiological studies pointed to a severalfold increased risk of premature CHD in persons with Lp(a) elevations in the upper 25% range (Dahlen et al., 1986; Rhoads et al., 1986). Angiographically documented CHD and saphenous vein bypass graft narrowing was markedly higher in individuals with elevated Lp(a) (Dahlen et al., 1986; Hoff et al., 1988).

Although some similarity in the atherogenicity of Lp(a) and LDL has been implied by comparable increase in the risk of CHD with elevated levels of these lipoproteins and the finding of both proteins in atheromatous lesions (Cushing et al., 1989; Rath et al., 1989; Smith et al., 1990; Smith and Cochran, 1990; Walton et al., 1974), it appears that the correlation between Lp(a) levels and lipid and lipoprotein levels is weak or absent (Boerwinkle et al., 1989a).

The **lipid** and **protein composition of Lp(a)** is quite similar to that of LDL, although Lp(a) has another protein, apo(a), attached to apo B via a cysteine residue (Fless et al., 1986; Gaubatz et al., 1987; Utermann and Weber, 1983). The combined lipoprotein particle varies widely in size among individuals, ranging between 400 and 700 kDa. The inter- and intraindividual **density heterogeneity** is reflected in electrophoretic mobility of the variants and at least six different Lp(a) protein species have been identified, using SDS-PAGE and immunoblotting, that are inherited in a codominant manner (Utermann et al., 1987, 1988 a,b).

The **size heterogeneity** has been related to variation in the number of kringle IV-like units in the apo(a) molecule (Kratzin et al., 1987; McLean et al., 1987). The apo(a) size polymorphism displays varying association with the Lp(a) concentration in different ethnic groups (Sandholzer et al., 1991). Although no effects of diet on Lp(a) level can be demonstrated (Pfaffinger et al., 1991; Van Biervliet et al., 1991), improved glycemic control in insulin-dependent diabetes mellitus decreases the Lp(a) level (Haffner et al., 1991). The decrease in Lp(a) concentration in the blood is not mediated via its removal by LDL receptors (Knight et al., 1991).

2.1.4.6.1 Lp(a) and Plasminogen

The primary amino acid and DNA sequences of apo(a) have been established, and it showed the presence of 1 kringle V-like domain, a serine protease domain, and between 15 and 40 tandem repeats resembling the kringle IV domain of plasminogen (McLean et al., 1987; Kratzin et al., 1987). The gene and protein have a striking resemblance to plasminogen and different apo (a) alleles code for proteins with a different number of kringle IV repeats (Breslow, 1989). Although there appears to be an inverse correlation between the level of Lp(a) in plasma and the size of the apo(a) protein in a given individual (Utermann et al., 1987, 1988a,b), the role of Lp(a) heterogeneity and its relationship to atherogenicity are not yet clear (Scanu and Fless, 1990). In fact, using specific DNA probes for the kringle IV and V domains of apo(a) cDNA in quantitative Southern blot analysis, a **ratio of kringle IV to kringle V** could be established (Gavish et al., 1989c). The assessment of this ratio is proposed as an **independent DNA marker for atherosclerosis risk associated with Lp(a)** (Gavish et al., 1989c).

The striking degree of homology between apo(a) and plasminogen raised the possibility that the genes encoding both proteins may be linked in the genome. In fact, by segregation analysis of the loci for plasminogen and apo(a) it has been established that the structural gene for the Lp(a) glycoprotein [apo(a)] is closely linked to the plasminogen locus on chromosome 6 (Lindahl et al., 1989).

The **similarity of apo(a) to plasminogen**, on the one hand, and an association with the atherogenicity of LDL, raised a number of questions about the role of Lp(a) in humans, and the nature of its own atherogenicity. It has been argued that although the amount of atherogenic lipids associated with Lp(a) is substantially lower than that found in LDL particles, and because Lp(a) is poorly, if at all, recognized by LDL receptors, that the lipid associated with Lp(a) may become available at sites of rapid cell degradation, active membrane biogenesis, or acute inflammatory process (Scanu and Fless, 1990). In this way, Lp(a) could contribute to atherogenesis. On the other hand, based on the structural similarity to plasminogen, it has been proposed that Lp(a) might have both **proatherosclerotic** and **prothrombotic** properties (Scanu, 1988).

Although **unmodified Lp(a)** is poorly taken up by blood-derived macrophages, treatment of the lysine residues with malonyldialdehyde markedly enhances Lp(a) uptake and degradation (Haberland et al., 1989). The more efficient uptake of **modified Lp(a)** particles is similar to the higher atherogenicity of acetylated or oxidized LDL particles. Some of the conclusions that can be drawn from the observed accumulation of Lp(a) in coronary arteries (Rath et al., 1989) and coronary artery bypass vein grafts (Cushing et al., 1989) suggest that Lp(a) may cross the endothelium of blood vessels by non-receptor-mediated mechanisms, possibly due to a direct action of Lp(a) on endothelial functions (Scanu and Fless, 1990).

One of the unresolved properties of Lp(a) is the **independence of its plasma concentration of the influence of environmental factors** such as diet and drugs, which are known to affect other lipoproteins markedly (Guraker et al., 1985; Carlson et al., 1989). The finding of a relationship between the kringle IV-to-kringle V ratio and apo(a) size and plasma Lp(a) level (Gavish et al., 1989c), the serine protease property of the neighboring amino acid sequences in apo(a) (Breslow, 1989; Scanu and Fless, 1990), as well as the ability of Lp(a) to compete *in vitro* with the binding of plasminogen to fibrinogen on fibrin monomers (Harpel et al., 1989; Loscalzo et al., 1990), with the streptokinase-mediated activation of human plasminogen (Edelberg et al., 1989; Karadi et al., 1988), and the tissue plasminogen activator-mediated lysis of fibrin clots (Loscalzo et al., 1990) — all these observations centered the attention on the **potential thrombotic role of Lp(a) in atherogenesis**.

While some researchers observed the binding of Lp(a) to insolubilized fibrin (Harpel et al., 1989; Loscalzo et al., 1988), other groups found no significant difference in the incorporation of Lp(a) and LDL into plasma clots (Kluft et al., 1989), while still others failed to demonstrate binding of Lp(a) to fibrin (Eaton and Tomlinson, 1989). Lp(a) has been found to bind reversibly to **tetranectin**, a plasma protein that itself binds reversibly to plasminogen and enhances plasminogen activation by tPA (Clemmensen et al., 1986; Kluft et al., 1989).

In clinical practice, however, even drastically elevated levels of Lp(a) were found not to affect the recanalization of coronaries in MI patient produced by prourokinase (Armstrong et al., 1989) or alteplase (rt-PA) (Tranchesi et al., 1990) thrombolytic therapy. With respect to Lp(a) interaction with cells, Lp(a) is able to compete for the binding of plasminogen to the plasminogen receptor on endothelial cells and macrophages at Lp(a) concentrations comparable to those found in the circulation (Hajjar et al., 1989; Gonzalez-Gronow et al., 1989; Miles et al., 1989).

The magnitude of affinities and capacities of endothelial and monocytoid blood cells for Lp(a), and, more specifically, for apo(a), suggests that at Lp(a) levels of 0.25 to 0.4 mg/ml, which are associated *in vivo* with increased risk of CHD, approximately 16 to 24% of plasminogen-binding sites can be occupied by Lp(a). By blocking plasminogen binding to vascular endothelial cells, **Lp(a) inhibits fibrinolysis and promotes surface coagulation** (e.g., Hajjar et al., 1989; Gonzalez-Gronow et al., 1989; Miles et al., 1989; Tranchesi et al., 1990).

Thus, it appears that Lp(a) may have both **atherogenic** and **thrombogenic** activities. In fact, the chronic formation and stabilization of mural thrombi is thought to play a significant role in atherogenesis, the more so because acute thrombotic events are frequently seen at vessel bifurcations and on the fissures of atherosclerotic plaques (Davies and Thomas, 1985; Schwartz et al., 1988). Although isolated reports of Lp(a) expressing protease activity appear (e.g., Salonen et al., 1989), at this point *the coincidental nature of the similarity of apo(a) and plasminogen cannot be fully excluded* (Miles and Plow, 1990). Lp(a) has been convincingly shown to regulate endothelial synthesis of a major fibrinolytic protein, **plasminogen activator inhibitor 1 (PAI-1)** (Etingin et al., 1991). By increasing PAI-1 expression, Lp(a) may support a specific prothrombotic endothelial cell phenotype (Etingin et al., 1991). Noteworthy, roles for plasminogen (Ichinose et al., 1991; Petersen et al., 1990) and fibrinogen (Moller and Kristensen, 1991) are being increasingly implicated among the risk factors for CHD.

2.1.4.7 LP(a) IN CORONARY HEART DISEASE

The atherogenic role of Lp(a) has been explored in FH. Its serum concentrations and phenotypes have been studied in FH heterozygotes and compared with those in patients with and without CHD (Seed et al., 1990). The frequencies of specific apo(a) phenotypes determined by immunoblotting with a monoclonal antibody were different between patients with and without CHD. The allele Lp^{S2} that is associated with high Lp(a) levels was almost *three times* as frequent among patients with CHD. On the contrary, the allele Lp^{S4} that is associated with low Lp(a) levels was almost twice as frequent among patients without CHD (Seed et al., 1990). Likewise, apo(a) levels in FH patients with CHD were found to be higher than in patients without CHD (Wiklund et al., 1990).

Thus, it appears that the level of Lp(a) and the phenotypic pattern of Lp(a) in an individual are **inherited characteristics predisposing to CHD** and are not affected by factors such as race, sex, diet, and age. Phenotypic inherited patterns are explainable, in part, in light of the documented inheritable quantitative DNA variations at the apo(a) gene level. Among these variations are polymorphisms involving the 2-kb DNA fragment detectable by the restriction enzyme *Msp*I (Kondo and Berg, 1990). Polymorphism of the apo(a) glycoprotein has been estimated to account for 41.9 and 9.6% of the observed variability in Lp(a) and total cholesterol levels, respectively (Boerwinkle et al., 1989a).

Attempts to find effective **pharmacological agents** capable of reducing the Lp(a) level in plasma are being made. Limited success has been reported with **neomycin** and **niacin**, which interfere with Lp(a) synthesis (Carlson et al., 1989; Guraker et al., 1985). A marked reduction of Lp(a) level, to the extent not seen with any other treatment, was achieved in two patients receiving *N*-acetylcysteine (NAC) (Gavish and Breslow, 1991). Clinical trials will have to be performed to assess the potential utility of this approach. Interference of NAC in the measurement of Lp(a) has been suggested, however, (Scanu, 1991), and caution in considering NAC as an efficient Lp(a)-lowering agent was advocated (Stalenhoef et al., 1991).

In a clinical setting, the **measurement of Lp(a)** is only beginning in selected laboratories. The heterogeneity of Lp(a) combined with the present scarcity of diagnostic methods, which mostly rely on immunological detection of the lipoprotein complex, require significant efforts to be directed toward rigorous standardization of the procedure. The wide spectrum of outstanding problems in the evaluation of Lp(a) has been reviewed (Albers et al., 1990).

2.1.5 GENETIC DEFECTS IN REVERSE CHOLESTEROL TRANSPORT

Transfer of cholesterol from peripheral tissues back to the liver for excretion constitutes the third pathway of lipoprotein transport (Breslow, 1989). The main lipoprotein particle involved in this transport cycle of cholesterol is **HDL**, which is primarily synthesized in the liver. **Apo A-I** is the principal protein of the nascent HDL that, when released into the circulation, attracts free cholesterol from extrahepatic cells. **Lecithin-cholesterol acyltransferase** esterifies this cholesterol in plasma, and apo A-I acts as a cofactor to the enzyme. Esterification of cholesterol induces changes in the initially discoid HDL particle, transforming it into a sphere designated as **small HDL** or HDL_3. Through further action of LCAT and LPL the particles enlarge to become HDL_2. An opposing effect of cholesterol ester transfer protein (**CETP**) and hepatic triglyceride lipase (**HTGL**) reduces the size of the HDL_2 particle to HDL_3. CETP serves to transport part of the HDL cholesterol to the apo B-containing particles (VLDL, LDL, and IDL), which participate in the disposal of cholesterol through the LDL receptor pathway, primarily in the liver (Breslow, 1989).

The recognized **genetic defects** of reverse cholesterol transport are infrequent. A homozygous form of apo A-I deficiency resulting in defective HDL production occurs at a frequency of less than 1 in 1 million in the population. Apparently related to the low HDL level, the affected individuals suffer from severe premature CHD and corneal opacities. Heterozygotes have half-normal HDL levels (Norum et al., 1989; Schaefer et al., 1982). Breslow (1989) distinguishes **three types of mutations** associated with this condition.

In **type I (apo A-I/C-III)** both apo A-I and C-III are deficient, resulting from a rearrangement at the apolipoprotein gene complex (*apo* A-I/C-III/A-IV) on **chromosome 11** that inactivates these two adjacent genes (Karathanasis, 1985; Karathanasis et al., 1983a,b, 1986). Deletion of the entire locus leads to **type II mutation**, in which apo A-I, C-III, and C-IV are all deficient (Ordovas et al., 1989). In **type III**, apo A-I is deficient due to a small insertion in the *apo* A-I gene that alters the reading frame and leads to premature termination of translation of the *apo* A-I mRNA (Breslow, 1989).

In addition to these mutations, **defective HDL metabolism** has been described due to a structural mutation in the *apo* A-I gene in the disorder called **apo A-I$_{Milano}$**, described in a local group of Italians. In spite of a reduced HDL level (33% of normal) and apo A-I (60% of normal) these individuals did not have increased risk of CHD (reviewed in Breslow, 1989).

Another case of **complete absence of HDL** associated with half-normal LCAT activity has been described in a 42-year-old patient suffering from massive corneal opacities but not CHD (Funke et al., 1991). Family studies suggested that the patient was homozygous for a gene defect in which the heterozygous phenotype is characterized by half-normal plasma HDL cholesterol and apo A-I concentrations. The gene defect proved to be a frameshift mutation in the *apo* A-I gene, as determined by direct sequencing of the PCR-amplified DNA segments containing exons of the candidate genes. The LCAT sequences were unaffected. The *apo* A-I mutation predicted an extensive alteration of the COOH-terminal portion of the defective apo A-I protein. The syndrome produced by this mutation was clinically and biochemically similar to **fish-eye disease** (Funke et al., 1991).

A number of phenotypic conditions have been described in which the **processing of HDL particles** appears to be affected. Among these is **Tangier disease**, which involves rapid clearance of HDL from circulation. The exact underlying genetic alteration has not been established so far (Breslow, 1989). Other conditions include **LPL** and **apo C-II deficiencies** (references in Breslow, 1989), as well as **apo A-II deficiency**, which is due to a single base substitution of the third intron of the *apo* A-II gene on **chromosome 1q21-q23** (Deeb et al., 1990). Defects in the *apo* C-II gene result in marked elevation of triglycerides (Connelly et al., 1987a,b; Fojo et al., 1984, 1988; Humphries et al., 1984).

No good classification of genetic defects affecting reverse cholesterol transport is available at present. The genetic variation of HDL cholesterol level in the population cannot be explained by the above-cited instances of mutations in the *apo* A-I/C-III/A-IV gene complex. The other approach to understanding the phenotypic variations of HDL is seen in studies of the related genes, and association of their polymorphisms with CHD (Breslow, 1989).

A **variant allele (*S2*)** of the *apo* A-I/C-III/A-IV complex is associated with high plasma levels in some populations, particularly in patient groups with premature CHD (Tas, 1991). Molecular genetic analysis indicates that a genetic predisposition to develop elevated plasma levels of apo C-III, alone or in combination with elevated plasma levels of apo A-IV, constitutes the primary defect responsible for the association of the *S2* allele with hyperlipidemia and premature CHD (Shoulders et al., 1991).

Frameshift (Funke et al., 1991) and **nonsense** (Matsunaga et al., 1991) **mutations** leading to **apo A-I deficiency** and premature atherosclerotic changes have been described in the *apo* A-I gene. The other gene of this complex, apo A-IV, does not appear to significantly contribute to hyperlidemias (Hanis et al., 1991a). In hemodialysis patients, the accumulation of triglyceride-rich lipoproteins appears to be related to abnormal concentrations of apo C-II, C-III, and E (Alsayed and Rebourcet, 1991).

In patients with **peripheral vascular disease**, within-individual variations in serum cholesterol seem to be associated with DNA polymorphisms at the *apo* B and A-I/C-III/A-IV loci (Monsalve et al., 1991). All the DNA markers tested so far seem to lack sufficient specificity to be clinically useful at the present time for CHD risk management. On the other hand, in **familial combined hyperlipidemia** (FCHL), a relatively common disorder, an *Xmn*I RFLP within the A-I/C-III/A-IV gene complex is associated with this condition: a linkage disequilibrium between FCHL and the 6.6-kb allele of the *Xmn*I RFLP can be unequivocally demonstrated (Wojciechowski et al., 1991)

Discussion of individual RFLPs in the A-I/C-III/A-IV complex associated with CHD has been given (Fisher et al., 1989; Humphries, 1988). In a population study in Austria, six RFLPs in the *apo* A-I/C-III/A-IV gene region demonstrated that the variation associated with some of these RFLPs was related to the lipid levels in patients and controls, but the **RFLPs themselves were deemed useless as markers for increased risk of CHD** (Paulweber et al., 1988). Alleles at this locus, or closely linked gene sequences, are thought to have major effects on lipid levels in an adolescent population and in all members of the general population (Anderson et al., 1989). At this point, only general trends toward RFLP allele-lipid level association can be observed in population studies (Anderson et al., 1989; Stocks et al., 1987).

A number of observations linking specific RFLPs with alterations in lipid metabolism associated with, or potentially leading to, CHD have been cast in the context of the possibility of the observed association being influenced by other factors contributing to the development of atherosclerosis and CHD. Thus, a specific *Sst*I RFLP in the C-III gene may be linked to a gene defect with a minor or subtle phenotypic effect that can enhance "the expression of a coexistent major monogenic defect of lipoprotein transport" (Henderson et al., 1987). In fact, a later study in 713 men related the increased CHD to specific RFLPs in the A-I/C-III/A-IV gene cluster (Price et al., 1989). While some haplotypes were independently and significantly related to the increased risk of CHD before age 60, others failed to show an association.

The selective association of some marker haplotypes in the A-I/C-III/A-IV region with significantly higher frequencies of familial CHD morbidity suggests that **predisposition to CHD may be enhanced by variants of one or more of these genes or of their regulators** (Price et al., 1990). In contrast, no association could be definitively established by analyzing allele frequencies of eight different RFLPs, within or adjacent to the A-I/C-III/A-IV gene complex, in CHD patients with confirmed narrowing of coronary vessels (Ordovas et al., 1991).

Some of the polymorphisms associated with reduced plasma levels of apo A-I can be explained by an **increased apo A-I fractional catabolic rate** when the rate of synthesis is normal (Roma et al., 1990). This observation supports the important **association of some of the observed RFLPs with the regulatory effects on gene expression**. Certain physiological aspects of HDL particles, such as their role as cofactors in LCAT activity and in the transformation occurring between HDL particles ($HDL_3 \rightarrow HDL_2 \rightarrow HDL_3$), may be reflected in the observed RFLPs (Borecki et al., 1988).

Polymorphisms affecting regulatory elements of the A-I/C-III/A-IV region account for **hyperalphalipoproteinemia** (Pagani et al., 1990). Specifically, polymorphism of the A→G transition in the *apo* **A-I gene promoter sequence** that affects transcription of the gene may influence the HDL cholesterol level. On the other hand, an increased HDL level and CETP deficiency have been related to a point mutation (G→A), in intron 14 of the *CETP* gene, that prevented normal processing of the *CETP* mRNA and yielded a null phenotype (Brown et al., 1989). The strong **antiatherosclerotic potential** of this type of HDL alteration was subsequently demonstrated in other patients (Inazu et al., 1990). **Polymorphisms in the gene coding for CETP, particularly the *Taq*I polymorphisms, appear to be related to plasma HDL cholesterol and transfer protein activity** (Freeman et al., 1990).

A more intimate role in HDL metabolism appears to emerge for a group of **steroid hormones** since the finding of regulation of the apo A-I gene by a novel member of the steroid receptor superfamily (Ladias and Karathanasis, 1991). In transfection experiments this protein, **ARP-1**, down regulated the apo A-I gene expression (Ladias and Karathanasis, 1991).

Modulation of the lipoprotein gene expression by **environmental influences** has been addressed at the gene level. In an animal study, two major loci and genotype-loci interactions were identified in baboons fed normal and atherogenic diets (Blangero et al., 1990). While most genotypes responded to the high cholesterol, saturated fat diet with an increase in apo A-I, two genotypes were identified that showed a decrease, and genetic analysis traced this decrease to the effect of one of the major loci.

An alternative explanation to the suggested absence of an association between the *S2* allele and CHD visualized the possibility that this allele (or a linked mutation) may produce mild phenotypic effects, but when combined with additional environmental or genetic factors, typical of different ethnic groups, may lead to CHD (Tas, 1991). In fact, apo C-III genotyping in Arab patients with and without MI confirmed the high frequency of this allele in the MI patients. Dietary excesses such as overconsumption of sugar and sweets among the studied ethnic groups were thought to contribute to the strength of the association of the *S2* allele with CHD (Tas, 1991).

In humans, the controversial association of a mutation in the 3′-untranslated region of the *apo* C-III gene with hypertriglyceridemia and CHD in Caucasians (Ferns et al., 1985; Ito et al., 1990; Rees et al., 1985; Tas, 1989) has not been confirmed in other studies (Aalto-Setala et al., 1987; Paulweber et al., 1988; Price et al., 1986).

Significant variability of the above-discussed observations is due, in part, to a marked variation of the allele frequency of many of the RFLPs so far examined in different populations with significant racial and ethnic heterogeneity. The multifactorial nature of atherosclerosis and CHD resulting from it may be masked in such studies by the effects of allele frequencies characteristic of contributing subpopulations, races, and ethnicities. Moreover, clinical definitions of CHD, the criteria used in sample selections, and methodological differences in many cases contribute to the inconclusive nature and limited clinical usefulness of the molecular genetic evidence accumulated so far.

2.1.6 OPTIONS IN PREDICTIVE MOLECULAR GENETIC TESTING FOR CORONARY HEART DISEASE

Mounting molecular biological evidence convincingly proves, on the whole, the role of genetic factors in the etiology of CHD; these factors appear to interact with environmental influences such as diet and life style, and with other contributing factors (diabetes, hypertension, etc.). The involvement of genetic factors in predisposing to CHD on a much broader scale than previously thought is now fully recognized. Some lipoprotein disorders such as FH and FDB have been well defined in molecular genetic terms, and identification of the related gene alterations is possible in affected individuals. This enables the specific therapy to be started earlier in the life of the patient, to at least mitigate the consequences of an unfavorable genetic make-up.

At this time, the use of predictive testing at the gene level may be limited to some RFLPs displaying strong associations with CHD (e.g., Ordovas et al., 1986), especially in family studies. Although testing for Lp(a) and apo E isoforms appears to provide a useful addition to evaluation of total, HDL, and LDL cholesterol, triglycerides, and apo A and B, *it is not clear at this time that more clinically useful information can be obtained by evaluating the respective genes*.

Undoubtedly the emerging understanding of the role of regulatory factors in modulating gene expression in lipid metabolism, the development of simpler molecular tools, and accumulation of relevant genetic data on lipoprotein disorders will make gene probe methodologies a helpful adjunct in wide predictive testing for CHD. Recent discussions of genetic aspects of predictive CHD testing have strengthened the importance of the "candidate gene" approach and the role of "variability genes" in determining the risk of an individual for CHD (Berg, 1989a,b; Chan and Dresel, 1990; Humphries, 1986; Lusis, 1988; Mahley et al., 1991; Motulsky, 1989).

Gastrointestinal Diseases

2.2.1 INTRODUCTION

Identification of either **occult** or **minor foci** of malignancies in an otherwise benign or borderline gastrointestinal lesion, or definitive recognition of a lesion at the **earliest stages** of malignancy, frequently presents a challenge. Additional objective criteria helpful in patient management and prognosis are needed to supplement histopathological assessment of lesions. **Molecular cytogenetic approaches** hold the promise of developing the least invasive means of objective assessment of gastrointestinal diseases. **Family studies** of patients with colorectal cancer indicate that in over 90% of patients, a first-degree relative is affected (Orrom et al., 1990). Coordinated efforts at establishing appropriate surveillance programs utilizing colonoscopy and biomarkers have been advocated (Lynch, 1990).

Several **models of colorectal tumorigenesis** based on the evidence accumulated in molecular genetics have been proposed (Astrin and Costanzi, 1989; Bruce, 1987; Fearon and Vogelstein, 1990; Guillem et al., 1988; Solomon, 1990), suggesting some practical approaches to the management of colorectal tumors. Although numerous DNA ploidy studies have been reported, *no unified conceptual scheme of gastric malignancies has been proposed* (Korenaga et al., 1986; Sasaki et al., 1988; Sowa et al., 1988; Tsushima et al., 1987; Yonemura et al., 1988).

In this discussion, the information on DNA content in gastrointestinal lesions, and the correlations with conventional histopathologic criteria, will be reviewed followed by evidence accumulated on chromosomal alterations and the role of oncogenes and tumor suppressor genes, growth factors and other markers.

2.2.2 DNA PLOIDY IN COLORECTAL TUMORS

Morphometric evaluation of epithelium in normal colorectal mucosa and adenocarcinoma performed by semiautomated image analysis on resection specimens reduced the number of discriminatory, informative variables in colorectal tumors to just two characteristics — **the mean nuclear cytoplasmic (N/C) ratio** and **the coefficient of variation of the nucleus to cell apex distance** (Hamilton et al., 1987). The N/C ratio was almost twice as high in adenocarcinoma cells compared to normal epithelium (39.7 ± 7.0 vs. 20.4 ± 2.0) accompanied by eccentricity of nuclei in adenocarcinoma cells (47.8 ± 9.1 vs. 19.2 ± 7.5). This approach was suggested for the more objective assessment of epithelial changes in chronic inflammatory diseases (Hamilton et al., 1987).

The same set of criteria fails to distinguish between **regenerative changes** and low-grade **dysplasia** as well as between these two groups and the normal and tumor tissues (Allen et al., 1987). All the high-grade dysplastic changes were allocated to the tumor category based on these criteria. Although this study demonstrated reproducibility of morphometric assessment of changes encountered in colorectal mucosa, operator bias in area selection and the need for visual evaluation of borderline cases, such as low-grade dysplasia with or without inflammatory component, underscore the utility of the morphometric approach as only a complementary tool in histopathologic evaluation.

Since progression from inflammatory changes through dysplasia to carcinoma is the accepted view of the developmental sequence of colorectal cancers in **chronic inflammatory bowel disease**, the objective differentiation of observed morphologic changes acquires practical significance. No general consensus exists among pathologists on some diagnostic criteria and precise evaluation remains a highly subjective process, although efforts are being made to standardize it (Riddell et al., 1983).

Adenomas are considered to be precursors of carcinoma of the large intestine, and the characterization of transitional stages can be of value for patient management. Cytofluorometry of the free cell nuclei isolated from formalin-fixed, paraffin-embedded tissue revealed no polyploid cells, no aneuploid cell clones, and low **proliferative activity** in normal epithelium at any age in either sex (Hamada et al., 1987). Increased proliferative activity in adenomas was correlated with the degree of **cellular atypia**, which was arbitrarily defined as low, moderate, and high. The proliferative activity, regardless of atypia, was higher in **adenomas adjacent to carcinomas**; however, similar to normal epithelium, neither polyploidy nor aneuploidy could be detected in the adenomas. The identification of aneuploidy in otherwise benign-appearing lesions is, therefore, considered as suggestive of colorectal carcinoma (Hamada et al., 1987).

An important **comparison of DNA FCM with cytogenetic analysis** of colorectal adenocarcinomas indicates that aneuploidy detected by FCM accurately reflects the chromosome content of the tumor over a broad range of variation from DNA diploid to hypertetraploid tumors (Remvikos et al., 1988). Because **polyploidization** appeared to be part of the progression of colorectal tumors with respect to malignancy and worse prognosis, DNA aneuploidy, apparently accurately reflecting chromosomal changes, may be seen as indicative of malignization (Remvikos et al., 1988). To what extent a certain degree of aneuploidy observed in adenomas (Goh and Jass, 1986; Quirke et al., 1986; Van den Ingh et al., 1985; Weiss et al., 1985) may reflect the precursor relationship of adenomas to carcinomas is, in this author's opinion, virtually impossible to infer. The molecular cytogenetic evidence discussed below suggests a much more complex correlation between gene level changes, diagnosis, and prognosis in colorectal carcinomas.

To investigate the possibility of the development of colorectal carcinoma *de novo*, rather than as a progression from the preexisting adenoma, the DNA ploidy patterns of colorectal carcinomas and mucosa with severe dysplasia have been studied and correlated with their macroscopic morphology (polypoid or crater-shaped lesions) (Hamada et al., 1988). Seventy-seven of the crater-shaped carcinomas were aneuploid as opposed to 17% of the polypoid tumors of the large intestine. This difference is interpreted to mean that approximately 60% of the crater-shaped invasive carcinomas may develop from small, nonpolypoid carcinomas.

Total nuclear DNA content as measured by flow cytometry (FCM) or image cytometry (ICM) is clearly a gross measure of the eventual accumulation of changes occurring at the chromosomal level, and is **insensitive to fine alterations at the gene level**. Nevertheless, attempts continue to be made to obtain as much practically useful information as possible from these techniques, which can be automated and hopefully made sufficiently objective.

Numerous studies have been reported on the use of **DNA content** as an objective criterion for differential diagnosis and/or prognosis of colorectal neoplasias. Cytophotometric evaluation of colonic mucosa of patients with longstanding **ulcerative colitis** (UC) yielded histogram patterns that identified low-grade dysplasia with a broad unimodal pattern of **hypertetraploid** cells (27%), and areas of high-grade dysplasia with 62.5% of cases being **aneuploid** (Cuvelier et al., 1987). DNA content histograms in adenocarcinomas of various degrees of differentiation were different. Correlation with histopathologic evaluation suggested that DNA content can be used for prognostic purposes, because aneuploidy, especially in dysplastic lesions, was frequently found in (although it could not be predictive of) invasive cases.

Polyploidy or aneuploidy is frequently (in 77% of cases) found in dysplastic tissue in UC (Suzuki et al., 1990). A good correlation exists between the frequency of polyploid cells and the grade of dysplasia as determined by **microspectrophotometric DNA analysis**.

Furthermore, a better prognosis could be entertained with **polypoid DNA content** in carcinomas to those with aneuploid pattern. Although there are benefits in quantitative assessment of epithelial changes for the diagnosis and prognosis of colorectal malignancies, **the finding of aneuploidy per se should not be overinterpreted to indicate or predict invasion.** Unfortunately, no distinction could be seen by this method between low- and high-grade dysplastic lesions, or between low- and high-grade malignant tumors (Cuvelier et al., 1987).

In order **to increase the informative content of the observed nuclear DNA aneuploidy** in colorectal carcinomas, the DNA content data have been subdivided into near diploid, hyperdiploid, hypotetraploid, tetraploid, and hypertetraploid according to the **DNA index** (Zaloudik et al., 1988). Several other parameters were correlated with these ploidy subdivisions such as **histologic grade**, vascular and serosal *invasion*, and the expression of **secretory component** (SC); as a marker of epithelial differentiation, the **carcinoembryonic antigen (CEA)** level and the expression of the **MHC class II antigens** were also measured (Zaloudik et al., 1988). The method could detect as few as 10% of cells with aneuploid nuclear DNA, but reproducibility on replicate samples was poor (discordance of up to 20%), resulting from either **sampling imperfection** or the **intrinsic heterogeneity** of the tumors. The dominant population of cells (over 50%) in all tumors has always been near-diploid, suggesting that a variable, but substantial, number of tumors must be diploid. The near-diploid population was heterogeneous, comprising hyperdiploid, diploid, and hyperdiploid cells unresolvable by this technique. Heterogeneity was also observed in the aneuploid population of cells. A correlation between serosal invasion and the tentatively "more dynamic hypotetraploid, tetraploid and hypertetraploid tumors" was established (Zaloudik et al., 1988).

Another statistically significant correlation could be seen between the **near-diploid DNA content** and the **secretory component**, as well as between the **HLA class II antigen expression** and that of the secretory component. However, no clearcut correlations of practical value could be established except to emphasize the heterogeneous nature of diploid and aneuploid nuclear DNA populations. **Multiple sampling** is strongly advocated to reduce the effects of marked heterogeneity of ploidy characteristics in individual samples from colorectal carcinomas (Koha et al., 1990).

A **correlation of ploidy with prognosis** of patients with colorectal carcinomas has been repeatedly demonstrated, particularly in cases with **Dukes' C stage** cancers (Armitage et al., 1985; Kokal et al., 1986; Melamed et al., 1986; Schutte et al., 1987; Wolley et al., 1982) as well as **in stage A** and **B** cases (Albe et al., 1990; Baretton et al., 1991a; Hood et al., 1990; Quirke et al., 1987; Scott et al., 1989). Although DNA FCM analysis is insensitive to minor chromosomal aberrations, this modality can be of practical value inasmuch as it provides the clinician with information on abnormal DNA content statistically correlated with poor survival (Schutte et al., 1987).

Extensive studies of colorectal adenocarcinomas by **FCM of fresh whole-cell suspensions** (Hood et al., 1990), or by **ICM of fresh samples** (Albe et al., 1990), emphasize the **lack of correlation** between ploidy and prognosis within individual Dukes' stages. Although tumors of higher histologic grade tended to be aneuploid, this correlation was not statistically significant, and DNA aneuploidy did not correlate with Dukes' stage (Hood et al., 1990). Overall survival analysis, taking into account the size of the tumor, CEA levels, sex, age, histologic differentiation, Dukes' stage, and ploidy **did not reveal independent prognostic significance of ploidy when all Dukes' stages were considered** (Albe et al., 1990). Ploidy as well as Dukes' stage and the degree of differentiation were, however, significantly **correlated with survival in patients who had had surgery**. Thus, ploidy status could be viewed as "an independent predictor of overall and disease-free survival for patients with early stages of disease" (Albe et al., 1990).

The FCM data on DNA ploidy and cell proliferation may be, however, no better than Dukes' staging for prognosis, especially in **rectal cancer** (Quirke et al., 1987; Scott et al., 1989). In fact, in a large prospective study, DNA and RNA indices as well as the proliferative characteristics derived from FCM analysis failed to be of independent prognostic value (Enker et al., 1991).

Others, however, maintain that the prognostic indicators in **colonic adenocarcinoma**, in particular DNA ploidy, seem to differ from those of **adenocarcinoma of the rectum** (Robey-Cafferty et al., 1990). Aneuploidy is a significant predictor of poor overall and disease-free survival in patients with rectal malignancy (Kouri et al., 1990; Robey-Cafferty et al., 1990), but not in those with colonic tumors (Robey-Cafferty et al., 1990).

A large morphometric study of DNA content in colorectal cancer conducted on excised specimens from 416 patients indicated that **pathological stage, local tumor extension**, and **DNA ploidy** were all independent prognostic indicators of patient survival following surgery (Armitage et al., 1990). The **histological grade, tumor size,** and **assessment of "curability"** were not; however, the strength of the pathological grade as predictor was greater than that of DNA ploidy or local tumor extension. Among the various factors related to the differing biologic behavior of the tumor in these two locations, the variability in DNA measurements in multiple tumor samples of colonic adenocarcinoma may contribute to the reported discrepant prognostic relationships (Wersto et al., 1991).

Substantially wider areas of involvement of **mucosa adjacent to carcinoma** of the large bowel than previously thought have been suggested by the finding of a high frequency of aneuploidy in mucosa up to 10 cm from a colorectal cancer (Ngoi et al., 1990). Interestingly, only a diploid pattern was found in mucosa adjacent to cancers with diploid DNA pattern. Ploidy patterns did not show any correlation with histologic abnormalities in the adjacent mucosa. Confirming previous reports, cells displaying abnormal proliferative properties were observed in the superficial component of colorectal mucosa. It is unclear to what extent methodologic aspects of the study contributed to this finding, but widespread occurrence of aberrations suggest that "the large bowel mucosa behaves as a unit in carcinogenesis", justifying close observation of patients undergoing surgical treatment for colorectal cancer (Ngoi et al., 1990).

Significant heterogeneity (difference in DNA ploidy pattern) is observed between the **primary lesions** in colorectal carcinomas and liver **metastases** (Yamaguchi et al., 1990). The DNA ploidy pattern of the metastatic lesions did not correlate with the metastasis period, the number of lesions, or extent of their spread; however, overall prognosis in diploid tumors was much better than in aneuploid tumors.

Although summary evaluation of ploidy is claimed to reflect accurately the chromosomal complement of the cells (Remvikos et al., 1988) other, more fine characteristics of chromatin DNA have also been evaluated. One such study correlated the degree of **susceptibility of chromatin DNA to acid denaturation** with different stages of colorectal tumors (Kunicka et al., 1987). Indeed, DNA denaturability *in situ*, assessed by FCM, was considered to be a sensitive probe for discriminating **tumors of different aggressiveness**. However, as will be discussed below, substantially more specific probes reflecting alterations at the gene level are available, making evaluation of the relatively gross physicochemical characteristics of chromatin DNA an unlikely candidate as a practically useful approach in managing patients with colorectal and, for that matter, other types of neoplasias.

2.2.3 DNA PLOIDY IN GASTRIC TUMORS

Only a very loose correlation could be observed between the nuclear DNA content of gastric cancers (**papillary adenocarcinoma, tubular adenocarcinoma** of various degrees of **differentiation**, and **mucinous adenocarcinomas**) and the clinical and histologic parameters (Sowa et al., 1988). A significant positive correlation was noted, however, between polyploidy and metastases to lymph nodes or liver.

The **histologic type of gastric cancer** did not correlate with DNA ploidy and proliferative activity measured by bromodeoxyuridine (BrdU) labeling *in vivo* in 129 cases of primary gastric cancers (Yonemura et al., 1988). Again, cases with metastatic spread tended to have a higher incidence of aneuploidy, and survival rate was higher in patients with diploid tumors. The BrdU labeling indices of early and advanced gastric carcinomas were not significantly different (8.1 vs. 11.9%, respectively), whereas that of diploid cancers was 6% (Yonemura et al., 1988).

The DNA ploidy pattern as determined by FCM also appears to display a significant correlation with survival and tumor grade in **gastric smooth muscle tumors** (**leiomyomas**, **leiomyosarcomas**, and **leiomyoblastomas**) (Tsushima et al., 1987). The larger tumors tended to be tetraploid/polyploid. This regularity was not absolute, however, because some of the histologically and biologically benign gastric smooth muscle tumors were aneuploid. On the other hand, some higher grade leiomyosarcomas with distinct malignant behavior were diploid. Thus, although DNA ploidy cannot be used to differentiate between benign and malignant gastric smooth muscle tumors, it can still be used as a prognostic tool.

DNA ploidy assessment has some prognostic value in gastric carcinomas only in advanced tumors with lymph node metastases (Baretton et al., 1991b; Kimura and Yonemura, 1991). Comparison of DNA ploidy in **primary gastric carcinomas** and **metastatic lesions** revealed that in 59% a change in DNA ploidy occurs in the metastases (Baba et al., 1990). The wide heterogeneity of ploidy in the primary lesions becomes markedly reduced in metastatic tumors, and the tumor cells with a lower DNA ploidy tend to metastasize to lymph nodes, particularly in cases of a differentiated carcinoma.

Multiple samples from a tumor specimen must be evaluated for the accurate determination of DNA ploidy in gastric cancers, as emphasized by the FCM finding of **intratumoral regional heterogeneity** of DNA ploidy parameters (Sasaki et al., 1988). This heterogeneity appears to be substantially higher in gastric carcinomas (40% of cases) than in colorectal carcinomas (7.4% of cases), varying from 60% for a single sample to 80% for multiple samples. Substantially lower DNA aneuploidy is reported in gastric adenocarcinomas (50 to 56%) (Hamada et al., 1988). Nevertheless, stability of the DNA ploidy pattern is characteristic of a given gastric carcinoma, both primary and recurrent (Korenaga et al., 1986).

2.2.4 CHROMOSOME ABERRATIONS

A loss of specific chromosomal regions in colorectal adenomas and carcinomas has been well documented and related to alterations in **tumor suppressor genes** and **oncogenes** implicated in the development of neoplasias (for a review, see Fearon and Vogelstein, 1990). The gene responsible for the inherited predisposition to the development of multiple colorectal adenomas, **familial adenomatous polyposis** (**FAP**), has been identified and assigned to chromosome 5q21-22 by linkage analysis (Bodmer et al., 1987; Leppert et al., 1987; Okamoto et al., 1988, 1990; Solomon et al., 1987). Although **loss of heterozygosity** (**LOH**) of 5q in colorectal adenomas is rare, it is observed in up to 50% of colorectal carcinomas (references in Fearon and Vogelstein, 1990). A very high percentage (22%) of family kindreds in cohort studies of FAP have been found to have spontaneous mutations as well as a higher incidence of colorectal carcinoma and extracolonic manifestations of FAP (Rustin et al., 1990).

A gene showing tumor-specific LOH in at least 50% of informative cases of colorectal carcinoma encodes the **colony-stimulating factor 1 receptor**, and has been mapped to **5q21-22**, close to the *FAP* locus (Gope et al., 1990b). Its role appears not to be limited to the development of hundreds of polyps in the intestines of affected people. Mutations at, or very near, the *FAP* locus predispose people to a lower grade of intestinal polyposis as well (Leppert et al., 1990). Furthermore, a possibility exists that alleles of the *APC* locus (responsible for FAP) may be associated with susceptibility to colon cancer not necessarily accompanied by polyposis. At this time, no firm practical conclusions can be drawn concerning the prognosis of patients demonstrating allelic changes in the *FAP* locus, and a more detailed understanding of the role of mutations at or near this locus on chromosome 5q is needed before specific clinical decisions can be made (Leppert et al., 1990).

Specific genetic changes seem to occur in benign adenomas that involve the **D5S43 locus on chromosome 5**. They appear to have a role in malignization of the epithelium, as suggested by the finding of precancerous changes only in those adenomas having this loss (Rees et al., 1989). In fact, the C11p11 probe as well as two other closely linked polymorphic DNA probes, **pi227** and **YN5.48**, which flank the *FAP* locus, allow the establishment of a presymptomatic diagnosis of FAP and can be used for prenatal diagnosis with higher than 99.9% reliability in the majority of families (Tops et al., 1989).

The original probe (**C11p11**) used to establish linkage to the *FAP* locus had limited utility for family studies because of low heterozygosity and significant distance from the *FAP* gene (Paul et al., 1990). The other probes useful in linkage analysis, **cKK5.33** and **pi227**, substantially enhance risk calculation, allowing predictive testing in presymptomatic FAP kindred, surveillance, genetic counseling, and development of treatment regimens.

Another chromosome loss, at **17p**, while infrequent in adenomas, is also present in colorectal carcinomas, appears to be involved in the progression of an adenoma to malignancy, and significantly is linked to the *p53* **gene** located on 17p (Baker et al., 1989; Sasaki et al., 1989; also see references in Fearon and Vogelstein, 1990). The loss of the *p53* activity associated with point mutations of this gene, and loss of the wild-type allele, is considered to impart a selective growth advantage to affected tissues leading to tumor progression (Fearon and Vogelstein, 1990) (see below).

Chromosome 18q is lost in more than 70% of colorectal carcinomas, and in up to 50% of late adenomas (Fearon and Vogelstein, 1990). The work of Vogelstein's group had identified a gene, termed *DCC* (for **d**eleted in **c**olon **c**ancer), that was assigned to this region. The gene encodes a protein that displays a significant homology to the **adhesion family of molecules**, it is expressed in normal colonic mucosa, and its expression is reduced or absent in the majority of colorectal carcinomas (Fearon and Vogelstein, 1990).

Losses of alleles from other chromosomes have also been observed in colorectal carcinomas: **1q, 4p, 6p, 6q, 8p, 9q,** and **22q** are lost in 25 to 50% of tumors, and the remaining chromosome arms are lost in 7 to 24% of the tumors (Vogelstein et al., 1989). Frequent chromosome 1p deletions, almost always involving band **1p35**, have been observed in colorectal carcinomas, particularly late in tumor development (Leister et al., 1990). The frequency of these deletions was higher in metastatic than in primary tumors. **Fragile 8q22** (Shabtai et al., 1988), **loss of 12q** (Okamoto et al., 1988), and clonal **allele loss of 17p** (Fey et al., 1989) have also been reported.

A certain regularity in the **order of appearance of cytogenetic changes** along the progression of the **adenoma-carcinoma sequence** has been observed in patients with FAP (Miyaki et al., 1990). Loss of heterozygosity affecting 5q, found in less than 2% of moderate adenomas, increased to 20% in severe adenomas. Carcinomas limited to the mucosa had LOHs on 5q (26%) and 17p (38%), whereas invasive carcinomas demonstrated multiple LOHs, on 5q (52%), 17p (56%), 18q (46%), and 22q (33%). Increase in the frequency of Ki-*ras* mutations was also noted in the more atypical adenomas (11% in moderate vs 36% in severe adenomas) (Miyaki et al., 1990). *In vitro* models of tumor progression from adenoma to carcinoma have been established and reviewed (Paraskeva et al., 1990).

The following progression of events is suggested: the heterozygous mutant/wild-type condition at the *APC* gene causes development of mild to moderate adenomas, while the loss of the normal allele in the *APC* gene accounts for a transition from moderate to severe adenoma, as does the Ki-*ras* mutation. Conversion of the latter to intramucosal carcinoma occurs following the LOH on chromosome 17p, and further progression to invasiveness involves additional LOHs on chromosomes 18q and 22q.

Other chromosomal aberrations have also been consistently observed. A **reduction in length of the telomeric repeat sequences** is noted in chromosomes from colorectal carcinomas compared to that of normal mucosa from the same patient (Hastie et al., 1990). A reduction in the telomere arrays and in the average telomere length also accumulates with age in blood and colon mucosa.

Although the role of **terminal repeat arrays** (**TRAs**) is not known, the length of TRAs in adenomas is reduced to the same extent as that found in the corresponding carcinomas, suggesting that **TRA reduction may be involved in malignization**. One explanation of this finding is that reduction in TRAs in colorectal carcinomas can be due to **accumulated progressive losses** occurring during cell division with age, which is consistent with the known age distribution of colorectal malignancies. It would be interesting to correlate the consistent allelic losses with the TRA reduction in colorectal carcinomas (Vogelstein et al., 1989). In either case, evaluation of chromosome aberrations by specific probes designed to identify character-istic loss of chromosome material may provide a practical molecular tool for better management of patients with colorectal tumors. The availability of specific oligonucleotide probes makes this objective quite realistic.

There appears to be a difference between gastric and colorectal cancer with respect to consistent chromosomal losses and loci involved, with **gastric cancers** frequently having LOHs at 1q and 12q (Fey et al., 1989). A highly specific chromosomal region at **11p13-15** harboring nonrandom rearrangements has been identified in gastric and esophageal adenocarcinomas, which could prove to be of diagnostic value (Rodriguez et al., 1990)

The genetic and molecular cytogenetic studies provide compelling evidence supporting the role of specific molecular events in the development of colorectal carcinomas. Several models that attempt to establish interrelationships between individual molecular events and suggest plausible diagnostic molecular tools have been proposed. These probes can be also used for patient management as alternatives to frequent colonoscopies and as reasonable guides for selecting optimum surgical strategies. Before discussing these models, studies dealing with oncogenes, TSGs, growth factors, and other tumor markers in colorectal neoplasias will be briefly discussed.

2.2.5 *RAS* FAMILY ONCOGENES IN COLORECTAL AND GASTRIC NEOPLASIAS

Mutations in all three members of the *ras* family of protooncogenes (Ha-*ras*, Ki-*ras*, and N-*ras*) have been studied with respect to their prevalence, temporal occurrence, association with specific histopathologic characteristics, and their prognostic value in gastric and colorectal neoplasias. Significant disparity exists in the results obtained by various investigators (Astrin and Costanzi, 1989).

Since activation of *ras* oncogenes by point mutations is known to produce a potentially transforming gene, and activation of *ras* alleles has been frequently observed in primary colorectal tumors and derivative cell lines (Astrin and Costanzi, 1989), highly discriminatory tools applied to a large sampling are needed to resolve existing controversies, and to establish specific clinicopathologic correlates of potential practical value.

Using an amplification-ASO assay, mutations of the *ras* gene were identified in 11 of 27 colorectal cancers examined (Bos et al., 1987). Mutations occurred predominantly (10 out of 11) in the **c-Ki-*ras* gene**, mostly at **codon 12**. One case of a mutation of codon 12 of the N-*ras* gene was also found. Importantly, the same mutation could be frequently identified in both the benign (adenomatous) and the malignant (carcinomatous) portions of the same tumor, implying at least the temporal relationship of the mutations to malignization. A similar high prevalence of mutations at codon 12 of the Ki-*ras* oncogene has been detected by other workers, both in adenomas and in carcinomas presumably originating from these adenomas (Forrester et al., 1987). The simultaneous presence of mutated Ki-*ras* and N-*ras* in both a colon adenocarcinoma and a premalignant villous adenoma was noted in the same series. No correlation could be established between the *ras* oncogene mutations and the degree of invasiveness or progression of the tumors.

Likewise, although *ras* mutations are found much more frequently in **larger adenomas** (58%, in tumors over 1 cm) than in the smaller ones (9%, in tumors less than 1 cm), **no significant correlation with the clinicopathologic grade** (location, invasion, degree of differentiation) could be established (Vogelstein et al., 1988). In contrast, in differentiating clonal cell lines derived from the colon carcinoma cell line HT29, the Ha-*ras* expression was found to be highest in the most differentiated cells of the colon (Augenlicht et al., 1987). Moreover, its expression was markedly diminished in metastatic lesions compared to primary colon tumors, suggesting a role for the Ha-*ras* oncogene in differentiation. Although this oncogene could not be directly implicated in transformation or tumorigenicity, there was a greater than a fivefold increase in Ha-*ras*, and a somewhat smaller increase in Ki-*ras*, protooncogene expression in the **mucin-secreting** cells. Again, no such increase could be detected in cells that were not mucin-secreting, pointing to **some association of the *ras* oncogene expression with a specific histologic type of differentiation of colonic epithelial cells**.

Consistent with the presence of mutated *ras* oncogenes in adenomas (Bos et al., 1987), a high percentage of adenomas (32.4%) show detectable levels of *ras* p21 product expression identified by the **monoclonal antibody RAP-5** (Lanza, 1988). The degree of p21 expression correlates with the degree of epithelial dysplasia and the size of adenoma, but not the histologic type. Adenocarcinomas show an even greater presence of *ras* p21 (77.1%), more intense staining, and staining of a greater number of cells than in adenomas. No consistent pattern could be observed in metastatic lesions, suggesting that *ras* p21 product, although playing a role in colon tumorigenesis, is not associated with metastatic tumor progression (Lanza,1988).

The expression of the **Ha-*ras* p21 product** is stronger in **deeply invasive** than in superficially invasive colorectal carcinomas chemically induced in laboratory animals; p21 is expressed in all subserosal tumors (Yasui et al., 1987). This observation suggested a role for Ha-*ras* p21 in tumor proliferation, invasion, as well as in promoting metastases, where this protein was also increased.

In contrast, evaluation of the level of *ras* gene product p21 in Dukes' B and C **rectal adenocarcinomas** by the RAP-5 monoclonal antibody indicated that it **correlates with survival** independent of Dukes' stage (Michelassi et al., 1988). Since high levels of p21 correlate with a more aggressive behavior of the tumor, accompanied by more frequent recurrences and worse prognosis, **measuring p21 may be recommended for the identification of patients at high risk for recurrent disease** (Michelassi et al., 1990).

The *ras* p21 product, revealed by the RAP-5 monoclonal antibody as absent from normal gastric mucosa or benign gastric lesions was, however, strongly expressed in gastric carcinomas as well as in colonic carcinomas irrespective of the degree of tumor differentiation (Czerniak et al., 1987). This agrees with other observations of overexpression, but not activation, of the *ras* genes in gastric carcinomas (Fujita et al., 1987). The reason for the discrepancies observed in *ras* family oncogene mutations in esophageal and gastric carcinomas is not clear (Jiang et al., 1989).

The high incidence of **colorectal carcinomas** developing **in patients with familial polyposis coli**, and the high frequency of point mutations of the *ras* family oncogenes, particularly Ki-*ras*, can be interpreted as **suggestive of a cause-and-effect relationship** reflected in the pattern of oncogene alterations. However, the frequency and the sites where Ki-*ras* gene mutations occur (codons 12 and 13) are similar in colorectal carcinomas arising in FAP patients, and in sporadic colorectal carcinoma, suggesting these mutations are **not the initial event** in colorectal tumorigenesis (Sasaki et al., 1990).

On the other hand, in addition to Ki-*ras* mutations identified by PCR in both adenomas and carcinomas, in diploid as well as aneuploid cells, these mutations could also be detected in the **histologically benign mucosa** adjacent to carcinoma (Burmer and Loeb, 1989). These findings favor the view that mutations in Ki-*ras* **occur early** in the malignization process, before changes in ploidy take place (Burmer and Loeb, 1989).

Analysis of Ki-*ras* protooncogene mutations by PCR in dysplasia or carcinoma associated with **chronic ulcerative colitis (UC)** and in **sporadic colon carcinomas** showed that mutations in codon 12 (52%) frequently occur in sporadic cases and only occasionally in UC-related cases (Burmer et al., 1990). Thus, genetic pathways of tumorigenesis in these two categories of colorectal carcinomas appear to be different. Similar conclusions are reached by other investigators in **dysplastic gastrointestinal lesions** and in cancers presumed to arise from dysplasia, although they failed to identify by PCR any point mutations in the second exon of Ki-*ras*, or in and around **codons 12, 13**, and **61** of **N-*ras*** and **Ha-*ras*** (Meltzer et al., 1990). In the meantime, the unsurpassed diagnostic potential of PCR in a rapid, nonradioactive, large-scale format has been used to demonstrate the presence of mutations in codons 12 and 13 of the Ki-*ras* gene in samples of colorectal, but not esophageal or gastric, carcinomas (Jiang et al., 1989).

An activated N-*ras* oncogene mutation has been detected in a case of **poorly differentiated adenocarcinoma** by an *in vivo* **selection assay** in nude mice transfected with NIH373 cells (Nishida et al., 1987). A more sensitive and specific assay using PCR amplification has also revealed point mutations in codon 12 of Ki-*ras* in approximately 20% of **gastric adenocarcinomas** (Miki et al., 1990). These were present only in tumors with the intestinal type of mucosal changes, not in other histologic types of tumor.

2.2.6 C-MYC

The level of expression of the *c-myc* gene in normal gastric and colonic mucosa is relatively low, and increased levels become detectable by the **monoclonal antibody 1-6E10** when inflammatory, metaplastic and dysplastic changes occur in chronic inflammatory bowel conditions, such as in **Crohn's disease** and **ulcerative colitis** (Ciclitira et al., 1987). Using a different antibody (**R5452**), the expression of **p62$^{c\text{-}myc}$** was undetectable in early gastric carcinomas, but 36.4% of advanced carcinomas were positive (Yamamoto et al., 1987). Stromal cellular components also demonstrated positivity for p62$^{c\text{-}myc}$, which was correlated with the better prognosis. Evaluation of the relative amount as well as the distribution of p62$^{c\text{-}myc}$ by immunohistochemical methods revealed an abnormal pattern of distribution in the mucosa of patients with FAP (Sundaresan et al., 1987). The staining patterns in polyps and carcinoma were markedly different from those in immediately adjacent, visibly uninvolved mucosa.

Contrasting with these observations is the uniform expression of the *c-myc* throughout the entire thickness of the colonic epithelium demonstrated by ISH and immunohistochemistry using a polyclonal rabbit anti-*c-myc* antiserum, arguing against the elevated expression of the *c-myc* gene in proliferating cells (ten Kate et al., 1989). Nevertheless, an enhanced *c-myc* expression has been repeatedly observed in gastrointestinal carcinomas, and in the majority of cases this did not involve aberrations of the *c-myc* gene (for a review, see Astrin and Costanzi, 1989).

Marked amplification of the *c-myc* gene has been found in mucinous (53.8% of cases) and poorly differentiated tumors (42.8% of cases), suggesting a **correlation between amplification of a member of the *c-myc* gene family with particularly aggressive subtypes of colorectal**

carcinomas (Heerdt et al., 1990). Supporting the involvement of chromosomal aberrations and enhanced c-*myc* expression, cell lines demonstrating numerous **double minute chromosomes**, loss of the **13 chromosome** and **Y chromosome**, as well as a 16- to 32-fold amplification of the c-*myc* oncogene have been successfully established from human **signet ring cell gastric carcinoma**, known for the worst clinical prognosis of all types of gastric cancer (Yanagihara et al., 1991).

Elevated expression of the c-*myc* gene appears to correlate with the **degree of differentiation** of colorectal neoplasms. c-*Myc* expression declines in human colon carcinoma cell lines induced to differentiate by either *N*-**methylformamide** (Chatterjee et al., 1989) or **transforming growth factor (TGF)-β** (Mulder et al., 1988; Mulder and Brattain, 1988). Nevertheless, a greater degree of c-*myc* mRNA overexpression is present in moderately and poorly differentiated carcinomas compared to well-differentiated tumors, and it seems to play an important role in the progression of colorectal carcinomas (Klimpfinger et al., 1990). In contrast, c-*fos* mRNA expression is higher in nonneoplastic tissue than in colorectal tumors and appears to be related to cell differentiation (Klimpfinger et al., 1990; Untawale and Blick, 1988). Overexpression of c-*myc* mRNA significantly correlates with chromosome 5q deletions apparently being associated with the hyperproliferative state of neoplastic colorectal tumors induced by chromosome 5q allelic deletion (Maestro et al., 1991). The level of expression of c-*myc* may be related to **polyamines**, which, in turn, are linked to **ornithine decarboxylase**, a rate-limiting enzyme in polyamine biosynthesis that is markedly increased in cells stimulated to grow (Celano et al., 1988).

The other *myc* protooncogene, **L-*myc*, failed to show any correlation** with conventional characteristics of colorectal cancers (Dukes' stage, metastases, or the degree of differentiation) (Ikeda et al., 1988), in contrast to the finding that L-*myc* RFLPs are related to the progression of human colorectal and gastric cancer (Kawashima et al., 1987). A low level of the N-*myc* - and L-*myc*-specific transcripts can be detected in normal colonic mucosa. These genes are frequently, if only modestly, overexpressed in colorectal polyps and carcinomas (Finley et al., 1989). L-*myc* overexpression occurs more frequently in adenomatous polyps than in tumors; it accompanies the c-*myc* overexpression and amplification, although the level of c-*myc* amplification does not correlate with the level of L-*myc* expression (Finley et al., 1989).

2.2.7 *ERB*A AND *ERB*B GENES

In normal human colonic mucosa, one *erb*A-β and two *erb*A-α-related transcripts of thyroid hormone receptors are expressed (Markowitz et al., 1989). These transcripts are also present in premalignant polyps and in adenomatous epithelium in cell culture. A marked and selective decrease of the *erb*A-β, but not of the *erb*A-α transcripts, is observed in colon carcinoma. A 6-kb transcript was markedly less expressed in eight of eight patients, the majority of whom already had widely metastasized tumors. Although no correlation with histopathological and clinical features was apparent, this decrease was not due to an alteration of the *erb*A-β gene, and was viewed as a novel marker of malignant transformation in the colon. Since *erb*A-β expression is characteristic of normal colonic epithelium, its loss in malignancy apparently reflects a specific transformation event whose mechanism remains unclear (Markowitz et al., 1989).

c-*erb*B-1, the **epidermal growth factor (EGF) receptor gene**, is amplified 16- to 32-fold along with a severalfold amplification of c-*myc* and a novel activation, at codon 13, of the Ha-*ras*-1 oncogene in human oral squamous cell carcinoma cell lines (Tadokoro et al., 1989). Coamplification of the c-*erb*B-1 (30-fold) and c-*myc* (5-fold) has been seen in a human gastric cancer representing simultaneous amplification of protooncogenes encoding the growth factor receptor (c-*erb*B-1) and the nuclear proteins (c-*myc*) (Nomura et al., 1986).

The other protooncogene of the *erb* family, c-*erb*B-2, also known as *neu* (*HER2*), encodes a transmembrane glycoprotein with **tyrosine-specific protein kinase** activity (Fukushige et al., 1986; Slamon et al., 1989; Yamamoto et al., 1986; Zhang et al., 1989). Prominent expression of the *neu* gene product, **p185^{neu}**, has been observed in the mucosal epithelium of stomach, small intestine (predominantly villi), colon, liver parenchyma, the exocrine and endocrine pancreas, and the salivary gland (Cohen et al., 1989). Normal mucosa shows positive immunostaining only on the luminal surface, with no expression in the crypts.

This protein is more abundant in adenomatous and preneoplastic polyps than in adjacent normal colonic epithelium, or areas of malignant degeneration. Its expression is also decreased in cell lines derived from the more anaplastic colonic tumors (see also D'Emilia et al., 1989). Although a role for the *neu* gene and the p185^{neu} product, which resembles the growth factor receptor in the regulation of cell growth and differentiation, appears highly probable, correlations between *neu* gene alterations, the p185^{neu} product expression, and colorectal tumor characteristics remain to be fully defined.

The expression of an **aberrant HER2/neu polypeptide** with a molecular weight of 190,000 (exceeding that of the normal HER2/*neu* protein by 5000) has been observed in colon adenocarcinomas (Tal et al., 1988). A rearrangement in the 3′ region of the *HER2/neu* gene appears to be responsible for this aberrant product. Amplification of the *HER2/neu* gene does not necessarily lead to aberrations in the EGF receptor.

The level of *neu* expression in **gastric cell lines** is comparable to that in colorectal carcinoma cell lines (Park et al., 1990). A more than fivefold amplification of the *neu* gene has been observed in 21% of gastric and esophageal adenocarcinomas (Houldsworth et al., 1990). A comparable frequency of the expression of p185^{neu} has been demonstrated in gastric adenocarcinomas by a polyclonal antibody assay of formalin-fixed, paraffin-embedded tissue (Falck and Gullick, 1989).

The ***erb*B-2 gene** is amplified in 25% of metastatic, but not early primary, tumors (regardless of the histologic type) with occasional coamplification of *hst*-1 and *int*-2 genes, suggesting that amplification of multiple oncogenes occurs during progression and metastatic spread of human gastric carcinomas (Tsujino et al., 1990).

*erb*B-2 protein expression is not associated with histologic type or venous invasion of **gastric cancer** (Yonemura et al., 1991). However, it is significantly associated with serosal invasion, lymph node metastases, lymphatic invasion, and the high number of lymph node metastases. In addition to higher risk of lymph node recurrences, patients with *erb*B-2/*neu*-positive tumors had a fivefold greater relative risk of death compared to *neu*-negative tumors. The high malignant potential of the *neu* positive tumors is apparently associated with the very high potential for lymph node metastasis. Thus, **erbB-2/*neu* protein expression can serve as an independent prognostic factor in gastric carcinomas.**

2.2.8 TUMOR SUPPRESSOR GENES

2.2.8.1 *p53*

The frequent loss of chromosome 17p alleles and the location of the *p53* gene in that area (17p13.1) (Baker et al., 1989) suggested the involvement of this gene in colorectal carcinomas. In conformity with the notion of suppressor genes, one observes frequent point mutations in one allele of the *p53* gene associated with the loss of the wild-type allele in gastric and colorectal tumors, similar to that in lung, breast, and brain tumors (Baker et al., 1989; Davidoff et al., 1991; Eliyahu et al., 1989; Finlay et al., 1989; Friedman et al. 1990; Hollstein et al., 1991; Iggo et al., 1990; Kim et al., 1990b; Levine et al., 1991; Nigro et al., 1989; Takahashi et al., 1989). While deletion of one allele of the *p53* gene is present in 75% of colon cancers, mutation of the remaining *p53* allele has been detected in 10 out of 10 colon cancers examined (Markowitz et al., 1990). **Transfection studies** with the mutant and wild-type alleles in a nontransformed adenoma cell line harboring a mutant Ki-*ras* oncogene showed that expression of the wild-type *p53* is compatible with normal growth of these cells. The transforming activity appears on joint expression of mutant *ras* and *p53* genes (Markowitz et al., 1990).

Analysis of *p53* gene mutations by sequencing cloned PCR products, and evaluation of the allelic losses by RFLPs, revealed the **low frequency of *p53* gene mutations in adenomas** irrespective of the size and the fact that the adenomas occurred in FAP patients (Baker et al., 1990). In carcinomas and adenomas, the *p53* gene mutations were infrequent (17%) when both copies of chromosome 17p were present, while the frequency rose to 86% if one copy of 17p was lost. These changes usually occurred near the point of transition from benign to malignant morphology, pointing to the causal role of the *p53* gene in malignization.

Using an **RNase protection assay** and **Northern and Southern analysis** to study abnormalities in the *p53* gene and its expression, the **incidence of *p53* gene abnormalities** was found to be similar in gastric and colorectal carcinoma cell lines, being somewhat higher in the cell lines than in the tumors themselves (Kim et al., 1990b). Aberrant *p53* alleles have also been identified in one third of human esophageal tumors as well (Hollstein et al., 1990; Meltzer et al., 1991). Amplification by PCR combined with direct sequencing revealed the presence of *p53* mutations in several exons, suggesting that mutations at this locus are common genetic events in the pathogenesis of **squamous cell carcinomas of the esophagus**. In hepatocellular carcinomas the pattern of mutations of the *p53* gene is somewhat different from that found in the lung, colon, esophogus, and breast, but the mutant *p53* protein is involved in selective clonal proliferation of hepatocytes in HCC (Bressac et al., 1990; Hsu et al., 1991). Asymmetric PCR-based assays can reliably and rapidly identify *p53* mutations as a diagnostic and prognostic molecular tool (Bartek et al., 1991). Assessment of **p53 protein** expression by immunohistochemistry showed positive tumors were significantly more frequent in the distal colon and had a higher rate of cell proliferation (Scott et al., 1991). No correlation with tumor grade, Dukes' stage, DNA ploidy, or patient survival has been established. This study, however, points to some biological distinctions between proximal and distal tumors of the digestive tract.

2.2.8.2 *RB*

The majority of colorectal carcinomas also have been found to overexpress compared to adjacent normal mucosa concomitantly the retinoblastoma (*RB*) and *p53* mRNA (Gope et al., 1990a, c). This increase in the expression of both genes was present in the primary tumors as well as in the derivative cell lines. A nearly proportional coincidental increase of the **histone *H3* gene** in the same cells, which is limited to the S phase of the cell cycle and reflects proliferative activity, suggests that the increased expression of *RB* and *p53* in colorectal carcinomas may be associated with the increased number of cycling cells in tumor tissue.

2.2.8.3 *DCC*

The other tumor suppressor gene, termed **DCC** (see Section 2.2.4), has been identified at **chromosome 18q** (Fearon et al., 1990). *DCC* is expressed in normal colonic mucosa, and **its expression is reduced or absent in the majority of colorectal carcinomas**. The homology of the *DCC* gene product to the cell adhesion family of molecules, and a decrease in its expression in cancer are believed to be related to the altered adhesion properties of mucosal cells and derangement of growth-restricting controls (Fearon and Vogelstein, 1990).

A strong confirmation for the role of *DCC* and *APC* genes as tumor suppressor genes in colonic tumors comes from the finding that introduction of normal human chromosome 5 (with the *APC* gene) and 18 (with the *DCC* gene) into a human colon carcinoma cell line, COKFu, through microcell hybridization, completely suppresses the tumorigenicity of the hybrid cells and normalizes their growth in culture (Tanaka et al., 1991). Similar results have been obtained on introduction of chromosome 5 into DT cells (Hoshino et al., 1991).

Yet another putative tumor suppressor gene operating in colon cancer appears to be the **protein kinase C** (*PKC*) gene when it is expressed at high levels. This is suggested by studies in cultured colon cancer cells stimulated by phorbol esters (Choi et al., 1990).

2.2.8.4 *MCC*

The *MCC* gene (**m**utated in **c**olorectal **c**ancer) is located at **chromosome 5q21** and encodes an 829-amino acid protein showing a similarity to the G protein-coupled m3 muscarinic acetylcholine receptor (Kinzler et al., 1991a). The presence of rearrangements and somatically acquired point mutations in *MCC* resulting in amino acid substitutions in sporadic colorectal carcinomas makes this a candidate for the putative colorectal tumor suppressor gene. *MCC* is present in normal colonic mucosa and, like the *APC* gene, is also expressed in a wide variety of tissues. *APC* is expessed in fetal tissues, lung carcinoma, bladder carcinoma, osteosarcoma, fibrosarcoma, and normal fibroblasts (Kinzler et al., 1991b). These two genes are located in close proximity to each other and encode proteins predicted to contain α-helical coiled-coil regions (Kinzler et al., 1991b). Both *MCC* and *APC* genes are somatically altered in tumors from colorectal cancer patients, and it appears that mutations in the *APC* gene may contribute to both FAP and Gardner's syndrome (Nishisho et al., 1991).

2.2.9 OTHER GENES ASSOCIATED WITH TUMORS OF THE DIGESTIVE TRACT

Understanding the role of histocompatibility antigens in human tumors has become increasingly important and may hold promise as yet another molecular tool in the diagnosis and management of patients with neoplasia (Eisenbach et al., 1983a,b; Festenstein and Garrido, 1986; Festenstein and Schmidt, 1981; Ljunggren and Karre, 1986; Smith et al., 1988). HLA class I gene expression is completely lost in some cases in colorectal, gastric, and laryngeal carcinomas, whereas a selective loss of HLA-B antigens and a different pattern of expression of class I antigens is found in primary tumors and metastases (Lopez-Nevot et al., 1989). Only laryngeal, but not gastric or colorectal, carcinomas displayed a strong correlation between the degree of differentiation and HLA class I gene expression. This is in contrast to the earlier findings of a correlation between the loss of HLA class I expression and the degree of differentiation of colorectal carcinoma. However, Southern analysis of the HLA class I genes in several tumors reveals no loss or rearrangement of DNA (Lopez-Nevot et al., 1989). The search for possible molecular genetic markers helpful in the earliest detection of tissue changes predicting subsequent neoplasias continues and a number of cloned DNA sequences are being tested for possible nonrandom association with specific abnormal phenotypes (Augenlicht et al., 1991; Dunlop et al., 1991; Hirohashi and Sugimura, 1991). The higher level of expression of the *MDR*1 gene appears to correlate with the higher degree of differentiation of colorectal and gastric carcinomas (Mizoguchi et al., 1990). Yet another oncogenic rearrangement (TPR-MET) has been detected even in precursor lesions of gastric carcinoma (Soman et al., 1991).

2.2.9.1 ORNITHINE DECARBOXYLASE

Ornithine decarboxylase is a rate-limiting enzyme catalyzing the conversion of ornithine to putrescine, a member of the family of low molecular weight polyamine bases, which together with spermine and spermidine are known to be essential for cellular proliferation and differentiation (Heby, 1981; Pegg and McCann, 1982). Their levels are elevated in rapidly growing tissues, in precancerous polyps (Welch and Welch, 1976), and in cancer (Janne et al., 1978), including colorectal cancer (Fenoglio and Pascal, 1982; Kingsnorth et al., 1984; La Muraglia et al., 1986; Moorehead et al., 1986; Porter et al., 1987).

Two genes encoding ODC, *ODC*1 and *ODC*2, have been localized to **chromosomes 2p25** and **7q31-qter**, respectively (Radford et al., 1990). Although the activity of *ODC* is elevated in colorectal polyps and carcinomas compared to that of adjacent uninvolved and control mucosa, no amplification of the loci could be detected, suggesting that the regulation of *ODC* in colorectal neoplasia occurs at the posttranscriptional level.

2.2.9.2 LAMININ

Laminin is one of the major components of basement membranes and has been implicated in numerous cellular functions, including attachment, spreading, migration, cell growth, differentiation, mitogenesis, and binding to extracellular matrix components (reviewed in Mafune et al., 1990). Some of these functions are essential for tumor invasion and growth, and differential expression of laminin has been observed in cell lines having different metastatic potential. Using a cDNA clone encoding a human **laminin-binding protein**, a significant increase in this protein mRNA expression has been observed in colorectal carcinomas (Mafune et al., 1990). This can be used as a marker for tumor progression and aggressiveness in clinical practice, especially because the increased expression of laminin receptor is specifically associated with the cancer cells and displays a correlation with the Dukes' classification and the invasive phenotype of colon carcinoma (Cioce et al., 1991).

2.2.9.3 *c-src*

The 10- to 30-fold increased activity of the product of the c-*src* protooncogene, $pp60^{c-src}$, is observed in cultured human colon carcinoma cell lines (Bolen et al., 1987a). The increased activity is related to the rate of phosphate turnover from tyrosine residues in the carboxy-terminal portion of $pp60^{c-src}$ (Bolen et al., 1987b), due to activated protein kinases (Cartwright et al., 1989), rather than to increased synthesis of the c-*src* encoded protein.

2.2.9.4 *trk*

Yet another protooncogene, *trk* (**t**ropomyosin **r**eceptor **k**inase), which encodes a hybrid molecule that contains parts resembling nonmuscle tropomyosin combined with tyrosine kinase (similar to that found in retroviral transforming genes), has been identified in colon carcinomas (for a review, see Fenoglio-Preiser et al., 1991). This 790-amino acid protein product has several features characteristic of cell surface receptors, including a 32-amino acid putative signal peptide, an N-glycosylation area, a transmembrane domain, a kinase catalytic region highly related to other tyrosine kinases, and a very short carboxy-terminal tail (Martin-Zanca et al., 1989). The product of the human *trk* protooncogene encodes a 140,000-Da glycoprotein, designated **gp140$^{proto-trk}$**. The primary product of *trk* is a 110,000-Da protein that is immediately glycosylated to yield the mature gp140$^{proto-trk}$. This product is a novel tyrosine kinase cell surface receptor whose ligand has not yet been identified.

The hybrid nature of gene products in colon carcinomas is not limited to gp140$^{proto-trk}$, and is also observed in a fusion product of **ubiquitin** and the **ribosomal protein S27a**, encoded by a novel **ubiquitin hybrid protein gene** (Wong et al., 1990). Its expression in cultured cancer cells and the kinetics of induction of this gene by serum growth factors show a pattern almost identical to that of protooncogenes c-*jun* or c-*fos*, both known as early growth response genes. This novel gene is overexpressed in tumor tissues compared to adjacent normal mucosa as determined by Northern blot in 10 of 13 patients. A role for this gene in the pathogenesis of human colon carcinoma is also suggested by an observed correlation between the level of its expression and clinical tumor staging (Wong et al., 1990).

2.2.9.5 HUMAN PAPILLOMAVIRUS

In addition to the previously described presence of HPV type 6 in **laryngeal papilloma**, and its putative role in the progression to carcinoma (Zarod et al., 1988), an intriguing, although still unexplained, association of human papillomavirus (**HPV**) infection and colon carcinomas has been demonstrated by immunohistochemistry and ISH (Kirgan et al., 1990a,b). A correlation was observed between the viral burden and the degree of malignant transformation of the glandular epithelium (Kirgan et al., 1990b). The viral distribution demonstrated by ISH was not limited to the distal parts of the colon, and was uniform throughout the large bowel. Moreover, analysis of human colon adenocarcinoma cells revealed the central clearing of chromatin, openings of the nuclear envelope characteristic of a lytic virus, and the presence of 45 to 55-nm viral particles consistent with the size of HPV, all within the cell nuclei. Further studies are certainly warranted to confirm these preliminary observations by more rigorous molecular biologic techniques (Southern blot, PCR) (Kirgan et al., 1990b).

2.2.9.6 CARCINOEMBRYONIC ANTIGEN

The well-known tumor marker, carcinoembryonic antigen (**CEA**), is a cell surface glyco-protein whose expression greatly increases in human colon carcinomas. With the recent cloning of the *CEA* gene, it has become possible to establish some aspects of the biological role this glycoprotein has in colon cancer (Beauchemin et al., 1987; Kamarck et al., 1987; Oikawa et al., 1987a,b; W. Zimmerman et al., 1987, 1988). CEA has a strong similarity to members of the immunoglobin superfamily that includes **intercellular adhesion molecules (ICAM)** (Benchimol et al., 1989). Interestingly, analysis of the various characteristics of CEA expression by immunocytochemistry in relation to those of oncogene products showed that CEA and T antigen were expressed most frequently, raising the possibility that these are expressed at an earlier stage in the neoplastic process than the *ras* gene products p21 and p21^{Ser} (mutated form) (McKenzie et al., 1987). CEA mRNA expression is much stronger in carcinomas than in adenomas, and is not present in noninvaded tissue (Higashide et al., 1990). The possibility of CEA having a role in carcinogenesis and in metastasis through its involvement in tissue architecture is very intriguing. In an experimental system, exogenously added CEA induced metastases, suggesting a direct or indirect role of CEA in the metastatic process (H. E. Wagner et al., 1990).

2.2.9.7 K-*sam*

The K-*sam* gene, encoding a heparin-binding growth factor receptor with a tyrosine kinase activity, has been isolated from stomach cancer cells, where it is markedly amplified (Hattori et al., 1990). This amplification, however, is restricted to poorly differentiated types of stomach cancer. The biological significance of this gene remains to be established.

2.2.9.8 *nm*23

The *nm*23 gene encodes a potential tumor metastasis suppressor, it is expressed in normal colonic mucosa, and its expression rises early in colon carcinogenesis. It remains increased in metastatic colon cancers, suggesting that, in the colon, a dissociation of *nm*23 gene expression from loss of tumor metastatic competence occurs in a tissue-specific manner (Haut et al., 1991). Using analysis of the *nm*23 allelic deletions with the *nm*23-H1 probe, a prognostic evaluation of colorectal carcinoma with respect to the development of distant metastases becomes possible (Cohn et al., 1991).

2.2.10 GROWTH FACTORS

The emerging understanding of the role of certain protooncogenes in the development and/ or progression of gastrointestinal tumors suggests that evaluation of protooncogenes may be of practical significance in patient management. Less is known about the utility of assessing **growth factor (GF)** genes and their products for diagnostic purposes. Activated cellular protooncogenes may disturb signal transduction pathways involved in growth regulation and thereby lead to loss of control over growth and differentiation (Weinstein, 1989).

Tumor suppressor genes, known to be consistently altered or deleted in various neoplasias, including those of the gastrointestinal tract, also encode products essential for the maintenance of normal **growth controls**. It appears that pleiotropic alterations of protooncogenes, tumor suppressor genes, and of their products are responsible for the development of malignancies in the digestive tract (see Section 2.2.13; Fearon and Vogelstein, 1990). The importance of evaluating GFs and their receptors stems from the realization that it is the aberrant expression of these that is, in part, responsible for autonomic growth of tumors through **autocrine stimulation** (Sporn and Roberts, 1985).

Coamplification of the *hst*-1 and *int*-2 genes occurs in approximately 50% of the primary squamous carcinomas of the esophagus and in 100% of the metastatic tumors (Tsuda et al., 1988, 1989a,b). The *hst*-1 gene product, first isolated from a gastric adenocarcinoma, bears a 40% homology to the mouse *int*-2 gene product and fibroblast growth factor (FGF) (Sakamoto et al., 1986; Yoshida et al., 1987, 1988). It is a heparin-binding GF for mouse fibroblasts and human vascular endothelial cells (Miyagawa et al., 1988). In spite of their amplification, these genes do not appear to be expressed in esophageal carcinomas. A positive correlation has been observed, however, between **coamplification** of the *hst*-1 and *int*-2 genes and the clinical stage or the presence of metastases (Miyagawa et al., 1988).

Amplification and overexpression of the **EGFR gene** occur in 71% of squamous cell carcinomas (SCCs) of the esophagus (Hollstein et al., 1988; Ozawa et al., 1987), and a good correlation exists between the EGF expression and the depth of tumor invasion and prognosis (Tahara, 1990).

Normal human gastrointestinal mucosa contains **TGF-α** (up to 4776 pg/wet weight of tissue) and **urogastrone epidermal growth factor** (**URO-EGF**) (0 to 216 pg/wet weight) (Cartlidge and Elder, 1989). Higher levels of expression of TGF-α and EGFR (c-*erb*B-1) genes are present in gastric carcinomas than in normal gastric mucosa (Yoshida et al., 1990), In one third of cases, the EGF gene is also expressed. About 10% of gastric carcinomas show amplification of the EGFR (c-*erb*B-1), c-*myc*, c-*yes*, and c-Ki-*ras* genes (Tahara, 1990; K. Yoshida et al., 1989). **c-erbB-2** is frequently found in adenocarcinoma of the stomach, and even more frequently, in metastatic lesions of any histologic type (Tsujino et al., 1989; Yokota et al., 1988b). A positive correlation has been observed between the depth of tumor invasion, poor prognosis, and coexpression of the EGF and EGFR genes (Yasui et al., 1988). Another correlation has been found between the degree of malignancy or the presence of metastases of gastric carcinomas and coexpression of TGF-α and c-Ha-*ras* p21 (Yamamoto et al., 1988).

Other GFs expressed in gastric carcinomas are platelet-derived growth factor (**PDGF**) and **TGF-β**, frequently simultaneously, suggesting a possible role for PDGF in the production of **scirrhous changes** (Tahara, 1990). In addition to TGF-β, the *sam* gene amplification is present only in **scirrhous carcinoma** and **poorly differentiated gastric adenocarcinoma** (T. Yoshida et al., 1989). In contrast to the situation observed in breast cancers, the synchronous expression of the estrogen receptor and EGFR portends a worse prognosis than that of the ER-negative tumors (Yokozaki et al., 1988).

In **colorectal carcinomas**, the genes for TGF-α, EGFR, EGF, PDGF-A, PDGF-B, and PDGF-R are expressed, some of these simultaneously (Tahara, 1990). Almost 50% of colorectal cancers show the presence of mRNA for **tumor necrosis factor (TNF)**, whereas only 6 of 26 matched normal areas do (Naylor et al., 1990). The level of TNF gene expression does not correlate with the stage of disease, degree of lymphocytic infiltration, or tumor necrosis. Analysis of tissue distribution of TNF mRNA by ISH revealed its presence only in a small fraction of macrophages and predominantly in stromal cells (Naylor et al., 1990).

An attempt to develop a practically useful marker for the presence of colorectal carcinomas based on the determination of EGF in the urine (**urogastrone**) although revealing a statistically significant difference between patients and control, could not unequivocally demonstrate its clinical utility (Sweetenham et al., 1990). In the case of breast cancer patients, no difference between the patients and controls was observed (Sweetenham et al., 1990).

Gastrin promotes the growth of gastrointestinal epithelial cells, and thereby acts as a GF (Tahara, 1990). Gastrin positivity is found in 17% of **gastric adenocarcinomas**, and in 50% of

scirrhous gastric carcinomas (Tahara, 1988). High-affinity **gastrin receptors** have also been identified in 65% of colorectal carcinomas, correlating with a better prognosis (Upp et al., 1989). Specific interaction appears to exist between gastrin and EGF, and possibly other **tyrosine protein kinases** (Tahara, 1990); however, the clinical usefulness of these observations for patient management at this time is not clear.

2.2.11 DNA METHYLATION

Changes in gene expression associated with the degree of **methylation of cytosines** at the 5′ position by a methyltransferase enzyme specific for the **CpG** sequence has been recognized as an important characteristic of neoplasia (Silverman et al., 1989). Although the predominantly observed change in carcinomas is either **total genomic hypomethylation** or **selective hypomethylation** of specific genes, hypermethylation patterns have also been observed. While hypermethylation appears to decrease consistently gene expression, hypomethylation has a variable effect — an increase in gene expression may depend on multiple other cellular factors.

In lung cancers, lymphomas, and acute myeloid malignancies, extensive hypermethylation of several genes on the short arm of **chromosome 11**, including the **calcitonin gene**, has been observed (Baylin et al., 1986, 1987; de Bustros et al., 1988). A specific pattern of changes in DNA methylation in the calcitonin gene suggests that even in the presence of widespread genomic hypomethylation the consistency of hypermethylation patterns in colon carcinoma may be associated with progression through various stages of malignization (Silverman et al., 1989). Thus, the novel pattern of calcitonin gene methylation is a feature of the majority of colonic adenomas and established colon carcinoma cell lines, but not of colon carcinomas.

2.2.12 GENOMIC IMPRINTING

The phenomenon of genetic imprinting (GI), implying the preferential expression of maternally, or paternally, derived genes, may in part be mediated through variations in DNA methylation (Cattenach and Kirk, 1985; see Section 1.1.6). Fine analysis of differences in the degree of methylation of specific alleles has been described in the Ha-*ras* gene, using restriction enzyme digestion, electrophoresis, and recovery of the bands from gels on ion-exchange paper (Bedford and van Helden, 1990). This technique can simplify and make the analysis of specific oncogene methylation patterns in clinical material much easier and reproducible. It remains to be seen whether the phenomenon of GI, which frequently amounts to the functional suppression of expression of certain genes (Sapienza, 1991), plays a role in the inheritance pattern of gastrointestinal neoplasias.

2.2.13 CONCLUDING REMARKS: CURRENT HYPOTHETICAL MODELS OF COLORECTAL TUMORIGENESIS

Confronted with a problem as widespread and complex as the development of colorectal tumors in humans, one is tempted to organize the accumulated evidence in the hope of finding certain reproducible features in the etiology and molecular mechanisms that would allow coherent analysis of clinical cases. Only a few such attempts by different research groups, relevant to events occurring at the gene level, will be discussed here.

The pros and cons of five hypotheses on the origin of colon cancer (Bruce, 1987) that are probably the farthest removed from the level of genetically operated mechanisms are as follows: (1) the role of **fecapentaenes** as potent mutagens produced in the colonic lumen, and acting as initiators of colon cancer, is not supported by current findings; (2) the effect of **3-ketosteroids** as cytotoxic and genotoxic substances initiating or promoting colorectal tumors, although tentatively implicated in *in vitro* studies, still remains far from established *in vivo*; (3) the significance of nuclear aberrations allegedly produced by **pyrolysis products** in food is far from clear, although theoretically possible; (4) the importance of **calcium deficiency** in enhancing the presence of free **bile** and **fatty acids**, which are toxic to colonic epithelium, is not supported by experimental data, which fail to demonstrate the protective effect of calcium; and last, (5) the impact of **high fecal pH**, leading to epithelial proliferation and to an increased sensitivity to carcinogens, could not be demonstrated unequivocally. In spite of that, Bruce (1987) appears to pursue the elucidation of the role of risk factors associated with diet in colonic tumorigenesis.

Turning to molecular genetics data, Astrin and Costanzi (1989) analyzed the role of oncogenes and tumor suppressor factors and proposed a model for colon carcinogenesis. They suggest that the disease has at least two forms. One is characterized by the loss of FAP gene function and deregulation of c-*myc* expression; the other involves an unidentified tumor suppressor gene locus that predisposes to cancer in the proximal (right sided) colon and does not involve c-*myc*. In both forms, the activation of the *ras* oncogene by point mutations plays a significant role.

Essentially two events are seen by Astrin and Costanzi (1989) as responsible and necessary for the neoplastic transformation of colonic epithelium: (1) disruption of the cell cycle control by deregulated c-*myc* coupled with (2) activation of the signal transduction as a result of point mutations in *ras*. The authors admit that multiple genetic events, occurring essentially at random without any particular order, are involved in tumor formation in the colon.

Offering an overview of the role of tumor suppressor genes in malignancy, Skuse and Rowley (1989) emphasize the relative imprecision of both the various terms used to define this class of genes as well as the limitations of current knowledge of their role in various types of tumors. In the case of colorectal carcinomas, the accumulated evidence points to losses involving multiple chromosomes. They argue that by the time the diagnosis is made, multiple mutational events are likely to have occurred involving both recessive and dominant oncogenes. They emphasize the importance of the study of chromosomal and molecular changes occurring in cases of Mendelian predisposition to tumorigenesis.

No "relevant alteration of the copy number and/or genomic structure" of N-*myc,* L-*myc* and *p53* genes has been detected by Dolcetti and coworkers (1988), leading them to believe that "at least in human colorectal malignancies, it is unlikely that nuclear oncogene structural alterations and/or amplification plays a major role in tumor induction or progression".

A **model of colorectal tumorigenesis** developed by the Vogelstein group, incorporating their own discoveries in molecular genetics studies as well as those of others, appears to offer the most comprehensive synthesis of the currently available evidence (Fearon and Vogelstein, 1990; Vogelstein et al., 1988, 1990). Their model comprises the following concepts and findings: (1) Tumorigenesis is a result of **accumulated changes** associated with mutations of at least four to six genes; (2) most of these mutations affect **tumor suppressor genes *p53*** and ***DCC*** (located at loci 17p and 18p, respectively), lost in over 75% of cases, and at least six additional loci, each lost in over 25% of the carcinomas; (3) *ras* is the only **consistently amplified oncogene** in colorectal carcinomas, although other oncogenes may be occasionally amplified as well. Mutations at codons 12 and 13 occur in over 50% of intermediate and late-stage adenomas and in carcinomas; (4) **hypomethylation** of DNA is an early event apparently promoting aberrant segregation leading to the loss of wild-type tumor suppressor genes; (5) although the *FAP* **gene** (mapped to 5q) is only infrequently lost in the sporadic adenomas of FAP patients, it is affected in over a third of cases; (6) a certain **order of genetic changes** can be recognized, with hypomethylation occurring very early, followed by *ras* gene mutations and 5q allelic losses, which precede 18q allelic losses, and the 17p allelic deletions completing the sequence. It is, however, the ***total accumulation of changes rather than the order of changes*** that is important. Based on this coherent model, the practical uses of the molecular genetic findings are envisaged (Vogelstein et al., 1990).

Identification of the defective genes in the blood of patients known to be predisposed to FAP or the Lynch syndrome can be performed early on, sparing regular colorectal examinations in individuals who do not have the affected genes. Mutant genes and their products, such as *p53* and *ras* genes, can be identified in the blood or feces, also by immunological methods. Prognostication based on the evaluation of the total accumulated allelic deletions in an individual patient may suggest the likelihood of potential metastases irrespective of the histopathologic features of the tumor at the time of evaluation. This may provide clinicians with helpful information for individualizing therapeutic regimens.

Hepatocellular Carcinoma

2.3.1 ANIMAL MODELS: CHEMICALLY INDUCED HEPATOCELLULAR CARCINOMA

2.3.1.1 PLOIDY ASSESSMENT

Extensive studies have been performed to elucidate the molecular mechanisms of chemical carcinogenesis in the livers of laboratory animals, mostly rats and mice. The earliest morphological events produced in the rat liver in the course of **chemically induced carcinogenesis** are the appearance of enzyme-altered foci (EAF), or altered hepatic foci (AHF) marked by enzyme elevations (Farber et al., 1987; Goldsworthy et al., 1986). These preneoplastic foci are thought to contain aneuploid cell populations, which have been studied for their **nuclear DNA content** by cytospectrophotometry (Wang et al., 1990b). The DNA content of **γ-glutamyltranspeptidase-positive foci** was measured to reveal that hepatocytes from control livers, and cells adjacent to the foci in treated livers, had the 2C, 4C, and 8C ploidy patterns, whereas in AHF the ploidy pattern was predominantly diploid, tetraploid, or heterogeneous. Because there are always cells with a normal DNA complement in tumor cell populations, **aneuploidy may not be the necessary condition for an irreversible commitment to malignancy** (Wang et al., 1990b). It constitutes an increased risk factor for carcinogenesis, reflecting **genomic instability** as a stage in the multistep process of carcinogenesis. Although the cellular DNA content in liver increases with the development of HCC, no such increase accompanies transition of chronic hepatitis to liver cirrhosis (Lin et al., 1990).

In **hepatoblastomas**, malignant hepatic tumors of childhood, the pattern of chromosomal aberrations (**trisomy 2q, trisomy 20**, and loss of heterozygosity [**LOH**] **at 11p**) is similar to that observed in pediatric embryonal rhabdomyosarcomas, suggesting a certain parallelism in the genetic pathways (Fletcher et al., 1991).

2.3.1.2 ONCOGENE STUDIES

Focal areas of hepatocytes marked by enzymatic elevations in the course of chemical carcinogenesis at the gene level reveal alterations of several oncogenes. These observations have led to the following conclusions (Beer and Pitot, 1989; Corral et al., 1988; Nagy et al., 1988):

1. One of the early events in chemical carcinogenesis is spontaneous mutation of Ha-*ras* at the first and second positions of **codon 61**.
2. The "**hot spots**" are genetically regulated at **codons 12** and **61** of Ha-*ras* playing a role in the susceptibility to chemical carcinogenesis (e.g., Buchmann et al., 1991). However, truncation of the 5′ exon of the Ha-*ras* gene, rather than hot spots, is thought by others to be involved (Duesberg, 1985).
3. The establishment of a causal relationship of activation of certain oncogenes to carcinogenesis in these models is complicated by the observation of an increase in Ha-*ras*, Ki-*ras*, c-*myc*, and c-*fos* protooncogene expression due to cell proliferation in the course of regenerative growth of affected liver parenchyma (Beer et al., 1986; Corcos et al., 1987;

Y. Li et al., 1988; Li and Lieberman, 1989; McMahon et al., 1986; Tashiro et al., 1986; Sinha et al., 1988).

4. Studies by the Pitot laboratory established that an increased expression of Ha-*ras* and c-*myc* is absent in the majority of preneoplastic foci; however, Ha-*ras* appears to be activated at the stage of progression (Beer and Pitot, 1989; Watatani et al., 1989). This is supported by *in situ* hybridization (ISH) studies failing to demonstrate any notable increase in the expression of Ha-*ras* and c-*myc*.

5. While mutational and transcriptional activation of protooncogenes in the rat model is observed at the stage of progression, in the mouse some protooncogenes appear to participate in carcinogenesis at earlier stages, but their specific role is not clear (Wiseman et al., 1986).

6. An alteration of the **c-*raf* mRNA levels** and the **gap junction protein** appears to be involved in the malignant development of preneoplastic foci (Beer et al., 1988).

7. Others believe that some of the early, and possibly critical, events in chemical hepatocarcinogenesis include transient elevation of c-*fos* (Kruijer et al., 1986), c-*myc* (Chandar et al., 1987, 1989), and Ha-*ras* (Galand et al., 1988). N-*myc* rises earlier than c-*myc*, and both protooncogenes are thought to be involved in the entry of hepatocytes into the cell cycle during the prereplicative stage of liver regeneration (Corral et al., 1988; Nagy et al., 1988).

8. Expression of the **α-fetoprotein (*AFP*)** genes detected by single-stranded RNA probes was present only 5 weeks after the initiation of hepatocarcinogenesis (Evarts et al., 1987). Interestingly, at later stages hepatocytes in the enzyme-altered foci and in neoplastic nodules were always negative for AFP. In fact, ISH analysis of *AFP* mRNA can be helpful in detecting preneoplastic cells in human liver cirrhosis that cannot be distinguished morphologically or by immunohistochemistry (Otsuru et al., 1988).

9. It appears that activation of **γ-glutamyltranspeptidase** and **glutathione transferase P** is induced by the activated *ras* protooncogene expression (Y. Li et al., 1988).

2.3.2 VIRALLY INDUCED HEPATOCELLULAR CARCINOMA IN ANIMAL MODELS

Enhanced expression of the c-*myc* oncogene is seen in **woodchuck hepatitis virus** (WHV)-induced HCC (Moroy et al., 1986). In addition, a high frequency of integration of WHV DNA into N-*myc* genes has been established (Fourel et al., 1990).

2.3.3 HUMAN HEPATOCELLULAR CARCINOMA

2.3.3.1 PLOIDY OF HUMAN HEPATOCELLULAR CARCINOMA

The larger the tumor, the lower the percentage of diploid cells detected in HCC (Ezaki et al., 1988). Tumors under 5 cm in diameter have 65% of cells showing a diploid DNA pattern and 35% having the nondiploid pattern. Those over 5 cm typically have 82% of cells featuring the nondiploid pattern. Although no correlation between ploidy and the presence of HBsAg or liver cirrhosis can be established, the DNA pattern does correlate with the morphologic grading and age of patients. The **ploidy pattern, however, fails to be of use in the prognosis of patients with HCC** (Ezaki et al., 1988).

2.3.3.2 CHROMOSOME ABERRATIONS

Allele loss on **chromosome 16q** is a common genetic alteration in **human HCC**, and a restriction fragment length polymorphism (RFLP) analysis of 70 surgically resected tumors with 15 polymorphic DNA markers revealed LOH in 52% of informative cases (Tsuda et al.,

1990). The most frequently affected region maps to an area between **16q22.1** and **16q22.3-q23.2**, and is detected in poorly differentiated larger tumors, and those with metastases, rather than at the earliest stage of hepatocarcinogenesis. These chromosome alterations are not associated with **hepatitis B virus (HBV)** DNA integration or **hepatitis C virus (HCV)** infection. They reflect, however, later stages of hepatocarcinogenesis associated with the enhancement of aggressiveness of HCC.

Frequently LOH has also been observed on **chromosome 4** (Buetow et al., 1989; Pasquinelli et al., 1988; Zhang et al., 1990), **11p** (Rogler et al., 1985; Wang and Rogler, 1988), **13q** (Wang and Rogler, 1988), and **17q** (Slagle et al., 1991).

In fact, a study of allelotype alterations in primary human HCC, using 44 RFLP markers, confirmed and emphasized the frequent occurrence of losses possibly involving **tumor suppressor genes** located on chromosomes **5q, 10q, 11p, 16q,** and **17p** (Fujimori et al., 1991). No significant correlation could be established between LOH and integration of HBV, in contrast to other studies (Pasquinelli et al., 1988; Rogler et al., 1985; Wang and Rogler, 1988). High levels of allele loss were also observed on **chromosome 16p** (83%) and **8q** (44% of cases) (Slagle et al., 1991), suggesting the presence of additional genes, the loss of which may contribute to the development of HCC. Consistent **rearrangements of DNA** occurring in about 10% of primary HCC have been confirmed by the isolation of a unique cellular DNA sequence adjacent to an HBV integration site (Blanquet et al., 1988). What specific role this particular DNA domain plays in the development of HCC is not yet clear.

2.3.3.3 ONCOGENES IN HUMAN HEPATOCELLULAR CARCINOMA

A study of expression of 12 cellular oncogenes in human hepatoma, and in human fetal liver, revealed an elevated expression of *erb*B, *erb* (A+B), H-*ras, fms, fos,* and *myc* (Zhang et al., 1987). On the other hand, Ki-*ras, rel, myb, sis, mos, src,* and *bas* did not show any clear changes in their expression in fetal liver during development, or in HCC. When the elevated level of c-*myc* mRNA was analyzed in a human hepatoma cell line, it appeared to be due to the increased c-*myc* gene transcription, rather than to stabilization of the mRNA (Huber and Thorgeirsson, 1987). This high constitutive expression of the c-*myc* gene in Hep G2 cells was not due to gene amplification, rearrangements or chromosomal translocations. The increased steady state levels of c-*myc* gene transcripts are coupled to expression of an activated N-*ras* gene (Huber et al., 1985).

c-*myc* expression appears to be higher in HCC than in cirrhotic liver tissue (Himeno et al., 1988). In contrast to the earlier findings (Zhang et al., 1987), no significant differences in mRNA levels of c-*fos*, N-*myc*, N-*ras*, Ha-*ras*, c-*erb*A, c-*erb*B, and c-*abl*, could be detected between patients with HCC, cirrhosis, and those in the normal-chronic hepatitis groups (Himeno et al., 1988). The discrepancies can be partially explained by differences in the fine topography of the distribution of altered oncogenes in different populations of cells involved and of those adjacent to the malignant areas.

Thus, analysis of the activated N-*ras* gene showed that it can be observed only in a small fraction of the tumor cells, and may possibly reflect tumor heterogeneity rather than an event critical to tumorigenesis (Takada and Koike, 1989). It appears, therefore, that to provide a meaningful assessment of the relative contributions of various oncogenes to tumor development a fine molecular genetic analysis must be correlated with the pathomorphological characteristics of involved or unaffected liver tissue in the course of hepatocarcinogenesis. If and when established, such changes in oncogene functions can be evaluated for their clinical utility.

Growth factors and **tumor suppressor genes** have also been implicated in liver cancer. Expression of **insulin-like growth factor II (IGF-II)** appears to serve as a marker of liver cell differentiation, and it is markedly enhanced in primary liver cancer (Cariani et al., 1990).

Analysis of the *p53* gene in HCC indicates that the loss of p53 expression associated with deletions or rearrangements, or aberrations in its product are frequently associated with HCC (Bressac et al., 1990). Specific point mutations in the *p53* gene in HCC differ from those found in other human cancers, and it is believed that the **mutant p53 protein is responsible for preferential clonal expansion of hepatocytes during carcinogenesis** (Hsu et al., 1991). Specific mutation at codon 249 of the *p53* gene is characteristic of HCC associated with high risk of exposure to **aflatoxins** (Ozturk and collaborators, 1991). This mutation can be used as marker for aflatoxin-related HCC. An association of tumor suppressor gene alterations with the development of HCC related to HBV integration is seen in the loss of one allele of the *p53* gene (chromosome 17p13) in 60% of patients with HCC (Slagle et al., 1991).

2.3.3.4 HEPATITIS B VIRUS DNA IN HEPATOCELLULAR CARCINOMA

2.3.3.4.1 Epidemiological Evidence

A strong association between **persistent HBV infection** and HCC has long been established (Beasley, 1982; Beasley et al., 1981; Beasley and Hwang, 1984; Blumberg and London, 1985; Chen, 1987; Kew and Popper, 1984; Nishioka, 1985; Munoz and Bosch, 1987; Szmuness, 1978). Most of the evidence has been based on the correlation between the incidence of, or mortality from, HCC and the prevalence of the HBsAg carrier status; however, exceptions to this general rule have been noted (Melbye et al., 1984). The relative risk of developing HCC in carriers has been estimated to be over 200-fold compared to noncarrier control populations (Munoz and Bosch, 1987). Although the presence of HBV DNA always precedes the development of HCC, other factors, such as **aflatoxins, tobacco smoking, alcohol**, and **oral contraceptives**, are involved in the onset of malignization (Henderson et al., 1983; Oshima et al., 1984; Peers et al., 1987; Trichopoulos et al., 1987; van Rensburg et al., 1985). Again, oncogenicity of HBV may be limited to specific tissues, since a number of extrahepatic sites of HBV DNA have been identified that are not prone to malignant transformation (di Bisceglie and Hoofnagle, 1990; Halpern et al., 1983; Lieberman et al., 1987; Lie-Injo et al., 1983; Tagawa et al., 1985; Yoffe et al., 1986, 1988, 1990). Chronicity of HBV infection also does not appear to be a prerequisite for the development of HCC, as shown in pediatric cases (Leuschner et al., 1988).

2.3.3.4.2 Oncogenesis Associated with Hepatitis B Virus

The potential mechanisms of hepatocarcinogenesis related to HBV infection range from the predisposition to malignancy associated with **cirrhosis** to **integration of HBV DNA** leading to dramatic changes in hepatocyte gene expression. **Chromosomal rearrangements, insertional mutagenesis**, as well as **trans-activation** of cellular genes by the **viral protein X** have been implicated (e.g., Dejean et al., 1983, 1984a,b; Hatada et al., 1988; Hino et al., 1989; Kekule et al., 1990; Nakamura et al., 1988; Matsumoto et al., 1988; Shafritz et al., 1981; Twu and Schloemer, 1987; Zahm et al., 1988; Wang et al., 1990a).

Important advances have been made in deciphering the molecular aspects of the biology of HBV and the events associated with HBV infection, including **viral DNA replication, gene expression**, and **integration** (see, e.g., Blum et al., 1983, 1988, 1989; Ganem and Varmus, 1987; Imai et al., 1987; Imazeki et al., 1988; Koshy et al., 1983; Mizusawa et al., 1985; Ogston et al., 1982; Pontisso et al., 1984; Romet-Lemonne et al., 1983; Shaul et al., 1984; Takada et al., 1990; Yaginuma et al., 1985; Yoffe et al., 1986, 1988). The presence of **direct 11-nucleotide repeat sequences** (DR1 and DR2), their role in HBV integration, and the functions of the four **open reading frames** (ORFs) in the course of infection and transformation have been studied in detail. Some findings include the realization that HBV can replicate and integrate into the genome of nonhepatic cells as well, and that the X protein of HBV is a trans-activating factor for viral enhancers possibly involved in hepatocarcinogenesis (see below). Although the molecular details of viral DNA integration are not yet fully elucidated, it appears that **HBV integration** is a **random event** with apparently specific predilection to the viral sequences near the direct repeats **DR1** and **DR2 as integration sites** (Ganem and Varmus, 1987; Takada et al., 1990).

The molecular mechanisms of hepatocarcinogenesis so far have not been firmly established, although several pertinent findings have been made. The role of deletions, translocations, and rearrangements accompanying HBV integration in tumorigenesis remains unclear, although the resulting loss of **tumor suppressor genes** has been invoked as one of the plausible modes of tumorigenesis by analogy with other tumors (retinoblastoma, colorectal cancer, breast cancer, etc.) (Blum et al., 1989).

Activation of a **steroid hormone receptor-related gene**, *hap*, resembling the c-*erb*A protooncogene, has been demonstrated in HCC (de The et al., 1987). The inappropriate expression of the *hap* gene product in the liver is thought to contribute to hepatocellular carcinogenesis. Activation of c-*myc* (Hsu et al., 1988; Moroy et al., 1986), and a **retinoic acid receptor** (Benbrook et al., 1988) have been identified in HCC. The transforming oncogene, *hst*-1, has been detected in HCC with integrated HBV genomes (Tokino et al., 1987). One of the four integrated HBV genomes was found in proximity to the *hst*-1 gene on **chromosome 11q13** (Hatada et al., 1988). The *hst*-1 and *int*-2 genes as well as the HBV DNA were coamplified, but not rearranged, in the tumor cells. To what extent the amplification of *hst*-1 and/or *int*-2 genes contributes to hepatocarcinogenesis is not clear, although the **fibroblast growth factor** properties of the *hst*-1 and *int*-2 homologs are known to induce transformation. The frequently observed invertedly duplicated structure of HBV genomes integrated in the vicinity of the **direct repeat 2** (DR2) is thought to induce overexpression of some oncogenes (Hatada et al., 1988).

Interruption of the normal functioning of the genes involved in the control of cell division is known to contribute to tumorigenesis. Recently, a case has been made for the possibility that HBV DNA integration within an intron of the **cyclin A** gene, identified at an early stage in human primary liver cancer cells, contributes to hepatocarcinogenesis (Wang et al., 1990a,b). In fact, these malignant tissues displayed an increased level of cyclin A mRNA compared with nontumor tissue.

Another important candidate for transformation inducing activities leading to HCC due to HBV DNA integration is the trans-activator functions of the viral gene products. The first **transcriptional trans-activation** has been ascribed to the X protein of HBV (Seto et al., 1988, 1990; Spandau and Lee, 1988; Twu and Schloemer, 1987). Transcriptional trans-activation implies the presence of specific cellular factors, because it is much higher in human cells and cells of higher primates than in rodent cells (Zahm et al., 1988). In fact, such factors, which bind to the HBV enhancer and core promoter domains, have been identified in liver cell nuclei (Karpen et al., 1988). In transgenic mice, HBx viral regulatory gene is definitely involved in liver carcinogenesis (Kim et al., 1991).

Yet another HBV gene sequence, that of the **pre-S/S region**, has been identified as a transcriptional trans-activator possibly associated with HBV-associated oncogenesis (Kekule et al., 1990). **Pre-S2/S** is the only HBV gene so far found to be almost always integrated in HBV-related HCC. The truncated pre-S2/S sequences in integrated HBV DNA of HCC encode a so far unidentified transcriptional trans-activation activity. Because truncation of this gene appears to be essential for trans-activation, it would be important to establish whether truncation is accidental or can be induced by external influences such as environmental factors, other viruses, and so on. One of the possible mechanisms leading to transformation by truncated **pre-S2/S** gene could be through its ability to stimulate the P2 promoter of the c-*myc* protooncogene (Kekule et al., 1990).

Furthermore, the **reverse transcriptase activity of HBV** has long been recognized (Bavand and Orgad, 1988; Miller et al., 1984; Seeger et al., 1986; Toh et al., 1983; Will et al., 1986). The **92-kDa gene product** of the HBV ORF P is believed to be the precursor of the reverse transcriptase of HBV (Mack et al., 1988). The details of potential carcinogenic activity of the reverse transcriptase activity of HBV associated with HCC remain to be defined.

2.3.4 CLINICAL RELEVANCE OF THE HEPATITIS B VIRUS DNA IDENTIFICATION IN LIVER DISEASE

Various methods of detecting the HBV nucleic acids are available. Comparison of the diagnostic efficiency of the HBV DNA detection by ISH with the identification of antibodies to HBsAg by immunohistochemistry revealed the superior qualities of ISH, even in routine clinical practice using biotinylated, commercially available DNA probes (Choi, 1990). ISH can be efficiently performed, even in an automated fashion on formalin-fixed, paraffin-embedded material, and various methodological improvements aimed at increasing the sensitivity of nonradioactive probes have been reported (e.g., Infantolino and Pinarello, 1987).

Since HBV DNA can be detected in the livers of patients negative for conventional serum markers (e.g., Brechot et al., 1985), the sensitivity of the HBV DNA detection may prove to be of critical importance in establishing the causal relationship between HBV and the observed pathology. In a series of 156 biopsies from cirrhotic patients suffering from various manifestations of long-term liver disease including HCC, the Southern blot analysis proved to be comparable in sensitivity to the detection of HBsAg in the serum (Y. S. White et al., 1990).

Obviously a much more sensitive detection is offered by polymerase chain reaction (**PCR**), which can be applied to detect HBV DNA and RNA sequences in primary liver carcinomas from patients negative for HBsAg (Paterlini et al., 1990). Clearly the issue of technical differences in the detection methods (e.g., Southern blotting vs. PCR, or the use of a different set of primers in the PCR amplification) acquires major importance in establishing the etiology of liver pathology. Five different sets of primers corresponding to the HBV genome regions coding for the **pre-S, S, C,** and **X genes** were used by Paterlini and coworkers (1990) and extensive precautions to avoid contamination have been taken. Two different patterns of HBV DNA molecules were identified in tumor cells of HBsAg-negative patients with HCC: those with deletions and those without deletions. The detection of HBV DNA in the tumors was strongly supported by the finding of HBV RNA molecules. Importantly, consistent with the lower number of HBV genomes per cell in HBsAg-negative tumors, a reduction in the expression of the viral genome is postulated to occur in the course of tumor development. This could be paralleled by masking of HBsAg in immune complexes. Analogous to the situation in other malignancies related to viral infection (e.g., human papilloma virus [HPV] 6b, Abelson murine leukemia virus, and bovine papillomavirus type 4), HBV infection may induce an initial event leading to transformation, followed by deletion of the viral sequences in some cases (Paterlini et al., 1991).

Although these findings support the etiological link between HBV and HCC, they do not unequivocally establish the causal relationship. The known variability of HBV may lead to underdiagnosing the infection due, for example, to the absence of S protein expression (Carman et al., 1990). The so-called **escape mutants** in an infant proved to have a different sequence from that of the isolate from the mother (Carman et al., 1990). This caused the highly antigenic α determinant of HBsAg, to which a large part of the vaccine-induced activity is directed, to be lost. The high genomic variability of HBV is well documented (e.g., Fiordalisi et al., 1990; Moriyama et al., 1991; Okamoto et al., 1987a–c), and it would be of interest to elucidate whether it is in any way related to the acquisition or manifestation of the carcinogenic potential of transformed hepatocytes by HBV alone or in association with other viruses or cellular factors. An important consideration that remains to be worked out is whether HCV, which may act as a cofactor, plays a role in hepatocarcinogenesis related to HBV. As discussed in Section 6.2.3, HCV has been detected in up to 60 to 70% of patients with HCC (see also Resnick and Antonioli, 1991).

2.3.5 CONCLUDING REMARKS

It is clear that before long the refinement of methodological aspects of PCR and other amplification methods will allow the identification of HBV nucleic acids in patient tissues on a routine basis in clinical laboratories. The practical utility of such determinations will be fully realized only when the firm etiological links between this virus and possibly other viruses (HCV, HDV) is established in the genesis of liver diseases, including HCC. In the meantime, the development of PCR procedures, and the use of other established molecular techniques, should be encouraged to expand our understanding of the natural history of HBV and its role in liver pathology.

Chapter 2.4

Pancreas

2.4.1 IMMUNOHISTOCHEMICAL MARKERS

Earlier studies using flow cytometry (FCM) demonstrated the feasibility of separating **endocrine pancreas** into individual cell populations by this technique (Fletcher et al., 1983; Pipeleers et al., 1985; Rabinovitch et al., 1982; van der Winkel et al., 1982). A controversy still exists concerning the type of precursor cell undergoing neoplastic transformation into a pancreatic carcinoma. The cell of origin in 29 primary, 5 metastatic **exocrine** pancreatic adenocarcinomas, and 5 **islet cell neoplasms** has been defined by immunohistochemical markers (Kim et al., 1990a). Polyclonal and monoclonal antibodies against **chromogranin A** and **somatostatin** were used to identify islet cells, secretory component, carbonic anhydrase II; **pancreatic cancer mucin Span-1**, to reveal ductal cells; and **trypsin** and **lipase**, to identify acinar cells. **Islet cell tumors** displayed a homogeneous expression of chromogranin A only, in contrast to the heterogeneous expression of markers in exocrine pancreatic adenocarcinomas. The results of this study supported the concept that the original transformed cell type in many exocrine pancreatic carcinomas belongs to **endodermal "stem cells"** capable of differentiating along multiple cell lineage pathways. This conclusion is in contrast to the often held view, based on morphologic characteristics, that the majority of pancreatic cancers derive from differentiated ductal epithelium cells (Kern et al., 1987; Kloppel and Fitzgerald, 1986).

The well-documented correlation of **serum tumor markers** of pancreatic tumors — **CA 19-9, CA-50, DU-PAN-2, CEA** — with malignancy and tumor load cannot be used for diagnosis or screening beyond providing a tentative indication or for monitoring of tumor recurrences (Beretta et al., 1987; Chung et al., 1987; Frebourg et al., 1988; Haglund et al., 1987; Harmenberg et al., 1988; Hayakawa et al., 1988; Ichihara et al., 1988; Paganuzzi et al., 1988; Takasaki et al., 1988; Suzuki et al., 1988). (For a review of medical and biological aspects of pancreatic carcinoma see Warshaw and Fernandez-del Castillo, 1992).

2.4.2 DNA PLOIDY

DNA ploidy and the fraction of cells in the S and G_2M phases have been estimated by FCM of formalin-fixed paraffin-embedded tissue of 47 **ductal adenocarcinomas** of the pancreas and 5 **adenocarcinomas of the ampulla of Vater** (Baisch et al., 1990). The fraction of cells with aneuploid DNA content, although being relatively low (15% of the tumors), was still significantly higher in carcinomas than in nonneoplastic tissue. While ploidy proved to be of no prognostic value for survival, aneuploidy was found predominantly in advanced tumors with lymph node metastases and in higher grade tumors. Statistical analysis indicated that the combined fraction of cells in $S + G_2M$ phases had an independent prognostic value, so that a high proportion of cells in $S + G_2M$ phases of the cell cycle portend a shorter survival time. It also appeared that tumors with the high percentage of cells in $S + G_2M$ were more **responsive to chemotherapy** (Baisch et al., 1990).

Comparative studies on fresh material and formalin-fixed paraffin-embedded gastric and pancreatic adenocarcinomas gave very close values for DNA ploidy (Baisch et al., 1990;

Macartney et al., 1986). The effect of the high content of connective tissue stroma in pancreatic adenocarcinomas could be an interfering factor in DNA estimates. A methodological study measured the effect of high stomal content on DNA ploidy assessment by **image cytometry (ICM)** of **Feulgen-stained formalin-fixed** and **paraffin-embedded** surgical specimens from 100 cases of ductal adenocarcinoma of the pancreas (Weger et al., 1989). The tumor cell nuclei in 76 of the 77 "readable" cases displayed an unequivocal "nondiploid" DNA distribution pattern. A judgment on the validity of such evaluations for diagnostic and prognostic purposes is reserved, however, suggesting that additional studies are needed to substantiate its use in a clinical setting (Weger et al., 1989). **Hyperploid DNA content has been observed to be the most frequent aneuploid DNA abnormality in the ductal-type pancreatic adenocarcinomas** (Wido et al., 1990). Although aneuploidy shows a significant association with decreased survival and DNA ploidy can be utilized to subdivide pancreatic cancer patients into different prognostic groups the value of detecting aneuploidy is rather limited since diploid pancreatic cancers are also generally rapidly fatal (Eskelinen et al., 1991).

A tentative and weak correlation of aneuploidy of **pancreatic neuroendocrine tumors** with a poor prognosis has been suggested in an FCM study of 17 pancreatic neuroendocrine tumors (Alanen et al., 1990). Caution in extrapolating ploidy data found in a particular type of tumor to diagnostic and prognostic conclusions is definitely warranted. In the case of **insulinoma**, for example, half of the 14 pancreatic insulinomas displayed aneuploidy, despite that 6 of the 7 patients were still alive and disease-free 2 to 5 years following surgery (Graeme-Cook et al., 1990). A similar survival pattern was observed in patients with normal DNA profiles, suggesting that **DNA ploidy should not be used for prognostic purposes in patients with insulinomas**.

2.4.3 CHROMOSOME STUDIES

Following the findings that normal pancreatic cells are stimulated by **epidermal growth factor (EGF)** (Korc et al., 1983) and have **EGF receptor (EGF-R)** encoded by a gene located on **chromosome 7** (Merlino et al., 1985), it was logical to see if alterations of chromosome 7 occur in pancreatic carcinomas. Indeed, in four human pancreatic carcinoma cell lines, numerical or structural abnormalities of chromosome 7 were found; however, no EGFR gene amplification was observed (Meltzer et al., 1988).

The gene for the MEN-1 syndrome has been assigned to **chromosome 11** with a loss of alleles in insulinomas, suggesting that unmasking of a recessive mutation at the *MEN-1* locus contributes to carcinogenesis (Larsson et al., 1988). Subsequently three unrelated cases of sporadic insulinoma were found to have an allele loss on both **11p** and **11q** (Patel et al., 1990).

2.4.4 SPECIFIC GENE EXPRESSION IN PANCREATIC TUMORS

Studies in **cultured pancreatic endocrine tumor cells** have identified the genes encoding various **DNA-binding proteins** specifically interacting with DNA regulatory elements in controlling the expression of the islet cell hormones (Habener et al., 1989). The unraveling of complex interactive relationships unique to individual genes, or sets of related genes, will bring us closer to understanding the genesis of tumors.

Preliminary observations so far offer no explanation for the role of the **parathyroid hormone-like peptide (*PLP*) gene** expression in normal and neoplastic **islets of Langerhans** (Drucker et al., 1989). Abundant expression of the *PLP* mRNA transcripts and peptide was observed in human pancreatic endocrine tumors not associated with hypercalcemia.

Still undisclosed interrelationships, possibly developmentally determined, appear to exist between the expression of various endocrine specific genes that become more apparent in neoplasia. Thus, the expression of the **transthyretin (*TTR*) gene**, normally identifiable in the pancreatic islet cells at a low level, is specifically enhanced in endocrine pancreatic tumors (**glucagonomas** and **insulinomas**) as well as in the intestinal **carcinoids** (Jacobsson et al., 1990). Interestingly, other endocrine tumors, **such as pheochromocytoma** and **paraganglioma** as well as adenocarcinoma of the pancreas, are *TTR* mRNA negative.

A step toward the identification of the gene encoding the pancreatic carcinoma marker, **DU-PAN2,** has been made and the cloning and sequencing of a cDNA for a human **pancreatic apomucin** has been accomplished (Lan et al., 1990). The cDNA and the deduced amino acid sequence of the pancreatic mucin proved to be over 99% homologous with a mucin cDNA derived from breast tumor mucin. The practical application of this gene probe awaits further studies needed to establish the range of expression of the mucin genes in the context of clinical presentation of pancreatic tumors.

2.4.4.1 ONCOGENES

Pancreatic beta cell replication is markedly stimulated in cultured cells into which the oncogenes *src* or a combination of *myc* and *ras* are introduced by electroporation (Welsh et al., 1988). Ki-*ras* mutations in codon 12 consisting predominantly of G→T transversions have been detected by PCR amplification in 28 of 30 patients with pancreatic adenocarcinoma (Smit et al., 1988). The most frequent point mutation detected by PCR in 75% of pancreatic tumors at the second position of codon 12 was G→A, followed by G→T and G→C (Mariyama et al., 1989). Similar observations are reported in Japanese patients, who differ slightly from the pattern of point mutations at codon 12 found in European countries (Nagata et al., 1990).

The combination of FNA technique and highly sensitive PCR amplification of specific gene sites offers a method to distinguish between normal and mutant *ras* oncogene (or other gene)-containing tumors, and the detection of such changes can be used for clinically useful diagnostic or therapeutic decisions. So far, **no specific correlations** have been established between the presence of c-Ki-*ras* mutations and tumor grade or tumor stage, sex, age, and survival of **pancreatic adenocarcinoma patients** (Grunewald et al., 1989). On the other hand, **in endocrine pancreatic tumors the level of overexpression of Ha-*ras* and Ki-*ras* is associated with tumor prognosis** (Hofler et al., 1988).

No mutations are detected by such sensitive techniques in **benign aspiration cytologies of the pancreas** (Shibata et al., 1990). The frequency of c-Ki-*ras* mutations at codon 12 in atypical cytologies is much lower (25%) than that found in FNA aspirates displaying malignant features (Shibata et al., 1990). The presence of mutant c-Ki-*ras* oncogenes does not significantly correlate with survival. In all cases of adenocarcinoma, where the mutations could be detected (72%), they were identified at initial presentation and no mutations were present in FNA specimens of patients without adenocarcinoma, suggesting **an important and early role of the c-Ki-*ras* oncogene in tumorigenesis**. The detection of c-Ki-*ras* oncogene mutations at codon 12 in pancreatic aspirates, however, cannot be considered a reliable diagnostic criterion of pancreatic carcinoma, nor can it be seen as an indicator of a benign disease (Shibata et al., 1990). Exactly the opposite findings also using PCR assays for the codon 12 of Ki-*ras* gene have been reported in pancreatic adenocarcinoma patients who showed the mutations in contrast to patients with chronic pancreatitis, none of whom displayed the characteristic Ki-*ras* mutations (Tada et al., 1991). No explanation for the discrepant findings with those of Shibata and coworkers (1990) has been given.

The **histological type** of pancreatic adenocarcinomas has been shown to vary depending on the **initiating transforming influence** (Sandgren et al., 1991). The most predominant, ductal type of pancreatic adenocarcinomas appears to derive from transformed acinar cells specifically modified by the oncogenic action of **c-*myc***, but not the tumor antigen of simian virus 40 (**SV40**) or *ras* (Sandgren et al., 1991). Recently the expression of **c-*erb*B-2** has also been demonstrated in pancretic adenocarcinomas (Williams et al., 1991).

Because there appears to be a **higher incidence of pancreatic cancers in families with a history of this disease** (6.7% of cases vs. 0.7% of controls) (Lynch et al., 1990), molecular genetic evaluations may also offer a powerful tool for the identification of persons at increased risk for developing this highly lethal malignancy.

It remains to be seen whether a combination of these markers with the evaluation of specific changes of oncogenes or other genes reported to be characteristic of endocrine or exocrine pancreatic tumors may provide a clinically useful test.

Chapter 2.5

Cystic Fibrosis

2.5.1 INTRODUCTION

Cystic fibrosis (**CF**) is the most frequent, fatal autosomal recessive disease affecting Caucasians with an incidence of 1:2000–2500. It is caused by a dysfunction of the fluid and electrolyte transport in exocrine epithelia (Quinton, 1990). The most serious clinical sequelae involving the respiratory tract lead to the median age of survival of about 26 years. The well-recognized increased concentration of **sweat NaCl** of CF patients (up to fivefold the normal level) has long served as a diagnostic marker for this disorder. The inherent defect of Cl⁻ transport, although not fatal in itself, leads to severe pulmonary infections, eventually causing death in 90% of CF homozygotes (Quinton, 1990). A set of **consensus guidelines** for CF patient management has been issued, and the diagnosis of CF at the gene level is believed to "contribute to confirmation of the diagnosis of CF in the future" (Cystic Fibrosis Foundation Center Committee and Guidelines Subcommittee, 1990).

The current understanding of the disturbances in electrolyte transport and its regulation with relation to the putative functions of the now identified *CF* gene product, the **cystic fibrosis transmembrane conductance regulator (CFTR)**, are discussed in detail elsewhere (Iannuzzi and Collins, 1990; Halley et al., 1990; Quinton, 1990; Welsh, 1990). In spite of intensive search for the responsible protein defect over some 40 years using "conventional" techniques, success has only recently crowned the "reverse genetics" approach in identifying the *CF* gene (Kerem et al., 1989; Riordan et al., 1989; Rommens et al., 1989).

Before discussing the currently available tools for the diagnosis of CF carrier state, the salient features of the molecular genetics of CF will be briefly summarized.

2.5.2 THE CYSTIC FIBROSIS GENE AND ITS PRODUCTS

Genetic studies first established a linkage between the ***CF* locus** and a polymorphic locus controlling the activity of the serum **aryl esterase paraoxanase (PON)** (Eiberg et al., 1985). Subsequently, the chromosomal location of *PON* was established on **chromosome 7** (Tsui et al., 1985). Another linkage was then established between the *CF* locus and the DNA marker **DOCRI-917** on 7q (Knowlton et al., 1985) and later also with an anonymous DNA probe, **pJ3.11**, that was assigned to chromosome 7 cen-q22 (Wainwright et al., 1985). Yet another linkage was established between the *CF* locus and the *met* oncogene locus, which placed the *CF* locus in the middle third of the long arm of chromosome 7, between bands q21 and q31 (Van der Hout et al., 1988; M. B. White et al., 1990; Zengerling et al., 1987). In fact, it has been speculated that the product of the *met* oncogene, a tyrosine kinase, could be involved in cAMP signal transmission pathways and/or regulation of ion transport (Dean et al., 1987).

The closest to the *CF* locus proved to be the *met* and J3.11 markers, each within 1 to 2 centimorgans (cM) of the *CF* gene (Beaudet et al., 1986). **Physical mapping** of the *CF* region, using **pulsed-field gel electrophoresis**, estimated the distance between two DNA markers (*met* and D7S8) to be 1300 to 1800 kb, corresponding to the distance of 1 to 2 cM determined by linkage analysis (Drumm et al., 1988; Poustka et al., 1988).

Two clones, **D7S122** and **D7S340**, identified by the technique called **saturation cloning**, were only 10 kb apart and 450 kb closer to the *CF* gene than *met* (Rommens et al., 1989). Other techniques, such as **chromosome jumping** (Collins and Weissman, 1984; Poustka and Lehrach, 1986), which identifies sites at large distances from each other, and **chromosome walking**, which can map contiguous DNA sequences, helped to accelerate cloning of DNA from the target region (Collins et al., 1987; Iannuzzi and Collins, 1990). Eventually, cDNA clones in sweat gland and tracheal cDNA libraries were identified, and a **6.5-kb mRNA transcript** was detected in tissues from CF patients (Kerem et al., 1989; Riordan et al., 1989; Rommens et al., 1989).

The gene size was estimated to be 250 kb, and the presence of a **3-bp deletion** affecting **phenylalanine at position 508** from the coding region identified the gene as the *CF* gene, present only in affected persons (Lemna et al., 1990; Rommens et al., 1989). This mutation affects about 75% of CF patients, and other mutations account for the defects in remaining CF patients (Kerem et al., 1989). A mutation causing **a two-nucleotide insertion** in the *CF* allele of the mother of the affected person, which introduced a termination codon in **exon 13** of the **CFTR gene** at residue 821, has been identified as leading to the production of a severely truncated nonfunctional protein (M. B. White et al., 1990).

While the 3-bp deletion mutation responsible for the omission of a phenylalanine residue at codon 508 (**ΔF508**) is almost exclusively associated with one haplotype, and is the **predominant mutation** in Caucasians, other mutations have already been identified (Cutting et al., 1990a,b). These occur in **exon 11**, three causing **amino acid substitutions** and one producing a termination codon. One of these mutations, **G551D**, is more frequent than other non-ΔF508 mutations. The location of these mutations affecting the nucleotide-binding fold, one of the most conserved positions, supports the assignment of the CF gene product as an **ATP-dependent transport protein** (Cutting et al., 1990b). Two **frameshift mutations** have been described in **exon 7** of the CFTR gene, one caused by a two-nucleotide insertion and the other caused by a deletion of one nucleotide, which introduce a termination codon UAA (Ochre) at residues 369 and 368, respectively (Iannuzzi et al., 1991). A compendium of the results of a worldwide survey of the most common mutations shows a marked variation in the prevalence of ΔF508 in different geographic populations (Cystic Fibrosis Genetic Analysis Consortium, 1990).

The *CF* gene product, **CFTR**, is 1480 amino acids long with a molecular mass of 168,138 Da, calculated from the gene sequence (Riordan et al., 1989). The predicted CFTR structure shows two amino acid sequence domains, resembling consensus nucleotide (ATP)-binding folds, and two repeated sequence motifs, suggesting structures characteristic of membrane proteins. Although the molecular function of CFTR is still not fully defined, the protein resembles the mammalian **multidrug resistance P-glycoprotein**, which acts as a pump removing toxins (drugs) from cells (Deuchars and Ling, 1989). In fact, a so-called **ATP-binding cassette (ABC)** superfamily of transport systems, which includes over 30 proteins sharing extensive sequence similarity and domain organization, has now been recognized (Higgins et al., 1990). These proteins, in addition to CFTR, include the **periplasmic binding protein-dependent uptake systems** of eukaryotes, **bacterial exporters**, and the P-glycoprotein product of the *MDR*1 **gene** among others.

Comparative analysis of these proteins allowed a tertiary structure model of the ATP-binding cassettes, characteristic of this class of transport systems, to be developed (Hyde et al., 1990). This model provides a framework for further elucidation of the specific transport characteristics of the CFTR, and analysis of the nature of the molecular defects affecting **chloride channel** activity in CF. Based on the presumed localization of CFTR in the epithelial membrane it is speculated that CFTR operates as an **unidirectional pump** to adjust the intracellular level of chloride channel inhibitors by regulating their transport across the cell membrane (Halley et al., 1990).

Sequences of the **CFTR open reading frame** have been cloned, and the product was identified and analyzed (Gregory et al., 1990). The CFTR is glycosylated and can associate with membranes. It can be phosphorylated by cAMP-dependent protein kinase, which is consistent with observations of the regulation of Cl⁻ channels by cAMP in normal, but not in CF, cell membranes. The CFTR function should not be equated, however, with that of the chloride channel activity, since in model cell lines no correlation could be established between the full range of CFTR mRNA levels and the distinctive depolarization-induced chloride channel expression (Ward et al., 1991).

Recent studies analyzing the effect of mutations of CFTR on the ion channel selectivity indicate that CFTR is a **cAMP-regulated chloride channel**, in which lysines at positions 95 and 335 determine anion selectivity (Anderson et al., 1991). At the intracellular level, defective CFTR function results in defective glycoprotein processing and ligand transport in the Golgi network, prelysosomes, and endosomes, leading to a phenotype similar to that of alkalinized cells and acidification-defective mutants (Barasch et al., 1991).

The expression of CFTR in heterologous (insect) nonepithelial cells is accompanied by a new cAMP-stimulated anion permeability (Kartner et al., 1991). This transport function strongly resembles that found in several CFTR-expressing human epithelial cells and supports the notion that CFTR may itself be a regulated anion channel. Analysis of the effect of mutations introduced into (CFTR at positions known to be altered in CF chromosomes) on the expression of mutant proteins in transfection experiments indicated that an incompletely glycosylated version of the protein was formed (Cheng et al., 1990). It is believed that the mutant versions of CFTR are recognized as abnormal and are incompletely processed in the endoplasmic reticulum, where they are subsequently degraded. It is argued that the absence of mature, completely glycosylated CFTR at the correct cellular location accounts for the molecular defects in most cases of CF (Cheng et al., 1990).

The **cellular functions of CFTR** in normal and affected persons are being actively studied at present. Elucidation of the molecular details of its operation is expected to provide clues to the possible restoration of the defective function, and to offer a therapeutic solution to CF. Correction studies performed in cultured CF airway epithelial cells have unequivocally demonstrated a causal relationship between mutations in the CFTR gene and defective Cl⁻ transport, when the expression of CFTR, but not of a mutant form of this glycoprotein, DF508, was able to correct the chloride channel defect (Halley et al., 1990; Rich et al., 1990; Scholte et al., 1989). Interestingly, truncation or even the absence of CFTR is not necessarily lethal, and appears to have a greater effect on pancreatic function than on lung function (Cutting et al., 1990a,b). The two major consequences of the ΔF508 mutation are an abnormal translocation of the CFTR protein affecting its incorporation into the cellular membrane and defective chloride ion transport (Dalemans et al., 1991).

The CF protein identifiable in elevated concentrations in the blood of CF carriers has been found to be immunologically indistinguishable from a complex of two proteins, **MRP-8** and **MRP-14 — calcium-binding proteins** expressed in the myeloid cells (Barthe et al., 1991). This observation may further help to define the exact molecular functions of the *CF* gene product. For a detailed review of CF studies see Tsui and Buchwald (1991).

2.5.3 TESTING FOR CYSTIC FIBROSIS CARRIERS

A hypothesis has been advanced, based on family studies combining the French, Spanish, Danish, and Greek data visualizing a unique ancestral mutation initially associated with a B (D1E2) restriction fragment length polymorphism (RFLP) haplotype with subsequent reassociation by cross-over with A, C, or D haplotypes (Serre et al., 1990). According to this hypothesis, the ancestral mutation is 3000 to 6000 years old. Be that as it may, an **estimated carrier frequency** of CF in a Caucasian population of northern European ancestry is high: 1 in 25 (Dawson et al., 1989) or even 1 in 15 in French Canadians suggesting the role of founder effect in elevated incidence in certain populations in Quebec (Rozen et al., 1992). According to

Mendelian genetics, parents carrying the *CF* mutation have a 1 in 4 chance of having either a normal or an affected child, and a 50% chance of the child being a carrier. The risk of CF in each subsequent child is 1 in 4. The child of a sibling of a CF patient has a 1 in 120 risk of CF, compared to a 1 in 1600–2000 risk of CF in the child of a couple having no relatives with CF (Amos et al., 1990; Dawson et al., 1989).

Indirect assessment of CF carrier status is possible by **linkage analysis** using DNA markers known to be linked to the *CF* locus. A number of probes have been developed that detect polymorphisms in strong linkage disequilibrium with the *CF* locus: e.g., **J3.11, met H, met D, 7C22, KM19, XV2c, B79a, CS7, E6, E7, pH131, W3D1.4, EG1.4, J44, T6/20, J32, CE1.0, and many other** (Beaudet and Buffone, 1987; Estivill et al., 1987a,b; Kerem et al., 1989; Tsui and Buchwald, 1991).

Application of linkage analysis to **prepregnancy family studies, prenatal diagnosis**, and **carrier testing** among relatives of an affected individual has also been extended to clarification of the status in asymptomatic persons with borderline or elevated **sweat chloride** concentrations (Amos et al., 1990). The results of such a study do not indicate that all asymptomatic individuals with a positive family history and borderline elevated sweat chloride levels are CF carriers. The elucidation of whether the majority of individuals having a borderline elevated or equivocal sweat chloride concentration are at greater risk for carrying a *CF* mutation can be answered only with a direct carrier test for the *CF* gene, rather than by linkage studies (Amos et al., 1990; Anwar et al., 1990). In fact, **direct gene diagnosis of CF** is now a possibility, as will be discussed below (Chong and Thibodeau, 1990; Prior et al., 1990; Strom et al., 1990; Wagner et al., 1990).

Since the cloning of the CFTR gene and identification of the ΔF508 mutation, the appropriateness of **wide population screening** has been debated (Brock, 1990; Gilbert, 1990; Schulman et al., 1990). When CF screening in white American couples with no family history of the disease has been offered, in cases when invasive fetal testing was performed for another indication, the acceptance rate was found to be high (Bick et al., 1990). However, it seems reasonable to postpone massive population screening until probes specific for all *CF* mutations become available in order to avoid a significant proportion of false-positive results (Gilbert, 1990). The problem is compounded by the logistical challenge posed by the necessity to provide genetic counselling and follow-up to approximately 300,000 couples in the United States alone, in which one or both partners would be a likely CF carrier (Gilbert, 1990). At present, it is recommended, therefore, that testing of CF status be provided to those with a family history of CF and to spouses of CF carriers or affecteds (Caskey et al., 1990).

Carrier screening at the present state of CF testing technology can be analyzed by various models to arrive at an acceptable level of efficiency in detecting mutations at a given prevalence of carriers in the population. In one such (somewhat simplified) model, the proportion of carriers in which the responsible mutation can be shown is expected to be more than 90.4% (ten Kate, 1990). Presently, this efficiency is only around 70% for CF carriers. It is emphasized also that the risk from **false-negative tests** in couples without previous children, who may be more interested in prenatal testing, can be more important than or equal to the mean risk from false-negative tests in the general population (ten Kate, 1990).

It has also been argued that because over 50 mutations were already identified within CFTR by 1990, any plans to conduct massive screenings for CF heterozygotes for all known mutations could be unrealistic (Anonymous, 1990). Relative risk calculations have been determined in various countries for the most common *CF* mutation (ΔF508), and its frequency is heterogeneous in different populations, ranging from 30.3% in Ashkenazi Jewish families (Lemna et al., 1990) to above 70% (Estivill et al., 1989; McIntosh et al., 1989a,b).

Using two **PCR protocols**, the frequency of the most common CF mutation, ΔF508, has been shown in 192 CF patients to occur in 72% of affected chromosomes (Highsmith et al., 1990). Chromosomes with the ΔF508 mutation tended to have haplotype B in 90.7%; this haplotype occurs less frequently (60.4%) for chromosomes with other mutations (Highsmith et al., 1990).

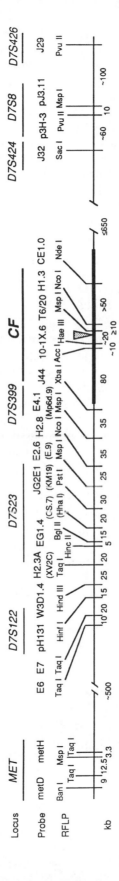

FIGURE 2.1. Map of the RFLP's closely linked to the CF locus. The inverted triangle indicates the location of the ΔF_{508} mutation. (Kerem et al. (1989). *Science 245*:1073–1080. With permission.)

Similar findings have been reported by other authors (Lemna et al., 1990), who observed the 96% frequency of occurrence of haplotype B on *CF* chromosomes carrying the ΔF508 mutation, whereas the same haplotype is encountered on 54% of chromosomes with non-ΔF508 mutations.

An interesting, and practically useful, observation has been made on the occurrence of a **4-bp tandem repeat (GATT)** at the 3' end of **intron 6** in the *CF* gene (Chehab et al., 1991). Screening of unaffected persons by PCR revealed that this repeat existed only in two polymorphic allelic forms, either as a hexamer or a heptamer, with a predicted heterozygote frequency of 41%. A strong linkage disequilibrium has been found for the hexamer allele linked to the ΔF508 deletion on all *CF* chromosomes investigated, which suggests ΔF508 had originated on the gene bearing the six-repeat allele. Furthermore, because this repeat marker is found in the *CF* gene, and is highly unlikely to recombine with the gene, the **GATT motif** can be a valuable DNA marker for haplotype analysis. The informative nature of the established DNA markers XV2c and KM19 is markedly improved in prenatal diagnosis when they are used with the repeat alleles; XV2c and KM19 alone fail (Chehab et al., 1991).

2.5.4 CYSTIC FIBROSIS WITH AND WITHOUT PANCREATIC INVOLVEMENT

Most CF patients develop **progressive pulmonary disease** and **pancreatic insufficiency with malabsorption**; however, marked variation in the disease expression is common (Shwachman, 1975; Wood et al., 1976). Approximately 10 to 15% of CF patients, who are **pancreas sufficient (PS),** have only mild pulmonary involvement. Higher genetic heterogeneity is found among PS patients than among **pancreas-insufficient (PI)** patients (Kerem et al., 1989). Comparison of the frequency of the ΔF508 mutation and haplotypes at the loci linked to the *CF* gene has been made in patients displaying very mild and severe lung involvement (Santis et al., 1990a,b). It appears that, in patients who are compound heterozygotes for the ΔF508 mutation (carrying the ΔF508 and other types of mutations), or in those who lack the mutation on both chromosomes, the as yet unidentified mutations influence the severity of lung involvement. The frequency of the ΔF508 mutation is similar in patients with mild and severe CF lung disease. **The variability in the severity of lung disease in patients homozygous for the ΔF508 mutation is apparently due to the influence of genes outside the *CF* locus** (Santis et al., 1990a).

More clarity exists on the role of genetic factors in determining the degree of pancreatic disease and the rate of its progression (Corey et al., 1989; Kerem et al., 1989). A study of the relationship between clinical phenotypes of CF and the prevalence of the ΔF508 allele has been conducted, and the variable clinical course was found to be at least in part attributable to specific genotypes of the *CF* gene (Kerem et al., 1990). Judging by the concordance of pancreatic and pulmonary status among siblings with CF, **the dominant role of genetic factors over nongenetic factors in determining the severity of pancreatic and lung disease has been established** (Santis et al., 1990b). In this study, the ΔF508 mutation was identified by allele-specific oligonucleotide (ASO) hybridization to genomic DNA sequences amplified by PCR. Genomic DNA lacking the ΔF508 mutation was screened by restriction analysis for the presence of four other mutations: **G551D, R553 Stop, S549N,** and **R347P**. No correlation was found between "the structural region of the protein affected by a mutation and the clinical phenotype" (Santis et al., 1990b). Because **pulmonary status can vary independently of the pancreatic status**, it is postulated that more than one genetic mechanism is at work. It is believed that caution should be exercised in the use of molecular genetic analysis in the presumptive diagnosis of pancreatic functional status and the decision on using enzyme therapy (Kopelman and Rozen, 1990).

This view is confirmed by the observation of a correlation between the severity of CF disease and the genotypes of the more distant flanking marker loci (Wulbrand et al., 1991). It appears therefore that, in addition to the mutations in the CFTR gene itself, **other sequences in the linkage group can exert influence on the manifestation of the disease and determine its clinical phenotype** (Wulbrand et al., 1991). Furthermore, analysis of the associations between subgroups of CF patients with different degrees of severity of the disease and the HLA haplotypes confirmed the validity of such subdivisions, where the C and D haplotypes were associated with lower age-corrected sweat sodium levels (Witt et al., 1991).

2.5.5 PRENATAL DIAGNOSIS OF CYSTIC FIBROSIS

As soon as the first probes demonstrating tight linkage with the *CF* locus became available (J3.11 and *met*), they were used for **prenatal diagnosis** on **chorionic villus DNA**; a summary of the earlier experience has been reported (Super et al., 1987). The results of prenatal testing were used by prospective parents in their decision to terminate or continue a pregnancy. Understandably, **linkage analysis** was not able to unequivocally resolve all the cases. Other markers surrounding D7S8 show strong linkage disequilibrium with the *CF* locus and have proved useful for prenatal diagnosis (Dean et al., 1990).

Comparison of **microvillar enzyme** and DNA analysis in prenatal diagnosis of CF showed the enzyme results were **falsely negative** in three cases (Buffone et al., 1988). In only one case was there a discrepancy between enzyme results and DNA RFLP analysis. The highest accuracy was observed at the 17th to 18th week of gestation. The false-negative rate of this diagnostic strategy may be over 10%.

One of the most useful probes for prenatal diagnosis by linkage analysis is **KM19**, a fact that was used to develop a **PCR-based prenatal diagnostic test** (Feldman et al., 1988), giving an adequate diagnosis in about 70% of families at risk for CF. A same-day prenatal diagnosis for CF based on PCR has been reported on trophoblast tissue with oligonucleotide primers specific for a sequence identifying the **CS7 locus** (Williams et al., 1988). Only about 50% of CF high-risk families were fully informative with CS7 alone, and the use of other markers was expected to improve informativeness of the test.

A combination of four closely linked DNA markers (XV2c, KM19, *met* H, and J3.11) permitted a fully informative pattern to be established in four pregnancies at the 1:4 risk of CF (Halley, 1989). Chorionic villus samples were retrieved at 10 to 11 weeks of gestation. PCR amplification of DNA from fetuses and relatives was followed by analysis of the amplified products with ASOs. Direct application of mutation analysis, in cases when both parents are heterozygous for the deletion, makes this method independent of the index patient. Thus, even if the material from the previous affected child is not available, this approach has a 50% chance of establishing a reliable prenatal diagnosis in the first trimester of future pregnancies (Halley, 1989). Independent identification of ΔF508 deletion by PCR amplification of DNA from chorionic villus samples has also been reported by other workers (McIntosh et al., 1989a,b).

In certain cases, the informativeness of the direct detection is inferior to the indirect RFLP approach. Comparison of RFLP analysis with direct detection of the ΔF508 deletion, in screening pregnancies at risk for CF in Italy and Spain, revealed that KM19, XV2c, and MP6d-9 probes used in RFLP analysis were informative in 86% of the families (Novelli et al., 1990). Direct analysis, on the other hand, was fully informative only in 30% of the cases, and partially informative in 34% of the cases. The use of ΔF508 deletion for the CF prenatal diagnosis and carrier screening in families of **South European ancestry** is of somewhat limited effectiveness, and the identification of other CF gene mutations is needed to enhance the predictive power of direct gene analysis in those populations. This study clearly demonstrated **regional variations in the predominance of major *CF* gene mutations**.

ΔF508 is much more informative (in up to 75% of cases) in **North American** (Kerem et al., 1989) and **North European populations** (Halley, 1989). In the **French population**, the results of testing 237 families showed the same linkage disequilibrium for haplotypes generated by XV2c and KM19 RFLPs as in Caucasian populations in North America (Chomel et al., 1989; Vidaud et al., 1989).

PCR amplification of the gene regions detected by the KM19 and CS7 probes for first-trimester prenatal diagnosis in 22 CF cases was informative in 86% in the **Italian populations** (Gasparini et al., 1989, 1991). In order to reach informativity for prenatal diagnosis in all the families tested, the results had to be confirmed by Southern blotting with a panel of probes.

In the **Belgian population**, prenatal diagnosis by RFLP analysis, using seven DNA markers in four families, allowed the accuracy of predictions to exceed 99%. However, as is the case with any linkage study, a major disadvantage of RFLP diagnosis is the need to have the parents and at least one living CF child available for haplotyping (Lissens et al., 1989).

As noted above, informativity of linked probes varies not only between **different geographic areas** but also in CF patients with and without pancreatic insufficiency, suggesting that pancreatic involvement may be determined by different mutations at the CF locus (Devoto et al., 1989). Alternative explanations can also be offered to account for the genotype/phenotype discrepancies found in patients with and without pancreatic involvement (Kupchik, 1991). Uncertainty in the indirect linkage analysis may also arise when the alleged father proves not to be the biologic father of the affected child, which fact could be established by the analysis of VNTR loci (Cahill et al., 1990).

A number of other **non-ΔF508 mutations** have been described in various ethnic groups and geographic locales (e.g., Gasparini et al., 1991; Zielenski et al., 1991a). Oligonucleotide primers have been designed for amplification of individual exons and the immediately flanking sequences in the introns for more efficient detection of CF-causing mutations (Zielenski et al., 1991b).

The definite advantage of DNA probes over the frequently inconclusive nature of the enzyme analysis performed on cell-free amniotic fluid has been reported in actual patients (Carey et al., 1990). The convenience of PCR amplification applies also to DNA extracted from Guthrie blood spots that had been in storage at room temperature for up to 7 years (Nelson et al., 1990).

Prenatal diagnosis of CF has been extended even to the level of **preconception** and **preimplantation diagnosis** for couples carrying the ΔF508 mutation (Strom et al., 1990). The test utilizes PCR amplification, removal of oocytes under ultrasonic guidance, removal of the first polar body before fertilization, and genetic analysis of the amplified polar body products.

2.5.6 OTHER ASPECTS OF PCR ASSAYS FOR CYSTIC FIBROSIS

Application of PCR for amplification of the known DNA sequences characteristic of the ΔF508 deletion directly and unambiguously identifies affected individuals in about 70% of cases (Chong and Thibodeau, 1990; Prior et al., 1990; Strom et al., 1990; Wagner et al., 1990). This test, although targeted for the most frequent specific mutation, cannot be used for the direct determination of carrier status because other as yet unidentified mutations will not be recognized without the use of **mutation-specific oligonucleotide primers**. In such cases, linkage analysis within families will still have to be used. However, simplification of the analytical process by the use of PCR for direct identification of the probably mutated sequences is a possibility, and will undoubtedly be used in developing **direct screening** procedures for CF.

In fact, marked improvements and modifications of the original PCR procedure for the direct detection of ΔF508 mutation have already been reported (Ferrari and Cremonesi, 1990; Friedman et al., 1990; Friedman and Stoerker, 1990; Rommens et al., 1990; Skogerboe et al., 1990). A novel approach, which is capable of detecting known DNA sequence variations down to a single nucleotide, uses the **primer extension strategy** and is based on the fidelity of the DNA polymerase (Kuppuswamy et al., 1991). Because this enzyme extends the synthesized strand by addition of only the correctly paired nucleotide onto the 3' terminus of the template-bound primer, the labeled nucleotides allow the identification of a mutant product in gel electrophoresis and autoradiography of the extended primer. This strategy, called **single-nucleotide primer extension (SNuPE)**, is also applicable to "carrier detection and prenatal diagnosis of every genetic disease with a known sequence variation" (Kuppuswamy et al., 1991).

To detect *CF* mutations when they do not create a useful restriction site, such a site can be generated by PCR and specially chosen primers. These provide restriction map alterations serving as a simple means of identifying small DNA abnormalities (Sorscher and Huang, 1991). Likewise, **PCR-mediated site-directed mutagenesis** has been used to create new allele-specific restriction sites, arising from a single-base mismatch specifically introduced into a PCR primer that abuts the mutation under study; this approach has been demonstrated for three of the more common *CF* mutations (Friedman et al., 1991).

A simplified PCR protocol has been described for the detection of the ΔF508 mutation (Dodson and Kant, 1991). Instead of the usual three-temperature cycling method, the new procedure uses a **two-temperature PCR**, which combines annealing and extension steps performed for 30 s at 73°C alternating with a 60-s denaturation step at 94°C. The identification of mutant alleles was performed by mixing aliquots of the amplified material with an equivalent amount of PCR-amplified material from known normal or ΔF508 homozygotes. The reaction products are eventually visualized in electrophoretic gels by ethidium bromide. The clinical utility of two-temperature PCR has been demonstrated by testing 105 individuals (Dodson and Kant, 1991). This method favorably compared to the more labor-intensive traditional approach that calls for hybridization of ASO probes to Southern blots prepared from PCR products amplified by a three-temperature method.

A promising adaptation of minisequencing of the PCR-amplified DNA using biotinylated primers and capture of the product on solid support suggests a potentially automatable protocol for large-scale quantitative detection of point mutations and deletions effectively demonstrated in the ΔF508 mutation (Jalanko et al., 1992).

To reduce the number of PCR tests in a mass screening program for the ΔF508 mutation, a **sample pooling strategy** has been proposed (Gille et al., 1991). For a **German population**, with an ΔF508 heterozygosity incidence of about 1:35, the optimum number of samples to be pooled in one assay is calculated to be 24. This approach allowed substantial economy to be achieved in analyzing large numbers of DNA samples, thereby reducing the number of PCR assays by as much as 77%, without reducing the efficiency of screening for this particular mutation.

Lung

2.6.1 SMALL CELL LUNG CARCINOMA

SCLC comprises up to 20% of all lung cancers, occurring predominantly in male smokers, that can be divided into three types: **oat cell**, **fusiform**, and **polygonal cell type** (Rosai, 1989, pp. 303–305). The latter type can be confused with other types of lung cancer, and ultrastructural studies may be needed for definitive diagnosis. In addition to characteristic **nuclear chromatin** and **nucleolar patterns**, the presence of **dense core neurosecretory-type granules** is found in at least some cells in most cases. Identification of the malignancy as belonging to the SCLC is also helped by **immunohistochemical markers** (neuron-specific enolase [NSE], neurofilaments, Leu-7), which, however, ca also be positive in other types of lung cancers, the finding that qualifies other types of cancer as accompanied "**with endocrine differentiation**" (references in Rosai, 1989, pp. 303–305). The serum level of **NSE** correlates with the clinical response of SCLC to treatment, and the pretreatment level of NSE has proved to be of prognostic significance (Harding et al., 1990).

Comparison of **synaptophysin, chromogranin A**, and **L-dopa decarboxylase** for defining neuroendocrine differentiation of lung cancer cells emphasizes the superior sensitivity and specificity of synaptophysin (Jensen et al., 1990). Others have emphasized the expression of chromogranin A mRNA, demonstrated by ISH as a specific marker for neuroendocrine differentiation in SCLC (Hamid et al., 1991). Another **pituitary polypeptide, 7B2**, which is selectively expressed in cells containing neurosecretory granules, is expressed in all SCLC cell lines of the classic type (Roebroek et al., 1989). Its specificity for neuroendocrine differentiation appears to be high, because it is virtually absent from variant cell lines as well as from NSCLC cells. Moreover, all the carcinoids expressed very high levels of 7B2 mRNA.

Additional immunohistochemical markers tested in formalin-fixed, paraffin-embedded tissue (antigens of the **membrane cluster-5** and **cluster-5A**) are found to be predominantly expressed in SCLC (Maier et al., 1989). This expression is not related to the morphologic subtypes of SCLC, it is not observed in normal lung tissue, and is insignificant in NSCLC, adenocarcinoma, squamous cell carcinoma (SCC), mesothelioma, or lung carcinoids.

Thus, it appears that gene level markers, if proven to be of higher specificity than the immunocytochemical markers mentioned above for a particular type of lung cancer, can be of significant practical use, particularly in light of the well-recognized differing response of SCLC and NSCLC to treatment (for a review, see Leonard, 1989).

2.6.1.1 CHROMOSOME ABERRATIONS

Analysis of chromosomes in cell cultures established from small cell and non-small cell lung cancers (**SCLC** and **NSCLC**) revealed structural abnormalities in virtually all chromosomes, the most consistent being a deletion of the short arm of **chromosome 3**, particularly in SCLC (93%) (Whang-Peng, 1989; Whang-Peng et al., 1982a,b). All SCLC patients had lost at least one codominant allele in the tumor tissue, which was also found, in addition to cell lines, in biopsy specimens, autopsy samples, and in an excised lymph node with metastatic tumor (Brauch et al., 1987). Some NSCLC specimens also demonstrated the loss at 3p. In fact, a consistent deletion at 3p21 has been demonstrated not only in SCLC but also in NSCLC, squamous cell carcinoma, and adenocarcinoma of the lung (Kok et al., 1987). Loss of alleles at 3p was observed before and after chemotherapy.

Various chromosomal mechanisms may account for the reported pleiotropy of genetic lesions observed in lung cancers, including **recombinatory events, nondisjunction** (Mori et al., 1989), **trisomy (of chromosome 7)** (Lee et al., 1987), and the formation of **isochromosomes** (Jin et al., 1988e). Consistent loss of DNA sequences from certain chromosomes in lung cancers, the known localization of the genes encoding **growth factor** or **tumor suppressor genes**, justifies precise mapping of these chromosomal alterations. A consistently found deletion of the short arm of chromosome 3 could play a role in lung cancer (Naylor et al., 1987). The specific mechanism is not yet clear but alterations of the receptor genes as shown for the **thyroid hormone receptor (3p24)** may be involved (Albertson et al., 1989). While confirming loss of heterozygosity (LOH) at 3p21, 3p24-25, and at 3p26 in SCLC, the pattern of loss of 3p alleles, was found to be different for NSCLC and SCLC (Brauch et al., 1990). The analysis of SCLC is of particular interest since the *RB* gene (chromosomal region 13q14.1) and the *p53* gene (chromosomal region 17p13) — the two genes with established tumor suppressor functions — appear to be involved in the development of this malignancy. Advances of molecular genetics in the understanding of the biology of some forms of lung cancer allow novel approaches to be devised for diagnosis, treatment, and prognosis. Reviews of molecular genetics of lung cancer have appeared (e.g., Bergh, 1990; Birrer and Minna, 1988; Viallet and Minna, 1990).

2.6.1.2 PLOIDY IN SMALL CELL LUNG CARCINOMA

Evaluation of various cytomorphonuclear parameters related to chromatin distribution and DNA ploidy has been performed on **cytospin preparations** of Feulgen-stained nuclei extracted from deparaffinized **typical** and **atypical carcinoids** and SCLC (Larsimont et al., 1990). Small cell lung carcinoma features a significantly higher DNA content than typical carcinoids, whereas atypical carcinoids show intermediate values. If these observations are confirmed in larger studies, this approach may offer an objective criterion for reproducible "scale" of differentiation helpful in differential diagnosis. Other studies, however, question the use of DNA ploidy as an independent criterion of malignant potential of carcinoid tumors (Yousem and Taylor, 1990), and of **atypical adenomatous hyperplasia** of the lung (Nakayama et al., 1990). A contrasting observation has been made by computer-assisted image cytometry (ICM) of cytospin preparations that allowed for differentiating between typical carcinoids and SCLC based on DNA content of cells (Larsimont et al., 1990). Atypical carcinoid constituted an intermediate group between these two entities.

2.6.1.3 ONCOGENES IN SMALL CELL LUNG CARCINOMA

2.6.1.3.1 The Retinoblastoma (*RB*) Gene

Structural abnormalities of the *RB* gene, aberrant *RB* mRNA expression, and absent or dysfunctional RB protein have all been observed in the majority of SCLC and only in 10 to 20% of NSCLC (Harbour et al., 1988; Hensel et al., 1990; Kratzke et al., 1990; Rygaard et al., 1990; Saksela et al., 1989; Yokota et al., 1988a).

Evaluation of 26 SCLC tumor and normal DNA samples with the *RB* cDNA clones **pRB4.5** and **p6NRO.5** revealed that tumor cells lost an allele of the *RB* gene (Hensel et al., 1990). RB transcripts were virtually absent in 75% of the cell lines tested, and it is likely that, similar to the situation in retinoblastoma, mutation of both alleles occurs in somatic cells in SCLC. The absence of *RB* gene rearrangements shown by Southern blots of normal cell DNA from 100 SCLC patients supports this contention (Hensel et al., 1990).

Apparently normal *RB* mRNA and the *RB* gene product, the M_r **110,000–116,000 phospho-protein**, are expressed in more than 50% of SCLC tumor cell lines and xenografts in nude mice

(Rygaard et al., 1990). Major structural abnormalities of the *RB* gene DNA could be detected in only 4 of 18 tumors investigated. No clear understanding has so far been gained of the specific functions of the *RB* gene product, and to what extent the degree of phosphorylation of this protein plays a role in SCLC. It appears, however, that a single amino acid change (**cysteine to phenylalanine**) resulting from a point mutation in exon 21 of the *RB* gene is sufficient to drastically alter its functional integrity (Kratzke et al., 1990). This substitution eliminates the ability of the *in vitro*-translated RB protein to bind **adenovirus EIA** and **simian virus 40 (SV40) large T** transforming proteins — the characteristic linked to the mechanism of tumorigenesis by the *RB* gene (DeCaprio et al., 1988; Whyte et al., 1988).

2.6.1.3.2 *p53*

Abnormalities of **chromosome 17p** (Mori et al., 1989; Weston et al., 1989; Yokota et al., 1987), the frequent mutations of *p53* in lung cancer (Takahashi et al., 1989), and assignment of the single copy of the *p53* gene to chromosome region **17p13** (Baker et al., 1989; Takahashi et al., 1989) strongly implicate a role for *p53* in the natural history of lung cancer. Indeed, a study of the *p53* gene structure and its expression in SCLC and NSCLC revealed DNA abnormalities, which included a **homozygous deletion, point** or **small mutations** occurring in the **open reading frame** of *p53*, as well as *p53* mRNA abnormalities (Minna et al., 1989). Small mutations were detected in patients' tumor tissue directly, rather than in cell lines established from such samples. These structural abnormalities were present in 17 of 30 cell lines representing all histologic subtypes of lung cancer as well as in a dozen primary NSCLC specimens (Takahashi et al., 1989). Using the **chemical mismatch cleavage technique** combined with preliminary amplification of target genomic DNAs by **PCR**, single base mutations in the *p53* gene have been efficiently identified (Curiel et al., 1990).

The transforming activity of *p53* appears to be enhanced by alterations in the gene structure or gene expression (for a review, see Lane and Benchimol, 1990). While the wild-type p53 proteins have a very short half-life, the mutant proteins have a much longer half-life (4–8 h vs. 6–20 min); this change accounts for the frequent finding of p53 proteins in cancer cells. In fact, Lane and Benchimol (1990) go so far as to suggest that "the ready detection of the protein is synonymous with mutation". The wide role of *p53* gene in tumorigenesis led to the suggestion that "alteration of *p53* expression by mutation is the most frequent known genetic change in human cancer" (Lane and Benchimol, 1990). Usually, *p53* appears to be involved in the later stages of tumor development interacting with other oncogenes such as c-*myc* and H-*ras* (Baker et al., 1989; Eliyahu et al., 1988; Finlay et al., 1989; Nigro et al., 1989).

Introduction of a mutated *p53* into **DMS53 SCLC**, which differentiates following v-H-*ras* insertion, led to changes in morphology of the cultured cells, a decrease in doubling time, increased cloning efficiency, and decreased calcitonin hormone production (Mabry et al., 1990). These observations pointed to a role for *p53* in SCLC **tumor progression** rather than initiation.

Analysis of exons 5–9 of the *p53* gene by PCR followed by sequencing of the amplified products revealed the presence of mutations in lung tumors of underground uranium miners (Vahakangas et al., 1992). The mutations detected differed from those ascribed to the effect of smoking or found at the usual sites reported in lung cancer suggesting specific association with exposure to radon. When the normal functions of *p53* in normal and tumor cells become more defined, this gene may prove to be of practical value in patient management. In the meantime, *p53* seems to blur the distinction between oncogenes and antioncogenes, because *p53* acts as an oncogene in its mutant form, and as an antioncogene, a tumor suppressor gene, in its wild-type form (Lane and Benchimol, 1990).

2.6.1.3.3 The *myc* Gene

Analysis of SCLC cell lines revealed amplification and expression of c-*myc*, N-*myc*, and L-*myc* protooncogenes (Brooks et al., 1987). While c-*myc* amplification could be readily demonstrated in cell lines derived from patients with SCLC, this amplification could not be established in direct tumor specimens taken at autopsy, suggesting that selection of tumor cells possessing amplified c-*myc* genes may have taken place in cell culture (Brooks et al., 1987). Using RNA-RNA ISH, however, the amplification of c-*myc* observed in cell lines could also be confirmed in surgical biopsy and autopsy material from patients with SCLC (Gu et al., 1988).

Amplification of the N-*myc* protooncogene has been readily demonstrated directly in tumor specimens taken from patients with noninvolved tissue showing no amplification (Brooks et al., 1987). **Restriction endonuclease mapping** and **DNA sequence analysis** of N-*myc* and c-*myc* genes revealed regions of homology with the protooncogene in the lung tumor, but otherwise sequences were unrelated to either the c-*myc* or N-*myc* genes, which Brooks and coworkers (1987) designated *L-myc* as first isolated from a lung tumor cell line. While the c-*myc* gene is assigned to human **chromosome 8** and N-*myc* is assigned to **chromosome 2**, the L-*myc* gene mapped to human **chromosome 1p** (Brooks et al., 1987). Retrospective analysis of the survival of the patients, from whom the cell lines had been derived, pointed to a **correlation between amplification of c-*myc* or N-*myc* with a shorter survival time** (Brooks et al., 1987). c-*myc* amplification also correlated with the more malignant behavior of the variant (as opposed to classic) histology of SCLC. An ISH study of SCLC cell lines and metastatic tumor obtained by surgical biopsy or from autopsy material, on the other hand, failed to reveal any morphologic difference between cells expressing the c-*myc*, N-*myc*, or L-*myc* oncogenes (Gu et al., 1988). Amplification of these three protooncogenes in SCLC apparently occurs through different regulatory mechanisms. While c-*myc* and L-*myc* amplification is due to a block in mRNA elongation during transcription, called **attenuation of transcription**, the regulation of transcription by a mutation occurs in c-*myc* gene expression, and is the sole mechanism of regulation of N-*myc* gene expression (Krystal et al., 1988). Somewhat discrepant observations are reported by other investigators on the amplification of various members of the *myc* gene family in SCLC (Gemma et al., 1988; Johnson et al., 1988). Thus, no difference in the gene frequency of the L-*myc* RFLP could be found between lung cancer patients and normal controls in a Norwegian population (Tefre et al., 1990).

c-*myc* amplification appears to play a more significant role in **malignant progression**, rather than in development of SCLC (Bergh, 1990; Noguchi et al., 1990). **Chemotherapeutic regimens** may affect the reported involvement of *myc* protooncogenes in lung cancer. In fact, analysis of tumors and cell lines derived from SCLC for *myc* family DNA amplification showed it to occur more frequently in specimens from treated than untreated patients (Brennan et al., 1991). Different chemotherapy regimens similarly affected the amplification demonstrable in tumor specimens and in tumor cell lines from the same patients.

Understanding the controlling mechanisms involved in protooncogene expression is essential to assessment of their role in clinical situations, where relative expression of specific oncogenes may prove of diagnostic and/or prognostic validity. **It does not appear at this point that assessment of alterations in the amplification and/or expression of the *myc* gene family may significantly contribute to management of patients with SCLC.**

2.6.1.3.4 The *jun* Gene

Attention to the *jun* protooncogene has been drawn because of its role as a mediator of growth factor or tumor promoter action on transcription (Ryder et al., 1988, 1989; Ryder and Nathans, 1988). In about 25% of cell lines derived from different types of lung cancer, the level of *jun*A expression has been found to be markedly elevated (Schutte et al., 1988). A related protooncogene, *jun*B, is expressed at high levels in all the tumor cell lines studied, and, like *jun*A, it is expressed to high levels in normal lung. No evidence of *jun* DNA amplification or rearrangement, or of changes in the *jun* mRNA half-life, could be found, suggesting that the high level of expression was due to transcriptional activation. While c-*jun* (*jun*A) does not transform cells alone, it causes transformation of rat embryo cells when cotransfected with an activated c-Ha-*ras* oncogene, but not c-*myc* or L-*myc* (Schutte et al., 1989).

Deregulated expression of the *jun* family transcription factors is thought to contribute to the development of lung cancer, although the precise sequence of events and the possible interaction of these transcription factors with other transcriptional regulators is not clear, especially since the high level of *jun* expression is found in normal lung (Minna et al., 1989). **It does not appear practical to consider evaluation of the *jun* family transcription factors for diagnostic or prognostic purposes at this time**, because similar levels of c-*jun* expression, with large variations between patient cases, have also been found by ISH in matched pairs of NSCLC and normal lung cells (Broers et al., 1990).

2.6.1.3.5 *erb*B, *ras,* and Growth Factors in Small Cell Lung Carcinoma

The amplification and rearrangements of 19 representative protooncogenes in 137 specimens of DNA from primary human lung carcinomas have been investigated and in 28% of the carcinomas amplifications of protooncogenes were detected (Shiraishi et al., 1989). The *myc, ras,* and *erb*B family genes constituted about 90% of the amplified protooncogenes. It was predominantly the *myc* family protooncogenes that contributed to the reported 28% of cases showing amplification, because Ki-*ras*-2 and N-*ras* were amplified only in 6 and 2 of the 137 tumors, respectively. The majority of the amplified *ras* genes were present in advanced stages of lung cancer. *erb*B-1 was amplified in 10 of 114 tumors, and *erb*B-2 (*HER*-2/*neu*) in 1 of 51 tumors. The *int*-2 gene was amplified in two squamous cell carcinomas. Loss of heterozygosity has been frequently observed at loci for various protooncogenes. Because members of the *myc, ras,* and *erb*B families are frequently amplified in a variety of human cancers, the alterations of these protooncogenes in human lung cancers may be related to the development of pulmonary malignancy (Shiraishi et al., 1989). Another gene, *src*, has been correlated with neuroendocrine differentiation, and reported in SCLC (Pahlman and Hammerling, 1990). Definitely more information is needed to make practical use of the observed oncogene alterations for patient management.

Along these lines the observation of differential expression of c-*erb*B-2 mRNA in SCLC is of interest. The highest level of c-*erb*B-2 expression has been found in four of four adenocarcinomas, with the expression of this protooncogene being different between NSCLC and SCLC (Schneider et al., 1989). Probing Southern blots of *Eco*RI-digested paired DNA samples as well as Northern blots and dot blots of paired RNA samples from SCLC, NSCLC, and normal lung tissues with human *neu* cDNA sequences indicated that the *neu* gene was not amplified (Schneider et al., 1990). No expression of the *neu* gene, or only minimal expression at best, could be detected in SCLC, whereas 38% of NSCLC specimens, in contrast, showed high *neu* expression both at early stages and in advanced tumors.

Immunohistochemical detection of the *ras* oncogene product, **p21^{ras}**, by the **monoclonal antibody RAP-5** reveals the higher frequency and intensity of p21^{ras} expression in squamous, **bronchoalveolar**, and neuroendocrine neoplasm, but not in SCLC (Lee et al., 1987). Most mesotheliomas do not immunostain for p21^{ras} either.

While p21^{ras} oncoprotein is overexpressed in adenocarcinomas compared to squamous cell carcinomas, overexpression of the c-*myc* gene product, **p62**, preferentially occurs in **poorly differentiated squamous cell carcinomas** compared to the **well** and **moderately differentiated** tumors (Spandidos et al., 1990). The role of N-*myc* and v-Ha-*ras* in the natural history of lung cancer is clearly demonstrated in cell culture studies by complete or partial conversion of SCLC to a more treatment-resistant (NSCLC) phenotype (Falco et al., 1990). Interaction of these two genes induces a specific change in cultured SCLC cells, including the expression of mRNA for three NSCLC-associated growth factors and receptors: **platelet-derived growth factor (PDGF) B chain, transforming growth factor (TFG)-α,** and **epidermal growth factor**

receptor (EGFR). These findings underscore the view that endocrine cells of the lung, imparting characteristic diagnostic features to this malignancy, are in fact linked to the other types of bronchial epithelial cells through a common progenitor cell (Falco et al., 1990). Similar conclusions have been reached by others, who demonstrated the expression of NSCLC proteins by SCLC cells transformed by c-*myc* and Ha-*ras* protooncogenes (Doyle et al., 1990). Apparently Ha-*ras*-1 gene is responsible, to a certain degree, for the development of the NSCLC phenotype. However, the earlier suggestion that the **a4 allele** of the Ha-*ras*-1 gene is found more frequently in NSCLC patients could not be confirmed by testing DNA from peripheral blood cells (G. R. M. White et al., 1990). The definite association of activated Ha-*ras*-1 gene with lung cancer can be shown by RFLP analysis with *Msp*I/*Hpa*II restriction digestion (Ryberg et al., 1990).

Patients with **Ki-*ras*** mutations have **significantly worse survival** than lung cancer patients without an activation, although these mutations are not associated with particular histological characteristics or specific clinical presentations of the adenocarcinomas (Rodenhuis and Slebos, 1990). *ras* oncogenes, but not the L-*myc* gene, are implicated in the development of **metastases** in lung adenocarcinomas (Reynolds et al., 1991).

The role of **growth factors** in determining the growth of normal and tumor lung cells is an active area of biological research (Barnes, 1990). Evaluation of **TGF-α** levels in lung carcinomas failed to produce clearcut information on its role in tumorigenesis, because it did not correlate with either histologic type, stage, grade, or degree of desmoplasia (Liu et al., 1990). Likewise, cell culture studies on a human SCLC cell line failed to produce unequivocal indications of the role of TGF-α, TFG-β or EGF in determining the phenotype (McLeod et al., 1990). Nevertheless, overexpression of the **growth factor receptors, EGFR** and **HER-2/neu**, is observed in tumors and normal bronchial epithelium of patients with lung cancer (Sozzi et al., 1991).

Some preliminary correlations are beginning to emerge between the **level of expression** of various growth factors (TGF-α, EGF) and growth factor receptors (EGFR) and survival of lung cancer patients (Tateishi et al., 1990). It is believed that TGF-α and PDGF may have both autocrine and paracrine roles in the biology and clinical behavior of NSCLC tumors and their surrounding stroma, and that progression of SCLC to the NSCLC phenotype may be influenced by these factors (Falco et al., 1990). The importance of **TFG-β** in regulating proliferation of normal lung tissue of mesenchymal and epithelial origin suggests a complex role in pulmonary neoplasia (Pelton and Moses, 1990).

The above-discussed findings suggest a role of activated protooncogenes and growth factors in lung carcinogenesis, although the specific functions of individual genes remain to be defined. The importance of these studies is underscored by the effective and selective inhibition of K-*ras* expression and associated tumorigenicity of lung cancer cells by antisense RNA (Mukhopadhyay et al., 1991). The practical use of evaluating the expression of various growth factors' genes still awaits further investigation.

2.6.2 NON-SMALL CELL LUNG CARCINOMA

Non-small cell lung carcinoma (NSCLC) includes a diverse group of lung carcinomas composing 75% of lung neoplasms. Its distinction from SCLC is, in part, justified by the difference in the response of these malignancies to treatment. While SCLC is predominantly treated by chemotherapy, surgery is used in NSCLC (Leonard, 1989; Mountain, 1977, 1983; Silverberg and Lubera, 1988).

2.6.2.1 CHROMOSOME ABERRATIONS
More than one chromosome alteration has been clearly demonstrated in 54 NSCLCs tested for LOH at 13 polymorphic genetic loci (Weston et al., 1989). The most frequent loss occurred

from the short arm of **chromosome 17**, more so in **squamous cell carcinomas** (8 of 9 cases) than in **adenocarcinomas** (2 of 11 cases). Loss of DNA sequences from **chromosome 11** accompanied the majority of chromosome 17 losses in squamous cell cancers. In almost 50% of cases losses were also noted in **chromosome 3**, including those areas previously shown to be lost consistently only in SCLC. Finally, LOH at the **13q** locus was observed in 2 of 21 informative cases. The pattern of DNA sequence losses is different not only between SCLC and NSCLC, but also between squamous cell carcinomas and adenocarcinomas, the latter showing similarity to NSCLC (Weston et al., 1989).

Cytogenetic analysis of NSCLC revealed all tumors (which included adenocarcinomas, squamous cell carcinomas, and large cell carcinomas) to be aneuploid and to display a **complex pattern of karyotypes** with multiple structural and numerical abnormalities (Lukeis et al., 1990). Loss of material from **chromosome 9p** occurred in 90% of cases, resulting from nonreciprocal translocations, deletion, or chromosome loss. Other **nonrandom alterations** involved **chromosomes 1p, 3p, 5q, 6p, 6q, 7p, 8p, 11q, 13p, 14p, 15p, 17p,** and **19p**. Clearly, loss of material from 9p is a potentially important alteration in NSCLC.

2.6.2.2 ONCOGENES AND GROWTH FACTORS

Evaluation of **protooncogenes** and **growth factors** in NSCLC revealed alterations of several protooncogenes (c-*myc*, N-*myc, ras, erb*B, and *neu*) (e.g., Cline and Battifora, 1987; Slebos et al., 1989), in addition to the above-mentioned deletions at 11p, 3p, and 17p (Ludwig et al., 1990). At this time, a still inconclusive picture emerges concerning the consistency of alterations of particular protooncogenes. Changes in c-*erb*B-1 and c-*erb*B-2 were found to correlate with the histologic type of tumor and were more common in advanced tumors (Cline and Battifora, 1987). Loss of c-Ha-*ras* or c-*myc* were frequently observed in primary tumors showing more aggressive behavior, such as recurrence or progression, following surgery (Cline and Battifora, 1987). Confirmation of these observations was provided in a study of 54 patients with operable NSCLC, in which changes in the copy number of Ha-*ras*, c-*myc*, and c-*raf*-1 were correlated with a relapse within 12 months, following complete surgical resection (Hajj et al., 1990). Amplification of protooncogenes, in particular c-*myc* and Ha-*ras*, appears to be a relatively rare event in NSCLC (Hajj et al., 1990; Slebos et al., 1989), although the correlation with relapses is quite strong (Hajj et al., 1990). In primary giant cell carcinoma of the lung a specific *in vivo* rearrangement of the c-*myc* gene has been detected about 6 kb upstream of the gene and implicated in tumorigenesis or tumor progression (Iizuka et al., 1990).

The amplification of c-*erb*B-1, which encodes the EGFR, commonly detected in squamous and large cell undifferentiated cancers in fresh tumor tissue (Cline and Battifora, 1987), can also be seen in paraffin-embedded sections. Others find that EGFR expression is more frequent in well-differentiated tumors (Dazzi et al., 1989). Opinions differ on the **frequency of protooncogene amplification**, ranging from 56% (Cline and Battifora, 1987) to rare (Slebos et al., 1989), and **no consensus exists on the utility of these measurements for diagnosis or prognosis**.

With respect to adenocarcinoma of the lung, mutations of Ki-*ras*, leading to its activation and 20-fold amplification, are thought to be among the early events in the development of NSCLC (Rodenhuis et al., 1987). Moreover, the detection of Ki-*ras* point mutations in **codon 12** by **PCR** appears to provide a useful indicator, defining a subgroup of lung adenocarcinoma patients with very poor prognosis and relapses, occurring despite a small tumor load and radical surgery (Slebos et al., 1990). These mutations are more frequent in **lung adenocarcinomas from smokers** than from similar tumors in nonsmokers, suggesting that **tobacco smoke contributes to the appearance of point mutations in Ki-*ras* in human lung adenocarcinomas** (Slebos et al., 1991). Tumors harboring the Ki-*ras* codon 12 mutations tended to be smaller in size and less differentiated than those without mutations and possibly have a different response to chemotherapy. A 30-fold amplification of N-*myc* has also been observed in adenocarcinoma of the lung (Saksela et al., 1986).

The role of the **tumor suppressor genes** in NSCLC is suggested by the finding that 11% of the cases studied have functional and structural abnormalities of the ***RB*** **gene**, ranging from the absence or alteration of *RB* mRNA to deletions of the *RB* gene (Reissmann et al., 1990). DNA from the adjacent uninvolved lung tissue did not show these alterations. Correlations with clinical behavior are still being investigated. The importance of studying **recessive oncogenes** in lung cancer (Kaye et al., 1990) is further emphasized by the finding that *p53*, which is known to be involved in the genesis of SCLC (see above), also plays a role in the pathogenesis of early-stage NSCLC (Chiba et al., 1990).

Clearly, **more work is needed before these gene-related characteristics may be used for patient management in lung cancer, especially in view of the observed conversion of SCLC cells to the NSCLC phenotype in culture** (Falco et al., 1990; Mabry et al., 1990). The cellular and molecular mechanisms underlying **transition between lung cancer phenotypes** is an active field of research undoubtedly bound to influence patient management strategies in the near future (Mabry et al., 1991). A **combination of molecular markers** is certainly preferable to reliance on a single prognosticator when devising treatment protocols. Evidence of the involvement of various genes in lung tumors includes the expression of the **atrial natriuretic factor gene** in SCLC (Bliss et al., 1990), **endothelin** mRNA (Giaid et al., 1990), and the expression of **blood group antigen A** in NSCLC tumor cells; these can be used as important favorable prognostic factors (Lee et al., 1991).

2.6.3 PLEURAL MESOTHELIOMA AND PULMONARY FIBROSIS

2.6.3.1 CHROMOSOME ABERRATIONS

Although clonal **chromosomal abnormalities** have been detected in 25 of 38 patients studied, no chromosomal aberration specific to mesothelioma could be demonstrated (Tiainen et al., 1989, and references therein). On the other hand, the number of copies of the **chromosome 7** short arm proved to be inversely correlated with survival, suggesting this numerical chromosome change may be used as a possible prognostic factor.

An attempt to develop a discriminating objective criterion to distinguish between **malignant** and **benign mesothelial tissue** has been made in a study measuring the expression of **EGFR** (Dazzi et al., 1990). No correlation could be established between different histologic subgroups of malignant mesothelioma and EGFR expression. Again, this characteristic could not be used as an independent prognostic factor for survival.

2.6.3.2 ONCOGENES AND GROWTH FACTORS

Meaningful correlations between growth factor(s) expression and pulmonary disease can certainly be of potential practical significance, although much is to be learned before such correlations can be used for patient management. Examples of such correlations abound. An association between the expression of c-*sis* protooncogene, encoding the B chain of PDGF, and pulmonary fibrosis has been established (Martinet et al., 1987). Alveolar macrophages recovered from the lungs of patients with **idiopathic pulmonary fibrosis** released four times more PDGF than did alveolar macrophages from normal persons. Pulmonary fibrosis induced by **silica** has also been linked to growth factor(s) such as **tumor necrosis factor (TNF)-α** in laboratory animals (Piguet et al., 1990). The expression of **intercellular adhesion molecule-1 (ICAM-1)** has been strongly implicated in the pathogenesis of asthma and possibly in a number of other diseases characterized by airway inflammation (chronic bronchitis, emphysema, and idiopathic pulmonary fibrosis) (Wegner et al., 1990).

2.6.4 NUCLEAR DNA CONTENT AND RESPONSE TO TREATMENT OF LUNG CANCERS

The few studies of **ploidy of lung carcinomas** favor the correlation of better prognosis with diploid DNA content (Volm et al., 1985; P. V. Zimmerman et al., 1987), although an opposite view has also been expressed (Bunn et al., 1983). In an attempt to exclude the influence of multiple variables (such as advanced disease) on the correlations sought, a group of 100 surgically treated patients was retrospectively evaluated (P. V. Zimmerman et al., 1987). A significantly shorter survival was observed in patients with aneuploid tumors determined by flow cytometry (FCM) of archival paraffin blocks. Although ploidy did not show a reproducible correlation with the clinical and pathological characteristics (adenocarcinoma vs. SCC), it proved to be the most important determinant of survival. Based on these observations, ploidy is recommended for use in deciding whether adjunct therapy can be of benefit — as in the case of aneuploid tumors. On the other hand, patients with diploid tumors, especially without lymph node involvement, have such a good prognosis with appropriate surgery that adjunct therapy can hardly improve it (P. V. Zimmerman et al., 1987).

In fact, the **nuclear DNA content** evaluated on specimens obtained by **bronchial washings** appears to predict the response of patients to **chemotherapy** in SCLC (Abe et al., 1987). The hyperdiploid DNA pattern was associated with a better response to combined Cytoxan plus Adriamycin plus Vincristine therapy than was the near-diploid pattern. On the other hand, patients with a near-diploid pattern of SCLC tumor had a better prognosis than did patients with a hyperploid pattern (Abe et al., 1985). As mentioned above, others have noted no difference, however, in survival characteristics between patients with near-diploid and aneuploid tumors (Bunn et al., 1983).

Because the majority of lung cancers are associated with **smoking**, the search for predisposing factors at the gene level acquires particular significance. A review of various aspects of lung metabolism, presumably related to predisposition to lung cancer, and which can be traced to genetic factors (Law et al., 1989), leads to the conclusion that "there is some variation between individuals in genetically determined lung cancer, but as yet no evidence for substantially increased risk in a minority of the population" (Law, 1990). It is felt that only molecular genetics can offer an approach that would allow the circumvention of various biases affecting population, metabolic, and clinical studies (Law, 1990).

An interesting approach to evaluating **multidrug resistance** (MDR) in lung cancer has been demonstrated by an FCM analysis of **Hoechst 33342 dye uptake** (Morgan et al., 1989). This dye represents a model chemotherapeutic agent whose accumulation parallels that of a number of cytotoxic drugs affected by the MDR phenotype. Measurement of the nuclear binding of this fluorescent compound by cells of different histologic types of lung cancer (SCLC vs. NSCLC) suggested that different mechanisms of drug resistance operate in lung tumors of different tissue type (Morgan et al., 1989). Expression of the major polycyclic aromatic hydrocarbon inducible-cytochrome P4501A1 gene (**CYP1A1**), implicated in lung carcinogenesis through modification of procarcinogens in tobacco smoke into potent carcinogens by the cytochrome-dependent monooxygenase, has been shown to increase in smokers and lung cancers (McLemore et al., 1990).

Tumors of the Head and Neck

2.7.1 DNA PLOIDY

The **squamous cell carcinomas (SCCs)** of the **oral cavity** display a high percentage of **aneuploidy** (48%) that correlates with the size of the tumor, low histologic grade, and the presence of lymph node metastases (Tytor et al., 1987a). Likewise, polyploidy is higher in poorly differentiated tumors and in metastases. The **proliferative index** does not reach statistically significant correlation with tumor size or duration of the disease, although it is **inversely correlated** with the degree of differentiation. This finding confirmed earlier reports (Johnson et al., 1985) and may, in part, be explained by the fact that the estimation of the SPF in nondiploid tumors is relatively inaccurate (Tytor et al., 1987a). These observations were further confirmed when **multifactorial malignancy grading systems** were applied to these tumors, and found to be significantly correlated with the presence of lymph node metastases and DNA ploidy (Tytor et al., 1987b). While the **prognosis** was worse for patients with high cumulative malignancy scores than for those with low scores, **the difference in prognosis between patients with diploid and those with nondiploid tumors was not statistically significant** (Tytor et al., 1987b).

In **nasopharyngeal cancer (NPC),** the **highest survival** following radiation therapy was observed in 55 patients with **anaplastic carcinoma** (75%), followed by **lymphoepithelioma** (60%), and the worst survival was in patients with SCC (30%) (Costello et al., 1990). **DNA ploidy analysis** in tumors taken before the initiation of therapy has also been correlated with 5-year survival, and the majority (60%) of all the tumors analyzed were aneuploid. No statistically significant difference in 5-year survival could be observed between patients with diploid and aneuploid tumors, even within the same histologic subgroup. Likewise, **there was no significant correlation between survival and the DNA index**. The most important predictor of survival was the degree of local invasion.

The majority (62%) of **adenoid cystic carcinomas** of the submandibular gland appear to be diploid (Luna et al., 1990). Aneuploid tumors are more frequently encountered in advanced clinical stages and their histology is different, displaying a solid pattern of cellular growth, intravascular extension, and invasion of nerves larger than 0.25 mm. It seems, therefore that **DNA ploidy measurements may be an effective adjunct to the clinicopathologic assessment in this type of tumors**.

Similar to NPC, the majority (57.1%) of SCCs of the **larynx** are aneuploid or tetraploid (Rua et al., 1991). Survival rates correlate with ploidy among tumors with a similar degree of differentiation, the lower survival observed in nondiploid cases compared to diploid (27.7 vs. 41.7%, respectively). **These relationships appear to offer practically useful guidance to the clinician, particularly at a time when other prognosticators are not yet available**.

Ploidy is significantly related to prognosis not only in early stages of SCC of the head and neck but also in advanced cases, such as those entering chemotherapy for **end-stage disease** (Cooke et al., 1990). In a large series of 172 patients with SCCs of the head and neck region, DNA aneuploidy and increasing DNA index were significantly correlated with both relapse-free and overall survival (Kearsley et al., 1991). Almost a threefold increase in death rate was observed in patients with aneuploid tumors, and a 6.6-fold higher death rate was observed in patients with hypertetraploidy DI over 2.11.

2.7.2 CHROMOSOME ABERRATIONS

Chromosomal abnormalities in the predominant type of cancer (SCC) of the head and neck region have been studied on a relatively limited basis (Teyssier, 1987; Jin et al., 1988a–e, 1990a–c). **Multiple unrelated structural clonal chromosome aberrations** have been a consistent finding. A **clustering** of chromosomal **breakpoints** seems to center nonrandomly around bands **1p22** and **11p13**, found to be rearranged in the majority of tumors (Jin et al., 1990c). **Homogeneously staining regions** and **double minutes**, indicating gene amplification areas, were found in some tumors. The **candidate genes** potentially affected in those regions were *bcl-1*, *int-2*, *hst-1* and N-*ras*. In fact, all of these protooncogenes have been found to be amplified in head and neck carcinomas (see below).

2.7.3 ONCOGENES AND GROWTH FACTORS

2.7.3.1 *ras*

PCR amplification combined with mutation-specific oligonucleotide probing revealed the presence of a **codon 61** mutation in Ki-*ras* in SCC of the oral cavity (Chang et al., 1989). Interestingly, the signal was stronger in the blood of the affected patient than in the DNA from the neoplasm or the salivary gland, although the patient had no manifestations of hematologic malignancy. The expression of the Ha-*ras* protooncogene in SCC of the oral cavity has been localized to the tumor as demonstrated by ISH (Hoellering and Shuler, 1989). The Ha-*ras* mRNA was expressed more in areas with higher proliferative activity and invasion, suggesting that the **evaluation of the Ha-*ras* expression pattern may be helpful in predicting the development and progression of oral SCC.**

In fact, another series of oral SCCs featured heterozygosity at the c-Ha-*ras* locus and loss of heterozygosity (LOH) at **11p**, which may also be related to the involvement of **tumor suppression genes** in oral SCC (Howell et al., 1989). Confirmation of LOH of one of the c-Ha-*ras* alleles as well as of other genes has been found among 25 specimens of primary SCC of the head and neck in 22% of heterozygous patients for this locus (Sheng et al., 1990). Analysis of c-Ha-*ras* gene expression in primary tumors and metastatic lesions suggested this gene is not as strongly involved at advanced stages of head and neck SCC as it appears to be at earlier clinical stages.

2.7.3.2 *myc*

The expression of c-*myc* in 44 SCCs of the head and neck failed to correlate with various clinicopathological parameters, including histopathological differentiation, lymph node invasion, extracapsular extension of the tumor, TNM staging, and so on (Field et al., 1989). On the other hand, the survival of patients with tumors having elevated levels of the c-*myc* oncoprotein was significantly shorter than those with lower levels of c-*myc* expression. It appears, therefore, that **measuring c-*myc* expression can be used as an effective prognostic indicator in head and neck SCC** (Field et al., 1989).

Using Southern analysis, the amplification of several protooncogenes has been evaluated, including c-*myc*, N-*myc*, L-*myc*, N-*ras*, Ha-*ras*, Ki-*ras*, erb-B, erbB-2, raf, and int-2 (Merritt et al., 1990). Frequent (11 out of 21 tumors) amplification of *int-2* and c-*myc* could be observed. Other oncogenes were not amplified in any of the tumors of the head and neck studied in that series. The same RFLP pattern of the L-*myc* gene can be demonstrated in **peripheral blood cells** in patients with oral SCC as in the tumor tissue (Saranath et al., 1990). The presence, rather than the absence, of the smaller, 6.6-kb **S fragment** was correlated with the likelihood of developing a poor to moderately differentiated tumor.

A **variable pattern** of oncogene amplification and deletions has been demonstrated in SCC cell lines and primary tumors of the head and neck (Yin et al., 1990). c-*myc* and **epidermal growth factor receptor (EGFR)** were observed in stage 3 tumors, whereas N-*myc* and *neu* amplifications, as well as *p53* deletion, correlated with recurrence. Although the expression of *neu* is observed in normal salivary glands, none of the 14 SCC of the oral and laryngeal mucosa displayed an enhanced *neu* transcription level (Riviere et al., 1991). In contrast, parotid **pleiomorphic adenomas** and a salivary gland adenocarcinoma featured enhanced *neu* expression (Riviere et al., 1991).

A strong positive correlation has been observed between smoking history and drinking, on the one hand, and the overexpression of *p53* gene in the head-and-neck SCC (Field et al., 1991, 1992).

The expression of the *neu* protooncogene is detectable in normal salivary gland tissue and it is variably expressed in pleomorphic adenoma, in the majority of cases without concomitant amplification of the gene (Kahn et al., 1992).

2.7.3.3 EPIDERMAL GROWTH FACTOR RECEPTOR

Conflicting findings have been reported concerning the expression of the EGFR gene in head and neck tumors. While only a single SCC cell line was found to show enhancement of the EGFR mRNA (Eisbruch et al., 1987), 19% of SCCs in other studies had **amplified** EGFR gene, with 53% of the tumors demonstrating **overexpression** of EGFR (Ishitoya et al., 1989). Other types of tumors of the head and neck failed to show amplification of the gene.

2.7.4 THE AMYLASE GENE

No evidence of structural rearrangements has been obtained by Southern analysis of the amylase gene in a **Warthin tumor**, an **adenoid cystic carcinoma**, and a **mucoepidermoid carcinoma of the parotid gland** (Morley and Hodes, 1988). **Pleiomorphic adenoma**, although positive for the amylase protein, failed to show an elevation of the gene expression. Warthin tumors were negative for both the amylase protein and amylase mRNA. One of the consistent findings in the salivary glands of Sjogren's syndrome patients is the presence of Epstein-Barr virus (EBV) detectable by immunocytochemistray, ISH, and PCR (e.g., Keinanen et al., 1989; Deacon et al., 1991; for a review of the genetic control of salivary proteins, see Karn, 1991).

2.7.5 EPSTEIN-BARR VIRUS

Epstein-Barr viurs (EBV) can be found in nasopharyngeal carcinoma (NPC) which can be detected in FNA aspirates by PCR amplification of EBV genomes (Feinmesser et al., 1992). The importance of definitive and early identification of EBV in NPC, particularly in tumors with squamous cell differentiation, underscores the need for prevention of the development of advanced lesions usually presented for diagnosis and treatment (Jacobs and Pinto, 1992).

References

Aalto-Setala, K. et al. (1987). DNA polymorphisms of apolipoprotein AI/CIII and insulin genes in familial hypertriglyceridemia and coronary heart disease. Atherosclerosis *66*:145–152.

Aalto-Setala, K. et al. (1988). *Xba*I and c/g polymorphisms of the apolipoprotein B gene locus are associated with serum cholesterol and LDL-cholesterol levels in Finland. Atherosclerosis *74*:47–54.

Aalto-Setala, K. et al. (1989). Genetic polymorphism of the apolipoprotein B gene locus influences serum LDL cholesterol level in familial hypercholesterolemia. Hum. Genet. *82*:305–307.

Abe, S. et al. (1985). Prognostic significance of nuclear DNA content in small cell carcinoma of the lung. Cancer *56*:2025–2030.

Abe, S. et al. (1987). Nuclear DNA content as an indicator of chemosensitivity in small-cell carcinoma of the lung. Anal. Quant. Cytol. Histol. *9*:425–428.

Aburatani, H. et al. (1988). A study of DNA polymorphism in the apolipoprotein B gene in a Japanese population. Atherosclerosis *72*:71–76.

Alanen, K. A. et al. (1990). DNA ploidy in pancreatic neuroendocrine tumors. Am. J. Clin. Pathol. *93*:784–788.

Albe, X. et al. (1990). Independent prognostic value of ploidy in colorectal cancer. A prospective study using image cytometry. Cancer *66*:1168–1175.

Albers, J. J. et al. (1990). The unique lipoprotein (a): properties and immunological measurement. Clin. Chem. *36*:2019–2026.

Albertson, D. G. et al. (1989). Localization of polymorphic DNA probes frequently deleted in lung carcinoma. Hum. Genet. *83*:127–132.

Allen, D. C. et al. (1987). Morphometrical analysis in ulcerative colitis with dysplasia and carcinoma. Histopathology *11*:913–926.

Alsayed, N. and Rebourcet, R. (1991). Abnormal concentrations of C-II,C-III, and E apolipoproteins among apolipoprotein B-containing, B-free, and A-I-containing lipoprotein particles in hemodialysis patients. Clin. Chem. *37*:387–393.

Amos, J. A. et al. (1990). DNA analysis of cystic fibrosis genotypes in relatives with equivocal sweat test results. Clin. Invest. Med. *13*:1–5.

Anderson, M. P. et al. (1991). Demonstration that CFTR is a chloride channel by alteration of its anion selectivity. Science *253*:202–207.

Anderson, R. A. et al. (1989). Restriction fragment length polymorphisms associated with abnormal lipid levels in an adolescent population. Atherosclerosis *77*:227–237.

Anonymous (1990). Cystic fibrosis: closing the gap [editorial]. Lancet *336*:539–540.

Anwar, R. et al. (1990). DNA sequence analysis of the KM-19 locus linked to cystic fibrosis. Design of new oligonucleotides to remove non-specific PCR products. Hum. Genet. *85*:319–323.

Armitage, N. C. et al. (1985). The influence of tumor cell DNA abnormalities on survival in colorectal cancer. Br. J. Surg. *72*:828–830.

Armitage, N. C. et al. (1990). The influence of tumor cell DNA content on survival in colorectal cancer: a detailed analysis. Br. J. Cancer *62*:852–856.

Armstrong, V. W. et al. (1989). Effect of HELP-LDL-apheresis on serum concentrations of human lipoprotein (a): kinetic analysis of the posttreatment return to baseline levels. Eur. J. Clin. Invest. *19*:235–240.

Arnheim, N. et al. (1991). Genetic mapping by single sperm typing. Animal Genet. *22*:105–115.

Astrin, S. M. and Costanzi, C. (1989). The molecular genetics of colon cancer. Semin. Oncol. *16*:138–147.

Augenlicht, L. H. et al. (1987). Overexpression of *ras* in mucus-secreting human colon carcinoma cells of low tumorigenicity. Cancer Res. *47*:3763–3765.

Augenlicht, L. H. et al. (1991). Patterns of gene expression that characterize the colonic mucosa in patients at genetic risk for colonic cancer. Proc. Natl. Acad. Sci. U.S.A. *88*:3286–3289.

Baba, H. et al. (1990). Comparison of DNA content in gastric cancer cells between primary lesions and lymph node metastases. Cancer *66*:1775–1780.

Baisch, H. et al. (1990). DNA ploidy and cell-cycle analysis in pancreatic and ampullary carcinoma: flow cytometric study of formalin-fixed paraffin-embedded tissue. Virchow's Arch. A: Pathol. Anat. *417*:145–150.

Baker, S. J. et al. (1989). Chromosome 17 deletions and p53 gene mutations in colorectal carcinomas. Science *244*:217–221.

Baker, S. J. et al. (1990). p53 gene mutations occur in combination with 17p allelic deletions as late events in colorectal tumorigenesis. Cancer Res. *50*:7717–7722.

Barasch, J. et al. (1991). Defective acidification of intracellular organelles in cystic fibrosis. Nature (London) *352*:70–73.

Baretton, G. et al. (1991a). Flow-cytometric analysis of the DNA content in paraffin-embedded tissue from colorectal carcinomas and its prognostic significance. Virchow's Arch. B: Cell Pathol. *60*:123–131.

Baretton, G. et al. (1991b). DNA ploidy and survival in gastric carcinomas: a flow cytometric study. Virchow's Arch. A: Pathol. Anat. *418*:301–309.

Barnes, P. J. (1990). Molecular biology of receptors: implications for lung disease. Thorax *45*:482–488.

Barthe, C. et al. (1991). Identification of "cystic fibrosis protein" as a complex of two calcium-binding proteins present in human cells of myeloid origin. Biochim. Biophys. Acta *1096*:175–177.

Bartek, J. et al. (1991). Asymmetric PCR-based strategy for gluetic analysis of the p53 tumor suppressor gene in cell lines and tumor tissues. Neoplasma *38*:93–99.

Bavand, M. R. and Orgad, L. (1988). Two proteins with reverse transcriptase activities associated with hepatitis B virus-like particles. J. Virol. *62*:626–628.

Baylin, S. B. et al. (1986). DNA methylation patterns of the calcitonin gene in human lung cancers and lymphomas. Cancer Res. *46*:2917–2922.

Baylin, S. B. et al. (1987). Hypermethylation of the 5' region of the calcitonin gene as a property of human lymphoid and acute myeloid malignancies. Blood *70*:412–417.

Beasley, R. P. (1982). Hepatitis B virus as the etiologic agent in hepatocellular carcinoma — epidemiologic considerations. Hepatology *2*:S21–26.

Beasley, R. P. and Hwang, L. Y. (1984). Hepatocellular carcinoma and hepatitis B virus. Semin. Liver Dis. *4*:113–121.

Beasley, R. P. et al. (1981). Hepatocellular carcinoma and hepatitis B virus: a prospective study of 22,707 men in Taiwan. Lancet *2*:1129–1133.

Beauchemin, N. et al. (1987). Isolation and characterization of full length functional cDNA clones for human carcinoembryonic antigen. Mol. Cell. Biol. *7*:3221–3230.

Beaudet, A. L. and Buffone, G. J. (1987). Prenatal diagnosis of cystic fibrosis. J. Pediatr. *111*:630–633.

Beaudet, A. et al. (1986). Linkage of cystic fibrosis to two tightly linked DNA markers: joint report from a collaborative study. Am. J. Hum. Genet. *39*:681–693.

Bedford, M. T. and van Helden, P. D. (1990). A method to analyze allele-specific methylation. BioTechniques *9*:744–748.

Beer, D. G. and Pitot, H. C. (1989). Proto-oncogene activation during chemically induced hepatocarcinogenesis in rodents. Mutat. Res. *220*:1–10.

Beer, D. G. et al. (1986). Expression of H-*ras* and c-*myc* protooncogenes in isolated γ-glutamyltranspeptidase-positive rat hepatocytes and in hepatocellular carcinomas induced by diethylnitrosamine. Cancer Res. *46*:2435–2441.

Beer, D. G. et al. (1988). Expression of the c-*raf* protooncogene, γ-glutamyltranspeptidase, and gap junction protein in rat liver neoplasms. Cancer Res. *48*:1610–1617.

Benbrook, D. et al. (1988). A new retinoic acid receptor identified from a hepatocellular carcinoma. Nature (London) *333*: 669–672.

Benchimol, S. et al. (1989). Carcinoembryonic antigen, a human tumor marker, functions as an intercellular adhesion molecule. Cell (Cambridge Mass) *57*:327–334.

Beretta, E. et al. (1987). Serum CA 19-9 in the postsurgical follow-up of patients with pancreatic cancer. Cancer *60*:2428–2431.

Berg, K. (1963). A new serum type system in man — the Lp system. Acta Pathol. Microbiol. Scand. *59*:369–382.

Berg, K. (1968). The Lp system (review). Ser. Hematol. *1*:111–136.

Berg, K. (1989a). Predictive genetic testing to control coronary heart disease and hyperlipidemia. Arteriosclerosis *9* (Suppl. I):50–58.

Berg, K. (1989b). Impact of medical genetics on research and practices in the area of cardiovascular disease. Clin. Genet. *36*:299–312.

Bergh, J. C. S. (1990). Gene amplification in human lung cancer. The *myc* family genes and other proto-oncogenes and growth factor genes. Am. Rev. Respir. Dis. *142*:S20–S26.

Bick, D. et al. (1990). Prenatal screening for F508 mutation in population not selected for cystic fibrosis. Lancet *336*:1324–1325.

Birrer, M. J. and Minna, J. D. (1988). Molecular genetics of lung cancer. Semin. Oncol. 15:226–235.

Blangero, J. et al. (1990). Genetic analysis of apolipoprotein A-I in two dietary environments. Am. J. Hum. Genet. *47*:414–428.

Blankenhorn, D. H. (1990). Can atherosclerotic lesions regress? Angiographic evidence in humans. Am. J. Cardiol. *65*:41F-43F.

Blanquet, V. et al. (1988). Regional mapping to 4q 32. 1 by *in situ* hybridization of a DNA domain rearranged in human liver cancer. Hum. Genet. *80*:274–276.

Bliss, D. P., Jr. et al. (1990). Expression of the atrial natriuretic factor gene in small lung cancer tumors and tumor cell lines. J. Natl. Cancer Inst. *82*:305–310.

Blum, H. E. et al. (1983). Detection of hepatitis B virus DNA in hepatocytes, bile duct epithelium and vascular elements by *in situ* hybridization. Proc. Natl. Acad. Sci. U.S.A. *80*:6685–6688.

Blum, H. E. et al. (1988). Latent hepatitis B virus infection with full-length viral genome in a patient serologically immune to hepatitis B virus infection. Liver *8*:307–316.

Blum, H. E. et al. (1989). The molecular biology of hepatitis B virus. Trends Genet. *5*:154–158.

Blumberg, B. S. and London, W. T. (1985). Hepatitis B virus and prevention of primary cancer of the liver. J. Natl. Cancer Inst. *74*:267–273.

Bodmer, W. F. et al. (1987). Localization of the gene for familial adenomatous polyposis on chromosome 5. Nature (London) *328*:614–616.

Boerwinkle, E. and Chan, L. (1989). A three codon insertion/deletion polymorphism in the signal peptide region of the human apolipoprotein B (APOB) gene directly typed by the polymerase chain reaction. Nucleic Acids Res. *17*:4003.

Boerwinkle, E. and Utermann, G. (1988). Simultaneous effects of the apolipoprotein E polymorphism on apolipoprotein E, apolipoprotein B, and cholesterol metabolism. Am. J. Hum. Genet. *42*:104–112.

Boerwinkle, E. et al. (1989a). Genetics of the quantitative Lp(a) lipoprotein trait. III. Contribution of Lp(a) glycoprotein phenotypes to normal lipid variation. Hum. Genet. *82*:73–78.

Boerwinkle, E. et al. (1989b). Rapid typing of tandemly repeated hypervariable loci by the polymerase chain reaction: application to the apolipoprotein B 3' hypervariable region. Proc. Natl. Acad. Sci. U.S.A. *86*:212–216.

Boerwinkle, E. et al. (1990). Rapid typing of apolipoprotein B DNA polymorphisms by DNA amplification. Association between Ag epitopes of human apolipoprotein B100, a signal peptide insertion/deletion polymorphism, and a 3' flanking DNA variable number of tandem repeats polymorphism of the apolipoprotein B gene. Atherosclerosis *81*:225–232.

Bolen, J. B. et al. (1987a). Activation of pp60c-src protein kinase activity in human colon carcinoma. Proc. Natl. Acad. Sci. U.S.A. *84*:2251–2255.

Bolen, J. B. et al. (1987b). Analysis of pp60c-src in human colon carcinoma and normal human colon mucosal cells. Oncogene Res. *1*:149–168.

Borecki, I. B. et al. (1988). Genetic factors influencing apolipoprotein AI and AII levels in a kindred with premature coronary heart disease. Genet. Epidemiol. *5*:393–406.

Bos, J. L. et al. (1987). Prevalence of *ras* gene mutations in human colorectal cancers. Nature (London) *327*:293–297.

Brauch, H. et al. (1987). Molecular analysis of the short arm of chromosome 3 in small cell and non-small cell carcinoma of the lung. N. Engl. J. Med. *317*:1109–1113.

Brauch, H. et al. (1990). Molecular mapping of the chromosome 3p deletion sites in lung cancer. Proc. Am. Assn. Cancer Res. *31*:320A.

Brechot, C. et al. (1985). Hepatitis B virus DNA in patients with chronic liver disease and negative tests for hepatitis B surface antigen. N. Engl. J. Med. *312*:270–276.

Brennan, J. et al. (1991). *myc* family DNA amplification in 107 tumors and tumor cell lines from patients with small cell lung cancer treated with different combination chemotherapy regimens. Cancer Res. *51*:1708–1712.

Brenninkmeijer, B. J. et al. (1984). Apo E polymorphism and lipoproteins in coronary artery disease and peripheral vascular disease [abstract]. Arteriosclerosis *4*:542A.

Breslow, J. L. (1985). Human apolipoprotein molecular biology and genetic variation. Annu. Rev. Biochem. *54*:699–727.

Breslow, J. L. (1987). Lipoprotein genetics and molecular biology. In: Plasma Lipoproteins. Gotto, A. M., Jr. (ed.). Elsevier, New York, pp. 359–397.

Breslow, J. L. (1988). Apolipoprotein genetic variation and human disease. Physiol. Rev. *68*:85–132.

Breslow, J. L. (1989). Genetics basis of lipoprotein disorders. J. Clin. Invest. *84*:373–380.

Bressac, B. et al. (1990). Abnormal structure and expression of p53 gene in human hepatocellular carcinoma. Proc. Natl. Acad. Sci. U.S.A. *87*:1973–1977.

Brink, P. A. et al. (1987). Familial hypercholesterolemia in South African Afrikaners. *Pvu*II and *Stu*I DNA polymorphisms in the LDL-receptor gene consistent with a predominanting founder gene effect. Hum. Genet. *77*:32–35.

Brock, D. (1990). Population screening for cystic fibrosis. Am. J. Hum. Genet. *47*:164–165.

Broers, J. L. V. et al. (1990). Unique patters of mRNA distribution of major surfactant associated protein, c-*myc* and c-*jun* in neoplastic and nonneoplastic human lung tissue using *in situ* hybridization. Proc. Am. Assn. Cancer Res. *31*:311A.

Brooks, B. J., Jr. et al. (1987). Amplification and expression of the *myc* gene in small cell lung cancer. Adv. Virol. Oncol. *7*:155–172.

Brown, M. L. et al. (1989). Molecular basis of lipid transfer protein deficiency in a family with increased high density lipoproteins. Nature (London) *342*:448–451.

Brown, M. S. and Goldstein, J. L. (1974). Expression of the familial hypercholesterolemia gene in heterozygotes: mechanism for a dominant disorder in man. Science *185*:61–63.

Brown, M. S. and Goldstein, J. L. (1986). A receptor-mediated pathway for cholesterol homeostasis. Science *232*:34–47.

Bruce, W. R. (1987). Recent hypotheses for the origin of colon cancer. Cancer Res. *47*:4237–4242.

Buchmann, A. et al. (1991). Mutational activation of the c-Ha-*ras* gene in liver tumors of different rodent strains: correlation with susceptibility to hepatocarcinogenesis. Proc. Natl. Acad. Sci. U.S.A. *88*:911–915.

Buetow, K. H. et al. (1989). Loss of heterozygosity suggests tumor suppressor gene responsible for primary hepatocellular carcinoma. Proc. Natl. Acad. Sci. U.S.A. *86*:8852–8856.

Buffone, G. J. et al. (1988). Prenatal diagnosis of cystic fibrosis: microvillar enzymes and DNA analysis compared. Clin. Chem. *34*:933–937.

Bunn, P. A. et al. (1983). Diagnostic and biological implications of flow cytometric DNA content analysis in lung cancer. Cancer Res. *43*:5026–5032.

Burmer, G. C. and Loeb, L. A. (1989). Mutations in the KRAS2 oncogene during progressive stages of human colon carcinoma. Proc. Natl. Acad. Sci. U.S.A. *86*:2403–2407.

Burmer, G. C. et al. (1990). c-Ki-*ras* mutations in chronic ulcerative colitis and sporadic colon carcinoma. Gastroenterology *99*:416–420.

Cahill, T. C. et al. (1990). Resolution of DNA linkage discrepancies through analysis of a VNTR locus in a family study of cystic fibrosis. Prenatal Diagn. *10*:795–799.

Camejo, G. et al. (1988). Identification of apoB-100 segments mediating the interaction of low density lipoproteins with arterial proteoglycans. Arteriosclerosis *8*:368–377.

Carey, W. F. et al. (1990). Cystic fibrosis prenatal diagnosis: confirmation of an equivocal microvillar enzyme result by direct analysis of the common gene mutation. Prenatal Diagn. *10*:613–616.

Cariani, E. et al. (1990). Expression of insulin-like growth factor II (IGF-II) in human primary liver cancer: mRNA and protein analysis. J. Hepatol. *11*:226–231.

Carlson, L. A. et al. (1989). Pronounced lowering of serum levels of lipoprotein Lp(a) in hyperlipidemic subjects treated with nicotine acid. J. Intern. Med. *226*:271–276.

Carman, W. F. et al. (1990). Vaccine-induced escape mutant of hepatitis B virus. Lancet *336*:325–329.

Cartlidge, S. A. and Elder, J. B. (1989). Transforming growth factor α and epidermal growth factor levels in normal human gastrointestinal mucosa. Br. J. Cancer *60*:657–660.

Cartwright, C. A. et al. (1989). pp60c-src activation in human colon cancer. J. Clin. Invest. *83*:2025–2033.

Caskey, C. T. et al. (1990). The American Society of Human Genetics statement on cystic fibrosis screening. Am. J. Hum. Genet. *46*:393.

Cattenach, B. M. and Kirk, M. (1985). Differential activity of maternally and paternally derived chromosome regions in mice. Nature (London) *315*:496–498.

Cavelier, C. A. et al. (1987). The DNA content in cancer and dysplasia in chronic ulcerative colitis. Histopathology *11*:927–939.

Celano, P. et al. (1988). Effect of polyamine depletion on c-*myc* expression in human colon carcinoma cells. J. Biol. Chem. *263*:5491–5494.

Chan, L. (1989). The apolipoprotein multigene family: structure, expression, evolution, and molecular genetics. Klin. Wochenschr. *67*:225–237.

Chan, L. and Dresel, H. A. (1990). Genetic factors influencing lipoprotein structure: implication for atherosclerosis. Lab. Invest. *62*:522–537.

Chandar, N. et al. (1987). Analysis of *ras* genes and linked viral sequences in rat hepatocarcinogenesis. Am. J. Pathol. *129*:232–241.

Chandar, N. et al. (1989). c-*myc* gene amplification during hepatocarcinogenesis by a choline-devoid diet. Proc. Natl. Acad. Sci. U.S.A. *86*:2703–2707.

Chang, S. E. et al. (1989). Novel Ki-*ras* codon 61 mutation in infiltrating leukocytes of oral squamous cell carcinoma. Lancet *i*:1014.

Chatterjee, D. et al. (1989). Reversible suppression of c-*myc* expression in a human colon carcinoma line by the anticancer agent *N*-methylformamide. Cancer Res. *49*:3910–3916.

Chehab, F. F. et al. (1991). A dimorphic 4-bp repeat in the cystic fibrosis gene is in absolute linkage disequilibrium with the F508 mutation: implications for prenatal diagnosis and mutation origin. Am. J. Hum. Genet. *48*:223–226.

Chen, D. S. (1987). Hepatitis B virus infection, its sequelae, and prevention in Taiwan. In: Neoplasmas of the Liver. Okuda, K. and Ishak, K. (eds.). Springer-Verlag, Tokyo, pp. 71–80.

Cheng, S. H. et al. (1990). Defective intracellular transport and processing of CFTR is the molecular basis of most cystic fibrosis. Cell (Cambridge, Mass) *63*:827–834.

Chiba, I. et al. (for the Lung Cancer Study Group) (1990). Mutations in the p53 gene are frequent in primary, resected non-small cell lung cancer. Oncogene *5*:1603–1610.

Choi, P. M. et al. (1990). Protein kinase C (PKC) can function as a tumor suppressor in HT29 colon cancer cells. Proc. Am. Assn. Cancer Res. *31*:320A.

Choi, Y. J. (1990). *In situ* hybridization using a biotinylated DNA probe on formalin-fixed liver biopsies with hepatitis B virus infections: *in situ* hybridization superior to immunohistochemistry. Mod. Pathol. *3*:343–347.

Chomel, J. C. et al. (1989). Usefulness of linkage disequilibrium of KM-19 and XV-2c DNA probes for genetic counselling in a high-risk CF family. Prenatal Diagn. *9*:297–300.

Chong, G. L. and Thibodeau, S. N. (1990). A simple assay for the screening of the cystic fibrosis allele in carriers of the Phe 508 deletion mutation. Mayo Clin. Proc. *65*:1072–1076.

Chung, Y. S. et al. (1987). The detection of human pancreatic cancer-associated antigen in the serum of cancer patients. Cancer *60*:1636–1643.

Ciclitira, P. J. et al. (1987). Expression of c-*myc* in non-malignant and premalignant gastrointestinal disorders. J. Pathol. *151*:293–296.

Cioce, V. et al. (1991). Increased expression of the laminin receptor in human colon cancer. J. Natl. Cancer Inst. *83*:29–36.

Clemmensen, I. et al. (1986). Purification and characterization of a novel, oligomeric, plasminogen kringle 4 binding protein from human plasma: tetranectin. Eur. J. Biochem. *156*:327–333.

Cline, M. J. and Battifora, H. (1987). Abnormalities of protooncogenes in non-small cell lung cancer. Correlations with tumor type and clinical characteristics. Cancer *60*:2669–2674.

Cohen, J. A. et al. (1989). Expression pattern of the *neu* (NGL) gene-encoded growth factor receptor protein (p185neu) in normal and transformed epithelial tissues of the digestive tract. Oncogene *4*:81–88.

Cohn, K. H. et al. (1991). Association of nm-23-H1 allelic deletions with distant metastases in colorectal carcinoma. Lancet *338*:722–724.

Collins, F. S. and Weissman, S. M. (1984). Directional cloning of DNA fragments at a large distance from an initial probe: a circularization method. Proc. Natl. Acad. Sci. U.S.A. *81*:6812–6816.

Collins, F. S. et al. (1987). Construction of a general chromosome jumping library, with application to cystic fibrosis. Science *235*:1046–1049.

Connelly, P. W. et al. (1987a). Structure of apolipoprotein CII$_{Toronto}$, a nonfunctional human apolipoprotein. Proc. Natl. Acad. Sci. U.S.A. *84*:270–273.

Connelly, P. W. et al. (1987b). Apolipoprotein CII$_{St. Michael}$, familial apolipoprotein CII deficiency associated with premature vascular disease. J. Clin. Invest. *80*:1597–1606.

Cooper, C. et al. (1991). An investigation of ring and dicentric chromosomes found in three Turner's syndrome patients using DNA analysis and *in situ* hybridisation with X and Y chromosome specific probes. J. Med. Genet. *28*:6–9.

Cooke, L. D. et al. (1990). Ploidy as a prognostic indicator in end stage squamous cell carcinoma of the head and neck region treated with cisplatinum. Br. J. Cancer *61*:759–762.

Corcos, D. et al. (1987). Expression of c-*myc* is under dietary control in rat liver. Oncogene Res. *1*:193–199.

Corey, M. et al. (1989). Familial concordance of pancreatic function in cystic fibrosis. J. Pediatr. *115*:274–277.

Corral, M. et al. (1988). Increased expression of the N-*myc* gene during normal and neoplastic rat liver growth. Exp. Cell Res. *174*:107–115.

Costello, F. et al. (1990). A clinical and flow cytometric analysis of patients with nasopharyngeal cancer. Cancer *66*:1789–1795.

Crespo, P. et al. (1990). Induction of apolipoprotein E gene expression in human and experimental atherosclerotic lesions. Biochem. Biophys. Res. Commun. *168*:733–740.

Cumming, A. M. and Robertson, F. (1984). Polymorphism at the apo E locus in relation to risk of coronary disease. Clin. Genet. *25*:310–313.

Curiel, D. T. et al. (1990). A chemical mismatch cleavage method usefull for the detection of point mutations in the p53 gene in lung cancer. Am. J. Respir. Cell Mol. Biol. *3*:405–411.

Cushing, G. L. et al. (1989). Quuantitation and localization of apolipoproteins (a) and B in coronary artery bypass veins grafts resected at re-operation. Atherosclerosis *9*:593–603.

Cutting, G. R. et al. (1990a). Two patients with cystic fibrosis, nonsense mutations in each cystic fibrosis gene, and mild pulmonary disease. N. Engl. J. Med. *323*:1685–1689.

Cutting, G. R. et al. (1990b). A cluster of cystic fibrosis mutations in the first nucleotide-binding fold of the cystic fibrosis conductance regulator protein. Nature (London) *346*:366–369.

Cystic Fibrosis Foundation Center Committee and Guidelines Subcommittee (1990). Cystic Fibrosis Foundation guidelines for patient services, evaluation, and monitoring in cystic fibrosis centers. Am. J. Dis. Child. *144*:1311–1312.

Cystic Fibrosis Genetic Analysis Consortium (1990). Worldwide survey of the ΔF508 mutation — report from the Cystic Fibrosis Genetic Analysis Consortium. Am. J. Hum. Genet. *47*:354–359.

Czerniak, B. et al. (1987). *ras* oncogene p21 as a tumor marker in the cytodiagnosis of gastric and colonic carcinomas. Cancer *60*:2432–2436.

D'Emilia, J. et al. (1989). Expression of the *erb*B-2 gene product (p185) at different stages of neoplastic progression in the colon. Oncogene *4*:1233–1239.

Daga, A. et al. (1988). *Pvu*II polymorphism of low density lipoprotein receptor gene and familial hypercholesterolemia. Study in Italians. Arteriosclerosis *8*:845–850.

Dahlen, G. H. et al. (1986). Association of levels of lipoprotein Lp(a), plasma lipids, and other lipoproteins with coronary artery disease documented by angiography. Circulation *74*:758–765.

Dalemans, W. et al. (1991). Altered chloride ion channel kinetics associated with the ΔF508 cystic fibrosis mutation. Nature *354*:526–528.

Dallongeville, J. et al. (1991). Apolipoprotein E polymorphism association with lipoprotein profile and endogenous hypertriglyceridemia and familial hypercholesterolemia. Arteriosl. Thromb. *11*:272–278.

Darnfors, C. et al. (1989). Lack of correlation between the apolipoprotein B *Xba*I polymorphism and blood lipid levels in a Swedish population. Atherosclerosis *75*:183–188.

Davidoff, A. M. et al. (1991). Genetic basis for p53 overexpression in human breast cancer. Proc. Natl. Acad. Sci. U.S.A. *88*:5006–5010.

Davies, M. J. and Thomas, A. C. (1985). Plaque fissuring: the cause of acute myocardinal infarction, sudden ischemic death, and crescendo angina. Br. Heart J. *53*:363.

Davignon, J. et al. (1988). Apolipoprotein E polymorphism and atherosclerosis. Atherosclerosis *8*:1–21.

Dawson, D. B. et al. (1989). Carrier identification of cystic fibrosis by recombinant DNA techniques. Mayo Clin. Proc. *64*:325–334.

Dazzi, H. et al. (1989). Expression of epidermal growth factor receptor (EGF-R) in non-small cell lung cancer. Use of archival tissue and correlation of EGF-R with histology, tumor size, mode status and survival. Br. J. Cancer *59*:746–749.

Dazzi, H. et al. (1990). Malignant pleural mesothelioma and epidermal growth factor receptor (EGF-R). Relationship of EGF-R with histology and survival using fixed paraffin-embedded tissue and the F4, monoclonal antibody. Br. J. Cancer *61*:924–926.

de Bustros, A. et al. (1988). The short arm of chromosome 11 is a "hot spot" for hypermethylation in human neoplasia. Proc. Natl. Acad. Sci. U.S.A. *85*:5693–5697.

de Knijff, P. et al. (1990). Influence of apo E polymorphism on the response to simvastatin treatment in patients with heterozygous familial hypercholesterolemia. Atherosclerosis *83*:89–97.

de The, H. et al. (1987). A novel steroid thyroid hormone receptor-related gene inappropriately expressed in human hepatocellular carcinoma. Nature (London) *330*:667–670.

Deacon, E. M. et al. (1991). Detection of Epstein-Barr virus antigens and DNA in major and minor salivary glands using immunocytochemistry and polymerase chain reaction: possible relationship with Sjogren's syndrome. J. Pathol. *163*:351–360.

Dean, M. et al. (1987). The *met* oncogene: a tyrosine kinase and a marker for cystic fibrosis. In: Molecular Mechanisms in the Regulation of Cell Behavior. Alan R. Liss, New York, pp. 107–112.

Dean, M. et al. (1990). Prenatal diagnosis and linkage disequilibrium with cystic fibrosis for markers surrounding D7S8. Hum. Genet. *85*:275–278.

DeCaprio, J. A. et al. (1988). SV40 large tumor antigen forms a specific complex with the product of the retinoblastoma susceptibility gene. Cell (Cambridge, Mass) *54*:275–283.

Deeb, S. S. et al. (1990). A splice-junction mutation responsible for familial apolipoprotein AII deficiency. Am. J. Hum. Genet. *46*:822–827.

Dejean, A. et al. (1983). Characterization of integrated hepatitis B viral DNA cloned from a human hepatoma and hepatoma-derived cell line PLC/PRF/5. Proc. Natl. Acad. Sci. U.S.A. *50*:2505–2509.

Dejean, A. et al. (1984a). Detection of hepatitis B virus DNA in pancreas, kidney and skin of two human carriers of the virus. J. Gen. Virol. *65*:651–655.

Dejean, A. et al. (1984b). Specific hepatitis B virus integration in hepatocellular carcinoma DNA through a viral 11-base-pair direct repeat. Proc. Natl. Acad. Sci. U.S.A. *81*:5350–5354.

Deuchars, K. L. and Ling, V. (1989). P-glycoprotein and multidrug resistance in cancer chemotherapy. Semin. Oncol. *16*:156–165.

Devoto, M. et al. (1989). Haplotypes in cystic fibrosis patients with or without pancreatic insufficiency from four European populations. Genomics *5*:894–898.

di Bisceglie, A. M. and Hoofnagle, J. H. (1990). Hepatitis B virus replication within the human spleen. J. Clin. Microbiol. *28*:2850–2852.

Dichek, H. L. et al. (1991). Identification of two separate allelic mutations in the lipoprotein lipase gene of a patient with the familial hyperchylomicronemia syndrome. J. Biol. Chem. *266*:473–477.

Dodson, L. A. and Kant, J. A. (1991). Two-temperature PCR and heteroduplex detection: application to rapid cystic fibrosis screening. Mol. Cell. Probes *5*:21–25.

Dolcetti, R. et al. (1988). Nuclear oncogene amplification or rearrangement is not involved in human colorectal malignancies. Eur. J. Cancer Clin. Oncol. *24*:1321–1328.

Doyle, L. A. et al. (1990). Oncogene activation induces human small cell lung cancer cells to secrete non-small cell lung cancer proteins. Proc. Am. Assn. Cancer Res. *31*:313A.

Drucker, D. J. et al. (1989). The parathyroid hormone-like peptide gene is expressed in the normal and neoplastic human endocrine pancreas. Mol. Endocrinol. *3*:1589–1595.

Drumm, M. L. et al. (1988). Physical mapping of the cystic fibrosis region by pulsed field gel electrophoresis. Genomics *2*:346–354.

Duesberg, P. H. (1985). Activated proto-oncogenes: sufficient or necessary for cancer? Science *228*:669–677.

Dunlop, M. G. et al. (1991). Linked DNA markers for presymptomatic diagnosis of familial adenomatous polyposis. Lancet *337*:313–316.

Dunning, A. M. et al. (1988). Relationships between DNA and protein polymorphism of apolipoprotein B. Hum. Genet. *78*:325–329.

Eaton, D. and Tomlinson, J. (1989). Does Lp(a) alter fibrinolysis? Circulation *80* (Suppl. II):467.

Edelberg, J. M. et al. (1989). Lipoprotein (a) inhibits streptokinase-mediated acctivation of human plasminogen. Biochemistry 28:2370–2374.

Ehnholm, C. et al. (1986). Apolipoprotein E polymorphism in the Finnish population: gene frequencies and relation to lipoprotein concentrations. J. Lipid Res. 27:227–235.

Eiberg, H. et al. (1985). Linkage relationships of paraoxonase (PON) with other markers: indication of PON-cystic fibrosis synteny. Clin. Genet. 78:265–271.

Eisbruch, A. et al. (1987). Analysis of the epidermal growth factor receptor gene in fresh human head and neck tumors. Cancer Res. 47:3603–3605.

Eisenbach, L. et al. (1983a). MHC imbalance and metastatic spread in Lewis lung carcinoma clones. Int. J. Cancer 32:113–120.

Eisenbach, L. et al. (1983b). Immunogenetic control of metastatic competence of cloned tumor cell populations. Symp. Fundam. Cancer Res. 36:101–121.

Eisenbach, L. et al. (1985). The differential expression of class I major histocompatibility complex antigens controls the metastatic properties of tumor cells. Transplant. Proc. 17:729–734.

Eliyahu, D. et al. (1988). Met A fibrosarcoma cells express two transforming mutant p53 species. Oncogene 3:313–321.

Eliyahu, D. et al. (1989). Wild-type p53 can inhibit oncogene mediated focus formation. Proc. Natl. Acad. Sci. U.S.A. 86:8763–8767.

Emi, M. et al. (1988). Genotyping and sequence analysis of apolipoprotein E isoforms. Genomics 3:373–379.

Enker, W. E. et al. (1991). DNA/RNA content and proliferative fractions of colorectal carcinomas: a five-year prospective study relating flow cytometry to survival. J. Natl. Cancer Inst. 83:701–707.

Eskelinen, M. et al. (1991). Relationship between DNA ploidy and survival in patients with exocrine pancreatic cancer. Pancreas 6:90–95.

Esser, V. and Russell, D. W. (1988). Transport-deficient mutations in the low density lipoprotein receptor. Alterations in the cysteine-rich and cysteine-poor regions of the protein block intracellular transport. J. Biol. Chem. 263:13276–13281.

Esser, V. et al. (1988). Mutational analysis of the ligand binding domain of the low density lipoprotein receptor. J. Biol. Chem. 263:13282–13290.

Estivill, X. et al. (1987a). A candidate for the cystic fibrosis locus isolated by selection for methylation-free islands. Nature (London) 326:840–845.

Estivill, X. et al. (1987b). Patterns of polymorphisms and linkage disequilibrium for cystic fibrosis. Genomics 1:257–263.

Estivill, X. et al. (1989). F508 gene deletion in cystic fibrosis in Southern Europe [letter]. Lancet ii:1404.

Etingin, O. R. et al. (1991). Lipoprotein (a) regulates plasminogen activator inhibitor-1 expression in endothelial cells. A potential mechanism in thrombogenesis. J. Biol. Chem. 266:2459–2465.

Eto, M. et al. (1989). Increased frequencies of apolipoprotein ε2 and ε4 alleles in patients with ischemic heart disease. Clin. Genet. 36:183–188.

Evarts, R. P. et al. (1987). In situ hybridization studies on expression of albumin and α-fetoprotein during the early stage of neoplastic transformation in rat liver. Cancer Res. 47:5469–5475.

Ezaki, T. et al. (1988). DNA analysis of hepatocellular carcinoma and clinicopathologic implications. Cancer 61:106–109.

Falck, V. G. and Gullick, W. J. (1989). c-erbB-2 oncogene product staining in gastric adenocarcinoma. An immuno-histochemical study. J. Pathol. 159:107–111.

Falco, J. P. et al. (1990). v-rasH induces non-small cell phenotype, with associated growth factors and receptors, in a small cell lung cancer cell line. J. Clin. Invest. 85:1740–1745.

Farber, E. et al. (1987). Cancer development as a multistep process: experimental studies in animals. In: Mechanisms of Environmental Carcinogenesis, Vol. 2. Barrett, J. C. (ed.). CRC Press, Boca Raton, FL, p. 1.

Fazio, S. et al. (1991). A form of familial hypobetalipoproteinemia not due to a mutation in the apolipoprotein B gene. J. Int. Med. 229:41–47.

Fearon, E. R. and Vogelstein, B. (1990). A genetic model for colorectal tumorigenesis. Cell (Cambridge, Mass) 61:759–767.

Fearon, E. R. et al. (1987). Clonal analysis of human colorectal tumors. Science 238:193–197.

Fearon, E. R. et al. (1990). Identification of a chromosome 18q gene that is altered in colorectal cancers. Science 247:49–56.

Feinmesser, R. et al. (1992). Diagnosis of nasopharyngeal carcinoma by DNA amplification of tissue obtained by fine-needle aspiration. N. Engl. J. Med. 326:17–21.

Feldman, G. L. et al. (1988). Prenatal diagnosis of cystic fibrosis by DNA amplification for detection of KM-19 polymorphism. Lancet ii:102.

Fenoglio, C. M. and Pascal, R. R. (1982). Colorectal adenomas and cancer. Cancer 50:2601–2608.

Fenoglio-Preiser, C. M. et al. (1991). Oncogenes and tumor suppressor genes in solid tumors: gastrointestinal tract. In: Molecular Diagnostics in Pathology. Fenoglio-Preiser, C. M. and Willman, C. L. (eds.). William & Wilkins, Baltimore, pp 189–218.

Ferns, G. A. A. et al. (1985). Genetic polymorphisms of apolipoprotein CIII and insulin in survivors of myocardial infaction. Lancet *ii*:300–303.

Ferns, G. A. A. et al. (1988). DNA haplotypes of the human apoprotein B gene in coronary atherosclerosis. Hum. Genet. *81*:76–80.

Ferrari, M. and Cremonesi, L. (1990). Letter to Editor. Clin. Chem. *36*:1703.

Festenstein, H. and Garrido, F. (1986). MHC antigens and malignancy. Nature (London) *322*:502–503.

Festenstein, H. and Schmidt, V. (1981). Variations in MHC antigen profiles of tumor cells and its biological effects. Immunol. Rev. *60*:85–127.

Fey, M. F. et al. (1989). Clonal allele loss in gastrointestinal cancers. Br. J. Cancer *59*:750–754.

Field, J. K. et al. (1989). Elevated expression of the c-*myc* oncoprotein correlates with poor prognosis in head and neck squamous cell carcinoma. Oncogene *4*:1463–1468.

Field, J. K. et al. (1991). Elevated p53 expression correlates with a history of heavy smoking in squamous cell carcinoma of the head and neck. Br. J. Cancer *64*:573–577.

Field, J. K. et al. (1992). Overexpression of p53 gene in head and neck cancer, linked with heavy smoking and drinking. Lancet *339*:502–503.

Finlay, C. A. et al. (1989). p53 proto-oncogene can act as a suppressor of transformation. Cell (Cambridge, Mass) *57*:1083–1093.

Finley, G. G. et al. (1989). Expression of the *myc* gene family in different stages of human colorectal cancer. Oncogene *4*:963–971.

Fiordalisi, G. et al. (1990). High genomic variability in the pre-C region of hepatitis B virus in anti-HBe, HBV DNA-positive chronic hepatitis. J. Med. Virol. *31*:297–300.

Fisher, E. A. et al. (1989). Gene polymorphisms and variability of human apolipoproteins. Annu. Rev. Nutr. *9*:139–160.

Fless, G. M. et al. (1986). Physiological properties of apolipoprotein (a) and lipoprotein (a–) derived from the dissocation of human plasma lipoprotein (a). J. Biol. Chem. *261*:8712–8718.

Fletcher, D. J. et al. (1983). Hormonal release by islet B cell-enriched and A and D cell-enriched populations prepared by flow cytometry. Endocrinology *113*:1791–1798.

Fletcher, J. A. et al. (1991). Consistent cytogenetic aberrations in hepatoblastoma: a common pathway of genetic alterations in embryonal liver and skeletal muscle malignancies? Genes Chromo. Cancer *3*:37–43.

Fojo, S. S. et al. (1984). Analysis of the apo CII gene in apo CII deficient patients. Biochem. Biophys. Res. Commun. *134*:308–313.

Fojo, S. S. et al. (1988). Donor splice site mutation in the apolipoprotein (Apo) CII gene (Apo C-II$_{Hamburg}$) of a patient with Apo CII deficiency. J. Clin. Invest. *82*:1489–1494.

Forrester, K. et al. (1987). Detection of high incidence of K-*ras* oncogenes during human colon tumorigenesis. Nature (London) *327*:298–303.

Fourel, G. et al. (1990). Frequent activation of N-*myc* genes by hepadna virus insertion in woodchuck liver tumours. Nature (London) *347*:294–298.

Frebourg, T. et al. (1988). The evaluation of CA-19-9 antigen level in the early detection of pancreatic cancer. A prospective study of 866 patients. Cancer *62*:2287–2290.

Freeman, D. J. et al. (1990). Polymorphisms in the gene coding for cholesteryl ester transfer protein are related to plasma high-density lipoprotein cholesterol and transfer protein activity. Clin. Sci. *79*:575–581.

Friedl, W. et al. (1990). Hypervariability in a minisatellite 3′ of the apolipoprotein B gene in patients with coronary heart disease compared with normal controls. J. Lipid Res. *31*:659–665.

Friedl, W. et al. (1991). Apolipoprotein B gene mutations in Austrian subjects with heart disease and their kindred. Arterioscl. Thromb. *11*:371–378.

Friedman, P. N. et al. (1990). Wild-type, but not mutant, human p53 proteins inhibit the replication activities of strain virus 40 large tumor antigen. Proc. Natl. Acad. Sci. U.S.A. *87*:9275–9279.

Friedman, K. and Stoerker, J. (1990). More on detection of cystic fibrosis by polymerase chain reaction. Clin. Chem. *36*:1702–1703.

Friedman, K. J. et al. (1990). Cystic fibrosis deletion mutation detected by PCR-mediated site-directed mutagenesis. Clin. Chem. *36*:695–696.

Friedman, K. J. et al. (1991). Detecting multiple cystic fibrosis mutations by polymerase chain reaction-mediated site-directed mutagenesis. Clin. Chem. *37*:753–755.

Fujimori, M. et al. (1991). Allelotype study of primary hepatocellular carcinoma. Cancer Res. *51*:89–93.

Fujita, K. et al. (1987). Frequent overexpression, but not activation by point mutation, of *ras* genes in primary human gastic cancers. Gastroenterology *93*:1339–1345.

Fukushige, S. I. et al. (1986). Localization of a novel v-*erb* B related gene, c-*erb*B-2, on human chromosome 17 and its amplification in a gastric cancer cell line. Mol. Cell. Biol. *6*:955–958.

Funahashi, T. et al. (1988). Mutations of the low density lipoprotein receptor in Japanese kindreds with familial hypercholesterolemia. Hum. Genet. *79*:103–108.

Funke, H. et al. (1991). A frameshift mutation in the human apolipoprotein AI gene causes high density lipoprotein deficiency, partial lecithin:cholesteryl-acyltransferase deficiency, and corneal opacities. J. Clin. Invest. *87*:371–376.

Galand, P. et al. (1988). Immunohistochemical detection of c-Ha-*ras* oncogene p21 product in pre-neoplastic and neoplastic lesions during hepatocarcinogenesis in rats. Int. J. Cancer *41*:155–161.

Ganem, D. and Varmus, H. E. (1987). The molecular biology of hepatitis B viruses. Annu. Rev. Biochem. *56*:651–693.

Gasparini, P. et al. (1989). First-trimester prenatal diagnosis of cystic fibrosis using the polymerase chain reaction: report of eight cases. Prenatal Diagn. *9*:349–355.

Gasparini, P. et al. (1991). The search for South European cystic fibrosis mutations: identification of two new mutations, four variants, and intronic sequences. Genomics *10*:193–200.

Gaubatz, J. W. et al. (1982). Human plasma lipoprotein (a): structural properties. J. Biol. Chem. *258*:4582–4589.

Gaubatz, J. W. et al. (1987). Isolation and characterization of the two major apoproteins in human lipoprotein (a). J. Lipid Res. *28*:69–79.

Gavish, D. and Breslow, J. L. (1991). Lipoprotein (a) reduction by *N*-acetylcysteine. Lancet *337*:203–304.

Gavish, D. et al. (1989a). Heritable allele-specific differences in the amounts of apo B and low density lipoproteins in plasma. Science *244*:72–76.

Gavish, D. et al. (1989b). Frequency of allele-specific differences in the amount of apo B in plasma [abstract]. Clin. Res. *37*:374a.

Gavish, D. et al. (1989c). Plasma Lp(a) concentration is inversely correlated with the ratio of kringle IV/kringle V encoding domains in the apo (a) gene. J. Clin. Invest. *84*:2021–2027.

Gemma, A. et al. (1988). *myc* family gene abnormality in lung cancers and its relation to renotransplantability. Cancer Res. *48*:6025–6028.

Genest, J. J., Jr. et al. (1990). DNA polymorphisms of the apolipoprotein B gene in patients with premature coronary artery disease. Atherosclerosis *82*:7–17.

Giaid, A. et al. (1990). Detection of endothelin immunoreactivity and mRNA in pulmonary tumors. J. Pathol. *162*:15–22.

Gilbert, F. (1990). Is population screening for cystic fibrosis appropriate now? Am. J. Hum. Genet. *46*:394–395.

Gille, C. et al. (1991). A pooling strategy for heterozygote screening of the ΔF508 cystic fibrosis mutation. Hum. Genet. *86*:289–291.

Goh, H. S. and Jass, J. R. (1986). DNA content and the adenoma-carcinoma sequence in the colorectum. J. Clin. Pathol. *39*:387–392.

Goldstein, J. L. and Brown, M. S. (1984). Progress in understanding the LDL receptor and HMG CoA reductase, two membrane proteins that regulate the plasma cholesterol. J. Lipid Res. *25*:1450–1461.

Goldstein, J. L. et al. (1985). Receptor-mediated endocytosis: concepts emerging from the LDL receptor system. Annu. Rev. Cell Biol. *1*:1–39.

Goldsworthy, T. L. et al. (1986). Models of hepatocarcinogenesis in the rat — contrasts and comparisons. CRC Crit. Rev. Toxicol. *17*:61–89.

Gonzalez-Gronow, M. et al. (1989). Further characterization of the cellular plasminogen binding site: evidence that plasminogen 2 and lipoprotein (a) compete for the same site. Biochemistry *28*:2374–2377.

Gope, R. et al. (1990a). Growth-regulated expression of the retinoblastoma and p53 genes in human colorectal cancers and cultured fibroblasts. Proc. Am. Assn. Cancer Res. *31*:320A.

Gope, R. et al. (1990b). Loss of heterozygosity at the colony-stimulating factor-1 receptor locus on chromosome 5 in human colorectal cancers. Proc. Am. Assn. Cancer Res. *31*:312A.

Gope, R. et al. (1990c). Increased expression of the retinoblastoma gene in human colorectal carcinomas relative to normal colonic mucosa. J. Natl. Cancer Inst. *82*:310–314.

Gordon, D. and Rifkind, B. M. (1989). Current concepts: high density lipoproteins — the clinical implications of recent studies. N. Engl. J. Med. *321*:1311–1115.

Graeme-Cook, F. et al. (1990). Aneuploidy in pancreatic insulinomas does not predict malignancy. Cancer *66*:2365–2368.

Gregg, R. E. et al. (1983). Apolipoprotein E-Bethesda: a new variant of apolipoprotein E associated with type III hyperlipoproteinemia. J. Clin. Endocrinol. Metab. *57*:969–974.

Gregory, R. J. et al. (1990). Expression and characterization of the cystic fibrosis transmembrane conductance regulator. Nature (London) *347*:382–386.

Grunewald, K. et al. (1989). High frequency of Ki-*ras* codon 12 mutations in pancreatic adenocarcinomas. Int. J. Cancer *43*:1037–1041.

Gu, J. et al. (1988). A study of *myc*-related gene expression in small cell lung cancer by *in situ* hybridization. Am. J. Pathol. *132*:13–17.

Gueguen, R. et al. (1989). An analysis of genotype effects and their interactions by using the apolipoprotein E polymorphism and longitudinal data. Am. J. Hum. Genet. *45*:793–802.

Guillem, J. G. et al. (1988). Changes in expression of oncogenes and endogenous retroviral-like sequences during colon carcinogenesis. Cancer Res. *48*:3964–3971.

Guraker, A. et al. (1985). Levels of lipoprotein Lp(a) decline with neomycin and niacin treatment. Atherosclerosis *57*:293–301.

Gylling, H. et al. (1989). Apolipoprotein E phenotype, and cholesterol metabolism in familial hypercholesterolemia. Atherosclerosis *80*:27–32.

Habener, J. F. et al. (1989). Factors that determine cell-specific gene expression in pancreatic endocrine tumor cells. Horm. Res. *32*:61–66.

Haberland, M. E. et al. (1989). Modification of Lp(a) by malondialdehyde leads to avid uptake by human monocyte-macrophages. Circulation *80*:163.

Haffner, S. M. et al. (1991). Decrease of lipoprotein (a) with improved glycemic control in IDDM subjects. Diabetes Care *14*:302–307.

Haglund, C. et al. (1987). Serum CA 50 as a tumor marker in pancreatic cancer: a comparison with CA 19-9. Int. J. Cancer *39*:477–481.

Hajj, C. et al. (1990). DNA alterations at proto-oncogene loci and their clinical significance in operable non-small cell lung cancer. Cancer *66*:733–739.

Hajjar, K. A. et al. (1989). Lipoprotein (a) modulation of endothelial cell surface fibrinolysis and its potential role in atherosclerosis. Nature (London) *339*:303–305.

Halley, D. J. J. (1989). Prenatal detection of major cystic fibrosis mutation. Lancet *ii*:972.

Halley, D. J. J. et al. (1990). The cystic fibrosis defect approached from different angles — new perspectives on the gene, the chloride channel, diagnosis and therapy. Eur. J. Pediatr. *149*:670–677.

Halpern, M. S. et al. (1983). Viral nucleic acid synthesis and antigen accumulation in pancreas and kidney of Peking ducks infected with duck hepatitis B virus. Proc. Natl. Acad. Sci. U.S.A. *80*:4865–4869.

Hamada, S. et al. (1987). Characteristics of colorectal epithelia and adenomas as revealed by DNA cytofluorometry. Jpn. J. Cancer Res. (Gann.) *78*:826–832.

Hamada, S. et al. (1988). The possibility of nonpolyploid carcinogenesis in the large intestine as inferred from frequencies of DNA aneuploidy of polyploid and crater-shaped carcinomas. Cancer *62*:1503–1510.

Hamid, Q. et al. (1991). Expression of chromogranin A mRNA in small cell carcinoma of the lung. J. Pathol. *163*:293–297.

Hamilton, P. W. et al. (1987). Classification of normal colorectal mucosa and adenocarcinoma by morphometry. Histopathology *11*:901–911.

Hanis, C. L. et al. (1991a). Apolipoprotein A-IV protein polymorphism: frequency and effects on lipids, lipoproteins, and apolipoproteins among Mexican-Americans in Starr County, Texas. Hum. Genet. *86*:323–325.

Hanis, C. L. et al. (1991b). Effects of the apolipoprotein E polymorphism on levels of lipids, lipoproteins, and apolipoproteins among Mexican-Americans in Starr County, Texas. Arterioscl. Thromb. *11*:362–370.

Harbour, J. W. et al. (1988). Abnormalities in structure and expression of the human retinoblastoma gene in SCLC. Science *241*:353–357.

Harding, M. et al. (1990). Neurone specific enolase (NSE) in small cell lung cancer: a tumor marker of prognostic significance? Br. J. Cancer *61*:605–607.

Harmenberg, U. et al. (1988). Tumor markers carbohydrate antigens CA 19-9 and CA 50 and carcinoembryonic antigen, in pancreatic cancer and benign diseases of the pancreatobiliary tract. Cancer Res. *48*:1985–1988.

Harpel, P. C. et al. (1989). Plasmin catalyzes binding of lipoprotein (a) to immobilized fibrinogen and fibrin. Proc. Natl. Acad. Sci. U.S.A. *86*:3847–3851.

Hastie, N. D. et al. (1990). Telomere reduction in human colorectal carcinoma and with ageing. Nature (London) *346*:866–868.

Hatada, I. et al. (1988). Co-amplification of integrated hepatitis B virus DNA and transforming gene *hst*-1 in a hepatocellular carcinoma. Oncogene *3*:537–540.

Hattori, Y. et al. (1990). K-*ram*, an amplified gene in stomach cancer, is a member of the heparin binding growth factor receptor genes. Proc. Natl. Acad. Sci. U.S.A. *87*:5983–5987.

Haut, M. et al. (1991). Induction of *nm*23 gene expression in human colonic neoplasms and equal expression in colon tumors of high and low metastatic potential. J. Natl. Cancer Inst *83*:712–716.

Havekes, L. et al. (1986). Apolipoprotein E3-Leiden. A new variant of human apolipoprotein E associated with familial type III hyperlipoproteinemia. Hum. Genet. *73*:157–163.

Hayakawa, T. et al. (1988). Sensitive serum markers for detecting pancreatic cancer. Cancer *61*:1827–1831.

Heby, O. (1981). Role of polyamines in the control of cell proliferation and differentiation. Differentiation *19*:1–20.

Heerdt, B. G. et al. (1990). Aggressive subtypes of human colorectal tumors frequently exhibit amplification of the c-*myc* gene. Proc. Am. Assn. Cancer Res. *31*:313A.

Hegele, R. A. et al. (1986). Apolipoprotein B gene DNA polymorphisms associated with myocardial infarction. N. Engl. J. Med. *315*:1509–1515.

Heizmann, C. et al. (1991). DNA polymorphism haplotypes of the human lipoprotein lipase gene: possible association with high density lipoprotein levels. Hum. Genet. 86:578–584.

Henderson, B. E. et al. (1983). Hepatocellular carcinoma and oral contraceptives. Br. J. Cancer *48*:437–440.

Henderson, H. E. et al. (1987). Association of a DNA polymorphism in the apolipoprotein CIII gene with diverse hyperlipidenic phenotypes. Hum. Genet. *75*:62–65.

Henderson, H. E. et al. (1990). Frameshift mutation in exon 3 of the lipoprotein lipase gene causes a premature stop codon and lipoprotein lipase deficiency. Mol. Biol. Med. 7:511–517.

Hensel, C. H. et al. (1990). Altered structure and expression of the human retinoblastoma susceptibility gene in small cell lung cancer. Cancer Res. *50*:3067–3072.

Higashide, T. et al. (1990). Detection of mRNAs of carcinoembryonic antigen and nonspecific cross-reacting antigen genes in colorectal adenomas and carcinomas by *in situ* hybridization. Jpn. J. Cancer Res. *81*:1149–1154.

Higgins, C. F. et al. (1990). Binding protein-dependent transport systems. J. Bioenerg. Biomembr. *22*:571–591.

Highsmith, W. E., Jr. et al. (1990). Frequency of the ΔPhe508 mutation and correlation with XV. 2c/KM-19 haplotypes in an American population of cystic fibrosis patients: results of a collaborative study. Clin. Chem. *36*:1741–1746.

Higuchi, K. et al. (1987). The human apo B-100 gene: apo B-100 is encoded by a single copy gene in the human genome. Biochem. Biophys. Res. Commun. *144*:1332–1339.

Himeno, Y. et al. (1988). Expression of oncogenes in human liver disease. Liver *8*:208–212.

Hino, O. et al. (1989). Features of two hepatitis B virus (HBV) DNA integrations suggest mechanisms of HBV integration. J. Virol. *63*:2638–2643.

Hirohashi, S. and Sugimura, T. (1991). Genetic alterations in human gastric cancer. Cancer Cells *3*:49–52.

Hirose, N. et al. (1987). Isolation and characterization of four heparin-binding cyanogen bromide peptides of human plasma apolipoprotein B. Biochemistry *26*:5505–5512.

Hixson, J. E. and Vernier, D. T. (1990). Restriction isotyping of human apolipoprotein E by gene amplification and clearage with *Hha* I. J. Lipid Res. *31*:545–548.

Hobbs, H. H. et al. (1987). Deletion in LDL receptor gene occurs in majority of French Canadians with familial hypercholesterolemia. N. Engl. J. Med. *317*:734–737.

Hobbs, H. H. et al. (1989). Evidence for a dominant gene that suppresses hypercholesterolemia in a family with defective low density lipoprotein receptors. J. Clin. Invest. *84*:656–664.

Hobbs, H. H. et al. (1990). The LDL receptor locus in familial hypercholesterolemia: mutational analysis of a membrane protein. Annu. Rev. Genet. *42*:133–170.

Hoellering, J. and Shuler, C. F. (1989). Localization of H-*ras* mRNA in oral squamous cell carcinomas. J. Oral Pathol. Med. *18*:74–78.

Hoff, H. F. et al. (1988). Serum Lp(a) level as a predictor of vein graft stenosis after coronary artery bypass surgery in patients. Circulation *77*:1238–1244.

Hofler, H. et al. (1988). Oncogene expression in endocrine pancreatic tumors. Virchow's Arch. B: Cell Pathol. *55*:355–361.

Hollstein, M. C. et al. (1988). Amplification of epidermal growth factor receptor gene but no evidence of *ras* mutations in primary human esophageal cancers. Cancer Res. *48*:5119–6123.

Hollstein, M. C. et al. (1990). Frequent mutation of the p53 gene in human esophageal cancer. Proc. Natl. Acad. Sci. U.S.A. *87*:9958–9961.

Hollstein, M. et al. (1991). p53 mutations in human cancers. Science *253*:49–53.

Hood, D. L. et al. (1990). Deoxyribonucleic acid ploidy and cell cycle analysis of colorectal carcinoma by flow cytometry. A prospective study of 137 cases using fresh whole cell suspensions. Am. J. Clin. Pathol. *93*:615–620.

Horsthemke, B. et al. (1986). Unequal crossing-over between two *Alu*-repetitive DNA sequences in the low density lipoprotein receptor gene. A possible mechanism for the defect in a patient with familial hypercholesterolemia. Eur. J. Biochem. *164*:77–81.

Hoshino, Y. et al. (1991). Normal human chromosome 5, on which a familial adenomatous polyposis gene is located, has tumor suppressive activity. Biochem. Biophys. Res. Commun. *174*:298–304.

Houldsworth, J. et al. (1990). Gene amplification in gastic and esophageal adenocarcinomas. Cancer Res. *50*:6417–6422.

Houlston, R. S. et al. (1989). Apolipoprotein (apo) E genotypes by polymerase chain reaction and allele-specific oligonucleotide probes: no detectable linkage disequilibrium between apo E and apo CII. Hum. Genet. *83*:364–368.

Houlston, R. S. et al. (1990). Genetic epidemiology of differences in low density lipoprotein (LDL) cholesterol concentration: possible involvement of variation at the apolipoprotein B gene locus in LDL kinetics. Genet. Epidemiol. *7*:199–210.

Howell, R. E. et al. (1989). Loss of Harvey *ras* heterozygosity in oral squamous carcinoma. J. Oral Pathol. Med. *18*:79–83.

Hsu, I. C. et al. (1991). Mutational hotspot in the p53 gene in human hepatocellular carcinomas. Nature (London) *350*:427–431.

Hsu, T. Y. et al. (1988). Activation of c-*myc* by woodchuck hepatitis virus insertion in hepatocellular carcinoma. Cell (Cambridge, Mass) *55*:627–635.

Huang, L. S. and Breslow, J. L. (1987). A unique AT-rich hypervariable minisatellite 3′ to the apo B gene defines a high information restriction fragment length polymorphism. J. Biol. Chem. *262*:8952–8955.

Huang, L. S. et al. (1990a). Exclusion of linkage between the human apolipoprotein B gene and abetalipoproteinemia. Am. J. Hum. Genet. *46*:1141–1148.

Huang, L. S. et al. (1990b). Molecular basis of five apolipoprotein B gene polymorphisms in noncoding regions. J. Lipid Res. *31*:71–77.

Huber, B. E. and Thorgeirsson, S. S. (1987). Analysis of c-*myc* expression in a human hepatoma cell line. Cancer Res. *47*:3414–3420.

Huber, B. E. et al. (1985). Tumorigenicity and transcriptional modulation of c-*myc* and N-*ras* oncogenes in a human hepatoma cell line. Cancer Res. *45*:4322–4329.

Humphries, S. E. (1986). Familial hypercholesterolemia as an example of early diagnosis of coronary artery disease risk by DNA techniques. Br. Heart J. *56*:201–205.

Humphries, S. E. (1988). DNA polymorphisms of the apolipoprotein genes — their use in the investigation of the genetic component of hyperlipidemia and atherosclerosis. Atherosclerosis 72:89–108.

Humphries, S. E. et al. (1984). Familial apolipoprotein CII deficiency: a preliminary analysis of the gene defect in two independent families. Hum. Genet. *67*:151–155.

Hyde, S. C. et al. (1990). Structural model of ATP-binding proteins associated with cystic fibrosis, multidrug resistance and bacterial transport. Nature (London) *346*:362–365.

Iannuzzi, M. C. and Collins, F. S. (1990). Reverse genetics and cystic fibrosis. Am. J. Respir. Cell Mol. Biol. 2:309–316.

Iannuzzi, M. C. et al. (1991). Two frameshift mutations in the cystic fibrosis gene. Am. J. Hum. Genet. *48*:227–231.

Ichihara, T. et al. (1988). Immunohistochemical localization of CA 19-9 and CEA in pancreatic carcinoma and associated diseases. Cancer *61*:324–333.

Ichinose, A. et al. (1991). Two types of abnormal genes for plasminogen in families with a predisposition for thrombosis. Proc. Natl. Acad. Sci. U.S.A. *88*:115–119.

Iizuka, M. et al. (1990). Joining of the c-myc gene and a line 1 family member on chromosome 8 in a human primary giant cell carcinoma of the lung. Cancer Res. *50*:3345–3350.

Iggo, R. et al. (1990). Increased expression of mutant forms of p53 oncogene in primary lung cancer. Lancet *335*:675–679.

Ikeda, I. et al. (1988). No correlation between L-*myc* restriction fragment length polymorphism and malignancy of human colorectal cancers. Jpn. J. Cancer Res. (Gann.) 79:674–676.

Imai, M. et al. (1987). Free and integrated forms of hepatitis B virus DNA in human hepatocellular carcinoma cells (PLC/342) propagated in nude mice. J. Virol. *61*:3555–3560.

Imazeki, F. et al. (1988). Integrated structures of duck hepatitis B virus DNA in hepatocellular carcinoma. J. Virol. *62*:861–865.

Inazu, A. et al. (1990). Increased high density lipoprotein levels caused by a common cholesterol-ester transfer protein gene mutation. N. Engl. J. Med. *323*:1234–1238.

Infantolino, D. and Pinarello, A. (1987). Detection of hepatitis B virus DNA in the liver by *in situ* hybridization: methodological approach to improve sensitivity on formalin-fixed, paraffin-embedded material. Basic Appl. Histochem. *31*:207–209.

Innerarity, T. L. et al. (1983). The receptor-binding domain of human apolipoprotein E. J. Biol. Chem. *258*:12341–12347.

Innerarity, T. L. et al. (1988). Isolation of defective receptor-binding low density lipoproteins from subjects with familial defective apolipoprotein B100 [abstract]. Arteriosclerosis *8*:551A.

Innerarity, T. L. et al. (1990). Familial defective apolipoprotein B-100: a mutation of apolipoprotein B that causes hypercholesterolemia. J. Lipid Res. *31*:1337–1349.

Ishitoya, J. et al. (1989). Gene amplification and overexpression of EGF receptor in squamous cell carcinomas of the head and neck. Br. J. Cancer *59*:559–562.

Ito, Y. et al. (1990). Hypertriglyceridemia as a result of human apo CIII gene expression in transgenic mice. Science *249*:790–793.

Jacobs, C. D. and Pinto, H. A. (1992). Head and neck cancer with an occult primary tumor. N. Engl. J. Med. *326*:58–59.

Jacobsson, B. et al. (1990). Transthyretin messenger ribonucleic acid expression in the pancreas and in endocrine tumors of the pancreas and gut. J. Clin. Endocrinol. Metab. *71*:875–880.

Jalanko, A. et al. (1992). Screening for defined cystic fibrosis mutations by solid-phase minisequencing. Clin. Chem. *38*:39–43.

Janne, J. et al. (1978). Polyamines in rapid growth and cancer. Biochem. Biophys. Acta *473*:241–293.

Jensen, S. M. et al. (1990). A comparison of synaptophysin, chromogranin, and L-dopa decarboxylase as markers for neuroendocrine differentiation in lung cancer cell lines. Cancer Res. *50*:6068–6074.

Jiang, W. et al. (1989). Rapid detection of *ras* oncogenes in human tumors: applications to colon, esophageal, and gastric cancer. Oncogene *4*:923–928.

Jin, Y. et al. (1988a). Isochromosomes i(8q) or i(9q) in three adenocarcinomas of the lung. Cancer Genet. Cytogenet. *33*:11–17.

Jin, Y. et al. (1988b). Unique karyotype abnormalities in a squamous cell carcinoma of the larynx. Cancer Genet. Cytogenet. *30*:177–179.

Jin, Y. et al. (1988c). t(6,7)(q23;p22) as the sole chromosomal anomaly in a vocal cord carcinoma. Cancer Genet. Cytogenet. *32*:305–307.

Jin, Y. et al. (1988d). Multiple apparently unrelated clonal chromosome abnormalities in a squamous cell carcinoma of the tongue. Cancer Genet. Cytogenet. *32*:93–101.

Jin, Y. et al. (1988e). Inversion inv(4)(p15q26) in a squamous cell carcinoma of the hypopharynx. Cancer Genet. Cytogenet. *36*:233–234.

Jin, Y. et al. (1990a). Multiple clonal chromosome aberrations in squamous cell carcinomas of the larynx. Cancer Genet. Cytogenet. *44*:209–216.

Jin, Y. et al. (1990b). Unrelated clonal chromosome aberrations in carcinomas of the oral cavity. Genes Chrom. Cancer *1*:209–215.

Jin, Y. et al. (1990c). Frequent rearrangement of chromosomal bands 1p22 and 11q13 in squamous cell carcinomas of the head and neck. Genes Chrom. Cancer *2*:198–204.

Johnson, B. E. et al. (1988). *myc* family DNA amplification in small cell lung cancer patients' tumors and corresponding cell lines. Cancer Res. *48*:5163–5166.

Johnson, T. S. et al. (1985). Flow cytometric analysis of head and neck carcinoma DNA index and S-fraction from paraffin-embedded sections: comparison with malignancy grading. Cytometry *6*:461–470.

Kahn, H. J. et al. (1992). Expression and amplification of *neu* oncogene in pleomorphic adenomas of salivary gland. Arch. Pathol. Lab. Med. *116*:80–83.

Kalin, M. F. and Zumoff, B. (1990). Sex hormones and coronary disease: a review of the clinical studies. Steroids *55*:331–352.

Kamarck, M. E. et al. (1987). Carcinoembryonic antigen family: expression in a mouse L-cell transfectant and characterization of a partial cDNA in a bacteriophage λ gt11. Proc. Natl. Acad. Sci. U.S.A. *84*:5350–5354.

Kamboh, M. I. et al. (1989). Genetic studies of human apolipoproteins. VI. Common polymorphism oof apolipoprotein E in blacks. Dis. Markers *7*:49–55.

Kamboh, M. I. et al. (1990). Genetic studies of human apolipoproteins. XIV. A simple agarose isoelectric focusing gel method for apolipoprotein E phenotyping. Electrophoresis *11*:314–318.

Kameda, K. et al. (1984). Increased frequency of lipoprotein disorders similar to type III hyperlipoproteinemia in survivors of myocardial infarction in Japan. Atherosclerosis *51*:241–249.

Karadi, I. et al. (1988). Lipoprotein (a) and plasminogen are immunologically related. Biochem. Biophys. Acta *960*:91–97.

Karathanasis, S. K. (1985). Apolipoprotein multigene family: tandem organization of human apolipoprotein AI, CIII, and CIV genes. Proc. Natl. Acad. Sci. U.S.A. *82*:6374–6378.

Karathanasis, S. K. et al. (1983a). Isolation and characterization of the human apolipoprotein AI gene. Proc. Natl. Acad. Sci. U.S.A. *80*:6147–6151.

Karathanasis, S. K. et al. (1983b). Linkage of human apolipoprotein AI and CIII genes. Nature (London) *304*:371–373.

Karathanasis, S. K. et al. (1986). Structure, evolution, and polymorphisms of the human apolipoprotein A4 gene (Apo A4). Proc. Natl. Acad. Sci. U.S.A. *83*:8457–8461.

Karn, R. C. (1991). Genetic control of mammalian salivary proteins. Front. Oral Physiol. *8*:117–140.

Karpen, S. et al. (1988). Identification of protein binding sites in the hepatitis B virus enhancer and core promoter domains. Mol. Cell. Biol. *8*:5159–5165.

Kartner, N. et al. (1991). Expression of the cystic fibrosis gene in non-epithelial invertebrate cells produces a regulated anion conductance. Cell (Cambridge, Mass) *64*:681–691.

Kawashima, K. et al. (1987). Restriction fragment length polymorphism (RFLP) of L-*myc* is related to the progression of human colon and stomach cancers. Proc. Jpn. Acad. *63*(B):300–303.

Kaye, F. J. et al. (1990). Recessive oncogenes in lung cancer. Am. Rev. Respir. Dis. *142*:S44–S47.

Kearsley, J. H. et al. (1991). Prognostic importance of cellular DNA content in head-and-neck squamous-cell cancers. A comparison of retrospective and prospective series. Int. J. Cancer *47*:31–37.

Keinanen, M. et al. (1990). Use of polymerase chain reaction to detect heterozygous familial hypercholesterolemia. Clin. Chem. *36*:900–903.

Keinanen, R. et al. (1989). Contemplation of virally induced changes in salivary glands in Sjogren's syndrome: on the use of *in situ* hybridization in such studies. J. Autoimmunity *2*:569–578.

Kekule, A. S. et al. (1990). The pre S2/S region of integrated hepatitis B virus DNA encodes a transcriptional transactivator. Nature (London) *343*:457–461.

Kerem, B. S. et al. (1989). Identification of the cystic fibrosis gene: genetic analysis. Science *245*:1073–1080.

Kerem, E. et al. (1990). The relation between genotype and phenotype in cystic fibrosis — analysis of the most common mutation (F508). N. Engl. J. Med. *323*:1517–1522.

Kern, H. F. et al. (1987). Fine structure of three major grades of malignancy of human pancreatic adenocarcinomas. Pancreas *2*:2–13.

Kew, M. C. and Popper, H. (1984). Relationship between hepatocellular carcinoma and cirrhosis. Semin. Liver Dis. *4*:136–146.

Kim, C. M. et al. (1991). HBx gene of hepatitis B virus induces liver cancer in transgenic mice. Nature (London) *351*: 317–320.

Kim, J. H. et al. (1990a). Cell lineage markers in human pancreatic cancer. Cancer *66*:2134–2143.

Kim, J. H. et al. (1990b). Occurrence of p53 gene abnormalities in gastric carcinoma tumors and cell lines. Proc. Am. Assn. Cancer Res. *31*:315A.

Kimura, H. and Yonemura, Y. (1991). Flow cytometric analysis of nuclear DNA content in advanced gastric cancer and its relationship with prognosis. Cancer 67:2588–2593.

Kingsnorth, A. N. et al. (1984). Polyamines in colorectal cancer. Br. J. Surg. 71:791–794.

Kinzler, K. W. et al. (1991a). Identification of a gene located at chromosome 5q21 that is mutated in colorectal cancers. Science 251:1366–1370.

Kinzler, K. W. et al. (1991b). Identification of FAP locus genes from chrosonone 5q21. Science 253:661–665.

Kirgan, D. et al. (1990a). Immunohistochemical demonstration of human papillomavirus antigen in human colon neoplasms. J. Surg. Res. 48:397–402.

Kirgan, D. et al. (1990b). Association of human papillomavirus and colon neoplasms. Arch. Surg. 125:862–865.

Klasen, E. C. et al. (1987). Apolipoprotein E phenotype and gene distribution in the Netherlands. Hum. Hered. 37:340–344.

Klimpfinger, M. et al. (1990). Expression of c-myc and c-fos mRNA in colorectal carcinoma in man. Virchow's Arch. B: Cell Pathol. 59:165–171.

Kloppel, G. and Fitzgerald, P. J. (1986). Pathology of nonendocrine pancreatic tumors. In: The Exocrine Pancreas: Biology, Pathobiology and Diseases. Go, V. L. W. et al. (eds.). Raven Press, New York, pp. 649–674.

Kluft, C. et al. (1989). Functional analogy between lipoprotein (a) and plasminogen in the binding to the kringle 4 binding protein, tetranectin. Biochim. Biophys. Res. Commun. 161:427–433.

Knight, B. L. et al. (1991). Catabolism of lipoprotein (a) in familial hypercholesterolemic subjects. Atherosclerosis 87:227–237.

Knowlton, R. G. et al. (1985). A polymorphic DNA marker linked to cystic fibrosis is located on chromosome 7. Nature (London) 318:380–382.

Koha, M. et al. (1990). Heterogeneity of DNA distribution pattern in colorectal carcinoma. A microspectrophometric study of fine needle aspirates. Anal. Quant. Cytol. Histol. 12:348–351.

Kok, K. et al. (1987). Deletion of a DNA sequence at the chromosomal region 3p21 in all major types of lung cancer. Nature (London) 330:578–581.

Kokal, W. et al. (1986). Tumor DNA content in the prognosis of colorectal carcinoma. JAMA 255:3123.

Kondo, I. and Berg, K. (1990). Inherited quantitative DNA variation in the LPA ("apolipoprotein (a)") gene. Clin. Genet. 37:132–140.

Kontula, K. et al. (1990). Apolipoprotein E polymorphism determined by restriction enzyme analysis of DNA amplified polymerase chain reaction: convenient alternative to phenotyping by isoelectric focusing. Clin. Chem. 36:2087–2092.

Kopelman, H. and Rozen, R. (1990). Genetic analysis and pancreatic function in cystic fibrosis. Lancet 335:1601.

Korc, M. et al. (1983). Binding of epidermal growth factor in rat pancreatic acini. Biochem. Biophys. Res. Commun. 111:1066–1073.

Korenaga, D. et al. (1986). Consistency of DNA ploidy between primary and recurrent gastric carcinomas. Cancer Res. 46:1544–1546.

Koshy, R. et al. (1983). Integration of hepatitis B virus DNA: evidence for integration in the single-stranded gap. Cell (Cambridge, Mass) 34:215–223.

Kouri, M. et al. (1990). The prognostic value of DNA-ploidy in colorectal carcinoma: a prospective study. Br. J. Cancer 62:976–981.

Kratzin, H. et al. (1987). Structural relationship of an apolipoprotein (a) phenotype (570 kDa) to plasminogen: homologous kringle domains are linked by carbohydrate-rich regions. Hoppe-Seylers Z. Biol. Chem. 368:1533–1544.

Kratzke, R. et al. (1990). Characterization of a mutant retinoblastoma protein from a small cell lung cancer line reveals a single amino acid change. Proc. Am. Assn. Cancer Res. 31:320A.

Kruijer, W. et al. (1986). Proto-oncogene expression in regenerating liver is simulated in cultures of primary adult rat hepatocytes. J. Biol. Chem. 261:7929–7933.

Krystal, G. et al. (1988). Multiple mechanisms for transcriptional regulation of the myc gene family in small-cell lung cancer. Mol. Cell. Biol. 8:3373–3381.

Kunicka, J. E. et al. (1987). DNA in situ sensitivity to denaturation: a new parameter for flow cytometry of normal human colonic epithelium and colon carcinoma. Cancer Res. 47:3942–3947.

Kupchik, G. (1991). Genotype/phenotype discrepancies in cystic fibrosis. Lancet 337:121.

Kuppuswamy, M. N. et al. (1991). Single nucleotide primer extension to detect genetic diseases: experimental application to hemophilia B (factor IX) and cystic fibrosis genes. Proc. Natl. Acad. Sci. U.S.A. 88:1143–1147.

Kuusi, T. et al. (1989). Apoprotein E polymorphism and coronary artery disease. Increased prevalence of apolipoprotein E4 in angiographically verified coronary patients. Arteriosclerosis 9:237–241.

La Muraglia, G. M. et al. (1986). High ornithine decarboxylase activity and polyamine levels in human colorectal neoplasia. Ann. Surg. 204:89–93.

Ladias, J. A. A. and Karathanasis, S. K. (1991). Regulation of the apolipoprotein AI gene by ARP-1, a novel member of the steroid receptor superfamily. Nature (London) 251:561–564.

Lalazar, A. et al. (1988). Site-specific mutagenesis of human apolipoprotein E: receptor binding activity of variants with single amino acid substitutions. J. Biol. Chem. *263*:3542–3545.

Lan, M. S. et al. (1990). Cloning and sequencing of a human pancreatic tumor mucin cDNA. J. Biol. Chem. *265*:15294–15299.

Lane, D. P. and Benchimaol, S. (1990). p53: oncogene or anti-oncogene? Genes Dev. *4*:1–8.

Langlois, S. et al. (1988). Characterization of six partial deletions in the low-density-lipoprotein (LDL) receptor gene causing familial hypercholesterolemia (FH). Am. J. Hum. Genet. *43*:60–68.

Lanza, G., Jr. (1988). *ras*^{p21} oncoprotein expression in human colonic neoplasia — an immunohistochemical study with monoclonal antibody RAP-5. Histopathology *12*:595–609.

Larsimont, D. et al. (1990). Characterization of the morphonuclear features and DNA ploidy of typical and atypical carcinoids and small cell carcinomas of the lung. Am. J. Clin. Pathol. *94*:378–383.

Larsson, C. et al. (1988). Multiple endocrine neoplasia type 1 gene maps to chromosome 11 and is lost in insulinoma. Nature (London) *332*:85–87.

Law, A. et al. (1986). Common DNA polymorphism within coding sequence of apolipoprotein B gene associated with altered lipid levels. Lancet *i*:1301–1303.

Law, M. R. (1990). Genetic predisposition to lung cancer. Br. J. Cancer *61*:195–206.

Law, M. R. et al. (1989). Debrisoquine metabolism and genetic predisposition to lung cancer. Br. J. Cancer *59*:686–687.

Lee, I. et al. (1987). Immunohistochemical evaluation of *ras* oncogene expression in pulmonary and pleural neoplasms. Virchow's Arch. B: Cell Pathol. *53*:146–152.

Lee, J. S. et al. (1991). Expression of blood-group antigen A — a favorable prognostic factor in non-small-cell lung cancer. N. Engl. J. Med. *324*:1084–1090.

Lehrman, M. A. et al. (1985). Internalization-defective LDL receptors produced by genes with nonsense and frameshift mutations that truncate the cytoplasmic domain. Cell (Cambridge, Mass) *41*:735–743.

Lehrman, M. A. et al. (1987). The Lebanese allele at the low density lipoprotein receptor locus. Nonsense mutation produces truncated receptor that is retained in endoplasmic reticulum. J. Biol. Chem. *262*:401–410.

Leister, I. et al. (1990). Human colorectal cancer: high frequency of deletions at chromosome 1p35. Cancer Res. *50*:7232–7235.

Leitersdorf, E. et al. (1990). Common low-density lipoprotein receptor mutations in the French Canadian population. J. Clin. Invest. *85*:1014–1023.

Lemna, W. K. et al. (1990). Mutation analysis for heterozygote defect and prenatal diagnosis of cystic fibrosis. N. Engl. J. Med. *322*:291–296.

Leonard, R. C. F. (1989). Small cell lung cancer. Br. J. Cancer *59*:487–490.

Leppert, M. et al. (1987). The gene for familial polyposis coli maps to the long arm of chromosome 5. Science *238*:1411–1413.

Leppert, M. et al. (1990). Genetic analysis of an inherited predisposition to colon cancer in a family with a variable number of adenomatous polyps. N. Engl. J. Med. *322*:904–908.

Leren, T. P. et al. (1988). Further evidence for an association between the *Xba*I polymorphism at the apolipoprotein B locus and lipoprotein level. Clin. Genet. *34*:347–351.

Leuschner, I. et al. (1988). The association of hepatocellular carcinoma in childhood with hepatitis B virus infection. Cancer *62*:2363–2369.

Levine, A. J. et al. (1991). The p53 tumor suppressor gene. Nature *351*:453–456.

Li, W. H. et al. (1988). The apolipoprotein multigene family: biosynthesis structure, structure-function relationships, and evolution. J. Lipid Res. *29*:245–271.

Li, Y. and Lieberman, M. W. (1989). Two genes associated with liver cancer are regulated by different mechanisms in *ras* T24 transformed liver epithelial cells. Oncogene *4*:795–798.

Li, Y. et al. (1988). MT *ras* T24, a metallothionein-*ras* fusion gene, modulates expression in cultured rat liver cells of two genes associated with in vitro liver cancer. Proc. Natl. Acad. Sci. U.S.A. *85*:344–348.

Lie-Injo, L. E. et al. (1983). Hepatitis B virus DNA in liver and white blood cells of patients with hepatoma. DNA *2*:301–308.

Lieberman, H. M. et al. (1987). Spenic replication of hepatitis B virus in the chimpanzee chronic carrier. J. Med. Virol. *21*:347–359.

Lin, H. H. et al. (1990). DNA measurements in chronic hepatitis, cirrhosis and hepatocellular carcinoma. Liver *10*:313–318.

Lindahl, G. et al. (1989). The gene for the Lp(a) -specific glycoprotein is closely linked to the gene for plasminogen on chromosome 6. Hum. Genet. *81*:149–152.

Lindgren, V. et al. (1985). Human genes involved in cholesterol metabolism: chromosomal mapping of the loci for the low density lipoprotein receptor and 3-hydroxy-3-methylglutaryl-coenzyme A reductase with cDNA probes. Proc. Natl. Acad. Sci. U.S.A. *82*:8567–8571.

Lissens, W. et al. (1989). Prenatal diagnosis of cystic fibrosis using closely linked DNA probes. J. Inherited Metab. Dis. *12* (Suppl.):308–310.

Liu, C. et al. (1990). Expression of transforming growth factor-α in primary human colon and lung carcinomas. Br. J. Cancer *62*:425–429.

Ljunggren, H. G. and Karre, K. (1986). Experimental strategies and interpretation in the analysis of changes in MHC gene expression during tumor progression. Opposing influences of T cell and natural killer mediated resistance? J. Immunogenet. *13*:141–151.

Loader, S. et al. (1991). Prenatal screening for hemoglobinopathies. II. Evaluation of counseling. Am. J. Hum. Genet. *48*:447–451.

Lopez-Nevot, M. A. et al. (1989). HLA class I gene expression on human primary tumors and autologous metastases: demonstration of selective losses of HLA antigens on colorectal, gastric and laryngeal carcinomas. Br. J. Cancer *59*:221–226.

Loscalzo, J. (1990). Lipoprotein (a). A unique risk factor for atherothrombotic disease. Atherosclerosis *10*:672–679.

Loscalzo, J. et al. (1988). Lipoprotein (a) inhibits fibrin-dependent enhancement of tissue plasminogen activator activity. Blood *72* (Suppl. I):370A.

Loscalzo, J. et al. (1990). Lipoprotein (a), fibrin binding and plasminogen activation. Arteriosclerosis *10*:240–245.

Ludwig, Ch. U. et al. (1990). DNA sequence deletions from chromosome 11, 3 and 17 in human non-small-cell lung carcinoma (NSCLC). Proc. Am. Assn. Cancer Res. *31*:319A.

Lukeis, R. et al. (1990). Cytogenetics on non-small cell lung cancer: analysis of consistent non-random abnormalities. Genes Chrom. Cancer *2*:116–124.

Luna, M. A. et al. (1990). Flow cytometric DNA content of adenoid cystic carcinoma of submandibular gland. Arch. Otolaryngol. *116*:1291–1296.

Lusis, A. J. (1988). Genetic factors affecting blood lipoproteins: the candidate gene approach. J. Lipid Res. *29*:397–429.

Lusis, A. J. and Sparkes, R. S. (1989). Genetic Factors in Atherosclerosis: Approaches and Model Systems. Karger, Basel, p. 227.

Lynch, H. T. (1990). The surgeon and colorectal cancer genetics. Case identification, surveillance, and management strategies. Arch. Surg. *125*:698–701.

Lynch, H. T. et al. (1990). Familial pancreatic cancer: clinicopathologic study of 18 nuclear families. Am. J. Gastroenterol. *85*:54–60.

Ma, Y. et al. (1987). Two DNA restriction fragment length polymorphisms associated with Ag (t/z) and Ag (g/c) antigenic sites of human apolipoprotein B. Arteriosclerosis *7*:301–305.

Ma, Y. et al. (1988). Apolipoprotein B gene haplotypes. Association between Ag and DNA polymorphisms. Arteriosclerosis *8*:521–524.

Ma, Y. et al. (1989). *Bsp* 1286 I restriction fragment length polymorphism detects Ag (c/g) locus of human apolipoprotein B in all 17 persons studied. Arteriosclerosis *9*:242–246.

Mabry, M. et al. (1990). Introduction of a mutated p53 gene results in increased growth and reduced hormone production in human small cell lung cancer. Proc. Am. Assn. Cancer Res. *31*:313A.

Mabry, M. et al. (1991). Transitions between lung cancer phenotypes — implications for tumor progression. Cancer Cells *3*:53–59.

Macartney, J. C. et al. (1986). DNA flow cytometry of histological material from human gastric cancer. J. Pathol. *148*:273–277.

Mack, D. H. et al. (1988). Hepatitis B virus particles contain a polypeptide encoded by the largest open reading frame: a putative reverse transcriptase. J. Virol. *62*:4786–4790.

Maestro, R. et al. (1991). Correlation between chromosome 5q deletions and different mechanisms of c-*myc*-overexpression in human colorectal cancer. Br. J. Cancer *63*:185–186.

Mafune, K.-i. et al. (1990). Expression of a M_r 32,000 laminin-binding protein messenger RNA in human colon carcinoma correlates with disease progression. Cancer Res. *50*:3888–3891.

Mahley, R. W. et al. (1991). Genetic defects in lipoprotein metabolism. Elevation of atherogenic lipoproteins caused by impaired catabolism. JAMA *265*:78–83.

Maier, A. et al. (1989). Expression of the small cell carcinoma antigens of cluster-5 and cluster-5A in primary lung tumors. Br. J. Cancer *59*:692–695.

Main, B. F. et al. (1991). Apolipoprotein E genotyping using the polymerase chain reaction and allele-specific oligonucleotide primers. J. Lipid Res. *32*:183–187.

Manttari, M. et al. (1991). Apolipoprotein E polymorphism influences the serum cholesterol response to dietary intervention. Metabolism 40:217–221.

Mariyama, M. et al. (1989). Frequency and types of point mutations at the 12th codon of the c-Ki-*ras* gene found in pancreatic cancers from Japanese patients. Jpn. J. Cancer Res. *80*:622–626.

Markowitz, S. et al. (1989). Expression of the erbA β class of thyroid hormone receptors is selectively lost in human colon carcinoma. J. Clin. Invest. *84*:1683–1687.

Markowitz, S. et al. (1990). Transforming activity of mutant p53 in colon adenoma cell lines. Proc. Am. Assn. Cancer Res. *31*:310A.

Martin-Zanca, D. et al. (1989). Molecular and biochemical characterization of the human *trk* protooncogene. Mol. Cell. Biol. *9*:24–33.

Martinet, Y. et al. (1987). Exaggerated spontaneous release of platelet-derived growth factor by alveolar macrophages from patients with idiopathic pulmonary fibrosis. N. Engl. J. Med. *317*:202–209.

Matsumoto, H. et al. (1988). Analysis of integrated hepatitis B virus DNA and cellular flanking sequences cloned from a hepatocellular carcinoma. Int. J. Cancer *42*:1–6.

Matsunaga, T. et al. (1991). Apolipoprotein A-I deficiency due to a codon 84 nonsense mutation of the apoliprotein A-I gene. Proc. Natl. Acad. Sci. U.S.A. *88*:2793–2797.

McClean, J. W. et al. (1983). Rat apolipoprotein E mRNA. Cloning and sequencing of a double stranded cDNA. J. Biol. Chem. *258*:8993–9000.

McIntosh, I. et al. (1989a). First-trimester prenatal diagnosis of cystic fibrosis by direct gene probing. Lancet *ii*:972–973.

McIntosh, I. et al. (1989b). Frequency of ΔF508 mutation on cystic fibrosis chromosomes in UK [letter]. Lancet *ii*:1404–1405.

McKenzie, K. J. et al. (1987). Expression of carcinoembryonic antigen, T-antigen, and oncogene products as markers of neoplastic and preneoplastic colonic mucosa. Hum. Pathol. *18*:1282–1286.

McLean, J. W. et al. (1987). cDNA sequence of human apolipoprotein (a) is homologous to plasminogen. Nature (London) *330*:132–137.

McLemore, T. L. et al. (1990). Expression of CYP1A1 gene in patients with lung cancer: evidence for cigarette smoke-induced gene expression in normal lung tissue and for altered gene regulation in primary pulmonary carcinomas. J. Natl. Cancer Inst. *82*:1333–1339.

McLeod, C. et al. (1990). The anchorage-dependent and -independent growth of a human SCC cell line: the roles of TGF-α/EGF and TGF-β. Br. J. Cancer *61*:267–269.

McMahon, G. et al. (1986). Identification of an activated c-Ki-ras oncogene in rat liver tumors induced by aflatoxin B1. Proc. Natl. Acad. Sci. U.S.A. *83*:9418–9422.

Melamed, M. R. et al. (1986). Flow cytometry of colorectal carcinoma with three year follow-up. Dis. Colon Rectum *29*:184–186.

Melbye, M. et al. (1984). Virus associated cancers in Greenland: frequent hepatitis B virus infection but low primary hepatocellular carcinoma incidence. J. Natl. Cancer Inst. *73*:1267–1272.

Meltzer, P. S. et al. (1988). Augmented expression of epidermal growth factor receptor in human pancreatic carcinoma cells exhibiting alterations of chromosome 7. Cellular and Molecular Biology of Tumors and Potential Clinical Applications. Alan R. Liss, New York, pp. 179–183.

Meltzer, S. J. et al. (1990). Activation of c-Ki-*ras* in human gastrointestinal dysplasias determined by direct sequencing of polymerase chain reaction products. Cancer Res. *50*:3627–3630.

Meltzer, S. J. et al. (1991). Reduction to homozygosity involving p53 in esophageal cancers demonstrated by the polymerase chain reaction. Proc. Natl. Acad. Sci. U.S.A. *88*:4976–4980.

Mendis, S. et al. (1991). Restriction fragment length polymorphisms in the Apo B gene in relation to coronary heart disease in a Southern Asian population. Clin. Chim. Acta *196*:107–118.

Menzel, H. J. et al. (1983). Apolipoprotein E polymorphism and coronary artery disease. Arteriosclerosis *3*:310–315.

Merlino, G. T. et al. (1985). Structure and localization of genes encoding aberrant and normal epidermal growth factor receptor RNAs from A431 human carcinoma cells. Mol. Cell. Biol. *5*:1722–1734.

Merritt, W. D. et al. (1990). Oncogene amplification in squamous cell carcinoma of the head and neck. Arch. Otolaryngol. *116*:1394–1398.

Michelassi, F. et al. (1988). *ras* oncogene expression as a prognostic indicator in rectal adenocarcinoma. J. Surg. Res. *45*:15–20.

Michelassi, F. et al. (1990). Relationship between *ras* oncogene expression and clinical and pathological features of colonic carcinoma. Hepatogastroenterol. *37*:513–516.

Miki, H. et al. (1990). K-*ras* activation in gastric epithelial tumors. Proc. Am. Assn. Cancer Res. *31*:316A.

Miles, L. A. and Plow, E. F. (1990). Lp(a): an interloper into the fibrinolytic system? Thromb. Hemostasis *63*:331–335.

Miles, L. A. et al. (1989). A potential basis for the thrombotic risks associated with lipoprotein (a). Nature (London) *339*:301–303.

Miller, R. H. et al. (1984). Hepatitis B viral DNA-RNA hybrid molecules in particles from infected liver are converted to viral DNA molecules during an endogenous DNA polymerase reaction. Virology *139*:64–72.

Minna, J. D. et al. (1989). Transcription factors and recessive oncogenes in the pathogenesis of human lung cancer. Int. J. Cancer (Suppl.):*4*:32–34.

Miyagawa, K. et al. (1988). hst-1 transforming protein: expression in silkworm cells and characterization as a novel heparin binding growth factor. Oncogene *3*:383–389.

Miyaki, M. et al. (1990). Genetic changes and histopathologic types in colorectal tumors from patients with familial adenomatous polyposis. Cancer Res. *50*:7166–7173.

Mizoguchi, T. et al. (1990). Expression of the MDR1 gene in human gastric and colorectal carcinomas. J. Natl. Cancer Inst. *82*:1679–1683.

Mizusawa, H. et al. (1985). Inversely repeating integrated hepatitis B virus DNA and cellular flanking sequences in the human hepatoma derived cell line huSP. Proc. Natl. Acad. Sci. U.S.A. *82*:208–212.

Moller, L. and Kristensen, T. S. (1991). Plasma fibrinogen and ischemic heart disease risk factors. Arterioscl. Thromb. *11*:344–350.

Monsalve, M. V. et al. (1988). DNA polymorphisms of the gene for apolipoprotein B in patients with peripheral arterial disease. Atherosclerosis *70*:123–129.

Monsalve, M. V. et al. (1991). Within-individual variation in serum cholesterol levels: association with DNA polymorphisms at the apolipoprotein B and A-I/C-III/A-IV loci in patients with peripheral arterial disease. Clin. Genet. *39*:260–273.

Moorehead, R. J. et al. (1986). A study of ornithine decarboxylase activity as a marker for colorectal neoplasia. Br. J. Surg. *73*:1042A.

Morgan, S. A. et al. (1989). Flow cytometric analysis of Hoechst 33342 uptake as an indicator of multi-drug resistance in human lung cancer. Br. J. Cancer *60*:282–287.

Mori, N. et al. (1989). Concordant deletions of chromosomes 3p and loss of heterozygosity for chromosomes 13 and 17 in small cell lung carcinoma. Cancer Res. *49*:5130–5135.

Moriyama, K. et al. (1991). Immunoselected hepatitis B virus mutant. Lancet *337*:125.

Morley, D. J. and Hodes, M. E. (1988). Amylase expression in human parotid neoplasms: evidence by *in situ* hybridization for lack of transcription of the amylase gene. J. Histochem. Cytochem. *36*:487–491.

Moroy, T. et al. (1986). Rearrangement and enhanced expression of c-*myc* in hepatocellular carcinoma of hepatitis virus infected woodchucks. Nature (London) *324*:276–279.

Morrisett, J. D. et al. (1987). Lipoprotein (a): structure, metabolism and epidemiology. In: Plasma Lipoproteins. Gotto, A. M., Jr. (ed.). Elsevier, Amsterdam, pp. 129–152.

Motti, C. et al. (1991). Using mutagenic polymerase chain reaction primers to detect carriers of familial defective apolipoprotein B-100. Clin. Chem. *37*:1762–1766.

Motulsky, A. G. (1989). Genetic aspects of familial hypercholesterolemia and its diagnosis. Arteriosclerosis *9* (Suppl. I):3–7.

Mountain, C. F. (1977). Assessment of the role of surgery for control of lung cancer. Ann. Thorac. Surg. *24*:365–373.

Mountain, C. F. (1983). Biological, physiologic and technical determinants in surgical therapy for lung cancer. In: Lung Cancer. Straus, M. (ed.). Grune & Stratton, New York, pp. 245–260.

Mukhopadhyay, T. et al. (1991). Specific inhibition of K-*ras* expression and turmorigenicity of lung cancer cells by antisense RNA. Cancer Res. *51*:1744–1748.

Mulder, K. M. and Brattain, M. G. (1988). Alterations in c-*myc* expression in relation to maturational status of human colon carcinoma cells. Int. J. Cancer *42*:64–70.

Mulder, K. M. et al. (1988). Modulation of c-*myc* by transforming growth factor β in human colon carcinoma cells. Biochem. Biophys. Res. Commun. *150*:711–716.

Munoz, N. and Bosch, F. X. (1987). Epidemiology of hepatocellular carcinoma. In: Neoplasmas of the Liver. Okuda, K. and Ishak, K. G. (eds.). Springer-Verlag, Tokyo, pp. 3–19.

Myklebost, O. et al. (1990). Association analysis of lipid levels and apolipoprotein restriction fragment length polymorphisms. Hum. Genet. *86*:209–214.

Nagata, Y. et al. (1990). Frequent glycine-to-aspartic acid mutations at codon 12 of c-Ki-*ras* gene in human pancreatic cancer in Japanese. Jpn. J. Cancer Res. *81*:135–140.

Nagy, P. et al. (1988). Cellular distribution of c-*myc* transcripts during chemical hepatocarcinogenesis in rats. Cancer Res. *48*:5522–5527.

Nakamura, T. et al. (1988). Microdeletion associated with the integration process of hepatitis B virus DNA. Nucleic Acids Res. *16*:4865–4873.

Nakayama, H. et al. (1990). Clonal growth of atypical adenomatous hyperplasia of the lung: cytofluorometric analysis of nuclear DNA content. Mod. Pathol. *3*:314–320.

Naylor, M. S. et al. (1990). Investigation of cytokine gene expression in human colorectal cancer. Cancer Res. *50*:4436–4440.

Naylor, S. L. et al. (1987). Loss of heterozygosity of chromosome 3p markers in small-cell lung cancer. Nature (London) *329*:451–454.

Nelson, P. V. et al. (1990). Gene amplification directly from Guthrie blood spots. Lancet *336*:1451–1452.

Ngoi, S. S. et al. (1990). Abnormal DNA ploidy and proliferative patterns in superficial colonic epithelium adjacent to colorectal cancer. Cancer *66*:953–959.

Nigro, J. M. et al. (1989). Mutations in p53 gene occur in diverse human tumor types. Nature (London) *342*:705–708.

Nishida, J. et al. (1987). A point mutation at codon 13 of the N-*ras* oncogene in a human stomach cancer. Biochem. Biophys. Res. Commun. *146*:247–252.

Nishioka, K. (1985). Hepatitis B virus and hepatocellular carcinoma: postulates for an etiological relationship. Adv. Viral Oncol. *5*:173–199.

Nishisho, I. et al. (1991). Mutations of chromosome 5q21 genes in FAP and colorectal cancer patients. Science *253*:665–669.

Noguchi, M. et al. (1990). Heterogenous amplification of *myc* family oncogenes in small cell lung carcinoma. Cancer *66*:2053–2058.

Nomura, N. et al. (1986). DNA amplification of the c-*myc* and c-*erb*B-1 genes in a human stomach cancer. Jpn. J. Cancer Res. (Gann.) *77*:1188–1192.

Norum, K. R. et al. (1989). Familial lecithin-cholesteryl acyltransferase deficiency; including fish eye disease. In: The Metabolic Basis of Inherited Diseases. Scriver, C. R. et al. (eds.). McGraw-Hill, New York, pp. 1181–1194

Novelli, G. et al. (1990). F508 gene deletion and prenatal diagnosis of cystic fibrosis in Italian and Spanish families. Prenatal Diagn. *10*:413–416.

Ogston, C. W. et al. (1982). Cloning and structural analysis of integrated woodchuck hepatitis virus sequences from hepatocellular carcinomas of woodchucks. Cell (Cambridge, Mass) *29*:385–394.

Oikawa, S. et al. (1987a). Primary structure of human carcinoembryonic antigen (CEA) deduced from cDNA sequence. Biochem. Biophys. Res. Commun. *142*:511–518.

Oikawa, S. et al. (1987b). The carcinoembryonic antigen (CEA) contains multiple immunoglobin-like domains. Biochem. Biophys. Res. Commun. *144*:634–642.

Okamoto, H. et al. (1987a). Defective mutant of hepatitis B virus in the circulation of symptom-free carriers. Jpn. J. Exp. Med. *57*:217–221.

Okamoto, H. et al. (1987b). Genomic heterogeneity of hepatitis B virus in a 54 year old woman who contracted the infection through materno-fetal transmission. Jpn. J. Exp. Med. *57*:231–236.

Okamoto, H. et al. (1987c). Point mutation in the S gene of hepatitis B virus for a d/y or w/r subtypic change in two blood donors carrying a surface antigen of compound subtype adyr or adwr. J. Virol. *61*:3030–3034.

Okamoto, M. et al. (1988). Loss of constitutional heterozygosity in colon carcinoma from patients with familial polyposis coli. Nature (London) *331*:273–277.

Okamoto, M. et al. (1990). Molecular nature of chromosome 5q loss in colorectal tumors and desmoids from patients with familial adenomatous polyposis. Hum. Genet. *85*:595–599.

Ordovas, J. M. et al. (1986). Apolipoprotein A-I gene polymorphism associated with premature coronary artery disease and familial hypoalphalipoproteinemia. N. Engl. J. Med. *314*:671–677.

Ordovas, J. M. et al. (1989). Familial apolipoprotein AI, CIII, and AIV deficiency and premature atherosclerosis due to deletion of a gene complex on chromosome 11. J. Biol. Chem. *264*:16339–16342.

Ordovas, J. M. et al. (1991). Restriction fragment length polymorphisms of the apolipoprotein A-I, C-III, A-IV gene locus. Relationships with lipids, apolipoproteins, and premature coronary artery disease. Atherosclerosis *87*:75–86.

Orrom, W. J. et al. (1990). Heredity and colorectal cancer. A prospective, community-based, endoscopic study. Dis. Colon Rectum *33*:490–493.

Oshima, A. et al. (1984). Follow-up study of HBsAg positive blood donors with special reference to effect of drinking and smoking on development of liver cancer. Int. J. Cancer *34*:775–779.

Otsuru, A. et al. (1988). Analysis of α-fetoprotein gene expression in hepatocellular carcinoma and liver cirrhosis by *in situ* hybridization. Cancer *62*:1105–1112.

Ozawa, S. et al. (1987). High incidence of EGF receptor hyperproduction in esophageal squamous cell carcinomas. Int. J. Cancer *39*:333–337.

Ozturk, M. et al. (1991). p53 mutation in hepatocellular carcinoma after aflatoxin exposure. Lancet *338*:1356–1359.

Pagani, F. et al. (1990). Human apolipoprotein AI gene promoter polymorphism: association with hyperalphalipoproteinemia. J. Lipid Res. *31*:1371–1377.

Paganuzzi, M. et al. (1988). CA 19-9 and CA 50 in benign and malignant pancreatic and biliary diseases. Cancer *61*:2100–2108.

Pahlman, S. and Hammerling, U. (1990). *src* expression in small-cell lung carcinoma and other neuroendocrine malignancies. Am. Rev. Respir. Dis. 142:S54–S56.

Paraskeva, C. et al. (1990). Colorectal carcinogenesis: sequential steps in the *in vitro* immortalization and transformation of human colonic epithelial cells (review). Anticancer Res. *10*:1189–1200.

Park, J. G. et al. (1990). Characterizations of cell lines established from human gastric carcinoma. Cancer Res. *50*:2773–2780.

Pasquinelli, C. et al. (1988). Rearrangement of a common cellular DNA domain on chromosome 4 in human primary liver tumors. J. Virol. *62*:629–632.

Patel, P. et al. (1990). Chromosome 11 allele loss in sporadic insulinoma. J. Clin. Pathol. *43*:377–378.

Paterlini, P. et al. (1990). Polymerase chain reaction to detect hepatitis B virus DNA and RNA sequences in primary liver cancers from patients negative for hepatitis B surface antigen. N. Engl. J. Med. *323*:80–85.

Paterlini, P. et al. (1991). Letter to Editor. N. Engl. J. Med. *324*:127.

Paul, P. et al. (1990). Evaluation of polymorphic genetic markers for linkage to the familial adenomatous polyposis locus on chromosome 5. Dis. Colon Rectum *33*:740–744.

Paulweber, B. et al. (1988). Genetic variation in the apolipoprotein AI-CIII-AIV gene cluster and coronary heart disease. Atherosclerosis *73*:125–133.

Peers, F. et al. (1987). Aflatoxin exposure, hepatitis B virus infection and liver cancer in Swaziland. Int. J. Cancer *39*:545–553.

Pegg, A. E. and McCann, P. P. (1982). Polyanine metabolism and function. Am. J. Physiol. *243*:C212–C221.

Pelton, R. W. and Moses, H. L. (1990). The β-type transforming growth factor. Mediators of cell regulation in the lung. Am. Rev. Respir. Dis. 142:S31–S35.

Penn, A. (1990). Mutational events in the etiology of arteriosclerotic plaques. Mutat. Res. *239*:149–162.

Pessah, M. et al. (1991). Anderson's disease: genetic exclusion of the apolipoprotein B gene in two families. J. Clin. Invest. *87*:367–370.

Petersen, T. E. et al. (1990). Characterization of the gene for human plasminogen, a key proenzyme in the fibrinolytic system. J. Biol. Chem. *265*:6104–6111.

Pfaffinger, D. et al. (1991). Relationship between apo(a) isoforms and Lp(a) density in subjects with different apo(a) phenotype: a study before and after a fatty meal. J. Lipid Res. *32*:679–683.

Piguet, P. F. et al. (1990). Requirement of tumor necrosis factor for development of silica-induced pulmonary fibrosis. Nature (London) *344*:245–247.

Pipeleers, D. G. et al. (1985). A new *in vitro* model for the study of pancreatic A and B cells. Endocrinology *117*:806–815.

Pontisso, P. et al. (1984). Detection of hepatitis B virus DNA in mononuclear blood cells. Br. Med. J. *288*:1563–1566.

Porter, C. W. et al. (1987). Polyamine biosynthetic activity in normal and neoplastic human colorectal tissues. Cancer *60*:1275–1281.

Poustka, A. and Lehrach, H. (1986). Jumping libraries and linking libraries: the next generation of molecular tools in mammalian genetics. Trends Genet. *2*:174–179.

Poustka, A. M. et al. (1988). A long range restriction map encompassing the cystic fibrosis locus and its closely linked genetic markers. Genomics *2*:337–345.

Price, J. O. et al. (1991). Prenatal diagnosis with fetal cells isolated from maternal blood by multiparameter flow cytometry. Am. J. Obstet. Gynecol. *165*:1731–1737.

Price, W. H. et al. (1986). Apolipoprotein CIII polymorphism and coronary heart disease. Lancet *ii*:1041.

Price, W. H. et al. (1989). DNA restriction fragment length polymorphisms as markers of familial coronary heart disease. Lancet *i*:1407–1411.

Price, W. H. et al. (1990). Genetic markers of familial coronary heart disease. Lancet *336*:629.

Prior, T. W. et al. (1990). A model for molecular screening of newborns: simultaneous detection of Duchenne/Becker muscular dystrophies and cystic fibrosis. Clin. Chem. *36*:1756–1759.

Quinton, P. M. (1990). Cystic fibrosis: a disease in electrolyte transport. FASEB J. *4*:2709–2717.

Quirke, P. et al. (1986). DNA aneuploidy in colorectal adenomas. Br. J. Cancer *53*:477–481.

Quirke, P. et al. (1987). Prognostic significance of DNA aneuploidy and cell proliferation in rectal adenocarcinomas. J. Pathol. *151*:285–291.

Rabinovitch, A. et al. (1982). Preparation of rat islet B-cell-enriched fractions by light-scatter flow cytometry. Diabetes *31*:939–943.

Radford, D. M. et al. (1990). Two chromosomal locations for human ornithine decarboxylase gene sequences and elevated expression in colorectal neoplasia. Cancer Res. *50*:6146–6153.

Rajput-Williams, J. et al. (1988). Variation of apolipoprotein-B gene is associated with obesity, high blood cholesterol levels, and increased risk of coronary heart disease. Lancet *ii*:1442–1446.

Rall, S. C. et al. (1983). Identification of a new structural variant of human apolipoprotein E (E2 Lys-146→Gln), in a type III hyperlipoproteinemic subject with the E3-2 phenotype. J. Clin. Invest. *72*:1288–1297.

Rath, M. et al. (1989). Detection and quantitation of lipoprotein (a) in the arterial wall of 107 coronary bypass patients. Arteriosclerosis *9*:579–592.

Reardon, M. F. et al. (1985). Lipoprotein predictors of the severity of coronary artery disease in men and women. Circulation *71*:881–888.

Rees, A. et al. (1985). Deoxyribonucleic acid polymorphism in the apolipoprotein AI, CIII gene cluster. Association with hypertriglyceridemia. J. Clin. Invest. *76*:1090–1095.

Rees, M. et al. (1989). Chromosome 5 allele loss in familial and sporadic colorectal adenomas. Br. J. Cancer *59*:361–365.

Reissmann, P. T. et al. and the Lung Cancer Study Group (1990). Inactivation of the retinoblastoma gene in non-small cell lung cancer. Proc. Am. Assn. Cancer Res. *31*:318A.

Remvikos, Y. et al. (1988). DNA content and genetic evolution of human colorectal adenocarcinoma. A study by flow cytometry and cytogenetic analysis. Int. J. Cancer *42*:539–543.

Resnick, R. H. and Antonioli, D. (1991). Letter to Editor. N. Engl. J. Med. *324*:126–127.

Reynolds, S. H. et al. (1991). Activated protooncogenes in human lung tumors from smokers. Proc. Natl. Acad. Sci. U.S.A. *88*:1085–1089.

Rhoads, G. G. et al. (1986). Lp(a) lipoprotein as a risk factor for myocardial infarction. JAMA *256*:2540–2544.

Rich, D. P. et al. (1990). Expression of cystic fibrosis transmembrane conductance regulator corrects defective chloride channel regulation in cystic fibrosis airway epithelial cells. Nature (London) *347*:358–363.

Riddell, R. H. et al. (1983). Dysplasia in inflammatory bowel disease: standardized classification with provisional clinical applications. Hum. Pathol. *14*:931–968.

Riordan, J. R. et al. (1989). Identification of the cystic fibrosis gene: cloning and characterization of complementary DNA. Science *245*:1066–1072.

Riviere, A. et al. (1991). Comparative investigation of c-*erb*B-2/*neu* expression in head and neck tumors and mammary cancer. Cancer *67*:2142–2149.

Robey-Cafferty, S. S. et al. (1990). Histologic parameters and DNA ploidy as predictors of survival in stage B adenocarcinoma of colon and rectum. Mod. Pathol. *3*:261–266.

Rodenhuis, S. and Slebos, R. J. C. (1990). The *ras* oncogenes in human lung cancer. Am. Rev. Respir. Dis. *142*:S27–S30.

Rodenhuis, S. et al. (1987). Mutational activation of the K-*ras* oncogene. A possible pathogenetic factor in adenocarcinoma of the lung. N. Engl. J. Med. *317*:929–935.

Rodriguez, E. et al. (1990). 11p13-15 is a specific region of chromosomal rearrangement in gastric and esophageal adenocarcinomas. Cancer Res. *50*:6410–6416.

Roebroek, A. J. M. et al. (1989). Differential expression of the gene encoding the novel pituitary polypeptide 7B2 in human lung cancer cells. Cancer Res. *49*:4154–4158.

Rogler, C. E. et al. (1985). Deletion in chromosome 11p associated with hepatitis B integration site in hepatocellular carcinoma. Science *230*:319–322.

Roma, P. et al. (1990). Apolipoprotein AI metabolism in subjects with a *Pst* I restriction fragment length polymorphism of the apo AI gene and familial hypoalphalipoproteinemia. J. Lipid Res. *31*:1753–1760.

Romet-Lemonne, J. L. et al. (1983). Hepatitis B virus infection in cultured human lymphoblastoid cells. Science *221*:667–669.

Rommens, J. M. et al. (1989). Identification of the cystic fibrosis gene: chromosome walking and jumping. Science *245*:1059–1065.

Rommens, J. et al. (1990). Rapid nonradioactive detection of the major cystic fibrosis mutation. Am. J. Hum. Genet. *46*:396–397.

Rosai J. (ed.). (1989). Ackerman's surgical pathology, 7th Ed. C. V. Mosby, St. Louis.

Rozen, R. et al. (1992). Cystic fibrosis mutations in French Canadians: three CFTR mutations are relatively frequent in a Quebec population with an elevated incidence of cystic fibrosis. Am J. Med. Genet. *42*:360–364.

Rua, S. et al. (1991). Relationship between histologic features, DNA flow cytometry, and clinical behavior of squamous cell carcinomas of the larynx. Cancer *67*:141–149.

Russell, D. W. et al. (1983). cDNA cloning of the bovine low density lipoprotein receptor: feedback regulation of a receptor in RNA. Proc. Natl. Acad. Sci. U.S.A. *80*:7501–7505.

Russell, D. W. et al. (1989). Molecular basis of familial hypercholesterolemia. Arteriosclerosis 9 (Suppl. I):8–13.

Rustin, R. B. et al. (1990). Spontaneous mutation in familial adenomatous polyposis. Dis. Colon Rectum *33*:52–55.

Ryberg, D. et al. (1990). Ha-*ras*-1 alleles in Norwegian lung cancer patients. Hum. Genet. *86*:40–44.

Ryder, K. and Nathans, D. (1988). Induction of proto-oncogene c-*jun* by serum growth factors. Proc. Nathl. Acad. Sci. U.S.A. *85*:8464–8467.

Ryder, K. et al. (1988). A gene activated by growth factors is related to the oncogene v-*jun*. Proc. Natl. Acad. Sci. U.S.A. *85*:1487–1491.

Ryder, K. et al. (1989). jun-D: a third member of the *jun* gene family. Proc. Natl. Acad. Sci. U.S.A. *86*:1500–1503.

Rygaard, K. et al. (1990). Abnormalities in structure and expression of the retinoblastoma gene in small cell lung cancer cell lines and xenografts in nude mice. Cancer Res. 50:5312–5317.

Sakamoto, H. et al. (1986). Transforming gene from human stomach cancers and a noncancerous portion of stomach mucosa. Proc. Natl. Acad. Sci. U.S.A. *83*:3997–4001.

Saksela, K. et al. (1986). Amplification of the N-*myc* oncogene in an adenocarcinoma of the lung. J. Cell. Biochem. *31*:297–304.

Saksela, K. et al. (1989). Oncogene expression in small cell lung cancer cell lines and a testicular germ-cell tumor: activation of the N-*myc* gene and decreased RB mRNA. Int. J. Cancer *44*:182–185.

Salonen, E. M. et al. (1989). Lipoprotein (a) binds to fibronectin and has serine proteinase activity capable of cleaving it. EMBO J. 8:4035–4040.

Sandgren, E. P. et al. (1991). Pancreatic tumor pathogenesis reflects the causative genetic lesion. Proc. Natl. Acad. Sci. U.S.A. *88*:93–97.

Sandholzer, C. et al. (1991). Effects of the apolipoprotein (a) size polymorphism on the lipoprotein (a) concentration in 7 ethnic groups. Hum. Genet. *86*:607–614.

Santis, G. et al. (1990a). Linked marker haplotypes and the F508 mutation in adults with mild pulmonary disease and cystic fibrosis. Lancet *335*:1426–1429.

Santis, G. et al. (1990b). Independent genetic determinants of pancreatic and pulmonary status in cystic fibrosis. Lancet *336*:1081–1084.

Sapienza, C. (1991). Genome imprinting and carcinogenesis. Biochim. Biophys. Acta *1072*:51–61.

Saranath, D. et al. (1990). Restriction fragment length polymorphism of the L-*myc* gene in oral cancer patients. Br. J. Cancer *61*:530–533.

Sasaki, K. et al. (1988). Intratumoral regional differences in DNA ploidy of gastrointestinal carcinomas. Cancer *62*:2569–2575.

Sasaki, M. et al. (1989). Loss of constitutional heterozygosity in colorectal tumors from patients with familial polyposis coli and those with nonpolyposis colorectal carcinoma. Cancer Res. *49*:4402–4406.

Sasaki, M. et al. (1990). K-*ras* activation in colorectal tumors from patients with familial polyposis coli. Cancer *65*:2576–2579.

Scanu, A. M. (1988). Lp(a): a potential bridge between the fields of atherosclerosis and thrombosis. Arch. Pathol. Lab. Med. *112*:1045–1047.

Scanu, A. M. (1991). *N*-Acetylcysteine and immunoreactivity of lipoprotein (a). Lancet *337*:1159.

Scanu, A. M. and Fless, G. M. (1990). Lipoprotein (a). Heterogeneity and biological relevance. J. Clin. Invest. *85*:1709–1715.

Schaefer, E. J. et al. (1982). Plasma apolipoprotein AI absence associated with a marked reduction of high density lipoproteins and premature coronary artery disease. Atherosclerosis 2:16–26.

Schneider, P. M. et al. (1989). Differential expression of the c-*erb*B-2 gene in human small cell and non-small cell lung cancer. Cancer Res. *49*:4968–4971.

Schneider, P. M. et al. (1990). Comparison of *neu* (c-*erb*B-2) gene expression in small cell lung cancer (SCLC), non-small cell lung cancer (NSCLC), and normal lung. Proc. Am. Assn. Cancer Res. *31*:312A.

Scholte, B. J. et al. (1989). Immortalization of nasal polyp epithelial cells from cystic fibrosis patients. Exp. Cell Res. *182*:559–571.

Schulman, J. D. et al. (1990). Screening for cystic fibrosis carriers. Am. J. Hum. Genet. *47*:740.

Schuster, H. et al. (1989). Four DNA polymorphisms in the LDL-receptor gene and their use in diagnosis of familial hypercholesterolemia. Hum. Genet. *82*:69–72.

Schuster, H. et al. (1990). Association of DNA haplotypes in the human LDL-receptor gene with normal serum cholesterol levels. Clin. Genet. *38*:401–109.

Schutte, B. et al. (1987). Retrospective analysis of the prognostic significance of DNA content and proliferative activity in large bowel carcinoma. Cancer Res. *47*:5494–5496.

Schutte, J. et al. (1988). Constitutive expression of multiple mRNA forms of the c-*jun* oncogene in human lung cancer cell line. Proc. Am. Assn. Cancer Res. *29*:455.

Schutte, J. et al. (1989). Deregulated expression of human c-*jun* transforms primary rat embryo cells in cooperation with an activated c-Ha-*ras* gene and transforms rat-1a cells as a single gene. Proc. Natl. Acad. Sci. U.S.A. *86*:2257–2261.

Schwartz, C. J. et al. (1988). Thrombosis and the development of atherosclerosis: Rokitansky revisited. Semin. Thromb. Hemostasis *14*:189–195.

Scott, J. (1990). RNA editing: a novel mechanism for the regulation of dietary cholesterol absorption. The Humphrey Davy Rolleston lecture 1989. J. Royal Coll. Phys. London *24*:101–106.

Scott, N. A. et al. (1989). Carcinoma of the anal canal and flow cytometric DNA analysis. Br. J. Cancer *60*:56–58.

Scott, N. et al. (1991). p53 in colorectal cancer: clinicopathological correlation and prognostic significance. Br. J. Cancer *63*:317–319.

Seed, M. et al. (1990). Relation of serum lipoprotein (a) concentration and apolipoprotein (a) phenotype to coronary heart disease in patients with familial hypercholesterolemia. N. Engl. J. Med. *322*:1494–1499.

Seeger, C. et al. (1986). Biochemical and genetic evidence for the hepatitis B virus replication strategy. Science *232*:477–484.

Serre, J. L. et al. (1990). Studies of RFLP closely linked to the cystic fibrosis locus throughout Europe lead to new considerations in population genetics. Hum. Genet. *84*:449–454.

Seto, E. et al. (1988). Transactivation of the human immunodeficiency virus long terminal repeat by the hepatitis B virus X protein. Proc. Natl. Acad. Sci. U.S.A. *85*:8286–8290.

Seto, E. et al. (1990). Transactivation by the hepatitis B virus X protein depends on AP-2 and other transcription factors. Nature (London) *344*:72–74.

Shabtai, F. et al. (1988). Familial fragile 8q22 involved as a cancer breakpoint in cells of a large bowel tumor. Cancer Genet. Cytogenet. *31*:113–118.

Shafritz, D. A. et al. (1981). Integration of hepatitis B virus DNA into the genome of liver cells in chronic liver disease and hepatocellular carcinoma: studies in percutaneous liver biopsies and post-mortem tissue specimens. N. Engl. J. Med. *305*:1067–1073.

Shaul, Y. et al. (1984). Cloning and analysis of integrated hepatitis virus sequences from a human hepatoma cell line. J. Virol. *51*:776–787.

Sheng, Z. M. et al. (1990). Analysis of the c-Ha-*ras* 1 gene for deletion, mutation, amplification and expression in lymph node metastases of human head and neck carcinomas. Br. J. Cancer *62*:398–404.

Shibata, D. et al. (1990). Detection of c-K-*ras* mutations in fine needle aspirates from human pancreatic adenocarcinomas. Cancer Res. *50*:1279–1283.

Shiraishi, M. et al. (1989). Amplification of protooncogenes in surgical specimens of human lung carcinomas. Cancer Res. *49*:6474–6479.

Shoulders, C. C. et al. (1991). Variation at the apo A-I/C-III/A-IV gene complex is associated with elevated plasma levels of apo C-III. Atherosclerosis *87*:239–247.

Shwachman, H. (1975). Gastrointestinal manifestation of cystic fibrosis. Pediatr. Clin. N. Am. *22*:787–805.

Silverberg, E. and Lubera, J. (1988). Cancer statistics 1988. CA 1988: *38*:5–22.

Silverman, A. L. et al. (1989). Abnormal methylation of the calcitonin gene in human colonic neoplasms. Cancer Res. *49*:3468–3473.

Sing, C. F. and Moll, P. P. (1990). Genetics of atherosclerosis. Annu. Rev. Genet. *24*:171–187.

Sinha, S. et al. (1988). Activiation of *ras* oncogene in aflatoxin-induced rat liver carcinogenesis. Proc. Natl. Acad. Sci. U.S.A. *85*:3673–3677.

Skogerboe, K. J. et al. (1990). Development and evaluation of a simplified dot blot method for detecting the ΔF508 mutation in cystic fibrosis. Clin. Chem. *36*:1984–1986.

Skuse, G. R. and Rowley, P. T. (1989). Tumor suppressor genes and inherited predisposition to malignancy. Semin. Oncol. *16*:128–137.

Slack, J. (1979). Inheritance of familial hypercholesterolemia. Atheroscl. Rev. *5*:35–66.

Slagle, B. L. et al. (1991). Hepatitis B virus integration event in human chromosome 17p near the p53 gene identifies the region of the chromosome commonly deleted in virus positive hepatocellular carcinomas. Cancer Res. *51*:49–54.

Slamon, D. J. et al. (1989). Studies of *HER-2/neu* proto-oncogene in human breast and ovarian cancer. Science *244*:707–712.

Slebos, R. J. C. et al. (1989). Cellular protooncogenes are infrequently amplified in untreated non-small cell lung cancer. Br. J. Cancer *59*:76–80.

Slebos, R. J. C. et al. (1990). K-*ras* oncogene activation as a prognostic marker in adenocarcinoma of the lung. N. Engl. J. Med. *323*:561–565.

Slebos, R. J. C. et al. (1991). Relationship between K-*ras* oncogene activation and smoking in adenocarcinoma of the human lung. J. Natl. Cancer Inst. *83*:1024–1027.

Smeets, H. J. M. et al. (1988). Identification of apolipoprotein E polymorphism by using synthetic oligonucleotides. J. Lipid Res. *29*:1231–1237.

Smit, V. T. H. B. M. et al. (1988). KRAS codon 12 mutations occur very frequently in pancreatic adenocarcinomas. Nucleic Acids Res. *16*:7773–7783.

Smith, E. B. and Cochran, S. (1990). Factors influencing the accumulation in fibrous plaques of lipid derived from low density lipoprotein. II. Preferential immobilization of lipoprotein (a) (Lp(a)). Atherosclerosis *84*:173–181.

Smith, E. B. et al. (1990). Factors influencing the accumulation in fibrous plaques of lipid derived from low density lipoprotein. I. Relation between fibrin and immobilization of Apo B-containing lipoprotein. Atherosclerosis *84*:165–171.

Smith, M. E. F. et al. (1988). Selective loss of HLA-A, B, C locus products in colorectal adenocarcinoma [letter]. Lancet *i*:823–824.

Solomon, E. (1990). Colorectal cancer genes. Nature (London) *343*:412–414.

Solomon, E. et al. (1987). Chromosome 5 allele loss in human colorectal carcinomas. Nature (London) *328*:316–319.

Soman, N. R. et al. (1991). The TPR-MET oncogenic rearrangment is present and expressed in human gastric carcinoma and precursor lesions. Proc. Natl. Acad. Sci. U.S.A. *88*:4892–4896.

Soria, L. F. et al. (1989). Association between a specific apolipoprotein B mutation and familial defective apolipoprotein B 100. Proc. Natl. Acad. Sci. U.S.A. *86*:587–591.

Sorscher, E. J. and Huang, Z. (1991). Diagnosis of genetic disease by primer-specified restriction map modification, with application to cystic fibrosis and retinitis pigmentosa. Lancet *337*:1115–1118.

Sowa, M. et al. (1988). An analysis of the DNA ploidy patterns of gastric cancer. Cancer *62*:1325–1330.

Sozzi, G. et al. (1991). Cytogenetic abnormalities and over-expression of receptors for growth factors in normal bronchial epithelium and tumor samples of lung cancer patients. Cancer Res. *51*:400–404.

Spandau, D. F. and Lee, C. H. (1988). Trans-activation of viral enhances by the hepatitis B virus X protein. J. Virol. *62*:427–434.

Spandidos, D. A. et al. (1990). Expression of *ras* p21 and *myc* p62 oncoproteins in small cell and non small cell carcinoma of the lung. Anticancer Res. *10*:1105–1114.

Sporn, M. B. and Roberts, A. B. (1985). Autocrine growth factors and cancer. Nature (London) *313*:745–747.

Stalenhoef, A. F. H. et al. (1981). Combined deficiency of apolipoprotein CII and lipoprotein lipase in familial hyperchylomicronemia. Metabolism *30*: 919–926.

Stalenhoef, A. F. H. et al. (1991). *N*-Acetylcysteine and lipoprotein. Lancet *337*:491.

Steinmetz, A. (1991). Clinical implications of the apolipoprotein E polymorphism and genetic variants: current methods for apo E phenotyping. Ann. Biol. Clin. *49*:1–8.

Stocks, J. et al. (1987). Haplotypes identified by DNA restriction fragment length polymorphisms in the AI-CIII-AIV gene region and hypertriglyceridemia. Am. J. Hum. Genet. *41*:106–118.

Strom, C. M. et al. (1990). Preconception genetic diagnosis of cystic fibrosis. Lancet *336*:306–307.

Stuyt, P. M. J. et al. (1984). Serum lipids, lipoproteins, and apolipoprotein E phenotypes in relatives of patients with type III hyperlipoproteinemia. Eur. J. Clin. Invest. *14*:219–226.

Sudhof, T. C. et al. (1987). 42 bp element from LDL receptor gene confers end-product repression by sterols when inserted into viral TK promoter. Cell (Cambridge, Mass) *48*:1061–1069.

Sundaresan, V. et al. (1987). Abnormal distribution of c-*myc* oncogene product in familial adenomatous polyposis. J. Clin. Pathol. *40*:1274–1281.

Super, M. et al. (1987). Clinical experience of prenatal diagnosis of cystic fibrosis by use of linked DNA probes. Lancet *ii*:782–784.

Suzuki, K. et al. (1990). Microspectrophotometric DNA analysis in ulcerative colitis with special reference to its application in diagnosis of carcinoma and dysplasia. Gut *31*:1266–1270.

Suzuki, Y. et al. (1988). High serum levels of DU-PAN-2 antigen and CA 19-9 in pancreatic cancer: correlation with immunocytochemical localization of antigens in cancer cells. Hepatogastroenterology *35*:128–135.

Sweetenham, J. W. et al. (1990). Urinary epidermal growth factor (hEGF) levels in patients with carcinomas of the breast, colon and rectum. Br. J. Cancer *62*:459–461.

Syvanen, A. C. et al. (1989). Direct sequencing of affinity-captured amplified human DNA: application to the detection of apolipoprotein E polymorphism. FEBS Lett. *258*:71–74.

Syvanen, A. C. et al. (1990). A primer-guided nucleotide incorporation assay in the genotyping of apolipoprotein E. Genomics *8*:684–692.

Szmuness, W. (1978). Hepatocellular carcinoma and the hepatitis B virus: evidence for a causal association. Progr. Med. Virol. *24*:40–69.

Tada, M. et al. (1991). Clinical application of *ras* gene mutation for diagnosis of pancreatic adenocarcinoma. Gastroenterology *100*:233–238.

Tadokoro, K. et al. (1989). Activation of oncogenes in human oral cancer cells: a novel codon 13 mutation of c-H-*ras*-1 and concurrent amplifications of c-*erb*B-1 and c-*myc*. Oncogenes *4*:499–505.

Tagawa, M. et al. (1985). Early events in duck hepatitis B virus infection: sequential appearance of viral deoxyribonucleic acid in the liver, pancreas, kidney, and spleen. Gastroenterology *89*:1224–1229.

Tahara, E. (1988). Endocrine tumors of the gastrointestinal tract: classification, function and biological behavior. In: Digestive Disease Pathology, Vol. 1. Watanabe, S. et al. (eds.). Field and Wood, Philadelphia, pp. 121–147.

Tahara, E. (1990). Growth factors and oncogenes in human gastrointestinal carcinomas. J. Cancer Res. Clin. Oncol. *116*:121–131.

Tajima, S. et al. (1988). Analysis of apolipoprotein E5 gene from a patient with hyperlipoproteinemia. J. Biochem. *104*:48–52.

Takada, S. and Koike, K. (1989). Activated N-*ras* gene was found in human hepatoma tissue but only in a small fraction of the tumor cells. Oncogene *4*:189–193.

Takada, S. et al. (1990). Structural rearrangement of integrated hepatitis B virus DNA as well as cellular flanking DNA is present in chronically infected hepatic tissues. J. Virol. *64*:822–828.

Takahashi, T. et al. (1989). p53: a frequent target for genetic abnormalities in lung cancer. Science *246*:491–494.

Takasaki, H. et al. (1988). Correlative study on expression of CA 19-9 and DU-PAN-2 in tumor tissue and in serum of pancreatic cancer patients. Cancer Res. *48*:1435–1438.

Tal, M. et al. (1988). Sporadic amplification of the HER2/*neu* protooncogene in adenocarcinomas of various tissues. Cancer Res. *48*:1517–1520.

Talmud, P. J. et al. (1987). Apolipoprotein B. gene variants are involved in the determination of serum cholesterol levels: a study in normo- and hyperlipidemic individuals. Atherosclerosis *67*:81–89.

Talmud, P. J. et al. (1988). Genetic evidence from two families that the apolipoprotein B gene is not involved in abetalipoproteinemia. J. Clin. Invest. *82*:1803–1806.

Talmud, P. et al. (1989). The molecular basis of truncated forms of apolipoprotein B in a kindred with compound heterozygous hypobetalipoproteinemia. J. Lipid Res. *30*:1773–1779.

Tanaka, K. et al. (1991). Suppression of tumorigenicity in human colon carcinoma cells by introduction of normal chromosome 5 or 18. Nature (London) *349*:340–342.

Tas, S. (1989). Strong association of a single nucleotide substitution in the 3′-untranslated region of the apolipoprotein-CIII gene with hypertriglyceridemia in Arabs. Clin. Chem. *35*:256–259.

Tas, S. (1991). Genetic predisposition to coronary heart disease and gene for apolipoprotein C-III. Lancet *337*:113–114.

Tashiro, F. et al. (1986). Expression of the c-Ha-*ras* and c-*myc* genes in aflatoxin B1-induced hepatocellular carcinomas. Biochem. Biophys. Res. Commun. *138*:858–864.

Tateishi, M. et al. (1990). Immunohistochemical evidence of autocrine growth factors in adenocarcinoma of the human lung. Cancer Res. *50*:7077–7080.

Tefre, T. et al. (1990). Studies of the L-*myc* DNA polymorphism and relation to metastasis in Norwegian lung cancer patients. Br. J. Cancer *61*:809–812.

ten Kate, J. et al. (1989). Expression of c-*myc* proto-oncogene in normal human intestinal epithelium. J. Histochem. Cytochem. *37*:541–545.

ten Kate, L. P. (1990). Carrier screening for cystic fibrosis and other autosomal recessive diseases [letter to the Editor]. Am. J. Hum Genet. *47*:359–361.

Teyssier, J. R. (1987). Nonrandom chromosomal changes in human solid tumors: application of an improved culture method. JNCI *79*:1189–1198.

Thompson, G. R. et al. (1989). Genotypic and phenotypic variation in familial hypercholesterolemia. Arteriosclerosis *9* (Suppl. I):75–80.

Tiainen, M. et al. (1989). Chromosomal abnormalities and their correlations with asbestos exposure and survival in patients with mesothelioma. Br. J. Cancer *60*:618–626.

Tikkanen, M. J. et al. (1990). Apolipoprotein E4 homozygosity predisposes to serum cholesterol elevation during high fat diet. Arteriosclerosis *10*:285–288.

Toh, H. et al. (1983). Sequence homology between retroviral reverse transcriptase and putative polymerases of hepatitis B virus and cauliflower mosaic virus. Nature (London) *305*:827–829.

Tokino, T. et al. (1987). Chromosomal translocation and inverted duplication associated with integrated hepatitis B virus in hepatocellular carcinomas. J. Virol. *61*:3848–3854.

Top, B. et al. (1990). Rearrangements in the LDL receptor gene in Dutch familial hypercholesterolemic patients and the presence of a common 4kb deletion. Atherosclerosis *83*:127–136.

Tops, C. M. J. et al. (1989). Presymptomatic diagnosis of familial adenomatous polyposis by bridging DNA markers. Lancet *i*:1361–1363.

Tranchesi, B. et al. (1990). Lack of association between raised serum lipoprotein (a) and thrombolysis. Lancet *336*:1587–1588.

Trichopoulos, D. et al. (1987). Hepatitis B virus, tobacco smoking and ethanol consumption in the etiology of hepatocellular carcinoma. Int. J. Cancer *39*:45–49.

Trieu, V. N. and McConathy, W. J. (1990). Lipoprotein (a) binding to other lipoprotein B containing lipoproteins. Biochemistry *29*:5919–5924.

Trieu, V. N. et al. (1991). Interaction of apolipoprotein (a) with apolipoprotein B-containing lipoproteins. J. Biol. Chem. *266*:5480–5485.

Tsuda, H. et al. (1990). Allele loss on chromosome 16 associated with progression of human hepatocellular carcinoma. Proc. Natl. Acad. Sci. U.S.A. *87*:6791–6794.

Tsuda, T. et al. (1988). Amplification of the *hst-1* gene in human esophageal carcinomas. Jpn. J. Cancer Res. (Gann.) *79*:584–588.

Tsuda, T. et al. (1989a). HST1 and INT2 gene coamplification in a squamous cell carcinoma of gallbladder. Jpn. J. Clin. Oncol. *19*:26–29.

Tsuda, T. et al. (1989b). High incidence of coamplification of HST1 and INT2 genes in human esophageal carcinomas. Cancer Res. *49*:5505–5508.

Tsui, L. C. et al. (1985). Cystic fibrosis defined by a genetically linked polymorphic DNA marker. Science *230*:1054–1057.

Tsui, L. C. and Buchwald, M. (1991). Biochemical and molecular genetics of cystic fibrosis. In: Advances in Human Genetics, Vol. 20. Harris, H. and Hirschhorn, K. (eds.). Plenum Press, New York, pp. 153–266.

Tsujino, T. et al. (1989). Oncogene amplification and metastasis in gastric and esophageal carcinomas. In: Proc. Jpn. Cancer Assn., 48th Annual Meeting, Tokyo, Nishizuka, Y. et al. (eds.), p. 25.

Tsujino, T. et al. (1990). Alterations of oncogenes in metastatic tumors of human gastric carcinomas. Br. J. Cancer *62*:226–230.

Tsushima, K. et al. (1987). Leiomyosarcomas and benign smooth muscle tumors of the stomach: nuclear DNA patterns studied by flow cytometry. Mayo Clin. Proc. *62*:275–280.

Twu, J. S. and Schloemer, R. H. (1987). Transcriptional trans activating function of hepatitis B virus. J. Virol. *61*:3448–3453.

Tybjaerg-Hansen, A. et al. (1990). Familial defective apolipoprotein B-100: detection in the United Kingdom and Scandinavia and clinical characteristics of ten cases. Atherosclerosis *80*:235–242.

Tytor, M. et al. (1987a). DNA pattern in oral cavity carcinomas in relation to clinical stage and histologic grading. Pathol. Res. Pract. *182*:202–206.

Tytor, M. et al. (1987b). DNA content, malignancy grading and prognosis in T1 and T2 oral cavity carcinomas. Br. J. Cancer *56*:647–652.

Untawale, S. and Blick, M. (1988). Oncogene expression in adenocarcinomas of the colon and in colon tumor-derived cell lines. Anticancer Res. *8*:1–8.

Upp, J. R., Jr. et al. (1989). Clinical significance of gastrin receptors in human colon cancers. Cancer Res. *49*:488–492.

Utermann, G. (1989). The mysteries of lipoprotein (a). Science *246*:904–910.

Utermann, G. and Weber, W. (1983). Protein composition of Lp(a) lipoprotein from human plasma. FEBS Lett. *154*:357–361.

Utermann, G. et al. (1984). Apo E phenotypes in patients with myocardial infarction. Hum. Genet. *65*:237–241.

Utermann, G. et al. (1987). Lp(a) glycoprotein phenotypes: inheritance and relation to Lp(a) — lipoprotein concentrations in plasma. J. Clin. Invest. *80*:458–465.

Utermann, G. et al. (1988a). Genetics of the quantitative Lp(a) lipoprotein trait. I. Relation of Lp(a) glycoprotein phenotypes to Lp(a) lipoprotein concentrations in plasma. Hum. Genet. *78*:41–46.

Utermann, G. et al. (1988b). Genetics of the quantitative Lp(a) lipoprotein trait. II. Inheritance of Lp(a) glycoprotein phenotypes. Hum. Genet. *78*:47–50.

Vanakangas, K. H. et al. (1992). Mutations of *p53* and *ras* genes in radon-associated lung cancer from uranium miners. Lancet *339*:576–580.

Van Biervliet, J. P. et al. (1991). Lipoprotein (a) profiles and evolution in newborns. Atherosclerosis *86*:173–181.

van der Winkel, M. et al. (1982). Islet cell analysis and purification by light scatter and autofluorescence. Biochem. Biophys. Res. Commun. *107*:525–532.

van den Ingh, H. F. et al. (1985). Flow cytometric detection of aneuploidy in colorectal adenomas. Cancer Res. *45*:3392–3397.

van der Hout, A. H. et al. (1988). Localization of DNA probes with tight linkage to the cystic fibrosis locus by *in situ* hybridization using fibroblasts with a 7q22 deletion. Hum. Genet. *80*:161–164.

van Rensburg, S. J. et al. (1985). Hepatocellular carcinoma and dietary aflatoxin in Mozambique and Transkei. Br. J. Cancer *51*:713–726.

Viallet, J. and Minna, J. D. (1990). Dominant oncogenes and tumor suppressor genes in the pathogenesis of lung cancer. Am. J. Respir. Cell Mol. Biol. *2*:225–232.

Vidaud, M. et al. (1989). Confirmation of linkage disequilibrium between haplotype B (XV-2c:allele 1, KM-19:allele 2) and cystic fibrosis allele in the French population. Hum. Genet. *81*:183–184.

Vogelstein, B. et al. (1988). Genetic alterations during colorectal-tumor development. N. Engl. J. Med. *319*:525–532.

Vogelstein, B. et al. (1989). Allelo type of colorectal carcinomas. Science *244*:207–211.

Vogelstein, B. et al. (1990). Clinical implications of colorectal tumor mutations. Proc. Am. Assn. Cancer Res. *31*:455–456.

Volm, M. et al. (1985). Prognostic significance of DNA patterns and resistive-predictive tests in non-small cell lung carcinoma. Cancer *56*:1396–1403.

Wagner, H. E. et al. (1990). Characterization of the tumorigenic and metastatic potential of a poorly differentiated human colon cancer cell line. Inv. Metast. *10*:253–266.

Wagner, M. et al. (1990). Direct gene diagnosis of cystic fibrosis by allele-specific polymerase chain reactions. Mol. Biol. Med. *7*:359–364.

Wainwright, B. J. et al. (1985). Localization of the cystic fibrosis locus to human chromosome 7cen-q22. Nature (London) *318*:384–385.

Wakabayashi, K. (1990). Animal studies suggesting involvement of mutagen/carcinogen exposure in atherosclerosis. Mutat. Res. 239:181–187.

Walton, K. W. et al. (1974). A study of methods of identification and estimation of Lp(a) lipoprotein and of its significance in health, hypenlipidemia and atherosclerosis. Atherosclerosis *20*:323–346.

Wang, H. P. and Rogler, C. E. (1988). Deletions in human chromosome arms 11p and 13q in primary hepatocellular carcinomas. Cytogenet. Cell Genet. *48*:72–78.

Wang, J. et al. (1990a). Hepatitis B virus integration in a cyclin A gene in a hepatocellular carcinoma. Nature (London) *343*:555–557.

Wang, J. H. et al. (1990b). Nuclear DNA content of altered hepatic foci in a rat liver carcinogenesis model. Cancer Res. *50*:7571–7576.

Ward, C. L. et al. (1991). Cystic fibrosis gene expression is not correlated with rectifying Cl⁻ channels. Proc. Natl. Acad. Sci. U.S.A. *88*:5277–5281.

Wardell, M. R. et al. (1982). Genetic variation in human apolipoprotein E. J. Lipid Res. *23*: 1174–1182.

Wardell, M. R. et al. (1987). Apolipoprotein E2-Christchurch (136 Arg-Ser). New variant of human apolipoprotein E in a patient with type III hyperlipoproteinemia. J. Clin. Invest. *80*:483–490.

Warshaw, A. L. and Fernandez-del Castillo, C. (1992). Pancreatic carcinoma. N. Engl. J. Med. *326*:455–465.

Watatani, M. et al. (1989). Infrequent activation of K-*ras*, H-*ras*, and other oncogenes in hepatocellular neoplasms initiated by methyl (acetoxymethyl) nitrosamine, a methylating agent, and promoted by phenobarbital in F344 rats. Cancer Res. *49*:1103–1109.

Weger, A. R. et al. (1989). Methodological aspects of DNA image cytometry in formalin-fixed paraffin-embedded material from pancreatic adenocarcinoma. Pathol. Res. Pract. *185*:752–754.

Wegner, C. D. et al. (1990). Intercellular adhesion molecule-1 (ICAM-1) in the pathogenesis of asthma. Science *247*:456–459.

Weinstein, I. B. (1989). The origins of human cancer: molecular mechanisms of carcinogenesis and their implications for cancer prevention and treatment — 27th G. H. A. Clowes Memorial Award Lecture. Cancer Res. *48*:4135–4143.

Weisgraber, K. H. and Rall, S. C., Jr. (1987). Human apolipoprotein B-100 heparin-binding sites. J. Biol. Chem. *262*:11097–11103.

Weisgraber, K. H. et al. (1981). Human E apoprotein heterogeneity. Cysteine-arginine interchanges in amino acid sequence of the apo E isoforms. J. Biol. Chem. *256*:9077–9083.

Weisgraber, K. H. et al. (1983). The receptor binding domain of human apolipoprotein E: monoclonal antibody inhibition of binding. J. Biol. Chem. *258*:12348–12354.

Weisgraber, K. H. et al. (1988). Apolipoprotein E genotyping using the polymerase chain reaction and allele-specific oligonucleotide probes. Biochem. Biophys. Res. Commun. *157*:1212–1217.

Weiss, H. et al. (1985). Characterization of human adenomatous polyps of the colorectal bowel by means of DNA distribution patterns. Oncology *42*:33–41.

Welch, J. P. and Welch, C. E. (1976). Villous adenomas of the colorectum. Am. J. Surg. *131*:185–191.

Welsh, M. J. (1990). Abnormal regulation of ion channels in cystic fibrosis epithelia. FASEB J. *4*:2718–2725.

Welsh, M. et al. (1988). Stimulation of pancreatic islet cell β-cell replication by oncogenes. Proc. Natl. Acad. Sci. U.S.A. *85*:116–120.

Wenham, P. R. et al. (1991a). Analysis of apolipoprotein E genotypes by the amplification refractory mutation system. Clin. Chem *37*:241–244.

Wenham, P. R. et al. (1991b). Apolipoprotein E genotyping by one-stage PCR. Lancet *337*:1158–1159.

Wenham, P. R. et al. (1991c). Rapid diagnosis of familial defective apolipoprotein B-100 by amplification refractory mutation system. Clin. Chem. *37*:1983–1987.

Wersto, R. P. et al. (1991). Variability in DNA measurements in multiple tumor samples of human colonic carcinoma. Cancer *67*:106–115.

Weston, A. et al. (1989). Differential DNA sequence deletions from chromosomes 3, 11, 13 and 17 in squamous cell carcinoma, large cell carcinoma and adenocarcinoma of the human lung. Proc. Natl. Acad. Sci. U.S.A. *86*:5099–5103.

Whang-Peng, J. (1989). 3p deletion in small cell lung carcinoma. Mayo Clin. Proc. *64*:256–260.

Whang-Peng, J. et al. (1982a). A non-random chromosomal abnormality, del 3p (14–23) in human small cell lung cancer (SCLC). Cancer Genet. Cytogenet. *6*:119–134.

Whang-Peng, J. et al. (1982b). Specific chromosome defect associated with human small cell lung cancer. Deletion 3p(14–23). Science *215*:181–182.

White, G. R. M. et al. (1990). Constitutional frequencies of c-Ha-*ras* alleles in patients with different types of lung cancer. Br. J. Cancer *61*:186.

White, M. B. et al. (1990). A frame-shift mutation in the cystic fibrosis gene. Nature (London) *344*:665–667.

White, R. et al. (1985). A closely linked genetic marker for cystic fibrosis. Nature (London) *318*:382–384.

White, Y. S. et al. (1990). Frequency of hepatic HBV DNA in patients with cirrhosis and hepatocellular carcinoma: relation to serum HBV markers. Br. J. Cancer *61*:909–912.

Whyte, P. et al. (1988). Association between an oncogene and an antioncogene: the adenovirus E1A proteins bind to the retinoblastoma gene product. Nature (London) *334*:124–129.

Wido, T. M. et al. (1990). Aneuploid DNA content in pancreatic adenocarcinoma. Int. J. Pancreatol. 129–134.

Wiklund, O. et al. (1989). *Xba*I restriction fragment length polymorphism of apolipoprotein B in Swedish myocardial infarction patients. Eur. J. Clin. Invest. *19*:255–258.

Wiklund, O. et al. (1990). Apolipoprotein (a) and ischemic heart disease in familial hypercholesterolemia. Lancet *335*:1360–1363.

Will, H. et al. (1986). Putative reverse transcriptase intermediates of human hepatitis B virus in primary liver carcinomas. Science *231*:594–596.

Williams, C. et al. (1988). Same day, first-trimester antenatal diagnosis for cystic fibrosis by gene amplificiation. Lancet *ii*:102–103.

Williams, T. M. et al. (1991). Expression of c-*erb*B-2 in human pancreatic adenocarcinomas. Pathobiology *59*:46–52.

Wiseman, R. W. et al. (1986). Activating mutations of the c-Ha-*ras* protooncogenes in chemically induced hepatomas of the male B6C3 F1 mouse. Proc. Natl. Acad. Sci. U.S.A. *83*:5825–5829.

Wissler, R. W. and Vesselinovitch, D. (1990). Can atherosclerotic plaques regress? Anatomic and biochemical evidence from nonhuman animal models. Am. J. Cardiol. *65*:33F-40F.

Witt, M. et al. (1991). Correlation of phenotypic and genetic heterogeneity in cystic fibrosis: variability in sweat electrolyte levels contributes to heterogeneity and is increased with the XV-2c/KM 19 B haplotype. Am. J. Med. Genet. *39*:137–143.

Wojciechowski, A. P. et al. (1991). Familial combined hyperlipidemia linked to the apolipoprotein AI-CIII-AIV gene cluster on chromosome 11q23-q24. Nature (London) *349*:161–164.

Wolley, R. C. et al. (1982). DNA distribution in human colon carcinomas and its relationship to clinical behavior. J. Natl. Cancer Inst. *69*:15–22.

Wong, J. M. et al. (1990). Ubiquitin hybrid protein gene is an early response gene and overexpressed in human colon carcinomas. Proc. Am. Assn. Cancer Res. *31*:306A.

Wood, R. E. et al. (1976). Cystic fibrosis. Ann. Rev. Respir. Dis. *113*:833–878.

Wulbrand, U. et al. (1991). Genetic determinants in cystic fibrosis. Lancet *337*:623.

Xu, C. F. et al. (1990). Variation at the apolipoprotein (apo) AI-CIII-AIV gene cluster and apo B gene loci is associated with lipoprotein and apolipoprotein levels in Italian children. Am. J. Hum. Genet. *47*:429–439.

Yaginuma, K. et al. (1985). Hepatitis B virus integration in hepatocellular carcinoma DNA: duplication of cellular flanking sequences at the integration site. Proc. Natl. Acad. Sci. U.S.A. *82*:4458–4462.

Yamaguchi, A. et al. (1990). Flow cytometric analysis of colorectal cancer with hepatic metastases and its relationship to metastatic characteristics and prognosis. Oncology *47*:478–482.

Yamamoto, T. et al. (1984). The human LDL receptor: a cysteine-rich protein with multiple *Alu* sequences in its mRNA. Cell (Cambridge, Mass) *39*:27–38.

Yamamoto, T. et al. (1986). Similarity of protein encoded by the human c-*erb*B-2 gene to epidermal growth factor receptor. Nature (London) *319*:230–234.

Yamamoto, T. et al. (1987). Immunohistochemical detection of c-*myc* oncogene product in human gastric carcinomas: expression in tumor cells and stromal cells. Jpn. J. Cancer Res. (Gann.) *78*:1169–1174.

Yamamoto, T. et al. (1988). Interaction between transforming growth factor-α and c-Ha-*ras* p21 in progression of human gastric carcinoma. Pathol. Res. Pract. *183*:663–669.

Yamamura, T. et al. (1984a). New mutants of apolipoprotein E associated with atherosclerotic diseases but not to type III hyperlipoproteinemia. J. Clin. Invest. *74*: 1229–1237.

Yamamura, T. et al. (1984b). A new isoform of apolipoprotein E — E5 — associated with hyperlipidemia and atherosclerosis. Atherosclerosis *50*:159–172.

Yanagihara, K. et al. (1991). Establishment and characterization of human signet ring cell gastric carcinoma cell lines with amplification of the c-*myc* oncogene. Cancer Res. *51*:381–386.

Yasui, W. et al. (1987). Expression of Ha-*ras* oncogene product in rat gastrointestinal carcinomas induced by chemical carcinogens. Acta Pathol. Jpn. *37*:1731–1741.

Yasui, W. et al. (1988). Interaction between epidermal growth factor and its receptor in progression of human gastric carcinoma. Int. J. Cancer *41*:211–217.

Yin, X. Y. et al. (1990). Gene dosage in head and neck squamous cell cancers. Proc. Am. Assn. Cancer Res. *31*:306A.

Yoffe, B. et al. (1986). Hepatitis B virus DNA in mononuclear cells and analysis of cell subsets for the presence of replicative intermediates of viral DNA. J. Infect. Dis. *153*:471–477.

Yoffe, B. et al. (1988). Demonstration of HBV DNA in brain in patients with AIDS. Hepatology *8*:1323A.

Yoffe, B. et al. (1990). Extrahepatic hepatitis B virus DNA sequences in patients with acute hepatitis B infection. Hepatology *12*:187–192.

Yokota, J. et al. (1987). Loss of heterozygosity on chromosomes 3, 13, and 17 in small-cell carcinoma and on chromosome 3 in adenocarcinoma of the lung. Proc. Natl. Acad. Sci. U.S.A. *84*:9252–9256.

Yokota, J. et al. (1988a). Altered expression of the retinoblastoma (RB) gene in small-cell carcinoma of the lung. Oncogene *3*:471–475.

Yokota, J. et al. (1988b). Genetic alterations of the c-*erb*B-2 oncogene occur frequently in tubular adenocarcinoma of the stomach and are often accompanied by amplification of the v-*erb*A homologue. Oncogene *2*:283–287.

Yokozaki, H. et al. (1988). Estrogen receptors in gastric adenocarcinoma: a retrospective immunohistochemical analysis. Virchow's Arch. A: Pathol. Anat. *413*:297–302.

Yonemura, Y. et al. (1988). Correlation of DNA ploidy and proliferative activity in human gastric cancer. Cancer *62*:1497–1502.

Yonemura, Y. et al. (1991). Evaluation of immunoreactivity for erbB-2 protein as a marker of poor short term prognosis in gastric cancer. Cancer Res. *51*:1034–1038.

Yoshida, K. et al. (1989). Amplification of epidermal growth factor receptor (EGFR) gene and oncogenes in human gastric carcinomas. Virchow's Arch. B: Cell Pathol. *57*:285–290.

Yoshida, K. et al. (1990). Expression of epidermal growth factor, transforming growth factor-α and their receptor genes in human gastric carcinomas; implication for autocrine growth. Jpn. J. Cancer Res. *81*:43–51.

Yoshida, M. C. et al. (1988). Human HST1 (HSTF1) gene maps to chromosome band 11q13 and coamplifies with INT2 gene in human cancer. Proc. Natl. Acad. Sci. U.S.A. *85*:4861–4864.

Yoshida, T. et al. (1987). Genomic sequences of *hst*, a transforming gene encoding a protein homologous to fibroblast growth factors and the int-2 encoded protein. Proc. Natl. Acad. Sci. U.S.A. *84*:7305–7309.

Yoshida, T. et al. (1989). Amplified gene in gastric cancer, *sam*, codes. Proc. Am. Assn. Cancer Res. *30*:783A.

Young, S. G. et al. (1988). Low plasma cholesterol levels caused by a short deleltion in the apolipoprotein B gene. Science *241*:591–593.

Young, S. G. et al. (1990). Familial hypobetalipoproteinemia caused by a mutation in the apolipoprotein B gene that results in a truncated species of apolipoprotein B (B-31). A unique mutation that helps to define the portion of the apolipoprotein B molecule required for the formation of buoyant, triglyceride-rich lipoproteins. J. Clin. Invest. *85*:933–942.

Yousem, S. A. and Taylor, S. R. (1990). Typical and atypical carcinoid tumors of lung: a clinicopathological and DNA analysis of 20 tumors. Mod. Pathol. 3:502–507.

Zahm, P. et al. (1988). The HBV X-orf encodes a transactivator: a potential factor in viral hepatocarcinogensis. Oncogene *3*:169–177.

Zaloudik, J. et al. (1988). The DNA content of colorectal carcinomas: an analysis of the heterogeneity of aneuploidy and correlation with immunopathological parameters. Neoplasma *35*:389–401.

Zannis, V. I. (1986). Genetic polymorphism in human apolipoprotein E. In: Methods in Enzymology, Vol. 128. Segrest, J. P. and Albers, J. J. (eds.). Academic Press, Orlando, pp. 823–851.

Zannis, V. I. and Breslow, J. L. (1981). Human very-low-density lipoprotein apolipoprotein E isoprotein polymorphism is explained by genetic variation and posttranslational modification. Biochemistry *20*:1033–1041.

Zarod, A. P. et al. (1988). Malignant progression of laryngeal papilloma associated with human papillomavirus type 6 (HPV-6) DNA. J. Clin. Pathol. *41*:280–283.

Zengerling, S. et al. (1987). Mapping of DNA markers linked to the cystic fibrosis locus on the long arm of chromosome 7. Am. J. Hum. Genet. *40*:228–236.

Zhang, W. et al. (1990). Frequent loss of heterozygosity on chromosomes 16 and 4 in human hepatocellular carcinoma. Jpn. J. Cancer Res. *81*:108–111.

Zhang, X. K. et al. (1987). The expression of oncogenes in human developing liver and hepatomas. Biochem. Biophys. Res. Commun. *142*:932–938.

Zhang, X. et al. (1989). Amplification and rearrangement of c-*erb*-B protooncogenes in cancer of human female genital tract. Oncogene *4*:985–989.

Zielenski, J. et al. (1991a). Identification of mutations in exons 1 through 8 of the cystic fibrosis transmembrane conductance regulator (CFTR) gene. Genomics *10*:229–235.

Zielenski, J. et al. (1991b). Genomic DNA sequence of the cystic fibrosis transmembrane conductance regulator (CFTR) gene. Genomics *10*:214–228.

Ziemer, M. et al. (1985). Sequence of hepatitis B virus DNA incorporated into the genome of a human hepatoma cell line. J. Virol. *53*:885–892.

Zimmerman, P. V. et al. (1987). Ploidy as a prognostic determinant in surgically treated lung cancer. Lancet *5*:530–533.

Zimmerman, W. et al. (1987). Isolation and characterization of cDNA clones encoding the human carcinoembryonic antigen reveal a highly conserved repeating structure. Proc. Natl. Acad. Sci. U.S.A. *84*:2960–2964.

Zimmerman, W. et al. (1988). Chromosomal localization of the carcinoembryonic antigen gene family and differential expression in various tumors. Cancer Res. *48*:2443–2450.

Part 3
Gene Level Evaluation of Selected Breast, Gynecologic, and Urologic Tumors

Chapter 3.1

Breast

3.1.1 INTRODUCTION

Breast cancer, one of the most frequent malignancies, is a heterogeneous disease with a variable and frequently unpredictable clinical course affecting 1 in 12 women in the Western world (Millis and Girling, 1989). The biology of various forms of breast cancer is poorly understood at present. Coordinate evaluation of the clinical and mammographic findings, histologic patterns, ploidy data, as well as biochemical and immunocytochemical analyses of the receptor status presents a major challenge. Efforts to create a systematic approach to **management of patients** with breast cancer have included the development of quality assurance guidelines for specifying the type of information to be included in the pathologist's report, and the format of its presentation (Hutter, 1990; National Cancer Institute, 1988). Assessment of appropriate strategy in patients with breast cancer has been the subject of numerous clinical trials. These were aimed at selecting the most informative diagnostic criteria for the prediction of a clinical course in patients with different forms of localized or disseminated disease.

TNM clinical and pathological staging (**T**, the extent of the primary tumor; **N**, absence or presence and extent of regional lymph node metastases; **M**, the absence or presence of distant metastases) provides a reference framework for assessment of cases, allowing uniform management strategies to be developed and compared on an intra- and intercase basis. The **new TNM** staging system of the American Joint Committee on Cancer/Union Internationale Contre le Cancer (AJCC/UICC) is given in Table 3.1.

A review of relative contributions of **fine needle aspirate (FNA) biopsy, frozen section** evaluation, **microscopic assessment** of histological characteristics in breast cancer management has been presented (Hutter, 1990). Morphological analysis of breast tumors follows the currently accepted World Health Organization **(WHO) histological classification** (Table 3.2).

This discussion will summarize developments in the chromosomal and molecular biological studies of breast cancer and their relevance to breast cancer patient management (for additional reviews, see also Adami et al., 1990; Donovan-Peluso et al., 1991; Koster et al., 1991; van de Vijver and Nusse, 1991).

3.1.2 PLOIDY ANALYSIS

The decision on the extent of appropriate **surgical intervention** as well as on the choice and administration of **systemic adjuvant therapy** in the treatmnent of breast cancer relies on the predictive clinical behavior of a given form of cancer and the evaluation of its histopathological parameters. A particular challenge is presented by node-negative disease. The **proliferative activity** of tumor cells (estimated by the labeling of replicating cells with thymidine) (Hery et al., 1987; Meyer and Province, 1988; Silvestrini et al., 1986; Tubiana et al., 1984, 1989) and **DNA content** of individual cells determined by **static cytomorphometry** can complement and correlate with a histopathological evaluation of the tumor by microscopy. Evaluation of cycling cells by the **Ki67 monoclonal antibody** and by flow cytometry-derived **S-phase fraction (SPF)** showed significant correlations with **nuclear grading** and **mitotic index**, the two methods being

TABLE 3.1
Guidelines for Breast Specimen Evaluation

Breast Task Force-CAP/ACR Patterns of Care Study

I. Type of specimen received
    A. Aspiration: smear(s)/biopsy
    B. Trochar biopsy
    C. Incisional biopsy
    D. Excisional biopsy
    E. Mastectomy (specify type as labeled in operating room)
II. Gross description (note tissue included and indicate whether or not completely present or undetermined)
    A. Laterality and type of specimen
        1. Skin
        2. Nipple
        3. Breast
        4. Pectoralis major
        5. Pectoralis minor
        6. Axillary contents
        7. Other
    B. Tumor(s)
        1. Size (three largest dimensions, in cm)
        2. Location
            a. Quadrant(s)
            b. Depth (from skin and distance from pectoralis fascia)
            c. Pattern of invasion should be noted (e.g., circumscribed or stellate)
            d. Operating room consultation report (frozen section)
            e. Tissue taken for estrogen and progesterone binding or other markers
    C. Presence or absence of gross tumor in margin of resection
III. Axillary lymph nodes
    A. Location (level if specified)
    B. Total number
    C. Number with metastases
    D. When possible, indicate highest lymph node present in dissection and whether or not free of
        tumor
    E. Statement in reference to capsular violation
IV. Histologic classification
    A. Use the revised WHO classification of breast tumors
    B. Classify by dominant pattern
V. Additional histologic factors
    A. Histologic grade with degree of differentiation
        1. Grade 1 — Well differentiated
        2. Grade 2 — Moderately differentiated
        3. Grade 3 — Poorly differentiated
        4. Grade 4 — Undifferentiated
    B. Presence or absence of microscopic tumor in margin of resection
    C. Involvement of adjacent structures (e.g., skin, nipple, pectoralis muscle, chest wall)
    D. Character of breast tissue remote from primary cancer (e.g., atrophic, fibrocystic changes, *in situ*
        cancer, other invasive cancer)

From Hutter, R. V. P. (1990). Cancer 66:1363–1372. With permission.

complementary in the information they provide, which may be further correlated with prognosis and stage of the disease (Vielh et al., 1991).

Reevaluation of the **thymidine-labeling index** as a prognostic indicator in a series of 190 patients with primary breast cancer over a period of 6 years **failed to confirm the validity of this technique for prognosis** (Winstanley et al., 1990). On the other hand, **nuclear DNA content** of tumor cells **significantly correlated with prognosis** in 409 breast adenocarcinomas followed for up to 13 years (Fallenius et al., 1988a). Nuclear DNA content estimated by static cytophotometry, however, did not correlate with the axillary node status, but strongly correlated

TABLE 3.2
WHO Histologic Classification of Breast Tumors

Epithelial tumors	ICD-B/SNOMED fields
A. Benign	
1. Intraductal papilloma	8503/0
2. Adenoma of the nipple	8506/0
3. Adenoma	8140/0
a. Tubular	8211/0
b. Lactating[a]	
4. Others	
B. Malignant	
1. Noninvasive	
a. Intraductal carcinoma	8500/2
b. Lobular carcinoma *in situ*	8520/2
2. Invasive	
a. Invasive ductal carcinoma	8500/3
b. Invasive ductal carcinoma with a predominant intraductal component	8500/3
c. Invasive lobular carcinoma with a predominant *in situ* component	8520/3
d. Mucinous carcinoma	8480/3
e. Medullary carcinoma	8510/3
f. Papillary carcinoma	8503/3
g. Tubular carcinoma	8211/3
h. Adenoid cystic carcinoma	8200/3
i. Secretory (juvenile) carcinoma	8502/3
j. Apocrine carcinoma	8401/3
k. Carcinoma with metaplasia	
i. Squamous type	8570/3
ii. Spindle-cell type	8572/3
iii. Cartilaginous and osseous type	8571/3
iv. Mixed type	
l. Others	
3. Paget's disease of the nipple	8540/3[b]

[a] No specific code is available for lactating adenoma.

[b] Paget's disease and invasive ductal carcinoma in coded 8541/3.

From Hutter, R. V. P. (1990). Cancer 66:1363–1372. With permission.

with the histopathological grading of ductal carcinomas (Fallenius et al., 1988b). Additional prognostic value of the nuclear DNA content was observed in both node-negative and node-positive patient groups serving as an objective biological **marker of aggressiveness.**

Although rigorous **standardization of technical aspects** of tumor tissue evaluation presents a problem, and affects the results of ploidy evaluation by static cytometry and flow cytometry (FCM) (Auer et al., 1987; Bocking et al., 1989a,b; Fallenius et al., 1987), a number of earlier studies emphasized the value of the analysis of DNA ploidy as a useful adjunct in the assessment of prognosis for breast cancer patients (Auer et al., 1984; Berryman et al., 1987; Cornelisse et al., 1987; Hedley et al., 1985; Kallioniemi et al., 1987; McGuire et al., 1985; McGuire and Dressler, 1985).

The **predictive value of aneuploidy of primary tumors** decreases in more advanced disease (Ewers et al., 1984). Tetraploid and near-tetraploid tumors appear to comprise a distinct group of endocrine-responsive breast tumors within the group of aneuploid tumors. Subdivision of DNA content data into diploid and aneuploid patterns may be masking important information relevant to therapy (Baildam et al., 1987a,b). A significantly better prognostic value of the combined evaluation of **ploidy** and **SPF** has been amply demonstrated (e.g., Clark et al., 1989;

Clark and McGuire, 1988; Dressler et al., 1987, 1988; Fallenius et al., 1987; Kallioniemi et al., 1988a; McGuire, 1987; Moran et al., 1984).

Ploidy, however, proved not to have significant influence on **long-term survival** (minimum follow-up of 22 years) in 115 patients with postsurgical stage $T_{1-2}N_0$ breast cancer (Toikkanen et al., 1990). The SPF had an independent prognostic value (Toikkanen et al., 1989) and although patients with a **DNA index (DI)** of less than 1.2 had a more favorable prognosis than those with a DI above 1.2, the most important independent prognostic variable for long-term survival was **lymphatic vessel invasion** of cancer cells (Toikkanen et al., 1990).

Mitotic activity of tumor cells in invasive ductal breast cancer is capable of predicting favorable and unfavorable prognosis among patients with diploid and tetraploid, but not aneuploid, tumors (Uyterlinde et al., 1988). When node-negative breast cancer patients are evaluated following surgical resection of such cancers, the most valuable prognostic information is afforded by an estimate of cells in S phase (Sigurdsson et al., 1990a). This parameter was able to identify patients in whom the risk of recurrence was sufficiently low so that adjuvant chemotherapy could be avoided. Although aneuploidy correlates with **invasiveness** of ductal breast carcinomas (Aasmundstad and Haugen, 1990), ploidy does not predict survival in this group of patients (Aasmundstad and Haugen, 1990; Blanco et al., 1990; Isola et al., 1990; O'Reilley et al., 1990a,b; Winchester et al., 1990). High SPF, on the other hand, was significantly associated with early relapse (Winchester et al., 1990).

In the case of **cystosarcoma phyllodes** of the breast, the DNA ploidy evaluation appears to offer a useful, objective adjunct to clinicopathological assessment of the neoplasm (El-Naggar et al., 1990b). Efforts to enhance the informative value of ploidy and SPF estimates have been made by **two-color multiparametric FCM analysis**, in which the cells are labeled by **cytokeratin monoclonal antibody** and **propidium iodide** (Visscher et al., 1990). Cytokeratin-gated DNA evaluation allows exclusion of contaminating stromal and inflammatory cells, and markedly improves the degree of correlation with the clinicopathological factors of known prognostic significance. Even in this format, SPF appears to be of greater prognostic value than ploidy. A **combination** of DI and SPF afforded the distinction of three types of DNA histograms (Kallioniemi et al., 1988a). DNA diploidy with low (<7%) SPF was associated with very favorable prognosis (type I DNA histogram). On the other hand, DNA aneuploidy with high DI (>2.20) or high SPF (>12%) (type III DNA histogram) increased the relative risk of death eightfold, and was associated with the worst prognosis. Further enhancement of the prognostic value of SPF may come from subdividing the SPF into **three prognostic categories** (<7, 7 to 11.9, and >12%) (Sigurdsson et al., 1990b). It should be emphasized that the predictive power of aneuploidy and SPF estimates may be significantly influenced by the very treatment the patients receive (Kute et al., 1990).

Increasingly, FNA combined with FCM is used for the evaluation of breast tumors and also for DNA content assessment, particularly in the course of treatment (Briffod et al., 1989; Mullen and Miller, 1990a; Remvikos et al., 1988, 1989; Spyratos et al., 1987). Substantial variations in cell cycle distribution and DI were observed in the same tumors, particularly aneuploid tumors, and in sequential FNAs in the course of treatment; similar variations were noted without treatment, or in clinical failure to treatment (Mullen and Miller, 1990b). A mathematical method of multisurface pattern separation has been successfully applied to the discriminatory analysis of breast FNA samples to distinguish between benign and malignant specimens (Wolberg and Mangasarian, 1990).

The **extreme DNA heterogeneity** of breast cancer is emphasized in an FCM study of primary and metastatic lesions (Beerman et al., 1991). This heterogeneity is thought to account for the disagreement in the findings on the prognostic value of DNA ploidy determinations, although some earlier studies refuted intratumoral heterogeneity as contributing to the discrepant observations made by different laboratories (Askensten et al., 1989). No less than four samples are now recommended for the reliable determination of the DNA ploidy status of primary breast tumors by FCM.

3.1.3 CHROMOSOMAL ABNORMALITIES IN BREAST CANCER

No single systematic study has been performed to define chromosomal alterations predisposing to the development of various forms of breast cancer (and/or benign lesions), those associated with the progression of the tumor, and, finally, those contributing to the metastatic spread. Numerous reviews of chromosomal aberrations in breast cancer have emphasized various aspects of genomic alterations associated with this pathology (e.g., Callahan et al., 1990; Callahan and Campbell, 1989; Mars and Saunders, 1990; Trent, 1985). The ultimate objective of chromosomal analysis is the identification of the gene(s) responsible for predisposition to, initiation, progression, and spread of a tumor. A combination of karyotype analysis, direct genomic DNA probing with defined sequences of known functional significance, as well as linkage analysis in family studies of breast cancer patients is needed to produce a coherent picture of an interplay of genetic factors responsible for the malignancy.

For the sake of simplicity, the genes associated with breast cancer (as well as other neoplasias) can be subdivided into those involved in the **initiation** of malignancy and those promoting **progression** of the disease (Mars and Saunders, 1990). Analysis of cytogenetic alterations in primary breast tumors by direct karyotype analysis as well as by cytogenetic analyses of breast cancer cell cultures reveals that the most frequently involved areas are **1p32-36, 1q21**, and **1p13** (Chen et al., 1989, 1991; Cruciger et al., 1976; Ferti-Passantonopoulou and Panani, 1987; Gebhart et al., 1986; Genuardi et al., 1989; Merlo et al., 1989; Mitchell and Santibanez-Koref, 1990; Rodgers et al., 1984, 1985; Zhang et al., 1989). Analysis of 113 primary breast cancers showed that **1p13** was affected in 20% of the tumors (Mitchell and Santibanez-Koref, 1990). While **1p** alteration may be associated with **early stages** of breast cancer (Anderson et al., 1985; Ferrell et al., 1989), changes on **1q** can be related to **metastatic spread** of the disease (Cruciger et al., 1976; Bello and Rey, 1989; Zhang et al., 1989).

Other chromosomal abnormalities in breast carcinomas include deletions of **17p** and **3p** (Ali et al., 1989; Devilee et al., 1989; Mackay et al., 1988a,b; Mitchell and Santibanez-Koref, 1990; Nigro et al., 1989; Zhang et al., 1989), and abnormalities on **chromosomes 6, 8, 11**, and **13** (Adnane et al., 1989; Ali et al., 1987, 1989; Devilee et al., 1989; Ferti-Passantonopoulou and Panani, 1987; Garcia et al., 1989; Guerin et al., 1988; Hill et al., 1987; C. Lundberg et al., 1987; Mackay et al., 1988a,b; Rodgers et al., 1984, 1985; T'Ang et al., 1988; Theillet et al., 1986; Varley et al., 1989). The frequent LOH on **17p** has been defined in the p13-3 region and associated with overexpression of *p53* mRNA pointing to a gene some 20 megabases telomeric of *p53* that may regulate its expression in the majority of breast cancers (Coles et al., 1990) (see below). Cytogenetic changes on **11p13-15** detected in pleural effusions suggested an **association with progression** of breast carcinoma (Bello and Rey, 1989). On the other hand, changes on **chromosomes 1** and **3** appear to be associated with an **early event** (Dutrillaux et al., 1990; Mars and Saunders, 1990). Other workers, however, believe that chromosome 1p36 abnormalities may be related to later stages of tumor progression, whereas deletion or rearrangements noted on **chromosome 16q22** may be important in the **early stages** of carcinogenesis (Hainsworth et al., 1990a). In new breast tumor cell lines established from pleural effusions of patients with confirmed breast cancer no common marker chromosome could be found (Sasi et al., 1991a). Several breakpoints involving, for example, **1q11, 3q11, 7p11, 9q11**, and **13q11** could frequently be detected. Oncogene probes revealed amplification of *neu* and c-*myc*.

Linkage analysis of pedigrees with early onset breast cancer using the markers **GH, CMM86, nm23, 42D6**, and **D17S250** places the responsible locus on **chromosome 17q** somewhere closer to D17S250 (Bishop, 1992, personal communication). Other researchers have linked familial breast cancer to the D17S74 locus on chromosome 17q (Narod et al., 1991). For the largest family with a hereditary predisposition to cancer of the breast and ovary, the lod score was 2.72 for the chromosomal region 17q12-q23 previously shown to harbor a gene for early-onset breast cancer (Hall et al., 1990). Up to 60 other polymorphic DNA markers scattered throughout the genome have been tested for linkage to breast-ovarian cancer syndrome and all failed to provide informative associations. A large statistical genetic analysis conducted on 4700 breast cancer patients and 4688 controls suggests the effect of genotype on the risk of breast cancer to be related to age and to be the greatest at young age (Claus et al., 1991).

3.1.4 PROTOONCOGENES AND TUMOR SUPPRESSOR GENES IN BREAST CANCER

Studies on the development of mammary glands in laboratory animals using the **mouse mammary tumor virus** (**MMTV**) demonstrated the activation by insertional mutagenesis of the expression of *int*-1, *int*-2, *int*-3, *int*-4, and *hst* cellular genes, which prior to infection remained silent (Callahan and Campbell, 1989). The relatedness of these genes to the **fibroblast growth factor** (FGF) family of genes suggests that abnormal expression of these genes may delay **senescence** of mammary epithelium or provide a stimulus for **angiogenesis** in the tumor.

Other genes (c-*myc* and Ha-*ras*), inserted in transgenic mice, do not interfere with normal mammary gland development, whereas in the presence of mutated, but not normal, *neu* oncogene, normal mammary gland development does not occur (Callahan and Campbell, 1989). Abnormal expression of these genes has been observed in primary and metastatic human breast carcinomas (see below). The chromosomal studies discussed above suggested that some of the well-known suppressor genes (*RB*-1, *p53*, *DCC*) may also be involved. In this section, a discussion will be given of pertinent findings that can potentially be used for clinical purposes.

3.1.4.1 *erb*B-2/*HER*-2/*neu*

3.1.4.1.1 Amplification of the c-*erb*B-2/*HER*-2/*neu* Gene

The c-*erb*B-2 protooncogene encodes a membrane receptor with protein kinase activity and, although related to **epidermal growth factor receptor** (EGFR), its amplification in human breast tumors can be identified and distinguished from the complex pattern of *EGFR* (King et al., 1985). Another member of the *EGFR* family, *erb*B-3, has been identified, mapped to **chromosome 12q13**, and found to be overexpressed in a subset of human mammary tumors (Kraus et al., 1989).

The c-*erb*B-2 protooncogene, also termed *neu* (because it had been known to be activated in neuro- and glioblastomas), has 50% of its amino acid sequence identical to that of the EGFR (Bargmann et al., 1986). Over 80% of the amino acids in the tyrosine kinase domain are identical, further enhancing the role of the *neu* gene product as a growth factor. A comparative analysis of the *neu* and *EGFR* gene products has been given (Maihle and Kung, 1988).

A point mutation in the *neu* protooncogene converts it into an oncogene and the ***neu* gene product**, **p185**^{neu}, is oncogenic, whereas the product of c-*neu*, p185^{c-neu}, is not (Weiner et al., 1989). An amino acid substitution in the oncogenic form of the product is associated with an aggregated form of the glycoprotein. The induction of an aggregated form may mimic the effects of ligand-induced receptor aggregation resulting in enzymatic activation that leads to cellular transformation (Weiner et al., 1989). In addition to point mutations in the kinase domain, which alone can impart sarcomagenic potential to the *neu* product, other modifications leading to activation of the transforming potential include carboxyl-terminal truncation and internal deletion (Pelley et al., 1989).

A study by Slamon and coworkers (1987) reported a 2- to over 20-fold amplification of the *neu* gene in 30% of human breast cancers, which had greater prognostic value than other prognostic factors including hormonal status. This observation has been made in lymph node-positive cases. Since that publication a large number of reports have appeared either supporting and extending this observation to node-negative cases, or refuting the prognostic value of *neu* amplification entirely, or admitting its usefulness with certain limitations.

Following the earlier observation on association of the c-*erb*B-2 amplification with advanced stages of breast cancer and metastasis to regional lymph nodes (Zhou et al., 1987), subsequent studies confirmed this correlation (Slamon et al., 1989; Tavassoli et al., 1989). Coamplification of c-*erb*B-2 (*neu*) and c-*erb*A oncogenes in breast cancers has been considered to be either fortuitous (van de Vijver et al., 1987), or to be indicative of lymph node metastasis (Tavassoli et al., 1989). In a retrospective analysis of 176 cases, the importance of the extent of *neu* amplification was found to be of inferior magnitude than that of the histological grade (Tsuda et al., 1990a,b). Both parameters, however, were strongly associated with the **aggressiveness** of the tumor, rather than its **metastatic spread**. The *neu* gene amplification has been more frequently found in aneuploid tumors, and in those with poor nuclear grade in node-negative tumors (Bacus et al., 1990; Ro et al., 1989).

Amplification of the *neu* gene is thought to be an **early event** in breast carcinogenesis (Barnes et al., 1988; Iglehart et al., 1990). Moreover, because *neu* alterations are present in all clinical stages, maintained during metastatic spread, present *in situ* as well as in the infiltrating component, and are homogeneously distributed throughout the tumor tissue, these alterations are considered to be important in the **initiation of breast cancer** (Iglehart et al., 1990). Amplification of the *neu* oncogene (usually in 20 to 30% of cases) can be accompanied by **rearrangements** or **translocations** (7%) of the gene (Borgen et al., 1990). The high proportion of cases with **amplified *neu*** (88%) are node-positive, although the average size of the tumor in such cases is significantly smaller than in nonamplified tumors (Borgen et al., 1990). This finding is interpreted as indicative of an association between *neu* amplification and early metastatic spread of the tumor.

In *neu*-negative patients, aneuploidy is associated with a significantly **shorter relapse-free survival** (Yuan et al., 1990). This observation can be used to predict the aggressiveness of breast tumors in patients with node-negative or *neu*-negative tumors. In lymph node-positive patients DNA ploidy did not influence either survival or relapse-free survival. Amplification of the *neu* oncogene observed in tumors with aneuploidy, poor nuclear grade (Ro et al., 1989), and early metastatic spread (Borgen et al., 1990) suggests and, in fact, has been documented to predict **poor prognosis of node-positive patients** (Borg et al., 1990; Cline et al., 1987; Slamon et al., 1987; Varley et al., 1987a; Wiltschke et al., 1990). In a large series of node-negative cases (704 patients) followed for up to 16 years, 18% of the node-negative patients with amplified *neu* gene had relapsed compared to only 5% of the patients who had remained in remission (Paterson et al., 1991). The occurrence of *neu* **amplification proved to be a significant predictor of a short disease-free interval and of poor prognosis**. It is strongly suggested that amplification of *neu* may contribute to the pathogenesis of some forms of node-negative breast cancer, and can be used to identify a subset of high-risk patients. Other studies offer conflicting observations on the importance on *neu* amplification.

That *neu* amplification can at least identify a subset of node-negative patients at higher risk for poor clinical outcome is acknowledged by investigators who tend to deemphasize the prognostic value of the *neu* amplification (McGuire, 1990). However, analysis of *neu* **alone is not considered to be of value**, because no significant correlation could be observed between its amplification and recurrence of the 157 primary and 14 metastatic breast cancers (Zhou et al., 1989).

More frequent early recurrence of tumors with detectable **amplification of any of the five protooncogenes** (c-Ki-*ras*, c-*myc*, c-*myb*, *neu*, *int*-2) was not considered to offer additional prognostic information, because all of these were stage III tumors that are already known to have a high recurrence rate. In a series of 362 tumors from patients with primary breast cancer (both node positive and node negative) the overall *neu* amplification rate was 33% (Clark and McGuire, 1991). **Amplification of *neu* did not correlate with either disease-free or overall survival**. The absence of correlation was particularly unambiguous for patients with node-negative disease. Similar conclusions were drawn from a series of 728 primary breast tumor specimens in which the *neu* gene product was estimated (Tandon et al., 1989), a series of 189 tumors (van de Vijver et al., 1988a), and yet another of 120 specimens of benign and malignant breast lesions (Mizukami et al., 1990).

neu amplification, although present in 14% of **estrogen-positive** (ER+) patients, is twice as frequent (28%) among **estrogen-negative** (ER−) patients (Zeillinger et al., 1989). Similarly, it is more frequent in **progesterone-negative** (PR−) compared to **progesterone-positive** (PR+) patients (22 vs. 16%). One possibility accounting for this observation could be a more stringent control by the *neu* protooncogene when hormone dependency is lost. Consistent with this notion was the finding of lower ER and PR levels in patients with amplified *neu* gene than in those with a single copy of the protooncogene. The degree of *neu* gene amplification is suggested as an additional parameter to be considered in clinical and laboratory assessment of a specific subset of patients that can benefit from adjuvant therapy (Zeillinger et al., 1989).

3.1.4.1.2 Expression of the *neu* Gene

A number of studies have evaluated the **degree of expression** of the *neu* gene, its association with specific histologic characteristics of breast cancer, a correlation between **overexpression** and **amplification** of the gene, as well as its utility for prognosis. The ***neu* gene product**, a 185-kDa protein, can be readily identified in frozen sections and paraffin-embedded tissue, and routine antibody staining technique can be used to reveal overexpression of the protein (Akiyama et al., 1986; Kraus et al., 1987; van de Vijver et al., 1988a,b). A systematic methodological study has addressed the issues of specific **cross-reactivity** of the anti-neu monoclonal antibodies with EGFR, and the specificity and sensitivity of the molecular and immunological probes for neu were established in cell lines genetically engineered to overexpress human neu or EGFR (Naber et al., 1990). In that study, the *neu* DNA amplification and *neu* mRNA overexpression were measured by slot-blot analysis, combined with *neu* mRNA *in situ* hybridization (ISH), and detection of the neu protein product (p185) with TA1 monoclonal antibody by immunocytochemistry. Twenty-nine percent of breast carcinomas had amplified *neu* DNA and mRNA overexpression closely correlated with the p185 expression. Weak staining of nonmalignant epithelium for p185 was also noted.

Comparison of the *neu* copy numbers and its expression, measured by immunohistological staining and by Western blotting performed on primary breast cancer tumors, demonstrated **an association of the amplification of the *neu* protooncogene with increased levels of expression of the neu protein** (Reid et al., 1990; Venter et al., 1987). Similar findings have also been made in cultured breast carcinoma cells: *neu* amplification and the corresponding *neu* mRNA overexpression have been shown in SKBR-3 cells, a mammary carcinoma cell line (van de Vijver et al., 1987). Demonstration of overexpression of the gene accompanying its amplification is important because, for example, the *erb*A gene is amplified in breast cancer, but not expressed (van de Vijver et al., 1987). In a large series of 668 breast cancer specimens, **the *neu* DNA amplification was paralleled in all cases by overexpression of *neu* mRNA and neu protein** (Slamon et al., 1989).

However, overexpression has not always been found to be linked to amplification of the *neu* gene, although **combined amplification and overexpression of the gene are significantly associated with poor prognosis** (Guerin et al., 1988). Other researchers found the highest level of *neu* overexpression to be associated with gene amplification, and that ***neu* amplification, but not *neu* mRNA expression**, was correlated with poor prognosis and significantly shorter time to treatment failure (King et al., 1989). In spite of that, there was no association of either *neu* expression, or the gene amplification, with the clinical stage of breast carcinoma or axillary lymph node involvement.

Atypical lobular hyperplasia or **lobular carcinoma** *in situ* associated with **ductal carcinoma** *in situ* (DCIS) or atypical ductal hyperplasia **does not show any expression of the** *neu* **gene product** (Lodato et al., 1990). No staining for neu oncoprotein has been detected in **papillary**, **cribriform**, or other *in situ* carcinomas; **only tumors of solid/comedo (large cell) type showed** *neu* **expression** (Keatings et al., 1990). Although positive staining for the *neu* gene product was also observed in **benign fibroadenomas** (see also Hanna et al., 1990), it was not correlated with the expression of estrogen receptors. In malignant tumors, however, strong immunoreactivity for the *neu* gene product has been associated with lymph node metastases but not the **histological grade** of breast carcinomas (Tauchi et al., 1989), suggesting that the finding of *neu* **overexpression** can be used as an adjunct prognostic criterion for preoperative prediction of the clinical course (Tauchi et al., 1989). Amplification and/or overexpression of c-*erb*B-2 viewed as a marker of advanced malignancy suggests that specific parts of the c-*erb*B-2 gene product should be targeted for immunodiagnostic and immunotherapeutic modalities (Gullick, 1990; Maguire et al., 1990; Saga et al., 1991).

Detailed analysis of the expression of the *neu* gene product has been performed in **invasive** and **noninvasive ductal breast carcinomas**, nonneoplastic lesions of the breast, normal adult and fetal breasts, as well as kidneys, small and large intestine, skin, endocervix, endometrium, thyroid, and hepatocytes (de Potter et al., 1989). The characteristic staining pattern was detected with two types of monoclonal antibodies: one, 3B5, was raised again a synthetic peptide corresponding to a portion of the *neu* gene product, and the other, 9G6, was raised against neu protein present on intact cells transfected with human *neu* DNA. Positive, strong staining observed along the cytoplasmic membrane was specific for malignancy, and was noted in 29% of the breast invasive and *in situ* carcinomas. **Estrogen receptor-negative tumors displayed higher expression of p185**. In contrast, a diffuse intracytoplasmic granular pattern of staining was observed in all the other tissues as well as in breast tissue that did not have the strong cytoplasmic membrane staining. Western blot experiments demonstrated that intracytoplasmic staining was produced by the cytoplasmic protein with a molecular weight of 155 kDa, whereas the membrane protein was the 185-kDa neu protein. Whether the intracytoplasmic 155-kDa protein is a *neu*-like cross-reacting protein, or a related product of the *neu* gene, remains to be established (de Potter et al., 1989).

Another study reported positive staining for the neu protein in 33% of **invasive ductal carcinomas** and foci of *in situ* **ductal** and **lobular carcinoma** (Hanna et al., 1990). Amplification of the *neu* oncogene was present, however, only in 25% of cases. With the polyclonal antibody used in this study the pattern of staining of tumor cells was cytoplasmic, with plasma membrane accentuation. Focal positive staining was also noted in normal breast epithelium and benign lesions of fibrocystic disease and fibroadenoma, as well as in myoepithelial cells and smooth muscles of blood vessels. One explanation of the cytoplasmic staining may be that the antibody was produced to the tyrosine kinase domain of the Neu oncoprotein, which is usually located within the peripheral portion of the cell cytoplasm. The strongest staining was observed in all *in situ* comedo, solid, and papillary carcinomas, but not in the cribriform patterns or lobulated type (Hanna et al., 1990).

Again, **overexpression** of this oncoprotein was related to **aggressiveness of the carcinomas**, and could be detected in 17% of breast carcinomas by the affinity-purified 21N polyclonal antibody specific for the 185-kDa protein. In contrast, only relatively rare positivity was detected in other malignancies, such as lung (1%), colon (4%), and bladder (2%) (McCann et al., 1990a). In the prostate, skin, or small cell lung carcinoma (SCLC), no positive staining could be observed at all.

Interrelationship between *neu* gene expression and sex steroid hormones and antihormones has been studied in MCF-7 and T47D cells, human breast cancer cell lines (Read et al., 1990). *neu* gene expression is modulated by estrogens, but not by progestins. Although estrogens stimulated the growth of carcinoma cells, and enhanced their aggressiveness, there was no increase in the expression of *neu*. It appears, therefore, that **aggressiveness of breast cancer, correlated with high levels of *neu* mRNA and protein, is uncoupled from estrogen-stimulated proliferation** (Read et al., 1990).

Significant correlation between **overexpression of *neu* and ER status** has been confirmed in subsets of ER+ patients characterized by a high risk (ER+ [R2]) and a lower risk (ER+ [R1]) of early relapse (May et al., 1990). *neu* mRNA overexpression was a significant predictor of early relapse, as significant as ER– negativity and ER+ (R2). ER– patients with no overexpression of *neu* constitute a group with a relatively good prognosis (see also McCann et al., 1990b). Overexpression of *neu* also correlated with lymph node status and the presence of inflammatory tumors. Similar association of *neu* overexpression with ER negativity and the presence of dense lymphocytic infiltration in inflammatory breast carcinomas displaying higher malignancy was also observed in other studies (Garcia et al., 1989; Lacroix et al., 1989; Tang et al., 1990).

It would be helpful if the **degree of *neu* overexpression** could be **quantified** and used in patient management. However, using **digital image processing** for the morphometric evaluation of neu protein expression in invasive carcinomas of 82 patients, the strong overexpression of the protein was found to be "prognostically overshadowed" by the morphological features (Baak et al., 1991). Immunohistochemical staining of the cell membrane for Neu oncoprotein in breast carcinomas is inversely correlated with the ER and PR status, and does not correlate with lymph node involvement (de Potter et al., 1990). It can have value, however, as an independent risk factor in predicting patients at risk for **hematogenous spread** and as a correlate with unfavorable prognosis (de Potter et al., 1990; Nesland et al., 1991). Likewise, strong staining fully correlates with amplification of the *neu* gene, and is associated with poor overall and disease-free survival (Berger et al., 1988; McCann et al., 1990b; Tsuda et al., 1990a,b; Winstanley et al., 1991; C. Wright et al., 1989, 1990). Others argue that although *neu* amplification is associated with an increased tumor growth rate, it is **not directly linked to its metastasizing potential** (Borg et al., 1991b).

Clearly, as seen from the above-cited observations, **a great deal of heterogeneity exists in the reported correlations between *neu* gene amplification/expression and prognosis**. To add to the confusing picture, still others report a definite trend toward poorer prognosis in patients demonstrating ***neu* mRNA overexpression regardless of amplification** (Parkes et al., 1990). Again, while some find immunoreactivity for the neu protein to be an independent indicator of poor short-term prognosis (Gullick et al., 1991; King et al., 1989; Lovekin et al., 1991; O'Reilly et al., 1990b, 1991; Perren, 1991; Walker et al., 1989), others fail to observe any association with early disease recurrence (Parkes et al., 1990), and still others correlate *neu* gene amplification with poor long-term survival (Dykin et al., 1990; Tsuda et al., 1990a,b; Winstanley et al., 1991). Yet some believe that although c-*erb*B-2 amplification is related to an increased tumor growth rate, it is not directly linked to the metastasizing potential of the tumor (Borg et al., 1991b). It should be used clinically as a prognostic factor identifying a high-risk subgroup of breast cancer in general considered to have good prognosis (Borg et al., 1991b). As an independent indicator of a shorter relapse-free survival and long-term survival, the overexpression of Neu oncoprotein cannot be solely attributed to more rapid proliferation of the primary tumor as measured by SPF evaluated by FCM (O'Reilly et al., 1991b).

One reason for the **apparent variability of Neu oncoprotein expression** may be the failure to correlate it with cellular growth rate or proliferation index (Benz et al., 1989). Support for this view comes from tissue culture studies. The level of expression of c-*myc*, c-Ha-*ras*-1, and c-*erb*B-2 (*neu*) increases dramatically in normal breast tissue in short-term cultures, reaching levels similar to those found in breast carcinoma. It appears that **dysregulation of protooncogene expression**, **rather than overexpression per se**, should be regarded as the contributing factor in breast carcinogenesis.

3.1.4.2 *myc*

Similar controversy exists concerning the role of *myc* family protooncogenes in breast carcinoma. An interrelationship between **c-*myc*** and **growth factors** in human breast carcinoma cell line has long been recognized (Fernandez-Pol et al., 1987a,b). Estrogen stimulation of breast cancer cell growth has been shown to induce c-*myc* oncogene expression (Dubik et al., 1987) at the transcription level (Dubik and Shiu, 1988), and also leads to posttranscriptional modulation of its expression (Santos et al., 1988). Tumor-specific rearrangement of one *myc* locus and amplification of the other *myc* locus have been detected in a **ductal adenocarcinoma**; these were absent in normal breast tissue from the same patient (Morse et al., 1988). Amplifications and rearrangements of the c-*myc* gene have been reported in 56% of the breast tumors studied (Bonilla et al., 1988). These genomic alterations were **not** correlated with the aggressive behavior of breast tumors, but rather seemed to be associated with development of breast carcinoma.

On the other hand, a **rearrangement** detected in the c-*myc* gene from an aggressive breast carcinoma was thought to be a contributing factor to the **tumor aggressiveness** (Machotka et al., 1989; Varley et al., 1987b). In transfection experiments, a transfected c-*myc* gene did not suppress endogenous c-*myc* levels, unlike the situation in lymphoma cells (Liu et al., 1989). Interestingly, elevated c-*myc* mRNA levels were found to be associated with slower *in vitro* growth rates. The influence of **augmented *myc* expression** on breast cancer was considered to be complex, and may not necessarily induce more malignant patterns of growth as seen in other human cancers. **Amplification of c-*myc*** is significantly associated with PR-negative status, and is prevalent in high-grade tumors (Adnane et al., 1989).

A correlation exists between **tumor stage** and c-*myc* expression, as shown by quantitating c-*myc* protein by ELISA (Spandidos et al., 1989). No correlation, however, could be observed between survival and elevated c-*myc* expression. Similar observations have been reported by others (Locker et al., 1989; Walker et al., 1990). The relationship of c-*myc* oncoprotein levels to tumor grade is explained by the role of c-*myc* in cellular differentiation, because high levels of p62^{c-myc} protein were associated with well-differentiated tumors (Locker et al., 1989). In contrast, the number of c-*myc*-positive tumor cells has been found to correlate with breast cancer invasiveness in another study (Pavelic et al., 1990). Only weak staining of the oncogene product could be observed in some ductal and lobular epithelial cells of normal breast samples.

Another protooncogene of the *myc* family, **L-*myc***, has been related to a significantly shorter survival period after relapse, because loss of heterozygosity (LOH) at L-*myc* was observed in these patients (Lidereau et al., 1990).

3.1.4.3 *ras* PROTOONCOGENES AND c-*fos*

Point mutations at position 12 were found to activate the Ki-*ras*-2 oncogene in mammary tumor cell lines (Prosperi et al., 1987). Although Ha-*ras* has been shown to enhance interactions of breast cancer cells with the basement membrane (Albini et al., 1986), and to be capable of inducing malignant phenotypes in immortalized human breast epithelial cells (Basolo et al., 1991), the elevated expression of Ha-*ras* protooncogene does not correlate with acquisition of metastatic potential in laboratory animals (Waghorne et al., 1987). An additional influence of **transforming growth factor-α (TGF-α)** was suggested as necessary for malignization and the enhanced growth rate in *ras*-transformed mammary epithelial cells (Ciardiello et al., 1988).

In primary and metastatic human breast cancers, qualitative and quantitative changes in *ras* gene expression appear to be independent events taking place during initiation as well as progression of breast carcinomas (Spandidos, 1987). In chemically induced mammary tumors in mice, the characteristic **codon 61 mutation** was observed accompanied by the expression of p21^{Ha-ras} product (Cardiff et al., 1988). However, although Ha-*ras*-1 activation apparently plays an important role in neoplastic progression, its role in the **initiation** of malignancy, or in the **maintenance** of the mammary hyperplasias, is questionable.

In order to test the hypothesis that *ras* activation may be involved in the **final stages** of breast cancer progression, breast tumors from invasive lesions, metastases, and established breast cancer cell lines have been studied (Rochlitz et al., 1989). DNA fragments from Ki-, Ha-, and N-*ras* codons 12, 13, and 61 were amplified by polymerase chain reaction (PCR), and then probed with allele-specific oligonucleotides (ASOs) to detect point mutations. A very low number of activating mutations could be detected, with more present in lymph node metastases and skin. **Activating *ras* mutations appear to be only rarely involved in either initiation or metastatic progression of human breast cancer** (Rochlitz et al., 1989).

Likewise, the **expression of p21 product** shows **no correlation** with tumor size or histological grade, although as a group tumors with lymph node metastases expressed higher levels of p21^{ras} than did nonmetastasizing tumors (Czerniak et al., 1989). The **wide scatter of the observed values p21^{ras} makes this parameter unsuitable for predicting the behavior of breast carcinomas**. p21^{ras} expression appears, however, to be quantitatively enhanced in a variety of tumors (stomach, lung, colon, and bladder) as well as in breast carcinomas (De Biasi et al., 1989), and in mammary and extramammary **Paget's disease** (Mori et al., 1990). Thus, although Ha-*ras* seems to be the predominantly transcribed gene of the *ras* gene family in breast cancer, its diagnostic and prognostic value in patient management is very low (Biunno et al., 1988). This view is strengthened by the finding of equal levels of expression of c-Myc (and c-Fos) protein in tumor and benign breast lesions (Walker et al., 1990).

Although c-*fos* appears to be the most constantly and significantly expressed nuclear oncogene in surgical specimens of breast carcinomas, according to some authors (Biunno et al., 1988), **its expression is similar to the level found in benign controls** in other studies (Walker et al., 1990). Even when a greater degree of expression is observed in the carcinomas than in the benign lesions or normal tissue, no distinct relationship can be established with proliferation or differentiation (Walker and Cowl, 1991).

3.1.4.4 *int*

Implication of protooncogenes of the ***int*** family in human breast carcinoma stems from the studies on the mechanism of oncogenesis produced by a retrovirus, the **mouse mammary tumor virus (MMTV)** (Brown et al., 1986). The two sites, termed *int*-1 (Nusse and Varmus, 1982) and *int*-2 (Peters et al., 1983), were identified and shown to map to different chromosomes (Nusse et al., 1984). *int*-2 has been detected in some human tumors, including breast tumors and squamous carcinomas of the head and neck (Zhou et al., 1988b). This protooncogene is one of several integration sites for MMTV (Nusse et al., 1984).

No retroviral components have been detected in human breast cancers so far (Hallam et al., 1990), in contrast to an earlier report (Al-Sumidaie et al., 1988). *int*-2 has been mapped to human **chromosome 11q13** (Brookes et al., 1989; Casey et al., 1986) in the region where another protooncogene, *HST*-1, is located (Adelaide et al., 1988; Yoshida et al., 1987, 1988). Both protooncogenes encode members of the fibroblast growth factor family and are situated only 35 kb apart (Wada et al., 1988).

The importance of these protooncogenes in human carcinogenesis is suggested by observations of frequent amplification of the chromosome 11q13 region harboring both protooncogene sites in a variety of malignancies, in addition to the already mentioned breast, head, and neck cancers (Berenson et al., 1989; Zhou et al., 1988b). These include hepatocellular carcinoma (Hatada et al., 1988), stomach cancer (Yoshida et al., 1988), esophageal and bladder carcinoma (Theillet et al., 1989; Tsuda et al., 1988; Tsutsumi et al., 1988), and melanomas (Adelaide et al., 1988; Theillet et al., 1989).

Amplification of *hst/int*-2 is always accompanied by amplification of ***bcl*-1**; the latter, however can be amplified solely (Theillet et al., 1990). It is suggested that there is a genetic element between *hst/int*-2 and *bcl*-1 that can be important in the development of a subset of breast tumors. Amplification of *hst/int*-2, usually on the order of 10 to 20% of cases, is frequently accompanied by an **alteration** of the *int*-2 gene (Varley et al., 1988), and its amplification is found predominantly in the more differentiated low-grade carcinomas (Adnane et al., 1989). The latter observation is at variance with other studies in which *hst/int*-2 amplification was associated with distant metastases and/or local recurrences (Lidereau et al., 1988a; Machotka et al., 1989). Amplification of *int*-2 is accompanied by the relatively low abundance of *int*-2 mRNA and their levels do not seem to correlate with the level of gene amplification (Liscia et al., 1989). In some tumors (colon), no expression of *int*-2 mRNA could be detected in the presence of unequivocal amplification of the gene.

An extensive study of 311 invasive human breast carcinomas demonstrated **coamplification** of the *hst/int*-1 genes in 9% of cases (Borg et al., 1991a). Amplification was not correlated to either tumor size, axillary lymph node status, patient age, menopausal status, or the stage of disease. Patients with *hst*-2/*int*-1 amplification had a significantly shorter disease-free survival in node-negative patients. However, no significant correlation to overall survival was observed. Although amplification of *hst*-2/*int*-1 seems to identify a subset of aggressive tumors, this characteristic is quite different from the breast tumor prognostic factors identified by *her*-2/*neu* amplification (Borg et al., 1991a). The third gene of the *int* family, *int*-3, is also related to *EGF*, but **apparently only *int*-2 may have prognostic significance in human breast cancer** (Dickson et al., 1990).

3.1.5 GROWTH FACTORS

Aside from c-*erb*B-2 (*neu*), which encodes a protein related to EGFR, a number of growth factors involved in breast carcinogenesis have been described. The *EGFR* gene amplification is accompanied by overexpression of the protein product less frequently (4% of cases) than overexpression of c-ErbB-2 (Neu) protein in the presence of its gene amplification (~16% of cases) (Lacroix et al., 1989). The expression of c-*erb*B-2 and *EGFR* does not occur simultaneously, and follows different paths (Lacroix et al., 1989, Zeillinger et al., 1989), although some findings support the role of both growth factors in the initiation of breast carcinomas (see above). No correlation between amplification of *HER*-2/*neu* gene with EGFR and androgen receptor content has been observed (Zeillinger et al., 1989).

The close structural homology of the *EGFR* and the c-*erb*B-2/Neu gene product extends also to a similarity in their enzymatic activities as **autophosphorylating tyrosine kinases** (King et al., 1988). The c-ErbB-2/Neu protein is a substrate of EGFR, and binding of EGF to its receptor (EGFR) causes a rapid increase in tyrosine phosphorylation of the c-ErbB-2/Neu protein (King et al., 1988). This suggests the possibility of **communication** between the c-ErbB-2/Neu protein and EGFR early in the signal transduction process.

The expression of *EGFR*, like that of *neu*, is not correlated with tumor size and is reciprocally correlated with the expression of ER (Tauchi et al., 1989). A highly sensitive **immunoenzymetric assay (IEMA)** for measuring EGFR in breast tumor tissue has been developed (Grimaux et al., 1989). The assay has adequate sensitivity to detect generally moderate, and often low (3 to 30 fmol/mg membrane protein), levels of EGFR found in breast cancers.

Epidermal growth factor receptors are modulated not only by their respective ligands but also by other growth factors such as **platelet-derived growth factor (PDGF), α-interferon,** and **TGF-β** (Assoian, 1985; Collins et al., 1983; Welch et al., 1990; Zoon et al., 1986). Experiments in cultured breast carcinoma cells demonstrated that the antiproliferative activity of TGF-β is modulated via its interaction with EGFR, and thus can amplify the EGF-induced inhibitory response of the cells (Fernandez-Pol et al., 1987a,b). Interestingly, the expression of

EGF by human breast carcinoma cell lines is enhanced by **progestins**, and is not affected by **estradiol, dexamethasone,** or **dihydrotestosterone** (Murphy et al., 1988).

In breast carcinoma biopsy samples, the level of EGF mRNA expression was correlated with the ER/PR status, being higher in ER+ and PR+ tumors (Dotzlaw et al., 1990). **The prognostic value of EGF and EGFR in breast carcinoma still remains to be established.** Along with TGF-α- and -β mRNA, the expression of EGFR in relation to the relapse-free and overall survival of patients with breast cancer cannot predict the outcome of these patients (Barrett-Lee et al., 1990). Others, however, emphasize the prognostic value of EGFR in predicting early recurrence (Locker et al., 1989; Sainsbury et al., 1987). **EGFR negativity is associated with good responses to endocrine therapy in ER positive patients** (Nicholson et al., 1990).

The other **protein marker, Ki67,** seemed to correlate with the response to therapy in EGFR-positive, ER-positive cases. Again, when comparing the prognostic value of EGFR with that of ER status, some conclude that the ER (but not EGFR) activity is related to prognosis (Hawkins et al., 1990), while others emphasize EGFR-1 staining by immunohistochemical techniques as showing the strongest association among existing prognostic factors (Hainsworth et al., 1990b). The clinical relevance of these findings may be emerging from the view that elevated endogenous growth factor expression may be one of the mechanisms associated with increased resistance of human breast cancer cells to the growth-suppressing agents, such as progestins and **antiestrogens,** employed in adjuvant therapies (Lippman and Mihich, 1990).

Among the other growth factors expressed by cultured breast carcinoma cells are those typically related to the growth of mesenchymal cells — **PDGF A chain, PDGF B chain, TGF-α,** and **insulin-like growth factor II** (IGF-II) (Peres et al., 1987) and **IGF-I** (Brunner, 1990; Peyrat et al., 1988a,b, 1989; Yee et al., 1989). Insulin-like growth factor I appears to be an independent prognostic factor and the **presence of IGF-I receptor is correlated with better overall and relapse-free survival,** as shown in 297 patients with lobular and ductal breast carcinomas (Bonneterre et al., 1990).

3.1.6 TUMOR SUPPRESSOR GENES

3.1.6.1 *p53*

The frequently observed LOH on the **chromosome 17p** region in breast carcinomas pointed to the possible involvement of the *p53* gene, which maps to the 13.1 region of 17p (Mackay et al., 1988b). Using 39 **restriction fragment length polymorphism (RFLP) markers**, including 25 **variable number of tandem repeat (VNTR)** probes, the frequent LOH in 79 primary breast carcinomas was found on **13q** (21% of cases), **16p** (45%), and **17q** (56%) (Sato et al., 1990). A combination of LOHs on both 13q and 17p was associated with more malignant histological features, while tumors with LOH on 16q presented with frequent lymph node metastases. The deletion mapping implied the involvement of a **tumor suppressor gene** (TSG) distal to the *p53* **gene** as well as the *p53* gene itself in primary breast cancer (Sato et al., 1990; see below).

In an attempt to characterize genomic alterations involved in initiation and progression of breast carcinomas, 52 tumors were studied for the occurrence of allele losses (Larsson et al., 1990). Different LOHs were most frequently noted in **lobular (chromosome 22)** and **ductal (chromosome 17)** breast carcinomas. A higher frequency of LOH was noted in primary tumors characterized by advanced disease developing after many years of mild clinical course, compared to patients with a continually mild form of the disease.

The involvement of a TSG in the development of breast carcinomas has been implied in numerous studies. The frequent LOH on 17p, particularly in band p13-1, suggested the **involvement of the *p53* suppressor gene** (Coles et al., 1990; Cropp et al., 1990a,b; Sato et al., 1990). In addition to LOH in this region, a significantly higher frequency of LOH at 17p13-3 was observed that was correlated with an overexpression of *p53* mRNA (Coles et al., 1990).

Another TSG has been identified 20 Mb telomeric to the *p53* gene (Coles et al., 1990). It is argued that **this regulatory gene is involved in the majority of breast cancers** (Coles et al., 1990). Further studies are needed to demonstrate to what extent *p53* and/or neighboring or related TSGs play a clinically important role in the development of breast carcinomas. Confirmation of the suggested role of loss or inactivation of gene(s) on chromosome 17p that differ from *p53*, or the presence of different mechanisms of *p53* gene inactivation in LOH of 17p, that is definitely associated with high SPF and DNA aneuploidy comes from the findings of rare mutations in the *p53* gene in breast cancers (Chen et al., 1991).

Apparently, TSGs involvement in breast carcinogenesis is not limited to the *p53* gene, as suggested by the findings of LOH on **chromosome 18** (between 18q21.3 and 18q23), with a possible candidate being gene **DCC** (**d**eleted in **c**olorectal **c**arcinomas) (Cropp et al., 1990a,b). While LOH on 17q is correlated with ER-negative cancers, LOH on 18q is associated with histopathological grade III cancers. Loss of heterozygosity on both chromosomes 17q and 18q seems to be associated with more aggressive tumors, although neither chromosomal change by itself is predictive of tumor stage or patient prognosis (Cropp et al., 1990b). Although LOH on 17p was frequent, no association with clinicopathological parameters was observed by some investigators (Borresen et al., 1990; Cropp et al., 1990b), while this chromosome change was found to be associated with more **aggressive tumor behavior** in other studies (Cattoretti et al., 1988; A. M. Thompson et al., 1990).

It is hypothesized that different subsets of mutations may make comparable contributions to the malignant phenotype, and that **there are subsets of tumors characterized by the particular mutations they contain** (Callahan et al., 1990; Cropp et al., 1990b). Using a panel of **anti-p53 antibodies**, the expression of the p53 protein was evaluated in normal breast tissue and in benign and malignant breast lesions (Bartek et al., 1990b). Over 50% of malignant lesions strongly expressed p53, whereas no expression was detectable in benign lesions. Since overexpression of p53 is associated with mutation of the *p53* gene (Bartek et al., 1990a), there is a possibility that alteration of this TSG plays a role in breast carcinogenesis.

Immunohistochemical evaluation of 200 primary breast cancers with anti-p53 monoclonal antibodies failed to establish a correlation between the expression of p53 and age of the patient, lymph node involvement, tumor size, ploidy, or labeling index scores (Cattoretti et al., 1988). However, its expression was associated with ER-negative, growth factor receptor-positive, high-grade tumors and, consequently, can be of prognostic value. Subsequently, the expression of *p53* mRNA was evaluated in 76 primary breast tumors with a cDNA probe and was found to be elevated in tumors with the low ER protein content, but no correlation was observed with other clinical parameters (A. M. Thompson et al., 1990). In contrast, the overexpression of *p53* alone was slightly correlated with lower overall survival, but, when combined with *neu* overexpression it was closely associated with larger tumor size, negative hormone receptor status, and reduced overall survival (Wiltschke et al., 1990). Some 30% of sporadic breast tumors have a *p53* mutation (Prosser et al., 1991).

Because more than 60% of breast cancers have LOH for markers in the 17p region (Mackay et al., 1988b; A. M. Thompson et al., 1990), and *p53* may behave as a TSG in sporadic breast tumors (Prosser et al., 1991), the possibility that mutations in *p53* occur in heritable breast cancer has been explored (Prosser et al., 1991). Using PCR amplification of all 11 exons of the *p53* gene, followed by oligonucleotide probing for possible mutations, samples from 5 pedigrees with frequent occurrence of breast cancer were tested. No structural abnormalities in the *p53* genes were noted, **discounting the possibility that structural abnormalities of the *p53* gene contribute to the heritable predisposition for the development of breast cancers**. Nevertheless, direct sequencing of *p53* exons 5–9 PCR amplification products obtained from touch preparations of human breast carcinoma allows the detection and analysis of point mutations in primary adenocarcinomas of the breast (Kovach et al., 1991).

In highly susceptible strains of rats, the *COP* **gene**, a **mammary cancer suppressor gene**, has been identified with exclusive specificity for the mammary gland, but not the kidneys or bladder, as shown by its failure to suppress renal or bladder chemically induced carcinomas (Isaacs, 1991).

3.1.6.2 *RB* GENE

The **retinoblastoma susceptibility (*RB*) gene**, is a recessive cancer gene whose mutational inactivation has been implicated in the development of retinoblastoma tumors and other neoplasms. Inactivation of *RB* has been demonstrated in some human breast cancer cell lines and the *RB* gene product, **pp110^{RB}**, could not be detected immunologically with specific antibodies (Lee et al., 1988). Analysis of *RB*-1 gene in breast cancer patients revealed an alteration at the 5′ end in a **mucoid carcinoma**, but no alterations of *RB*-1 were found in ductal adenocarcinomas (Bowcock et al., 1990). In addition, linkage analysis conducted for 14 loci on **chromosome 13q** has been performed in extended families, in which breast cancer demonstrates an autosomal dominant pattern of inheritance. Both the linkage data and structural analysis of *RB*-1 indicate that changes in the *RB*-1 locus either occur by chance, or represent secondary alterations associated with progression of some tumors (Bowcock et al., 1990). An association of frequent LOH on **chromosome 7q31** in primary breast tumors with a short metastasis-free and overall survival points to a possible site of a breast tumor or metastasis suppressor gene (Bieche et al., 1992). No significant association of this genetic aberration could be established with the usual prognostic characteristics such as tumor size, histopathologic grade, and lymph-node or steroid receptor status.

3.1.7 OTHER LINKAGE STUDIES IN BREAST CANCER

A collaborative study of human breast cancers by American and French researchers summarized its findings on linkage of breast cancers with specific gene alterations (Lidereau et al., 1988b) (see also Section 3.1.3). A role for deletions on **chromosome 11p** resulting in hemizygosity of **c-Ha-*ras*, β-globin, PTH, calcitonin,** and **catalase genes** in the aggressiveness of breast tumors has been suggested. c-Ha-*ras* RFLP data were not conclusive, but **c-*mos* RFLPs** could be detected only in patients with breast cancers or other types of tumors.

Testing of **genetic linkage of breast cancer susceptibility** to 9 oncogenes (Ha-*ras*, Ki-*ras*, N-*ras*, *myc*, *myb*, *erb*A-2, *int*-2, *raf*-1, and *mos*) in 12 extended families strongly suggested that these oncogenes are not the sites of primary alterations leading to breast cancer (Hall et al., 1989). However, alterations in one or more of these oncogenes may be associated with tumor progression. On the other hand, a significant association has been observed between *int*-2 RFLPs and **the presence of more than three positive lymph nodes** (Meyers et al., 1990). The observed low frequency of *int*-1, *int*-2, *neu,* and c-*myc* amplification limits their usefulness as clinical predictors of disease recurrence.

Fingerprinting analysis using *Hae*III enzyme digestion of tumor DNA from 22 patients with a variety of benign, *in situ*, and invasive lesions was able to detect minimal DNA changes that could allegedly identify the fine borderline between normal and malignant cells (O'Rourke et al., 1990). It remains to be seen whether this line of studies can offer a clinically useful discriminatory tool.

3.1.8 CATHEPSIN D

Cathepsin D (CD) is the most recently evaluated prognostic factor in breast cancer. According to some authors, overexpression of CD is correlated with **aggressive tumor behavior** and, consequently, shortened disease-free and overall survival (Tandon et al., 1990). Others disagree because aggressive behavior is not a prerequisite for its expression; however, a high proportion of cancers detected at mass population screening express CD, consistent with

ER positivity and better prognosis (Cowan et al., 1990). Cathepsin D expression is frequent, present in 61% of tumors as shown by yet another study of 171 cases of primary breast cancer (Henry et al., 1990). **Positive CD staining was associated with significantly superior overall survival, particularly demonstrable in the subgroup of patients with lymph node metastases.** Furthermore, there was a correlation between the intensity of CD staining and overall survival (Henry et al., 1990).

In node-negative patients, however, a relationship between CD elevation and worse survival appears to be more apparent (Lippman and Mihich, 1990). No explanation has been proposed to account for such controversial findings on this lysosomal protease in breast carcinomas.

3.1.9 *NM*23, *WDNM*1, *WDNM*2, *PGM*21, AND GLA PROTEIN

The *nm*23 gene has been isolated by differential hybridization from murine melanoma cell lines with different metastatic potentials (Steeg et al., 1988). It was expressed more in cells with low metastasis potential, than in clones with high metastasis potential. In the mouse, the *nm*23 gene product is an M_r 17,000 peptide sharing 96% homology with its counterpart in humans. In preliminary studies, allelic deletions of *nm*23 were found in a large number of colon, renal, and small cell cancers (Lippman and Mihich, 1990). Reintroduction of the gene into a metastatic cell line results in reduction of both spontaneous metastases, and those developed following intravenous injection. In a series of 47 invasive ductal, one medullary, and one lobular carcinoma, *nm*23 mRNA levels were determined, and a low level of *nm*23 gene expression was found in 86% of lymph node-positive patients (Hennessy et al., 1990a). A similarly low level of *nm*23 gene expression in node-negative patients was detected only in 40% of cases. *In situ* hybridization confirms the higher *nm*23 mRNA levels in adenomas and carcinomas from node-negative patients than those found in tumors from node-positive patients (Bevilacqua et al., 1989). The usefulness of *nm*23 as an indicator of the metastatic potential of human cancers certainly deserves further study.

In another series of 69 human primary breast cancers, the lymph node status, **relapse-free survival (RFS)**, and overall survival were evaluated with respect to the level of the *nm*23 mRNA expression (Hennessy et al., 1990b). Seventy-three percent of tumors from lymph node-positive patients demonstrated a low level of expression, whereas similarly low levels were found in only 33% of tumors from node-negative patients. Better RFS and overall survival were significantly correlated with higher levels of *nm*23 mRNA expression.

A **gene associated with an effect apparently opposite to that of *nm*23** has been isolated by differential hybridization and termed *pGM*21 (Phillips et al., 1990). The enhanced expression of *pGM*21 mRNA in rat mammary adenocarcinoma cell lines with high metastatic potential was not associated with gene rearrangements or amplification. In a highly metastatic DMBA8 ascites cell line, the *pGM*21 gene overexpression was up to 25-fold higher than in the less metastatic cell lines. Other genes that are augmented in less metastatic rat mammary cell lines are **WDNM1** and **WDNM2** (Dear et al., 1988, 1989). Their products, however, have not been characterized so far. A review of metastasis suppressor genes has been given (Sobel, 1990).

A specific expression of the matrix **Gla protein gene** is detected by differential cDNA hybridization in breast carcinoma cell lines (Chen et al., 1990). Since the Gla protein is vitamin K dependent and vitamin K antagonists are known to inhibit metastases in experimental models, this gene may be involved in breast cancer metastasis.

3.1.10 OTHER MARKERS OF BREAST CARCINOMA

3.1.10.1 *pS*2-*pNR*-2

The human *pS*2-*pNR*-2 **gene** was originally isolated from estrogen-dependent breast carcinoma cell lines (Masiakowski et al., 1982; May and Westley, 1988; Prud'homme et al.,

1985; Westley et al., 1989), but it was not detected in breast cancer cell lines unresponsive to estrogen. The gene product, pS2/pNR-2, has been also identified in normal stomach mucosa, although its function is not apparent (Rio et al., 1988). Prediction of the gene product sequence suggests a resemblance to the growth factor IGF-I, and shows homology to porcine pancreatic spasmolytic polypeptide (Rio et al., 1988). Comparison of the levels of estrogen receptor mRNA and the *pNR*-2 mRNA in 96 breast tumor samples demonstrated the expression of *pNR* mRNA only in a subset of ER-positive tumors (Henry et al., 1989). Furthermore, the level of the *pNR*-2 gene product was positively correlated with response to primary tamoxifen therapy.

Significantly better survival of ER+ patients with *pS2/pNR*-2-positive tumors was particularly pronounced in patients also positive for PR (Foekens et al., 1990). Among node-positive as well as node-negative patients, *pS2* status was used to discriminate between a good and bad prognostic group. So far, no definite role for the protein has been established (Chambon and Rio, 1990).

3.1.10.2 p24 AND p29 PROTEINS

These estrogen-regulated proteins have been described in human breast cancers that show **some correlation with ER status**. The **p24** protein is more frequent in ER-positive tumors (Seymour et al., 1990). The **p29** protein shows a variable relation to ER status and no relationship to PR levels; no association with lymph node status has been observed (Mora et al., 1989).

3.1.10.3 c-*myb*

The protooncogene **c-*myb*** is detected in 64% of breast carcinomas, and **its expression is associated with good prognosis**, allowing it to be used for the characterization of a new class of estrogen-dependent tumors (Guerin et al., 1990).

3.1.10.4 STRESS-RESPONSE PROTEIN

Srp-27 is a 27,000-Da heat-shock or stress-response protein found to be **an independent prognostic indicator for disease-free survival only in patients having an advanced breast cancer** with one to three positive lymph nodes (Thor et al., 1991). A significant correlation has been observed in such patients between Srp-27 overexpression and ER content, pS2 protein expression, advanced stage of the disease, and lymphatic and vascular invasion.

3.1.11 CONCLUDING REMARKS

The significant diversity of observations on gene markers and their correlates with various forms of breast tumors reflect not only differences in the methodological approaches, structures of the studies undertaken, and complex interrelationships with therapeutic interventions, but appear to reflect heterogeneity of molecular and morphological aspects of breast tumors as well. Combinations of most informative gene level markers will eventually find their way into routine patient management. Among these, the *neu* and *nm*23 gene expression and fingerprinting analysis appear to hold the greatest promise at this time. As in other tumors, the evaluation of P-glycoprotein expression may serve to predict the course of the disease and its response to treatment (Verrelle et al., 1991).

Therapeutic implications of the molecular biology of breast cancer may follow a number of strategies based on targeted interaction with specific molecules to produce antitumor effects (Lippman and Mihich, 1990). Among those could be potentiation of the overexpression of **ErbB-2**, the use of inhibitory ligands such as **mammostatin, anti-receptor antibodies** to EGFR and **ErbB-2 receptor, anti-growth factor antibodies**, and so on. Use of these targets for antitumor drug development and their potential value as diagnostic and prognostic factors in breast cancer patient management underscore the importance of wide exploration of the practical value of the gene level testing discussed in this chapter.

Tumors of the Uterine Cervix and Endometrium

3.2.1 INTRODUCTION

The sex steroid receptor proteins have been studied in endometrial cancer, and correlated with survival data and response to hormonal and/or cytotoxic chemotherapy (for a review, see Chambers, 1988). In essence, the level of **estrogen** and **progestin receptor** protein (ER and PR) in the tumor is inversely proportional to the grade of the tumor, that is, the more differentiated tumors have higher levels of ER and PR protein. This parallels the survival data — a favorable prognosis for survival is expected with high levels of the sex steroid receptor protein. The receptor status, however, does not correlate with the depth of myometrial invasion or peritoneal cytology findings.

The receptor status may even offer a better prognostic tool than histologic grading (Chambers, 1988). An improvement of technical aspects of receptor protein assessment is needed before this parameter may be used clinically for prognosis and appropriate selection of hormonal and/or chemotherapy. Technical deficiencies of receptor evaluation plague this approach when competitive binding of specific radioactive and nonradioactive ligands is used.

3.2.2 DNA PLOIDY OF CERVICAL TUMORS

In an experimental animal model of **squamous cell carcinoma (SCC) of the uterine cervix**, the tumor can be produced in mice by application of 3,4-benzopyrene, leading to progressively increasing cellular atypia accompanied by a parallel increase in and heterogeneity of cellular DNA content (Naslund et al., 1987). Flow cytometry (FCM) analysis of cervical cells from patients with **cervical intraepithelial neoplasia (CIN)**, and human papilloma virus (HPV) lesions, revealed a significantly higher percentage of dividing cells than in women with normal cervices (Hughes et al., 1987). On the other hand, DNA aneuploidy varied insignificantly between different stages of HPV lesions.

In CIN III, with or without accompanying invasive SCC, very significant differences can be observed in the pattern of DNA ploidy in the affected cells between patients of two age groups (35 years old and younger vs. 50 years old and older) (Hanselaar et al., 1988). In the **older women**, about 80% of CIN III lesions are aneuploid, irrespective of the coexisting invasive lesions. On the other hand, in the **younger patients**, the majority (60%) of CIN III lesions with concomitant invasive cancer are diploid. The DNA ploidy pattern is the same in invasive lesions, and the adjacent areas of CIN, supporting the notion that the two lesions are related.

Progressive cytological signs of viral (HPV) effects and dysplasia positively correlate with the proportion of cells having abnormal DNA content measured by FCM (De Vita et al., 1990), or by microphotometric DNA analysis (Kashyap et al., 1990). An FCM study of cervical adenocarcinoma in 125 patients revealed 31% of the tumors to be aneuploid (Leminen et al., 1990). Aneuploidy and the high percentage of cells in the **S phase** were correlated with tumor size, histological grade, and clinical stage, confirming the earlier observations (Rutgers et al.,

1986, 1987; Strang et al., 1987). The patients with triploid tumors had an especially poor prognosis, because triploidy is associated with an aggressive disease. Clinical stage and histological grade failed to perform as independent prognosticators. Thus, it appears that **tumors with diploid DNA content and a low percentage of cells in the S phase have the best prognosis in cervical adenocarcinoma**.

3.2.3 CHROMOSOME ABERRATIONS IN CERVICAL CARCINOMA

When 19 different chromosomes were screened with 34 polymorphic markers by Southern blot analysis of fresh tumors, loss of heterozygosity (LOH) at the D3S2 locus was consistently observed on **chromosome 3p** in all of the patients evaluated (Yokota et al., 1989). These findings suggest the **involvement of recessive genetic changes** in the development of cervical carcinomas. While no amplification of any of the 13 oncogenes tested, including **c-*myc*** and **Ha-*ras***, could be detected, HPV16 was present in the majority of tumors.

In a series of 148 patients with invasive cervical carcinoma, abnormalities of **chromosome 1** proved to be the most frequent chromosomal aberration detected by G banding (Sreekantaiah et al., 1988). Multiple locations on chromosome 1 were affected, and the two preferentially affected regions were 1p11-p13 and 1q21-q32; however, these have never been the only chromosomal abnormalities.

3.2.4 DNA PLOIDY OF ENDOMETRIAL TUMORS

DNA FCM of hyperplastic endometrium, known to be the precursor of endometrial adenocarcinoma, revealed a higher percentage of nondiploid cells (20%) than in normal endometrium (6%), approaching that in well-differentiated adenocarcinoma (29%) (Lindahl and Alm, 1991). The high frequency of nondiploid cells persisted even after abrasio alone or in combination with high-dose gestagen treatment. Importantly, these cells were detected in areas without histologically demonstrable hyperplasia.

DNA FCM of normal endometrium and endometrial carcinomas revealed that the majority (57%) of the malignant tumors had no differences compared to the ploidy characteristics of benign endometrium (Lindahl et al., 1987a). Although FCM was of no value for screening purposes, **the degree of aneuploidy correlated with histopathological findings**. Aneuploidy was more frequently observed in poorly differentiated tumors than in highly or moderately differentiated ones (62 vs. 29%). With respect to staging, aneuploidy was more frequent in FIGO stage IV tumors compared to stages I and II. DNA ploidy had a significantly better prognostic value for 2-year recurrence-free survival than the degree of tumor differentiation, range of myometrial invasion, or ER/PR concentrations (Lindahl et al., 1987b). The S-phase fraction (SPF) value determined by FCM failed to correlate with prognosis. The **range of myometrial invasion, however, is the strongest prognostic characteristic, superseding even the DNA parameters** (Lindahl and Gullberg, 1991).

Other workers reported better 5 to 7 year survival of patients with stage I endometrial carcinoma if their tumors were PR positive (van der Putten et al., 1989). A combination of the mean shortest nuclear axis, DNA index (DI), and depth of myometrial invasion offered the best prognosticator. Analysis of the DNA content of archival paraffin-embedded specimens of endometrial adenocarcinoma by FCM failed to detect aneuploidy in low-grade carcinomas (Newbury et al., 1990). Although its overall occurrence in all tumors was only 18% in this series, aneuploidy was associated with adverse histological type (such as serous papillary carcinoma), high grade and depth of myometrial invasion. A DI over 1.5 strongly predicted poor survival within 8.7 years. Again, proliferative activity of diploid tumors could not predict the clinical outcome. It appears, therefore, that **DNA FCM of endometrial neoplasias can be of help in selecting a subgroup of patients who have low-stage, high-grade endometrial cancers, and who may benefit from adjuvant therapy**.

DNA FCM of tumor cells performed on paraffin-embedded tissue of stage I carcinoma of endometrium obtained from 203 patients revealed 84% of the specimens to be diploid (Moberger et al., 1990). The nondiploid (aneuploid and tetraploid) cases accounted for 50% of all the relapses. The survival estimates were higher in diploid than nondiploid cases (92 vs. 63% of progression-free 5-year survival). The higher grades of tumor, stages III and IV, received postoperative adjuvant therapy and, therefore, were not included in this study, which aimed at defining better criteria for prognosis and recommendations for therapy.

Multivariate analysis of the effects of histological grade, subtypes, and DNA ploidy on progression-free survival pointed to an **independent influence of DNA ploidy and histological subtype**. In contrast, **histological grade was not significantly correlated with prognosis**. Thus, DNA ploidy is suggested as an adjunct to other traditional modes of tumor assessment. An effort has been reported to correlate tumor DNA content of endometrial adenocarcinoma cells with the pattern of tumor growth beyond the commonly estimated depth of invasion (Moberger et al., 1990). Indeed, it was found that the **pattern of infiltration** (contiguous vs. discontinuous growth), which can be assessed only postoperatively in hysterectomy specimens, shows a strong correlation with the tumor cell DNA content in curetted diagnostic material obtained prior to treatment. This study was performed on material gathered from 21 endometrial carcinoma patients (who died from their disease), and 23 patients selected at random from 307 survivors beyond 5 years postoperatively. Patients with stage IV tumors were not included, because no survival could be found to serve as matching comparison. A **contiguous growth pattern** appears to be a sign of a low-grade malignant (mostly diploid) tumor with delayed spread and fewer metastases before therapy is instituted. On the other hand, tumors with a **discontinuous pattern of growth** had a predominantly aneuploid nuclear DNA pattern. An automated image analysis system operating on cell dispersion and monolayer smear processing of tissue has been described and its application to mass screening for endometrial cancer tested (Tanaka et al., 1987 a,b). The results of that study have not been published at the time of this writing.

3.2.5 CHROMOSOME ABNORMALITIES IN UTERINE CANCER

Chromosomal aberrations in endometrial carcinomas most frequently feature **chromosome 1** (Couturier et al., 1988a,b; Fujita et al., 1985; Gibas and Rubin, 1987). In addition to heterogeneous chromosome profiles a number of complex aberrations such as minutes, rings, homogeneously stained regions, premature chromosome condensations, premature centromere division, and telomere joinings have been described in endometrial carcinomas (Simon et al., 1990). The most important abnormality appeared to involve **chromosome 10**, which was found in increased numbers (50 to 70) in several cases.

The possible role of **tumor suppressor genes** in the development of endometrial adenocarcinoma is suggested by the complete suppression of tumorigenicity of human endometrial carcinoma tumor cells (HHUA) on introduction via microcell hybrids of **chromosomes 1, 6, or 9** (Yamada et al., 1990). The introduction of **chromosome 11** had the same effect in some, but not all, cases.

Smooth muscle tumors of the uterus (uterine leiomyomas, parametrial leiomyomas) show an identical abnormal stemline pattern in different tumors from the same patient with each leiomyoma arising as a separate clone (Mark et al., 1991).

3.2.6 ONCOGENES IN CERVICAL TUMORS

With respect to cervical cancer the prognosis of the disease is not always predictable based on the involvement of the lymph nodes. Again, more accessible, noninvasive, and reproducible markers are sought in the pattern of alterations consistently present in tumor tissues. In this context, oncogene expression and the presence of their mutations have been looked into. Thus, the loss of one **c-Ha-*ras*** allele in tumor cells from cervical cancers at different clinical stages has been detected (Riou, 1988). It was not, however, correlated with the aggressiveness of cervical cancer, because it was found with essentially similar frequency at different stages of cervical cancer. Point mutations of the c-Ha-*ras* oncogene have been observed *in situ* in 24% of stage III and IV cervical carcinomas, emphasizing their association with poor prognosis. The mutant oncogene was detected in 40% of cervical carcinomas accompanied by a deletion of the c-Ha-*ras* gene on the other allele (Riou, 1988).

The **c-*myc*** protooncogene is overexpressed (4 to 20 times the level in normal tissue) in stage I and II SCCs of the uterine cervix (Riou et al., 1987). This finding is highly correlated with the eightfold higher incidence of early recurrences, and the 18-month relapse-free survival rates of 49 and 90% in the patients with c-*myc* overexpression and normal controls, respectively.

A high frequency (almost 90%) of marked amplification and/or rearrangement of the c-*myc* gene has been reported in stage II cervical carcinomas (Ocadiz et al., 1987). Some of the tumors showed up to 60-fold amplification of c-*myc*. **Concordant interaction of several oncogenes** in the development of cancer was again suggested in the finding of activation of the c-*myc* protooncogene in all tumors that contained a mutated c-Ha-*ras* gene.

The level of c-*myc* gene transcripts has been evaluated in tumors obtained from 150 untreated patients with primary invasive squamous cell carcinomas of the uterine cervix at different clinical stages (Riou, 1988). The **c-*mos*** oncogene was used as a control for chromosome 8, and no amplification of this oncogene was found in any of the cases (in 21% of tumors), whereas c-*myc* amplification was 3 to 30 times the control levels.

A significantly higher rate of c-*myc* amplification is found in stage III and IV carcinomas (49%), compared with only 6% in stage I and II tumors, suggesting a role for c-*myc* protooncogene in cervical **cancer progression** (Riou, 1988). Again, the majority of amplified c-*myc* genes are also overexpressed in what apparently constitutes an early event in the natural history of cervical cancer, whereas c-*ras* gene amplification seems to arise only in the most advanced cancers (Riou, 1988). c-*myc* expression has also been observed in eutopic and ectopic **endometriotic tissue** removed during surgery, suggesting that expression of this oncogene plays a role in regulating cell proliferation in endometriotic tissue as well (Schenken et al., 1991).

In a study of the **prognostic significance of c-*myc* gene expression**, the risk of relapse was found to be significantly associated with c-*myc* overexpression, nodal status, and geographical origin of the patients. The **major prognostic factor was c-*myc* overexpression**. Thus, the 18-month relapse-free survival of patients with c-*myc* overexpression was 49% in contrast to 90% for patients without c-*myc* overexpression. The presence of **c-*myc* gene activation outweighed nodal status as a prognostic factor in cervical cancer patients**. Overexpression of c-*myc* is correlated not only with the advanced stage of cervical carcinoma, but shows a significant value as a prognosticator of 3-year disease-free survival (Riou et al., 1990). **Overexpression of c-*myc* is associated with a markedly reduced survival rate**, almost half that of controls without c-*myc* amplification (51 vs. 93%, respectively). Amplification of all three oncogenes — Ha-*ras*, c-*myc*, and *erb*B-2 — has been found in CIN III and invasive cancer compared with normal cervix and CIN I (Pinion et al., 1991). It appears, therefore, given these observations are confirmed in larger studies, that **evaluation of c-*myc* mRNA levels can serve as a practical clinical tool for identifying patients with cervical cancer at high risk for early recurrence, thus facilitating the customized therapeutic approach on an individual patient basis**.

3.2.7 ONCOGENES AND GROWTH FACTORS IN ENDOMETRIAL TUMORS

Very little is known so far about the role of protooncogenes in endometrial neoplasias. In the rat uterus, estrogen is known to produce transient activation of **c-*fos*** and c-*myc* protooncogene expression, which precedes increases in DNA synthesis and cell proliferation (Weisz et al., 1990). Although the function of the c-*fos* oncogene product is not fully clarified, it appears that this gene is involved in controlling tissue differentiation (Distel et al., 1987; Miller et al., 1984; Muller and Wagner, 1984). In this context, changes of c-*fos* oncogene expression in uterine malignancies may offer a diagnostically useful tool.

In addition, **c-*jun*** gene transcription is increased three- to fourfold upon estrogen stimulation and constitutes a primary response to the hormone. The involvement of c-*jun* in the formation of the composite transcription factor Jun/AP-1 points to the possibility that this protooncogene, as well as c-*fos*, which is regulated by the same growth-promoting factors, may be activated in human endometrial neoplasias. I am not aware at this point of direct studies demonstrating this phenomenon in patients, which may provide a sensitive marker of early proliferative activity in the endometrium (Weisz et al., 1990).

In order to better understand the aggressive biological behavior of **serous papillary adenocarcinoma** of the endometrium, a thorough study of an advanced case has been performed (Sasano et al., 1990c). Amplification of three protooncogenes, ***int*-2**, **c-*erb*B-2**, and c-*myc*, was estimated by Southern blot hybridization, DNA ploidy was evaluated by FCM, and the expression of ER and PR, keratin, vimentin, and carcinoembryonic antigen was determined by immunohistochemical and biochemical methods. Of the three oncogenes, only c-*myc* amplification was observed, in the tumor itself (10-fold) and in its metastases to the omentum (5-fold). Amplification of c-*myc* in a serous adenocarcinoma endometrium parallels that detected in serous carcinoma of the ovary (Zhou et al., 1988a). Amplification of c-*myc* in the serous variant, but not in classical endometroid carcinoma of the endometrium, may be related to the particularly aggressive behavior of this type of carcinoma.

Based on immunocytochemical studies, Sasano and coworkers (1990c) argue that there are differences in the cells comprising the papillary structures of this type of carcinoma, and the solid sheets of poorly differentiated areas in the same tumor. These histological variants apparently did not contribute to the characteristics of aneuploid peaks detected by FCM. The similarity of c-*myc* amplification in endometrial and ovarian serous epithelial neoplasms appears to emphasize the **possible association of c-*myc* amplification with serous papillary morphology**. Estrogens appear to exert their control on the growth of all three uterine cell types (epithelial, stromal, and myometrial) through several oncogenes: N-*myc*, c-*myc*, c-Ha-*ras*, *erb*B, and c-*fos*, at least (Lingham et al., 1988; Loose-Mitchell et al., 1988; Murphy et al., 1987; Travers and Knowler, 1987).

Among various tumors of the female reproductive tract, including endometrial carcinomas and cell line derivatives, SCC of the uterine cervix, and a **malignant melanoma** of the vagina and ovarian tumors, the activation of **Ki-*ras*** in endometrial cancer is frequently detected by polymerase chain reaction (PCR) amplification of *ras* oncogene sequences (Enomoto et al., 1990a,b). In contrast, cervical and ovarian tumors do not show *ras* activation.

Analysis of the effects of **epidermal growth factor** (EGF) and **transforming growth factors** α and β (TGF-α and TGF-β) on RL95-2 human endometrial carcinoma cells demonstrated a complex pattern of inhibitory and stimulatory activities (Korc et al., 1987). Cellular responses varied depending on the cell density and the concentrations of the growth factors. The inhibitory activity of TGF-β_1 appears to be inversely correlated with the level of expression of TGF-β_1 mRNA by the cells (Boyd and Kaufman, 1990). The physiological implications of these activities are still far from being fully understood in terms of their relevance to endometrial neoplasia, the more so because evidence of new growth factor activities appears to emerge all the time.

One such example is the finding of a new activity of EGF that has been demonstrated to produce uterine contractions in laboratory animals (Gardner et al., 1987). The role of EGF in endometrial growth control is implicated by the progressive decrease in the concentration of **EGF receptor** (EGFR) in endometrial cancers of increasing grade (Reynolds et al., 1990). This inverse relationship with grade points to receptor regulation by hormones and/or growth factors. Some of the proposed mechanisms of involvement of EGF in the control of growth of normal and neoplastic endometrium included the down regulation of EGFR by increased levels of EGF, loss of EGFR from transformed cells, and the binding of EGFR by **basic fibroblast growth factor** (FGF). While the role of EGFR in endometrial adenocarcinoma may be closely related to estrogen influences, this relationship is not as evident in sarcomas, which also demonstrate a 72% decrease in EGFR (Reynolds et al., 1990). At this time, the diagnostic utility of these observations for endometrial cancer management is low and further research and clinical correlations are needed.

The ectopic expression of the **placental alkaline phosphatase** gene has been demonstrated using a specific cDNA probe in a human endometrial adenocarcinoma cell line (Albert et al., 1990). Estrogen was shown to stimulate this activity at the level of transcription of the gene, which could be interrupted by a new antiestrogen, ICI 164,384. Much effort is going into unraveling the fine details of hormonally regulated cascades of mitogenic activity related to normal and neoplastic endometrial growth. It is believed that assessment of the oncogenes and growth factor interactions at the gene level will soon allow not only efficient interference in the process of endometrial tumorigenesis but also its **earliest detection**.

3.2.8 UTERINE SMOOTH MUSCLE TUMORS

The majority of **leiomyomas** display a DNA diploid pattern (Tsushima et al., 1988). On the other hand, **leiomyosarcomas** exhibit a DNA tetraploid/polyploid or aneuploid pattern. Although FCM evaluation of smooth muscle tumors is not recommended as a diagnostic tool for differentiation between benign and malignant tumors, its use for prognosis of clinical behavior of these tumors may be of practical value. At least in the tumors of the gastrointestinal tract, a combination of DI and tumor size proved to offer good discrimination between benign and malignant smooth muscle tumors, allowing the separation of 79% of the benign and 97% of the malignant smooth muscle tumors (Federspiel et al., 1987). A combination of several histopathological characteristics usually improves the diagnostic accuracy and/or predictive value of FCM evaluations. No studies have been presented so far that combine morphological, cytogenetic, and molecular characteristics of uterine smooth muscle tumors to afford maximum diagnostic power.

Cytogenetic analysis of uterine leiomyomas, the most frequent benign smooth muscle tumor of the uterus, indicated that the most frequent finding in uterine leiomyomas is **translocation t(12,14)**, although the specific gene rearrangement responsible for the tumor is not yet known (Nilbert et al., 1988). In contrast, there is little similarity from case to case in complex rearrangements observed in another malignant smooth muscle tumor, **angioleiomyoma**. Still another type of malignant smooth muscle tumor, **rhabdomyosarcoma**, has been found to contain the characteristic reciprocal **translocation t(2,13) (q37,q14)** in its embryonal, undifferentiated, and alveolar histological subtypes (Nilbert et al., 1988). Normal karyotypes were found in 145 of 224 leiomyomas, and clonal chromosomal abnormalities were detected in 44 tumors (Nilbert et al., 1990). These included **del(7) (q21.2q31.2), trisomy 12,** and **t(12;14) (q14-15;q23-24).** Multiple aberrations, including deletions, inversions, and translocations, were noted on **chromosome 6p**.

Essentially consonant with these observations are the cytogenetic studies on 78 short-term cultures of human uterine leiomyomas (Mark et al., 1990). Leiomyomas appear to contain a number of normal and abnormal stemlines with variant cells, carrying one or more specific chromosomal deviations (simple numerical changes, particularly losses of various chromosome types) (Mark et al., 1990). These abnormalities were noted particularly in **chromosomes 2, 6, 9, 10, 11, 16, 17, 18, 21,** and **22.** The multitude of chromosome aberrations appears to suggest that several genetic changes are necessary for the development of uterine leiomyomas.

Cytogenetic studies on a **low-grade metastatic uterine leiomyosarcoma** and on a large degenerating uterine leiomyoma revealed multiple and consistent chromosomal aberrations (Fletcher et al., 1990). Leiomyosarcoma demonstrated an extreme cytogenetic instability in contrast to leiomyoma, which was characterized by cytogenetic stability. A grading system of practical diagnostic value has been proposed for the evaluation of cytogenetic instability (Fletcher et al., 1990). Identification of **high-grade cytogenetic instability** in uterine smooth muscle tumors is likely to predict aggressive clinical behavior of the tumor.

Specific molecular characteristics are sought to allow better differentiation between histologically similar patterns. In fact, malignant smooth muscle tumor leiomyosarcoma is known to express, among other things, the gene encoding **insulin-like growth factor II** (IGF-II) at levels markedly higher than are observed in other, normal tissues (Daughaday et al., 1988). Thus, specific gene expression may eventually be defined in these tumors to help in differential diagnosis. Further molecular studies are needed to identify the specific cytogenetic changes in these benign tumors that may be used to develop discriminating diagnostic probes.

Chapter 3.3

Ovarian Neoplasms

3.3.1 DNA PLOIDY OF OVARIAN NEOPLASMS

Flow cytometry (FCM) studies of ovarian tumors point to a significant correlation between tumor stage, degree of differentiation, and ploidy. A few more recent representative studies will be cited here.

Although morphological studies (typing and staging) of **primary ovarian carcinomas** of the **FIGO I stage** are prognostically important, they tend to be poorly reproducible (Baak et al., 1987). DNA ploidy and DNA cytometric studies offer a more objective approach, and add additional prognostic value to morphological evaluation of ovarian carcinomas. A prospective study of 112 fresh tumor samples collected from 83 patients with ovarian tumors and analyzed by FCM showed all the benign and semimalignant tumors as well as 48% of malignant tumors to be diploid (Iversen and Skaarland, 1987). Aneuploidy was more frequent in the advanced stages of the disease, in tumors of low degree of differentiation, and in older patients. In contrast to the other studies (Erba et al., 1989), aneuploid tumors were smaller (Iversen and Skaarland, 1987), and tumors found predominantly in postmenopausal patients had a higher percentage of cells in the S phase. No association was found with tumor type in that series, which included 50 ovarian carcinomas, 6 of borderline malignancy (5 serous and 1 mucinous), and 24 benign tumors. The latter consisted of serous cystadenomas or cystadenofibromas, mucinous cystadenomas, fibromas, teratomas, and one Brenner tumor. All the metastases of diploid primary tumors were also diploid, whereas the majority of metastases of aneuploid tumors were correspondingly aneuploid (Iversen and Skaarland, 1987).

The **DNA index (DI)** determined by FCM appears to predict mainly the **degree of differentiation** of ovarian tumors (Rutgers et al., 1987). This is reflected in a correlation of the DI with prognosis. The majority of low-grade tumors are diploid, and subgroups with a favorable prognosis can be defined by a combination of age ($\leq$40 years), grade of tumor, as well as by tumor stage (I or II). The presence of over 85% of the cells in the G_1 phase of the cell cycle added an additional favorable factor to prognosis. Subgroups of patients in the prognostically poor category were defined by age ($>$60), by tumor grade (2 or 3), and by stage (clinically advanced). The worst prognosis reflected a combination of grade 2 or 3, or a clinically advanced stage, with G_1 cell fraction $\leq$85%.

Better performance of the gross DNA data (ploidy and S-phase fraction [SPF]) evaluated by FCM has been reported for a series of 69 patients with stage I ovarian cancer (Punnonen et al., 1989). In addition to a DNA histogram, the FIGO stage and histological type appeared to be the best prognostic parameters in a 5-year survival assessment, whereas the age and type of treatment had little prognostic data. Stepwise discriminant analysis in stage Ia ovarian cancer was accurate in 82% of prognoses, and was based on the combination of histological type and DNA histogram type, which incorporates the SPF as well as clonality of the cells tested. Again, patients with diploid or near-diploid untreated ovarian carcinomas (DI $\leq$1.25) have markedly better survival, as do those with tumors characterized by a high proportion of G_0- to G_1-phase cells (Volm et al., 1989). No significant correlation was found between cytometric data and stage. Other parameters, such as the type of surgical intervention, cytostatic treatment, histological tumor type, and degree of differentiation, had no significant effects on the length of survival, underscoring the practical importance of cytometric prognostic factors in a 5-year follow-up study. An important aspect of such an analysis should be the consideration of **intra-** and **interlesional heterogeneity** of cellular DNA content in ovarian carcinomas affecting the sampling and averaging of the data obtained (Baker et al., 1987).

Erba and coworkers (1989) agree that the **DI does not have good prognostic power in ovarian carcinoma**. The somewhat longer survival of patients with DNA diploid tumors was not statistically significantly different compared to the survival of patients with aneuploid tumors. Furthermore, no statistical significance could be established between the survival data and the SPF. Although tumor size does not show any significant prognostic value in sample sizes of the magnitude used in this study (155 samples, 101 patients), the size of tumor seems to be of prognostic significance in markedly larger series (over 500 patients). Although this study included serous, endometrioid, mucinous, and undifferentiated types of ovarian carcinoma, the ploidy analysis was related only to the well, moderately, and poorly differentiated tumor classification. Such a generalized evaluation is bound to level off finer distinctions of various histological types of tumors. The overall conclusion is that **the DI cannot play a major role as a biological prognostic factor in ovarian cancer**. Use of the DI appears, however, to be justified in some cases, as shown in a study limited to the same histological type, serous ovarian tumors, in which the DNA content was estimated in a series of benign, borderline, and malignant tumors (Klemi et al., 1988). Although DNA aneuploidy was observed in both benign and borderline malignant serous ovarian tumors, **the DI of over 1.3 proved to be the most effective parameter in predicting inferior prognosis**.

High cellular DNA content of advanced ovarian carcinomas correlates with a poor prognosis (Rodenburg et al., 1987). It appears that **complementation of FCM data on DNA ploidy by DNA image cytometry (ICM) significantly improves the predictive power of either modality used separately**. After adjustment for stage, residual tumor mass, histopathological type, treatment, and age of patients, the DNA ploidy was able to predict the risk of death to be twofold higher in DNA aneuploid tumors and sixfold higher in DNA multiploid tumors as compared to that in patients with DNA diploid tumors (Kallioniemi et al., 1988b). **The SPF increased the prognostic value of DNA ploidy evaluations**, particularly in DNA diploid tumors.

Flow cytometry analysis of paraffin-embedded tumor samples from 128 previously untreated patients with FIGO stage III and IV malignant common epithelial ovarian cancers showed 73% of the tumors to be aneuploid, the rest diploid (Friedlander et al., 1988). The DNA content and FIGO stage proved to be the only significant prognostic variables. Irrespective of ploidy, the prognosis was poor for stage IV cancers that spread beyond the peritoneal cavity, or those with liver metastases. A few borderline tumors turned out to be diploid and, despite spread to the perifoveal cavity, had an unusually good prognosis defying conventional histological criteria, emphasizing the practical value of DNA FCM as a prognostic indicator.

It is debatable to what extent DNA ploidy data can influence the prognostic value of histopathological analysis of **borderline ovarian tumors**. These comprise up to 15 to 20% of all common epithelial ovarian tumors (Friedlander et al., 1988), and **pathologists differ widely in their acceptance of DNA ploidy as a diagnostic aid**. Moreover, opinions vary on the definition of the histological criteria of borderline tumors, especially those that have spread beyond the ovary. Improved resolution of FCM in differentiating ploidy populations of tumor cells, which should have less than 2% CV on fresh tissue, is critical. The practical benefit of such discriminatory finesse is also seen in diploid tumors having a relatively good prognosis independent of the type of treatment and, therefore, justifying a milder regimen of chemotherapy. Therefore, staging should be supplemented by FCM analysis of ovarian tumors, when deciding on the specific therapeutic modalities in a clinical setting (Friedlander et al., 1988).

Although a specific histological variant of ovarian tumor, **granulosa cell tumor**, presents as a heterogeneous group with variable clinical behavior, a number of commonly accepted criteria (clinical stage, diffuse growth pattern, numerous mitotic figures, cellula atypia, and age) have been routinely used for prognosis; their significance, however, is less certain than the extraovarian spread of tumor at the time of diagnosis (Klemi et al., 1988, 1990). Flow cytometry on paraffin-embedded material from 23 tumors showed 17 diploid, 1 tetraploid, and 5 aneuploid tumors, the latter having poorer survival. An SPF over 6% was associated with poor survival, even for patients assigned to the same stage. This correlates with the histological criterion of finding over 10 mitoses per 10 high-power fields, which has always been predictive of the final outcome.

A contrasting conclusion was reached in another, smaller series of **granulosa cell tumors** of the ovary (Suh et al., 1990). The majority of the tumors from 10 patients with primary lesions, metastases, and recurrences were found to be diploid. The clinical course of granulosa cell tumors could not be predicted by DNA FCM, although DNA ploidy correlated well with cellular anaplasia and high mitotic activity, which usually have prognostic value for granulosa cell tumors.

In a series of 115 patients with **epithelial ovarian cancer**, the parameters that correlated best with survival included stage, SPF, and size of largest metastases. DNA FCM, although also showing a correlation with survival, did not significantly improve the prognostic value of these characteristics (Barnabei et al., 1990). An interesting use of DNA FCM has been demonstrated by the assessment of DNA content in multiple tumors of the female genital tract, predominantly bilateral ovarian tumors (Smit et al., 1990). Since two independent tumors are deemed unlikely to show a near-identical aneuploidy pattern, this finding suggests metastatic spread of the disease.

A good correlation has been detected between the DI and FIGO stages. The more advanced stages III and IV showed a preponderance of aneuploid tumors, while those at stages I and II were more likely to have DNA diploid tumors; this is supported by a corresponding degree of the tumor. DNA diploid tumors are smaller in size, and contain a significantly lower percentage of cells in the S phase than in DNA aneuploid tumors. To what extent normal cells with a low S-phase index contribute to overshadowing the population of cells in S phase in diploid tumors remains to be elucidated. Understandably, the degree of malignancy correlates well with the S-phase values, the latter being higher in poorly differentiated tumors (Erba et al., 1989).

An interesting reversal of the situation with DNA ploidy prognosis is found in **partial hydatidiform moles** (Lage et al., 1991). All patients followed for partial moles, who had developed persistent gestational trophoblastic tumors and had triploid partial moles, achieved complete remission with one course of single-agent chemotherapy. Diploid partial moles with persistent tumor were less sensitive to the same chemotherapy.

Chemotherapeutic agents used in treating ovarian tumors markedly affect DNA and cell kinetic characteristics of the tumors. A total of 155 ovarian carcinoma samples, both primary tumors and metastases, retrieved from 101 patients either receiving **cisplatin**, **adriamycin**, and **cyclophosphamide** or not on chemotherapy have been evaluated for DNA content and SPF (Erba et al., 1989). Thirty-four samples (22%) were DNA diploid tumors, the rest (78%) were aneuploid, in conformity with the earlier findings. In a few cases tumor multiclonality was observed showing cell populations with a different DI. Interestingly, the DI proved to be stable over time when assessed in the primary tumors, metastases, and ascitic fluids of 25 patients.

A large study designed to identify factors, including treatment, that can predict survival has been conducted over 10 years and includes almost a thousand ovarian cancer patients (Marsoni et al., 1990). The **residual tumor size** was found to be the major determinant of survival. Within the subgroup with tumor size less than 2 cm, other variables (age, stage, and histology) further add to the identification of patients with a better prognosis. When ploidy and proliferation indices, chemoresistance testing, and hormone receptor levels were compared for their predictive value, only the chemoresistance testing at initiation of therapy, and second-look status at a later point, retained prognostic significance (Schneider et al., 1990).

Some studies have attempted to **correlate FCM data with other biochemical characteristics** of tumor cells, such as **cyclic 3',5'-guanosine monophosphate** (cGMP) excreted in the urine of patients with epithelial ovarian cancer (Redman et al., 1990). There was a **significant correlation between the proliferative index (PI) and cGMP. No variable, however, predicted the response of patients to cisplatin therapy**. The level of excretion of cGMP was of prognostic value, although of little clinical relevance, even when the stage and residual disease were accounted for in the predictive model. It remains to be seen whether this biochemical marker proves to have any clinical utility.

3.3.2 CHROMOSOMAL ABNORMALITIES IN OVARIAN NEOPLASMS

Using Southern analysis with polymorphic DNA probes to a variety of chromosomal loci, LOH has been observed in 64% of human ovarian carcinomas on **chromosome 6q** at the estrogen (ER) locus (Lee et al., 1990). Additional LOH was observed on **chromosome 17p** at two loci and an allelic loss at the Ha-*ras*-1 gene locus on chromosome 11p. **The LOH at chromosome 6q appears to be specific to ovarian carcinomas**. Losses on 11p and 17p, also noted in other cancers, may be related to the presence of tumor suppressor genes in those regions. A more detailed study of chromosome 17 in epithelial ovarian carcinomas confirmed the high frequency of **allele loss on 17q** (77%), and a slightly lower frequency on 17p (69%), indicating **loss of tumor suppressor genes is involved in the genesis or progression of epithelial ovarian carcinomas** (Eccles et al., 1990; Russell et al., 1990). In a study of 30 patients with sporadic ovarian cancer, *Pvu II* digests probed with three **VNTR probes** (CMM86-17q22; LCN5A2-17q23, and pTHH59-17q23-25.3) used in linkage studies of familial breast-ovarian cancer revealed that in sporadic ovarian cancer the allele loss at CMM86 was present in 65% of informative tumors compared to only 9% in sporadic breast cancer (Foulks et al., 1991). It appears that a gene near this locus is involved in most ovarian cancers which when identified will certainly be helpful in screening a select group of patients.

Other chromosomal aberrations in ovarian cancers are the **translocation t(15;20) (p11;q11)** (Murty et al., 1990), **chromosome 1q heterochromatin asymmetry**, and the increased incidence of **C-band inversion** (Kopf et al., 1990).

3.3.3 DNA FINGERPRINTING IN OVARIAN TUMORS

The most common trophoblastic tumor, **complete mole**, is highly prevalent, ranging from 1:100 to 1:1200 pregnancies (Nobunaga et al., 1990). Appropriate diagnostic procedures, including macro- and microscopic evaluation, sometimes fail to distinguish complete mole from the hydropic abortus. Definitive diagnosis assumes added importance because of the relatively increased (up to 4000 times) risk of choriocarcinoma associated with molar pregnancy. RFLPs of the **minisatellite DNA probe 33.15 allow the determination of androgenesis of the complete mole** (Nobunaga et al., 1990). The **DNA fingerprints** were obtained by *Hinf*I digestion of DNA from molar tissues and their parental mononuclear cells. Comparison of parental fingerprints with those of the moles established that **specific bands were inherited exclusively from the fathers**. This unequivocally established that complete mole developed when the fertilization of an anuclear, empty ovum by a single sperm took place. This was followed by duplication of its haploid chromosomal complement. Alternatively, two sperms may have fertilized such an ovum. In two cases, however, various polymorphic fragments, banding in the 4- to 20-kb size range, were not abnormal and could be traced to the parent, thus establishing parental contribution to the genome of hydropic placenta that resulted from normal fertilization (Nobunaga et al., 1990).

This study exemplifies an approach to using DNA fingerprinting analysis in a histologically challenging problem. Certainly **karotyping** with chromosome marking can distinguish **com-**

plete mole usually having a 46,XX set of chromosomes, and occasionally 46,XY. Special banding techniques, with a lengthy tissue culture procedure (several weeks), make this approach unacceptable for resolving patient management problems requiring shorter time intervals. An RFLP analysis, on the other hand, can be completed within a few days. A complicated situation involving hydatidiform moles for DNA fingerprint analysis occurs in distinguishing the **partial mole, which is typically triploid (69,XXX)**. However, since **partial mole is not associated with choriocarcinoma**, the differentiation of partial mole from hydropic abortus is of lesser clinical significance (Vassilakos et al., 1977). DNA fingerprints in this case will show the polymorphic fragments transmitted from their parents as in normal pregnancy or hydropic abortus, because partial mole results from the fertilization of a normal haploid ovum by two normal sperms (Szulman and Surti, 1984).

Using **quinacrine-fluorescent heteromorphism** and **RFLP analysis** on a fresh tumor sample in a case of **choriocarcinoma** (rather than on established cell lines as performed previously), the origin of the tumor from the product of the same fertilization that led to the previous pregnancy could be established (Chaganti et al., 1990). In order to interpret the findings properly, the host, her spouse, and the son (previous pregnancy) were tested by restriction analysis to elucidate whether maternal markers were present in the tumor.

DNA fingerprinting can also reveal specific DNA alterations in **epithelial ovarian tumors**. The **satellite probes 33.15, 228S,** and **216S** identify somatic changes in 70% of ovarian tumors, the most common being a deletion or reduction in intensity of a band suggesting LOH (Boltz et al., 1990). In one patient, both the primary tumor and metastasis displayed the same DNA fingerprinting pattern, with only minor variations detectable in DNA taken from different sites, suggesting that different cell populations were derived from a common stem line.

3.3.4 PROTOONCOGENES IN OVARIAN NEOPLASMS

A substantially finer degree of analysis at the gene level is offered by the evaluation of **oncogene amplification** in ovarian neoplasm (Sasano et al., 1990a,b). Twenty-four ovarian tumors, including 16 carcinomas, were examined by DNA FCM, cell cycle analysis, and ER and PR analysis, and for amplification of the protooncogenes c-*myc*, *int*-2, and c-*erb*B-2. Earlier, an elevation of the c-*myc* oncoprotein, p62^{c-myc}, was found to precede the development of aneuploidy in papillary serous ovarian adenocarcinoma (Watson et al., 1987). **No protooncogene amplification was detected in benign ovarian neoplasms, or in ovarian carcinomas with low malignant potential**. c-*erb*B-2 amplification was not observed, only a single case of *int*-1 amplification was present, whereas 12 cases showed c-*myc* amplification in invasive carcinomas. No correlation between ploidy and oncogene amplification was established, however, the latter being frequently associated with morphological nuclear aplasia and high mitotic count. Importantly, a **correlation exists between protooncogene amplification and the serous type or some degree of serous differentiation of the tumors**, while none of the purely mucinous carcinomas showed any evidence of amplification. **Protooncogene amplification, particularly c-*myc*, apparently may be used as a marker of aggressiveness of epithelial tumors of the ovary** (Sasano et al., 1990a,b). The more aggressive nature of serous adenocarcinoma of the ovary correlates with c-*myc* amplification. Nuclear atypia correlates to a greater extent than mitotic activity and histological grade with protooncogene amplification, which offers a more objective criterion than morphological grading. Assessing protooncogene amplifications offers substantial benefits compared to DNA FCM, because FCM reflects only major changes in DNA brought about by abnormalities of chromosomal replication. Gene amplification, on the other hand, discloses relatively minor DNA changes. c-*myc* amplification may thus serve as an earlier marker of subsequent aggressive tumor behavior than aneuploidy (Sasano et al., 1990a,b).

The absence of protooncogene amplification in **ovarian tumors with low malignant potential** supports the earlier contention, expressed by Friedlander and coworkers (1988), that tumors showing morphological evidence of low malignant potential should be separated from invasive ovarian carcinomas.

Previous studies pointed to the involvement of the Ki-*ras*, Ha-*ras*, and c-*myc* genes in **serous cystadenoma of the ovary** (e.g., Feig et al., 1984; Slamon et al., 1984; Yokota et al., 1986; van't Veer et al., 1988; Zhou et al., 1988a), while others provided conflicting evidence (Haas et al., 1987; Rodenburg et al., 1988). A 10- to 20-fold amplification of Ki-*ras* in ascitic cells from an ovarian carcinoma does not change over several months, irrespective of chemotherapy and clinical progression of the disease (Filmus et al., 1986). Ki-*ras* amplification is accompanied by an abnormal banding region on **chromosome 6**. Multiple oncogenes, which are altered in approximately 20 to 30% of ovarian carcinomas, include c-Ki-*ras*, *neu*, c-*jun*, c-*myc*, and c-*fos* (Enomoto et al., 1990a,b; Garcia et al., 1990). Up to 40% of the tumors with altered oncogenes have multiple aberrant transcripts of these oncogenes.

A panel of probes for 17 oncogenes has been tested by *in situ* hybridization (ISH) on paraffin sections of paraformaldehyde-fixed specimens of ovarian benign and malignant specimens (Kacinski et al., 1989). Ovarian neoplastic epithelial cells were found to express *neu, fes, fms, trk,* c-*myc,* and *PDGF*-A significantly. Of all the 17 oncogenes studied, only *fms* correlated strongly with high FIGO clinical stage, and/or historic grade indicative of aggressive tumor behavior and poor outcome. *sis* expression was noted in mixed Müllerian neoplasms as opposed to adenocarcinomas.

Levels of *fes* transcripts (usually observed in rapidly proliferating cells) were characteristic of these aggressive neoplasms, distinguishing them from benign or borderline cases. The consistency of the correlations between levels of c-*fos* and c-*myc* could be expected, because both genes are known to be expressed in rapidly proliferating cells (Kacinski et al., 1989). Again, correlated with these two genes, the expression of the *src* family oncogenes *erb*B, *neu, fms,* and *ros* was also biologically reasonable, suggesting that these kinases and their ligands play a role in the control of ovarian epithelial cell proliferation. It appears that the observed **coordinate expression of multiple growth factor receptors** (ErbB, Neu, Fms, Ros, and Trk) is not an artifact of cross-hybridization.

The *fms* oncogene, which is expressed in adenocarcinomas of ovary, codes for the receptor for **macrophage colony-stimulating factor** (CSF-1), a mutagen, chemoattractant, and activator of tissue macrophages and trophoblast cells (Kacinski et al., 1989). The protein has a role in wound healing, immune response, implantation, and development of the human placenta. No significant genomic rearrangements have been detected in c-*fms*, and Northern blot, cDNA PCR, and immunological and immunohistochemical studies suggest the expression of c-*fms* mRNA and protein by ovarian adenocarcinoma is normal or near normal. Thus, it appears that a mutant *fms* gene is not responsible, all by itself, for uncontrolled proliferation of adenocarcinoma cells.

Similar to the situation in breast cancer, the expression of **epidermal growth factor receptor** (*EGFR*) and *neu* genes, in epithelial ovarian cancer, correlates with poor prognosis (Berchuck et al., 1990 a,b). Likewise, overexpression of *HER*-2/*neu* oncogenes in ovarian cancer is associated with resistance to the cytotoxic effect of **tumor necrosis factor** (TNF) (Lichtenstein et al., 1990). The cell lines that did not express the oncogene showed sensitivity to TNF and were sensitive to lymphokine-activated killer cells.

Overexpression of *neu* appears to confer on tumor cells resistance to several different cytotoxic influences and, thereby, a proliferative advantage over normal cells (Lichtenstein et al., 1990). The resistance of ovarian carcinoma cells related to the expression of **multidrug-resistant protein P** can be readily detected in routinely processed tissues, and helps in predicting failure to respond to standard chemotherapeutic treatments (Rutledge et al., 1990).

The *neu* oncogene, also implicated in breast and ovarian malignancies (de Potter et al., 1989; Parkes et al., 1990; Slamon et al., 1989a,b; Tsuda et al., 1990a,b; van de Vijver et al., 1988) and in primary tumors of the peripheral nervous system (Perantoni et al., 1987), does not show significantly different levels of expression in benign and borderline neoplasms from those found in invasive lesions; there is, however, a statistically significant correlation between levels of *neu* and *fms* expression in both ovarian and breast neoplasms (Kacinski et al., 1989).

A homozygous deletion of the **RB gene** in an endometrioid tumor of the ovary of low malignant potential has been implicated as playing a role in its aggressive biological behavior (Sasano et al., 1990a). The aggressive nature of this tumor was not reflected in any other markers of poor prognosis such as nuclear atypia, DNA aneuploidy, and protooncogene amplifications. Allelic loss at the *RB* locus has also been detected in 30% of the informative cases of other ovarian neoplasms, which included leiomyosarcoma, sex cord-stromal tumor, epithelial carcinomas (mucinous, endometrioid, mixed Müllerian, and serous), and their metastases (Li et al., 1991).

Two genes, *PSGGA* and *PSGGB*, that code for the human **pregnancy-specific β glycoprotein** have been cloned (Leslie et al., 1990). Oligonucleotide probes specific for *PSGGB* revealed that this gene encodes a 1.7-kb mRNA in **hydatidiform mole** tissues, and a 2.0-kb mRNA in term placental tissues. A diagnostically useful aspect of the *PSGGB*-specific probe is its preferential hybridization to mRNA from trophoblastic tissue, suggesting a practical way for diagnosing gestational trophoblastic disease at the gene level.

3.3.5 CONCLUDING REMARKS

Another aspect of oncogene amplification in ovarian malignancies, as well as in other tissues, deserves particular attention in terms of its practical use in diagnosis and prognosis. Namely, the observed correlation between oncogene amplification and nuclear atypia may have a specific molecular explanation. Although the mechanisms accounting for the tissue-specific gene expression have not been elucidated completely, several lines of investigations point to potentially important practical outcomes.

One line of research concerns the **role of nuclear structure** in the control of specific gene expression. Alterations in nuclear structure are noted during mitogenic stimulation (Bladou et al., 1988) and the induction of differentiation (Stuurman et al., 1989, 1990). The **nuclear matrix** has different protein compositions in different cell lines, and also differs between transformed and nontransformed cells of similar origin (Fey and Penman, 1988). Characteristic alterations occur in its protein composition in response to normal stimulation (Getzenberg and Coffey, 1990). It has been proposed that tissue specificity of the nuclear matrix is also responsible for arranging the DNA in a unique conformation, and, thereby, it eventually participates in the **tissue-specific patterns of gene expression** (Getzenberg and Coffey, 1990).

One of the plausible consequences of protooncogene amplification and overexpression could be alterations in the nuclear matrix. If these alterations do indeed lead to modification of the tissue-specific gene expression, the amplification and/or overexpression of certain protooncogenes that results in nuclear atypia reflecting nuclear matrix alterations should be more quantitatively evaluated. At this time, quantitative estimates of nuclear spatial characteristics are far from being easy, accurate, or reproducible. It is unclear whether physical and/or optical means could soon be developed to detect specific structural rearrangements characteristic of a given type of neoplasia. In the meantime, gene level evaluation of ovarian, as well as other, malignancies offers the most helpful adjuncts in evaluating a given condition.

Chapter 3.4

Prostate Neoplasms

3.4.1 INTRODUCTION

Evaluation of prostate hyperplasia for possible foci of malignancy has been traditionally performed by light microscopy examination of tissue sections on the basis of morphological characteristics grouped into diagnostic clusters reflecting our understanding of the natural history of prostatic cancers. Histologic grading is also used for prognosis of the disease. These generalizations, however, frequently fail, because patients with the same pathologic stage and histologic grade of prostatic adenocarcinoma may experience a different course of the disease. Therefore, an important challenge facing prostatic cancer evaluation, is the development of individualized prognosis to supplement conventional assessment by grade and stage (Gittes, 1991; Lieber, 1989; Oesterling, 1991).

Naturally, because the ploidy assessment has largely been correlated with the prognosis of tumor development in various organs, this approach has been applied to prostate adenocarcinoma as well. Only a few representative relevant studies will be discussed here.

3.4.2 PLOIDY IN PROSTATE NEOPLASMS

DNA flow cytometry (FCM) of 91 samples from patients with advanced stages (D1) of prostate adenocarcinoma, having metastases in pelvic lymph nodes, showed 42% of the tumors to be diploid, 45% tetraploid, and 13% aneuploid (Winkler et al., 1988). Differences in tumor behavior were strongly correlated with ploidy, so that 75% of tetra- or aneuploid tumors were found to progress subsequently in patients followed for 5 to 19 years; at the same time, only 15% of DNA diploid tumors progressed locally or systemically. None of the patients with diploid tumors died during the period of observation. In contrast, over 40% of those having tetra- or aneuploid tumors at the time of resection had died of prostate cancer 10 years after prostatectomy. DNA FCM appears to offer highly significant and objective prognostic information in patients with advanced D1 stage prostatic adenocarcinoma (Winkler et al., 1988).

The vast experience gained at the Mayo Clinic through the study of archival paraffin-embedded prostate cancer samples by FCM suggests "that tumor nuclear DNA pattern is perhaps the most important variable now known for forecasting the pace of tumor progression for patients with adenocarcinoma of the prostate" (Lieber, 1989). The management of such patients must take into account their tumor ploidy data. Again, the observation that even metastasized diploid tumors grow very slowly, and are unlikely to kill the patient, is emphasized. **Ploidy pattern** thus does not serve as an index of metastasizing potential of prostate adenocarcinoma, but rather reflects the rate of **tumor progression**. Indeed, given the progressive introduction of early screening for prostate cancer by ultrasound with simultaneous sampling of the questionable lesions at the time of ultrasonography, the DNA ploidy evaluation may become an important modality in diagnosis and management of prostatic neoplasias.

Determination of the nuclear DNA content in prostatic needle biopsy specimens, and the prediction of disease progression, has been evaluated in patients with stage A or B prostatic cancer (Peters et al., 1990). Image analysis (ICM) of Feulgen-stained nuclei was performed on specimens retrieved from 44 patients, and suggested that **nuclear DNA quantitation on needle biopsy specimens allows better estimates of prostatic cancer progression than those based on tumor grade alone**. Although DNA analysis by ICM is less sensitive than FCM estimates of ploidy, the resolution of ICM, combined with the convenience of selective evaluation of areas of particular interest, offers definite advantages for routine patient management, even in cases of localized adenocarcinoma of the prostate (Peters et al., 1990). Comparison of the advantages and drawbacks of evaluating prostatic carcinoma in smears and tissue sections led to the conclusion that tissue histograms were difficult to interpret and **cytologic preparations should be preferred for ICM evaluations** (Epstein et al., 1990).

To find better correlates between DNA ploidy and the clinical behavior of prostatic cancer, quantitative cut-off values were imposed on the ploidy measurements performed on Feulgen-stained smears in a series of 213 patients (Forsslund and Zetterberg, 1990). The diploid/tetraploid tumors had less than 30% aberrant tumor cells, whereas aneuploid (high-grade) tumors had over 50% nondiploid and nontetraploid tumor cells. All patients assigned an aneuploid tumor type died within 5 years, whereas all patients in the diploid/tetraploid group lived over 5 years.

A substantially **higher rate of progression** has been observed in nondiploid tumors, with all (100%) DNA aneuploid tumors progressing toward higher stages in 349 patients following radical retropubic prostatectomy for pathological stage B adenocarcinoma of the prostate (Montgomery et al., 1990). Again, the analysis of ploidy by flow or static cytometry is suggested as an "essential tool" for patient management in localized prostate cancer (Montgomery et al., 1990).

Initial ploidy evaluation in another study correlated with the cytological grade, but not tumor staging, **prostatic acid phosphatase** activity, or patient age, and had prognostic value as a single predictor and in combination with other parameters (Adolfsson et al., 1990). A serial follow-up evaluation of DNA ploidy on prostate needle biopsy specimens from untreated patients confirmed the notion that a process of **gradual dedifferentiation** occurs in the course of natural progression of prostate cancer (Adolfsson and Tribukait, 1990).

A largely **contrasting conclusion** has been reached by other urologists, who also studied DNA content of prostatic adenocarcinoma by FCM at Stanford University (E. C. Jones et al., 1990). Formalin-fixed, paraffin-embedded tissue from 57 radical prostatectomies was analyzed for ploidy patterns which were compared with the grade, volume, and pathological stage, as an aid in selecting the best candidates for surgical treatment. Aneuploidy was present in 46% of specimens, almost all of them larger than 4 cm^3. Aneuploidy was more predominant in advanced stages of the disease and less differentiated tumors. However, diploid tumors were also found in this category, and **the relationship of aneuploidy vs. pathological stage and grade was not significant**. Sensitivity and specificity of DNA content analysis was low (50 to 72%), and only in the larger tumors (over 4 cm^3) did it reach 91%. This suggested that aneuploidy is associated with tumor progression, but is not a requirement for progression to occur. **DNA content analysis is, therefore, discounted as an independent predictor of the clinical course of prostatic adenocarcinoma** (E. C. Jones et al., 1990).

It is argued that, although FCM established an empirical correlation between aneuploidy and prognosis of various carcinomas, no molecular or structural basis for such correlations has been developed beyond the notion that an abnormal DNA content implies **genetic instability** (E. C. Jones et al., 1990). One of the possible reasons for failure of DNA aneuploidy determined by FCM to be of prognostic value could be the discrepancy reaching up to 30% between the estimates made by FCM and the corresponding karyotypic analysis (Remvikos et al., 1988). Jones and coworkers (1990) admit that with refinement of DNA FCM analysis, this modality may prove clinically useful on biopsy material. Because E. C. Jones and coworkers (1990) screened each case before operation as well as by intraoperative lymph node examination, to exclude incurable carcinomas, they believe that the observations presented in their series are the most representative, and inclusive of cases ordinarily considered for definitive therapy.

DNA FCM and replication rate estimates performed on deparaffinized specimens from patients with localized prostate cancer failed to enhance significantly the ability of standard histopathological grading to predict disease recurrence (Ring et al., 1990). Other studies can be cited **questioning the validity of aneuploidy as a prognostic factor** (S. Lundberg et al., 1987; Seppelt et al., 1986; Stephenson, 1988). It appears that a finer (than flow cytometric) level of evaluation of changes at the gene, rather than "bulk" DNA, level may provide clues to reliable, objective, and reproducible means of prostate cancer evaluation on biopsy material. This may lead to the development of treatment alternatives to radical prostatectomy.

3.4.3 CHROMOSOMAL ABERRATIONS IN PROSTATE CANCER

If more advanced malignancies do tend to be aneuploid and less differentiated, then at least numerical chromosomal aberrations are to be expected in the karyotype of virtually all prostate adenocarcinoma samples. Such cytogenetic analyses have been reported on primary tumors and their metastases (Ahmann et al., 1986; Atkin and Baker, 1985; Gibas et al., 1985), and on prostatic adenocarcinoma cell lines (Gibas et al., 1984; Iizumi et al., 1987; Ohnuki et al., 1980; Konig et al., 1987, 1988; Pittman et al., 1987). A detailed cytogenetic analysis of a xenografted human prostatic adenocarcinoma cell line PC-82 has compared affected chromosomes in this line with all data available on prostatic carcinoma and identified **chromosomes 1, 2, 3, 6, 7, 10, and 15** as involved in rearrangements in over 50% of all prostatic carcinomas. Chromosomes 1 (90% of cases), 7 (80%), and 10 (70%) appear to be consistently involved in all the primary tumors studied.

A survey of previous cytogenetic data on human prostate adenocarcinoma, considered in conjunction with cytogenetic evaluation of primary prostatic cancer cell cultures, indicated that the most common chromosomal changes included the **loss of chromosomes 1, 2, 5,** and **Y**, the **gain of chromosomes 7, 14, 20,** and **22**, and **rearrangements involving 2p, 7q,** and **10q** (Brothman et al., 1990). Although the appearance of any clonal cytogenetic abnormality in general correlates with a poorly differentiated state of prostate cancer, **a single chromosomal aberration cannot be used as a marker for early-stage prostate cancer** (Brothman et al., 1990).

The role of **tumor suppressor genes** in prostate cancer has been suggested by the frequent LOH (30% of all tumors) affecting predominantly **chromosomes 16q** and **10q** (Carter et al., 1990a,b). Fifty-four percent of the localized cancers, and 100% of metastasized tumors, displayed LOH on at least one chromosome.

3.4.4 DNA FINGERPRINTING OF PROSTATIC CANCERS

Another approach can be seen in the **DNA fingerprinting** analysis of prostatic adenocarcinoma and benign prostatic hyperplasia (BPH) specimens (White et al., 1990). This technique was used to scan a large number of independent genetic loci for possible alterations in these two conditions, which frequently present diagnostic dilemmas, especially on biopsy specimens. Two DNA probes, containing multiple repeats with variant core sequences, and capable of detecting two different subsets of **hypervariable loci**, were used to generate DNA fingerprints after digestion with *Hae*III or *Hin*fI restriction enzymes of matched tumor and blood DNAs. There is an infinitely small probability (1:9 billion) of two unrelated individuals having the same DNA fingerprint pattern (White et al., 1990).

Among the 26 individuals with prostatic neoplasms, **altered band intensities** were observed in DNA fingerprints prepared from tumor DNA in 12 of 14 prostatic adenocarcinoma cases, as well as in 5 of 12 BPH cases, compared to constitutive DNA (White et al., 1990). **Decreased band intensities** were significantly more common in adenocarcinoma than in BPH specimens. **Increased band intensities** were noted in both. One of the two probes (**33.15**) proved more informative than the other (**33.6**) in revealing distinctions between adenocarcinoma and BPH. It was not established whether decreases or increases of band intensities or the new bands observed in the DNA fingerprints represented consistent and characteristic deletions, rearrangements, or amplifications at a single polymorphic locus or reflect independent patient-specific events occurring at different chromosomal locations. They were, nevertheless, consistently correlated with the type of neoplasia.

Because Southern hybridization analysis of total genomic DNA is unable to detect deletions occurring in less than 20% of the cells present, the finding of band alterations in some BPH patients is unlikely to be explained by minimal volume, undiagnosed adenocarcinomas. A note has been made of a reported correlation between the extent of **hypomethylation** and prostatic tumor grade (Bedford and van Helden, 1987). Again, because these changes in BPH specimens were detected predominantly with one of the two restriction enzymes (*Hin*fI), which is known to be methylation sensitive, these observations could be partially due to changes in methylation of bases adjacent to the minisatellite regions. Methylation changes may not account for all the observed band intensity variations, also because these were observed when the other restrictase, *Hae*III insensitive to methylation, had been used (White et al., 1990). DNA fingerprinting appears to offer an extraordinarily helpful approach to diagnosis of prostate neoplasia and more studies will hopefully follow to confirm these initial observations.

3.4.5 GROWTH FACTORS AND ONCOGENES IN PROSTATE TUMORS

A number of growth factors and their receptors have been identified in the prostate gland: the **fibroblast growth factor** (FGF) and **transforming growth factor β** (TGF-β) families, **TGF-α**, and **epidermal growth factor (EGF)**. Elevated protooncogene and activated oncogene activities (elevated expression of *myc*, and the expression of activated *ras*) have also been detected in prostate cancer (for a review, see T. C. Thompson, 1990).

Within the FGF family of growth factors, the overexpression of *int*-2 gene, **basic fibroblast growth factor** (bFGF) mRNA, and **acidic fibroblast growth factor** (aFGF) mRNA has been studied in animal models, and their role in human prostate tumors is being actively investigated (Table 3.3). Members of the TGF-β family appear to be expressed at higher levels in the poorly differentiated prostatic adenocarcinomas (T. C. Thompson, 1990). **Epidermal growth factor (EGF) receptor** (EGFR) binding and its mRNA levels appears to be elevated in human prostate cancer. The level of EGFR in human BPH is higher than that in prostate cancer, and the well-differentiated tumors demonstrate higher expression of EGFR (Maddy et al., 1989).

Although the diverse range of metabolic activities in the prostate gland is under the control of **androgens**, the changes in androgen responsiveness do not appear to be associated with detectable alterations in the structure of the androgen receptor gene, at least in prostate carcinoma cell lines (Tilley et al., 1990). The regulatory pathways involving androgen control of genes in the human prostate gland are far from clear at present, and the interrelationship with protooncogenes and growth factors is the subject of intense research (Bussemakers et al., 1990; Sonnenschein et al., 1989; T. C. Thompson, 1990).

Activation of *ras* genes by point mutations has been amply documented in the human prostate tumors (Cooke et al., 1988a,b; T. C. Thompson, 1990; Varma et al., 1989). The general consensus, however, is that the frequency of *ras* gene mutations in prostate tumors is low (Carter et al., 1990a,b; Gumerlock et al., 1991). It appears that these mutations may play a role in the progression of prostate cancer or the development of the rare ductal variant of prostatic

TABLE 3.3

Growth Factor and Oncogene Expression in Benign and Malignant Prostatic Tissues[a]

Growth factor/oncogene	Species	Tissue/cell line	Material assayed	Comments
bFGF	Human	Benign prostate hyperplasia was compared to normal prostate	mRNA	Elevated bFGF mRNA in benign prostatic hyperplasia Limited number of control, normal prostate specimens
aFGF	Rat	Transplantable rat prostate adenocarcinoma and derived cell lines	mRNA	aFGF elevated compared to normal mature adult prostate aFGF expressed in epithelial cells of normal immature prostate and in mesenchymal cells of adenocarcinoma
TGF-β_1	Rat	Lineage of transplantable rat prostate adenocarcinomas	mRNA	Elevated TGF-β_1 mRNA in all adenocarcinomas compared to normal adult prostate Increased TGF-β, mRNA in less differentiated adenocarcinomas
TGF-β_1	Mouse	Poorly differentiated primary adenocarcinomas induced by *ras* + *myc* using the mouse prostate reconstitution model system	mRNA	Elevated TGF-β, mRNA in all adenocarcinomas compared to normal and hyperplastic prostate
TGF-β_2	Human	Benign prostatic hyperplasia was compared to normal prostate	mRNA	Elevated TGF-β, mRNA in benign prostatic hyperplasia Limited number of normal control prostate specimens
Ki-*ras*	Human	Adenocarcinoma	Cellular DNA	NIH-3T3 focus assay and Southern blotting indicated transforming Ki-*ras* oncogene in 1 of 8 prostatic adenocarcinomas
Ha-*ras*	Human	Adenocarcinoma	Cellular DNA	A $\rightarrow$ G transition in codon 61 of the Ha-*ras* gene detected by oligonucleotide hybridization Low-frequency occurrence, only 1 of 24 adenocarcinomas examined Cancer with *ras* mutation was rare ductal variant

[a] This table is not a complete listing but it represents a selected group of studies that have demonstrated clear differences in specific growth factor activities or the presence of dominant transforming oncogenes in hyperplastic/malignant prostate tissue vs. normal prostate.

From Thompson, T. C. (1990). Cancer Cells 2:345–354 With permission.

adenocarcinoma (Carter et al., 1990a,b). Furthermore, the level of methylation of c-Ha-*ras* is the same in prostate cancer and the normal gland (Menegazzi et al., 1988).

Among other protooncogenes whose levels are elevated in prostate cancer are *myc, p53,* and *sis* (Cooke et al., 1988a; T. C. Thompson, 1990). Variable levels of *fos* mRNA have been detected (Cooke et al., 1988a), and these are involved in the cascade of gene activity leading to prostatic cell death (Buttyan et al., 1988). An interesting finding of marked phenotypic changes of cultured human prostate carcinoma cells, following the introduction of a normal *RB* gene, points to the role of this tumor suppressor gene in the genesis of prostate cancer (Bookstein et al., 1990).

The widely recognized high association of prostatic adenocarcinoma with **osteoblastic metastases** (Jacobs et al., 1979) has stimulated the search for mitogens with specificities for bone (Koutsilieris et al., 1987). While hyperplastic prostate tissue contained two peptides with apparent molecular masses of 10,000 and 13,000 Da, only a single polypeptide of 10,000 Da could be identified in the adenocarcinoma specimens. Analysis of the growth factor activities recoverable from the prostatic carcinoma cell line conditioned medium detected a **unique mitogen**, an M_r 26,000 to 30,000 basic protein, that has a higher specificity for human osteoblasts than for human fibroblasts (Perkel et al., 1990). It is believed that the production of this unique growth factor plays a role in the intense osteoblast-specific stimulation that is observed at the sites of prostatic adenocarcinoma bone metastases.

3.4.6 PROSTATE-SPECIFIC MARKERS: PROSTATE-SPECIFIC ANTIGEN AND HUMAN GLANDULAR KALLIKREIN-I

Attempts to find objective molecular distinctions between BPH and adenocarcinoma (AC) have drawn attention to possible differences at the gene expression level. One such effort has been to analyze the pattern of expression of two genes encoding closely related protein products — **prostate-specific antigen** (PSA) and **human glandular kallikrein-I** (hGK-I) (Gittes, 1991; Henttu et al., 1990; Oesterling, 1991). Northern and slot-blot hybridizations were used to evaluate mRNA species transcribed from both genes in BPH and CA specimens.

Prostate-specific antigen is a 33-kDa glycoprotein that has been extensively used for the evaluation of prostate involvement with hyperplasia and/or cancer. Although a significant range of PSA values determined in the blood is found in pathological conditions, most people do not exceed a certain low level of PSA if the prostate is normal and has not been manipulated. The serum concentration of PSA has been found roughly proportional to the mass of either benign hyperplastic tissue or cancer, and it is especially useful in monitoring the effects of specific therapy (Barak et al., 1989; Killian and Chu, 1990; Rainwater et al., 1990; Stamey et al., 1987, 1989) as well as in screening for prostate cancer in conjunction with rectal examination (Catalona et al., 1991).

Both PSA and hGK-I are **serine proteases** showing similarities in structure and at the respective gene level (Henttu et al., 1990; Klobeck et al., 1989). The determination of PSA and hGK-I mRNA species in BPH and CA revealed that there were **no significant differences in the pattern of PSA gene expression between BPH and CA**. Interestingly, while the total amounts of PSA mRNAs in all BPH samples were fairly similar, there was a threefold range of intersample variation in PSA mRNA levels in CA patients. Although the hGK-I mRNA variation in BPH samples was greater than that of the PSA mRNA, **the overall correlation between the expression of the two genes in BPH was good. Differences were noted in the expression of these two genes in the CA samples**. Because the histological distribution of PSA gene expression was not defined, the wide scatter of PSA mRNA in cancer tissue could be due either to a higher level of transcription in some cells, or to the expression of this gene in a wider range of carcinoma cells.

Prostate-specific antigen has been found to occur in complex with α_1-**antitrypsin**, in addition to its existence in a monomeric form (Stenman et al., 1991). The level of the complexed form progressively increases parallel to the elevation of PSA concentration, reaching over 85% at PSA levels above 1000 µg/l. Because the proportion of **PSA-α_1-anti-chymotrypsin** was higher in prostate cancer than in BPH patients, assaying for **this complex has a higher sensitivity for cancer than does an assay for total PSA immunoreactivity** alone. A combination of PSA with α_2-**macroglobulin** and **inter-α-trypsin inhibitor** has also been identified; however, its occurrence was similar in BPH and cancer patients. The search for discriminatory markers in prostate diseases continues. High levels of **5'-nucleotidase** have been found in human prostates without cancer, and this enzyme activity decreases in prostatic carcinoma (Rackley et al., 1989).

3.4.7 CONCLUDING REMARKS

Thus it appears that, in spite of the existence of a number of rather specific markers correlated with pathological conditions of the human prostate, at this time no clinically useful gene level probe can be reliably selected for patient management.

It seems that the informative practical value of gene level probes would have been significantly enhanced if the levels of the gene transcripts (e.g., PSA mRNA), determined biochemically in total tissue specimens, could be related to the specific epithelial and/or stromal markers, representing **a structural and/or functional unit of the tissue mass**. This marker could be either a protein or a gene dose such as actin. The **specific expression of a gene per unit target tissue** could reveal the finer and more consistent distinctions between benign hyperplasia and cancer.

This approach would have been of particular practical significance in the study of Henttu and coworkers (1990), because some aberrant PSA mRNA species were observed. It appears that CA samples contained somewhat larger amounts of these mRNAs, but whether these were expressed into variant PSA proteins, and how efficiently and to what extent these were involved in malignancy, is at present unknown. It is tempting to speculate that such aberrant PSA mRNAs may reflect an altered state of the putative steroid response elements (Riegman et al., 1989), and provide a clue to the role of androgens in prostate cancer. Investigations along these lines will point to distinctions in tissue- and tumor-specific genes that can be exploited for gene level diagnosis or monitoring of prostate pathology.

Chapter 3.5

Kidney and Urinary Bladder

3.5.1 RENAL NEOPLASMS

The traditional classification of renal neoplasms recognizes renal cell carcinoma (adenocarcinoma), transitional cell carcinoma and squamous cell carcinoma of the renal pelvis, and sarcomas of the kidney and nephroblastoma (Silva and Childers, 1989). For a discussion of trends in the diagnosis and treatment of renal, upper urinary tract, and bladder carcinomas see Stenzl and de Kernion (1991). It appears that even certain cases of **familial nephrolithiasis** can benefit from molecular diagnostic approach as shown in an extensive pedigree that had an X-linked recessive nephrolithiasis with renal failure (Frymoyer et al., (1991).

3.5.1.1 RENAL CELL CARCINOMA (ADENOCARCINOMA)

3.5.1.1.1 DNA Ploidy

DNA cytometry has been applied to differentiation between **renal cell carcinoma** (RCC) and **renal adenoma**, which frequently presents a challenge when the diagnosis is to be based solely on histology and size (conventionally, tumors less than 3 mm in size are considered adenomas) (Silva and Childers, 1989). The size criterion frequently fails, and so does eosinophilia of tumors, referred to as **oncocytomas** and **tubular adenomas**. When 6 adenomas and 15 adenocarcinomas were retrospectively evaluated by DNA image cytometry (ICM), the tumor histograms differed substantially from controls (Bennington and Mayall, 1983). Statistical analysis established a correlation between the cytometric data and tumor grade. And although the cytometric features accurately predicted the favorable and poor outcomes, no clear distinctions could be made, either between renal adenomas and grade I adenocarcinomas, or between oncocytic and clear cell tumors of the same grade by these criteria.

A significant correlation appears to exist between the overall survival of the patient, the spread of metastases, nuclear elongation, nuclear crowding, mitotic density, and tumor grade (Bibbo et al., 1987). Likewise, a strong correlation has been observed between these nuclear parameters and histological grades of RCC, confirming the earlier observations (Baisch et al., 1982; Ljungberg et al., 1985, 1986; Otto et al., 1984; Schwabe et al., 1983). A high proportion (up to 75%) of large RCCs contain aneuploid population of cells (Ljungberg et al., 1985). A retrospective ICM study of RCC has correlated the size of the tumor and DNA content with its predominant cellular population, and demonstrated that **abnormal DNA content and heterogeneous cell populations begin to appear in tumors between 2.5 and 5.0 cm in size** (Banner et al., 1990). All tumors over 5.0 cm contained nondiploid cell populations. The appearance of nondiploid stem lines and heterogeneous DNA content seems to parallel tumor growth and indicates more aggressive behavior of the RCC. Quantitative criteria have even been proposed, such as a mean nuclear area of 32 mm^2 as the decision threshold for a 5-year survival rate after surgery (Tosi et al., 1986). High-speed nuclear morphometry may possibly be considered as an adjunct to the evaluation of RCC (Bibbo et al., 1987).

Flow cytometry (FCM) of **renal tumors of childhood** has been viewed as a helpful supportive tool for determining prognosis (Douglas et al., 1986; Schmidt et al., 1986). On the other hand, although 75% of Wilms' tumors, showing pleiomorphic histology, were aneuploid, "the population index was found to be of no prognostic value", and no correlation was observed in these tumors between ploidy and staging in any of the histological types of Wilms' tumor studied (Kumar et al., 1989). (See also Chapter 3.6).

Thus, it appears that the **summary evaluation of DNA content** in RCC, even when related to fine morphological characterization, leaves room for better diagnostic and prognostic tools, both in terms of specificity, convenience, objectivity, and better correlation with the biology of RCC. An interesting use of **RNA content** estimated by acridine orange FCM of RCC indicated that although overall RNA level did not correlate with clinical stage of the tumor, within a set of diploid RCCs, high RNA content was correlated with high nuclear grade, large tumor size, advanced clinical stage, and cytoplasmic granularity (El-Naggar et al., 1990a). **RNA content assessment appears to be a valuable objective and quantitative parameter helpful in the clinicopathological evaluation of RCC.** As will be shown below, molecular cytogenetic and oncogene evaluation seems to offer a substantially better approach to clinicopathological assessment of RCC patients and their prognosis.

3.5.1.1.2 Molecular Cytogenetics of RCC

Successful establishment of cell lines from RCC has been reported, and the more aggressively behaving tumors proved easier to establish in culture (Ebert et al., 1990). Availability of cell culture derivatives of tumors retaining the characteristics of the source tumors offers a convenient way to study the cytogenetic and molecular mechanisms responsible for specific characteristics of the tumor under study, as demonstrated, for example, in lung tumor studies. Renal cancer tumor cell lines as well as cell lines established from normal kidney have shown so far a significant cellular heterogeneity (Ebert et al., 1990).

In RCC the **chromosome 3** rearrangement (3p11-p21) has been repeatedly observed (Berger et al., 1986; Carroll et al., 1987; Cohen et al., 1979; Dal Cin et al., 1989; de Jong et al., 1988; Kovacs et al., 1987; Kovacs and Frisch, 1989; Miles et al., 1988; Pathak et al., 1982; Szucs et al., 1987; Wang and Perkins, 1984; Yoshida et al., 1986). Loss of heterozygosity (LOH) in the 3p region is a nonrandom alteration observed in all **nonhereditary RCCs** studied, and appears to be important in the origin or progression of RCC (Zbar et al., 1987). Which specific gene may be lost from that region, or what associated changes (such as a null mutation) may occur at one 3p allele related to the loss of the wild-type allele, is not known yet. It is of interest, however, that the molecular probe that identified allele loss in RCC has been used to demonstrate allele loss in small cell lung carcinoma (SCLC), suggesting the involvement of a recessive oncogene located on 3p (Zbar et al., 1987). The frequency of LOH on 3p is 53% in **sporadic RCC**, predominantly associated with the **clear cell RCC** (in 75% of cases) (Ogawa et al., 1991). Distinct association of LOH on 3p with a particular histological type of RCC is suggested by the absence of this chromosome abnormality in mixed cell-type tumors consisting predominantly of granular cell components.

In support of the role of genetic loss in the development of RCC, similar to the situation in retinoblastoma and Wilms' tumor, the loss of the putative "RCC gene" located near the most frequently lost region on 3p, D3F15S2, either severely reduces the level of expression or leads to the complete absence of expression of a gene located near this site, as observed in primary RCC (Erlandsson et al., 1990). This gene has been provisionally designated as "*RIK*", and found to be fully expressed in normal kidney. The same gene has been independently isolated by Naylor and coworkers (1989).

The primary role of the chromosome 3p loss in RCC, over that of the activation of **H-*ras*** oncogenes and **chromosome 11p** allele loss (Anglard et al., 1989; Fujita et al., 1987, 1988), is suggested by the failure of chromosome 11 introduced into a human RCC cell to affect tumorigenicity or cell growth (Shimizu et al., 1990). On the other hand, the introduction of a normal chromosome 3p modulated the tumorigenicity of the same cells. Although the majority of RCC cases have LOH on 3p, as shown both microscopically and at the molecular level, the location of the putative tumor suppressive gene on 3p has not been precisely defined.

A family has been identified with a balanced **constitutional t(3;8) translocation** in eight of the family members who developed renal cancer with the breakpoint either at 3p21 or 3p14.2 (Cohen et al., 1979; Wang and Perkins, 1984). Three probes (pH3E4/D3S48, pHF12-32/D3S2, and pMS1-37/D3S3) have been used to localize the breakpoint by *in situ* hybridization (ISH) (van der Hout et al., 1991). The first two probes located in the 3p21 region, whereas the third probe hybridized at 3p14. Distortion of the higher order spatial conformation of chromosome 3p in the region of the breakpoint is thought to account for the failure of ISH, which shows only a weak signal at this site for the probe pMS1-37/D3S3 (van der Hout et al., 1991).

Characteristic histopathological features of four types of renal cortical tumors have been correlated with cytogenetic and RFLP data. The four histological categories were (1) clear cell, nonpapillary RCC, (2) nonclear cell, nonpapillary RCC, (3) papillary RCC, and (4) oncocytic tumors. Only 28 tumors (56%) were successfully karyotyped. Unbalanced translocations and terminal or interstitial deletions were noted on chromosome 3p. The majority of abnormalities of **chromosome 5** were due to unbalanced translocation between 3p and 5q. Trisomy or more copies of **chromosomes 17, 7, and 12** were observed. Also, there was a relative deficiency of chromosomes 17p and 18, in both cases accompanied by deletion of 3p. The majority of 3p deletions established by RFLP and cytogenetic analysis were in clear cell, nonpapillary RCC. Oncocytic and nonclear cell type, nonpapillary RCC, on the other hand, did not have 3p deletions by either technique. **No association with a particular histological type could be observed in the loss of alleles on chromosomes 17p and 18**. An increased dosage of gene(s) on chromosomes 5q and 7 suggested these phenomena were primarily related to **progression** of the disease (Presti et al., 1991).

Another study of 38 RCCs by RFLP analysis confirmed the high prevalence of LOH on 3p encountered in 64% of the tumors (Morita et al., 1991). The possibility of the involvement of more than one tumor suppressor gene in the development of RCC was suggested by the relatively high occurrence (30% of cases) of LOH at informative loci also on six other chromosomes (**5q, 6q, 10q, 11q, 17p,** and **19p**). Evidence supporting the role of a tumor susceptibility locus in markedly enhancing susceptibility to chemical carcinogenesis and the development of renal tumors with cell-type specific predilection to carcinogenesis has been provided in an animal model carrying a single germ line mutation at this locus (Walker et al., 1992).

Although the frequency of **familial cases of RCC** is relatively low (1%), the accumulated clinical and molecular genetic evidence allows the earliest recognition of this condition (Maher and Yates, 1991). The most frequent cause of inherited RCC is **von Hippel-Lindau** (VHL) disease, a rare autosomal dominant syndrome with a heterozygote prevalence of 1 in 50,000 individuals, with some 700 cases reported so far. RCC in VHL disease is frequently bilateral and multicentric. Patients at risk for familial RCC should be screened annually from age 20 as in VHL disease to ensure the earliest detection of the tumor that may significantly improve prognosis (Maher and Yates, 1991).

3.5.1.1.3 Papillary and Nonpapillary RCC

A distinction between papillary and nonpapillary RCC, as well as between tubulopapillary adenomas and papillary carcinomas, has been claimed to be of practical value and supported by molecular cytogenetic characteristics of these types of RCC (Kovacs, 1989b). Hereditary, nonpapillary RCC features a breakpoint on **chromosome 3** (3p13-14.2), and functional inactivation of one allele of a putative suppressor gene in each germ-line cell (Kovacs, 1989b). In VHL disease, renal involvement (multiple cysts and/or nonpapillary RCC) is accompanied by loss of the chromosome 3p segment. It is not yet known whether it is the same gene that is responsible for the initiation of familial nonpapillary RCC and the development of different tumors, including RCC in VHL patients.

Kovacs (1990) believes that a gene, or cluster of genes, localized to 3p is responsible in both cases. This area of chromosome 3p appear to be also involved in the initiation of sporadic nonpapillary RCC. RFLP analysis of normal and tumor tissues form RCC patients pointed to a loss of a presumed tumor suppressor gene from 3p and the LOH for loci distal to the 3p13 breakpoint has been demonstrated nonpapillary, but never in papillary, RCC (Kovacs, 1990). **Loss of 3p segment is even proposed as a diagnostic marker for nonpapillary RCC.** Another chromosomal aberration frequently accompanying a loss of 3p segment is a **trisomy of 5q22-qter** observed in about 50% of nonpapillary RCCs. Because growth factor or receptor genes are localized to this segment, aberrations in this chromosomal area are possibly involved in proliferation of tubular cells (Kovacs, 1990).

The next most common chromosomal aberration observed in about 50% of RCC cases is the loss of **chromosome 14** (Kovacs and Frisch, 1989). Among the putative control mechanisms participating in the development of RCC, the loss of chromosome 14 may point to the c-*fos* protooncogene assigned to 14q.

Papillary RCC, although not universally recognized as a separate category of tumors, arguably deserves to be considered as such, based on a better survival, distinctive histologic characteristic (tendency to massive necrosis and cystic changes), and characteristic cytogenetic features (Kovacs, 1990). In contrast to nonpapillary RCC, **no aberration of chromosome 3p or 5q is reported in papillary RCC**, whereas about 70% of these have **trisomy 17**, and frequently **trisomy 7** (Kovacs, 1990). Confronted with the challenge of differentiating papillary adenomas from RCC, a suggestion has been made to consider papillary tumors of the kidney parenchyma showing only trisomy 7 and 17 as benign renal adenomas (Kovacs, 1990). The hypothetical molecular mechanisms involved in progression of renal adenomas to RCC may involve c-*erb*B-1 (*HER*-1) (chromosome 7) and c-*erb*B-2 (*HER*-2) (chromosome 17) (Kovacs, 1990).

3.5.1.1.4 Oncogenes and Growth Factors in RCC

Evaluation of the expression of three groups of protooncogenes by Northern blot (nuclear oncogenes c-*myc*, N-*myc,* and c-*fos*; **G proteins** encoded by c-Ha-*ras*, c-Ki-*ras,* and N-*ras*; and **protein kinases** encoded by c-*abl,* c-*fes* c-*fms,* c-*raf,* and c-*erb*B-2) revealed consistent overexpression of c-*myc* (73% of cases) and c-*erb*B-1 (47% of cases) (Yao et al., 1988). This overexpression was not due to gene amplification or rearrangement in the tumors as shown by Southern analysis. The concomitant overexpression of these two protooncogenes is thought to reflect their possible interaction and involvement in RCC tumorigenesis.

c-*myc* expression is not sufficient by itself to transform cells (Stern et al., 1986). In fact, in an animal model, a marked increase of c-*myc* and c-Ha-*ras* expression was observed in a normal kidney within hours of removal of the contralateral kidney (Bailey et al., 1990). A similar increase in the overexpression of these protooncogenes was also detected in the livers of the nephrectomized animals, suggesting this overexpression was an early, nonspecific response to the regenerative stimulus, and other signals must contribute to tissue specificity necessary for regeneration and cellular proliferation. By the same token, c-*myc* overexpression in RCC may reflect either reactive behavior of the cells of the uninvolved renal parenchyma, or constitute a part of the proliferative pattern of tumor cells (Bailey et al., 1990).

In any case, a correlation is observed between the degree of c-*myc* expression in the nuclei of cells in RCC, and the grade of the tumor reflected in nuclear pleiomorphism (Kinouchi et al., 1989). Positive staining for the c-*myc* gene product by monoclonal antibody was more frequent in grade 3 (100%) than in grade 1 (12%) tumors. A similar correlation was also observed in tissues from metastatic tumors of different grades, consistent with the notion that c-*myc* expression is not associated with the metastatic potential of RCC.

On the other hand, a correlation has been established between the frequency of distant organ metastases in RCC patients and the RFLP pattern of another protooncogene of the *myc* family,

L-*myc* (Kakehi and Yoshida, 1989). It appears that the **L-*myc* RFLP can be widely used as a molecular tool to predict prognosis in RCC patients**. Expression of the L-*myc* gene is also present in markedly metastatic SCLC, suggesting it is a feature of metastatic disease, involving lung among other organs (Kawashima et al., 1988). No correlation had been found between L-*myc* RFLP and colorectal tumors spread to various organs but sparing lungs and brain (Ikeda et al., 1988) and, clearly, more studies should be done on metastasized malignancies before the specificity of L-*myc* expression for lung involvement can be confirmed and become practically useful.

An interesting and important observation has been made on the differential expression of the genes for the precursors of **epidermal growth factor** (pro-EGF), **transforming growth factor α** (pro-TGF-α), and for the **EGF receptor** (EGFR) in tissue specimens from RCC and uninvolved kidney (Petrides et al., 1990). **For the first time in primary tumors in humans the gene for a growth factor (pro-EGF) has been found to be underexpressed**. The genes for pro-TGF-α and EGFR were **overexpressed** in virtually all tumor specimens compared to normal renal tissue. This overexpression was not due to amplification of the respective genes. Although almost all the RCC specimens overexpressed the gene for EGFR, confirming earlier reports (Mydlo et al., 1989; Yao et al., 1988), in a few of these there was also a coexpression of the gene for pro-TGF-α. The observed pattern of expression of the genes for EGF, TGF, and EGFR apparently plays a critical role in the development of RCC although the specific histological phenotype may be imparted by alterations of some additional genes (Petrides et al., 1990). The reciprocal relationship between the expression of the *EGFR* and *HER-2/neu* genes has been observed (Weidner et al., 1990). Again, the high expression of the *EGFR* gene was confirmed in 73% of cases. In the majority of cases (93%) the *HER-2/neu* gene displayed low expression. The reverse holds true for the normal kidney: high expression of the *HER-2/neu* gene and low expression of the *EGFR* gene.

Other oncogenes also showed a correlation with the degree of differentiation of RCC (Weidner et al., 1990) — **the c-*myc* expression was enhanced and c-*fos* oncogene expression was concomitantly decreased with progressively malignant (poorly differentiated) phenotype**. The decrease in c-*fos* expression can be related to the reported frequent cytogenetic aberration in RCC — a deletion in a region of **chromosome 14 (del(14) (q22-qter))** (Kovacs and Frisch, 1989), to which the c-*fos* gene has been mapped. The progressive decrease of c-*fos* expression with increasing malignancy can be related to the loss of one or possibly both c-*fos* alleles (Weidner et al., 1990). If confirmed, this observation may provide a basis for another practical, objective measure of malignancy in RCC.

Testing for the polymorphism in codon 72 of the *p53* gene by polymerase chain reaction (PCR) offers a rapid and sensitive tool to screen a large number of routinely processed RCC or bladder cancer specimens (Oka et al., 1991). Following microscopic evaluation of frozen sections, DNA can be isolated from areas with higher abundance of tumor cells. Loss of heterozygosity in the *p53* gene was 60% in RCC and 73% in bladder cancer specimens, as determined by this method (Oka et al., 1991).

3.5.1.2 RENAL CORTICAL ADENOMAS

In the meantime, cytogenetic analysis of renal cortical adenomas revealed a consistent pattern of chromosome abnormalities: a **tetrasomy 7(+7,+7), a trisomy 17(+17),** and **loss of Y chromosome** as determined by G banding of metaphases in short-term cell cultures established from these tumors (Dal Cin et al., 1989). Additional findings included a **trisomy 12(+12)** and a **monosomy 21**. The authors suggest that the specific combination of +7, +7, +17, and −Y defines a subgroup of renal cell tumors, which they called **giant cortical adenomas**. A combination of these four cytogenetic abnormalities is considered unusual suggesting a revision of other cases previously classified as RCC, which possibly included renal cell adenomas, because some of these changes were also noted among RCCs (Berger et al., 1986; Dal Cin et al., 1989; Yoshida et al., 1986).

3.5.1.3 RENAL ONCOCYTOMA

Renal oncocytomas, typically having a distinctive mosaic pattern of cells with normal and aberrant karyotypes, differ also in this respect from other renal tumors. They may constitute a **mitochondrial disease**, rather than be related to deletion of a tumor suppressor gene (Kovacs, 1990). The **distinctive restriction pattern** observed only in oncocytomas, and not encountered in papillary or nonpapillary RCC, may serve as a **differential diagnostic tool** in neoplasms showing histological similarities (Kovacs et al., 1989a).

Image cytometry of oncocytomas, oncocytic-granular RCC, and pure RCC as well as mixed tumors revealed that ICM is most efficient in identifying the heterogeneity of ploidy patterns in mixed cell type tumors (Veloso et al., 1992). Ploidy patterns correlated with the stage of the disease and were able to predict overall prognosis. Since both oncocytic tumors and oncocytic granular cell tumors are diploid and cannot be differentiated by this criterion, finer distinctions of ploidy patterns revealed by ICM may be of additional benefit over simple morphologic evaluation (Veloso et al., 1992).

3.5.2 POLYCYSTIC KIDNEY DISEASE

Autosomal dominant polycystic kidney disease (PKD) is most commonly (95% of cases) caused by a mutation at **chromosome 16p13** (Sutherland et al., 1987) locus *PKD*-1, the region extensively studied by numerous tightly linked genetic markers (Germino et al., 1990). The eventual aim is to clone the *PKD*-1 gene to enable more informative linkage studies in families, and to facilitate detection of carriers of mutations in members of a linked family. So far, attempts to isolate candidate genes from a number of CpG islands in regions expected to harbor the disease mutations suggested a possible role for a mutated proton channel in the pathogenesis of PKD (Gillespie et al., 1991). No differences could be found, however, in the cDNAs (AJ1) corresponding to both alleles of an affected individual, and the transcript size as well as the level of the putative gene expression were not different from those in the normal kidney (Gillespie et al., 1991).

The established linkage of genetic markers on chromosome 16 can be used for prenatal and presymptomatic diagnosis of PKD. However, some of the earlier markers displayed 5% recombination with the *PKD*-1 gene, and genetic heterogeneity of PKD exists in some families (Kimberling et al., 1988; Romeo et al., 1988). Additional paired flanking markers have been described, which increase the diagnostic utility of RFLP analysis in PKD (Breuning et al., 1989). Because families exist with PKD-carrying genes of unknown location, it is important to perform careful family studies in order for diagnostic DNA studies to be informative (Dalgaard and Norby, 1989).

However, because in about 4% of the families the disease is caused by mutations not occurring at the *PKD*-1 locus, linkage studies in that case are of little value (Parfrey et al., 1990). In the *PKD*-1 families, the presence of the disease established by ultrasonography could be confirmed by linkage studies. The disorder was found to cosegregate with polymorphic DNA markers flanking the *PKD*-1 locus in some families, whereas in other cases linkage studies were not informative. A correlation between linkage data and clinical variation of the disease suggests that, in people with negative ultrasonography in early adult life, there is a small likelihood of inheriting a *PKD*-1 mutation (Parfrey et al., 1990). In those who inherit a non-*PKD*-1 mutation for the disease, renal failure is likely to occur late in life.

By combining family studies in kindreds with PKD and the analysis of chromosome 16 from cell lines containing rearranged chromosomes, probes from 16 polymorphic loci were mapped to 16p13.1-pter (Breuning et al., 1990a). **A linkage map for the distal part of 16p has been constructed that incorporates newly defined RFLPs valuable for presymptomatic and prenatal diagnosis of *PKD*-1.** In fact, a two-step procedure for the diagnosis of *PKD*-1 by family studies utilizing these new markers has been described (Breuning et al., 1990b). In nonrecombinants, which comprise 90% of family members, the accuracy of diagnosis using these new markers exceeds 99%. Because autosomal dominant PKD is known to be genetically heterogeneous, the prenatal diagnosis of PKD on chorionic villi samples is recommended only after completion of haplotyping of the index family (Breuning et al., 1990b). Families are to be sufficiently large to rule out the rare form of PKD not caused be a mutation at 16p13.1-pter.

While linkage studies for the *PDK*-1 locus, particularly with the probe **CMM65**, can be used for presymptomatic and prenatal diagnosis, they have to be performed within each family with markers from both sides of the locus for accurate diagnosis to be made (Harris et al., 1991). A novel, promising approach uses polymorphic variations in the length of **microsatellites** — sequences composed of simple sequence repeats, such as dinucleotides $(CA)_n$, which can be detected by PCR and electrophoresis of the amplified products (Harris et al., 1991). Two microsatellite markers, in particular, have been identified that lie closer to the *PKD*-1 locus than any of the previously described ones: **SM7** (locus D16S283) identifies repeat sequence $(AC)_{19}$; and **SM5B** (locus D16S284) for the sequence $(A)_7$ GTT $(ATT)_5$ $(ATTT)_2$ $(GT)_{16}$ (Harris et al., 1991). The SM7 marker is particularly useful revealing 14 alleles with the observed heterozygosity in Caucasians of 62.7%. The use of ethidium bromide staining of the PCR-amplified products in nondenaturing acrylamide gels allows this linkage analysis to be completed in informative cases within one day.

A study of the **renin-angiotensin-aldosterone system** in PKD showed that the increase in renin release is apparently due to real ischemia caused by pressure from the expanding renal cysts, thus contributing to the early development of hypertension in PKD (Chapman et al., 1990). However, the role of renin and angiotensin II may not be so straightforward, because most patients with hypertension have normal or low circulating concentrations of renin and angiotensin II (Leckie, 1990). This problem has been addressed in transgenic animals, in which **the renin gene (*Ren*-1)** load has been studied, and the development of high blood pressure was linked to an extra gene for renin (Mullins et al., 1990). However, other candidate genes for hypertension are being investigated as well (Leckie, 1990; Morris, 1991; Soubrier et al., 1990).

A model of PKD developed in laboratory animals has been investigated with respect to **c-*myc*** expression, and an elevated and potentially abnormal c-*myc* expression was implicated in the pathogenesis of PKD (Cowley et al., 1987).

3.5.3 BLADDER CANCER

Early diagnosis of urinary bladder neoplasias presents a challenge when tumor size, location, and histopathological characteristics are not conspicuous enough to diagnose the malignancy unequivocally. Particularly difficult is to diagnose *in situ* carcinoma of the bladder, and various noninvasive diagnostic techniques have been devised to provide an accurate and objective diagnostic tool (Koss, 1979).

The most widely used approach is to evaluate exfoliated cells collected as bladder irrigation specimens. Flow cytometry of bladder irrigation specimens allows for differentiating urothelial cells from granulocytes and squamous cells by the estimates of the relative content of nucleic acids stained by the fluorescent dye, **acridine orange** (Colliste et al., 1980). Virtually all (27 out of 28) patients with papillary carcinoma, carcinoma *in situ*, and invasive carcinoma of the bladder could be identified. A good, but not complete, agreement was found between the FCM results of biopsy specimens and those of irrigation specimens (Devonec et al., 1982). No examples of positive conventional cytology and negative FCM findings could be found (Devonec et al., 1982). After conservative treatment of low-stage tumors, FCM of bladder irrigation specimens may provide a more sensitive tool than conventional cytology (Devonec et al., 1982).

Differential staining of blood and urothelial cells in FCM analysis of bladder irrigation specimens, using **propidium iodide** staining for DNA, can mimic aneuploidy or contribute to an increased hyperdiploid fraction (Wheeless et al., 1991). By using slit-scan enhancement of morphological features to enrich for urothelial cells based on the ratio of nuclear diameter to cell diameter for identification of urothelial cells, a significantly better discrimination of the method was achieved. In addition, the specificity increased from 61 to 77%.

By selectively sampling cells from, and around, **transitional cell carcinoma** (TCC) of the bladder at the time of cystoscopy, a correlation has been established between the WHO grading of TCC and the **fraction of cells in the S and G$_2$ phases** determined by high-resolution FCM measurement of cellular DNA content (Farsund et al., 1984a). Rather fine subdivision of grade 2 tumors into diploid and aneuploid could be achieved, taking into consideration also the extent of involvement of the surrounding, normal-appearing mucosa. In this way, based on the type of DNA aberrations, a reliable objective measure of the extent of bladder involvement with malignancy, and a biologic subclassification could be easily and rapidly achieved.

Image analysis performed on voided urine sediment proved to be a sensitive technique for the recognition of abnormal DNA patterns, especially in cases of prostatic disease in the absence of significant morphological abnormalities of the urothelial cells (Koss et al., 1987).

A comparison of **microspectrofluorimetry** (SFM), semiautomated **quantitative fluorescence image analysis** (QFIA), routine **Papanicolaou (Pap) cytopathology**, and **histopathology** has been performed on 272 samples from 67 symptomatic patients with TCC (Bass et al., 1987). Absolute nuclear fluorescence intensity of individual acridine orange-stained cells was able to detect recurrent and precancerous bladder lesions as well as kidney, ureter, and prostate lesions. The method is based on sensitive identification of abnormal DNA content, and is thought to be suitable for use in screening for early malignancy of the urinary tract. Actually, the evaluation of aneuploidy by FCM as a way to monitor intravesical chemotherapy, the success of which is estimated by the degree of elimination of aneuploid cells, has been reported in patients with urothelial cancer (Farsund et al., 1984b).

A retrospective evaluation of paraffin-embedded specimens from patients with primary superficial bladder carcinoma followed for up to 4 years has been performed by selective nuclear morphometry and FCM (Blomjous et al., 1989). **The quantitative assessment of morphology was found to be of value as a predictor of tumor recurrence.** The size of nuclei was important: only 62.1% of the cases with large nuclei remained free of progressive recurrence, whereas 92.2% of those with small nuclei did not have recurrences. DNA ploidy was selected as the best discriminator. None of the classic prognosticators, including histological grade or morphometry, added prognostic value. These observations suggest a more aggressive treatment approach to patients presenting with superficial carcinoma with large nuclei (nuclear area over 95 μm^2), or aneuploid DNA content.

An attempt to establish a new morphological indicator for the classification of bladder cancers has been made at the **electron microscopic** (EM) level to define reproducible characteristics of nucleoli (Shimazui et al., 1990). A number of morphometric parameters, such as the nucleolar volumes, nucleolus/nucleus volume ratio, volume densities of granular and fibrillar nucleolar components, as well as the number of **fibrillar centers** (FCs), were significantly changed with an increase in tumor grade. **Fibrillar centers significantly increase in number in bladder tumors compared to normal urothelial cells**, and this characteristic is considered a valuable diagnostic adjunct in defining malignancy of bladder tumors. **Nucleolar organizer regions** (NORs), related to regulation of protein synthesis and cell proliferation, are more numerous in bladder cancer samples and their number appears to be significantly related to survival (Lipponen and Eskelinen, 1991). In spite of that, **only a weak correlation** has been observed with lymph node involvement and distant metastases, necessitating further correlative studies before this parameter can be recommended for practical application (Lipponen and Eskelinen, 1991).

An enhanced discriminatory power of FCM is obtained when this methodology is combined with the selective power of **monoclonal antibodies** (MAbs). T16 MAb is able to differentiate normal, metaplastic, and neoplastic urothelium, to which it binds, from the inflammatory cells, fibroblasts, smooth muscle cells, and other nonepithelial cells found in a bladder irrigation specimen (Bretton et al., 1989). It appears that reevaluation of hyperploidy data previously obtained with acridine orange staining that failed to exclude inflammatory cells may be in order.

Another study utilizing immunoselection of bladder tumor cells by the monoclonal antibody DU 83.21 also eliminates the interference from inflammatory cells in the DNA FCM of bladder carcinoma cells (T24) propagated in culture (G. L. Wright et al., 1990). Following incubation of the T24 cells with the MAb DU 83.21 and propidium iodide (DNA stain), a good separation from buffy coat cells could be achieved.

3.5.3.1 CHROMOSOME ABERRATIONS, ONCOGENES, AND GROWTH FACTORS IN BLADDER CANCER

Chromosomal analysis of TCC of the bladder demonstrated a frequent loss of **11p** sequences, which in its frequency of occurrence approached that reported for Wilms' tumor (42 vs. 55%) (Fearon et al., 1985). Other chromosomal abnormalities involved **chromosomes 1, 5, 8, 9,** and **13** (Fearon et al., 1985). Analysis of 40 TCC of the urinary bladder by ISH with probes specific to **chromosomes 1, 7, 9,** and **11** revealed the occurrence of monosomy for chromosome 9 and trisomy for chromosomes 1, 7, and 11 (Hopman et al., 1991). The **loss of chromosome 9** was the most frequent chromosomal aberration in low-grade TCC of the bladder, and appears to be one of the **primary genetic events in TCC oncogenesis.** Secondary events, such as tetraploidization accompanied by underpresentation of chromosome 9 in comparison to chromosomes 1, 7, and 11, correlate with tumor progression.

Although no mutation activating *ras* oncogenes has been detected in human bladder SCC, an enhanced expression of the *ras*-21 proteins was demonstrated in four of nine studied **schistosomiasis patients** (Fujita et al., 1987). The human bladder carcinoma cell lines T24 and EJ are known for the presence of a point mutation within codon 12 of the c-Ha-*ras* gene; however, their transforming activity, if any, requires additional factors and is dependent on culture condition (Senger et al., 1988). *ras* oncogenes are either activated by point mutations or overexpress their product, p21, as a result of chemical carcinogenesis of urinary bladders in laboratory animals (Fujita et al., 1988). However, in experimental animals and *in vitro* cell culture models, *ras* gene activation (Knowles et al., 1987) and expression (Ward et al., 1988) are not necessary steps in urothelial transformation produced by chemical carcinogens.

Point mutations in codons 12, 13, and 61 of the *ras* gene have been frequently observed in TCC of the bladder, but less common point mutations can be identified by **PCR** in **codons 59, 61,** and **63** in paraffin-embedded tissue from the tumors of patients (Grimmond et al., 1990). In Japanese patients with **urothelial cancer**, no significant increase in the frequency of rare and of the three common alleles of Ha-*ras* could be detected (Ishikawa et al., 1987). Some of the genetic alterations in Ha-*ras* were not accompanied by increased expression of this protooncogene. Therefore, although evaluation of patients by RFLP for the Ha-*ras* gene is not a helpful modality in urothelial cancer, **deletions** of this gene may have a role in the development of these tumors (Ishikawa et al., 1987).

Substitution of **valine for glycine (G → T mutation)** at codon 12 of the Ha-*ras* gene has been detected by PCR in 12 of 33 tumor samples of human urinary bladder carcinomas studied (Czerniak et al., 1990). The frequency of this mutation as detected by PCR is higher than other estimates obtained by transfection assays and restriction analysis. Because this mutation was frequently found in the more aggressive urothelial tumors, the **performance of PCR amplification for the Ha-*ras* gene on voided urine specimens may serve as a helpful adjunct for identifying the aggressive variants of urothelial carcinomas** (Czerniak et al., 1990). A review of the earlier observations on the activation of *ras* and c-*myc* oncogenes in bladder cancer, their relevance to prognosis, as well as newer developments in the management of bladder cancer has appeared (Raghavan et al., 1990).

Comparison of chromosomal aberrations between **high-grade** and **low-grade TCC of the bladder** revealed the frequent occurrence of allelic loss of **chromosome 17p** (Olumi et al., 1990). Because this portion of chromosome 17 harbors the *p53* TSG often mutated in other cancers, this particular chromosomal abnormality may be associated with molecular mechanisms distinguishing low-grade from high-grade TCC (Olumi et al., 1990).

The loss of a putative **TSG** from **chromosome 11p**, one of the possible factors involved in bladder cancer (Fearon et al., 1985), may be complemented by the observed loss of the genetic region **3p14-21** in RCC, which also contains putative tumor suppressor genes (Klingelhutz et al., 1990). In a significant proportion of primary TCCs (16 out of 44), **c-*erb*B-2** overexpression could be demonstrated using an immunohistochemical technique (C. Wright et al., 1990). Analysis of the expression of **c-*erb*B-2** and a specific, codon 282 mutation of the *p53* **gene** in a patient with a large, poorly differentiated invasive renal pelvis TCC, ureteric TCC, and bladder TCC helped to establish that the bladder cancer arose from a distal spread of the renal pelvis tumor (Lunec et al., 1992).

The analysis of molecular genetic evidence related to bladder cancer convincingly points to the role of TSGs in addition to oncogene activation by point mutations, gene amplification, or deregulation of gene expression (Perucca et al., 1990).

The presence of a high concentration of **EGFR** is correlated with poor differentiation and invasiveness of human bladder carcinomas and, therefore, with poor prognosis (Hader et al., 1987; Neal et al., 1985). A study in a urothelial culture model demonstrated reduced requirements of transformed urothelium for the growth factor mitogens usually required for growth of normal urothelial cells, and lack of growth inhibition of the transformed cells by **TGF-β**, also contrasting with the behavior of normal cells (Jones et al., 1990).

As discussed elsewhere in this book, **fingerprinting** of malignant tumors emerges as a valuable tool for the evaluation of various events related to tumorigenesis, and/or the appearance of specific biologic characteristics of a tumor that may have consequences to the host. Multiple highly polymorphic loci can be analyzed using **minisatellite probing**. Following changes in the intensity of the bands as well as changes in the fine aspects of the fingerprint pattern may reflect, as in the case of bladder cancer, various events related to chromosomal loss or rearrangements unique to a specific tumor and associated with changes in the tumor behavior or aggressiveness (Uchida et al., 1990).

3.5.4 CONCLUDING REMARKS

At the time of this writing it appears that no definitive consensus has yet been achieved on the gene level diagnostics that may permit specific evaluation of bladder carcinoma patients, although certain modalities such the PCR assay for the Ha-*ras* gene and fingerprinting show much promise. There is little doubt that further analysis of molecular characteristics of the consistently involved oncogenes (Ha-*ras*, c-*myc*) and growth factors (e.g., EGFR) may provide such practically useful diagnostic tools in the near future. These could be combined with either FCM, ICM, or PCR detection of the characteristic gene aberrations specific for a particular malignant phenotype revealed in fingerprint patterns.

Wilms' Tumor

3.6.1 INTRODUCTION

The group of **small round cell tumors** also includes, in addition to neuroblastoma, Wilms' tumor (WT) (nephroblastoma), Ewing's sarcoma, rhabdomyosarcoma, and malignant lymphoma. Wilms' tumor is a common renal tumor of childhood, although extrarenal locations including retroperitoneum, mediastinum, sacrococcygeal region, testis, and inguinal canal have been reported (reviewed, e.g., in Rosai, 1989; Silverberg, 1990). The majority of cases are sporadic, although a minority are hereditary; the latter are often bilateral and display an autosomal dominant trait with variable penetrance. WT has been reported in association with other malignancies including retinoblastoma, botryoid rhabdomyosarcoma, and osteosarcoma (Rosai, 1989). WT showing a predominance of the blastematous component resembles other small round cell tumors, including neuroblastoma.

Variability of histological expression of WT may present a challenge for differential diagnosis, which can be aided by immunohistochemical panels. Characteristically, one finds positivity for vimentin (focally positive in the blastematous elements), keratin, and various lectins for the epithelial elements. Lack of specificity of these stains is well recognized. In fact, the presence of both cytokeratin and vimentin positivity is seen in aneuploid RCC particularly related to a tubulo-papillary and pseudosarcomatous growth patterns that can be used as an additional prognostic indicator (Dierick et al. 1991). The mesenchymal elements show variable positivity depending on the constituent components, such as desmin and myoglobin in rhabdomyoblastomatous foci. When neural elements are present positive reactivity for the S-100 protein, neuron-specific enolase, and glial fibrillary acidic protein can be observed (Rosai, 1989).

In the elaborate evaluation of histopathological and clinical characteristics of WT for predicting the disease course and selection of postsurgical treatment, so far only little, if any, use has been made of the gene related information.

3.6.2 CHROMOSOME ABERRATIONS

Some progress had been made in defining potential candidates for the WT gene (Call et al., 1989; Gessler et al., 1990). The limited success of DNA ploidy analysis for prognosis of WT could in part be accounted for by the variability of cytogenetic features expressed in the occurrence of normal karyotypes in WT (Solis et al., 1988; Wang-Wuu et al., 1990), although 80% of the karyotypes in one study were abnormal, including pseudodiploid and hyperdiploid patterns (Wang-Wuu et al., 1990). Nonrandom structural abnormalities have been observed in **1p/1q, 11p, 7p/7q, 16p/16q, 12q,** and **17q**. Abnormalities of **chromosome 1** seem to be the most recent secondary structural change, with 80% of the breakpoints occurring in **1q** and 20% in **1p34** regions (Wang-Wuu et al., 1990). **Trisomy 12** was the **most common abnormality**, detected in 52% of all tumors, being even more frequent in hyperdiploid tumors (81%), followed by trisomy 18 (28% in the authors' series). Many of the same secondary chromosome abnormalities, especially numerical (**+12, +18, +6, +8**), are also observed in other solid tumors and in leukemias (Wang-Wuu et al., 1990). The **11p13** region has been implicated in patients with the **WT-aniridia syndrome** (Francke et al., 1979; Riccardi et al., 1978, 1980), as well as **WT without aniridia** (Kaneko et al., 1981, 1983; Slater et al., 1985).

It appears that although deletions from **11p13** and **11p15** are the **primary lesions** in WT, the chromosomal analysis frequently misses these, because only in 31% of the total 93 reviewed cases could it be determined. **Loss of heterozygosity (LOH)** at chromosome 11p persists in all lines established from the differentiated tissues present in WT (Brown et al., 1989). Interestingly, pulsed field gel electrophoresis of WT DNA shows that tumor-specific deletions of the 11p13 region are very rare, and no hemizygous deletions can be detected (Royer-Pokora et al., 1991). Further analysis of the deletion area revealed that the predominant alteration in sporadic WT is produced by subtle changes such as point mutations, alterations in the methylation patterns, or due to very small deletions. The role of epigenetic changes such as **methylation-related imprinting** in the development of WT tumors and developmental abnormalities appears to be substantial. This phenomenon may explain the fact that deletions in the WT gene itself may not always be found (Royer-Pokora et al., 1991). A role for **genomic imprinting** in the pathogenesis of WT is again demonstrated by the preferential LOH of the maternal alleles from 11p (Pal et al., 1990).

3.6.3 TUMOR SUPPRESSOR GENES IN WT

Intensive research is currently under way to elucidate the role of **TSGs** in a number of malignancies. LOH at 11p13 and 11p15 suggests both regions may be involved in WT (Dao et al., 1987; Grundy et al., 1988; Koufos et al., 1989; Mannens et al., 1988, 1990; Raizis et al., 1985; Reeve et al., 1989). Suppression of the tumorigenic phenotype of WT by **introduction of a human chromosome 11** into the tumor cells (Weismann et al., 1987) strengthened the belief that TSGs may be located on chromosome 11 (Mannens et al., 1990). Further support for this notion was provided by the finding that the fibroblasts of aniridia patients with deletions from 11p13 can be morphologically transformed to a malignant phenotype by the early region of human **polyoma BK virus** (deRonde et al., 1988). Although providing no information on the specific location of these genes, the consistent LOH at 11p emphasizes the major role for chromosome 11p in WT development and apparent noninvolvement of other TSGs (Mannens et al., 1990).

3.6.3.1 *WT1* GENE

A **candidate gene** for the *WT* locus has been identified and sequences mapping in the *WT* area of band 11p13 have been isolated by two groups of researchers (Call et al., 1989, 1990; Gessler et al., 1990). The **WT1 gene** has a restricted pattern of expression in certain embryonal tissues, but not in adult kidney (Haber et al., 1990; Pritchard-Jones et al., 1990). Although at least three loci have been described as involved in the genesis of WT (11q13, 11q15, and neither of these in familial forms of WT), **only one candidate gene for the 11p13 locus has been identified**. Interactions between mutant alleles at 11p13 and 11p15 are possible, and various modes of their interdependence in the development of WT have been proposed (Haber et al., 1990).

In contrast to retinoblastoma, in which the absence of a single gene appears to cause the malignancy, the study of individuals with WT pointed to an **autosomal microdeletion**, that removed more than one locus, leading to a constellation of disorders that are otherwise known as single-gene, dominant disorders (Francke, 1990). Cytogenetic studies on WT patients as well as those with the **Beckwith-Wiedemann syndrome** pointed to, but did not limit, the involvement of 11p13 and 11p15.5. It appears the **at least three genes** exist that, when mutated, predispose to or cause WT (Francke, 1990). In contrast to the current views on TSGs as responsible only for suppression functions, a wider role for these is suggested in the differentiation process where the absence of these genes may cause an arrest in the development of the cells at a **stage of undifferentiation**, a notion fully consistent with the known high growth rate and aggressiveness of malignant tumors (Francke, 1990).

Gessler and coworkers (1990) have isolated a 3-kb clone, designated **LK15**. The product of this clone appears to be a **DNA-binding protein** that could be a **translational regulator**. In testing a panel of DNAs from 65 WTs they detected homozygous deletions in two tumors. The low incidence of homozygous deletions, identified within 50 kb of the *WT* locus, and the presence of an apparently normal-sized transcript for LK15 in 12 tumors, is in contrast with the known frequent deletions and internal rearrangements characteristic of the **retinoblastoma gene *RB*-1**. It is possible that another important gene in this region or at other sites of the genome may have a larger role in sporadic WT than previously believed (Gessler et al., 1990). Judging by the expression pattern of the LK15 transcript, its corresponding gene may play a part in normal kidney development (Gessler et al., 1990). In support of this view is the fact that the genitourinary anomaly component of the WAGR syndrome maps to this genomic region (Gessler et al., 1990).

Another group of researchers has isolated a series of genomic and cDNA clones that map within the constitutional and tumor deletions on chromosome **11p13 — the *WT* locus** (Call et al., 1990; Rose et al., 1990). A transcript spanning over approximately 50 kb within the interval that must contain the *WT* genes has been defined. Four cDNA clones, two from human embryonic kidney (*WT*4, *WT*2), one from human adult kidney (*WT*22), and one from a pre-B cell line (*WT*33), were studied in detail and the longest one, *WT*33, proved to be 2313 bp long. Variations in mRNA sequences in the 5' segment of the coding region were observed using RNA polymerase chain reaction (PCR), suggesting **alternative splicing patterns** among various tissue types. The product of this gene was found to contain four "**zinc finger**" **domains** known to function as transcription regulators through specific binding to nucleic acids. A 51% similarity has been found between amino acid sequences of the zing finger regions of the two identified human **early-growth response genes (*EGR*1 and *EGR*2)** and other polypeptides related to WT33. These genes are involved in pathways controlling cell proliferations (Call et al., 1990; Rose et al., 1990).

These findings are compared with the known functions of the **retinoblastoma gene *RB*** and its product **p105^{RB}** (Call et al., 1990; Rose et al., 1990). While the *RB* gene product appears to be expressed at significant levels in almost all types of normal dividing cells, the **WT33 transcript** appears to be limited predominantly to the kidney and to a subset of hematopoietic cells. Inactivation of the *RB* gene accordingly occurs in a wide range of tumors, whereas this does not seem to be the case for the *WT* putative gene. Furthermore, the difference extends to the function of the respective gene products: while **p105^{RB}** is a nuclear protein, it does not appear to bind to DNA specifically, and exerts control over growth indirectly through an interaction with other nuclear proteins (Call et al., 1990).

The presence of the zinc finger motifs on the predicted WT33 protein suggests its direct binding to DNA, or possibly RNA, making it a **likely direct transcription regulator**. Based on the genetic location of *WT*33 gene, its tissue-specific expression, and the function predicted from its sequence the authors suggest that *WT*33 represents the 11p13 *WT* gene (Call et al., 1990). Analysis of premalignant renal lesions in patients with known hereditary predisposition to WT may suggest that **inactivation of the 11p13 *WT* gene may be a necessary, but not a sufficient, event for the development of malignancy**. These findings open practical avenues for specific diagnostic (and prognostic) evaluation of patients using probes that can be developed from the described *WT* candidate genes.

Further confirmation that the candidate *WT* gene is indeed the *WT* gene comes from mRNA *in situ* hybridization (ISH) on sections of human embryos and *WT* (Pritchard-Jones et al., 1990). This gene (*WT*2-1) was found to be expressed specifically in the condensed mesenchyme, renal vesicle, and glomerular epithelium of the developing kidney, in mesonephric glomeruli, and in cells resembling these structures in tumors. The genital ridge, fetal gonads, and mesothelium also showed expression of this gene. Significantly, there was an almost 500-fold variation in *WT* gene expression so that the tumors that were predominantly **blastemal**, or showed epithelial differentiation, expressed high levels of the gene, whereas tumors with **stromal** predominance had low levels.

Comparison of *WT2-1* expression in developing kidney and tumors revealed that it was expressed only in the malignant counterparts of the cell types that express the gene during normal kidney development. It is possible that the **WT2-1 gene product acts as a transcriptional regulator** during normal kidney development to switch off genes involved in maintaining cell proliferation (Pritchard-Jones et al., 1990). In tumors, mutations apparently alter the ability of the gene product to regulate downstream target gene(s), which leads to uncontrolled proliferation with aberrant differentiation and eventual tumor formation. Constitutional mutations in the *WT-1* gene have been identified in two patients with WT and genital abnormalities supporting its role as a recessive oncogene that operates in mammalian development (Pelletier et al., 1991).

In any case, it appears that the **cloned sequences of the *WT* gene may now find practical application for resolving differential diagnostic challenges and, potentially, be of prognostic value**, if correlations between the level of its expression and clinical course of this malignancy are established.

3.6.3.2 N-*myc*

In the meantime, the pattern of expression of various markers at the gene level has been used to solve a challenging diagnostic problem presented by an unusual small round cell tumor arising from a fallopian tube in a 15-year-old girl (Bendit et al., 1990). Histological studies failed to differentiate between a neuroectodermal neoplasm and WT. **N-*myc* proved to be overexpressed in the tumor, but not in normal tissues**. This was not due to amplification of the gene. Extensive **overexpression of the N-*myc* gene without amplification, similar to that in neuroblastoma, is characteristic of WT** (Nisen et al., 1986). The electron microscopy (EM) picture and the pattern of immunohistochemical markers supported the WT diagnosis. Likewise, N-*myc* DNA probes allowed a diagnosis of **intrarenal neuroblastoma** (Nisen et al., 1988). Measurement of the N-*myc* gene amplification predicted poor prognosis, subsequently confirmed by the clinical course.

References

Aasmundstad, T. A. and Haugen, O. A. (1990). DNA ploidy in intraductal breast carcinomas. Eur. J. Cancer 26:956–959.

Adami, H. O. et al. (1990). Breast cancer etiology. Int. J. Cancer Suppl. 5:22–39.

Adelaide, J. et al. (1988). Chromosomal localization of the HST oncogene and its co-amplification with the *INT*-2 oncogene in a human melanoma. Oncogene 2:413–416.

Adnane, J. et al. (1989). Proto-oncogene amplification and human breast tumor phenotype. Oncogene 4:1389–1395.

Adolfsson, J. and Tribukait, B. (1990). Evaluation of tumor progression by repeated fine needle biopsies in prostate adenocarcinoma: modal deoxyribonucleic acid value and cytological differentiation. J. Urol. 144:1408–1410.

Adolfsson, J. et al. (1990). The prognostic value of model deoxyribonucleic acid in low grade, low stage untreated prostate cancer. J. Urol. 144:1404–1407.

Ahmann, F. R. et al. (1986). Growth in semi-solid agar of prostate cancer cells obtained from bone marrow aspirates. Cancer Res. 46:3560–3564.

Akiyama, T. et al. (1986). The product of the human c-*erb* B-2 gene: a 185 kilodalton glycoprotein with tyrosine kinase activity. Science 232:1644–1646.

Al-Sumidaie, A. M. et al. (1988). Particles with properties of retroviruses in monocytes from patients with breast cancer. Lancet i:5–9.

Albert, J. L. et al. (1990). Estrogen regulation of placental alkaline phosphatase gene expression in a human endometrial adenocarcinoma cell line. Cancer Res. 50:3306–3310.

Albini, A. et al. (1986). 17 β-estradiol regulates and v-Ha-*ras* transfection constitutively enhances MCF7 breast cancer cell interactions with basement membrane. Proc. Natl. Acad. Sci. U.S.A. 83:8182–8186.

Ali, I. U. et al. (1987). Reduction to homozygosity of genes on chromosome 11 in human breast neoplasia. Science 238:185–188.

Ali, I. U. et al. (1989). Presence of two members of c-*erb*A receptor gene family (c-*erb*A-β and c-*erb*A-2) in smallest region of somatic homozygosity in chromosome 3p21-p25 in human breast carcinoma. J. Natl. Cancer Inst. 81:1815–1820.

Anderson, D. E. et al. (1985). A linkage study of human breast cancer. Cytogenet. Cell Genet. 40:568–569.

Anglard, P. et al. (1989). Loss of alleles at loci on chromosome 11 in human renal cell carcinoma [meeting abstract]. J. Urol. 141:296 A.

Askensten, U. G. et al. (1989). Intratumoral variations in DNA distribution patterns in mammary adenocarcinomas. Cytometry 10:326–333.

Assoian, R. K. (1985). Biphasic effects of type β transforming growth factor on epidermal growth factor receptors in NRK fibroblasts. J. Biol. Chem. 260:9613–9617.

Atkin, N. B. and Baker, M. C. (1985). Chromosome study of five cancers of the prostate. Hum. Genet. 70:359–364.

Auer, G. et al. (1984). Prognostic significance of nuclear DNA content in mammary adenocarcinomas in humans. Cancer Res. 44:394–396.

Auer, G. U. et al. (1987). Comparison between slide and flow cytophotometric DNA measurements in breast tumors. Anal. Quant. Cytol. Histol. 9:138–146.

Baak, J. P. A. et al. (1987). Evaluation of the prognostic value of morphometric features and cellular DNA content in FIGO I ovarian cancer patients. Anal. Quant. Cytol. Histol. 9:287–290.

Baak, J. P. A. et al. (1991). Comparative long term prognostic value of quantitative HER-2/neu protein expression, DNA ploidy, and morphometric and clinical features in paraffin-embedded invasive breast cancer. Lab. Invest. 64:215–223.

Bacus, S. S. et al. (1990). HER-2/neu oncogene expression and DNA ploidy analysis in breast cancer. Arch. Pathol. Lab. Med. 114:164–169.

Baildam, A. D. et al. (1987a). DNA analysis by flow cytometry, response to endocrine treatment and prognosis in advanced carcinoma of the breast. Br. J. Cancer 55:553–559.

Baildam, A. D. et al. (1987b). Effect of tamoxifen upon cell DNA analysis by flow cytometry in primary carcinoma of the breast. Br. J. Cancer 55:561–566.

Bailey, A. et al. (1990). Stimulation of renal and hepatic c-*myc* and c-Ha-*ras* expression by unilateral nephrectomy. Oncogene Res. 5:287–293.

Baisch, H. et al. (1982). DNA content of human kidney carcinoma cells in relation to histological grading. Br. J. Cancer 45:878–886.

Baker, V. V. et al. (1987). Cellular DNA content and CA-125 antigen expression in ovarian carcinomas. Oncology 44:283–286.

Banner, B. F. et al. (1990). DNA analysis of multiple synchronous renal cell carcinomas. Cancer 66:2180–2185.

Barak, M. et al. (1989). Evaluation of prostate-specific antigen as a marker for adenocarcinoma of the prostate. J. Lab. Clin. Med. *113*:598–603.

Bargmann, C. I. et al. (1986). The *neu* oncogene encodes an epidermal growth factor receptor-related protein. Nature (London) *319*:226–230.

Barnabei, V. M. et al. (1990). Flow cytometric evaluation of epithelial ovarian cancer. Am. J. Obstet. Gynecol. *162*:1584–1592.

Barnes, D. M. et al. (1988). An immunohistochemical evaluation of c-*erb*B-2 expression in human breast carcinoma. Br. J. Cancer *58*:448–452.

Barrett-Lee, P. et al. (1990). Transcripts for transforming growth factors in human breast cancer: clinical correlates. Br. J. Cancer *61*:612–617.

Bartek, J. et al. (1990a). Genetic and immunochemical analysis of mutant p53 in human breast cancer cell lines. Oncogene *5*:893–899.

Bartek, J. et al. (1990b). Patterns of expression of the p53 tumor suppressor in human breast tissue and tumors *in situ* and *in vitro*. Int. J. Cancer *46*:839–844.

Basolo, F. et al. (1991). Tranformation of human breast epithelial cells by c-Ha-*ras* oncogene. Mol. Carcinogenesis *4*:25–35.

Bass, R. A. et al. (1987). DNA cytometry and cytology by quantitative fluorescence image analysis in symptomatic bladder cancer patients. Int. J. Cancer *40*:698–705.

Bedford, M. T. and van Helden, P. D. (1987). Hypomethylation of DNA in pathological conditions of the human prostate. Cancer Res. *47*:5274–5276.

Beerman, H. et al. (1991). Flow cytometric analysis of DNA stemline heterogeneity in primary and metastatic breast cancer. Cytometry *12*:147–154.

Bello, M. J. and Rey, J. A. (1989). Cytogenetic analysis of metastatic effusions from breast tumors. Neoplasma *35*:71–81.

Bendit, I. et al. (1990). Molecular phenotype of a pediatric small round cell tumor. Cancer *66*:1534–1538.

Bennington, J. L. and Mayall, B. II. (1983). DNA cytometry on four-micrometer sections of paraffin-embedded human renal adenocarcinomas and adenomas. Cytometry *4*:31–39.

Benz, C. C. et al. (1989). Expression of c-*myc*, c-Ha-*ras*-1, and c-*erb*B-2 protooncogenes in normal and malignant human breast epthelial cells. J. Natl. Cancer Inst. *81*:1704–1709.

Berchuck, A. et al. (1990a). Epidermal growth factor receptor and *neu* expression in advanced epithelial ovarian cancer. Proc. Am. Assn. Cancer Res. *31*:314A.

Berchuck, A. et al. (1990b). Overexpression of *HER*-2/*neu* is associated with poor survival in advanced epithelial ovarian cancer. Cancer Res. *50*:4087–4091.

Berenson, J. R. et al. (1989). Frequent amplification of the *bcl*-1 locus in head and neck squamous cell carcinomas. Oncogene *4*:1111–1116.

Berger, C. S. et al. (1986). Chromosomes in kidney, ureter, and bladder cancer. Cancer Genet. Cytogenet. *23*:1–24.

Berger, M. S. et al. (1988). Correlation of c-*erb*B-2 gene amplification and protein expression in human breast carcinoma with nodal status and nuclear grading. Cancer Res. *48*:1238–1243.

Berryman, I. L. et al. (1987). The nuclear DNA content of human breast carcinoma. Associations with clinical stage, axillary lymph node status, estrogen receptor status and outcome. Anal. Quant. Cytol. Histol. *9*:429–434.

Bevilacqua, G. et al. (1989). Association of low *nm*23 RNA levels in human primary infiltrating ductal breast carcinomas with lymph node involvement and other histopathological indicators of high metastatic potential. Cancer Res. *49*:5185–5190.

Bibbo, M. et al. (1987). Karyometry and histometry of renal-cell carcinoma. Anal. Quant. Cytol. Histol. *9*:182–187.

Bieche, I. et al. (1992). Loss of heterozygosity on chromosome 7q and aggressive primary breast cancer. Lancet *339*:139–143.

Bishop, T. (1992). Personal communication.

Biunno, I. et al. (1988). Structure and expression of oncogenes in surgical specimens of human breast carcinomas. Br. J. Cancer *57*:464–468.

Bladou, T. et al. (1988). Changes in structure and protein composition of bovine lymphocyte nuclear matrix during concanavalin-A-induced mitogenesis. Biochem. Cell Biol. *66*:40–53.

Blanco, G. et al. (1990). Prognostic factors in recurrent breast cancer: relationships to site of recurrence, disease-free interval, female sex steroid receptors, ploidy and histological malignancy grading. Br. J. Cancer *62*:142–146.

Blomjous, E. C. M. et al. (1989). The value of morphometry and DNA flow cytometry in addition to classic prognosticators in superficial urinary bladder carcinoma. Am. J. Clin. Pathol. *91*:243–248.

Bocking, A. et al. (1989a). DNA grading of malignancy in breast cancer. Prognostic validity, reproducibility and comparison with other classifications. Anal. Quant. Cytol. Histol. *11*:73–80.

Bocking, A. et al. (1989b). Representativity and reproducibility of DNA malignancy grading in different carcinomas. Anal. Quant. Cytol. Histol. *11*:81–86.

Boltz, E. M. et al. (1990). Demonstration of somatic rearrangements and genomic heterogeneity in human ovarian cancer by DNA fingerprinting. Br. J. Cancer *62*:23–27.

Bonilla, M. et al. (1988). *In vivo* amplification and rearrangement of c-*myc* oncogene in human breast tumors. J. Natl. Cancer Inst. *80*:665–671.

Bonneterre, J. et al. (1990). Prognostic significance of insulin-like growth factor I receptors in human breast cancer. Cancer Res. *50*:6931–6935.

Bookstein, R. et al. (1990). Suppression of tumorigenicity of human prostate carcinoma cells by replacing a mutated RB gene. Science *247*:712–715.

Borg, A. et al. (1990). *HER*-2/*neu* amplification predicts poor survival in node-positive breast cancer. Cancer Res. *50*:4332–4337.

Borg, A. et al. (1991a). Association of *INT*2/*HST*1 coamplification in primary breast cancer with hormone-dependent phenotype and poor prognosis. Br. J. Cancer *63*:136–142.

Borg, A. et al. (1991b). *ERB*B2 amplification in breast cancer with a high rate of proliferation. Oncogene *6*:137–143.

Borgen, P. I. et al. (1990). Amplification of *HER*-2/*neu* oncogene is associated with early metastasis in human breast cancer. Proc. Am. Assn. Cancer Res. *31*:313A.

Borresen, A. L. et al. (1990). Amplification and protein over-expression of the *neu*/*HER*-2/c-*erb*B-2 protooncogene in human breast carcinomas: relationship to loss of gene sequences on chromosome 17, family history and prognosis. Br. J. Cancer *62*:585–590.

Bowcock, A. M. et al. (1990). Exclusion of the retinoblastoma gene and chromosome 13q as the site of a primary lesion for human breast cancer. Am. J. Hum. Genet. *46*:12–17.

Boyd, J. A. and Kaufman, D. G. (1990). Expression of transforming growth factor β 1 by human endometrial carcinoma cell lines: inverse correlation with effects on growth rate and morphology. Cancer Res. *50*:3394–3399.

Bretton, P. R. et al. (1989). Initial evaluation of a new epithelial antigen (T16) for bivariate flow cytometry of bladder irrigation specimens. Cytometry *10*:339–344.

Breuning, M. H. et al. (1989). Characterization of new probes for diagnosis of polycystic kidney disease. In: Genetics of Kidney Disease. Alan R. Liss, New York, pp. 69–75.

Breuning, M. H. et al. (1990a). Map of 16 polymorphic loci on the short arm of chromosome 16 close to the polycystic kidney disease gene (*PKD*1). J. Med. Genet. *27*:603–613.

Breuning, M. H. et al. (1990b). Two step procedure for early diagnosis of polycystic kidney disease with polymorphic DNA markers on both sides of the gene. J. Med. Genet. *27*:614–617.

Briffod, M. et al. (1989). Sequential cytopunctures during preoperative chemotherapy for primary breast cancer: cytomorphologic changes, initial tumor ploidy and tumor regression. Cancer *63*:631.

Britton, L. C. et al. (1989). Flow cytometric DNA analysis of stage I endometrial carcinoma. Gynecol. Oncol. *34*:317–322.

Brookes, S. et al. (1989). Sequence organization of the human *int*-2 gene and its expression in teratocarcinoma cells. Oncogene *4*:429-.

Brothman, A. R. et al. (1990). Frequency and pattern of karyotypic abnormalities in human prostate cancer. Cancer Res. *50*:3795–3803.

Brown, A. M. C. et al. (1986). A retrovirus vector expressing the putative mammary oncogene *int*-1 causes partial transformation of a mammary epithelial cell line. Cell *46*:1001–1009.

Brown, K. W. et al. (1989). Loss of chromosome 11p alleles in cultured cells derived from Wilms' tumors. Br. J. Cancer *60*:25–29.

Brunner N. (1990). Human breast cancer growth and progression: role of secreted polypeptide growth factors. Int. J. Cancer Suppl *5*:62–66.

Bussemakers, M. J. G. et al. (1990). Differential hybridization analysis reveals theree genes that are overexpressed in progressionally advanced rat prostatic cancer lines. Proc. Am. Assn. Cancer Res. *31*:305A.

Buttyan, R. et al. (1988). Cascade induction of c-fos, c-myc, and heat shock 70K transcripts during regression of the rat ventral prostate gland. Mol. Endocrinol. *2*:650–657.

Call, K. M. et al. (1989). Isolation of a cDNA in the WAGR region: a candidate gene for Wilms' tumor [abstract]. Am. J. Hum. Genet. *45*:A179.

Call, K. M. et al. (1990). Isolation and characterization of a zinc finger polypeptide gene at the human chromosome 11 Wilms' tumor locus. Cell (Cambridge, Mass) *16*:509–520.

Callahan, R. and Campbell, G. (1989). Mutations in human breast cancer: an overview. J. Natl. Cancer Inst. *81*:1780–1786.

Callahan, R. et al. (1990). Mutations and breast cancer. Proc. Am. Assn. Cancer Res. *31*:465–466.

Cardiff, R. D. et al. (1988). c-H-*ras*-1 expression in 7,12-dimethyl benzanthracene-induced Balb/c mouse mammary hyperplasias and their tumors. Oncogene *3*:205–213.

Carroll, P. R. et al. (1987). Abnormalities at chromosome region 3p 12-14 characterize clear cell renal carcinoma. Cancer Genet. Cytogenet. *26*:253–259.

Carter, B. S. et al. (1990a). Allelic loss of chromosomes 16q and 10q in human prostate cancer. Proc. Natl. Acad. Sci. U.S.A. *87*:8751–8755.

Carter, B. S. et al. (1990b). *ras* gene mutations in human prostate cancer. Cancer Res. *50*:6830–6832.

Casey, G. et al. (1986). Characterization and chromosome assignment of the human homolog of *int*-2, a potential proto-oncogene. Mol. Cell. Biol. *6*:502–510.

Catalona, W. J. et al. (1991). Measurement of prostate-specific antigen in serum as a screening test for prostate cancer. N. Engl. J. Med. *324*:1156–1161.

Cattoretti, G. et al. (1988). *p53* expression in breast cancer. Int. J. Cancer *41*:178–183.

Chaganti, R. S. K. et al. (1990). Genetic origin of a trophoblastic choriocarcinoma. Cancer Res. *50*: 6330–6333.

Chambers, J. T. (1988). Sex steroid receptors in endometrial cancer. Yale J. Biol. Med. *61*:339–350.

Chambon, P. and Rio, M. C. (1990). Expression of the estrogen-responsive pS2 gene in breast cancer: a review. Proc. Am. Assn. Cancer Res. *31*:466–468.

Chapman, A. B. et al. (1990). The renin-angiotensin-aldosterone system and autosomal dominant polycystic kidney disease. N. Engl. J. Med. *323*:1091–1096.

Chen, L. C. et al. (1989). Loss of heterozygosity on chromosome 1q in human breast cancer. Proc. Natl. Acad. Sci. U.S.A. *86*:7204–7207.

Chen, L. C. et al. (1990). Overexpression of matrix Gla protein mRNA in malignant human breast cells: isolation by differential cDNA hybridization. Oncogene *5*:1391–1395.

Chen, L. C. et al. (1991). Loss of heterozygosity on the short arm of chromosome 17 is associated with high proliferative capacity and DNA aneuploidy in primary human breast cancer. Proc. Natl. Acad. Sci. U.S.A. *88*:3847–3851.

Ciardiello, F. et al. (1988). Induction of transforming growth factor α expression in mouse mammary epithelial cells after transformation with a point-mutated c-Ha-*ras* protooncogene. Mol. Endocrinol. *2*:1202–1216.

Clark, G. M. and McGuire, W. L. (1988). Steroid receptors and other prognostic factors in primary breast cancers. Semin. Oncol. *15*:20–25.

Clark, G. M. and McGuire, W. L. (1991). Follow-up study of *HER-2/neu* amplification in primary breast cancer. Cancer Res. *51*:944–948.

Clark, G. M. et al. (1989). Prediction of relapse or survival in patients with node-negative breast cancer by DNA flow cytometry. N. Engl. J. Med. *320*:627–633.

Claus, E. B. et al. (1991). Genetic analysis of breast cancer in the cancer and steroid hormone study. Am J. Hum. Genet. *48*:232–242.

Cline, M. J. et al. (1987). Proto-oncogene abnormalities in human breast cancer: correlations with anatomic features and clinical course of disease. J. Clin. Oncol. *5*:999–1006.

Cohen, A. F. et al. (1979). Hereditary renal cell carcinoma associated with a chromosomal translocation. N. Engl. J. Med. *301*:592–595.

Coles, C. et al. (1990). Evidence implicating at least two genes on chromosome 17p in breast carcinogenesis. Lancet *336*:761–763.

Collins, M. K. L. et al. (1983). Platelet-derived growth factor treatment decreases the affinity of the epidermal growth factor receptors of Swiss 3T3 cells. J. Biol. Chem. *258*:11689–11693.

Colliste, L. G. et al. (1980). Bladder cancer diagnosis by flow cytometry. Correlation between cell samples from biopsy and bladder irrigation fluid. Cancer *45*:2389–2394.

Cooke, D. B. et al. (1988a). Oncogene expression in prostate cancer: Dunning R3327 rat dorsal prostatic adenocarcinoma system. Prostate *13*:263–272.

Cooke, D. B. et al. (1988b). Expression of ras proto-oncogenes in the Dunning R3327 rat prostatic adenocarcinoma system. Prostate *13*:273–287.

Cornelisse, C. J. et al. (1987). DNA ploidy and survival in breast cancer patients. Cytometry *8*:225–234.

Couturier, J. et al. (1988a). Trisomy and tetrasomy for long arm of chromosome 1 in near-diploid human endometrial adenocarcinomas. Int. J. Cancer *38*:17–19.

Couturier, J. et al. (1988b). Chromosome imbalance in endometrial adenocarcinoma. Cancer Genet. Cytogenet. *33*:67–76.

Cowan, W. K. et al. (1990). Characteristics of breast carcinomas detected by mass population screening compared to those presenting clinically: expression of cathepsin D, c-erbB-2 oncoprotein and epidermal growth factor receptor determined by immunohistochemistry. Br. J. Cancer *62* (Suppl. XII):42A.

Cowley, B. D., Jr. et al. (1987). Elevated c-*myc* protooncogene expression in autosomal recessive polycystic kidney disease. Proc. Natl. Acad. Sci. U.S.A. *84*:8394–8398.

Cropp, C. S. et al. (1990a). Loss of heterozygosity of chromosomes 17 and 18 in primary breast cancers. Proc. Am. Assn. Cancer Res. *31*:317A.

Cropp, C. S. et al. (1990b). Loss of heterozygosity on chromosomes 17 and 18 in breast carcinoma: two additional regions identified. Proc. Natl. Acad. Sci. U.S.A. *87*:7737–7741.

Cruciger, Q. V. et al. (1976). Human breast carcinomas: marker chromosomes involving 1q in seven cases. Cytogenet. Cell Genet. *17*:231–235.

Czerniak, B. et al. (1989). Expression of *ras* oncogene p21 protein in relation to regional spread of human breast carcinomas. Cancer *63*:2008–2013.

Czerniak, B. et al. (1990). Ha-*ras* gene codon 12 mutation and DNA ploidy in urinary bladder carcinoma. Br. J. Cancer *62*:762–763.

Dal Cin, P. et al. (1989). Renal cortical tumors. Cytogenetic characterization. Am. J. Clin. Pathol. *92*:408–414.

Dalgaard, O. Z. and Norby, S. (1989). Autosomal dominant polycystic kidney disease in the 1980's. Clin. Genet. *36*:320–325.

Dao, D. D. et al. (1987). Genetic mechanisms of tumor-specific loss of 11p DNA sequences in Wilms' tumor. Am. J. Hum. Genet. *41*:202–217.

Daughaday, W. H. et al. (1988). Synthesis and secretion of insulin-like growth factor II by a leiomyosarcoma with associated hypoglycemia. N. Engl. J. Med. *319*:1434–1440.

De Biasi, F. et al. (1989). Evidence of enhancement of the *ras* oncogene protein product (p21) in a spectrum of human tumors. Int. J. Cancer *43*:431–435.

Dierick, A.-M. et al. (1991). Vimentin expression of renal cell carcinoma in relation to DNA content and histologic grading: a combined light microscopic, immunocytochemical and cytophotometrical analysis. Histopathology *18*:315–322.

de Jong, B. et al. (1988). Cytogenetics of 12 cases of renal adenocarcinomas. Cancer Genet. Cytogenet. *30*:53–61.

de Potter, C. R. et al. (1989). The expression of the neu oncogene product in breast lesions and in normal fetal and adult human tissues. Histopathology *15*:351–362.

de Potter, C. R. et al. (1990). The *neu*-oncogene protein as a predictive factor for hematogenous metastases in breast cancer patients. Int. J. Cancer *45*:55–58.

de Ronde, A. et al. (1988). Morphologic transformation by early region human polyomavirus BK DNA of human fibroblasts with deletions in the short arm of one chromosome 11. J. Gen. Virol. *69*:467–471.

De Vita, R. et al. (1990). Flow cytometric DNA analysis of the human cervix affected by human papillomavirus and/ or intraepithelial neoplasia. Anal. Quant. Cytol. Histol. *12*:306–313.

Dear, T. N. et al. (1988). Differential expression of a novel gene, *WDNM*1, in nonmetastatic rat mammary adenocarcinoma cells. Cancer Res. *48*:5203–5209.

Dear, T. N. et al. (1989). Transcriptional down-regulation of a rat gene, *WDNM*2, in metastatic DMBA-8 cells. Cancer Res. *49*:5323–5328.

Devilee, P. et al. (1989). At least four different chromosomal regions are involved in loss of heterozygosity in human breast carcinoma. Genomics *5*:554–560.

Devonec, M. et al. (1982). Flow cytometry of low stage bladder tumors: correclation with cytologic and cystoscopic diagnosis. Cancer *49*:109–118.

Dickson, R. B. et al. (1990). UCLA colloquium: new insights into breast cancer: the molecular biochemical and cellular biology of breast cancer [meeting report]. Cancer Res. *50*:4446–4447.

Distel, R. J. et al. (1987). Nucleoprotein complexes that regulate gene expression in adipocyte differentiation: direct participation of c-*fos*. Cell (Cambridge, Mass) *49*:835–844.

Donovan-Peluso, M. et al. (1991). Oncogene amplification in breast cancer. Am. J. Pathol. *138*:835–845.

Dotzlaw, H. et al. (1990). Epidermal growth factor gene expression in human breast cancer biopsy samples: relationship to estrogen and progesterone receptor gene expression. Cancer Res. *50*:4204–4208.

Douglas, E. C. et al. (1986). Hyperploidy and chromosomal rearrangements define the anaplastic variants of Wilms' tumor. J. Clin. Oncol. *4*:975–981.

Dressler, L. G. et al. (1987). Evaluation of a modeling system for S-phase estimation in breast cancer by flow cytometry. Cancer Res. *47*:5294–5302.

Dressler, L. G. et al. (1988). DNA flow cytometry and prognostic factors in 1331 frozen breast cancer specimens. Cancer *61*:420–427.

Dubik, D. and Shiu, R. P. C. (1988). Transcriptional regulation of c-*myc* oncogene expression by estrogen in hormone-responsive human breast cancer cells. J. Biol. Chem. *263*:12705–12708.

Dubik, D. et al. (1987). Stimulation of c-*myc* oncogene expression associated with estrogen-induced proliferation of human breast cancer cells. Cancer Res. *47*:6517–6521.

Dutrillaux, B. et al. (1990). Characterization of chromosomal anomalies in human breast cancer. A comparison of 30 paradiploid cases with few chromosome changes. Cancer Genet. Cytogenet. *49*:203–217.

Dykin, R. et al. (1990). Long term survival in breast cancer related to overexpression of the c-erbB-2 oncoprotein: an immunohistochemical study using monoclonal antibody NCL-CB11. Br. J. Cancer *62* (Suppl. XII):25A.

Ebert, T. et al. (1990). Establishment and characterization of human renal cancer and normal kidney cell lines. Cancer Res. *50*:5531–5536.

Eccles, D. M. et al. (1990). Allele losses on chromosome 17 in human epithelial ovarian carcinoma. Oncogene *5*:1599–1601.

El-Naggar, A. K. et al. (1990a). Acridine orange flow cytometric analysis of renal cell carcinoma. Clinicopathologic implications of RNA content. Am. J. Pathol. *137*:275–280.

El-Naggar, A. K. et al. (1990b). DNA content and proliferative activity of cystosarcoma phyllodes of the breast. Am. J. Clin. Pathol. *93*:480–485.

Enomoto, T. et al. (1990a). K-*ras* activation in neoplasms of the human female reproductive tract. Cancer Res. *50*:6139–6145.

Enomoto, T. et al. (1990b). K-*ras* activation in neoplasms of the human female reproductive tract. Proc. Am. Assn. Cancer Res. *31*:314A.

Epstein, J. I. et al. (1990). Comparison of DNA ploidy and nuclear size, shape and chromatin irregularity in tissue sections and smears of prostatic carcinoma. Anal. Quant. Cytol. Histol. *12*:352–358.

Erba, E. et al. (1989). Flow cytometric analysis of DNA content in human ovarian cancers. Br. J. Cancer *60*:45–50.

Erlandsson, R. et al. (1990). A gene near the D3F15S2 site on 3p is expressed in normal human kidney but not or only at a severely reduced level in 11 of 15 primary renal cell carcinomas. Oncogene *5*:1207–1211.

Ewers, S. B. et al. (1984). Flow-cytometric DNA analysis in primary breast carcinomas and clinicopathologic correlations. Cytometry *5*:408–419.

Fallenius, A. G. et al. (1987). The reliability of microspectrophotometric and flow cytometric nuclear DNA measurements in adenocarcinomas of the breast. Cytometry *8*:260–266.

Fallenius, A. G. et al. (1988a). Prognostic significance of DNA measurements in 409 consecutive breast cancer patients. Cancer *62*:331–341.

Fallenius, A. G. et al. (1988b). Predictive value of nuclear DNA content in breast cancer in relation to clinical and morphologic factors. A retrospective study of 227 consecutive cases. Cancer *62*:521–530.

Farsund, T. et al. (1984a). Relation between flow cytometric DNA distribution and pathology in human bladder cancer. A report of 69 cases. Cancer *54*:1771–1777.

Farsund, T. et al. (1984b). Local chemotherapeutic effects in bladder cancer demonstrated by selective sampling and flow cytometry. J. Urol. *131*:22–32.

Fearon, E. R. et al. (1985). Loss of genes on the short arm of chromosome 11 in bladder cancer. Nature (London) *318*:377–380.

Federspiel, B. H. et al. (1987). Morphometry and cytophotometric assessment of DNA in smooth-muscle tumors (leiomyomas and leiomyosarcomas) of the gastrointestinal tract. Anal. Quant. Cytol. Histol. *9*:105–114.

Feig, L. A. et al. (1984). Somatic activation of *ras*^K gene in a human ovarian carcinoma. Science *223*:698–701.

Fernandez-Pol, J. A. et al. (1987a). Modulation of epidermal growth factor receptor gene expression by transforming growth factor β in a human breast carconima cell line. Cancer Res. *47*:4260–4265.

Fernandez-Pol, J. A. et al. (1987b). Suppression of the EGF-dependent induction of c-*myc* proto-oncogene expression by transforming growth factor β in a human breast carcinoma cell line. Biochem. Biophys. Res. Commun. *144*:1197–1205.

Ferrell, R. E. et al. (1989). A genetic linkage study of familial breast-ovarian cancer. Cancer Genet. Cytogenet. *38*:241–248.

Ferti-Passantonopoulou, A. D. and Panani, A. D. (1987). Common cytogenetic findings in primary breast cancer. Cancer Genet. Cytogenet. *27*:289–298.

Fey, E. G. and Penman, S. (1988). Nuclear matrix proteins reflect cell type of origin in cultured human cells. Proc. Natl. Acad. Sci. U.S.A. *85*:121–125.

Filmus, J. et al. (1986). A cell line from a human ovarian carcinoma with amplification of the K-*ras* gene. Cancer Res. *46*:5179–5182.

Fletcher, J. A. et al. (1990). Chromosome aberrations in uterine smooth muscle tumors: potential diagnostic relevance of cytogenetic instability. Cancer Res. *50*:4092–4097.

Foekens, J. A. et al. (1990). Prediction of relapse and survival in breast cancer patients by pS2 protein status. Cancer Res. *50*:3832–3837.

Forsslund, G. and Zetterberg, A. (1990). Ploidy level determinations in high-grade and low-grade malignant variants of prostatic carcinoma. Cancer Res. *50*:4281–4285.

Foulkes, W. et al. (1991). Allele loss on chromosome 17q in sporadic ovarian cancer. Lancet *338*:444–445.

Francke, U. (1990). A gene for Wilms' tumor? Nature (London) *343*:692–694.

Francke, U. et al. (1979). Aniridia-Wilms' tumor association: evidence for specific deletion of 11p13. Cytogenet. Cell Genet. *24*:185–192.

Friedlander, M. L. et al. (1988). Prediction of long-term survival by flow cytometric analysis of cellular DNA content in patients with advanced ovarian cancer. J. Clin. Oncol. *6*:282–290.

Frymoyer, P. A. et al. (1991). X-linked recessive nephrolithiasis with renal failure. N. Engl. J. Med. *325*:681–686.

Fujita, H. et al. (1985). Marker chromosomes of the long arm of chromosome 1 in endometrial carcinoma. Cancer Genet. Cytogenet. *18*:183–193.

Fujita, J. et al. (1987). Frequency of active ras oncogenes in human bladder cancers associated with schistosomiasis. Jpn. J. Cancer Res. (Gann) *78*:915–920.

Fujita, J. et al. (1988). Activation of H-*ras* oncogene in rat bladder tumors induced by *N*-butyl-*N*-(4-hydroxybutyl)-nitrosamine. J. Natl. Cancer Inst. *80*:37–43.

Garcia, I. et al. (1989). Genetic alterations of c-*myc*, c-*erb*B-2, and c-Ha-*ras* protooncogenes and clinical associations in human breast carcinomas. Cancer Res. *49*:6675–6679.

Garcia, M. et al. (1990). Protooncogene alterations in human ovarian carcinomas. Proc. Am. Assn. Cancer Res. *31*:308A.

Gardner, R. M. et al. (1987). Contractions of the isolated uterus stimulated by epidermal growth factor. FASEB J. *1*:224–228.

Gebhart, E. et al. (1986). Cytogenetic studies on human breast carcinomas. Breast Cancer Res. Treat. *8*:125–138.

Genuardi, M. et al. (1989). Distal deletion of chromosome 1p in ductal carcinoma of the breast. Am. J. Hum. Genet. *45*:73–82.

Germino, G. G. et al. (1990). Identification of a locus which shows no genetic recombination with the autosomal dominant polycystic kidney disease gene on chromosome 16. Am. J. Hum. Genet. *46*:925–933.

Gessler, M. et al. (1990). Homozygous deletion in Wilms' tumors of a zinc-finger gene identified by chromosome jumping. Nature (London) *343*:774–778.

Getzenberg, R. H. and Coffey, D. S. (1990). Tissue specificity of the hormonal response in sex accessory tissues is associated with nuclear matrix protein patterns. Mol. Endocrinol. *4*:1336–1342.

Gibas, Z. and Rubin, S. C. (1987). Well-differentiated adenocarcinoma of endometrium with simple karyotypic change: a case report. Cancer Genet. Cytogenet. *25*:21–26.

Gibas, Z. et al. (1984). A high-resolution study of chromosome changes in a human prostatic carcinoma cell line (LNCaP). Cancer Genet. Cytogenet. *11*:399–404.

Gibas, Z. et al. (1985). Chromosome rearrangements in a metastatic adenocarcinoma of the prostate. Cancer Genet. Cytogenet. *16*:301–304.

Gillespie, G. A. J. et al. (1991). CpG island in the region of an autosomal dominant polycystic kidney disease locus defines the 5' end of a gene encoding a putative proton channel. Proc. Natl. Acad. Sci. U.S.A. *88*: 4289–4293.

Gittes, R. F. (1991). Carcinoma of the prostate. N. Engl. J. Med. *324*:236–245.

Grimaux, M. et al. (1989). Immunoenzymetric assay of epidermal growth factor receptor. Application to breast tumor samples. Tumor Biol. *10*:215–224.

Grimmond, S. M. et al. (1990). Detection of a rare point mutation of K-*ras* in a human bladder cancer xenograft by polymerase chain reaction and direct sequencing. Proc. Am. Assn. Cancer Res. *31*:311A.

Grundy, P. et al. (1988). Familial predisposition to Wilms' tumor does not map to the short arm of chromosome 11. Nature (London) *336*:374–376.

Guerin, M. et al. (1988). Overexpression of either c-*myc* or c-*erb*B-2/*neu* protooncogenes in human breast carcinomas: correlation with poor prognosis. Oncogene Res. *3*:21–31.

Guerin, M. et al. (1990). Strong association between c-*myb* and estrogen-receptor expression in human breast cancer. Oncogene *5*:131–135.

Gullick, W. J. (1990). The role of the epidermal growth factor and c-erb B-2 protein in breast cancer. Int. J. Cancer Suppl. *5*:55–61.

Gullick, W. J. et al. (1991). c-erbB-2 protein overexpression in breast cancer is a risk factor in patient with involved and uninvolved lymph nodes. Br. J. Cancer *63*:434–438.

Gumerlock, P. H. et al. (1991). Activated *ras* alleles in human carcinoma of the prostate are rare. Cancer Res. *51*:1632–1637.

Haas, M. et al. (1987). Evidence against *ras* activation in human ovarian carcinomas. Mol. Biol. Med. *4*:265–275.

Haber, D. A. et al. (1990). An internal deletion within an 11p13 zinc finger gene contributes to the development of Wilms' tumor. Cell (Cambridge, Mass) *61*:1257–1269.

Hader, M. et al. (1987). Epidermal growth factor receptor expression, proliferation, and colony stimulating activity production in the urinary bladder carcinoma cell line 5637. J. Cancer Res. Clin. Oncol. *113*:579–585.

Hainsworth, P. J. et al. (1990a). Chromosome 1p and 16q abnormalities in primary breast cancer. Br. J. Cancer *62* (Suppl. XII):4A.

Hainsworth, P. J. et al. (1990b). Neu and "group"-ras oncoprotein and epidermal growth factor receptor (EGFR) immunohistochemistry in primary breast cancer. Br. J. Cancer *62* (Suppl. XII):12A.

Hall, J. M. et al. (1989). Oncogenes and human breast cancer. Am. J. Hum. Genet. *44*:577–584.

Hall, J. M. et al. (1990). Linkage of early-onset familial breast cancer to chromosome 17q21. Science *250*:1684–1689.

Hallam, N. et al. (1990). Absence of reverse transcriptase activity in monocyte cultures from patients with breast cancer. Lancet *336*:1079.

Hanna, W. et al. (1990). Distribution and patterns of staining of *neu* oncogene product in benign and malignant breast diseases. Mod. Pathol. *3*:455–461.

Hanselaar, A. G. J. M. et al. (1988). DNA ploidy patterns in cervical intraepithelial neoplasia grade III, with and without synchronous invasive squamous cell carcinoma. Cancer *62*:2537–2545.

Harris, P. C. et al. (1991). Rapid genetic analysis of families with polycystic kidney disease 1 by means of a microsatellite marker. Lancet *338*:1484–1487.

Hatada, I. et al. (1988). Co-amplification of integrated hepatitis B virus DNA and transforming gene *hst*-1 in a hepatocellular carcinoma. Oncogene *3*:537–540.

Hawkins, R. A. et al. (1990). The prognostic value of receptors for epidermal growth factor and estrogen in patients with breast cancer. Br. J. Cancer *62* (Suppl. XII):12A.

Hedley, D. W. et al. (1985). Application of DNA flow cytometry to paraffin-embedded archival material for the study of aneuploidy and its clinical significance. Cytometry *6*:327–333.

Hennessy, C. et al. (1990a). Expression of the anti-metastatic gene *nm*23 in human primary breast cancers and its correlation to lymph node status. Proc. Am. Assn. Cancer Res. *31*:319A.

Hennessy, C. et al. (1990b). Expression of the metastatic gene NM23 in human breast cancers related to lymph node status and survival. Br. J. Cancer *62* (Suppl. XII):5A.

Henry, J. A. et al. (1989). Expression of the estrogen regulated pNR-2 mRNA in human breast cancer: relation to estrogen receptor mRNA levels and response to tamoxifen therapy. Br. J. Cancer *61*:32–38.

Henry, J. A. et al. (1990). The prognostic significance of the lysosomal protease cathepsin D in human breast cancer: an immunohistochemical study. Br. J. Cancer *62* (Suppl. XII):24A.

Henttu, P. et al. (1990). Expression of the gene coding for human prostate-specific antigen and related hGK-1 in benign and malignant tumors of the human prostate. Int. J. Cancer *45*:654–660.

Hery, M. et al. (1987). The DNA labeling index: a prognostic factor in node-negative breast cancer. Breast Cancer Res. Treat. *9*:207–211.

Hill, S. et al. (1987). Cytogenetic analysis in human breast carcinoma. II. Seven cases in the triploid/tetraploid range investigated using direct preparations. Cancer Genet. Cytogenet. *24*:45–62.

Hopman, A. H. N. et al. (1991). Numerical chromosome 1,7,9 and 11 aberrations in bladder cancer detected by *in situ* hybridization. Cancer Res. *51*:644–651.

Hughes, R. G. et al. (1987). Nuclear DNA analysis of koilocytic and premalignant lesions of the uterine cervix. Br. Med. J. *294*:267–269.

Hutter, R. V. P. (1990). The role of the pathologist in breast cancer management. Cancer *66*:1363–1372.

Iglehart, J. D. et al. (1990). Increased *erb*B-2 gene copies and expression in multiple stages of breast cancer. Cancer Res. *50*:6701–6707.

Iizumi, T. et al. (1987). Establishment of a new prostatic carcinoma cell line (TSU-PR1). J. Urol. *137*:1304–1306.

Ikeda, I. et al. (1988). No correlation between L-*myc* restriction fragment length polymorphism and malignancy of human colorectal cancers. Gann *79*:674–676.

Isaacs, J. T. (1991). A mammary cancer suppressor gene and its site of action in the rat. Cancer Res. *51*:1591–1595.

Ishikawa, J. et al. (1987). Lack of correlation between rare Ha-*ras* alleles and urothelial cancer in Japan. Int. J. Cancer *40*:474–478.

Isola, J. J. et al. (1990). Evaluation of cell proliferation in breast carcinoma. Comparison of Ki-67 immunohistochemical study, DNA flow cytometric analysis, and mitotic count. Cancer *65*:1180–1184.

Iversen, O. E. and Skaarland, E. (1987). Ploidy assessment of benign and malignant ovarian tumors by flow cytometry. A clinicopathologic study. Cancer *60*:82–87.

Jacobs, S. C. et al. (1979). Prostatic osteoblastic factor. Invest. Urol. *17*:195–198.

Johnson, B. E. et al. (1988). myc family DNA amplification in small cell lung cancer patients' tumors and corresponding cell lines. Cancer Res. *48*:5163–5166.

Jones, E. C. et al. (1990). DNA content in prostatic adenocarcinoma. A flow cytometry study of the predictive value of aneuploidy for tumor volume, percentage, Gleason grade 4 and 5 and lymph node metastates. Cancer *66*:752–757.

Jones, R. F. et al. (1990). Altered responses to TGF-β and mitogenic growth factors of oncogene-expressing cell cultures derived from experimental bladder carcinoma. Proc. Am. Assn. Cancer Res. *31*:312A.

Kacinski, B. M. et al. (1989). Oncogene expression *in vivo* by ovarian adenocarcinomas and mixed-mullerian tumors. Yale J. Biol. Med. *62*:379–392.

Kakehi, Y. and Yoshida, O. (1989). Restriction fragment length polymorphism of the L-*myc* gene and susceptibility to metastasis in renal cancer patients. Int. J. Cancer *43*:391–394.

Kallioniemi, O. P. et al. (1987). Tumor DNA ploidy as an independent prognostic factor in breast cancer. Br. J. Cancer *56*:637–642.

Kallioniemi, O. P. et al. (1988a). Improving the prognostic value of DNA flow cytometry in breast cancer by combining DNA index and S-phase fraction. A proposed classification of DNA histograms in breast cancer. Cancer *62*:2183–2190.

Kallioniemi, O. P. et al. (1988b). Prognostic significance of DNA index, multiploidy, and S-phase fraction in ovarian cancer. Cancer *61*:334–339.

Kaneko, Y. et al. (1981). Interstitial deletion of the short arm of chromosome 11 limited to Wilms' tumor cells in a patient without aniridia. Cancer Res. *41*:4577–4578.

Kaneko, Y. et al. (1983). Further chromosome studies on Wilms' tumor cells of patients without aniridia. Cancer Genet. Cytogenet. *10*:191–198.

Kashyap, V. et al. (1990). Microphotometric nuclear DNA analysis in cervical dysplasia of the uterine cervix: its relation to the progression to malignancy and regression to normalcy. Neoplasma *37*:497–500.

Kawashima, K. et al. (1988). Close correlation between restriction fragment length polymorphism of the L-MYC gene and metastasis of human lung cancer to the lymph nodes and other organs. Proc. Natl. Acad. Sci. U.S.A. *85*:2353–2356.

Keatings, L. et al. (1990). c-*erb*B-2 oncoprotein expression in mammary and extramammary Paget's disease: an immunohistochemical study. Histopathology *17*:243–247.

Killian, C. S. and Chu, T. M. (1990). Prostate-specific antigen: questions often asked. Cancer Invest. *8*:27–37.

Kimberling, W. V. et al. (1988). Linkage heterogeneity of autosomal dominant polycystic kidney disease. N. Engl. J. Med. *319*:913–918.

King, C. R. et al. (1985). Amplification of a novel v-*erb*-B-related gene in a human mammary carcinoma. Science *229*:974–978.

King, C. R. et al. (1988). EGF binding to its receptor triggers a rapid tyrosine phosphotylation of the erbB-2 protein in the mammary tumor cell line SK-BR-3. EMBO J. *7*:1647–1651.

King, C. R. et al. (1989). Heterogeneous expression of *erb*B-2 messenger RNA in human breast cancer. Cancer Res. *49*:4185–4191.

Kinouchi, T. et al. (1989). Correlation of c-*myc* expression with nuclear pleomorphism in human renal cell carcinoma. Cancer Res. *49*:3627–3630.

Klemi, P. J. et al. (1988). Clinical significance of abnormal nuclear DNA content in serous ovarian tumors. Cancer *62*:2005–2010.

Klemi, P. J. et al. (1990). Prognostic value of flow cytometric DNA content analysis in granulosa cell tumor of the ovary. Cancer *65*:1189–1193.

Klingelhutz, A. J. et al. (1990). Loss of putative 3p suppressor gene in human uroepithelial cell carcinogenesis. Proc. Am. Assn. Cancer Res. *31*:318A.

Klobeck, H. G. et al. (1989). Genomic sequence of human prostate specific antigen (PSA). Nucleic Acids Res. *17*:3981.

Knowles, M. A. et al. (1987). *N*-methyl-*N*-nitrosourea-induced transformation of rat urothelial cells *in vitro* is not mediated by activation of *ras* oncogenes. Oncogene *1*:143–148.

Konig, J. J. et al. (1987). Cytogenetics of prostatic carcinoma: studies on two human tumor cell lines. Cancer Genet. Cytogenet. *28*:43.

Konig, J. J. et al. (1988). Cytogenetic characterization of an established xenografted prostatic adenocarcinoma cell line (PC-82). Cancer Genet. Cytogenet. *34*:91–99.

Kopf, I. et al. (1990). Heterochromatin variants in 109 ovarian cancer patients and 192 healthy subjects. Hereditas *113*:7–16.

Korc, M. et al. (1987). Divergent effects of epidermal growth factor and transforming growth factors on a human endometrial carcinoma cell line. Cancer Res. *47*:4909–4914.

Koss, L. G. (ed.) (1979). Diagnostic Cytology and its Histopathologic Bases, Vol. 2. Lippincott, Philadelphia, pp. 767–768.

Koss, L. G. et al. (1987). DNA cytophotometry of voided urine sediment. Comparison with results of cytologic diagnosis and image analysis. Anal. Quant. Cytol. Histol. *9*:398–404.

Koster, A. et al. (1991). Expression of oncogenes in human breast cancer specimens. Anticancer Res. *11*:193–202.

Koufos, A. et al. (1989). Familial Wiedemann-Beckwith syndrome and a second Wilms' tumor locus both map to 11p15. 5. Am. J. Hum. Genet. *44*:711–719.

Koutsilieris, M. et al. (1987). Characteristics of prostate-derived growth factors for cells of the osteoblast phenotype. J. Clin. Invest. *80*:941–946.

Kovach, J. S. et al. (1991). Direct sequencing from touch preparations of human carcinomas: analysis of p53 mutations in breast carcinomas. J. Natl. Cancer Inst. *83*:1004–1009.

Kovacs, G. and Frisch, S. (1989). Clonal chromosome abnormalities in tumor cells from patients with sporadic renal cell carcinomas. Cancer Res. *49*:651–659.

Kovacs, G. et al. (1987). Specific chromosome aberration in human renal cell carcinoma. Int. J. Cancer *40*:171–178.

Kovacs, G. et al. (1989a). Renal oncocytoma. A phenotypic, and genotypic entity of renal parenchymal tumors. Am. J. Pathol. *134*:967–971.

Kovacs, G. (1989b). Papillary renal cell carcinoma. A morphologic and cytogenetic study of 11 cases. Am. J. Pathol. *134*:27–34.

Kovacs, G. (1990). Application of molecular cytogenetic techniques to the evaluation of renal parenchymal tumors. J. Cancer Res. Clin. Oncol. *116*:318–323.

Kraus, M. H. et al. (1987). Overexpression of the EGF receptor-related protooncogene *erb*B-2 in human mammary tumor cell lines by different molecular mechanisms. EMBO J. *6*:605–610.

Kraus, M. H. et al. (1989). Isolation and characterization of ERBB3, a third member of the ERBB/epidermal growth factor receptor family: evidence for overexpression in a subset of human mammary tumors. Proc. Natl. Acad. Sci. U.S.A. *86*:9193–9197.

Kumar, S. et al. (1989). Prognostic relevance of DNA content in childhood renal tumors. Br. J. Cancer *59*:291–295.

Kute, T. E. et al. (1990). The use of flow cytometry for the prognosis of stage II adjuvant treated breast cancer patients. Cancer *66*:1810–1816.

Lacroix, H. et al. (1989). Overexpression of erbB-2 or EGF receptor proteins present in early stage mammary carcinoma is detected simultaneously in matched primary tumors and regional metastases. Oncogene *4*:145–151.

Lage, J. M. et al. (1991). Flow cytometric analysis of DNA content in partial hydatidiform moles with persistent gestational trophoblastic tumor. Obstet. Gynecol. *77*:111–115.

Larsson, C. et al. (1990). Genomic alterations in human breast carcinomas. Genes Chrom. Cancer *2*:191–197.

Leckie, B. (1990). Stress on the renin gene. Nature (London) *344*:487–488.

Lee, E. Y. H. P. et al. (1988). Inactivation of the retinoblastoma susceptibility gene in human breast cancers. Science *241*:218–221.

Lee, J. H. et al. (1990). Frequent loss of heterozygosity on chromosomes 6q, 11, and 17 in human ovarian carcinomas. Cancer Res. *50*:2724–2728.

Leminen, A. et al. (1990). Deoxyribonucleic acid flow cytometric analysis of cervical adenocarcinoma: prognostic significance of deoxyribonucleic acid ploidy and S-phase fraction. Am. J. Obstet. Gynecol. *162*:848–853.

Leslie, K. K. et al. (1990). Linkage of two human pregnancy-specific β 1-glycoprotein genes: one is associated with hydatidiform mole. Proc. Natl. Acad. Sci. U.S.A. *87*:5822–5826.

Li, S. B. et al. (1991). Allele loss at the retinoblastoma locus in human ovarian cancer. J. Natl. Cancer Inst. *83*: 637–640.

Lichtenstein, A. et al. (1990). Resistance of human ovarian cancer cells to tumor necrosis factor and lymphokine-activated killer cells: correlation with expression of *HER*2/*neu* oncogenes. Cancer Res. *50*:7364–7370.

Lidereau, R. et al. (1988a). Amplification of the *int*-2 gene in primary human breast tumors. Oncogene Res. *2*:285–291.

Lidereau, R. et al. (1988b). Genetic variability of proto-oncogenes for breast cancer risk and prognosis. Biochimie *70*:951–959.

Lidereau, R. et al. (1990). Loss of heterozygosity of the l-*myc* proto-oncogene in human breast tumors. Proc. Am. Assn. Cancer Res. *31*:315A.

Lieber, M. M. (1989). Prostatic dysplasia: significance in relation to nuclear DNA ploidy studies of prostate adenocarcinoma. Urology *34* (Suppl.):43–48.

Lindahl, B. and Alm, P. (1991). Flow cytometrical DNA measurement in endometrial hyperplasias. A prospective follow-up study after abrasio only or additional high-dose gestagen treatment. Anticancer Res. *11*:391–396.

Lindahl, B. and Gullberg, B. (1991). Flow cytometrical DNA- and clinical parameters in the prediction of prognosis in stage I-II endometrial carcinoma. Anticancer Res. *11*:397–402.

Lindahl, B. et al. (1987a). Flow cytometric DNA analysis of normal and cancerous human endometrium and cytological-histopathological correlations. Anticancer Res. *7*:781–790.

Lindahl, B. et al. (1987b). Prognostic value of flow cytometrical DNA measurements in stage I-II endometrial carcinoma: correlations with steroid receptor concentration, tumor myometrial invasion, and degree of differentiation. Anticancer Res. *7*:791–798.

Lingham, R. et al. (1988). Estrogen induction of the epidermal growth factor receptor in RNA. Mol. Endocrinol. *2*:230–235.

Lippman, M. and Mihich, E. (1990). Meeting report. Second annual Pezcoller symposium: therapeutic implications of the molecular biology of breast cancer. Cancer Res. *50*:7402–7404.

Lipponen, P. K. and Eskelinen, M. J. (1991). Nucleolar organizer regions (NORs) in bladder cancer; relation to histological grade, clinical stage and prognosis. Anticancer Res. *11*:75–80.

Liscia, D. S. et al. (1989). Expression of *int*-2 mRNA in human tumors amplified at the int-2 locus. Oncogene *4*:1219–1224.

Liu, E. et al. (1989). Effects of c-*myc* overexpression on the growth characteristics of MCF-7 human breast cancer cells. Oncogene *4*:979–984.

Ljungberg, B. et al. (1985). DNA content in renal cell carcinoma with reference to tumor heterogeneity. Cancer *56*:503–508.

Ljungberg, B. et al. (1986). Prognostic significance of the DNA content in renal cell carcinoma. J. Urol. *135*:422–426.

Locker, A. P. et al. (1989). c-*myc* oncogene product expression and prognosis in operable breast cancer. Br. J. Cancer *60*:669–672.

Lodato, R. F. et al. (1990). Immunohistochemical evaluation of c-*erb*B-2 oncogene expression in ductal carcinoma *in situ* and atypical ductal hyperplasia of the breast. Mod. Pathol. *3*:449–454.

Loose-Mitchell, D. S. et al. (1988). Estrogen regulation of c-*fos* messenger ribonucleic acid. Mol. Endocrinol. *2*:946–951.

Lovekin, C. et al. (1991). c-*erb*B-2 oncoprotein expression in primary and advanced breast cancer. Br. J. Cancer *63*:439–443.

Lundberg, C. et al. (1987). Loss of heterozygosity in human ductal breast tumors indicates a recessive mutation on chromosome 13. Proc. Natl. Acad. Sci. U.S.A. *84*:2372–2376.

Lundberg, S. et al. (1987). DNA flow cytometry and histopathologic grading of paraffin-embedded prostate biopsy specimens in a survival study. Cancer Res. *47*:1973–1977.

Lunec, J. et al. (1992). c-*erb*B-2 amplification and identical p53 mutations in concomitant transitional carcinomas of renal pelvis and urinary bladder. Lancet *339*:439–440.

Machotka, S. V. et al. (1989). Amplification of the proto-oncogenes *int*-2, c-*erb*B-2 and c-*myc* in human breast cancer. Clin. Chim. Acta *184*:207–218.

Mackay, J. et al. (1988a). Partial deletion of chromosome 11p in breast cancer correlates with size of primary tumor and estrogen receptor level. Br. J. Cancer *58*:710–714.

Mackay, J. et al. (1988b). Allele loss on short arm of chromosome 17 in breast cancers. Lancet *ii*:1384–1385.

Maddy, S. Q. et al. (1989). Epidermal growth factor receptors in human prostate cancer: correlation with histological differentiation of the tumor. Br. J. Cancer *60*:41–44.

Maguire, H. C., Jr. et al. (1990). *neu* (c-*erb*B-2), a tumor marker in carcinoma of the female breast. Pathobiology *58*:297–303.

Maher, E. R. and Yates, J. R. W. (1991). Familial renal cell carcinoma: clinical and molecular genetic aspects. Br. J. Cancer *63*:176–179.

Maihle, N. J. and Kung, H. J. (1988). c-*erb* B and the epidermal growth-factor receptor: a molecule with dual identity. Biochem. Biophys. Acta *948*:287–304.

Mannens, M. et al. (1988). Molecular nature of genetic changes resulting in loss of heterozygosity of chromosome 11 in Wilms' tumors. Hum. Genet. *81*:41–48.

Mannens, M. et al. (1990). Loss of heterozygosity in Wilms' tumors, studied for six putative tumor suppressor regions, is limited to chromosome 11. Cancer Res. *50*:3279–3283.

Mark, J. et al. (1990). Chromosomal patterns in human benign uterine leiomyomas. Cancer Genet. Cytogenet. *44*:1–13.

Mark, J. et al. (1991). Cytogenetics of multiple uterine leiomyomas, parametrial leiomyoma and disseminated peritoneal leiomyomatosis. Anticancer Res. *11*:33–40.

Mars, W. M. and Saunders, G. F. (1990). Chromosomal abnormalities in human breast cancer. Cancer Metast. Rev. *9*:35–43.

Marsoni, S. et al. (Gruppo Interregionale Cooperativo di Oncologia Ginecologica [GICOG]). (1990). Prognostic factors in advanced epithelial ovarian cancer. Br. J. Cancer *62*:444–450.

Masiakowski, P. et al. (1982). Cloning of cDNA sequences of hormone-regulated genes from the MCF-7 human breast cancer cell line. Nucleic Acids Res. *10*:7895–7903.

May, E. et al. (1990). Human breast cancer: identification of populations with a high risk of early relapse in relation to both estrogen receptor status and c-*erb*B-2 overexpression. Br. J. Cancer *62*:430–435.

May, F. E. B. and Westley, B. R. (1988). Identification and characterization of estrogen regulated RNAs in human breast cancer cells. J. Biol. Chem. *263*:12901–12908.

McCann, A. H. et al. (1990a). c-erbB-2 oncoprotein expression in primary human tumors. Cancer *65*:88–92.

McCann, A. H. et al. (1990b). The combined significance of c-*erb*B-2 and estrogen receptor status in human breast cancer. Br. J. Cancer *62* (Suppl. XII):12.

McGuire, W. L. (1987). Prognostic factors for recurrence and survival in human breast cancer. Breast Cancer Res. Treat. *10*:5–9.

McGuire, W. L. (1990). Prognostic factors in axillary node-negative breast cancer patients. Proc. Am. Assn. Cancer Res. *31*:469–470.

McGuire, W. L. and Dressler, L. G. (1985). The emerging impact of flow cytometry in predicting recurrence and survival in breast cancer: a review. J. Natl. Canc. Inst. *75*:405–410.

McGuire, W. L. et al. (1985). Impact of flow cytometry in predicting recurrence and survival in breast cancer patients. Breast Cancer Res. Treat. *5*:117–128.

Menegazzi, M. et al. (1988). Analysis of the methylation pattern of c-Ha-*ras* oncogene in human prostate cancer. Ital. J. Biochem. *37*:104–110.

Merlo, G. R. et al. (1989). Frequent alteration of the DF3 tumor-associated antigen gene in primary human breast carcinomas. Cancer Res. *49*:6966–6971.

Meyer, J. S. and Province, M. (1988). Proliferative index of breast carcinoma by thymidine labeling: prognostic power independent of stage, estrogen and progestrone receptors. Breast Cancer Res. Treat. *12*:191–204.

Meyers, S. L. et al. (1990). Analysis of the *int* 1, *int* 2, c-*myc* and *neu* oncogenes in human breast carcinomas. Cancer Res. *50*:5911–5918.

Miles, J. et al. (1988). Genomic defects in non-familial renal cell carcinoma: a possible specific chromosome change. Cancer Genet. Cytogenet. *34*:135–142.

Miller, A. D. et al. (1984). c-*fos* protein can induce cellular transformation: a novel mechanism of activation of a cellular oncogene. Cell (Cambridge, Mass) *36*:51–60.

Millis, R. R. and Girling, A. C. (1989). The breast. In: Diagnostic Surgical Pathology. Sternberg, S. S. (ed.). Raven Press, New York, pp. 253–313.

Mitchell, E. L. D. and Santibanez-Koref, M. F. (1990). 1p13 is the most frequently involved band in structural chromosomal rearrangements in human breast cancer. Genes Chromo. Cancer *2*:278–289.

Mizukami, Y. et al. (1990). Immunohistochemical demonstration of growth factors, TGF α, TGF β, IGF-I, and *neu* oncogene product in benign and malignant human breast tissues. Anticancer Res. *10*:1115–1126.

Moberger, B. et al. (1990). The prognostic significance of growth pattern and its relation to tumor cell nuclear DNA content in endometrial carcinoma. Acta Oncol. *29*:17–23.

Montgomery, B. T. et al. (1990). Stage B prostate adenocarcinoma. Flow cytometric nuclear DNA ploidy analysis. Arch. Surg. *125*:327–331.

Mora, A. A. et al. (1989). p29 protein in breast cancer: relation between estrogen and progestron receptors. Tumori *75*:113–116.

Moran, R. E. et al. (1984). Correlation of cell cycle kinetics, hormone receptors, histopathology, and nodal status in human breast cancer. Cancer *54*:1586–1590.

Mori, O. et al. (1990). Expression of *ras* p21 in mammary and extramammary Paget's disease. Arch. Pathol. Lab Med. *114*:858–861.

Morita, R. et al. (1991). Allelotype of renal cell carcinoma. Cancer Res. *51*:820–823.

Morris, B. J. (1991). Mollecular genetics links renin to hypertension. Mol. Cell. Endocrinol. *75*:C13-C18.

Morse, B. et al. (1988). Insertional mutagenesis of the *myc* locus by a LINE-1 sequence in a human breast carcinoma. Nature (London) *333*:87–90.

Mullen, P. and Miller, W. R. (1990a). Variations associated with the DNA analysis of multiple fine needle aspirates obtained from breast cancer patients. Br. J. Cancer *59*:688–691.

Mullen, P. and Miller, W. R. (1990b). DNA flow cytometry of breast cancer fine needle aspirates. A response. Br. J. Cancer *61*:343–344.

Muller, R. and Wagner, E. F. (1984). Differentiation of F9 teratocarcinoma stem cells after transfer of c-*fos* proto-oncogenes. Nature (London) *311*:438–442.

Mullins, J. J. et al. (1990). Fulminant hypertension in transgenic rats harboring the mouse *Ren*-2 gene. Nature (London) *344*:541–544.

Murphy, L. C. et al. (1988). Epidermal growth factor gene expression in human breast cancer cells: regulation of expression by progestins. Cancer Res. *48*:4555–4560.

Murphy, L. J. et al. (1987). Estrogen induction of N-*myc* and c-*myc* protooncogene expression in the rat uterus. Endocrinology *120*:1882–1888.

Murty, V. V. V. S. et al. (1990). Nonrandom chromosome abnormalities in testicular and ovarian germ cell tumor cell lines. Cancer Genet. Cytogenet. *50*:67–73.

Mydlo, J. H. et al. (1989). Expression of transforming growth factor α and epidermal growth factor receptor messenger RNA in neoplastic and nonneoplastic human kidney. Cancer Res. *49*:3407–3411.

Naber, S. P. et al. (1990). Strategies for the analysis of oncogene overexpression. Studies of the *neu* oncogene in breast carcinoma. Am. J. Clin. Pathol. *94*:125–136.

Narod, S. A. et al. (1991). Familial breast-ovarian cancer locus on chromosome 17q12-q23. Lancet *338*:82–83.

Naslund, I. et al. (1987). Nuclear DNA changes during pathogenesis of squamous carcinoma of the cervix in 3,4-benzopyrene treated mice. Anal. Quant. Cytol. Histol. *9*:411–418.

National Cancer Institute (1988). Clinical alert from the National Cancer Institute, May 15, 1988. Breast Cancer Res. Treat. *12*:3–5.

Naylor, S. L. et al. (1989). The DNF 1552 locus at 3p21 is transcribed in normal lung and small lung cancer. Genomics *4*:335–361.

Neal, D. E. et al. (1985). Epidermal growth factor receptors in human bladder cancer: comparison of invasive and superficial tumors. Lancet *i*:366–368.

Nesland, J. M. et al. (1991). The c-erbB-2 protein in primary and metastatic breast carcinomas. Ultrastr. Pathol. *15*:281–289.

Newbury, R. et al. (1990). DNA content as a prognostic factor in endometrial carcinoma. Obstet. Gynecol. *76*:251–257.

Nicholson, R. I. et al. (1990). Multiple antibodies predict hormone sensitivity/hormone insensitivity in breast cancer. Br. J. Cancer *62* (Suppl. XII):18A.

Nigro, J. M. et al. (1989). Mutations in the *p53* gene occur in diverse human tumor types. Nature (London) *342*:705–708.

Nilbert, M. et al. (1988). Karyotypic rearrangements in 20 uterine leiomyomas. Cytogenet. Cell Genet. *49*:300–304.

Nilbert, M. et al. (1990). Characteristic chromosome abnormalities, including rearrangements of 6p, del(7q), +12, and t(12,14), in 44 uterine leiomyomas. Hum. Genet. *85*:605–611.

Nisen, P. D. et al. (1986). Enhanced expression of the N-*myc* gene in Wilms' tumor. Cancer Res. *46*:6217–6222.

Nisen, P. D. et al. (1988). N-*myc* oncogene expression in histopathologically unrelated bilateral pediatric renal tumors. Cancer *61*:1821–1826.

Nobunaga, T. et al. (1990). Differential diagnosis between complete mole and hydropic abortus by deoxyribonucleic acid fingerprints. Am. J. Obstet. Gynecol. *163*:634–638.

Nusse, R. and Varmus, H. E. (1982). Many tumors induced by the mouse mammary tumor virus contain a provirus integrated in the same region of the host genome. Cell (Cambridge, Mass) *31*:99–109.

Nusse, R. et al. (1984). Mode of proviral activation of a putative mammary oncogene (*int*-1) on mouse chromosome 15. Nature (London) *307*:131–136.

O'Reilly, S. M. et al. (1990a). DNA index, S-phase fraction, histologic grade and prognosis in breast cancer. Br. J. Cancer *61*:671–674.

O'Reilly, S. M. et al. (1990b). The relationship between c-*erb*B-2 expression, S-phase fraction (SPF) and prognosis in breast cancer. Br. J. Cancer *62* (Suppl. XII):25A.

O'Rourke, S. et al. (1990). Analysis of breast lesions by genetic fingerprinting. Br. J. Cancer *62* (Suppl. XII):16A.

Ocadiz, R. et al. (1987). High correlation between molecular alterations of the c-*myc* oncogene and carcinoma of the uterine cervix. Cancer Res. *47*:4173–4177.

Oesterling, J. E. (1991). Prostate specific antigen: a critical assessment of the most useful tumor marker for adenocarcinoma of the prostate. J. Urol. *145*:907–923.

Ogawa, O. et al. (1991). Allelic loss at chromosome 3p characterizes clear cell phenotype of renal cell carcinoma. Cancer Res. *51*:949–953.

Ohnuki, Y. et al. (1980). Chromosomal analysis of human prostatic adenocarcinoma cell lines. Cancer Res. *40*:524–534.

Oka, K. et al. (1991). Detection of loss of heterozygosity in the p53 gene in renal cell carcinoma and bladder cancer using the polymerase chain reaction. Mol. Carcinogenesis *4*:10–13.

Olumi, A. F. et al. (1990). Allelic loss of chromosome 17p distinguishes high grade from low grade transitional cell carcinomas of the bladder. Cancer Res. *50*:7081–7083.

Otto, U. et al. (1984). Tumor cell deoxyribonucleic acid content and prognosis in human renal cell carcinoma. J. Urol. *132*:237–239.

Pal, N. et al. (1990). Preferential loss of maternal alleles in sporadic Wilms' tumor. Oncogene *5*:1665–1668.

Parfrey, P. S. et al. (1990). The diagnosis and prognosis of autosomal dominant polycystic kidney disease. N. Engl. J. Med. *323*:1085–1090.

Parkes, H. C. et al. (1990). C-*erb*B-2 mRNA expression in human breast tumors: comparison with c-*erb*B-2 DNA amplification and correlation with prognosis. Br. J. Cancer *61*:39–45.

Paterson, M. C. et al. (1991). Correlation between c-*erb* B-2 amplification and risk of recurrent disease in node-negative breast cancer. Cancer Res. *51*:556–567.

Pathak, S. et al. (1982). Familial renal cell carcinoma with a 3;11 chromosome translocation limited to tumor cells. Science *217*:939–941.

Pavelic, Z. P. et al. (1990). Detection of the c-*myc* proto-oncogene in human breast cancer. Proc. Am. Assn. Cancer Res. *31*:329A.

Pelletier, J. et al. (1991). WT1 mutations contribute to abnormal genital system development and hereditary Wilms' tumor. Nature *353*:431–434.

Pelley, R. J. et al. (1989). Disease tropism of c-*erb* B: effects of carboxyl-terminal tyrosine and internal mutations on tissue-specific transformation. Proc. Natl. Acad. Sci. U.S.A. *86*:7164–7168.

Perantoni, A. O. et al. (1987). Activated *neu* oncogene sequences in primary tumors of the peripheral nervous system induced in rats by transplacental exposure to ethylnitrosourea. Proc. Natl. Acad Sci. U.S.A. *84*:6317–6321.

Peres, R. et al. (1987). Frequent expression of growth factors for mesenchymal cells in human mammary carcinoma cell lines. Cancer Res. *47*:3425–3429.

Perkel, V. S. et al. (1990). Human prostatic cancer cells, PC3, elaborate mitogenic activity which selectively stimulates human bone cells. Cancer Res. *50*:6902–6907.

Perren, T. J. (1991). c-*erb*B-2 oncogene as a prognostic marker in breast cancer. Br. J. Cancer *63*:328–332.

Perucca, D. et al. (1990). Molecular genetics of human bladder carcinomas. Cancer Genet. Cytogenet. *49*:143–156.

Peters, G. et al. (1983). Tumorigenesis by mouse mammary tumor virus: evidence for a common region for provirus integration in mammary tumors. Cell (Cambridge, Mass) *33*:369–377.

Peters, J. M. et al. (1990). Prognostic significance of the nuclear DNA content in localized prostatic adenocarcinoma. Anal. Quant. Cytol. Histol. *12*:359–365.

Petrides, P. E. et al. (1990). Modulation of pro-epidermal growth factor, pro-transforming growth factor α and epidermal growth factor receptor gene expression in human renal carcinomas. Cancer Res. *50*:3934–3939.

Peyrat, J. P. et al. (1988a). Presence and characterization of insulin-like growth factor 1 receptors in human benign breast disease. Eur. J. Cancer Clin. Oncol. *24*:1425–1431.

Peyrat, J. P. et al. (1988b). Insulin-like growth factor I receptor (IGF 1-R) in human breast cancer. Relation to estradiol and progesterone receptors. Cancer Res. *48*:6429–6433.

Peyrat, J. P. et al. (1989). Characterization of insulin-like growth factor I receptor (IGF I-R) in human breast cancer cell lines. Bull. Cancer *76*:311–319.

Phillips, S. M. et al. (1990). Isolation of gene associated with high metastatic potential in rat mammary adenocarcinomas. J. Natl. Cancer Inst. *82*:199–203.

Pinion, S. B. et al. (1991). Oncogene expression in cervical intraepithelial neoplasia and invasive cancer of cervix. Lancet *337*:819–820.

Pittman, S. et al. (1987). Flow cytometric and karyotypic analysis of a primary small cell carcinoma of the prostate: a xenografted cell line. Cancer Genet. Cytogenet. *26*:165–169.

Presti, J. C., Jr. et al. (1991). Histopathological, cytogenetic, and molecular characterization of renal cortical tumors. Cancer Res. *51*:1544–1552.

Pritchard-Jones, K. et al. (1990). The candidate Wilms' tumor gene is involved in genitourinary development. Nature (London) *346*:194–197.

Prosperi, M. T. et al. (1987). Two adjacent mutations at position 12 activate the K-*ras* 2 oncogene of a human mammary tumor cell line. Oncogene Res. *1*:121–128.

Prosser, J. et al. (1991). Mutations in *p53* do not account for heritable breast cancer: a study in five affected families. Br. J. Cancer *63*:181–184.

Prud'homme, J. F. et al. (1985). Cloning of a gene expressed in human breast cancer and regulated by estrogen in MCF-7 cells. DNA *4*:11–21.

Punnonen, R. et al. (1989). Prognostic assessment in stage I ovarian cancer using a discriminant analysis with clinicopathological and DNA flow cytometric data. Gynecol. Obstet. Invest. *27*:213–216.

Rackley, R. R. et al. (1989). 5'-Nucleotidase activity in prostatic carcinoma and benign prostatic hyperplasia. Cancer Res. *49*:3702–3707.

Raghavan, D. et al. (1990). Biology and management of bladder cancer. N. Engl. J. Med. *322*:1129–1138.

Rainwater, L. M. et al. (1990). Prostate-specific antigen testing in untreated and treated prostatic adenocarcinoma. Mayo Clin. Proc. *65*:1118–1126.

Raizis, A. M. et al. (1985). A mitotic recombination in Wilms' tumor occurs between the parathyroid hormone locus and 11p13. Hum. Genet. *70*:344–346.

Read, L. D. et al. (1990). Hormonal modulation of *HER*-2/*neu* protooncogene messenger ribonucleic acid and p185 protein expression in human breast cancer cell lines. Cancer Res. *50*:3947–3951.

Redman, C. W. E. et al. (1990). Tumor cell activity markers in epithelial ovarian cancer: are biochemical and cytometric indices complementary? Br. J. Cancer *61*:755–758.

Reeve, A. E. et al. (1989). Loss of allelic heterozygosity at a second locus on chromosome 11 in sporadic Wilms' tumor cells. Mol. Cell. Biol. *9*:1799–1803.

Reid, C. P. et al. (1990). c-*erb* B-2 activity in screen detected breast cancers. Br. J. Cancer *62* (Suppl. XII):22A.

Remvikos, Y. et al. (1988). Relevance of DNA ploidy as cytometry and cytogenetics in 25 cases of human breast cancer. Cytometry *9*:612–618.

Remvikos, Y. et al. (1989). Pretreatment proliferative activity of breast cancer correlates with the response to cytotoxic chemotherapy. J. Natl. Cancer Inst. *81*:1383.

Reynolds, R. K. et al. (1990). Characterization of epidermal growth factor receptor in normal and neoplastic human endometrium. Cancer *66*:1967–1974.

Riccardi, V. M. et al. (1978). Chromosomal inbalance in the aniridia-Wilms' tumor association: 11p interstitial deletion. Pediatrics *61*:604–610.

Riccardi, V. M. et al. (1980). The aniridia-Wilms' tumor association: the critical role of chromosomal band 11p13. Cancer Genet. Cytogenet. *2*:131–137.

Riegman, P. H. J. et al. (1989). Characterization of the prostatic-specific antigen gene: a novel human kallikrein-like gene. Biochem. Biophys. Res. Commun. *159*:95–102.

Ring, K. S. et al. (1990). Flow cytometric analysis of localized adenocarcinoma of the prostate: the use of archival DNA analysis in conjunction with pathological grading to predict clinical outcome following radical retropubic prostatectomy. Prostate *17*:155–164.

Rio, M. C. et al. (1988). Breast cancer-associated pS2 protein: synthesis and secretion by normal stomach mucosa. Science *241*:705–708.

Riou, G. F. (1988). Proto-oncogenes and prognosis in early carcinoma of the uterine cervix. Cancer Surv. *7*:441–456.

Riou, G. et al. (1987). c-*myc* proto-oncogene expression and prognosis in early carcinoma of the uterine cervix. Lancet *i*:761–763.

Riou, G. F. et al. (1990). The c-*myc* proto-oncogene in invasive carcinomas of the uterine cervix: clinical relevance of overexpression in early stages of the cancer. Anticancer Res. *10*:1225–1232.

Ro, J. et al. (1989). c-*erb*B-2 amplification in node-negative human breast cancer. Cancer Res. *49*:6941–6944.

Rochlitz, C. F. et al. (1989). Incidence of activating *ras* oncogene mutations associated with primary and metastatic human breast cancer. Cancer Res. *49*:357–360.

Rodenburg, C. J. et al. (1987). Use of DNA image cytometry in addition to flow cytometry for the study of patients with advanced ovarian cancer. Cancer Res. *47*:3938–3941.

Rodenburg, C. J. et al. (1988). Immunohistochemical detection of the *ras* oncogene product p21 in advanced ovarian cancer. Arch. Pathol. Lab. Med. *112*:151–154.

Rodgers, C. S. et al. (1984). Cytogenetic analysis in human breast carcinoma. I. Nine cases in the diploid range investigated using direct preparations. Cancer Genet. Cytogenet. *13*:95–119.

Rodgers, C. S. et al. (1985). Cytogenetic analysis in a case of cancer of the male breast. Cancer Genet. Cytogenet. *15*:113–117.

Romeo, G. et al. (1988). A second genetic locus for autosomal polycystic kidney disease. Lancet *ii*:8–10.

Rosai, J. (1989). Wilms' tumor. In: Ackerman's Surgical Pathology. C. V. Mosby, St. Louis, pp. 860–864.

Rose, E. A. et al. (1990). Complete physical map of the WAGR region of 11p13 localizes a candidate Wilms' tumor gene. Cell (Cambridge, Mass) *60*:495–508.

Royer-Pokora, B. et al. (1991). Direct pulsed field gel electrophoresis of Wilms' tumors shows that DNA deletions in 11p13 are rare. Genes, Chromosomes and Cancer *3*:89–100.

Rutgers, D. H. et al. (1986). DNA flow cytometry of squamous cell carcinomas from the human uterine cervix: the identification of prognostically different subgroups. Radiother. Oncol. *7*:249–258.

Russell, S. E. H. et al. (1990). Allele loss from chromosome 17 in ovarian cancer. Oncogene *5*:1581–1583.

Rutgers, D. H. et al. (1987). DNA flow cytometry, histological grade, stage, and age as prognostic factors in human epithelial ovarian carcinomas. Pathol. Res. Pract. *182*:207–213.

Rutledge, M. L. et al. (1990). Monoclonal antibody C219 detection of the multidrug-resistant protein P-glycoprotein in routinely processed tissues: a study of 36 cases of ovarian carcinoma. Mod. Pathol. *3*:298–301.

Saga, T. et al. (1991). Scintigraphic detection of overexpressed c-*erb*-B-2 protooncogene products by a class-switched murine anti-c-Erb B-2 protein monoclonal antibody. Cancer Res. *51*:990–994.

Sainsbury, J. R. C. et al. (1987). Epidermal growth factor receptor status as predictor of early recurrence and of death from breast cancer. Lancet *i*:1398–1402.

Santos, G. F. et al. (1988). Estrogen-induced post-transcriptional modulation of c-*myc* proto-oncogene expression in human breast cancer cells. J. Biol. Chem. *263*:9565–9568.

Sasano, H. et al. (1990a). An analysis of abnormalities of the retinoblastoma gene in human ovarian and endometrial carcinoma. Cancer *66*:2150–2154.

Sasano, H. et al. (1990b). Protoncogene amplification and tumor ploidy in human ovarian neoplasms. Hum. Pathol. *21*:382–391.

Sasano, H. et al. (1990c). Serous papillary adenocarcinoma of the endometrium. Analysis of proto-oncogene amplification, flow cytometry, estrogen and progesterone receptors, and immunohistochemistry. Cancer *65*:1545–1551.

Sasi, R. et al. (1991a). Chromosome aberrations and oncogene alterations in two new breast tumor cell lines. Cancer Genet. Cytogenet. *51*:239–254.

Sasi, R. et al. (1991b). Prenatal sexing and detection of ZFY gene sequences in sex chromosome disorders by polymerase chain reaction. J. Clin. Lab. Anal. *5*:193–196.

Sato, T. et al. (1990). Allelotype of breast cancer: cumulative allele losses promote tumor progression in primary breast cancer. Cancer Res. *50*:7184–7189.

Schenken, R. S. et al. (1991). c-*myc* protooncogene polypeptide expression in endometriosis. Am. J. Obstet. Gynecol. *164*:1031–1037.

Schmidt, D. et al. (1986). Flow cytometric analysis of nephroblastomas and related neoplasms. Cancer *58*:2494–2500.

Schneider, J. et al. (1990). DNA analysis, chemoresistance testing and hormone receptor levels as prognostic factors in advanced ovarian carcinoma. Arch. Gynecol. Obstet. *248*:45–52.

Schwabe, H. W. et al. (1983). Flow-cytophotometric studies in renal carcinoma. Urol. Res. *11*:121–125.

Senger, D. R. et al. (1988). T24 human bladder carcinoma cells with activated Ha-*ras* protooncogene: nontumorigenic cells susceptible to malignant transformation with carcinogen. Proc. Natl. Acad. Sci. U.S.A. *85*:5107–5111.

Seppelt, U. et al. (1986). Investigation of automated DNA diagnosis and grading of prostate cancer. Anal. Quant. Cytol. Histol. *8*:152–157.

Seymour, L. et al. (1990). Detection of p24 protein in human breast cancer: influence of receptor status and estrogen exposure. Br. J. Cancer *61*:886–890.

Shimazui, T. et al. (1990). Morphometry of nucleoli as an indicator for grade of malignancy of bladder tumors. Virchow's Arch. B: Cell Pathol. *59*:179–183.

Shimizu, M. et al. (1990). Introduction of normal chromosome 3p modulates the tumorigenicity of a human renal cell carcinoma line Ycr. Oncogene *5*:185–194.

Sigurdsson, H. et al. (1990a). Flow cytometry in primary breast cancer: improving the prognostic value of the fraction of cells in the S-phase by optimal categorization of cut-off levels. Br. J. Cancer *62*:786–790.

Sigurdsson, H. et al. (1990b). Indicators of prognosis in node-negative breast cancer. N. Engl. J. Med. *322*:1045–1053.

Silva, F. G. and Childers, J. H. (1989). Adult renal diseases. In: Diagnostic Surgical Pathology, Vol. 2. Sternberg, S. S. (ed.). Raven Press, New York, pp. 1255–1330.

Silverberg, S. G. (ed.) (1990). Wilms' tumor. In: Principles and practice of Surgical Pathology, 2nd Ed. Churchill Livingstone, New York, pp. 1468–1471.

Silvestrini, R. et al. (1986). Prognostic implication of labelling index versus estrogen receptors and tumor size in node-negative breast cancer. Breast Cancer Res. Treat. *7*:161–169.

Simon, D. et al. (1990). Is chromosome 10 a primary chromosomal abnormality in endometrial adenocarcinoma? Cancer Genet. Cytogenet. *47*:155–162.

Slamon, D. J. et al. (1984). Expression of cellular oncogenes in human malignancies. Science *224*:256–262.

Slamon, D. J. et al. (1987). Human breast cancer: correlation of relapse and survival with amplification of the *HER*-2/*neu* oncogene. Science *235*:177–182.

Slamon, D. J. et al. (1989). Studies of the *HER*-2/*neu* proto-oncogene in human breast and ovarian cancer. Science *244*:707–712.

Slater, R. M. et al. (1985). A cytogenetic study of Wilms' tumor. Cancer Genet. Cytogenet. *14*:95–109.

Slave, I. et al. (1990). *myc* gene amplification and expression in primary human neuroblastoma. Cancer Res. *50*:1459–1463.

Smit, V. T. H. B. M. et al. (1990). Flow cytometric DNA ploidy analysis of synchronously occurring multiple malignant tumors of the female genital tract. Cancer *66*:1843–1849.

Sobel, M. E. (1990). Metastasis suppressor genes. J. Natl. Cancer Inst. *82*:267–276.

Solis, V. et al. (1988). Cytogenetic changes in Wilms' tumors. Cancer Genet. Cytogenet. *34*:223–234.

Sonnenschein, C. et al. (1989). Negative controls of cell proliferation: human prostate cancer cells and androgens. Cancer Res. *49*:3474–3481.

Soubrier, F. et al. (1990). Similar frequencies of renin restriction fragment length polymorphisms in hypertensive and normotensive subjects. Hypertension *16*:712–717.

Spandidos, D. A. (1987). Oncogene activation in malignant transformation: a study of H-*ras* in human breast cancer. Anticancer Res. *7*:991–996.

Spandidos, D. A. et al. (1989). High levels of c-myc protein in human breast tumors determined by a sensitive ELISA technique. Anticancer Res. *9*:821–826.

Spyratos, F. et al. (1987). Flow cytometric study of DNA distribution in cytopunctures of benign and malignant breast lesions. Anal. Quant. Cytol. Histol. *9*:485–494.

Sreekantaiah, C. et al. (1988). Chromosome 1 abnormalities in cervical carcinoma. Cancer *62*:1317–1324.

Stamey, T. A. et al. (1987). Prostate-specific antigen as a serum marker for adenocarcinoma of the prostate. N. Engl. J. Med. *317*:909–916.

Stamey, T. A. et al. (1989). Prostate-specific antigen in the diagnosis and treatment of adenocarcinoma of the prostate. IV. Anti-androgen-treated patients. J. Urol. *141*:1088–1090.

Steeg, P. S. et al. (1988). Evidence for a novel gene associated with low tumor metastic potential. J. Natl. Cancer Inst. *80*:200–204.

Stenman, U. H. et al. (1991). A complex between prostate-specific antigen and α-1-antichymotrypsin is the major form of prostate-specific antigen in serum of patients with prostatic cancer: assay of the complex improves clinical sensitivity for cancer. Cancer Res. *51*:222–226.

Stenzl, A. and de Kernion, J. B. (1991). Current trends in diagnosis and treatment of renal, upper urinary tract, and bladder carcinomas. In: Endocrine Dependent Tumors. Voigt, K. D. and Knabbe, C. (eds.). Raven Press, New York, pp. 215–240.

Stephenson, R. A. (1988). Flow cytometry in genitourinary malignancies using paraffin-embedded material. Semin. Urol. *6*:46–52.

Stern, D. F. et al. (1986). Differential responsiveness of *myc*- and *ras*-transfected cells to growth factors: selective stimulation of *myc*-transfected cells by epidermal growth factor. Mol. Cell. Biol. *6*:870–877.

Strang, P. et al. (1987). S-phase rate as a predictor of early recurrences in carcinoma of the uterine cervix. Anticancer Res. *7*:807–810.

Stuurman, N. et al. (1989). The protein composition of the nuclear matrix of murine P19 embryonal carcinomal cells is differentiation-stage dependent. Exp. Cell Res. *180*:460–466.

Stuurman, N. et al. (1990). The nuclear matrix from cells of different origin. Evidence for a common set of matrix proteins. J. Biol. Chem. *265*:5460.

Suh, K. S. et al. (1990). Granulosa cell tumor of the ovary. Histopathologic and flow cytometric analysis with clinical correlation. Arch. Pathol. Lab. Med. *114*:496–501.

Sutherland, G. R. et al. (1987). Molecular genetics of human chromosome 16. J. Med. Genet. *24*:451–456.

Szucs, S. et al. (1987). Deletion 3p: the only chromosome loss in a primary renal cell carcinoma. Cancer Genet. Cytogenet. *26*:369–373.

Szulman, A. E. and Surti, U. (1984). The syndrome of partial and complete molar gestation. Clin. Obstet. Gynecol. *27*:172–180.

T'Ang, A. et al. (1988). Structural rearrangement of the retinoblastoma gene in human breast carcinoma. Science *242*:263–266.

Tanaka, N. et al. (1987a). CYBEST-CDMS Model 2. Automated cell dispersion and monolayer smearing system for CYBEST. Anal. Quant. Cytol. Histol. *9*:445–448.

Tanaka, N. et al. (1987b). CYBEST Model 4. Automated system for uterine cancer utilizing image analysis processing. Anal. Quant. Cytol. Histol. *9*:449–454.

Tandon, A. K. et al. (1989). *HER*-2/*neu* oncogene protein and prognosis in breast cancer. J. Clin. Oncol. *7*:1120–1128.

Tandon, A. K. et al. (1990). Cathepsin D and prognosis in breast cancer. N. Engl. J. Med. *322*:297–302.

Tang, R. et al. (1990). Oncogene amplification correlates with dense lymphocytic infiltration in human breast cancers: a role for hematopoietic growth factor release by tumor cells? J. Cell. Biochem. *44*:189–198.

Tauchi, K. et al. (1989). Immunohistochemical studies on oncogene products (c-*erb*B-2, *EGFR*, c-*myc*) and estrogen receptor in benign and malignant breast lesions. With special reference to their prognostic significance in carcinoma. Virchow's Arch. A: Path. Anat. *416*:65–73.

Tavassoli, M. et al. (1989). c-*erb*B-2/c-*erb*A co-amplification indicative of lymph node metastasis, and c-*myc* amplification of high tumor grade, in human breast carcinoma. Br. J. Cancer *60*:505–510.

Theillet, C. et al. (1986). Loss of c-H-*ras* allele and aggressive human primary breats carcinomas. Cancer Res. *46*:4776–4781.

Theillet, C. et al. (1989). Amplification of FGF-related genes in human tumors: possible involvement of HST in breast carcinomas. Oncogene *4*:915–922.

Theillet, C. et al. (1990). BCL-1 participates in the 11q13 amplification found in breast cancer. Oncogene *5*:147–149.

Thompson, A. M. et al. (1990). *p53* gene mRNA expression and chromosome 17p allele loss in breast cancer. Br. J. Cancer *61*:74–78.

Thompson, T. C. (1990). Growth factors and oncogenes in prostate cancer. Cancer Cells *2*:345–354.

Thor, A. et al. (1991). Stress response protein (Srp-27) determination in primary human breast carcinomas: clinical, histologic and prognostic correlations. J. Natl. Cancer Inst. *83*:170–178.

Tilley, W. D. et al. (1990). Endrogen receptor gene expression in human prostate carcinoma cell lines. Cancer Res. *50*:5382–5386.

Toikkanen, S. et al. (1989). The prognostic signficance of nuclear DNA content in invasive breast cancer — a study with long-term follow-up. Br. J. Cancer *60*:693–700.

Toikkanen, S. et al. (1990). Nuclear DNA content as a prognostic factor in $T_{1-2}N_0$ breast cancer. Am. J. Clin. Pathol. *93*:471–479.

Tosi, P. et al. (1986). Nuclear morphometry as an important prognostic factor in stage I renal cell carcinoma. Cancer *58*:2512–2518.

Travers, M. T. and Knowler, J. T. (1987). Estrogen-induced expression of oncogenes in the immature rat uterus. FEBS Lett. *211*:27–30.

Trent, J. (1985). Cytogenetic and molecular biologic alterations in human breast cancer: a review. Breast Cancer Res. Treat. *5*:221–229.

Tsuda, H. et al. (1990a). Correlation between histologic grade of malignancy and copy number of c-*erb*B-2 gene in breast carcinoma. A retrospective analysis of 176 cases. Cancer *65*:1794–1800.

Tsuda, H. et al. (1990b). Immunohistochemical study on overexpression of c-ErbB-2 protein in human breast cancer: its correlation with gene amplification and long-term survival of patients. Jpn. J. Cancer Res. *81*:327–332.

Tsuda, T. et al. (1987). Analysis of N-*myc* amplification in relation to disease stage and histologic types in human neuroblastomas. Cancer *60*:820–826.

Tsuda, T. et al. (1988). Amplification of the *hst*-1 gene in human esophageal carcinomas. Jpn. J. Cancer Res. *79*:584–588.

Tsushima, K. et al. (1988). Uterine leiomyosarcomas and benign smooth muscle tumors: usefulness of nuclear DNA patterns studied by flow cytometry. Mayo Clin. Proc. *63*:248–255.

Tsutsumi, M. et al. (1988). Co-amplification of the *hst*-1 and *int*-2 genes in human cancers. Jpn. J. Cancer Res. *79*:428–432.

Tubiana, M. et al. (1984). The long-term prognostic significance of the thymidine labelling index in breast cancer. Int. J. Cancer *33*:441–445.

Tubiana, M. et al. (1989). Growth rate, kinetics of tumor cell proliferation and long-term outcome in human breast cancer. Int. J. Cancer *44*:17–22.

Uchida, T. et al. (1990). DNA minisatellites demonstrate somatic DNA changes in most human bladder cancers. Cytogenet. Cell Genet. *53*:61–63.

Uyterlinde, A. M. et al. (1988). Limited prognostic value of cellular DNA content to classical and morphometrical parameters in invasive ductal breast cancer. Am. J. Clin. Pathol. *89*:301–307.

van de Vijver, M. J. and Nusse, R. (1991). The molecular biology of breast cancer. Biochim. Biophys. Acta *1072*:33–50.

van de Vijver, M. J. et al. (1987). Amplification of the neu (c-*erb*B-2) oncogene in human mammary tumors is relatively frequent and is often accompanied by amplification of the linked c-*erb*A oncogene. Mol. Cell. Biol. *7*:2019–2023.

van de Vijver, M. J. et al. (1988a). Immunohistochemical detection of the Neu protein in tissue sections of human breast tumors with amplified *neu* DNA. Oncogene *2*:175–178.

van de Vijver, M. J. et al. (1988b). Neu-protein overexpression in breast cancer. Association with comedo-type ductal carcinoma *in situ* and limited prognostic value in stage II breast cancer. N. Engl. J. Med. *319*:1239–1245.

van der Hout, A. H. et al. (1991). Localization by *in situ* hybridization of three 3p probes with respect to the breakpoint in a t(3;8) in hereditary renal cell carcinoma. Cancer Genet. Cytogenet. *51*:121–124.

van der Putten, H. W. H. M. et al. (1989). Prognostic value of quantitative pathologic features and DNA content in individual patients with stage I endometrial adenocarcinoma. Cancer *63*:1378–1387.

van't Veer, L. J. et al. (1988). *ras* oncogene activation in human ovarian carcinoma. Oncogene *2*:157–165.

Varley, J. M. et al. (1987a). Alterations to either c-*erb*B-2 (*neu*) or c-*myc* protooncogenes in breast carcinomas correlate with poor short-term prognosis. Oncogene *1*:423–430.

Varley, J. M. et al. (1987b). An unusual alteration in c-*myc* in tissue from a primary breast carcinoma. Oncogene *1*:431–438.

Varley, J. M. et al. (1988). A common alteration to the *int*-2 proto-oncogene in DNA from primary breast carcinomas. Oncogene *3*:87–91.

Varley, J. M. et al. (1989). The retinoblastoma gene is frequently altered leading to loss of expression in primary breast tumors. Oncogene *4*:725–729.

Varma, V. A. et al. (1989). Antibodies to *ras* oncogene p21 proteins lack immunohistochemical specificity for neoplastic epithelium in human prostate tissue. Arch. Pathol. Lab. Med. *113*:16–19.

Vassilakos, P. et al. (1977). Hydatidiform mole: two entities. Am. J. Obstet. Gynecol. *127*:167–170.

Veloso, J. D. et al. (1992). DNA ploidy of oncocytic-granular renal cell carcinomas and renal oncocytomas by image analysis. Arch. Pathol. Lab. Med. *116*:154–158.

Venter, D. J. et al. (1987). Overexpression of the c-ErbB-2 oncoprotein in human breast carcinomas: immunohistological assessment correlates with gene amplification. Lancet *ii*:69–72.

Verrelle, P. et al. (1991). Clinical relevance of immunohistochemical detection of multidrug resistance P-glycoprotein in breast carcinoma. J. Natl. Cancer Inst. *83*:111–116.

Vielh, P. et al. (1991). Ki 67 index and S-phase fraction in human breast carcinomas. Comparison and correlations with prognostic factors. Am. J. Clin. Pathol. *94*:681–686.

Visscher, D. W. et al. (1990). Multiparametric deoxyribonucleic acid and cell cycle analysis of breast carcinomas by flow cytometry. Clinicopathologic correlations. Lab. Invest. *62*:370–378.

Volm, M. et al. (1989). Flow-cytometric prognostic factors for the survival of patients with ovarian carcinoma: a 5 year follow-up study. Gynecol. Oncol. *35*:84–89.

Wada, A. et al. (1988). The homologous oncogenes, *HST*1 and *INT*2, are closely located in human genome. Biochem. Biophys. Res. Commun. *157*:828–835.

Waghorne, C. et al. (1987). Metastatic potential of SP1 mouse mammary adenocarcinoma cells is differentially induced by activated and normal forms of c-H-*ras*. Oncogene *1*:149–155.

Walker, C. et al. (1992). Predisposition to a renal cell carcinoma due to alteration of a cancer susceptibility gene. Science *255*:1693–1695.

Walker, R. A. et al. (1989). An evaluation of immunoreactivity for c-ErbB-2 protein as a marker of poor short-term prognosis in breast cancer. Br. J. Cancer *60*:426–429.

Walker, R. et al. (1990). Oestrogen receptor, EGF receptor and oncogene expression in cancer-containing breasts. Br. J. Cancer *62* (Suppl. XII):28A.

Walker, R. A. and Cowl, J. (1991). The expression of *c-fos* protein in human breast. J. Pathol. *163*:323–327.

Wang, N. and Perkins, K. L. (1984). Involvement of band 3p14 in t(3;8) hereditary renal carcinoma. Cancer Genet. Cytogenet. *11*:479–481.

Wang-Wuu, S. et al. (1990). Chromosome analysis of 31 Wilms' tumors. Cancer Res. *50*:2786–2793.

Ward, J. M. et al. (1988). H-*ras* p21 and peanut lectin immunoreactivity of hyperplastic, preneoplastic and neoplastic urinary bladder lesions in rats. Jpn. J. Cancer Res. *79*:152–155.

Watson, J. V. et al. (1987). Oncogene expression in ovarian cancer: a pilot study of c-*myc* oncoprotein in serous papillary ovarian cancer. Gynecol. Oncol. *28*:137–150.

Weidner, U. et al. (1990). Inverse relationship of epidermal growth factor receptor and *HER*2/*neu* gene expression in human renal cell carcinoma. Cancer Res. *50*:4504–4509.

Weiner, D. B. et al. (1989). A point mutation in the *neu* oncogene mimics ligand induction of receptor aggregation. Nature (London) *339*:230–231.

Weismann, B. E. et al. (1987). Introduction of a normal human chromosome 11 into a Wilms' tumor cell line controls its tumorigenic expression. Science *236*:175–180.

Weisz, A. et al. (1990). Estrogen stimulates transcription of c-*jun* proto-oncogene. Mol. Endocrinol. *4*:1041–1050.

Welch, D. R. A. et al. (1990). Transforming growth factor β stimulates mammary adenocarcinoma cell invasion and metastatic potential. Proc. Natl. Acad. Sci. U.S.A. *87*:7678–7682.

Westley, B. R. et al. (1989). Effects of estrogen and the antiestrogens, tamoxifen and LY117018, on four estrogen regulated RNAs in the EFM-19 breast cancer cell lines. J. Steroid Biochem. *32*:365–372.

Wheeless, L. L. et al. (1991). DNA slit-scan flow cytometry of bladder irrigation specimens and the importance of recognizing urothelial cells. Cytometry *12*:140–146.

White, J. J. et al. (1990). DNA alterations in prostatic adenocarcinoma and benign prostatic hyperplasia: detection of DNA fingerprint analyses. Mutat. Res. *237*:37–43.

Wiltschke, C. et al. (1990). Expression of the *HER*-2/*neu* and the *p53* oncogene-protein in breast cancer patients; correlation to clinical parameters. Proc. Am. Assn. Cancer Res. *31*:315A.

Winchester, D. J. et al. (1990). The importance of DNA flow cytometry in node-negative breast cancer. Arch. Surg. *125*:886–889.

Winkler, H. Z. et al. (1988). Stage D1 prostatic adenocarcinoma: significance of nuclear DNA ploidy patterns studied by flow cytometry. Mayo Clin. Proc. *63*:103–112.

Winstanley, J. H. R. et al. (1990). The role of thymidine labelling as a prognostic factor in the long term survival in primary breast cancer. Br. J. Cancer *62* (Suppl. XII):26A.

Winstanley, J. et al. (1991). The long term prognostic significance of c-*erb*B-2 in primary breast cancer. Br. J. Cancer *63*:447–450.

Wolberg, W. H. and Mangasarian, O. L. (1990). Multisurface method of pattern separation for medical diagnosis applied to breast cytology. Proc. Natl. Acad. Sci. U.S.A. *87*:9193–9196.

Wright, C. et al. (1989). Expression of c-ErbB-2 oncoprotein: a prognostic indicator in human breast cancer. Cancer Res. *49*:2087–2090.

Wright, C. et al. (1990). Expression of c-ErbB-2 protein product in bladder cancer. Br. J. Cancer *62*:764–765.

Wright, G. L., Jr. et al. (1990). Flow cytometric DNA analysis after immunoselection of bladder tumor cells with monoclonal antibody DU 83.21. Cancer *66*:1242–1252.

Yamada, H. et al. (1990). Identification of multiple chromsomes carrying tumor suppressor activity for a uterine endometrial carcinoma cell line. Proc. Am. Assn. Cancer Res. *31*:329A.

Yao, M. et al. (1988). Enhanced expression of c-*myc* and epidermal growth factor receptor (c-*erb*B-1) genes in primary human renal cancer. Cancer Res. *48*:6753–6757.

Yee, D. et al. (1989). The insulin-like growth factor binding protein BP-25 is expressed by human breast cancer cells. Biochem. Biophys. Res. Commun. *158*:38–44.

Yokota, J. et al. (1986). Alterations of *myc, myb, ras*^{Ha} proto-oncogenes in cancers are frequent and show clinical correlations. Science *231*:261–265.

Yokota, J. et al. (1989). Loss of heterozygosity on the short arm of chromosome 3 in carcinoma of the uterine cervix. Cancer Res. *49*:3598–3601.

Yoshida, M. A. et al. (1986). Cytogenetic studies of tumor tissue from patients with non-familial renal cell carcinoma. Cancer Res. *46*:2139–2147.

Yoshida, M. C. et al. (1988). Human *HST*-1 (*HSTF*1) gene maps to chromosome band 11q13 and co-amplifies with the *INT*2 gene in human cancer. Proc. Natl. Acad. Sci. U.S.A. *85*:4861–4864.

Yoshida, T. et al. (1987). Genomic sequence of *hst*, a transforming gene encoding a protein homologous to fibroblast growth factors and the *int*-2 encoded protein. Proc. Natl. Acad. Sci. U.S.A. *84*:7305–7309.

Yuan, J. et al. (1990). DNA ploidy in human breast cancer: its relationship to other indicators of poor prognosis and to long-term survival. Br. J. Cancer *62* (Suppl. XII):6A.

Zbar, B. et al. (1987). Loss of alleles in loci on the short arm of chromosome 3 in renal cell carcinoma. Nature (London) *327*:721–724.

Zeillinger, R. et al. (1989). *HER*-2 amplification, steroid receptors and epidermal growth factor receptor in primary breast cancer. Oncogene *4*:109–114.

Zhang, R. et al. (1989). Rare clonal karyotype variants in primary cultures of human breast carcinoma cells. Cancer Res. *49*:444–449.

Zhou, D. et al. (1987). Association of multiple copies of the c-*erb*B-2 oncogene with spread of breast cancer. Cancer Res. *47*:6123–6125.

Zhou, D. J. et al. (1988a). A unique pattern of proto-oncogene abnormalities in ovarian adenocarcinomas. Cancer *62*:1573–1576.

Zhou, D. J. et al. (1988b). Amplification of human *int*-2 in breast cancers and squamous carcinomas. Oncogene *2*:279–282.

Zhou, D. J. et al. (1989). Proto-oncogene abnormalities in human breast cancer: c-*erb*B-2 amplification does not correlate with recurrence of disease. Oncogene *4*:105–108.

Zoon, K. C. et al. (1986). Modulation of epidermal growth factor receptors by human α interferon. Proc. Natl. Acad. Sci. U.S.A. *83*:8226–8230.

Diabetes Mellitus

4.1.1 INTRODUCTION

The elucidation of molecular genetic aspects of **non-insulin-dependent diabetes mellitus (NIDDM)** and **insulin-dependent diabetes mellitus (IDDM)** has stimulated attempts to define approaches capable of identifying predisposition of individuals to DM. The literature dealing with molecular genetic underpinnings of susceptibility and resistance to DM is enormous, conflicting observations abound, and a number of hypotheses have been advanced attempting to sort out the vast amount of data (e.g., Nepom, 1990; Rich, 1990; Wassmuth and Lernmark, 1989). The weight of the accumulated evidence clearly points to an association of DM susceptibility with a certain pattern of arrangement of **HLA class II genes**. The complexity of interactions of various alleles already defined and those still to be identified, on the one hand, and, at this time, the rather obscure role of environmental factors (e.g., viral, chemical, and cultural) dictated a somewhat simplified approach in describing the major findings pertinent to laboratory diagnosis of DM susceptibility at the gene level.

Susceptibility to IDDM and NIDDM will be **discussed only as an example** of a variety of conditions in which the role of HLA is becoming progressively recognized. The use of newer molecular biological techniques, such as polymerase chain reaction (PCR)-based analysis of polymorphism patterns and specific alleles of interest, will markedly enhance the speed of accumulation and finesse of analysis of relevant associations between diseases and underlying genetic factors at the gene level. DNA typing is likely to become a common approach to obtain additional and important information when working up common cases such as DM.

Important advances have also been made in the identification of the details of synthesis and function of **insulin, insulin receptor (IR)**, and **glucokinase (GK)** at the gene level. A brief overview of these will be given to better appreciate the emerging potential for evaluating clinical problems in diabetology at the gene level. In view of the intense efforts made to develop practical approaches to gene therapy of DM, gene level diagnostics may acquire additional practical significance much sooner than is generally thought.

4.1.2 SYNOPSIS OF MOLECULAR PATHOLOGY OF DIABETES MELLITUS

4.1.2.1 INSULIN-DEPENDENT DM (IDDM)

IDDM is characterized by onset in adolescence, predominantly in males, associated with an excess morbidity and mortality. At the advanced stages of the disease, featuring hyperglycemia, hypoinsulinemia, ketoacidosis, complications tend to develop in micro- and macrovasculature that eventually contribute to the shortened life expectancy of IDDM patients. Over the last two decades, the key role of **autoimmune responses** resulting in the destruction of the β cells has been defined and a working hypothesis of the etiology of IDDM was proposed (Nerup, 1981; Nerup et al., 1984). It envisioned **genetic predisposition (susceptibility) to IDDM** determined by two genes on **chromosome 6**, one associated with **HLA-D/DR3**, the other with **HLA-D/DR4** (see below). Susceptible individuals develop an abnormal response to the damage produced in β cells by a cytotoxic virus or a chemical. An immunologically-driven destructive process destroys the β cells, or the β cells fail to regenerate after the damage. The complex chain of events emphasizing the immunological aspects of the β cell damage in IDDM has been reviewed (Campbell and Harrison, 1990; Eisenbarth, 1986; Mandrup-Poulsen, 1988; Nerup et al., 1984).

The characteristic **mononuclear cell infiltrate** and **β-cell destruction** (**insulitis**) observed in recent onset IDDM cases with appropriate clinical findings reflect an already advanced stage of the destructive process because approximately 90% of the β-cell mass must be destroyed before clinical symptoms become evident. Insulitis (13 to 75% of the islets destroyed) is more severe in younger patients, and is markedly reduced in long-standing IDDM. Substantial evidence has been provided to implicate **cell-mediated immunity** in the process. Animal studies and findings in IDDM patients suggest that **cytotoxic T cells** apparently participate in the destruction of the β cells at the later stages, rather than during an initial assault. Although the predominant infiltrative component in the islets is T lymphocyte, monocytes and macrophages are also present. The specificity of recognition of β-cell **autoantigens** occurring early in the immune response is apparently mediated by the **activated autoreactive T cells** recruiting natural killer (NK) and/or macrophages (Mandrup-Poulsen, 1988).

Humoral autoimmune reactions complement the cell-mediated immunity in the development of IDDM. The presence of islet cell antibodies, islet cell surface antibodies, as well as other antibodies, reacting with cytoplasmic and surface components of the islet and nonislet tissues, has been observed in IDDM. None of these demonstrate absolute specificity for the β-cells. The clinical significance of these antibodies, the **anti-64-kDa protein antibody** and **insulin autoantibodies**, as markers of IDDM remains to be elucidated. **Islet cells autoantibodies** are thought to result from **polyclonal B cell activation** following the liberation of antigens from destroyed β cells, and their role in β cell destruction remains to be demonstrated (Mandrup-Poulsen, 1988).

The molecular details of β cell destruction have not been conclusively defined; however, the soluble products liberated from the infiltrating immune cells (**cytokines**) have been implicated in delivery of the lethal hit. Studies with crude cytokine preparations suggested that **interleukin 1** (IL-1) is one of the cytokines with islet cytotoxic activity (Mandrup-Poulsen, 1988). Depending on the dose and length of exposure to IL-1, the changes observed in isolated islets can be reversible or cause irreversible cell damage. Degenerative changes in the form of autophagic vacuoles and the accumulation of β granules have been observed as a result of IL-1 action on the islets. The effect of IL-1 appears to be selective, but not exclusive, in specifically destroying β cells. The mechanism of IL-1 action is not entirely clear although the formation of damaging **free radicals** in response to overstimulation of β cell function by high local doses of IL-1 has been observed (Mandrup-Poulsen, 1988).

The molecular events leading to islet cell destruction seem to begin early on with the **hyperexpression of MHC class I molecules** on islet cells and the **induction of MHC class II on β cells** (reviewed in Campbell and Harrison, 1990). **Interferon α** (IFN-α) is implicated in mediating the MHC class I molecules of β cells; however, the molecular mechanism of induction of MHC class II molecules remains unknown. In addition to promoting autoimmune destruction of β cells, the hyperexpression of MHC molecules may constitute a nonimmune component of β cell destruction via a **transgene activation mechanism**. Furthermore, expression of the MHC class I and II molecules by β cells is not sufficient for antigen presentation, and experimental evidence suggests that cytokine products of immunoinflammatory cells mediate "**phenotype switching**" of the β cell, leading to the induction of many properties of antigen-presenting cells (Campbell and Harrison, 1990). In cells from IDDM patients, the expression of MHC class I is decreased, and T cell responses to self antigens are defective, contributing to aberrant antigen presentation (Faustman et al., 1991).

It appears that **interleukin 6** (IL-6) is produced by the β cells of the islets and acts as a costimulator (along with IL-1) in T lymphocyte activation. Associated with this process is lymphocyte adherence, which is mediated on the lymphocyte surface by the **lymphocyte function-associated antigen 1** (LFA-1) and by the **intercellular adhesion molecule 1** (ICAM-1) on the target cell surface. The expression of ICAM-1 has indeed been observed also on the surface of the majority of islet cells exposed to **IFN-γ** or **IFN-α** (Campbell and Harrison, 1990). The cytokine-induced ICAM-1 expression at sites of acute or chronic islet inflammation is thought to play a role in attracting lymphocytes to β cells. In addition to inducing the hyperexpression of MHC, ICAM-1, and IL-6, cytokines (IL-1, IFN-α, and IFN-γ) directly affect the β cell function.

4.1.2.2 NON-INSULIN DEPENDENT DM (NIDDM)

NIDDM is usually defined as **maturity-onset diabetes**. The insulin level in the basal state is comparable to that found in unaffected persons; however, after a glucose challenge a mild hyperglycemia develops followed by late hyperinsulinemia. Both obesity and diabetes independently affect basal insulin levels (Porte, 1991). In addition to glucose, a number of other compounds (**glucagon**, **gastrointestinal inhibitory peptide** [GIP], **cholecystokinin** [CCK], **secretin**, and **amino acids**, in particular **arginine**), may produce rapid insulin responses. Insulin response to nonglucose stimulants is markedly reduced in NIDDM. Because glucose regulates β cell function by producing a direct release of insulin and by modulating the response to other islet stimulants, it appears that the hyperglycemia of NIDDM may represent a way of compensating for islet dysfunction (Porte, 1991). **Insulin resistance**, present in virtually all NIDDM patients, interacts with deficient islet function and leads to **compensatory hyperglycemia**, the degree of which is proportional to the extent of the β **cell dysfunction**. A combination of a reduction in the number of β cells and a defect in the function of the entire β cell mass leads to the islet cell dysfunction in NIDDM. There appears to be a generalized defect in the ability of glucose to regulate islet function. (For a review see Moller and Flier, 1991).

4.1.2.2.1 Islet Amyloid Polypeptide (IAPP)

Relative hyperproinsulinemia and the presence of amyloid deposits in the islets of NIDDM patients have been related to the inappropriate release of unprocessed **prohormone** and its intermediates, which contributes to the development of hyperglycemia (Porte, 1991). Interestingly, the **amyloid deposits** characteristic of NIDDM also occur in insulinomas. The constituent of the islet amyloid is a new peptide, called **islet amyloid polypeptide (IAPP)** or **amylin** (Cooper et al., 1987). This 37-amino acid neuropeptide-like molecule has a 50% homology with the **neuropeptide calcitonin gene-related peptide (CGRP)** (Johnson et al., 1989; Westermark et al., 1987). The two peptides, however, are encoded by different genes located on different chromosomes — the *CGRP* genes on **chromosome 11** and that for IAPP on **chromosome 12** (Mosselman et al., 1988, 1989).

Similarity in the excess of **proinsulin** in NIDDM and insulinoma led to the suggestion that the processing of proinsulin and pro-IAPP may be abnormal and in some way also contributes to amyloid deposition (Porte and Kahn, 1989). IAPP is a normally secreted peptide produced along with insulin in β cells from a larger precursor, and the same enzymes that convert proinsulin to insulin are involved in pro-IAPP processing (Sanke et al., 1988).

IAPP is present in significant amounts in normal islets but constitutes only a very small fraction of the level of insulin in the β-cells (Kanatsuka et al., 1989; Mosselman et al., 1988; Nakazato et al., 1989). The level of IAPP markedly increases on glucose stimulation, with higher levels observed in obesity, however, evidence suggesting an increase in IAPP in NIDDM is lacking (Butler et al., 1990; Hartter et al., 1990; Lukvik et al., 1990).

Although the **biological role of IAPP** has not been established several hypotheses have been advanced (Johnson et al., 1991; Steiner et al., 1991). IAPP does not appear to affect the secretion of insulin (Nagamatsu et al., 1990) or other hormones (Silvestre et al., 1990). A restricting effect on the blood flow in acinar cells is a possibility (Steiner et al., 1991). Structural similarity with

CGRP points to possible modulating effects of IAPP on **adenylate cyclase**. The overproduction of IAPP and the deposition of amyloid in NIDDM may be related to the overproduction of insulin in NIDDM (Cooper et al., 1989; Reaven, 1988; Westermark and Johnson, 1988). The other possibility is a defective release of secretory granules in NIDDM due to impaired glucose responsiveness. Sequencing of the *IAPP* genes from NIDDM patients suggests no abnormality of either IAPP or its precursors (Nishi et al., 1990). Because IAPP is also deposited in normal individuals with aging, **the peptide *per se* is apparently not a diabetogenic molecule but is deposited as a consequence of disordered β cell function associated with diabetes and aging** (Steiner et al., 1991). Most likely, IAPP contributes to NIDDM by its accumulation as amyloid in β cells, which in turn leads to a reduced β cell mass (Porte, 1991). The possibility of IAPP inducing the peripheral insulin resistance has not been demonstrated. An important observation of the markedly greater responsiveness of IAPP secretion relative to insulin in severely hyperglycemic states points to a yet undisclosed regulatory function IAPP can play (Johnson et al., 1991).

4.1.2.2.2 Other Factors in NIDDM

Addressing the glucose transport function in NIDDM, the profound underexpression of **GLUT-2**, the high-K_m facilitative **glucose transporter** on β cells, was found not to be secondary to hyperglycemia, but rather to play a causal role, at least in an animal model (Orci et al., 1990). At this stage, the evidence strongly favors the view that **NIDDM is a heterogeneous disease syndrome characterized by both impaired islet function and insulin resistance** (Porte, 1991). The etiology of NIDDM is probably related to hereditary risk factors modulated by changes in the environment. An intriguing finding has pointed to the possibility of persistent cytomegalovirus (CMV) infection selectively affecting β cells of the islets of Langerhans in persons with NIDDM (Lohr and Oldstone, 1990), although subsequent studies failed to confirm this finding (Hattersley et al., 1992).

In fact, one form of NIDDM, the so-called **maturity-onset diabetes of the young** (MODY), has an autosomal dominant mode of inheritance, and the gene responsible for MODY in one well-characterized pedigree is tightly linked to the marker **adenosine deaminase gene** (*ADA*) on chromosome 20q (Bell et al., 1991).

4.1.2.3 REGULATION OF INSULIN GENE EXPRESSION

The **insulin gene** is located on the short arm of **chromosome 11** (Owerbach et al., 1981). The transcription of the insulin gene occurs in the β cells of the Langerhans islets where the mRNA transcripts are processed (Selden et al., 1987). Translation of the mature insulin mRNA yields **preproinsulin**, which is proteolytically cleaved to produce mature insulin and the **prepeptide** and **C peptide** sequences (Chan et al., 1976; Steiner et al., 1967). Insulin and C peptide are secreted together in response to physiologic stimuli. A cell type-specific enhancer regulatory sequence has been identified in the 5′-flanking region of the insulin gene (Bell et al., 1982; Nir et al., 1986; Ohlsson and Edlund, 1986). Experiments in transgenic mice suggest that introduction of human insulin gene serves to down regulate endogenous mouse insulin gene expression to maintain insulin mRNA and protein levels at a physiological set point (Selden et al., 1987). A significant role in both insulin gene regulation and the modulation of insulin action in target tissues is played by **guanine nucleotide-binding proteins (G proteins)** (for a review, see Robertson et al., 1991).

4.1.2.4 GLUCOKINASE

Stimulation of insulin gene expression by glucose appears to involve **glucokinase** as **glucose sensor**, on the surface of β cells, that serves to generate cell type-specific signals (Matshinsky, 1990). The enzyme is also found in hepatic cells. The glucokinase gene has been characterized and shown to have differently regulated sequences in the liver and pancreas (Iynedjian et al.,

1986; Schwab and Wilson, 1989). At least three glucokinase isoforms are generated by differential RNA processing of the glucokinase gene transcripts (Magnuson and Shelton, 1989). It appears that 5′-noncoding sequences of the glucokinase gene differ in the liver, and β cells may play a role in regulating the expression of glucokinase mRNA (Magnuson, 1990).

Glucokinase gene expression may be involved in the determination of glucose homeostasis, including the dysregulation characteristic of diabetes, because even a slight diminution of β cell glucokinase activity can significantly reduce the glucose sensitivity of the pancreatic β cell (Meglasson and Matschinsky, 1986). Analysis of the glucokinase gene restriction fragment length polymorphisms (RFLPs) and those of its tissue-specific promoter regions in NIDDM pedigrees suggests that primary glucose refractoriness may involve glucose-sensor molecules in β cells and hepatocytes (e.g., **glucokinase**, **phosphorylase *a*, glucose-6-phosphatase**) and account for the symptomatology of NIDDM (Matschinsky, 1990). In fact, a close linkage could be established between the glucokinase locus on chromosome 7p and diabetes in MODY patients (Froguel et al., 1992). Some 45 to 95% of the families studied showed linkage to glucokinase, supporting the view that a gene involved in glucose metabolism can be responsible for the development of NIDDM (Froguel et al., 1992).

4.1.2.5 GLUCOSE TRANSPORTER

Metabolic defects potentially involved in the pathogenesis of NIDDM include impaired **insulin binding** and **receptor kinase activity** (Arner et al., 1987; Caro et al., 1987; Olefsky, 1990; Sinha et al., 1987), **defective glucose transport** (Ciaraldi et al., 1982; Garvey et al., 1988), and **abnormal glycogen synthesis** (Freymond et al., 1988). **Insulin deficiency**, however, also contributes to the metabolic disturbances of NIDDM (Garvey et al., 1985).

Because glucose transport is mediated by a family of tissue-specific membrane glycoproteins, these **glucose transporters** (GTs) become likely candidates for a highly specific genetic defect predisposing to insulin resistance (Mueckler, 1990). Genetic studies, however, so far provided conflicting results: some favoring an association between RFLPs at the Hep G2 **glucose transporter locus** (*GLUT*) (Shows et al., 1987) and NIDDM (Li et al., 1988), whereas others demonstrated a lack of such an association (Kaku et al., 1990; Matsutani et al., 1990; Xiang et al., 1989). Studies, in mouse models and in humans, however, support the view that deficiencies in the transporter locus may indeed exist (Faustman et al., 1991).

4.1.2.6 INSULIN RECEPTOR

Interaction of insulin with the cell begins with IR specifically binding the hormone and initiating the physiological response of the cell, thereby **modulating the cell surface signal into a specific intracellular response**. The structure of the receptor has been elucidated (for reviews, see Houslay and Siddle, 1989; Olefsky, 1990; Rosen, 1987, 1989). The processed receptor is a tetramer composed of two α subunits that bind insulin outside the cell, and two β subunits that span the thickness of the plasma membrane and have protein tyrosine kinase activity in their cytosolic domains. Each domain of the tetrameric protein can function independently of alterations in other parts of the molecule (Rosen, 1989). Binding of insulin to IR leads to the appearance of the insulin-dependent protein tyrosine kinase activity of the IR that is essential, but not sufficient, for insulin signal transduction.

The **amino acid sequence of IR** and the **gene encoding IR** have been identified (Ebina et al., 1985; Ullrich et al., 1985). The *IR* gene has been partially sequenced and assigned to **chromosome 19** (Seino et al., 1989, 1990a). The different functional domains of IR are encoded by discrete exons. It is believed that the multifunctional IR arose through assembly of the existing DNA sequences in a modular fashion to encode required protein functions (Olefsky, 1990). Following ligand binding, IR is internalized and specific comformational changes of IR domains appear to be necessary for internalization.

A specific **insulin-degrading enzyme** (IDE) has been identified, cDNA produced, and the *IDE* gene mapped to human **chromosome 10** (Affholter et al., 1990). Specific structural sites on the IR molecule also determine other events such as recycling, retroendocytosis, receptor turnover rate, down regulation, and ligand degradation (McClain et al., 1988). For example, exon 11 exerts a modulating effect on the affinity of the receptor for insulin (McClain et al., 1988). Mutations of tyrosine residues 1162 and 1163 of the IR disrupt the normal process of internalization, but do not affect the process of receptor recycling after internalization (Reynet et al., 1990).

A possibility exists that altered domains of one insulin receptor can interfere with the function of the same domains in other IR molecules, thereby amplifying the functional consequence of an altered protein structure (Olefsky, 1990). In fact, a study has demonstrated that insulin receptors *per se* are capable of intermolecular phosphorylation *in vitro* that could constitute an initial step in the **amplification of the insulin-binding signal** *in vivo* (Hayes et al., 1991).

Several **mutations in the *IR* gene** have been described that result in expression of structurally abnormal proteins (Seino et al., 1990a,b). These mutations are associated with extreme **insulin resistance** (Accili et al., 1989; Kadowski et al., 1988; Klinkhamer et al., 1989; Moller et al., 1990; Moller and Flier, 1988; Odawara et al., 1989; Shimada, 1990; Taira et al., 1989; Yoshimasa et al., 1988). Furthermore, linkage studies suggest that there is also genetic variation in the *IR* gene that is associated with an increased (McClain et al., 1988) or decreased (Xiang et al., 1989) risk for NIDDM with a less pronounced insulin resistance than in the extreme resistance syndromes.

Characterization of the *IR* gene provides an opportunity to amplify individual exons of the gene by PCR and **relate the specific sequence alterations to the symptomatology of the NIDDM patients**, subdivided with respect to their insulin resistance (Seino et al., 1990b). PCR amplification and sequencing of exons 17 to 21, which code for the tyrosine kinase domain of the IR, are suggested to be the first priority. In the less resistant patients, exons 2 and 3, coding for the domains responsible for insulin binding, should be analyzed (Seino et al., 1990b). Genomic DNA encompassing exons 1 and 2 has been cloned and shown to contain many repetitive sequences homologous to the *Alu* repeat family surrounding the promoter region (McKeon et al., 1990). Sequences demonstrating enhancer functions are found in the promoter region as well, appearing to contribute to tissue specificity of *IR* gene expression.

So far, a linkage relationship between a specific RFLP and NIDDM failed to reveal any abnormality in the *IR* gene in a patient homozygous for this RFLP (Kusari et al., 1991). It appears, therefore, that insulin resistance seen in this NIDDM patient is not related to a structural alteration in the IR itself. The heterogeneous nature of NIDDM makes it possible that this patient is an exception and other NIDDM subjects may have IR abnormalities accounting for their insulin resistance (for a review on cellular regulatory mechanisms related to IR expression and function see, e.g., Knutson, 1991).

4.1.3 SUSCEPTIBILITY TO DM

The contribution of **genetic factors** to the development of DM has long been appreciated, and the advances of molecular genetics have allowed direct testing for the postulated role of alterations at the gene level. One of the first associations to be studied was that between certain **HLA allele frequencies** in patients with IDDM and controls (Barbosa and Rich, 1988). As discussed above, associations between the candidate gene involvement in predisposing to, or accounting for, the symptomatology of IDDM or NIDDM were not always clear-cut, particularly in the case of NIDDM. Disease associations have been studied predominantly by **linkage analysis** using RFLP markers for **candidate susceptibility genes**, including the **HLA locus** (Baisch et al., 1990; Erlich, 1989; Erlich et al., 1990; Owerbach et al., 1987–1990a,b; Todd et al., 1987), the **insulin gene** (Bell et al., 1984; Cox et al., 1988c; Cox and Bell, 1989; Hitman et al., 1985; Owerbach and Nerup, 1982), the *IR* **gene** (Elbein et al., 1986; McClain et al., 1988;

Morgan et al., 1990; Xiang et al., 1989), the **glucose transporter gene** (Cox et al., 1988a; Li et al., 1988; Matsutani et al., 1990), the **T lymphocyte receptor (TCR) β chain gene** (Hoover et al., 1986; Millward et al., 1987; Spielman et al., 1987), as well as **apolipoprotein genes** (Buraczynska et al., 1985; Xiang et al., 1989). The following discussion aims only to illustrate the salient features of, and reported approaches to, the analyses of susceptibility to DM that may provide useful predictive tools within the foreseeable future (for reviews, see Barbosa and Rich, 1988; Krolewski and Warram, 1985; Rotter et al., 1990).

The specific genetic factors accounting for susceptibility to IDDM and NIDDM have not yet been defined conclusively, although numerous correlations between certain genetic loci and the frequency of occurrence of these conditions have been observed. An insightful analysis of the currently entertained models of the role of genetic factors in determining susceptibility to IDDM and NIDDM has been given (Rich, 1990; Wassmuth and Lernmark, 1989). Family and **monozygotic (MZ) twin** studies show the MZ concordance rate for NIDDM to be close to 100%, suggesting an autosomal recessive inheritance (Newman et al., 1987; O'Rahilly et al., 1988). As mentioned above, the mode of inheritance of a NIDDM variant, a maturity-onset diabetes of the young (MODY), appears to be autosomal dominant. The **most powerful predictor of risk for diabetes** (either NIDDM or IDDM) is being a monozygotic (MZ) twin of a diabetic person (Rich, 1990). In contrast to NIDDM, the MZ twin concordance rate for IDDM is much lower than 100%, in the range of 25 to 55% (Barnett et al., 1981; Tattersall and Pyke, 1975). This implies an important contribution by nongenetic, **environmental factors** such as "triggering" influences — chemical damage, viruses, as well as cultural traditions, diets, and so on. Estimated risks of DM for distant relatives of MZ twins vary widely in contrast to the relatively uniform predictions proposed for siblings of diabetic persons (Rich, 1990). A number of hypotheses have been advanced to account for the nonlinear decrease for IDDM and NIDDM risk observed with progressive distance in relationship (Rich, 1990).

While the underlying pathology of NIDDM appears to be highly heterogeneous in nature (see above), the principal defect accounting for IDDM is the specific destruction of β cells, in which an **autoimmune mechanism** appears to play a key role (see above; also see Nepom, 1989a,b; Riley, 1989). Among the first **candidate genes** responsible for susceptibility to IDDM were the HLA genes or alleles, initially **HLA class I**, and later the **class II HLA-*DR* locus** (Thomsen et al., 1985). The structure of the major histocompatibility complex (MHC), its products, and the diverse functions they play have been thoroughly reviewed (e.g., Auffray and Strominger, 1986; Kappes and Strominger, 1988; Nepom, 1988, 1989a,b; Rask et al., 1985; Tiwari and Terasaki, 1985; Trowsdale, 1987). Analysis of maternal effects, association of specific *DQA1* and *DQB1* alleles with susceptibility to IDDM, and the possible involvement of *DQA2* has been presented (Rubinstein, 1991).

Most of the evidence on associations between diseases and the MHC genes has been obtained using **serological techniques** relying on complement-dependent lymphocytotoxicity assays (Richiardi and Curtoni, 1986; Schreuder and van Leuwen, 1986). The cellular methods for typing class II molecules performed in **mixed lymphocyte cultures (MLC)** evaluate the proliferative responses of T lymphocytes to define recognition epitopes (Bach et al., 1986; Duquesnoy and Zeevi, 1986). On the other hand, molecular biological techniques, in particular **RFLP** analysis, **PCR** amplification, and **sequencing** of specific gene regions as well as generation of **allele-specific oligonucleotide (ASO) probes** allow us to define in molecular genetic terms all of the specificities previously characterized serologically and by cellular methods (Andersson et al., 1984; Bidwell et al., 1988; Bugawan et al., 1988; Cox et al., 1988b,c; Gyllensten and Erlich, 1988; Holbeck and Nepom, 1986; Hyldig-Nielsen et al., 1987; Michelsen and Lernmark, 1987; Monos et al., 1987; Owerbach et al., 1987, 1988; Rich et al., 1985; Robinson et al., 1989; Scharf et al., 1986; Tilanus et al., 1986). Using **sequence-specific oligonucleotides (SSO)** conjugated to horseradish peroxidase for probing PCR-amplified DRB alleles one can distinguish 31 of 34 HLA-DRB1 alleles (Scharf et al., 1991).

Numerous reports describe an association between the **risk for IDDM and the frequency of specific HLA alleles** (for detailed discussions see Nepom, 1989b; Rich, 1990; Wassmuth and Lernmark, 1989). Initially, a significant association was observed between **HLA-*B15*, HLA-*B8*,** and IDDM (Tiwari and Terasaki, 1985). The association of susceptibility to IDDM with the **HLA-*D/DR*** genes, in particular **HLA-*DR3*** and ***DR4*** (on the average twice as frequent among IDDM patients as in controls), has been amply documented (Bach et al., 1985; Monos et al., 1987; Nepom, 1988; Nepom et al., 1986; Sheehy et al., 1985; Thomson et al., 1988; Tiwari and Terasaki, 1985). Particularly strong association has been observed between **insulin autoimmune syndrome** and HLA-DR4, pointing to a strong genetic predisposition toward the development of this condition (Uchigata et al., 1992). While family studies usually fail to confirm linkage association between increased risk of IDDM and the variable number of tandem repeats (VNTRs) in the 5' region of the insulin gene (INS) on **chromosome 11p**, the analysis of polymorphisms of INS and neighboring loci demonstrated that HLA-DR4-positive diabetic patients display such an association (Julier et al., 1991). The existence of a gene(s) affecting HLA-DR4 IDDM susceptibility in a 19-kb region of INS-IGF-2 (the third intron of the gene for **insulin-like growth factor II**) is strongly suggested by this finding (Julier et al., 1991).

The alleles at the **HLA-*DQ*** locus, in particular ***DQw8***, were shown to occur in over 95% of DR4-positive IDDM patients compared to 60 to 70% of healthy DR4- positive controls (Aparicio et al., 1988; Baisch et al., 1990; Carrier et al., 1989; Michelsen and Lernmark, 1987; Nepom et al., 1986; Owerbach et al., 1983; Rowe et al., 1990; Schreuder et al., 1986; Sterkers et al., 1988). Restriction polymorphism defined by the *Bam*HI 3.7-kb fragment occurs more frequently in unaffected persons (37%) than in IDDM patients (0 to 2%) (Owerbach et al., 1983).

Several studies have described **very specific amino acid deletions** (aspartic acid at position 57 of the HLA-*DQ*β chain) to be **associated with protection against IDDM** (Horn et al., 1988; Morel et al., 1988; Ronningen et al., 1989; Todd et al., 1987, 1988). Amplification of *DQ*β regions by the PCR revealed a correlation between the presence of an aspartic acid residue at position 57 and low susceptibility to IDDM, on the one hand, and the presence of an **alanine** (*DR4*), **valine** (*DRw6*), or **serine** (*DR2*) residue at this position and higher susceptibility, on the other (Erlich et al., 1990). These associations were not as strong in Chinese IDDM patients. The overall conclusion was reached that the **aspartic acid** at position 57 does not confer complete resistance or protection; instead, **rather specific combinations of *DQ*β and *DR*β sequences play a major role in determining susceptibility to IDDM** (Erlich et al., 1990).

Likewise, exceptions to the role of **aspartic acid** at position 57 have been observed in Japanese patients, and accurate *DQ* genotyping was recommended to obtain valuable information on disease susceptibility to IDDM (Baisch et al., 1990). In fact, the presence or absence of **aspartic acid** at position 57 does not by itself determine the risk for IDDM. Therefore, **when testing for IDDM susceptibility, a combination of dot-blot analysis using ASO probes not only for the HLA-*DQ* gene region, but also for the HLA-*DR* region, including also the α chains, should be advocated** (Baisch et al., 1990). Testing with ASO probes for MHC class II *DQA1* and *DQB1* genes in IDDM patients confirmed that the DQw1.18 allele of the *DQB1* gene appears to be associated with inherited protection whereas the A3 allele of the *DQA1* gene and DQw2 allele of the *DQB1* gene are associated with susceptibility to IDDM (Jenkins et al., 1991). Comparison of subtypes of HLA-*DR4* emphasizes the primary MHC association with IDDM to reside in the *DQ* region in most *DR4* patients (Owerbach et al., 1989). **Higher susceptibility to IDDM has also been observed in *DR3*-associated haplotypes**, and among Caucasians 35 to 45% of IDDM patients have been shown to be *DR3/4* heterozygotes, compared to 2 to 5% of healthy controls (reviewed in Nepom, 1990; Wassmuth and Lernmark, 1989). Other *DR* specificities also have associations with IDDM susceptibility (Thomson et al., 1988).

The numerous genetic associations suggesting the existence of susceptibility and resistance haplotypes exhibiting cis- and trans-encoded protective, and/or synergistic, effects can be united within the framework of a concept visualizing a hierarchy of affinities determining the strength of interactions between a diabetogenic peptide and different HLA class II molecules (Nepom, 1990). This hypothesis emphasizes the competition of different class II molecules for a single peptide, for example, a **class II susceptibility gene product** such as **DQ3.2**. Those individuals in whom DQβ3.2 has the highest affinity for a diabetogenic peptide develop a predisposition to diabetes. If other genes (*DR*, or other *DQ* genes) generate products with higher affinity than DQ3.2 for the same ligand peptide, then their competitive advantage in this example confers a relative resistance to IDDM. This model envisions variation of quantitative levels of the susceptibility alleles relative to other class II genes that may explain, in part, the variation in the incidence of diabetes among genetically at risk individuals. Furthermore, non-MHC genes may generate products competing with the diabetogenic peptides for binding to class II susceptibility genes (Nepom, 1990).

Studies of genetic markers of NIDDM (Matsutani et al., 1990; Morgan et al., 1990; Surwit et al., 1991) have so far failed to identify a single major gene, and the proposed risk models visualize an **interplay of a few genes with moderate effects** (Rich, 1990). The heterogeneous nature of NIDDM emphasizes the importance of either the involvement of multiple loci with small effects interacting with the environmental factors, or genetic heterogeneity (Rich, 1990).

A genetic association between an RFLP in the polymorphic region flanking the 5′-end of the insulin gene and atherosclerosis detected by *Bg*lII digestion has been reported in adult (Mandrup-Poulsen et al., 1984; Owerbach et al., 1982) and pediatric populations (Amos et al., 1989). In the Bogalusa Heart Study, four alleles were observed at this locus and the class 3 allele, in particular the presence of two copies of this allele, was found to be significantly associated with risk for DM and CAD that can be observed in childhood (Amos et al., 1989). Further family studies are obviously needed to establish a quantitative relationship between DM phenotypes and the alleles of the flanking regions of the insulin gene that, if confirmed, may offer a tool for identifying individuals at risk for NIDDM at an early age.

4.1.4 CONCLUDING REMARKS

The identification of precise molecular characteristics controlling the structure and function of the key factors in expression of the genes for insulin, IR, glucose transporter, glucokinase, and IAPP makes it possible to specifically relate a diabetes-associated pathology to alterations at the respective gene levels. Advances in molecular immunology and molecular immunogenetics, in particular, offer direct diagnostic possibilities for identifying individuals at increased risk for developing DM well before the symptoms of the disease appear.

In the meantime, a combination of amplification of specific HLA alleles by PCR or other techniques combined with ASO probes, and direct sequencing of the amplified products can be used to accumulate evidence on specific disease associations in informative pedigrees. Simplification of these procedures makes large-scale testing a practical possibility. In fact, Rich (1990) suggests that NIDDM, although being clinically heterogeneous with a number of factors contributing to the disease, may require alterations to occur only at a few loci to provide a significant risk to relatives. This makes the challenge of mapping a number of single genes with moderate effect on NIDDM susceptibility a manageable task. Linkage analysis strategies in defining genetically complex traits such as NIDDM have been described (Risch, 1990a–c).

Much is yet to be learned before predictive DM testing becomes a practical tool in the clinical laboratory. The enormous advances in our understanding of the fine molecular events associated with specific clinical variants of DM give hope that such diagnostic tools may become available in the foreseeable future.

Thyroid and Parathyroid Disorders

4.2.1 THYROID GLAND

4.2.1.1 DNA Ploidy

Although the histopathology of **thyroid neoplasms** has been well developed and presents little problems in clear-cut cases, diagnostic challenges can be encountered in combinations of two or more histological patterns, in poorly differentiated tumors, and in some **Hürthle tumors**, as well as when distinctions are to be made between **benign** and **malignant follicular thyroid nodules**.

The determination of **DNA content** by **image cytometry (ICM)** or **flow cytometry (FCM)** as an aid to differential diagnosis over the past few years has yielded conflicting data. In contrast to some reports suggesting that **ploidy** of thyroid tumors can define their biological behavior (Greenbaum et al., 1985), other authors showed that **aneuploidy occurs in normal glands, adenomas, and nodular goiter**, and fails to distinguish between benign and malignant Hürthle cell tumors (Joensuu et al., 1986; Johannessen et al., 1981, 1982). Ploidy determinations can be of value as an adjunct to histological assessment in **papillary, follicular, medullary,** and **anaplastic thyroid carcinomas** as well as in **Hürthle cell tumors**, **metastatic lesions,** and **malignant lymphoma of the thyroid.** In general, aneuploidy has been correlated with poor survival in papillary and follicular carcinomas. Even benign Hürthle cell tumors, however, showed aneuploid DNA patterns not associated with malignant behavior. Aneuploidy in medullary carcinoma, in contrast, was correlated with aggressive growth pattern (Joensuu et al., 1986; Johannessen et al., 1981, 1982).

In retrospective analysis, the **predictive value** of DNA aneuploidy, age, tumor grade, and extent and size of tumor proved to be significant with respect to mortality from papillary thyroid carcinoma, allowing an assessment of an individual patient's risk of dying from this disease (Smith et al., 1988). A nationwide study of **medullary thyroid carcinoma** has been conducted in Sweden to establish the relation between nuclear DNA content determined by ICM and survival in 211 patients, with complete follow-up for up to 27 years (Ekman et al., 1990). The percentage of cells having other than diploid DNA content can be used as a continuous variable that increases the information on the relative risk. These data essentially confirmed the earlier observations that aneuploidy is associated with a worse prognosis in medullary thyroid carcinoma, all **anaplastic carcinomas**, known for their poor prognosis, are aneuploid (e.g., Joensuu et al., 1986; Schroder et al., 1988).

Again, conflicting observations are reported for **metastatic thyroid carcinomas**, in which the primary tumor and metastases may have different ploidy patterns. Barring artifactual variability in the reported observations as due to specimen handling and source (fresh tissue vs. formalin fixed, paraffin embedded), ploidy evaluation may be used as an adjunct in diagnostic pathology, although it is inferior to the prognostic value of age, tumor size, and extrathyroidal extension (Joensuu et al., 1986; Schroder et al., 1988).

A DNA FCM study of medullary thyroid carcinoma confirmed that DNA content as well as age, hereditary background, calcitonin immunoreactivity, and type of surgery were strong predictors of clinical outcome (El-Naggar et al., 1990). However, in multivariate analysis none of these factors, or patient's gender, clinical stage, histological subtype, and amyloid content of the tumor proved to be of significant predictive value with respect to the 5-year survival of any individual patient.

4.2.1.2 CHROMOSOME STUDIES

If DNA ploidy of thyroid neoplasms does not offer a superior diagnostic tool, could chromosome characteristics be identified and correlated with specific histopathological types? This question was asked in a cytogenetic study of 12 thyroid carcinomas: 7 papillary, 3 follicular, and 2 anaplastic (Jenkins et al., 1990). Using new culture and harvesting techniques chromosomally abnormal clones were found in nine tumors. Although the sampling was too small for meaningful correlations to be established, some potentially useful associations have been noted. An anomaly of the **chromosome 10q** arm, represented by inversions or insertions, balanced or imbalanced translocation, was consistently observed in cytogenetically abnormal papillary adenocarcinomas. No cytogenetic abnormalities could be detected in **chromosomes 2, 11,** and **12** (Jenkins et al., 1990). On the other hand, chromosome 10 appears to be also involved in nonpapillary thyroid cancers in association with other chromosome aberrations, consistent with the observations of other investigators.

The importance of the involvement of chromosome 10 may be viewed in the context of several other possibly related observations. Thus, the gene for **multiple endocrine neoplasia type II (MEN IIA)**, which involves medullary thyroid carcinoma, has been mapped to **10q11.2-10q21**. Furthermore, the *PTC* oncogene (for **p**apillary **t**hyroid **c**arcinoma) has also been mapped to 10q11-q12 (Donghi et al., 1989; Sobol et al., 1989). The conventional way of screening for this neoplasm consists of **pentagastrin challenge** leading to **calcitonin stimulation**. The calcitonin gene expression has been monitored by nonradioactive *in situ* hybridization (ISH) (Denijn et al., 1990). An elevated level indicates **C cell hyperplasia** and thyroidectomy may prevent cancerous transformation of the gland. Regular screening of relatives of affected persons may last for many years, because it is not known who may develop the disease, or when.

Employing polymorphic **DNA probe RBP3** for the chromosome 10q1 region, some 130 members of 11 families of European and North African origin were analyzed. For 63% of the families, linkage information obtained was adequate to offer genetic counseling. The allele associated with the susceptibility gene could be identified in several cases and the carrier state for MEN IIA could be predicted (Sobol et al., 1989). One drawback of this approach is that the small size of the families and the lack of informative matings limit the power of RFLP analysis. In those cases, screening with pentagastrin challenge must be used.

Mapping for **medullary thyroid carcinoma** (MTC) without **pheochromocytoma** was found not to be precise enough to distinguish it conclusively from MEN IIA. Only a few families with a sufficient number of surviving relatives are readily available in MEN IIB groups, therefore linkage studies for MEN IIB and MEN I have so far not been brought to the stage of informativity developed for MEN IIA (Sobol et al., 1989). The involvement of chromosome 10 in MTC, particularly in association with the MEN II syndrome, does not appear to be high (Nelkin et al., 1989). This supports the view that the development of MTC is different from the sequence of events operating in retinoblastoma and some other inherited tumors (Nelkin et al., 1989).

Other chromosomal abnormalities of interest involve **chromosome 3**, where the β **thyroid hormone receptor gene** (*erb*A-2) maps at **3p25-p21**. The prognostic utility of these findings will be substantially clarified by studies on the established cell lines for thyroid tumors (Pfragner et al., 1990). In fact, the predominant findings in the MTC cell line MTC-SK, were terminal chromosomal rearrangements most frequently in **chromosome 11p**, which is the region harboring the H-*ras* oncogene, and the locus of the **calcitonin** and **calcitonin gene-related peptide genes**. Other aberrations included a characteristic instability of the centromeric region of **chromosome 16** and somatic pairing of the homologous chromosomes 16 (Pfragner et al., 1990).

4.2.1.3 ANALYSIS OF SPECIFIC GENES IN THYROID DISEASE

4.2.1.3.1 Thyroid-Stimulating Hormone β Gene

In efforts to establish better diagnostic and prognostic correlates in thyroid neoplasia, analysis of specific genes and their expression has been undertaken. Studies at the gene level allowed the identification of a specific thyroid-stimulating hormone (*TSH-β*) **gene** defect in members of five Japanese families with inherited **TSH deficiency** (Hayashizaki et al., 1990). Single base substitutions detected in **codon 29** of the gene accounted of the production of a defective β polypeptide, which was unable to form a functional hormone through its association with the α subunit.

This analysis exemplifies an approach that will be progressively employed for the delineation of specific gene defects in a number of human diseases. In fact, in another study of **hereditary hypothyroidism** in members of two Greek families, a nonsense mutation in exon 2 of the *TSH-β* subunit gene has been identified (Dacou-Voutetakis et al., 1990). This led to the production of a truncated peptide incapable of functional association with the TSH-α subunit.

4.2.1.3.2 Thyroid Peroxidase Gene

Advances in the study of the pathogenesis of **thyroid autoimmune disorders** led to the identification of the thyroid microsomal-microvillar antigen as **thyroid peroxidase (TPO)** (for a review, see Mariotti et al., 1989). A recombinant DNA approach provided new insights not only into the mechanism of thyroid damage in such disorders as **Hashimoto's thyroiditis**, but also pointed to another cause of **congenital hypothyroidism** related to defects in the **organification of iodide**. DNA polymorphism in the *TPO* gene has been used to analyze hereditary hypothyroidism cases in the Netherlands (de Vijlder et al., 1990). A 0.7-kb *TPO* cDNA probe was able to detect an *Eco*RI variable number of tandem repeats (VNTR) polymorphism identifying at least nine alleles. Another probe, a 3.0-kb *TPO* cDNA, revealed a different set of *Eco*RI restriction fragment length polymorphisms (RFLPs). These proved to be informative markers in families with a total absence of iodide organification, the most frequent cause of congenital hypothyroidism in the Dutch population (de Vijlder et al., 1990). Analysis of the *TPO* gene products by immunohistochemistry has been reported to be a good marker of malignancy in thyroid tumors as well (De Micco et al., 1990).

4.2.1.3.3 HLA Associations

A further elaboration of the details of the autoimmune basis of thyroid disorders comes from molecular studies utilizing DNA probes in order to establish specific disease associations with a given pattern of RFLP. These studies provide essentially a new basis for the concept of genetic susceptibility in **Hashimoto's thyroiditis** (Badenhoop et al., 1990). The strongest relative risk has been associated with **DQw7**, found to occur in 56% of affected persons and in only 21% of controls. Variations in DNA sequences in codons 45 and 57 of the *DQB1* **gene** are the critical points distinguishing *DQw7* from other *DQw* specificities. Significant associations have also been established with the adjacent *DQA1* **genes**. These findings provide further support to the notion that gene-level analysis is rapidly moving into the realm of specific disease associations that may prove helpful in clinical management of thyroid patients.

4.2.1.3.4 Insulin Growth Factors

One line of inquiry pursued the notion of "**autocrine secretion**" observed in cancer cells, which allows these cells to establish a degree of autonomy independent of "external" growth factors. Rats of the Wag/Rij strain, which spontaneously develop MTC, were used to study the expression of **insulin growth factors I** and **II** (IGF-I and IGF-II), which occurs in most differentiated MTCs (Hoppener et al., 1987). Virtually no transcription of these genes occurs in anaplastic medullary carcinoma. Thus, a differential pattern of specific genes has been observed in two types of MTC. Because the gene for IGF-II is not known to be expressed in adult tissues, its consistent transcription in differentiated rat MTC may reflect tumor-specific activation or enhanced expression in this type of tumor (Hoppener et al., 1987).

4.2.1.3.5 *PTC, ras,* and *TRK* Genes

Activation of a specific oncogene has been observed in **papillary thyroid carcinoma**; the gene was designated *PTC* (Fusco et al., 1987) and assigned to the q arm of **chromosome 10** (Donghi et al., 1989). This was the first oncogene specifically associated with thyroid cancer. Activation of another oncogene, *ras*, has subsequently been shown to occur in high frequency in microfollicular adenomas, follicular carcinomas, and undifferentiated carcinomas, and with a low frequency in papillary thyroid carcinomas (Lemoine et al., 1988, 1989; Suarez et al., 1988).

Employing transfection assays in NIH3T3 cells, Bongarzone and coworkers (1989) analyzed oncogene activation in 16 cases. The *PTC* oncogene was found to be activated in four cases, *TRK* **oncogene** was activated in four cases, and activated **N-*ras*** oncogene accounted for transforming activity in two cases. Structural alterations of these oncogenes detected in the transformants were identical to those found in the original tumors. Because both *PTC* and *TRK* oncogene products display tyrosine kinase activity, the activation of this class of oncogenes may be specifically associated with papillary thyroid carcinoma (Bongarzone et al., 1989). The frequency of N-*ras* activation in follicular neoplasms is only 53 vs. 20% in papillary cancers (Bongarzone et al., 1989; Lemoine et al., 1988, 1989). N-*ras* mutations are believed to occur only rarely in primary medullary thyroid carcinomas (Moley et al., 1991).

Subsequent analysis of point mutation of **Ki-*ras*** revealed that all three mutations detected in **papillary** cancers are different (Wright et al., 1989). In contrast, the combined evidence from **follicular** thyroid cancer cases suggested a similarity of point mutations of this gene. **The pattern of point mutations of the Ki-*ras* gene appears to correlate significantly with the type of thyroid malignancy**; it is not clear, however, to what extent this distinction may be correlated with the known difference in their biological behavior (Wright et al., 1989). The other *ras* oncogene, **Ha-*ras***, is activated roughly in the same proportion of benign and malignant thyroid neoplasms, suggesting that it may constitute an **early tumorigenic event** (Namba et al., 1990).

4.2.1.3.6 Receptors

Cytoplasmic and **nuclear receptors** are involved in the control of gene expression. The molecular mechanisms responsible for the receptor-binding selectivity of this process, transmission of the signal to the targeted genes, and their role in switching gene expression on and off are being actively studied. Interaction of the **DNA-binding proteins** with specific nucleotide sequences is one of the mechanisms controlling transcription. Analysis of human **steroid receptor genes** revealed a very close similarity with the v-*erb*A (avian erythroblastosis virus) oncogene. Similarities of **thyroid hormone receptors** with steroid hormone receptors are seen not only in their mode of action and regulation of gene expression. At the structural level, the cDNA sequence of human **c-*erb*A** indicates that the protein encoded by the gene is related to the family of steroid hormone receptors (Sap et al., 1986; Weinberger et al., 1986). The expression of c-*erb*A oncogene is markedly modulated by the degree of **methylation** (Bahn et al., 1988) and is directly controlled by the thyroid hormone at the transcription level (Glass et al., 1987; Lazar and Chin, 1988).

The similarity between v-*erb*A, which is required for the maintenance of the fully transformed phenotype of avian erythroblasts, and c-*erb*A suggests that alterations of the hormone-binding region of this receptor may enable it to play a modulatory role in transcriptional activation similar to that observed in malignancy (Weinberger et al., 1986). Subsequent studies revealed that high-affinity thyroid hormone-binding proteins are found in a wide range of organs. The affinity of thyroid hormone binding to the c-*erb*A product appears to be defined by the highly conserved C-terminal 37 to 40 residues (Schueler et al., 1990).

In reconstruction experiments with cells known to be receptor-deficient, the introduction of the thyroid hormone receptor resulted in specific gene expression of target genes. Thus, c-*erb*A products, acting as thyroid hormone receptor, stimulated the expression of **growth hormone** and **prolactin genes** in 235-1 cells (Forman et al., 1988). Interestingly, these products acted in a hormone-independent fashion, much like the v-*erb*A oncogene products causing transformation.

FAO hepatoma cells characteristically having no specific nuclear binding of T_3, when infected with a retrovirus construct containing the chicken c-*erb*A gene, display T_3-binding activity similar to that found in normal liver cells and start expressing several hormone-dependent genes in response to T_3 (Munoz et al., 1990). Fine analysis of the two c-*erb*A genes, c-*erb*A2 and c-*erb*A3, demonstrated that the so-called phenomenon of **generalized thyroid hormone resistance** is associated with the gene that is tightly linked to the **c-*erb*AB locus on chromosome 3** (Usala et al., 1988). The importance of this linkage suggests that c-*erb*AB codes for a thyroid receptor in humans, and if abnormal it may account for the mutant phenotype of c-*erb*AB involved in various metabolic and growth abnormalities (Degroot, 1989; Goldberg et al., 1989; Usala et al., 1988). While the structure, function, and possible role of thyroid hormone receptors in the pathology of various organs is being elucidated, their role in **benign thyroid** conditions or neoplasia remains largely unknown. It is conceivable that some form of c-*erb*A derangement may have a role in the autocrine control of thyroid malignancy. We are not aware at this time, however, of specific findings to indicate this.

The role of **TSH** and **LH/CG** (luteinizing hormone/choriogonadotropin) receptors in thyroid pathology may be important. Some of the critical aspects of the TSH receptor (TSHR) function include (1) control of thyroid cell metabolism, (2) control of the growth and development of the gland itself (Field, 1986), and (3) interaction with autoantibodies to TSHR, which are responsible for the pathogenesis and hyperthyroidism of **Graves' disease** (Morris et al., 1988; Rees-Smith et al., 1988). All this underscores the practical significance of the cloned TSHR cDNAs (Frazier et al., 1990). Unexpectedly, the cloning thyroid library contained several cDNAs encoding the human LH/CG receptor formerly thought to be present only in gonadal cells. Analysis of the genes for LH/CG showed similarity to the genes for TSH. However, the majority of LH/CG receptor mRNA in the thyroid was incompletely spliced, suggesting that **tissue-specific splicing** may be an important link in the control of glycoprotein hormone metabolism (Frazier et al., 1990). It remains to be seen to what extent the interplay of hormonal receptiveness demonstrated in the thyroid gland between TSH and LH/CG may play a role in thyroid pathology. At any rate, the presence of the GH/CG receptor genes in the thyroid may, in part, account for the observed elevation of T_3 and T_4 in early pregnancy, when TSH is not elevated (Frazier et al., 1990).

Important insights may be expected from studies of the **TSHR gene**, which has been assigned by ISH to **chromosome 14q31** (Rousseau-Merck et al., 1990). The association between this gene and that of immunoglobulin G heavy chain haplotypes, suggested by the autoimmune nature of Graves' disease, appears to be more complex than that emerging from chromosomal linkage studies (Rousseau-Merck et al., 1990).

4.2.1.3.7 Nuclear Oncogenes

The involvement of the "**nuclear**" family oncogenes (**c-*myc*, N-*myc*, L-*myc*, *fos*, *myb*,** and ***p53***) assessed by Southern blot analysis failed to reveal any evidence of arrangement or amplification of any member of the nuclear oncogenes in thyroid follicular cell neoplasia

(Wyllie et al., 1989). Changes in the expression of the oncogenes were thought to be the secondary consequences of the tumor phenotype. On the other hand, **N-*myc*** expression assessed by ISH is viewed not only as a differentiation marker of C cell tumors, but is interpreted as indicative of the role N-*myc* activation plays in neuroendocrine neoplasia (Boultwood et al., 1988). On the other hand, direct sequencing of mutations in the ***p53* gene** following PCR amplification of exons 5 and 8 revealed the absence of any mutations in these exons in differentiated papillary adenocarcinomas whereas almost all of the undifferentiated tumors harbored specific mutations at codons 135, 141, 178, 213, 248, and 273 (Ito et al., 1992). It appears, therefore, that specific mutations of the *p53* gene are involved in the progression of differentiated carcinomas to the undifferentiated state.

The **c-*fos*** gene expression is reduced in inverse correlation to the degree of differentiation in follicular carcinoma (Wyllie et al., 1989). The role of c-*fos* in thyroid differentiation is supported by the finding of the induction of the c-*fos* oncogene in response to thyrotropin action on the thyroid (Colletta et al., 1986; Foti et al., 1990). Furthermore, the elevation of the level of c-*fos* antisense transcript (Foti et al., 1990) is consistent with the degree of growth inhibition in cell culture.

4.2.2 PARATHYROID GLAND

Differentiation between benign and malignant neoplasia of the parathyroid gland may present a diagnostic dilemma, and a combination of morphological and clinical features, including the presence or absence of metastases, is needed to recognize a malignancy.

Cellular DNA content and synthesis has been assessed in parathyroid adenomas, carcinomas, primary and secondary hyperplasia, and in normal glands by DNA FCM (Roth et al., 1990). **Hyperplastic glands** showed aneuploidy in 22% and up to 1.5% of cells were in the S phase. The **S-phase fraction (SPF)** in **adenomas** constituted up to 3.8%, whereas in **carcinomas** this value could be as high as 14.1%. All types of parathyroid neoplasia, but not normal parathyroid gland, demonstrated aneuploidy and tetraploidy and, thus, **adenomas cannot be distinguished from primary hyperplasia by DNA FCM**, although **an increase in the proliferative activity above 3.8 to 4% is highly suggestive of malignancy**.

According to other authors parathyroid carcinomas are more likely to be aneuploid than adenomas (Obara et al., 1990). This, however, can justify the use of FCM in the evaluation of parathyroid diseases only as an adjunct to the histopathological diagnosis.

Adrenal Gland and Multiple Endocrine Neoplasias 1 and 2

4.3.1 ADRENAL GLAND

4.3.1.1 DNA PLOIDY

Adrenocortical neoplasms present certain diagnostic problems and efforts have been made to define histological criteria helpful in diagnosing borderline cases, and in predicting their clinical behavior (Cutler and Laue, 1990; Hough et al., 1979; Weiss, 1984). One of the characteristic features of **adrenocortical carcinoma** is its usually large size, exceeding 100 g. Hough and coworkers (1979) and Weiss (1984) developed histological criteria that are usually applied to predict malignant behavior of adrenal tumors. In one such case, a thorough analysis of a tumor, initially diagnosed as benign, subsequently convincingly demonstrated the presence of a small adrenocortical carcinoma when these criteria were applied (Gandour and Grizzle, 1986).

Analysis of the nuclear DNA content of 22 paraffin-embedded adrenal cortical tumors (16 adenomas and 6 carcinomas) by flow cytometry (FCM) showed all the adenomas to be diploid and 83.3% of the carcinomas (as well as all the metastasizing tumors) to be aneuploid (Bowlby et al., 1986). DNA FCM proved to offer prognostic information. Likewise, another FCM study of adrenocortical benign and malignant neoplasms suggested that aneuploid tumors were more likely to be larger (over 50 g) and display several histological features of carcinoma (Amberson et al., 1987). Again, neoplasms, which recurred or metastasized, were more apt to be aneuploid. This study, however, emphasized the **inconsistency** of the FCM criteria, and discounted its greater prognostic value over that of size and histology.

Among the 11 histopathological parameters such as nuclear grade, mitotic rate, architecture of the tumor, and invasive behavior, only one of these, the **mitotic rate**, proved to be a reliable predictor of the outcome in 42 cases of adrenocortical carcinoma (Weiss et al., 1989). It was proposed that adrenocortical carcinomas with 20 or more mitoses should be considered **high grade**, and those with fewer mitoses as **low-grade** neoplasms (Weiss et al., 1989). Essentially the same conclusion was reached in a large series of 43 adrenocortical benign and malignant tumors analyzed by FCM and histopathological criteria with respect to the diagnostic value of the cellular DNA content and its predictive value as a prognostic indicator (Cibas et al., 1990). **Because ploidy showed low sensitivity and specificity for prognosis and poor correlation with the outcome, assessment of the biological behavior of adrenocortical tumors by FCM is not advised.**

Particular caution should be exercised in making a diagnosis of an individual patient solely on the results of DNA ploidy assay. In a study limited to adenomas only, aneuploidy was found in 25% or more of 164 cases, depending on the organ site (Joensuu and Klemi, 1988). In an adult with **Wiedemann-Beckwith syndrome (WBS),** a disorder that includes congenital exomphalos, macroglossia, visceromegaly, and gigantism, an **adrenal adenoma** showed **loss of heterozygosity (LOH) on chromosome 11p**, suggesting that LOH in a region harboring the c-Ha-*ras*-

1 locus may be involved early in the tumorigenesis of an adrenal neoplasm (Hayward et al., 1988).

4.3.1.2 CONGENITAL ADRENAL HYPERPLASIA

Congenital adrenal hyperplasia (CAH) is a manifestation of a constellation of metabolic derangements largely due to steroid **21-hydroxylase (21-OH) deficiency**. This enzyme of the **glucocorticoid pathway** converts 17-hydroxyprogesterone to 11-deoxycortisol, and then eventually to cortisol. In the **mineralocorticoid pathway**, leading to the production of aldosterone, 21-OH converts progesterone to deoxycorticosterone. Deficient activity of 21-OH results in the **accumulation of precursors**, particularly **17-hydroxyprogesterone**, which stimulates the production of adrenal androgens clinically manifested as **virilization**. The other consequence of 21-OH deficiency is the defective production of aldosterone leading to the syndrome designated as "**salt wasting**". Excellent reviews of the clinical and molecular pathology of CAH have been published (Morel and Miller, 1991; Strachan, 1990).

This condition is associated with aberrations in the two genes for 21-OH located on **chromosome 6** within the class III region of the HLA complex. They are flanked on both sides by class II (HLA-*DR*) and class I HLA genes (HLA-*A*, HLA-*B*, HLA-*C*). Thanks to the high degree of polymorphism of the flanking genes, these markers can be used to identify affected individuals and their parents. Of the two 21-OH genes, only one, the *CYP21B* gene, is functional and encodes cytochrome **P450c21**, which mediates 21-hydroxylation (Strachan, 1990; Urabe et al., 1990). The other gene, *CYP21A*, located about 30 kb upstream from *CYP21B*, is a closely related pseudogene defective in gene expression due to several mutations resulting in the lack of its transcriptional activity.

The location of *CYP21A* and *CYP21B* in close proximity to the **complement *C4A*** and ***C4B* genes**, respectively, is functionally significant. In addition, there is a high degree of sequence homology between genes for 21-OH and *C4* genes. As Strachan (1990) notes, the tandemly repeated organization of the 21-OH and *C4* genes makes the unequal crossover events in this region more probable. Thus, some of the resulting recombination events may end up with either a gene deletion haplotype (a **missing 21-OH + *C4* gene unit**), or gene addition haplotype (an **extra 21-OH + *C4* gene unit**). Gene deletion, in which the single residual 21-OH gene has retained enough *CYP21* gene sequences to code for functional 21-OH activity, will not be manifested in adrenal insufficiency. As established by RFLP with specific DNA probes to Southern blots of *Taq*I-digested genomic DNA, disease-associated gene deletion appears to include elimination of markers of both a *CYP21B* gene and also a *C4B* gene (Strachan, 1990).

The **variable degree of clinical manifestations** of 21-OH deficiency depends on the specific mutation in a given case. A single amino acid substitution (Val-281 → Leu) in the mild "**nonclassical**" form leads to the enzyme having 20 to 50% of normal activity, whereas another mutation (Ile-172 → Asn) results in a "**simple virilizing**" form with an enzyme having less than 2% of normal activity (Tusia-Luna et al., 1990). Still another cluster mutation results in an enzyme without any detectable activity. A large proportion (up to one third) of nonclassic, late-onset CAH cases in Israeli Jewish women have been traced to secondary adrenal biosynthetic defects (Eldar-Geva et al., 1990).

Pulsed-field gel electrophoresis capable of resolving larger fragments of nucleic acids helped to identify approximately 37% of disease haplotypes as lacking the 21-OH + *C4* gene unit, suggesting that the **21-OH deficiency is attributable to gene deletion in about one in three disease haplotypes** (Strachan, 1990). The presence of additional 21-OH genes in disease haplotypes of patients with 21-OH deficiency is consistent with a pathological point mutation in the *CYP21B* gene. The appearance of pathological point mutations in the *CYP21B* locus is due to the so-called phenomenon of **gene conversion**, when normal sequences of the *CYP21B* genes are replaced with the corresponding, but nonfunctional sequences of the *CYP21A* gene (Strachan, 1990). Nonclassical forms of 21-OH deficiency, and various gradations of the severity of this condition appear to have correlates in particular mutation classes resulting in different clinical phenotypes (Strachan, 1990). Indeed, yet another mutant *CYP21B* gene with a C→T change, leading to a termination codon TAG in exon 8, has been identified in a patient with 21-OH deficiency (Urabe et al., 1990).

4.3.1.2.1 Prenatal Diagnosis of CAH

The laboratory identification of underlying biochemical and genetic aberration in CAH (also in **prenatal diagnosis**) begins with the measurement of steroid precursor levels. **Indirect identification** of affected individuals relies on linkage analysis of polymorphisms at the gene loci mapping in the vicinity of the 21-OH locus, DNA-based HLA and *C4* gene polymorphisms, and serologically defined HLA polymorphisms. For these methods to be informative definitive diagnosis of parental disease and nondisease haplotypes is required (Raux-Demay et al., 1989; Strachan, 1990).

DNA probes applied to **chorionic villi biopsy** as early as gestation weeks 8 to 10 allow an earlier diagnosis to be established. Prenatal diagnosis of **classic variants** of 21-OH deficiency can be accomplished by the determination of **17-hydroxyprogesterone (17-OHP)** values in amniotic fluid as well as by direct and indirect DNA methods (Raux-Demay et al., 1989). Elevation of 17-OHP is diagnostic in the first trimester of pregnancy. DNA extracted from chorionic villi can be analyzed for CAH **indirectly** by highly polymorphic HLA class I and class II probes as well as by the **direct method** using the cDNA **probe pc21/3c** for the 21-OH gene. The lack of informativeness of the direct 21-OH probe assay in all the cases studied indicates that this probe cannot be used routinely. The indirect method, on the other hand, can be used for all couples at risk, when DNA of the index case is available (Raux-Demay et al., 1989).

Although direct 21-OH gene probes can, theoretically, identify pathological mutations in the *CYP21B* genes, such mutations account for only about 40% of cases, and only about 15% of classic patients will have two haplotypes of this kind (Strachan, 1990). This is not sufficient for definitive prenatal diagnosis. The remaining 60% of cases harbor elusive point mutations partly still unknown, and technically challenging to demonstrate. Thus, **indirect linkage analysis should be used to infer the inheritance of pathological point mutations** (Strachan, 1990). An optimum strategem for prenatal diagnosis, at this time, is a combination of a 21-OH gene probe, a CYB probe, an HLA-*B* probe, and an HLA-*DR* β probe, which together identify over 95% of families with a surviving affected child who is informative for at least one of these probes (Strachan, 1990).

4.3.1.2.2 Identification of CAH Carriers

Identification of carriers still relies on HLA typing involving the siblings or close relatives of a known 21-OH-deficient patient. DNA-based probes identify approximately 40% of the affected patients displaying either deletions in the *CYP21B* gene or large-scale *CYP21B* → *CYP21A* conversion haplotypes (Speiser et al., 1988; Strachan, 1990).

A typical approach involved the construction of an oligonucleotide probe derived from the *CYP*21B gene that had been isolated from a patient homologous for HLA-*B14*, *DR1* who suffered from "nonclassic" 21-OH deficiency (Speiser et al., 1988). The probe could identify the Val-281 mutation in *Taq*I digests of this gene in eight patients (Speiser et al., 1988). The codon-281 mutation has been found to be consistently associated with 21-OH deficiency of the nonclassic type displaying the HLA-*B14*, *DR1* haplotype. This mutation has never been detected in normal controls. The opposite, however, is not true: there are people with nonclassic 21-OH deficiency, who do not have the codon-281 mutation (Speiser et al., 1988).

The close physical proximity of the 21-OH and *C4* genes allows the use of *C4* cDNA probes for the prenatal diagnosis of 21-OH deficiency on chorionic villi at 10 weeks of gestation (Nakura et al., 1990). The use of the *C4* cDNA probe appears preferable to that of HLA complex probes because of the shorter distance (60 kb) between the *C4B* gene, where *Taq*I RFLPs are located, and the 21-OH B gene compared to the distance between the latter and the HLA-*DR* and HLA-*B* genes, which are about 400 kb and 800 kb away, respectively.

4.3.1.2.3 Polymerase Chain Reaction in CAH Diagnosis

The recently demonstrated direct characterization of the *CYP21B* and *CYP21A* genes by polymerase chain reaction (PCR) (Owerbach et al., 1990b) offers a new approach to diagnosing salt-wasting conditions in CAH by clinical laboratories. Evaluation of the *CYP21B* gene by Southern analysis and PCR in the leukocyte DNA of 30 CAH patients revealed the major abnormalities of this gene detected in 22 of the 26 CAH patients to be deletion of the *CYP21B* gene, gene conversion of the entire *CYP21B* gene to *CYP21A* (its inactive counterpart), frameshift mutations in exon 3, an intron 2 mutation that causes abnormal RNA splicing, and a mutation leading to a stop codon in exon 8 (Owerbach et al., 1990b). DNA from patients with the salt-wasting variant of CAH as well as from the simple virilizing form, in which aldosterone production is largely unaffected, contained the **intron 2-G mutation**, suggesting that other causes account for the variable clinical presentations. **Drastic simplification of laboratory diagnosis of CAH is due to the availability of specific primers for *CYP21B* and *CYP21A*, automation of the PCR reaction, and analysis of the PCR reaction products by gel electrophoresis.**

4.3.1.2.4 Oncogenes in Adrenal Disorders

No amplification of N-*myc*, L-*myc*, or *erb*B-1 could be detected by Southern analysis and specific DNA probes in **pheochromocytomas** of the adrenal medulla (Moley et al., 1990). In mice, the overexpression of c-*mos* has been found in lens and brain of the animals developing pheochromocytomas, and in medullary thyroid cancer simulating the human MEN 2 syndrome (Schulz et al., 1990). In malignant Y-1 mouse adrenal cells, the protooncogene c-Ki-*ras* is amplified, whereas c-*fos* does not appear to be amplified (Kimura and Armelin, 1988).

4.3.2 MULTIPLE ENDOCRINE NEOPLASIAS 1 AND 2

4.3.2.1 MULTIPLE ENDOCRINE NEOPLASIA 1

MEN 1 syndrome is an autosomal dominant condition predisposing to neuroendocrine neoplasms of the parathyroid glands and tumors of the pancreatic islet cell and anterior pituitary gland, with the genetic defect assigned to **chromosome 11q13** (Bale et al., 1989; Larsson et al., 1988; Majewski and Wilson, 1979; Nakamura et al., 1989). It has been proposed (Larsson et al., 1988) that the development of MEN 1 conforms to **Knudson's two-hit hypothesis**, whereby the recessive unmasking of an inherited mutation occurs when a somatic mutation or deletion affects the other normal allele (Knudson, 1971). This hypothesis has been tested in application to MEN 1, and evidence supporting the gene deletion phenomenon has been provided (Bystrom et al., 1990). Interestingly, **sporadic primary hyperparathyroidism** appears to share the same mechanisms. LOH on chromosome 11q13 in MEN 1 is quite high, reaching 82% as estimated by restriction fragment length polymorphisms (RFLPs) in tumor and constitutional tissues (Radford et al., 1990).

4.3.2.2 MULTIPLE ENDOCRINE NEOPLASIA 2A

MEN 2A is an autosomal dominant syndrome that predisposes to the development of medullary thyroid carcinoma, pheochromocytoma, and hyperparathyroidism, with the responsible gene assigned to the pericentric region of **chromosome 10p11.2-q11.2** (Mathew et al., 1987a,b; Simpson et al., 1987). Centromeric probes **D10Z1** (Wu et al., 1990) and **D10S94** (Goodfellow et al., 1990) have been described, and a marked sex difference was noted in recombination frequencies in this pericentromeric region (Wu et al., 1990). This phenomenon should be taken into account when determining the genotypes at the D10Z1 locus. **Presymptomatic diagnosis for MEN 2A syndrome is possible with pericentromeric DNA markers** (Mathew et al., 1991). In families at risk for this syndrome, even in the absence of biochemical evidence of thyroid C cell hyperplasia (the stimulated calcitonin assay), a combination of four DNA markers substantially altered the carrier risks of most of the individuals tested, suggesting that DNA probe analysis should be a part of routine screening for MEN 2A syndrome. Furthermore, the **availability of informative markers for MEN 2A makes prenatal diagnosis a possibility**. An additional marker for MEN 2A is presented by the gene for mannose-binding protein, which maps to the pericentromeric region of chromosome 10 (Schuffenecker et al., 1991). While protooncogenes N-*myc*, L-*myc*, or *erbB*-1 were not amplified in both familial (multiple endocrine neoplasia 2A [MEN 2A]) and sporadic pheochromocytomas, LOH on chromosome 1p is always present in the tumors associated with MEN 2A, but does not appear to be a part of sporadic pheochromocytomas (Moley et al., 1990).

4.3.2.3 MULTIPLE ENDOCRINE NEOPLASIA 2B

In addition to the medullary carcinoma of the thyroid gland, pheochromocytoma, the MEN 2B syndrome, also an autosomal dominant disorder, includes neuromas of the mucous membranes and skeletal abnormalities, including the Marfanoid habitus with arachnodactyly. Similar to MEN 2A, the gene for MEN 2B has been assigned by linkage studies to the centromere of chromosome 10 by the D10Z1 probe (Jackson et al., 1988; Norum et al., 1990). Subsequent studies of patients with MEN 2A and MEN 2B syndromes demonstrated that a locus for familial medullary thyroid carcinoma and a locus for MEN 2B map to the same pericentromeric region of chromosome 10 as the locus for MEN 2A (Lairmore et al., 1991).

Although at this stage the evidence strongly supports the notion of genetic homogeneity of the MEN 2A syndrome, the genetic heterogeneity of MTC and the MEN 2B syndrome cannot be excluded until more studies are performed and the gene(s) for these familial disorders are fully characterized (Lairmore et al., 1991).

4.3.3 SELECTED ANALYTICAL TECHNIQUES USEFUL IN GENE LEVEL DIAGNOSIS OF ENDOCRINE DISORDERS

This brief discussion of some molecular biological techniques useful in the evaluation of patients with various endocrine disorders will include examples from the recent literature. Detailed reviews of molecular biology of the endocrine system have appeared (Melmed and Prager, 1991; Shapiro, 1991). Unquestionably, the choice of appropriate probes for diagnostic use on target tissues draws attention to the relative advantages and deficiencies of available options. Some of the features of nucleic acid probes used for *in situ* tissue probing are briefly summarized here.

Diagnostic application of **ISH** in endocrinology has been reviewed (Lloyd, 1987, 1990a,b; Lloyd et al., 1990a,b). Among the advantages of ISH over immunohistochemistry, particularly for the diagnosis of endocrine diseases, are the ability of ISH to (1) distinguish *de novo* synthesis from uptake, (2) identify specific genes that are actively expressed, even if their final products are not detectable for a number of reasons (degradation, utilization, release and dissipation in tissue, etc.), (3) identify the status of a specific gene without relying on identification of the final product, which can be misidentified by immunological techniques due to posttranslational modification (e.g., Lloyd et al., 1989). Moreover, one mRNA species can be processed into various protein products. At least two well-known examples of this phenomenon are (1) the **proopiomelanocortin mRNA** giving rise to messages translated into **adrenocorticotropin, β-endorphin, melanocyte-stimulating hormone, lipotropins,** and other bioactive peptides (Nakanishi et al., 1979) and (2) **preproglucagon**, which, in addition to glucagon, is processed into a number of related peptides (Hamid et al., 1986).

Lloyd (1990a,b) and Lloyd and coworkers (1990a,b) illustrate a similar situation with a family of chromogranin-related derivatives. Thus, posttranslational modification of the parent molecules by proteolytic cleavage produces several **chromogranin A** and **B** peptides distributed in various tissues. Chromogranin A can be identified either immunologically by antibodies to its different portions (**pancreastatin** and **betagranin**), or by ISH detecting the intact chromogranin A mRNA coding for both pancreastatin and betagranin.

Lloyd and colleagues (1989) have identified chromogranin A and B mRNA in a variety of tissues and neoplasms. Although chromogranin A mRNA is abundantly present in small cell lung carcinoma (SCLC), the determination of the neuroendocrine origin of this neoplasm by immunological techniques may be unreliable, because only small amounts of **immunoreactive chromogranin A** are usually expressed (Lloyd et al., 1989). Chromogranin A and B have been detected by ISH in bronchial, rectal, and midgut **carcinoid tumors** (Funa et al., 1991). In **pituitary adenomas** causing acromegaly, the identification of mRNA for **growth hormone** by ISH is the better diagnostic tool, because the stored hormone could not be identified by immunostaining in spite of large amounts of growth hormone secreted by the tumor (Lloyd et al., 1989). In clinically silent cases, the growth hormone mRNA is detectable in the tumors, whereas the product is either dysfunctional or not efficiently produced (Lloyd et al., 1989).

Correlative studies performed on **thyroid neoplasms** provided direct indication that ISH, combined with morphological grading, can offer an objective, if not yet quantifiable, estimate of malignancy. Given observations on **thyroglobulin gene expression** in well-differentiated papillary and follicular neoplasms (Berge-LeFranc et al., 1985) are confirmed, then thyroglobin mRNA levels could be used for finer diagnosis and monitoring of thyroid tumors, especially in the context of the tissue-specific features of thyroglobulin gene expression (Sinclair et al., 1990). Calcitonin mRNAs can be visualized by ISH using biotinylated probes (Lloyd, 1990a).

The use of recombinant DNA techniques in identifying the molecular basis of endocrine disorders and in diagnostic applications have been reviewed (Jameson and Arnold, 1990). Selected examples include Southern blot analysis that is used to identify the presence or absence of the **functional (GH1)** or **variant (GH2) growth hormone gene**, which contains inactivating mutations. MEN I and IIa syndromes are transmitted in an autosomal dominant pattern, and identification of the specific genes responsible for the syndrome can present definite diagnostic benefits (see above). Unfortunately, the identity of the gene for MEN, and of its products, remains unknown at this time beyond the assignment of the responsible genes to chromosomes 11 (MEN I) and 10 (MEN IIa) (see above). A strategy of genetic mapping performed on XX males and XY females that allowed the assignment of the testis-determining factor to a specific location on the short arm of the Y chromosome and the H-Y antigen location to the long arm of the Y chromosome has been described in detail (Jameson and Arnold, 1990). Given the need for predicting the gender of a child prenatally, in cases when the genetic anomaly is predominantly expressed in one sex, the delineation of fine genetic assignments on sex chromosomes may shift from the domain of fundamental research into clinical practice (Jameson and Arnold, 1990).

Identification of point mutations by PCR can be seen as a useful diagnostic tool in cases with variable clinical presentations related to genetic abnormalities. A case in point is the **vitamin D receptor gene analysis** (Jameson and Arnold, 1990). An approach to evaluation of gene expression in endocrine tissues has been elaborated in a model system using the pheochromocytoma cell line, PC12 (Tsutsumi et al., 1990). In response to a number of stimulators such as nerve growth factor, dexamethasone, forskolin, and lithium, a cumulative increase in neurotensin production is observed that can be demonstrated by radiolabeled **oligonucleotide probes** complementary to neurotensin mRNA. Stimulated, but not unstimulated, cells showed two mRNA bands (10 and 15 kb) on Northern blots. ISH revealed approximately 30% of the cells to be stimulated and strongly positive for neurotensin mRNA. Substantially higher sensitivity of detection of neurotensin gene expression was obtained by PCR, which detected some level of neurotension gene expression in unstimulated cells as well.

Oligonucleotide probes offer numerous **advantages** over the immunohistochemical approach. In various endocrine tissues, some cells may only appear to be secretion depleted, failing to reveal the tested gene products because of their **rapid turnover**. This may account for little or no peptide immunoreactivity even in the stimulated cells. Examples of such false-negative assays can be seen in the failure to detect ectopic production of adrenocorticotropic hormone (ACTH) by SCLC (Osamura et al., 1984), and glucagon synthesis by malignant islet cell tumor (Boidi et al., 1981).

The multipurpose use of synthetic oligonucleotides for ISH, Northern blotting, and PCR makes such probes particularly appealing for diagnostic use in neuroendocrine pathology and beyond (Tsutsumi et al., 1990). Detailed techniques for the preparation of oligonucleotide probes for use in hybridization histochemistry, demonstrated in neuroendocrine tissues, have been described (Penschow et al., 1989).

Among the **advantages of using oligonucleotide probes over larger fragments of DNA or RNA** are a high degree of consistency of tissue assessment over time helped by the strictly defined nature of the probe, easily controlled concentration and specific activity used in each case. Further practical advantages are the ease of customized synthesis (which can include common or unique sequences), absence of subcloning, no requirement for denaturation steps (as needed with double-stranded cDNA), reduced needs for pretreatment of target tissue (Triton X-100, proteinase K, and acetic anhydride treatments can be omitted), long shelf life at 4°C, and so on.

Among the **disadvantages of oligonucleotides** for ISH is the lower sensitivity than that achieved with cRNA probes, because of the relatively low specific activity and lower binding of DNA-RNA hybrids than RNA-RNA hybrids (De Lellis and Wolfe, 1987; Singer et al., 1986). Although **single-stranded cDNAs** seem to give a high background signal due to entrapment of nonhomologous bacterial DNA sequences, they do not self-hybridize, unlike the double-stranded probes. The latter must be unwound prior to and during the hybridization process and the self-annealing and hybridization to the target sequences are in competition as a result (Giaid et al., 1989).

The **reverse transcription (RT)-PCR sequencing** has been used previously for a variety of research objectives such as cloning genes (e.g., Berchtold, 1989; Frohman et al., 1988), detection of dystrophin gene transcripts (e.g., Chelly et al., 1988), and detection of chromosomal translocations in leukemia (e.g., Kawasaki et al., 1988). A combination of **RT** of mRNA into cDNA followed by PCR amplification proved to be more sensitive than Northern blotting and ISH for detecting specific mRNAs in endocrine pathology (Tsutsumi et al., 1990). Again, there is no need to use poly(A) selection of mRNAs by oligo-dT chromatography, when PCR is used for the detection of specific sequences.

Cloned RNA probes are appealing, because they are usually longer than cDNA and have higher thermal stability, allowing for higher stringency conditions to be used; also beneficial are the high affinity of RNA-RNA hybrids, constant probe size, and absence of vector sequences in the probe. An additional advantage is in the preparation of a **sense probe** as the control, which has a sequence identical to the target mRNA and, therefore, is unable to hybridize with the target (Giaid et al., 1989).

An ingenious extension of *in situ* evaluation of gene expression has been the development of ***in situ* transcription (IST)** (Altar et al., 1989). IST allows a specific cDNA to be synthesized *in situ* by the enzyme **reverse transcriptase** following the addition of appropriate primers. The cDNA thus produced can be isolated from the tissue section and analyzed by a variety of techniques. Identification of individual cells harboring the targeted mRNA molecules is difficult by this technique; its main advantage is its higher sensitivity compared to that of conventional ISH. The quantitation of mRNA levels within a tissue could be one of the applications of IST. It remains to be explored to what extent this approach can be used for diagnostic purposes in a clinical setting, especially when compared with PCR.

Some of the **advantages of IST** appear to be the relative ease of **quantitation of mRNAs**, because the labeled cDNA, produced in the tissue section and isolated for analysis, can be of high specific activity, and its synthesis can be quantitatively related to the target mRNA. Besides, the morphology of the tissue section does not have to be destroyed for isolation of the nucleic acid sequences, as required for conventional PCR. On the other hand, PCR offers unparalleled sensitivity of detection. Moreover, when nonradioactively labeled primers are used combined with automation of the process, PCR is an appealing approach for diagnostic gene probing.

References

Accili, D. et al. (1989). A mutation in the insulin receptor gene that impairs transport of the receptor to the plasma membrane and causes insulin resistant diabetes. EMBO J. *8*:2509–2517.

Affholter, J. A. et al. (1990). Insulin-degrading enzyme: stable expression of the human complementary DNA, characterization of its protein product, and chromosomal mapping of the human and mouse genes. Mol. Endocrinol. *4*:1125–1135.

Altar, C. A. et al. (1989). *In situ* mRNA hybridization: standard procedures and novel approaches. In: Gene Probes. Conn, P. M. (ed.). Academic Press, New York, pp. 238–281.

Amberson, J. B. et al. (1987). Flow cytometric analysis of nuclear DNA from adrenocortical neoplasms. A retrospective study using paraffin-embedded tissue. Cancer *59*:2091–2095.

Amos, C. I. et al. (1989). Polymorphism in the 5′-flanking region of the insulin gene and its potential relation to cardiovascular disease risk: observations in a biracial community. The Bogalusa Heart Study. Atherosclerosis *79*:51–57.

Andersson, M. et al. (1984). Genomic hybridization with class II transplantation antigen cDNA probes as a complementary technique in tissue typing. Hum. Immunol. *11*:57–67.

Aparicio, J. M. et al. (1988). HLA-DQ system and insulin-dependent diabetes mellitus in Japanese: does it contribute to the development of IDDM as it does in Caucasians? Immunogenetics *28*:240–246.

Arner, P. et al. (1987). Defective insulin receptor tyrosine kinase in human skeletal muscle in obesity and type 2 (non-insulin-dependent) diabetes mellitus. Diabetologia *30*:437–440.

Auffray, C. and Strominger, J. L. (1986). Molecular genetics of the human major histocompatibility complex. Adv. Hum. Genet *15*:197–247.

Bach, F. H. et al. (1985). Insulin-dependent diabetes-associated HLA-D region encoded determinants. Hum. Immunol. *12*:59–64.

Bach, F. H. et al. (1986). Cellular detection of HLA class II-encoded determinants: subtype polymorphisms of HLA-D. In: HLA Class II Antigens. Solheim, B. G. et al. (eds.). Springer-Verlag, Berlin/Heidelberg, pp. 249–265.

Badenhoop, K. et al. (1990). Susceptibility to thyroid autoimmune disease: molecular analysis of HLA-D region genes identifies new markers from goitrous Hashimoto's thyroiditis. J. Clin. Endocrinol. Metab. *71*:1131–1137.

Bahn, R. S. et al. (1988). *n*-Butyrate increases c-*erb* A oncogene expression in human colon fibroblasts. Biochem. Biophys. Res. Commun. *150*:259–262.

Baisch, J. M. et al. (1990). Analysis of HLA-DQ genotypes and susceptibility in insulin-dependent diabetes mellitus. N. Engl. J. Med. *322*:1836–1841.

Bale, S. J. et al. (1989). Linkage analysis of multiple endocrine neoplasia type 1 with *INT*2 and other markers on chromosome 11. Genomics *4*:320–322.

Barbosa, J. and Rich, S. S. (1988). Genetics of insulin-dependent diabetes. In: Immunogenetics of Endocrine Disorders. Farid, N. R. (ed.). Alan R. Liss, New York, pp. 163–202.

Barnett, A. H. et al. (1981). Diabetes in identical twins. Diabetologia *20*:87–93.

Bell, G. I. et al. (1982). The highly polymorphic region near the human insulin gene is composed of simple tandemly repeating sequences. Nature (London) *295*:31–35.

Bell, G. I. et al. (1984). A polymorphic locus near the human insulin gene is associated with insulin-dependent diabetes mellitus. Diabetes *33*:176–183.

Bell, G. I. et al. (1991). Gene for non-insulin-dependent diabetes mellitus (maturity-onset diabetes of the young subtype) is linked to DNA polymorphism on human chromosome 20q. Proc. Natl. Acad. Sci. U.S.A. *88*:1484–1488.

Berchtold, M. W. (1989). A simple method for direct cloning and sequencing of cDNA by the use of a single specific oligonucleotide and oligo(dT) in a polymerase chain reaction (PCR). Nucleic Acids Res. *17*:453.

Berge-LeFranc, J. L. et al. (1985). Quantification of thyroglobulin messenger RNA by *in situ* hybridization in differentiated thyroid cancers. Cancer *56*:345–350.

Bidwell, J. L. et al. (1988). A DNA RFLP typing system that positively identifies serologically well-defined and ill-defined HLA-DR and DQ alleles, including DRw10. Transplantation *45*:640–646.

Bloss, J. D. et al. (1990). The use of molecular probes to distinguish new primary tumors from recurrent tumors in gynecological malignancies. Am. J. Clin. Pathol. *94*:432–434.

Boidi, C. et al. (1981). The morphology of glucagonomas. In: Glucagon, Physiology, Pathophysiology, and Morphology of the Pancreatic A-cells. Unger, R. H. and Orci, L. (eds.). Elsevier North Holland, New York, pp. 399–411.

Bongarzone, I. et al. (1989). High frequency of activation of tyrosine kinase oncogenes in human papillary thyroid carcinoma. Oncogene *4*:1457–1462.

Boultwood, J. et al. (1988). N-*myc* expression in neoplasia of human thyroid C-cells. Cancer Res. *48*:4073–4077.

Bowlby, L. S. et al. (1986). Flow cytometric analysis of adrenal cortical tumor DNA. Cancer *58*:1499–1505.

373

Bugawan, T. L. et al. (1988). Analysis of HLA-DP allelic sequences polymorphism using the *in vitro* enzymatic amplification of DP-α and DP-β loci. J. Immunol. *141*:4024–4030.

Buraczynska, M. et al. (1985). Apolipoprotein A-I gene polymorphism and susceptibility of non-insulin-dependent diabetes mellitus. Am. J. Hum. Genet. *37*:1129–1137.

Butler, P. C. et al. (1990). Effects of meal ingestion on plasma amylin concentration in NIDDM and nondiabetic humans. Diabetes *29*:752–756.

Bystrom, C. et al. (1990). Localization of the MEN1 gene to a small region within chromosome 11q13 by deletion mapping in tumors. Proc. Natl. Acad. Sci. U.S.A. *87*:1968–1972.

Campbell, I. L. and Harrison, L. C. (1990). Molecular pathology of type I diabetes. Mol. Biol. Med. *7*:299–309.

Caro, J. F. et al. (1987). Insulin receptor kinase in human skeletal muscle from obese subjects with and without non-insulin dependent diabetes. J. Clin. Invest. *79*:1330–1337.

Carrier, C. M. et al. (1989). Definition of IDDM-associated HLA-DQ and DX RFLPs by segregation analysis of multiplex sibships. Hum. Immunol. *24*:51–63.

Chan, S. J. et al. (1976). Cell-free synthesis of rat pre-proinsulins: characterization and partial amino acid sequence determination. Proc. Natl. Acad. Sci. U.S.A. *73*:1964–1968.

Chelly, J. et al. (1988). Transcription of the dystrophin gene in human muscle and non-muscle tissues. Nature (London) *333*:858–860.

Ciaraldi, T. P. et al. (1982). Role of glucose transport in the post-receptor defect of non-insulin-dependent diabetes mellitus. Diabetes *31*:1016–1022.

Cibas, E. S. et al. (1990). Cellular DNA profiles of benign and malignant adrenocortical tumors. Am. J. Surg. Pathol. *14*:948–955.

Colletta, G. et al. (1986). Induction of the c-*fos* oncogene by thyrotropic hormone in rat thyroid cells in culture. Science *233*:458–460.

Cooper, G. J. S. et al. (1987). Purification and characterization of a peptide from amyloid-rich pancreas of type 2 diabetic patients. Proc. Natl. Acad. Sci. U.S.A. *84*:8628–8632.

Cooper, G. J. S. et al. (1989). Amylin and the amylin gene: structure, function and relationship to islet amyloid and to diabetes mellitus. Biochem. Biophys. Acta *1014*:247–258.

Cox, N. J. and Bell, G. I. (1989). Disease associations. Chance, artifact, or susceptibility genes? Diabetes *38*:947–950.

Cox, N. J. et al. (1988a). Glucose transporter gene and non-insulin-dependent diabetes. Lancet *ii*:793–794.

Cox, N. J. et al. (1988b). HLA-DR typing at the DNA level: RFLPs and subtypes detected with a DR-β cDNA probe. Am. J. Hum. Genet. *43*:954–963.

Cox, N. J. et al. (1988c). Insulin-gene sharing in sib pairs with insulin-dependent diabetes mellitus: no evidence for linkage. Am. J. Hum. Genet. 42:167–172.

Cutler, G. B., Jr. and Laue, L. (1990). Congenital adrenal hyperplasia due to 21-hydroxylase deficiency. N. Engl. J. Med. *323*:1806–1813.

Dacou-Voutetakis, C. et al. (1990). Familial hypothyroidism caused by a nonsense mutation in the thyroid-stimulating hormone β subunit gene. Am. J. Hum. Genet. *46*:988–993.

De Lellis, R. A. and Wolfe, H. J. (1987). New techniques in gene product analysis. Arch. Pathol. Lab. Med. *111*:620–627.

de Micco, C. et al. (1990). *TPO* as marker of malignancy in thyroid tumors: immunohistochemical study. In: Thyroperoxidase and Thyroid Autoimmunity, Vol. 207. Carayon, P. and Ruf, J. (eds.). Colloque INSERM/John Libbey Eurotext, pp. 133–136.

de Vijlder, J. J. M. et al. (1990). DNA polymorphisms in the *TPO* gene and hereditary defects in iodide organification. In: Thyroperoxidase and Thyroid Autoimmunity, Vol. 207. Carayon, P. and Ruf, J. (eds.). Colloque INSERM/John Libbey Eurotext, pp. 149–155.

Degroot, L. J. (1989). Thyroid hormone nuclear receptors and their role in the metabolic action of the hormone. Biochimie *71*:269–277.

Denijn, M. et al. (1990). Detection of calcitonin-encoding mRNA by radioactive and non-radioactive *in situ* hybridization: improved colorimetric detection and cellular localization of mRNA in thyroid sections. J. Histochem. Cytochem. *38*:351–358.

Donghi, R. et al. (1989). The oncogene associated with human papillary thyroid carcinoma (PTC) is assigned to chromosome 10q11-q12 in the same region as multiple endocrine neoplasia type 2A (MEN 2A). Oncogene *4*:521–523.

Duquesnoy, R. J. and Zeevi, A. (1986). Cellular detection of human class II MHC antigens by alloreactive T cell clones. In: HLA Class II Antigens. Solheim, B. G. et al. (eds.). Springer-Verlag, Berlin/Heidelberg, pp. 266–280.

Ebina, Y. et al. (1985). The human insulin receptor cDNA: the structural basis for hormone-activated transmembrane signalling. Cell (Cambridge, Mass) *40*:747–758.

Eisenbarth, G. S. (1986). Type I diabetes mellitus. A chronic autoimmune disease. N. Engl. J. Med. *314*:1360–1368.

Ekman, E. T. et al. and the Swedish Medullary Thyroid Cancer Study Group. (1990). Nuclear DNA content and survival in medullary thyroid carcinoma. Cancer *65*:511–517.

El-Naggar, A. K. et al. (1990). Clinicopathologic and flow cytometric DNA study of medullary thyroid carcinoma. Surgery *108*:981–985.

Elbein, S. C. et al. (1986). Multiple restriction fragment length polymorphisms at the insulin receptor locus: a highly informative marker for linkage analysis. Proc. Natl. Acad. Sci. U.S.A. *83*:5223–5227.

Eldar-Geva, T. et al. (1990). Secondary biosynthetic defects in women with late-onset congenital adrenal hyperplasia. N. Engl. J. Med. *323*:855–863.

Erlich, H. A. (1989). HLA and insulin-dependent diabetes mellitus. Nature (London) *337*:415.

Erlich, H. A. et al. (1990). HLA-DQ β sequence polymorphism and genetic susceptibility to IDDM. Diabetes *39*:96–103.

Faustman, D. et al. (1991). Linkage of faulty major histocompatibility complex class I to autoimmune diabetes. Science *254*:1756–1761.

Field, J. B. (1986). Mechanism of action of thyrotropin. In: The Thyroid. Ingbar, S. H. and Braverman, L. E. (eds.). J. B. Lippincott, Philadelphia, pp. 288–303.

Forman, B. M. et al. (1988). c-*erb* A protooncogenes mediate thyroid hormone-dependent and independent regulation of the rat growth hormone and prolactive genes. Mol. Endocrinol. *2*:902–911.

Foti, D. et al. (1990). Studies on the role of c-*fos* in TSH-stimulated thyroid cell proliferation. Cell Mol. Biol. *36*:363–373.

Frazier, A. L. et al. (1990). Isolation of TSH and LH/CG receptor cDNAs from human thyroid: regulation by tissue specific binding. Mol. Endocrinol. *4*:1264–1276.

Freymond, D. et al. (1988). Impaired insulin-stimulated muscle glycogen synthase activation *in vivo* in man is related to low fasting glycogen synthase phosphatase activity. J. Clin. Invest. *82*:1503–1509.

Froguel, Ph. et al. (1992). Close linkage of glucokinase locus on chromosome 7p to early-onset non-insulin-dependent diabetes mellitus. Nature (London) *356*:162–164.

Frohman, M. A. et al. (1988). Rapid production of full-length cDNAs from rare transcripts: amplification using a single gene-specific oligonucleotide primer. Proc. Natl. Acad. Sci. U.S.A. *85*:8998–9002.

Funa K. et al. (1991). *In situ* hybridization study of chromogranin A and B mRNA in carcinoid tumors. Histochemistry *95*:555–559.

Fusco, A. et al. (1987). A new oncogene in human thyroid papillary carcinomas and their lymph-nodal metastases. Nature (London) *328*:170–172.

Gandour, M. J. and Grizzle, W. E. (1986). A small adrenocortical carcinoma with aggressive behavior: an evaluation of criteria for malignancy. Arch. Pathol. Lab. Med. *110*:1076–1079.

Garvey, W. T. et al. (1985). The effect of insulin treatment on insulin secretion and insulin action in type II diabetes mellitus. Diabetes *34*:222–232.

Garvey, W. T. et al. (1988). Role of glucose transporters in the cellular insulin resistance of type II non-insulin-dependent diabetes mellitus. J. Clin. Invest. *81*:1528–1536.

Giaid, A. et al. (1989). *In situ* detection of peptide messenger RNA using complementary RNA probes. In: Gene Probes. Conn, P. M. (ed.). Academic Press, New York, pp. 150–163.

Glass, C. K. et al. (1987). A c-*erb*-A binding site in rat growth hormone gene mediates trans-activation by thyroid hormone. Nature (London) *329*:738–741.

Goldberg, Y. et al. (1989). Thyroid hormone action and the *erb*-A oncogene family. Biochimie *71*:279–291.

Goodfellow, P. J. et al. (1990). A new DNA marker (D10S94) very tightly linked to the multiple endocrine neoplasia type 2A (MEN2A) locus. Am. J. Hum. Genet. *47*:952–956.

Greenbaum, E. et al. (1985). The diagnostic value of flow cytometric DNA measurement in follicular tumors of the thyroid gland. Cancer *56*:2011–2018.

Gyllensten, U. B. and Erlich, H. A. (1988). Generation of single-stranded DNA by the polymerase chain reaction and its application to direct sequencing of the HLA-DQA locus. Proc. Natl. Acad. Sci. U.S.A. *85*:7652–7656.

Hamid, Q. A. et al. (1986). Immunocytochemical characterization of 10 pancreatic tumors associated with the glucagonoma syndrome, using antibodies to separate regions of the pro-glucagon molecule and other neuroendocrine markers. Histopathology *10*:119–133.

Hartter, E. et al. (1990). Reduced islet-amyloid polypeptide in insulin-dependent diabetes mellitus. Lancet *i*:854.

Hattersley, A. T. et al. (1992). Failure to detect cytomegalovirus DNA in pancreas in type 2 diabetes. Lancet *339*:459–460.

Hayashizaki, Y. et al. (1990). Deoxyribonucleic acid analyses of five families with familial inherited thyroid stimulating hormone deficiency. J. Clin. Endocrinol. Metab. *71*:792–796.

Hayes, G. R. et al. (1991). Intermolecular phosphorylation of insulin receptor as possible mechanism for amplification of binding signal. Diabetes *40*:300–303.

Hayward, N. K. et al. (1988). Generation of homozygosity at the c-Ha-*ras* 1 locus on chromosome 11p in an adrenal adenoma from an adult with Wiedemann-Beckwith syndrome. Cancer Genet. Cytogenet. *30*:127–132.

Hitman, G. A. et al. (1985). Type 1 (insulin-dependent) diabetes and a highly variable locus close to the insulin gene on chromosome 11. Diabetologia *28*:218–222.

Holbeck, S. L. and Nepom, G. T. (1986). Exon-specific oligonucleotide probes localize HLA-DQ β allelic polymorphisms. Immunogenetics *24*:251–258.

Hoover, M. L. et al. (1986). HLA-DQ and T cell receptor genes in insulin dependent diabetes mellitus. Cold Spring Harbor Symp. Quant. Biol. *51*:803–809.

Hoppener, J. W. M. et al. (1987). Expression of insulin-like growth factor I and II genes in rat medullary thyroid carcinoma. FEBS Lett. *215*:122–126.

Horn, G. T. et al. (1988). Allelic sequence variation of the HLA-DQ loci: relationship to serology and to insulin-dependent diabetes susceptibility. Proc. Natl. Acad. Sci. U.S.A. *85*:6012–6016.

Hough, A. J. et al. (1979). Prognostic factors in adrenal cortical tunors. A mathematical analysis of clinical and morphologic data. Am. J. Clin. Pathol. *72*:390–400.

Houslay, M. D. and Siddle, K. (1989). Molecular basis of insulin receptor function. Br. Med. Bull. *45*:264–284.

Hyldig-Nielsen, J. J. et al. (1987). Restriction fragment length polymorphism of the HLA-DP subregion and correlations to HLA-DP phenotypes. Proc. Natl. Acad. Sci. U.S.A. *84*:1644–1648.

Ito, T. et al. (1992). Unique association of p53 mutations with undifferentiated but not with differentiated carcinomas of the thyroid gland. Cancer Res. *52*:1369–1371.

Iynedjian, P. B. et al. (1986). Tissue-specific expression of glucokinase: identification of the gene product in liver and pancreatic islets. Proc. Natl. Acad. Sci. U.S.A. *83*:1998–2001.

Jackson, C. E. et al. (1988). Linkage between MEN2B and chromosome 10 markers linked to MEN2A. Am. J. Hum. Genet. *43*:154A.

Jameson, J. L. and Arnold, A. (1990). Clinical review 5. Recombinant DNA strategies for determining the molecular basis of endocrine disorders. J. Clin. Endocrinol. Metabol. *70*:301–307.

Jenkins, D. et al. (1991). Allele-specific gene probing supports the DQ molecule as a determinant of inherited susceptibility to type 1 (insulin-dependent) diabetes mellitus. Diabetologia *34*:109–113.

Jenkins, R. B. et al. (1990). Frequent occurrence of cytogenetic abnormalities in sporadic non-medullary thyroid carcinoma. Cancer *66*:1213–1220.

Joensuu, H. and Klemi, P. J. (1988). DNA aneuploidy in adenomas of endocrine organs. Am. J. Pathol. *132*:145–151.

Joensuu, H. et al. (1986). Influence of cellular DNA content on survival in differentiated thyroid cancer. Cancer *58*:2462–2467.

Johannessen J et al. (1981). A flow cytometric deoxyribonucleic acid analysis of papillary thyroid carcinoma. Lab. Invest. *45*:336–341.

Johannessen, J. et al. (1982). The diagnostic value of flow cytometric DNA measurements in selected disorders of the human thyroid. Am. J. Clin. Pathol. *77*:20–25.

Johnson, K. E. et al. (1989). Islet amyloid, islet amyloid polypeptide, and diabetes mellitus. N. Engl. J. Med. *321*:513–518.

Johnson, K. H. et al. (1991). Newly identified pancreatic protein islet amyloid polypeptide. What is its relationship to diabetes? Diabetes *40*:310–314.

Julier, C. et al. (1991). Insulin-IGF-2 region on chromosome 11p encodes a gene implicated in HLA-DR4-dependent diabetes susceptibility. Nature (London) *354*:155–159.

Kadowski, T. et al. (1988). Two mutant alleles of the insulin receptor gene in a patient with extreme insulin resistance. Science *240*:787–790.

Kaku, K. et al. (1990). Polymorphisms of Hep G2/erythrocyte glucose-transporter gene. Linkage relationships and implications for genetic analysis of NIDDM. Diabetes *39*:49–56.

Kanatsuka, A. et al. (1989). Secretion of islet amyloid polypeptide in response to glucose. FEBS Lett. *259*:199–201.

Kappes, D. and Strominger, J. L. (1988). Human class II major histocompatibility complex genes and proteins. Annu. Rev. Biochem. *57*:991–1028.

Kawasaki, E. S. et al. (1988). Diagnosis of chronic myeloid and acute lymphocytic leukemias by detection of leukemia-specific mRNA sequences amplified *in vitro*. Proc. Natl. Acad. Sci. U.S.A. *85*:5698–5702.

Kimura, E. and Armelin, H. A. (1988). Role of proto-oncogene c-Ki-*ras* amplification and overexpression in the malignancy of Y-1 adrenocortical tumor cells. Braz. J. Med. Biol. Res. *21*:189–201.

Klinkhamer, M. P. et al. (1989). A leucine-to-proline mutation in the insulin receptor in a family with insulin resistance. EMBO J. *8*:2503–2507.

Knudson, A. G. (1971). Mutation and cancer: statistical study of retinoblastoma. Proc. Natl. Acad. Sci. U.S.A. *68*:820–823.

Knutson, V. P. (1991). Cellular trafficking and processing of the insulin receptor. FASEB J. *5*:2130–2138.

Krolewski, A. S. and Warram, J. H. (1985). Epidemiology of diabetes mellitus. In: Joslin's Diabetes Mellitus, 12th Ed. Marble, A. et al. (eds.). Lea & Febiger, Philadelphia, pp. 12–42.

Kusari, J. et al. (1991). Insulin-receptor cDNA sequence in NIDDM patient homozygous for insulin-receptor gene RFLP. Diabetes *40*:249–254.

Lairmore, T. C. et al. (1991). Familial medullary thyroid carcinoma and multiple endocrine neoplasia type 2B map to the same region of chromosome 10 as multiple endocrine neoplasia type 2A. Genomics *9*:181–192.

Larsson, C. et al. (1988). Multiple endocrine neoplasia type 1 gene maps to chromosome 11 and is lost in insulinoma. Nature (London) *332*:85–87.

Lazar, M. A. and Chin, W. W. (1988). Regulation of two c-*erb* A messenger ribonucleic acids in rat GH3 cells by thyroid hormone. Mol. Endocrinol. *2*:479–484.

Lemoine, N. R. et al. (1988). Activated *ras* oncogenes in human thyroid cancers. Cancer Res. *48*:4459–4463.

Lemoine, N. R. et al. (1989). High frequency of *ras* oncogene activation in all stages of human thyroid tumorigenesis. Oncogene *4*:159–164.

Li, S. R. et al. (1988). Association of genetic variant of the glucose transporter with non-insulin-dependent diabetes mellitus. Lancet *2*:368–370.

Lloyd, R. V. (1987). Use of molecular probes in the study of endocrine diseases. Hum. Pathol. *18*:1199–1211.

Lloyd, R. V. (1990a). Molecular probes and endocrine diseases. Am. J. Surg. Path. *14* (Suppl. 1):34–44.

Lloyd, R. V. (1990b). Endocrine Pathology. Springer-Verlag, New York.

Lloyd, R. V. et al. (1989). Chromogranin A and B messenger ribonucleic acids in pituitary and other normal and neoplastic human endocrine tissues. Lab. Invest. *60*:548–556.

Lloyd, R. V. et al. (1990a). Detection of chromogranins A and B in endocrine tissues with radioactive and biotinylated oligonucleotide probes. Am. J. Surg. Path. *14*:35–43.

Lloyd, R. V. et al. (1990b). Analysis of endocrine active and clinically silent corticotropic adenomas by *in situ* hybridization. Am. J. Pathol. *137*:479–488.

Lohr, J. M. and Oldstone, M. B. A. (1990). Detection of cytomegalovirus nucleic acid sequences in pancreas in type 2 diabetes. Lancet *336*:644–648.

Lukvik, B. et al. (1990). Basal and stimulated plasma amylin levels in diabetes mellitus [abstract]. Diabetologia *33* (Suppl.):39A.

Magnuson, M. A. (1990). Glucokinase gene structure. Functional implications of molecular genetic studies. Diabetes *39*:523–527.

Magnuson, M. A. and Shelton, K. D. (1989). An alternate promoter in the glucokinase gene is active in the pancreatic β-cell. J. Biol. Chem. *264*:15936–15942.

Majewski, J. T. and Wilson, S. D. (1979). The MEN1 syndrome: an all or none phenomenon? Surgery *86*:476–480.

Mandrup-Poulsen, T. (1988). On the pathogenesis of insulin-dependent diabetes mellitus. Dan. Med. Bull. *35*:438–460.

Mandrup-Poulsen, T. et al. (1984). DNA sequences flanking the insulin gene on chromosome 11 confer risk of atherosclerosis. Lancet *i*:250–252.

Mariotti, S. et al. (1989). Recent advances in the understanding of humoral and cellular mechanisms implicated in thyroid autoimmune disorders. Clin. Immunol. Immunopathol. *50*:S73-S84.

Mathew, C. G. P. et al. (1987a). Deletion of genes on chromosome 1 in endocrine neoplasia. Nature (London) *328*:524–526.

Mathew, C. G. P. et al. (1987b). A linked genetic marker for multiple endocrine neoplasia type 2A on chromosome 10. Nature (London) *328*:527–528.

Mathew, C. G. P. et al. and the MEN2A International Collaborative Group (1991). Presymptomatic screening for multiple endocrine neoplasia type 2A with linked DNA markers. Lancet *337*:7–11.

Matschinsky, F. M. (1990). Glucokinase as glucose sensor and metabolic signal generator in pancreatic β-cells and hepatocytes. Diabetes *39*:647–652.

Matsutani, A. et al. (1990). Polymorphisms of *GLUT2* and *GLUT4* genes. Use of evaluation of genetic susceptibility to NIDDM in blacks. Diabetes *39*:1534–1542.

McClain, D. A. et al. (1988). Restriction fragment length polymorphism in insulin receptor gene and insulin resistance in NIDDM. Diabetes *37*:1071–1075.

McKeon, C. et al. (1990). Structural and functional analysis of the insulin receptor promoter. Mol. Endocrinol. *4*:647–656.

Meglasson, M. D. and Matschinsky, F. M. (1986). Pancreatic islet glucose metabolism and regulation of insulin secretion. Diabetes/Metab. Rev. *2*:163–214.

Melmed, S. and Prager, D. (1991). Molecular biology of the endocrine system. In: Functional Endocrine Pathology. Kovacs, K. and Asa, S. L. (eds.). Blackwell Scientific, Boston, pp. 124–142.

Mueckler, M. (1990). Family of glucose-transporter genes. Implications for glucose homeostasis and diabetes. Diabetes *39*:6–11.

Michelsen, B. and Lernmark, A. (1987). Molecular cloning of a polymorphic DNA endonuclease fragment associates insulin-dependent diabetes mellitus with HLA-DQ. J. Clin. Invest. *79*:1144–1151.

Millward, B. A. et al. (1987). T cell receptor β chain polymorphisms are associated with insulin dependent diabetes. Clin. Exp. Immunol. *70*:152–157.

Moley, J. F. et al. (1990). Chromosome deletions in pheochromocytomas from patients with the MEN2A syndrome. Proc. Am. Assn. Cancer Res. *31*:319A.

Moley, J. F. et al. (1991). Low frequency of *ras* gene mutations in neuroblastomas, pheochromocytomas, and medullary thyroid cancers. Cancer Res. *51*:1596–1599.

Moller, D. E. et al. (1990). Functional properties of a naturally occurring Trp1200 → Ser1200 mutation of the insulin receptor. Mol. Endocrinol. *4*:1183–1191.

Moller, D. E. and Flier, J. S. (1988). Detection of an alteration in the insulin receptor gene in a patient with insulin resistance, acanthosis nigricans, and the polycystic ovary syndrome (type A insulin resistance). N. Engl. J. Med. *319*:1526–1529.

Moller, D. E. and Flier, J. S. (1991). Insulin resistance mechanisms, syndromes, and implications. N. Engl. J. Med. *325*:938–948.

Monos, D. S. et al. (1987). The HLA-DQw3.2 allele of the DR4 haplotype is associated with insulin dependent diabetes: correlation between DQ-β restriction fragments and DQ-β chain variation. Immunogenetics *26*:299–303.

Morel, P. A. et al. (1988). Aspartic acid at position 57 of the HLA-DQ-β chain protects against type I diabetes: a family study. Proc. Natl. Acad. Sci. U.S.A. *85*:8111–8115.

Morgan, R. et al. (1990). Allelic variants at insulin-receptor and insulin gene loci and susceptibility to NIDDM in Welsh population. Diabetes *39*:1479–1484.

Morris, J. C., III et al. (1988). Clinical utility of thyrotropin-receptor antibody assays: comparison of radioreceptor and bioassay methods. Mayo Clin. Proc. *63*:707–717.

Mosselman, S. et al. (1988). Islet amyloid polypeptide: identification and chromosomal localization of the human gene. FEBS Lett *239*:227–232.

Mosselman, S. et al. (1989). The complete amyloid polypeptide precursor by proteolytic processing. FEBS Lett. *247*:154–158.

Munoz, A. et al. (1990). The chicken c-*erb* A-α product induces expression of thyroid-hormone-responsive genes in 3, 5, 3′-triiodothyronine receptor deficient rat hepatoma cells. Mol. Endocrinol. *4*:312–320.

Nagamatsu, S. et al. (1990). Lack of islet amyloid polypeptide regulation of insulin biosynthesis or secretion in normal rat islets. Diabetes *39*:871–874.

Nakamura, Y. et al. (1989). Localization of the genetic defect of multiple endocrine neoplasia type 1 within a small region of chromosome 11. Am. J. Hum. Genet. *44*:751–755.

Nakanishi, S. et al. (1979). Nucleotide sequence of cloned cDNA for bovine corticotropin B-lipotropin precursor. Nature (London) *278*:423–427.

Nakazato, M. et al. (1989). Establishment of radioimmunoassay for human islet amyloid polypeptide and its tissue content and plasma concentration. Biochem. Biophys. Res. Commun. *164*:394–399.

Nakura, J. et al. (1990). A case of first trimester prenatal diagnosis of 21-hydroxylase deficiency with human complement C4 cDNA probe. Endocrinol. Jpn. *37*:615–618.

Namba, H. et al. (1990). Point mutations of *ras* oncogenes are an early event in thyroid tumorigenesis. Mol. Endocrinol. *4*:1474–1479.

Nelkin, B. D. et al. (1989). Low incidence of loss of chromosome 10 in sporadic and hereditary human medullary thyroid carcinoma. Cancer Res. *49*:4114–4119.

Nepom, B. S. et al. (1986). Specific genomic markers for the HLA-DQ subregion discriminate between DR4+ insulin-dependent diabetes mellitus and DR4+ seropositive juvenile rheumatoid arthritis. J. Exp. Med. *164*:345–350.

Nepom, G. T. (1988). Immunogenetics of HLA-associated diseases. Concepts Immunopathol. *5*:80–105.

Nepom, G. T. (1989a). Genetic structure and functions of the human major histocompatibility complex. In: A Textbook of Rheumatology. McCarthy, D. J. (ed.). Lea & Febiger, Philadelphia, pp. 440–452.

Nepom, G. T. (1989b). Determinants of genetic susceptibility in HLA-associated autoimmune disease. Clin. Immunol. Immunopathol. *53*:S53-S62.

Nepom, G. T. (1990). A unified hypothesis for the complex genetics of HLA associations with IDDM. Diabetes *39*:1153–1157.

Nerup, J. (1981). Etiology and pathogenesis of insulin-dependent diabetes mellitus: present views and future developments. In: Etiology and Pathogenesis of Insulin-dependent Diabetes Mellitus. Martin, J. M. et al. (eds.). Raven Press, New York, pp. 275–288.

Nerup, J. et al. (1984). Autoimmunity. In: Immunology of Diabetes in Experimental Animals and Man. Gupta, S. (ed.). Plenum Press, New York and London, pp. 351–367.

Newman, B. et al. (1987). Concordance for type 2 (non-insulin-dependent) diabetes mellitus in male twins. Diabetologia *30*:763–768.

Nir, U. et al. (1986). Regulation of rat insulin 1 gene expression: evidence for negative regulation in nonpancreatic cells. Proc. Natl. Acad. Sci. U.S.A. *83*:3180–3184.

Nishi, M. et al. (1990). Islet amyloid polypeptide (amylin): no evidence of an abnormal precursor sequence in 25 type 2 (non-insulin-dependent) diabetic patients. Diabetologia *33*:628–630.

Norum, R. A. et al. (1990). Linkage of the multiple endocrine neoplasia type 2B gene (MEN2B) to chromosome 10 markers linked to MEN2A. Genomics *8*:313–317.

O'Rahilly, S. O. et al. (1988). Type 2 (non-insulin-dependent) diabetes mellitus: new genetics for old nightmares. Diabetologia *31*:407–414.

Obara, T. et al. (1990). Flow cytometric DNA analysis of parathyroid tumors. Implication of aneuploidy for pathologic and biologic classification. Cancer *66*:1555–1562.

Odawara, M. et al. (1989). Human diabetes associated with a mutation in the tyrosine kinase domain of the insulin receptor. Science *245*:66–68.

Ohlsson, H. and Edlund, T. (1986). Sequence-specific interactions of nuclear factors with the insulin gene enhancer. Cell (Cambridge, Mass) *45*:35–44.

Olefsky, J. M. (1990). The insulin receptor. A multifunctional protein. Diabetes *39*:1009–1016.

Orci, L. et al. (1990). Evidence that down-regulation of β-cell glucose transporters in non-insulin-dependent diabetes may be the cause of diabetic hyperglycemia. Proc. Natl. Acad. Sci. U.S.A. *87*:9953–9957.

Osamura, R. Y. et al. (1984). Light and electron microscopic localization of ACTH and pro-opiomelanocortin-derived peptides in human developmental and neoplastic cells. J. Histochem. Cytochem. *32*:885–893.

Owerbach, D. and Nerup, J. (1982). Restriction fragment length polymorphism of the insulin gene in diabetes mellitus. Diabetes *31*:275–277.

Owerbach, D. et al. (1981). The insulin gene is located on the short arm of chromosome 11 in humans. Diabetes *30*:267–270.

Owerbach, D. et al. (1982). Possible association between DNA sequences flanking the insulin gene and atherosclerosis. Lancet *ii*:1291–1293.

Owerbach, D. et al. (1983). HLA-D region β-chain DNA endonuclease fragments differ between HLA-DR identical healthy and insulin-dependent diabetic individuals. Nature (London) *303*:815–817.

Owerbach, D. et al. (1987). Molecular biology of the HLA system in insulin-dependent diabetes mellitus. Diabetes Metab. Rev. *3*:819–833.

Owerbach, D. et al. (1988). Oligonucleotide probes for HLA-DQA and DQB genes define susceptibility to type 1 (insulin-dependent) diabetes mellitus. Diabetologia *31*:751–757.

Owerbach, D. et al. (1989). Primary association of HLA-DQw8 with type I diabetes in DR4 patients. Diabetes *38*:942–945.

Owerbach, D. et al. (1990a). Association of HRAS 1 polymorphism with HLA-DR3, DQw2/DR4, DQw8. Diabetes *39*:1504–1509.

Owerbach, D. et al. (1990b). Direct analysis of CYP21B genes in 21-hydroxylase deficiency using polymerase chain reaction amplification. Mol. Endocrinol. *4*:125–131.

Penschow, J. D. et al. (1989). Location of gene expression in tissue sections by hybridization histochemistry using oligodeoxyribonucleotide probes. In: Gene Probes, Vol. 1 (Methods in Neurosciences). Conn, Q. M. (ed.). Academic Press, New York, pp. 222–238.

Pfragner, R. et al. (1990). Establishment and characterization of continuous cell line MTC-SK derived from a human medullary thyroid carcinoma. Cancer Res. *50*:4160–4166.

Porte, D., Jr. (1991). β-Cells in type II diabetes mellitus. Diabetes *40*:166–180.

Porte, D., Jr. and Kahn, S. E. (1989). Hyperproinsulinemia and amyloid in NIDDM: clues to etiology of islet β cell dysfunction? Diabetes *38*:1333–1336.

Radford, D. M. et al. (1990). Loss of heterozygosity of markers on chromosome 11 in tumors from patients with multiple endocrine neoplasia syndrome type 1. Cancer Res. *50*:6529–6533.

Rask, L. et al. (1985). Generation of class II antigen polymorphism. Immunol. Rev. *84*:123–143.

Raux-Demay, M. et al. (1989). Early prenatal diagnosis of 21-hydroxylase deficiency using amniotic fluid 17-hydroxyprogesterone determination and DNA probes. Prenatal Diagn. *9*:457–466.

Reaven, G. M. (1988). Role of insulin resistance in human disease. Diabetes *37*:1595–1607.

Rees-Smith, B. et al. (1988). Autoantibodies to the thyrotropin receptor. Endocrinol. Rev. *9*:106–121.

Reynet, C. et al. (1990). Mutation of tyrosine residues 1162 and 1163 of the insulin receptor affects hormone and receptor internalization. Mol. Endocrinol. *4*:304–311.

Rich, S. S. (1990). Mapping genes in diabetes. Genetic epidemiological perspective. Diabetes *39*:1315–1319.

Rich, S. et al. (1985). Complement and HLA: further definition of high risk haplotypes in insulin-dependent diabetes. Diabetes *34*:504–509.

Richiardi, P. and Curtoni, E. S. (1986). Serology of HLA class II antigens: methodological aspects. In: HLA Class II Antigens. Solheim, B. G. et al. (eds.). Springer-Verlag, Berlin/Heidelberg, pp. 169–186.

Riley, W. J. (1989). Insulin dependent diabetes mellitus, an autoimmune disorder. Clin. Immunol. Immunopathol. *53*:S92-S98.

Risch, N. (1990a). Linkage strategies for genetically complex traits. I. Multilocus models. Am. J. Hum. Genet. *46*:222–228.

Risch, N. (1990b). Linkage strategies for genetically complex traits. II. The power of affected relative pairs. Am. J. Hum. Genet. *46*:229–241.

Risch, N. (1990c). Linkage strategies for genetically complex traits. III. The effect of marker polymorphism on analysis of affected relative pairs. Am. J. Hum. Genet. *46*:242–253.

Robertson, R. P. et al. (1991). G proteins and modulation of insulin secretion. Diabetes *40*:1–6.

Robinson, D. M. et al. (1989). HLA class II typing using oligonucleotide probes. Genet. Epidemiol. *6*:27–30.

Ronningen, K. S. et al. (1989). The amino acid at position 57 of the HLA-DQ β chain and susceptibility to develop insulin-dependent diabetes mellitus. Hum. Immunol. *26*:215–225.

Rosen, O. M. (1987). After insulin binds. Science *237*:1452–1458.

Rosen, O. M. (1989). Structure and function of insulin receptors. Diabetes *38*:1508–1511.

Roth, S. I. et al. (1990). Flow cytometric DNA analysis of the parathyroids. Mod. Pathol. *3*:86A.

Rotter, J. I. et al. (1990). Genetics of diabetes mellitus. In: Diabetes Mellitus. Theory and Practice, 4th Ed. Rifkin, H. and Porte, D., Jr. (eds.). Elsevier, New York, pp. 378–413.

Rousseau-Merck, M. F. et al. (1990). Assignment of the human thyroid stimulating hormone receptor (TSHR) gene to chromosome 14q31. Genomics *8*:233–236.

Rowe, J. R. et al. (1990). HLA-DQA2 (DX α) polymorphism and insulin dependent diabetes. Hum. Immunol. *29*:256–262.

Rubinstein, P. (1991). HLA and IDDM: facts and speculations on the disease gene and its mode of inheritance. Hum. Immunol. *30*:270–277.

Sanke, T. et al. (1988). An islet amyloid peptide is derived from an 89-amino acid precursor by proteolytic processing. J. Biol. Chem. *263*:17243–17246.

Sap, J. et al. (1986). The c-*erb*-A protein is a high-affinity receptor for thyroid hormone. Nature (London) *324*:635–640.

Scharf, S. J. et al. (1986). Direct cloning and sequence analysis of enzymatically amplified genomic sequences. Science *233*:1076–1078.

Scharf, S. J. et al. (1991). Rapid typing of DNA sequence polymorphism at the HLA-DRB1 locus using the polymerase chain reaction and nonradioactive oligonucleotide probes. Hum. Immunol. *30*:190–201.

Schreuder, G. M. T. and van Leuwen, A. (1986). Alloantisera against human class II antigens: definition and methodology. In: HLA Class II Antigens. Solheim, B. G. et al. (eds.). Springer-Verlag, Berlin/Heidelberg, pp. 187–203.

Schreuder, G. M. T. et al. (1986). HLA-DQ polymorphism associated with resistance to type I diabetes detected with monoclonal antibodies, isoelectric point differences, and restriction fragment length polymorphism. J. Exp. Med. *164*:938–943.

Schroder, S. et al. (1988). Prognostic factors in medullary thyroid carcinomas: survival in relation to age, sex, stage, histology, immunochemistry, and DNA content. Cancer *61*:806–816.

Schueler, P. A. et al. (1990). Binding of 3,5,3′-triiodothyronine (T3) and its analogs to the *in vitro* translational products of c-*erb* A protooncogenes: differences in the affinity of the α- and β-forms for the acetic acid analog and failure of the human testis and kidney α-2 products to bind T3. Mol. Endocrinol. *4*:227–234.

Schuffenecker, I. et al. (1991). The gene for mannose-binding protein maps to chromosome 10 and is a marker for multiple endocrine neoplasia type 2. Cytogenet. Cell. Genet. *56*:99–102.

Schulz, N. et al. (1990). Pheochromocytomas and C-cell thyroid neoplasms in transgenic c-*mos* mice, a model for the human MEN2 syndrome. Proc. Am. Assn. Cancer Res. *31*:325A.

Schwab, D. A. and Wilson, J. E. (1989). Complete amino acid sequence of rat brain hexokinase, deduced from the cloned cDNA and proposed structure of a mammalian hexokinase. Proc. Natl. Acad. Sci. U.S.A. *86*:2563–2567.

Seino, S. et al. (1989). Structure of the human insulin receptor gene and characterization of its promoter. Proc. Natl. Acad. Sci. U.S.A. *86*:114–118.

Seino, S. et al. (1990a). Human insulin-receptor gene. Partial sequence and amplification of exons by polymerase chain reaction. Diabetes *39*:123–128.

Seino, S. et al. (1990b). Human insulin-receptor gene. Diabetes *39*:129–133.

Selden, R. F. et al. (1987). Regulation of insulin-gene expression. Implications for gene therapy. N. Engl. J. Med. *317*:1067–1076.

Shapiro, L. R. (1991). Cytogenetics. In: Functional endocrine pathology. Kovacs, K. and Asa, S. L. (eds.). Blackwell Scientific, Boston, pp. 143–153.

Sheehy, M. J. et al. (1985). A particular subset of HLA-DR4 accounts for all or most of the DR4 association in type I diabetes. Diabetes *24*:942–944.

Shimada, F. (1990). Insulin with partial deletion of insulin-receptor gene. Lancet *335*:1179–1181.

Shows, T. B. et al. (1987). Polymorphic human glucose transporter gene (*GLUT*) is on chromosome 1p31.3-p35. Diabetes *36*:546–549.

Silvestre, R. A. et al. (1990). Rat amylin inhibits insulin release without affecting glucagon and somatostatin output in the rat pancreas [abstract]. Diabetologia *33* (Suppl.):39A.

Simpson, N. E. et al. (1987). Assignment of multiple endocrine neoplasia type 2A to chromosome 10 by linkage. Nature (London) *328*:528–530.

Sinclair, A. J. et al. (1990). The tissue-specific expression of the thyroglobulin gene requires interaction between thyroid-specific and ubiquitous factors. Eur. J. Biochem. *193*:311–318.

Singer, R. H. et al. (1986). Optimization of *in situ* hybridization using isotopic and non-isotopic detection methods. BioTechniques *4*:230–244.

Sinha, M. K. et al. (1987). Insulin-receptor kinase activity of adipose tissue from morbidity obese humans with and without NIDDM. Diabetes *36*:620–625.

Smith, S. A. et al. (1988). Mortality from papillary thyroid carcinoma. A case-control study of 56 lethal cases. Cancer *62*:1381–1388.

Sobol, H. et al. and the GETC (1989). Genetic screening of endocrine tumor syndromes with DNA probes: the example of medullary thyroid carcinoma. Hormone Res. *32*:34–40.

Speiser, P. W. et al. (1988). Molecular genetic analysis of nonclassic steroid 21-hydroxylase deficiency associated with HLA-B14, DR1. N. Engl. J. Med. *319*:19–23.

Spielman, R. S. et al. (1987). Segregation of T-cell receptor β chain RFLPs in sibs with insulin-dependent diabetes mellitus (IDDM) [abstract]. Am. J. Hum. Genet. *41*:186A.

Steiner, D. F. et al. (1967). Insulin biosynthesis: evidence for a precursor. Science *157*:697–700.

Steiner, D. F. et al. (1991). Is islet amyloid polypeptide a significant factor in pathogenesis or pathophysiology of diabetes? Diabetes *40*:305–309.

Sterkers, G. et al. (1988). HLA-DQ rather than HLA-DR region might be involved in dominant nonsusceptibility to diabetes. Proc. Natl. Acad. Sci. U.S.A. *85*:6473–6477.

Strachan, T. (1990). Molecular pathology of congenital adrenal hyperplasia. Clin. Endocrinol. *32*:373–393.

Suarez, H. G. et al. (1988). Detection of activated *ras* oncogenes in human thyroid carcinomas. Oncogene *2*:403–406.

Surwit, R. S. et al. (1991). Control of expression of insulin resistance and hyperglycemia by different genetic factors in diabetic C57BL/6J mice. Diabetes *40*:82–87.

Taira, M. et al. (1989). Human diabetes associated with a deletion of the tyrosine kinase domain of the insulin receptor. Science *245*:63–66.

Tattersall, R. B. and Pyke, D. A. (1975). Diabetes in identical twins. Lancet *2*:1120–1125.

Thomsen, M. et al. (1985). MLC typing in juvenile diabetes and idiopathic Addison's disease. Transplant Proc. *22*:125–147.

Thomson, G. et al. (1988). Genetic heterogeneity, modes of inheritance, and risk estimates for a joint study of Caucasians with insulin-dependent diabetes mellitus. Am. J. Hum. Genet. *43*:799–816.

Tilanus, M. G. J. et al. (1986). An overview of the restriction fragment length polymorphism of the HLA-D region: its application to individual D-DR-typing by computerized analyses. Tissue Antigens *28*:218–227.

Tiwari, J. L. and Terasaki, P. I. (1985). Endocrinology: juvenile diabetes mellitus. In: HLA and Disease Associations. Tiwari, J. L. and Terasaki, P. I. (eds.). Springer-Verlag, New York/Berlin, pp. 185–212.

Todd, J. A. et al. (1987). HLA-DQ β gene contributes to susceptibility and resistance to insulin-dependent diabetes mellitus. Nature (London) *329*:599–604.

Todd, J. A. et al. (1988). A molecular basis for MHC class II associated autoimmunity. Science *24*:1003–1009.

Trowsdale, J. (1987). Genetics and polymorphism: class II antigens. Br. Med. Bull. *43*:15–36.

Tsutsumi, Y. et al. (1990). Demonstration of neurotensin messenger RNA in PC12 pheochromocytoma cells using synthetic oligonucleotides for hybridization and polymerase chain reaction. Biomed. Res. *11*:1–9.

Tusia-Luna, M. T. et al. (1990). Determination of functional effects of mutations in the steroid 21-hydroxylase gene (CYP21) using recombinant vaccinia virus. J. Biol. Chem. *265*:20916–20922.

Uchigata, Y. et al. (1992). Strong association of insulin autoimmune syndrome with HLA-DR4. Lancet *339*:393–394.

Ullrich, A. et al. (1985). Human insulin receptor and its relationship to the tyrosine kinase family of oncogenes. Nature (London) *313*:756–761.

Urabe, K. et al. (1990). Gene conversion in steroid 21-hydroxylase genes. Am. J. Hum. Genet. *46*:1178–1186.

Usala, S. J. et al. (1988). Tight linkage between the syndrome of generalized thyroid hormone resistance and the human c-*erb*A β gene. Mol. Endocrinol. *2*:1217–1220.

Wassmuth, R. and Lernmark, A. (1989). The genetics of susceptibility to diabetes [review]. Clin. Immunol. Immunopathol. *53*:358–399.

Weinberger, C. et al. (1986). The c-*erb* A gene encodes a thyroid hormone receptor. Nature (London) *324*:641–646.

Weiss, L. M. (1984). Comparative histologic study of 43 metastasizing and nonmetastasizing adrenocortical tumors. Am. J. Surg. Pathol. *8*:163–169.

Weiss, L. M. et al. (1989). Pathologic features of prognostic significance in adrenocortical carcinoma. Am. J. Surg. Pathol. *13*:202–206.

Westermark, P. and Johnson, K. H. (1988). The pathogenesis of maturity-onset diabetes mellitus: is there a link to islet amyloid polypeptide? Bioassays *9*:30–33.

Westermark, P. et al. (1987). Amyloid fibrils in human insulinoma and islets of Langerhans of the diabetic cat are derived from a neuropeptide-like protein also present in normal islets. Proc. Natl. Acad. Sci. U.S.A. *84*:3881–3885.

Wright, P. A. et al. (1989). Papillary and follicular thyroid carcinomas show a different pattern of *ras* oncogene mutation. Br. J. Cancer *60*:576–577.

Wu, J. et al. (1990). The genetic defect in multiple endocrine neoplasia type 2A maps next to the centromere of chromosome 10. Am. J. Hum. Genet. *46*:624–630.

Wyllie, F. S. et al. (1989). Structure and expression of nuclear oncogenes in multistage thyroid tumorigenesis. Br. J. Cancer *60*:561–565.

Xiang, K. S. et al. (1989). Insulin receptor and apolipoprotein genes contribute to the development of NIDDM in Chinese Americans. Diabetes *38*:17–23.

Yoshimasa, Y. et al. (1988). Insulin resistant diabetes due to a point mutation that prevents insulin receptor processing. Science *240*:784–787.

Part 5
Gene Level Evaluation of Selected Neurologic and Connective Tissue Disorders

Muscle, Bone, and Skin Disorders

5.1.1 DUCHENNE AND BECKER MUSCULAR DYSTROPHIES: OVERVIEW OF MOLECULAR BIOLOGICAL FINDINGS

The triumphant discoveries of the gene and its product, **dystrophin**, responsible for **Duchenne muscular dystrophy (DMD)** and a related milder, allelic form, **Becker muscular dystrophy (BMD)**, illustrate the power of the formerly-called "reverse genetics" in deciphering the long-recognized inherited conditions, the biochemical basis of which could not be otherwise identified. The history of the research efforts eventually crowned by the successful development of molecular tools allowing the unequivocal differentiation of these muscular dystrophies from a number of clinically similar diseases has been described (Darras, 1990; Dubowitz, 1989; Harper, 1988; Hodgson and Bobrow, 1989; Hoffman and Kunkel, 1989; Kunkel, 1989; Kunkel and Hoffman, 1989; Lachmann, 1989; Love et al., 1989; Oexle and Schliwa, 1989; Prior, 1991; Rowland, 1989; Witkowski, 1989).

A brief summary of the major findings related to the molecular biology of DMD/BMD will be given here only to emphasize the relative importance of specific diagnostic approaches of current and future clinical value. The diagnostic tools and protocols currently in use will be undoubtedly modified already in the near future as more detailed information becomes available on the fine structure of the exons of the *DMD* gene, and molecular tools such as appropriate primers for the polymerase chain reaction (PCR) analysis of the dystrophin gene alterations are developed.

Essentially three major groups of muscular dystrophies are recognized: X-linked, autosomal recessive, and autosomal dominant. The sex-linked muscle disorders include DMD, BMD, and **Emery-Dreifuss dystrophy.** The **autosomal recessive childhood muscular dystrophy, adult limb girdle dystrophy**, and **congenital muscular dystrophy** compose the autosomal recessive conditions. Among the **autosomal dominant** diseases are **myotonic dystrophy, fascioscapulohumeral dystrophy**, and **oculopharyngeal muscular dystrophy**. Although each of these entities can be described by specific clinicopathological features, the advances of molecular diagnostics necessitate some revision of the formerly held distinctions.

DMD/BMD is the first genetic disorder whose biochemical basis has remained unknown, and in which the identification of the gene led to the discovery of its principal gene product, although the pathobiology of the disease at the level of muscle still remains to be elucidated. DMD and BMD should be considered together because they share the same genetic locus, and share comparable molecular pathology. Clinical differences in clear-cut cases justify distinction between these two diseases, however, that require different management and genetic counseling approaches (Harper, 1988).

DMD, manifested predominantly in boys, with an incidence of 1 in 3000, is characterized by an **early onset** usually before the age of 3, incapacitating muscle weakness before age 12, frequently accompanied by mental retardation, cardiac involvement, and a dramatic increase in creatine kinase (CK; about 20,000 to 30,000 IU/ml). The death, usually following pulmonary complications, occurs at the age of 15 to 25. The milder form, known as **BMD** (incidence approximately 1 in 30,000), has a **variable time of onset**, often as late as early adult life, loss of ambulation occurring much later than in DMD patients, frequently in the third and fourth decade. Cardiac involvement and mental retardation are infrequent, CK elevations are more pronounced in childhood, and calf hypertrophy is often striking, more so than in DMD. Death rarely occurs before 30 years of age. **Up to one third of DMD/BMD cases are represented by new mutations** (Moser, 1984).

Uncommonly, carriers can manifest DMD/BMD. Differentiating such patients from those with autosomal recessive limb-girdle muscular dystrophy represents a diagnostic challenge (Norman and Harper, 1989). An estimated manifesting carrier frequency in Wales was found to be 2.5% with a prevalence of 1 in 100,000 of the female population, comparable to that of autosomal recessive limb-girdle muscular dystrophy. One of the proposed mechanisms of disease expression in the affected female is **uneven lyonization (X chromosome inactivation)** as demonstrated in a twin study on the basis of methylation differences of the paternal and maternal chromosome (Richards et al., 1990). Biochemical and ultrastructural studies performed over the past 30 years and aimed at identifying the primary defect responsible for the characteristic pathology failed to detect an abnormal muscle protein (for a review, see Rowland et al., 1989).

Application of retriction fragment length polymorphism (RFLP) analysis to X chromosomes isolated by flow cytometry (FCM), using several markers on unusual patients, led to the finding of deletions and translocations of the middle of the short arm of the **X (Xp21) chromosome** (Boyd and Buckle, 1986; Francke et al., 1985). Cytogenetic studies performed on deletion syndrome patients, in which muscular dystrophy was accompanied by mental retardation, enzyme deficiencies, and other concomitant disorders, narrowed the location of the *DMD* gene to Xp21, at least for cases of "simple" DMD (Kunkel et al., 1986).

Genetic and structural studies demonstrated that **both the *DMD* and *BMD* loci map at Xp21**, establishing that these two diseases represent allelic conditions (Kingston et al., 1984). **Identification of deletions** in DMD/BMD cases offers a definitive diagnosis in 65% of cases, which can be performed prenatally in pregnancies at risk. This offers a significantly better evaluation than probabilistic estimates based on linkage studies (Harper and Thomas, 1986).

Testing random cloned segments from the Xp21 region on the DNA from a particularly informative patient, derived by a special procedure called the **phenol emulsion reassociation technique**, allowed Kunkel and coworkers eventually to identify potential coding sequences for *DMD* (Monaco et al., 1986); confirmation from other researchers followed (Burghes et al., 1987). When the complete cDNA for *DMD* was cloned (Koenig et al., 1987), a 14-kb transcript, encoded by over 70 small exons spread over nearly 2500 kb of genomic DNA, was identified (Burmeister et al., 1988; Koenig et al., 1987; van Ommen et al., 1986). This proved to be the **largest known gene**, almost 10 times the upper limit of the average gene, its large size accounting for the high mutation rate (Beggs and Kunkel, 1990a,b). All of the exons have been tested with segments of cDNA as probes by Southern analysis, showing that about 65% of DMD and BMD patients had deletions or duplications of one or more exons of this gene (Forrest et al., 1987; Hu et al., 1988).

A survey for the presence of duplications conducted in 72 unrelated nondeletion patients analyzed by Southern blot with clones representing the entire *DMD* cDNA, and quantitative estimates of hybridization band intensity, revealed 10 cases with a duplication of part of the gene (Hu et al., 1990). This represented 6% of all cases. Restriction analysis identified a shift of the reading frames in some DMD cases, but not in BMD patients. Analysis of 258 independent deletions at the *DMD/BMD* locus revealed **a correlation between the phenotype and the type of deletion mutation**, confirming the "reading frame" theory in 92% of cases (Koenig et al., 1989). Many "in-frame" deletions of the dystrophin gene may not be detected because they fail to produce typical DMD/BMD phenotypes. Phenotypic characteristics of BMD have been related to the presence of minor in-frame alternatively spliced mRNA species giving rise to truncated dystrophin molecules (Chelly et al., 1990a,b). Some cases have also been identified among DMD patients who had slightly reduced amounts of in-frame truncated mRNA but no detectable dystrophin (Chelly et al., 1990b).

PCR-amplified dystrophin transcripts from muscle specimens of DMD and BMD patients demonstrated that **alternative splicing patterns correlated with the observed clinical phenotypes**. Although no precise correlation between clinical severity and specific deletion pattern was observed in 49% of families, a common pattern of deleted exons (0.5, 1.5, and 10 kb) has been related to a mild phenotype (Norman et al., 1990). Using a number of probes (**Cf23a, Cf56a**, and **Cf115**) to screen for deletions in the dystrophin gene, the pattern of deletions fails to predict the precise clinical course of the disease; **no correlation has been observed between the extent of the gene deletions and the severity of the disease** (Passos-Bueno et al., 1990). Duplications appear to originate more frequently in males than in females, and unequal sister-chromatid exchange is the most common mechanism responsible for the appearance of duplications in DMD and BMD.

Segregation analysis of almost 2000 cases from 4 continents performed on DMD patients diagnosed by clinical criteria, electromyography (EMG), biochemical studies, and muscle biopsy, but no DNA probes or dystrophin analysis, revealed a **significantly lower than theoretically predicted frequency of sporadic cases** (Barbujani et al., 1990). The simplest explanation of these findings visualizes the contribution of germinal mosaicism in the mothers that has been incorrectly attributed to segregation, unequal mutation rates in male and female gametes, and imprecision of diagnosis. It appears that more meaningful genetic analysis would be possible only, if the testing protocols developed by Kunkel and coworkers are adopted for definitive diagnosis in proband families (Beggs and Kunkel, 1990b).

What makes this possible is the eventual identification of the *DMD* **gene product** — a protein of 3685 amino acids with calculated molecular mass of 427 kDa, named "**dystrophin**" (Hoffman et al., 1987). This protein, located at the inner surface of the myofiber cell membrane, is related to other **cytoskeletal proteins, α-actinin** and **spectrin**, as determined by sequence analysis (Beggs and Kunkel, 1990b). Most significantly, evaluation of dystrophin presence and molecular weight in muscle biopsies from **patients with DMD consistently demonstrated the absence of detectable dystrophin**, whereas **patients with BMD invariably had dystrophin, which was altered either in size or quantity** (Bonilla et al., 1988; Hoffman et al., 1988, 1989). Dystrophin constitutes a major component of the sarcolemmal cytoskeleton in skeletal muscle (Ohlendieck and Campbell, 1991).

Tissue-specific expression of the dystrophin gene is also confirmed in studies with fibroblasts from DMD patients and controls (Brown et al., 1990). The comparative spreading properties on glass of monensin-treated fibroblasts (with inhibited glycoprotein synthesis responsible for adhesiveness) from DMD and normal individuals proved to be the same. This points to the myofibers as the predominant target cells in DMD/BMD, rather than other cell types. Molecular studies addressing the tissue-specific pattern of **dystrophin gene expression in muscle tissue and brain**, using PCR primed by oligonucleotides specific for the brain and muscle promoters, reveal that the **brain promoter is active only in neurons, whereas the muscle promoter operates in skeletal, cardiac, and smooth muscle as well as in glial and neuronal cells** (Chelly et al., 1990a).

Further studies of the subcellular localization and function of dystrophin in neurons will help define the role of this protein in the brains of some DMD patients with mental retardation. Indeed, analysis of the dystrophin gene promoter in the brain suggested that in certain patients a **deletion of either brain-specific or muscle-specific promoter might result in reduced dystrophin expression selective to brain or muscle** (Boyce et al., 1991). A BMD patient has been identified with a specific deletion of the dystrophin muscle promoter, and the authors predict that specific loss of the brain promoter may account for some cases of X-linked mental retardation. In the central nervous system (CNS), dystrophin is localized at postsynaptic membrane sites, and alterations in its level at those sites may be the basis for the cognitive impairment observed in DMD (Lidov et al., 1990). Important observations related to the development of genetic engineering and myoblast implantation for therapy of DMD point to a **severely reduced replicative life-span of DMD myoblasts** compared to normal controls (Webster and Blau, 1990). At age 7, DMD myoblasts were rarely capable of 10 doublings compared to 56 doublings from a 5-year old normal control, apparently reflecting the extensive division of DMD myoblasts in an attempt to regenerate degenerating myofibers (Webster and Blau, 1990).

Computer-generated reconstruction of dystrophin based on the protein sequence suggests a rodlike protein consisting of a long central domain that separates an **N-terminal actin-binding domain** from a **cysteine-rich domain** and a **C-terminal domain** (Koenig et al., 1988). It appears that deletions or alterations involving the cysteine-rich and C-terminal domains have **the most severe pathological consequences** (Koenig et al., 1989; England et al., 1990). On the other hand, mutations affecting the spectrin-like structure of the dystrophin molecule (but which are not involved in interactions with other muscle proteins, such as actin) and extending to 46% of the molecule produce only **very mild muscular dystrophy** (England et al., 1990).

Molecular analysis of the dystrophin complex revealed the presence of four glycoproteins (**156K**, **50K**, **43K**, and **35K**), integral components of the dystrophin oligomeric complex (~18S), and the concentration of one of these is markedly reduced in DMD patients (Ervasti et al., 1990). **The reduction of the 156K glycoprotein may be the first step in the molecular pathogenesis of DMD**.

The exact **biological function of dystrophin** remains unknown, although a functional, full-length (14 kb) cDNA for the protein has been constructed and shown to be expressed as a dystrophin molecule indistinguishable from normal dystrophin, being localized in the cell membrane (C. C. Lee et al., 1991). This creates an opportunity not only to study better the function of dystrophin, but also to consider approaches to gene therapy (C. C. Lee et al., 1991). In fact, experimental transfection, in which dystrophin DNA constructs are introduced into *mdx* mice, appears to hold promise, as the injected genes show surprising longevity and functionality (Acsadi et al., 1991).

Further evolution of our understanding of the molecular biological events responsible for DMD pathology, in particular the susceptibility of muscle fibers to necrosis comes from the deciphering from cDNAs of the primary sequences of two components of the dystrophin-glycoprotein complex (Ibraghimov-Beskrovnaya et al., 1992). The 43K and 156K **dystrophin-associated glycoproteins** (DAGs) are encoded by a single mRNA that is expressed at normal amounts in **mdx** and DMD skeletal muscle. Both glycoproteins, particularly the 43K DAG, however, are greatly reduced in dystrophin-deficient muscle. The 156K DAG binds **laminin** and thus links the sarcolemmal cytoskeleton to the extracellular matrix. The 43-156K DAG is termed **dystroglycan** (Ibraghimov-Beskrovnaya et al., 1992). It is believed that absence of dystrophin in DMD leads to the loss of dystroglycan binding muscle fibers to the extracellular matrix; this disruption underlies the various physiological processes resulting in eventual muscle cell necrosis. The normal production of dystroglycan in dystrophic muscle may be important for potential DMD therapies (Ibraghimov-Beskrovnaya et al., 1992).

5.1.1.1 STRATEGIES FOR DUCHENNE AND BECKER MUSCULAR DYSTROPHY TESTING

Importantly, the vast majority of patients with muscular disorders other than DMD and BMD have, as do normal controls, unaltered levels of normally appearing dystrophin. Essentially two groups of diagnostic approaches are used for DMD/BMD: (1) assessment of the gene product, dystrophin, by Western blot and immunofluorescence, on the one hand; and (2) evaluation of the gene by Southern blotting with cDNA probes, PCR, or RFLP-based linkage studies, on the other hand (Beggs and Kunkel, 1990b).

5.1.1.1.1 Immunoblotting

For **dystrophin immunoblotting** only 10 to 15 mg of tissue is needed, so needle or open biopsy can be used provided the tissue is immediately transferred to −70°C. Following sodium dodecyl sulfate (SDS) electrophoresis and transfer to the nitrocellulose membrane, the protein is detected by specific antibodies and visualized by an immunoperoxidase technique (Hoffman et al., 1988). The **complete absence of dystrophin is predictive of DMD with over 99% accuracy**, whereas the **reduction of dystrophin** compared to normal levels, and/or **the detection of larger or smaller sized molecules** of the protein by SDS electrophoresis, **establishes a diagnosis of BMD with over 95% accuracy** (Hoffman et al., 1988, 1989). An intermediate phenotype has been recognized with dystrophin levels under 20% of normal levels; the patients (so-called "outliers") become wheelchair bound between the ages of 13 and 20 years (Brooke et al., 1983). In some cases, testing with multiple antisera is needed to establish the diagnosis of BMD (Bulman et al., 1989).

There is a need for immunological tools capable of specifically identifying the integrity of the cysteine-rich and C-terminal domains of dystrophin. Because cross-reaction of polyclonal antisera raised against the C-terminal portion of dystrophin with unidentified, less abundant proteins of the same molecular weight has been observed the need for monoclonal antibodies is evident.

A cDNA fragment coding for the last 485 amino acids (which include the cysteine-rich domain and the entire C-terminal domain) was used to derive a recombinant fusion protein (Ellis et al., 1990). It was then used to immunize mice for the production of a **monoclonal antibody** (MAb). Two classes of MAbs were obtained. One stained dystrophin in normal muscle extracts and, similar to polyclonal antisera, cross-reacted in dystrophic muscle with a lower abundance protein of the same size as dystrophin (400 kDa). Also isolated were 17 MAbs that were dystrophin-specific and stained nothing in the dystrophic muscle. Six of the dystrophin-specific MAbs also strongly stained dystrophin on frozen sections from normal, but not Duchenne, muscle. MAbs against the central rod region of the dystrophin molecule have also been developed (Nguyen Thi Man et al., 1990).

Sensitivity of the immunoblotting assay for dystrophin using polyclonal antisera is sufficient to diagnose DMD and BMD in boys, but it is not capable of detecting the smaller reductions in dystrophin level in most DMD carrier females (Beggs and Kunkel, 1990b). It is not yet known whether MAbs, such as those developed by Ellis and coworkers (1990), can provide the required sensitivity for immunoblotting detection of DMD carrier females.

Alternatively, culturing of myoblasts from suspected carriers can be performed to allow dystrophin expression on clonal populations of cells that have been induced to fuse into myotubes (Hurko et al., 1989; Oronzi-Scott et al., 1988). In spite of the length and expense of this method, Beggs and Kunkel (1990b) believe it "may be the most sensitive approach to carrier detection for DMD and BMD". Undoubtedly, more information will become available on immunological testing for DMD and BMD, so that such specialized testing becomes an established modality in the major clinical centers. At present, at least one commercial laboratory offers its service for comprehensive dystrophin evaluation (Genica Pharmaceuticals Corporation, The Neuromuscular Disease Company, Worcester, MA). Association of the leading researchers in the field with this company ensures competent evaluation of the submitted cases. A large number of laboratories already offers testing for dystrophin gene mutations.

5.1.1.1.2 Immunofluorescence

Dystrophin visualization in biopsy material or frozen sections from normal individuals reveals the sarcolemmal protein present as a homogeneous ring around the periphery of all muscle fibers (Arahata et al., 1988; Bonilla et al., 1988; Ellis et al., 1990; Zubrzycka-Gaarn et al., 1988). Muscle fibers from DMD patients show **no immunofluorescent staining**, whereas in specimens from BMD patients the staining pattern varies from normal to patchy and is significantly lighter (Arahata et al., 1989a). In DMD carriers, patches of negative fibers are scattered among positive fibers. The pattern of expression of dystrophin in muscle biopsies from symptomatic carriers of DMD can be identified by a distinct mosaic staining of surface membranes of fibers when 2 to 8% of fibers show partial immunofluorescence (Arahata et al., 1989b). Dystrophin staining pattern can be successfully analyzed even in freeze-dried tissue (Khurana et al., 1991).

Sampling errors, or nonrandom X chromosome inactivation with selective loss of dystrophin-negative fibers, may produce false-negative results in which all the fibers stain positively. At this time, immunofluorescence does not identify BMD carriers (Beggs and Kunkel, 1990b).

5.1.1.1.3 Detection of Dystrophin Gene Mutations

At present, two methods are used to detect dystrophin gene mutation: Southern blotting and PCR, which identify alterations of the dystrophin gene. DNA can be amplified by **multiplex PCR from Guthrie spots** (McCabe et al., 1990). Alternatively, DNA isolated from peripheral blood is probed with cDNA on **Southern blots** and, for the 65% of DMD/BMD patients who have detectable mutations (duplications or deletions), **this test is definitive**; it is also capable of carrier detection and prenatal diagnosis, the latter on either chorionic villi or amniocentesis-derived cells. Southern analysis with cDNA probes allows almost quantitative assessment of the missing or duplicated exons, permitting an **estimate of the severity of a patient's disease** in about 92% of cases. If the mutation introduces a frameshift in the protein translation, the disease phenotype is that of DMD with little or no functional dystrophin produced. In BMD, mutations usually occur "in-frame", producing an internally deleted or duplicated, partially functioning protein (Koenig et al., 1989; Malhotra et al., 1988).

Because dystrophin gene deletions preferentially cluster at the "hot spots" — the 5' end and at 7 to 8 kb on the cDNA — a rapid screening of the deletions can be performed by multiplex PCR analysis with primers for only a subset of exons (Beggs and Kunkel, 1990b). Two regions within the hot-spot region of the dystrophin gene have been tested by PCR in 42 DMD or BMD patients (Hentemann et al., 1990). Deletions were found in 16.6% of patients. This, essentially a methodological study, compared the suitability and reliability of PCR in deletion screening and prenatal diagnosis. Oligonucleotide primers specific for exon sequences were used in multiplex DNA amplification and Southern blotting. The simultaneous amplification of several deletion-prone exons in the dystrophin gene had been previously developed in the effort to detect the majority of the DMD deletions (Chamberlain et al., 1988).

Confirming earlier reports of interference of maternal cells in prenatal screening of deletions by PCR, as could occur in chorionic villi samples, the highest level of allowable maternal DNA contamination was shown to be under 2% when amplification was limited to 25 cycles (Hentemann et al., 1990). If more than 30 cycles were performed, as low as 0.2% of contaminating DNA affected the results, producing a visible amplification product. The screening for mutations in the dystrophin gene should be confined to coding regions of the dystrophin gene, simultaneously amplifying several exon sequences in a rapid screen (Speer et al., 1989).

Southern analysis of the dystrophin gene using cDNA requires up to seven different subcloned probes to examine at least two different restriction digests of genomic DNA (Forrest et al., 1988; Hu et al., 1988; Koenig et al., 1987). Since most of the deletions cluster in two hot spots, a subset of approximately 70 exons is sufficient to identify the majority of deletions. The PCR multiplex system developed by Chamberlain and coworkers (1988) amplified six and, subsequently, nine exons predominantly deleted in the majority of DMD/BMD patients. Another nine exons can be amplified with other primers to cover all of the known deletions in BMD, and over 97% of deletions in DMD patients (Beggs et al., 1990). Thus, PCR-based detection of carriers, prenatal diagnosis, and mutation detection can be performed substantially faster than the conventional Southern blot analysis. Some of the advantages of the PCR system include the ability of PCR to work even on highly degraded DNA unsuitable for Southern blotting. The entire procedure can be performed in 1 or 2 days with a sensitivity of up to 98%, thus virtually eliminating the need for Southern blotting in the 65% of patients with deletions (Beggs et al., 1990).

The assay can be run on fixed, or even embedded, tissue specimens such as those from deceased patients and relatives. The PCR does not eliminate the need for linkage studies on the remaining 35% of DMD/BMD patients, for whom polymorphic markers must be used. Several dystrophin gene polymorphisms have already been developed (Beggs and Kunkel, 1990b; Roberts et al., 1989), and further assays along these lines will allow the entire dystrophin gene analysis to be conducted using the PCR technique.

For diagnostic purposes, however, the more reliable Southern blot analysis should be performed at this time. The importance of performing diagnostic DNA studies in cases of undetermined muscle diseases is underscored by the finding of mutations within the dystrophin gene in what is clinically described as chronic limb-girdle syndrome, "atypical" spinal muscular atrophy, and other conditions (Clarke et al., 1989; Lunt et al., 1989: Norman et al., 1989a,b). Analysis of molecular deletions in DMD and BMD patients, using the **entire cDNA for the *DMD* gene as hybridization probes**, has been combined with segregation analysis for prenatal and carrier diagnosis in 17 DMD Japanese families (Sugino et al., 1989). An estimated diagnostic rate proved to be 63% of that actually confirmed in the population tested.

Determination of **carrier status in DMD/BMD** by Southern blot analysis relies on the assessment of the relative abundance/decrease in specific bands detected by an array of cDNA probes. Efforts have been made to quantitate this technique through **densitometry of the autoradiographic bands** (Prior et al., 1989), and the use of **multiple probes** (Mao and Cremer, 1989). Given the difficulty of reproducing high-quality Southern blots as well as the intrinsic limitations of the technique, when new deletions occur in families for which probes may not yet be available, **quantitative PCR** (Chamberlain et al., 1988) using allele-specific oligonucleotide (ASO) primers (Prior et al., 1990) offers several advantages. Not the least of these is its speed and the possibility of automation.

In order to meet the challenge of identifying the full spectrum of DNA sequence alterations possible in diseases with high new mutation rates, **a combination of three PCR-based techniques is advocated** (Grompe et al., 1990). The first technique, **multiplex amplification**, allows the rapid detection of alterations in a large-sized gene such as the *DMD* gene. The second method, **chemical mismatch cleavage**, is used for screening DNA fragments up to 1.5 kb in size. The full spectrum of mutations is detected by a third approach, **direct automated sequencing**, applicable to DNA pieces up to 350 bp. Although currently applied to a few specific genetic conditions (DMD is evaluated by the first technique, ornithine transcarbamylase [OTC] deficiency by the second, and Lesch-Nyhan syndrome by the third), a combination of all three techniques is recommended for routine, rapid diagnosis of any genetic conditions for which the defective gene had been cloned. Grompe and coworkers (1990) offer **an algorithm for diagnostic strategies in new mutation disorders** (Figure 5.1). Extending this general approach, a **multiplex PCR-based linkage analysis of DMD families** has also been developed (Grompe et al., 1990).

A **nonradioactive method** for the detection of deletions or duplications in the dystrophin gene has been proposed based on **peripheral blood lymphocyte mRNA** (Roberts et al., 1990). This method avoids the complication of carrier detection by gene dosage estimation from Southern blots. The entire 11 kb coding region of the dystrophin mRNA was amplified with a set of 10 reverse transcription and overlapping nested PCR reactions, each covering an average size of 1.1 kb segment of the gene. The amplified products were directly visualized on acrylamide minigels by ethidium staining, and **major mutations were identified by the appearance of a band of different size than that of the wild type**. In addition to offering a

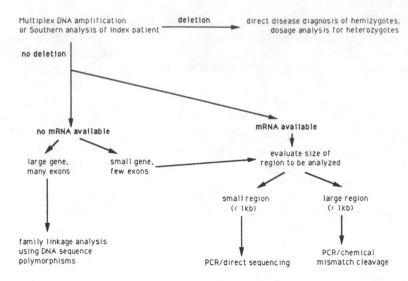

FIGURE 5.1. Strategies for diagnosis in new mutation diseases. (From Grompe et al. In: Biotechnology and Human Genetic Predisposition to Disease. Wiley-Liss, New York, p. 56. Copyright ©1990. Reprinted by permission of Wiley-Liss, a division of John Wiley & Sons, Inc.)

convenient carrier detection method, the described procedure allows the production of a sufficient amount of amplified products for sequencing.

This approach may help to further define the relationship between the effect of deletions or duplication on disease severity, since not all of the observed cases can be adequately explained by the **frameshift mutation hypothesis** (Monaco et al., 1988). As mentioned above, it stipulates that the phenotype severity is determined by whether the gene alteration affects the **open reading frame (ORF)**: if it does, a truncated dystrophin is produced and the DMD phenotype ensues; when the mutation does not affect the ORF, a functionally defective protein with an interstitial deletion or duplication leads to a mild (Becker) phenotype. Deletions in exons 45 to 47 have been identified by PCR analysis in lymphocytes of a BMD patient and a heterozygous carrier (Schloesser et al., 1990).

A specific deletion, removing 52 bp out of 88 bp of exon 19 of the dystrophin gene, has been detected in a Japanese patient whose mother and sister have similar allelic defects (Matsuo et al., 1990). This frameshift mutation introduces a termination codon in exon 20, and is predicted to result in a truncated protein. The smallest deletion of the *DMD* gene reported so far is designated DMD-Kobe, classified as a new type of *DMD* gene abnormality (Matsuo et al., 1990). It appears that **alternative splicing may account for the previously unexplained phenotypic variants in some patients that do not fit the frameshift hypothesis** (Roberts et al., 1990).

An interesting approach to screening for DMD diagnosis is offered by a combination of PCR amplification of the dystrophin gene exon 45 with the **single strand "conformational" polymorphism (SSCP)** analysis of the amplified product (Zietkiewicz et al., 1992). The absence of amplification denoted the presence of the characteristic deletion, whereas SSCP data helped to resolve the carrier status of the patient's mother and his sister both of whom showed a heterozygous pattern. This approach appears to provide the only qualitative positive identification of non-carrier status of mothers of sporadic cases of dystrophin gene deletions (Zietkiewicz et al., 1992).

Yet another approach to performing linkage analysis in DMD families is the detection of CA repeats which can be amplified by PCR; their location at the extreme 5′ terminus of the dystrophin gene coupled with their high polymorphism can be easily utilized (Feener et al., 1991).

5.1.1.1.4 Linkage Analysis of DMD/BMD

As noted above, in 65% of patients with DMD or BMD the alteration (deletion or duplication) of the dystrophin gene affects the exons (Beggs and Kunkel, 1990b; den Dunnen et al., 1989). In other cases, point mutations or other alterations occur in noncoding sequences, and may require indirect identification methods. It requires, however, an appropriate number of individuals from a family with a positive history of the disease, and extensive testing of multiple DNA samples by restriction analysis (Bushby et al., 1991; Greenberg et al., 1991; Laing et al., 1991). **Amino- and carboxy-terminal antisera specific for dystrophin have been found inadequate for differential diagnosis of disease severity based solely on dystrophin quantitation** (Bulman et al., 1991).

At present, **19 intragenic RFLPs** within the cDNA have been identified using genomic probes (Darras and Francke, 1988; Forrest et al., 1988; Koenig et al., 1987; Laing et al., 1991; Liechti-Gallati et al., 1990; Walker et al., 1988). A systematic survey of DMD and BMD patients with intronic probes, and with cDNA probes covering three fourths of the coding sequence, identified 45 molecular deletions within the dystrophin gene (Gilgenkrantz et al., 1989). Forty-two percent of the breakpoints were located in the intronic sequence identified by probe P20; the rest of the deletions were scattered in the more proximal part of the gene. In BMD patients, most of the deletions were in the intronic sequences around P20. Surprisingly, disruption of the reading frame was also identified by pulsed-field gel electrophoresis in three BMD patients.

Responding to requests of 53 pregnant women for **prenatal testing for DMD/BMD carrier status**, DNA RFLP was performed and proved informative in 51 cases (Cole et al., 1988). Prenatal diagnosis utilizing dystrophin analysis for establishing DMD in the fetus proved of value in a case of doubtful maternal carrier status as assessed by RFLP (Boelter et al., 1990). A number of important diagnostic pitfalls including nonpaternity, karyotypic anomalies, and gonadal mosaicism have been found to complicate carrier detection and prenatal diagnosis for DMD and BMD with DNA analysis (Bakker et al., 1989).

The use of **intragenic RFLP analysis** for carrier risk estimates on 100 possible female carriers enabled 78% of possible carriers to be assigned high or low risks (Kelly et al., 1990; Yoshioka et al., 1990). Multiple probes are to be used to generate probabilistic estimates of the likelihood of inheritance of a mutation. A variety of PCR-based assays are being developed to allow for analysis of polymorphisms informative in constructing linkage relationships (Grompe et al., 1990; Roberts et al., 1989; Weber and May, 1989).

New methodologies being developed around the world will undoubtedly modify the currently advocated approaches for analysis of DMD/BMD as well as other conditions resulting from gene alterations. Clinical evaluation of DMD/BMD families should include a genetic register for X-linked muscular dystrophies, banking of DNA samples, thorough carrier testing by creatine kinase analysis, deletion testing, linkage prediction, and counseling (Harper, 1988).

An algorithm combining linkage studies and direct detection of gene alterations by PCR technique currently employed by Beggs and Kunkel (1990b; Beggs et al., 1990) in their evaluation of clinical cells is shown in Figure 5.2. Dosage analysis has proved to be a suitable alternative, and in some cases even more efficient than linkage, to determine carrier status of female relatives of DMD patients known to have a deletion within the *DMD* gene (Bejjani et al., 1991).

One of the difficulties interfering with genetic counselling and prenatal diagnosis of DMD is the high frequency of recombinational and mutational events in the relatively large dystrophin gene (Shomrat et al., 1992). In order to overcome these difficulties it is possible to combine genomic and cDNA probes for this gene. This allows to make distinctions between mutant and normal dystrophin alleles in families with single cases and normal CK levels in all females as well as in families seeking prenatal diagnosis when the affected member may not be alive (Shomrat et al., 1992).

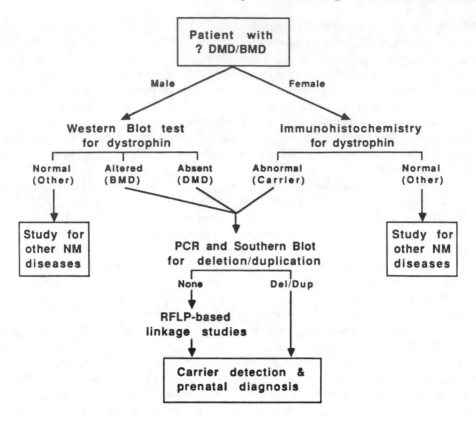

FIGURE 5.2. Flow diagram illustrating the protocol recommended to establish a "molecular diagnosis" of DMD or BMD. (Reproduced from the *Journal of Clinical Investigation*, 1990, Vol. 85, pp. 613-619, by copyright permission of the American Society for Clinical Investigation.)

The detection of dystrophin gene deletions by amplification of mRNA using RT-PCR on lyophilized biopsy tissue offers a higher level of resolution than Southern or multiplex PCR analysis of DNA (Anderson et al., 1992).

5.1.2 MYOTONIC DYSTROPHY

Myotonic dystrophy (DM) is the most common form of adult muscular dystrophy with a variable age of onset and clinical severity (Harper, 1979). This autosomal dominant disorder varies in its manifestation from a congenital form that is lethal in the newborn period to an asymptomatic condition with normal life span (Streib et al., 1987). In affected cases, multisystem involvement is common. The incidence varies from 2 to 14 cases per 100,000 to 1 in 617 among French-Canadians (Korneluk et al., 1989).

Linkage studies in this disorder point to the proximal long arm of chromosome 19 in the proximity of the apolipoprotein CII gene (*apo*-CII) and the enzyme peptidase D gene (for a review, see Harper, 1988). Earlier studies on chromosome 19, and the methodological aspects of the characterization of the closest possible approximation of the *DM* gene, have been reviewed (Milunsky et al., 1991; Roses and Bartlett, 1989; Shaw and Harper, 1989).

Subsequent analysis of 421 persons in 53 DM families, using 7 **DM-linked DNA markers** combined with somatic cell hybrid studies, assigned the *DM* gene locus to a position more distal on the **chromosome 19** long arm than previously assumed (Korneluk et al., 1989). Furthermore, having unequivocally identified DNA markers flanking the disease locus, a 10-cM region of chromosome 19 was established that contains the *DM* locus. These findings may prove helpful in identifying the causative gene and the molecular diagnosis of DM. In fact, a number of new probes have been assigned close to the *DM* locus using human/hamster hybrid cell lines containing defined breakpoints in the region close to the **gene for muscle-specific creatine kinase (*CKMM*)** (Johnson et al., 1989, 1990; Norman et al., 1989a).

Using **probes for *CKMM*, *apo* CII** and the repair genes ***ERCC*1** and ***ERCC*2**, and separation of large-sized DNA restriction fragments by **field inversion gel electrophoresis**, a long-range restriction map of the 19q13 region has been constructed (Smeets et al., 1990). The *DM* locus is contained in this region. The *CKMM* gene and both DNA repair genes are physically linked within a 250-kb segment of the 19q13 region that does not include the *apo* CII gene.

In the meantime, using the genetic markers for *CKMM* and *apo* CII, four families have been tested by linkage analysis for DM (Speer et al., 1990). **Genetic linkage studies in DM as in other diseases can establish the carrier status, but may not predict the severity and progression of the disease**, especially because DM is known for its variable expressivity and age-dependent penetrance. A novel **chromosome 19** marker, *ERCC*1, which shows close linkage to DM, has been characterized (Shutler et al., 1991). **PCR-based protocols** have been developed for the direct identification of *CKMM* haplotypes (Lavedan et al., 1990).

The variable nature of severity and age of onset of MD has been related to the recently identified novel restriction fragment of larger than normal length present in affected individuals (Harley et al., 1992). The **unstable fragment** of DNA characteristic of DM individuals has been identified and confirmed by several research teams (Brook et al., 1992; Buxton et al., 1992; Harley et al., 1992) and the putative defect has been mapped (Aslanidis et al., 1992). The characteristic **CTG triplet repeat** that is found in normal individuals in 5 to 27 copies is expanded to at least 50 repeats in minimally affected MD patients. In severely affected individuals the triplet is amplified up to several kilobase pairs (Brook et al., 1992). The CTG repeat is located in the 3′ untranslated region of an mRNA that encodes a polypeptide of the protein kinase family expressed in tissues of MD patients (Brooks et al., 1992; Mahadevan et al., 1992). These proteins are known to modulate the activity of excitable cells by affecting ion channels, glycogen and lipid metabolism and by modifying gene expression through phosphorylation (Brooks et al., 1992; Mahadevan et al., 1992).

By defining the physical map location and the characteristic RFLPs, the **molecular diagnosis of DM becomes possible**. Moreover, **the larger the unstable genomic fragment the more severe the disease symptomatology** (Aslanidis et al., 1992). The amplification of the unstable portion of the DM gene is found also in Japanese patients which, similar to Caucasians, display an amplification from a few hundred bp to 5 kb, and again the increase in size correlates with the severity of symptoms (Yamagata et al., 1992).

5.1.3 CHRONIC SPINAL MUSCULAR ATROPHY AND X-LINKED SPINAL AND BULBAR ATROPHY

Spinal muscular atrophy (SMA) is a group of heritable degenerative diseases that selectively affects the α-**motor neurons**. SMAs with onset in childhood are second to cystic fibrosis among autosomal recessive disorders and are the leading cause of heritable infant mortality (Brzustowicz et al., 1990). Linkage analysis of 13 clinically homogeneous SMA families indicates that "chronic" childhood-onset SMA including its variants (intermediate, type II, and Kugelberg-Welander, type III SMAs) is genetically homogeneous, with the causative gene mapping to chromosomal region **5q11.2-13.3** (Brzustowicz et al., 1990; Gilliam et al., 1990).

Research on the most common and severe form, type I (Werdnig-Hoffmann), unfortunately is hampered by the lack of pedigrees of sufficiently large size. Linkage analysis of the chronic forms of the disease (types II and III) in 24 multiplex families led to the assignment of the locus for the chronic forms to the long arm of chromosome 5 (5q12-q14) (Melki et al., 1990a). Again, no genetic heterogeneity was found between type II and III forms of the disease, based on the use of the **anonymous DNA marker D5S39,** suggesting that these are allelic disorders.

When the same marker was used for linkage analysis of 25 families with the acute type I form of SMA, the mutant gene was found to be tightly linked to the D5S39 locus. Thus, it appears **that all three forms of SMA, formerly distinguished on the basis of onset and phenotypic severity, are due to different mutations at a single locus on chromosome 5** (Melki et al., 1990b).

It can be expected that, although the biochemical defect underlying this disorder is still unknown, the strategies so successfully used in deciphering DMD/BMD molecular pathology can also be applied to this cluster of allelic conditions in the near future.

Another condition — **X-linked spinal and bulbar muscular atrophy (Kennedy's disease)**, which is an adult-onset form of motoneuron disease has been shown to be related to androgen insensitivity, and the **androgen receptor (AR) gene** appears to be a candidate gene for this disease (La Spada et al., 1991). By analogy with fragile X syndrome and DM, this gene located at the proximal arm of the X chromosome has been shown to harbor mutations consisting of an increase in the size of a **polymorphic tandem CAG repeat** in the coding region (La Spada et al., 1991) .

A very strict correlation is observed between the number of these repeats and the presence of the disease: the affected persons have enlarged segments (ca. 500 bp), whereas their unaffected siblings have fragment sizes of normal lengths (405–450 bp). Carrier females (obligate heterozygotes) have both an enlarged fragment and a normal fragment (La Spada et al., 1991). It is believed that these repeats encode a **polyglutamine motif** that may prevent the AR from functioning normally in motoneurons (where ARs are normally located) thereby leading to degeneration of these cells — a characteristic feature of this disease (La Spada et al., 1991).

5.1.4 FACIOSCAPULOHUMERAL MUSCULAR DYSTROPHY (LANDOUZY-DEJERINE DISEASE)

Facioscapulohumeral muscular dystrophy is an autosomal dominant disorder, with a prevalence of 1 in 20,000; it manifests at the end of the first decade, or in the second decade, with **facial or shoulder girdle weakness**. The biochemical defect has been traced to deficiency of complex III of the mitochondrial respiratory chain that does not appear to result from major deletions or rearrangements (Slipetz et al., 1991). The rate of progression of the disease varies within families (Lunt, 1989; Lunt et al., 1989). Following extensive linkage studies that excluded almost 90% of the genome, and erroneous assignment to chromosome 14 (Harper, 1988), the use of **microsatellite marker Mfd22 (D4S171)** allowed assignment of the causative gene to **chromosome 4** (Wijmenga et al., 1990). The microsatellites are evenly distributed all over the genome, which makes them highly useful for linkage analysis. **This study represents the first successful attempt to locate a human disease gene with the help of this type of marker**. The positive linkage results argue against genetic heterogeneity being responsible for the phenotypic heterogeneity of this disease, which is frequently associated with other symptoms such as hearing loss and retinal vasculopathy (Wijmenga et al., 1990).

Using the microsatellite DNA marker D4S171, which maps to chromosome 4, and **probe D4S139** on the distal region of 4q, **presymptomatic detection** and **prenatal diagnosis** of facioscapulohumeral disease have been shown to be clinically useful (Lumley et al., 1990; Wijmenga et al., 1991).

5.1.5 EMERY-DREIFUSS MUSCULAR DYSTROPHY

The gene for this X-linked disorder, characterized by relatively **benign muscle wasting** of the humeral and peroneal muscle groups as well as limb girdle muscle and involving the heart at an early stage, has been firmly assigned to the distal arm of the X chromosome (Harper, 1988).

Linkage analysis with seven Xq28 marker loci suggested that the candidate gene for this condition is located in distal Xq28, distal to DXS305 (Consalez et al., 1991).

5.1.6 MARFAN SYNDROME

This serious connective tissue disorder is inherited as an autosomal dominant trait, with the diagnosis usually made clinically during childhood and sometimes in adulthood (Tsipouras, 1990). The **heterogeneous manifestations** of the syndrome, affecting a variety of organ systems, may have more than one cause and more than one mutant gene locus. In spite of over 30 years of intensive research, and the long-held view that it represents a prototypical heritable disorder of the connective tissue, and is a **prime example of gene pleiotropy** (Tsipouras, 1990), more recent progress has dramatically changed the course of research.

Hollister and collaborators (Godfrey et al., 1990a,b) have concentrated their efforts on the **microfibrillar fibers** accumulating in Marfan skin and produced by cultured Marfan syndrome fibroblasts. Using several monoclonal antibodies (MAbs) against **fibrillin**, a major structural protein of microfibrils, tissues from affected individuals were examined by indirect immunofluorescence. There was a consistent deficient accumulation, and/or production and assembly, of immunoreactive fibrous components (Hollister et al., 1990). In a single-blind study (performed without knowledge) of clinical findings on 27 patients with Marfan syndrome, 24 were correctly identified by the decreased content of microfibrillar fibers in their skin, cultured fibroblasts, or both. On the other hand, all 13 normal subjects as well as 19 of 25 patients with other heritable connective tissue disorders were correctly identified as "non-Marfan" cases.

Topical alterations of the microfibrillar fibers account for the unilateral manifestations of the Marfan syndrome as demonstrated by marked reduction in microfibrillar antigen in tissues taken from the affected side (Godfrey et al., 1990b). Assuming that abnormalities in the microfibrillar component of elastic fiber systems play a role in the etiology and pathogenesis of the Marfan syndrome, an analysis of the segregation pattern of this specific structural deficiency has been performed in affected families (Godfrey et al., 1990a). In all cases, immunofluorescent abnormalities cosegregated with the Marfan phenotype, while **all nonaffected family members did not show fibrillin abnormalities**.

The next step will be cloning of the fibrillin gene, and testing of typical and variant Marfan syndrome cases. This direction of research is promising, although a few unanswered questions remain (Tsipouras, 1990). Thus, **fibrillin cannot account for all mutations in the pleiotropic phenotype of Marfan syndrome** as demonstrated by the immunofluorescent studies. Also, it appears that **fibrillin immunofluorescence is not absolutely specific for Marfan syndrome**, because false-positive results have been observed with other connective tissue disorders—**cutis laxa and homocystinuria** (Hollister et al., 1990). If not related to technical artifacts, the different immunofluorescent patterns between the skin and fibroblast cultures of Marfan patient samples could possibly be explained by allelic heterogeneity (Tsipouras, 1990).

In the meantime, an important observation has been made in a linkage analysis of five families with Marfan syndrome that allowed the assignment of the defective gene to **chromosome 15** at **D15S45** (Kainulainen et al., 1990). One may notice that chromosome 15 contains the genes coding for **type I collagen receptor, chondroitin sulfate proteoglycan I core protein**, and **cardiac muscle X-actin**, which could be viewed as candidate genes for the mutation in Marfan syndrome (Kainulainen et al., 1990). Further molecular studies suggest that the Marfan syndrome locus on chromosome 15 is apparently flanked by D15S28 and D15S49 (Tsipouras et al., 1991). Moreover, a **partial sequence of a candidate gene for the Marfan syndrome has been defined** (Maslen et al., 1991), *de novo* missense mutations have been detected in the fibrillin gene in some Marfan syndrome cases (Dietz et al., 1991), and DNA polymorphism in the fibrillin gene has been closely linked to sporadic cases of Marfan syndrome (Dietz et al., 1991; B. Lee et al., 1991). How these findings will be reconciled with the impressive demonstration of the fibrillin association with the disease will be seen only in detailed molecular studies on the tentative gene(s) structure and function.

Using specific markers for the **fibrillin genes** — a $(TAAAA)_n$ for the fibrillin gene on **chromosome 15 (fibrillin 15)** and a $(GT)_n$ for the fibrillin gene on **chromosome 5 (fibrillin 5)** — and primers specific for these genes, PCR amplification studies were performed in 28 families with the Marfan syndrome and 8 families with other phenotypically related disorders — **congenital contractural arachnodactyly, ectopia lentis, mitral-valve prolapse syndrome**, and **annuloaortic ectasia** (Tsipouras et al., 1992). Genetic linkage analysis established an association between the Marfan syndrome and only the fibrillin 15 suggesting that **this syndrome is produced by mutations in a single fibrillin gene on chromosome 15**. Consequently, the diagnosis of the Marfan syndrome by genetic linkage and molecular analysis becomes possible in many families (Tsipouras et al., 1992).

5.1.7 OSTEOSARCOMA

Osteosarcoma (OS) is the most common malignant tumor of the bone, excluding plasma cell myelomas, arising most frequently in the second decade of life and affecting the knee area, particularly in the distal femoral metaphysis (Rosai, 1989).

5.1.7.1 CHROMOSOME ABERRATIONS

It has long been recognized that among the genetic factors responsible for, or contributing to, the development of primary OSs are **alterations of the retinoblastoma (*RB*) gene at 13q14** in a high percentage of cases (Dryja et al., 1986; Hansen et al., 1985). The frequent association of OS with RB has been thought to proceed according to the **two-hit model** proposed by Knudson (1975). The **primary event** is seen in the alteration of one of the alleles at 13q14, so that **subsequent chromosomal rearrangements** or mutations affecting the other allele result in tumorigenesis. Similar to RB, loss of heterozygosity (LOH) at various loci on chromosome 13 is also present in OS (Dryja et al., 1986; Ford et al., 1990; Hansen et al., 1985).

5.1.7.2 THE *RB* GENE

As discussed in the section on RB, evidence has confirmed the hypothetical model of Knudson (1971) by demonstrating structural and functional aberrations of the *RB* gene (Benedict et al., 1988; Friend et al., 1986; Fung et al., 1987; Lee et al., 1987a,b). Thirty cases of OS have been studied, using cDNA probes for the *RB* gene, to see if structural abnormalities of the *RB* gene are associated with this malignancy (Toguchida et al., 1988). A total or partial deletion or rearrangement of the *RB* gene was found in 43% of the cases; seven cases presented with homozygous deletions, and six with hemizygous deletions or rearrangements. The LOH on chromosome 13 by either homozygosity or hemizygosity was found in 64% of all informative cases, including those without detectable structural alteration of the *RB* gene. The majority of these tumors showed LOH also at other loci on chromosome 13. The highest frequency (77%) of LOH on other chromosomes was detected particularly on **chromosome 17**. This does not appear to be a coincidental finding, and these two chromosomes may share a common role in the development of OS (Toguchida et al., 1988). The *RB* gene is altered in the high percentage of sporadic OS, and deletions in the *RB* gene correlate with the grade of OS: these are found in 40% of high-grade OS but never in low-grade tumors (Wunder et al., 1991).

5.1.7.3 THE *p53* GENE

In this context, the involvement of the ***p53* gene**, which maps to the short arm of chromosome 17 in OSs, may prove to be coordinated with *RB* gene activity in the pathogenesis of OSs (Masuda et al., 1987; Romano et al., 1989). While changes in the *RB* gene in RB patients could be detected by specific *RB* probes in the fibroblasts cultured from these patients (Benedict et al., 1988), similar findings are not presently available for OS.

A parallel study of the two **tumor suppressor genes**, *RB* and *p53*, with respect to their structural alterations and/or aberrant expression, has been done in a murine model of OS and in human OS cell lines (Chandar et al., 1990). The *p53* gene was found to be altered in the majority of the human OS cell lines, whereas the *RB* gene appeared unchanged, at least at the level of detection employed in that study. It is possible that inactivation of the *p53* gene plays a greater role in the development of OS compared to that of the *RB* gene (Chandar et al., 1990).

Cloning and sequencing of the *p53* gene in a human OS cell line (HOS-SL) demonstrated that it harbored a mutation (Romano et al., 1989). Interestingly, although similar OS cell lines proved to have the same mutations in the *p53* gene, the **wild-type** form of *p53* appears to be **lost** from the tumor cell lines. Moreover, the half-life of the *p53* gene product in tumor cells is prolonged, apparently due to its association with heat shock proteins of the 70-kDa family (Romano et al., 1989). Again, in addition to the multiple forms of p53 protein found in HOS-SL cells, apparently arising from posttranslational modification, a population of p53 molecules associated with heat shock proteins (slower migrating in gel electrophoresis) has been detected (Romano et al., 1989). Thus, the lack of wild-type p53 in these cell lines, coupled with the complete lack of wild-type sequence at codon 156 in each of these lines, pointed to the causative role of the **p53 gene as a recessive cancer susceptibility gene** in the development of OS (Romano et al., 1989). In soft tissue sarcomas (including leiomyosarcomas, rhabdomyosarcomas, and malignant fibrous histiocytomas) evidence supports the notion of concurrent abnormalities of the *RB*1 and *p53* TSGs (Stratton et al., 1990).

5.1.7.4 PROTOONCOGENES

Among other genes characteristically associated with OS some, at least in cell cultures, are the protooncogenes **c-*sis*** (Graves et al., 1984), ***met*** (Cooper et al., 1984; Park et al., 1986), ***PDGF*** (platelet-derived growth factor) (Graves et al., 1984; Heldin et al., 1986), **c-*myc*** (Bogenmann et al., 1987; Sturm et al., 1990), and **v-Ki-*ras*** (Carloni et al., 1988). Amplification of the **c-*myc*** DNA and overexpression of the oncogene could be demonstrated, both *in vivo* and *in vitro*, in the cell lines established from the tumor and metastases of a pediatric patient before and after chemotherapy (Bogenmann et al., 1987).

In contrast to the heterogeneous expression of **N-*myc*** in human neuroblastoma, c-*myc* is uniformly expressed. The c-*myc* gene is apparently involved in the control of osteogenic differentiation of transformed cells. This is indicated by the abnormally high expression of c-*myc* in some radiation-induced murine OSs that correlated with an early stage of osteogenic differentiation (Sturm et al., 1990). In some tumors, rearrangements could be identified in the c-*myc* gene; however, the expressed c-*myc* transcripts were of normal size. The elevated c-*myc* mRNA level in some tumors was accompanied by a low expression of **osteopontin (bone sialoprotein)** and of **bone gla protein**, another marker of highly differentiated osteogenic cells. In the radiation-induced murine OS, N-*myc*, c-*sis*, c-*mos*, and c-*fos* are not amplified and appear normal structurally (Sturm et al., 1990). Whether c-*myc* amplification occurs early in the development of the primary tumors or at a later stage of growth is not established; however, its pattern of amplification in transplantable tumors reproduces that found in primary tumor.

Three groups or types of OS tumors are distinguished (Sturm et al., 1990). The first type shows no apparent alteration of either the gene structure or its expression, suggesting that c-*myc* is probably not involved in tumorigenesis. The second includes malignancies with enhanced c-*myc* mRNA expression, apparently due to amplification or rearrangements in the c-*myc* gene; c-*myc* transcripts contribute to the enhanced growth of the tumor cells. The third group consists of tumors in which the c-*myc* mRNA transcripts are increased, but in which the osteogenic markers osteopontin and bone gla protein are at low levels. This lowered expression of osteogenic differentiation markers may be correlated with the constitutive c-*myc* expression resulting in a block of cell differentiation (Freytag, 1988).

A similar correlation has been made in osteogenic cell lines established from a spontaneous murine OS (Schmidt et al., 1989). Studies of c-*myc* involvement in OS so far have not addressed the possible interaction of c-*myc* with the *p53* and *RB* genes. In view of the established role of the latter in OS such an interaction may be highly probable. If the above-discussed findings are extensively confirmed in human OS, the degree of c-*myc* gene expression or alteration may be used in **evaluating the level of dedifferentiation of osteogenic tumors** in clinical situations.

5.1.7.5 DIFFERENTIAL DIAGNOSIS

The variability of the microscopic patterns encountered in osteogenic malignancies necessitates consideration of a broad range of benign and malignant lesions (Enzinger and Weiss, 1988; Rosai, 1989). An aid to diagnosis is the ultrastructural similarity of the better differentiated tumor cells with normal osteoblasts (Rosai, 1989), but the few studies performed on extraskeletal OS point to features essentially indistinguishable from those reported for primary OS of bone (Enzinger and Weiss, 1988).

Positive staining for **vimentin intracellularly** and **type I collagen extracellularly**, as well as focal positivity for **S-100 protein** and **type II collagen**, are characteristic immunohistochemical features (Rosai, 1989). The observed stong positive reaction of **osteonectin** (Rosai, 1989) would be interesting to compare with that for **osteopontin** and **bone gla protein**. The latter two osteogenic markers appear to be reduced in inverse relationship to the degree of differentiation of OS (Sturm et al., 1990). It seems that, at this stage, analysis of the ***RB, p53*** and c-*myc* genes and their products, combined with the osteogenic markers osteopontin, osteonectin, and bone gla protein, may offer a practical approach to resolving some of the diagnostic and prognostic challenges presented by OS. Although discussion of collagen disorders is not included here, a lucid review of mutations in collagen genes by Prockop (1992) should not be missed.

5.1.8 SKIN

5.1.8.1 MALIGNANT MELANOMA

5.1.8.1.1 DNA Ploidy

The increasing incidence of **cutaneous malignant melanoma (MM)** is related to **solar ultraviolet radiation** (Koh, 1991; Tucker, 1988). The major precursor of MM is considered to be the **dysplastic nevi** (Lever and Schaumburg-Lever, 1990). Some cases (8 to 12%) can be related to familial predisposition and the presence of dysplastic nevi (Greene et al., 1985; Greene and Fraumeni, 1979). **MM** is often recognized at early stages of tumorigenesis, and that accounts for the significant success of curative surgery. However, it is estimated that approximately 20% of all cases show a relapse within 5 years of treatment (Bartkowiak et al., 1991). Analysis of nuclear DNA content of 804 primary MM by FCM revealed that 57% of the cases were diploid, one abnormal cell population was encountered in 32%, and 11% of the cases showed multiclonality. An S-phase fraction (SPF) of over 15% has bad prognostic significance in MM. Aneuploidy and multiclonality and the occurrence of hypertetraploidy were other significant prognostic factors of a poor outcome (Bartkowiak et al., 1991). Comparing the prognostic power of DNA ploidy with other criteria conventionally used in evaluating stage I MM, DNA content, although significantly correlating with these as well as with disease-free survival, ranked fifth behind growth pattern, ulceration, thickness, and pathologic stage (Gattuso et al., 1990).

In stage II MM with **metastatic spread** to lymph nodes, aneuploidy reaches 86%, the rest of the cells being tetraploid (Zeng et al., 1990). The development of multiclonality is associated with tumor progression, which is also accompanied by a wide spread of nuclear DNA values.

These features quantified by digital imaging systems are considered to provide useful prognostic information.

5.1.8.1.2 Chromosome Studies

The most significant findings related to events at the gene level in skin diseases have been made in MM. The progressive nature of the disease has been linked to the **accumulation** of **karyotypic abnormalities**. Tumor progression from karyotypically normal melanocytes to primary melanoma is characterized by the appearance of frequent random karyotypic changes and abnormalities of **chromosomes 1, 6**, and **7**, which appear at increasing frequencies in metastatic MM lesions (Balaban et al., 1986; Herlyn et al., 1985; Richmond et al., 1986).

Monoclonality of the metastic lesions has been confirmed by RFLPs at 61 autosomal and 7 sex-linked loci in all lines established from 6 different metastases of a single melanoma patient (Dracopoli et al., 1987). The LOH at different loci varied, with the majority of segregations (66.7%) being due to loss of the same allelic fragments in all six metastases on **chromosome 9** and the **X chromosome**. All cells derived from the metastases were aneuploid, and rearranged chromosomes could be identified in each cell line derived from the metastases. Frequent and random LOHs in metastatic MM cells contrast with LOHs at specific loci, presumed to be closely linked to recessive tumor genes, in other human primary tumors (RB, Wilms' tumor [WT], etc.). This phenomenon is thought to be due to differences between primary tumors and the allelic segregation occurring during tumor progression and expression in metastases (Dracopoli et al., 1987).

Following earlier indications of the localization of the gene associated with the development of MM to **chromosome 1p**, a multipoint linkage analysis of 26 polymorphic loci on chromosome 1p in 6 families with MM has been performed (Bale et al., 1989). The gene for susceptibility to melanoma-dysplastic nevus syndrome (MM-DNS) was assigned to a location between an anonymous DNA marker (**D1S47**) and the gene locus for **pronatrodilatin** (**PND**), a commonly used reference gene, in chromosome band **1p36**.

Using the same reference points for linkage analysis in 63 members of 5 Dutch families, in which 39 subjects had cutaneous MM-DNS, this locus could not be proved (Gruis et al., 1990). The cause of these discrepancies is not clear; however, clinical and genetic heterogeneity between the Dutch families and American patients analyzed in these two studies is invoked (Bale et al., 1990). Admittedly, other loci for the DNS can be found that are distinct from those identified for cutaneous MM.

Another study in three Utah kindreds with multiple cases of MM has been conducted (Cannon-Albright et al., 1990) using the same two markers, *PND* and D1S47, reported to be most closely linked by Bale and coworkers (1989). Again, no evidence for linkage could be found and using multipoint linkage analysis, the **cutaneous melanoma-nevus syndrome locus** was excluded from an area of 55 cM containing the *PND*-D1S47 region.

A **high-resolution chromosome banding** study has been accomplished on 62 patients with cutaneous MM (Trent et al., 1990a,b). Structural chromosomal abnormalities that occurred 10 or more times were correlated with patient **survival**. The most frequent sites of abnormalities were found on **chromosomes 1** (31 times), **6** (15 times), **2** (14 times), **7** and **9** (11 times each), and **11, 21**, and **3** (10 times each). No correlation could be found between specific chromosomal changes and histopathological parameters of the diagnostic biopsy samples; however, patients with structural abnormalities of chromosome 7 or 11 had a **significantly shorter survival**.

Nonrandom alterations of chromosome 6 in patients with MM appear to have relevance to malignancy as demonstrated by cell culture studies (Trent et al., 1990a,b). When a normal chromosome 6 is introduced into melanoma cell lines by **microcell hybridization technique**, the resulting microcell hybrids drastically change their malignant phenotypic features and fail to induce tumors in nude mice. This change is reversible, as the loss of chromosome 6 from melanoma microcell hybrids makes the cells tumorigenic again. It thus appears that **one or more genes on chromosome 6 may be involved in the expression of the malignant characteristics of MM** (Trent et al., 1990a,b).

In uveal MM, cytogenetic and DNA polymorphism analysis reveal monosomy of **chromosome 3** and multiplication of **chromosome 8q** material defining a subgroup of uveal tumors with poor prognosis (Prescher et al., 1992).

So far, all the cytogenetic and chromosomal studies cited have succeeded in providing only tentative evidence on a rather wide range of correlations between chromosomal aberrations and biology of cutaneous MM. **No specific, diagnostically helpful changes at the chromosome level can be used at present in differential diagnosis or patient management in malignant melanoma.**

5.1.8.1.3 *ras* Protooncogene in Malignant Melanoma

Analysis of particular ***Bam*HI** Ha-*ras* protooncogene alleles in members of two families with **hereditary melanoma (HM)-DNS** failed to show any linkage with melanoma occurrence (Sutherland et al., 1986). Likewise, no correlation could be found between MM and rare Ha-*ras*-1 alleles defined by ***Msp*I/*Hpa*II** digestion (Radice et al., 1987). Further analysis of the same DNA for a different polymorphism using ***Taq*I** sites identified a group of allele variants in the variable number of tandem repeats (VNTR) region of **Ha-*ras*-1**, named **Tp**, which were significantly more frequent in patients with MM than in normal controls.

Combining DNA-mediated gene transfer and tumorigenicity in nude mice, activated *ras* oncogenes mutated at **codons 12**, **13**, and **61** have been looked for in sporadic human MM (Raybaud et al., 1988). In 13 surgical specimens of primary tumors and metastases, neither Ha-*ras* nor **Ki-*ras*** oncogenes could be detected by this assay. An **N-*ras*** oncogene mutated at codon 61 was found in only one specimen. This assay does not give information on activation of *ras* genes by mechanisms other than mutation, such as amplification or deletion. Mutations of the Ki-*ras* gene could be missed, due to the large size of this gene, for technical reasons (Raybaud et al., 1988).

On the contrary, a correlation between *in vivo* exposure to **solar radiation** and N-*ras* mutation at codons 12 and 13 has been found in 7 of 37 patients with MM (van't Veer et al., 1989). The same mutation was present in the primary site and the metastases, suggesting the activating mutation was an early event in malignization. Primary tumors with activated N-*ras* were localized exclusively in continuously sun-exposed body areas. Furthermore, support for the etiological role of solar radiation in MM comes from the finding of the affected sequences opposite the codons of the N-*ras* oncogene, all of which contained **pyrimidine dimers (TT or CC)**, known to be produced by UV irradiation (van't Veer et al., 1989).

Activation of the N-*ras* gene was later confirmed in a thorough investigation of patients' specimens and cultured MM cells using the PCR technique (Albino et al., 1989). Moreover, a correlation of these with specific phenotypic characteristics of tumors and precursor lesions, the expression of *ras*-encoded p21 proteins, and their quantitative evaluation by FCM has also been performed (Albino et al., 1989). Almost a quarter of cultured MM cell lines had activated *ras* genes, N-*ras* occurring 10 times as frequently as Ha-*ras,* and always containing a mutation at codon 61. Cell surface characteristics of all the melanomas having an activated *ras* gene were similar. Interestingly, only 5 to 6% of noncultured primary and metastatic MMs had mutated *ras* genes. No *ras* mutations could be found in normal nevi or dysplastic nevi considered to be precursors of melanomas. Evaluation of p21 expression in tissue sections of tumors revealed no correlation with tumor progression, or with any growth characteristics of melanomas with or without *ras* mutations. The overall impression was that "the role of *ras* mutations may be limited to an indirect involvement in the transformation of a subset of melanomas" (Albino et al., 1989). *ras* mutations apparently reflect the genetic instability in melanomas, and are manifestations of the high mutation rates characteristic of cancer cells in general, in which *ras* **mutations seem to be the result, rather than the cause, of transformation.**

These findings contrast sharply with the **immunohistochemical demonstration** of a strong p21 protein presence in cutaneous MM, **which correlated with the degree of malignancy** (Yasuda et al., 1989). Thus, the anti-p21^{ras} reactivity, was higher in nodular melanoma, epithelial type melanoma, and deeply invading tumors than in any other types of MM. No reactivity could be observed in melanocytic nevi with functional activity nor in nevus cells in the epidermis in compound nevi. On the other hand, dermally located nevus cells reacted strongly. The p21 protein expression was more pronounced in **immature melanocytes** and could not be viewed as a marker of malignant transformation, but rather only of the tumor cell differentiation.

5.1.8.1.4 Other Genes in Malignant Melanoma

In two MM cell lines, DNA sequences with partial homology to the **human papillomavirus (HPV) 9** and **Epstein-Barr virus nuclear antigen (EBNA) 1** genomes have been detected (Klingel et al., 1987). No presence of previously reported HPV38 in any of the MM samples could be confirmed, although HPV9 is known to be closely related to HPV38.

In an effort to correlate the frequent occurrence of changes involving chromosomes 1, 6, and 7, thirty cell lines derived from primary and metastatic melanomas of 28 patients were analyzed by Southern blotting for 28 various genes known to map near frequent chromosomal breakpoints in melanoma samples (Linnenbach et al., 1988). An alteration in *myb* protooncogene correlated with a 6p22 chromosomal abnormality; a tumor-specific deletion was found in the gene encoding α-**type protein kinase C (PKC)**. In addition, polymorphic alleles for the gene coding for **epidermal growth factor receptor (EGFR)** and α-type PKC were detected.

Later, normal human melanocytes were found not to express *PKC*-α, -β, or -γ (Becker et al., 1990). In contrast, *PKC*-α was expressed in primary and metastatic MM. It is unclear to what extent this gene is involved in malignancy, because antisense oligonucleotides against different regions of the *PKC* gene failed to produce any growth-inhibitory effects. Similar observations were also made on the expression of **cAMP-dependent protein kinase** (Becker et al., 1990).

The presence of activated cellular protooncogenes has been sought in two cell lines (NH and HM1) derived from patients with metastatic MM (Husain et al., 1990). **Consistent overexpression** (9- to 14-fold) of the c-*myc* gene was detected, whereas **no detectable level of expression** of c-*fms*, **c-abl**, v-*src*, **c-erbA-1**, **c-erbB**, v-*mos*, **TGF-β**, and **c-myb** could be observed. A similar elevation of c-*myc* expression was also observed in long-term cultured melanoma cell lines. The level of expression of **c-Ha-ras**, **N-ras**, **c-fos**, and **c-sis** was the same in normal human melanocytes and metastatic MM.

The level of expression of 21 different oncogenes, **tumor suppressor genes**, **growth factors**, and **proteases** has been evaluated in cultured melanocytes and 17 melanoma cell lines (Chenevix-Trench et al., 1990). Significant intercorrelations were observed between c-*myc*, *p53*, and c-*src*-1 levels, and between *p53* and **c-erbB-2**. In at least some cell lines, the expression of **c-mel**, **c-erbB-2**, **c-myc**, **c-src-1**, *p53*, **PDGF** A chain, *gro*, **TGF-α**, **EGFR**, and **tissue plasminogen activator** genes has been detected. No specific differences between melanocytes and melanoma cells could be established, because the levels of expression of all the genes studied fell within the wide range observed in cell lines derived from metastases. Further studies are needed to unravel the importance, if any, in MM tumorigenesis of the correlations in expression of *p53*, c-*myc*, c-*src*-1, and c-*erb*B-2. A preliminary observation has been reported on the possible role of **c-jun** and **c-fos** in the transformation of cultured human primary melanocytes to MM melanoma (Yamanishi et al., 1990).

*Taq*I RFLP analysis of the human *TGF*-α locus in melanoma biopsy tissue, melanoma cell lines, and matched controls revealed a significantly higher frequency of a 2.7-kb allele in MM (Hayward et al., 1988). It is yet to be shown whether this correlation is relevant to the biology of MM. It is possible that TGF-α plays a role in MM similar to that in other tumors, supporting their autocrine growth or contributing to the neovascularization associated with solid tumors (Hayward et al., 1988).

Increased expression of closely related growth factors (TGF and EGF) has been observed in surgical specimens of MM (TGF-α, TGF-β, and EGFR mRNA) as well as in cultured melanoma cells (TGF-α mRNA) (Derynck et al., 1987). Release of a **low molecular weight human TGF in the urine** of patients with MM, and its transforming potential in a fibroblast culture, have been demonstrated (Kim et al., 1985).

Urinary TGF levels appear to parallel **tumor burden** as demonstrated in patients with **breast cancers** (Stromberg et al., 1987). In fact, the release of TGF-α in the urine appears not only to reflect the tumor burden in a MM patient declining to levels undetectable by Western blots following surgical removal of the tumor, but may also serve as an **early marker** for melanoma (Ellis et al., 1987). Likewise, the expression of **EGFR** in various **hyperproliferative skin lesions** accompanying MM in a patient who also had acanthosis nigricans, the sign of Leser-Trélat (appearance or increase in the number of seborrheic keratoses in a patient with an internal malignant disease), and multiple acrochordons markedly declined following removal of melanoma (Ellis et al., 1987).

Association of the **process of melanosis** with the development of tumors was investigated in transgenic mice. The transgene consisted of the **tyrosinase promoter** fused to the simian virus 40 (SV40) early-region oncogenic sequences (Bradl et al., 1991; Klein-Szanto et al., 1991). The interrelationships between specific differentiation and the development of tumors in such animal models will help in understanding the molecular pathogenesis of MM. The processes of melanization and proliferation appear to follow independent and possibly alternative pathways, as demonstrated by the response of cultured melanoma cells to **prostaglandin** (Abdel-Malek et al., 1987).

Some **53 new human genes** have been found in human MM cells by the technique of molecular subtraction in an effort to identify specific MM tumor markers (Hutchins et al., 1991). Their characterization has not been completed at the time of this writing.

A very sensitive procedure for the detection of MM by assaying peripheral blood for the presence of melanoma cells is based upon identification of the gene for **tyrosinase**, a tissue-specific gene in melanocytes (Smith et al., 1991). Using RT-PCR for tyrosinase mRNA (normally absent from peripheral blood), a single melanoma cell can be detected in 2 ml of blood.

Another gene specific for melanoma cells encodes the **antigen MZ2-E** that is recognized by cytotoxic T lymphocytes; it can be used as a target either for specific detection of MM cells or for direct immunotherapy (van der Bruggen et al., 1991).

5.1.8.2 SKIN DISEASES OTHER THAN MM

A coherent picture of key molecular events in skin diseases involving the epidermis is not available at present. The diversity of skin pathologies and lack of concentrated effort lead to only fragmentary observations on the biology of keratinocytes affected by some skin disorders.

In chemically induced transformation of mouse epidermal tissue, a definite role for activated Ha-*ras* oncogene has been shown (Roop et al., 1986). Although mouse skin **papillomas** and **squamous cell carcinomas (SCC)**, produced by a combined action of urethane and phorbol esters, contain an activated Ha-*ras* gene, specific differences in the mutation pattern of the protooncogene between these two forms of tumors are discernible (Harper et al., 1987). While primary papillomas and carcinomas contain a point mutation in codon 61 of the Ha-*ras* gene, producing a p21 protein identifiable by SDS-polyacrylamide gel electrophoresis (PAGE), the papilloma cell lines lack both the gene alteration and, consequently, an altered p21 product. Risk of SCC appears to be significantly associated with HLA-B mismatching and HLA-DR homozygosity as observed in renal-transplant recipients (Bouwes Bavinck et al., 1991).

Activated Ha-*ras* genes appear to be a common aberration not restricted to a specific type of cancer. In addition to its occurrence in MM and chemically induced tumors, activation of Ha-*ras* also occurs in **nonmelanotic human skin cancers** originating in sun-exposed body sites (Ananthaswamy et al., 1988). While Ha-*ras* is implicated in the **initiation** of skin carcinogenesis, the *fos* oncogene appears to be involved in the malignant **progression** of premalignant skin cell lines. Introduction of v-*fos* alone results in normal or hyperplastic skin (Greenhalgh et al., 1990).

Likewise, introduction of c-Ha-*ras* alone into **cultured human keratinocytes** markedly affects the growth potential of the cells, but no direct correlation between high levels of expression of its gene product, p21, and malignant growth can be observed (Boukamp et al., 1990).

Coinfection with v-Ha-*ras* and v-*fos* transforms keratinocytes grown in nude mice into SCC. Because *fos* acts as a transcription regulator of other genes, its role in the malignization of epidermal cells appears to be related to the function of the *fos*-encoded protein (Greenhalgh et al., 1990). The induction of c-*fos* expression during chemical tumorigenesis in mouse skin cell cultures appears to be a relatively early event (Cochran et al., 1984; Greenberg and Ziff, 1984; Kruijer et al., 1984), preceding the expression of c-*myc* (Mueller et al., 1984). In contrast, the c-*myc* mRNA levels decrease during terminal differentiation of various cells, consistent with its role in cell cycle control (Campisi et al., 1984; Dotto et al., 1986). A selective and specific induction of c-*fos* by various phorbol esters, preceding that of c-*myc* and **ornithine decarboxylase gene (*ODC*)**, has also been confirmed in mouse models of skin carcinogenesis (Rose-John et al., 1988).

Other transcriptional regulators are also involved in epidermal tumorigenesis. Thus, v-*jun* oncogene appears to operate as a dominant negative regulator in mouse epidermal tumor cells, as shown by transfection studies in mouse epidermal papilloma cell lines (Tsang et al., 1990).

In the more "distal" aspect of cell growth regulation, the role of EGF and EGFR has been demonstrated by Southern analysis in **epidermoid carcinoma** AY31 cells (King and Sartorelli, 1989). Expanding on the earlier findings of the ability of EGF to suppress the terminal differentiation of malignant keratinocytes (Boonstra et al., 1985; Fuchs and Green, 1981; King et al., 1986), a decreased level of EGFR could be correlated with the higher capacity of A431 cells than parental cells to form cornified envelopes (King and Sartorelli, 1989). The level of the EGFR protein is regulated at the gene level, since relatively fewer copies of the *EGFR* genes were found in the cells exhibiting a higher capacity to form cornified envelopes (King and Sartorelli, 1989). The mechanisms by which suprabasal cells regulate the expression of *EGFR* genes, and how these cells enter the terminal differentiation program, remains unknown, although the involvement of this gene in the differentiation of keratinocytes is beyond doubt (Maguire et al., 1989).

A related growth factor, **TGF-α**, appears to be expressed at higher levels in a nonmalignant, proliferative skin disorder, **psoriasis**, than in normal skin (Elder et al., 1989). Epidermal hyperplasia in this condition is positively stimulated by this growth factor, rather than being released from under the constraints of a growth-inhibiting factor, **TGF-β_1**, since the latter is present essentially at the same levels in normal, uninvolved, and lesional psoriatic epidermis (Elder et al., 1989).

Application of molecular biological techniques for the study of psoriasis and the development of gene level diagnostic markers encounters a number of obstacles (Meyer, 1990). To start with, it is not clear whether all cases of psoriasis represent a single disease; its mode of inheritance is compatible with either a polygenic pattern or autosomal dominant inheritance with incomplete penetrance (Meyer, 1990). Linkage analysis has so far assigned the four RFLPs informative in a large kindred to a region about 30 cM wide on **chromosome 6p**, with a probable translocation defect in that area (Wuepper et al., 1990).

The pattern of expression of a number of protooncogenes in psoriasis suggests that overexpression is more characteristic of acute cellular responses not observed in the steady state hyperplasia characteristic of psoriasis (Elder et al., 1990). In fact, **no increase** in the expression of c-*myc*, c-Ha-*ras*, c-*erb*B (*EGFR*), c-*jun*, or *TGF*-β transcripts has been detected in psoriatic lesions by RNA blot hybridization. The expression of c-*fos* was lower, while that of *TGF*-α was markedly increased in psoriatic lesions compared to normal skin (Elder et al., 1990). In cultured cells, the EGFR tyrosine kinase is activated by TGF-α, suggesting a mechanism that may be important in the pathogenesis of psoriatic epidermal hyperplasia (Elder et al., 1990). Another characteristic feature of psoriatic lesions, as well as those of **pemphigus** and **bullous pemphigoid**, is the elevated *in situ* expression of **tissue-type plasminogen activator (tPA)** (Baird et al., 1990; Jensen et al., 1990).

Elevated oncogene expression is not only related to malignization of the epidermis, but appears to play a role in conferring an unusual **radioresistance** to skin fibroblasts (Chang et al., 1987). An indirectly related analysis of noncancerous skin fibroblasts from members of a cancer-prone family with the **Li-Fraumeni syndrome** showed that the level of c-*myc* expression was three- to eightfold that of controls accompanied by an apparent activation of c-*raf*-1 gene (Chang et al., 1987).

Significant insights have been gained in understanding molecular events responsible for another condition with striking inherited sunlight sensitivity and multiple manifestations involving skin — **xeroderma pigmentosum**. This disease exemplifies the role of **defective DNA repair** mechanisms in the development of neoplasia and is discussed in a separate section (see Section 1.5.7.1).

Basal cell carcinoma (BCC) is the most common type of human skin cancer for which very limited cytogenetic data exists. A study on short-term cultures from 33 BCCs (Mertens et al., 1991) revealed the majority of tumors (23) had a very high level of **nonclonal structural rearrangements**. In eight tumors, **clonal chromosomal aberrations**, all of them different, were detected. It appears that BCC has a **multicellular origin** reflecting **field cancerization** (Mertens et al., 1991). The importance of environmental mutagenic factors, such as UV light, in leading to lifelong accumulation of chromosomal rearrangements is challenged by, among other observations, the similarity of cytogenetic characteristics in BCCs from sun-exposed and unexposed areas of the body. The field cancerization phenomenon is viewed as the occurrence of "a high frequency of rearrangements in an area of epithelial cells that are not yet fully neoplastic but that also are no longer completely normal" (Mertens et al., 1991, p. 956). Earlier observations also point to the coexistence of clonal and nonclonal structural chromosomal changes in BCC (Aledo et al., 1989; Scappaticci et al., 1989). The gene for the **nevoid-BCC (the Gorlin syndrome)** has been assigned by linkage analysis to **chromosome 9q22.3-q31** between DNA markers D9S12 and D9S53 thus allowing identification of presymptomatic patients for surveillance of the developing skin cancer (Farndon et al., 1992).

Erythema multiforme, in particular its recurrences, has been associated with **herpes simplex virus (HSV)**. The presence of the virus has been demonstrated by serological, immunofluorescent, electron microscopic, and PCR analysis (Darragh et al., 1991). On the other hand, extensive analysis of the **HPV DNA** of various types performed by PCR and Southern blot hybridization has shown that known HPV types apparently play only a minor role in the development of skin cancer in the general population (Kawashima et al., 1990).

An interplay of numerous **immunologic factors** involved in the generation of local responses of the skin to the action of environmental "noxious" agents is currently a field of intense and promising research. One of the theories visualizes keratinocytes as "signal transducers" converting external stimuli into a cascade of interaction between cytokines, adhesion molecules, and chemotactic factors responsible for the initiation of "antigen-independent" cutaneous inflammation (Barker et al., 1991).

5.1.8.3 CHEMICAL CARCINOGENESIS OF THE SKIN

Activation of the c-Ha-*ras* gene is thought to constitute one of the early events in malignization of the **chemically induced** skin **papillomas** in the mouse (Harper et al., 1986; Roop et al., 1986). In fact, activated *ras* genes can replace chemical carcinogens and initiate transformation of mouse skin on their own (Brown et al., 1986). A specific mutation of codon 61 of the c-Ha-*ras*-1 gene appears to be a critical step in the development of mouse skin tumors (Bizub et al., 1986,

1987). Analysis of the cytogenetic characteristics of mouse skin tumors and derivative cell lines containing activated *ras* genes did not show any karyotypic abnormalities (Aldaz et al., 1988). The activated Ha-*ras* gene is apparently capable of generating **benign lesions** without the involvement of major chromosomal alterations. While activated *ras* genes can generate benign tumors alone, in the chemically induced tumors progression from papilloma to carcinoma may involve the amplification of mutated **Ha-*ras*** alleles (Quintanilla et al., 1988). In **virally induced carcinomas** the level of expression of viral Ha-*ras* is particularly high. The other protooncogene, amplified and expressed in a chemically induced **epidermoid carcinoma** cell line derived from hamster cheek pouch, is c-*erb*B (Wong and Biswas, 1987).

Chapter 5.2
Neurological and Mental Disorders

5.2.1 HUNTINGTON'S DISEASE

Huntington's disease (HD) is an autosomal dominant, incurable disorder affecting the central nervous system, the onset of which usually occurs when the patient is in the fourth or fifth decade of life. In the advanced stages of the disease, the characteristic choreic movements, cognitive impairment, and behavioral changes eventually are superseded by physical debilitation, resulting from premature nerve cell death throughout the brain, being most prominent in the striatum.

The focus of this discussion is on the current status of Huntington's disease **diagnosis at the gene level**. Reviews of the biology of the disease and spectacular discoveries, which enormously accelerated the search for the *HD* gene, are available (Gilliam and Gusella, 1988; Gusella and Gilliam, 1989). Several laboratories in the United States and elsewhere have experienced numerous ups and downs in their concentrated efforts to identify definitively the *HD* gene (Roberts, 1990). The inevitable psychological, ethical, and legal issues accompanying the availability of a highly predictive test for an incurable disease have been superbly summarized by N. S. Wexler (1989). The practical use of the first DNA probe (G8), showing a markedly improved degree of predictability in linkage analysis, initially had to be tempered by the still not excluded possibility of allelic or nonallelic heterogeneity (Gusella, 1986; Watt et al., 1986).

Although the biochemical defect responsible for the neurological manifestations of HD still awaits identification of the gene and its product, indications are that the premature death of selected medium-sized spiny γ-aminobutyric acid (GABA)-containing neurons may be due to **defective repair** (Gilliam and Gusella, 1988). An attractive hypothesis attempting to unify neuroanatomical, positron emission tomography, and neurotransmitter data points to the possible role of **derangement of glucose metabolism** as the leading cause of neuronal death (Carter, 1986).

In analyzing the **linkage association** of an anonymous **DNA locus D4S10**, identified by the **DNA probe G8**, with the disease, Gusella and coworkers (1983) succeeded in assigning it to **chromosome 4**. The linkage data between D4S10 and the Huntington's disease gene, assessed in the large Venezuela disease pedigree, one American black family, another of Germanic descent, as well as in several families from Wales, established that this locus displays about 4% recombination with the Huntington's disease gene (Gilliam and Gusella, 1988). The positive linkage to D4S10 observed in all of the pedigrees studied indicated that the causative mutations occur at the same locus, excluding the likelihood of nonallelic heterogeneity in HD. Subsequent family studies suggested that over 95% of HD cases are caused by a mutation on chromosome 4. Interestingly, in contrast to the usually observed dependence of the phenotypic severity of dominant human disorders on the gene dosage (symptoms being more severe in homozygotes than heterozygotes), the motor and cognitive dysfunction as well as psychiatric symptoms in HD are not exaggerated by a double dose of the defective gene, nor are they ameliorated by the presence of the normal allele (Gilliam and Gusella, 1988).

Further refinement of the *HD* gene location came from studies performed on chromosomes from **Wolf-Hirschhorn syndrome** patients, suffering from a congenital anomaly involving heterozygous deletion of part of the short arm of chromosome 4, most commonly involving the **4p16** band (Gusella et al., 1985). *In situ* hybridization (ISH) of G8 to metaphase chromosomes provided further support (Gilliam and Gusella, 1988). To narrow down the minimal chromosomal areas harboring the *HD* gene, the search has centered on the **flanking markers** (Gilliam et al., 1987a,b; MacDonald et al., 1987, 1989; Pritchard et al., 1989; Richards et al., 1988; Smith et al., 1988; Wasmuth et al., 1988; Whaley et al., 1988; Youngman et al., 1988, 1989).

Since the approximate target area still remains relatively large (2 to 6×10^6 bp), other molecular technologies are also being employed. **Long-range restriction mapping** using a **pulsed-field gradient gel electrophoretic system** (PFGGE) is capable of separating fragments from 10 to 2500 kb, and helps in constructing physical maps of the target chromosome region. These can then be compared with genetic maps of the same area, allowing optimal selection of DNA sequences for cloning. In fact, PFGGE is also used to isolate, and clone large DNA restriction fragments of interest, such as those identified by G8. Chromosome analysis by the **"jumping library"** (Richards et al., 1988) and **linking library techniques** allow detailed physical maps to be constructed (Dogget et al., 1989; Pohl et al., 1988; Poustka et al., 1987). **Patterns of inheritance** in relation to the age of onset of the HD symptomatology suggest an effect of **paternal imprinting** seemingly resulting in a major change in gene expression — the early onset/rigid variant (Ridley et al., 1991).

Analysis of recombination events in HD families moved the location of the *HD* gene farther toward the telomere from the most distal tested **marker D5 (D4S90)**, which is removed from the end of the chromosome by about 300 kb (Bucan et al., 1990). To delineate and clone the telomeric candidate region, the telomere of the short arm of chromosome 4 has been cloned in the yeast *Saccharomyces cerevisiae* (YAC clones) and analyzed by PFGGE to map approximately 200 kb distal to D4S90 (Bates et al., 1990). It appears that the *HD* mutation may be distal even to this locus. At this time it is clear that the rate of recombination, complicating the exact identification of the *HD* gene, is not continuously increased with progress toward the telomere of chromosome 4p (Allitto et al., 1991).

5.2.2 TESTING FOR HUNTINGTON'S DISEASE

In spite of the limitations of the presently available DNA probes for HD, the presymptomatic and prenatal identification of probable gene carriers can now be established in informative families with a high degree of probability (2 to 98%) (Brandt et al., 1989; Brock et al., 1989; Craufurd et al., 1989; Fahy et al., 1989; Farrer et al., 1988; Meissen et al., 1988; Quarrell et al., 1987).

At this stage the test is based on linkage studies, which identify the allele, or form of the DNA marker (D4S10 or others), that is genetically linked to the *HD* gene. To this end, the segregation of the DNA marker is determined among affected and unaffected persons in a pedigree; many people must cooperate for the result to be informative (Conneally et al., 1984). Early experience gained in a predictive testing program conducted in 47 persons at 50% risk of inheriting HD revealed that some people "in the early stages of HD may seek presymptomatic testing rather than neurologic evaluation" (Meissen et al., 1988). Expected psychological reactions to positive testing such as depression were observed, although no suicidal tendencies were observed.

Imperfection of the probes available for linkage analysis of HD carriers at this time necessitates a number of considerations to be incorporated into the evaluation of risk, including such variables as the age at onset (Farrer et al., 1988). A detailed protocol incorporating linkage data and age at onset adjusted for censored observations, sex of affected parent, and familial correlation has been developed into a **computer program**, **MLINK**, for calculating the risk of having HD (Farrer et al., 1988). Also recommended is pre- and posttest counseling, psychological testing, and paternity testing as part of a formal presymptomatic testing protocol (Brandt et al., 1989; Farrer et al., 1988).

Interestingly, **neurological**, **psychiatric**, **psychological**, and **social variables** evaluated in 55 persons at the point of entry as participants in a presymptomatic testing program failed to predict reliably the genetically affected individuals (Brandt et al., 1989; Strauss and Brandt, 1990). Lack of definitiveness of linkage testing and absence of cure or treatment for HD certainly reduced the number of at-risk individuals willing to be tested.

Still unresolved issues of possible **discrimination** by insurance companies and employers against persons testing positive for HD, or other untreatable conditions with delayed onset diagnosed genetically, are of major concern. The highest acceptance of the test is understandably demonstrated by individuals who seek testing themselves, compared to those who are invited (Craufurd et al., 1989). A multitude of problems ranging from inappropriate use of the test prior to or after evaluation to loss of samples in transit to the laboratory have been addressed (Morris et al., 1989).

Analysis of the **psychological factors** affecting attitude and the decision to undergo presymptomatic testing for HD revealed that fewer people request genetic testing than expected, and those who decide to be tested have a more favorable attitude at both the beginning and the end of the educational sessions (Quaid et al., 1989). Some authors argue that there is no point in offering the test unless the couple is committed to termination of pregnancy if the result is positive (Quinn, 1989).

Testing of minor children is deemed inappropriate for a number of reasons (Bloch and Hayden, 1990). Among these is the inability to obtain an informed decision from the person to be tested, when the request is actually made by a third party or by parent(s). Because no treatment is currently available to improve the condition of the person affected with HD, the effects of such testing on the child's relationship with peers and parents, as well as on subsequent adjustments to the challenges of life during the long presymptomatic period, are not predictable (Brock et al., 1990; Dalby, 1990; Harper et al., 1990). **Prenatal testing** can be performed when a parent is affected or is at increased risk for HD, and when the fetus has a 50% chance of inheriting the *HD* gene (Bloch and Hayden, 1990).

With the newer markers **D4S43** and **D4S95**, in addition to D4S10, virtually all families became fully informative in the Edinburgh predictive testing program; however, even with these markers the test can still be "less than ideal" (Brock et al., 1989). The most efficient strategy in presymptomatic, prenatal, and exclusion testing for HD incorporates the use of seven closely linked DNA markers, D4S10, D4S81, D4S95, D4S43, D4S111, D4S125, and D4S115 (Skraastad et al., 1991). A combination of exclusion prenatal testing and definitive testing of a fetus, when a parent is determined to be at increased risk of carrying the *HD* gene, or when he/she is already affected with HD, is currently under study in British Columbia, Canada (Fahy et al., 1989).

Analysis of the genetic linkage between Japanese HD and the DNA markers D4S10 and **D4S43/S127** firmly established the identity of the Japanese *HD* gene with the Western gene, in spite of the lower prevalence of HD in Japan (Kanazawa et al., 1990). Restriction fragment length polymorphism (RFLP) analysis of HD in Finland, where the prevalence of the disease is very low (~5 in 1 million), demonstrated a genetic homogeneity of HD (Ikonen et al., 1990). Linkage disequilibrium of RFLP haplotypes of D4S10 and D4S43 loci polymorphisms was observed in this population in contrast to those from other countries.

Clearly, **problems arising in presymptomatic testing** for Huntington's disease are numerous and not easily resolvable. Efforts are being made to address these complex issues (Bloch and Hayden, 1990; Tyler and Morris, 1990). Several discussions address the complicated **ethical** and **legal issues** associated with predictive testing for HD (Chapman, 1990; Huggins et al., 1990)

5.2.3 NEUROFIBROMATOSIS

Neurofibromatosis is a degenerative disease of neural crest origin characterized by predominant hyperplasia of Schwann cells. This single-gene disorder occurs in two forms: **peripheral neurofibromatosis (NF1)**, also known as **von Recklinghausen's disease (VRNF)**, and **central** or **bilateral acoustic neurofibromatosis (NF2)** (Martuza, 1984; Riccardi, 1981).

5.2.3.1 PERIPHERAL NEUROFIBROMATOSIS

NF1 is one of the most common autosomal dominant disorders (with an incidence of 1 in 3000) (Crowe et al., 1956), manifested as multiple **benign neurofibromas** and **cafe-au-lait spots**. Cell culture studies suggest that the degenerating nerve cell appears to release a stimulant acting on Schwann cells, leading to the formation of neurofibromas (Ratner et al., 1988). The clinical diagnosis is established when two or more of the criteria proposed by the **NIH Health Consensus Development Conference on Neurofibromatosis** (1987) are met. These include: six or more cafe-au-lait macules, optic glioma, two or more Lisch nodules (iris hamartomas), two or more neurofibromas of any type or one plexiform neurofibroma, a first-degree relative with NF1, and so on. Infrequently, benign neurofibromas develop into malignant nerve sheath tumors; NF1 patients also have an increased risk of developing other malignancies, including leukemias (Sorensen et al., 1986). All these features suggest NF1 is similar to other hereditary cancer syndromes such as **retinoblastoma** and **Wilms' tumor** (Glover et al., 1991).

Genetic linkage studies (Barker et al., 1987; Goldgar et al., 1989; Seizinger et al., 1987a,b) and analysis of rearrangements of **chromosome 17** in NF1 patients (Schmidt et al., 1987) mapped the *NF1* **gene** to **17q11.2** (Fain et al., 1989; Fountain et al., 1989; O'Connell et al., 1989). Although in some respects NF1 resembles retinoblastoma (RB), in contrast to RB all NF1 cases result from the inheritance of a mutant allele, and the rate of mutation at the *NF1* locus is very high (around 10^{-4}), apparently resulting from the large size of the locus (Marshall, 1991). Cloning of the candidate NF1 gene (Cawthon et al., 1990; Viskochil et al., 1990; Wallace et al., 1990) showed that it encodes an 11- to 13-kb transcript ubiquitously expressed and denoted **NF1 LT** (Cawthon et al., 1990). It is frequently disrupted in NF1 patients due to mutant alleles in which a premature termination or an insertion is encoded (Cawthon et al., 1990; Viskochil et al., 1990; Wallace et al., 1990; Xu et al., 1990a,b).

Like *RB*, *NF1* **is expressed in all tissues studied so far,** although a high frequency of neoplasia is expressed only in some types of tissues. **Sequence analysis** of the encoded protein points to a homology with the catalytic domain of the human *ras* **GTPase activating protein (GAP)**, and an even closer homology to the yeast *IRA* **gene** product (Ballester et al., 1990; Buchberg et al., 1990; G. A. Martin et al., 1990; Xu et al., 1990a). It appears that NF1 may act as a significant regulator of p21^{ras} activity (G. A. Martin et al., 1990).

Similarity of the *NF1* gene to a **tumor suppressor gene** is strongly suggested by the frequent disruption of the *NF1* gene in VRNF patients and the autosomal dominant mode of inheritance (Buchberg et al., 1990). Members of the NF1 family have been previously shown to act as **transcription regulators** and **initiation factors** (Jones et al., 1987; Rosenfeld and Kelly, 1986; Rossi et al., 1988). The possible functions of the *NF1* gene are suggested also by its close homology to the *IRA*-1 and *IRA*-2 genes of the yeast *S. cerevisiae*, which negatively regulate the Ras-cAMP pathway (Buchberg et al., 1990). The **negative regulatory role** of the NF1 family of proteins has been further demonstrated in their interaction with the CRE1 element of the human proenkephalin-inducible promoter (Chu et al., 1991). Some of the possible results of the disruption of normal NF1 function may lead to growth stimulation and interference with a differentiation pathway (Stanbridge and Nowell, 1990).

No explanation for the role of the *NF1* gene in **benign neurofibromas** and **malignant neurofibrosarcomas** has been given so far, one of the possible scenarios visualizes a similarity with familial adenomatous polyposis (FAP), second allele inactivation in benign colorectal adenomas that eventually leads to malignization (Marshall, 1991). Likewise, inactivation of the second *NF1* allele may be needed for **neurofibrosarcomas** to develop. It appears that other genes (e.g., *p53*) are involved in malignant tumors of the Schwann cells (Menon et al., 1990).

Now that the *NF1* gene has been cloned and sequenced, **direct testing for this gene is possible**. It is not yet clear at present how *NF1* contributes to neoplasia, being ubiquitously expressed in all tissues. In the meantime, a number of probe-restriction enzyme combinations allow informative linkage analysis in clinical carriers and prenatal diagnosis (Ward et al., 1990). Using PCR-based analysis of an intragenic *Alu*-repeat polymorphism, a prenatal diagnosis has been made in a family with only one affected person (the mother) (Lazaro et al., 1992). Since the majority of NF1 mutations are different, the identification of the specific mutation in each patient or family may prove difficult, and the *Alu* repeat polymorphism markedly improves the efficiency of NF1 diagnosis (Lazaro et al., 1992). An excellent discussion of molecular genetics of NF1 has been given (Wallace and Collins, 1991).

5.2.3.2 BILATERAL ACOUSTIC NEUROFIBROMATOSIS

The development of **unilateral lesions** in sporadic cases, and invariably **bilateral acoustic neuromas** in familial **NF2** suggests a similarity to the clinical presentations of **RB** and **Wilms' tumor** (Gusella and Gilliam, 1989). The dominant pattern of inheritance in this disorder is due to the presence of one already defective allele at the *NF2* locus in familial NF2 cases. As in RB, one more hit appears to be enough to destroy the remaining functional allele and for tumor formation in NF2 (Gusella and Gilliam, 1989). In fact, loss of heterozygosity (LOH) in DNA markers on **chromosome 22** has been found in acoustic neuromas of NF2 patients as well as in leukocytes of patients with sporadic unilateral acoustic neuromas (Rouleau et al., 1987; Seizinger et al., 1986, 1987a,b; Wertelecki et al., 1988), and in the **meningioma** of a patient with central neurofibromatosis (Wullich et al., 1989). These findings suggest that inactivation of a **tumor suppressor gene** is involved in NF2. By bracketing the NF2 locus between flanking polymorphic DNA markers on chromosome 22, a 13-cM region has been defined (Rouleau et al., 1990).

Although the incidence of NF2 is substantially lower (1 in 100,000) than that of VRNF, it is highly penetrant and the slow growing tumors that probably begin in childhood do not manifest themselves until after puberty, leading to deafness by age 30 (Gusella and Gilliam, 1989). Furthermore, other **central nervous system tumors**, which can be lethal, are common in patients with NF2 (Parry, 1990).

Because early diagnosis of NF2 is complicated, to a greater degree than NF1, by the lack of diagnosable manifestations, presymptomatic or even prenatal diagnosis by genetic markers assumes particular importance. Early surgical intervention can preserve hearing and facial nerve functions (Gusella and Gilliam, 1989). Flanking markers moving progressively closer to the *NF2* gene can be used for accurate presymptomatic and prenatal diagnosis before the defective gene(s) is isolated and direct probes are constructed (Rouleau et al., 1990).

5.2.3.3 RETINOBLASTOMA

5.2.3.3.1 General Features

Retinoblastoma (RB) is a malignant tumor affecting the retina of predominantly very young children. Essentially two forms are known: hereditary and sporadic. Both eyes are involved, when the disease is inherited. When only one eye is affected the disease is usually considered a nonhereditary, sporadic from. This is an autosomally dominant condition with approximately 90% penetrance, so there is a 50% chance that a child of affected parents will also develop the tumor (Judisch and Patil, 1985; Knudson, 1975; Vogel, 1979). For the malignancy to appear, two events must occur: (1) the presence of a mutated, tumor-predisposing allele, and (2) either a somatic mutation, leading to loss of the normal *RB* allele, or some other chromosomal alteration (for reviews, see Benedict et al., 1988; Buchanan and Cavenee, 1987; Cowell, 1991; Freeman et al., 1990; Ponder, 1988).

Eighty-five percent of patients with only one affected eye have not inherited the *RB* gene. Children with an inherited mutant *RB* allele are also prone to develop **osteosarcoma** and **soft-tissue sarcoma** with a markedly increased frequency. **Melanoma** and **bladder cancer** were been associated with RB in 23 cases among 117 relatives, who were known carriers of the *RB* mutation, in a series of 1438 parents and 2663 other relatives of RB patients (Sanders et al., 1989). For relatives who do not carry the mutation, there appears to be no excess risk of developing cancer. It is unclear so far to what extent **environmental factors**, such as exposure to X-rays, play a role in the development of the nonfamilial, sporadic form of RB (Bunin et al., 1989).

5.2.3.3.2 Chromosome Aberrations

Differentiation between **heritable** and **sporadic forms** of RB based only on clinical data may be difficult. Because the predisposition to the development of tumors in survivors of hereditary RB is 1000-fold higher than in the general population, the evaluation of **chromosomal characteristics** in the tumor may be of great prognostic significance. The cytogenetic detection of multiple unrelated clones in the tumor has been helpful in a clinical case in which one clone of tumor cells displayed completely different markers, suggesting a hereditary, multifocal origin of the tumor (Tien et al., 1989).

Cytogenetic evidence pointed to chromosomal region **13q14** as the area in some way involved in the primary and secondary events leading to the development of RB. The presence of a **homozygous deletion** has been established in region 13q14 by a **high-resolution banding technique** of RB tumor cells in primary culture and in early subcultures (Lemieux et al., 1989). **Genomic imprinting** may play a role in this disease as shown by the **paternal origin** of affected chromosome 13, carrying a reduction of homozygosity in both bilateral and unilateral tumors (Leach et al., 1990).

5.2.3.3.3 The *RB* Gene

Genomic organization of the human *RB* **susceptibility locus** has been described (McGee et al., 1989). This gene, located within **chromosome 13q14**, spans approximately 200 kb. The wild-type gene contains 27 exons and 26 introns. Genetic changes at 13q14 have been identified not only in RB, but also in tumors of various tissues (see below). Analysis of chromosomal aberrations in this region can be helpful in elucidating whether *RB*1 or a closely linked gene(s) is involved in a wide range of malignancies.

A long-range, physical map of the *RB*1 region with overlapping areas around the gene has been developed (Ford et al., 1990). A number of **CpG islands**, frequent sites of **methylation**, have been identified in that region. Because they mark the start of many genes, the putative genes in the vicinity of *RB*1 may be relevant to the development of many malignant tumors, if they harbor deletions and/or translocations (Ford et al., 1990). RFLP analysis of DNA from leukocytes of *RB* carriers shows LOH of chromosome 13 markers. Frequently, LOH is also observed in **osteosarcomas** from patients with and without a history of retinoblastoma (Ford et al., 1990).

The *RB* gene 4.7R on chromosome 13q14 contains, in some cases, homozygous internal deletions yielding corresponding **truncated transcripts** (Friend et al., 1986; Fung et al., 1987; Lee et al., 1987a,b). The majority (60%) of tumors with no identifiable structural changes showed either the absence of an *RB* transcript, or abnormal transcripts, that were larger or smaller than normal, indicating the **loss of a splice junction**. The **Y79 retinoblastoma cell line** most commonly used consistently shows a reduced amount of a truncated *RB* transcript, although no gross DNA changes can be detected by cDNA probes. The high frequency of mutations in 4.7R supported its identification as the *RB* gene (Dunn et al., 1988).

The **PCR** and **RNase protection assays** show that inactivation of both alleles of the *RB*1 gene is produced by small deletions, duplications, and point mutations leading to splice alterations and, consequently, loss of an exon from the mature mRNA (Dunn et al., 1989). The mutant gene has variable expression regulated at the level of transcription and manifested in the absence of a functional *RB*1 gene product. This has been demonstrated by the absence of the mutant allele expression in lymphoblasts from affected patients (Dunn et al., 1989).

Somatic mutations at the *RB*1 locus may be involved in the formation and development of tumors in RB patients (Greger et al., 1990). In a patient with a hereditary extraocular ectodermal form of retinoblastoma — **sinonasal undifferentiated carcinoma**, a deletion at the *RB*1 locus could be detected in **metastatic carcinoma cells**, but not in normal unaffected tissues of that patient (Greger et al., 1990).

An interesting **animal model of heritable retinoblastoma** has been described, where a specific association between **p105^{RB}** and **simian virus 40 (SV40) T antigen** could be demonstrated in the RB tumor cells (Windle et al., 1990). The expression of this viral oncogene — SV40 T antigen — in the retina of **transgenic mice** produces heritable ocular tumors identical to human RB in their histological, ultrastructural, and immunohistochemical characteristics. Inactivation of the RB protein, p105^{RB}, in this case through the binding of a viral oncoprotein, T antigen, has been unequivocally proven to be the key factor in the development of the tumors (Windle et al., 1990).

5.2.3.3.4 The RB Protein

The *RB* gene encodes a protein of 928 amino acids, called **p105** or **p110**, which is localized to the nucleus and displays DNA-binding activity (Lee et al., 1987a,b; Xu et al., 1989). This is a ubiquitous protein identified in all vertebrate species examined, but **conspicuously absent from many tumor cell lines** (Huang et al., 1988). Interaction of the RB protein with oncogene products causes its **inactivation** and transformation of cultured calls. The truncated forms of RB protein are not phosphorylated and fail to bind to the **SV40 T antigen** (Shew et al., 1990). The crucial portion of the RB product in this interaction is its C-terminal domain (Shew et al., 1990). Binding to **adenovirus E1A antigen** is also abrogated (Kaelin et al., 1990).

The extent of the **RB protein phosphorylation** is important in the control of its transforming potential. The phosphorylated RB protein lacks the growth supporting function, and it is the **dephosphorylated product that apparently prevents tumors from developing** (Benedict et al., 1990a,b; Freeman et al., 1990; Green, 1989). The state of phosphorylation of p105^{RB} varies throughout the **cell cycle**, so that between the end of G_2 and the beginning of G_1 the protein is completely dephosphorylated (Cooper and Whyte, 1989; Ludlow et al., 1990). It also changes markedly during **cell differentiation**, irrespective of the method triggering differentiation (e.g., tumor promoter or dimethylsulfoxide [DMSO]). Induction of the differentiation state in human hematopoietic cells in culture is associated with the predominance of the dephosphorylated form of the RB protein (Akiyama and Toyoshima, 1990).

While **RB phosphorylation** appears, in general, to be linked to the ability of a cell to synthesize DNA, there are **cell-specific** or **lineage-specific** differences in both the regulation of RB phosphorylation and *RB*-1 gene expression at the mRNA level (Furukawa et al., 1990). Not only phosphorylation of the RB product, but even its total content per cell varies with the state of differentiation, decreasing with progression toward the differentiated state (Yen et al., 1990). Phosphorylation of the RB protein is an intrinsic property of a normal RB protein and mutant protein are unable to become hyperphosphorelated and fail to bind to nuclear structures and are also defective in binding to the E1A oncoprotein — another characteristic feature of wild-type RB protein (Templeton et al., 1991). The state of phosphorylation of the RB protein varies in the cell cycle so that it is phosphorylated from S to M phase and dephosphorylated in G_1 (Lin et al., 1991). One of the protein kinases essential for transitions through the cell cycle — **cdc2** — appears to be a specific enzyme that phosphorylates the RB protein accounting for the cell-cycle dependent phosphorylation of the RB protein (Lin et al., 1991).

The **effect of aging** on the RB protein has been studied in cultured cells. Although the extent of phosphorylation of the RB protein in young and aging human diploid fibroblasts, and in the immortalized but untransformed mouse fibroblasts, is not known, it is clear that a change in the regulation of *RB* transcription is not part of senescence and that *RB* activity is regulated primarily **post transcriptionally** (Shigeoka and Yang, 1991). In Syrian hamster embryo senescent cells, that RB protein is found entirely in a hypophosphorylated form (Futreal and Barrett, 1990). In addition to the *RB* gene product, two ***RB*-associated proteins** have been identified that could be important in *RB* gene regulation or interact with it in a dependent fashion (Xu et al., 1989).

Another issue of practical significance is that **specific antibodies** developed against *RB* gene products and associated proteins are also capable of identifying their truncated forms (Xu et al., 1989). Consequently, analysis of tumor explants with such antibodies may provide clinically useful information, especially if differences in the biological behavior, aggressiveness, or resistance to certain therapeutic modalities are found in **RB+** and **RB– tumors.** A very important observation has been made regarding the ability of a cellular protein (RbAP46) to specifically associate with the RB protein (Huang et al., 1991). This protein displays RB-binding properties very similar to the well-characterized specific RB ligand SV40 T antigen. The details of RbAP46 function in the cell are not yet known but its competitive interaction with the products of the oncogenic virus may lead to the disclosure of important regulatory roles of the RB protein.

5.2.3.3.5 The *RB* Gene in Other Tumors

The *RB* gene represents a **prototype of a recessive gene**, in which absence of both alleles, or the loss of function affecting both alleles, results in tumorigenesis. This appears to apply not only to affected retinal cells, but to other human tumors as well. Using cDNA *RB* probes, analysis of the expression of the *RB* gene is possible irrespective of the presence of structural aberrations in the gene itself (Benedict et al., 1988).

The *RB* gene is inactivated in two of nine human **breast cancer cell lines** examined (Lee et al., 1988). The *RB* mRNA is lengthened in one cell line, reflecting internal duplication of a 5-kb region. Consequently, translation of the message is probably prematurely terminated due to a shifted reading frame. The opposite is found in the other cell line, which has a homozygous deletion of the *RB* gene. As a result, the entire gene is removed beyond exon 2. The *RB* gene product, p105^{RB}, cannot be detected with specific antibodies in either cell line (Lee et al., 1988).

Analysis of *RB* gene expression by a highly sensitive **immunological assay** for the *RB*-encoded p105 protein confirmed that **inactivation of the RB product is universal** in RB cells and is nearly as frequent in **small cell lung carcinoma (SCLC)** (Horowitz et al., 1990). In addition, one third of the **bladder carcinoma cell lines** tested carried altered or absent p105^{RB}. On the other hand, the **majority of other pulmonary neoplasms have readily detectable levels of intact *RB* mRNA and p105^{RB}** (Horowitz et al., 1990).

RB mutations could be confirmed in SCLC and bladder carcinoma cells by **PCR amplification of *RB* mRNA**, and the demonstration of **splicing mutations resulting in a truncated product** that fails to bind adenovirus E1A and SV40 large T antigen (Horowitz et al., 1990). Only infrequent *RB* inactivation is detected by this particular assay in human **colon carcinoma, breast carcinoma,** and **melanoma**. The binding assay, however, may not exclude the possibility of subtle structural alterations interfering with the function of an apparently normal p105^{RB} (Horowitz et al., 1990).

Loss of *RB* gene expression was also observed in 90% of **SCLC** by others (Hensel et al., 1990). Although no evidence of familial predisposition of patients to SCLC is known, it remains to be seen whether alteration of *RB* gene expression in this type of lung tumor is related to their neuroendocrine characteristics (Hensel et al., 1990).

Structural alterations of the *RB*-1 gene have been observed in 33% of **soft-tissue sarcomas** especially in high-grade (Wunder et al., 1991), and particularly in **leiomyosarcomas**, which sometimes have homozygous deletions of this gene (Stratton et al., 1989). The expression of $p105^{RB}$ is normal and homogeneous in all tumor cells in 30% of primary sarcomas whereas tumors with metastatic spread have altered expression of the RB protein (Cance et al., 1990). Expression of the RB protein appears to correlate with survival. In patients whose tumors had homogeneous RB expression the survival was markedly increased, compared to those with either heterogeneous expression or no expression at all (Cance et al., 1990).

The critical role of the *RB* gene in some forms of tumors is supported by **transfection experiments**, in which a cloned *RB* gene is introduced via retroviral-mediated gene transfer into retinoblastoma WERI-Rb27 or osteosarcoma Saos-2 cell lines, which have partial deletions of endogenous *RB* genes (Huang et al., 1988). Indeed, introduction of the *RB* gene drastically affects cell morphology, growth rate, soft agar colony formation, and tumorigenicity of cultured cells in nude mice. Some differences are noted, however, in the response of the retinoblastoma and osteosarcoma cell lines to the RB protein replacement (Huang et al., 1988). This is apparently due to alternative pathways in osteosarcoma genesis, which may not involve *RB* gene inactivation.

The other possibility is that *RB* **gene inactivation, if present, acts in concert with other mechanisms, such as oncogene amplification**. In fact, N-*myc* amplification is known to occur in retinoblastoma and osteosarcoma cells in RB patients, but appears to be an event independent of *RB* gene expression (Sakai et al., 1988; Squire et al., 1986). Again, yet another tumor suppressor gene, *p53*, is altered in osteosarcoma cell lines (Romano et al., 1989). Thus, **replacement of suppressor genes** in tumor cells opens a practical approach to gene therapy of clinical malignancy, at least in selected cases.

Oncogene expression (N-*myc*, c-*myc,* and *sis*) in pediatric cancers such as **rhabdomyosarcoma**, **Wilms' tumor**, **neuroblastoma**, **osteosarcoma**, and **retinoblastoma** does not seem to be under control of the putative tumor suppressor genes (Pasquale et al., 1988; Squire et al., 1986). It is possible that, although tumorigenicity behaves as a recessive genetic trait in human **pediatric** cancer, different genetic elements may be affected during tumor development in **adults** (Pasquale et al., 1988; Squire et al., 1986).

5.2.3.3.6 Detection of *RB* Gene Abnormalities

Diagnosis of RB or other malignancies, in which the *RB* gene is suspected to contribute to tumorigenesis, has relied on RFLP analysis using probes from the cloned gene (Wiggs et al., 1988). In 10 to 20% of all families, alteration of the gene produces an easily recognizable restriction fragment that is diagnostic for the mutant *RB* allele, whereas 10% of all families are not informative for any of the RFLPs (Yandell and Dryja, 1989). Those falling between these two extremes have been diagnosed by an RFLP-based analysis using informative RFLPs closely linked to or within the disease locus.

An **intragenic *Bam*HI RFLP** within the 5' end of the *RB*-1 gene appears to be superior to other polymorphic probes within *RB*-1 for genetic counseling of familial and nonfamilial RB patients and their relatives (Goddard et al., 1990). Intragenic probes derived from within the *RB* predisposition gene are also used for **prenatal screening** for the inheritance of mutant alleles (Onadim et al., 1990). Using densitometric analysis of Southern blots, qualitative and quantitative estimates of the *RB*-1 gene deletions can be made in patient samples (Kloss et al., 1991).

A different **screening strategy** has been developed using **PCR amplification** of the *RB* gene regions followed by **direct DNA sequencing** (Yandell et al., 1989; Yandell and Dryja, 1989). An ***Xba*I RFLP** within the human *RB* susceptibility gene (Wiggs et al., 1988) can be used for analysis by PCR (McGee et al., 1990). Another pair of primers has been reported for the analysis of a polymorphic *Bam*HI site in intron 1 of the human *RB* gene (Bookstein et al., 1988, 1990).

Normal allelic variations in 13 regions of the *RB* locus have been analyzed, and four DNA sequence polymorphisms were found. The higher informative power of some of these polymorphisms was demonstrated, as no other known RFLPs at this locus proved informative (Yandell et al., 1989; Yandell and Dryja, 1989). Combining PCR with direct sequencing of well-characterized loci offers a viable diagnostic approach for detecting useful DNA polymorphisms that complement RFLP analysis, denaturing gradient gel electrophoresis, and RNase mismatch methods (see above) (Yandell and Dryja, 1989). Moreover, it is procedurally convenient, rapid, and capable of detecting essentially all polymorphic sites. Combining reverse transcription (RT)-PCR with single strand conformation polymorphism (SSCP) analysis of the amplified product, the identification of abnormal *RB* gene transcripts allows a highly sensitive DNA analysis for possible nucleotide substitutions, deletions, etc. (Murakami et al., 1991).

5.2.4 NEUROBLASTOMA

Neuroblastomas (NBs) are highly malignant neoplasms, predominantly of infancy and early childhood (under 3 years), that usually arise in the **adrenal medulla** and **sympathetic ganglia** of the retroperitoneum and posterior mediastinum (Lloyd, 1990). Histologically, the tumor is composed of **small round to fusiform cells**, usually with large hyperchromatic nuclei.

5.2.4.1 DIFFERENTIAL DIAGNOSIS

In the differential diagnosis of NB the other **small round cell tumors** are considered, including **Ewing's sarcoma**, **Wilms' tumor**, **malignant lymphoma**, and **rhabdomyosarcoma**. In cases of poorly differentiated neoplasms, light microscopy must be complemented by immunohistochemical and/or ultrastructural studies. **Neuron-specific enolase (NSE)** and **synaptophysin** staining is usually positive, and **chromogranin A** displays focal positivity (Lloyd, 1990). The serum level of NSE can be used for monitoring childhood NB (Cooper et al., 1987), **Merkel cell tumor** (primary neuroendocrine tumor of skin) (Plowman, 1990a), and **medulloblastoma** (Plowman, 1990b).

Additional markers (**PAS, cytokeratin, leukocyte common antigen** and **lymphocyte markers, myoglobin**, and **S-100 protein**) can also be used to differentiate NBs from similar appearing tumors. Alternatively, staining for the **adhesion molecules L1** and for the **highly sialylated neural cell adhesion molecule isoforms** helps in differentiating NB from other tumors (Figarella-Branger, 1990).

Differentiation of NB into **ganglio-NB** and **ganglioneuroma** may occur either spontaneously or following chemotherapy. Gene level markers, particularly evaluation of the amplification of N-*myc* vs. c-*myc*, can help in differentiating NB from tumors with similar morphology (Yeger et al., 1990). A cell line, **NUB-20**, established from a histopathologically diagnosed case of metastatic NB was classified as a peripheral **neuroepithelioma** on the basis of the phenotypic characteristics, a **t(11,22) translocation**, as well as amplification and overexpression of the **c-*myc*** gene rather than N-*myc*. Multiparameter analysis of cellular and molecular characteristics is particularly emphasized in the classification of the heterogeneous group of small round cell tumors (Yeger et al., 1990) in support of the earlier proposed use of the pattern of N-*myc* amplification and expression for classification of small round cell tumors (Thiele et al., 1987a,b, 1988a; Triche et al., 1988). While N-*myc* may not be as helpful for differential diagnosis of other types of tumors, its amplification and/or high expression have been reported in **retinoblastomas** (Lee et al., 1984), **SCLC** (Ibson et al., 1987; Johnson et al., 1988; Kiefer et al., 1987; Nau et al., 1986), **Wilms' tumor** (Nisen et al., 1986, 1988), **rhabdomyosarcoma** (Garson et al., 1986; Garvin and Sens, 1986), **astrocytoma** (Garson et al., 1985), and **thyroid C cell tumors** (Boultwood et al., 1988).

5.2.4.2 PROGNOSIS

Although the determination of **total catecholamines** is marginally the best single test to diagnose NB (Worthington et al., 1988), comparison of a number of parameters demonstrated that **clinical staging** and **DNA content** could be used as independent prognostic indicators, particularly in stage III and IV patients (Taylor et al., 1988).

In a series of 58 patients with NB or ganglio-NB, the **age of patients** under 1.5 years was the best prognostic indicator, followed by the **degree of tumor differentiation** (Oppedal et al., 1988). Immunohistochemical pattern was of no additional help. **Aneuploidy is associated with a better prognosis**, possibly due to a better response to treatment (Gansler et al., 1986; Hayashi et al., 1988; Kaneko et al., 1987, 1988; Look et al., 1984; Oppedal et al., 1988; Taylor et al., 1988).

5.2.4.3 N-*myc*

5.2.4.3.1 N-*myc* and Catecholamine Metabolism

Molecular biological analysis of NBs has repeatedly demonstrated N-*myc* amplification and expression (Kanda et al., 1987; reviewed in Thiele et al., 1988a; Thiele and Israel, 1988). An **inverse relationship** exists between N-*myc* **amplification** and **catecholamine metabolism** in patients with advanced NB (Nakagawara et al., 1990a,b). Nevertheless, **mass screening based on urinary catecholamine metabolites, particularly if performed only at 6 months of age, may be inadequate for the detection of high-risk NBs** (Kaneko et al., 1990). Based on chromosome and N-*myc* amplification studies, a mass screening program conducted at 6 and 12 months of age is recommended for the early detection of high-risk NBs (Kaneko et al., 1990).

5.2.4.3.2 N-*myc* and DNA Ploidy

Comparison of **nuclear DNA content** and N-*myc* **amplification** in NBs demonstrated that tumors with N-*myc* amplification represent a subset of nonaneuploid tumors (Taylor and Locker, 1990). The **degree of amplification** (20 to 1500 copies) was not strongly correlated with ploidy, and independently predicted poor prognosis regardless of the histological signs of differentiation (Oppedal et al., 1989). Failure to correlate DNA ploidy with prognosis and N-*myc* amplification may, in part, be due to the fact that only one allele of the N-*myc* gene appears to be amplified (Yamada et al., 1988).

5.2.4.3.3 Mechanism of N-*myc* Amplification

Analysis of the **mechanism of N-*myc* amplification** by Southern blot and PCR analysis using somatic cell hybrids revealed that one of the **initial steps** leading to amplification is excision of the N-*myc* locus from the vast majority of the **chromosomes 2** (Hunt et al., 1990). This finding is consistent with the **rare recombination models** visualizing the deletion of N-*myc* (or any other gene, for that matter) as being part of the amplification process, but alternative mechanisms cannot yet be excluded as important elements of other amplification systems (Hunt et al., 1990).

Elucidation of the molecular events associated with other chromosomal abnormalities found in NB, including the most frequent deletion of the short arm of **chromosome 1**, will shed additional light on the molecular mechanism of NB (Brodeur et al., 1981, 1988; Gilbert et al., 1984; Kaneko et al., 1987). As new facts emerge, the **role of N-*myc* amplification** in NB becomes progressively complicated. Although N-*myc* amplification is a recognized event in the development of human NB, other genetic lesions appear to contribute to NB aggressiveness *in vivo*, and to its growth *in vitro*. This was suggested in a newly established NB cell line, in which loss of N-*myc* amplification occurred during the cloning process (Cornaglia-Ferraris et al., 1990b). On the other hand, judging from the changes induced by human **interferon γ** in the

morphology of cultured NB cells and alteration in N-*myc* gene expression, these two characteristics of the tumor appear to be related (Watanabe et al., 1989).

Because N-*myc* amplification appears to play such a prominent role in determining the course
of NB, the characterization of the N-*myc* **amplification units** is important in understanding
the role of this oncogene in the disease. Cell lines and primary NB tumors have been studied
with respect to rearrangements, homogeneity of N-myc amplification units within individual
tumors, and conservation of the amplified sequences among different tumors (Zehnbauer et al.,
1988). Relatively few rearrangements within the N-*myc* amplification unit have been found.
Moreover, there was a significant homogeneity among the amplification units within one tumor,
and there was a great similarity between the amplicons in different tumors (Zehnbauer et al.,
1988).

Based on these findings **gene probes for N-*myc*** can be developed that are especially useful
when measuring low levels of amplification in bone marrow samples during remission
(Zehnbauer et al., 1988). **A single major amplicon can possibly be defined for each NB
tumor**, in contrast to the situation with many drug-resistant cell lines, in which these amplification units are extremely heterogeneous. N-*myc* amplification leads (as does any gene
amplification, for that matter) to a perturbation of the nucleosomal structure of chromatin
containing the amplified gene, as determined by micrococcal nuclease digestion (Telford and
Stewart, 1990).

The **stability** of N-*myc* amplification, its **uniformity** within a tumor, and the **similarity**
of its characteristics observed in different patients make this marker an attractive tool
for diagnosing, monitoring, and predicting the course of disease in a clinical setting. **Consistency** of N-*myc* gene copy numbers within a tumor is observed not only at different tumor sites,
but also at different times *in vivo* (Brodeur et al., 1987). Not only is amplification of the oncogene a constant feature in human NBs, but the enhanced N-*myc* expression is also supported
by the **increased stability of the N-*myc* mRNA**; this is controlled by a posttranscriptional
mechanism that is similar to that described for c-*myc* in other cell types (Amy and Bartholomew,
1987).

5.2.4.3.4 N-*myc* and Prognosis

Amplification of N-*myc* in NBs is associated with an **advanced stage** of the disease and **poor
prognosis** (Brodeur et al., 1984). Multiple studies, using Southern blot and ISH, repeatedly
established **a correlation between N-*myc* amplification and poor outcome** (Brodeur et al.,
1986, 1987; Cohen et al., 1988; Garvin et al., 1990; Grady-Leopardi et al., 1986; Nakagawara
et al., 1987; Tsuda et al., 1987). In the majority of known cases, the induction of a differentiated
state in cultured hybrid NB cells by **dibutyryl cAMP** leads to a four- to eightfold increase in the
level of N-*myc* mRNA, compared to that in untreated, undifferentiated controls (Berman et al.,
1989). N-*myc* expression contributes to malignant progression possibly by stimulating the
expression of **autocrine growth factor(s)** (Schweigerer et al., 1990). The elevated expression
of N-*myc* also contributes to the acquisition of an **enhanced malignant phenotype** as shown
by transfection experiments in cultured cells and by the induction of tumors in nude mice
(Schweigerer et al., 1990). Neutralization of this autocrine growth factor activity may offer an
avenue for the treatment of human NBs (Schweigerer et al., 1990).

The significance of the N-*myc* **gene load** for **prognosis** of NB is emphasized by the
significantly better, progression-free survival of patients with single-copy tumors compared to
those having multicopy tumors, who typically experienced rapid tumor progression (Cohn et al.,
1988). Despite correlation of the **level of expression** of the N-*myc* gene with prognosis, the
relationship between the degree of NB differentiation and malignancy is not a simple one
(Hammerling et al., 1987). Moreover, a highly tumorigenic NB cell line has been described that
has undetectable N-*myc* expression (Cornaglia-Ferraris et al., 1990a).

5.2.4.3.5 N-*myc* and DNA Index

Although N-*myc* amplification occurs in **advanced stages** of disease, N-*myc* evaluation alone is probably insufficient to predict the course of disease in all cases and, therefore, the determination of **DNA content** or **chromosomal analysis** was suggested to provide more useful information (Taylor and Locker, 1990). On the other hand, **no correlation** was found between **DNA index** (DI) and N-*myc* gene amplification or expression (Slavc et al., 1990). One cannot evaluate prognosis based solely on the expression of N-*myc* and c-*myc*, which are usually expressed in primary untreated NBs. In the absence of N-*myc* amplification, the expression of these genes does not correlate with disease progression (Slavc et al., 1990). Also, the **DI of tumors determined by flow cytometry (FCM) does not correlate with disease behavior**. Patients with nonaneuploid tumors have shown a good response to therapy (Slavc et al., 1990). Thus, in contrast to Taylor and Locker (1990), Slavc and coworkers (1990) believe that "N-*myc* amplification remains the best molecular genetic marker of disease progression in patients with NB".

Consequently, a **combination of DI** and N-*myc* amplification level is advocated for routine management of NB patients, because **only these two parameters in combination proved to be significantly associated with a high risk of relapse independent of the age of patients** (Bourhis et al., 1991). The use of N-*myc* amplification for patient management requires its evaluation prior to administration of any treatment modalities; not only may the histology of tumors be modified by treatment, but the assessment of N-*myc* amplification may also be inaccurate (Combaret et al., 1989).

5.2.4.3.6 N-*myc* and Drug Resistance of NB

Evaluation of N-*myc* gene amplification and its expression in NBs helps to unravel the mechanism(s) underlying the **resistance** of these neoplasms to therapy. Tumors with amplified N-*myc* are initially sensitive to chemotherapy; later, however, they acquire resistance to therapy and tend to recur. **Recurrent NBs with N-*myc* amplification fail to respond to chemotherapy** (Nakagawara et al., 1988). One of the culprits of inherent or acquired drug resistance could be the ***mdr*-1 gene**, known to be responsible for multidrug resistance in various cell lines (Roninson et al., 1986). This gene appeared to be expressed not only in normal adrenal gland, but was also found to be overexpressed in some tumors, although not in NB cell lines (Fojo et al., 1987).

Because amplification and overexpression of the N-*myc* oncogene in NB is associated with poor prognosis, it is important to determine whether the *mdr* gene expression contributes to the observed poor outcome. Interestingly, an **inverse correlation** between the expression of N-*myc* and *mdr*-1 has been found in the same tumors: *mdr*-1 mRNA was expressed at a **lower level** in the advanced stage tumors and in the histologically undifferentiated NBs (Nakagawara et al., 1990a,b). Thus, the acquisition of drug resistance by these neoplasms is not associated with an increase in *mdr*-1 expression.

These findings contrast with the reported increase in the level of *mdr*-1 expression in human NBs in response to chemotherapy (Bourhis et al., 1989). The reason for these conflicting observations is not clear at this time and other mechanisms known to be involved in drug resistance, such as the expression of **epidermal growth factor** and the increased activity of **topoisomerase II** or **glutathione *S*-transferase**, are being investigated (Nakagawara et al., 1990a).

In a retrospective study, the level of P-glycoprotein has been evaluated by immunohisto-chemistry in NB samples and the patients found negative for P-glycoprotein had significantly longer relapse-free and overall survival than the P-glycoprotein-positive children (Chan et al., 1991). It appears, therefore, that measuring the P-glycoprotein expression prior to treatment may be of value in predicting the success or failure of therapy for nonlocalized NB. Furthermore, attempts to combine chemotherapy with reduction or reversal of the P-glycoprotein-mediated multidrug resistance may be of clinical benefit (Chan et al., 1991).

Likewise, the established association of N-*myc* amplification with arrest of differentiation can be exploited to develop **specific antitumor agents**, as demonstrated in the effective growth inhibition and change of phenotype characteristics of cultured neuroepithelioma cells on addition of N-*myc* **antisense oligomers** (Rosolen et al., 1990).

5.2.4.3.7 Other Oncogenes Related to Prognosis of NB

Knowledge of the molecular mechanisms underlying the **developmental progression** of a tumor allows selection of specific markers tightly associated with individual differentiation stages. When NB cell lines are induced to differentiate by the addition of **retinoic acid**, a marked decrease in the expression of N-*myc*, **c-*myb***, and **c-Ha-*ras*** is observed (Thiele and Israel, 1988). The expression of two other protooncogenes, **c-*ets*-1** and **c-*fos***, increases during NB differentiation, whereas the expression of c-*fos* parallels development of an extensive network of neuritic processes. The high level of expression of the N-*myc* gene, while not related to the cell growth characteristics, appears to prevent NB cells from differentiating (Thiele and Israel, 1988).

The other protooncogene that displays a close correlation with the clinical course of NB is Ha-*ras* (Ireland, 1989; Tanaka et al., 1988). In contrast to N-*myc*, the expression of the Ha-*ras* gene is **positively correlated with good prognosis**, so that the amount of the Ha-*ras* p21 protein, detected by a simple immunohistochemical procedure, can be of clinical importance. These observations are strengthened by the previously established association of Ha-*ras* p21 expression with the differentiation of neural tissues (Bar-Sagi and Feramisco, 1985; Hagag et al., 1986; Noda et al., 1985; Tanaka et al., 1986).

Thus it appears that N-*myc* amplification and expression, marking the **arrest stage** in the differentiation of NB, and the mutations in the Ha-*ras* gene directly evaluated by PCR (Ireland, 1989) and taken as a measure of differentiation, may together provide a practical tool for the prognostic assessment of NB. Yet another potential candidate for a prognostic marker in NB is the **c-*myb*** protooncogene, whose expression progressively declines during differentiation (Thiele et al., 1988b).

The expression of the protooncogene **c-*src***, in particular the neuronal product of the gene, **pp60**$^{c-srcN}$, that appears during normal development of neuroblasts and neurons is associated with **good prognosis**, especially in the evaluation of infants (Bjelfman et al., 1990). The expression of **pp60**$^{c-srcN}$ is limited to NBs and RBs, while **pp60**$^{c-src}$ is found in most tumors studied.

Fine analysis of **developmental regulation** of the **neuropeptide Y gene** in normal adrenal gland and NBs indicated that the expression of neuropeptide Y mRNA in NB tumors may be associated with a **better prognosis** among patients under 1 year of age at the time of diagnosis (Cohen et al., 1990). Similarly, the expression of neuropeptide Y mRNA in **pheochromocytoma** was also associated with an improved prognosis (Helman et al., 1989). These findings suggest that benign and malignant pheochromocytomas may arise from subpopulations of adrenal medullary cells that differ in their expression of neuropeptide Y, much like NBs at stages IVS and IV. It is the possible **programmed death** of neuroblastic islets, which share common biochemical (neuropeptide Y mRNA expression) and biological characteristics (disappearance/regression) with NB stage IVS, that explains the better clinical prognosis of tumors expressing neuropeptide Y (Cohen et al., 1990).

An **oncoprotein**, designated **Op18**, has been described in acute lymphocytic leukemia and some solid tumors, including NB (Hailat et al., 1990). This protein appears to be similar, if not identical, to the previously described phosphoprotein **prosolin**, and another phosphoprotein called **stathmin**, all of these proteins being related to the state of differentiation. In NB the **state of phosphorylation of Op18** is inversely related to the level of N-*myc* amplification whereas the level of Op18 phosphorylation progressively increases with differentiation of NB (Hailat et al., 1990).

The *smg* **p25A** (a p21^{ras}-like GTP-binding protein) **gene** is expressed in neural crest-derived tumor cell lines and NB tissues (Sano et al., 1990). The level of expression of the smg p25A mRNA is closely related to the neuronal differentiation state of tumors derived from the neural crest.

Yet another marker of differentiated NB cells is **β$_2$-microglobulin**, which is not associated with N-*myc* expression (Cooper et al., 1990). No association has been observed between N-*myc* expression and diminished levels of MHC class I antigen expression in NB.

Nerve growth factor receptor (NGFR) is a biological marker for neuroepithelioma, whereas its expression is heterogeneous in NB cell lines (Baker et al., 1989). The *NGFR* gene is apparently normal in NB cell lines, but there are multiple distinct defects in the NGF/NGFR pathway that may play an important role in the initiation or maintenance of the undifferentiated state of NB (Azar et al., 1990). The known **tumor suppressor genes**, such as *RB*-1 and *p53*, have so far failed to correlate with tumorigenic expression in NB hybrid cell lines, and the search for yet undisclosed tumor suppressor activities and their utility for the diagnosis and/or prognosis of NBs and sarcomas continues (Chandar et al., 1990; Futreal and Barrett, 1990; Weissman et al., 1990).

In NB cell lines, but not in non-NB tumor cell lines, a significant expression of the *ret* protooncogene can be observed, although the role it plays in NB cells has not been defined yet (Ikeda et al., 1990).

5.2.5 OTHER SELECTED TUMORS OF THE NERVOUS SYSTEM

Although some of the tumors of the nervous system are relatively frequent, such as **gliomas** (Black, 1991a,b), the accumulated molecular genetic evidence has so far failed to implicate specific etiological factors. Studies on protooncogenes in experimental animals, cell lines, and surgical tumor specimens revealed certain regularities in the spatial and temporal **expression of protooncogenes** in normal neural tissues (for reviews, see Fenoglio-Preiser et al., 1991; Sudol, 1988) and tumors of neural origin (reviewed in Bigner et al., 1989; Bigner and Vogelstein, 1990; Black, 1991a,b; McDonald and Dohrmann, 1988; Rutka et al., 1990). Here, only a brief summary of selected findings will be given.

5.2.5.1 CHROMOSOMAL ABNORMALITIES

Cytogenetic studies have established the structural alterations of **chromosomes 7, 10, 22,** and the **sex chromosomes** as the most common in **malignant gliomas** (Rutka et al., 1990). Other chromosomal abnormalities in glial tumors involve the short arm of **chromosome 9, 3,** and **11** as well as **chromosomes 1, 6,** and **19** (Pantazis et al., 1985); **double minutes (DMs)** are present in about 50% of gliomas, suggesting DNA amplification of potentially diagnostic significance (see below). LOH has been repeatedly demonstrated in **meningiomas** in the CNS, and in peripheral neural tumors, **schwannomas,** and **neurofibromas,** pointing to the possible involvement of tumor suppressor genes by analogy with the loss of 13q in RB and 11p in Wilms' tumor.

Glioblastoma multiforme (GBM) displays a characteristic overrepresentation of chromosomes 7 and 22, the chromosomes harboring genes for the α and β **chains of platelet-derived growth factors (PDGF)** (Scheck and Shapiro, 1990). A lower grade malignant tumor, **anaplastic astrocytoma,** shows focal areas of overexpression of the genes for PDGF α chain, which is substantially more pronounced in GMB (Scheck and Shapiro, 1990). This marker could potentially be used for differentiating the histologically equivocal cases.

RFLP analysis of malignant astrocytomas revealed LOH on **chromosomes 10** and **17p,** suggesting that tumor suppressor genes in astrocytoma tumorigenesis may be located on two different chromosomes (Fults et al., 1990). In fact, extensive analysis of all chromosomes in 41 malignant astrocytoma patients detected LOHs on all autosomes except chromosome 21 (Fults et al., 1990). Furthermore, the number of chromosomes with LOH correlated with tumor histopathology. A particularly frequent event, especially in GBM, was loss of broad regions of chromosome 10. Comparison of the patterns of LOH in malignant astrocytomas and colorectal carcinomas suggested that different sets of tumor suppressor genes are involved in the tumorigenesis of these two cancers (Fults et al., 1990).

Detailed analysis of **chromosome 22** in 81 unrelated meningioma patients demonstrated the retention of the constitutional genotype along chromosome 22 in 37% of cases, monosomy 22 in the tumor DNA in 52%, and partial LOH at 22q in 11% of the cases (Dumanski et al., 1990). A meningioma locus is tentatively mapped, based on the deletion data, within **22q 12.3-qter**. Consistent with the hormonal effects in the pathogenesis of meningioma (Black, 1991), the male cases showed a higher percentage of tumors with no detectable aberrations on chromosome 22 (Dumanski et al., 1990).

5.2.5.2 ONCOGENES AND GROWTH FACTORS

Table 5.1, modified from Rutka and colleagues (1990), and supplemented with later findings, summarizes the **protooncogenes** and **growth factors** identified so far in tumors of the central and peripheral nervous systems. Comparison of **pediatric brain tumors** with adult tumors suggested that oncogene amplification is a relatively uncommon mechanism of oncogene activation in the pediatric group (Wasson et al., 1990). Moreover, the data suggest that even within one group of tumors, **medulloblastoma**, mechanisms of oncogene activation may vary. One of the consistently reported findings is the difference in the expression of oncogenes in primary tumors and the derived cell lines. This was noted in medulloblastomas (Wasson et al., 1990) and in astrocytomas (Wu and Chikaraishi, 1990).

In **astrocytomas**, a positive correlation has been observed between the level of **EGFR expression**, the phosphotyrosine kinase activity associated with the EGFR, and the degree of malignancy (Baugnet-Mahieu et al., 1990). Taking precautions to preserve fastly degrading **N-*myc*** protein by immediate freezing of tissue, an enhanced expression of this protein has been demonstrated in medulloblastoma tumors (Garson et al., 1989). N-*myc* amplification has been reported in primitive neuroectodermal tumors of the central nervous system by other workers as well (Rouah et al., 1989). The significance of this observation lies in the observed positive correlation between N-*myc* expression and poor prognosis, which can be used in patient management.

Other oncogenes noted in various brain tumors include the ***ROS*-1** gene rearrangement observed in **glioblastoma** cells (Birchmeier et al., 1987), **Ki-*ras*** gene expression in **meningiomas** (Carstens et al., 1988), rare **Ha-*ras*** and **c-*mos*** alleles in **intracranial tumors** (Diedrich et al., 1988), **N-*ras*** and ***EGFR*** genes in glioblastomas (Gerosa et al., 1989; Wong et al., 1987), **c-*myc*** amplification in medulloblastoma cell lines and transplants (Friedman et al., 1988) and glioblastoma (Trent et al., 1986), **v-*sis*, v-*myc*,** and **v-*fos*** in primary brain tumors of neuroectodermal origin (Fujimoto et al., 1988), rearrangement of the ***abl*** gene in glioblastoma (Heisterkamp et al., 1990), coexpression of the genes for **PDGF-β** and **PDGR-R** in glioblastoma (Hermansson et al., 1988) and **TGF-α** in malignant glioma cell lines (Nister et al., 1988; Nister and Westermark, 1986), and human **malignant gliomas** (Rutka et al., 1984). Analysis of low grade astrocytomas and high grade glioblastomas for the presence of the *p53* gene mutations revealed that histologically identifiable progression of brain tumors is associated with clonal expansion of those cells which had previously acquired a mutation in the *p53* gene (Sidransky et al., 1992).

5.2.6 SELECTED MENTAL DISORDERS

5.2.6.1 INTRODUCTION

Practical application of gene level testing for predisposition to or definitive diagnosis of psychiatric disorders is **still a matter of the distant future**. Nevertheless, despite the enormous complexity of the problem, an optimistic view suggests that, at least on a limited basis, useful

TABLE 5.1
Oncogenes and Growth Factors in Tumors of the Nervous System

Tumor	Probe	Source-DNA/RNA	Result[a]	Ref.
Astrocytoma	c-*myc*	Tumor-DNA	A, RA, EE	Sauceda et al. (1988)
	N-*myc*	Tumor-DNA	A, EE	Garson et al. (1985)
	N-*ras*	Tumor-DNA/RNA	A	
	Ha-*ras*	Tumor-DNA	Rare alleles	Diedrich et al. (1988)
	c-*mos*	Tumor-DNA	Rare alleles	
	v-*sis*	Tumor-RNA	EE	Fujimoto et al. (1988)
	v-*fos*	Tumor-RNA	EE	
	v-*myc*	Tumor-RNA	Minimal expression	
	c-*ros*	Tumor-RNA	A	Wu and Chikaraishi (1990)
Glioblastoma	c-*myc*	Tumor-DNA	Minimal Expression	Sauceda et al. (1988)
	abl	Cell line-DNA	RA	Heisterkamp et al. (1990)
		Cell line-DNA/RNA	A, RA, EE	Trent et al. (1986)
	v-*myc*	Tumor-RNA	EE	Fujimoto et al. (1988)
	N-*ras*	Tumor-DNA/RNA	A, EE	Sauceda et al. (1988); Gerosa et al. (1989)
	Ha-*ras*	Tumor-DNA	Rare alleles	Diedrich et al. (1988)
	c-*mos*	Tumor-DNA	Rare alleles	
	PDGF	Cell lines-RNA	EE	Nister and Westermark (1986); Nister et al. (1988)
		Tumor-RNA	EE (regional)	Hermansson et al. (1988)
	EGFR	Tumor-DNA/RNA	A, EE	Gerosa et al. (1989); Libermann et al. (1985)
		Cell lines-DNA/RNA	A, RA, EE	Nister and Westermark (1986); Nister et al. (1988; Wonu et al. (1987)
	TGF-α	Cell lines-RNA	EE	Nister et al, (1988); Rutka et al. (1984)
	c-*ros*	Cell lines-DNA/RNA	A	Birchmeier et al. (1987); Wu and Chikeraishi (1990)
Meningioma	c-*myc*	Tumor-DNA	A	Sauceda et al. (1988)
	N-*myc*	Tumor-DNA/RNA	A, RA	
	Ha-*ras*	Tumor-DNA	Rare alleles	Diedrich et al. (1988)
	c-*mos*	Tumor-DNA	Rare alleles	
	Ki-*ras*	Tumor-RNA	A	Carstens et al. (1988)
Medulloblastoma	v-*myc*	Tumor-RNA	EE	Fujimoto et al. (1988)
	N-*myc*	Tumor-protein	EE	Garson et al. (1989)
	N-*ras*	Tumor-DNA/RNA	A	Sauceda et al. (1988)
	*erb*Bl	Cell line-DNA	A	Wasson et al. (1990)
	N-*myc*	Cell line-DNA	A	Wasson et al. (1990)
	N-*myc*	Tumor-DNA	A	Nisen et al. (1986); Rouah et al. (1989)
	c-*myc*	Cell line-DNA	A	
		Tumor-DNA	EE	Bigner et al. (1989); Bigner and Vogelstein (1990); Friedman et al. (1988)
Neuroblastoma	N-*myc*	Cell lines-RNA	A, EE	Cohen et al. (1988); Seeger et al. (1984); Thielo et al. (1987a)
	c-*ets*	Cell lines-RNA	Minimal expression	Thiele et al. (1987a)
	c-*myb*	Cell lines-RNA	Minimal expression	Thielo et al. (1987a)
Adenoma	c-*myc*	Tumor-DNA/RNA	A, RA	Sauceda et al. (1988)
	N-*myc*	Tumor-DNA/RNA	A, EE	
	N-*ras*	Tumor-DNA/RNA	A, EE	
	Ha-*ras*	Tumor-DNA	Rare alleles	Diedrich et al. (1988)
	c-*mos*	Tumor-DNA	Rare alleles	

TABLE 5.1 (continued)
Oncogenes and Growth Factors in Tumors of the Nervous System

Tumor	Probe	Source-DNA/RNA	Result[a]	Ref.
Oligodendroglioma	c-*mos*	Tumor-DNA	Rare alleles	Diedrich et al. (1988)
	Ha-*ras*	Tumor-DNA	Rare alleles	
Ependymoma	c-*mos*	Tumor-DNA	Rare alleles	Diedrich et al. (1988)
	Ha-*ras*	Tumor-DNA	Rare alleles	
Neuroepithelioma	c-*sis*	Cell line-DNA	RA	Thiele et al. (1987b)

Abbreviations: A, amplification; EE, enhanced expression; RA, rearrangement.

Adapted and expanded from Rutka, J. T. et al. (1990). *Cancer Invest.* 8:425–438. With permission.

information can be gained at the molecular genetic level in the foreseeable future. A compendium of papers dealing with all aspects of research at the gene level in the field of mental activity has been published recently (McHugh and McKusick, 1991).

Specific genetic linkages are few and reflect the complexity both at the clinical and genetic levels (e.g., Baron et al., 1990; Baron, 1991; Edwards, 1990; Gershon et al., 1990; Gurling, 1990; Kidd et al., 1989; McGuffin et al., 1990; Risch, 1990a,b). Although the characterization of **brain-specific gene expression** has been attempted (e.g., Heilig and Widerlov, 1990; Milner et al., 1986; Sapienza and St. Jacques, 1986), **no disease-specific associations have been firmly established so far**.

Among the relatively reliable associations of a subset of bipolar affective disorders are those pointing to **X-linked transmission** (Baron, 1991; Baron et al., 1987, 1990; Schwartz et al., 1990; Stevenson et al., 1990). Other chromosomes implicated in mental disorders are **chromosomes 11** (see Baron, 1991; St. Clair et al., 1990), **6** (Weitkamp et al., 1981), and **16** (Wilkie et al., 1990a,b). The once suspected association of susceptibility to schizophrenia with the **chromosome 5q11-q13** region had to be reevaluated (Gelernter and Kidd, 1991; McGuffin et al., 1990).

New linkage data continue to appear on various genetic diseases, bringing the possibility of developing clinically useful tests for diagnosis and monitoring of neurological diseases closer everyday. Examples include the demonstrated linkage between the gene causing **familial amyotrophic lateral sclerosis** (Lou Gehrig disease) and four DNA markers on the long arm of **chromosome 21** (Siddique et al., 1991). Assignment of the gene for **tryptophan oxygenase** to **4q31** allows further exploration of the role of this gene in **alcoholism** and certain **depression** conditions (Comings et al., 1991; Goodwin, 1991).

Cloning of the D_1 (Sunahara et al., 1990; Zhou et al., 1990) and D_3 (Sokoloff et al., 1990) **dopamine receptors** opens new possibilities for unraveling drug interactions in **Parkinson's disease** and **schizophrenia** as well as shedding light on the molecular aspects of **drug addiction** and alcoholism (Golbe, 1990; Seeman and Niznik, 1990). The amplification of **cholinesterase-encoding genes** noted in hematopoietic disorders, ovarian carcinomas, and malignant gliomas points to the role of neurotransmitters in malignancy (for a comprehensive review of the subject, see Soreq and Zakut, 1990).

Positive DNA diagnosis has been possible in the evaluation of 49 members of **familial amyloid polyneuropathy** (FAP) pedigrees (Harada et al., 1989). The variant **prealbumin (transthyretin) gene** was found to be closely linked to clinical manifestations of the disease in both sporadic cases and familial nonsymptomatic gene carriers. (For a discussion of the molecular biology of FAP, see Herbert, 1989).

Detailed analysis of molecular biological research in the field of neurological diseases is outside the scope of this book. Excellent coverage of the topic, summarizing developments of recent years, has been presented (e.g., Boulton et al., 1990; McHugh and McKusick, 1991;

Rosenberg and Harding, 1988; Rowland et al., 1989). Here, only **selected neurological conditions** will be briefly summarized to illustrate the most recent progress made toward clinically useful gene level probing.

5.2.6.2 FRAGILE X SYNDROME

Fragile X syndrome, also known as **Martin-Bell syndrome**, is the most common form of mental retardation after Down syndrome, afflicting 1 in 1250 males and 1 in 2000 females. Affected males characteristically are mentally retarded (Freund and Reiss, 1991), have long faces and large ears and testes (Fryns, 1989). Although the pattern of inheritance follows that of an X chromosome-linked mutation, it has a number of unusual features, described as **variable penetrance**.

The syndrome is not expressed in at least 20% of males that carry a fragile X chromosome; daughters to whom they pass the trait are also asymptomatic (Sherman et al., 1985). Daughters of transmitting males are obligate carriers of the fragile X chromosome, and **mental retardation** is first expressed among the grandchildren of transmitting males (Laird, 1987). Because affected males or females always inherit the fragile X chromosome from their mothers, it has been proposed that two independent events are required for the syndrome to be expressed (Laird, 1987). First, the fragile X mutation occurs (it is known to have a high mutation rate — 4 to 5 $\times$ 10^{-4}; Sherman et al., 1984); next, the X chromosome is **inactivated** in pre-oogonial cells. The syndrome is not expressed as long as the X chromosome remains inactivated. Normal **reactivation** of the X chromosome prior to oogenesis is blocked by the mutation. It is this **block to reactivation** that leads to mental retardation in affected progeny (Laird, 1987; Laird et al., 1990; Rocchi et al., 1990; Sved and Laird, 1990).

The **fragile site** in this syndrome **(FRAXA)** has been located in the **long arm of the X chromosome** at the band position **q27.3**, and can be revealed by folate deprivation (Sood et al., 1990; Suthers et al., 1990; Vincent ct al., 1989). **Methylation of the fragile site** in affected individuals, leading to a reduction in transcriptional activity in methylated areas compared to normal chromosomal regions, has been proposed (Laird, 1987) and later demonstrated to occur (Bell et al., 1991). Contrasting findings failing to observe any differences in DNA methylation in the vicinity of the FRAXA on active and inactive X chromosomes and in fragile X males have also been reported (Khalifa et al., 1990).

These observations, although not refuting **Laird's imprinting hypothesis**, suggest that mechanisms other than methylation are at work to allow an imprint to persist through generations. However, pulsed-field gel mapping studies on DNA from fragile X patients with probes close to the fragile X locus revealed that **rare-cutter restriction sites** (the so-called **CpG islands**) were resistant to digestion in fragile X males due to **hypermethylation** (Bell et al., 1991; Vincent et al., 1991). Further analysis defined a single CpG island containing a cluster of rare-cutter sites between markers **DXS463** and **DXS465** (Heitz et al., 1991).

Cytogenetic analysis detects only 55% of carrier females, and even fewer (30%) of the mentally normal ones (Sherman et al., 1985); it usually fails to identify phenotypically normal males, therefore diagnostic probes for this frequent condition are of significant practical value. Some of the better markers **(DXS304)** lie within 5 cM of FRAXA (Vincent et al., 1989).

CpG islands are usually situated at the 5' end of genes, and indeed a gene, designated *FMR-1* (for fragile X mental retardation) has been identified from the complementary DNA library of the fragile X region (Verkerk et al., 1991). The *FMR-1* gene is highly conserved in evolution, and shows a strong similarity to a family of late-onset neurodegeneration genes, *deg-1* and *mec-4*, of the nematode *Caenorhabditis elegans* (Verkerk et al., 1991). These genes encode proteins termed "**degenerins**" (Driscoll and Chalfie, 1991). Because *mec*-4 appears to cross-hybridize with human DNA, it was suggested that related human genes may be involved in the process of neurodegeneration and cell death in various human neurodegenerative conditions (e.g., **Huntington's disease** and **amyotrophic lateral sclerosis**) (Driscoll and Chalfie, 1991).

The isolated DNA region that spans the fragile site appears to be critically involved in the expression of the clinical phenotype. Moreover, immediately distal to the CpG island a 7-kb *Eag*I-*Bgl*II fragment has been identified, designated **St B 12** (Oberlé et al., 1991). The relationship between the breakpoints, the CpG island, and the fragile site has been defined, and shown by ISH with cloned DNA from the 1.8-Mb region, containing the entire fragile X region, to coincide (Dietrich et al., 1991).

While in affected males the CG-containing restriction sites surrounding the CpG island are predominantly methylated, these remain largely **unmethylated in normal males** and **unaffected transmitting males**. The latter group was found to have a **DNA insertion** (up to 400 bp) that is passed on to their daughters (Oberlé et al., 1991; Verkerk el al., 1991; Yu et al., 1991). This insertion is of the same size, or only slightly larger in the daughter, but becomes significantly increased (up to 1.5 to 3.5 kb) in their children expressing the phenotype. The probes **St X21E** or **St A22** applied to blots generated by *Ban*I digestion reveal a distinctive pattern in fragile X families that allows to distinguish normal males (who have a 1.15-kb fragment) and normal females (featuring also an additional 1.7-kb fragment) from almost all carrier females who show abnormal patterns regardless of whether they have any manifestation of the characteristic phenotype (Oberlé et al., 1991). All daughters of **normal transmitting males (NTMs)** have an abnormal band located between 1.9 to 2.2 kb. This band is much larger in females with X expression. Abnormal fragments with these two probes (St X21E and St A22) in *Ban*I digests can be observed only if this restriction site centromeric to the **CpG island is methylated** and, therefore, it cannot be detected in NTMs (Oberlé et al., 1991). The diagnostic specificity of the combined use of these two probes is 100% with a sensitivity of 98%. In contrast to the cytogenetic analysis that carries a false negative rate of 50%, this molecular genetic diagnostic test displays almost total reliability and specificity (Oberlé et al., 1991).

Analysis of the fragile X pedigrees with *Pst*I restriction revealed a **region of instability** localized to a trinucleotide repeat **p(CCG)$_n$** (Kremer et al., 1991). Characteristically, in both normal and affected persons the sequences flanking these repeats could be observed in a nonhomologous host and after **PCR** amplification. This observation points to the molecular mechanism accounting for the instability at the fragile site arising from variation in the p(CCG)$_n$ repeat copy number (Kremer et al., 1991). These important findings emphasize the superior diagnostic value of DNA probing for fragile X over the conventional cytogenetic methods, allowing direct diagnosis of the fragile X condition (Davies, 1991; Oberlé et al., 1991; Yu et al., 1991).

The three nucleotide repeats [**p(CCG)$_n$** in the *Pst*I fragment (Kremer et al., 1991) or **(CGG)$_n$** after *Eco*RI, *Hind*II, or *Bgl*II digestion and probing with **Ox1.9** (Nakahori et al., 1991)] have been shown to offer direct diagnostic tests. While normal individuals contain 2 to 50 copies of CGG, NTMs have 55 to 125 copies, the affected males generally containing over 125 repeats (Hirst et al., 1991). The use of the Ox1.9 probe on chorionic villi allows a prenatal evaluation of the fetus; however, a number of fine details still remain to be worked out for the definitive prediction to be possible (Hirst et al., 1991).

A better diagnostic resolution appears to be afforded by a combined use of *Eco*RI and *Eag*I digests which allow to unambiguously distinguish **in a single test** between the normal genotype, the premutation, and the full mutation (Rousseau et al., 1991). This approach can be used for prenatal, after birth, and for genetic counseling evaluations.

Another combination of restriction enzymes and probes [*Pst*I digests probed with the **pfxa3 probe** (Oncor, Gaithersburg, MD), and *Taq*I blots probed with **VK 21A**, and the **pS8 control probe**] has been used for direct diagnosis of fragile X syndrome by detecting prenatally the unstable sequences in DNA due to p(CCG)$_n$ repeat amplification (Sutherland et al., 1991). The **chorionic villus** DNA restricted with *Eco*RI and *Sac*II can be probed with the pfxa3 probe to demonstrate the characteristic pattern of methylation thus providing a direct diagnostic test for the fragile X genotype (Sutherland et al., 1991). Restriction analysis using *Eco*RI and *Eag*I enzymes and the St B12.3 probe can be accomplished also on **amniotic fluid cultures** (Dobkin et al., 1991).

Further development of the diagnostic test for the fragile X genotype comes from the use of **PCR amplification of the CGG repeats** (Pergolizzi et al., 1992). The results of this PCR-based assay fully compare with those using Southern analysis with the probe St B12.3. PCR blots can be probes with an oligonucleotide probe homologous to the CGG repeat — **[5′(CGG)₅3′]**. Affected males have markedly increased numbers of the repeat (deltas) ranging from about 500 to over 6000 (Pergolizzi et al., 1992).

Parenthetically, complementation of conventional cytogenetic analysis by DNA probing is also of diagnostic value in unraveling the cause of **familial premature ovarian failure** in women with a **partial deletion of the long arm of the X chromosome** (Krauss et al., 1987). DNA hybridization studies demonstrated that the deletion was interstitial, not terminal, and the same in all patients with ovarian failure syndrome tested.

Significant progress has also been made in defining the putative molecular mechanisms responsible for selective X chromosome inactivation (Brown et al., 1991a,b; Fisher et al., 1990). The *XIST* gene located in the **X-inactivation center** on the human X chromosome is implicated at least in some aspects of X inactivation (Brown et al., 1991a,b). An intriguing observation has been made on changes in common fragile sites on chromosomes from peripheral blood lymphocytes during different phases of the **menstrual cycle** (Furuya et al., 1991). The highest frequency of **spontaneous breaks** was noted at the luteal phase, affecting primarily bands 2q32, 3q27, 6q26, and 16q23, whereas sister chromatid exchanges, while showing marked variation, did not correlate with the menstrual cycle. To what extent this observation will be practically relevant to detecting the higher occurrence of chromosomal abnormalities in patients with various cancers such as breast cancer, in which hormonal influences may affect chromosomal fragility, still remains to be seen.

5.2.6.3 DOWN SYNDROME

Down syndrome (**DS**) is the most common chromosomal disorder in humans, occurring in 1 in 600 to 1000 live births. This is also the most common genetically defined cause of **mental retardation**, and a major cause of **congenital heart defects** (Patterson et al., 1990). The early recognition of the association of DS with **trisomy 21** has been subsequently refined to suggest that it is the **trisomy of only specific genes**, not any sets of genes, that accounts for the Down phenotype. Somatic cell hybrid studies and molecular mapping have led to the construction of a detailed map of chromosome 21, pointing to **parts of bands 21q22.1, q22.2, and q22.3** as responsible for certain phenotypic characteristics of DS (Korenberg, 1990; Patterson et al., 1990).

Routine diagnosis of trisomy 21 can be accomplished by conventional **cytogenetic banding** of metaphase chromosomes. This technique, however, fails to recognize some cases of DS caused by a **translocation at 21q22-21qter**. Trisomy of only this subregion is sufficient to produce a Down phenotype (Lichter et al., 1990). Using specific DNA probes, chromosome 21 abnormalities can also be identified by ISH (Fuscoe et al., 1989; Lichter et al., 1988, 1990; Pinkel et al., 1988; van Camp et al., 1989). In fact, using a single **cosmid clone probe (c519)**, the **quantitation of trisomy 21** in interphase cells is possible (Lichter et al., 1990). An unequivocal diagnosis can be established on a relatively small sample of fetal tissue, due to the high resolution of the nonradioactive, fluorescence detection system. Furthermore, there is a diagnostic approach carrying an even lower risk, if specific probes are combined with PCR amplification on maternally derived fetal reticulocytes (Bianchi et al., 1990).

The **parental origin** of the supernumerary 21 chromosome can be established using **DNA polymorphism** by scoring the polymorphic alleles in the parents, the proband, and siblings, if available (e.g., Antonarakis et al., 1991; Rudd et al., 1988; Stewart et al., 1988; Warren et al., 1987). A study in 200 families established that the extra chromosome 21 was **maternal in origin** in about 95% of the cases, and **paternal** in only about 5% — much less than conventionally estimated by cytogenetic methods (Antonarakis et al., 1991).

As an alternative to either RFLP studies or Southern analysis, which use specific and reference probes, the estimation of the gene dosage is possible with a **slot-blot method** for the quantification of the copy numbers of unique chromosome 21 sequences (Blouin et al., 1990). Intensities of the signals generated on blots probed by the chromosome 21 DNA sequences (**D21S11** and **D21S17**) are quantified by **densitometric scanning**. Comparison of the signal from an unknown individual to that of a control can identify the presence of three vs. two copies of a given chromosome 21 sequence. The accuracy of the method is 100%. This technique can also be used for efficient identification of chromosome 21 rearrangements given the appropriate probes are available.

A number of specific **chromosomal structural rearrangements** have been mapped (Munke et al., 1988). By applying PCR amplification of unique sequences on chromosome 21 in a **recombination-based assay (RBA)**, a **prenatal diagnosis of DS** as well as analysis of genes encoded by chromosome 21 becomes possible (Jankowski et al., 1990; Sacchi, 1990).

Of particular clinical interest is the study of **specific chromosomal assignments** of various features constituting the DS phenotype performed at the molecular level (Korenberg, 1990; Korenberg et al., 1990; McCormick et al., 1989; Rahmani et al., 1989). The minimal region on chromosome 21 that, when duplicated, accounts for the majority of phenotypic characteristics composing the DS phenotype has been defined in molecular terms from the analysis of a three-generation family with four affected members (Korenberg et al., 1990). So far, **no single defect accounting for the DS phenotype has been identified** (Korenberg, 1990).

5.2.6.4 ALZHEIMER'S DISEASE

Alzheimer's disease (AD) is the most common form of dementia in older people, characterized by degenerative neurological changes associated with progressive impairment of memory, reasoning, orientation, and judgment. While generally AD occurs as a sporadic disorder of late onset, some cases, designated **familial Alzheimer's disease (FAD),** are transmitted in an autosomal dominant fashion suggesting a genetic defect (Davies, 1986). FAD has a somewhat earlier onset than the sporadic form, although phenotypically the two forms are indistinguishable.

The **characteristic histopathological feature** of AD is the presence of numerous **neuritic plaques** and tangles that, on postmortem examination of the brains, appear to correlate in their frequency of occurrence with the degree of intellectual impairment of the patients. The neuritic plaques represent aggregates of altered axons and dendrites that surround extracellular deposits of **amyloid filaments** (Selkoe et al., 1987). Similar amyloid filaments are recognized not only in normal aging human brain, but also in the neuritic plaques and cerebrovascular amyloid deposits in **Down syndrome** patients, and in **aged mammals** of some other species (Selkoe et al., 1987).

The principal protein of the amyloid filaments has been isolated and called β-**protein** or β-**amyloid**, a 4.2-kDa polypeptide, and the amyloid β-protein gene has been isolated (Tanzi et al., 1987a,b). The locus encoding the β amyloid gene maps to **chromosome 21** in the vicinity of the genetic defect causing FAD (Griffin, 1988; St. George-Hyslop et al., 1987b; Tanzi et al., 1987a,b).

The β/**A4 protein**, the principal component of the amyloid fibril, is derived from a large membrane-bound protein, **Alzheimer amyloid protein precursor (APP)** (Glenner and Wong, 1984; Kang et al., 1987; Masters et al., 1985). The deposition of amyloid is thought to result from **defective processing** of APP, particularly due to self-aggregation of its C-terminal portion, which is an inherent characteristic of the C-terminal peptide of APP (Maruyama et al., 1990).

The presence of AD-like neuropathological changes in Down syndrome raised the possibility that AD could be caused by overexpression or duplication of one or more genes on chromosome 21. This possibility was studied by RFLP and Southern analysis in both FAD and its sporadic form cases, and no evidence of increased DNA dosage could be observed in these patients (Podlisny et al., 1987; Morrison, 1988; St. George-Hyslop et al., 1987a; Tanzi et al., 1987b).

Searching for **candidate genes for FAD**, a physical map of the proximal long arm of chromosome 21 has been constructed using pulsed-field gel electrophoresis and somatic cell hybrids (Owen et al., 1990). In a region spanning 10 Mb, several **CpG islands** were identified close to the four most centromeric loci, suggesting the possible location of genes. Several probes map to within 1.5 Mb of one another, providing the most tightly linked probes to the *AD* locus (Owen et al., 1990).

A study of the **inheritance mode** of five polymorphic DNA markers from the proximal arm of chromosome 21 (**D21S16, D21S13, D21S52, D21S1,** and **D21S11**) in a large unselected series of 48 pedigrees with FAD, suggested that AD is not a single entity "but rather results from genetic defects on chromosome 21 and from other genetic or nongenetic factors" (St. George-Hyslop et al., 1990). However, linkage analysis of FAD using chromosome 21 markers in 48 kindreds including Volga German families thought to be genetically homozygous for FAD failed to support the view that a chromosome 21 gene is responsible for late-onset FAD and even some forms of early-onset FAD (Schellenberg et al., 1991).

What other **interactive possibilities** may play a role is not clear at this point, although some suggestions have been advanced (Breitner et al., 1991; G. M. Martin et al., 1990). According to studies in two families with AD, mutation within the *APP* gene can account for at least some cases of early-onset AD (Goate et al., 1991). In two unrelated kindreds studied, a mutation causes **Val-Ile substitution** close to the carboxyl terminal of the β-amyloid peptide. This mutation could not be identified, however, in nine families with late-onset AD. If the reports of recombination events between the *APP* and *FAD* genes are not artifactual (Goate et al., 1991), they suggest that there may be another gene on chromosome 21 predisposing to FAD (Wright et al., 1991).

The **clinical implication** of this mutation is open to speculation at this time. One of the possibilities is that isoleucine substitution alters the anchoring of APP in the membrane (Goate et al., 1991). An **alternative view** holds that, in patients heterozygous for this mutation, the control of translation of APP is deregulated, leading to overproduction of the protein (Tanzi and Hyman, 1991). It is argued that **interference with the translational regulation** may open possibilities for both environmental and genetic factors to affect APP production, which opens specific routes of therapeutic intervention (Tanzi and Hyman, 1991). A mutation at codon 717 of the β-amyloid precursor protein gene, Val→Gly substitution and yet another change Val→Phe in the same codon have been found to segregate with FAD (Chartier-Harlin et al., 1991).

Similarity between the histopathological changes found typically in AD and those observed in normal aged brains suggested that AD may be a variant of **aging**. This possibility is being actively explored. In cultured fibroblasts, senescence was found to be accompanied by an increased gene expression of β-amyloid precursor protein (Adler et al., 1991). The similarity to a physiological process is strengthened by the reversible induction of this gene expression in fibroblasts made quiescent by serum deprivation.

Other investigators claim that, although the similarity of AD to an aging process-related phenomenon is valid in general, it is the **aberration of the normal aging process** that accounts for the AD phenomenology (Bosman et al., 1991). In particular, attention should be drawn to the **cellular membranes** and, in fact, a selective alteration of band 3 in red blood cells (RBC) suggests that the normal aging process of these cells from AD patients is disturbed. Future research will shed light on the validity of these hypotheses. It appears intuitively possible that a link exists between specific membrane alterations, such as those observed in band 3 of RBCs (Bosman et al., 1991) and the β amyloid protein precursor (APP), the defective processing of which leads to the accumulation of amyloid (Maruyama et al., 1990). It is clear that gene level testing for predisposition to AD at this time is a matter of future research. Fine molecular analysis of βAPP processing by lysosomal proteases suggests that aberrant endosomal-lysomal processing may produce amyloidogenic peptides and play a critical role in the pathogenesis of AD (Estus et al., 1992; Golde et al., 1992). (For reviews see also Selkoe, 1991; Yanker and Mesulam, 1991.)

5.2.6.5 MITOCHONDRIAL ENCEPHALOPATHIES

Yet another group of neurological diseases, the so-called **mitochondrial encephalopathies**, can be diagnosed now at the gene level. Mutations in **mitochondrial DNA (mtDNA)** have been described in a number of neurological conditions, including **chronic progressive external ophthalmoplegia (CPEO), Kearns-Sayre syndrome, myoclonus epilepsy with ragged red fibers (MERRF),** and in **mitochondrial encephalopathy with active acidosis and strokelike episodes (MELAS)** (DiMauro et al., 1989; Goto et al., 1990; Ino et al., 1991; Johns et al., 1989; Shoffner et al., 1990). Another mitochondrial disease has been described that in addition to dementia, features ataxia, neurogenic muscle weakness, and retinitis pigmentosa in a maternally inherited condition (Holt et al., 1990). Depletion of mtDNA can be found in affected tissues correlated with aberrations in the respiratory chain function (Moraes et al., 1991). The role of mtDNA deletions appears to be complex since in some cases rearrangements of the mtDNA are observed in healthy individuals (Cormier et al., 1991).

One of the most frequently detected mutations is an A→G transition at position 3234 in the **mitochondrial gene for tRNA^{Leu}** (Goto et al., 1990; Ino et al., 1991; Johns and Hurko, 1991). This mutation is present in MELAS, Kearns-Sayre syndrome, and CPEO patients throughout the world (Johns and Hurko, 1991). It appears that **mtRNA dysfunction** is a key feature of mitochondrial encephalopathies (Johns and Hurko, 1991). Importantly, this mutation can be **detected noninvasively** in blood samples and even urine. Yet another adult onset myopathy and cardiomyopathy, without the involvement of the nervous system, that is maternally inherited has been identified and found to be genetically determined through a mutation (A→G^{3260}) of the tRNA$^{leu(UUR)}$ (Zeviani et al., 1991).

PCR amplification of the region harboring the mutation at position 3234, which creates an *Apa*I restriction site, as well as the MERRF mutation at position 8344, can be efficiently applied for the molecular genetic diagnosis of mitochondrial encephalopathies from blood samples, even when abnormal histological findings in muscle biopsy specimens are absent (Hammans et al., 1991; Johns and Hurko, 1991). This rapid, reliable, and inexpensive screening test is advocated in patients with **myoclonic epilepsy, ataxia**, other undiagnosed encephalopathies, in **early** or **atypical strokes** for accurate diagnosis, and genetic counseling (Hammans et al., 1991).

References

Abdel-Malek, Z. A. et al. (1987). *In vivo* modulation of proliferation and melanization of S91 melanoma cells by prostaglandins. Cancer Res. *47*:3141–3146.

Acsadi, G. et al. (1991). Human dystrophin expression in *mdx* mice after intramuscular injection of DNA constructs. Nature (London) *352*:815–818.

Adler, M. J. et al. (1991). Increased gene expression of Alzheimer disease β amyloid precursor protein in senescent cultured fibroblasts. Proc. Natl. Acad. Sci. U.S.A. *88*:16–20.

Akiyama, T. and Toyoshima, K. (1990). Marked alteration in phosphorylation of the RB protein during differentiation of human promyelocytic HL60 cells. Oncogene *5*:179–183.

Albino, A. P. et al. (1989). Analysis of *ras* oncogenes in malignant melanoma and precursor lesions: correlation of point mutations with differentiation phenotype. Oncogene *4*:1363–1374.

Aldaz, C. M. et al. (1988). Cytogenetic profile of mouse skin tumors induced by the viral Harvey-*ras* gene. Carcinogenesis *9*:1503–1505.

Aledo, R. et al. (1989). Cytogenetic study on eleven cutaneous neoplasms and two pre-tumoral lesions from xeroderma pigmentosum patients. Int. J. Cancer *44*:79–83.

Allitto, B. A. et al. (1991). Increased recombination adjacent to the Huntington disease-linked D4S10 marker. Genomics *9*:104–112.

Amy, C. M. and Bartholomew, J. C. (1987). Regulation of N-*myc* transcript stability in human neuroblastoma and retinoblastoma cells. Cancer Res. *47*:6310–6314.

Ananthaswamy, H. N. et al. (1988). Detection and identification of activated oncogenes in human skin cancers occurring on sun-exposed body sites. Cancer Res. *48*:3341–3346.

Anderson, M. D. S. et al. (1992). Dystrophin mRNA in lyophilized tissue. Nature (London) *355*:778.

Antonarakis, S. E. and the Down Syndrome Collaborative Group (1991). Parental origin of the extra chromosome in trisomy 21 as indicated by analysis of DNA polymorphisms. N. Engl. J. Med. *324*:872–876.

Arahata, K. et al. (1988). Immunostaining of skeletal and cardiac muscle surface membrane with antibody against Duchenne muscular dystrophy peptide. Nature (London) *333*:861–863.

Arahata, K. et al. (1989a). Dystrophin diagnosis: comparison of dystrophin abnormalities by immunofluorescent and immunoblot analyses. Proc. Natl. Acad. Sci. U.S.A. *86*:7154–7158.

Arahata, K. et al. (1989b). Mosaic expression of dystrophin in symptomatic carriers of Duchenne's muscular dystrophy. N. Engl. J. Med. *320*:138–142.

Aslanidis, C. et al. (1992). Cloning of the essential myotonic dystrophy region and mapping of the putative defect. Nature (London) *355*:548–551.

Azar, C. G. et al. (1990). Multiple defects of the nerve growth factor receptor in human neuroblastomas. Cell Growth Differ. *1*:421–428.

Baird, J. et al. (1990). mRNA for tissue-type plasminogen activator is present in lesional epidermis from patients with psoriasis, pemphigus, or bullous pemphigoid, but is not detected in normal epidermis. J. Invest. Dermatol. *95*:548–552.

Baker, D. L. et al. (1989). Analysis of nerve growth factor receptor expression in human neuroblastoma and neuroepithelioma cell lines. Cancer Res. *49*:4142–4146.

Bakker, E. et al. (1989). Prenatal diagnosis of Duchenne muscular dystrophy: a three-year experience in a rapidly evolving field. J. Inherited Metab. Dis. *12* (Suppl. 1):174–190.

Balaban, G. B. et al. (1986). Karyotypic evolution in human malignant melanoma. Cancer Genet. Cytogenet. *19*:113–122.

Bale, S. J. et al. (1989). Mapping the gene for hereditary cutaneous malignant melanoma-dysplastic nevus to chromosome 1p. N. Engl. J. Med. *320*:1367–1372.

Bale, S. J. et al. (1990). A reply. N. Engl. J. Med. *322*:854.

Ballester, R. et al. (1990). The NF1 locus encodes a protein functionally related to mamalian GAP and yeast IRA proteins. Cell (Cambridge, Mass) *63*:851–859.

Bar-Sagi, D. and Feramisco, J. R. (1985). Microinjection of the *ras* oncogene protein into PC12 cells induces morphological differentiation. Cell (Cambridge, Mass) *42*:841–848.

Barbujani, G. et al. (1990). Segregation analysis of 1885 DMD families: significant departure from the expected proportion of sporadic cases. Hum. Genet. *84*:522–526.

Barker, D. et al. (1987). Gene for von Recklinghausen neurofibromatosis is in the pericentromeric region of chromosome 17. Science *236*:1100–1102.

Barker, J. N. W. N. et al. (1991). Keratinocytes as initiators of inflammation. Lancet *337*:211–214.

Baron, M. (1991). Genetics of manic depressive illness: current status and evolving concepts. In: Genes, Brain, and Behavior. McHugh, P. R. and McKusick, V. A. (eds.). Raven Press, New York, pp. 153–164.

Baron, M. et al. (1987). Genetic linkage between X-chromosome markers and bipolar affective illness. Nature (London) *326*:289–292.

Baron, M. et al. (1990). The impact of phenotypic variation on genetic analysis: application to X-linkage in manic-depressive illness. Acta Psychiatr. Scand. *82*:196–203.

Bartkowiak, D. et al. (1991). DNA flow cytometry in the prognosis of primary malignant melanoma. Oncology *48*:39–43.

Bates, G. P. et al. (1990). A yeast artificial chromosome telomere clone spanning a possible location of the Huntington disease gene. Am. J. Hum. Genet. *46*:762–775.

Baugnet-Mahieu, L. et al. (1990). Epidermal growth factor receptors in human tumors of the central nervous system. Anticancer Res. *10*:1275–1280.

Becker, D. et al. (1990). Differential expression of protein kinase C and cAMP-dependent protein kinase in normal human melanocytes and malignant melanomas. Oncogene *5*:1133–1139.

Beggs, A. H. and Kunkel, L. M. (1990a). A polymorphic CACA repeat in the 3′ untranslated region of dystrophin. Nucleic Acids Res. *18*:1931.

Beggs, A. H. and Kunkel, L. M. (1990b). Improved diagnosis of Duchenne/Becker muscular dystrophy. J. Clin. Invest. *85*:613–619.

Beggs, A. H. et al. (1990). Detection of 98% of DMD/BMD gene deletions by polymerase chain reaction. Hum. Genet. *86*:45–48.

Bejjani, B. et al. (1991). The value of deletion analysis for carrier detection in Duchenne muscular dystrophy (DMD). Clin. Genet. *39*:245–252.

Bell, M. V. et al. (1991). Physical mapping across the fragile X: hypermethylation and clinical expression of the fragile X syndrome. Cell (Cambridge, Mass) *64*:861–866.

Bendit, I. et al. (1990). Molecular phenotype of a pediatric small round cell tumor. Cancer *66*:1534–1538.

Benedict, W. F. et al. (1988). The gene responsible for the development of retinoblastoma and osteosarcoma. Cancer *62* (Suppl.):1691–1694.

Benedict, W. F. et al. (1990a). Role of the retinoblastoma gene in the initiation and progression of human cancer. J. Clin. Invest. *85*:988–993.

Benedict, W. F. et al. (1990b). The retinoblastoma gene: its role in human malignancies. Cancer Invest. *8*:535–540.

Berman, S. A. et al. (1989). Increased N-*myc* mRNA expression associated with dibutyryl cyclic AMP induced neuroblastoma differentiation. J. Neurogenet. *6*:75–86.

Bianchi, D. W. et al. (1990). Isolation of fetal DNA from nucleated erythrocytes in maternal blood. Proc. Natl. Acad. Sci. U.S.A. *87*:3279–3283.

Bigner, S. H. and Vogelstein, B. (1990). Cytogenetic and molecular genetics of malignant gliomas and medulloblastoma. Brain Pathol. *1*:12–18.

Bigner, S. H. et al. (1989). Cytogenetics and gene amplification in malignant brain tumors. Proceedings of the Third International Workshop on Chromosomes in Solid Tumors. Cancer Genet. Cytogenet. *41*:226.

Birchmeier, C. et al. (1987). Expression and rearrangement of the ROS1 gene in human glioblastoma cells. Proc. Natl. Acad. Sci. U.S.A. *84*:9270–9274.

Bizub, D. et al. (1986). Mutagenesis of the Ha-*ras* oncogene in mouse skin tumors induced by polycyclic aromatic hydrocarbons. Proc. Natl. Acad. Sci. U.S.A. *83*:6048–6052.

Bizub, D. et al. (1987). Antisera to the variable region of *ras* oncogene proteins, and specific detection of H-*ras* expression in an experimental model of chemical carcinogenesis. Oncogene *1*:131–142.

Bjelfman, C. et al. (1990). Expression of the neuronal form of pp60 c-*src* in neuroblastoma in relation to clinical stage and prognosis. Cancer Res. *50*:6908–6914.

Black, P. McL. (1991a). Brain tumors. Part I. N. Engl. J. Med. *324*:1471–1476.

Black, P. McL. (1991b). Brain tumors. Part 2. N. Engl. J. Med. *324*:1555–1564.

Bloch, M. and Hayden, M. R. (1990). Opinion: predictive testing for Huntington disease in childhood: challenges and implications. Am. J. Hum. Genet. *46*:1–4.

Blouin, J. L. et al. (1990). Slot blot method for the quantification of DNA sequences and mapping of chromosome rearrangements: application to chromosome 21. Am J. Hum Genet. *46*:518–526.

Boelter, W. D. et al. (1990). Dystrophin protein and RFLP analysis for fetal diagnosis and carrier confirmation of Duchenne muscular dystrophy. Prenatal Diagn. *10*:703–715.

Bogenmann, E. et al. (1987). c-*myc* amplification and expression in newly established human osteosarcoma cell lines. Cancer Res. *47*:3808–3814.

Bonilla, E. et al. (1988). Duchenne muscular dystrophy: deficiency of dystrophin at the muscle cell surface. Cell (Cambridge, Mass) *54*:447–452.

Bookstein, R. et al. (1988). Human retinoblastoma susceptibility gene: genomic organization and analysis of heterozygous intragenic deletion mutants. Proc. Natl. Acad. Sci. U.S.A. *85*:2210–2214.

Bookstein, R. et al. (1990). PCR-based detection of a polymorphic *Bam*HI site in intron 1 of the human retinoblastoma (RB) gene. Nucleic Acids Res. *18*:1666.

Boonstra, J. et al. (1985). Epidermal growth factor receptor expression related to differentiation capacity in normal and transformed keratinocytes. Exp. Cell. Res. *161*:421–433.

Bosman, G. J. C. G. M. et al. (1991). Alzheimer's disease and cellular aging: membrane-related events as clues to primary mechanisms. Gerontology *37*:95–112.

Boukamp, P. et al. (1990). c-Ha-*ras* oncogene expression in immortalized human keratinocytes (Ha Ca T) alters growth potential *in vivo* but lacks correlation with malignancy. Cancer Res. *50*:2840–2847.

Boulton, A. A. et al. (1990). Molecular Neurobiological Techniques. Humana Press, Clifton, NJ.

Boultwood, J. et al. (1988). *myc* expression in neoplasia of human thyroid C-cells. Cancer Res. *48*:4073–4077.

Bourhis, J. et al. (1989). Correlation of *MDR*1 gene expression with chemotherapy in neuroblastoma. J. Natl. Cancer Invest. *81*:1401–1405.

Bourhis, J. et al. (1991). Combined analysis of DNA ploidy index and N-*myc* genomic content in neuroblastoma. Cancer Res. *51*:33–36.

Bouwes Bavinck, J. N. et al. (1991). Relation between skin cancer and HLA antigens in renal-transplant recipients. N. Engl. J. Med. *325*:843–848.

Boyce, F. M. et al. (1991). Dystrophin is transcribed in brain from a distant upstream promoter. Proc. Natl. Acad. Sci. U.S.A. *88*:1276–1280.

Boyd, Y. and Buckle, V. J. (1986). Cytogenetic heterogeneity of translocations associated with Duchenne muscular dystrophy. Clin. Genet. *29*:108–115.

Bradl, M. et al. (1991). Malignant melanoma in transgenic mice. Proc. Natl. Acad. Sci. U.S.A. *88*:164–168.

Brandt, J. et al. (1989). Presymptomatic diagnosis of delayed onset disease with linked DNA markers. The experience in Huntington's disease. JAMA *261*:3108–3114.

Breitner, J. C. S. et al. (1991). Case-control studies of environmental influences in diseases with genetic determinants, with an application to Alzheimer's disease. Am. J. Epidemiol. *133*:246–256.

Brock, D. J. H. et al. (1989). Predictive testing for Huntington's disease with linked DNA markers. Lancet *ii*:463–466.

Brock, D. J. H. et al. (1990). Options for prenatal testing for Huntington's disease using linked DNA probes. J. Med. Genet. *27*:68–69.

Brodeur, G. M. et al. (1981). Cytogenetic features of human neuroblastomas and cell lines. Cancer Res. *41*:4678–4686.

Brodeur, G. M. et al. (1984). Amplification of N-*myc* in untreated human neuroblastomas correlates with advanced disease stage. Science *224*:1121–1124.

Brodeur, G. M. et al. (1986). Clinical implications of oncogene activation in human neuroblastomas. Cancer *58*:541–545.

Brodeur, G. M. et al. (1987). Consistent N-*myc* copy number in simultaneous and consecutive neuroblastoma samples from sixty individual patients. Cancer Res. *47*:4248–4253.

Brodeur, G. M. et al. (1988). Molecular analysis and clinical significance of N-*myc* amplification and chromosome 1p monosomy in human neuroblastomas. In: Advances in Neuroblastoma Research 2. Evans, A. E. et al. (eds.). Alan R. Liss, New York, pp. 3–16.

Brooke, J. D. et al. (1992). Molecular basis of myotonic dystrophy: expansion of a trinucleotide (CTG) repeat at the 3′ end of a transcript encoding a protein kinase family member. Cell (Cambridge, Mass.) *68*:799–808.

Brooke, M. et al. and the CIDD Group (1983). Clinical investigations in Duchenne dystrophy 2. Determination of the "power" of therapeutic trials based on the natural history. Muscle Nerve *6*:91–103.

Brown, C. J. et al. (1991a). A gene from the region of the human X inactivation center is expressed exclusively from the inactive X chromosome. Nature (London) *349*:38–44.

Brown, C. J. et al. (1991b). Localization of the X inactivation center on the human X chromosome in Xq13. Nature (London) *349*:82–84.

Brown, K. et al. (1986). v-*ras* genes from Harvey and Balb murine sarcoma viruses can act as initiators of two-stage mouse skin carcinogenesis. Cell (Cambridge, Mass) *46*:447–456.

Brown, S. C. et al. (1990). Monensin does not consistently inhibit the spreading of normal or Duchenne fibroblasts on glass. J. Cell Sci. *97*:149–156.

Brzustowicz, L. M. et al. (1990). Genetic mapping of chronic childhood-onset spinal muscular atrophy to chromosome 5q 11.2-13.3. Nature (London) *344*:540–541.

Bucan, M. et al. (1990). Physical maps of 4p16.3, the area expected to contain the Huntington disease mutation. Genomics *6*:1–15.

Buchanan, J. A. and Cavenee, W. K. (1987). Genetic markers for assessment of retinoblastoma predisposition. Dis. Markers *5*:141–152.

Buchberg, A. M. et al. (1990). Sequence homology shared by neurofibromatosis type 1 gene and IRA-1 and IRA-2 negative regulators of the RAS cyclic AMP pathway. Nature (London) *347*:291–294.

Bulman, D. E. et al. (1989). Characterization of Duchenne muscular dystrophy patients with antisera which recognize the amino and carboxy-terminal regions of dystrophin. Am. J. Hum. Genet. *45* (Suppl.):179A.

Bulman, D. E. et al. (1991). Differentiation of Duchenne and Becker muscular dystrophy phenotypes with amino- and carboxy-terminal antisera specific for dystrophin. Am. J. Hum. Genet. *48*:295–304.

Bunin, G. R. et al. (1989). Pre- and postconception factors associated with sporadic heritable and nonheritable retinoblastoma. Cancer Res. *49*:5730–5735.

Burghes, A. H. M. et al. (1987). A cDNA clone from the Duchenne/Becker muscular dystrophy gene. Nature (London) *328*:434–437.

Burmeister, M. et al. (1988). A 10-megabase physical map of human Xp21, including the Duchenne muscular dystrophy gene. Genomics *2*:189–202.

Bushby, K. M. D. et al. (1991). Prevalence and incidence of Becker muscular dystrophy. Lancet *337*:1022–1024.

Buxton, J. et al. (1992). Detection of an unstable fragment of DNA specific to individuals with myotonic dystrophy. Nature (London) *355*:547–548.

Campisi, J. et al. (1984). Cell-cycle control of c-*myc* but not of c-*ras* expression is lost following chemical transformation. Cell (Cambridge, Mass) *36*:241–243.

Cance, W. G. et al. (1990). Altered expression of the retinoblastoma gene product in human sarcomas. N. Engl. J. Med. *323*:1457–1462.

Cannon-Albright, L. A. et al. (1990). Evidence against the reported linkage of the cutaneous melanoma-dysplastic nevus syndrome locus to chromosome 1p36. Am. J. Hum. Genet. *46*:912–918.

Carloni, G. et al. (1988). Integration and loss of a single v-Ki-*ras* gene affects tumorigenic potential of human osteosarcoma cells. FEBS Lett. *229*:333–339.

Carstens, C. et al. (1988). Human KRAS oncogene expression in meningioma. Cancer Lett. *43*:37–41.

Carter, C. J. (1986). The biochemical pathology of Huntington's disease. Rev. Neurosci. *1*:1–34.

Cawthon, R. M. et al. (1990). A major segment of the neurofibromatosis type 1 gene: cDNA sequence, genomic structure, and point mutations. Cell (Cambridge, Mass) *62*:193–201.

Chamberlain, J. S. et al. (1988). Deletion screening of the Duchenne muscular dystrophy locus via multiplex DNA amplification. Nucleic Acids Res. *16*:11141–11156.

Chan, H. S. L. et al. (1991). P-glycoprotein expression as a predictor of the outcome of therapy for neuroblastoma. N. Engl. J. Med. *325*:1608–1614.

Chandar, N. et al. (1990). Analysis of tumor suppressor genes *Rb* and *p53* in a transplantable murine osteosarcoma. Proc. Am. Assn. Cancer Res. *31*:316A.

Chang, E. H. et al. (1987). Oncogenes in radioresistant, noncancerous skin fibroblasts from a cancer-prone family. Science *237*:1036–1039.

Chapman, M. A. (1990). Invited editorial: predictive testing for adult-onset genetic disease: ethical and legal implications of the use of linkage analysis for Huntington disease. Am. J. Hum. Genet. *47*:1–3.

Chartier-Harlin, M.-C. et al. (1991). Early-onset Alzheimer's disease caused by mutations at codon 717 of the β-amyloid precursor protein gene. Nature (London) *353*:844–846.

Chelly, J. et al. (1990a). Dystrophin gene transcribed from different promoters in neuronal and glial cells. Nature (London) *344*:64–65.

Chelly, J. et al. (1990b). Effect of dystrophin gene deletions on mRNA levels and processing in Duchenne and Becker muscular dystrophies. Cell (Cambridge, Mass) *63*:1239–1248.

Chenevix-Trench, G. et al. (1990). Gene expression in melanoma cell lines and cultured melanocytes: correlation between levels of c-*src* 1, c-*myc* and *p53*. Oncogene *5*:1187–1193.

Chu, H. M. et al. (1991). NF-1 proteins from brain interact with the proenkephalin cAMP inducible enhancer. Nucleic Acids Res. *19*:2721–2728.

Clarke, A. et al. (1989). Xp21 DNA probe in diagnosis of muscular dystrophy and spinal muscular atrophy. Lancet *i*:443.

Cochran, B. H. et al. (1984). Expression of the c-*fos* gene and of a *fos*-related gene is stimulated by platelet derived growth factor. Science *226*:1080–1082.

Cohen, P. S. et al. (1988). Detection of N-*myc* gene expression in neuroblastoma tumors by *in situ* hybridization. Am. J. Pathol. *131*:391–397.

Cohen, P. S. et al. (1990). Neuropeptide Y expression in the developing adrenal gland and in childhood neuroblastoma tumors. Cancer Res. *50*:6055–6061.

Cohn, S. L. et al. (1988). Single copies of the N-*myc* oncogene in neuroblastomas from children presenting with the syndrome of opsoclonus-myoclonus. Cancer *62*:723–726.

Cole, C. G. et al. (1988). Prenatal testing for Duchenne and Becker muscular dystrophy. Lancet *i*:262–265.

Combaret, V. et al. (1989). Clinical value of N-*myc* oncogene amplification in 52 patients with neuroblastoma included in recent therapeutic protocols. Eur. J. Cancer Clin. Oncol. *25*:1607–1612.

Comings, D. E. et al. (1991). Human tryptophan oxygenase localized to 4q31: possible implications for alcoholism and other behavioral disorders. Genomics *9*:301–308.

Conneally, P. M. et al. (1984). Huntington disease: estimation of heterozygote status using linked genetic markers. Genet. Epidemiol. *1*:81–88.

Consalez, G. G. et al. (1991). Assignment of Emery-Dreifuss muscular dystrophy to the distal region of Xq28: the results of a collaborative study. Am. J. Hum. Genet. *48*:468–480.

Cooper, C. S. et al. (1984). Molecular cloning of a new transforming gene from a chemically-transformed human cell line. Nature (London) *311*:29–33.

Cooper, J. A. and Whyte, P. (1989). RB and the cell cycle: entrance or exit? Cell (Cambridge, Mass) *58*:1009–1011.

Cooper, E. H. et al. (1987). Serum neuron-specific enolase in children's tumor. Br. J. Cancer *56*:65–67.

Cooper, M. J. et al. (1990). β_2-Microglobulin expression in human embryonal neuroblastoma reflects its developmental regulation. Cancer Res. *50*:3694–3700.

Cormier, V. et al. (1991). Autosomal dominant deletions of the mitochondrial genome in a case of progressive encephalomyopathy. Am. J. Hum. Genet. *48*:643–648.

Cornaglia-Ferraris, P. et al. (1990a). A new human highly tumorigenic neuroblastoma cell line with undetectable expression of N-*myc*. Pediatr. Res. *27*:1–6.

Cornaglia-Ferraris, P. et al. (1990b). Evidence of loss of N-*myc* amplification during the establishment of a human neuroblastoma cell line [letter to the Editor]. Int. J. Cancer *45*:578–579.

Cowell, J. K. (1991). The genetics of retinoblastoma. Br. J. Cancer *63*:333–336.

Craufurd, D. et al. (1989). Uptake of presymptomatic predictive testing for Huntington's disease. Lancet *ii*:603–605.

Crowe, F. W. et al. (1956) A Clinical, Pathological, and Genetic Study of Multiple Neurofibromatosis. Charles C Thomas, Springfield, IL.

Dalby, S. (1990). Prenatal exclusion testing for Huntington's disease [letter]. J. Med. Genet. *27*:407–408.

Darragh, T. M. et al. (1991). Identification of herpes simplex virus DNA in lesions of erythema multiforme by the polymerase chain reaction. J. Am. Acad. Dermatol. *24*:23–26.

Darras, B. T. (1990). Molecular genetics of Duchenne and Becker muscular dystrophy. J. Pediatr. *117*:1–15.

Darras, B. T. and Francke, U. (1988). Normal human genomic restriction fragment patterns and polymorphisms revealed by hybridization with the entire dystrophin cDNA. Am. J. Hum. Genet. *43*:612–619.

Davies, K. (1991). Breaking the fragile X. Nature (London) *351*:439–440.

Davies, P. (1986). The genetics of Alzheimer's disease: a review and a discussion of the implications. Neurobiol. Aging *7*:459–466.

den Dunnen, J. T. et al. (1989). Topography of the Duchenne muscular dystrophy (*DMD*) gene: FIGE and cDNA analysis of 194 cases reveals 115 deletions and 13 duplications. Am. J. Hum. Genet. *45*:835–847.

Derynck, R. et al. (1987). Synthesis of messenger RNAs for transforming growth factors α and β and the epidermal growth factor receptor by human tumors. Cancer Res. *47*:707–712.

Diedrich, U. et al. (1988). Rare Ha-*ras* and c-*mos* alleles in patients with intracranial tumors. Neurology *38*:587–589.

Dietrich, A. et al. (1991). Molecular cloning and analysis of the fragile X region in man. Nucleic Acids Res. *19*:2567–2572.

Dietz, H. C. et al. (1991). Marfan syndrome caused by a recurrent *de novo* missense mutation in the fibrillin gene. Nature (London) *352*:337–340.

DiMauro, S. et al. (1989). Mitochondrial diseases. In: Molecular Genetics in Diseases of Brain, Nerve, and Muscle. Rowland, L. P. et al. (eds.). Oxford University Press, New York, pp. 285–298.

Dobkin, C. S. et al. (1991). Prenatal diagnosis of fragile X chromosome. Lancet *338*:957–958.

Doggett, N. A. et al. (1989). The Huntington disease locus is most likely within 325 kilobases of the chromosome 4p telomere. Proc. Natl. Acad. Sci. U.S.A. *86*:10011–10014.

Dotto, G. P. et al. (1986). c-*myc* and c-*fos* expression in differentiating mouse primary keratinocytes. EMBO J. *5*:2853–2857.

Dracopoli, N. C. et al. (1987). Loss of heterozygosity at autosomal and X-linked loci during tumor progression in a patient with melanoma. Cancer Res. *47*:3995–4000.

Driscoll, M. and Chalfie, M. (1991). The *mec*-4 gene is a member of a family of *Caenorhabditis elegans* genes that can mutate to induce neuronal degeneration. Nature (London) *349*:588–593.

Dryja, T. P. et al. (1986). Chromosomal 13 homozygosity in osteosarcoma without retinoblastoma. Am. J. Hum. Genet. *38*:59–66.

Dubowitz, V. (1989). The Duchenne dystrophy story: from phenotype to gene and potential treatment. J. Child. Neurol. *4*:240–250.

Dumanski, J. P. et al. (1990). Molecular genetic analysis of chromosome 22 in 81 cases of meningioma. Cancer Res. *50*:5863–5867.

Dunn, J. M. et al. (1988). Identification of germline and somatic mutations affecting the retinoblastoma gene. Science *241*:1797–1800.

Dunn, J. M. et al. (1989). Mutations in the *RB1* gene and their effects on transcription. Mol. Cell. Biol. *9*:4596–4604.

Edwards, J. H. (1990). Genetic linkage and psychiatric disease [letter]. Nature (London) *344*:298–299.

Elder, J. T. et al. (1989). Overexpression of transforming growth factor α in psoriatic epidermus. Science *243*:811–814.

Elder, J. T. et al. (1990). Growth factor and proto-oncogene expression in psoriasis. J. Invest. Dermatol. *95*:7S-9S.

Ellis, D. L. et al. (1987). Melanoma, growth factors, acanthosis nigricans, the sign of Leser-Trelat, and multiple acrochordons. A possible role for α-transforming growth factor in cutaneous paraneoplastic syndromes. N. Engl. J. Med. *317*:1582–1587.

Ellis, J. M. et al. (1990). Specificity of dystrophin analysis improved with monoclonal antibodies. Lancet *336*:881–882.

England, S. B. et al. (1990). Very mild muscular dystrophy associated with the deletion of 46% of dystrophin. Nature (London) *343*:180–182.

Enzinger, F. M. and Weiss, S. W. (1988). Soft Tissue Tumors, 2nd Ed. C. V. Mosby, St. Louis.

Ervasti, J. M. et al. (1990). Deficiency of a glycoprotein component of the dystrophin complex in dystrophic muscle. Nature (London) *345*:315–319.

Estus, S. et al. (1992). Potentially amyloidogenic, carboxyl-terminal derivatives of the amyloid protein precursor. Science *255*:726–728.

Fahy, M. et al. (1989). Different options for prenatal testing for Huntington's disease using DNA probes. J. Med. Genet. *26*:353–357.

Fain, P. R. et al. (1989). Refined physical and genetic mapping of the NF1 region on chromosome 17. Am. J. Hum. Genet. *45*:721–728.

Farrer, L. A. et al. (1988). Considerations in using linkage analysis as a presymptomatic test for Huntington's disease. J. Med. Genet. *25*:577–588.

Farndon, P. A. et al. (1992). Location of gene for Gorlin syndrome. Lancet *339*:581–582.

Feener, C. A. et al. (1991). Rapid detection of CA polymorphisms in cloned DNA: application to the 5' region of the dystrophin gene. Am. J. Hum. Genet. *48*:621–627.

Fenoglio-Preiser, C. M. et al. (1991). Oncogenes and tumor suppressor genes in solid tumors: neural tumors. In: Molecular Diagnostics in Pathology. Fenoglio-Preiser, C. M. and Willman, C. L. (eds.). Williams & Wilkins, Baltimore, pp. 167–187.

Figarella-Branger, D. F. et al. (1990). Differential spectrum of expression of neural cell adhesion molecule isoforms and L1 adhesion molecules on human neuroectodermal tumors. Cancer Res. *50*:6364–6370.

Fisher, E. M. C. et al. (1990). Homologous ribosomal protein genes on the human X and Y chromosomes: escape from X inactivation and possible implications for Turner syndrome. Cell (Cambridge, Mass) *63*:1205–1218.

Fojo, A. T. et al. (1987). Expression of a multidrug-resistance gene in human tumors and tissues. Proc. Natl. Acad. Sci. U.S.A. *84*:265–269.

Ford, G. M. et al. (1990). A physical map around the retinoblastoma gene. Genomics *6*:284–292.

Forrest, S. M. et al. (1987). Preferential deletion of exons in Duchenne and Becker muscular dystrophies. Nature (London) *329*:638–640.

Forrest, S. M. et al. (1988). Further studies of gene deletions that cause Duchenne and Becker dystrophies. Genomics *2*:109–114.

Fountain, J. W. et al. (1989). Physical mapping of the von Recklinghausen neurofibromatosis region on chromosome 17. Am. J. Hum. Genet. *44*:58–67.

Francke, U. et al. (1985). Minor Xp21 chromosome deletion in a male associated with expression of Duchenne muscular dystrophy, chronic granulomatous disease, retinitis pigmentosa, and McLeod syndrome. Am. J. Hum. Genet. *37*:250–267.

Freeman, C. S. et al. (1990). An overview of tumor biology. Cancer Invest. *8*:71–90.

Freund, L. S. and Reiss, A. L. (1991). Cognitive profiles associated with the Fra (X) syndrome in males and females. Am. J. Med. Genet. *38*:542–547.

Freytag, S. O. (1988). Enforced expression of the c-*myc* oncogene inhibits cell differentiation by precluding entry into a distinct predifferentiation state in G_0/G_1. Mol. Cell. Biol. *8*:1614–1624.

Friedman, H. S. et al. (1988). Phenotypic and genotypic analysis of a human medulloblastoma cell line and transplanted xenograft (D341 Med) demonstrating amplification of c-*myc*. Am. J. Pathol. *130*:472–484.

Friend, S. H. et al. (1986). A human DNA segment with properties of the gene that predisposes to retinoblastoma and osteosarcoma. Nature (London) *323*:643–646.

Fryns, J. P. (1989). X-linked mental retardation and the fragile X syndrome: a clinical approach. In: The Fragile X Syndrome. Davies, K. E. (ed.). Oxford University Press, New York, pp. 1–39.

Fuchs, E. and Green, H. (1981). Regulation of terminal differentiation of cultured human keratinocytes by vitamin A. Cell (Cambridge, Mass) *25*:617–625.

Fujimoto, M. et al. (1988). Expression of three viral oncogenes (v-*sis*, v-*myc*, v-*fos*) in primary brain tumors of neuroectodermal origin. Neurology *38*:289–293.

Fults, D. et al. (1990). Allelotype of human malignant astrocytoma. Cancer Res. *50*:5784–5789.

Fung, Y. K. T. et al. (1987). Structural evidence for the authenticity of the human retinoblastoma gene. Science *236*:1657–1661.

Furukawa, Y. et al. (1990). Expression and state of phosphorylation of the retinoblastoma susceptibility gene product in cycling and noncycling human hematopoietic cells. Proc. Natl. Acad. Sci. U.S.A. *87*:2770–2774.

Furuya, T. et al. (1991). Changes of common fragile sites of chromosomes according to the menstrual cycle. Hum. Genet. *86*:471–474.

Fuscoe, J. C. et al. (1989). An efficient method for selecting unique-sequence clones from DNA libraries and its application to fluorescent staining of human chromosome 21 using *in situ* hybridization. Genomics *5*:100–109.

Futreal, P. A. and Barrett, J. C. (1990). Decreased phosphorylation of the retinoblastoma gene product correlates with cellular senescence. Proc. Am. Assn. Cancer Res. *31*:316A.

Gansler, T. et al. (1986). Flow cytometric DNA analysis of neuroblastoma: correlation with histology and clinical outcome. Cancer *58*:2453–2458.

Garson, J. A. et al. (1985). N-*myc* amplification in malignant astrocytoma. Cancer *2*:718–719.

Garson, J. A. et al. (1986). N-*myc* oncogene amplification in rhabdomyosarcoma at relapse. Lancet *1*:1496.

Garson, J. A. et al. (1989). N-*myc* gene expression and oncoprotein characterization in medulloblastoma. Br. J. Cancer *59*:889–894.

Garvin, A. J. and Sens, D. A. (1986). The *in vitro* growth and differentiation of a human rhabdomyosarcoma cell line. Lab. Invest. *56*:26A.

Garvin, J., Jr. et al. (1990). N-*myc* oncogene expression and amplification in metastatic lesions of stage IV-S neuroblastoma. Cancer *65*:2572–2575.

Gattuso, P. et al. (1990). Is DNA ploidy of prognostic significance in stage I malignant melanoma? Surgery *108*:702–708.

Gelernter, J. and Kidd, K. K. (1991). The current status of linkage studies in schizophrenia. In: Genes, Brain and Behavior. McHugh, P. R. and McKusick, V. A. (eds.). Raven Press, New York, pp. 137–152.

Gerosa, M. A. et al. (1989). Overexpression of N-*ras* oncogene and epidermal growth factor receptor gene in human glioblastomas. J. Natl. Cancer Inst. *81*:63–67.

Gershon, E. S. et al. (1990). Genetic mapping of common diseases: the challenges of manic-depressive illness and schizophrenia. TIG *6*:282–287.

Gilbert, F. et al. (1984). Human neuroblastomas and abnormalities of chromosomes 1 and 17. Cancer Res. *44*:5444–5449.

Gilgenkrantz, H. et al. (1989). Analysis of molecular deletions with cDNA probes in patients with Duchenne and Becker muscular dystrophies. Genomics *5*:574–580.

Gilliam, T. C. and Gusella, J. F. (1988). Huntington's disease. In: The Molecular Biology of Neurological Disease. Rosenberg, R. N. and Harding, A. E. (eds.). Butterworths, London, pp. 163–182.

Gilliam, T. C. et al. (1987a). Localization of the Huntington's disease gene to a small segment of chromosome 4 flanked by D4S10 and the telomere. Cell (Cambridge, Mass) *50*:565–571.

Gilliam, T. C. et al. (1987b). A DNA segment encoding two genes very tightly linked to Huntington's disease. Science *238*:950–952.

Gilliam, T. C. et al. (1990). Genetic homogeneity between acute and chronic forms of spinal muscular atrophy. Nature (London) *345*:823–825.

Glenner, G. G. and Wong, C. W. (1984). Alzheimer's disease: initial report of the purification and characterization of a novel cerebrovascular amyloid protein. Biochem. Biophys. Res. Commun. *120*:885–890.

Glover, T. W. et al. (1991). Molecular and cytogenetic analysis of tumors in von Recklinghausen neurofibromatosis. Genes, Chromosomes Cancer *3*:62–70.

Goate, A. et al. (1991). Segregation of a missense mutation in the amyloid precursor protein gene with familial Alzheimer's disease. Nature (London) *349*:704–706.

Goddard, A. D. et al. (1990). Use of the *RB*1 cDNA as a diagnostic probe in retinoblastoma families. Clin. Genet. *37*:117–126.

Godfrey, M. et al. (1990a). Cosegregation of elastin-associated microfibrillar abnormalities with the Marfan phenotype in families. Am. J. Hum. Genet. *46*:652–660.

Godfrey, M. et al. (1990b). Unilateral microfibrillar abnormalities in a case of asymmetric Marfan syndrome. Am. J. Hum. Genet. *46*:661–671.

Golbe, L. I. (1990). The genetics of Parkinson's disease: a reconsideration. Neurology *40* (Suppl 3):7–14.

Golde, T. E. et al. (1992). Processing of the amyloid protein precursor to potentially amyloidogenic derivatives. Science *255*:728–730.

Goldgar, D. E. et al. (1989). Multipoint linkage analysis in neurofibromatosis type 1: an international collaboration. Am. J. Hum. Genet. *44*:6–12.

Goodwin, D. W. (1991). The genetics of alcoholism. In: Genes, Brain and Behavior. McHugh, P. R. and McKusick, V. A. (eds.). Raven Press, New York, pp. 219–226.

Goto, Y. et al. (1990). A mutation in the tRNA$^{Leu\ (UUR)}$ gene associated with the MELAS subgroup of mitochondrial encephalomyopathies. Nature (London) *348*:651–653.

Grady-Leopardi, E. F. et al. (1986). Detection of N-*myc* oncogene expression in human neuroblastoma by *in situ* hybridization and blot analysis: relationship to clinical outcome. Cancer Res. *46*:3196–3199.

Graves, D. T. et al. (1984). Detection of c-*sis* transcripts and synthesis of PDGF-like proteins by human osteosarcoma cells. Science *226*:972–974.

Green, M. R. (1989). When the products of oncogenes and anti-oncogenes meet. Cell (Cambridge, Mass) *56*:1–3.

Greenberg, C. R. et al. (1991). Three years' experience with neonatal screening for Duchenne/Becker muscular dystrophy: gene analysis, gene expression, and phenotype prediction. Am. J. Med. Genet. *39*:68–75.

Greenberg, M. E. and Ziff, E. B. (1984). Stimulation of 3T3 cells induces transcription of the c-*fos* protooncogene. Nature (London) *311*:433–438.

Greene, M. H. and Fraumeni, J. F., Jr. (1979). The hereditary variant of malignant melanoma. In: Human Malignant Melanoma, Clark, W. H., Jr. et al. (eds.). Grune & Stratton, New York, pp. 139–166.

Greene, M. H. et al. (1985). High risk of malignant melanoma in melanoma-prone families with dysplastic nevi. Ann. Intern. Med. *102*:458–465.

Greenhalgh, D. A. et al. (1990). Two oncogenes, v-fos and v-ras, cooperate to convert normal keratinocytes to squamous cell carcinoma. Proc. Natl. Acad. Sci. U.S.A. *87*:643–647.

Greger, V. et al. (1990). Possible involvement of the retinoblastoma gene in undifferentiated sinonasal carcinoma. Cancer *66*:1954–1959.

Griffin, S. (1988). *In situ* hybridization: visualizing brain messenger RNA. In: The Molecular Biology of Neurological Disease. Rosenberg, R. N. and Harding, A. E. (eds.). Butterworths, London, pp. 35–43.

Grompe, M. et al. (1990). Simplified diagnosis of new mutation X-linked disease. In: Biotechnology and Human Genetic Predisposition to Disease. Wiley-Liss, New York, pp. 47–59.

Gruis, N. A. et al. (1990). Locus for susceptibility to melanoma on chromosome 1p [letter to the Editor]. N. Engl. J. Med. *322*:853–854.

Gurling, H. (1990). Genetic linkage and psychiatric disease [letter]. Nature (London) *344*:298.

Gusella, J. F. (1986). Probes in Huntington's chorea. A reply. Nature (London) *320*:21–22.

Gusella, J. F. and Gilliam, T. C. (1989). DNA markers in dominant neurogenetic diseases. In: Molecular Genetics in Diseases of Brain, Nerve and Muscle. Rowland, L. P. et al. (eds.). Oxford University Press, New York, pp. 327–337.

Gusella, J. F. et al. (1983). A polymorphic DNA marker genetically linked to Huntington's disease. Nature (London) *306*:234–238.

Gusella, J. F. et al. (1985). Deletion of Huntington's disease-linked G8 (D4S10) locus in Wolf-Hirschhorn syndrome. Nature (London) *318*:75–78.

Hagag, N. et al. (1986). Inhibition of growth factor-induced differentiation of PC12 cells by microinjection of antibody to *ras* p21. Nature (London) *319*:680–682.

Hailat, N. et al. (1990). N-*myc* gene amplification in neuroblastoma is associated with altered phosphorylation of a proliferation related polypeptide (Op 18). Oncogene *5*:1615–1618.

Hammans, S. R. et al. (1991). Mitochondrial encephalopathies: molecular genetic diagnosis from blood samples. Lancet *337*:1311–1313.

Hammerling, U. et al. (1987). Different regulation of N- and c-*myc* expression during phorbol ester-induced maturation of human SH-S454 neuroblastoma cells. Oncogene *2*:73–77.

Hansen, M. F. et al. (1985). Osteosarcoma and retinoblastoma: a shared chromosomal mechanism revealing recessive predisposition. Proc. Natl. Acad. Sci. U.S.A. *82*:6216–6220.

Harada, T. et al. (1989). Clinical features and diagnosis by recombinant DNA techniques of familial amyloid polyneuropathy in Japan. Res. Commun. Chem. Pathol. Pharmacol. *65*:237–244.

Harley, H. G. et al. (1992). Expansion of an unstable DNA region and phenotypic variation in myotonic dystrophy. Nature (London) *355*:545–546.

Harper, J. R. et al. (1986). Transfection of the EJ *ras*-Ha gene into keratinocytes derived from carcinogen-induced mouse papillomas causes malignant progression. Mol. Cell. Biol. *6*:3144–3149.

Harper, J. R. et al. (1987). Analysis of the *ras*^{H} oncogene and its p21 product in chemically induced skin tumors and tumor-derived cell lines. Carcinogenesis *8*:1821–1825.

Harper, P. S. (1979). Myotonic dystrophy. Saunders, Philadelphia.

Harper, P. S. (1988). Molecular genetics and muscular dystrophy. In: The Molecular Biology of Neurological Disease. Rosenberg, R. N. and Harding, A. E. (eds.). Butterworths, London, pp. 183–198.

Harper, P. S. and Thomas, N. S. T. (1986). A molecular approach to genetic counselling in the X-linked muscular dystrophies. Am. J. Med. Genet. *25*:687–702.

Harper, P. S. et al. (1990). Genetic testing for Huntington's disease. Br. Med. J. *300*:1089–1090.

Hayashi, Y. et al. (1988). Chromosome findings and prognosis in 15 patients with neuroblastoma found by VMA mass screening. J. Pediatr. *112*:567–571.

Hayward, N. et al. (1988). A *Taq*I RFLP of the human TGF-α gene is significantly associated with cutaneous malignant melanoma. Int. J. Cancer *42*:558–561.

Heilig, M. and Widerlov, E. (1990). Neuropeptide Y: an overview of central distribution functional aspects, and possible involvement in neuropsychiatric illness. Acta Psychiatr. Scand. *82*:95–114.

Heisterkamp, N. et al. (1990). Rearrangement of the human ABL oncogene in a glioblastoma. Cancer Res. *50*:3429–3434.

Heitz, D. et al. (1991). Isolation of sequences that span the fragile X and identification of a fragile X-related CpG island. Science *251*:1236–1239.

Heldin, C. H. et al. (1986). A human osteosarcoma cell line secretes a growth factor structurally related to a homodimer of PDGF A-chains. Nature (London) *319*:511–514.

Helman, L. J. et al. (1989). Neuropeptide Y expression distinguishes malignant from benign pheochromocytoma. J. Clin. Oncol. *7*:1720–1725.

Hensel, C. et al. (1990). Altered structure and expression of the human retinoblastoma gene in small cell lung cancer. Cancer Res. *50*:3067–3072.

Hentemann, M. et al. (1990). Rapid detection of deletions in the Duchenne muscular dystrophy gene by PCR amplification of deletion-prone exon sequences. Hum. Genet. *84*:228–232.

Herbert, J. (1989). Familial amyloidotic polyneuropathies. In: Molecular Genetics in Diseases of Brain, Nerve, and Muscle. Rowland, L. P. et al. (eds.). Oxford University Press, New York, pp. 299–325.

Herlyn, M. et al. (1985). Characteristics of cultured human melanocytes isolated from different stages of tumor progression. Cancer Res. *45*:5670–5676.

Hermansson, M. et al. (1988). Endothelial cell hyperplasia in human glioblastoma: coexpression of mRNA for PDGF B chain and PDGF receptor suggests autocrine growth stimulation. Proc. Natl. Acad. Sci. U.S.A. *85*:7748–7752.

Hirst, M. et al. (1991). Prenatal diagnosis of fragile X syndrome. Lancet *338*:956–957.

Hodgson, S. V. and Bobrow, M. (1989). Carrier detection and prenatal diagnosis in Duchenne and Becker muscular dystrophy. Br. Med. Bull. *45*:719–744.

Hoffman, E. P. and Kunkel, L. M. (1989). Dystrophin abnormalities in Duchenne/Becker muscular dystrophy. Neuron *2*:1019–1029.

Hoffman, E. P. et al. (1987). Dystrophin: the protein product of the Duchenne muscular dystrophy locus. Cell (Cambridge, Mass) *51*:919–928.

Hoffman, E. P. et al. (1988). Characterization of dystrophin in muscle biopsy specimens from patients with Duchenne's or Becker's muscular dystrophy. N. Engl. J. Med. *318*:1363–1368.

Hoffman, E. P. et al. (1989). Improved diagnosis of Becker muscular dystrophy via dystrophin testing. Neurology *39*:1011–1017.

Hollister, D. W. et al. (1990). Immunohistologic abnormalities of the microfibrillar-fiber system in the Marfan syndrome. N. Engl. J. Med. *323*:152–159.

Holt, I. J. et al. (1990). A new mitochondrial disease associated with mitochondrial DNA heteroplasmy. Am. J. Hum. Genet. *46*:428–433.

Horowitz, J. M. et al. (1990). Frequent inactivation of the retinoblastoma antioncogene restricted to a subset of human tumor cells. Proc. Natl. Acad. Sci. U.S.A. *87*:2775–2779.

Hu, X. et al. (1988). Partial gene duplication in Duchenne and Becker muscular dystrophy. J. Med. Genet. *25*:369–376.

Hu, X. et al. (1990). Duplicational mutation at the Duchenne muscular dystrophy locus: its frequency, distribution, origin and phenotype-genotype correlation. Am. J. Hum. Genet. *46*:682–695.

Huang, H. J. S. et al. (1988). Suppression of the neoplastic phenotype by replacement of the *RB* gene in human cancer cells. Science *242*:1563–1566.

Huang, S. et al. (1991). A cellular protein that competes with SV40 T antigen for binding to the retinoblastoma gene product. Nature (London) *350*:160–162.

Huggins, M. et al. (1990). Ethical and legal dilemmas arising during predictive testing for adult-onset disease: the experience of Huntington disease. Am. J. Hum. Genet. *47*:4–12.

Hunt, J. D. et al. (1990). Excision of N-*myc* from chromosome 2 in human neuroblastoma cells containing amplified N-*myc* sequences. Mol. Cell. Biol. *10*:823–829.

Hurko, O. et al. (1989). Dystrophin analysis in clonal myoblasts derived from a Duchenne muscular dystrophy carrier. Am. J. Hum. Genet. *44*:820–826.

Husain, Z. et al. (1990). Comparison of cellular protooncogene activation and transformation-related activity of human melanocytes and metastatic melanoma. J. Invest. Dermatol. *95*:571–575.

Hutchins, J. T. et al. (1991). Novel gene sequences expressed by human melanoma cells identified by molecular subtraction. Cancer Res. *51*:1418–1425.

Ibraghimov-Berskrovnaya, O. et al. (1992). Primary structure of dystrophin-associated glycoproteins linking dystrophin to the extracellular matrix. Nature (London) *355*:696–702.

Ibson, J. M. et al. (1987). Oncogene amplification and chromosomal abnormalities in small cell lung cancer. J. Cell Biochem. *33*:267–288.

Ikeda, I. et al. (1990). Specific expression of the *ret* proto-oncogene in human neuroblastoma cell lines. Oncogene *5*:1291–1296.

Ikonen, E. et al. (1990). Huntington disease in Finland: linkage disequilibrium of chromosome 4 RFLP haplotypes and exclusion of a tight linkage between the disease and D4S43 locus. Am. J. Hum. Genet. *46*:5–11.

Ino, H. et al. (1991). Mitochondrial leucine tRNA mutation in a mitochondrial encephalomyopathy. Lancet *337*:234–235.

Ireland, C. M. (1989). Activated N-*ras* oncogenes in human neuroblastomas. Cancer Res. *49*:5530–5533.

Jankowski, S. et al. (1990). Molecular approaches to trisomy 21. In: Molecular Genetics of Chromosome 21 and Down Syndrome. Wiley-Liss, New York, pp. 79–88.

Jensen, P. J. et al. (1990). Tissue plasminogen activator in psoriasis. J. Invest. Dermatol. *95*:13S–14S.

Johns, D. R. and Hurko, O. (1991). Mitochondrial leucine tRNA mutation in neurological diseases. Lancet *337*:927–928.

Johns, D. R. et al. (1989). Directly repeated sequences associated with pathogenic mitochondrial deletions. Proc. Natl. Acad. Sci. U.S.A. *86*:8059–8062.

Johnson, B. E. et al. (1988). *myc* family DNA amplification in small cell lung cancer patients' tumors and corresponding cell lines. Cancer Res. *48*:5163–5166.

Johnson, K. et al. (1989). Recombination events that locate myotonic dystrophy distal to APOC2 on 19q. Genomics *5*:746–751.

Johnson, K. et al. (1990). A new polymorphic probe which defines the region of chromosome 19 containing the myotonic dystrophy locus. Am. J. Hum. Genet. *46*:1073–1081.

Jones, K. A. et al. (1987). A cellular DNA-binding protein that activates eukaryotic transcription and DNA replication. Cell (Cambridge, Mass) *48*:79–89.

Judisch, G. F. and Patil, S. R. (1985). Retinoblastoma, genetics and genetic councelling. In: Retinoblastoma. Blodi, F. C. (ed.). Churchhill Livingstone, New York, pp. 151–162.

Kaelin, W. G. et al. (1990). Definition of the minimal Simian virus 40 large T antigen and adenovirus E1A-binding domain in the retinoblastoma gene product. Mol. Cell. Biol. *10*:3761–3769.

Kainulainen, K. et al. (1990). Location on chromosome 15 of the gene defect causing Marfan syndrome. N. Engl. J. Med. *323*:935–939.

Kanazawa, I. et al. (1990). Studies on DNA markers (D4S10 and D4S43/S127) genetically linked to Huntington's disease in Japanese families. Hum. Genet. *85*:257–260.

Kanda, N. et al. (1987). Amplification of IMR-32 clones 8, G21 and N-*myc* in human neuroblastoma xenografts. Cancer Res. *47*:3291–3295.

Kaneko, Y. et al. (1987). Different karyotypic patterns in early and advanced stage neuroblastomas. Cancer Res. *47*:311–318.

Kaneko, Y. et al. (1988). Chromosomes and screening for neuroblastoma. Lancet *i*:174–175.

Kaneko, Y. et al. (1990). Current urinary mass screening for catecholamine metabolites at 6 months of age may be detecting only a small portion of high-risk neuroblastomas: a chromosome and N-*myc* amplification study. J. Clin. Oncol. *8*:2005–2013.

Kang, J. et al. (1987). The precursor of Alzheimer's disease amyloid A4 protein resembles a cell-surface receptor. Nature (London) *325*:733–736.

Kawashima, M. et al. (1990). Premalignant lesions and cancers of the skin in the general population: evaluation of the role of human papillomaviruses. J. Invest. Dermatol. *95*:537–542.

Kelly, E. D. et al. (1990). Carrier estimations in Duchenne muscular dystrophy families in Northern Ireland using RFLP analysis. J. Med. Genet. *27*:101–104.

Khalifa, M. M. et al. (1990). Methylation status of genes flanking the fragile site in males with the fragile X syndrome: a test of the imprinting hypothesis. Am. J. Hum. Genet. *46*:744–753.

Khurana, T. S. et al. (1991). Dystrophin detection in freeze-dried tissue. Lancet *338*:448.

Kidd, K. K. et al. (1989). Applications of molecular genetic methods to affective disorders. In: Molecular Probes: Technology and Medical Applications. Albertini, A. (ed.). Raven Press, New York, pp. 61–68.

Kiefer, P. E. et al. (1987). Amplification and expression of protooncogenes in human small cell lung cancer cell lines. Cancer Res. *47*:6236–6242.

Kim, M. K. et al. (1985). Purification and characterization of a low molecular weight transforming growth factor from the urine of melanoma patients. J. Biol. Chem. *260*:9237–9243.

King, I. and Sartorelli, A. C. (1989). Epidermal growth factor receptor gene expression, protein kinase activity, and terminal differentiation of human malignant epidermal cells. Cancer Res. *49*:5677–5681.

King, I. et al. (1986). A sensitive method to quantify the terminal differentiation of cultured epidermal cells. Exp. Cell Res. *167*:252–256.

Kingston, H. M. et al. (1984). Localization of the Becker muscular dystrophy gene on the short arm of the X chromosome by linkage to cloned DNA sequences. Hum. Genet. *67*:6–17.

Klein-Szanto, A. et al. (1991). Melanosis and associated tumors in transgenic mice. Proc. Natl. Acad. Sci. U.S.A. *88*:169–173.

Klingel, R. et al. (1987). An amplification unit in human melanoma cells showing partial homology with sequences of human papillomavirus type 9 and with nuclear antigen 1 of the Epstein-Barr virus. Cancer Res. *47*:4485–4492.

Kloss, K. et al. (1991). Characterization of deletions at the retinoblastoma locus in patients with bilateral retinoblastoma. Am. J. Med. Genet. *39*:196–200.

Knudson, A. G., Jr. (1971). Mutation and cancer: statistical study of retinoblastoma. Proc. Natl. Acad. Sci. U.S.A. *68*:820–823.

Knudson, A. G. (1975) The genetics of childhood cancer. Cancer *35*:1022–1026.

Koenig, M. et al. (1987). Complete cloning of the Duchenne muscular dystrophy (DMD) cDNA and preliminary genomic organization of the DMD gene in normal and affected individuals. Cell (Cambridge, Mass) *50*:509–517.

Koenig, M. et al. (1988). The complete sequence of dystrophin predicts a rod-shaped cytoskeletal protein. Cell (Cambridge, Mass) *53*:219–228.

Koenig, M. et al. (1989). The molecular basis for Duchenne versus Becker muscular dystrophy: correlation of severity with type of deletion. Am. J. Hum. Genet. *45*:498–506.

Koh, H. K. (1991). Cutaneous melanoma. N. Engl. J. Med. *325*:171–182.

Korenberg, J. R. (1990). Molecular mapping of the Down syndrome phenotype. In: Molecular Genetics of Chromosome 21 and Down Syndrome. Wiley-Liss, New York, pp. 105–115.

Korenberg, J. R. et al. (1990). Molecular definition of a region of chromosome 21 that causes features of the Down syndrome phenotype. Am. J. Hum. Genet. *47*:236–246.

Korneluk, R. G. et al. (1989). A reordering of human chromosome 19 long-arm DNA markers and identification of markers flanking the myotonic dystrophy locus. Genomics *5*:596–604.

Krauss, C. M. et al. (1987). Familial premature ovarian failure due to an interstitial deletion of the long arm of the X chromosome. N. Engl. J. Med. *317*:125–131.

Kremer, E. J. et al. (1991). Mapping of DNA instability at the fragile X to a trinucleotide repeat sequence $p(CCG)_n$. Science *252*:1711–1714.

Kruijer, W. et al. (1984). Platelet derived growth factor induces a rapid but transient expression of the c-*fos* gene and protein. Nature (London) *312*:711–716.

Kunkel, L. M. (1989). Duchenne muscular dystrophy: identification of the gene. In: Molecular Genetics in Diseases of Brain, Nerve and Muscle. Rowland, L. P. et al. (eds.). Oxford University Press, New York, pp. 370–379.

Kunkel, L. M. and Hoffman, E. P. (1989). Duchenne/Becker muscular dystrophy: a short overview of the gene, the protein, and current diagnostics. Br. Med. Bull. *45*:630–643.

Kunkel, L. M. et al. (1986). Analysis of deletions in the DNA of patients with Becker and Duchenne muscular dystrophy. Nature (London) *322*:73–77.

Lachmann, P. J. (1989). Molecular genetics of muscle disease — Duchenne and other dystrophies. Future prospects. Br. Med. Bull. *45*:819–824.

Laing, N. G. et al. (1991). The diagnosis of Duchenne and Becker muscular dystrophies: two years' experience in a comprehensive carrier screening and prenatal diagnostic laboratory. Med. J. Aust. *154*:14–18.

Laird, C. D. (1987). Proposed mechanism of inheritance and expression of the human fragile-X syndrome of mental retardation. Genetics *117*:587–599.

Laird, C. D. et al. (1990). Two progenitor cells for human oogonia inferred from pedigree data and the X-inactivation imprinting model of the fragile X syndrome. Am. J. Hum. Genet. *46*:696–719.

La Spada, A. R. et al. (1991). Androgen receptor gene mutations in X-linked spinal and bulbar muscular atrophy. Nature (London) *352*:77–79.

Lavedan, C. et al. (1990). Direct haplotyping by double digestion of PCR-amplified creatine kinase (CKMM): application to myotonic dystrophy diagnosis. Genomics *8*:739–740.

Lazaro, C. et al. (1992). Prenatal diagnosis of sporadic neurofibromatosis 1. Lancet *339*:119–120.

Leach, R. J. et al. (1990). Preferential retention of paternal alleles in human retinoblastoma: evidence for genomic imprinting. Cell Growth Differ. *1*:401–406.

Lee, B. et al. (1991). Linkage of Marfan syndrome and a phenotypically related disorder to two different fibrillin genes. Nature (London) *352*:330–334.

Lee, C. C. et al. (1991). Expression of recombinant dystrophin and its localization to the cell membrane. Nature (London) *349*:334–336.

Lee, E. Y. H. P. et al. (1988). Inactivation of the retinoblastoma susceptibility gene in human breast cancers. Science *241*:218–221.

Lee, W. H. et al. (1984). Expression and amplification of the N-*myc* gene in primary retinoblastoma. Nature (London) *309*:458–460.

Lee, W. H. et al. (1987a). Human retinoblastoma susceptibility gene: cloning, identification, and sequence. Science *235*:1394–1399.

Lee, W. H. et al. (1987b). The retinoblastoma susceptibility gene encodes a nuclear phosphoprotein associated with DNA binding activity. Nature (London) *329*:642–645.

Lemieux, N. et al. (1989). First cytogenetic evidence of homozygosity for the retinoblastoma deletion in chromosome 13. Cancer Genet. Cytogenet. *43*:73–78.

Lever, W. F. and Schaumburg-Lever, G. (1990). Histopathology of the Skin, 7th Ed. J. B. Lippincott, Philadelphia, pp. 756–805.

Libermann, T. A. et al. (1985). Amplification and overexpression of the EGF receptor gene in primary human glioblastomas. J. Cell Sci. *3* (Suppl):161–172.

Lichter, P. et al. (1988). Rapid detection of human chromosome 21 aberrations by *in situ* hybridization. Proc. Natl. Acad. Sci. U.S.A. *85*:9664–9668.

Lichter, P. et al. (1990). Detection of Down syndrome by *in situ* hybridization with chromosome 21 specific DNA probes. In: Molecular Genetics of Chromosome 21 and Down Syndrome. Wiley-Liss, New York, pp. 69–78.

Lidov, H. G. W. et al. (1990). Localization of dystrophin to postsynaptic regions of central nervous system cortical neurons. Nature (London) *348*:725–728.

Liechti-Gallati, S. et al. (1990). RFLPs for Duchenne muscular dystrophy cDNA clones 9 and 10. Am. J. Hum. Genet. *46*:1090–1094.

Lin, B. T. Y. et al. (1991). Retinoblastoma cancer suppressor gene product is a substrate of the cell cycle regulator cdc2 kinase. EMBO J. *10*:857–864.

Linnenbach, A. J. et al. (1988). Structural alteration in the *myb* protooncogene and deletion within the gene encoding α-type protein kinase C in human melanoma cell lines. Proc. Natl. Acad. Sci. U.S.A. *85*:74–78.

Lloyd, R. V. (1990). Molecular probes and endocrine diseases. Am. J. Surg. Pathol. *14* (Suppl 1):34–44.

Look, A. T. et al. (1984). Cellular DNA content as a predictor of response to chemotherapy in infants with unresectable neuroblastoma. N. Engl. J. Med. *311*:231–325.

Love, D. R. et al. (1989). Molecular analysis of Duchenne and Becker muscular dystrophies. Br. Med. Bull. *45*:659–680.

Ludlow, J. W. et al. (1990). The retinoblastoma susceptibility gene product undergoes cell cycle-dependent dephsphorylation and binding to and release from SV40 large T. Cell (Cambridge, Mass) *60*:387–396.

Lumley, M. et al. (1990). DNA marker applicable to presymptomatic and prenatal diagnosis of fascioscapulohumeral disease. Lancet *336*:1320–1321.

Lunt, P. W. (1989). A workshop on fascioscapulohumeral (Landouzy-Dejerine) disease. J. Med. Genet. *26*:535–537.

Lunt, P. W. et al. (1989). DNA probes in differential diagnosis of Becker muscular dystrophy and spinal muscular atrophy. Lancet *i*:46–47.

MacDonald, M. E. et al. (1987). A somatic hybrid panel for localizing DNA segments near the Huntington disease gene. Genomics *1*:29–34.

MacDonald, M. et al. (1989). Recombination events suggest two possible locations for the Huntington's disease gene. Neuron *3*:183–190.

Maguire, H. C., Jr. et al. (1989). Distribution of neu (c-erb B-2) protein in human skin. J. Invest. Dermatol. *89*:786–790.

Mahadevan, M. et al. (1992). Myotonic dystrophy mutation: an unstable CTG repeat in the 3′ untranslated region of the gene. Science *255*:1253–1258.

Malhotra, S. B. et al. (1988). Frame-shift deletions in patients with Duchenne and Becker muscular dystrophy. Science *242*:755–759.

Mao, Y. and Cremer, M. (1989). Detection of Duchenne muscular dystrophy carriers by dosage analysis using the DMD cDNA clone 8. Hum. Genet. *81*:193–195.

Marshall, C. J. (1991). Tumor suppressor genes. Cell (Cambridge, Mass) *64*:313–326.

Martin, G. A. et al. (1990). The GAP-related domain of the neurofibromatosis type 1 gene product interacts with *ras* p21. Cell (Cambridge, Mass) *63*:843–849.

Martin, G. M. et al. (1990). Dominant susceptibility genes. Nature (London) *347*:124.

Martuza, R. L. (1984). Genetic neurooncology. Clin. Neurosurg. *31*:417–440.

Maruyama, K. et al. (1990). Formation of amyloid-like fibrils in COS cells overexpressing part of the Alzheimer amyloid protein precursor. Nature (London) *347*:566–569.

Maslen, C. L. et al. (1991). Partial sequence of a candidate gene for the Marfan syndrome. Nature (London) *352*:334–337.

Masters, C. L. et al. (1985). Amyloid plaque core protein in Alzheimer disease and Down syndrome. Proc. Natl. Acad. Sci. U.S.A. *82*:4245–4249.

Masuda, H. et al. (1987). Rearrangement of the p53 gene in human osteogenic sarcomas. Proc. Natl. Acad. Sci. U.S.A. *84*:7716–7719.

Matsuo, M. et al. (1990). A very small frame-shifting deletion within exon 19 of the Duchenne muscular dystrophy gene. Biochem. Biophys. Res. Commun. *170*:963–967.

McCabe, E. R. B. et al. (1990). DNA from Guthrie spots for diagnosis of DMD by multiplex PCR. Biochem. Med. Metab. Biol. *44*:294–295.

McCormick, M. K. et al. (1989). Molecular genetic approach to the characterization of "Down syndrome region" of chromosome 21. Genomics *5*:325–331.

McDonald, J. D. and Dohrmann, G. J. (1988). Molecular biology of brain tumors. Neurosurgery *23*:537–544.

McGee, T. L. et al. (1989). Structure and partial genomic sequence of the human retinoblastoma susceptibility gene. Gene *80*:119–128.

McGee, T. L. et al (1990). Detection of the *Xba*I RFLP within the retinoblastoma locus by PCR. Nucleic Acids Res. *18*:207.

McGuffin, P. et al. (1990). Exclusion of a schizophrenia susceptibility gene from the chromosome 5q11-q13 region: new data and a reanalysis of previous reports. Am. J. Hum. Genet. *47*:524–535.

McHugh, P. R. and McKusick, V. A. (eds.) (1991). Genes, Brain, and Behavior. Raven Press, New York.

Meissen, G. J. et al. (1988). Predictive testing for Huntington's disease with use of a linked DNA marker. N. Engl. J. Med. *318*:533–542.

Melki, J. et al. (1990a). Gene for chronic proximal spinal muscular atrophies maps to chromosome 5q. Nature (London) *344*:767–768.

Melki, J. et al. and the French Spinal Muscular Atrophy Investigators (1990b). Mapping of acute (type I) spinal muscular atrophy to chromosome 5q12-q14. Lancet *336*:271–273.

Menon, A. G. et al. (1990). Chromosome 17p deletions and *p53* gene mutations associated with the formation of malignant neurofibrosarcomas in von Recklinghausen neurofibromatosis. Proc. Natl. Acad. Sci. U.S.A. *87*:5435–5439.

Mertens, F. et al. (1991). Cytogenetic analysis of 33 basal cell carcinomas. Cancer Res. *51*:954–957.

Meyer, L. J. (1990). Psoriasis: the application of genetic technology and mapping. J. Invest. Dermatol. *95*:5S-6S.

Milner, R. J. et al. (1986). Brain-specific gene expression. Biochem. Soc. Symp. *52*:107–111.

Milunsky, A. et al. (1991). Prenatal diagnosis of myotonic muscular dystrophy with linked deoxyribonucleic acid probes. Am. J. Obstet. Gynecol. *164*:751–755.

Monaco, A. P. et al. (1986). Isolation of candidate cDNAs for portions of the Duchenne muscular dystrophy gene. Nature (London) *323*:646–650.

Monaco, A. P. et al. (1988). An explanation for the phenotypic differences between patients bearing partial deletions of the *DMD* locus. Genomics *2*:90–95.

Moraes, C. T. et al. (1991). mtDNA depletion with variable tissue expression: a novel genetic abnormality in mitochondrial diseases. Am. J. Hum. Genet. *48*:492–501.

Morel, Y. and Miller, W. L. (1991). Clinical and molecular genetics of congenital adrenal hyperplasia due to 21-hydroxylase deficiency. In: Advances in Human Genetics. Harris, H. and Hirschhorn, K. (eds.). Plenum Press, New York, pp. 1–68.

Morris, M. J. et al. (1989). Problems in genetic prediction for Huntington's disease. Lancet *ii*:601–603.

Morrison, M. R. (1988). Messenger RNA levels in neurological disease. In: The Molecular Biology of Neurological Disease. Rosenberg, R. N. and Harding, A. E. (eds.). Butterworths, London, pp. 135–152.

Moser, H. (1984). Duchenne muscular dystrophy: pathogenic aspects and genetic prevention. Hum. Genet. *66*:17–40.

Mueller, R. et al. (1984). Induction of c-*fos* gene and protein by growth factors preceds activation of c-*myc*. Nature (London) *312*:716–720.

Munke, M. et al. (1988). Regional assignments of six polymorphic DNA sequences on chromosome 21 by *in situ* hybridization to normal and rearranged chromosomes. Am. J. Hum. Genet. *42*:542–549.

Murakami, Y. et al. (1991). Inactivation of the retinoblastoma gene in a human lung carcinoma cell line detected by single-strand conformation polymorphism analysis of the polymerase chain reaction product of cDNA. Oncogene *6*:37–42.

Nakagawara, A. et al. (1987). N-*myc* oncogene amplification and prognostic factors of neuroblastoma in children. J. Pediatr. Surg. *22*:895–898.

Nakagawara, A. et al. (1988). Surgical aspects of N-*myc* oncogene amplification of neuroblastoma. Surgery *104*:34–40.

Nakagawara, A. et al. (1990a). Inverse correlation between expression of multidrug resistance gene and N-*myc* oncogene in human neuroblastomas. Cancer Res. *50*:3043–3047.

Nakagawara, A. et al. (1990b). Inverse correlation between N-*myc* amplification and catecholamine metabolism in children with advanced neuroblastoma. Surgery *107*:43–49.

National Institutes of Health (NIH) (1987). Consensus Development Conference Statement. Neurofibromatosis *6*:1–7.

Nau, M. M. et al. (1986). Human small-cell lung cancers show amplification and expression of the N-*myc* gene. Proc. Natl. Acad. Sci. U.S.A. *83*:1092–1096.

Nguyen thi Man et al. (1990). Monoclonal antibodies against defined regions of the muscular dystrophy protein, dystrophin. FEBS Lett. *262*:237–240.

Nisen, P. D. et al. (1986). Enhanced expression of the N-*myc* gene in Wilms' tumors. Cancer Res. *46*:6217–6222.

Nisen, P. D. et al. (1988). N-*myc* oncogene expression in histopathologically unrelated bilateral pediatric renal tumors. Cancer *61*:1821–1826.

Nister, M. and Westermark, B. (1986). Clonal variations in the production of a PDGF-like protein and expression of corresponding receptors in human malignant gliomas. Cancer Res. *46*:332–340.

Nister, M. et al. (1988). Expression of messenger RNAs for PDGF and TGF α and their receptors in human malignant glioma cell lines. Cancer Res. *48*:3910–3918.

Noda, M. et al. (1985). Sarcoma viruses carrying *ras* oncogenes induce differentiation-associated properties in neural cell lines. Nature (London) *318*:73–75.

Norman, A. and Harper, P. (1989). A survey of manifesting carriers of Duchenne and Becker muscular dystrophy in Wales. Clin. Genet. *36*:31–37.

Norman, A. M. et al. (1989a). Presymptomatic detection and prenatal diagnosis for myotonic dystrophy by means of linked DNA markers. J. Med. Genet. *26*:750–754.

Norman, A. et al. (1989b). Distinction of Becker from limb-girdle muscular dystrophy by means of dystrophin cDNA probes. Lancet *i*:466–468.

Norman, A. M. et al. (1990). Becker muscular dystrophy: correlation of deletion type with clinical severity. J. Med. Genet. *27*:236–239.

Oberlé, I. et al. (1991). Instability of a 550-base pair DNA segment and abnormal methylation in fragile X syndrome. Science *252*:1097–1102.

O'Connell, P. et al. (1989). Fine structure DNA mapping studies of the chromosomal region harboring the genetic defect in neurofibromatosis type 1. Am. J. Hum. Genet. *44*:51–57.

Oexle, K. and Schliwa, M. (1989). Dystrophin under scrutiny. Biol. Cell *67*:239–241.

Ohlendieck, K. and Campbell, K. P. (1991). Dystrophin constitutes 5% of membrane cytoskeleton in skeletal muscle. FEBS Lett. *283*:230–234.

Onadim, Z. O. et al. (1990). Application of intragenic DNA probes in prenatal screening for retinoblastoma gene carriers in United Kingdom. Arch. Dis. Child. *65*:651–656.

Oppedal, B. R. et al. (1988). Prognostic factors in neuroblastoma: clinical, histopathologic, and immunohistochemical features and DNA ploidy in relation to prognosis. Cancer *62*:772–780.

Oppedal, B. R. et al. (1989). N-*myc* amplification in neuroblastomas: histopathological, DNA ploidy, and clinical variables. J. Clin. Pathol. *42*:1148–1152.

Oronzi-Scott, M. et al. (1988). Duchenne muscular dystrophy gene expression in normal and diseased human muscle. Science *239*:1418–1420.

Owen, M. J. et al. (1990). Physical mapping around the Alzheimer disease locus on the proximal long arm of chromosome 21. Am. J. Hum. Genet. *46*:316–322.

Pantazis, P. et al. (1985). Synthesis and secretion of proteins resembling platelet-derived growth factor by human glioblastoma and fibrosarcoma cells in culture. Proc. Natl. Acad. Sci. U.S.A. *82*:2404–2408.

Park, M. et al. (1986). Mechanism of *met* oncogene activation. Cell (Cambridge, Mass) *45*:895–904.

Parry, D. M. (1990). Gene mapping and tumor genetics. Neurofibromatosis 1 (Recklinghausen disease) and neurofibromatosis 2 (bilateral acoustic neurofibromatosis): an update. Mulvihill, J. J. (moderator) Ann. Intern. Med. *113*:39–52.

Pasquale, S. R. et al. (1988). Tumorigenicity and oncogene expression in pediatric cancers. Cancer Res. *48*:2715–2719.

Passos-Bueno, M. R. et al. (1990). Screening of deletions in the dystrophin gene with the cDNA probes Cf23a, Cf56a, and Cf115. J. Med. Genet. *27*:145–150.

Patterson, D. et al. (1990). Translocations and rearrangements involving chromosome 21. In: Molecular Genetics of Chromosome 21 and Down Syndrome. Wiley-Liss, New York, pp. 27–37.

Pergolizzi, R. G. et al. (1992). Detection of full fragile X mutation. Lancet *339*:271–272.

Pinkel, D. et al. (1988). Fluorescence *in situ* hybridization with human chromosome-specific libraries: detection of trisomy 21 and translocations of chromosome 4. Proc. Natl. Acad. Sci. U.S.A. *85*:9138–9142.

Plowman, P. N. (1990a). Serum marker for Merkel cell tumor. Clin. Radiol. *40*:542.

Plowman, P. N. (1990b). Neuron-specific enolase as a marker for medulloblastoma. Lancet *ii*:1388.

Podlisny, M. B. et al. (1987). Gene dosage of the amyloid B precursor protein in Alzheimer's disease. Science *238*:669–671.

Pohl, T. M. et al. (1988). Construction of a *Not*I linking library and isolation of new markers close to the Huntington's disease gene. Nucleic Acids Res. *16*:9185–9198.

Ponder, B. (1988). Gene losses in human tumors. Nature (London) *335*:400–402.

Poustka, A. et al. (1987). Construction and use of human chromosome jumping libraries from *Not*I digested DNA. Nature (London) *325*:353–355.

Prescher, G. et al. (1992). Chromosomal aberrations defining uveal melanoma of poor prognosis. Lancet *339*:691–692.

Prior, T. W. (1991). Duchenne muscular dystrophy gene. Arch. Pathol. Lab. Med. *115*:984–990.

Prior, T. W. et al. (1989). Detection of Duchenne/Becker muscular dystrophy carrier by densitometric scanning. Clin. Chem. *35*:1256–1257.

Prior, T. W. et al. (1990). Determination of carrier status in Duchenne and Becker muscular dystrophies by quantitative polymerase chain reaction and allele-specific oligonucleotides. Clin. Chem. *36*:2113–2117.

Pritchard, C. A. et al. (1989). Isolation and field inversion gel electrophoresis analysis of DNA markers located close to the Huntington disease gene. Genomics *4*:408–418.

Prockop, D. J. (1992). Mutations in collagen genes as a cause of connective tissue diseases. N. Engl. J. Med. *326*:540–546.

Quaid, K. A. et al. (1989). Knowledge, attitude, and the decision to be tested for Huntington's disease. Clin. Genet. *36*:431–438.

Quarrell, O. W. J. et al. (1987). Exclusion testing for Huntington's disease in pregnancy with a closely linked DNA marker. Lancet *ii*:1281–1283.

Quinn, N. (1989). Predictive testing for Huntington's disease with linked DNA markers [letter]. Lancet *ii*:921.

Quintanilla, M. et al. (1988). Oncogene mutation and amplification during initiation and progression stages of mouse skin carcinogenesis. In: Growth Factors, Tumor Promoters, and Cancer Genes, Alan R. Liss, New York, pp. 257–266.

Radice, P. et al. (1987). HRAS1 proto-oncogene polymorphisms in human malignant melanoma: *Taq*I defined alleles significantly associated with the disease. Oncogene 2:91–95.

Rahmani, Z. et al. (1989). Critical role of the D21S55 region on chromosome 21 in the pathogenesis of Down syndrome. Proc. Natl. Acad. Sci. U.S.A. *86*:5958–5962.

Ratner, N. et al. (1988). The neuronal cell-surface molecule mitogenic for Schwann cells is a heparin-binding protein. Proc. Natl. Acad. Sci. U.S.A. *85*:6992–6996.

Raybaud, F. et al. (1988). Detection of a low frequency of activated *ras* genes in human melanomas using a tumorigenicity assay. Cancer Res. *48*:950–953.

Riccardi, V. M. (1981). Von Recklinghausen neurofibromatosis. N. Engl. J. Med. *305*:1617–1627.

Richards, C. S. et al. (1990). Skewed X inactivation in a female MZ twin results in Duchenne muscular dystrophy. Am. J. Hum. Genet. *46*:672–681.

Richards, J. E. et al. (1988). Chromosome jumping from D4S10 (G8) toward the Huntington's disease gene. Proc. Natl. Acad. Sci. U.S.A. *85*:6437–6441.

Richmond, A. et al. (1986). Growth factor and cytogenetic abnormalities in cultured nevi and malignant melanomas. J. Invest. Dermatol. *86*:295–302.

Ridley, R. M. et al. (1991). Patterns of inheritance of the symptoms of Huntington's disease suggestive of an effect of genomic imprinting. J. Med. Genet. *28*:224–231.

Risch, N. (1990a). Genetic linkage and complex diseases, with special reference to psychiatric disorders. Genet. Epidemiol. *7*:3–16.

Risch, N. (1990b). Genetic linkage and complex diseases: a response. Genet. Epidemiol. *7*:41–45.

Roberts, L. (1990). Huntington's gene: so near, yet so far. Science *247*:624–627.

Roberts, R. G. et al. (1989). Rapid carrier and prenatal diagnosis of Duchenne and Becker muscular dystrophy. Nucleic Acids Res. *17*:811.

Roberts, R. G. et al. (1990). Direct diagnosis of carriers of Duchenne and Becker muscular dystrophy by amplification of lymphocyte RNA. Lancet *336*:1523–1526.

Rocchi, M. et al. (1990). Mental retardation in heterozygotes for the fragile X mutation: evidence in favor of an X inactivation-dependent effect. Am. J. Hum. Genet. *46*:738–743.

Romano, J. W. et al. (1989). Identification and characterization of a *p53* gene mutation in a human osteosarcoma cell line. Oncogene *4*:1483–1488.

Roninson, I. B. et al. (1986). Isolation of the human *mdr* DNA sequences amplified in multidrug-resistant KB carcinoma cells. Proc. Natl. Acad. Sci. U.S.A. *83*:4538–4542.

Roop, D. R. et al. (1986). An activated Harvey *ras* oncogene produces benign tumors on mouse epidermal tissue. Nature (London) *323*:822–824.

Rosai, J. (1989). Wilms' tumor. In: Ackerman's Surgical Pathology. C. V. Mosby, St. Louis, pp. 860–864.

Rose-John, S. et al. (1988). Differential effects of phorbol esters on c-*fos* and c-*myc* and ornithine decarboxylase gene expression in mouse skin *in vivo*. Carcinogenesis *9*:831–835.

Rosenberg, R. N. and Harding, A. E. (eds.) (1988). The molecular biology of neurological disease. Butterworths, London.

Rosenfeld, P. J. and Kelly, T. J. (1986). Purification of nuclear factor I by DNA recognition site affinity chromatography. J. Biol. Chem. *261*:1398–1408.

Roses, A. D. and Bartlett, R. J. (1989). Myotonic muscular dystrophy. In: Molecular Genetics in Diseases of Brain, Nerve, and Muscle. Rowland, L. P. et al. (eds.). Oxford University Press, New York, pp. 338–348.

Rosolen, A. et al. (1990). Antisense inhibition of single copy N-*myc* expression results in decreased cell growth without reduction of c-*myc* protein in a neuroepithelioma cell line. Cancer Res. *50*:6316–6322.

Rossi, P. et al. (1988). A nuclear factor 1 binding site mediates the transcriptional activation of a type I collagen promoter by transforming growth factor-β. Cell (Cambridge, Mass) *52*:405–414.

Rouah, E. et al. (1989). N-*myc* amplification and neuronal differentiation in human primitive neuroectodermal tumors of the central nervous system. Cancer Res. *49*:1797–1801.

Rouleau, G. A. et al. (1987). Genetic linkage of bilateral acoustic neurofibromatosis to a DNA marker on chromosome 22. Nature (London) *329*:246–248.

Rouleau, G. A. et al. (1990). Flanking markers bracket the neurofibromatosis type 2 (NF2) gene on chromosome 22. Am. J. Hum. Genet. *46*:323–328.

Rousseau, F. et al. (1991). Direct diagnosis by DNA analysis of the fragile X syndrome of mental retardation. N. Engl. J. Med. *325*:1673–1681.

Rowland, L. P. (1989). The transformation of clinical concepts and clinical practice by molecular genetics. In: Molecular Genetics in Diseases of Brain, Nerve, and Muscle. Rowland, L. P. et al. (eds.). Oxford University Press, New York, pp. 8–23.

Rowland, L. P. et al. (eds.) (1989). Molecular Genetics in Diseases of Brain, Nerve, and Muscle. Oxford University Press, New York.

Rudd, N. L. et al. (1988). The use of DNA probes to establish parental origin in Down syndrome. Hum. Genet. *78*:175–178.

Rutka, J. T. et al. (1984). Isolation and partial purification of growth factors with TGF-like activity from human malignant gliomas. JNS *71*:875–883.

Rutka, J. T. et al. (1990). Molecular probes in neuro-oncology: a review. Cancer Invest. *8*:425–438.

Sacchi, N. (1990). Genes on chromosome 21 and cancer. In: Molecular Genetics of Chromosome 21 and Down Syndrome. Wiley-Liss, New York, pp. 169–185.

Sakai, K. et al. (1988). Increase in copy number of N-*myc* in retinoblastomas in comparison with chromosome abnormality. Cancer Genet. Cytogenet. *30*:119–126.

Sanders, B. M. et al. (1989). Non-ocular cancer in relatives of retinoblastoma patients. Br. J. Cancer *60*:358–365.

Sano, K. et al. (1990). Expression of the *smg* p25A (a *ras* p21-like GTP-binding protein) gene in human neuroblastoma cell lines and tumor tissues. Cancer Res. *50*:7242–7245.

Sapienza, C. and St-Jacques, B. (1986). "Brain-specific" transcription and evolution of the identifier sequence. Nature (London) *319*:418–420.

Sauceda, R. et al. (1988). Novel combination of c-*myc*, N-*myc* and N-*ras* oncogene alteration in brain tumors. Mol. Brain Res. *3*:123–132.

Scappaticci, S. et al. (1989). Multiple clonal chromosome abnormalities in a superficial basal cell epithelioma. Cancer Genet. Cytogenet. *42*:309–311.

Scheck, A. C. and Shapiro, J. R. (1990). Regional heterogeneity in human malignant gliomas occurs in anaplastic astrocytomas as well as glioblastomas multiforme. Proc. Am. Assn. Cancer Res. *31*:306A.

Schellenberg, G. D. et al. (1991). Linkage analysis of familial Alzheimer disease, using chromosome 21 markers. Am. J. Hum. Genet. *48*:563–583.

Schloesser, M. et al. (1990). Characterization of pathological dystrophin transcripts from the lymphocytes of a muscular dystrophy carrier. Mol. Biol. Med. *7*:519–523.

Schmidt, J. et al. (1989). Establishment and characterization of osteogenic cell lines from a spontaneous murine osteosarcoma. Differentiation *39*:151–160.

Schmidt, M. A. et al. (1987). Cases of neurofibromatosis with rearrangements of chromosome 17 involving band 17q11.2. Am. J. Med. Genet. *28*:771–777.

Schwartz, C. E. et al. (1990). Allan-Herndon syndrome. II. Linkage to DNA markers in Xq21. Am. J. Hum. Genet. *47*:454–458.

Schweigerer, L. et al. (1990). Augmented MYCN expression advances the malignant phenotype of human neuroblastoma cells: evidence for induction of autocrine growth factor activity. Cancer Res. *50*:4411–4416.

Seeger, R. C. et al. (1984). Association of multiple copies of N-*myc* oncogene with rapid progression of neuroblastoma. N. Engl. J. Med. *313*:1111–1116.

Seeman, P. and Niznik, H. B. (1990). Dopamine receptors and transporters in Parkinson's disease and schizophrenia. FASEB J. *4*:2737–2744.

Seizinger, B. R. et al. (1986). Loss of genes on chromosome 22 in tumorigenesis of human acoustic neuroma. Nature (London) *322*:644–647.

Seizinger, B. R. et al. (1987a). Common pathogenetic mechanism for three tumor types in bilateral acoustic neurofibromatosis. Science *236*:317–319.

Seizinger, B. R. et al. (1987b). Genetic linkage of von Recklinghausen neurofibromatosis to the nerve growth factor receptor gene. Cell (Cambridge, Mass) *49*:589–594.

Selkoe, D. J. (1991). The molecular pathology of Alzheimer's disease. Neuron *6*:487–498.

Selkoe, D. J. et al. (1987). Conservation of brain amyloid proteins in aged mammals and humans with Alzheimer's disease. Science *235*:873–877.

Shaw, D. J. and Harper, P. S. (1989). Myotonic dystrophy: developments in molecular genetics. Br. Med. Bull. *45*:745–759.

Sherman, S. L. et al. (1984). The marker (X) syndrome: a cytogenetic and genetic analysis. Ann. Hum. Genet. *48*:21–37.

Sherman, S. et al. (1985). Further segregation analysis of the fragile X syndrome with special reference to transmitting males. Hum. Genet. *69*:289–299.

Shew, J. Y. et al (1990). C-terminal truncation of the retinoblastoma gene product leads to functional inactivation. Proc. Natl. Acad. Sci. U.S.A. *87*:6–10.

Shigeoka, H. and Yang, H. C. (1991). Regulation of the retinoblastoma anti-oncogene in aging human diploid fibroblasts. Mech. Ageing Dev. *57*:63–70.

Shoffner, J. M. et al. (1990). Myoclonic epilepsy and ragged-red fiber disease (MERRF) is associated with a mitochondrial DNA tRNA[Lys] mutation. Cell (Cambridge, Mass) *61*:931–937.

Shomrat, R. et al. (1992). Use of dystrophin genomic and cDNA probes for solving difficulties in carrier detection and prenatal diagnosis of Duchenne muscular dystrophy. Am. J. Med. Genet. *42*:281–287.

Shutler, G. et al. (1991). Physical and genetic mapping of a novel chromosome 19 ERCC1 marker showing close linkage with myotonic dystrophy. Genomics *9*:500–504.

Siddique, T. et al. (1991). Linkage of a gene causing familial amyotrophic lateral sclerosis to chromosome 21 and evidence of genetic locus heterogeneity. N. Engl. J. Med. *324*:1381–1384.

Sidransky, D. et al. (1992). Clonal expansion of *p53* mutant cells is associated with brain tumor progression. Nature (London) *355*:846–847.

Skraastad, M. I. et al. (1991). Presymptomatic, prenatal, and exclusion testing for Huntington disease using seven closely linked DNA markers. Am. J. Med. Genet. *38*:217–222.

Slavc, I. et al. (1990). *myc* gene amplification and expression in primary human neuroblastoma. Cancer Res. *50*:1459–1463.

Slipetz, D. M. et al. (1991). Deficiency of complex III of the mitochondrial respiratory chain in a patient with facioscapulohumeral disease. Am. J. Hum. Genet. *48*:502–510.

Smeets, H. et al. (1990). A long-range restriction map of the human chromosome 19q13 region: close physical linkage between CKMM and the ERCC1 and ERCC2 genes. Am. J. Hum. Genet. *46*:492–501.

Smith, B. et al. (1988). Isolation of DNA markers in the direction of the Huntington disease gene from the G8 locus. Am. J. Hum. Genet. *42*:335–344.

Smith, B. et al. (1991). Detection of melanoma cells in peripheral blood by means of reverse transcriptase and polymerase chain reaction. Lancet *338*:1227–1229.

Sokoloff, P. et al. (1990). Molecular cloning and characterization of a novel dopamine receptor (D_3) as a target for neuroleptics. Nature (London) *347*:146–156.

Sood, R. et al. (1990). Genetic mapping of two new DNA markers in Xq26-q28 relative to the fragile X syndrome locus. Am. J. Hum. Genet. *47*:395–402.

Sorensen, S. A. et al. (1986). Long-term follow-up of von Recklinghausen neurofibromatosis: survival and malignant neoplasms. N. Engl. J. Med. *314*:1010–1015.

Soreq, H. and Zakut, H. (1990). Cholinesterase Genes: Multileveled Regulation. Karger, Basel.

Speer, A. et al. (1989). DNA amplification of a further exon of Duchenne muscular dystrophy locus increases possibilities for deletion screening. Nucleic Acids Res. *17*:4892.

Speer, M. C. et al. (1990). Presymptomatic and prenatal diagnosis in myotonic dystrophy by genetic linkage studies. Neurology *40*:671–676.

Squire, J. et al. (1986). Tumor indication by the retinoblastoma mutation is independent of N-*myc* expression. Nature (London) *322*:555–557.

St. Clair, D. et al. (1990). Association within a family of a balanced autosomal translocation with major mental illness. Lancet *336*:13–16.

St. George-Hyslop, P. H. et al. (1987a). Absence of duplication of chromosome 21 genes in familial and sporadic Alzheimer's disease. Science *238*:664–666.

St. George-Hyslop, P. H. et al. (1987b). The genetic defect causing familial Alzheimer's disease maps on chromosome 21. Science *235*:885–890.

St. George-Hyslop, P. H. and other members of the FAD Collaborative Study Group (1990). Genetic linkage studies suggest that Alzheimer's disease is not a single homogeneous disorder. Nature (London) *347*:194–197.

Stanbridge, E. J. and Nowell, P. C. (1990). Origins of human cancer revisited. Cell (Cambridge, Mass) *63*:867–874.

Stevenson, R. E. et al. (1990). Allan-Herndon syndrome. I. Clinical studies. Am. J. Hum. Genet. *47*:446–453.

Stewart, G. D. et al. (1988). Trisomy 21 (Down syndrome): studying nondisjunction and meiotic recombination by using cytogenetic and molecular polymorphisms that span chromosome 21. Am. J. Hum. Genet. *42*:227–236.

Stratton, M. R. et al. (1989). Structural alterations of the *RB1* gene in human soft tissue tumors. Br. J. Cancer *60*:202–205.

Stratton, M. R. et al. (1990). Mutation of the p53 gene in human soft tissue sarcomas: association with abnormalities of the RB1 gene. Oncogene *5*:1297–1301.

Strauss, M. E. and Brandt, J. (1990). Are there neuropsychological manifestations of the gene for Huntington's disease in asymptomatic, at-risk individuals? Arch. Neurol. *47*:905–908.

Streib, E. W. et al. (1987). Myotonic dystrophy sine myotonia: normal EMG in two obligate gene-carriers of advanced age. Electromyogr. Clin. Neurophysiol. *27*:443–446.

Stromberg, K. et al. (1987). Urinary TGFs in neoplasia: immunoreactive TGF-α in the urine of patients with disseminated breast cancer. Biochem. Biophys. Res. Commun. *144*:1059–1068.

Sturm, S. A. et al. (1990). Amplification and rearrangement of c-*myc* in radiation-induced murine osteosarcomas. Cancer Res. *50*:4146–4153.

Sudol, M. (1988). Expression of proto-oncogenes in neural tissues. Brain Res. Rev. *13*:391–403.

Sugino, S. et al. (1989). Molecular genetic study of Duchenne and Becker muscular dystrophies: deletion analyses of 45 Japanese patients and segregation analyses in their families with RFLPs based on the data from normal Japanese females. Am. J. Med. Genet. *34*:555–561.

Sunahara, R. K. et al. (1990). Human dopamine D_1 receptor encoded by an intronless gene on chromosome 5. Nature (London) *347*:80–83.

Sutherland, C. et al. (1986). Harvey-*ras* oncogene restriction fragment alleles in familial melanoma kindreds. Br. J. Cancer *54*:787–790.

Sutherland, G. R. et al. (1991). Prenatal diagnosis of fragile X syndrome by direct detection of the unstable DNA sequence. N. Engl. J. Med. *325*:1720–1722.

Suthers, G. K. et al. (1990). Physical mapping of new DNA probes near the fragile X mutation (FRAXA) by using a panel of cell lines. Am. J. Hum. Genet. *47*:187–195.

Sved, J. A. and Laird, C. D. (1990). Population genetic consequences of the fragile X syndrome, based on the X-inactivation imprinting model. Am. J. Hum. Genet. *46*:443–451.

Tanaka, T. et al. (1986). Organ specific expression of *ras* oncoproteins during growth and development of the rat. Mol. Cell. Biochem. *70*:97–104.

Tanaka, T. et al. (1988). Expression of Ha-*ras* oncogene products in human neuroblastomas and the significant correlation with a patient's prognosis. Cancer Res. *48*:1030–1034.

Tanzi, R. E. and Hyman, B. T. (1991). Alzheimer's mutation. Nature (London) *350*:564.

Tanzi, R. E. et al. (1987a). Amyloid β protein gene: cDNA, mRNA distribution, and genetic linkage near the Alzheimer locus. Science *235*:880–884.

Tanzi, R. E. et al. (1987b). The amyloid β protein gene is not duplicated in brains from patients with Alzheimer's disease. Science *238*:666–669.

Taylor, S. R. and Locker, J. (1990). A comparative analysis of nuclear DNA content and N-*myc* gene amplification in neuroblastoma. Cancer *65*:1360–1366.

Taylor, S. R. et al. (1988). Flow cytometric DNA analysis of neuroblastoma and ganglioneuroma: a 10 year retrospective study. Cancer *62*:749–754.

Telford, D. J. and Stewart, B. W. (1990). Structural features of the amplified N-*myc* oncogene detected by micrococcal nuclease digestion of neuroblastoma cell nuclei. Biochem. Int. *21*:1025–1032.

Templeton, D. J. et al. (1991). Nonfunctional mutants of the retinoblastoma protein are characterized by defects in phosphorylation, viral oncoprotein association, and nuclear tethering. Proc. Natl. Acad. Sci. U.S.A. *88*:3033–3037.

Thiele, C. J. and Israel, M. A. (1988). Regulation of N-*myc* expression is a critical event controlling the ability of human neuroblasts to differentiate. Exp. Cell Biol. *56*:321–333.

Thiele, C. J. et al. (1987a). Differential protooncogene expression characterizes histopathologically indistinguishable tumors of the peripheral nervous system. J. Clin. Invest. *80*:804–811.

Thiele, C. J. et al. (1987b). Translocation of c-*sis* protooncogene in peripheral neuroepithelioma. Cancer Genet. Cytogenet. *24*:119–128.

Thiele, C. J. et al. (1988a). Patterns of proto-oncogene expression: a tool to subtype histopathologically similar solid tumors. In: Cellular and Molecular Biology of Tumors and Potential Clinical Applications. Alan R. Liss, New York, pp. 301–305.

Thiele, C. J. et al. (1988b). Regulation of c-*myb* expression in human neuroblastoma cells during retinoic acid-induced differentiation. Mol. Cell. Biol. *8*:1677–1683.

Tien, H. F. et al. (1989). Cytogenetic evidence of multifocal origin of a unilateral retinoblastoma. A help in genetic counseling. Cancer Genet. Cytogenet. *42*:203–208.

Toguchida, J. et al. (1988). Chromosomal reorganization for the expression of recessive mutation of retinoblastoma susceptibility gene in the development of osteosarcoma. Cancer Res. *48*:3939–3943.

Trent, J. et al. (1986). Evidence for rearrangement, amplification, and expression of c-*myc* in a human glioblastoma. Proc. Natl. Acad. Sci. U.S.A. *83*:470–473.

Trent, J. et al. (1990a). Tumorigenicity in human melanoma cell lines is controlled by the introduction of human chromosome 6. Proc. Am. Assn. Cancer Res. *31*:321A.

Trent, J. M. et al. (1990b). Relation of cytogenetic abnormalities and clinical outcome in metastatic melanoma. N. Engl. J. Med. *322*:1508–1511.

Triche, T. J. et al. (1988). N-*myc* protein expression in small round cell tumors. Adv. Neuroblastoma Res. *2*:475–485.

Tsang, T. et al. (1990). Viral *jun* oncogene acts as a dominant negative regulator in mouse epidermal tumor cells. Proc. Am. Assn. Cancer Res. *31*:323A.

Tsipouras, P. (1990). Invited editorial: Marfan syndrome: light at the end of the tunnel? Am. J. Hum. Genet. *46*:643–645.

Tsipouras, P. et al. (1992). Genetic linkage of the Marfan syndrome, ectopia lentis, and congenital contractural arachnodactyly to the fibrillin genes on chromosomes 15 and 5. N. Engl. J. Med. *326*:905–909.

Tsipouras, P. et al. (1991). Marfan syndrome is closely linked to a marker on chromosome 15q1.5-22.1. Proc. Natl. Acad. Sci. U.S.A. *88*:4486–4488.

Tsuda, T. et al. (1987). Analysis of N-*myc* amplification in relation to disease stage and histologic types in human neuroblastomas. Cancer *60*:820–826.

Tucker, M. A. (1988). Individuals at high risk of melanoma. Pigm. Cell *9*:95–109.

Tyler, A. and Morris, M. (1990). National symposium on problems of presymptomatic testing for Huntington's disease, Cardiff. J. Med. Ethics *16*:41–42.

van Camp, G. et al. (1989). Selection of human chromosome 21 specific DNA probes for genetic analysis in Alzheimer's dementia and Down syndrome. Hum. Genet. *83*:58–60.

van der Bruggen, P. et al. (1991). A gene encoding an antigen recognized by cytolytic T lymphocytes on a human melanoma. Science *254*:1643–1647.

van Ommen, G. J. B. et al. (1986). A physical map of 4 million bp around the Duchenne muscular dystrophy gene on the human X-chromosome. Cell (Cambridge, Mass) *47*:499–504.

van't Veer, L. J. et al. (1989). N-*ras* mutations in human cutaneous melanoma from sun-exposed body sites. Mol. Cell. Biol. *9*:3114–3116.

Verkerk, A. J. M. H. et al. (1991). Identification of a gene (*FMR*-1) containing a CGG repeat coincident with a breakpoint cluster region exhibiting length variation in fragile X syndrome. Cell (Cambridge, Mass) *65*:905–914.

Vincent, A. et al. (1989). The polymorphic marker DXS304 is within 5 centimorgans of the fragile X locus. Genomics *5*:797–801.

Vincent, A. et al. (1991). Abnormal pattern detected in fragile-X patients by pulsed-field gel electrophoresis. Nature (London) *349*:624–626.

Viskochil, D. et al. (1990). Deletions and a translocation interrupt a cloned gene at the neurofibromatosis type 1 locus. Cell (Cambridge, Mass) *62*:187–192.

Vogel, F. (1979). Genetics of retinoblastoma. Hum. Genet. *52*:1–54.

Walker, A. P. et al. (1988). RFLP for Duchenne muscular dystrophy cDNA clone 30-2. Nucleic Acids Res. *16*:9072.

Wallace, M. R. et al. (1990). Type 1 neurofibromatosis gene: identification of a large transcript disrupted in three NF1 patients. Science *249*:181–186.

Wallace, M. R. and Collins, F. S. (1991). Molecular genetics of von Recklinghausen neurofibromatosis. In: Advances in Human Genetics 20. Harris, H. and Hirschhorn, K. (eds.). Plenum Press, New York, pp. 267–307.

Ward, K. et al. (1990). Diagnosis of neurofibromatosis I by using tightly linked, flanking DNA markers. Am. J. Hum. Genet. *46*:943–949.

Warren, A. C. et al. (1987). Evidence for reduced recombination on the nondisjoined chromosomes 21 in Down syndrome. Science *237*:652–654.

Wasmuth, J. J. et al. (1988). A highly polymorphic locus very tightly linked to the Huntington's disease gene. Nature (London) *322*:734–736.

Wasson, J. C. et al. (1990). Oncogene amplification in pediatric brain tumors. Cancer Res. *50*:2987–2990.

Watanabe, H. et al. (1989). Effect of human interferons on morphological differentiation and suppression of N-*myc* gene expression in human neuroblastoma cells. Jpn. J. Cancer Res. *80*:1072–1076.

Watt, D. C. et al. (1986). Probes in Huntington's chorea. A letter. Nature (London) *320*:21.

Weber, J. L. and May, P. E. (1989). Abundant class of human DNA polymorphisms which can be typed using the polymerase chain reaction. Am. J. Hum. Genet. *44*:388–396.

Webster, C. and Blau, H. M. (1990). Accelerated age-related decline in replicative life-span of Duchenne muscular dystrophy myoblasts: implications for cell and gene therapy. Somatic Cell. Mol. Genet. *16*:557–565.

Weissman, B. E. et al. (1990). A common genetic defect in the control of tumorigenicity in human sarcomas and neuroblastomas. Proc. Am. Assn. Cancer Res. *31*:316A.

Weitkamp, L. R. et al. (1981). Depressive disorders and HLA: a gene on chromosome 6 that can affect behavior. N. Engl. J. Med. *305*:1301–1306.

Wertelecki, W. et al. (1988). Neurofibromatosis 2: clinical and DNA linkage studies of a large kindred. N. Engl. J. Med. *319*:278–283.

Wexler, N. S. (1989). The oracle of DNA. In: Molecular Genetics in Diseases of Brain, Nerve, and Muscle. Rowland, L. P. et al. (eds.). Oxford University Press, New York, pp. 429–442.

Whaley, W. L. et al. (1988). Mapping of D4S98/S114/S113 confines the Huntington's defect to a reduced physical region at the telomere of chromosome 4. Nucleic Acids Res. *16*:11769–11780.

Wiggs, J. et al. (1988). Prediction of the risk of hereditary retinoblastoma, using DNA polymorphisms within the retinoblastoma gene. N. Engl. J. Med. *318*:151–157.

Wijmenga, C. et al. (1990). Location of fascioscapulohumeral muscular dystrophy gene on chromosome 4. Lancet *336*:651–653.

Wijmenga, C. et al. (1991). Mapping of fascioscapulohumeral muscular dystrophy gene to chromosome 4q35-qter by multipoint linkage analysis and *in situ* hybridization. Genomics *9*:570–575.

Wilkie, A. O. M. et al. (1990a). Clinical features and molecular analysis of the α thalassemia/mental retardaton syndromes. I. Cases due to deletions involving chromosome band 16p13.3. Am. J. Hum. Genet. *46*:1112–1126.

Wilkie, A. O. M. et al. (1990b). Clinical features and molecular analysis of the α thalassemia/mental retardation syndromes. II. Cases without detectable abnormality of the α globin complex. Am. J. Hum. Genet. *46*:1127–1140.

Windle, J. J. et al. (1990). Retinoblastoma in transgenic mice. Nature (London) *343*:665–669.

Witkowski, J. A. (1989). Dystrophin-related muscular dystrophies. J. Child. Neurol. *4*:251–271.

Wong, A. J. et al. (1987). Increased expression of the epidermal growth factor receptor gene in malignant gliomas is invariably associated with gene amplification. Proc. Natl. Acad. Sci. U.S.A. *84*:6899–6903.

Wong, D. T. W. and Biswas, D. K. (1987). Expression of c-*erb* B proto-oncogene during dimethylbenzanthracene-induced tumorigenesis in hamster cheek pouch. Oncogene 2:67–72.

Worthington, D. J. et al. (1988). Neuroblastoma — when are urinary catecholamines and their metabolites "normal"? Ann. Clin. Biochem. 25:620–626.

Wright, A. F. et al. (1991). β amyloid resurrected. Nature (London) 349:653–654.

Wu, J. K. and Chikaraishi, D. M. (1990). Differential expression of *ros* oncogene in primary human astrocytomas and astrocytoma cell lines. Cancer Res. 50:3032–3035.

Wuepper, K. D. et al. (1990). Psoriasis vulgaris: a genetic approach. J. Invest. Dermatol. 95:2S–4S.

Wullich, B. et al. (1989). Cytogenetic and *in situ* DNA hybridization studies in intracranial tumors of a patient with central neurofibromatosis. Hum. Genet. 82:31–34.

Wunder, J. S. et al. (1991). Analysis of alterations in the retinoblastoma gene and tumor grade in bone and soft-tissue sarcomas. J. Natl. Cancer Inst. 83:194–200.

Xu, G. et al. (1990a). The neurofibromatosis type 1 gene encodes a protein related to GAP. Cell (Cambridge, Mass) 62:599–608.

Xu, G. et al. (1990b). The catalytic domain of the neurofibromatosis type 1 gene product stimulates *ras* GTPase and complements IRA mutants of *S. cerevisiae*. Cell (Cambridge, Mass) 63:835–841.

Xu, H. J. et al. (1989). The retinoblastoma susceptibility gene product: a characteristic pattern in normal cells and abnormal expression in malignant cells. Oncogene 4:807–812.

Yamada, M. et al. (1988). Amplified allele of the human N-*myc* oncogene in neuroblastomas. Jpn. J. Cancer Res. 79:670–673.

Yamagata, H. et al. (1992). Expansion of unstable DNA region in Japanese myotonic dystrophy patients. Lancet 339:692.

Yamanishi, D. T. et al. (1990). Expression of c-*jun* and c-*fos* oncogenes in human primary melanocytes and melanomas. Proc. Am. Assn. Cancer Res. 31:310A.

Yandell, D. W. and Dryja, T. P. (1989). Detection of DNA sequence polymorphisms by enzymatic amplification and direct genomic sequencing. Am. J. Hum. Genet. 45:547–555.

Yandell, D. W. et al. (1989). Oncogenic point mutations in human retinoblastoma gene: their application to genetic counseling. N. Engl. J. Med. 321:1689–1695.

Yanker, B. A. and Mesulam, M.-M. (1991). β-Amyloid and the pathogenesis of Alzheimer's disease. N. Engl. J. Med. 325:1849–1857.

Yasuda, H. et al. (1989). Differential expression of *ras* oncogene products among the types of human melanomas and melanocytic nevi. J. Invest. Dermatol. 93:54–59.

Yeger, H. et al. (1990). Importance of phenotypic and molecular characterization for identification of a neuroepithelionia tumor cell line, NUB-20. Cancer Res. 50:2794–2802.

Yen, A. et al. (1990). Regulation of the Rb gene product by retinoic acid and 1,25-(OH)$_2$ vitamin D3 during cell differentiation: role as a "status quo" gene. Proc. Am. Assn. Cancer Res. 31:328A.

Yoshioka, M. et al. (1990). An isolated case of Duchenne muscular dystrophy (DMD) in a female with a deletion of DMD cDNA. Clin. Genet. 38:474–478.

Youngman, S. et al. (1988). A DNA probe, D5 (D4S90), mapping to human chromosome 4p16. 3. Nucleic Acids Res. 16:1648.

Youngman, S. et al. (1989). A new DNA marker (D4S90) is located terminally on the short arm of chromosome 4, close to the Huntington disease gene. Genomics 5:802–809.

Yu, S. et al. (1991). Fragile X genotype characterized by an unstable region of DNA. Science 252:1179–1181.

Zehnbauer, B. A. et al. (1988). Characterization of N-*myc* amplification units in human neuroblastoma cells. Mol. Cell. Biol. 8:522–530.

Zeng, Q. S. et al. (1990). Nuclear DNA measurements of metastatic melanoma by a computerized digital imaging system. Hum. Pathol. 21:1112–1116.

Zeviani, M. et al. (1991). Maternally inherited myopathy and cardiomyopathy: association with mutation in mitochondrial DNA tRNA$^{leu(UUR)}$. Lancet 338:143–147.

Zhou, Q. Y. et al. (1990). Cloning and expression of human and rat D$_1$ dopamine receptors. Nature (London) 347:76–80.

Zietkiewicz, E. et al. (1992). Carrier status diagnosis in Duchenne muscular dystrophy with "conformational" DNA polymorphism. Lancet 339:134.

Zubrzycka-Gaarn, E. E. et al. (1988). The Duchenne muscular dystrophy gene product is localized in the sarcolemma of human skeletal muscle fibers. Nature (London) 333:466–469.

Molecular Biological Tools for Analysis of Infections

6.1.1 INTRODUCTION

Evaluation of infectious agents at the gene level frequently offers the fastest, most accurate and sensitive method of diagnosis and epidemiological characterization of known or yet undisclosed etiological agents. **Transfer of molecular biological technology** to the identification of viruses, bacteria, and parasites has translated into a continuously expanding arsenal of **nucleic acid probes**, adaptations of various analytical and preparatory techniques, and the accumulation of **sequence data** allowing a vast range of infectious agents to be reliably identified in the **environment, vectors**, and **patients**. Although the number of commercially available gene-level diagnostics is rather limited at present, the development of progressively **user-friendly procedures**, the introduction of **nonradioactive labels**, refinements in the specificity and sensitivity of the available **kits**, the continuing research in virtually all areas of infectious diseases, and adaptation of amplification methodologies, mostly the **polymerase chain reaction (PCR)** and the anticipated introduction of ligase-based amplifications, self-sustained sequence replication (3SR), and other amplification systems in commercially adapted products (see Part 1), ensure that diagnostic laboratories are able to benefit from this technology.

While the advantages of nucleic acid probes may be apparent in many cases, this technology most likely will complement the **conventional immunological, culture**, and **microscopic techniques** in selected routine applications (Yolken, 1988). The power of gene level diagnostics, while desirable in some situations (e.g., in diagnosing hepatitis B virus [HBV] and human immunodeficiency virus [HIV]), may pose **new dilemmas** in determining what is clinically relevant, to what extent, and under what circumstances (e.g., in the identification of human papilloma virus [HPV] infections). The **superior discriminatory power** of **genotyping** microorganisms allows heretofore unavailable precision in the epidemiological characterization of isolates, sometimes redefining the concept of patient management. The role of gene level diagnostics in establishing the clonality of microorganisms and their potential for epidemiological studies has been discussed (Eisenstein, 1990).

6.1.2 IDENTIFICATION OF PATHOGENS IN CLINICAL SPECIMENS

The challenge of identifying a pathogen in clinical material can be met in a number of ways. The probes can be applied directly without prior disaggregation of the specimen, isolation, concentration and enrichment of the pathogen, its partial or complete disruption, and, finally, isolation of the nucleic acids from the pathogen. Alternatively, molecular biological techniques may be used to partially disrupt the tissue (e.g., by protease digestion), and the pathogen can be identified directly with or without **amplification** (e.g., the PCR) of its nucleic acids using *in situ* hybridization (ISH). The relative merits of various approaches for the identification and characterization of pathogens will be discussed in appropriate subsections. The superior sensitivity and specificity of amplification methods is particularly attractive for the detection of low concentrations of infectious agents, as in latent conditions, or for the fastest identification of the pathogen without laborious isolation and/or purification of target nucleic acids. Enumeration of the reported nucleic acid probes developed for genomic or plasmid sequences is not the objective of this review. Attempts to catalogue the continuously appearing communications in this field may be a frustrating endeavor indeed, albeit gratifying in witnessing the expansion in the application of the molecular biological technology to clinical practice.

In earlier sections (Part 1), a description was given of representative molecular biological approaches to characterization of nucleic acids for the purpose of diagnosis. Understandably, most of these techniques can also be applied to the characterization of infectious agents. Mentioned here are only a few studies indicative of the approaches being taken in the development of practically useful nucleic acid probes for general microbiological practice. For detailed reviews a number of excellent discussions can be recommended (e.g., Eisenstein, 1990; Macario and Conway de Macario, 1990; Tompkins, 1989).

Adaptation of molecular biological techniques to the clinical laboratory environment with the maximum utilization of automated and semiautomated equipment motivates researchers to devise hybridization protocols that can be run in a manner similar to the more familiar immunological techniques — for long a mainstay of diagnostic clinical laboratories. Some of the examples will be cited here. One such adaptation can be seen in the development of a simple **dot hybridization** and **fluorimetric hybridization** method in **microdilution plates** for rapid and routine genotyping of **viridans group streptococci** (Ezaki et al., 1988). Almost 100% sensitive and specific **digoxigenin-labeled oligonucleotide probes** have been constructed for the identification of *Streptococcus pyogenes* (Podbielski et al., 1990).

The identification of certain types of the **M protein** that are associated with the development of dangerous sequelae of streptococcal infections such as rheumatic fever and acute glomerulonephritis is routinely performed by serological means (**M antigen typing**). Due to the technically demanding process of generating specific typing antisera, this procedure is performed in only a few laboratories. Besides up to 50% of all group A streptococcal isolates remain untypable (Podbielski et al., 1990). Furthermore, unequivocal results are difficult to obtain using **ELISA** procedure on certain isolates. Considering the specificity, speed, and use of a nonradioactive label, the **oligonucleotide method** of typing group A streptococcal isolates is certainly preferable to the serological approach (Podbielski et al., 1990).

Difficulties are commonly encountered in the identification of **pneumococcal** isolates showing atypical responses in the traditional battery of tests (Fenoll et al., 1990). **Specific DNA probes** have been derived from *Streptococcus oralis, S. mitis, S. sanguis* (Schmidhuber et al., 1987), and *S. pneumoniae* (Pozzi et al., 1989). Among clinical pneumococcal isolates, a large number of strains cannot be identified as pneumococci by serotyping and antibiotic susceptibility testing (Fenoll et al., 1990). Therefore, a new specific DNA probe containing a 0.65-kb fragment coding for a position of the **major pneumococcal autolysin (amidase)** has been constructed (Fenoll et al., 1990). This new probe was designed such that it reduced the possibility of the DNA hybridizing with other genes encoding proteins that could contain homologous choline-binding domains, reducing the specificity of recognition of *S. pneumoniae*.

DNA probes for culture confirmation of *S. agalactiae, Hemophilus influenzae,* and *Enterococcus* spp. are commercially available (Accuprobe; Gen-Probe, Inc.), eliminating the need for **subculturing** (Daly et al., 1991). The 40-min **chemoluminescent assay** with only 10 min of hands-on time was 100% specific, identifying 325 of 327 isolates (Daly et al., 1991).

Comparison of **restriction endonuclease digestion patterns** probed with **rRNA gene-specific probes (ribotyping)**, with restriction endonuclease fingerprinting of total chromosomal DNA in the typing of **coagulase-negative staphylococci**, revealed that, although ribotyping patterns were easier to interpret, they were slightly less discriminating (Bialkowska-Hobrzanska et al., 1990).

By isolating extra bands on DNA restriction digests observed in **methicillin-resistant** *Staphylococcus aureus* strains, a specific probe for the detection of methicillin-resistant *S. aureus* has been constructed (Fluit et al., 1990). Restriction endonuclease fingerprinting has been successfully used in the analysis of the genetic relatedness of *Actinobacillus pleuropneumonia* to selected members of the family Pasteurellaceae (Borr et al., 1991).

An example of direct application of restriction endonuclease analysis to patient management can be seen in a study of *Pseudomonas aeruginosa*, for which traditional typing systems may fail due to instituted therapy. Analysis of **Southern blot** patterns of *P. aeruginosa* prior to and after antibiotic treatment of infected patients helped to establish that failure to eradicate the organism was due to the development of drug resistance rather than to reinfection (Ogle et al., 1988). The application of nucleic acid probes to the identification of drug-resistant factors is discussed elsewhere in this book (see Chapter 1.6). A comprehensive coverage of the detection of **TEM β-lactamase genes** using **DNA probes** has been presented (Towner, 1990).

6.1.2.1 *IN SITU* HYBRIDIZATION

Application of ISH for viral diagnosis in tissues has long been recognized as an invaluable technique, particularly for the demonstration of low levels of viral burden in specific cell types (Haase et al., 1984; Macario and Conway de Macario, 1990; McDougall, 1988; Wolber et al., 1988). Detailed protocols for the analysis of various viral particles by ISH have been described, which include the criteria for probe selection and labeling, specific features of tissue preparation, hybridization conditions, and analysis of data (also see Part 1). One of the advantages of ISH for virus identification is the possibility to **semiquantify** the viral load in individual cell populations. By combining **ISH with electron microscopy (EM)**, the analysis of ultrastructural features of viral replication has been possible in cytomegalovirus (CMV)-infected fibroblasts (Wolber et al., 1988). An appropriate combination of detergents and proteases was shown to allow good **penetration** of DNA probes and gold label into formalin-fixed cells affording a high degree of preservation of structural detail of the cells (Wolber et al., 1988). Identification of HPV in tissue sections from cervical condylomata revealed a progressive increase in viral burden with epithelial differentiation (Schwartz et al., 1985). ISH has been particularly helpful in identifying CNS involvement with slow viruses and HIV (reviewed in Haase et al., 1984; Grody et al., 1987; McDougall, 1988).

6.1.2.2 rRNA GENE-SPECIFIC PROBES

In an effort to improve the **sensitivity** of nucleic acid probes for the detection of bacteria, a variety of methods have been devised targeting the highly **conserved RNA sequences** (Yehle et al., 1987). A combination of **rRNA-DNA solution hybridization** with **immunodetection** of the hybrids has been adapted to a latex bead format, allowing the hybridization to be completed within 10 to 15 minutes rather than hours (Miller et al., 1988). Using alkaline phosphatase as the reporter molecule, **quantitative estimates** become possible if the entire complex is collected on filters and the enzyme concentration is measured. This method is capable of detecting a broad range of bacterial species with a sensitivity of 500 cells per assay (Miller et al., 1988). The utility of this assay has been demonstrated in screening urine specimens for bacteriuria, for which the sensitivity of the method, 96.2%, is comparable to that of the conventional culture (Miller et al., 1988).

A **quantitative slot-blot** method has been described that is capable of detecting 3 pg rRNA in its radioactively labeled version, and 50 pg rRNA when the digoxigenin-labeling system is used (Cox et al., 1990). Digoxigenin-labeled probes were developed for rapid biotyping of *Campylobacter* (*Helicobacter*) spp. in clinical isolates.

rRNA can also be used to probe restriction fragment length polymorphism (RFLP) patterns as demonstrated in a epidemiological study of a **nosocomial infection** with *Corynebacterium* *jeikeium* where patterns of antibiotic susceptibility, which appeared not to be plasmid determined, correlated with RFLP profiles of *C. jeikeium* strains (Pitcher et al., 1990).

Fluorescent oligonucleotide probes constructed to identify 16S rRNA sequences can be directly applied to nonfractionated material from the intestine for bacterial strain identification using fluorescent microscopy (Amann et al., 1990).

6.1.2.3 POLYMERASE CHAIN REACTION

Amplification of the nucleic acid material of pathogens for the diagnosis and epidemiological studies of infectious disease has become a **method of choice** for a large number of organisms. The advantages of speed, specificity, low cost, and the possibility of identifying a pathogen without isolation and purification complement and, in many cases, supersede the practical usefulness of culture methods and immunological approaches. Furthermore, many organisms cannot be cultured with ease at the present time. Additional benefits of the PCR are seen in retrospective epidemiological studies on fixed, or archival, material (e.g., Impraim et al., 1987; D. P. Jackson et al., 1989) and in monitoring the therapeutic efficiency of pharmacological interventions when the pathogenic organisms may not be viable, but still present (Stoker, 1990). The availability of automated cyclers and optimization of incubation conditions have substantially reduced the hands-on time and make PCR diagnosis an attractive alternative in a clinical laboratory setting.

Standardization of amplification protocols, and the ability to detect the products of amplification without the use of radioactive labels, will make PCR a method of choice for clinical diagnostic laboratories for a wide range of applications. Further refinement of PCR systems allowing the **simultaneous amplification** of several target sequences of different nature (e.g., viral and human) in a **multiplex PCR** format will significantly enhance the efficiency and economy of diagnostic laboratories. In fact, a judicious selection of primers, shortening of incubation times, and appropriate adjustment of temperature regimes allows 100% specific and 100% reproducible fast multiplex PCR to be performed for a number of microbiological applications (Vandenvelde et al., 1990).

Some of the **shortcomings of PCR** for infectious disease evaluation include the danger of **contamination** and "**carry over**" and the **lack of quantification**, particularly pronounced when applied to clinical material without purification of target DNA (see Part 1; also Stoker, 1990). Some reappraisal of the old concepts concerning the prevalence of various pathogens, of what constitutes a latent infection, and of what should be considered a "**gold standard**" will have to come from epidemiological studies using this new amplification technology (Stoker, 1990). A number of established PCR protocols for the identification of HIV, HBV, CMV, HPV, enteroviruses, and bacterial pathogens in the environment have been published (e.g., Innis et al., 1990) and new protocols continuously appear in periodicals. The list of infectious agents that have been successfully detected and/or characterized by PCR amplification is growing exponentially. A sampling of reported studies at the time of this writing is given in Table 6.1.

6.1.2.4 EPIDEMIOLOGICAL ANALYSIS

Several approaches have been applied to the study of **clonality of pathogens**, which can be arbitrarily separated into phenotyping and genotyping methods (for a review, see e.g., Eisenstein, 1990). The former rely on the pattern of the pathogen **response to a variety of substrates**, **interaction with antisera** and **chemical challenges** (e.g., drugs), as well as **multilocus enzyme electrophoresis**. Characterization based on the nucleic acid specificity of an organism has relied on **plasmid** or **chromosome restriction patterns**. While plasmid-related analysis suffers from the intrinsic variability of the plasmid profiles due to frequent, and often extensive, loss or gain of DNA material by the plasmid (Eisenstein, 1990), **chromosome fingerprinting** is a substantially more permanent characteristic, and the specific differentiation of a bacterial species does not require the expression of a phenotype (Stull et al., 1988).

TABLE 6.1
Selected Assays for Infectious Agents

Pathogens	Ref.
Viral pathogens	
Cytomegalovirus	Demmler et al. (1988)
Epstein-Barr virus	T. Saito et al. (1989)
Herpes simplex virus	Brice et al. (1989)
Human herpes virus 6	Qavi et al. (1989)
Human papilloma viruses	Shibata et al. (1988); Wagatsuma et al. (1990); Griffin et al. (1990); McNichol and Dodd (1990)
Parvoviruses	Salimans et al. (1989a,b)
Rotaviruses	Xu et al. (1990); Wilde et al. (1990, 1991); Gouvea et al. (1990)
HIV-1	Greenberg et al. (1989); Shibata et al. (1989); Loche and Mach (1988); Hart et al. (1988); Laure et al. (1988); Hufert et al. (1989); Abbott et al. (1988); Hewlett et al. (1989, 1990); Carman and Kidd (1989); Lai-Goldman et al. (1988); Albert and Fenyo (1990); Holodniy et al. (1991); Ou et al. (1988); Dahlen et al. (1991); Kellogg et al. (1990); Clarke et al. (1990); Cassol et al. (1991); Dickover et al. (1990); Conway et al. (1990); Wages et al. (1991); Horsburgh et al. (1990); Lifson et al. (1990)
HIV-2	Rayfield et al. (1988)
HTLV-I	Abbott et al. (1988)
HTLV-II	Ehrlich et al. (1989)
Enteroviruses	Rotbart (1990a,b)
Rhinoviruses	Gama et al. (1988a,b)
Hepatitis A virus	Jansen et al. (1990)
Hepatitis B virus	Larzul et al. (1988, 1989); Sumazaki et al. (1989); S. Kaneko et al. (1989a,b, 1990a); Matsuda et al. (1989)
Hepatitis C virus	Weiner et al. (1990a,b); S. Kaneko et al. (1990b)
Influenza viruses	Bressoud et al. (1990)
Mumps virus	Forsey et al. (1990)
Adenoviruses	Allard et al. (1990)
Bacterial pathogens	
Chlamydia spp.	Dean et al. (1990); Pollard et al. (1989)
Mycoplasma pneumoniae	Bernet et al. (1989); Jensen et al. (1989)
Rickettsia rickettsii	Tzianabos et al. (1989)
Mycobacterium leprae	Hartskeerl et al. (1989)
Mycobacterium tuberculosis	Brisson-Noel et al. (1989); Shankar et al. (1989)
Borrelia burgdorferi	Rosa and Schwan (1989)
Treponema pallidum	Noordhoek et al. (1990, 1991); Burstain et al. (1991)
Enterotoxigenic *Escherichia coli*	Olive (1989); Pollard et al. (1990)
Shigella	Frankel et al. (1989)
Bordetella pertussis	Houard et al. (1989)
Clostridium difficile	Wren et al. (1990); Kato et al. (1991)
Leptospira	van Eys et al. (1989)
Helicobacter pylori	Valentine et al. (1991)
Listeria monocytogenes	Deneer and Boychuk (1991)
Yersinia enterocolitica	Wren and Tabaqchali (1990)
Coliform bacteria in water supply	Bej et al. (1990)
Naegleria fowleri	McLaughlin et al. (1991)
Legionella pneumophila	Bej et al. (1991); Mahbubani et al. (1990)
Fungi	
Cryptococcus spp.	Vilgalys and Hester (1990)
Protozoal pathogens	
Toxoplasma gondii	Savva et al. (1990)
Drug-resistant *Plasmodium falciparum*	Zolg et al. (1990)

TABLE 6.1 (continued).
Selected Assays for Infectious Agents

Pathogens	Ref.
Parasites	
Echinococcus multilocularis	Gottstein and Mowatt (1991)
Pneumocystis carinii	Wakefield et al. (1990)
Trypanosoma cruzi	Moser et al. (1989)

Adapted and expanded from Stoker, N. G (1990). Trans. Roy. Soc. Trop. Med. Hyg. *84*:755–756. With permission.

Restriction digests of **total chromosomal DNA** have the disadvantage of being too complex (about 1000 bands in an average fingerprint), making multiple comparisons cumbersome and poorly reproducible. Therefore, **restriction patterns of selected genetic regions** have been successfully used for species differentiation. This approach, however, suffers from the relatively narrow range of differentiation largely limited to species-specific or strain-specific identification (Stull et al., 1988). A substantially broader spectrum of hybridization specificities is afforded by probing chromosomal DNA with **rRNA** (Grimont and Grimont, 1986). **Ribosomal DNA (rDNA)** typing has been successfully used for subtyping strains isolated in sporadic outbreaks as well as in monitoring the incidence of various isolates in both sporadic and epidemic cases (Altwegg and Mayer, 1989; Rappuoli et al., 1988; Stull et al., 1988).

Restriction analysis of genomic DNA using a variety of probes, with or without the aid of PCR amplification, has been successfully used for molecular analysis of viridans streptococci in a case of myocarditis (Stein and Libertin, 1989), *P. aeruginosa* strains in sputum samples of cystic fibrosis patients (Hjelm et al., 1990) and in immunocompromised children (Boukadida et al., 1991), *Listeria monocytogenes* in epidemic outbreaks (Nocera et al., 1990), *Vibrio cholerae* strains (Yam et al., 1991), *Shigella* strains (Litwin et al., 1991), *Salmonella* strains (Nastasi et al., 1990), *Acanthamoeba* strains (Kilvington et al., 1991), *Leptospira* strains (Van Eys et al., 1991), *Candida* species (Pfaller et al., 1990; Stevens et al., 1990; Clemons et al., 1991; Carruba et al., 1991; Reagan et al., 1990), penicillin resistance of *Neisseria meningitidis* strains (Zhang et al., 1990) and human adenovirus type 3 isolated from ocular lesions (Itakura et al., 1990) — to name just a few published examples.

An important assumption in taxonomic, epidemiological, and diagnostic studies involving nucleic acid hybridization is the expected specificity of the hybrid formation with the probe. A rigorous study analyzing various parameters resulting in **experimental error in DNA-DNA pairing** has been presented (Hartford and Sneath, 1990).

6.1.2.5 PULSED-FIELD GEL ELECTROPHORESIS

A method of bacterial **strain identification** has been developed based on pulsed-field gel electrophoresis (PFGE) of genomic restriction digests (Le Bourgeois et al., 1989; McClelland et al., 1987; Tanskanen et al., 1990; Yan et al., 1991). In fact, in the case of *Campylobacter*, PFGE restriction profiles using *Sma*I-restricted genomic DNA allows a reliable differentiation between *C. jejuni* and *C. coli*, offering a better practical alternative for epidemiological studies than combining conventional DNA fingerprinting with RNA hybridization procedures (Yan et al., 1991).

Chapter 6.2

Enteric Pathogens

6.2.1 ENTEROVIRUSES

Human enteroviruses include **polio viruses, Coxsackie viruses, echoviruses**, and **hepatitis A virus**, which cause a wide spectrum of diseases affecting the nervous, respiratory, cardiovascular, and gastrointestinal systems (Cherry, 1987; Melnick, 1990). An excellent update on the biology of pathogens causing acute viral gastroenteritis has recently appeared (Blacklow and Greenberg, 1991). The importance of developing fast, specific, and highly sensitive diagnostic methods complementing the existing "gold standard" of cell culture is underscored by the severity of diseases caused by these viruses particularly in children with frequently nonspecific symptomatology (Rotbart, 1991). The lack of adequate speed, sensitivity, and specificity in traditional diagnostic methods for enterovirus identification prompted the development of nucleic acid-based techniques. Detailed discussions of the earlier methods designed for the detection of enteroviral nucleic acids by **cRNA** (e.g., Cova et al., 1988), **cDNA** and **oligonucleotide probes** (e.g., Bruce et al., 1989; Rotbart et al., 1988), as well as later **amplification methods** (e.g., Rotbart, 1990a,b) have been presented (Hyypia, 1989; Rotbart, 1990a,b, 1991; Rotbart and Levin, 1989). The earlier cDNA technique, when directly applied to clinical specimens of cerebrospinal fluid (CSF) or stool, had low sensitivity (33%) (Rotbart, 1991). Subsequent improvements in **target retention**, the use of **single-stranded RNA probes**, and, finally, **target amplification** by PCR led to the development of clinically useful assays (Rotbart, 1990a,b, 1991).

6.2.1.1 ENTERIC ADENOVIRUSES
Certain types of **enteric adenovirus EA (types 40** and **41)** cause gastroenteritis in infants and their identification by gene probes directly in clinical specimens has been developed in a **direct spot test** format using radioactive and nonradioactive probes (Hammond et al., 1987; Kidd et al., 1985; Niel et al., 1986; Stalhandske et al., 1983, 1985; Takiff et al., 1985). The high reproducibility of DNA hybridization and the ability to detect EA directly in stool specimens, although with a slightly lower sensitivity than that available with **electron microscopy (EM) methods** following **ultracentrifugation** of specimens, makes this modality a useful diagnostic test for patient management (Hammond et al., 1987). An even higher level of diagnostic ability on diluted stool samples is offered by a PCR assay that compared favorably with **cell culturing** in routine diagnostic laboratories and with **restriction endonuclease analysis (REA)** of viral DNA (Allard et al., 1990).

6.2.1.2 ROTAVIRUSES
The worldwide distribution of rotaviruses (RV) causing gastrointestinal infections in infants and children has long been recognized and rapid diagnosis has traditionally been made by **immunological methods** (Haikala et al., 1983; Hammond et al., 1982; Hughes et al., 1984; Miotti et al., 1985), visualization of the virus by **EM**, and the viral RNA detection by **polyacrylamide gel electrophoresis** and silver staining (Herring et al., 1982; Rodger et al., 1981; Spencer et al., 1983; Theil et al., 1981). Some fastidious nongroup A RV cannot be cultivated in tissue culture and are not detectable by conventional immunological techniques

(Bellinzoni et al., 1989). Earlier hybridization methods utilized a **dot-blot** format with radioactively labeled probes (e.g., Dimitrov et al., 1985; Estes and Tanaka, 1989; J. H. Lin et al., 1987). Later, **biotinylated single-stranded (ssRNA)** probes were constructed and applied to dot-blot and **Northern analysis** of RV double-stranded RNA (dsRNA) genomes (Bellinzoni et al., 1989).

Alkaline phosphatase-conjugated oligonucleotide probes have also been used to characterize extracted viral RNA from stool samples and it compared favorably with polyacrylamide gel electrophoretic and immunological (**ELISA**) analysis of RV (Arens and Swierkosz, 1989; Olive and Sethi, 1989; Yamakawa et al., 1989). Several ELISA systems e.g., **EIA**, International Diagnostic Laboratories; **Pathfinder**, Kallestad Laboratories; **Rotaclone**, Cambridge BioScience, Inc., and a nonradioactive oligonucleotide assay kit (**SNAP**, Molecular BioSystems, Inc.) are available for RV diagnosis. Comparison of these systems on stool specimens demonstrated the acceptable sensitivity and specificity of all these diagnostics (Arens and Swierkosz, 1989). The oligonucleotide-based system required an overnight incubation for maximum sensitivity and involved laborious isolation of viral nucleic acids, necessitating further improvements to make it more appealing to clinical diagnostic laboratories.

Development of effective anti-RV vaccines is under way and this effort requires accurate determination of RV serotypes circulating in the community (Sethabutr et al., 1990). Immunological assays are suitable only for the cultivable virus strains and monoclonal antibody systems can identify about 70% of field isolates. Alternative approaches rely on characterization of RV serotypes by **sequencing** the genes encoding serotype-specific VP7 glycosylated products of genes 7, 8, or 9 (Green et al., 1988), or by **RNA segmental oligonucleotide mapping** (Desselberger et al., 1986). Hybridization with oligonucleotide probes (Sethabutr et al., 1990) and **solution RNA-RNA hybridization** (Nakagomi et al., 1989; A. Ohshima et al., 1990) have been used for molecular characterization of RV isolates. Fractionation of isolated RV RNA into the 11 RNA segments assigned to 4 groups with concurrent subgrouping and serotyping by monoclonal antibodies has proved to be effective in analyzing **epidemic outbreaks** (Pipittajan et al., 1991).

A new amplification typing approach (**PCR typing**) complements grouping of RV by **electrophenotyping**, particularly in defining RV-positive specimens untypable by serological methods (Gouvea et al., 1990). Identification of RV by PCR in clinical specimens can be hampered by **inhibitory substances** present in unpurified fecal samples (Wilde et al., 1990). Processing of the specimens through chromatographic cellulose fiber powder (CF11 powder) to remove substances interfering with PCR allowed the virus to be detected by PCR at dilutions beyond the detection limits of standard techniques (ELISA or the direct visualization of RNA following polyacrylamide gel electrophoresis) (Wilde et al., 1990). PCR is helpful in determining **RV shedding** by children, substantially improving RV detection compared to ELISA (58 vs. 36% positivity, respectively) (Wilde et al., 1991). Whether the degree of RV shedding detectable only by PCR, but not by enzyme immunoassay, can contribute to spreading wild-type virus must be clinically determined.

6.2.2 HEPATITIS B VIRUS

Molecular probes have enabled scientists and clinicians to make certain observations on the **biology of hepatitis B virus (HBV)** and **pathogenesis of HBV infection** that have been helpful in patient management. A brief synopsis of pertinent findings is given here to illustrate this point, emphasizing the techniques used. No attempt is made to cover the subject comprehensively.

6.2.2.1 HEPATITIS B VIRUS IN SERUM

One of the traditional **biochemical markers** of liver involvement in HBV infection is the **level of transaminases**, in particular, **alanine aminotransferase (ALT)**. The precise relationship of a rise in ALT to structural and functional alterations of hepatocytes in the course of HBV infection has not been defined yet. Dot-blot hybridization with a ^{32}P-labeled probe established that the peak of serum HBV DNA was reached shortly before or simultaneously with the maximum elevation of ALT in over 90% of the acute **exacerbations** of chronic HBV hepatitis (Liaw et al., 1988). Elevation of the viral DNA level preceded hepatocytolysis in acute exacerbations.

Serological evidence of the replicative phase of HBV infection has usually been obtained by demonstrating the presence of **HBeAg** and a rising level of the homologous antibody (**anti-HBe**), which is assumed to follow clearance of HBeAg and herald transition to the inactive phase of the infection. A number of discordant reports have appeared to imply that HBV replication can occur in the presence of anti-HBe (Carloni et al., 1987; Morace et al., 1985; Negro et al., 1984). Liver histology and serological and biochemical markers have been correlated with HBV DNA detection, and HBV DNA was found in the serum of anti-HBe carriers suffering from **major liver damage**, but not in anti-HBe carriers having normal or only minimally compromised liver (Bonino et al., 1986).

In contrast to the situation involving exacerbation of **chronic active hepatitis (CAH)**, the HBV genome was detectable in blood over prolonged periods of time, indicating that HBV replication was an integral part of the patients' serological profile. The atypical **concomitant presence of HBV DNA** and **anti-HBe** in serum correlated with histologically demonstrated **severe periportal hepatitis** leading to cirrhosis in a number of cases. Furthermore, a correlation was found between HBV DNA and an unusual type of intracytoplasmic assembly of **HBcAg**. This is in contrast to its nuclear localization, prominent in carriers with "classical" HBV replication serologically expressed in HBe antigenemia. It thus appears that the **positive HBV DNA test in patients with positive HBsAg and anti-HBe serology is significant in predicting an unfavorable outcome** (Bonino et al., 1986).

A simplified **spot hybridization** test was used to demonstrate that HBV DNA in the sera of HBeAg-negative patients with severe chronic liver disease correlated with elevated ALT and **anti-HBc IgM**. The presence of HBV DNA not only identified potentially infectious subjects but also indicated **chronic progression** of hepatitis (Carloni et al., 1987). During acute clinical exacerbations of HBV CAH in patients positive for anti-HBe, the serum levels of HBV DNA rose in 88% of patients (Lai et al., 1988a,b). In addition, in 83% of cases, the liver tissues obtained during clinical exacerbations showed **free replicative forms** of HBV DNA demonstrated by Southern blot. In a longitudinal study, HBV DNA was found to be a better **index of impending chronicity** than HBeAg (Moestrup et al., 1985). Based of the HBV DNA hybridization studies, the **infectivity** in acute hepatitis B is characteristic of the **presymptomatic** and **early symptomatic** period.

HBV DNA is present in the plasma of 85.7% of people positive for both HBsAg and HBeAg, and in 11.8% of those positive for HBsAg and negative for HBeAg (Peng et al., 1988). Interestingly, free HBV DNA was found in 14.3% of the leukocyte samples of the first group. An integrated form of HBV DNA was also detected in clinically healthy controls. In Sardinia, highly endemic for HBV, a very high percentage (69%) of the 71 **blood donors** negative for all serological HBV markers tested positive for HBV DNA in serum by dot-blot hybridization whereas the remaining 31% were positive for anti-HBc (Lai et al., 1989).

Substantially higher estimates of the HBV DNA prevalence are made using PCR. Thus, **PCR** analysis of the prevalence of HBV DNA in 116 HBsAg carriers demonstrated that 100% of the HBeAg-positive individuals had HBV DNA in their blood (Shih et al., 1990). Among the HBeAg-negative carriers, those with CAH were 90% positive for HBV DNA, the healthy carriers were 80% positive, and patients with cirrhosis had the lowest rate of HBV DNA presence — 69.2%. It appears that viral replication is most active in carriers with CAH **despite seroconversion** to anti-Be status. It is anticipated that simplified PCR assays using nonradioactive probes will prove indispensable in mass screenings for HBV DNA (Shih et al., 1990).

Evaluation of HBV DNA in serum has also been used in **monitoring steroid treatment** of patients with hepatitis (Nair et al., 1986; Tong et al., 1987; Wood et al., 1987), for establishing **maternal-infant HBV transmission** (Lee et al., 1986), as well as in **screening** medical workers (Hofmann et al., 1988). **Restriction analysis** and **sequencing** of HBV DNA from the sera of HBsAg-positive persons allows the identification of **epidemiological relationships** between cases of HBV infection (Lin et al., 1990). Evidence of intrafamilial transmission has been presented at the HBV genome level using probes for specific regions of the viral genome (Lin et al., 1990).

In summary, the above-cited studies as well as numerous others confirm the superior **diagnostic potential** of molecular hybridization and PCR performed on serum specimens compared to serology and biochemical markers in establishing infectivity and predicting the development of HBV infection (Berris et al., 1987; Brechot et al., 1985; Burnett et al., 1985, 1986; Feinman et al., 1988; Govindarajan et al., 1986; Liaw et al., 1987; Matsuyama et al., 1985; Moestrup et al., 1985; Moller et al., 1987; Shih et al., 1990).

6.2.2.2 HEPATITIS B VIRUS DNA IN WHITE BLOOD CELLS

HBV DNA has been detected in **bone marrow** cells (Elfassi et al., 1984) and peripheral blood **mononuclear cells** (PBMCs) of homosexual men, hemophiliacs, and intravenous drug users (Lie-Injo et al., 1983; Pontisso et al., 1984). HBV DNA was detected by Southern blot in peripheral blood **lymphocytes** as well as lymphocytes from the bone marrow, semen, and lymph nodes patients with AIDS (Laure et al., 1985). The restriction DNA pattern was consistent with integration of HBV DNA. In addition, free, but not replicative, viral forms were identified in these cells. Similar observations were reported by others in 4 out of 23 AIDS patients (Lie-Injo et al., 1985).

However, PBMCs have been shown by Southern analysis to contain replicative forms of HBV DNA in 14 carriers positive for HBsAg (Yoffe et al., 1986). Moreover, considering the presence of **1.0- to 3.2-kbp fragments** generated by *Eco*RI digestion as indicative of the **replicative forms**, HBV DNA could be detected in the sera of only those HBsAg carriers whose PBMCs contained replicative forms. In patients positive for HBeAg, the viral DNA was found in an **extrachromosomal** state. The replicating forms were detected only in monocytes or B cells, whereas T cells were devoid of these replicating forms (Yoffe et al., 1986).

HBV DNA replicates in PBMCs, as shown by the expression of HBV antigens as well as the presence of HBV DNA in PBMCs and hepatocytes in 104 CAH patients (Parvaz et al., 1987). Moreover, the **pattern of expression** of serological markers of HBV replication in PBMCs showed an excellent correlation with the expression of the HBV gene products in hepatocytes. Both **replication** and **integration** of HBV DNA could occur in **T lymphocytes** as well as in **non-T lymphocytes**. Although HBV **replication** occurs in both hepatocytes and PBMCs, it is more successful in hepatocytes, whereas **integration** occurs more readily in PBMCs (Parvaz et al., 1987). Integration of HBV DNA into PBMCs allowed the construction of a human B cell hybridoma that carried HBV DNA in an episomal state and expressed the major viral transcripts as well as HBsAg, HBcAg, and HBeAg (Colucci et al., 1988). Although HBV could not complete its life cycle in the hybridoma, it was able to transmit episomal HBV DNA to normal lymphocytes in cocultures.

Confirmation of HBV DNA **transcription** in PBMCs was provided, for example, by **ISH** demonstrating RNA production in HBsAg-positive and -negative patients with liver diseases (Hadchouel et al., 1988). Although no definite conclusions have been drawn regarding the contribution of replicative HBV DNA forms found in PBMCs to transmission of the infection (Lobbiani et al., 1990), it appears that persistence of the virus in peripheral blood constitutes a part of the infectious process at an early stage (Harrison, 1990).

6.2.2.3 HEPATITIS B VIRUS DNA IN THE LIVER

As mentioned above, HBV DNA has been repeatedly demonstrated in the liver in hepatitis. ISH is the method of choice for histological assignment of HBV. A biotinylated probe has been used to examine **formalin-fixed** liver sections obtained from 45 patients with various chronic liver diseases (Rijntjes et al., 1985). The results were positively correlated with HBV serological markers and histopathological features in 41 out of 45 patients. HBV DNA has always been detectable in the liver of chronic hepatitis patients, regardless of the serological status. On the other hand, no HBV DNA could be identified in sections showing a normal histological pattern, despite positive anti-HBc and anti-HBs.

Although the presence of HBV DNA may not correlate with serological markers of HBV infection, the relationship between **HBV gene expression**, **liver disease**, and HBV **RNA** shows a more consistent correlation (Yokosuka et al., 1986). Using the detection of RNA with a ^{32}P-labeled DNA probe, liver FNA biopsies have been evaluated from 30 chronic liver disease patients both seropositive and seronegative for HBsAg. HBV-specific RNA was found in all seropositive but not seronegative individuals. The 29S and 21S RNAs were found to be mRNAs for HBcAg and HBsAg, respectively. Free HBV DNA and its products were causally related to the activity of liver disease (Yokosuka et al., 1986). Absence of transcription in seronegative individuals is apparently due to the quiescent stage of the disease following integration of HBV DNA. The **integrated DNA** has a 21S RNA transcript and HBsAg as its sole expression, whereas **free DNA** gives rise to a 29S RNA transcript and expression of HBcAg in the liver and of HBeAg in the serum. **Extrachromosomal monomeric viral DNA** seems to persist in patients serologically immune to HBV infection, apparently due to a restriction of the viral expression the via methylation, mutation, gene rearrangements, or other mechanisms (Blum et al., 1987).

The presence of **free replicative forms** of HBV DNA in the liver of patients with CAH has been repeatedly correlated with **greater tissue damage** (Tozuka et al., 1989). In contrast, **integration of HBV DNA**, which occurs early in CAH, was found predominantly in livers showing **no necrosis** or **inflammatory activity** in patients positive for anti-HBe. Areas of **regenerating hepatocytes** were frequently seen in livers with HBV DNA integration. Interestingly, similar to the situation with hepatocellular carcinoma (see below), the clonal growth of hepatocytes was found to occur also in noncancerous disease. These regenerating hepatocytes were thought to have clonal integration of HBV DNA, suggesting that hepatocarcinogenesis is preceded by integration of HBV DNA (Tozuka et al., 1989).

Evaluation of the HBV DNA presence in the blood and liver is used as a guide in devising treatment protocols. Important correlations have been established. For example, in the course of **antiviral or immunomodulatory therapy**, the amount of HBV DNA in the liver estimated by dot-blot and Southern hybridization was found to correlate with the expression of HBeAg: HBV DNA in the liver was significantly reduced in patients demonstrating an absence of HBeAg; conversely, the amount of DNA did not change appreciably in HBeAg-positive patients (di Bisceglie et al., 1987). Evaluation of the **state of integration** and expression of HBV DNA in the hepatocyte and the serum has been used to develop **optimal therapeutic protocols**. Thus, antiviral therapy can be advised for HBsAg-positive CAH patients with episomal HBV DNA irrespective of the presence of integrated sequences (Heijtink et al., 1987). The selection and evaluation of patients for antiviral therapy should involve consideration of the HBV-DNA state (**episomal vs. integrated**) found in liver biopsy, because the **existence of HBV DNA in the episomal state in the liver is not always reflected by the presence of HBV DNA in the serum**. It should be noted, however, that HBV DNA can no longer be detected by liver biopsy in children receiving interferon α treatment, because they lose circulating HBV DNA from their blood (Bartolome et al., 1990a). Parallel disappearance of HBV DNA from the liver, PBMCs, and serum has been observed also in adults receiving interferon therapy (Bartolome et al., 1990b).

A difference has been found among the molecular forms of HBV DNA in PBMCs and the liver, and the HBV DNA form in PBMCs did not correlate with serum HBV markers (Bouffard et al., 1990).

6.2.2.4 EXTRAHEPATIC LOCALIZATION OF HBV

The role of HBV gene expression in determining the various manifestations of fulminant hepatitis may be better understood when the physical state and levels of expression of viral DNA during acute infection are evaluated in various body tissues. Aside from the liver and white blood cells (WBCs) (Lie-Injo et al., 1983; Pontisso et al., 1984; Romet-Lemonne et al., 1983; Yoffe et al., 1986), the extrahepatic distribution of HBV has been described in the **brain, saliva, pancreas, kidney, skin,** and **vessels** of animals and humans (Blum et al., 1983; Davison et al., 1987; Dejean et al., 1984; Halpern et al., 1983, 1986; Lieberman et al., 1983; Tagawa et al., 1987; Yoffe et al., 1986). HBV DNA and HBV-related RNA have been studied in **patients who have died from fulminant** and **resolving hepatitis** (Yoffe et al., 1990). HBV nucleic acids were identified in **lymph nodes, spleen, gonads, thyroid gland, kidneys, pancreas,** and **adrenal glands**. The strongest signal was detected in the lymph nodes. Interestingly, little or no HBV nucleic acid was detected in the sera or livers in these patients. In contrast, HBV was predominantly found in the livers of HBV carriers, although extrahepatic HBV DNA was also detectable. No specific histological changes could be correlated with the extrahepatic HBV DNA presence. Replicative intermediate forms of HBV DNA were also identified by other workers in the spleens of patients with chronic hepatitis (di Bisceglie and Hoofnagle, 1990).

6.2.2.5 HEPATOCELLULAR CARCINOMA

ISH with biotinylated probes to HBV DNA in formalin-fixed, paraffin-embedded liver biopsies from patients with hepatocellular carcinoma (HCC), combined with Southern blot, showed that the **number of HBV DNA-positive cells varied from tumor to tumor** (Brambilla et al., 1986). HBV DNA **nuclear staining** did not correlate either with tumor localization, the gene expression markers (HBsAg and HBcAg) or with the type or degree of differentiation of the tumor. Southern blot analysis of **total tumor genomic DNA** obtained from mice injected with PLC/PRF/5 cells showed **multiple integration events**. ISH was thought to be more sensitive than the Southern blot technique when the viral DNA level was low. ISH showed HBV DNA was present in hepatocytes **adjacent** to HCC foci, suggesting that **integration of viral DNA precedes tumor development**. Integration of HBV DNA, although detected by Southern analysis in the majority of HBsAg-positive cases of HCC (e.g., Hada et al., 1986; Imazeki et al., 1986; Lai et al., 1988a,b), in a large number of HCC cases **could not be** convincingly shown (Hada et al., 1986). Moreover, no free viral DNA could be seen in neoplastic areas (Imazeki et al., 1986).

Fine analysis of **HBV DNA integration into specific sites** on the host chromosome has been performed in **HCCs in children** (Giacchino et al., 1987; Yaginuma et al., 1987). There appear to be **multiple integration sites**, the location of which may be related to carcinogenesis, if such integration affects the expression of **oncogenes** (Yaginuma et al., 1987). In contrast, others believe there is a **single integration site** accounting for clonal proliferation of such infected cells (Giacchino et al., 1987).

The overwhelming evidence, however, indicates that various types of rearrangements (**deletions, duplication, insertions**) occur both in integrated HBV sequences and in the cellular flanking sequences (e.g., Hino et al., 1986; Rogler et al., 1985; Yaginuma et al., 1985). Although **integration of the virus occurs at random** in the host genome, it appears that HBV DNA preferentially integrates in areas with **abundant repetitive sequences** (Shaul et al., 1984, 1986). **Amplification of integrated HBV** and its **flanking cellular sequences** has been detected in HCC (Matsumoto et al., 1988). Again, the relationship of this amplification to that of oncogenes triggered by the viral integration cannot be excluded as contributing to carcinogenesis. No final word has been pronounced on this as yet. When studying **noncancerous** regions of the liver by Southern analysis, **heterogeneous sites** of HBV DNA integration into hepatocyte DNA were found (Tanaka et al., 1988). These appeared to **precede hepatocarcinogenesis** and lead to **multifocal clonal cell proliferations** stemming from the hepatocytes harboring integrated HBV DNA (Lai et al., 1988a,b; Tanaka et al., 1988). The role of HBV integration in HCC does not seem to be as direct as that of viruses with oncogenes or that of viruses acting by a "promoter-insertion" mechanism (Robinson et al., 1990). Other factors, such as hepatocyte injury (Shafritz, 1991) and regeneration (Robinson et al., 1990) play an important role in hepatocarcinogenesis related to HBV.

The **expression of the HBV genome** in HCC and chronic hepatitis was studied by REA (Bowyer et al., 1987). In **chronic hepatitis, hypomethylation** of the HBV genome was found in both the virus and in infected tissues. **HBV DNA copy number** low in 17 out of 29 tumor tissues, preventing evaluation of the role of methylation. However, **hypermethylation** was detected in six of the tumor tissues tested as well as in the PLC/PRF/5 cell line, suggesting it plays a role in HBV gene expression. Methylation per se cannot be an absolute determinant of the HBV genome expression in human HCC, and additional factors must contribute to the HBV genome expression (Bowyer et al., 1987).

Detailed understanding of the role of HBV in hepatocarcinogenesis will certainly influence the use of HBV DNA probes in a clinical setting. It is apparent that a combination of morphological evaluation of liver biopsy with a sensitive evaluation of the HBV DNA status may offer substantially more precise assessment of HBV infection and its sequelae. There is little doubt these molecular biological techniques will soon be used in routine patient management.

6.2.2.6 HYBRIDIZATION METHODS

The search for diagnostic tools better than serological assays for the detection of HBV, the identification of its physical form in a patient, and, consequently, prognosis of the infection resulted in the development of DNA probes in the early 1980s (e.g., Berninger et al., 1982; Bonino et al., 1981; Scotto et al., 1983; Weller et al., 1982). **Dot-blot** hybridization of serum samples has been performed either following **phenol-chloroform extraction**, or using **direct application** of radioactively labeled DNA probes. Relatively small samples of serum (less than 500 µl) were sufficient to generate visible spots by autoradiography when **cloned HBV DNA** was used as a probe.

Evaluation of the positive results could be **semiquantitative** at best. Although a large number of samples could be processed simultaneously, the methodological aspects of this technique with regard to sample preparation, construction of the probe, and evaluation of results have not been standardized between different laboratories. **Reproducibility** of the method even in the same laboratory has never been thoroughly studied. In my experience, dot-blot hybridization with some of the commercially developed kits suffered from poor reproducibility, beginning with the collection of serum samples on membranes: parallel samples took different times to filter and wash even within the same run, the autoradiographic spots showed sometimes uneven distribution of radioactivity, and quantitation of the signals was rather subjective. A serial dilution of a standard HBV DNA-positive sample can be used, against which the test spots are compared to give an **indirect approximation** of the number of viral particles. The sensitivity of the dot-blot method was 1×10^5 DNA-containing particles per milliliter of serum (Berninger et al., 1982). This method, although detecting 0.1 pg or less of pure HBV DNA, could not be used to quantitate the level of Dane particles or HBV DNA in serum (Lieberman et al., 1983). Again, a negative result could be due to the possible presence of HBV at very low levels.

A **spot hybridization** test cannot identify the **physical nature** or **size** of the DNA being tested. Therefore the simplicity and the high throughput of specimens available with spot hybridization can be used only for the fast identification of patients with relatively high levels of HBV in their blood. In spite of these limitations molecular hybridization in a dot-blot or spot-blot format has provided significant information on the natural history of HBV infection. The **superior sensitivity** of HBV DNA hybridization over that of conventional serological methods was already evident in these earlier studies. The hybridization method proved to be more sensitive than the DNA polymerase and HBeAg assays in detecting complete viral particles. This proved useful in determining the **level of infectivity** in patients and in monitoring **response to therapy** (Weller et al., 1982). It was firmly established that HBV DNA could be reliably detected in individuals positive for **anti-HBe**, suggesting continuing infectivity (Lieberman et al., 1983; Scotto et al., 1983).

As with any hybridization assay, the **purity** of the cloned fragments used as a probe is crucial in determining the specificity of identification of targeted sequences. **False-positive** results have been observed in HBV DNA dot hybridization assays due to the **residual portions of the plasmid** used to construct the probe. It hybridized to the contaminating plasmid-containing bacteria in the samples (Diegutis et al., 1986).

A thorough, quantitative study of the **HBV genome titers** in sera of infected individuals distinguished **virus carriers** with high ($>5 \times 10^7$ HBV genomes), moderate (10^5 to 5×10^7), and low ($<10^5$) infectivity (Zyzik et al., 1986). The detection limit of the method, based on comparison with a series of reference dilutions, was 0.3 pg DNA per milliliter, equal to 10^5 HBV genomes. In **symptomatic carriers**, low or moderate HBV DNA and HBeAg titers were seen during clinical **exacerbations** in the most severe cases (Zyzik et al., 1986). In addition, quantitative assays for HBV DNA detected the beginning of **HBV elimination** from the blood earlier than any other marker of chronic HBV infection, although failing to detect HBV DNA in healthy HBsAg carriers who were anti-HBe positive (Zyzik et al., 1986).

Improvements in the molecular assays for HBV DNA essentially follow the same pattern as that found in the diagnosis of other infections by nucleic acid probes — introduction of **oligonucleotide** probes, **nonradioactive** labels, **solution hybridization** (Kuhns et al., 1989), and the **PCR** technique. A 21-mer **oligonucleotide ^{32}P-labeled probe** present in all known HBV genomes has been tested on 988 HBsAg-positive or -negative serum samples and showed a 98% correlation with the results of HBV DNA probe hybridization (J. H. Lin et al., 1987). The assay time was significantly shortened (1 to 2 vs. 16 h), the sensitivity did not suffer, and the background signal was significantly reduced. Due to **nonspecific binding**, however, the probe could not be used either for membrane-bound deproteinized serum samples or for purified nucleic acids applied to the membrane without serum. Although the probe appeared to be universal, no true universality can be expected from the 21-mer probe used because nucleotide sequences of HBV in different carriers may vary to some extent (J. H. Lin et al., 1987).

A combination of **biotinylated** DNA probe with the detection system using **avidin-β-galactosidase complex** detected as little as a few picograms of target DNA (Yokota et al., 1986). The method was comparable to the conventional ^{32}P-labeled probe method. A **streptavidin-alkaline phosphatase conjugate** has been used as a detector system with biotinylated DNA probe in a dot-blot hybridization format (Quibriac et al., 1987, 1989). Comparison with the ^{32}P-labeled probe showed the nonradioactive counterpart to be sensitive and specific (96% correlation with the results of the radioactive probe tested on 150 specimens). Sensitivity of the assay was 10 pg, and no interference of related or unrelated material (HeLa DNA, CMV, HSV1, HSV2, Adv2) was noted. In addition, no false-positive signals due to the plasmid sequences were detected. Following removal of the ^{32}P-labeled DNA probe from the filters, reprobing HBV DNA in a dot-blot hybridization assay with biotinylated DNA revealed that the sensitivity of the radioactive probe was 0.25 pg, whereas that of the biotin-labeled probe was 1 pg (Saldanha et al., 1987). The biotin-labeled probes could be successfully stored at –20°C for over 1 year without loss of sensitivity or specificity.

Another nonradioactive HBV DNA probe used **2-acetylaminofluorene (AAF)** as a reporter molecule (Larzul et al., 1987). Using the same DNA probe labeled either with ^{32}P or AAF, a correlation of 96% between the two probes was observed on 328 sera, the AAF test detecting 86% of 65 HBV DNA-positive patients. The sensitivity limit of the AAF-modified probe, which used an **anti-AAF monoclonal antibody** detection system, was found to be 5×10^6 viral particles per milliliter of serum compared to 5×10^5 particles per milliliter for the radioactive probe.

Three nonradioactive HBV DNA probes (labeled with **biotin**, **photobiotin**, and **digoxygenin**) have been compared (Casacuberta et al., 1988). The sensitivities for these probes tested on serial dilutions of plasmid DNA processed as serum samples were 1–5 pg with ^{32}P and the digoxygenin-labeled probe, 5–10 pg with the biotin-labeled probe, and 50 pg with the photobiotin-labeled DNA. Unfortunately, a high rate of 23% **false-positive** results was noted with the biotin-labeled probes. However, phenol extraction followed with ethanol precipitation proved to completely eliminate this nonspecific hybridization. **Digoxigenin-labeled riboprobes** have been developed for the detection of HBV DNA in serum, showing a detection limit of 0.5 to 1.0 pg HBV DNA and a high specificity (Gerritzen and Scholt, 1990).

Chemoluminescent substrate for the **alkaline phosphatase** linked to a DNA probe for HBV significantly improved sensitivity of the assay, which detected 1.18×10^6 copies of the "core antigen" HBV$_c$ plasmid DNA in 30 min (Bronstein et al., 1989). This favorably compares with a similar assay (sensitivity 9.8×10^7 copies of DNA) employing a **tetrazolium-based** detection system.

A **commercial** HBV DNA **solution hybridization** system introduced by Abbott Laboratories for research use in this country (**Abbott Genostics**^{TM} **hepatitis B viral DNA** package insert; Abbott Laboratories, 1987) is based on the assay developed by Kuhns and coworkers (1988). The kit has been tested in Europe and compared with spot hybridization in chronic hepatitis B carriers undergoing viral therapy (Zarski et al., 1989). A good correlation was found between the two methods, with the Abbott Genostics system displaying a slightly higher sensitivity than the spot test. The Abbott kit uses an ^{125}**I-labeled DNA probe** and is well standardized, providing a quantitative evaluation of the HBV DNA level in serum. The cut-off level of sensitivity for a positive result was equal to or greater than 1.5 pg HBV DNA per milliliter. This is essentially an overnight procedure that uses column separation of hybridized and free probes.

My experience with the early versions of this assay was encouraging; its simplicity and the excellent support provided by the Abbott staff and local representatives certainly help to make it a suitable system for wide use even in routine clinical laboratories. Certain technical deficiencies were encountered at the early stages of the development of this kit (poor uniformity of the separation columns, less than desirable reproducibility, fast progressive deterioration of radioactivity of the label), but the practical utility of this convenient testing system appears to be high given the necessary improvements are made. Comparison of the ^{125}I-solution hybridization method (Abbott Laboratories) with the dot-blot hybridization procedure revealed the solution hybridization gave positive results in 82% of all dot-blot positive sera (Makris et al., 1991). The correspondence between the two methods was higher (97%) in strongly positive dot-blot cases. The solution hybridization although proving less sensitive than the dot-blot assay still is capable of identifying the majority of replicating HBV offering a means for monitoring antiviral treatment (Makris et al., 1991).

Evaluation of the deleterious effects of an infection as organ-specific as HBV cannot be complete without establishing correlations between the **morphological alterations** and the presence and extent of tissue involvement with the virus. In this context, the **ISH** approach offers advantages unavailable in serum assay systems (Blum et al., 1983; Gowans et al., 1981, 1983;

see above). The cytoplasmic HBV DNA visualized by ISH in hepatocytes positive for HBcAg correlates with the level of cytoplasmic HBcAg, but not with the presence of nuclear HBcAg (Gowans et al., 1985). In **persistent acute hepatitis** at least the major part of the **replication** of viral DNA occurs in the cytoplasm of hepatocytes, as revealed by ISH with radioactive and biotin-labeled probes (Gowans et al., 1985; Herrmann and Hubner, 1987). ISH shows that HBV DNA is **diffusely distributed** in the liver during the early stage of chronic liver disease, whereas the site of HBV replication becomes localized in the advanced stage of the disease (Michitaka et al., 1988).

6.2.2.7 PCR IN HBV DETECTION

The PCR allows detection of HBV early in an infection, when it is present at the level of only 4×10^4 particles per milliliter serum (Larzul et al., 1988, 1989). Another PCR assay had a somewhat greater sensitivity for HBV DNA, identifying as little as 0.4 fg viral DNA, corresponding to about 130 genome equivalents (Sumazaki et al., 1989). The extent of HBV infection can now be estimated, when a definitive serological diagnosis (anti-HBc positivity) is impossible, and when even direct hybridization assays fail to detect HBV DNA.

An even finer PCR system for HBV DNA has been developed using an **AAF**-labeled DNA probe (Larzul et al., 1989). In an earlier version of the assay, dot-blot hybridization with this probe on the PCR-amplified products had a detection limit of 3 to 30 particles. Comparison of this nonradioactive assay with the classical radioactive test without enzymatic amplification revealed an acceptable correlation between these two techniques. **Southern analysis** of the amplified products allowed the resolution of a single HBV DNA. **Thermal treatment** of the sample for 5 min at 115°C to disrupt purified particles enhanced the detection so that a single viral particle could be identified. When applied to **crude serum** the detection limit was approximately 2×10^5 viral particles after 40 amplification cycles. The decrease in sensitivity of the PCR on crude serum samples was probably due to either **entrapment** of viral particles in the clot formed during heating, or the presence of **inhibitors** of the PCR in serum. The **need for very high stringency** conditions to minimize false-positive results is emphasized. In the present form the assay is not perfectly reliable and should be used as a good **qualitative**, but not quantitative, test (Larzul et al., 1989). Interestingly, the removal of nucleases and PCR inhibitors present in sera can be accomplished by the phenol/chloroform extraction of sera (Nicholson et al., 1992). Treatment with **heparinase** may also substantially enhance the efficiency of PCR detection as shown in HIV-1 evaluation (Imai et al., 1992).

Because rigorous purification of viral DNA is not required for PCR amplification, a release of HBV DNA from viral particles appears to suffice. Kaneko and coworkers (1989a,b) used a somewhat different approach from that employed by Larzul and coworkers (1989). They incubated serum with 0.1 M **NaOH** for 1 h at 37°C. The HBV DNA sequences amplified by the PCR for 30 cycles are then detected by 2% (w/w) **agarose gel electrophoresis** and **ethidium bromide** staining. The detection limit of that method is 10^{-5} pg HBV DNA, it can be completed in 1 day, and it does not require the use of radiolabeled reagents. A marked increase in sensitivity of the assay was attained by a **second round of amplification** of the products of the first PCR cycle. **Combining PCR with Southern analysis** of the amplified material using radiolabeled probes allows a 10^4-fold enhancement in sensitivity over that of a slot-blot assay of unamplified sample (Kaneko et al., 1989b). As few as three HBV DNA molecules can be identified.

A systemic study of **primers for optimal PCR** amplification of HBV DNA concluded that (1) primers specific for the single- or double-stranded region of the HBV genome are equally effective, (2) the maximal separation of the primers on the viral genome should be no more than 500 nucleotides, and (3) up to 14% mismatch can be tolerated with the retention of effective amplification of HBV DNA (Kaneko and Miller, 1990). PCR protocols have been described for typing HBV genomes without (Norder et al., 1990) and with REA of the amplified products (Shih et al., 1991). Effective PCR amplification has been demonstrated on DNA isolated from **formalin-fixed, paraffin-embedded** liver tissues (Lampertico et al., 1990).

To simplify the most sensitive HBV DNA detection protocol that uses PCR followed by Southern hybridization (S. Kaneko et al., 1989a,b, 1990a), a **microtiter sandwich hybridization assay** has been developed for the detection of the PCR-amplified product (Keller et al., 1990). Using an **ELISA-like format** with **biotin-labeled probes** a sensitivity of five copies of HBV DNA, equal to that of Southern hybridization, was obtained with sandwich hybridization. The specificity of that assay was 95% compared to Southern analysis (1 discordant result out of 18). The sandwich assay can have the additional advantage of yielding **quantitative** estimates of the HBV DNA amplified product.

An **automated PCR** procedure for HBV detection, using only **two thermal steps** and employing three primer pairs, performed efficiently under highly stringent conditions and detected a **single HBV DNA molecule** with the **X region-specific primers** (Larzul et al., 1990). The X antigen is a protein of 154 amino acids that is expressed during HBV infection and whose main function is transcriptional transactivation (for a review see Rossner, 1992). Using primers for the **S** and **C** regions of the HBV genome, HBV DNA has been detected by PCR in sera from patients found negative for HBV DNA by conventional hybridization technique (Chemin et al., 1991). It appears that HBV DNA detected by PCR is associated with the presence of pre-S1 envelope proteins and reflects the presence of complete virions. Evaluation of symptomatic and asymptomatic HbsAg carriers indicates that the clearance of infectious particles occurs at a slower rate than assumed previously (Chemin et al., 1991).

The presence of HBV DNA has been shown by the PCR not only in serum but also in the **colostral whey** and **colostral cells** of mothers known to be HBsAg carriers (Mitsuda et al., 1989). Southern blot hybridization of PCR-amplified product allowed a sensitivity of 3.6×10^{-15} g and better to be achieved. PCR analysis of **polyadenylated RNA** in HBV-infected **PBMCs** definitively establishes the occurrence of transcription of viral DNA in peripheral blood cells (Baginski et al., 1991). The presence of HBV DNA in serum detectable by PCR appears to be a good marker of viremia that correlates with the presence of HBsAg in serum and aminotransferase levels (Baker et al., 1991).

6.2.3 HEPATITIS C VIRUS

6.2.3.1 IMMUNOLOGICAL TESTS

The etiological agent of **non-A, non-B (NANB) hepatitis**, first proposed as a cause of posttransfusion hepatitis in 1975 (Feinstone et al., 1975), has been identified and referred to as **hepatitis C virus (HCV)** (Choo et al., 1989). The triumphant success of the Chiron Corporation and CDC scientists resulted in the identification of one **clone (5-1-1)** reactive with chimpanzee serum known to be positive for NANB hepatitis antibody. The viral nucleic acid was shown to be a **single-stranded RNA** molecule up to 10,000 nucleotides long and apparently enclosed in a lipid envelope. These characteristics suggested that HCV is similar to **togaviruses** and **flaviviruses**. HCV is considered an unusual virus that is most related to the pestiviruses (Choo et al., 1991). The HCV genome has been thoroughly characterized and significant genetic diversity, particularly within the putative 5' structural gene region of different HCV isolates, has been described (Choo et al., 1991; Houghton et al., 1991; Nobuyuki Kato et al., 1990a,b; Okamato, H. et al., 1991).

On the basis of nucleotide sequence variation HCV has been classified into four types (**PT, K1, F2a** and **K2b**) (Enomoto et al., 1990; Nakao et al., 1991). The prevalence of each HCV type has been evaluated by RT-PCR amplification of the HCV RNA genome encoding a portion of the NS5 domain followed by genotyping of the amplified product with type-specific probes or by restriction analysis (Takada et al., 1992). It turns out that HCV-PT type is found in 70% of cases in the U.S., in 42% in Europe, and 36% in Brazil, whereas HCV-K2 is predominantly found in the Oriental countries (Japan, 78% K1 and 17% K2a; China, 44% K1 and 50% K2a). There is an almost equal distribution of the PT and K1 HCV types in Europe and Brazil. Thus, except for the U.S., the prevalence of HCV K1 is high everywhere (Takada et al., 1992).

On the basis of variation in the nucleotide sequences within a region of the C (**core**) gene of HCV four groups of HCV, designated I through IV, have been identified (Okamoto et al., 1992). Using specific primers for a C gene sequence and a mixture of four type-specific primers the typing of HCV can be accomplished. With respect to the typing developed by Enomoto and coworkers (1990), type I corresponds to PT, type II to K1, type III to K2a, and type IV to K2b. Introduction of HCV typing by PCR will significantly improve the diagnostic and epidemiologic analysis of HCV infections. The development of a testing antibody followed the production of **clone C100**, comprising the four initially isolated **clones 5-1-1, 32, 36,** and **81**. The C100 clone was also fused with the human gene encoding **superoxide dismutase** and expressed in recombinant yeast. The resultant **fusion polypeptide, C-100-3**, containing 363 viral amino acids, was used first in a **radioimmunoassay**, and then in a commercially developed **ELISA** offered by Ortho Diagnostics (Raritan, NJ; Kuo et al., 1989).

This assay has been widely tested and was able to demonstrate that about 80% of patients worldwide clinically diagnosed to have NANB hepatitis are positive for the antibodies to HCV (Alter et al., 1989; Kuo et al., 1989; van der Poel et al., 1989–1991). The overall prevalence of HCV antibodies in blood donors is very low. It has been found to be 0.1% in Germany, 0.3–0.7% in the United Kingdom, 0.7% in France, 0.9% in Italy, and 1.2–1.4% in the United States and Japan (Choo et al., 1990; Contreras and Barbara, 1989). Certain groups at increased risk for HCV, such as intravenous drug abusers, hemodialysis patients, and hemophiliacs, have a high frequency of seropositivity. Health-care workers have an occupational risk of HCV infection. Anti-HCV is more than 10 times as frequent among dentists in New York City as among controls (1.75% vs. 0.14%), similar to that among oral surgeons (9.3% vs. 0.97%) (Klein et al., 1991). Cases of vertical transmission of HCV during the perinatal period have been reported (Giovanninni et al., 1990; Reesink et al., 1990; Thaler et al., 1991) as well as horizontal transmission among family members of hemodialysis patients (e.g., Calabrese et al., 1991; Kiyosawa et al., 1991). Hemophiliacs who have received untreated blood products showed a 64% level of anti-HCV positivity in Spain (Esteban et al., 1989) and 85% in the United Kingdom (Ludlam et al., 1989). In contrast, hemophiliacs given only dry-heated factor VIII apparently are not infected (Skidmore et al., 1990). The prevalence of anti-HCV seropositivity among intravenous drug abusers is equally high, ranging in different countries from 48% in Germany (Roggendorf et al., 1989) to 81% in the United Kingdom (Mortimer et al., 1989). As will be discussed below, patients with HCV are reported to have a high frequency of anti-HCV (Colombo et al., 1989; Hasan et al., 1990; Saito et al., 1990).

6.2.3.2 LIMITATIONS OF SEROLOGICAL HCV TESTING

Since the widespread introduction of **immunological assessment** of HCV infection into clinical practice, a number of limitations have become apparent. In the first place, as with any test relying on the immunological response of the host to infection, there is a "window" of nonreactivity. In the case of HCV, antibody production may take on average 15 weeks (range, 4 to 32 weeks) after onset of hepatitis, or even up to 1 year, and may remain absent in up to 20% of patients (Alter et al., 1989; Kuo et al., 1989). Thus, it is prudent to refrain, based on that assay, from excluding HCV infection in antibody-negative patients up to 6 months after the onset of symptoms. The **HCV antibody assay is apparently not sensitive enough**, because **PCR** amplification confirms the presence of the virus in anti-HCV-negative and anti-HCV-positive patients (Naoya Kato et al., 1990; Ulrich et al., 1990; Weiner et al., 1987, 1990a).

Another limitation of the **first-generation** anti-HCV test is that it measures **IgG antibody** (Martin, 1990). The dynamics of appearance and decline in IgG antibody formation, particularly in the resolution phase of infection, may therefore affect the test results. A third limitation is the concern about the **specificity** of the initially introduced immunological tests. Apparently false-positive results were observed in patients with active autoimmune chronic liver disease (Esteban et al., 1989; Gray et al., 1990; McFarlane et al., 1990) or the so-called type 2 autoimmune hepatitis (Lenzi et al., 1990). It was suggested that the source of nonspecificity in the first-generation anti-HCV test was the presence of the superoxide dismutase gene product used in the Ortho ELISA antigen (Ikeda et al., 1990). A substantial **increase in the specificity** of the original ELISA procedure, which eliminates false-positive results in patients with chronic liver disease and rheumatoid arthritis, can be achieved by an **8 *M* urea wash** (Goeser et al., 1990; Gray et al., 1990). This procedure dissociates low-affinity antibodies from the antigen or removes nonspecific binding to the plastic well, thereby increasing specificity without altering the sensitivity.

The **next generation of immunological anti-HCV test** was based on the **recombinant immunoblot assay (RIBA)** developed by Chiron and Ortho Diagnostics (Skidmore, 1990). The test appears to be more specific and strongly predictive of infectivity (Ebeling et al., 1990). The RIBA assay is designed against the two antigens 5-1-1 and C100, and when it is used to assess infectivity the reactivity even to one of the antigens correlates with infectivity (Bellobuono et al., 1990). Superiority of the RIBA assay over the Ortho ELISA was demonstrated by the single positive result by RIBA compared to the 33 Ortho ELISA reactive specimens (van der Poel et al., 1990). Another assay, the **peptide test,** developed from synthetic peptides derived from immunodominant regions of both capsid and nonstructural proteins, allows earlier detection (by 4 to 10 weeks) of HCV antibody (Hosein et al., 1991). Newer generations of RIBA assays — **RIBA-2**, and then **4-RIBA** (using four antigens) — have been developed and appraised (e.g., Bassetti et al., 1991; Ebeling et al., 1991; Leon et al., 1991; van der Poel et al., 1991). It is suggested to use PCR testing when RIBA-2 assays yield indeterminate results, although in general there is a good correlation between the two (Follett et al., 1991). On the other hand, a substantial number of PCR-negative samples react with only Ortho RIBA C22, suggestive of the existence of divergent HCV variants and the need for exclusion of such "indeterminate" cases from blood donations (Chan et al., 1991).

6.2.3.3 IMMUNOLOGICAL TESTING OF BLOOD PRODUCTS FOR HEPATITIS C VIRUS

Some debate has been generated by the differing positions of the **European community** and that proposed by FDA in the U.S. toward screening of sources of plasma for HCV. The **French** workers argue for exclusion of antibody-positive donors from contributing to the plasma pools used in the preparation of blood products such as albumin (Habibi and Garretta, 1990), despite the use of apparently effective procedures for inactivation of HCV. The risk of presence of HCV and other viruses in the pooled plasma appears to outweigh the tenuous benefits of the neutralizing potential of anti-HCV, the coagulation factors, and immunoglobulins. The **FDA** proposed, however, to conduct a scientific experiment, arguing that exclusion of anti-HCV-positive donors may substantially lower the amount of antibody (Finlayson and Tankersley, 1990). Support for the latter position has been voiced (Thomas, 1990) based on the overwhelming evidence that the current techniques of preparing immunoglobulins ensure adequate safety of the blood-derived products. It is apparent, however, that the current inactivation procedures may be inadequate for total elimination of HCV, as shown by transmission of the virus that sometimes occurs with use of factor VIII concentrate purified by monoclonal antibody adsorption followed by exposure to 60°C for 30 h (Berntorp et al., 1990). Also, concern has been voiced about the validity of FDA experimental findings in chimpanzees (Cash, 1990), because products failing to transmit NANB hepatitis to chimpanzees proved ineffective in humans (Colombo et al., 1985). Organic solvent/detergent treatment of blood preparations is being widely introduced to reduce or eliminate the HIV, HBV, and HCV from these products.

Up to 52% of **intravenous immunoglobulin** (IVIG) preparation lots have been found to be positive for anti-HCV, as tested by the Ortho Diagnostics ELISA (Quinti et al., 1990). In fact, depending on the brand and the country of origin of the donor plasma, the prevalence of anti-HCV antibody varied from 0 to 100% of the lots. The development of chronic hepatitis has been documented in several patients who had received IVIG preparations with high titers of anti-HCV antibodies (Quinti et al., 1990). The prevalence of anti-HCV positivity of plasma pools from United Kingdom sources has been reported to be 0.4%, whereas that from United States sources reached up to 100%, apparently due to the use of paid donors (Minor et al., 1990).

A study of NANB transfusion-associated hepatitis in Spain revealed anti-HCV positivity of donor specimens to be 1.5% (Esteban et al., 1990). The seroconversion rate reached 88% in recipients given anti-HCV-positive blood when followed up to 12 months following blood transfusion. The overall occurrence of transfusion-associated hepatitis was 9.6% in 280 transfusion recipients. Using PCR assays, HCV RNA has been detected in all batches of **commercially** available factor VIII specimens, and even at low concentrations in some pools of plasma donations from volunteers (Simmonds et al., 1990). Interestingly, **factor VIII concentrate produced from plasma donations by volunteers was uniformly negative for HCV RNA by PCR analysis** (Simmonds et al., 1990). A 5-year follow-up of hemophiliacs receiving preparations that were dry heat treated or exposed to wet heat or solvent-detergent treatments revealed that virucidal treatments of concentrates are effective in preventing HCV transmission (Pistello et al., 1991).

6.2.3.4 IMMUNOLOGICAL HEPATITIS C VIRUS TESTING OF LOW-PREVALENCE GROUPS

Anti-HCV positivity was detected in the relatives of anti-HCV-positive patients enrolled in an interferon trial study, whereas none of the relatives of anti-HCV-negative patients showed the presence of the antibody (Ideo et al., 1990). The relatively increased prevalence of NANB hepatitis among health-care workers has long been associated with **occupational transmission**; health-care workers constitute about 5% of U.S. patients with NANB hepatitis (Alter and Sampliner, 1989). In fact, a documented case of needlestick HCV seroconversion has been reported in a surgeon (Vaglia et al., 1990) and a nurse (Cariani et al., 1991). Caution should also be exercised in dental practice, because HCV RNA has been detected in the saliva, although at lower levels than in the blood, of HCV carriers (Takamatsu et al., 1990; Wang et al., 1991) (see also Section 6.2.3.1).

While screening a population with a presumably relatively low prevalence of HCV, a surprising **discrepancy between assay results obtained by the test kits of three different manufacturers** (Ortho Diagnostics, Abbott and Chiron) was found (Wong et al., 1990). All three tests use target antigen from the same region of the HCV genome (Wong et al., 1990). The unexpectedly high prevalence of anti-HCV positivity (77%) found among members of the Kwaio tribe on Malaita in the Solomon Islands is comparable to that of the high-risk groups, such as intravenous drug abusers, hemophiliacs, or patients with hepatocellular carcinoma. The apparent false positivity of these findings has not been satisfactorily explained, and the use of caution was advocated in interpreting the results of the Ortho Diagnostics ELISA test. The prevalence of anti-HCV in recipients of blood products before transfusion is comparable to that found in the blood products (less than 1%) (1/383 vs. 6/5150) (van der Poel et al., 1990).

6.2.3.5 IMMUNOLOGICAL HCV TESTING IN LIVER DISEASE

As mentioned above, anti-HCV testing in autoimmune CAH by the Ortho ELISA test showed a **high degree of false positivity** (65%) (McFarlane et al., 1990). None of the patients with primary biliary cirrhosis tested positive. Histological changes in the liver of hemophiliac patients characteristic of posttransfusion hepatitis are associated invariably with anti-HCV positivity of the patients' sera (Makris et al., 1990). On the other hand, the absence of histological features consistent with hepatitis does not exclude anti-HCV positivity, as almost 50% of such patients had anti-HCV antibodies. Using a test developed against antigen expressed by cloned HCV cDNA, the anti-HCV antibody was detected in 88.2% of posttransfusion hepatitis patients who had multiple elevations of ALT (Miyamura et al., 1990). Once detected, the anti-HCV positivity persisted for years.

Retrospective immunological analysis of patients with HCV who were HBsAg negative and had no evidence of alcoholic liver disease, primary biliary cirrhosis, autoimmune hepatitis, hemochromatosis, or α_1-antitrypsin deficiency revealed anti-HCV positivity in 34% of cases (Hasan et al., 1990). These patients were negative for HCV core antibody, but all the patients had underlying cirrhosis.

It appears that **HCV is a common causative agent of HCC**, although immunological detection fails to identify the virus in a large number of HCC patients. A similar study in 253 HCC patients in Japan revealed a high prevalence (68.6%) of HCV antibody in patients, who had no evidence of a previous or current HBV infection (I. Saito et al., 1990). The strong association of HCV infection with the development of HCC in patients negative for HBV infection suggests a significant role of HCV as a causative factor. Moreover, modes of **transmission of HCV other than blood transfusion** are suggested, because only about 30% of the HCV antibody-positive HCC patients had a history of blood transfusions. Furthermore, a high prevalence of HCV antibody positivity was found in patients in whom HCC had been ascribed to very high alcohol abuse. The prevalence of HCV infection in sexually promiscuous groups (female prostitutes, 8.97%; clients of prostitutes, 16.36%; homosexual men, 5.48%) is markedly higher than in voluntary blood donors (0.48%), suggesting that HCV can be transmitted by sexual intercourse (Sanchez-Quijano et al., 1990). The association of anti-HCV with HCC is found to be much more pronounced in persons positive for HBsAg than for HBsAg-negative individuals suggesting an infeaction between HBV and HCV in the pathogenesis of HCC (Kaklamani et al., 1991).

A novel UBI anti-HCV assay (United Biomedical Lake Success, NY) based entirely on synthetic HCV peptides from both the structural and nonstructural regions of the HCV genome differs from the assay system based on recombinant antigens (Hosein et al., 1991). When compared with Ortho HCV ELISA on British and Italian populations, a certain discrepancy in the results arose suggesting that seropositivity by the Ortho HCV ELISA in U.K. patients could represent a false-positive result (Lenzi et al., 1991). It appears that genuine geographic and genetic differences exist in autoimmune liver disease among different populations. The importance of definitive diagnosis of HCV infection is critical for an appropriate choice between corticosteroid or interferon therapies. Because difficulties in interpretation of anti-HCV seropositivity arise not only from the test used but may vary with the country of origin and the treatment efficiency may be different in different populations, all patients with CAH should be treated with corticosteroids and the alpha interferon treatment reserved for those who remain anti-HCV positive despite therapy (Lenzi et al., 1991; Omata et al., 1991). For a comparative analysis of the first and second-generation HCV assays see Aach and coworkers (1991).

6.2.3.6 HEPATITIS C VIRUS RNA ASSAYS

Reverse transcription of HCV RNA followed by **PCR amplification (RT-PCR)** of the derived cDNA has been used to compare the viral isolates obtained from a chronically infected chimpanzee (used in the original development of HCV tests) and a Japanese blood donor (Kubo et al., 1989). The human cDNA displayed 79.8% homology at the nucleotide level. In spite of this relatively low homology with the prototype HCV cDNA sequences, the amino acid sequences are highly conserved (92.2% homology). This accounts for the ability of the recombinant C100-3 antibody (RIBA) assay derived from the prototype HCV cDNA to detect most cases of posttransfusion NANB hepatitis in the U.S., Japan, and other countries. Significantly greater variation is observed between the core and envelope domains at the polypeptide level (97 vs. 75%, respectively), whereas the 5′-untranslated and putative nucleocapsid (core) protein regions are highly conserved (99 and 91% nucleotide identities, respectively)(Takeuchi et al., 1990). Molecular cloning of the HCV genome from Japanese patients (HCV-J) with NANB hepatitis revealed that HCV-J has a 22.6% difference in nucleotide sequence, and a 15.1% difference in amino acid sequence, compared to the American isolate (Nobuyuki Kato et al., 1990a,b) (see also Section 6.2.3.1).

A **direct assay** for HCV virions based on the identification of HCV RNA is clearly desirable to eliminate false-positive results due to nonspecificity, false-negative results arising from immunologically low responses to infection or low titer of the antibody as well as due to silent or latent infection. **PCR amplification** offers an approach to **direct HCV RNA identification** as demonstrated by the earlier assays (S. Kaneko et al., 1990b; Weiner et al., 1990a,b). Amplification of the viral sequences is the preferred method of viral nucleic acid identification because, due to the low level of virion particles in the tested samples, the **hybridization signal is not sensitive enough** to detect HCV RNA. Besides, HCV RNA nucleotide sequences show **significant variability** among different isolates, a known property of single-stranded RNA viruses, and thus some HCV genomes fail to be detected.

Improvements in the **selection of primer sequences** have been continuously made by different researchers. The detection rate of a newer assay using **"nested" PCR** with primers for sequences of the newly derived clones (S. Kaneko et al., 1990b), was significantly increased compared to that of the original assay described by Chiron Corporation. However, when the primers described by Weiner and coworkers (1990a,b) and S. Kaneko and coworkers (1990b) were used on 16 samples in Britain, they did not perform as well as a new combination of primers (Clewley, 1990). One of the possible explanations for the observed differences was **primer self-annealing** and **extension**. It is proposed that in addition to the use of **nested PCR** to increase sensitivity in the detection of HCV RNA, a **sequence-specific hybridization** appears to be essential to ensure the bands of amplified material visualized by **ethidium bromide** are true positives, and not **primer artifacts** (Clewley, 1990).

Other RT-PCR protocols with different primer sequences for the detection of HCV RNA have been described (Garson et al., 1990a). This assay also used the nested PCR amplification approach, first using sets of "outer" and then "inner" primers. Although the system is still considered preliminary, the superior diagnostic potential of the PCR-based approach was fully demonstrated in establishing the carrier status of some blood donors over a long period of time. The same assay system has also been applied to the detection of **HCV in clotting factor concentrates** (Garson et al., 1990b). **All the unheated concentrates except one were found to contain HCV RNA**. No HCV sequences were identified in any of the "superheated" (80°C, 72 h) batches, whereas "less heated" concentrates (60°C, 32 h) did have the viral sequences.

In order to improve the PCR assay, which may have reduced sensitivity due to HCV RNA sequence variations, a **new set of primers** (outer primers NCR1 and NCR2, and inner primers NCR3 and NCR4) has been designed and synthesized (Garson et al., 1990c). The latest set of primers was based on the **highly conserved** (>99%) 5'-noncoding region. When tested on serum samples from 20 hemophiliacs treated repeatedly with unheated commercial factor VIII concentrate, the original set of primers detected HCV DNA in 35% of samples, whereas the new set detected the viral sequences in 85%.

Using the originally described set of primers described by Weiner and colleagues (1990a), the **existence of symptom-free HCV carriers has been demonstrated in the absence of detectable anti-HCV** (Zanetti et al., 1990). HCV RNA has also been detected by PCR amplification in the saliva of patients with posttransfusion HCV infection (Wang et al., 1991). Current cDNA/PCR assays for HCV RNA are **likely to underestimate** the true incidence of the virus, because they may not be as sensitive as PCR assays in detecting DNA genomes (Weiner et al., 1990b). Furthermore, the methods for specimen retrieval, storage, and preparation for the assay may affect the number of viral genomes available for cDNA analysis (Weiner et al., 1990b). Nevertheless, the **PCR assay for HCV RNA by far exceeds the detection capabilities of immunological techniques** and shows that the majority of anti-HCV-positive patients **are carriers of HCV** (Naoya Kato et al., 1990). Improvements in the primer selection for PCR HCV assays are being continuously reported (e.g., Garson et al., 1991).

Accurate and precise identification of HCV RNA will certainly continue to improve, because unequivocal diagnosis of its presence is crucial in eliminating the risk of transfusion-related

HCV hepatitis as well as in the selection of appropriate treatment of patients with hepatitis. Failure to identify the virus may lead to treatments that have been demonstrated to be contraindicated in patients with HCV hepatitis (Czaja et al., 1990, 1991; Martin, 1990). Thus, it has been demonstrated that the liver of a patient with chronic HCV infection, in fact, deteriorates as a result of corticosteroid treatment (Davis et al., 1989; Di Bisceglie et al., 1989).

Likewise, administration of **recombinant α-interferon** to patients with hepatic disease with non-HCV related etiology leads to their deterioration (Vento et al., 1989). On the contrary, recombinant α-interferon treatment for HCV infection leads to significant improvements in the patient's condition (e.g., Di Bisceglie et al., 1990; Ferenci et al., 1990; Gomez-Rubio et al., 1990; Saracco et al., 1990; Weiland et al., 1990). These observations underscore the importance of accurate diagnosis of HCV, which is currently available only by PCR-based assays.

6.2.4 HEPATITIS DELTA VIRUS

6.2.4.1 THE BIOLOGY OF HEPATITIS DELTA VIRUS

The **delta hepatitis antigen** was first described by Rizzetto and coworkers in 1977 in HBV carriers (Rizzetto et al., 1977). The infection is endemic in the Mediterranean area (Rizzetto et al., 1980a,b) and appears to play a role in the development of HBV-associated disease in other parts of the world, e.g., within the African continent (Roingeard et al., 1992). Interestingly, HDV infection is uncommon in Southeast Asia and China where HBsAg is highly prevalent (10–20%).

Epidemiology of the HDV infection has been well studied and shows marked variation its prevalence in different parts of the world. It is rare in the Far East, but extremely frequent in the Middle East, Brazil, parts of Africa, and in the South Pacific islands, where the percentage of HDV infection in carriers of HBsAg with liver disease ranges from 30 to 90%. In Europe and the U.S., the infection is widespread, predominantly among high-risk groups such as **intravenous drug abusers** and **hemophiliacs**. The route of transmission is similar to that of HBV (Fay et al., 1987; Jilg et al., 1985; Lettau et al., 1987; Novick et al., 1988; Ponzetto et al., 1985; Saracco et al., 1987; Tamura et al., 1987).

Hepatitis delta virus (HDV) is a defective RNA virus displaying hepatotropism and depending exclusively on HBV. It appears that assembly of HDV requires the coat proteins of HBV or other related hepadnaviruses (Ryu et al., 1992). In contrast to HBV, HDV is invariably pathogenic and produces the primary or complicating forms of severe HBsAg-positive acute and chronic liver disorders (Purcell et al., 1984; Rizzetto et al., 1984).

HDV RNA is circular, single stranded, and composed of 1678 nucleotides; its structure, sequence, and expression have been extensively characterized (Kuo et al., 1988; Saldanha et al., 1989; Wang et al., 1986). Hybridization studies with HBV DNA and HCV RNA revealed a lack of sequence homology between HDV RNA and those viruses (Weiner et al., 1987). Many **open reading frames (ORFs)** were found in the HDV genome, and the one responsible for the synthesis of a polypeptide recognized by the sera of HDV-infected individuals has been identified as a candidate for the sequence encoding an HDV antigen. HDV proteins in the serum appear to be the same size as those in the liver (Roggendorf et al., 1987). These proteins can be recognized by a human monoclonal antibody (Pohl et al., 1987). Significant variation in HDV RNA sequences among different strains apparently accounts for differences in the severity of HDV symptomatology (Chao et al., 1990, 1991).

The virus produces an **antigen, HDAg**, and antibody to this antigen (**anti-HD**) can be detected in infected persons. Infection produced by HDV appears to occur only when the HBV load becomes infectious. Although superinfection by HDV is rare in acute HBV cases, it is more prevalent in chronic HBsAg carriers and acquires a critical epidemiological significance (Ponzetto et al., 1985). Initially it appeared that HDV interfered with the replication of HBV (Govindarajan and Valinluck, 1985). Later studies, however, found no mutual inhibition between HDV and HBV replication as deduced from the assessment of HDV replication markers and their relationship to HBV replication in chronic carriers of HBsAg with superinfection of HDV (Bas et al., 1988).

It appears that **superinfection** with HDV results in a fulminant form of hepatitis (Govindarajan et al., 1985; Lettau et al., 1987; Saracco et al., 1987). The load of HDV is related to the incubation time, rather than to the severity and duration of hepatitis (Ponzetto et al., 1987). It has been suggested that superinfection by HDV may interfere with the progression of HBV infection and even result in clearance of HBsAg in HBV carriers (Ichimura et al., 1988; Govindarajan and Valinluck, 1985). However, high-level HBV replication can continue for several years in coinfected persons (Bas et al., 1988; Buti et al., 1988; Farci et al., 1988; Saldanha et al., 1989).

HDV superinfection appears to induce the clearance of HBV-associated antigens as demonstrated in 1029 HBV carriers in Japan by the elimination rate of HBeAg and HBsAg in HDV-superinfected carriers (Ichimura et al., 1988). Although HDV causes an acute self-limited infection in Japan with predominantly heterosexual transmission (Ichimura et al., 1988), this virus causes chronic liver disease according to other authors (Negro et al., 1988). These workers compared the expression of HDAg with the morphological characteristics of HDV infection and the outcome of liver disease as studied in 101 patients followed for an average of 12 years. The highest degree of **tissue damage** was found before the elimination of the virus. The **immune response** to HDV replication appears to have a major role in determining the outcome of the infection. It is the response of the host to HDV infection that contributes to the different effects of HDV coinfection and superinfection of HBV replication as determined by serological markers and HBV DNA levels (Farci et al., 1988).

Histological changes in the livers of patients infected with HDV seem to **correlate with the HDV load** (Negro et al., 1988). The highest degree of inflammation and intrahepatic expression of HDAg occurs before the elimination of the virus. The outcome of HDV disease, however, is not directly related to the severity of the initial histological changes and suggests that the **immune response of an individual** plays an important part in the pathogenesis of HDV hepatitis.

6.2.4.2 IMMUNOLOGICAL METHODS

Initially, **total delta antibody evaluation** was performed by a **solid-phase blocking radioimmunoassay (RIA)** as described by Rizzetto (Rizzetto et al., 1980b). Subsequently **commercially available kits** for blocking RIA assay of anti-HD became available (Abbott Laboratories, North Chicago, IL; Deltassay, Noctoch, Dublin, Ireland).

6.2.4.3 HEPATITIS DELTA VIRUS RNA DETERMINATION

A number of **hybridization protocols** have been developed over the years by various groups for the identification of HDV RNA in the serum and liver tissue (Buti et al., 1988; Denniston et al., 1986; Farci et al., 1988; Saldanha et al., 1989; Weiner et al., 1987). A detailed study of HDV RNA in acute delta infection has been reported by Buti and coworkers (1988) suggesting the possibility that evaluation of HDV RNA in serum may be useful in clinical situations following infection, avoiding the necessity to perform liver biopsy for this purpose. A **spot hybridization** technique can be used for detecting HDV RNA in serum using a cDNA of HDV RNA as a probe.

Another fine methodological study of HDV infection using nucleic acid probes was reported by Saldanha and coworkers (1989), who described in detail a **slot-blot assay** for the evaluation of serum HDV RNA. Cloning of HDV RNA has been successful (e.g., Kuo et al., 1988), suggesting that a widely available diagnostic probe can be developed for wide clinical use. By using **Northern blot hybridization** with a **riboprobe** derived from a cDNA fragment, HDV RNA can be demonstrated in the patient's serum in 92% of those who had HDAg in their liver biopsy specimens (Smedile et al., 1986).

When the sera of patients positive for HDAg in their liver tissue were tested only 76% were positive for HDV RNA, as indicated by the ^{32}P-labeled RNA of the SP6 Riboprobe system, which appears to be one of the most sensitive and specific probes for detecting small amounts of RNA (Rasshofer et al., 1988). A **riboprobe** derived from a cDNA fragment of a 650-bp HDV RNA sequence has been developed for **spot hybridization** (Rasshofer et al., 1988). This approach offers greater ease in performing large batches of tests than the earlier-described **Northern blot** hybridization assay using a cDNA of HDV RNA (Smedile et al., 1986). The spot hybridization procedure was subsequently confirmed to offer a rapid and sensitive approach to the detection of HDV RNA that correlated well with the results of Northern blotting (Gupta et al., 1989). Substantial **enhancement in the sensitivity** of HDV RNA detection offered by the DNA hybridization assay is achieved in a riboprobe assay (83 vs. 63%) (Smedile et al., 1990). It also offers significant savings in time compared to the Northern blot using a cDNA probe. When applied to the detection of **HDV viremia**, the riboprobe method proved to be the most sensitive of the hybridization-based methods described so far (Smedile et al., 1990).

A novel **ISH** method applicable to the detection of HDV RNA in **formalin-fixed, paraffin-embedded** liver biopsies complements the earlier ISH assays adapted for **frozen sections** (Negro et al., 1989). A **RT/PCR**-based method amplifies a region of about 1200 bp that includes the **entire HDAg ORF** (Zignego et al., 1990). The product of amplification corresponds to the expected size of the gene and shows up to 98.6% sequence identity with previously published HDV sequences. This method allows detection of HDV RNA in 10-pl samples of serum or in less than 0.1 pg of liver RNA. Comparing PCR with slot-blot hybridization on serial dilutions of known concentrations of HDV RNA, the PCR proved to be 10,000 times more sensitive (Madejon et al., 1990).

6.2.5 HEPATITIS A VIRUS

ssRNA probes have been constructed with radioactive (^{32}P) and nonradioactive (biotin) labels for **dot-blot** hybridization to detect hepatitis A virus (HAV) (Jiang et al., 1987). Using **differential hybridization** with probes for positive and negative sense strands (compared with HAV genomic RNA) applied to clinical and field samples, a positive to negative ratio was used to develop a **semiquantitative expression** of HAV presence. The same ssRNA probes can be used for **ISH** evaluation of infected cells (Jiang et al., 1989). **Genomic fingerprinting** of HAV isolates has been reported and revealed only a minor (9%) variability of the viral genome (Fiore et al., 1990).

6.2.6 OTHER GASTROINTESTINAL PATHOGENS

6.2.6.1 *ESCHERICHIA COLI*

Escherichia coli strains causing gastrointestinal disease have been classified into four major groups depending on the predominant pathogenetic route by which these bacteria cause diarrhea: **enterotoxigenic (ETEC), enteroinvasive (EIEC), enteropathogenic (EPEC),** and **enterohemorrhagic (EHEC)** strains (Levine, 1987). Nucleic acid probes have long since been developed for identification of these organisms in **pure culture** and subsequently **directly in clinical specimens** (for a detailed review of earlier studies, see Echeverria et al., 1990). Comprehensive coverage of the variety of probes constructed for the detection of plasmid genes encoding enterotoxin (**heat labile, *LT*; heat stable, *ST***), genes encoding invasiveness factors and adherence, and chromosomally encoded genes for **Shiga-like enterotoxin (*SLT*)** is outside the scope of this discussion. Only selected studies primarily aimed at making gene level diagnostics a practical alternative for clinical laboratories and representing the latest developments in this field, will be cited here.

Although the earlier probes for the **pInv** (**plasmid encoding invasion factors**) were more effective than serotyping techniques (Bohnert et al., 1988; Gomes et al., 1987) they were more effective in colony hybridization assays than in direct stool blots (Taylor et al., 1988). Nevertheless, **DNA hybridization using synthetic oligonucleotides**, particularly with respect to analysis of ST-ETEC strains, is the method of choice, in spite of their somewhat lower sensitivity (78.8%) compared to **ELISA** and the **infant mouse assay** (Cravioto et al., 1988). Further improvement came from the construction of a **single hybrid RNA probe** containing sequences complementary to the *ST* and *LT* alleles (Saez-Llorens et al., 1989). The constructs, labeled either radioactively or with biotin, showed a 100% correlation with the biological assay for both the *LT* and *ST* toxins. Strains of *E. coli* producing SLT or verotoxin (VTEC) often include representatives of both EPEC and EHEC. VTEC has been implicated in outbreaks of **hemorrhagic colitis**, **hemolytic uremic syndrome**, and **thrombotic thrombocytopenic purpura** (Karmali, 1989). The earlier probes to *SLT*-related genes (e.g., Brazil et al., 1988; Newland and Neill, 1988; Venkatesan et al., 1988) have been supplemented with PCR-based assays (Pollard et al., 1990b). **Oligonucleotide probes** for the **EPEC adherence factor** proved to be more sensitive and specific, producing clearer results on hybridization than the corresponding DNA sequences (Jerse et al., 1990).

Nonradioactively labeled probes are particularly suitable for diagnostic clinical laboratories, and a number of reporter molecules have been used to label probes for various genes of *E. coli*. As in other gene probe systems, **biotin** (e.g., Carter et al., 1987; Danbara, 1990; Gicquelais et al., 1990; Kumar et al., 1988), **alkaline phosphatase** (e.g., Nishibuchi et al., 1988), **horseradish peroxidase** (e.g., Abe et al., 1990), and **digoxigenin** (e.g., Riley and Caffrey, 1990) have been used. Of these, digoxigenin-labeled DNA fragments proved to be more sensitive and produced less **background noise** than biotinylated probes for *ETEC* genes (Riley and Caffrey, 1990). Pretreatment of colonies with **sodium dodecyl sulfate** and **proteinase K** was reported to significantly reduce nonspecific hybridization of horseradish peroxidase-labeled trivalent polynucleotide probe directed against *LTh, STIa,* and *STIb* genes of ETEC cells (Abe et al., 1990). This probe appears to be suitable for routine use in clinical laboratories for the detection of ETEC.

A PCR-based assay for LT-producing *E. coli* has been reported to amplify target genes when amplifications were done directly on *E. coli* colonies in a relatively pure culture (Victor et al., 1991). Routine diagnosis in clinical isolates was sensitive and specific enough not to require posthybridization with radioactively or alkaline phosphatase-labeled oligonucleotide probes. A comprehensive coverage of DNA probes for *E. coli* causing extraintestinal infections has been given (Williams, 1990).

6.2.6.2 *CAMPYLOBACTER* AND *HELICOBACTER* SPP.

The importance of *Campylobacter* species as major gastrointestinal pathogens has led to the early development of gene-level probes that have been thoroughly discussed elsewhere (Terpstra et al., 1990; Wetherall and Johnson, 1990). Only a few reports will be cited here. As in other bacterial identifications, ribosomal gene-specific probes have been constructed for *Campylobacter* spp. analysis (Cox et al., 1990; Moureau et al., 1989). A superior commercial detection system for the identification of (but not differentiation between) three *Campylobacter* species (*C. jejuni, C. coli,* and *C. laridis*) has been developed by Gen-Probe (San Diego, CA) and proven to offer 100% accuracy in an evaluation study of 214 *Campylobacter* isolates and 210 other bacterial isolates (Tenover et al., 1990). The test is fast (45 min), easy to perform, but positive results exclude just those three species. Therefore, conventional laboratory biochemical and microscopic evaluations should be performed on organisms negative with the probe, but suggestive of campylobacters by **morphology**, **motility**, and **oxidase positivity** (Tenover et al., 1990).

Another commercially available system using alkaline phosphatase as label (**SNAP**; Molecular BioSystems, Inc., San Diego, CA) is specific for *C. jejuni* and *C. coli*, but not for other *Campylobacter* and *Helicobacter pylori* (Olive et al., 1990). This kit showed excellent performance on isolated bacterial colony isolates with a reduced sensitivity (82.6%) on direct stool samples.

Differentiation between *C. jejuni*, causing the majority of human gastroenteritis cases, and *C. coli*, implicated in only 2 to 5% of cases, can be accomplished by the probes using **digoxigenin** label with an alkaline phosphatase detection system (Taylor and Hiratsuka, 1990). These probes, when applied to fecal specimens, had a sensitivity of 89 and 93% and a specificity of at least 84.3%. Chromosomal DNA fragments have been cloned from *C. hyointestinalis*, the predominantly porcine pathogen that can also cause infection in humans (Gebhart et al., 1989). These highly specific probes had a detection limit of 104 *Campylobacter* organisms, when applied to bacterial spot blotting. On the basis of 16S rRNA sequence data, oligonucleotide probes have been constructed that are capable of distinguishing *C. fetus* and *C. hyointestinalis* (Wesley et al., 1991). Other ribosomal genes probes have been developed for the identification and subtyping of aerotolerant *Campylobacter* species by restriction fragment length polymorphism (RFLP) analysis (Kiehlbauch et al., 1991).

The identification of *H. pylori* by DNA and RNA probe assays complements culturing, histological, serological, and biochemical tests (Clayton et al., 1989; Morotomi et al., 1989; Vanderberg et al., 1989; Wetherall et al., 1988). A **PCR** amplification assay detects *H. pylori* directly in gastric biopsy and aspirate specimens proved positive in 93% of positive tissues and 85% of aspirate specimens positive by microbiological culture and histological examination (Valentine et al., 1991). All the attractive features of PCR amplification assays using nonradioactive primers and detection systems, combined with the ability to evaluate archival material retrospectively, make this fast and specific assay an attractive alternative to conventional methods. As in other organisms, the analysis of rRNA gene patterns of *H. pylori* generated by *Hind*III digestion of the DNA from gastric biopsies of patients with duodenal ulcers allowed an epidemiological analysis to be carried out (Owen et al., 1992).

6.2.6.3 *CLOSTRIDIUM* SPECIES

Clostridium difficile is the etiological agent of **pseudomembranous colitis**, and may be in part responsible for antibiotic-associated diarrhea (e.g., Gerding et al., 1986). Epidemiological studies have successfully used **whole-cell DNA fingerprinting** (Kuijper et al., 1987) and **species-specific oligonucleotide probes** for rRNA (Wilson et al., 1988) to distinguish between various isolates and related species. The diagnosis of *C. difficile* is confirmed by the isolation of toxigenic *C. difficile* from stool or the detection of toxins A (**enterotoxin**) and/or B (**cytotoxin**). Other methods include **cell culture assay** (Walker et al., 1986), **latex agglutination** (Borriello et al., 1987; Kohno, 1986), **ELISA** (Walker et al., 1986), and **counterimmunoelectrophoresis** (Wu and Fung, 1983).

A **PCR** amplification assay has been described that used primers for a fragment of the gene encoding toxin A in *C. difficile* (Wren et al., 1990). Although highly sensitive, that assay showed some cross-reactivity with *Clostridium sordellii*. To avoid this, primers to both repetitive and unique sequences of the *C. difficile* toxin A gene have been constructed and a PCR assay with these primers proved to be highly specific for *C. difficile* (Kato et al., 1991). Although the specific PCR assay for *C. difficile* has been effective on purified clinical material, it appears that only minimal processing of stool samples may be needed for specific amplification of toxin A gene sequences (Kato et al., 1991). These findings also support previous observations that nontoxigenic *C. difficile* does not contain the entire toxin A gene. Consequently, this PCR assay may be used to reduce the labor involved in conventional cell toxicity and biochemical assays for toxigenic *C. difficile*.

C. perfrigens causes food poisoning primarily due to the toxin it releases in the intestine (**CPE**) during sporulation (Granum et al., 1984). Identification of CPE-producing strains of *C. perfrigens* is difficult and four oligonucleotide probes for the *CPE* gene sequences have been developed (Van Damme-Jongsten et al., 1989). Because *C. perfrigens* appears to produce only one enterotoxin, and *C. perfrigens* strains normally colonize intestine without causing a disease, positive hybridization testifies to the presence of toxigenic *C. perfrigens* strains (Wernars and Notermans, 1990). Using these probes for the analysis of 245 isolates from 186 separate **food poisoning** outbreaks, 59% of the cases proved to give hybridization signals with the probes for the *CPE* gene (Van Damme-Jongsten et al., 1990). Interestingly, although toxigenic strains were seen only infrequently among nontypable isolates, no correlation between serotype and the presence of the enterotoxin (*CPE*) gene could be found (Van Damme-Jongsten et al., 1990).

C. botulinum is identified by its ability to produce **neurotoxin** and studies are in progress to clone the gene encoding the toxin B and other *C. botulinum* genes for the construction of appropriate probes (Wernars and Notermans, 1990).

6.2.6.4 *YERSINIA* SPECIES

Yersinieae are inherently virulent; however, they express full virulence only when they possess the 67- to 72-kb virulence plasmid (Portnoy et al., 1981; Robins-Browne et al., 1989). *Y. enterocolitica* is the most frequently encountered enteric pathogen implicated in a variety of diseases from diarrhea to hepatosplenic abscesses and septicemia (Bottone, 1977). DNA and oligonucleotide probes have long since been constructed for the identification of *Y. enterocolitica* in foods (Hill et al., 1983; Jagow and Hill, 1986, 1988; Miliotis et al., 1989). A **PCR**-based assay has been developed for *Y. enterocolitica* (Wren and Tabaqchali, 1990).

6.2.6.5 *SALMONELLA* SPECIES

Interest in developing gene probes for the identification of *Salmonella* spp. stems from the discrepancy between the need for a clinician to have positive identification of the pathogen within hours of infection, and the ability of presently available methods to produce results no sooner than within a few days. Furthermore, epidemiological evaluation of *Salmonella* infections requires a simple, fast, and definitive identification of the organism (Rubin, 1990; Rubin et al., 1988). A **DNA probe** constructed for the **via B antigen locus** of *Citrobacter freundii* identifies bacteria that synthesize **Vi antigen**, the capsular antigen of *S. typhi* (Rubin et al., 1985). This probe has been tested in two populations in Peru and Indonesia by *in situ* **colony hybridization** (Rubin et al., 1988). Although the probe was cloned from *C. freundii*, because only rare strains of this organism are capable of synthesizing Vi antigen, no false hybridization results have been observed even with 46 strains of *C. freundii*. This **8.6-kb Vi DNA probe** proved to be highly specific and sensitive in studies using freshly isolated colonies from clinical specimens from these two geographic areas (Rubin et al., 1988).

Because **blood culture** is the standard method for diagnosis of typhoid fever, detecting 40 to 70% of *S. typhi*-infected patients (Hoffman et al., 1984, 1986), the development of a more sensitive probe for *S. typhi* in the blood would be desirable. The main problem with the use of a DNA probe on blood samples would be the low bacteremia in typhoid patients (15 *S. typhi* organisms per milliliter of blood) compared to the sensitivity of the probe (500 bacteria) (Rubin, 1990; Rubin et al., 1989). A method has been devised to **concentrate bacteria** from blood samples by lysis and centrifugation of blood with Isolator tubes followed by **growth amplification**, resulting in a detection level by the DNA probe roughly comparable to that of the culture method (Rubin et al., 1989). That probe identified 14 of 33 patients with culture-confirmed typhoid fever in a concentrated specimen equivalent to 2.5 ml of blood, whereas the culture method identified 17 of the same 33 using 8 ml of blood. Hybridization of the probe to specimens from 4 patients, from whom *S. typhi* had not been isolated, indicated that those were either false positives, or, in fact, the probe identified *S. typhi* not detectable by culture technique (Rubin et al., 1989).

A different DNA probe for *Salmonella* spp., **SAL6**, has been tested in a **rapid DNA hybridization** procedure (Scholl et al., 1990). The probe was capable of identifying all 70 *Salmonella* serotypes and was not reactive with 101 stock strains of other enteric bacteria. This radioactively labeled probe has been tested only on cultured isolates treated by a novel, so-called **wicking procedure**. It appears that amplification techniques such as **PCR** will soon be developed to detect *Salmonella* spp. directly in clinical specimens.

6.2.6.6 *LISTERIA MONOCYTOGENES*

Recovery of *Listeria monocytogenes* from food and environmental samples is a laborious procedure and nucleic acid probes can significantly simplify this process. A number of **DNA probes** have been developed targeted to **rRNA** sequences, the **β-hemolysin gene** and the **DTH (delayed type hypersensitivity) factor** (for a review, see Wernars and Notermans, 1990). The DTH probe identifies only *L. monocytogenes*, and nonpathogenic *Listeria* spp. do not interfere. As in other organisms, **DNA fingerprinting** proved valuable in epidemiological studies of **neonatal listeriosis** traceable to cross-infection confirming a case of nosocomial infection (Farber et al., 1991).

Newer probes, tagged with horseradish peroxidase, have been constructed for the detection of *L. monocytogenes* by **direct colony hybridization** on hydrophobic grid membrane filters amenable to quick automated counting (Peterkin et al., 1991). By constructing primers to the **listeriolysin O gene** the specific amplification of *L. monocytogenes* has been accomplished by **PCR**, detecting less than 50 organisms without the need for subsequent hybridization with labeled probes (Deneer and Boychuk, 1991).

6.2.6.7 LACTOCOCCI AND ENTEROCOCCI

Difficulty in applying colony hybridization of nucleic acid probes to gram-positive bacteria is due to their different **cell wall** structure (Datta et al., 1987). There is an increasing need to differentiate enterococci contaminating the starter lactococci cultures in dairy processes (Betzl et al., 1990). A set of oligomers to the **23 rRNA** sequences (naturally abundant in any cell — 10^4 copies per cell) has been constructed (Betzl et al., 1990). The probes not only effectively entered the organisms but also allowed specific identification of the cocci and their quantification.

6.2.6.8 COLIFORM BACTERIA IN THE ENVIRONMENT

Coliform bacteria are used for monitoring the bacteriological purity of **water supplies**; they are indicators of potential human fecal contamination and thus signal the presence of dangerous enteric pathogens (e.g., Bonde, 1977). A highly sensitive **PCR** amplification assay has been developed capable of detecting as little as 1 to 10 fg of genomic *E. coli* DNA and as few as one to five viable *E. coli* cells in 100 ml of water (Bej et al., 1990). PCR amplification using appropriate primers can offer not only a simple and specific tool for monitoring water quality, but can also identify the organisms without relying on indirect measures quantifying indicator organisms (Bej et al., 1990).

6.2.6.9 *GIARDIA LAMBLIA*

Giardia lamblia identification in fecal specimens by the presence of characteristic cysts is only partially (50%) successful (Kamath and Murugasu, 1974). At least two types of **DNA probes** have been reported: one is a **cloned fragment** (Butcher and Farthing, 1989), and the other is a **total genomic DNA probe** (Lewis et al., 1990) showing similar sensitivity and specificity to the cloned probe. Using trophozoite or cyst *G. lamblia* DNA, the sensitivity of the total genomic probe was found to be 10 ng of *G. lamblia* DNA, 10^5 trophosoites, and 10^4 guanine thiocyanate-treated cysts. The detection of *G. lamblia* cysts in water can be accomplished using dot-blot hybridization with a short cDNA probe from the small subunit of rRNA, following disruption of the cysts with glass beads in the presence of proteinase K. The method is capable of detecting as few as one to five cysts per milliliter of water sample concentrates (Abbaszadegan et al., 1991).

6.2.6.10 *ENTAMOEBA HISTOLYTICA*

Infection with *Entamoeba histolytica* is widespread, affecting about 500 million people worldwide but causing disease in only about one tenth of those infected (Guerrant, 1986). Identification of the pathogenic *E. histolytica* strains can help in defining variations in host responses and the biological and clinical behavior of the amoeba. Molecular biological analysis of extrachromosomal circular DNA molecules containing **rRNA** genes as well as clusters of **tandemly reiterated sequences** identified two types of sequences characterized by their restriction patterns and displaying specificity to pathogenic and nonpathogenic *E. histolytica* (Bracha et al., 1990; Garfinkel et al., 1989). In fact, **PCR** studies demonstrated that repetitive sequences characteristic of pathogenic *E. histolytica* are also present in low copy numbers in a nonpathogenic strain (Mirelman et al., 1990).

A **cDNA library** has been constructed on the basis of RNA isolated from an axenically grown strain of *E. histolytica* expressing a **pathogenic isoenzyme pattern (zymodeme)** (Burch et al., 1991). After testing a number of cDNA clones, **clone C2** was selected: it was capable of differentiating between pathogenic and nonpathogenic *E. histolytica* zymodemes and detect as few as only 100 amoebic trophosoites (Burch et al., 1991). This dot-blot format assay may be helpful in identifying pathogenic *E. histolytica*.

Yet another approach to differentiation between pathogenic and nonpathogenic *E. histolytica* is amplification of a 482-bp segment that contained identical 5′ and 3′ ends, but in which internal cleavage sites for restriction endonucleases differed between pathogenic and nonpathogenic strains (Tannich and Burchard, 1991). This was used as a basis for differentiating between the strains by **restriction analysis** of the amplified product. The **genotypic patterns** correlated closely with the clinical picture produced by the organisms in patients, confirming the notion that pathogenic and nonpathogenic strains of *E. histolytica* constitute distinct subspecies (Tannich and Burchard, 1991). Other enteropathogenic organisms identifiable with gene probes (*Vibrio* spp., *Staphylococcus aureus*, etc.) have been extensively discussed (Wernars and Notermans, 1990).

Chapter 6.3

Respiratory Pathogens

6.3.1 RESPIRATORY VIRUSES

6.3.1.1 INFLUENZA VIRUS, ADENOVIRUS, AND RESPIRATORY SYNCYTIAL VIRUS

Laboratory identification of **influenza A virus** is hampered by the need for cumbersome **cell culture** methods. This RNA virus causes significant morbidity due to its escape from immunosurveillance (Klenk and Rott, 1988). A combination of **reverse transcription** of a region of the virus genome encoding the relatively stable viral hemagglutinin and **polymerase chain reaction (PCR)** amplification of this 441-bp region allowed the detection of influenza virus H1 subtype in nasopharyngeal lavages (Bressoud et al., 1990). This approach makes cultivation of the virus on embryonated eggs or in cell culture unnecessary, whereas the construction of primers directed toward the relatively genetically stable region of the **hemagglutinin gene** markedly reduces the rate of false negativity (Bressoud et al., 1990).

A number of DNA clones as well as total adenovirus type 2 and 16 (Ad2 and Ad16) genomic DNA have been selected for the detection of adenoviruses in cells by **dot-blot** hybridization (Huang and Deibel, 1988). A commercial kit for **Ad2** (PathoGene II; Enzo Biochemical, Inc.) is also available that uses biotinylated total Ad2 DNA. Total Ad2 DNA and cloned radioactive probes displayed similar sensitivity, detecting 10 pg of adenovirus DNA. This compared with previously reported findings (Hyypia, 1985; Gomes et al., 1985; Lehtomaki et al., 1986; Virtanen et al., 1983). Elimination of cross-reactivities with other subgroups and further improvements in sensitivity are needed to apply DNA probes for respiratory adenoviruses directly to clinical material without preliminary amplification in cell culture.

Respiratory syncytial (RS) virus causes lower respiratory tract infections in infants and young children worldwide (Collins and Wertz, 1986). A relatively simple hybridization assay using **cDNA** probes specific for the subgroups A or B of RS virus detects the viral RNA in fixed, virally infected cells (Sullender el al., 1990). An **RT-PCR** with nested PCR can be used to detect RSV RNA by targeting the F glycoprotein of the virus in effusions collected from patients with otitis media (Okamoto et al., 1991).

6.3.1.2 RHINOVIRUSES

Human rhinoviruses (HRVs) are the major cause of the common cold. HRV infection can be diagnosed by **culturing** the virus, which is difficult to grow, or by an **ELISA** test for viral antigen (Dearden and Al-Nakib, 1987). The ELISA test is type specific and, because there are over 100 distinct serotypes, there is a need for a broad diagnostic assay. Using a cDNA to the 5′-noncoding region of the HRV RNA genome, which has up to 90% homology between different HRV serotypes, only homologous serotypes can be identified (Al-Nakib et al., 1986). To further narrow the region that is totally conserved among different serotypes, **synthetic oligonucleotide probes** capable of detecting HRVs in nasal wash samples have been constructed (Bruce et al., 1988). cDNA fragments have also been used for grouping HRV according to their **genetic relationships** (Auvinen et al., 1990) and for *in situ* hybridization **(ISH)** on nasal epithelial cells (Bruce et al., 1990). **cDNA probes**, however, are not adequately sensitive in detecting HRV in clinical specimens without prior preamplification of the virus in cell culture. The development of **PCR**-based methods for HRV identification (Gama et al., 1988a,b) may replace cell culture amplification of HRVs (Hyypia, 1989).

6.3.1.3 HERPES VIRUSES

6.3.1.3.1 Herpes Simplex Virus

Numerous hybridization assays have been developed for the diagnosis of herpes simplex virus (HSV) infections (e.g., Brautigam et al., 1980; Redfield et al., 1983; Stalhandske and Pettersson, 1982). Using cloned viral DNAs as probes, Schuster and coworkers (1986a) were able to detect at least 5 pg of homologous DNA. Even higher sensitivity (3 pg of DNA) was reported in their dot-blot hybridization assay, using *in vitro* synthesized radioactively labeled single-stranded RNA transcripts (Schuster et al., 1986b). These probes, however, did not type HSV.

Single-stranded, short (22 bp) **oligonucleotide probes**, homologous to unique regions of HSV types 1 and 2, have been constructed (Peterson et al., 1986). These radioactively labeled probes detected HSV in vesicle material and genital swabs with a sensitivity of 10^4 to 10^5 HSV infectious units, showing 99% agreement between hybridization and **monoclonal antibody typing**. Several reports described HSV probes for detecting the virus by **ISH** in genital lesions (Langenberg et al., 1988), eyes (Nago et al., 1988), and in cutaneous squamous cell carcinoma (SCC) (Claudy et al., 1989). All these probes used **biotin** labels. Diagnostic DNA hybridization procedures using **radioiodinated probes** have been developed for detection of acyclovir-resistant HSV isolates (Swierkosz et al., 1987).

The expression of **HSV type 1** genes during latent infections of human sensory ganglia was found (using ISH) to differ from that in an acute infection; instead of **"sense" transcripts** of this gene, **"antisense" transcripts** were produced (Croen et al., 1987). Production of antisense transcripts is believed to have a role in maintaining the latency of HSV by modulating its expression.

ISH with **dsDNA probes** labeled with **photobiotin** (Chantralita et al., 1989) or **alkaline phosphatase-labeled synthetic oligonucleotide probes** on paraffin-embedded tissue have been constructed (Bruner, 1990). **Commercially available HSV DNA probes** (Enzo Bio-chemical) demonstrate their practical utility for improving the diagnosis of HSV in clinical material (e.g., bronchoalveolar lavage), particularly when cytological diagnosis is equivocal (Crosby et al., 1991). The Enzo Biochemical **biotinylated** commercial kit for the detection of both HSV1 and 2 has been evaluated and found to have an overall sensitivity of 92% in comparison to tissue cultures (Langenberg et al., 1988). The convenience of using this probe for diagnosing HSV infection in paraffin-embedded tissue has been demonstrated (Claudy et al., 1989). HSV2 DNA could be identified in SCC by ISH with biotinylated probes and the cells were negative for HPV2, 5, 16 and 18 probes (Claudy et al., 1989). Comparison of the **biotinylated Enzo HSV DNA probe** with a **shell vial culture assay** enhanced by a **direct fluorescent (HSV-monoclonal) antibody** stain and conventional **tube cultures** has been conducted on 199 specimens (Seal et al., 1991). The **poor performance** of the DNA probe (sensitivity, 24.5%; specificity, 88.3%) argued against its routine use for HSV identification.

Among the **complicating morphological features** resolvable by HSV DNA probes are the similarity of HSV inclusions to those of cytomegalovirus (CMV) and adenovirus, dual infections with HSV and CMV, and multinucleation produced by viruses other than respiratory tract pathogens (Crosby et al., 1991). The **Ortho DNA probe** HSV confirmation kit (Ortho Diagnostic Systems, Inc., Raritan, NJ) has been evaluated for the detection of HSV in clinical specimens from various sources after inoculation of coverslips seeded with A-549 cells following centrifugation and overnight incubation (Woods and Yam, 1989). This method also was **not** considered to be a **suitable** substitute for conventional tube cell culture for detection of HSV in clinical specimens, showing a sensitivity of 82% and a specificity of 100%. *In situ* hybridization for HSV DNA with PathoGene DNA probe assay (Enzo Diagnostics, NY) can identify the virus in placenta in the absence of characteristic morphologic findings (Schwartz and Caldwell, 1991).

The extraordinary sensitivity of **PCR** amplification has proven to be of great clinical value for the diagnosis of HSV DNA in **chorioretinal inflammatory diseases** (Fox et al., 1991), for the early diagnosis of **HSV encephalitis** (Aurelius et al., 1991; Puchhammer-Stockl et al., 1990), in a case of **Mollaret's meningitis** (Yamamoto et al., 1991), as well as in **occult dermatological lesions** (Penneys et al., 1991). HSV type 1 has been reportedly detected in the brains of normal people as well as Alzheimer's disease patients (Jamieson et al., 1991). The PCR has been adapted not only to detect HSV but also to **type** it (Kimura et al., 1990).

6.3.1.3.2 Varicella-Zoster Virus

A related herpesvirus, varicella-zoster virus **(VZV)**, causes a variety of diseases in children and immunodeficient patients or patients on immunosuppressive therapy, for whom accurate and early diagnosis may be of critical importance. While traditional methods of diagnosing VZV rely on the **cytopathic phenomenology** in cell cultures and may take from 3 to 14 days, the other techniques (**immunofluorescence microscopy** and **EM**) either require highly specific sera or lack viral specificity (EM). **DNA probes** derived from the VZV genome, labeled by ^{32}P or **biotin**, proved to be sensitive in detecting VZV in infected human cells when used in a **dot-blot** hybridization assay (Cuthbertson and Grose, 1988). The sensitivity limit of the radioactive VZV DNA probe was approximately 5 to 10 pg of viral DNA, whereas that of the biotinylated probe was 10 to 15 pg of viral DNA; the latter in addition offered the results much faster than the radioactively labeled probe. A **tritium-labeled VZV DNA probe** has been used to detect the virus in human T lymphocytes during acute zoster or even after disappearance of rash (Gilden et al., 1988; Koropchak et al., 1989). **PCR** amplification demonstrated **latent VZV infection** in trigeminal and thoracic ganglia in a postmortem study (Mahalingam et al., 1990). PCR was equally efficient in identifying the virus in vesicle samples, crusts, and throat swabs from patients with clinical varicella (Kido et al., 1991), and in peripheral blood mononuclear cells from adult humans (the virus was absent from umbilical cord blood or from PBMC of young children) (Devlin et al., 1992).

6.3.1.4 CYTOMEGALOVIRUS

Human **CMV**, a typical member of the herpesvirus family (Griffiths and Grundy, 1987) causes debilitating conditions in **newborns** accompanied with growth retardation, microcephaly, and psychomotor disorders following primary infection in mothers during pregnancy; it has also been implicated in the pathogenesis of **type 1** (Pak et al., 1988) and **type 2** (Lohr and Oldstone, 1990) **diabetes**, in multiorgan infections in **immunocompromised patients** such as recipients of organ and bone marrow transplants, as well as in **patients with AIDS** (Grose and Weiner, 1990; Hillyer et al., 1990). The diagnosis of CMV has relied on the isolation of the virus in **tissue culture** and/or detection of virus-specific antibodies by **immunofluorescence** and **ELISA** (Apperley and Goldman, 1988). Interesting observations have been reported on the presence of CMV in the **vascular tree** of otherwise healthy individuals (Melnick et al., 1983; Petrie et al., 1987), similar to that of other herpesviruses (Yamashiroga et al., 1988), and its role in vascular damage and the formation of **atherosclerotic lesions** (Hendrix et al., 1989–1991), provide support for the existing view of the **role of viruses in atherosclerosis** (Benditt et al., 1983).

6.3.1.4.1 Immunological Assays

Rapid diagnosis of CMV infection has acquired clinical importance in part because of the possibility of counteracting the infection with such antiviral agents as ganciclovir (Erice et al., 1987; Keay et al., 1987). A whole range of patients, from the congenitally infected to AIDS patients with CMV viremia, transplant recipients on immunosuppressive treatment, and patients with malignancies, can benefit from the earliest possible diagnosis of CMV infection. A critique of **complement fixation**, **IgM**, **neutralizing antibody assays**, as well as **viral isolation** and **antigen detection** has been presented (Chou, 1990). While this discussion is focusing on **gene level** diagnostics, selected immunological techniques and their relative merits will be briefly summarized.

One of the standard approaches to the diagnosis of CMV infection has been the demonstration of specific IgM in the patient's serum. Interference from **anti-CMV IgG antibodies** significantly undermined the specificity of such assays. Removal of these interfering antibodies thus had to be accomplished by either **column chromatography** or **removal** of IgG by treatment of the specimens to be tested with goat **anti-human IgG** antibodies prior to anti-CMV IgM testing. Comparison of these methods and assessment of the use of anti-CMV IgM antibody determinations with the detection of CMV by the **shell vial assay** has been performed to explore the ability of the test to diagnose infection in liver tissue of transplant patients (Smith and Shelley, 1988).

The detection of **anti-CMV IgM** appeared to be **of little value** for rapid diagnosis of CMV infection even though this approach is easier than detection of the virus in cell cultures. Other, faster immunological methods such as the rapid shell viral assay should be preferred. Various antibody preparations from a number of manufactures can be used in this assay system. Comparison of the available reagents revealed that only the **Du Pont monoclonal antibody** showed 100% sensitivity and specificity and an absence of nonspecific reactions (Randazzo and Michalski, 1988). Again, such assessments stipulate that the immune system of the patients has not been compromised to the point of inability to mount an immune response to infection. The detection of CMV viremia by observing the specific **cytopathic effects** takes on average over 36 days. In a large number of cases with laboratory-proven CMV infections, no cytopathic effect can be observed (Schirm et al., 1987).

The **early antigen immunofluorescence (IF)** method for the detection of CMV in clinical specimens, using coverslip method of buffy coats, reduced the speed of virus identification from the usual 10 weeks to 4 to 6 days. The use of monoclonal antibodies allowed a specificity of 98.1% (Schirm et al., 1987). Monoclonal antibodies have been used for the **detection of early antigen fluorescent foci (DEAFF)** in cell cultures (Stirk and Griffiths, 1987). The cultures were incubated for 22 h after inoculation with the clinical specimen to be tested. The monolayer was then stained using a pool of seven monoclonal antibodies directed against the early antigens of CMV (strain AD 169) in an indirect IF test. Characteristic **nuclear fluorescence** was then detected microscopically. In comparison with conventional cell culture assays, which take several weeks, this test can be performed within 24 h. Unfortunately, false-negative results were observed due to a number of possible causes, including a low titer of the virus and lack of specificity of the monoclonals used. The **cell culture technique** also has its problems, including failure of fibroblasts to grow, variable susceptibility of the cells to the virus, and contamination. On balance, the DEAFF fails to produce results in 6.5% of cases, whereas cell culture fails in 10.5% of cases. The DEAFF assay was tested on urine, saliva, and blood samples.

Comparison of the relative merits of immunological techniques was done on 500 blood/urine donors tested for CMV infection; detection of early antigens was carried out by four serological assays (Lentz et al., 1988). Urine specimens were cultured on human foreskin fibroblasts, and sera were tested by **ELISA** (Litton Bionetics, Charleston, SC) and **latex agglutination** (Hynson, Wescott, and Dunning, Baltimore, MD). **CMV-induced IgM** was determined by an **indirect immunofluorescence assay** (IFA; Electro-Nucleonics, Inc., Columbia, MD) and by an **enzyme-labeled antigen assay (ELA)** (Hamburg, Germany). The results of that study indicated that the prevalence of CMV viruria in blood donors was low (0.6%), and that screening for CMV seropositivity for **early antigen antibody**, but not for CMV-specific IgM, can identify infected donors (Lentz et al., 1988).

Another comparison of different tests for the detection of anti-CMV antibodies in blood donors examined a CMV ELISA using a nuclear CMV antigen, **Vironostika CMV** anti-IgG micro-ELISA, **CMVSCAN**, a latex agglutination test, and an ELISA from Organon Teknika — the Vironostika CMV one-step micro-ELISA (Grillner, 1987). Sera from 419 blood donors were tested. The sensitivity was 100% for CMV ELISA and CMVSCAN, and 99 and 96% for the CMV anti-IgG and CMV one-step tests, respectively. The specificities were 100, 99, 94, and 100%, respectively. ELISA formats were considered convenient for large-scale screenings, although the sensitivity of the CMV one-step assay should be increased. The latex agglutination test has the advantage of being extremely rapid.

Although apparently a suitable modality for testing potential blood donors, these assays may have additional **disadvantages** when applied to the diagnosis of CMV infection in immunocompromised patients or monitoring those receiving specific antiviral therapy. Actually, testing of different early assay formats for the detection of antibodies to CMV has been performed in normal blood donors and immunocompromised patients (de Ory et al., 1987). **Enzyme immunoassay** (Enzygnost-CMV, Behringwerke, Germany), **fluoroimmunoassay** (FIAX; International Diagnostic Technology, Inc., U.S.), **latex agglutination** (Hynson, Wescott, and Dunning, Inc., U.S.) and a **complement fixation test** using a standard micromethod were evaluated on 490 serum samples, including 278 samples from immunocompromised patients. An **indirect immunofluorescence test** was used to resolve discrepancies. The enzyme immunoassay and the latex agglutination tests had the highest sensitivity. For **large screenings** the latex agglutination was considered preferable. In high-risk patients, the screening should be performed using highly specific methods having high predictive values (de Ory et al., 1987).

A more sophisticated, and at least sixfold more sensitive, method for the detection of antibodies to CMV is achieved by **flow cytometry (FCM)** (McHugh et al., 1988a). Human antiviral antibodies were detected with a **biotin-streptavidin amplification** procedure using **phycoerythrin** as the fluorescent label. **Microsphere-associated fluoresence** was quantitated by FCM. Although this technique can simultaneously detect several specific antibodies at sensitivities exceeding those of conventional assays, it is difficult to anticipate wide acceptance of this approach in routine clinical laboratories at this stage.

Direct identification of CMV in urine samples by monoclonal antibody has been adapted to a **dot immunoperoxidase assay** performed on nitrocellulose paper dotted with urine cell-free pellets (Zerbini et al., 1987). The complex of the monoclonal antibody to the capsid antigen was visualized by **immunoperoxidase staining**. The results agreed completely with the detection of CMV-induced antigens in cell cultures inoculated with clinical specimens, yielding results within 8 h. **Visual interpretation** of the results relying on identification of CMV in the urine allows for a rapid, practical assay for active CMV infections.

From the above summary, several features of the traditional, immunological methods for the CMV detection become apparent. These methods are relatively **lengthy**, **labor intensive**, and **unreliable**, when the virus is either not abundant or inactivated in transport to the testing facility or by the therapeutic measures, or when the patient fails to mount an immune response of adequate strength. Reviews describing the biology of CMV infection and the available methods for its diagnosis have been published (e.g., Apperley and Goldman, 1988). Significantly less attention has been directed toward comparison and analysis of the merits and deficiencies of the newer molecular biological approaches.

6.3.1.4.2 Blot Hybridization

Beginning with the work of Spector and Spector (1985), numerous reports have described DNA and RNA probes constructed for the identification of CMV in clinical isolates. Because the virus is shed in the urine, efforts have been made to develop a fast diagnostic method of applying **DNA** or **RNA probes** to samples of urine following the **isolation of nucleic acids**.

A probe for assays of clinical specimens was developed using two fragments (17 bp each) representing the short and the long **unique segments** constituting about 14% of the CMV genome, transcribed early and late, respectively, during CMV infection (Spector and Spector, 1985). No cross-hybridization to other herpesviruses was noted. The sensitivity of the probe to either of the two fragments was 1 pg of DNA. Hybridization was **more sensitive** than the standard tissue culture techniques (Spector and Spector, 1985). **Antiviral drugs**, excreted in high concentrations, were found to inhibit viral growth if evaluated by the culture technique, and the resulting low amount of virus excreted may spuriously reduce the apparent sensitivity of culture technique below detection limits. The DNA probes, however, allowed the detection of CMV DNA or mRNA under these conditions. In addition, the probes were able to detect CMV in sections of Kaposi's sarcoma by **ISH**, whereas the tissue culture technique failed to detect the virus.

Using an 11.7-kb *Hin*dIII CMV DNA L fragment (derived from the laboratory strain **AD169 CMV**) as a probe, as well as *in vitro* synthesized RNA transcripts, a **dot-blot** hybridization detecting 3.2 to 10 pg of CMV DNA has been developed (Schuster et al., 1986a,b). The sensitivity and specificity of that assay in comparison to the results of cell culture were 100 and 83%, respectively. **Viral isolation**, however, was possible only in 95 specimens out of 301 due to **microbial contamination**. A nonradioactive version of the assay using **horseradish peroxidase** as label had a markedly lower sensitivity of 80 pg of homologous DNA. A significantly more sensitive **alkaline phosphatase**-labeled probe detecting 10 pg of isolated CMV DNA, when tested directly on urine samples, proved to be at least 10 times less sensitive than the same radiolabeled probe (Buffone et al., 1988). The nonisotopic probe (cloned as a *Bam*HI 14.3-kb fragment of fragment E of the Towne strain of CMV) was also plagued by a high rate of **false-negative** reactions and various **nonspecific interferences** producing **false-positive** reactions. Another drawback was the length of the **extraction procedure**, 36 h, although the sensitivity and specificity of the probe were 92.2 and 97.9%, respectively.

An even worse performance (a sensitivity of 52% and a specificity of 87%) has been reported for a **biotinylated probe** derived from *Bam*HI restriction fragment B, which could detect 30 pg of CMV DNA (Lurain et al., 1986). A somewhat higher sensitivity and specificity (81 and 68.8%, respectively) in comparison with cell culture was achieved when fragments D and H were combined and used as probes. These fragments, however, showed a **cross-reactivity** with HSV types 1 and 2 (Lurain et al., 1986). Although there was a 70.5% agreement between hybridization and culture methods, no correlation was found between the amount of CPE in cell culture and the intensity of the color reaction of the probes. Technical problems were noted when using biotinylated probes, such as **interference** by substances found in urine samples.

Meyer and Enders (1988) used a biotinylated *Bam*HI AB DNA probe of CMV Towne strain coupled with **horseradish peroxidase-streptavidin conjugate** in an assay capable of identifying 60 pg of CMV DNA. Compared with tissue culture results, this assay displayed a sensitivity of 69.5% and a specificity of 60.3%. Another hybridization assay using the biotinylated *Hin*dIII L fragment of CMV AD169 strain and **¹²⁵I-labeled streptavidin conjugate** detected 30 pg of CMV DNA. The sensitivity of that assay was 84.8% and the specificity was 74.7% (Meyer and Enders, 1988). The early appearance of a cytopathic effect correlated with the virus concentration and with a positive early antigen fluorescent test; however, the **low sensitivity**, **length of performance**, and **the high cost** of the probe assay were disappointing.

³²P-Labeled fragments of CMV DNA (a *Hin*dIII L fragment [11.7 kb] of the coding sequence for the late viral polypeptides and an *Eco*RI J fragment [10.6 kb] containing the coding sequence for the major immediate early RNA, and other transcripts of the immediate-early region) have been used in other assays (Augustin et al., 1987). The purified probes were able to detect 500 to 750 fg of cloned CMV DNA after 4 to 12 h of autoradiography. Following different combinations of **urine sample preparation** involving extraction of the virus from **low-speed**

and **ultracentrifuged** and untreated specimens, CMV DNA in urine was detected as a **free virus particle** as well as in a **cell-associated form**. Because reactivity of the urine samples with the probes could be eliminated by DNase, but not RNase, the signals were thought to be due to CMV-specific DNA, rather than RNA.

Several general observations on the use of cloned probes have been made (Augustin et al., 1987). For instance, the use of whole insert-containing plasmids, rather than of isolated inserts, as probes may give false-positive results, which cannot be excluded by control hybridization of the vector alone. Dilution series of the vector as well as of the CMV-infected and uninfected cells were used to assess the specificities of the probes in each experiment. The other concern was to avoid having probes with sequences homologous to human sequences. Careful selection of restriction fragments minimized that problem. Another operational drawback was the need to perform laborious isolations involving ultracentrifugation. Even with all these efforts the very low amount of the virus present, and/or nonhomogeneous distribution of the virus in the sample, could lead to false-negative results. Despite all the deficiencies the probe assay was superior to serology and culture technique in detecting CMV DNA.

The presence of CMV DNA in peripheral blood monocytes (PBMCs) of seropositive donors has been established by spot hybridization with a ^{32}P-labeled 32-kb probe representing the *Xba*I fragment C of the Towne strain (Jackson et al., 1987). The probe was specific, did not cross-hybridize with other herpesviruses or human DNA, and could detect at least 10 pg of CMV. No difference in signal intensity was noted between the spots of DNA isolated from either neutrophils or mononuclear cells from seronegative donors and those from IgG or IgM-seropositive donors. Apparently the cells contained less than 1 CMV genome per 20 cells (the sensitivity limit of the assay) (Jackson et al., 1987), which was corroborated by ISH reports of the CMV RNA occurrence of 0.03 to 2.0% in peripheral blood mononuclear cells (Schrier et al., 1985). The utility of the **radiolabeled dot-blot hybridization** assay for CMV has been demonstrated by screening 281 consecutive urines from infants (Buffone et al., 1988). Comparing it to the standard tissue culture assay, a sensitivity of 100% and specificity of 98.9% could be observed.

Significant efforts have been expended in comparing the relative merits of various laboratory methods for the detection of CMV in clinical situations. **Comparable sensitivities** have been observed between the **enzyme immunofiltration** carried out on glass fiber filters in microplates using a monoclonal antibody and a peroxidase conjugate, and a radiolabeled probe dot-blot procedure using an *Eco*RI D fragment as probe (Rossier et al., 1987). The enzyme immunofiltration detected 1.82×10^3 cells although the former method was found to be faster and simpler. Both methods, however, were tested **not on clinical samples** but on laboratory-grown MRC-5 cells infected with the AD 169 strain of CMV. When five different methods for the diagnosis of CMV infection were compared in renal transplant and dialysis cases, the following levels of detection were found: DNA hybridization using radiolabeled probes, 31.9%; virus isolation in conventional tube cell cultures, 19.2%; immunofluorescence with monoclonal antibodies on centrifuged vial cultures, 25.2%; complement fixation test, 23.4%; and EM, 2.1% (Agha et al., 1989).

The mode of radiolabeling CMV DNA probes affects their performance. Thus, radiolabeled **nick-translated** probes are able to detect 1 to 10 pg of homologous DNA, or the equivalent of 10 to 50 CMV-infected cells (Kimpton et al., 1988). **Random hexanucleotide-primed DNA probes** were even more sensitive, with a detection limit of 0.1 to 0.5 pg, or one to five CMV-infected fibroblasts.

An even higher sensitivity in comparison with cell culture was shown by a **dot-blot** hybridization procedure using **digoxygenin** labeling of the probe in 516 urine samples from organ transplant recipients, leukemic patients, AIDS patients, children aged between 3 and 6 years, and the laboratory staff (Musiani et al., 1990a,b). The biology of CMV must be considered when comparing the efficiency of different techniques. Thus, dot-blot hybridization has been compared to early antigen (EA) detection in monitoring the appearance and disappearance of CMV infection (Stockl et al., 1988a). The dot-blot hybridization assay proved to be an **earlier marker** than EA detection. Some of the observed discrepant cases could be due to the difference in the time frame of the viral markers identifiable by the two techniques.

Currently, the evaluation of **antiviral activity** of various drugs against CMV is performed by **plaque reduction assays**. A rapid and reliable dot-blot hybridization assay was shown to be less tedious and to offer an excellent alternative to the plaque reduction assay (Dankner et al., 1990). **Dot-blot hybridization** has been adapted to produce **quantitative assessment** of CMV DNA in blood leukocytes of viremic patients by employing a **video densitometer-based technique** to monitor the effects of **antiviral therapy** (Saltzman et al., 1990). Up to 10 pg of CMV DNA could be detected even in the presence of microgram quantities of the host cell DNA.

6.3.1.4.3 *In Situ* Hybridization

ISH has been repeatedly tested in various tissues using radioactive and nonradioactive probes for the detection of CMV. Using probes that identified viral **RNA transcripts**, the OKT4 lymphocytes were identified as the cell population serving as a reservoir in natural CMV infection for maintaining infection as well as acting as a vehicle for its spread by blood transfusion (Schrier et al., 1985). ISH was shown to be more sensitive in detecting CMV infection than **routine light microscopy** in Kaposi's sarcoma cases (Grody et al., 1988). Only 20% of the tumors showed evidence of CMV involvement in AIDS patients. No evidence of CMV presence could be provided in elderly patients with "classic" Kaposi's sarcoma. The virus was apparently present at a copy number below the detection limit for the methods used unless the putative role of CMV in Kaposi's sarcoma must be reexamined (Grody et al., 1988).

In their elegant histochemical study using sequential detection of CMV DNA and antigens by ISH and **immunological staining**, Wolber and Lloyd (1988) used a biotin-labeled commercial CMV DNA probe (Enzo Biochemical, New York) and a two-color technique for the identification of the virus in formalin-fixed paraffin-embedded tissues. Most of the CMV DNA detected by the probe revealed **unencapsulated**, **intranuclear** viral DNA and not the DNA contained within complete **CMV nucleocapsids**. The CMV antigen **immunostaining** was, on the other hand, predominantly cytoplasmic. The detection of CMV infection by the probe proved to be a **more sensitive** technique than immunocytochemistry (Wolber and Lloyd, 1988).

Strictly methodological investigation of comparative values of **immunocytochemical staining** with DNA ISH using a biotin-labeled probe and **virus isolation** has been performed (Janssen et al., 1987). Detection of the early antigen by immunocytochemistry during the first 2 days of inoculation of a fibroblast culture with the virus was found to be the **most sensitive** method. After the 5th day, the sensitivities of the immunological method, conventional virus isolation, and ISH were found to be **equal**. In **quantitative terms**, however, ISH identified two to five times more CMV DNA than CMV early antigen-positive cells could have accounted for. It should be noted, however, that so far the sensitivity of biotin-labeled probes is inferior to that of radioactively labeled ones. A detailed technical study of this aspect of CMV DNA probe methods has been performed (Mifflin et al., 1987).

A similar comparison of ISH using biotinylated DNA probe with an **immunoperoxidase technique** employing a monoclonal antibody against CMV early antigen has been performed on sections of the colon (Robey et al., 1988). Contrary to Wolber and Lloyd (1988), DNA ISH was found to be less useful for diagnosis because it was usually positive in cells displaying **cytopathic changes**. According to that study, immunostaining for the early antigen added useful information to the evaluation of early or focal cases of CMV colitis.

When two versions of nonradioactive CMV DNA probe labeling — **biotinylation** and **direct linkage of enzymatically active horseradish peroxidase (HRP)** — were compared to **indirect fluorescent monoclonal antibody** (IFA) and **direct FA staining**, only IFA and ISH with direct horseradish peroxidase were able to give consistent and reliable CMV detection as early as 45 h postinoculation (McClintock et al., 1989). Biotin-label ISH (PathoGene II; Enzo Biochemical) performed **much worse** than HRP-DNA probe tests (Vista-Probe, Inc., and Digene Diagnostics, Inc.).

ISH using biotinylated probes with two different detection systems (**alkaline phosphatase** and **peroxidase**) (Enzo Biochemical) proved to be more difficult to perform, and in some instances more difficult to interpret, than immunohistochemistry for detection of CMV (HSV) in routinely processed tissues (Strickler et al., 1990). A **combination** of **digoxygenin** label with an **alkaline phosphatase**-labeled detection system using **anti-digoxygenin Fab fragments** has been found to be sensitive, specific, and a provider of good resolving power on both **frozen** and **paraffin-embedded** tissue sections for detection of CMV infection in AIDS patients (Musiani et al., 1990b).

CMV infection has been identified by ISH with a commercial **biotin-labeled CMV DNA probe** (Enzo Biochemical) in apparently normal cells in **renal allograft biopsies**, when clinical and serological evidence of CMV infection was lacking (Ulrich et al., 1986a,b). CMV could be found predominantly within the proximal tubular epithelial cells. ISH proved to be a useful diagnostic tool even in the absence of typical nuclear inclusions.

The **higher efficiency of ISH** using either biotinylated CMV DNA or ^{32}P DNA probes compared to serological, histological, and culture methods for CMV detection has also been shown in **liver biopsies** (Einsele et al., 1988; Naoumov et al., 1988). Radioactively labeled probe, but not the biotin-labeled one, was even capable of identifying **reactivation** of CMV infection in liver biopsy specimens from a large group of **bone marrow transplant** recipients (Einsele et al., 1988).

The differences noted between the results obtained with the **monoclonal detection system** and **ISH** were in large measure accounted for by the natural course of CMV infection. While the cells expressing early antigens are still morphologically normal and do not contain CMV DNA, the cells positive by ISH are associated with clusters of polymorphonuclear leukocytes (PMNs), lobular inflammation, and contain viral inclusions. Such cells support active CMV replication and CMV DNA can be found in their cytoplasm. ISH for CMV in **liver biopsies** using the biotinylated label in the PathoGene kit (Enzo Biochemical) proved to be less sensitive, although highly specific, when compared to immunostaining with a monoclonal antibody against an early CMV antigen (Paya et al., 1990).

In another study, the same Enzo Biochemical biotinylated CMV kit for ISH proved to be equally **sensitive and as fast as immunohistochemical evaluation** of immediate early antigen and both techniques were comparable or even better than the serological detection of CMV in autopsy tissues of recipients of **bone marrow transplants** (Rasing et al., 1990). Similar observations of **better resolution** and **sensitivity** of ISH with biotin labels have also been previously reported by others (Griffith et al., 1990; Grody et al., 1988; Keh and Gerber, 1988; Roberts et al., 1988; Wiley et al., 1986).

Contrasting findings on the relative diagnostic value of ISH for CMV in **PBMCs** compared to conventional assay systems (virus isolation, CMV-specific IgM and IgG serology, complement fixation) have been reported by other workers (Stockl et al., 1988b). In 8 of the 18 continually monitored patients who developed a productive CMV infection, ISH assays yielded positive results 1 to 2 weeks before the conventional assays did.

The Enzo CMV DNA probe has been used for detecting virus in **bronchoalveolar lavage** (BAL) specimens by ISH (Hilborne et al., 1987). When compared to the results of tissue culture a sensitivity of 90% with a specificity of 63% was obtained. Conventional **cytological examination** was markedly less efficient, revealing CMV inclusions in only 23%. The speed (24 h vs. 21 days by culture) and higher sensitivity of the probe method make it an attractive approach for CMV detection on BAL specimens. In assessing the efficiency of ISH for diagnosis of CMV infection in **BAL** samples, the Enzo PathoGene CMV kit with biotinylated label proved to have a high specificity but very poor sensitivity (62%) (Gleaves et al., 1989). It was only slightly better than histological and/or immunofluorescent techniques (56%), and substantially worse than the results of **centrifugation culture** (95%) and **standard culture** (87%).

The **consensus** is that, given a simplification of sample preparation for probe assay is achieved and that the label can be made nonradioactive, the molecular biological approach will assume the leading role in clinical laboratories for the diagnosis and monitoring of CMV infection in tissues. The relative sensitivity of the methods compared must be scrupulously established for the results to be clinically meaningful. This is particularly true of some findings using ISH. For example, the reported markedly lower sensitivity of ISH using biotinylated probes accounted for the delay in the diagnosis of CMV infection that could be established only at 5 days by the ISH assay (Scott et al., 1988). In comparison, the detection of early antigen by indirect immunofluorescence (**EA-IFA**) in shell vial cultures was able to diagnose CMV infection within 24 h.

6.3.1.4.4 Polymerase Chain Reaction

A PCR protocol for the identification of CMV in formalin-treated **peripheral blood PMNs** of patients with AIDS has been described (Shibata et al., 1988). PCR was able to detect CMV DNA in 20 μl of blood. The virus was more likely to be detected in the blood of patients with AIDS (52%) than in HIV-infected patients who did not have AIDS. Normal volunteers showed no CMV DNA in their blood. All specimens positive for CMV by culture were also positive for CMV DNA by PCR. Amplification of a portion of the CMV genome before amplification of the detection probe allowed detection of PCR products from as few as 1 plaque-forming unit or 10 CMV plasmid copies. The specificity of the PCR assay, which can be completed within 2 days, was determined by both the primer oligomers and the detection probes. The PCR system was twice as sensitive as the culture method in detecting CMV in blood from HIV-infected individuals (Shibata et al., 1988).

PCR has been used to determine the presence of HIV-1 and CMV viruses in **fixed autopsy tissues** from AIDS patients (Shibata and Klatt, 1989). Tissue slices (10 μm thick) were deparaffinized and **DNA was extracted**. Specific probes for CMV and HIV-1 confirmed the identity of the amplified products. CMV was detected in tissues from six of nine patients. In some cases, CMV was detected in all tissues tested, in others it was most often found in the adrenal gland and lung.

With primers specific for the **major immediate-early** and **late antigen genes** of the CMV genome, PCR was able to detect the virus in urine from newborns and in tissue culture of urine specimens (Demmler et al., 1988). The virus could be detected in 1 μl of urine supernatants without any pretreatment of the samples. Direct gel analysis identifying 400- and 435-bp bands was possible on a 5-μl aliquot of a 10^{-3} dilution of the tissue culture mixture. Dot-blot hybridization with ^{32}P-labeled homologous oligonucleotide probes was positive at a 10^{-4} dilution of the same culture. The specificity of CMV PCR was tested using other members of the herpes family of viruses (HSV, VZV, EBV) and no cross-amplification could be detected by direct gel analysis or by dot-blot hybridization of the PCR-amplified products.

Importantly, urine samples were found to contain an **inhibitor** substance(s), which interfered with the PCR if not sufficiently diluted (Demmler et al., 1988). Five- and even 2-μl aliquots of urine yielded negative PCR results. However, a smaller volume (1 μl) reduced the inhibitory effect of the urine, and allowed the amplification products to be detected by direct gel analysis. The overall sensitivity of the amplification procedure was 93% and the specificity was 100% relative to tissue culture results. Dot-blot assay of the amplified product increased the sensitivity to 100%, raising the predictive value of a negative result from 90 to 100% while keeping the specificity and predictive value of a positive result at 100% (Demmler et al., 1988). The advantages of the PCR assay over tissue culture detection are numerous — the test can be completed within 24–48 h vs. 2–28 days on a minute amount of urine. The virus does not have to be viable, nor does its DNA have to be intact. Again, no special treatment of the sample is required.

Optimal conditions for the CMV PCR using primers produced by Collaborative Research, Inc. for the conserved region of the CMV major immediate early gene have been developed (Olive et al., 1989b). By varying the reaction conditions, 0.15 fg of a plasmid containing the cloned target gene corresponding to almost 6 gene copies could be detected after 35 cycles of amplification. Five microliters of **preheated urine** could be used directly in the reaction without further treatment before amplification. The assay could be completed within 5 h and showed no interference with other herpesviruses. It was positive in all clinical specimens in which the virus had been identified either by culture, hybridization, or ELISA. Detection of the amplified products was made by direct gel observation. This PCR consistently detected CMV **directly in urine** specimens of renal transplant patients, whereas ELISA, nonradioactive DNA hybridization, and virus isolation were variable (Olive et al., 1989a).

The exquisite sensitivity of PCR amplification has been used to investigate whether there are **differences among strains of human CMV** by amplifying the **a-seq** region of the CMV genome (Zaia et al., 1990). The discriminatory power of the PCR assay could be demonstrated with some sets of primers to specific regions of the viral genome but not with others. Isolates from different patients proved to be different. Restriction enzyme analysis of the amplified products of the a-seq region confirmed these differences. At this time no clinically significant correlations have been provided to determine CMV strain differences. Technical improvements of the existing assays continue to increase the specificity and sensitivity of PCR amplifications.

Superior performance of a **nested PCR** assay allows a significant **reduction of the background** that is usually produced by false positivity due to amplification of fragmented DNA (Porter-Jordan et al., 1990). A PCR assay for CMV DNA allows the detection of as few as 3 copies of the viral genome, representing a **1000-fold increase in sensitivity** over conventional DNA hybridization assays (Shibata et al., 1990). Using 2 primer sets, including those for a CMV-specific **pp150 structural protein** not found in other herpesviruses, on 16 isolates from different patients, the reliability of that particular PCR protocol was found to be high (Shibata et al., 1990). By using PT-PCR and primers that flank a splice junction, viral mRNA could be differentiated from CMV DNA by the smaller amplified product for the mRNA template (Buffone et al., 1990). The expression of CMV mRNA could be detected as early as 6 h after infection. CMV genomes could be detected in the blood of children and immunocompromised patients. Incidentally, PCR amplification on **kidney** samples from patients with **IgA nephropathy** demonstrated that CMV is not involved in this immune disorder affecting mesangial matrix (Bene et al., 1990).

Somewhat discrepant findings on the prevalence of CMV among healthy donors are reported by different workers using PCR. The low frequency of detection of CMV DNA by PCR observed by some (e.g., Docke et al., 1991; Olive et al., 1989b) is contrasted with the detection of CMV DNA in most healthy donors by nested PCR on enriched mononuclear fractions (Taylor-Wiedeman et al., 1991). This raises the issue of specimen presentation to the amplification reaction. The **presence of various inhibitory substances** in the sample may in part account for the discrepant results. An ingenious method proposes to absorb DNA from urine on glass beads thereby allowing to elute protein and other substances prior to recovery of DNA for PCR (Buffone et al. 1991). The results obtained after such absorption fully correlate with those of tissue culture detection of CMV. **Growth of the virus, however, cannot always be used as a gold standard**. Thus, superior performance of PCR for the detection of CMV DNA proved to be more informative in monitoring the effects of antiviral ganciclovir therapy in bone-marrow transplant patients revealing that CMV persists after cessation of the therapy when viral cultures were negative (Einsele et al., 1991).

6.3.1.5 EPSTEIN-BARR VIRUS

Analysis of lymphoproliferative conditions for the presence of **EBV genomes** allows definitive confirmation of the involvement of this virus in the disease process. EBV-related lymphoproliferative disorders appear to be increasing in frequency (Magrath, 1989; Nalesnik et al., 1988; Wilkinson et al., 1989). Numerous viral antigens specific for different stages of the viral cycle are widely used for diagnosis. An interesting survey of some important developments in the molecular epidemiology of EBV infections has been presented (Miller et al., 1987). Interpretation of **serological findings** of EBV infection is particularly complicated in the presence of immunosuppression (M. Ho et al., 1985; Katz et al., 1989; Okano et al., 1988). As in other cases, phenotypic expression of viral markers may be distorted by the specific circumstances of the disease process, making detection of the viral genome an invaluable tool in establishing the pathogenetic role of a virus. The enormous potential and resolving power of **recombinant DNA techniques** using cloned subfragments of viral DNA as probes has been decisively shown.

Southern blot and ISH definitively established the causative role of EBV in producing **liver cell necrosis** in a pediatric case of fulminant **infectious mononucleosis** (Deutsch et al., 1986). By comparing the intensity of hybridization signals, the EBV genome content per average liver cell was estimated to be 2.5. Unfortunately, the EBV DNA signal could not be assigned to a specific cell type and the virus was believed to enter the liver within lymphocytes.

A **slot-blot** hybridization approach was used to test for the presence of EBV genomes in tissues of 21 cases of **Hodgkin's disease (HD)** (Weiss et al., 1987). The *Bam*HI-W probe containing EBV DNA from a **variably repeated region** present in the genome of different isolates was chosen. Another probe (*Eco*RI-B), which detected a **single-copy DNA** sequence in EBV DNA, was employed to **estimate the number of EBV genomes** present in positive cases. **Southern blot** hybridization performed on the positive cases ensured that the slot-blot analyses were detecting actual EBV genomes and not some cross-hybridizing DNA. An important finding contrasting with the previously held views was that the EBV DNA was detected both in cases **with** and **without clonal gene rearrangements** as well as in both the **nodular sclerosing** and **mixed cellularity**-type lymphomas (Weiss et al., 1987).

Some 29% of **HD patients** had EBV DNA detectable by Southern analysis in their lymph nodes (Staal et al., 1989). Again, both nodular sclerosis and mixed cellularity histological types were positive for the virus. Significantly fewer (only 8%) of the **diffuse large cell** lymphomas harbored EBV. It remains unresolved whether EBV is involved in the etiology of HD or infects the lymphoid tissue secondary to the immune deficiency accompanying HD (Staal et al., 1989). Further insight into the role of EBV in HD was gained in a study that used **tritium-labeled EBER-I antisense RNA** for **ISH** of archival formalin-fixed paraffin-embedded HD specimens (Wu et al., 1990). The majority of tumor specimens displayed an intense signal in virtually all of the tumor cells in tumors of mixed cellularity and nodular sclerosis histological types. The better resolution achieved in this study demonstrated the very large number of EBER transcripts in the target cells. This finding argues in favor of the view that "the virus is not merely a silent passenger in Hodgkin's disease", but has a role in growth regulation of Reed-Sternberg cells (Wu et al., 1990). Indeed, studies on EBV-encoded latent gene products suggest that EBV may be associated with more cases of HD than previously suspected (Pallesen et al., 1991).

Two types of biotinylated EBV DNA probes — one cloned from the *Bam*HI-V (W), **large internal repeat region** and the other, the *Not*I/*Pst*I synthetic, oligonucleotide probe from **tandem repeat regions** — demonstrated equal distribution of positivity in ISH studies (Montone et al., 1990). The **synthetic oligonucleotide probe**, however, gave a **stronger signal** confined to the nuclei of fewer lymphocytes. These ISH studies point to the **heterogeneous distribution** of gene amplification in EBV-infected cells (Montone et al., 1990).

To elucidate the extent of occurrence of EBV infection and its possible contributory role in the development of AIDS-related complex (ARC) and AIDS, the viral load in patients with these conditions was estimated by **dot-blot** hybridization in oropharyngeal specimens (Alsip et al., 1988). The probe used in this study was the 28-kb *Eco*RI A fragment of EBV. Although high levels of EBV DNA were found in 33% of the patients and asymptomatic HIV antibody-positive persons, it was noted that the hybridization assay may not consistently detect EBV DNA in specimens with fewer than 10^5 to 10^6 genome equivalents per milliliter, as is the case in ARC patients. Actually, the hybridization assay was inferior to the lymphocyte transformation assay for use in general screening for the presence of EBV. It could, however, be used for **gross approximation** of EBV load in specimens with more than 10^6 EBV genome equivalents per milliliter (Alsip et al., 1988).

Absence of EBV DNA sequences from lymph nodes in patients with HIV and those with **persistent generalized lymphadenopathy (PGL)** revealed that EBV infection is not directly involved in the pathogenesis of PGL at the cellular level (Uccini et al., 1989).

A study of the involvement of EBV in **T cell lymphomas** has been conducted (Jones et al., 1988). DNA from tumors was subjected to **restriction analysis** to reveal T cell receptor and immunoglobulin gene rearrangements, and ISH was performed with the biotinylated *Bam*HI-V(W) fragment of EBV genome as a probe. Biotinylated plasmid used to clone the EBV fragment was used as a negative control for the probe. Detection of EBV DNA was combined with the identification of cell-surface phenotype in the same tissue by exposing sections to the monoclonal antibody LN-3 and developing it with avidin-biotin peroxidase. Although EBV was present in T cell tumors in patients with mucosal responses to EBV, it is believed that examination of T cell lymphomas for EBV DNA seems to be indicated on a wider scale (Jones et al., 1988).

ISH on swabbed specimens from **oral hairy leukoplakia** has proved to have almost equal sensitivity to that of ISH on biopsy specimens for the detection of EBV DNA (De Souza et al., 1990). This alternative for early identification of EBV in suspicious lesions usually appearing before overt manifestations of AIDS avoids the often undesirable invasive biopsy procedure.

The etiological role of EBV in **nasopharyngeal carcinoma** has long been suspected (Old et al., 1966), and the notion has been strengthened with the demonstration by radioactive EBV probes of the presence of EBV DNA in malignant epithelial cells (Klein et al., 1974; Wolf et al., 1973; zur Hausen et al., 1970). An *in situ* **cytohybridization assay** has been developed using nonradioactive **digoxygenin-labeled** *Bam*HI fragments of EBV DNA as a probe (Permeen et al., 1990). The results are available within 96 h compared to 5 weeks for ^{3}H-labeled ISH probes used in the **autoradiographic** approach. EBV DNA is found in malignant epithelial cells of nasopharyngeal carcinoma as well as in the lymphocytes infiltrating the tumor and in some morphologically unaltered epithelial cells in the tumor biopsies. The specificity of the digoxygenin probe was matched by the added advantage of the label for ISH, that is, the **absence of background staining** usually unavoidable with any radioisotopic labels. This method has been tested on frozen sections and no observations have been made on paraffin-embedded tissue. Using **recombinant EBV proteins** the diagnosis of nasopharyngeal carcinoma is possible with substantially improved diagnostic probability (Littler et al., 1991).

Using **PCR** that detects a single EBV genome, EBV DNA sequences could be identified in 41 of 41 undifferentiated **nasopharyngeal carcinoma** cells, in 2 of 4 moderately differentiated tumors, and in 3 of 5 keratinized specimens (Chang et al., 1990). The higher number of EBV genome copies in undifferentiated tumors suggested that EBV replication interferes with the differentiation of nasopharyngeal carcinoma cells. The distinctive histological appearance of an undifferentiated lymphoepithelial gastric carcinoma also resembles that of the nasopharynx in that it harbors EBV DNA detectable by PCR, in contrast to adenocarcinomas with lymphoid stroma (Burke et al., 1990). Interestingly, up to 23% of **normal healthy adults** have been found by PCR to harbor EBV DNA in samples of **oropharyngeal cells** (Gopal et al., 1990).

Another technique has proven its applicability to formalin-fixed, paraffin-embedded tissues — **PCR** amplification of the **long internal direct repeat region** of the EBV genome (Peiper et al., 1990). The identification of the amplified products in Southern blots was achieved with ^{32}P-labeled oligonucleotides complementary to EBV sequences flanked by the amplimers. Tissue specimens retrieved from **infectious mononucleosis** patients, **renal** and **heart transplant** patients, and a **nasopharyngeal carcinoma** patient all showed the presence of EBV. Histologically normal spleen and lymph node tissue used as controls were negative for EBV sequences. This assay may also prove useful in differentiating cases of monomorphic proliferation of large lymphocytes simulating malignant lymphoma from hyperplasia associated with EBV by demonstrating absence of the virus in the lymphoma case (Peiper et al., 1990). This specific application, however, may have only limited practical utility because the diversity of lymphomas, including those harboring EBV, is quite significant (e.g., Jones et al., 1988).

PCR has been employed to identify EBV DNA in blood and tissue biopsies from **Sjögren's syndrome** (I. Saito et al., 1989), **HD** (Bignon et al., 1990), and **immunocompromised patients** (Fermand et al., 1990; Rogers et al., 1990). It appears that the superior sensitivity of PCR can help resolve the subtle variations in individual responses to various viral antigens, including those tentatively ascribed to EBV and constituting a part of the **chronic fatigue syndrome** (Miller, 1991).

An extensive study using PCR for the detection of EBV DNA was performed on DNA extracted from the blood of 25 **transplant** patients, 5 patients with **infectious mononucleosis**, biopsy or autopsy tissue from 29 patients with **lymphoproliferative disorders**, and 13 healthy controls (Telenti et al., 1990). Primers were constructed to be specific for the conserved regions of the EBV genome encoding **capsid protein gp220** and **Epstein-Barr nuclear antigen 1 (EBNA)**. The sensitivity of the method was 10 to 50 copies of the EBV genome. No EBV was identified in the blood specimens of 13 healthy adults, 11 of whom were seropositive. EBV was detected in all 10 immunocompromised patients, as well as in the blood of 11 of 25 transplant patients. EBV was present in the blood of 2 patients with EBV-positive lymphomas and in all 5 patients with infectious mononucleosis, but only in 1 of the 19 patients with other T or B cell lymphomas or leukemias. The lack of identification of EBV in seropositive individuals is explained by the fact that DNA material for the PCR in that study was extracted from approximately 15 to 75×10^3 leukocytes, whereas the published estimates hold that there is only 1 EBNA-positive lymphocyte per 10^7 or 10^8 circulating cells in the peripheral blood (Telenti et al., 1990). Moreover, it is estimated that only 1 to 10 per 10^6 cells are capable of developing into a continuous line *in vitro* as a result of EBV infection. Therefore, the chances of detecting EBV in seropositive individuals in this study would have been greater if the DNA sampling had been performed on a larger number of cells.

A number of lymphomatoid proliferative conditions have been analyzed with the aid of PCR assays for EBV DNA. **Lymphomatoid granulomatosis** showed a high (72.4%) rate of EBV DNA presence while primary, well-differentiated lymphocytic lymphoma of the lung as well as metastatic large cell lymphoma in the lung were negative (Katzenstein and Peiper, 1990). Lymph nodes in **angioimmunoblastic lymphadenopathy** have been found to harbor the entire EBV genome in a large number of cells (at least 1 viral copy per 15,000 cells) (Knecht et al., 1990).

The results of the detection of EBV DNA obtained by ISH, Southern analysis, and PCR have been compared in **lymphoid tumors**, **benign lymphadenopathy**, and normal **thymuses** (K. Ohshima et al., 1990). The presence of EBV DNA was not related to lymphoid malignancy, although in some neoplastic conditions its presence is enhanced. ISH was equal to Southern blot analysis, and both ISH and Southern analysis were inferior to PCR in their efficiency to detect EBV DNA (K. Ohshima et al., 1990). **Southern analysis**, however, may have a **greater discriminatory power** in assigning the presence of EBV DNA sequences to a specific cell type, as demonstrated by the presence of EBV infection in T cells rather than B cells in a case of infectious mononucleosis (Yoneda et al., 1990).

Further sophistication of a PCR assay for EBV DNA comes from the construction of **three pairs of primers** specific for genomic sequences coding for the two forms of **EBNA**, 2A and 2B, and for a DNA sequence from the *BamZ/BamR* region (Jilg et al., 1990). This ensures a reliable and rapid detection of types A and B viruses, which differ in their ability to transform B lymphocytes, but so far it is not clear whether the two forms have a different pathogenic potential in humans.

The use of **paraffin-embedded** tissue for EBV identification by PCR **reduced the sensitivity** of detection in a manner paralleled by the reduced signal from the control β-globin gene amplification in the same tissue (Telenti et al., 1990). Optimally, DNA extracted from **frozen** tissue should be used.

6.3.2 *MYCOPLASMA* SPECIES

These organisms, which lack a cell wall, number over 100 species (Tully, 1985) and cause a variety of diseases in humans, animals, and plants (Razin, 1981; Razin and Freundt, 1984). Deficiencies of cell culture methods and mycoplasma DNA staining for identification of these pathogens include poor reproducibility, ambiguity (~11%), and a significant proportion of false-positivity (~2%) and false negativity (0.7 to 2.4%) (Bolske, 1988). The detection of mycoplasmas by **DNA probes** directed to **total DNA** or **rRNA gene** sequences has been described (Frydenberg et al., 1987; Gobel et al., 1987; Hyman et al., 1987; Mattsson et al., 1991; Razin et al., 1987; Roberts et al., 1987a). A **commercially available** DNA probe assay (Gen-Probe, San Diego, CA) for *M. pneumoniae* favorably compares with culture methods (Tilton et al., 1988). Probes specific for *M. genitalium*, which appears to have evolved independently of other human mycoplasma species (Dallo et al., 1989), have been used to assess the prevalence of this species in urethral infections (Hooton et al., 1988).

Very low, if any, cross-hybridization could be noted with specific DNA probes for *M. pneumoniae* and *M. genitalium* using radioactive labels. The sensitivity of the same probes labeled by **biotinylation** or **sulfonation** processes was lower than that of the radioactively labeled counterparts by an order of magnitude (Hyman et al., 1987; Razin et al., 1987). A review of the earlier studies in mycoplasma diagnosis by gene probes, and the methodologies for constructing such probes, has been presented (Dular, 1990).

PCR amplification assays for *M. pneumoniae* offer a highly species-specific identification of the organism in BAL with a sensitivity of 10^2 to 10^3 organisms (Bernet et al., 1989) or even better, equivalent to about 35 genome copies per milliliter (Jensen et al., 1989). In fact, PCR assays revealed that some 11% of HIV-1-seropositive patients carry *M. fermentans,* however, no correlation with the HIV clinical stage could be found (Hawkins et al., 1992).

6.3.3 *LEGIONELLA* SPECIES

Although a number of *Legionella* species are recognized, one species, *Legionella pneumophila,* is responsible for over 90% of *Legionella* infections. Diagnosis is frequently based on the clinical symptoms and serological conversion (Brenner et al., 1984). **Cultivation** of the organism is laborious and identification requires the use of **direct fluorescent-antibody tests** or **gas chromatography**; however, **DNA probes** selected by subtractive hybridization have been developed (Grimont et al., 1985). These probes were tested in **colony** and **dot hybridization**, although they are also potentially useful for direct identification of *L. pneumophila* in clinical and water samples. The advantages of DNA probe techniques over conventional immunological and biochemical identification of suspected *Legionella* isolates have been stressed (Ezaki et al., 1988). A rapid and simple dot-blot hybridization assay using photobiotin-labeled DNA probes can be performed in a clinical laboratory. Among 500 isolates, 20 strains were identified as *L. pneumophila* where serological methods failed (Ezaki et al., 1988). The rapid diagnostic system for *Legionella* offered by Gen-Probe, Inc. (San Diego, CA) utilizes radioiodinated cDNA as a probe for rRNA sequences specific for members of the genus *Legionella*. The Gen-Probe system has been evaluated (Edelstein et al., 1987) vs. direct fluorescent antibody staining and isolation of *Legionella* in culture (Edelstein, 1986; Doebbeling et al., 1988; Wilkinson et al., 1986).

In a prospective evaluation, avoiding testing of frozen respiratory specimens, the DNA probe was considered a practical alternative to a direct fluorescent antibody assay for rapid diagnosis of *Legionella* infections (Doebbeling et al., 1988; Pasculle et al., 1989). False-negative results did occur, emphasizing the importance of testing multiple specimens and underscoring the role of culture in *Legionella* diagnosis. Another inconvenience of the Gen-Probe assay in the radioactive format is the requirement for scheduling and operational adjustments that must be made and that are not always realistic in a clinical laboratory setting. Undoubtedly a nonradio-active version of the assay will find wide appeal to testing laboratories.

The sensitivity and specificity of the Gen-Probe assay was found to be 93.9 and 97.4%, respectively, in an experimental system (Pasculle et al., 1989). The unacceptably low positive predictive value of 60.8% was found in that study. By raising the threshold value of positive results above those suggested by the manufacturer, the specificity of the probe assay was improved to 99.2%, making the performance of the probe and direct fluorescent antibody methods essentially equivalent.

In certain situations identification of *Legionella* at the species level becomes of practical importance and currently there are no commercially available methods for serologically speciating the isolates, although species-specific antisera have been developed by the Centers for Disease Control (Atlanta, GA). This led Ezaki and coworkers (1990) to elaborate a microplate technique by preparing microdilution plates with immobilized unlabeled reference DNAs and stocking them in a dehydrated state.

Restriction analysis has been useful in tracking down the source of various isolates linked to nosocomial *Legionella* infections and proved to be easier to perform than **alloenzyme typing** (Tompkins et al., 1987). Alternatively, pulsed-field gel electrophoresis (PFGE) using *Not*I and *Sfi*I digests of *L. pneumophila* DNA has been used for the detection of nosocomial *L. pneumophila* in water supplies on the basis of restriction patterns (Ott et al., 1991). DNA hybridization allowed the identification of a new isolate from cooling water tower that proved to be a new *Legionella* species, *Legionella adelaidensis* (Benson et al., 1991).

The detection of *L. pneumophila* in water by **PCR** amplification is possible (Mahbubani et al., 1990) even following biocide or elevated temperature treatment (Bej et al., 1991). **Nonviable** cells did not show PCR amplification, whereas culturable and **viable** noncultivable cells could be detected by PCR. By selecting primers to the **macrophage infectivity potentiator (*mip*) gene**, a good correlation was established with culture methods in the differentiation between viable and dead *L. pneumophila*.

6.3.4 *BORDETELLA* SPECIES

The identification of *Bordetella pertussis* and *Bordetella parapertussis* by a DNA hybridiza-tion assay without preculturing the organisms was only marginally successful (38% positives compared to serological assays) compared to the 50% sensitivity of culture methods (Reizenstein et al., 1990). A significantly higher sensitivity, reaching 69%, could be achieved following **preculture** of nasopharyngeal aspirates from patients with suspected pertussis for 24 to 72 h. This, however, was achieved at the expense of specificity.

6.3.5 *MORAXELLA (BRANHAMELLA) CATARRHALIS*

Moraxella (Branhamella) catarrhalis is a respiratory pathogen identified in patients with underlying pulmonary disease (Nicotra et al., 1986), in immunocompromised patients (Sweeney et al., 1985), as well as in children with otitis media and sinusitis (Bluestone, 1986). A **synthetic oligonucleotide DNA probe** derived from a random fragment of **chromosomal DNA** 100% specific to the *M. (B.) catarrhalis* species has been developed (Beaulieu et al., 1991). This gene probe will make *M. (B.) catarrhalis* testing definitive on pure cultures instead of only presumptive identification possible when using currently available commercial kits for *Neisseria* species. No reports on testing this probe on clinical material have been published at the time of this writing.

6.3.6 FUNGI

6.3.6.1 *HISTOPLASMA CAPSULATUM*

Using **differential hybridization**, several yeast phase-specific genes of *Histoplasma capsulatum* have been cloned, including *Yps*-3, the one apparently determining the pathogenicity of the organism (Keath et al., 1989). The specificity of this 1.85-kb nuclear DNA probe is helpful not only in classification studies, but may be used for the identification of *H. capsulatum* as well.

6.3.6.2 *CRYPTOCOCCUS* SPECIES

A detailed study of **restriction patterns** of rDNA sequences of several *Cryptococcus* spp. amplified by **PCR** has demonstrated the utility of this approach to **taxonomic** and **phylogenetic evaluation** (Vilgalys and Hester, 1990). In addition it also emphasized the power of fingerprint analysis for species- and strain-specific identification within *Cryptococcus* spp. (Vilgalys and Hester, 1990). Restriction patterns of the three separate regions of the **rDNA repeat (A, B,** and **C)** revealed that *C. neoformans* had no variation among the 19 strains tested. While in the four enzyme restriction patterns of *C. neoformans* only minor variations were present, considerable variability exists among PCR fingerprints of the other *Cryptococcus* and *Candida* strains examined.

6.3.6.3 *NOCARDIA ASTEROIDES*

Nocardia asteroides causes pulmonary nocardiosis with frequent dissemination to other organs. The prognosis for the patient is poor if the appropriate treatment is delayed (Brownell and Belcher, 1990). Because the nocardiae may produce infections indistinguishable from those caused by **mycobacteria** in both **immunocompetent** and **immunocompromised** hosts, a gene-level evaluation of this slow-growing organism may be helpful. The existing laboratory methods for diagnosis of *N. asteroides* may require detection of diagnostic **metabolites** of the organism frequently outside the capabilities of routine clinical laboratories. **HPLC** can be helpful, as are **serological** methods. Two **recombinant clones** have been constructed that identified 31% of the *N. asteroides* strains in a reference collection (Brownell and Belcher, 1990). Additional clones were then selected to provide pooled DNA probes capable of identifying all the strains tested. Efforts are being directed toward the development of an amplification procedure for *N. asteroides,* using some of the probes as primers (Brownell and Belcher, 1990).

6.3.7 *MYCOBACTERIUM TUBERCULOSIS*

The resurgence of **tuberculosis (TB)** infection in recent years adds to the challenge of finding the fastest methods of identification of specific mycobacterial infection. Although **acid-fast staining** of material obtained from the patient is a rapid method of diagnosis, Ziehl-Neelsen staining is insensitive and nonspecific, requiring over 10^4 organisms per milliliter to be present in the specimen (David, 1976). Only half of patients with pulmonary TB and a quarter of those with extrapulmonary disease are identified by the Ziehl-Neelsen and **rhodamine-auramine stains** (cited in Wilkins and Ivanyi, 1990). Moreover, over half of the estimated 8 million TB cases per year worldwide are smear-negative pulmonary and extrapulmonary diseases (Wilkins and Ivanyi, 1990). Immunological detection by **latex agglutination, radioimmunoassay**, and **enzyme-linked immunoasorbent assay** also suffers from insensitivity and low specificity (Kadival et al., 1986; Krambovitis et al., 1984; Yanez et al., 1986).

Poor specificity and sensitivity have been reported for **serological** methods of mycobacterial diagnosis (Daniel and Debanne, 1987; Narayanan et al., 1983; Nassau et al., 1976). Elegant studies by J. J. McFadden and colleagues (e.g., 1987) address the fine distinctions between isolates of *Mycobacterium avium* complex at the DNA level. Important observations have been made by the authors on *M. paratuberculosis* isolated from tissues exhibiting morphological signs of Crohn's disease. These studies also provide important insights into the mycobacteria affecting a high proportion of AIDS patients (McFadden et al., 1987).

6.3.7.1 DNA PROBES

Whole chromosomal DNA probes for rapid identification of *Mycobacterium tuberculosis* and *M. avium* complex have been developed (Roberts et al., 1987b). **Dot-blot** hybridization could be accomplished within 48 h and the results were significantly improved by lysing the cultures prior to application of the probes.

A general mycobacterial identification method that recognized not only *M. tuberculosis*, but also **mycobacteria other than tuberculosis (MOTT)**, uses a **biotin**-based detection system (Picken et al., 1988). One of these probes recognizes the closely related *M. tuberculosis, M. bovis,* and *M. microti,* but no other mycobacterial species. The second probe was specific for *M. malmoense,* and yet another probe hybridized, in addition to *M. tuberculosis,* to *M. intracellulare, M. malmoense, M. scrofulaceum, M. simiae, M. xenopi, M. avium, M. szulgai, M. kansasii,* and *M. hemophilum.*

Differentiation of *M. tuberculosis* from *M. bovis* can be performed by restriction analysis using *Pst*I and *Hae*III restriction fragment length polymorphisms (RFLPs) (e.g., Andrew and Boulnois, 1990). A random fragment of *M. bovis* DNA generated by sonication and cloned in the **M13 vector** can hybridize to 27 isolates of *M. bovis* and *M. tuberculosis* at high stringency conditions (0.1 × SSC, 65°C) (Andrew and Boulnois, 1990). No hybridization is observed to DNA from *M. avium, M. fortuitum,* or *M. phlei.* At low stringency, the probe also hybridized to nonmycobacterial species. In this way, certain DNA fragments could be used either as *M. tuberculosis* **complex-specific probes** or as **panmycobacterial probes**, depending on the stringency condition of hybridization and washing conditions. A detailed description of technical aspects of specimen preparation, construction, and labeling at earlier stages in the use of mycobacterial probes is available (Andrew and Boulnois, 1990). Convenient adaptation of quantitative microdilution plate hybridization with immobilized reference DNAs for species identification of mycobacteria uses colorimetric detection of hybrids (Kusunoki et al., 1991).

Analysis of the IS 6110 sequence reveals its significant conservation across strain and species lines, which allows **RFLP analysis** for this sequence to be used in fingerprinting various strains of the *M. tuberculosis* complex (Cave et al., 1991). Probes for the insertion sequence IS 6110 proved to be convenient strain-specific markers helpful in epidemiological studies of tuberculosis (Mazurek et al., 1991).

A combination of **sonication** of mycobacteria in chloroform followed by probing for the **rRNA genes** in **dot blots** of the lysate gave the highest sensitivity in detecting sequences homologous to the DNA probes (Rahav et al., 1990). The latter were generated by restriction digestion of the complete rRNA operon from *M. smegmatis.* The detection limit of that assay was 200 organisms, which is comparable to some PCR-based methods. Using a similar approach with probes developed for the repetitive sequences of the mycobacterial genome is one way of increasing the sensitivity of this assay.

A kit available from Gen-Probe uses ^{125}I-labeled *M. avium* and *M. intracellulare* DNA probes for hybridization with rRNA from a test organism. Musial and coworkers (1988) conducted an evaluation with emphasis on the specificity of the Gen-Probe system and found the kits to be fast, highly sensitive, and specific in identifying *M. avium-M. intracellulare* complex and *M. tuberculosis* grown in culture. This system has been used for analysis of multiple isolates from various locales in Japan and comparison of their drug susceptibility (H. Saito et al., 1989). In view of the differing resistance of *M. avium* and *M. intracellulare* to therapeutic agents, such a fast and reproducible differentiation in a routine laboratory by DNA probes justifies their use as the next generation of diagnostic tools in the evaluation of mycobacterial infections.

Although routine cultures are time consuming, the BACTEC system (Johnston Laboratories, Inc., Townson, MD) employing a radioactive estimate of the cultured mycobacteria growth reduces the diagnostic procedure to approximately 2 weeks. By combining the Gen-Probe method with the BACTEC radiometric technique, most isolates of *M. tuberculosis, M. avium,* and *M. intracellulare* can be identified within 4 weeks in contrast to the 9 to 11 weeks usually required with conventional media and biochemical identification (Ellner et al., 1988). Although this technology still requires cultivation of the organism, the lack of nonspecific reactions and the savings in time and labor make the DNA probe approach the method of choice.

Further improvement is offered by **lysing** and **concentrating** mycobacteria from blood specimens grown in BACTEC 13A bottles prior to testing with the Gen-Probe diagnostic system for *Mycobacterium avium* complex (Conville et al., 1989). Identification of the acid-fast bacilli directly from specimens grown in BACTEC 12B bottles has been performed by the Gen-Probe system (Peterson et al., 1989). In contrast to conventional cultures taking 4 to 6 weeks for final identification, the BACTEC system allows an adequate amount of mycobacteria to be grown within 7 to 8 days. A combination of BACTEC with the Gen-Probe assay allows over 83% of specimens containing clinically relevant mycobacteria to be identified within 7 days for *M. avium-M. intracellulare* isolates and within 18 days for *M. tuberculosis* isolates. Confirmation of the efficiency of Gen-Probe assays, particularly in combination with specimens cultured in the BACTEC system, has been provided (Body et al., 1990). **Parallel subculturing** of positive BACTEC bottles is suggested for review of mixed infections such as are encountered in AIDS patients positive for both *M. tuberculosis* and *M. avium* complex.

An intriguing observation has been made suggesting various mycobacterial strains (*M. avium*) isolated from many AIDS patients are distinct from strains isolated from other patients (Hampson et al., 1989). This raises the possibility that the role played by mycobacteria in suppressed patients is more than simply opportunistic, suggesting their participation in down regulation of the already compromised immunity of the AIDS patient. Interestingly, *M. paratuberculosis* could not be isolated either from these patients or from healthy controls, supporting the view that *M. paratuberculosis*, isolated only in a disease state and only from patients with Crohn's disease, can in fact be involved in the etiology of this inflammatory bowel disease.

The **superior diagnostic ability of the RFLP** technique over serotyping was demonstrated by the failure of the latter approach to establish the genetic identity of the mycobacteria in that study (Hampson et al., 1989). A more recently reported **immunological method** (Wilkins and Ivanyi, 1990) appears, however, to offer a somewhat higher sensitivity and specificity than previous methods taking advantage of the simplicity of the procedure. The relative procedural simplicity of **serological** assays still makes this modality appealing to routine clinical laboratories, especially in developing countries. Certain difficulties encountered in applying currently available gene probes for *M. tuberculosis* to clinical specimens directly (Daniel, 1989), particularly in field studies, still cause laboratories to favor the development of convenient serological methods.

One of the latest improvements in the line of existing **ELISA**-based techniques directed to **immunodominant specific antigens** of *M. tuberculosis* is the **sandwich ELISA** adaptation of the **solid-phase antibody competition test** (SACT-SE) (Wilkins and Ivanyi, 1990). The ability of the SACT-SE to distinguish TB patients from those presenting with a similar clinical picture using a combination of 4 murine **monoclonal antibodies** has been evaluated in 236 patients. In cases of extrapulmonary TB this test showed a positivity of 70 to 80% unrelated to the organ localization of the disease. Therefore the need for biopsy in the majority of cases may be eliminated, and the specific therapy can be initiated 12.5 days earlier (Wilkins and Ivanyi, 1990).

6.3.7.2 PCR DIAGNOSIS OF MYCOBACTERIA

The DNA and RNA probe methods, including the commercially available Gen-Probes (San Diego, CA), allow a specific and fast diagnosis of mycobacterial infection to be made. However, becasue reliable identification by this method requires the presence of at least 10^6 organisms in the test sample, the practical sensitivity of such methods is still too low to make them appealing in a clinical setting. They can be successfully used on cultured specimens for the fastest final identification, but not on a patient's sample.

The **time factor** becomes particularly important in the outcome of **tuberculosis meningitis** (Kennedy and Fallon, 1979). Specific primers and reaction conditions for the identification of the *M. tuberculosis* complex bacteria by PCR have been developed, making it the **method of choice** for the fastest, most specific, and most sensitive detection of the organism. The first reports on the use of PCR amplification of mycobacterial DNA in clinical material were made by Hance and coworkers (Brisson-Noel et al., 1989; Hance et al., 1989). Primers based on the evolutionarily conserved **65-kDa heat shock gene** were selected in those studies. Selective **oligonucleotide probes** specific for *M. tuberculosis-M. bovis, M. avium-M. paratuberculosis*, or *M. fortuitus* strains were used for specific identification of the amplified PCR product. Some PCR procedures developed for *M. tuberculosis* (Brisson-Noel et al., 1989; Shankar et al., 1989) allow the fastest identification of a mycobacterium strain on the basis of **agarose gel electrophoresis patterns** generated by the amplified products without the need for confirmation by hybridization. Among the several **repetitive sequences** in the genome of members of the *M. tuberculosis* complex (*M. tuberculosis*, 10 to 12 copies; *M. bovis*, 1 to 3 copies), the **insertion sequence IS 6110** has been the target of PCR amplification for diagnostic purposes (Thierry et al., 1990). Oligonucleotides derived from this sequence were used as primers for the **PCR** assay.

Subsequently, Hermans and coworkers (1990) identified an *M. tuberculosis* **DNA insert in a λgt11 gene** library that specifically hybridized with DNA from mycobacterial species belonging to the *M. tuberculosis* complex. Parts of this insert (**158 bp**), which was found to occur in many copies in the mycobacterial chromosome, were used as primers for the PCR. None of the 45 strains not belonging to the *M. tuberculosis* complex species contained the 158-bp fragment. On the other hand, all strains of the species *M. africanum, M. bovis,* and *M. microti* and 29 of the 34 *M. tuberculosis* strains tested were positive by PCR. The assays were performed on sputum samples and lung tumor biopsy from patients confirmed by culture to have *M. tuberculosis*. The detection limit for **chromosomal DNA isolated** from *M. tuberculosis* was 100 fg, corresponding to approximately 20 bacteria. When the PCR was tested on whole bacteria presented either as a suspension of *M. tuberculosis* cells in a buffer or in sputum from a healthy person, the sensitivity fell to 200 bacteria for cells suspended in buffer and to 1000 bacteria for cells suspended in sputum. Even the latter level of sensitivity represents an **improvement of three orders of magnitude** over the detection methods that use DNA or oligonucleotide probes for hybridization assays only (Hermans et al., 1990). Testing the PCR assay on DNA isolated from clinical samples of sputum, urine, tumor biopsies, pleural fluids, and bronchial washings demonstrated that all specimens positive for acid-fast stain were also positive by PCR. The amplification assay remained positive 16 weeks after specific **isoniazid-rifampin-pyrazinamide** treatment, at which time no bacteria could be cultured or detected by Ziehl-Neelsen staining (Hermans et al., 1990).

Another PCR method targets the nucleotide sequences of the gene for the 38-kDa protein antigen **b** (**Pab**) of *M. tuberculosis* (Sjobring et al., 1990). Serological methods also identify this antigen in *M. bovis*. Primers for the *Pab* DNA sequence are used to successfully amplify a 419-

bp DNA sequence on the DNA isolated from clinical specimens, confirming they harbor *M. tuberculosis* complex infection (Sjobring et al., 1990). Because the PCR sensitivity is affected by the condition of a target nucleotide sequence, some effort was put into improving the DNA isolation from clinical specimens. Lysis of mycobacterial cells is difficult and lengthy, and a combination of boiling and ultrasonication simultaneously yielded better DNA. No cross-reactivity of the reaction was noted with 27 other strains of mycobacteria, and sensitivity of the assay was less than 10 mycobacterial cells when assayed on serial dilutions of a culture containing approximately 4×10^6 *M. bovis* BCG cells per milliliter.

A promising approach to **direct** and **specific identification** of mycobacteria (although still applied to cultured bacteria) is based on the **amplification of rRNA** (Boddinghaus et al., 1990). rRNA sequences are characteristic for a given organism and can be used for fine taxonomic differentiation as well as for the identification of microorganisms, especially considering the abundance of rRNA in the cell (10^3 to 10^4 molecules per cell). Based on the 16S RNA sequences, a highly sensitive and specific amplification assay has been developed for distinguishing closely related taxa within the genus *Mycobacterium* (Boddinghaus et al., 1990). First, the rRNA sequences of interest were transcribed into cDNA by **reverse transcriptase**, followed by amplification of the DNA by PCR. The subsequent identification of the amplified DNA fragment was performed by **hybridization with oligonucleotides**, the specificity of which was defined at a genus, group, or species level. Fewer than 100 mycobacteria could be detected by this assay.

No evaluation of this assay has been carried out on clinical specimens, but the general idea of targeting rRNA sequences for enhancing the sensitivity of an assay has been used in Gen-Probe testing systems. In fact, the **Gen-Probe rapid diagnostic system** has been used to reassign various serovar strains of the *M. avium* complex (H. Saito et al., 1990).

PCR amplification has targeted repeated nucleotide sequences not only in rRNA (Boddinghaus et al., 1990), but also in mycobacterial DNA (Eisenbach et al., 1990). Using a segment of one of the cloned repeats of *M. tuberculosis*, a **PCR assay based on the repetitive 123-bp fragment** has been developed (Eisenbach et al., 1990). The detection limit of the assay was 1 fg of purified target DNA, which is equivalent to about one copy of the *M. tuberculosis* chromosome (~3000 kb). The product was identified by gel electrophoresis after 30 amplification cycles. Addition of the *M. avium* DNA to the reaction did not have any effect on the sensitivity and specificity of the assay; it was, however, greatly affected if applied to the patient's sample, rather than to purified DNA.

Yet another group of researchers has developed **genus-** and **species-specific DNA probes** for the identification of mycobacteria by PCR (Fries et al., 1990). Two **genomic DNA probes** for mycobacteria have been isolated: one genus specific and one species specific for *M. avium*, the latter of particular practical significance in immunocompromised patients. One cloned fragment (**pMAv 17**) was recognized by the three mycobacterial species (*M. tuberculosis, M. avium,* and *M. intracellulare*), and was selected as a genus-specific probe. The other two clones (**pMAv 22** and **pMAv 29**) hybridized predominantly to *M. avium*. The banding pattern in gel electrophoresis produced by the amplified product of the *M. tuberculosis* complex was distinct from that of other mycobacteria. A striking difference was observed in the **detectibility of the target DNA with and without PCR**: the amplification reduced the time of required X-ray film exposure from 24 h to 30 min. The sensitivity was 1 fg *M. avium* DNA by PCR, which corresponds to **less than a single genome** of mycobacterial DNA. In contrast, the direct assay, without PCR amplification, using only specific DNA probes shows a sensitivity of about 1 mg of DNA. This corresponds to 2.5×10^6 mycobacteria, the number of organisms that can be detected by light microscopy (Fries et al., 1990).

The next assay developed by that group of researchers was based on the PCR product generated at **low primer-annealing temperatures** that was unique for *M. tuberculosis* (Patel et al., 1990). Further modification of the PCR assay used a 240-bp fragment that encodes part of the **MPB 64 protein** and a 17-mer oligonucleotide complementary to the central portion of the amplified product was used as a probe (Shankar et al., 1991). Amplification of a **336-bp repetitive fragment** in the *M. tuberculosis* chromosome developed by other workers is capable of detecting the amount of DNA corresponding to **fewer than 10 organisms** (de Wit et al., 1990). Following isolation of the mycobacterial DNA from clinical cerebrospinal fluid (CSF) samples, pleural and pericardial fluid aspirates, and biopsies, as well as from open lung biopsy, this assay proved to be **at least as sensitive as conventional culture** detection methods on the same specimens.

The application of PCR to **uncultured clinical specimens** has been described (Pao et al., 1990). Two 24-bp synthetic oligonucleotides derived from a sequence coding for the 65-kDa antigen of *M. tuberculosis* were used as primers. The assay distinguishes *M. tuberculosis* and *M. bovis* BCG from all other species of mycobacteria. As few as 40 *M. tuberculosis* bacteria (0.1 pg of DNA) could be identified even in the presence of DNA equivalent to 10^6 human cells. The amplified nontuberculous bacterial DNA gave rise to shorter products than characteristic bands of mycobacterial DNA, which could be easily distinguished on ethidium bromide-stained gels. In addition, the PCR assay was also positive in patients receiving specific medications, whereas the same isolates failed to grow in culture. This discrepancy accounted for the apparent low specificity of the assay (62.6%) compared to the "gold standard" — *in vitro* culture of mycobacteria. The exquisite sensitivity of the PCR assay, even in the presence of vast excess amounts of human DNA, makes it a suitable modality for fast determination as well as large-scale evaluation of mycobacterial infection.

A **nested PCR** using two pairs of primers to the insertion sequence IS6110 has been applied for the detection of *M. tuberculosis* in a case of **tuberculous pericarditis** (Godfrey-Faussett et al., 1991). DNA had to be extracted from pericardial fluid, and the diagnostic band could be identified in the patient's specimen within 2 days.

In the case of **tuberculous meningitis (TBM)**, the PCR assay compared favorably with conventional bacteriology and ELISA for CSF antibodies (Shankar et al., 1991). In fact, it proved to be the most sensitive technique, exceeding by some 20% the efficiency of detection by culture (75 vs. 55%). The 20 to 30% false-positive rate of both antigen and antibody detection assays in high-prevalence areas is definitely inferior to the almost 100% specificity of PCR. **Yet another PCR assay** for TBM detected *M. tuberculosis* in CSF specimens from five or six clinically diagnosed patients (K. Kaneko et al., 1990). The primers were for the MTB 64 protein gene sequences.

Compared to the standard amplification protocol, the PCR assay for the 65-kDa antigen using a nested approach in a second set of reamplification was much more sensitive, yielding 13 positive results among 35 samples from tuberculosis patients that tested negative by culture (Pierre et al., 1991). The standard PCR procedure was positive in only one of six culture-negative specimens. Significant **reduction of false-negative** results and improved efficiency of amplification in specimens containing **inhibitors** of PCR can be achieved by pretreatment of clinical samples with **guanidium thiocyanate** (Brisson-Noel et al., 1991). A PCR procedure using genus-specific or *M. avium*-specific probes has been adapted to identify **whole mycobacteria** grown under different culture conditions without prior lysis or isolation of nucleic acids from the organisms (Fries et al., 1991).

6.3.8 *MYCOBACTERIUM PARATUBERCULOSIS*

The role for *M. paratuberculosis*, the etiological agent of Johne's disease, a chronic enteritis of ruminants, and in human inflammatory bowel disease, has caused some controversy. While some researchers did not find a specific relationship between *Mycobacterium* strain Linda and Crohn's disease (Yoshimura et al., 1987), others, using the RFLP technique, established that three fastidious mycobacteria isolated from three persons with Crohn's disease were not only indistinguishable from each other, but also from the type strain of *M. paratuberculosis* (McFadden et al., 1987).

The difficulty of separating *M. paratuberculosis* from *M. avium* complex organisms is due to the close homology of their DNAs; however, a clone, **pMB22**, containing an **insertion sequence IS900** that is a highly specific marker for *M. paratuberculosis* has been defined (Green et al., 1989). It is present in multiple copies in *M. paratuberculosis* strains, but is absent from *M. avium* DNA (Vary et al., 1990). A PCR assay for the identification of *M. paratuberculosis* targeted this insert, providing the first fast (a few hours) and accurate test (Vary et al., 1990) for the assessment of *M. paratuberculosis* in animal (Whipple et al., 1990) and human disease. Analysis of archival material of **tissues from Crohn's disease by PCR failed, however, to identify any mycobacteria, including** *M. paratuberculosis* (Wu et al., 1991).

6.3.9 *MYCOBACTERIUM LEPRAE*

In the case of *M. leprae* neither **monoclonal antibodies** nor **DNA probes** for hybridization offer the required sensitivity. On the other hand, the selection of appropriate primers for **PCR** amplification allowed the detection of a **single bacterium** (Hartskeerl et al., 1989). A nonradioactive assay for *M. leprae* has been developed using two outside and two inside **nested primers** in an amplification assay using a **two-step PCR** (Plikaytis et al., 1990). As little as 3 fg of *M. leprae* genomic DNA could be detected, which corresponds to a single bacillus. Moreover, the assay was capable of amplifying target sequences in **crude lysates** of *M. leprae* bacilli, identifying as few as 20 organisms in tissue biopsy specimens from infected animals and humans. **Another nested PCR assay** detects down to 40 *M. leprae* bacilli using DNA isolated from cultivated organisms (Hackel et al., 1990). Amplification of a 530-bp fragment of the gene encoding the proline-rich antigen of *M. leprae* can be performed on DNA from *M. leprae* in clinical **fixed** or **frozen biopsy** samples from leprosy patients (de Wit et al., 1991).

6.3.10 PARASITES

6.3.10.1 *PNEUMOCYSTIS CARINII*

Pneumocystis carinii has acquired particular notoriety as the major cause of fatal pneumonia in immunocompromised patients due to either **immunosuppressive therapy**, **cancer**, or **AIDS**. So far, the identification of the organism has relied essentially on **direct microscopy** of silver- or Giemsa-stained preparations of bronchoscopic lavage or sputum, the use of **monoclonal antibody reagents** (Linder et al., 1987), or **immunofluorescence** (Leigh et al., 1989). A highly specific and sensitive **PCR** assay for *P. carinii*, combined with oligoblotting for enhancing the specificity, has been described (Wakefield et al., 1990). **Oligoblotting** increased the sensitivity of this PCR assay 100-fold compared to that obtained with **ethidium bromide** band staining. Further clinical correlation studies are in progress to establish the clinical relevance of the extremely low levels of *P. carinii* infection in patients with different immunoprotection levels. An **ISH** protocol using biotinylated oligonucleotide probes for rRNA has demonstrated its utility in complementing conventional staining and immunohistochemical methods for the diagnosis of *P. carinii* (Hayashi et al., 1990).

6.3.10.2 *ECHINOCOCCUS MULTILOCULARIS*

Echinococcus multilocularis causes the most lethal zoonotic disease, alveolar echinococcosis (Schantz and Gottstein, 1986). Using differential antibody screening of cDNA clones of *E. multilocularis*, two potentially **immunodiagnostic antigen gene clones**, designated EM2 and EM4, have been identified (Hemmings and McManus, 1989). While the EM2 clone proved to be nonspecific, the EM4 native antigens are coded for by a single-copy gene showing no homology with other parasites (Hemmings and McManus, 1991). A highly specific **PCR** assay for *E. multilocularis* using two primer sets has been developed that is capable of detecting 50 pg of parasite DNA (Gottstein and Mowatt, 1991). This assay is capable of identifying the organism in a variety of sources, whereas oligonucleotides used in the assay can also be helpful in assessing strain variation, RFLPs, and other manifestations of genetic variation in *E. multilocularis* (Gottstein and Mowatt, 1991).

Chapter 6.4
Sexually Transmitted Diseases

6.4.1 *TRICHOMONAS VAGINALIS*

Trichomonas vaginalis causes one of the most widespread sexually transmitted diseases (STDs) in the world (reviewed in Honigberg, 1989). Although a number of diagnostic modalities exist for the diagnosis of *T. vaginalis*, for example, **microscopy** of wet mounts or stained preparations (Spence et al., 1980), *in vitro* **culture** (e.g., Philip et al., 1987), and **immunological methods** (e.g., Carney et al., 1988; Mason, 1979; Smith, 1986; Street et al., 1982; Watt et al., 1986), the organism is difficult to identify when a low number of *T. vaginalis* are present. To this end, a **2.3-kb *T. vaginalis* DNA fragment** has been cloned and used for the diagnosis of trichomoniasis by **dot-blot** hybridization on vaginal discharge specimens (Rubino et al., 1991). The sensitivity of this radioactively labeled probe was 200 *T. vaginalis* isolates. Of the 20 samples identified as positive by the probe, only 14 were positive by the standard wet mount method.

6.4.2 *HEMOPHILUS DUCREYI*

Conventional laboratory methods used in the identification of *Hemophilus ducreyi* in chancroid lesions are hindered in part by the **autoagglutination** of the organism in clumps when suspended in liquid, precluding **serological agglutination** tests and interfering with the interpretation of **fluorescent-antibody** tests (e.g., Schalla et al., 1986). The association of chancroid with an increased probability of heterosexual transmission of HIV, and the difficulties in conventional methods of identifying *H. ducreyi*, led to the introduction of a **DNA probe assay** for this organism (Parsons et al., 1989). DNA fragments from a virulent strain of *H. ducreyi* used as radiolabeled probes could identify the organism with the sensitivity of 103 colony-forming units (CFU) in **colony hybridization assays** in pure or mixed cultures. An even higher sensitivity must be attained, however, to detect the bacilli directly in a clinical specimen with an **amplification assay** (Parsons et al., 1989). The epidemiology of chancroid can be studied by ribotyping *H. ducreyi* using genomic fingerprints with rRNA from *Escherichia coli* as a probe (Sarafian et al., 1991).

6.4.3 *CHLAMYDIA* SPECIES

The range of diseases caused by chlamydiae, in particular *Chlamydia trachomatis*, an obligate intracellular parasite of eukaryotes, includes urethritis, cervicitis, salpingitis, female infertility, lymphogranuloma venereum, and trachoma (Barnes, 1989; Stamm and Holmes, 1984). A variety of laboratory techniques have been developed for the isolation and detection of chlamydiae, frequently presenting significant diagnostic challenges (for reviews, see Barnes, 1989; Dutilh et al., 1990). Although **automated fluorescence image cytometry** has been described for DNA quantitation and detection of chlamydiae (Lockett et al., 1991), the relative merits of various "conventional" methods will not be discussed here — the cited publications

509

offer detailed analysis of those. In the following, representative applications of gene-level diagnostics for *C. trachomatis* identification are reviewed.

Total **sulfonated** *C. trachomatis* **DNA** has been used as a probe for the **ISH** detection of the parasite on microscope slide-mounted experimentally infected cells as well as in clinical specimens (Dutilh et al., 1988). Visualization of the probe was done by **enzymatic sandwich immunodetection** with alkaline phosphatase as reporter molecule. **Intracellular inclusions** could be detected within 8 h of inoculation of cultured cells using a high probe concentration (10 μg/ml). The ISH results on clinical specimens correlated with the identification of *C. trachomatis* by culture.

Synthetic oligonucleotide probes have been constructed for the 7.4-kb plasmid of *C. trachomatis* (Clyne et al., 1989). Following **solution hybridization** with unlabeled probes, the probes-target duplex is captured onto a microtiter dish surface and, using a **novel signal amplification technique**, and an **alkaline phosphatase-labeled probe**, the organism can be quantitated. Eventually the **chemoluminescent reporter dioxetane substrate** is triggered by the enzyme and the signal is registered by a luminometer or by exposure to instant film. This 5-h assay is specific for all serovars of *Chlamydia*, detecting as few as 10,000 to 20,000 chlamydial elementary bodies (Clyne et al., 1989). The dioxetane detection system offers substantial advantages over colorimetric detection with other substrates, including **horseradish peroxidase.**

A commercial DNA probe for *C. trachomatis* (Gen-Probe PACE, Gen-Probe, Inc., San Diego, CA) uses **ssDNA** labeled with **acridinium ester**, complementary to *C. trachomatis* **rRNA**. Using culture positivity and direct fluorescent antibody detection as the "gold standard", the overall sensitivity and specificity of the probe ranged from 71 and 98%, respectively, to 80 and 98% (Woods et al., 1990). Based on the number of probe-positive results that could not be confirmed by culture, the Gen-Probe PACE was not recommended for use in screening for *C. trachomatis* in women with a low to moderate risk of infection. A somewhat better performance of the Gen-Probe probe was reported in a population with a 17.4% prevalence, in which the sensitivity and specificity were 82.8 and 99.4%, respectively, relative to cell culture (LeBar et al., 1989). A comparison of the Gen-Probe method with culture in **direct testing** of clinical specimens from 117 women and 50 men revealed a particularly diminished sensitivity (68%) and specificity (75%) in the male population when compared to the culture method (Gratton et al., 1990). The other disadvantage of the probe assay is that the quality of the specimen cannot be assessed. Continuing efforts are being put into further improvement of the commercially offered gene probes, and it appears that before long fully satisfactory labor-saving systems will be on the market for clinical laboratory users.

In the meantime, one of the most sensitive techniques available, **PCR** amplification of the target nucleic sequences, has been adapted for *Chlamydia* spp. and is capable of detecting one chlamydial DNA molecule in 105 cells. No cross-reactivity was observed when a variety of mammalian and bacterial DNAs were probed with the *Chlamydia*-specific primers (Pollard et al., 1989). An extremely fast (less than 1 h) and sensitive (10^{-18} g of a sequence of *C. trachomatis* plasmid DNA) PCR assay allowed the resolution of one bacterium equivalent with the bands clearly visible on ethidium bromide-stained gels (Welch et al., 1990). PCR amplification of DNA sequences from the chlamydial genome using **degenerate oligonucleotide primers** has been used to study regulation of *C. trachomatis* RNA polymerase production (Engel et al., 1990).

A **PCR-EIA protocol** that combines detection of the PCR-amplified DNA with an enzyme immunoassay (EIA) was found to be highly sensitive and specific when applied **directly on cervical specimens** (Bobo et al., 1990). This detection system, developed earlier (Coutlee et al., 1989), uses a monoclonal antibody to DNA-RNA duplexes to quantify hybrids formed between amplified DNA and the complementary RNA probes. The combination of solution hybridization, capture of the amplified product by an immunological method, absence of radioactive labels, and **quantifiable format** of the assay makes this PCR-EIA approach an attractive system for the evaluation of infectious processes in a clinical setting. This protocol has also been used to diagnose and monitor *C. trachomatis* eye infection (Bobo et al., 1991).

Another PCR assay for *C. trachomatis* detects one copy of the *Chlamydia* plasmid (Ostergaard et al., 1990). In comparison with cell culture assays, PCR had a sensitivity of 100% and a specificity of 93% when testing 228 clinical samples. However, compared to cell culture or IDEIA enzyme immunosorbent assay, a corrected specificity of the PCR assay proved to be 99%. Assessment of the clinical utility of PCR assays for *C. trachomatis* infection showed an almost complete agreement with the results of culture methods in evaluating the effect of deoxycycline treatment (Claas et al., 1991). The simplicity and speed of the PCR assay allows it to be used for monitoring *Chlamydia* infections. PCR amplification has also been combined with **restriction analysis** of the amplified *C. trachomatis* DNA for **genotyping** (Frost et al., 1991). *C. psittaci* has been specifically detected by a two-step PCR that did not cross-react with *C. trachomatis* and *C. pneumoniae* (Kaltenboeck et al., 1991).

6.4.4 *NEISSERIA* SPECIES

6.4.4.1 *NEISSERIA GONORRHOEA*

A number of cultural and immunological methods are available for the detection of *Neisseria gonorrhoea* in clinical isolates. An evaluation of five **rapid immunologic diagnostic systems** has been conducted on **quadFERM** (API Analytab Products, Plainview, NY); **RapID NH system** (Innovative Diagnostic Systems, Inc., Atlanta, GA); **Gonochek II** (E. I. du Pont de Nemours & Co., Inc., Wilmington, DE), Rim Series **RIM-N kit** (Austin Biological Laboratories, Inc., Austin, TX), and **Phadebact monoclonal GC OMNI test** (Pharmacia Diagnostics AB, Uppsala, Sweden) (Dolter et al., 1990). Of these, the best overall performance was found with the quadFERM test; however, two other kits, RIM-N and RapID NH, also were 100% sensitive and specific on clinical isolates.

A discussion of the earlier gene probes for *N. gonorrhoea* has been presented (Goltz et al., 1990). The simplicity of a **dot-blot** DNA-DNA hybridization format and **rRNA-derived DNA probes** specific for the less conserved regions of rRNA were combined in an assay system that could either exclusively react with *N. gonorrhoea* DNA or could detect other bacterial species (Rossau et al., 1989). rRNA-derived DNA probes were able to differentiate isolates of *Neisseria* at the species, and even at the subspecies, level without the use of Southern analysis (see also Tonjum et al., 1989).

Several groups have conducted comparative studies on the Gen-Probe **PACE** system for *N. gonorrhoea*. In a patient population with a gonococcal disease prevalence of 21%, the Gen-Probe assay demonstrated sensitivity, specificity, and positive and negative predictive values of 93, 99, 97, and 99%, respectively (Granato and Franz, 1989). A subsequent study by the same authors in a similar population obtained similar results (90, 99, 98, and 97%) (Granato and Franz, 1990). Modification of the Gen-Probe assay for *N. gonorrhoea* (**PACE2**) was intended to increase the sensitivity of the assay. In a comparative study of 469 specimens, with an overall disease prevalence of 6.1%, the PACE2 assay performed virtually identically to the culture method in both symptomatic and asymptomatic populations (Panke et al., 1991). The latest modification of the Gen-Probe assay is considered, therefore, suitable for the screening and diagnosis of gonorrheal genital infections in women (Panke et al., 1991).

A different **DNA digoxigenin probe assay** is designed to evaluate **cryptic plasmid-encoded segment gene** *cpp*B sequences, which are also found in the chromosome (Torres et al., 1991). The sensitivity of that assay is 25 pg of DNA or 500 CFU of *N. gonorrhoea*. All *N. gonorrhoea* strains, even those free of the plasmid, tested positive with this probe. The sensitivity of the assay was 95% and the specificity was 98% in 201 patients visiting an STD clinic. The small size of the probe (237 bp) allows short hybridization times (under 2 h), the entire assay taking less than

5 h (Torres et al., 1991). Other probe systems for the detection of *N. gonorrhoea* (Kolberg et al., 1989; Urdea et al., 1989), using genomic DNA sequences coding for the **gonococcal pilin**, showed cross-reactivity with *N. meningitidis*.

6.4.4.2 *NEISSERIA MENINGITIDIS*

A specific **DNA probe, pUS210**, containing a 2-kb insert specifically hybridizing to restriction bands from *N. meningitidis*, has been used for **epidemiological studies** of outbreaks of meningococcal disease in two distinct loci in the United Kingdom (Knight et al., 1990). A different DNA probe reacting with all the meningococci tested has also been described (Jones et al., 1990). The results obtained with these two different types of probes were discordant, emphasizing the need for careful analysis of RFLP data in epidemiological studies (Jones et al., 1990). A PCR assay for *N. meningitidis* using primers directed to the sequences flanking the **dihydropteroate synthase gene (dhps)** allows the earliest diagnosis of meningococcal meningitis when cultures of the CSF may be negative (Kristiansen et al., 1991).

6.4.5 *TREPONEMA PALLIDUM*

With the resurgence of syphilitic infection, better diagnostic methods than **dark-field microscopy, rabbit intratesticular inoculation**, and **serologic tests** (e.g., Hart, 1986; Tramont, 1990) may be welcome. A highly sensitive **PCR** amplification assay capable of detecting 0.01 pg of purified *Treponema pallidum* DNA and even a single treponeme in a suspension has been developed (Burstain et al., 1991). This assay detected *T. pallidum* DNA in serum, cerebrospinal fluid (CSF), and amniotic fluid from syphilis patients as well as in paraffin-embedded tissue. It can be expected that the PCR assay for syphilis may become a valuable addition to the existing battery of tests. The test is somewhat complicated by the need for **confirmatory DNA-DNA hybridization** with a **tpp 47-specific probe** for the enhancement of its sensitivity and specificity, and by its not relying solely on **ethidium bromide** identification of the amplified product on the gel. Compared with the rabbit infectivity test, the PCR test for *T. pallidum* is 100% specific, its sensitivity, however, varies in different types of clinical material, from 100% in amniotic fluid to lower values in sera and CSF, which usually have fewer organisms; the combined sensitivity for all clinical specimens is 78% (Grimpel et al., 1991). Nevertheless, the PCR assay directed to the DNA sequence of the 39-kDa *bmp* gene is capable of detecting *T. pallidum* DNA in CSF for up to 3 years following intravenous administration of penicillin (Noordhoek et al., 1991).

6.4.6 HUMAN IMMUNODEFICIENCY VIRUS 1

6.4.6.1 IMMUNOLOGICAL ASSAYS

The initial approach to the identification of HIV-1 infection has been to use **electro-immunoassay, ELISA, Western blot**, and **immunofluorescence** — all of which determine the presence of antibodies produced in the patient against various structural components of the virus. The detection of antibodies to HIV-1 proteins and the presence of HIV-1 antigens in serum provides valuable diagnostic information on prior exposure to the virus and disease progression (for reviews, see e.g., De Cock et al., 1990; Downie et al., 1989; Lombardo et al., 1989; McHugh et al., 1988). However, instances of misidentification in the case of HIV-2 proteins by Western blot analysis have been reported (Pau et al., 1991). More recently, **recombinant** or **synthetic peptides** have been used, instead of native proteins, in the development of serological tests. Not only is their preparation less hazardous, but it is expected that fewer false-positive results would be generated than in the case of testing sera with cellular components of the native viral extracts (for a review, see Naylor et al., 1990).

Specific antigen epitopes are targeted in the most informative assays designed to evaluate the status of infection, and, possibly, predict its future progression. Several **immunodominant proteins** have been the subject of close scrutiny in this regard: **p24 (gag), p66/p51 (pol), gp41 (transmembrane),** and **gp120 (envelope)** proteins and, more recently, **p17 (gag** or **shell** protein) (Cullen, 1991; Greene, 1991; Haseltine, 1991; Naylor et al., 1990; Rosenberg and Fauci, 1991). It appears that mutations in the envelope protein-determining sequences (*env*) may not be necessary for disease progression (McNearney et al., 1990). Peptides representing **HIV-1-p17 epitopes** have been used to develop novel assays to measure HIV-1-p17-related antibodies and antigens in patient's sera.

The genetic variability of HIV-1 detectable in HIV-seropositive patients leads, among other phenomena, to the variability in proviral ***gag* DNA epitope sequences** in the cytotoxic T lymphocytes of these patients (Phillips et al., 1991). Some of the emerging viral variants are not recognized by autologous T lymphocytes, and the accumulation of such variants is seen as leading to immune escape by HIV-1.

This variability affects the efficiency of detection of HIV-1 in the peripheral blood. Using PCR with primers directed to **the highly conserved sequences** flanking the 5′end of the **gp41** encoding region of the ENV-gene this virus could be identified in PBMCs of all the patients tested (Naher et al., 1991). In contrast, **when primers to variable sequences of the same gene were used some patients could not be diagnosed**.

The definitive test for an active HIV-1 infection is the recovery of HIV-1 or provirus DNA from the infected person. The methods for **HIV-1 cultivation** have been markedly improved, with coculture being the most sensitive in isolating the virus from patients (Jackson et al., 1988; Felber et al., 1990; Ulrich et al., 1988). **Quantitation** of HIV-1 in microculture of infected PBMCs has been described (Dimitrov et al., 1990; Ho et al., 1989; Re et al., 1989). An RNA-RNA hybridization assay has been developed for the semiquantitative estimation of resistance to zidovudine and dideoxyinosine by HIV-1 viral isolates grown in cultured PBMCs (Japour et al., 1991). Detection of HIV-1 and HIV-2 nucleic acids provides not only definitive evidence of viral infection, but it is also used to identify the viruses in blood products. The following discussion will concentrate on the approaches used to **identify** and **quantitate HIV-1** (and to a lesser extent HIV-2) **nucleic acids** in blood and other tissues.

6.4.6.2 HIV HYBRIDIZATION ASSAYS

6.4.6.2.1 Solution Hybridization

Because extraordinarily low concentrations of HIV-1, on the order of 1 infected cell in 100,000 PBMCs, have been documented to occur in asymptomatic infected individuals capable of transmitting the infection, exquisitely sensitive and specific hybridization assays are needed (Harper et al., 1986). The problem of detecting a hybridization signal at exceedingly low concentrations of the hybrids formed has generated a number of ingenious assays. One approach uses a probe that has a **replicatable reporter** — the part of the probe that generates the signal — when a specific hybrid is formed and must be identified (Lizardi et al., 1988). This assay uses the very high specificity of **QB replicase** for its target on the RNA attached to the HIV probe (see Part 1). Subsequent improvement of this assay using an amplifiable hybridization probe resulted in a very sensitive solution hybridization assay directed to a conserved region of the HIV-1/*pol* gene. **Quantitation of the viral molecules** in the sample is based on the kinetics of amplification of the replicable RNA in the probe (Lomeli et al., 1989). The **limit of detection** was about **10,000 target molecules**. Further improvement of the assay can significantly enhance the sensitivity of the assay, and introduction of automated procedures for kinetic measurements of the amplified signal can make the assay useful for clinical laboratories.

Techniques analogous to **sandwich hybridization**, except for the reversibility of the link between hybrid complexes and the affinity support, aim at improving the detection limit of hybridization assays and **quantitating** the target sequences (Gillespie et al., 1989). These procedures were called "**reversible target capture**" (Gillespie et al., 1989; Morrissey et al., 1989). Because the hybrids are released, better washing protocols can be used that reduce the assay noise and thereby increasing the sensitivity of the assay. **Subpicogram** quantities of HIV-1 *pol* gene RNA could be measured with a **signal-to-noise ratio of over 10** (Gillespie et al., 1989).

6.4.6.2.2 *In Situ* Hybridization

Initial reports on the application of **ISH** for the detection of HIV described virally infected cells in frozen brain tissue from AIDS patients (Shaw et al., 1985). Subsequently, infected lymphocytes in lymph nodes and peripheral blood were detected by ISH (Harper et al., 1986). Improvement in the early detection of infected lymphocytes could be achieved by ISH of cocultured cells, thus markedly reducing the speed of diagnosis by culture methods (Busch et al., 1987). Detection of HIV-1 by a **nonisotopic technique**, using **streptavidin** and an **alkaline phosphatase reporter system**, makes it better suited for clinical laboratories (Singer et al., 1989). A **fluorescent modification** of the HIV ISH technique is capable of detecting target sequences as small as 5.3 kb with a sensitivity allowing the identification of HIV proviral DNA when only 5% of the cells in a population are infected (Spadoro et al., 1990). ISH for HIV-1 nucleic acids applied to **bone marrow** examination in patients with AIDS and AIDS-related complex (ARC) showed the infection to involve predominantly mononuclear cells (Sun et al., 1989). The method has been adapted to **formalin-fixed**, **paraffin-embedded tissues**, **cell blocks**, and **smears** of cultured cells using radiolabeled DNA and RNA probes (Shapshak et al., 1990). Commercially available HIV-1 ISH kits are produced by several companies (e.g., Enzo Biochemical, Inc.; Applied Biotechnology, Inc.).

6.4.6.3 HIV POLYMERASE CHAIN REACTION

Exposure to HIV-1 can lead to a **latent form** of infection persisting for up to several years in the absence of active virus replication (e.g., Blattner, 1991). A small fraction (less than 1%) of both T lymphocytes and monocytes in blood has been found by PCR to carry a latent infection at all stages of HIV-1 infection (McElrath et al., 1991). **Immunological methods** may be inadequate for the diagnosis and monitoring of infection in such cases, because antibody response to HIV-1 may vary greatly. Quantitation of the viral load is also not always possible by immunological methods. **Culture methods** can be used but their efficiency varies with different isolates, sample handling variations, and the inherent variability of duration and magnitude of viremia at different stages of infection. Besides, viral cultures using periodic assays of the reverse transcriptase activity over time are lengthy (up to several weeks) and expensive. Although **antigen capture methods** are available their sensitivity may often be below that required for the detection of the low levels of virus present in most clinical specimens (Schneweis et al., 1989).

Conventional nucleic acid hybridization assays using ^{32}P-**labeled probes** detect 10^4 to 10^6 copies of HIV-1, and **nonradioactive labels** lower the assay sensitivity even further (Pellegrino et al., 1987; Richman et al., 1987). Thus, in spite of the relative simplicity and high specificity, the antigen detection and unamplified nucleic acid assays are only suitable for virus detection during the **active stage** of infection, and may fail to detect HIV-1 particularly in **asymptomatic infections** (Keller and Manak, 1990). The above-described ISH and hybridization assays using the **reversible target capture** approach are more sensitive and can detect as few as 10 to 100 infected cells per 10^6 uninfected cells. However, the presence of **integrated viral sequences** in the host genome in only one or a few copies may not be detected by either ISH or other hybridization assays.

6.4.6.3.1 Primers and Amplification Conditions

This challenge is better met by using amplification of viral sequences by **PCR**, which has been shown to possess the required sensitivity and specificity for the detection of HIV-1-related sequences in clinical blood specimens (e.g., Hart et al., 1988; Hufert et al., 1989; Li et al., 1990; Loche and Mach, 1988; Lucotte and Reveilleau, 1989; Psallidopoulos et al., 1989; Rolfs, 1990; Schnittman et al., 1989).

The **standard conditions** for performing PCR amplifications using the original set of primers have been published and widely used in clinical studies (Kwok et al., 1987; Ou et al., 1988). Significant effort has been given to the selection of optimal amplification buffers, annealing temperatures and time, chain elongation temperatures and time, the number of cycles, the amount of *Taq*I polymerase needed, the concentrations of primers, and the electrophoretic conditions for analysis of amplified products (Carman and Kidd, 1989). It was found that deoxynucleotide and primer concentrations and chain elongation time were not important so long as they were kept above a critical level. In **oligomer restriction technique**, in addition to the initial pair of primers, a third, radioactively end-labeled primer (such as K7 or SK03) is added to the amplified product. This **primer contains a restriction enzyme site**, which is then recognized by a corresponding enzyme that generates a restriction fragment of a specific size detectable on the gel (Carman and Kidd, 1989).

Under certain conditions the cycling of the reaction through **three temperatures can be avoided**, because both annealing and chain elongation can be done at 60°C, particularly when DNA free from genomic sequences is being amplified (Carman and Kidd, 1989). This situation, however, may be only infrequently encountered in testing clinical material.

Coamplification is a useful approach to simultaneously generate products to **more than one target sequence**, using a number of primer pairs to sequences of interest. This can be used with particular advantage when only a limited amount of clinical material is available, or when unrelated sequences are added as an internal control of the reaction (Keller and Manak, 1990). As few as 10 copies of the HIV-1 RNA and 10 to 20 copies of integrated proviral DNA can be detected using **simultaneous addition** of primer pairs to the *gag, env, tat,* and *nef* regions of the HIV-1 genome in the same reaction mixture, followed by **analysis of PCR products by liquid hybridization** with end-labeled oligonucleotides and **gel electrophoresis** (Hewlett et al., 1989). This approach can be used for the simultaneous detection of HIV-1 and HTLV-1, or of other unrelated sequences.

Refinements of reaction conditions and **primer selection** are continually being made. One such study has combined lysis of PBMCs directly in a buffer containing sodium dodecyl sulfate (SDS), Triton X-100, and proteinase K (Albert and Fenyo, 1990). This **crude lysate** was then directly amplified in a **two-step nested PCR** using two pairs of primers, first with outer primers, then with inner primers nested within the first primers. The amplified product was **directly visualized by ethidium bromide** staining of agarose gels. Again, the reaction used multiple sets of primers added simultaneously to amplify HIV-1 *gag*, *env*^{gp120}, *env*^{gp41}, and *pol* sequences. The specificity of the assay was proven by the absence of amplification products in samples from healthy controls, as well as in cells infected *in vitro* with HIV-2 and HTLV-1. Detection and quantitation of HIV RNA in the **cell-free portion** of the blood by PCR is described as a useful assay for evaluating disease progression or in monitoring antiviral therapy (Holodniy et al., 1991).

Addition of **several pairs of primers** also increases the **detection rate** compared to that observed with two primer pairs (86 vs. 75%) (Stoeckl et al., 1989). A combination of PCR amplification with the subsequent resolution of the amplified products by **capillary DNA chromatography** has been developed, aimed at the future automation of screening of blood for HIV-1 (Brownlee et al., 1990).

6.4.6.3.2 Amplified Product Analysis and Quantitation of HIV-1

Detection of the PCR-amplified product can be achieved by the **oligomer restriction** described above, **liquid hybridization, filter hybridization** (including **dot blot**, **slot blot**, and **spot blot**), and **direct nucleotide incorporation**. A comparison of these methods suggested that **quantitative estimates** of target sequences can be obtained with the help of **scanning densitometry** of slot-blot hybridization signals, and calibrated solution hybridization (Abbott et al., 1988). **Titration of HIV-1** and **quantitative analysis** of virus expression can be achieved with liquid RNA-RNA hybridization (Volsky et al., 1990), or subsequent to PCR amplification by capturing the products first in solution hybridization incorporating a **europium label**, and then onto streptavidin-coated microtitration wells (Dahlen et al., 1991). The europium label is then quantitated in a **time-resolved fluorometer**. Direct incorporation of radioactive labels for the quantitation of PCR-amplified products yielded ambiguous results. **Southern analysis** is cumbersome and not easily quantifiable. Solution hybridization may provide false-negative results due to sequence diversity of retroviruses. Therefore the recommended approach was to use multiple primer pairs with a comparative analysis of blot and liquid hybridization and, ideally, all positive samples should be confirmed from an independent DNA sample (Dahlen et al., 1991; Volsky et al., 1990).

Retention of the amplified products can be achieved by **affinity-based hybrid collection** (Harju et al., 1990), or by a similar procedure using a **DNA sandwich hybridization** format (Keller and Manak, 1990). The former technique uses 5′-**biotinylated primers**, and the amplified DNA fragments are detected by liquid hybridization using radiolabeled probes. For quantitation, the hybrids are collected on polystyrene microparticles or onto microtiter wells using the **biotin-avidin interaction** (Harju et al., 1990). In the latter approach, the capture of the amplified product is achieved via hybridization to the **immobilized portion** of the viral genome attached to a microtiter well (Keller and Manak, 1990). Both methods are highly sensitive, detecting as few as 10 to 30 molecules of DNA.

Quantitation of the PCR-amplified products is used to extrapolate the amount of target present in the assayed sample. The amount of DNA after amplification (Y) is related to the starting number of target copies (S) by the equation $Y = S(1 + e)^r$, where r is the number of cycles performed and e is the efficiency of the reaction (Dickover et al., 1990). **Variation in efficiency** from 0 (no amplification) to 1 (except doubling of the target sequence) depends on the reaction conditions and the amount of target material being amplified. This variability complicates quantitation by PCR assays. A promising approach allowing quantitation of PCR-amplified products has been described using electrochemiluminescence detection and shown by quantifying the HIV-1 *gag* gene (Blackburn et al., 1991).

Standardization of the assay by introducing either internal or external standards is aimed at offering an amplified product independent of target sequences to be used for the evaluation of assay efficiency and, thereby, the magnitude of amplification of the target. **Internal standards** use **parallel amplification** of a eukaryotic gene sequence amplified in the same reaction as the target. A region of the HLA locus, *DQα*, has been used in a coamplification assay with a portion of HIV-1 *gag* to achieve **simultaneous amplification** and **quantitation** of both target sequences (Kellogg et al., 1990). A less rigorous standardization still allowing quantitation by a PCR assay uses β-globin sequences amplified in a separate reaction alongside amplification of the HIV-1 sequences (Keller and Manak, 1990). Following amplification, the products can be analyzed by oligomer hybridization, dot blot, or sandwich hybridization using appropriate probes for the internal standard and the target.

In an **external standard approach**, known amounts of the target are **amplified in different tubes** (Dickover et al., 1990). One of the drawbacks of internal standardization is the relatively different efficiency of amplification of the target and standard sequences, because they may be present in different amounts. In fact, because of the disparity in the copy number of cellular HLA sequences and HIV-1 templates, an **attenuation in the efficiency** of the HLA amplification was required (Kellogg et al., 1990). Quantitation of the target HIV-1 DNA has been reported on the basis of a linear relationship observed between the common logarithms of the original amount of HIV-1 DNA and those of the radioactivity of amplified HIV-1 DNA generated by hybrids using ^{32}P-labeled probe and measured by a β scanner following electrophoresis (Oka et al., 1990).

6.4.6.3.3 Quality Control

One of the most potentially dangerous aspects of PCR assays with respect to production of erroneous results stems from the very nature of the reaction, that is, its ability to generate up to 10^6 copies of the target sequence. If contaminated, the reaction may produce a signal comparable to that expected in a positive sample but amplified from **spuriously introduced target** sequences. This **carryover** is the main problem that plagues PCR assays in general and various precautions have been suggested to avoid it (see Part 1). **Physical separation of facilities** in which isolation of the nucleic acids from the tested sample is performed from those where the amplification reaction is carried out has been advocated. Dedicated Pipettemen and other laboratory tools, segregated by separate stages of the PCR assay workflow, should be used. The use of **reagents** dispensed in small aliquots and avoidance of movements or procedures that are capable of **aerosolizing** solutions or specimens are among the important precautions. A summary of procedures designed to minimize PCR product carryover and considerations for judicious selection of controls is available (Kwok, 1990).

The use of **standardization** procedures and **multiple primers** to ensure quantitation and required sensitivity and specificity of the assay was discussed above. A set of appropriate **negative controls** is mandatory as well as **parallel testing** of related sequences to ensure maximum specificity. The stringency of hybridization conditions during identification of the amplified products affects the results. **Sequence variability** of the retroviral RNA necessitates that primers be selected for the **conserved regions of the viral genome**, when the assay is used for screening. The **heterogeneity** of HIV-1 has been well documented (Goodenow et al., 1989; Saag et al., 1988). One currently recommended pair of primers (**SK38/SK39**) is designed for the amplification of a 115-bp region of HIV-1 *gag* (Kellogg and Kwok, 1990). The other pair (**SK145** or **SK150/SK101**) amplifies a 130-bp region of *gag* that is conserved among the HIV-1 isolates and HIV-2 (isolate ROD). The confirmatory probes for the SK38-39 and SK145-101 products are **SK19** and **SK102**, respectively (Kellogg and Kwok, 1990).

The effects of **primer-template mismatches** on PCR, such as those that may occur in HIV-1 variants or in different isolates, have been investigated (Kwok et al., 1990). **Single internal mismatches** were found to have **no significant effect** on the PCR product yield. Mismatches that occurred at the 3′-terminal base had varied effects: A-G, G-A, and C-C **mismatches reduced overall PCR product yield** about 100-fold, and A-A mismatches about 20-fold. In spite of that, even the presence of two mismatches allowed significant amplification to proceed (Kwok et al., 1990). The detection efficiency of HIV-1 RNA by RT-PCR is markedly enhanced following the treatment of plasma samples with **heparinase** (Imai et al., 1992). This pretreatment increased the percentage of HIV-1-RNA-positive results from 26 to 68% in the same samples.

6.4.6.3.4 Comparison of HIV-1 Detection by PCR and Serological Methods

Virus isolation from anti-HIV antibody-positive hemophiliacs was more sensitive than **p24 antigen detection** or **decline of p24 antibody**. **PCR** detected the virus in 6 of 10 patients from whom virus could not be isolated (Schneweis et al., 1989). It appears that the PCR may detect latent virus.

While the detection of HIV-1 by the PCR in **seropositive** individuals (Hart et al., 1988; Kwok et al., 1987; Loche and Mach, 1988; Ou et al., 1988) raises little controversy, and is definitely more accurate than tissue culture isolation methods, particularly early in the infection (Wages et al., 1991), results of PCR assays in **seronegative** individuals raise some questions. Some authors detect HIV sequences in peripheral blood lymphocytes from patients seronegative for anti-HIV antibodies, and even in seronegative sexual partners of seropositive individuals (Loche and Mach, 1988). The results can be confirmed with a different set of primers. Negative PCR results in HIV-1 seropositive individuals do occur (Cassol et al., 1991; Mariotti et al., 1990a,b; Ou et al., 1988; Zazzi et al., 1992) and a **low viral titer** may be responsible for this failure (Zazzi et al., 1992), although PCR is capable of detecting HIV-1 proviral DNA in **needles** from injecting drug users when no visible blood is present (Heimer et al., 1992). Furthermore, **sexual partners of blood donors** with isolated and persistent HIV-1 core antibodies did not have HIV-1 DNA according to a PCR assay using one primer for the *gag* region and two primers for the *pol* region (Mariotti et al., 1990a). The latter findings were interpreted to mean that persons with isolated and **persistent core antibodies** (*anti-p24* or *p17*) are not infected with HIV-1. Similar results have been reported by others (Jackson et al., 1990).

PCR testing of HIV-1-**seronegative homosexual men** considered at high risk of HIV infection (over 100 sexual contacts with an HIV-1-seropositive partner without protection), using three sets of primers to the *gag, pol,* and long terminal repeat (**LTR**) **regions**, did not detect the virus (Lefrere et al., 1990a). Likewise, HIV-**seronegative hemophilia patients** tested by the PCR with the same three sets of primers showed no presence of HIV DNA (Lefrere et al., 1990b).

A large comparative study has been conducted in persons at high risk for HIV infection using PCR with the *gag* region primer pair SK38/39 and the *env* region primer pairs SK68/69 and CO71/72 (Horsburgh et al., 1990). While 92% of the HIV-seropositive individuals proved to be positive by the PCR, only 3.4% had HIV DNA detectable by PCR among HIV-seronegative persons. The controversy is not yet resolved because some of the previously HIV-seronegative persons on retesting were HIV seropositive, whereas others remained seronegative. Thus, the possibility of **false-positive PCR** results (3%) cannot be excluded. It appears that **neither PCR nor antibody testing is completely specific for detecting the presence of HIV infection** (Horsburgh et al., 1990). Testing of an **independent sample** and a **clinical follow-up** are advised when discordant results are obtained. The other possibility, that of the presence of **persistent infection without a demonstrable antibody** response reported earlier (Imagawa et al., 1989; Ranki et al., 1987), could not be confirmed by Horsburgh and coworkers (1990).

In another study, 98% of the HIV-1-seronegative patients were also PCR negative (Lifson et al., 1990). In two cases, **PCR positivity could not be reproduced on repeat testing**. PCR was positive in 97% of the seropositive individuals. On repeat testing the three previously PCR-negative antibody-positive men were PCR positive as well. Thus, a generally good agreement has been observed between serological tests and the PCR. It is believed that the **low concentration of the virus** in latent infection and the **effects of freezing and thawing** (the testing was performed largely on frozen samples) may have contributed to the observed discrepancies (Lifson et al., 1990).

Similar **concordance of serological and PCR results** has been observed using three pairs of primers in testing HIV-exposed seronegative neurophilia patients (Shoebridge et al., 1990). However, in a different study, a high proportion (11 out of 27) of sex partners of seropositive hemophiliacs were seronegative, but tested positive by the PCR, suggesting these persons had a latent infection (Hewlett et al., 1990). HIV-1 infection as detected by the PCR was transmitted to 60% of exposed children, including one child from a seronegative PCR-positive mother. Subsequent seroconversion of the PCR-positive, antibody-negative persons suggests that **amplification by PCR using multiple pairs of primers is capable of detecting latent infection**, as suggested earlier (Imagawa et al., 1989).

Detection of HIV-1 RNA or DNA by PCR amplification is of particular importance in **cases when immunological response cannot be expected**, as in the case of **newborn babies**. Besides, maternal antibodies to HIV may persist in the newborns for many months and clinical presentation of HIV-1 infection is not apparent. Although there is no single laboratory test capable of identifying all cases of HIV-infected infants (Rakusan et al., 1991), comparison of HIV detection by lymphocyte cultures with PCR amplification of viral RNA and DNA clearly established that HIV DNA detection by the PCR is far more sensitive for the early diagnosis of HIV infection in offspring of seropositive mothers (Escaich et al., 1991). The PCR assay provided evidence of infection of infants born to seropositive and, in some cases, seronegative mothers well before the development of any AIDS-related symptoms (Hewlett et al., 1990; Laure et al., 1988; Rogers et al., 1989; Rossi and Moschese, 1991). It is advocated that **prenatal screening** should include not only women at risk for having HIV infection but be extended to all pregnant women (Barbacci et al., 1991). Analysis of multiple HIV-1 sequences from the **V3 and V4-V5 regions of the envelope gene** conducted in several mother-infant pairs suggests that only a minor subset of maternal virus infects the infant judging by the limited variety of sequences in the infant compared to that in the mother (Wolinsky et al., 1992).

6.4.6.3.5 Detection of HIV by PCR in Various Tissues

As mentioned above, HIV-1 PCR has been extensively applied to the identification of the viral nucleic acids in peripheral blood (Busch et al., 1991; Hart et al., 1988; Kwok et al., 1987; Loche and Mach, 1988; Lucotte and Reveilleau, 1989; Ou et al., 1988; Psallidopoulos et al., 1989; Schnittman et al., 1989). **Proviral HIV DNA** has been predominantly detected in **CD4** lymphocytes, rather than **CD8** cells, so that every 100th CD4 cell may be infected with HIV-1 (Hufert et al., 1989). DNA amplification for the detection of HIV-1 proviral sequences is possible in **crude cell lysates** of PBMCs by **bracketed** and **nested PCR** protocols, using colorimetric and chemoluminescent substances for the identification of the amplified products in dot-blot hybridization or agarose electrophoresis and Southern blotting (Conway et al., 1990). Evaluation of **pooled PBMCs** from 76,500 blood donations made in San Francisco by viral culture and PCR revealed the very low probability (1 in 61,171) that a screened donor will be HIV-1 positive (Busch et al., 1991).

Application of PCR to **formalin-fixed, paraffin-embedded** tissues demonstrated the presence of HIV-1 DNA in the **lymph nodes, lungs** (Clarke et al., 1990), **periodontal lesions** (Murray et al., 1991), **dried blood spot specimens** (Cassol et al., 1991), and **aortas** of seropositive individuals (Lai-Goldman et al., 1988; Shibata et al., 1989).

Early involvement of the **central nervous system** in the course of HIV-1 infection presenting as meningitis or encephalitis is a well-recognized complication of this disease (Gabuzda and Hirsch, 1987; D. D. Ho et al., 1985; Hollander and Stringari, 1987; Merrill and Chen, 1991). Two patients presenting with meningitis, who tested negative for anti-HIV-1 antibodies in the serum at the most recent evaluation, but who at first evaluation were repeatedly serologically positive for HIV-1 in their **CSF** (but not in serum), proved to have the viral nucleic acids in CSF cells (Rolfs, 1990). PCR amplification with primers for different regions of the *env, gag,* and *pol* genes of the HIV-1 genome demonstrated that there may be a **primary intrathecal B cell immune reaction** against the virus even preceding the appearance of antibodies in the blood.

Analysis of HIV-1 DNA by a semiquantitative modification of PCR in the **brain** tissue and blood of AIDS patients and **autopsy samples** of HIV encephalitis patients revealed the presence of all three forms of retroviral DNA in AIDS patients: **unintegrated linear, unintegrated circular**, and **integrated (the provirus)** (Pang et al., 1990). In autopsy samples from patients with HIV encephalitis, a considerably increased proportion of unintegrated viral DNA was found, which form is thought to account for a higher frequency of cytopathic effects and superinfection (Pang et al., 1990).

The presence of HIV-1 proviral DNA in **urine** pellets containing mononuclear cells from HIV-1 seropositive individuals (Li et al., 1990) could be demonstrated by PCR amplification using only one pair of primers (SK38/39). Urine samples were also positive for the HIV-1 antibodies.

6.4.6.4 CONCLUDING REMARKS

Comparison of the results of serological and PCR amplification approaches indicates that for **screening purposes** the immunological tools may be adequate, whereas the appropriateness of the PCR for screening is not clear. Immunological and cultural methods have been effective in demonstrating the presence of HIV-1 in cool aerosols generated by some surgical instruments (Johnson and Robinson, 1991). Infectious virus and viral antigens could be detected only infrequently and at lower levels in the body fluids other than blood; however, HIV nucleic acids were **detectable in saliva and semen samples in the absence of infectious virus** (O'Shea et al., 1990). On the other hand, the evaluation of **latent forms** of HIV-1 infection as well as the resolution of some controversial findings in **low level infections** can benefit from PCR-based methodology in combination with other modalities.

As the role of HIV integration in the host genome becomes more clear the structural and functional aspects of HIV-1 replication may provide answers to the discrepancies observed in some HIV-1 PCR assays. The complex **topological arrangements of DNA strands** found in unintegrated HIV-1 DNA (for a review, see Kulkosky and Skalka, 1990) as well as the **extreme sequence variability** of the virus even in the conserved regions (e.g., Genesca et al., 1990; Goodenow et al., 1989; Goudsmit et al., 1991) may affect the identification and quantitation of the virus by PCR. Further improvements of amplification techniques will undoubtedly make this modality a **method of choice** for the diagnosis and monitoring of the disease, as well as for the identification of HIV-1 in blood products.

6.4.7 HUMAN PAPILLOMAVIRUSES

6.4.7.1 INTRODUCTION

Human papillomavirus (HPV) became a focus of attention among biologists and clinicians when its putative role as a causal agent of premalignant and malignant lesions seemed to gain support from **epidemiological** (e.g., Koutsky et al., 1988; Meanwell, 1988), **molecular biological** (e.g., Ostrow and Faras, 1987; zur Hausen, 1989a,b), and **clinical** observations (numerous reports in the above-cited references). The pioneering works of zur Hausen (reviewed in zur Hausen, 1989a) and Orth (Orth, 1986) set the stage for a subsequent avalanche of molecular biological evidence linking a **predominant HPV type** with a characteristic **anatomic site** or **lesion**. Based on fine evaluation of the HPV genotype by **high-stringency hybridization** over 60 HPV types and subtypes have been identified so far, with the clear understanding that many more will be characterized in the future (zur Hausen, 1989a). Table 6.2 lists currently recognized HPV types. Analysis of **archival specimens** dating from the 1920s to the 1980s by ISH, Southern blot and PCR indicated that no significant change in the prevalence of HPV DNA or the HPV types occurred over that period (Thompson et al., 1992). In fact, phylogenetic studies performed with PCR assays on HPV16 isolates from the four continents suggest the HPV16 age to be at least several centuries if not equal to the respective age of human races (Chan et al., 1992).

Numerous comprehensive **reviews** dealing with the biology of HPV, its epidemiology, its relation to anogenital and other forms of cancer, methods for its detection and characterization, as well as a description of new HPV types have been published (e.g., Johnson et al., 1991; Lancaster et al., 1986; Ostrow and Faras, 1987; Reid and Campion, 1988; Roman and Fife, 1989; K. J. Syrjanen, 1987, 1989; S. M. Syrjanen, 1987, 1990; zur Hausen, 1989a,b). **Books** and **symposia** dedicated to HPV (e.g., "Papillomaviruses", 1986; Reid, 1987; Syrjanen et al., 1987) leave virtually no aspect of HPV infection untouched, ranging from the molecular genetics to socioeconomic and ethical issues of widespread testing. A detailed discussion of HPV in this chapter, therefore, is not warranted and only **selected facets** of HPV research and clinical concerns will be briefly highlighted to emphasize the practical importance of evaluating HPV infections at the genome and, apparently, subgenome level.

TABLE 6.2
Currently Recognized Human Papillomavirus and Their Clinical Manifestations

HPV Type	Clinical manisfestations[a]
1a–c	Myrmecia-type plantar warts
2a–e	Common warts, filiform warts, mosaic-type plantar warts, palatal warts
3a, b	Flat warts, juvenile warts, mild forms of EV
4	Palmar and plantar warts of hyperkeratotic type
5a, b	Macular lesions of EV, squamous cell carcinomas (EV)
6a–f	Condylomata acuminata, CINI–III, VINI–III, laryngeal papillomas, Buschke-Löwenstein tumors
7	Common hand warts in meat handlers and butchers
8	Macular lesions of EV, squamous cell carcinomas (EV)
9	Macular and flat warts lesions of EV
10a, b	Flat warts
11a, b	Condylomata acuminata, CINI–III, laryngeal papillomas, conjunctival papillomas, inverted nasal papillomas
12	Macular and flat wart lesions of EV
13a, b	Focal epithelial hyperplasia in oral mucosa
14a, b	Macular and flat wart lesions of EV, squamous cell carcinomas (EV)
15	Macular and flat wart lesions of EV
16	Condylomata acuminata, CINI–III, VINI–III, bowenoid papulosis, carcinoma of the cervix, penis, bronchus, sinuses
17a, b	Macular and flat wart lesions of EV, squamous cell carcinomas (EV)
18	Condylomata acuminata, CINI–III, VINI–III, bowenoid papulosis, carcinoma of the cervix and penis
19	Macular and flat wart lesions of EV
20	Macular and flat wart lesions of EV, squamous cell carcinomas (EV)
21	Flat wart lesions of EV
22	Macular lesions of EV
23	Macular and flat wart lesions of EV
24	Macular and flat wart lesions of EV
25	Macular and flat wart lesions of EV
26	Cutaneous warts (immunosuppressed patient)
27	Cutaneous warts (immunosuppressed patient)
28	Cutaneous flat wart lessions
29	Cutaneous warts
30	Squamous cell carcinoma of larynx, CINI, II
31	CINI–III, carcinoma of the cervix
32	Focal epidthelial hyperplasia in oral mucosa, oral papilloma
33	Bowenoid papulosis, CINI–III, carcinoma of the cervix
34	Bowen's disease, CIN
35	CINI–III, carcinoma of the cervix
36	Actinic keratosis, benign EV lesions
37	Keratoacanthoma
38	Malignant melanoma
39	Carcinoma of the cervix, CINI, II, bowenoid papulosis
40	PIN, CIN
41	Multiple flat wart lesions of the skin, squamous cell cancer
42	Genital papillomas, bowenoid papulosis, flat condylomas, CIN
43	CIN
44	CIN
45	CIN, carcinoma of the cervix
46	Benign EV lesions
47	Macular and flat wart lesions of EV
48	Squamous cell carcinoma of the skin
49	Cutaneous flat wart lesions
50	Macular and flat wart lesions of EV
51	CINI, carcinoma of the cervix
52	CIN, carcinoma of the cervix
53	CIN

TABLE 6.2 (continued)
Currently Recognized Human Papillomavirus and Their Clinical Manifestations

HPV Type	Clinical manisfestations[a]
54	Condyloma acuminatum
55	Bowenoid papulosis
56	CIN, carcinoma of the cervix
57	CIN, skin warts, inverted nasal papillomas
58	CIN
59	CIN
60	Cutaneous warts

[a] CIN, Cervical intraepithelial neoplasia; VIN, vulvar intraepithelial neoplasia; EV, epidermodysplasia verruciformis; PIN, penile intraepithelial neoplasia.

From Syrjanen, K. J. (1989). APMIS 97:957–970. With permission.

Much is still to be learned about the causal role of HPV in cervical cancer; its mode of transmission, reservoir, and contributing factors. Several reviews dealing with the **epidemiology** of HPV infection have appeared (e.g., Koutsky et al., 1988; Meanwell, 1988; Reeves et al., 1989a,b; K. J. Syrjanen, 1990) along with reports of broad-based **screening studies** in various countries (e.g., Melchers et al., 1989a,b; Mitrani-Rosenbaum et al., 1988; Kulski et al., 1987; Munoz et al., 1988) and in women of different ages (e.g., Sadeghi et al., 1989; Celentano et al., 1988). Although significant indirect evidence points toward association of cervical cancer with predominantly "**high-risk**" HPV types 16, 18 and, to a lesser extent, types 31, 33, 35, 41–45, and 51–56, no firm cause-effect relationship has so far been unequivocally established. Dissenting arguments questioning even the indirect relationship mentioned above have been repeatedly voiced **advocating abandonment of mass screening for HPV** (Skrabanek, 1988a,b; Skrabanek and Jamieson, 1985). The preferred epidemiological approach appears to be a prospective longitudinal study, such as one under way in Birmingham (England) on 2000 young women (Meanwell, 1988). Stored cytological preparations will be examined for HPV, Epstein-Barr virus (EBV), and herpes simplex virus (HSV) DNA by various techniques from ISH to PCR.

Although studies of this type will better describe the prevalence of HPV infection in a normal population, the multiplicity of factors involved in the initiation and development of a malignancy tentatively related to a viral infection presents a formidable task to untangle in a large population sample. Only a combination of detailed evaluation of the archival material, using sensitive techniques such as the PCR, with prospective studies may narrow down the few cardinal factors and their temporal and causal relationships. Hopefully, confirmation of hypothetical mechanisms can be demonstrated *in vitro* on cultured cells, such as are being developed by zur Hausen and colleagues (e.g., zur Hausen, 1989a).

In the meantime, wider HPV typing correlated with tissue morphology should be conducted. Rigorously controlled laboratory procedures should be adequate in their sensitivity and specificity to the objectives of such studies. The introduction of nonradioactively labeled probes without a compromise in sensitivity will make such testing a realistic goal. It would appear appropriate, however, to combine such testing with **quantitative evaluation** of at least some tentatively corroborating factors such as hormonal status, oncogene activity, concomitant infections, and tobacco metabolites.

6.4.7.2 HUMAN PAPILLOMAVIRUS INFECTION IN MALES

The role of HPV infection in the natural history of cervical cancer is still not entirely clear. Even less understood is the importance of HPV infection of males in the etiology of either **penile cancer** or cervical lesions of their sexual partners. The significant association of the presence of HPV16 and HPV18 in penile lesions associated with dysplasia, more so than that of HPV6 and HPV11 (S. M. Syrjanen et al., 1987, and references therein), corroborates the earlier findings suggesting the need for detailed evaluation of **anogenital lesions in males**. The molecular biological approach appears to be particularly appropriate because neither location, gross appearance, nor cytological details of penile condylomata are indicative of the presence of HPV16 and HPV18 (S. M. Syrjanen et al., 1987).

HPV infection is **sexually transmissible** and evaluation of males either with signs of infection, and/or those whose **sexual partners** are known to have lesions or HPV, may contribute to the understanding of the natural history of HPV. Schneider and coworkers (1987a) found that 65% of male sexual partners of HPV DNA-positive women also harbored the virus. HPV16 and HPV18 were found in 27% of the couples examined, with infections in the males ranging from completely asymptomatic conditions to condylomata. Again, 58% of the West German and 70% of the U.S. women tested in that study had evidence of infection in the vagina, the vestibule, and the vulva, although most of the women were asymptomatic. Again, a high prevalence of HPV-associated lesions ranging from multiple condylomata to penile dysplasia was found in 65% of male sexual partners of 127 women with cervical intraepithelial neoplasia (Krebs and Schneider, 1987).

On the other hand, no HPV16 DNA has been found in samplings of condyloma acuminata in men, suggesting that "most of the lesions in men are perhaps not as important as the reservoirs of HPVs which cause cervical neoplasia in their female sexual partners" (Sakuma et al., 1988). It would be difficult to share this point of view considering the small size of their sample population (35 men), and the fact that other researchers convincingly detect HPV16 (Campion et al., 1985) and HPV18 (S. M. Syrjanen et al., 1987) in males. Besides, HPV6 was detected in a **primary carcinoma of the urethra** in men (Grussendorf-Conen et al., 1987).

In a series of 108 men with **genitoanal warts**, the high-risk HPV types were found in 11%, with HPV16 being most common in warts with some degree of **dysplasia** (von Krogh et al., 1987). Whether the presence of genitoanal warts is a risk factor in transmission of HPV infection to sexual partners is difficult to assess at present. However, the fact that HPV has been detected in the **semen** of males with severe chronic wart disease (HPV5 or HPV2) at least suggests a possibility that this mode of transmission can take place (Ostrow et al., 1986).

Colposcopic examinations have been combined with HPV typing of male sexual partners of women with flat condyloma or cervical intraepithelial neoplasia (CIN) before and after **5% acetic acid** application to the penis and the anogenital area (Barrasso et al., 1987). Various lesions were found in 64.4% of males with condyloma lesions and in 41.2% of partners of women with condyloma. Penile lesions with features of **penile intraepithelial neoplasia (PIN)** harboring HPV16 and 33 were seen in 32.8% of partners of women with **CIN**.

A very high percentage (93.5%) of the 214 male sexual partners of women with genital tract abnormalities had visible genital lesions, and 6.1% of those biopsied showed histological evidence of PIN (Kennedy et al., 1988). Infected males appear to represent "an important reservoir for female infection and reinfection with the sexually transmitted agent that is associated with CIN and cancer" (Kennedy et al., 1988). This study underscores the need for DNA typing methods in screening males, because in a high percentage of cases histological examination fails to identify the virus. Moreover, about 40% of flat lesions in men are missed if acetic acid application is not used.

In a separate study from the same group (Selvey et al., 1989), the most common finding in colposcopically identified penile lesions was HPV6/11 in the partners of patients with warty atypia or condylomata, whereas HPV16/18 were most commonly seen in partners of women with CIN. In this context it should be noted that in a prospective study of 508 women with HPV lesions of the cervix the "low-risk" types HPV6 and HPV11 were found as frequently as the "high-risk" types HPV16 and HPV18 in the subsequently developed carcinoma *in situ* (K. J. Syrjanen et al., 1988). HPV6 and 11 DNA can be revealed in penile lesions by PCR followed with ISH even in the absence of perinuclear halos and nuclear atypia (Nuovo et al., 1992). The virus, however, was localized to areas of relative hyperkeratosis and a thickened granular layer associated with epithelial crevices.

The importance of testing **symptomatic** as well as **nonsymptomatic males** for HPV infection is emphasized by numerous studies detecting HPV positivity in male partners of women with anogenital HPV-related lesions (Barrasso et al., 1987; Krebs and Schneider, 1987; Ostrow et al., 1986; Schneider et al., 1987a). Although obvious and subclinical HPV-related lesions in men can be **successfully treated** (Krebs, 1989), controlling the infection in men does not seem to influence the rate of failure in the treatment of cervical dysplasia in their female sexual partners (Krebs and Helmkamp, 1990).

6.4.7.3 TISSUE TROPISM OF HPVS

Association of different HPV types with particular kinds of **cell transformation** and **body sites** has been amply documented in a large number of studies thoroughly reviewed elsewhere (e.g., Papillomaviruses, 1986; Reid, 1987; Shah, 1990; K. J. Syrjanen et al., 1987). The major impressions emerging from these studies link HPV types 16 and 18 with the higher grade lesions (e.g., Crum et al., 1987; Lorincz et al., 1987; Macnab et al., 1986; Reid et al., 1987; von Krogh et al., 1988). **HPV18** is thought to play a role in the enhancement of **progression** of cervical cancer (Kurman et al., 1988; Walker et al., 1989), and tends to induce **adenodifferentiation**, whereas **HPV16** leads to **squamous maturation** (Wilczynski et al., 1988a,b). The **multicentric nature** of HPV infections of the anogenital tract is an established fact, with up to 81% of women with **cervical** disease also having **vulvar** disease (Spitzer et al., 1989). HPV16 and HPV18 have been repeatedly implicated in the development of invasive SCC of the vulva (van Sickle et al., 1990).

ISH with biotinylated DNA probes for HPV types 6, 11, 16, 18, and 31 detected HPV DNA in 19% of **anal cancers** overall, and in 32.9% of SCCs (Beckmann et al., 1989). Most of the lesions containing HPV16 displayed only subtle **koilocytotic atypia**. Filter ISH (FISH) and ISH identified HPV6 and HPV11, in equal proportion, in 55% of SCCs arising in condyloma acuminatum of the vulva (Downey et al., 1988). Condylomata arising on **oral mucosa** have been found to contain HPV6, 11 or related genomes in 85% of the lesions (Eversole et al., 1987). In juvenile **laryngeal papillomata**, all cases studied (10) contained HPV6 and/or HPV11 (Terry et al., 1987). A significant proportion of oral SCC (11.8%) and dysplastic lesions (28.6%) contain HPV DNA (S. M. Syrjanen et al., 1988). The most frequently detected HPV types were 6, 11, 16, and 18. HPV35 has also been detected in laryngeal SCC (Somers et al., 1990). Immortalization of **cultured nasal** and **nasopharyngeal cells** has been demonstrated for HPV16 (Debiec-Rychter et al., 1991). Using PCR assays for the *L1* gene of HPV, 71% of patients with **esophageal carcinoma** were found to have HPV DNA either in the tumor itself or in adjacent tissue (Williamson et al., 1991). Interestingly, 15% of patients without esophageal malignancy also demonstrated HPV DNA in the biopsy material. Half of these HPV-positive cases were affected with **clear cell acanthosis**.

Anatomic predisposition of certain HPV types, although not being exclusive of other types or location, remains an **enigma** at present. Some considerations related to the importance of **local hormonal microenvironment** appear to be supported by experimental evidence for at least cervical lesions (reviewed in Peto and zur Hausen, 1986). It is speculated that the **elevated level of progesterone**, as found in oral contraceptive users and in pregnant women, accounts for the high prevalence of HPV in these groups of patients, possibly related to the presence of a **hormone-responsive element** in the noncoding region of genital HPVs.

The widespread presence of HPV in **morphologically unaffected** cervices and progression of selected groups of patients to malignant transformation implicated additional factors as contributing to the process. Among these, in addition to hormonal influences, the role of **smoking, herpesviruses**, and **hygiene** has been implicated (zur Hausen, 1989a). Interestingly, tobacco products and their metabolites are incriminated as suspect cofactors of malignant transformation triggered or driven by HPV not only in the cervix, but also in the oral cavity in the development of **verrucous carcinoma** (S. M. Syrjanen, 1987). In this context, it would be of interest to try and overcome the stubborn resistance of cell lines to transformation by HPV through combined exposure to the viruses and tobacco metabolites.

6.4.7.4 MOLECULAR GENETIC ASPECTS OF HPV INFECTIONS

Following the formulation of a hypothesis linking cancer of the cervix to HPV infection (zur Hausen 1976, 1977), the identification of koilocytic changes in cervical smears as indicative of papillomavirus infection (Meisels and Fortin, 1976), and the demonstration of the genetic heterogeneity of HPVs (reviewed in zur Hausen, 1989a,b) intensive molecular biological studies on the **gene expression** of papillomaviruses started. Although the precise **mechanism of malignization** attributed to HPVs and the details of their interaction with the host cell and other factors contributing to tumorigenesis still remain to be established, sufficient molecular evidence has accumulated to justify specific HPV evaluation at the molecular genetic level in some cases of patient management. The salient features of the current understanding of HPV gene expression are summarized below.

Papillomaviruses (PV) belong to the **Papovaviridae** family of small **DNA viruses**, which have distinct **species, tissue type**, and **differentiation stage tropism**. They induce epithelial or fibroblastic tumors of the skin or mucosa of higher vertebrates (Table 6.3). The virus contains a **single, circular, double-stranded DNA** molecule, on the average 7800 bp long (Sousa et al., 1990). Three genomic regions are recognized in PVs: an **early coding (E)** region, composed of eight genes (*E1* through *E8*) and a **late coding region** (*L1* and *L2*), separated by a **noncoding segment** — the **long control region (LCR)** — which is 0.4 to 1.0 kb long. The LCR harbors a number of **regulatory elements**, including several RNA polymerase II promoters, constitutive and inducible transcriptional enhancers, binding sites for cellular transcription factors, and several copies of the **palindrome ACCN$_6$GGT** (Sousa et al., 1990).

6.4.7.4.1 Role of HPV Integration

HPV DNA can exist in the **extrachromosomal** form, as an **episome**, or **integrated** in the cellular genome. Some HPV types exits exclusively as episomes (e.g., HPV6 and HPV11), whereas others (e.g., HPV16 and HPV18) are found as free DNA molecules in intraepithelial neoplasms (Crum et al., 1985, 1987; Durst et al., 1985) and as integrated DNA sequences in invasive cervical cancers (Fukushima et al., 1990; Lehn et al., 1985; Matsukura et al., 1986). Integration of HPV16 DNA sequences apparently occurs as an **early event** in the pathogenesis of genital tumors (Schneider-Maunoury et al., 1987). HPV DNA is also integrated in the **cell lines** derived from cervical cancers (HeLa, SW756, SiHa, and C4-I) (Boshart et al., 1984; Popescu et al., 1987; Schwartz et al., 1985; Yee et al., 1985). In contrast to **premalignant** lesions containing the HPV DNA in episomal form, **most carcinomas appear to harbor viral DNA integrated into the host cell chromosomes** (reviewed by zur Hausen, 1989b).

Integrated HPV DNA occurs either as **single copies** (SiHa and C4-I — HPV16 and HPV18, respectively) or **amplified** up to 10- to 50-fold (as is HPV18 DNA in HeLa and SW756) (Durst et al., 1985; Pater and Pater, 1985; Schwartz et al., 1985; Yee et al., 1985). HPVs found as episomes can replicate, but in latently infected transformed cells the viral replication is nonpermissive for the production of mature virions. Integrated viral sequences replicate along with those of the cellular genome (Sousa et al., 1990). Interestingly, integration of HPV DNA is **not a random event** and **specific integration sites** are recognized, such as, for example, **chromosome 12** at band **q13** for HPV18 DNA (Popescu et al., 1987). The **opening sites** on circular viral DNA consistently occur within the **3' end** of the *E1* **open reading frame (ORF)**, or the **5' end** of the adjacent *E2* **ORF**. What the controlling event responsible for **disruption of the viral DNA** at that site is, and why this site is consistently selected for opening the circular molecule for subsequent integration, is not yet clear. Little is known concerning the **specificity**

of the site of disruption for different HPV DNA types and correlation of these with the development of carcinomas other than in the anogenital region.

Sometimes integration occurs near cellular **oncogenes**, as found in some cervical carcinomas (Durst et al., 1987). One of the important aspects of the transforming and immortalizing activities of PVs is their **cooperation with oncogenes**. In cell culture models such cooperation has been demonstrated for EJ-*ras* (Storey et al., 1988), v-*fos* (Crook et al., 1988), and c-*myc* (Crook et al., 1989). An important aspect of the transforming activity of HPVs that may lead to a better understanding of their role in anogenital cancers is the ability of certain HPV genes such as *E7* and *E6* to interact with the products of **tumor suppressor genes *RB1* and *p53***, and with simian virus 40 **(SV40) large T, polyoma large T, and adenovirus E1A proteins,** in addition to the product of activated *ras* oncogene (Chesters and McCance, 1989; Crook et al., 1988, 1989, 1991; Matlashewski et al., 1987; Phelps et al., 1988; Scheffner et al., 1990; Storey et al., 1988). The binding of pRB is not essential, however, for the immortalization of primary keratinocytes (Jewers et al., 1992).

It appears that the **flanking sequences** may have an effect on the expression of the viral genome (Shirasawa et al., 1989). In fact, the sites of incorporation of HPV16 and HPV18 DNA sequences into chromosomes of cultured cells as well as in a primary cervical carcinoma were found to be located near **oncogenes** c-*myc*, c-*src*-1, c-*raf*-1, *abl*, and *sis*, although the exact locations have not been determined (Durst et al., 1987; Popescu et al., 1987a,b). It is speculated that activation of the oncogenes located not far from the integration site may contribute to the malignant transformation of the cells harboring HPV16 and HPV18 DNA (Cannizzaro et al., 1988; Durst et al., 1987; Popescu et al., 1987a,b).

The formerly held belief that only the integrated form of the virus is found in cervical cancers should be reviewed in light of evidence that **both episomal and integrated forms of HPV16 are present in invasive cervical cancer** (Matsukura et al., 1989). Up to 70% of specimens showed only an episomal form of HPV16 DNA without integration, as testified to by the analysis of *Pst*I restriction patterns. Some specimens (30%) showed **integrated multimeric forms** of viral DNA either without the episomal form or with the concomitant episomal form. No correlation was found between the forms of viral DNA and the clinical stages of tumor; however, in **pelvic lymph nodes** the virus was present **only in the episomal form**, whereas primary cervical cancer showed both episomal and integrated forms of viral DNA.

Fuchs and coworkers (1989) examined HPV16 DNA-positive primary cancers and several lymph nodes to determine whether HPV16 DNA can be used as a diagnostic marker for the detection of **early node dissemination**. In the primary tumors DNA was integrated in 39%, was predominantly episomal in 36%, and was present in both integrated and extrachromosomal forms to a similar extent in 25% of the cases. Thirteen of the 16 involved lymph nodes contained HPV16 DNA. Integrated DNA showed the same pattern in primary tumors and the nodes, whereas **episomal DNA was reduced in the metastases**. Importantly, HPV16 DNA was detected in 18 out of 59 histologically negative lymph nodes. The prognostic significance of its presence, however, remains to be established. PCR amplification detected the same HPV DNA type in metastases as that found in the primary tumor (Claas et al., 1989). Actually, integration of the HPV genome can occur even in the precancerous stage (Lehn et al., 1988; Schneider-Maunoury et al., 1987).

6.4.7.4.2 Transcriptional Regulation of Human Papilloma Virus

The **nonrandom incorporation of HPV near or in the *E1/E2/E4/E5* regions**, disrupting or deleting these ORFs, is particularly important in determining the **transcription pattern of HPVs** reflecting complex regulatory events. As a result, the 5′ part of the early region (ORFs *E6* and *E7*) is joined to downstream host cell sequences (Sousa et al., 1990). This is reflected in part in the pattern of transcription characteristic of CIN, in which HPV16 DNA exists as an episome, on the one hand, and that observed in invasive carcinomas containing integrated HPV DNA, predominantly HPV16. **Only *E6* and *E7* transcripts are found in invasive tumors**, whereas in CIN, in addition to these transcripts, there are also transcripts from the *E1* and *E2/E4/E5* coding regions. Late region transcripts cannot be detected in either case (Shirasawa et al., 1988; Wilczynski et al., 1988).

It appears that both the *E6/E7* and the *E2/E4/E5* regions are required for the production of the **complete transformed phenotype** (Sousa et al., 1990). Furthermore, both *E6* and *E7* can induce the transformed phenotype (e.g., Bedell et al., 1987), *E7* producing a stronger effect, although in some cases both genes are required for the transformation to occur (Barbosa et al., 1991; Barbosa and Schlegel, 1989; Hudson et al., 1990; Sato et al., 1989; Watanabe et al., 1989). The ***E7* gene product** is abundantly present in cervical carcinomas and derived cell lines, whereas in lesions produced by the benign HPV types 1a, 6, and 11, no *E7* message could be identified (Baker et al., 1987; Bedell et al., 1989; Kanda et al., 1988a,b; Phelps et al., 1988; Schneider-Gaedicke and Schwarz, 1986; Smotkin and Wettstein, 1986; Vousden et al., 1988). Again, ***E7* transcripts are more abundant in cancers than *E6*** when compared to benign papillomas (Nasseri and Wettstein, 1984). Increased HPV18 anti-E7 antibody prevalence is observed in cervical cancer patients (Bleul et al., 1991). Both E6 and E7 proteins appear to have **zinc-binding activity** (commonly observed in regulatory proteins binding to specific DNA sequences) that is associated with, but not essential for, the transforming activity of E7 (e.g., Grossman et al., 1989; Grossman and Laimins, 1989). Consistent with the lower association of HPV6 with malignant tumors it was found that *in vitro* the E6 and E7 genes of HPV6 display only a weak immortalizing activity in human foreskin epithelial cells (Halbert et al., 1992). The *E7* gene product is a multifunctional phosphoprotein resembling the structure and function E1A proteins and the T antigens of other papovaviruses (Phelps et al., 1992).

Binding of the *E7* gene product to the product of the retinoblastoma tumor suppressor gene (pRB) accounts at least in part for its transforming potential and introduction of mutations into the *E7* gene abrogates some of the transforming and transactivating functions of its product (Phelps et al., 1992). As already mentioned, the binding to pRB is not an absolute requirement for the immortalization of cultured keratinocytes (Jewers et al., 1992). Introduction of mutations into a specific Cys-X-X Cys motif that is part of a zinc finger arrangement, however, completely eliminates the ability of HPV16 DNA to immortalize primary keratinocytes (Jewers et al., 1992).

The *E2* ORF plays a regulatory role in **controlling the expression of transformation** and **replication** functions (Sousa et al., 1990). This appears to be common to all PVs and **different versions of the *E2* gene product** are one of the most important regulatory elements in PV gene expression. Thus, while the entire *E2* gene product can produce trans-activation of target genes, a portion of the *E2* ORF encodes a protein termed **E2-TR** that specifically **represses trans-activation** (Chin et al., 1988; Giri and Yaniv, 1988; Hirochika et al., 1987). The E2-TR has been identified only in BPV1, although the described repressor activity of the C-terminal region of HPV16 and HPV11 E2 has also been demonstrated (Chin et al., 1988; Cripe et al., 1987; Lambert et al., 1987).

The **transcriptional regulation by E2** is mediated through direct binding to viral cis elements, specifically to the **short palindromic repeated sequences ACCN$_6$GGT** (for a review, see also Sousa et al., 1990). The minimum requirement for the E2-responsive enhancer is at least two E2-binding sites, although the precise mechanism of transcriptional activation by the *E2* gene products is not clear so far (Sousa et al., 1990).

Binding sites for the E2 protein are found in the noncoding, LCR region of the viral genome (e.g., Kasher and Roman, 1988; Wu and Mounts, 1988; see also Sousa et al., 1990). Interestingly, although E2-binding sites do occur outside of the LCR, in the *E2/E4/E5* as well as in the *E6* (Chin et al., 1989) coding regions, these binding sites are absent in oncogenic HPV types 16, 18, and 33, suggesting a different type of transcriptional regulation in these viruses (Sousa et al., 1990).

Depending on the location of the binding sites in relation to the promoter, a full-length **E2 protein can either activate or silence transcription**. The predominant function of the *E2* gene product appears to be transcriptional **repression** (Sousa et al., 1990).

This function of *E2* correlates with what is presently understood as the tentative **mechanism of malignization** by HPV. Because HPVs in premalignant lesions are in the episomal form, *E2* can efficiently suppress the transcription of the oncogenic ORFs *E6* and *E7*. On the other hand, because the integration frequently accompanying malignization has been shown to disrupt *E2*, the repressor function of the *E2* ORF is lost and the unimpeded transcription of ORFs *E6* and *E7* occurs (Sousa et al., 1990). Although this is an incomplete and obviously simplified model, it helps to explain at least some aspects of HPV involvement in cancers. In studies on primary human keratinocytes, the differential immortalization activities of HPV16 and HPV18 have been assigned to the regulatory region located upstream of the *E6* and *E7* genes (Romanczuk et al., 1991).

The two **late genes *L1*** and ***L2*** are **expressed only in terminally differentiated keratinocytes**. Moreover, *L1* appears to be the most highly conserved PV gene and codes for the major capsid protein carrying the major antigenic determinants for **group-specific cross-reactivity** among different groups of PVs (Browne et al., 1988; Jenison et al., 1989; Sousa et al., 1990; Strike et al., 1989; Thompson and Roman, 1987; Tomita et al., 1987). In fact, antibodies raised against the *L1* gene product are considered for the detection of specific HPVs in clinical lesions (Jenison et al., 1989), whereas, as discussed below, PCR amplification of the *L1* gene sequence is meant to provide the **most broad identification** of HPVs.

6.4.7.5 HPV AND HORMONES

Important features of LCRs are the **constitutive enhancer elements**, active only in some epithelial cells, and the presence of a **glucocorticoid response element (GRE)**, which confers strong inducibility by dexamethasone (e.g., Gloss et al., 1987) and which is apparently involved in the observed elevated susceptibility of pregnant women to HPV infection (Fife et al., 1987). The GRE can also function as the **progesterone response element (PRE)** (Chan et al., 1989). It is not clear whether antihormone influence can modulate HPV infection in the genital tract. It is, therefore, of interest to elucidate whether a relationship can be established between HPV infection and hormonal use. Only a slightly elevated risk of invasive cancer has been detected among women who had been using **oral contraceptives** for a long time (Brinton et al., 1990). No synergism between the presence of HPV16 and HPV18 and oral contraceptive use could be observed, pointing, on the other hand, to an independent contribution of each factor to risk.

A further question was then asked — whether oral contraceptives **interact** with other risk factors, including HPV infection and sexual behavior, leading to the development of preinvasive cervical lesions (Negrini et al., 1990). HPV type was determined by Southern blot analysis. Although oral contraceptive use was not found to correlate with **low-grade** squamous intraepithelial lesions, its use and the length of use were associated with an elevated risk of **high-grade** squamous intraepithelial lesions (Negrini et al., 1990). Finally, although HPV16 and HPV18 seemed to be more readily detected in oral contraceptive users, the etiological role of either factor could not be established.

The response of viral sequences, particularly *E6* and *E7*, to **hormonal stimulation** may vary in different cells (von Knebel Doeberitz et al., 1991). The origin of the cell line and the chromosomal integration site of the HPV DNA influence the rate of response to hormones. It appears that other mechanisms such as viral elements outside the LCR (or the upstream regulatory region, URR), viral-cellular junction fragments, or the flanking cellular sequences at the integration site of the viral DNA are able to override the response of the HPV promoter elements to glucocorticoid hormones (von Knebel Doeberitz et al., 1991). This may account for differences in viral gene expression and cell growth among cancers of independent clonal origin.

In situ studies in cervical carcinomas of various grades demonstrated a reduction in **estrogen receptor (ER) positivity** in cervical cells containing the carcinogenic HPV types 16 and 18, suggesting that the virus genes may suppress ER expression (Nonogaki et al., 1990). This suppression appears to be more pronounced by HPV DNA types 16 and 18 than by types 31, 33, and 35. Clearly, a number of aspects of HPV infection still remain to be elucidated, among them being the intricate regulatory networks reactive to some but not other influences, including hormones, as well as the surprising tropism of PVs for epithelial cells.

6.4.7.6 HPV-HOST CELL INTERACTION

As noted above, the characteristic occurrence of certain types of HPV in **specific anatomic locations**, such as HPV types 6, 11, 16, 18, 31, 33, and 45 in anogenital cancers, so far has not been explained in molecular terms. It is not clear to what extent the genotypic distinctions between different HPV types and relatedness of those found in a particular anatomic site account for specific structure-function relationships with the host cell regulatory systems. Finding both the integrated and episomal forms of HPV16 DNA in CIN suggests that the integration of HPV16 DNA, although frequently associated with malignant cells, is not necessary for cells to become malignant (Fukushima et al., 1990).

As epidemiological studies show (see below), **the sheer presence of HPVs is not sufficient for anogenital cancers to develop**. Interaction of HPVs with host cell gene expression is obviously important in tumorigenesis (for reviews, see also Matlashewski, 1989; Sousa et al., 1990; Spalholz and Howley, 1989). The presence of HPV16 DNA in AIDS-related Kaposi's sarcoma in homosexual men suggests that the role of HPV16 also in this cancer should be examined (Huang et al., 1992). Production of **mature HPV particles** is associated with the process of **host cell differentiation** and proceeds only in **highly differentiated keratinocytes**, although HPV DNA can be detected in **basal cells** as well (Schneider et al., 1985). In stratified cultured keratinocytes, the amplification of HPV31b was found to parallel epithelial differentiation (Bedell et al., 1991). Likewise, concentration of HPV33 mRNA increases in condyloma acuminata with the degree of differentiation of the keratinocytes (Beyer-Finkler et al., 1990). An indication of the role of **host cell factors** in controlling HPV transforming activity also comes from the finding of a specific translocation event involving **chromosome 11** in HPV-transformed cultured cells. Moreover, the introduction of a single copy of chromosome 11 suppressed the malignant phenotype of tumorigenic HeLa/fibroblast hybrid cells (Saxon et al., 1986).

The role of **viral-host interaction** in the course of HPV infection is an intriguing area under intensive study. The concept of an **intracellular surveillance system** controlling the expression of HPV in host cells, developed by zur Hausen (1987), although only partially substantiated by the available clinical and experimental evidence, is intellectually appealing. It may account for the observed latency of HPV infection and the epidemiological observations on the high prevalence of infection in a clinically normal population. zur Hausen (1987) and Spalholz and Howley (1989) summarize some of the ingenious cell culture experiments pointing to the existence of a putative **cellular interfering factor (CIF)** that controls HPV transcription in growing cells *in vivo*. The role of the ***E6-E7* transcripts** in mediating proliferative changes, and the operation of **cellular regulatory mechanisms** controlling early HPV transcription, suggest a coherent **model of a stepwise transforming process**, wherein HPV aided by various cofactors interacts with the host cell and cancer eventually occurs. Most of the arguments in favor of this hypothesis have been derived from experiments on *in vitro* cultured cell hybrids.

The relative abundance of E7 over E6 in cancers and the intricate interplay between these factors in producing malignization in some but not in other cells points to the importance of **cell-specific factors** (Sousa et al., 1990). Interestingly, the E7 nucleoprotein of HPV16 has been found to induce a **tumor-specific rejection response** in mice infected with HPV16 *E7* gene-containing cells (Chen et al., 1991).

Reviews of the cellular factors possibly involved in the control of HPV infection have been presented (Matlashewski, 1989; Spalholz and Howley 1989). An important role of cell-mediated immune responses in determining the fate of papillomavirus infection has been suggested by the observed association of **HLA-DQw3** and **HLA-DR5** and the biology of SCC of the cervix (Wank and Thomssen, 1991). HLA-DR5 is apparently associated with increased risk of this cancer and thus is accompanied also by a marked decrease in the frequency of **HLA-DRw6**. While some contend this finding (Glew et al., 1992), others were able to confirm the association although of a lower strength in a different population (Helland et al., 1992). In fact, analysis of wart progression or regression by RFLPs revealed **DRα *Eco*RI polymorphism** is associated with wart regression and an increased risk of eventual malignant transformation is associated with a **DQα*Pvu*II fragment**. If these preliminary findings are further confirmed also in humans, an additional tool for evaluating the potential development of HPV infection may become available (Han et al., 1992). Besides, the effect of population diversity with respect to the heterogeneity of HLA frequencies still remains to be defined before such associations may be used for evaluating the development of malignancies in patients.

6.4.7.7 RISK FACTORS

The literature dealing with the epidemiology of HPV infections is vast and frequently contradictory (e.g., Azocar et al., 1990; Herrero et al., 1990). Although continuing epidemiological research addresses a wide range of relevant issues, from the age at first intercourse to the prevalence of specific HPV types, the ultimate goal of such studies, i.e., practical recommendations for population screening, patient management, and disease prevention, still remains a rather distant objective. Detailed coverage of pertinent findings is outside the scope of this chapter (for reviews see, e.g., Biro and Hillard, 1990; Herrero et al., 1990; Meanwell, 1988; Munoz et al., 1988; Reeves et al., 1989a,b; K. J. Syrjanen, 1989). Although no general consensus has been reached on a number of issues, and emphatic appeals to "scientific continence" have been voiced (Skrabanek, 1988a,b), a few **general correlations** seem to have emerged. These do not suggest direct "cause-effect" relationships and a **constellation of factors** is certainly at play modulating the potentially oncogenic effects of HPV.

The **variability** of genital HPV infections, fluctuating from manifest to a subclinical or latent infection in the same individual examined at regular intervals, contributes to the markedly divergent prevalence estimates reported by various authors (reviewed in K. J. Syrjanen, 1989). It is also possible that **sampling differences** reflecting the known **heterogeneous distribution** of HPV DNA in epithelia were confounding the incidence data (Meanwell, 1988). Besides, **genomic variability of HPVs**, although recognized, has not been adequately used in epidemiological studies. The potential of this approach has been demonstrated by using sequence variants of HPV16 in evaluating infections with multiple variants in a single patient as well as tracing HPV infections in epidemiological studies (Ho et al., 1991).

Among the **risk factors** implicated in the development of anogenital cancers are an early **age at first intercourse** and **number of sexual partners**, suggesting the possibly higher vulnerability of cervical epithelium of adolescents to potentially transforming factors in semen (e.g., Herrero et al., 1990). The length of exposure to a transmissible agent is thought to be reflected in the increased risk with an increase in the number of steady sexual partners, particularly before age 30 (Herrero et al., 1990). A strong relation between the occurrence of HPV types 16 and 18 and risk of cervical cancer has been repeatedly established, as with **low socioeconomic level**, the **number of pregnancies**, **smoking**, certain **venereal diseases**, and patterns of **male sexual behavior** (reviewed in, e.g., Pfister, 1990; Reid, 1987; K. J. Syrjanen et al., 1987). The weight of epidemiological data, although indirectly suggestive of the role of HPV in anogenital cancers, fails to implicate HPVs as the only cause of cervical neoplasia, despite the experimental evidence demonstrating the oncogenic potential of some HPVs (Munoz et al., 1988).

An indication of the potential interaction between HPVs and **herpesviruses** comes from the findings of enhancement of BPV-induced transformation of NIH 3T3 cells during coinfection with cytomegalovirus (CMV). This is possibly related to an enhancement of host DNA synthesis, thereby augmenting the carcinogenic potential of PVs (Goldstein et al., 1987).

Overexpression of the host proteins has been consistently observed as a result of HSV2 infection in CINIII, but not in normal or abnormal smears of cervical cells (Davis et al., 1988). The same marker proteins, however, were also observed in cells transformed by agents other than HSV. A combination of HSV2 and HPV type 16 or 18 sequences leads to oncogenic transformation of immortalized Syrian hamster embryo cells (Iwasaka et al., 1988). HSV1 sequences as well as a **phorbol ester** have also been found to activate HPV18 gene expression in epithelial cells (Gius and Laimins, 1989).

Analysis of tissue specimens from patients with SCC of the vulva revealed concurrent infection with HPV and HSV2 in 41% of the cases (Kaufman et al., 1988). Likewise, about 30% of genital invasive carcinomas have been found to harbor HSV2 sequences, the concurrent HSV2 and HPV16 DNA presence detected in six of eight tumor samples (Di Luca et al., 1987, 1989). **HIV-related immune deficiency** appears to allow **reactivation** of HPV, eventually leading to epithelial abnormalities (Caussy et al., 1990).

No association of HPV16 infection of the vulva with cigarette smoking, the presence of HSV DNA sequences, and the use of contraceptive drugs could be found in cases of Bowen's type *in situ* SCC from patients with invasive disease (Pilotti et al., 1990). Earlier, grade 3 vulvar intra-epithelial neoplasia was found to display dual HPV and HSV2 infection (Bornstein et al., 1988).

The role of HPV in anogenital cancer is continually being refined, taking into account the multiple risk factors (Koutsky, 1991). Some reappraisal of epidemiological data and risk factor assessment should be made using the most sensitive detection methods available. A representative study of this kind has been performed on 467 women attending a university health service, where HPV has been assayed by the PCR (Ley et al., 1991). The key risk factors for cervical carcinoma (numbers of sexual partners, use of oral contraceptives, younger age, and black race) were strongly associated with genital HPV infection.

6.4.7.8 LABORATORY METHODS FOR THE EVALUATION OF HPV INFECTION

A whole range of molecular biological tools is available for both the researcher and the clinical laboratory to characterize in fine detail not only the type of virus present in the submitted specimen but also its physical form — its integrated or episomal status with respect to the host cell chromosome. A thorough and thoughtful review of these methods has been given by Roman and Fife (1989). In discussing the pros and cons of various techniques — Southern blot, dot blot, reverse blot, *in situ*, filter ISH, and PCR — the authors mention the specificity and sensitivity of the methods, where established, as well as their advantages and deficiencies. The interested reader is highly recommended to consult this review for an introduction to and critical appraisal of the comparative features of molecular biological methods currently available for HPV typing. An important factor affecting all epidemiological studies already alluded to above and frequently addressed in analysis of the vast literature on the subject, is the sensitivity and specificity of the techniques used in the evaluation of HPV infection (K. J. Syrjanen, 1989).

Understanding the **relative merits** and **limitations** of various cytological, histological, immunological, and molecular biological methods used over the years in the evaluation of HPV infections is essential for the analysis of the vast epidemiological data. The introduction of newer techniques, such as the PCR, which is capable of detecting a single viral particle, dramatically changes our ideas about the prevalence of HPV in clinically unaffected individuals. In fact, its exquisite sensitivity challenges the conventional notions of the epidemiological relevance of HPV presence and requires a reappraisal of the accumulated evidence. Clearly, the objective is the establishment of rigorous cause-effect relationships between HPV infection and anogenital cancer.

Detection of a **clinically manifest HPV infection** can be accomplished by any of the clinical diagnostic techniques: **colposcopy**, **Pap smear**, or **punch biopsy**. At this level of detection, the prevalence is about 2 to 3% in the general population (age range, 25 to 60 years) (K. J. Syrjanen, 1989). Although even in clinically manifest HPV infections the more sensitive techniques of DNA-DNA hybridization can fail to detect the virus, it is in the **subclinical** and/or **latent** infections that limitations of various techniques become more pronounced. The presence of HPV DNA may not be expressed in morphological alterations of the tissue and, therefore, will remain undetectable by colposcopy or Pap smears.

Absence of HPV infection by definition implies absence of the virus even when evaluated by the most sensitive technique — **PCR** amplification. This poses a practical dilemma that has not been resolved so far. Namely, because the relevance of a low-grade infection, detectable only at the PCR level, to the eventual development of anogenital cancer is not known, a database must also be obtained at this sensitive level for such a relationship to be analyzed. On the other hand, practical considerations of **mass screenings** of unselected populations favor the use of a significantly less sensitive method, for example, the Pap smear. Even with a follow-up with more sensitive techniques in selected cases, the bias introduced would affect meaningful extrapolations to general populations.

One of the **suggested compromises** is the use of ISH on the tissue material collected at mass screenings; this would also allow topographical tissue evaluation for the presence of HPV DNA (K. J. Syrjanen, 1989).

Availability of probes labeled with **nonradioactive** reporter molecules definitely contributes to their wider use in less sophisticated clinical laboratories. **Biotin-labeled** probes are known for their markedly lower sensitivity (100 to 800 copies per genome) compared to that of ^{3}H-labeled probes (1 to 10 copies per genome) in the DNA-DNA hybridization format (cited in Crum et al., 1988). Using high concentrations (up to 400 µg/ml) of biotinylated DNA, the sensitivity of detection of HPV16 and HPV18 DNA in cervical swabs was found to be comparable to that demonstrated by ^{32}P-labeled probes (Neumann et al., 1986). Significant enhancement of the biotinylated DNA probe sensitivity for a relatively low copy number of HPV DNA, such as in HeLa cells (fewer than 50 copies), can be obtained by a **three-step amplification technique** (a rabbit anti-biotin antibody, a biotinylated goat anti-rabbit antibody, and a complex of streptavidin-alkaline phosphatase or streptavidin-gold or streptavidin-fluorescein) (Guerin-Reverchon et al., 1989).

A **combination** of biotinylated capture probes and ^{32}P-labeled detector probes has been used in a **solution hybridization** method (Parkkinen et al., 1989). This rapid (about 4 h) **sandwich assay** designed for HPV16 DNA in cervical scrapes was positive with SiHa cells containing just one HPV16 DNA copy per cell; ISH failed to detect the viral sequences. The solution sandwich method is suitable for diagnosis of infections with a low copy number of HPV, whereas the *in situ* method is more suited for diagnosis of infections with a high copy number of HPV, even if present in only a few infected cells. A more detailed **summary of comparative features** of a selection of HPV diagnostic studies using filter hybridization methods is given in Table 6.3. In spite of the abundance of studies examining HPV infection in various tissues, only a few reports are available comparing different molecular diagnostic methods (see Table 6.4).

Although no comparison with the sensitivity of radioactively labeled probes is provided, a nonradioactive labeling of DNA by chemical insertion of a **sulfone group** on cytidine residues has been accomplished (Melki et al., 1988). The modified DNA was then detected in a dot-blot hybridization format by monoclonal antibody with high sensitivity and specificity, which can be easily performed in a clinical laboratory setting. HPV11 and HPV16 DNA was found in 92% of samples suspected by histology or colposcopic criteria to have HPV infection as well as in 31 samples that were normal by these criteria. A novel technology using **electrochemiluminescent (ECL) labels** allows a rapid (less than 0.5 h) quantitation of PCR-amplified products (Blackburn et al., 1991) as demonstrated by detecting HPV16 and 18 and Ha-*ras* oncogene (Kenten et al.,

TABLE 6.3
Comparison of Different Hybridization Techniques for Human Papillomavirus Diagnosis

Technique	Sensitivity[a] (copies detected per cell)	DNA needed	Label and testing time	Advantages	Disadvantages
Southern blot	0.1	10 µg	^{32}P, 4–5 d Biotin, 2 d	1. Detects new types 2. Integrated/episomal 3. High/low stringency	1. Very laborious 2. Not suitable for screening
Dot blot	1.0	500 ng	^{32}P, 4–5 d Biotin, 2 d	1. Rapid 2. Suitable for screening	1. False positive 2. High stringency only
Reverse blot	1.0	500 ng	^{32}P, 4–5 d Biotin, 2 d	1. High/low stringency 2. Several types can be identified at the same same	1. Requires individual labeling reactions
Tissue *in situ*	20–50	Few cells	^{35}S, 6–8 d Biotin, 1–2 d	1. Cellular localization 2. Correlation of HPV type with morphology 3. Use of routinely fixed tissue	1. Relatedness only
Filter *in situ*[b]	$1–5 \times 10^{4–5}$ HPV DNA molecules	Few cells	^{32}P, 1–2 d	1. No DNA extraction 2. Rapid	1. Background problems 2. Relatively high copy numbers per cell detected
Sandwich hybridization	$1–5 \times 10^5$ HPV DNA molecules	$10^5–10^6$ cells	^{32}P, 1 d Biotin, 6 h	1. Rapid 2. Detection with liquid scintillation	1. Only one type/sample can be analyzed
PCR	0.00001	10 pg	5 h	1. Highly sensitive 2. Semiautomated	1. Contamination 2. Relatedness only 3. Requires sequence information

[a] Sensitivities given are for radioactive labels.
[b] Only for exfoliated cells.

From Syrjanen, S. M. (1990). APMIS *98*:95–110. With permission.

1991). The ECL labels afford the sensitivity comparable to ^{32}P-based detection systems, reduce PCR contamination and offer a possibility to quantitate the amplified products.

An interesting application of **fingerprinting HPV DNA** has been demonstrated in identifying the later developed SCC of the genital tract as a secondary primary tumor, rather than as a result of the metastatic spread of the previously treated tumor (Bloss et al., 1990). Although HPV16 DNA could be detected in both tumors, Southern analysis with radiolabeled viral probes demonstrated the presence of the *E1* gene in the later tumor specimen. It had been deleted from the HPV DNA in the hysterectomy specimen retrieved 18 months earlier, suggesting that the later tumor was not related to the earlier malignancy. This distinction dictates selection of a different patient management strategem. The practical value of **genotyping HPV** infection is also supported by the findings of the better response of HPV6/11-positive lesions compared to those harboring HPV16/18 to treatment with systemic or topical **interferon** (Schneider et al., 1987b).

Clearly, a **combination** of adequate experimental systems allowing an estimate of oncogenically relevant viral loads to be made, on the one hand, and the use of voluminous data collected by one of the compromise combinations of histological and molecular biological methods of mass screening, on the other hand, will have to be used to generate a practically useful

TABLE 6.4
Comparative Features of Representative Molecular Human Papillomavirus Diagnostic Studies

Ref.	Tissue	Lable; HPV type; time	Sensitivity and specificity	Operational features	Findings	Comments
Henderson et al. (1987)	141 patients: cervical and anal scrapes	^{32}P; HPV6, 11, 16, 18		Radiolabeled *Alu* probes were used for relative quantitation of the signals	HPV DNA present in 16% of women without past or present history of lesions, in 24% of those without visible genital dysplasia, 30% with dysplasia of CINI	HPV group-specific antigen could be detected only in 60% of HPV DNA-positive samples
Webb et al. (1987)	66 patients: scrappings and smear biopsy	^{32}P; HPV6, 11, 16, 18; and 31	1 HPV genome per cell for HPV6 and HPV16	Reverse blotting	96% of 54 patients with condylomas or dysplasia had HPV DNA, mostly type 16	Suitable for detecting multiple defined sequences in small quantities of DNA
de Villiers et al. (1987)	9295 smears	HPV6, 11, 16, 18		FISH. Cytologically normal smears had HPV DNA in 10%, CINI–III had HPV DNA in 35–40%, mostly in women under 30 years	FISH underestimates the total rate of HPV infection by a factor of 2 to 3	
Bergeron et al. (1988)	28 endometrial samples	^{32}P; HPV6/11, 16, 18; 31		Southern blot	No HPV DNA was identified in endometrial samples, either normal, hyperplastic, or neoplastic	
Battista et al. (1988)	88 biopsies: cervix, uterus, vulva, larynx	^{32}P; HPV6, 11, 16, 18; 48 h to 2 weeks		Southern blot	Types 6/11 are present as episomes in benign lesions, types 16/18 are found as episomes and/or integrated in benign, premalignant, and malignant lesions	Analysis of the state of HPV DNA in lesions is suggested as an adjunct tool in monitoring HPV infection of anogenital areas

Reference	Probe	Sensitivity	Method	Results	Comments
Demeter et al. (1988)	19 patients: scrapings and biopsy	^{32}P; HPV6, 11, 16, 18; 1–3 days	24.8 ± 11.5 µg o^{-2} cell DNA; 1 pg HPV DNA	High correlation between scrapings and biopsy. 96-well manifold. Amount of DNA related to *Alu* DNA signals. Pepsin digestion is replaced by alkali detergent lysis of tissue sections. DNA neutralized for nitrocellulose binding. Alkaline DNA solution	No HPV DNA in histologically normal tissues. Formalin-fixed, paraffin-embedded cervical biopsies were suitable for HPV detection and typing by FISH. Alkaline DNA solution binds to nylon. Alkaline lysis improved filterability of samples	Can be used for retrospective studies also correlated with immunohistochemistry. Can be rehybridized with different probes. An alternative to ISH
Nuovo et al. (1988)	39 vaginal condylomata biopsies	^{32}P; HPV6, 11, 16, 18		Southern or dot-blot hybridization	63% of lesions with koilocytotic atypia contain HPV DNA; without koilocytotic changes, 11.7%	Lack of HPV DNA detection by these methods was interpreted as evidence of HPV-unrelated etiology of the lesions without koilocytotic atypia (see, however, PCR data below)
Caussy et al. (1988)	23 condyloma patients, 23 patients with cervical cancer, 33 patients after hysterectomy for nonneoplastic diseases	^{32}P; HPV6, 11, 16,18	Percentage detection: 82% — Southern, 62% — FISH, 72% — ISH; specificity of ISH: 72% for condylomas, 30% for invasive cancer. Overall relative sensitivities, 61–66%; relative specificities, 86%; agreement 77–80%. FISH, 0.1 pg HPV DNA	FISH modified to digest cellular proteins and mucus. ISH most vulnerable to sampling variation. 10 copies of HPV16 genome, detected in SiHa cells, not detectable by ISH	Prevalence of HPV DNA in cancer: 89%; FISH; 70% ISH. Majority of cancers in some lesions have less than 500 copies of the HPV genomes per call	Southern requires unfixed tissue, tedious. FISH is fast, but many nonspecific reactions. ISH works on paraffin-embedded tissue, low sensitivity in a large number of invasive cases. Suitable for retrospective studies on multiple sections

TABLE 6.4 (continued)
Comparative Features of Representative Molecular Human Papillomavirus Diagnostic Studies

Ref.	Tissue	Lable; HPV type; time	Sensitivity and specificity	Operational features	Findings	Comments
Cornelissen et al. (1988)	SiHa, Caski cell lines, cervical smears, biopsy	^{32}P; HPV16	Dot blot: 10^3 Caski cells, 10^5 SiHa cells; FISH, 50 pg HPV DNA; dot blot, 1–5 pg DNA	Dot blot is 10–50 times more sensitive than FISH. Binding capacity of filters saturated at 10^{-6} cells	5 pg of homologous DNA detectable in presence of 500 pg of heterologous DNA. Dot-blot finding coincided with Southern blot analysis for HPV16 DNA. Findings in smears are confirmed by biopsy. No correlation between HPV prevalence and severity of lesions	False positives on smears if vector probes are used. Prevalence by dot blot and Southern is twice that determined by FISH
Gerber-Huber et al. (1988)	Scrappings	^{32}P; HPV6, 11, 16, 18; quanti-tation by *Alu* hybridization, 6–7 days	10^4 HPV molecules	Slot blot is 100 times more sensitive than FISH or Southern blot. Southern analysis used to define HPV type when questionable	90% HPV positive in dysplasia outpatient clinic patients. 83% positive for HPV16/18 alone (42%) combined with other types (41%). 17% positive for HPV6/11	Possible cross-reaction with pBR vector. In slot blot the purified cellular DNA is loaded onto a small area concentrating it and leading to increased sensitivity
Duggan et al. (1989)	Colposcopy scrappings	Biotin (Enzo Biochemical) HPV6, 11, 16, 18, 33	1–10 pg viral DNA = 10^5–10^6 HPV DNA molecules. Similar to Southern blot	Dot blot. Limited cocktail of probes. Fails to identify unrecognized HPV types. Cervical scrapes — low cell yield for good hybrid signal. Dilution of infected cells by uninfected cells leads to false negative	Low-risk types infrequent (7%), high risk (41%), condyloma/CIN I (75% of cases). Typable HPV DNA in 60% cases of condyloma. Nondysplastic noncondylomatous changes suggest latent virus	Unexpectedly low frequency of HPV6/11 and high frequency of HPV16/18/33 in first time colposcopy

Reference	Specimen	Probe	Sensitivity	Method	Results	Comments
Nuovo and Richart (1989)	205 patients: biopsy, smears	ViraPap; Vira Type 6/11, 16/18, 3/33, 35			FISH is superior to ISH in equivocal cases or when other HPV types are present. ISH helpful in differentiating HPV6/11 from oncogenic types. Slot blot and Southern are highly concordant	ISH fails to identify HPV presence in equivocal histology
Melchers et al. (1989a,b)	1963 scraping samples	^{32}P; HPV6, 11, 16, 18, 31, 33	1 pg HPV DNA	FISH, alkaline denaturation, and neutralization prior to application onto the membrane increases sensitivity (5-fold for episomal HPV DNA, and 16-fold for integrated HPV DNA)	FISH and Southern analysis show a 100% correlation. 92.2% of all specimens tested by FISH showed no HPV6/11 or 16 confirmed by Southern analysis. 3.6% were equivocal. 3.2% had both HPV6/11 and 16	Reducing surface of sample application on the membrane 28 mm^2 vs. conventional 314 mm^2; 54 samples/membrane. Improved filterability of sample
Colgan et al. (1989)	30 patients: biopsies of cervical lesions and unaffected area	^{32}P; HPV 16; 10 days	1 pg HPV DNA = 0.1 HPV gene copy per cell; occasional cross-reaction with HPV 31	Southern analysis	28% of patients with CIN or condyloma had HPV16 DNA in lesions and adjacent epithelium. 57% were HPV DNA positive if invasive carcinoma was present	
Auvinen et al. (1989)	467 biopsy specimens	^{32}P; HPV6, 11, 16, 18	Single-stranded RNA probes are more specific than DNA in typing clinical specimens	Spot blot and Southern. PEG 6000-containing hybridization solution suitable for typing specimens, comparable to formamide. Increasing formamide decreases sensitivity, but specificity increases 10%. Dextran increases specificity without loss of sensitivity	Total percentage positive 33.5%; HPV16/18, 14.7%; HPV6/11, 18.2%. High SDS concentration is necessary to diminish background. 10–15% PEG 6000 + 7% SDS is optimal. In clinical samples superiority of PEG over dextran sulfate is not evident	

TABLE 6.4 (continued)
Comparative Features of Representative Molecular Human Papillomavirus Diagnostic Studies

Ref.	Tissue	Lable; HPV type; time	Sensitivity and specificity	Operational features	Findings	Comments
Duggan et al. (1990a,b)	119 patients; scrappings	Biotin; HPV6, 11, 16, 18, 33; 1–10 pg HPV DNA or 10^5–10^6 HPV DNA molecules	HPV DNA positivity rate almost equal for wood (30%) and plastic (32%) spatulas	Dot blot	Wood spatulas collect more cells (over 10^5 cells) than plastic, optimizing the sensitivity of the detection	Nonpurple dots in scrapes by wood spatulas do not affect detection rate
Duggan et al. (1990a,b)	401 patients: cervical scrapes	Biotin; HPV6, 11, 16, 18, 33		Dot blot	41.3% of CINI condylomas had HPV16/18/33; 65.4% in CINII and III	Possible underdiagnosis of HPV6/11 due to lower sensitivity of dot blot
Neumann et al. (1990)	18 patients: scrapings, biopsy	^{32}P; HPV16, 18		FISH less sensitive than ISH	ISH correlates with cytology and histology	No correlation established between the grade of differentiation, clinical stage and HPV positivity, HPV type, or copy number
Ikenberg et al. (1990)	18 patients: vagina, lymph node, metastases	^{32}P		Southern analysis	56% with invasive carcinoma of vagina had HPV DNA, mostly type 16, in 0.5 to 50 copies per cell, some lymph nodes were positive for HPV DNA. 56% of HPV-positive patients were alive without recurrence, 100% of HPV-negative patients died from disease within 13 months	
Nuovo et al. (1990)	100 patients: 130 biopsies	^{32}P; HPV6, 11, 16, 18, 31, 35, 51		Southern analysis	Absence of CIN has 34% positive for HPV. Presence of HPV DNA in cervical lesions lacking CIN characteristic histology predicts current or future CIN in such patients	ISH deemed insensitive for this type of study. Dot-blot technique may replace Southern analysis for this purpose

Ranki et al. (1990)	178 patients: smears, swabs	AffiProbe kit, ^{35}S; 1 day ViraPap/ViraType 6/11, 16, 18; 4 days to 2 weeks	AffiProbe and ViraPap have equal sensitivity cut-off: 1.5 times the mean of the background. 1.5×10^7 HPV DNA molecules	Solution hybridization. 10^9 unrelated HPV DNA molecules are needed for cross-reactivity	Internal positive standards define the cutoff values. Comparable to ViraPap/ViraType but faster, suitable for screenings	No purfication is needed for AffiProbe. RBCs and mucus do not interfere, no cross-reactions with unrelated bacteria or viruses
David et al. (1990)	Biopsy	^{32}P; HPV6, 11, 16; 18 days	1 ng to 0.01 pg: HPV6 and 11 cross-react. Slot blot is more sensitive than dot blot and Southern. FISH, 5×10^4 HPV DNA molecules; Southern, 10^7 HPV DNA molecules, and needs μg of DNA material, labor intensive	Slot blot is faster than Southern, increased stringency allows discrimination of HPV6 from HPV11	Detection rate overall, 35.2%. FISH: not quantitative, subjective evaluation. Southern allows highest specificity	
Bartholoma et al. (1991)	50 swabs, 11 biopsies	^{32}P-Oncor; HPV6, 11, 16; 18, 31, 33, 35; up to 2 weeks. Variability in exposure time needed for different specimens. Dot blot could be held at room temperature for up to 2 weeks, and transported without refrigeration	Semiquantitive	Southern blot (Oncor), dot blot (LTI) with the Probe Tech I instrument up to two membranes simultaneously (18 specimens) for Southern. 21 specimens per run of LTI dot blot with triplicate assays	Overall agreement 78.8%, 8 of 13 specimens identified by dot blot, but not by Southern. Blood-tinged specimens give false negatives. Overall prevalence, 50.8%. 78.6% HPV DNA positive in histologically abnormal specimens	Dot blot preferred for clinical specimens. Southern failed to identify in some swab specimens. Southern required freezing throughout. Unexplainable smearing of pattern in Oncor Southern analysis; difficult to interpret

database for the development of screening, treatment, and preventive protocols. In the mean-time, a discussion of the observations made with various techniques evaluating HPV infections at the gene level may be useful.

6.4.7.8.1 Commercial Kits for HPV Assays

The proliferation of commercial kits for the identification and typing of HPV by conventional hybridization procedures (e.g., Enzo Diagnostics, Inc., New York; Genemed Biotechnologies, Inc., San Francisco, CA; Digene Diagnostics, Inc., Silver Spring, MD), and the availability of such testing offered by reference laboratories (e.g., Roche Biomedical, Specialty Laboratories, SKBS, etc.) allow interested clinicians to accumulate sizable statistics on various clinicopathological correlations, including those at the molecular level.

In my experience, HPV typing on smears and scrapes using the **Life Technologies/Digene Diagnostics kit** proved to be a convenient, reproducible, and dependable tool, especially if combined with appropriate controls (e.g., for the DNA content of the material tested). We have also observed that a number of specimens taken from grossly uninvolved areas of the cervix revealed the presence of HPV DNA types 6, 11, 16, 18, 31, 33, and 35. Similar finding have been reported (e.g., Macnab et al., 1986; Schneider et al., 1987a; Wickenden et al., 1987).

The **mode of specimen collection** with the Life Technologies/Digene Diagnostics kit has a significant influence on the results of testing necessitating strict standardization of the collection procedure. Although sampling with a Cytobrush introduced blood into the specimen, the cellularity has always been adequate (see also Duggan et al., 1990a,b). The average yield of cells obtained from Cytobrush samples collected into a lytic collection buffer is about 10^6 cells, which is roughly twice the amount of squamous cells usually collected in ethanol, or that found in an average biopsy specimen (Hjerpe et al., 1990). When a spatula was used to obtain scrapings, a frequently inadequate sampling of cells was obtained.

The results of HPV typing by dot-blot hybridization on **scrapes vs. biopsy** specimen show a poor correlation between the two methods of specimen collection (Hallam et al., 1989). Although the overall prevalence of HPV in scrapes and biopsies was comparable, the scrapes detected more of the HPV18 and HPV31 DNA than did the biopsies. However, only the difference for HPV18 was statistically significant. The differences observed between the two sampling methods possibly reflected the **heterogeneity of distribution** of different HPV types within the cervix. HPV18 and HPV31 appear to be more superficial than HPV16 (Hallam et al., 1989). Further studies are needed to substantiate this assumption; however, rigorous standard-ization of specimen collection for HPV DNA typing is an obvious necessity. **Cervicovaginal lavage** is advocated as a more sensitive cell collection method than the scrape-swab technique for HPV DNA typing (Vermund et al., 1989).

6.4.7.8.2 ISH in HPV Diagnosis

ISH is best suited for the **intracellular "visualization"** of target nucleic acids and their **relationship to the morphology** of cells and tissues. Although **less sensitive** than filter hybridization assays, specific probes for ISH use relatively uncomplicated procedures on **conventional tissue sections** routinely processed in pathology laboratories. A number of findings clarifying various aspects of the pathogenesis of HPV infections have been obtained using ISH with radiolabeled and nonradioactive probes. A brief overview of representative ISH studies for HPV detection is given below, illustrating the various diagnostic possibilities of this method.

6.4.7.8.3 Sensitivity of ISH

Comparing the sensitivity of **biotin-** and **^{35}S-labeled RNA** probes for ISH of HPV16 RNA in **precancerous lesions**, Crum and coworkers (1988) found the ^{35}S-labeled probe displayed almost an order of magnitude higher sensitivity that, on longer exposures, approached that of tritiated probes. However, other reports indicate that both biotin- and ^{35}S-labeled probes can detect one to two copies of HPV16 per SiHa cell using ISH, the only advantage of radiolabeled probes being the possibility of quantitation of the autoradiographic signals (Unger et al., 1991). Colorimetric detection of biotinylated probes offers the advantage of speed and the potential for automation and improved localization of signal, with a sensitivity comparable to that of the radioactive method. Comparison of the Enzo Biochemical ISH kit using biotinylated HPV6/11 and HPV16 DNA probes with Southern blot hybridization demonstrated the sensitivity of ISH to be 88 to 89%, with a specificity of 99% (Todd et al., 1989). The markedly lower sensitivity of biotinylated (Enzo Biochemical) DNA probes for ISH, compared to that of tritium-labeled probes for HPV types 6, 11, and 16, has been demonstrated in vulvular, vaginal, and cervical carcinomas (Grussendorf-Conen et al., 1987; Grussendorf-Conen and Cremer, 1990).

Although the sensitivity of a given hybridization procedure is usually decreased by an order of magnitude when nonradioactive labels are used, ISH with **digoxygenin**-labeled probes for HPV types 11, 16, and 18 shows the **same sensitivity** as that of radiolabeled probes on paraffin-embedded sections. The **topographical localization** of hybrids is in fact even better with digoxygenin-labeled probes than with ^{35}S-labeled probes (Heino et al., 1989).

Fixation in **Bouin** solution markedly reduces the rate of detection of HPV DNA by ISH compared to that in tissue fixed in **buffered formalin** (Nuovo, 1991). This difference was particularly pronounced in the case of high-grade intraepithelial lesions, reaching 23% between the two fixatives. Analysis of the effects of **varying stringency conditions** on ISH using biotinylated DNA probes has been given for **frozen sections** and **paraffin-embedded, formalin-fixed** tissues (Guerin-Reverchon et al., 1990). In addition it was found that frozen sections were more frequently positive than paraffin sections, suggesting that fixation and processing of tissues may impede the penetration of probes to the viral DNA.

ISH with **biotinylated** DNA probes is strongly influenced by the **proteinase pretreatment**. An increase in proteinase concentration to 1–5 from 0.1 µg/ml increased the detection limit for HPV16 DNA from 30 to 40 copies per carcinoma cell to at least 20, at the expense of poor preservation of morphology (Walboomers et al., 1988). While ISH fails to detect HPV DNA in equivocal vulvar lesions, **Southern analysis** can be the method of choice to detect HPV DNA in such cases (Nuovo et al., 1989). The higher diagnostic power of Southern analysis cannot be accounted for solely by the higher sensitivity. In one study, the ISH sensitivity was 90.4%, compared to that of Southern analysis (98.1%), and the specificity of ISH was 86.2% (Garuti et al., 1989). ISH for HPV mRNA was negative in morphologically negative tissues. Although only 53% of morphologically positive cases proved to be antigen positive, 83% were positive for HPV mRNA, with two thirds being HPV6 in ISH assays (Wilbur et al., 1988). A substantially enhanced sensitivity of the detection of HPV6/11, 16, and 18 can be achieved on formalin-fixed, paraffin-embedded tissue sections by introducing the silver enhancement of the horseradish peroxidase-labeled probe (Higgs et al., 1990). This system performs markedly better than the streptavidin-alkaline phosphatase detection.

6.4.7.8.4 ISH in HPV Research

ISH using **single-stranded (ss) sense** and **antisense RNA** radiolabeled probes gives higher signals that correlate with the degree of cellular differentiation, the koilocytotic cells showing the strongest signal (Stoler and Broker, 1986). HPV mRNA can be detected even in relatively undifferentiated dysplastic cells and in invasive carcinomas (Stoler and Broker, 1986).

ISH with ssRNA probes for the **early** and **late regions of HPV genomes** allows evaluation of the HPV gene expression in tissues. The early regions are present even in mature epithelia, whereas the late regions in the oldest and most differentiated keratinocytes (Stoler et al., 1989). Using subgenomic riboprobes for HPV16 mRNA, a **strong antisense RNA signal** could be detected in some genital SCC, suggesting that transcription of antisense RNA could be a natural feature of HPV infection in some tumors (Higgins et al., 1991a).

An ISH study in which the sensitivity of the biotinylated DNA probes was comparable to that of ^{35}S-labeled RNA probes showed that the **intensity of signals** for HPV16 and HPV18 in the malignant lesions was **less** than in the dysplastic or benign lesions in the same patients, consistent with the suppression of replication of HPV DNA in the transformed cells (Fujita et al., 1990).

Studying the relationship of malignancy to **precursor lesions** by ISH with biotinylated probes confirmed the origin of cervical carcinoma from **areas of dysplasia** (Hara et al., 1990). Moreover, a decrease in the presence of HPV16 could be seen in progressively less differentiated carcinoma lesions. A direct correlation could be established between ISH and histopathological data; no such relationship was observed with the Southern analysis data (Hara et al., 1990). ISH with biotinylated probes helps to differentiate between the possible **origins** of malignancy: **cervical adenocarcinomas** (positive for HPV16/18) from **endometrial adenocarcinomas** (negative for HPV16/18) (A. L. Nielsen, 1990). ISH with RNA probes (HPV types 6, 11, 16, and 18) showed the presence of HPV in some inverted papillomas (Brandwein et al., 1989).

6.4.7.8.5 ISH in Clinical HPV Studies

ISH alone or in combination with other techniques helps to establish the source of infection. In a **longitudinal analysis** of a 17-year history of anogenital condylomata with an unusual bladder tumor, ISH analysis identified HPV6 and HPV11 in both types of lesions, suggesting a **common source** and **spread** (Wilson et al., 1990). **Conjunctival papillomas** of children and young adults, but not adult **conjunctival dysplasias**, were found to contain HPV6, or HPV11, apparently acquired during passage through an infected birth canal (McDonnell et al., 1987). **Condylomata acuminata in children** are associated with the same types of HPV found in anogenital lesions in adults, including types 16 and 18 (Vallejos et al., 1987).

In analyzing the relationship of **metastatic lesions** to the **primary tumor**, ISH offers certain advantages over Southern analysis. Specifically, although Southern blot hybridization can establish the presence of HPV in both cases, it requires fresh frozen tissue and cannot assign HPV to individual cell types. ISH helps to locate HPV precisely within tumor cells in formalin-fixed, paraffin-embedded tissues. Using ^{35}S-labeled HPV16 DNA probes the viral sequences could be identified both in the primary cervical squamous cell tumor and in its lymph node metastases (Lewandowski et al., 1990).

The suitability of a **combination of Pap smears with ISH** for the **screening** and **follow-up** of patients with cervical lesions has been demonstrated in the higher persistence or progression of HPV16/18 DNA and HPV31/35/51 positive lesions compared to HPV6/11 DNA-positive cell alterations (Schon et al., 1991). Although establishing the dependence of HPV ISH on the sampling method used, by correlating HPV ISH with Southern analysis the diagnostic efficiency of ISH was found to be 98.9% for biopsies and 96.9% for cell smears (Czerwenka et al., 1991). Interestingly, although **anal** condylomata and dysplasias harbor the same HPV types as the genital tract (HPV DNA was present in 73% of condylomata and 67% of anal dysplasias), no evidence of HPV DNA can be found in **anal SCC** (Duggan et al., 1989; see also Wolber et al., 1990).

6.4.7.8.6 HPV PCR

Decidedly, one of the most central issues is the consensus on what is to be defined as infection in a case of such a highly prevalent occurrence as one finds with HPV in a clinically normal population. The dilemma is further compounded by the progress in technology, which now allows detection of as few as one copy of viral genome per 10^5 cells. The full potential of implementation of the PCR for HPV detection and typing is still far from clear. Given the method does not lead to accumulation of medically irrelevant data due to inherent sensitivity of the method as well as to contaminations or carryovers, a much better understanding of the true epidemiology of HPV may be obtained. Because PCR can be applied to soluble material as well as to tissues, including archival material, accumulation of such data may not be a matter of the too distant future.

The superior analytical power of the PCR has been applied to identify HPV DNA where ISH had failed (Claas et al., 1989). Indeed, PCR confirmed the presence of the same HPV DNA type in cervical carcinoma as that found by ISH. HPV16 DNA could be detected only in histologically visible CIN lesions and/or koilocytes (Cornelissen et al., 1989). PCR offers unparalleled sensitivity, with the **specificity essentially determined by the oligonucleotide primers used to target sequences**. Therefore, detection of unknown types of HPV DNA would not be possible if the primer does not share common sequences with the target. This may presumably account for some of the false-negative results (Claas et al., 1989). To bypass this limitation, primers corresponding to **highly conserved HPV sequences** capable of detecting both known and unknown HPV DNA types have been constructed (Gregoire et al., 1989). The **universal primers** can be used to detect HPV in **metastasis** in women treated for cervical cancer. It is not clear at this stage, however, to what extent the identification of HPV DNA in metastatic lesions may have clinical significance that would contribute to improved treatment of patients.

Seventy percent of women with a positive cytology are found by PCR to have HPV DNA in their cervical scrapes, compared to 46% detected by the modified filter ISH and Southern analysis (Melchers et al., 1989a). In the same study, PCR detected HPV DNA in 5% of the control group women with normal smears and no history of cervical lesions. Again, it remains to be seen whether the **sheer presence** of HPV DNA in amounts detectable only by PCR in cervical samples and not by ISH or Southern analysis can provide valuable **practical** information necessitating appropriate clinical response, such as more rigorous surveillance or even therapeutic measures.

Epidemiological studies on the prevalence of HPV16 in cervical cancer biopsies and in cervical scrapes from women with normal cytology using a PCR assay revealed that the virus was present in about 80% of normal specimens and in 100% of cancer biopsies and in women after laser treatment for CIN (Tidy et al., 1989a; Young et al., 1989). A subsequent study by Tidy and coworkers (1989b) attempted to resolve the apparent controversy of finding the oncogenic HPV16 in such a high percentage of women with normal cytology. **Pitfalls in PCR application** for clinical diagnosis are numerous, among which contamination of the testing system by the previously generated PCR products is one source of potential misdiagnosis. An example of such a misleading finding is the erroneous identification by Tidy and coworkers (1989a) of an HPV16b subtype initially thought by the authors to explain the high prevalence of HPV16 in the normal population. Such findings could undoubtedly affect our views and understanding of the epidemiology and biology of cervical cancer. However, a **more rigorous assay** for HPV16b indicated that the high estimates of prevalence were probably due to accidental contamination (or carryover) to which PCR is vulnerable (Manos et al., 1990). Consequently, the high prevalence results had to be questioned and the **report retracted** (Tidy and Farrell, 1989).

Theoretically, PCR is capable of detecting **single-copy genes** and thus can give a "true" representation of target sequences. In the case of HPV infection, analysis of paraffin-embedded tissues, although 10 to 40 times less sensitive than analysis of cytological preparation, revealed a significant level of carriage of high- and low-risk HPV types in both normal and cancerous tissues (Griffin et al., 1990). A **further refinement of the HPV PCR assay** has been reported for the detection of HPV types 6, 11, 16, 18, and 33 in cervical scrapes that excluded the possibility of detecting cloned HPV plasmids (van den Brule et al., 1989). **The overall prevalence of HPV in cytologically normal scrapes was estimated to be 6%; in those showing cytologic dysplasia the HPV prevalence went up to 60%**. In biopsies of **SCC of the cervix**, the HPV prevalence was 90%, all cells containing only HPV16 and HPV18. These findings agree with the HPV16 PCR data reported by Manos and coworkers (1990).

A study of the reaction conditions using **genotype-specific primers** for HPV types 6b, 16, and 18 emphasized the importance of a particular set of parameters, other than selected primers, that affect the **sensitivity** and **specificity** of amplification (Chow et al., 1990). The expected diagnostic bands in gel electrophoresis were 110 and 154 bp long for HPV16 and HPV18, respectively, which could be confirmed by Southern hybridization. A fragment longer than expected was synthesized with primers for the LCR of HPV6b, possibly explainable by recombinational events in this region of the viral genome (Chow et al., 1990). In a **multiplex format of PCR**, amplified HPV6b products were obtained using both HPV6b and HPV16 primer pairs on a HPV6b template. HPV16 sequences, however, could not be amplified when all three sets of primers were pooled in the same reaction mixture using SiHa DNA with integrated HPV16 DNA as template (Chow et al., 1990). **Different annealing temperatures** were found to be **optimal for amplification** of HPV16 (50°C, but not 60°C) and HPV18 (60°C), whereas HPV6b primers could be annealed at either 50 or 60°C. These **empirical findings** coincided with the **calculated annealing temperatures** for the selected oligonucleotide primers.

A **multiplex PCR assay** designed for the simultaneous amplification of four HPV sequences specific for types 6b, 11, 16, and 18 in a single tube has also been described (Anceschi et al., 1990). That amplification system works on cervical scrapes, as well as on paraffin-embedded material. The HPV PCR amplification systems using **type-specific primer sets** identify respective viral DNA in the target by the product of a particular size and, although being highly specific, they depend on appropriate primer sets for each known viral type (Resnick et al., 1990). Therefore, this approach is unlikely to detect the presence of variant or novel viral types, which can be identified using **consensus primers** (Gregoire et al., 1989; Maki et al., 1991; Snijders et al., 1990; Ting and Manos, 1990; van den Brule et al., 1990a,b). The consensus primers designed for the late HPV region, *L1*, are complementary to the highly conserved viral sequences (see above). Amplification of a wide range of HPV types, including types 6, 11, 16, 18, 31, 33, 35, 39, 40, 42, 45, 51 to 59, as well as 25 other secondary or novel HPV types can be performed (Resnick et al., 1990).

Another consensus primer set designed for a 240-bp region of the **early *E6* gene** aids the detection of a gene consistently associated with HPV-related tumors (Resnick et al., 1990). The **positive-strand primers** selected for the *E6* region are complementary to sequences just upstream of the *E6* ORF, and contain sequences considered to be the target of the regulatory interaction with the ***E2* ORF product** (see above). The **negative-strand primers** are designed for the ***E6* coding region**. Other primers have been constructed that incorporated specificities for the C-X-X-C motifs related to **transcriptional activators**. For various primer sets some of the amplification reaction parameters had to be **empirically established**: the optimal primer ratios, $MgCl_2$ concentrations, and the cycling parameters. Simultaneous amplification of β-globin fragment served to assess DNA **integrity** and the relative **quantity of DNA** in the sample. Although bands of expected sizes could be visualized in the gels in the majority of tumors, in some cases a **high background** necessitated that Southern hybridization be performed to identify amplification products. **Disintegrated target DNA** and **nonspecific priming** were thought to be responsible for the background signals. In each case, the final estimates were positively confirmed by dot-blot hybridization with HPV type-specific probes (Resnick et al., 1990). An additional **advantage of the second set of consensus primers** for the *E6* region yielding a smaller amplification product is that they increase the likelihood of detecting HPV sequences of degraded DNA such as are found in archival material and paraffin-embedded tissues (Resnick et al., 1990).

A similar combination of sets of primers for the late (*L1*) (Snijders et al., 1990) and early (*E1*) regions of HPVs has been developed to expand the HPV types detectable by PCR (van den Brule et al., 1990a,b). Termed the **general primer-PCR (GP-PCR)**, this amplification system complemented the **type-specific (TS)-PCR** and increased the HPV detection from 50 to 80% in mild dysplasia, and from 67 to 88% in severe dysplasia. In cytologically normal scrapes, the prevalence of unsequenced HPV types was about 10%, whereas in abnormal scrapes and carcinoma *in situ* it was as high as 30%. These could be identified as HPV types 13, 30, 31, 45, 51, and other unknown HPV genotypes. Invasive cervical carcinomas harbored only HPV16 and

HPV18. The possibility of using GP/TS-PCR amplifications on **crude cell suspensions** makes this system applicable for large HPV-screening programs (van den Brule et al., 1990b).

6.4.7.8.7 Comparison of PCR with Other Modalities in HPV Diagnosis

PCR-based systems for HPV detection are continuously being improved to achieve the objectives of appropriate scope, specificity, and sensitivity required for a particular study, e.g., mass screening, analysis of archival material, or viral gene expression research. A few studies have addressed the comparative diagnostic features of PCR amplification, filter hybridization methods, and histopathological evaluation of anogenital HPV infections. When the assessment of invasive cervical carcinomas was limited to the detection of only HPV16 and HPV18, 51.8% of lesions proved to contain HPV16 DNA, 14.5% contained HPV18 DNA, and 6.0% contained both viral types detectable by dot-blot or Southern hybridization (Low et al., 1990). Some 91% of dot blot-negative specimens proved to be positive for HPV DNA by PCR, the majority containing HPV16 DNA sequences.

Testing a group of 109 women attending a sexually transmitted disease clinic, 51% of these proved to be positive for HPV DNA by PCR amplification using type-specific primers for HPV types 6, 11, 16, 33, and 18 (Morris et al., 1990). Dot-blot hybridization could identify only 44% of HPV-infected persons. Some 19% of specimens were dot blot negative but PCR positive. In a large number of women with **normal cytology** (33%), HPV DNA was detectable by either method. However, a relatively high concordance of both DNA diagnostic methods with cytology was noted (56 to 59%). It was concluded, therefore, that the PCR was "**slightly more sensitive** than DNA hybridization for detection of HPV" (Morris et al., 1990). This conclusion was at odds with the findings of others (Melchers et al., 1989a,b), and the discrepancies were attributed to the difference between the nature of the specimens tested: **cervicovaginal lavage** vs. **scrapings**. Unexpectedly, 12% of samples negative by the PCR were positive by dot-blot hybridization (Morris et al., 1990). **Primer failure**, the presence of **inhibitors** of PCR amplification and **false-positive dot-blot estimates**, frequently **subjective estimates** of positivity, and **cross-hybridizations** may account for the discrepancies.

A significant number of **histologically abnormal** biopsies **failed** to reveal HPV DNA by PCR (15%) or dot blot (21%), in agreement with earlier observations (McNichol et al., 1989). It is not clear what accounted for this discrepancy — latent infection, imperfections of the assays used or the yet unexplained absence of HPV DNA in a subset of tumors carrying a very poor prognosis (Riou et al., 1990). An intriguing observation has been made in 106 **early stage invasive carcinomas** of the **cervix** (Riou et al., 1990). While there was no difference in the risk of overall relapse associated with individual HPV types, patients with **no detectable HPV DNA** by either Southern analysis or PCR had a 2.6 times higher overall relapse rate and 4.5 times higher risk of distant metastases. It appears that **HPV-negative cervical carcinomas represent a biologically distinct subset of tumors carrying a worse prognosis than do HPV-positive cancers** (Riou et al., 1990). Dissenting opinions maintain that HPV DNA-negative tumors do not carry a worse prognosis and the presence or absence of HPV DNA in cervical carcinomas had little or no influence on clinical outcome (Rose et al., 1991). However, in a retrospective study of 212 patients followed for 6 years using radiolabeled **riboprobes** essentially two groups of patients could be discerned — a younger, HPV-RNA-positive group, with a better prognosis, and older patients, HPV-RNA-negative, who had poorer prognosis (Higgins et al., 1991b). The opposite finding of HPV DNA in otherwise healthy normal cervices was suggested to arise through the transfer of viral DNA with instruments used to examine other infected patients (Lou et al., 1991). Treating the instruments for 2 h in 2% glutaraldehyde was shown to destroy HPV DNA as tested by PCR (Lou et al., 1991).

A large comparative study carried out in 397 randomly selected women participating in an HPV prevalence study evaluated the commercially available ViraPap/ViraType (Digene Diagnostics, Inc.) expanded cocktail (detecting HPV types 6, 11, 16, 18, 31, 33, 35, 42, 43, 44,

45, 51, 52, and 56) and PCR (Gravitt et al., 1991). The primers were specific for the **consensus** **L1 sequences** and amplified products were also identified by **generic** and **type-specific** oligonucleotide probes as well as by a **novel restriction digest** scheme. Furthermore, the generic amplification positive products were then confirmed by PCR amplification using primers complementary to the *E5* ORFs specific for HPV types 6, 16, and 18. The **higher** **sensitivity of the PCR diagnostic assay** was in part limited by the narrow range of oligomer reporter probes used in hybridization rather than by the consensus amplimer pair used. Identification of amplified products by gel **electrophoresis** or the **restriction digest** protocol **broadened the range of detectable HPVs**, although both ViraPap and PCR yielded about a **10% rate of uninterpretable results**. The overall agreement between the two methods was 77.6%, with the PCR identifying 1.7 times as many positives as ViraPap. An important medical observation made in this study was that a **substantial proportion of the population harbors** **the less common HPV types** (Gravitt et al., 1991).

An even larger study conducted in 467 women seen at a university health service for a routine annual gynecological examination compared ViraPap with PCR (Bauer et al., 1991). Up to 46% of the persons tested were infected with HPV as detected by PCR, whereas the ViraPap was positive only in 11%. The majority of women (92%) with condylomatous atypia or dysplasia were HPV positive. The degenerate consensus primers capable of amplifying not only the known types but many unidentified HPV types as well revealed that up to 13% of women have yet unidentified HPV types (Bauer et al., 1991).

Methodologically, the identification of amplified products **solely by their sizes** in electro-phoretic gels proved to be **unreliable** because occasionally apparent human amplimers of molecular weight similar to that of the expected *L1*-amplified product were noted (Gravitt et al., 1991). Restriction analysis of the amplified product used in that study helped to resolve the occurrence of multiple HPV infections difficult to distinguish by hybridization methods alone. The recommendation that PCR be used to screen HPV infections should await further developments making the amplification method more inclusive and verifiable (Gravitt et al., 1991). A step in this direction is clearly seen in the introduction of **simultaneous amplifications** of the ***L1*** and ***E6* ORF consensus sequences** (Resnick et al., 1990).

Comparing ISH with radiolabeled probes, histological reexamination, and PCR assays for *E6-E7* ORFs of HPV types 6, 11, 16, and 18, the HPV DNA detection rate was found to be highest (50%) in the cervical biopsy specimens and lowest (28.6%) in the anal biopsy specimens in 109 women with normal Pap smears (S. Syrjanen et al., 1990). The PCR assay identified HPV DNA in 34.8% of the biopsy specimens found negative by ISH. A **reexamination of the concept** **of the existence of HPV DNA in "healthy women"** , that is, those with histologically normal squamous epithelium, is advocated because, based on this study, it was concluded that (1) **latent** infections of the anogenital tract are **extremely common**; (2) the only **adequately sensitive** method to diagnose such infections is PCR; (3) a **single Pap smear** is **grossly inadequate** in excluding the presence of clinical, subclinical, or latent HPV infection (or even CIN); (4) colposcopic examination with **acetowhite technique cannot distinguish HPV** infections from changes unrelated to HPV; and finally (5) **light microscopic criteria** currently used **fail** to predict subclinical and latent HPV infections adequately (S. Syrjanen et al., 1990).

The characteristic cytological feature, **koilocytotic atypia**, is found in condylomata of the **lower female genital tract**, over 90% of which contain HPV DNA (Nuovo, 1990; Nuovo et al., 1990a,b). Lesions lacking koilocytotic atypia are found, nevertheless, to harbor HPV DNA. The incidence of HPV infection in such cases is estimated to be 7% for vulvular and 3% for cervical lesions when ISH is used, rising to 19 and 36%, respectively, when Southern analysis is performed (Nuovo, 1990; Nuovo et al., 1990a,b). PCR assays gave estimates somewhat **lower** than those by Southern hybridization: 27 and 17% for vulvular and cervical lesions, respectively, the difference apparently being due to the presence of a relatively high proportion of "**novel**" **HPV types**.

The higher estimates obtained for vulvular lesions by PCR than ISH are attributed to the low copy number of viral DNA, apparently below the threshold of the ISH analysis used. In fact, the relatively high number of viral particles required for ISH to be informative can be advantageous. It can be used for discriminating **presumably infectious tissues**, lacking koilocytotic atypia, from tissues without "active" viral proliferation where HPV ISH is negative (Nuovo, 1990). On the contrary, cervical lesions that lack koilocytotic atypia are associated with a high risk of CIN if abnormality of Pap smears is noted at the time of biopsy, and ISH would then be misleading (Nuovo, 1990). These observations contrast with the earlier findings, in which HPV16 DNA could be detected by the PCR only in the cervical conization biopsy containing CIN lesions and/ or koilocytes (Cornelissen et al., 1989).

Likewise, although 96% of **penile condylomata** contain HPV DNA, mostly types 6 and 11, PCR detects the presence of HPV DNA in up to 20% of the lesions lacking koilocytotic atypia negative for HPV by ISH (Nuovo et al., 1990a,b). Interestingly, the majority of penile lesions identifiable only by the acetowhite method were found to lack HPV DNA by ISH and PCR assays. With respect to the pattern of HPV colonization of genital lesions it has been found that 2.4% of cervical and vulvular lesions contain **two** or **more different HPV types** detectable **by ISH** (Nuovo et al., 1991). A higher rate of occurrence (18%) of two different HPV types has been detected by a **type-specific PCR**, although the overall conclusion is that **concurrent active HPV infections** are not a frequent event. More research is needed to establish the precise nature of interaction between concurrent HPV infections.

Using highly sensitive type-specific PCR amplification with the *E6* ORF specific primers of HPV types 16 and 18, followed by dot-blot hybridization of the amplified product with radiolabeled HPV16- and HPV18-specific oligonucleotide probes, 88 **invasive-squamous carcinomas** have been evaluated (Kiyabu et al., 1989). HPV DNA sequences were abundantly present in anal (100%), vulvar (75%), vaginal (64%), penile (40%), and cervical (74%) tumors. A lower incidence of HPV DNA was observed in laryngeal (40%), buccal (38%), and glossal (29%) tumors. No HPV DNA could be detected in esophageal carcinomas. The predominant HPV type was HPV16 (93%) (Kiyabu et al., 1989). In **sinonasal carcinomas**, ISH failed to detect any HPV type, whereas PCR identified HPV6 in invasive papillary SCC of the maxillary sinus (Judd et al., 1991).

The exquisite sensitivity of PCR allowed the detection of HPV18 DNA sequences in **anal adenocarcinomas**, in addition to the well-established presence of HPV in SCC (Koulos et al., 1991). In contrast, **no HPV DNA** could be identified in any of the studied **adenocarcinomas** of the **rectum** and **colon**, or in the **adenomatous polyps** of the colon tested. ISH assays of the same lesions were positive in only one case of anal SCC and in one case of adenocarcinoma of the cervix, failing to detect the virus in any of the other tumors, indicating that, even if present, the HPV genome load did not exceed 10 copies per cell (Koulos et al., 1991). In **sinonasal carcinomas**, ISH failed to detect any HPV type, whereas PCR identified HPV6 in invasive papillary SCC of the maxillary sinus (Judd et al., 1991).

The high level of HPV DNA presence in anal invasive SCC (84.6%) has been confirmed by PCR in another study (Palefsky et al., 1991). Only in 50% of the cases could ISH detect the viral DNA. Whereas PCR was able to detect **herpesvirus (HSV)** in **advanced** cancers (38.5%) and in 75% of **high-grade** anal intraepithelial lesions, ISH failed to detect HSV DNA at all. It is speculated that HSV infection may also play a role in anal SCC progression (Palefsky et al., 1991).

Simplification of the evaluation of patients for HPV infection could be achieved by **skin tests** and efforts at developing such tests using the **HPV16 *E4*** and ***L1* ORF proteins** have been reported (Hopfl et al., 1991). Women with CIN had clearly positive reactions to the C-terminal part of the *L1* ORF protein. The usefulness of this recombinant HPV16 skin test is supported by its apparent specificity for the L1 protein, but not *E4*, reflecting an *in vivo* reaction to HPV by analogy with the tuberculin reaction (Hopfl et al., 1991).

The informative value of testing for HPV infection by the detection of the **major capsid protein** expression still remains to be established. For example, in a series of 20 patients, 10 were known to be HPV16 DNA positive by dot-blot hybridization; only half of these, however, were positive by ISH (Lacey et al., 1991). And only one of the HPV16 DNA positive patients displayed the L1 protein expression demonstrated by monoclonal antibodies to L1 protein, possibly reflecting a recent infection.

6.4.7.9 CONCLUDING REMARKS

A number of important issues in the biology of HPV infections, their prevalence, the role of various risk factors, as well as the choice of the most appropriate technology for population screenings and for individual patient management still remain unresolved. It is clear, however, that adequate molecular tools are available to the practicing physician for objective characterization of HPV infection. Precise recommendations for using a particular technique must be based on the specific objectives of diagnosis and treatment in a given clinical situation. Familiarity with the relative benefits and deficiencies of various diagnostic modalities in the HPV infection evaluation is imperative, and, hopefully, this chapter illustrates this point.

Chapter 6.5

Pathogens Causing
Multisystem Infections

6.5.1 *AEROMONAS HYDROPHILA*

Aeromonas hydrophila can cause both diarrheal and extraintestinal infections with hemolytic and soft-tissue necrosis, and with wound infections (Altwegg and Geiss, 1989; Janda and Duffey, 1988). Immunocompromised hosts may develop *Aeromonas* sepsis or meningitis. It is believed that the major toxin responsible for the broad symptomatology of *Aeromonas* infections is **aerolysin**, produced by some strains of *A. hydrophila*, which exhibits both hemolytic and cytolytic properties (Pollard et al., 1990a). A **PCR** assay targeting a 209-bp fragment of the ***aer*** gene sequence has been developed, and showed a 100% concordance with biological assays for the hemolytic, cytotoxic, and enterotoxin activity of *A. hydrophila* (Pollard et al., 1990a). Cross-reactivity with streptococcal fragments was observed, although the identical bands in gel electrophoresis could be differentiated by *Nci*I digestion. Furthermore, consistent amplification of the streptococcal sequences required 60 cycles with **1 μg** of nucleic acids, whereas only 30 cycles of amplification using **10 ng** of nucleic acids was adequate in the PCR for aerolysin-positive *A. hydrophila*. This assay, therefore, offers a rapid, species-specific virulence test, given that other hemolytic and cytotoxic species of *Aeromonas* and other enteric pathogens are excluded (Pollard et al., 1990a).

6.5.2 *CANDIDA ALBICANS*

Candida albicans produces an advanced multiorgan system failure; however, routine laboratory diagnosis of nosocomial candidemia and disseminated candidiasis is relatively slow (Buchman et al., 1990). Because specific antifungal therapy is delayed by definitive identification of the organism from clinical isolates, a rapid probe assay is definitely desirable. By targeting a gene responsible for the metabolic step (the conversion of lanosterol to ergosterol) that is inhibited by commonly used antifungal drugs (e.g., miconazole, ketonazole), a **PCR** amplification assay has been developed for *C. albicans* (Buchman et al., 1990).

DNA probe methodology can frequently complement the diagnosis and confirmation of other systemic mycoses as well such as caused by *Blastomyces dermatitidis, Coccidioides immitis* and *Histoplasma capsulatum* (Kaufman, 1992).

6.5.3 RUBELLA VIRUS

Rubella virus (RV) is a single-stranded RNA virus that usually produces an acute, mild systemic illness with fever and exanthema (German measles), but in some cases may result in acute and chronic rheumatologic, neurologic, and autoimmune disorders (Eggerding et al., 1991). Isolation of RV in chronic cases and identification by characteristic **cytopathic effects** or by an **indirect enterovirus interference assay** are slow, unreliable, and labor intensive. The

ISH detection of rubella RNA (Filipenko et al., 1988) and the detection of fetal rubella by **nucleic acid hybridization** (Ho-Terry et al., 1988) have been described.

Two **PCR**-based assays of reverse-transcribed RV RNA are capable of fetal diagnosis of rubella infection (Carmen et al., 1989; Ho-Terry et al., 1990). Yet another detailed study of a **nested PCR**-based assay for the detection of both acute and persistent RV infections has been designed to detect a segment of the RV gene that encodes the E1 membrane glycoprotein of RV (Eggerding et al., 1991). The assay was performed either on a single-stranded RV RNA extracted from infected cells or released from virions by **proteinase K digestion**. Two sets of primers were used either independently (nonnested) or sequentially (nested PCR).

Following **reverse transcription** and **amplification**, the DNA product was **sequenced** to confirm the specificity of the RV PCR, which can also be established by REA and Southern blot analysis. The nested RV PCR proved to reduce background amplification and improve the efficiency of amplification. The **sensitivity** of the nested RV PCR was 5 fg of total cytoplasmic RNA from RV-infected cells. Calculations indicate that the assay sensitivity was high enough to **detect one molecule of viral RNA** (Eggerding et al., 1991). By reducing the amount of cDNA in the initial amplification reaction and by using a nested PCR assay, most spurious bands in the amplified product could be eliminated. The **automated sequencing** of the amplified product is advocated as a useful method for confirming sequence information, and specificity of amplification, but may also identify strain differences (Eggerding et al., 1991).

6.5.4 B19 PARVOVIRUS

B19 parvovirus was discovered by accident in 1975 by Cossart and coworkers (1975) in serum from blood-bank donors. The virus causes a number of diverse conditions, such as the common childhood exanthema called fifth disease (Anderson et al., 1984), acute joint symptoms (Reid et al., 1985; White et al., 1985), absolute reticulocytopenia, and anemia due to cytotoxic effects on the erythroid precursors in the bone marrow (Mortimer et al., 1983; Ozawa et al., 1986; Young et al., 1984a,b). Infection during pregnancy may result in hydrops fetalis (Anand et al., 1987; Brown et al., 1984). A **dot-blot** hybridization assay for B19 parvovirus had been developed and used in screening of blood donors for B19 parvovirus (Mori et al., 1989). Screening by **countercurrent immunoelectrophoresis** and dot-blot hybridization has identified 1 positive donor among 24,000 donors tested (Cohen et al., 1990). In a case of chronic bone marrow failure due to persistent B19 parvovirus infection, the presence of B19 virus in the erythroid progenitor cells has been demonstrated by **ISH**, quantitated by dot-blot hybridization, and the replicating forms revealed by **restriction analysis** (Kurtzman et al., 1987).

A dot-blot hybridization assay for the detection of B19 virus in blood samples using **digoxigenin-labeled probes** demonstrates a high sensitivity and reproducibility, making it a suitable test for routine screening of B19 parvovirus DNA in clinical specimens (Azzi et al., 1990). Using this assay, 19 serum samples were found to be positive among 10,150 samples screened for B19 virus (Zerbini et al., 1990).

As mentioned above, intrauterine fetal infection with B19 parvovirus may result in hydrops fetalis (also see Clewley et al., 1987; Franciosi and Tattersall, 1988). A dot-blot hybridization assay is capable of detecting as little as 0.3 pg of viral DNA, when radiolabeled RNA probes are used (Salimans et al., 1989a,b). Sensitivity of the nonradiolabeled DNA probe is much lower (3 pg of viral DNA); however, the **PCR** assay is capable of detecting 100 fg of viral DNA without hybridization of the amplified product. After hybridization with a radioactively labeled probe up to 10 fg of B19 DNA can be detected (Salimans et al., 1989b).

Analysis of **placental tissues** from women diagnosed to have B19 infections during pregnancy, but who gave birth to healthy infants at term, revealed 83% of these to be positive for B19 DNA by PCR (Clewley, 1989). At the high sensitivity level of detection by PCR, these findings were interpreted as reflecting **"decay" of viral DNA** after the peak of viremia, and to have no clinical significance. Comparison of various primer pairs for PCR assays identified the most efficient combinations, which supported an amplification reaction that was 107 times more sensitive than dot-blot hybridization with an internal radiolabeled probe (Koch and Adler, 1990). Not all sets of probes detect all the isolates of B19 virus. Because PCR detects the virus only during the period of viremia, a more comprehensive evaluation should also include anti-B19 IgM (Koch and Adler, 1990). The PCR assay for B19 DNA is of particular benefit for the evaluation of **chronic infection** with the parvovirus in immunocompromised patients, who fail to mount IgG or IgM responses. Analysis of amniotic fluid, fetal serum and maternal serum for B19 DNA by PCR in 56 pregnancies in parallel with the evaluation of B19IgM and IgG allowed the prenatal evaluation of acute and prior infections to be performed in order to identify high-risk pregnancies (Torok et al., 1992).

6.5.5 HUMAN HERPESVIRUS-6

Human herpesvirus-6 **(HHV-6)**, also known as **human B-lymphotrophic virus (HBLV)**, was first isolated from patients suffering from AIDS and other lymphoproliferative disorders (Ablashi et al., 1988). HHV-6 differs from known human herpesviruses biologically, immunologically and by molecular analysis. Elevated antibody titers have been observed in patients with certain malignancies, Sjögren's syndrome, sarcoidosis, chronic fatigue syndrome, some B cell lymphomas, and is frequently associated with HIV-induced immunodeficiency (Ablashi et al., 1988). HHV-6 can lead to interstitial pneumonitis that produces the formerly unexplained lung disease in marrow-transplant recipients and other immunocompromised persons (Carrigan et al., 1991). For a review of HHV-6 see Oren and Sobel (1992).

A **PCR**-based assay for HHV-6 has been developed and demonstrated the prevalence of this virus among AIDS patients to be 83% (Buchbinder et al., 1988). **ISH** of tumors of B cell origin revealed the presence of HHV-6 in some of these. HHV-6 is also detectable by PCR in healthy people. Eighteen of 20 whole-saliva samples and the majority of peripheral blood samples tested positive for the virus (Jarrett et al., 1990). Although only the rare cell in the peripheral blood is infected, it appears that HHV-6 persists in the peripheral blood and oropharynx or salivary glands of most healthy individuals following primary infection (Jarrett et al., 1990). The salivary gland appears to be the site of replication of HHV-6 and a potential site for HHV-6 persistence (Fox et al., 1990). Although there is no correlation between the presence of HHV-6 and antibodies against the virus (Jarrett et al., 1990), the prevalence of the antibody among 200 normal blood donors was found to be 54%, and 63% in patients with non-Hodgkin's lymphoma, and 83% in 25 Hodgkin cases (Torelli et al., 1990).

HHV-6 has been detected in 10-day cultured PBMCs from patients with reactivation of the chronic fatigue syndrome (Josephs et al., 1991). The identity of the virus was established by **Southern analysis** with the **HHV-6-specific probe ZVH 14**. It is suggested that **short-term cultures** as used in this study may offer a reliable **test for reactivation** of HHV-6. HHV-6 is also frequently detected by **PCR** assays in the peripheral blood mononuclear cells from kidney transplant patients (Kikuta et al., 1991).

Different strains of HHV-6 are very closely related as determined by Southern analysis of PCR-amplified HHV-6 sequences of eight HHV-6 strains combined with **partial nucleotide sequencing** (Aubin et al., 1991). It appears that HHV-6 is a unique species with genetic polymorphism, and two groups of related strains with 96% homology can be distinguished on the basis of concordant differences in restriction patterns. Further studies of restriction patterns of various HHV-6 strains combined with PCR will allow **epidemiological tracing** of transmission and pathogenicity of this virus (Aubin et al., 1991).

6.5.6 PATHOGENS AFFECTING THE HEART

Coxsackie viruses, although in general causing common enteric infections and a number of clinically important diseases, particularly in children, may produce, especially the group B coxsackie viruses, acute pericarditis and myocarditis leading to fatal arrythmias (Easton and Eglin, 1988; Melnick, 1984). The presence of coxsackie B virus-specific RNA has been demonstrated in myocardial biopsies from patients with chronic dilated cardiomyopathy and myocarditis (Bowles et al., 1986).

An **ISH** procedure using radiolabeled coxsackie virus **B3 cDNA** has demonstrated the potential of this approach for an unequivocal diagnosis of enteroviral heart disease using athymic mice as a model system (Kandolf et al., 1987). In these animals, the myocardium was affected by the virus in a disseminated multifocal manner. ISH using a cDNA probe derived from coxsackie **B4** virus-infected cell RNA can reveal different enteroviruses, including **coxsackie A** and **B** viruses, **echoviruses**, and **poliovirus** (Easton and Eglin, 1988). Interestingly, the coxsackie virus RNA in cardiac biopsies from patients diagnosed to have the infection is localized in areas **other than histologically recognized sites of affected tissue** (Easton and Eglin, 1988). Unfortunately, only 46% of cases proved positive by ISH in that study of serologically diagnosed patients, suggesting a low sensitivity of the method.

A study in infected cell cultures and in mice has demonstrated the feasibility of ISH using **strand-specific probes** for enteroviral RNA of broad specificity (Hohenadl et al., 1991). This method may allow a more informative evaluation of the state of viral infection because total RNA of acutely infected cells contains about 100-fold more viral genomic **plus-strand RNA** than complementary **minus-strand RNA**. After the acute phase of infection is over the amount of plus-strand and minus-strand RNA becomes similar. Analysis of persistent enteroviral infection in dilated cardiomyopathies using strand-specific hybridization may help in elucidating the virus role in this pathology (Hohenadl et al., 1991).

A significantly more sensitive **PCR** assay has been able to detect the presence of enteroviruses in 5 of 48 patients with clinically suspected myocarditis or dilated cardiomyopathy (Jin et al., 1990). Primers were selected such that sequences of a broad range of enteroviruses would be amplified. Probing the PCR-positive myocardial biopsies by slot-blot hybridization for coxsackie viruses under stringent conditions failed to yield positive results, testifying to the inadequate sensitivity of slot-blot hybridization for the detection of this type of infection of the heart. Residual viral genomes were observed in tissue showing no histological signs of acute myocarditis, suggesting that persistent infection may eventually contribute to the development of dilated cardiomyopathy. Some of the reasons for the relatively low level of detection of enteroviruses in this study compared to other studies (Bowles et al., 1986; Kandolf, 1988) could be **sampling errors** and **degradation of the viral RNA** in biopsy specimens (Jin et al., 1990).

6.5.7 ORAL PATHOGENS

The significance of bacteria in initiating tooth decay due to the two most prevailing oral diseases, **dental caries** and **periodontal disease**, has long been recognized (Russell, 1991). No final agreement has so far been reached on whether particular etiological organisms or the entire bacterial complement of **dental plaque** is responsible. An organized plaque is richly populated and may yield isolates of **several hundred different species** from a wide range of genera (Russell, 1991). Precise identification of all the species colonizing the oral cavity is hampered by the lack of availability of easily determined phenotypic characteristics among some bacteria (e.g., the streptococci) and by the lack of physiological or serological tests for discrimination (Russell, 1991). The earlier stages in the application of the molecular genetic approach to the study of oral flora and its relationship to periodontal diseases, in particular, have been thoroughly discussed in a report covering findings up to 1989 (Russell, 1991).

Although the major emphasis has been placed on the study of **streptococci** in oral diseases (Russell, 1991), a number of other pathogens have also been analyzed at the gene level. In a random study of 87 young adults, a significant degree of **periodontitis** was observed in 18 individuals (van Steenbergen et al., 1991). The most prevalent species was found to be

Prevotella intermedia. **Restriction endonuclease analysis** revealed different digestion patterns among all isolates of *Actinobacillus actinomycetemcomitans, Prevotella gingivalis,* and *P. intermedia* suggesting that no cross-infection occurred among the subjects studied.

As for the **oral spirochetes,** four cultivable species, *Treponema denticola, Treponema socranskii, Treponema vincentii,* and *Treponema pectinovorum,* have been identified by **chromosomal DNA probes** isolated from representative strains within each species (DiRienzo et al., 1991). The probes were also capable of identifying *T. denticola* in uncultured plaque samples.

A battery of **oligonucleotide probes** has been constructed by sequencing the **16S rRNA** for each bacterium, identifying **hypervariable regions**, and chemically synthesizing **species-specific probes** (Dix et al., 1990). Four of these probes are useful for the rapid detection and identification of *Bacteroides gingivalis* in highly mixed samples from subgingival sulci (Moncla et al., 1990).

A large-scale study of DNA probe assays for a number of oral pathogens has been conducted to define optimum **sample collection sites** in periodontal disease for research and diagnostic purposes (Savitt et al., 1990). Comparing the efficiency of DNA probe assays with that of indirect **immunofluorescence,** the sensitivity of DNA probes was found to be an **order of magnitude higher** (Zappa et al., 1990). It was particularly better than immunofluorescence for *B. gingivalis* and *B. intermedius.* These probes, however, did not identify any samples as positive for *Actinobacillus actinomycetemcomitans.*

The importance of identifying oral pathogens, and the extraordinary sensitivity and specificity of gene probes, have led to the establishment of a diagnostic service for oral pathogens (BioTechnical Diagnostics, Cambridge, MA).

6.5.8 SOME OF THE PATHOGENS AFFECTING THE NERVOUS SYSTEM

6.5.8.1 HERPES SIMPLEX VIRUS

Herpes simplex virus (HSV) can cause severe **focal encephalitis** that, if not treated early, can be fatal (e.g., Whitley, 1988). Because **viral cultures** of cerebrospinal fluid (CSF) are only infrequently productive and **serological tests** are rarely positive in early stages of infection (Kahlon et al., 1987; Lakeman et al., 1987), a **PCR** assay developed for **HSV in CSF** samples can be life saving (Rowley et al., 1990). This assay, using the earlier constructed primers derived from a **highly conserved region** of the **DNA polymerase gene** of HSV (Rowley and Wolinsky, 1989), amplifies viral DNA isolated from CSF by proteinase K/chloroform-phenol extraction (Rowley et al., 1990). Additional information gained in this assay comes from the detection of viral DNA in the CSF supernatant of ultracentrifuged samples, indicating the presence of free viral DNA rather than intact viral particles. A set of primers to a different region of the HSV genome flanks a conserved region within the **glycoprotein D gene** (Powell et al., 1990). PCR and **isoelectric focusing with affinity immunoblotting (IEF)** are proposed as routine diagnostic modalities for adequate and timely recognition of HSV-encephalitis (Pohl-Koppe et al., 1992). Following acyclovir treatment PCR becomes negative, whereas IEF allows to follow an HSV-specific humoral immune response in CSF.

6.5.8.2 RABIES VIRUS

Infection with rabies virus produces eosinophilic cytoplasmic inclusions, called **Negri bodies,** which are not always detectable (Perl, 1975). Although **immunofluorescent techniques** and **immunoperoxidase staining** as well as **cell culture** techniques and **mouse inoculation** tests have been used for the diagnosis of rabies infection, the newly developed **ISH** procedure (A. C. Jackson et al., 1989) offers a substantially better tool because it can also be applied to paraffin-embedded tissues. This technique, using **minus-strand radiolabeled RNA**

probes, was developed in an experimental animal system, but it has the potential for clinical application as a diagnostic test for rabies.

6.5.8.3 MEASLES VIRUS

Subacute sclerosing panencephalitis (SSP) is a rare chronic inflammatory consequence of the involvement of the brain in persistent infection with the measles virus (Fournier et al., 1985). **ISH** using **cloned measles virus DNA** has demonstrated the presence of the measles viral RNA not only in peripheral blood mononuclear cells, but also in nerve cells and cells from the perivascular infiltrates in brain biopsies from a patient with SSP (Fournier et al., 1985).

6.5.8.4 EPSTEIN-BARR VIRUS

Epstein-Barr virus (**EBV**) has been identified so far in lymphoproliferative conditions such as Hodgkin's disease, lymphoepithelioma, infectious mononucleosis, posttransplant B cell lymphoproliferations (Niedobitek et al., 1989; Weiss et al., 1989a,b; Weiss and Movahed, 1989), hairy cell leukoplakia (De Souza et al., 1990), angioimmunoplastic lymphadenopathy (Knecht et al., 1990), lymphomatoid granulomatosis (Katzenstein and Peiper, 1990), lymphoepithelial carcinoma of the stomach (Burke et al., 1990), in the blood and oropharynx of healthy adults and HIV-positive individuals (Gopal et al., 1990), and nasopharyngeal carcinoma (Chang et al., 1990). The presence of EBV has been demonstrated by **ISH** with biotinylated probes on **formalin-fixed paraffin-embedded primary CNS lymphomas** from immunocompromised but not immunocompetent patients (Bashir et al., 1990). The extent of virus-positive cells in the tumors showed no correlation with survival. The significant heterogeneity in the topographic distribution of virally infected cells in the tumors suggested the possibility of lytic or secondary infection in a small number of the latently infected tumor cells.

In a different study, EBV sequences have been found in CNS lymphomas developed in two immunocompetent patients, although the majority of EBV-positive tumors were observed in immunocompromised patients secondary to renal transplantation, HIV infection, leukemia and Wiskott-Aldrich syndrome (Nakhleh et al., 1991). That study used **radiolabeled EBV probes** in ISH, and it is possible that the higher sensitivity of radiolabeled probes accounted for the detection of EBV sequences in lymphomas from immunocompetent patients. A comparative study using PCR assay would be more informative in resolving this issue.

6.5.8.5 *NAEGLERIA FOWLERI*

Naegleria fowleri is a ubiquitous free-living amoeba isolated from heated aquatic environments worldwide (Tydall et al., 1989). When the amoeba enters the nasopharyngeal space of humans, it can find its way into the brain, causing CNS disease (Martinez, 1985). Amoebic meningoencephalitis produced by *N. fowleri* is often fatal and because only this species is known to be pathogenic in humans, accurate and early diagnosis is essential. Using **mitochondrial DNA** sequences from *N. fowleri,* a **PCR** assay for the specific detection of this pathogenic amoeba has been developed and its efficiency demonstrated in tissue from mice experimentally infected with *N. fowleri* (McLaughlin et al., 1991).

6.5.9 OTHER VIRUSES

A high rate of infections, particularly viral infections, is observed in **heart-lung transplant recipients** undergoing immunosuppressive therapy (Weiss et al., 1990). Using **ISH** with biotinylated probes, some 313 lung and 164 heart biopsies from 20 heart-lung recipients have been tested for the presence of CMV, HSV, and adenovirus (Weiss et al., 1990). Although none of the specimens had evidence of HSV or adenoviral infection, 25 lung biopsies and 1 heart biopsy had detectable **CMV DNA**. These findings contrast sharply with the 100% rate of lasting (2.5 years) infection with CMV, adenovirus, and HSV in myocytes and endothelial cells in heart biopsy specimens from cardiac transplant patients found in the other study (Kemnitz et al., 1989).

Lassa virus causes an often fatal disease transmitted to humans by contact with persistently infected rodents in West and Central Africa (Lunkenheimer et al., 1990). Because early symptoms of Lassa fever are nonspecific and resemble influenza, the laboratory diagnosis is usually protracted (**cell culture**), the body's immune response is delayed, and, because the disease can be effectively treated, particularly at early stages, the need for the fastest diagnosis is apparent (Lunkenheimer et al,1990). To this end, a combination of **reverse transcription** of the Lassa virus **RNA sequences** followed by **PCR** has been developed (Lunkenheimer et al., 1990). The assay works on blood and urine specimens, offering a rapid alternative diagnostic for this dangerous systemic infection.

Chapter 6.6

Pathogens Transmitted by Arthropods

6.6.1 VIRUSES

6.6.1.1 ARBOVIRUSES. DENGUE VIRUS

Dengue virus (**DV**) causes the most important mosquito-borne viral infection of humans, affecting millions of people annually in tropical countries (Halstead, 1988). *Aedes* mosquitos are the common vectors of DV (Halstead, 1988). DV belongs to the Flavivirus genus that includes, among other members, pathogens causing yellow fever, Japanese encephalitis, and tick-borne encephalitis. The laboratory diagnosis of DV relies on **serology** and **virus isolation**, which are either lengthy or nonspecific, although **nucleic acid probes** for DV2 have been described (Henchal et al., 1987; Khan and Wright, 1987). In **spot hybridization assays** with RNA extracted from cells infected with 1 of 14 different flaviviruses or Semliki Forest virus, DV was detectable by DV-specific **photobiotin-labeled probes** both under low- and high-stringency conditions (Khan and Wright, 1987).

Radioactively labeled RNA probes have been constructed for **mixed phase** and **solution hybridization assays** for DV2 RNA in mosquito vectors (Olson et al., 1990). Solution hybridization was faster (2–3 vs. 16–20 h) and had a broader probe specificity than mixed phase hybridization. The utility of specific nucleic acid probes may be somewhat limited due to the need for specific probes corresponding to each serotype; besides, some samples display a very low level of viremia (Deubel et al., 1990). To this end, a combination of **PCR** amplification of the DV cDNA with type-specific probing of the amplified product has been developed (Deubel et al., 1990). The sensitivity of the PCR assay is 20 times higher than that of cell culture. Occasional false-negative PCRs in serologically diagnosed dengue fever cases have not been explained.

6.6.1.2 COLORADO TICK FEVER VIRUS

Colorado tick fever (**CTF**) virus is the only orbivirus that causes disease in humans, mainly in the western United States and western Canada. Its primary vector is the soft-bodied tick *Dermatocentor andersoni* (Emmons, 1981). **Molecular hybridization** studies have long been conducted for the genetic characterization of this virus and an **RNA-RNA dot-blot** format has been used to analyze the relatedness of various strains (Bodkin and Knudson, 1987). Although I am unaware of the existing nucleic acid probes for the **diagnosis** of CTF in a clinical setting, they may soon appear given the extent of information on the genomic relatedness among different species of the CTF virus.

6.6.1.3 DUGBE VIRUS

Dugbe (**DUG**) virus is transmitted by *Amblyomma variegatum* ticks; the virus infrequently causes Crimean-Congo hemorrhagic fever in humans, with a resulting mild, febrile disease (Ward et al., 1990). A **PCR**-based assay following **reverse transcription** of the viral RNA and combined with **dot-blot** analysis of the amplified product with a **DUG-specific cDNA probe**

557

is still inferior in sensitivity to the **intracerebral inoculation** of mice (Ward et al., 1990). The PCR assay, however, yields results within 48 h compared to at least 8 days needed for the inoculation test.

6.6.2 RICKETTSIAE

PCR amplification assays have been described for three members of the genus *Rickettsia*: *R. typhi*, the etiological agent of murine typhus (Webb et al., 1990), and *R. rickettsii* and *R. conorii*, the etiological agents of Rocky Mountain spotted fever (RMSF) and boutonneuse fever, respectively (Tzianabos et al., 1989). Both assays already display sufficient sensitivity and specificity to be regarded as supplementary to, if not **replacing, candidates for routine early diagnosis** of respective pathogens in a clinical setting. Failure of the *R. rickettsii* PCR to detect some cases may be due to the extremely low number of rickettsiae in the blood at the early stages of infection, because the sensitivity of the assay is around 50 rickettsiae (Tzianabos et al., 1989). Yet undefined interference may partly account for the occasional PCR failures. Compared to the **ELISA** test for *R. typhi* the PCR assay is more sensitive when evaluated on rickettsiae isolated from fleas (Webb et al., 1990). Further refinement in the construction of primer pairs will undoubtedly increase species specificity of the PCR assay.

6.6.3 SPIROCHETES

6.6.3.1 *BORRELIA BURGDORFERI*

Borrelia burgdorferi (**BB**) is the etiological agent of the tick-borne **Lyme borreliosis** affecting most commonly the skin, joints, heart, and nervous system (for a review, see Karlsson, 1990). The laboratory diagnosis of Lyme borreliosis can be accomplished by **IFA** and **antigen detection** (Hyde et al., 1989; Karlsson, 1990). **Serological evaluation** of Lyme disease patients includes **cultivation** of BB from CSF (Karlsson et al., 1990), **Western blot analysis** and **ELISA** (Karlsson et al., 1988, 1989). **Restriction endonuclease assay (REA)** and **DNA hybridization** have been used to characterize BB from different isolates in North America and Europe and to demonstrate genotypic heterogeneity within this genus and species (LeFebvre et al., 1989). It appears that REA is an accurate and reliable method for identifying the Lyme disease agent among different isolates.

In order to detect even a single BB spirochete **PCR-based assays** are optimal, given appropriate primers are constructed. To achieve this goal, the complete sequence of the insert of one genomic clone of chromosomal origin has been deciphered and a BB-specific PCR assay developed for this sequence (Rosa and Schwan, 1989). The assay proved to be sensitive to fewer than 5 copies of the BB genome, even in the presence of a 106-fold excess of eukaryotic DNA. **Another PCR assay** for BB, when **combined with dot hybridization** detection, recognizes 10 organisms per milliliter of blood or urine (Malloy et al., 1990). This BB-PCR protocol uses amplification of the *OSP*-**A gene** of the spirochete, similar to **yet another PCR assay** (Persing et al., 1990). The latter study analyzed the infection of *Ixodes dammini* tick with BB and demonstrated the advantages of the PCR assay over **direct fluorescent antibody staining**, which is diagnostic only with live ticks. A **different BB-PCR assay**, using sets of primers for the *OSP*-A gene combined with hybridization of the amplified product using radiolabeled oligonucleotides, has a sensitivity of fewer than 50 spirochetes without cross-reactivity to any other organisms tested (Nielsen et al., 1990). Direct comparison of the performance of *in vitro* culture and PCR assays for the detection of BB in spleen, kidney, and bladder from infected gerbils showed that the PCR assay was much faster and had at least a comparable diagnostic sensitivity (Lebech et al., 1991).

Chapter 6.7

Parasites

Molecular biological techniques have long been applied to the characterization and study of human parasites of medical importance, and the development of an antimalaria vaccine is being actively pursued (Anders and Smythe, 1989). A comprehensive coverage of this vast field is outside the scope of this chapter (for reviews, see Hide and Tait, 1991; Nantulya, 1991). Here some illustrative, diagnosis-oriented studies will be cited to demonstrate typical approaches to gene level detection of parasitic human pathogens.

6.7.1 *PLASMODIUM FALCIPARUM*

Specific **DNA probes** are capable of distinguishing between past and present *Plasmodium falciparum* infections in lysed blood specimens from infected patients spotted on nitrocellulose filters (Barker et al., 1986). Hybridization of the parasite DNA with a **species-specific repetitive sequences probe** distinguishes *P. falciparum* from other plasmodium species with a sensitivity of 10 pg of purified *P. falciparum* DNA, which is equivalent to 100 parasites. When used in a field setting this **dot-blot** test, requiring minimum equipment and sample handling, can detect approximately 40 parasites per microliter of blood (Barker et al., 1986). **Spot assays** detecting five parasites per microliter in a 10-µl sample of blood have also been reported (Pollack et al., 1985). However, nonspecific cross-reactivities were observed.

An oligonucleotide probe for the repetitive sequences used in a previous study (Franzen et al., 1984) has been constructed for detecting *P. falciparum* in clinical specimens (Mucenski et al., 1986). This markedly simplified assay was subsequently tested on specimens from 175 febrile patients in different countries (Indonesia, the Philippines, Peru, Kenya, Somalia, and Pakistan) and proved to be highly sensitive (70 parasites per microliter) and specific for *P. falciparum* (Buesing et al., 1987). Although the spot hybridization methods have become relatively simple to perform, so that they can be easily run in field studies, in addition to being at least semiquantitative, more sensitive techniques are needed to distinguish between different clinical isolates of *P. falciparum*. By constructing primers to sequences of the **polymorphic major merozoite surface antigen (*MSA* 1) gene**, analysis of the repetitive elements in the genome of *P. falciparum* isolates has been made possible (Scherf et al., 1991).

Identification of insects in field collections by a fast and relatively simple technique is important in ecological studies and insect control programs (Cockburn, 1990). Distinguishing insect species in some cases presents a problem, and **computer** and **polytene chromosome analysis** may be needed. Using **hybridization of squash blots** of *Anopheles quadrimaculatus* with **species-specific repetitive DNA sequences probes**, a simple and rapid identification of large numbers of individual mosquitoes can be accomplished in field conditions (Cockburn, 1990). Although described for *A. quadrimaculatus,* this technique can be adapted for rapid identification of other species in field collections.

Using PCR with primers targeted for a 206 bp *P. falciparum* DNA sequence as little as 0.01 picogram DNA equivalent to one-half of a parasite can be detected in blood without prior DNA extraction (Tirasophon et al., 1991). This procedure allows to conduct epidemiological studies assessing very low parasitemia levels.

6.7.2 *SCHISTOSOMA MANSONI*

A short, **0.64-kb DNA probe** containing 121-bp repeats is used in a **hybridization assay** of high **species specificity** for *Schistosoma mansoni* (Hamburger et al., 1991). Because this probe occupies at least 12% of the total *S. mansoni* genome, this repeat motif is highly representative, thereby ensuring a high detection sensitivity. Cross-reactivity with *S. haematobium* suggests the presence of this repeat also in this species. The specificity of the **pSm1-7 insert** for *S. mansoni* is over 1000 times higher than for *S. haematobium*, and even greater with respect to *S. magrebowiei*. The pSm1-7 O.64-kb probe could detect schistosomal DNA in 1 ng of infected snail DNA, allowing identification of *S. mansoni* in individual infected snails without prior purification of DNA.

6.7.3 *BABESIA* SPECIES

Babesiosis is a tick-transmitted infection caused by malaria-like parasites of the genus *Babesia*, which invade and destroy erythrocytes (Imes and Neafie, 1976). Of the 20 known species of *Babesia* that infect a great variety of mammals, including horses, sheep, cattle, goats, pigs, and so on, the infrequent infections in humans are caused mostly by *B. microti*, although *B. bovis* and *B. divergens* also have been reported in humans. Laboratory diagnosis is usually made by **microscopy** of Giemsa-stained thin blood smears. Species identification cannot be made morphologically but requires **serological tests** and **animal inoculation**. Significant improvement has been achieved in the diagnosis of *Babesia* spp. with the introduction of a **DNA probe** capable of detecting 100 pg of *B. bovis* DNA (McLaughlin et al., 1986). That probe, however, displayed cross-reactivity with other species, limiting its utility for species-specific diagnosis. To develop a better species-specific probe two *B. bovis* repetitive DNA sequences have been isolated, one of which (Bo25) was capable of distinguishing geographic isolates of *B. bovis* at a sensitivity of 100 pg for *B. bovis* Mexican isolate DNA, and of 1 ng for the Australian isolate DNA (Jasmer et al., 1990). **Repetitive DNA probes** for the other *Babesia* sp., *B. equi*, which affects horses, are çapable of identifying the parasite even when present at low levels (limit of detection, 0.49 and 0.97 ng *B. equi* DNA for the two probes, respectively) (Posnett and Ambrosio, 1989).

6.7.4 *LEISHMANIA* SPECIES

The conventional identification of *Leishmania* strains is performed by **isoenzyme analysis** (van Eys et al., 1989). Even better identification is possible by using **recombinant DNA probes** cloned from *Leishmania infantum* DNA that recognize characteristic patterns on Southern blots of different isolates of the organism (van Eys et al., 1989). This approach appears to offer a better tool than previously used **restriction analyses** and **probing with kinetoplast DNA** (references in van Eys et al., 1989; see also Laskay et al., 1991). By using **circular amplicons**, which hybridize with amplified DNA cloned from a tunicamycin-resistant strain of *Leishmania*, consistent identification of the drug-resistant isolates is possible (Katakura et al., 1991). This assay can be used not only for the analysis of the mechanism of drug-resistance of leishmania, but also for the identification of drug resistant strains in clinical situations. For a review of earlier findings see Barker (1987).

6.7.5 *TRYPANOSOMA* SPECIES

The detection of trypanosome infection in **tsetse flies** conventionally has been very laborious, requiring **dissection** and **examination of midgets**, **salivary glands**, and **proboscids** of the flies (Kukla et al., 1987). A marked simplification and relatively low-cost solution to the species- and subspecies-specific identification of trypanosomes has come with the introduction of **DNA probes** used in **slot-blot** hybridization formats (Kukla et al., 1987).

The detection of *Trypanosoma cruzi*, the etiological agent of Chagas' disease, can be accomplished by a variety of probes (Ashall et al., 1988; Gonzalez et al., 1984). Even more sensitive detection is possible using a **PCR**-based amplification assay that is capable of processing a large number of samples of the insect vector or mammalian blood (Moser et al., 1989). This assay, designed for the **repetitive motif** in the nuclear DNA of *T. cruzi*, the most abundant sequence in this organism, is sensitive enough to detect 1/200 of the DNA in a single parasite, when a radiolabeled probe is hybridized to the amplified product. A further refinement in the detection and classification of *T. cruzi* can be accomplished using **PCR amplification of the kinetoplast DNA** combined with **restriction analysis of the amplified product** (Avila et al., 1990).

6.7.6 FILARIAL SPECIES

A number of **DNA probes** have been developed so far to warrant a workshop on DNA diagnostics and filariasis and a symposium on filariasis and onchocerciasis (references in McReynolds et al., 1991). A summary of the more recent achievements in the field of gene level diagnostics of filarial diseases emphasized the **need for simplification of testing procedures** using nonradioactive probes, better transportation and preservation protocols for the material under study, and the need to use gene-level diagnostics in conjunction with the established morphologically based methods (McReynolds et al., 1991). Table 6.5 from McReynolds and coworkers (1991) lists some of the available filarial DNA probes.

6.7.7 *TOXOPLASMA GONDII*

Several **PCR** assays for *Toxoplasma* **DNA** in clinical specimens have been described. The first assay to be developed was based on the identification of the **repetitive *B1* gene**, and reportedly could detect the DNA of a **single organism** in a crude cell lysate (Burg et al., 1989). In purified DNA preparations this assay could detect as few as 10 parasites in the presence of 100,000 white blood cells, which is comparable to the **maximal pleocytosis** encountered in CSF of patients with toxoplasmic encephalitis (Burg et al., 1989).

In a study of 43 documented cases of **maternal toxoplasmosis**, a comparison of the PCR assay with the detection of **specific IgM** from fetal blood and inoculation of amniotic fluid into **tissue culture** (the two standard methods) has been made (Grover et al., 1990). PCR identified *Toxoplasma gondii* in five of five samples of amniotic fluid from proven cases of congenital toxoplasmosis. The other 2 methods correctly identified only 3 or 4 of the 10 positive samples. The other **indirect methods**, inoculation of amniotic fluid and fetal blood into mice, were able to detect 7 of 10 positive samples. The speed, sensitivity, and specificity of the PCR assay is clearly superior to the other modalities available for **prenatal diagnosis of congenital toxoplasmosis**.

A different PCR assay was developed based on the amplification of a part of the *P30* **gene** of *T. gondii* (Savva et al., 1990). Following amplification, the PCR product can be either **directly visualized** in the gel or by **Southern hybridization** with radioactive or nonradioactive probes. This assay has been used on brain biopsies of AIDS patients to confirm cerebral *Toxoplasma* infection (Holliman et al., 1990), and on specimens from pregnant women and their fetuses and infants (Johnson et al., 1990). The assay is capable of detecting a single *T. gondii* organism in the presence of over 10^6 human cells. If the level of infection is low, the expected amplified 914-bp fragment may not be visualized in the gel, and then a second amplification cycle using a different set of primers allows the product to be directly visualized by ethidium bromide (Savva and Holliman, 1990a,b). Unequivocal results can be obtained within 7 to 8 h.

TABLE 6.5
Specificity of Filarial DNA Probes

Genus	Probes	Specificity[a]	Size	Copy number (per haploid genome)	Organization
Brugia	BM 45 and BM 45B	*B. malayi*[b]	322 bp	30,000	Direct tandem
	pBM15	*B. malayi*	322 bp	30,000	Direct tandem
	BP41 and BP41B	*B. pahangi*	322 bp	3,000	Direct tandem
Loa	LL3	*Loa loa*[c]	356 bp	5,500	Interspersed
Wuchereria	3Wb34	*W. bancrofti*[c]	NK[d]	NK	NK
	IWb35	*W. bancrofti*	1.3 kb	500–1000	Interspersed
	Wb 1, 2, and 6	*W. bancrofti*	NK	2,000–8,000	Interspersed
	pWb 6 and 11	*W. bancrofti*	NK	2,000–8,000	Interspersed
Onchocerca	CIAl-2	*O. volvulus*	149 bp	4,000–5,000	Direct tandem
	pFS-I	*O. volvulus*[e]	149 bp	4,000–5,000	Direct tandem
	pSS-IBT	*Onchocerca*	149 bp	4,000–5,000	Direct tandem
	pOVS134	*Onchocerca*	149 bp	4,000–5,000	Direct tandem
	pOV2	*Onchocerca*	149 bp	4,000–5,000	Direct tandem
	pOV6	*Onchocerca*	149 bp	4,000–5,000	Direct tandem
	puOV3	*Onchocerca*	149 bp	4,000–5,000	Direct tandem
	pOA1	*O. armillata*	147 bp	NK	Tandem
	pOV5	Filariae	NK	NK	NK
	pOV8	Helminths	NK	100–1,000	Tandem
	pOV26	*Onchocerca*	NK	100–1,000	Interspersed

[a] The specificities presented in the table were determined from the results of the coded DNA filters. Each investigator used their own stringent hybridization conditions to achieve specificity. Information on the size of the repeat, copy number, and repeat organization was supplied by the individual investigators.

[b] These probes also hybridize to *B. timori* and *B. patei* DNA.

[c] These probes hybridize with unrelated DNAs as described in the text.

[d] NK, Not known.

[e] Under stringent hybridization conditions pFS-I is specific for forest isolates of *O. volvulus*.

From McReynolds, L. A. et al. (1991). Parasitol. Today 7:65–67. With permission.

Another PCR assay for the detection of *T. gondii* is also based on amplification of the *P30* gene coupled with the identification of the amplified product by Southern analysis with a **210-bp insert probe** (Dupouy-Camet et al., 1990). The occasional failure of this method to identify the parasite is ascribed to the low number of the organisms in some samples, such as fetal blood, so that the target nonrepetitive portion of the *T. gondii* genome could be absent in the blood samples tested. Other assays (Savva et al., 1990; Savva and Holliman, 1990a,b) seem to provide detection of congenital toxoplasmosis with a higher degree of certainty.

References

Aach, R. D. et al. (1991). Hepatitis C virus infection in post-transfusion hepatitis. N. Engl. J. Med. *325*:1325–1329.

Abbaszadegan, M. et al. (1991). Detection of *Giardia* cysts with a cDNA probe and applications to water samples. Appl. Environ. Microbiol. *57*:927–931.

Abbott Laboratories (1987). Abbott Genostics™ Hepatitis B viral DNA. A radiological molecular hybridization assay for the detection of hepatitis B viral DNA in serum. [package insert]. Abbott Laboratories, North Chicago, IL.

Abbott, M. A. et al. (1988). Enzymatic gene amplification: qualititative and quantitative methods for detecting proviral DNA amplified *in vitro*. J. Infect. Dis. *158*:1158–1169.

Abe, A. et al. (1990). Trivalent heat-labile and heat-stable enterotoxin probe conjugated with horseradish peroxidase for detection of enterotoxigenic *Escherichia coli* by hybridization. J. Clin. Microbiol. *28*:2616–2620.

Ablashi, D. V. et al. (1988). Human B-lymphotropic virus (human herpesvirus-6) J. Virol. Methods *21*:29–48.

Agha, S. A. et al. (1989). Early diagnosis of cytomegalovirus infection in renal transplant and dialysis patients by DNA-DNA hybridization assay. J. Med. Virol. *27*:252–257.

Al-Nakib, W. et al. (1986). The detection of human rhinoviruses and their relationship using cDNA probes. J. Med. Virol. *20*:289–296.

Albert, J. and Fenyo, E. M. (1990). Simple, sensitive, and specific detection of human immunodeficiency virus type 1 in clnical specimens by polymerase chain reaction with nested primers. J. Clin. Microb. *28*:1560–1564.

Allard, A. et al. (1990). Polymerase chain reaction for detection of adenoviruses in stool samples. J. Clin. Microbiol. *28*:2659–2667.

Alsip, G. R. et al. (1988). Increased Epstein-Barr virus DNA in oropharyngeal sections from patients with AIDS, AIDS-related complex, or asymptomatic human immunodeficiency virus infections. J. Infect. Dis. *157*:1072–1076.

Alter, H. J. et al. (1989). Detection of antibody to hepatitis C virus in prospectively followed transfusion recipients with acute and chronic non-A, non-B hepatitis. N. Engl. J. Med. *321*:1494–1500.

Alter, M. J. and Sampliner, R. E. (1989). Hepatitis C and miles to go before we sleep. N. Engl. J. Med. *321*:1538–1540.

Altwegg, M. and Geiss, H. K. (1989). *Aeromonas* as a human pathogen. CRC Crit. Rev. Microbiol. *16*:253–286.

Altwegg, M. and Mayer, L. W. (1989). Bacterial molecular epidemiology based on a non-radioactive probe complementary to ribosomal RNA. Res. Microbiol. *140*:325–333.

Amann, R. I. et al. (1990). Fluorescent-oligonucleotide probing of whole cells for determinative, phylogenetic, and environmental studies in microbiology. J. Bacteriol. *172*:762–770.

Anand, A. et al. (1987). Human parvovirus infection in pregnancy and hydrops fetalis. N. Engl. J. Med. *316*:183–187.

Anceschi, M. M. et al. (1990). Multiple primer pairs polymerase chain reaction for the detection of human papillomavirus types. J. Virol. Methods *28*:59–66.

Anders, R. F. and Smythe, J. A. (1989). Polymorphic antigens in *Plasmodium falciparum*. Blood *74*:1865–1875.

Anderson, M. J. et al. (1984). An outbreak of erythema infectiosum associated with human parvovirus infection. J. Hyg. (London) *93*:85–93.

Andrew, P. W. and Boulnois, G. J. (1990). Early days in the use of DNA probes for *Mycobacterium tuberculosis* and *Mycobacterium avium* complexes. In: Gene Probes for Bacteria. Macario, A. J. L. and de Marcario, E. C. (eds.). Academic Press, New York, pp. 179–203.

Apperley, J. F. and Goldman, J. M. (1988). Cytomegalovirus: biology, clinical features and methods for diagnosis. Bone Marrow Transplant *3*:252–264.

Arens, M. and Swierkosz, E. M. (1989). Detection of rotavirus by hybridization with a nonradioactive synthetic DNA probe and comparison with commercial enzyme immunoassays and silver-stained polyacrylamide gels. J. Clin. Microbiol. *27*:1277–1279.

Ashall, F. et al. (1988). Radiolabeled total parasite DNA probe specifically detects *Trypanosoma cruzi* in mammalian blood. J. Clin. Microbiol. *26*:576–578.

Aubin, J. T. et al. (1991). Several groups among human herpesvirus 6 stains can be distinguished by Southern blotting and polymerase chain reaction. J. Clin. Microbiol. *29*:367–372.

Augustin, S. et al. (1987). Problems in detection of cytomegalovirus in urine samples by dot-blot hybridization. J. Clin. Microbiol. *25*:1973–1977.

Aurelius, E. et al. (1991). Rapid diagnosis of herpes simplex encephalitis by nested polymerase chain reaction assay of cerebrospinal fluid. Lancet *337*:189–192.

Auvinen, E. et al. (1989). Polyethylene glycol increases specificity of hybridization typing of HPV specimens. Mol. Cell. Probes *3*:289–298.

Auvinen, P. et al. (1990). Identification of rhinoviruses by cDNA probes. J. Virol. Methods *27*:61–68.

Avila, H. et al. (1990). Schizodeme analysis of *Trypanosoma cruzi* stocks from South and Central America by analysis of PCR-amplified minicircle variable region sequences. Mol. Biochem. Parasitol. *42*:175–188.

Azocar, J. et al. (1990). Prevalence of cervical dysplasia and HPV infection according to sexual behavior. Int. J. Cancer *45*:622–625.

Azzi, A. et al. (1990). Detection of B19 parvovirus infections by a dot-blot hybridization assay using a digoxigenin-labeled probe. J. Virol. Methods *27*:125–134.

Baginski, I. et al. (1991). Detection of polyadenylated RNA in hepatitis B virus-infected peripheral blood mononuclear cells by polymerase chain reaction. J. Infect. Dis. *163*:996–1000.

Baker, B. L. et al. (1991). Determination of hepatitis B virus DNA in serum using the polymerase chain reaction: clinical significance and correlation with serological and biochemical markers. Hepatology *13*:632–636.

Baker, C. C. et al. (1987). Structural and transcriptional analysis of human papillomavirus type 16 sequences in cervical carcinoma cell lines. J. Virol. *61*:962–971.

Barbacci, M. et al. (1991). Routine prenatal screening for HIV infection. Lancet *337*:709–711.

Barbosa, M. S. and Schlegel, R. (1989). The E6 and E7 genes of HPV-18 are sufficient for inducing two-stage *in vitro* transformation of human keratinocytes. Oncogene *4*:1529–1532.

Barbosa, M. S. et al. (1991). *In vitro* biological activities of the E6 and E7 genes vary among human papillomaviruses of different oncogenic potential. J. Virol. *65*:292–298.

Barker, D. C. (1987). DNA diagnosis of human leishmaniasis. Parasitol. Today *3*:177–184.

Barker, R. H., Jr. et al. (1986). Specific DNA probe for the diagnosis of *Plasmodium falciparum* malaria. Science *231*:1434–1436.

Barnes, R. C. (1989). Laboratory diagnosis of human chlamydial infections. Clin. Microbiol. Rev. *2*:119–136.

Barrasso, R. et al. (1987). High prevalence of papillomavirus-associated penile intraepithelial neoplasia in sexual partners of women with cervical intraepithelial neoplasia. N. Engl. J. Med. *317*:916–923.

Bartholoma, N. Y. et al. (1991). Evaluation of two commercially available nucleic acid hybridization assays for the detection and typing of human papillomavirus in clinical specimens. Am. J. Clin. Pathol. *95*:21–29.

Bartolome, J. et al. (1990a). Hepatitis B virus DNA patterns in the liver of children with chronic hepatitis B. J. Med. Virol. *31*:195–199.

Bartolome, J. et al. (1990b). Hepatitis B virus DNA in liver and peripheral blood mononuclear cells during reduction in virus replication. Gastroenterology *99*:1745–1750.

Bas, C. et al. (1988). Assessment of hepatitis B virus DNA levels in chronic HBsAg carriers with or without hepatitis delta virus superinfection. J. Hepatol. *6*:208–213.

Bashir, R. M. et al. (1990). Variable expression of Epstein-Barr virus genome as demonstrated by *in situ* hybridization in central nervous system lymphomas in immunocompromised patients. Mod. Pathol. *3*:429–434.

Bassetti, D. et al. (1991). Second generation RIBA to confirm diagnosis of HCV infection [letter to Editor]. Lancet *337*:912.

Battista, C. et al. (1988). Presence of human papillomavirus types 16 and 18 in genital warts and cervical neoplasias. Med. Oncol. Tumor Pharmacother. *5*:1–9.

Bauer, H. M. et al. (1991). Genital human papillomavirus infection in female university students as determined by a PCR-based method. JAMA *265*:472–477.

Beaulieu, D. et al. (1991). Development of a species-specific DNA probe for (Branhamella) catarrhalis. Mol. Cell. Probes *5*:37–48.

Beckmann, A. M. et al. (1989). Human papillomavirus infection and anal cancer. Int. J. Cancer *43*:1042–1049.

Bedell, M. A. et al. (1987). The E6-E7 region of human papillomavirus type 18 is sufficient for transformation of NIH 3T3 and Rat-1 cells. J. Virol. *61*:3635–3640.

Bedell, M. A. et al. (1989). Identification of human papillomavirus type 18 transforming genes in immortalized and primary cells. J. Virol. *63*:1247–1255.

Bedell, M. A. et al. (1991). Amplification of human papillomavirus genomes *in vitro* is dependent on epithelial differentiation. J. Virol. *65*:2254–2260.

Bej, A. K. et al. (1990). Detection of coliform bacteria in water by polymerase chain reaction and gene probes. Appl. Environ. Microbiol. *56*:307–314.

Bej, A. K. et al. (1991). Detection of viable *Legionella pneumophila* in water by polymerase chain reaction and gene probe methods. Appl. Environ. Microbiol. *57*:597–600.

Bellinzoni, R. et al. (1989). Rotavirus gene detection with biotinylated single-stranded RNA probes. Mol. Cell. Probes *3*:233–244.

Bellobuono, A. et al. (1990). Infectivity of blood that is immunoblot intermediate reactive on hepatitis C virus antibody testing. Lancet *336*:309.

Benditt, E. A. et al. (1983). Viruses in the etiology of atherosclerosis. Proc. Natl. Acad. Sci. U.S.A. 6386–6389.

Bene, M. C. et al. (1990). Absence of cytomegalovirus DNA in kidneys in IgA nephropathy. Lancet *336*:868.

Benson, R. F. et al. (1991). *Legionella adelaidensis*, a new species isolated from cooling tower water. J. Clin. Microbiol. *29*:1004–1006.

Bergeron, C. et al. (1988). Search for human papillomaviruses in normal, hyperplastic, and neoplastic endometria. Obstet. Gynecol. *72*:383–387.

Bernet, C. et al. (1989). Detection of *Mycoplasma pneumoniae* by using the polymerase chain reaction. J. Clin. Microbiol. *27*:2492–2496.

Berninger, M. et al. (1982). An assay for the detection of the DNA genome by hepatitis B virus in serum. J. Med. Virol. *9*:57–68.

Berntorp, E. et al. (1990). Hepatitis C virus transmission by monoclonal antibody purified factor VIII concentrate. Lancet *335*:1531–1532.

Berris, B. et al. (1987). Hepatitis B virus DNA in asymptomatic HBsAg carriers: comparison with HBeAg/Anti-HBe status. J. Med. Virol. *23*:233–239.

Betzl, D. et al. (1990). Identification of lactococci and enterococci by colony hybridization with 23 rRNA-targeted oligonucleotide probes. Appl. Environ. Microbiol. *56*:2927–2929.

Beyer-Finkler, E. et al. (1990). Cell differentiation-related gene expression of human papillomavirus 33. Med. Microbiol. Immunol. *179*:185–192.

Bialkowska-Hobrzanska, H. et al. (1990). Typing of coagulase-negative staphylococci by Southern hybridization of chromosomal DNA fingerprints using a ribosomal RNA probe. Eur. J. Clin. Microbiol. Infect. Dis. *9*:588–594.

Bignon, V. J. et al. (1990). Detection of Epstein-Barr viral genomes in lymph nodes of Hodgkin's disease patients. Mol. Carcinogenesis *3*:9–11.

Bingen, E. et al. (1992). Analysis of DNA restriction fragment length polymorphism extends the evidence for breast milk transmission in *Streptococcus agalactiae* late-onset neonatal infection. J. Inf. Dis. *165*:569–573.

Biro, F. M. and Hillard, P. A. (1990). Genital human papillomavirus infection in adolescents. Med. Clin. N. Am. *74*:1235–1249.

Blackburn, G. F. et al. (1991). Electrochemiluminescence detection for development of immunoassays and DNA probe assays for clinical diagnostics. Clin. Chem. *37*:1534–1539.

Blacklow, N. R. and Greenberg, H. B. (1991). Viral gastroenteritis. N. Engl. J. Med. *325*:252–264.

Blattner, W. A. (1991). HIV epidemiology: past, present, and future. FASEB J. *5*:2340–2348.

Bleul, C. et al. (1991). Human papillomavirus type 18 E6 and E7 antibodies in human sera: increased anti-E7 prevalence in cervical cancer patients. J. Clin. Microbiol. *29*:1579–1588.

Bloss, J. D. et al. (1990). The use of molecular probes to distinguish new primary tumors from recurrent tumors in gynecological malignancies. Am. J. Clin. Pathol. *94*:432–434.

Bluestone, C. D. (1986). Otitis media and sinusitis in children. Role of *Branhamella catarrhalis*. Drugs *31* (Suppl 3):132–141.

Blum, H. E. et al. (1983). Detection of hepatitis B virus DNA in hepatocytes, bile duct epithelium, and vascular elements by *in situ* hybridization. Proc. Natl. Acad. Sci. U.S.A. *80*:6685–6688.

Blum, H. E. et al. (1987). Letter to the Editor. N. Engl. J. Med. *317*:116–117.

Bobo, L. et al. (1990). Diagnosis of *Chlamydia trachomatis* cervical infection by detection of amplified DNA with an enzyme immunoasssay. J. Clin. Microbiol. *28*:1968–1973.

Bobo, L. et al. (1991). Diagnosis of Chlamydia trachomatis eye infection in Tansania by polymerase chain reaction/ enzyme immunoassay. Lancet *338*:847–850.

Boddinghaus, B. et al. (1990). Detection and identification of mycobacteria by amplification of rRNA. J. Clin. Microbiol. *28*:1751–1759.

Bodkin, D. K. and Knudson, D. L. (1987). Genetic relatedness of Colorado tick fever virus isolates by RNA-RNA blot hybridization. J. Gen. Virol. *68*:1199–1204.

Body, B. A. et al. (1990). Use of Gen-Probe and Bactec for rapid isolation and identification of mycobacteria. Am. J. Clin. Pathol. *93*:415–420.

Bohnert, M. G. et al. (1988). Detection of enteric pathotypes of *Escherichia coli* by hybridization using six DNA probes. Ann. Inst. Pasteur/Microbiol. *139*:189–202.

Bolske, G. (1988). Survey of Mycoplasma infections in cell cultures and a comparison of detection methods. Zbl. Bakt. Hyg. A *269*:331–340.

Bonde, G. T. (1977). Bacterial indicators of water pollution. Adv. Aquat. Microbiol. *1*:273–364.

Bonino, F. et al. (1981). Hepatitis B virus DNA in the sera of HBsAg carriers: a marker of active hepatitis B virus replication in the liver. Hepatology *1*:386–391.

Bonino, F. et al. (1986). Chronic hepatitis in HBsAg carriers with serum HBV-DNA and anti-HBe. Gastroenterology *90*:1268–1273.

Bornstein, J. et al. (1988). Multicentric intraepithelial neoplasia involving the vulva. Clinical features and association with human papillomavirus and herpes simplex virus. Cancer *62*:1601–1604.

Borr, J. D. et al. (1991). Analysis of *Actinobacillus pleuropneumoniae* and related organisms by DNA-DNA hybridization and restriction endonuclease fingerprinting. Int. J. System. Bacteriol. *41*:121–129.

Borriello, S. P. et al. (1987). Analysis of latex agglutination test for *Clostridium difficile* toxin A (D-1) and differentiation between *C. difficile* toxins A and B and latex reactive protein. J. Clin. Pathol. *40*:573–580.

Boshart, M. et al. (1984). A new type of papillomavirus DNA; its presence in genital cancer biopsies and in cell lines derived from cervical cancer. EMBO J. *3*:1151–1157.

Bottone, E. J. (1977). *Yersinia enterocolitica*: a panoramic view of a charismatic microorganism. CRC Crit. Rev. Microbiol. *5*:211–241.

Bouffard, P. et al. (1990). Different forms of hepatitis B virus DNA and expression of HBV antigens in peripheral blood mononuclear cells in chronic hepatitis Br. J. Med. Virol. *31*:312–317.

Boukadida, J. et al. (1991). Outbreak of gut colonization by *Pseudomonas aeruginosa* in immunocompromised children undergoing total digestive decontamination: analysis by pulsed-field electrophoresis. J. Clin. Microbiol. *29*:2068–2071.

Bowles, N. E. et al. (1986). Detection of coxsackie B virus-specific RNA sequences in myocardial biopsies from cases of myocarditis and dilated cardiomyopathy. Lancet *i*:1120–1122.

Bowyer, S. M. et al. (1987). Expression of the hepatitis B virus genome in chronic hepatitis B carriers and patients with hepatocellular carcinoma. Proc. Natl. Acad. Sci. U.S.A. *84*:847–850.

Bracha, R. et al. (1990). Differentiation of clinical isolates of *Entamoeba histolytica* by using specific DNA probes. J. Clin. Microbiol. *28*:680–684.

Brambilla, C. et al. (1986). Varying nuclear staining intensity of hepatitis B virus DNA in human hepatocellular carcinoma. Lab. Invest. *55*:475–481.

Brandwein, M. et al. (1989). Human papillomavirus 6/11 and 16/18 in Schneiderian inverted papillomas. Cancer *63*:1708–1713.

Brautigam, A. R. et al. (1980). Rapid typing of herpes simplex virus isolates by deoxyribonucleic acid: deoxyribonucleic acid hybridization. J. Clin. Microbiol. *12*:226–234.

Brazil, G. M. et al. (1988). Development of DNA probes for cytotoxin and enterotoxin genes in enteric bacteria. Experientia *44*:848–853.

Brechot, C. et al. (1985). Hepatitis B virus DNA in patients with chronic liver disease and negative tests for hepatitis B surface antigen. N. Engl. J. Med. *312*:270–276.

Brenner, D. J. et al. (1984). Family VII, Legionellaceae Brenner, Steigerwalt and McDade 1979, 658. In: Bergey's Manual of Systematic Bacteriology, Vol 1. Krieg, N. R. and Hold, J. G. (eds.). Williams & Wilkins, Baltimore, pp. 279–288.

Bressoud, A. et al. (1990). Rapid detection of influenza virus H1 by the polymerase chain reaction. Biochem. Biophys. Res. Commun. *167*:425–430.

Brice, S. L. et al. (1989). Detection of herpes simplex virus DNA in cutaneous lesions of erythema multiforme. J. Invest. Dermatol. *93*:183–187.

Brinton, L. A. et al. (1990). Oral contraceptive use and risk of invasive cervical cancer. Int. J. Epidemiol. *19*:4–11.

Brisson-Noel, A. et al. (1989). Rapid diagnosis of tuberculosis by amplification of mycobacterial DNA in clinical samples. Lancet *ii*:1069–1071.

Brisson-Noel, A. et al. (1991). Diagnosis of tuberculosis by DNA amplification in clinical practice evaluation. Lancet *338*:364–366.

Bronstein, I. et al. (1989). A comparison of chemiluminescent and colorimetric substrates in a hepatitis B virus DNA hybridization assay. Anal. Biochem. *180*:95–98.

Brown, T. et al. (1984). Intrauterine parvovirus infection associated with hydrops fetalis. Lancet *ii*:1033–1034.

Browne, H. M. et al. (1988). Analysis of the L1 gene product of human papillomavirus type 16 by expression in a vaccinia virus recombinant. J. Gen. Virol. *69*:1263–1273.

Brownell, G. H. and Belcher, K. E. (1990). DNA probes for the identification of *Nocardia asteroides*. J. Clin. Microbiol. *28*:P2082–P2086.

Brownlee, R. G. et al. (1990). Application of capillary DNA chromatography to detect AIDS virus (HIV-1) DNA in blood. J. Chromatogr. *533*:87–96.

Bruce, C. B. et al. (1988). Synthetic oligonucleotides as diagnostic probes for rhinoviruses. Lancet *ii*:53.

Bruce, C. et al. (1989). Detection of enteroviruses using cDNA and synthetic oligonucleotide probes. J. Virol. Methods *25*:233–240.

Bruce, C. et al. (1990). Detection of rhinovirus RNA in nasal epithelial cells by *in situ* hybridization. J. Virol. Methods *30*:115–126.

Bruner, J. M. (1990). Oligonucleotide probe for herpes virus: use in paraffin sections. Mod. Pathol. *3*:635–638.

Buchbinder, A. et al. (1988). Polymerase chain reaction amplification and *in situ* hybridization for the detection of human B-lymphotropic virus. J. Virol. Methods *21*:191–197.

Buchman, T. G. et al. (1990). Detection of surgical pathogens by *in vitro* DNA amplification. I. Rapid identification of *Candida albicans* by *in vitro* amplification of a fungus-specific gene. Surgery *108*:338–345.

Buesing, M. et al. (1987). An oligonucleotide probe for detecting *Plasmodium falciparum*: an analysis of clinical specimens from six countries. J. Int. Dis. *155*:1315–1318.

Buffone, G. T. et al. (1988). DNA hybridization assay for congenital cytomegalovirus infection. J. Clin. Microbiol. *26*:2184–2186.

Buffone, G. T. et al. (1990). Detection of mRNA from the immediate early gene of human cytomegalovirus in infected cells by in vitro amplification. Mol. Cell. Probes *4*:143–151.

Buffone, G. T. et al. (1991). Improved amplification of cytomegalovirus DNA from urine after purification of DNA with glass beads. Clin. Chem. *37*:1945–1949.

Burch, D. J. et al. (1991). Isolation of a stain-specific *Entamoeba histolytica* cDNA clone. J. Clin. Microbiol. *29*:699–701.

Burg, J. L. et al. (1989). Direct and sensitive detection of a pathogenic protozoan, *Toxoplasma gondii* by polymerase chain reaction. J. Clin. Microbiol. *27*:1787–1792.

Burke, A. P. et al. (1990). Lymphoepithelial carcinoma of the stomach with Epstein-Barr virus demonstrated by polymerase chain reaction. Mod. Pathol. *3*:377–380.

Burnett, L. et al. (1985). Can potentially infectious specimens containing hepatitis B virus be identified on the basis of their biochemical profile? Clin. Biochem. *31*:1329–1330.

Burnett, L. et al. (1986). Are specimens with "at risk" biochemical profiles more likely to be infectious for hepatitis B virus? Clin. Chem. *32*:1951–1953.

Burstain, J. M. et al. (1991). Sensitive detection of *Treponema pallidum* by using the polymerase chain reaction. J. Clin. Microbiol. *29*:62–69.

Busch, M. P. et al. (1987). *In situ* hybridization and immunocytochemistry for improved assessment of human immunodeficiency virus cultures. Am. J. Clin. Pathol. *88*:673–680.

Busch, M. P. et al. (1991). Evaluation of screened blood donations for human immunodeficiency virus type 1 infection by culture and DNA amplification of pooled cells. N. Engl. J. Med. *325*:1–5.

Butcher, P. D. and Farthing, M. J. G. (1989). DNA probes for the fecal diagnosis of *Giardia lamblia* infections in man. Biochem. Soc. Trans. *17*:363–364.

Buti, M. et al. (1988). Hepatitis D virus RNA in acute delta infection: serological profile and correlation with other markers of hepatitis D virus infection. Hepatology *8*:1125–1129.

Calabrese, G. et al. (1991). Transmission of anti-HCV within the household of hemodialysis patients. Lancet *338*:1466.

Campion, M. J. et al. (1985). Increased risk of cervical neoplasia in consorts of men with penile condylomata acuminata. Lancet *i*:943–946.

Cannizzaro, L. A. et al. (1988). Regional chromosomal localization of human papillomavirus integration sites near fragile sites, oncogenes, and cancer chromosome breakpoints. Cancer Genet. Cytogenet. *33*:93–98.

Cariani, E. et al. (1991). Detection of HCV RNA and antibodies to HCV after needlestick injury. Lancet *337*:850.

Carloni, G. et al. (1987). Detection of HBV infectivity by spot hybridization in HBeAg negative chronic carriers: HBV DNA in sera from asymptomatic and symptomatic subjects. J. Med. Virol. *21*:15–23.

Carman, W. F. and Kidd, A. H. (1989). An assessment of optimal conditions for amplification of HIV cDNA using Thermus aquaticus polymerase. J. Virol. Methods *23*:277–290.

Carman, W. F. et al. (1989). Reverse transcription and subsequent DNA amplification of rubella virus RNA. J. Virol. Methods *25*:21–30.

Carney, J. A. et al. (1988). New rapid latex agglutination test for diagnosing *Trichomonas vaginalis* infection. J. Clin. Pathol. *41*:806–808.

Carrigan, D. R. et al. (1991). Interstitial pneumonitis associated with human herpesvirus-6 infection after marrow transplantation. Lancet *338*:147–149.

Carruba, G. et al. (1991). DNA fingerprinting and electrophoretic karyotype of environmental and clinical isolates of *Candida parapsilosis*. J. Clin. Microbiol. *29*:916–922.

Carter, G. I. et al. (1987). Detection of TEM β-lactamase genes by non-isotopic spot hybridization. Eur. J. Clin. Microbiol. *6*:406–409.

Casacuberta, J. M. et al. (1988). Comparison of different non-isotopic methods for hepatitis B virus detection in human serum. Nucleic Acids Res. *16*:11834.

Cash, J. D. (1990). Screening donors for hepatitis C virus antibody. Lancet *336*:309.

Cassol, S. et al. (1991). Use of dried blood spot specimens in the detection of human immunodeficiency virus type 1 by the polymerase chain reaction. J. Clin. Microbiol. *29*:667–671.

Caussy, D. et al. (1988). Evaluation of methods for detecting human papillomavirus deoxyribonucleotide sequences in clinical specimens. J. Clin. Microbiol. *26*:236–243.

Caussy, D. et al. (1990). Interaction of human immunodeficiency and papilloma viruses: association with anal epithelial abnormality in homosexual men. Int. J. Cancer *46*:214–219.

Cave, M. D. et al. (1991). IS6110: conservation of sequence in the *Mycobacterium tuberculosis* complex and its utilization in DNA fingerprinting. Mol. Cell. Probes *5*:73–80.

Celentano, D. D. et al. (1988). Cervical cancer screening practices among older women: results from the Maryland cervical cancer case-control study. J. Clin. Epidemiol. *41*:531–541.

Chan, S. W. et al. (1991). Serological responses to infection with three different types of hepatitis C virus. Lancet *338*:1391.

Chan, S. Y. et al. (1992). Molecular variants of human papillomavirus type 16 from four continents suggest ancient pandemic spread of the virus and its coevolution with humankind. J. Virol. *66*:2057–2066.

Chan, W. K. et al. (1989). Progesterone and glucocorticoid response elements occur in the long control regions of several human papillomaviruses involved in anogenital neoplasia. J. Virol. *63*:3261–3269.

Chang, Y. S. et al. (1990). Detection of Epstein-Barr virus DNA sequences in nasopharyngeal carcinoma cells by enzymatic DNA amplification. J. Clin. Microbiol. *28*:2398–2402.

Chantralita, W. et al. (1989). Rapid detection of herpes simplex virus DNA by *in situ* hybridization with photobiotin-labeled double-stranded DNA probes. Mol. Cell. Probes *3*:363–373.

Chao, Y.-C. et al. (1990). Sequence conservation and divergence of hepatitis delta (δ) virus RNA. Virology *178*:384–392.

Chao, Y.-C. et al. (1991). Molecular cloning and characterization of an isolate of hepatitis delta virus from Taiwan. Hepatology *13*:345–352.

Chemin, I. et al. (1991). Correlation between HBV DNA detection by polymerase chain reaction and pre-S1 antigenemia in symptomatic and asymptomatic hepatitis B virus infections. J. Med. Virol. *33*:51–57.

Chen, L. et al. (1991). Human papillomavirus type 16 nucleoprotein E7 is a tumor rejection antigen. Proc. Natl. Acad. Sci. U.S.A. *88*:110–114.

Cherry, J. D. (1987). Enteroviruses: polioviruses (poliomyelitis), coxsackieviruses, echoviruses, and enteroviruses. In: Textbook of Pediatric Infectious Diseases, 2nd Ed. Feigin, R. D. and Cherry, J. D. (eds.). W. B. Saunders, Philadelphia, pp. 1729–1841.

Chesters, P. and McCance, D. (1989). Human papillomavirus type 6 and 16 in cooperation with Ha-*ras* to transform secondary rat embryo fibroblasts. J. Gen. Virol. *70*:353–365.

Chin, M. T. et al. (1988). Regulation of human papillomavirus type 11 enhancer and E6 promoter by activating and repressing proteins from the E2 open reading frame: functional and biochemical studies. J. Virol. *62*:2994–3002.

Chin, M. T. et al. (1989). Identification of a novel constitutive enhancer element and an associated binding protein: implications for human papillomavirus type 11 enhancer regulation. J. Virol. *63*:2967–2976.

Choo, Q. L. et al. (1989). Isolation of a cDNA clone derived from a blood borne non-A, non-B viral hepatitis genome. Science *244*:359–362.

Choo, Q. L. et al. (1990). Hepatitis C virus: the major causative agent of viral non-A, non-B hepatitis. Br. Med. Bull. *46*:423–441.

Choo, Q. L. et al. (1991). Genetic organization and diversity of the hepatitis C virus. Proc. Natl. Acad. Sci. U.S.A. *88*:2451–2455.

Chou, S. (1990). Newer methods for diagnosis of cytomegalovirus infection. Rev. Infect. Dis. *12* (Suppl. 7):S727-S736.

Chow, V. T. K. et al. (1990). *Thermus aquaticus* DNA polymerase-catalysed chain reaction for the detection of human papillomaviruses. J. Virol. Methods *27*:101–112.

Claas, E. C. J. et al. (1989). Human papillomavirus detection in paraffin-embedded cervical carcinomas and metastases of the carcinomas by the polymerase chain reaction. Am. J. Pathol. *135*:703–709.

Claas, H. C. J. et al. (1991). Diagnostic value of the polymerase chain reaction for *Chlamydia* detection as determined in a follow-up study. J. Clin. Microbiol. *29*:42–45.

Clarke, J. R. et al. (1990). Detection of HIV-1 in human lung macrophages using the polymerase chain reaction. AIDS *4*:1133–1136.

Claudy, A. L. et al. (1989). Detection of herpes simplex virus DNA in a cutaneous squamous cell carcinoma by *in situ* hybridization. Arch. Dermatol. Res. *281*:333–335.

Clayton, C. L. et al. (1989). Molecular cloning and expression of *Campylobacter pylori* species-specific antigens in *Escherichia coli* K-12. Infect. Immunol. *57*:623–629.

Clemons, K. V. et al. (1991). Epidemiologic study by DNA typing of a *Candida albicans* outbreak in heroin addicts. J. Clin. Microbiol. *29*:205-207.

Clewley, J. P. (1989). Polymerase chain reaction assay of parvovirus B19 DNA in clinical specimens. J. Clin. Microbiol. *27*:2647–2651.

Clewley, J. P. (1990). Detection of hepatitis C virus RNA in serum. Lancet *336*:309–310.

Clewley, J. P. et al. (1987). Detection of parvovirus B19 DNA, antigen and particles in the human fetus. J. Med. Virol. *23*:367–376.

Clyne, J. M. et al. (1989). A rapid chemoluminescent DNA hybridization assay for the detection of *Chlamydia trachomatis*. J. Bioluminesc. Chemoluminesc. *4*:357–366.

Cockburn, A. F. (1990). A simple and rapid technique for identification of large numbers of individual mosquitoes using DNA hybridization. Arch. Insect Biochem. Physiol. *14*:191–199.

Cohen, B. J. et al. (1990). Blood donor screening for parvovirus B19. J. Virol. Methods *30*:233–238.

Colgan, T. J. et al. (1989). Human papillomavirus infection of morphologically normal cervical epithelium adjacent to squamous dysplasia and invasive carcinoma. Hum. Pathol. *20*:316–319.

Collins, P. and Wertz, G. W. (1986). Human respiratory syncytial virus genome and gene products. In: Concepts in Viral Pathogenesis II. Notkins, A. and Oldstone, M. (eds.). Springer-Verlag, New York, pp. 40–46.

Colombo, M. et al. (1985). Transmission of non-A, non-B hepatitis by heat treated factor VIII concentrate. Lancet *ii*:1–4.

Colombo, M. et al. (1989). Prevalence of antibodies to hepatitis C virus in Italian patients with hepatocellular carcinoma. Lancet *2*:1006–1008.

Colucci, G. et al. (1988). Production of hepatitis B virus infected human B cell hybridomas: transmission of the viral genome to normal lymphocytes in cocultures. Virology *164*:238–244.

Contreras, M. and Barbara, J. A. J. (1989). Screening for hepatitis C virus antibody. Lancet *ii*:505.

Conville, P. S. et al. (1989). Comparison of three techniques for concentrating positive BACTEC 14A bottles for mycobacterial DNA probe analysis. Diagn. Microbiol. Infect. Dis. *12*:309–313.

Conway, B. et al. (1990). Detection of HIV-1 DNA in crude cell lysates of peripheral blood mononuclear cells by the polymerase chain reaction and non-radioactive oligonucleotide probes. J. AIDS *3*:1059–1064.

Cornelissen, M. T. E. et al. (1988). Evaluation of different DNA-DNA hybridization techniques in detection of HPV16 DNA in cervical smears and biopsies. J. Med. Virol. *25*:105–114.

Cornelissen, M. T. E. et al. (1989). Localization of human papillomavirus type 16 DNA using the polymnerase chain reaction in the cervix uteri of women with cervical intraepithelial neoplasia. J. Gen. Virol. *70*:2555–2562.

Cossart, Y. E. et al. (1975). Parvovirus-like particles in human sera. Lancet *i*:72–73.

Coutlee, F. et al. (1989). Immunodetection of DNA with biotinylated RNA probes: a study of reactivity of a monoclonal antibody to DNA-RNA hybrids. Anal. Biochem. *181*:96–105.

Cova, L. et al. (1988). Use of cRNA probes for the detection of enteroviruses by molecular hybridization. J. Med. Virol. *24*:11–18.

Cox, N. et al. (1990). Characterization of an rRNA gene-specific cDNA probe: applications in bacterial identification. J. Gen. Microbiol. *136*:1639–1643.

Cravioto, A. et al. (1988). DNA hybridization with oligodeoxyribonucleotide probes for identifying enterotoxin-producing *Escherichia coli*. Mol. Cell. Probes *2*:125–130.

Cripe, T. P. et al. (1987). Transcriptional regulation of the human papillomavirus-16 E6-E7 promoter by a keratinocyte-dependent enhancer, and by viral E2 transactivator and repressor gene products: implications for cervical carcinogenesis. EMBO J. *6*:3745–3753.

Croen, K. D. et al. (1987). Latent herpes simplex virus in human trigeminal ganglia. Detection of an immediate early gene "anti-sense" transcript by *in situ* hybridization. N. Engl. J. Med. *317*:1427–1432.

Crook, T. et al. (1988). Human papillomavirus type 16 cooperates with activated *ras* and *fos* oncogenes in the hormone dependent transformation of primary mouse cells. Proc. Natl. Acad. Sci. U.S.A. *85*:8820–8824.

Crook, T. et al. (1989). Constitutive expression of c-*myc* oncogene confers hormones independence and enhanced growth-factor responsiveness on cells transformed by human papillomavirus type 16. Proc. Natl. Acad. Sci. U.S.A. *86*:5713–5717.

Crook, T. et al. (1991). Modulation of immortalizing properties of human papillomavirus type 16 E7 by p53 expression. J. Virol. *65*:505–510.

Crosby, J. H. et al. (1991). *In situ* DNA hybridization for confirmation of herpes simplex virus in bronchoalveolar lavage smears. Acta Cytol. *35*:248–250.

Crum, C. P. et al. (1985). Cervical papillomaviruses segregate within morphologically distinct precancerous lesions. J. Virol. *54*:675–681.

Crum, C. P. et al. (1987). Morphological correlates of genital HPV infection: viral replication, transcription and gene expression. In: Viruses and Human Cancer. Alan R. Liss, New York, pp. 355–369.

Crum, C. P. et al. (1988). A comparison of biotin and isotope-labeled ribonucleic acid probes for *in situ* detection of HPV-16 ribonucleic acid in genital precancers. Lab. Invest. *58*:354–359.

Cullen, B. R. (1991). Regulation of HIV-1 gene expression. FASEB J. *5*:2361–2368.

Cuthbertson, G. and Grose, C. (1988). Biotinylated and radioactive DNA probes for detection of varicella-zoster virus genome in infected human cells. Mol. Cell. Probes *2*:197–207.

Czaja, A. J. et al. (1990). Frequency and significance of antibody to hepatitis C virus in severe corticosteroid-treated cryptogenic chronic active hepatitis. Mayo Clin. Proc. *65*:1303–1313.

Czaja, A. J. et al. (1991). Frequency and significance of antibody to hepatitis C virus in severe corticosteroid-treated autoimmune chronic active hepatitis. Mayo Clin. Proc. *66*:572–582.

Czerwenka, K. F. et al. (1991). Reliability of *in situ* hybridization of smears and biopsies for papillomavirus genotyping of the uterine cervix. Eur. J. Clin. Chem. Clin. Biochem. *29*:139–145.

Dahlen, P. O. et al. (1991). Detection of human immunodeficiency virus type 1 by using the polymerase chain reaction and a time-resolved fluorescence-based hybridization assay. J. Clin. Microbiol. *29*:798–804.

Dallo, S. F. et al. (1989). DNA and protein sequence homologies between the adhesins of *Mycoplasma genitalium* and *Mycoplasma pneumoniae*. Infect. Immunol. *57*:1059–1065.

Daly, J. A. et al. (1991). Use of rapid, nonradioactive DNA probes in culture confirmation tests to detect *Streptococcus agalactiae, Haemophilus influenzae,* and *Enterococcus* spp. from pediatric patients with significant infections. J. Clin. Microbiol. *29*:80–82.

Danbara, H. (1990). Identification of enterotoxigenic *Escherichia coli* by colony hybridization using biotinylated LTIh, STIa and STIb enterotoxin probes. In: Gene Probes for Bacteria. Macario, A. J. L. and Conway de Macario, E. (eds.). Academic Press, San Diego, pp. 167–178.

Daniel, T. M. (1989). Rapid diagnosis of tuberculosis: laboratory techniques applicable in developing countries. Rev. Infect. Dis. *11* (Suppl. 2):471–478.

Daniel, T. M. and Debanne, S. M. (1987). The serodiagnosis of tuberculosis and other mycobacterial diseases by enzyme-linked immunosorbent assay. Am. Rev. Respir. Dis. *135*:1137–1151.

Dankner, W. M. et al. (1990). Rapid antiviral DNA-DNA hybridization assay for human cytomegalovirus. J. Virol. Methods *28*:293–298.

Datta, A. R. et al. (1987). Detection of hemolytic *Listeria monocytogenes* by using DNA colony hybridization. Appl. Environ. Microbiol. *53*:2256–2259.

David, F. et al. (1990). Detection and typing of human papillomavirus DNA from cervical biopsies by the slot-blot hybridization method. Mol. Cell. Probes *4*:53–61.

David, H. L. (1976). Bacteriology of the Mycobacterioses. U.S. Government Printing Office, Washington, D. C.

Davis, J. M. et al. (1988). Cellular polypeptides overexpressed after herpes simplex infection permit virus subtyping and may help diagnose cervical cancer. Genitourin. Med. *64*:321–326.

Davis, G. L. et al. (1989). Hepatitis interventional therapy group: treatment of chronic hepatitis C with recombinant interferon α: a multicenter randomized, controlled trial. N. Engl. J. Med. *321*:1501–1506.

Davison, F. et al. (1987). Detection of hepatitis B DNA in spermatozoa, urine, saliva and leucocytes, of chronic HbsAg carrier. A lack of relationship with serum markers of replication. J. Hepatol. *4*:37–44.

De Cock, K. M. et al. (1990). Rapid and specific diagnosis of HIV-1 and HIV-2 infections: an evaluation of testing strategies. AIDS *4*:875–878.

de Ory, F. et al. (1987). Comparison of four methods for screening of cytomegalovirus antibodies in normal donors and immunocompromised patients. Eur. J. Clin. Microbiol. *6*:402–405.

De Souza, Y. G. et al. (1990). Diagnosis of Epstein-Barr virus infection in hairy leukoplakia by using nucleic acid hybridization and noninvasive techniques. J. Clin. Microbiol. *28*:2775–2778.

de Villiers, E. M. et al. (1987). Human papillomavirus infections in women with and without abnormal cervical cytology. Lancet *ii*:703–706.

de Wit, D. et al. (1990). Direct detection of *Mycobacterium tuberculosis* in clinical specimens by DNA amplification. J. Clin. Microbiol. *28*:2437–2441.

de Wit, M. Y. L. et al. (1991). Application of a polymerase chain reaction for the detection of *Mycobacterium leprae* in skin tissues. J. Clin. Microbiol. *29*:90610.

Dean, D. et al. (1990). Improved sensitivity of a modified polymerase chain reaction amplified DNA probe in comparison with serial tissue culture passage for detection of *Chlamydia trachomatis* in conjunctival specimens from Nepal. Diagn. Microbiol. Infect. Dis. *12*:133–137.

Dearden, C. J. and Al-Nakib, W. (1987). Direct detection of rhinoviruses by an enzyme-linked immunosorbent assay. J. Med. Virol. *23*:179–189.

Debiec-Rychter, M. et al. (1991). Chormosomal characterization of human nasal and nasopharyngeal cells immortalized by human papilomavirus type 16 DNA. Cancer Genet. Cytogenet. *52*:51–61.

Dejean, A. et al. (1984). Detection of hepatitis B virus DNA in pancreas, kidney and skin of two human carriers of the virus. J. Gen. Virol. *65*:651–655.

Demeter, T. et al. (1988). Detection of human papillomavirus DNA in cell scrapes and formalin-fixed, paraffin-embedded tissue of the uterine cervix by filter *in situ* hybridization. J. Med. Virol. *26*:397–409.

Demmler, G. J. et al. (1988). Detection of cytomegalovirus in urine from newborns by using polymerase chain reaction DNA amplification. J. Infect. Dis. *158*:1177–1184.

Deneer, H. G. and Boychuk, I. (1991). Species-specific detection of *Listeria monocytogenes* by DNA amplification. Appl. Environ. Microbiol. *57*:606–609.

Denniston, K. J. et al. (1986). Cloned fragment of the hepatitis delta virus RNA genome: squence and diagnostic application. Science *232*:873–875.

Desselberger, U. et al. (1986). Genome analysis of human rotavirus by oligonucleotide mapping of isolated RNA segments. Virus Res. *4*:357–368.

Deubel, V. et al. (1990). Identification of dengue sequences by genomic amplification: rapid diagnosis of dengue virus serotypes in peripheral blood. J. Virol. Methods *30*:41–54.

Deutsch, J. et al. (1986). Demonstration of Epstein-Barr virus DNA in a previously healthy boy with fulminant hepatic failure. Eur. J. Pediatr. *145*:94–98.

Devlin, M. E. et al. (1992). Peripheral blood mononuclear cells of the elderly contain varicella-zoster virus DNA. J. Inf. Dis. *165*:619–622.

Di Bisceglie, A. M. and Hoofnagle, J. H. (1990). Hepatitis B replication within the human spleen. J. Clin. Microbiol. *28*:2850–2852.

Di Bisceglie, A. M. et al. (1987). Hepatitis B virus deoxyribonucleic acid in liver of chronic carriers. Correlation with serum markers and changes associated with loss of hepatitis Be antigen after antiviral therapy. Gastroenterology *93*:1236–1241.

Di Bisceglie, A. M. et al. (1989). Recombinant interferon α therapy for chronic hepatitis C: a randomized, double-blind, placebo-controlled trial. N. Engl. J. Med. *321*:1506–1510.

Di Bisceglie, A. M. et al. (1990). A randomized, double-blind, placebo-controlled trial of recombinant human α-interferon therapy for chronic non-A, non-B (type C) hepatitis. J. Hepatol. *11*:S36-S42.

Di Luca, D. et al. (1987). Simultaneous presence of herpes simplex and human papillomavirus sequences in human genital tumors. Int. J. Cancer *40*:763–768.

Di Luca, D. et al. (1989). Search for human papillomavirus, herpes simplex virus and c-*myc* oncogene in human genital tumors. Int. J. Cancer *43*:570–577.

Dickover, R. E. et al. (1990). Quantitation of human immunodeficiency virus DNA by using the polymerase chain reaction. J. Clin. Microbiol. *28*:2130–133.

Diegutis, P. S. et al. (1986). False-positive results with hepatitis B virus DNA dot-hybridization in hepatitis B surface antigen-negative specimens. J. Clin. Microbiol. *23*:797–799.

Dimitrov, D. H. et al. (1985). Detection of rotaviruses by nucleic acid hybridization with cloned DNA of simian rotavirus SA11 genes. J. Infect. Dis. *152*:293–300.

Dimitrov, D. H. et al. (1990). Isolation of HIV-1 and titration of infected peripheral blood mononuclear cells by microculture. In: Monogr. Virol. Human Immunodeficiency Virus: Innovative Techniques, Vol. 18. Khan, N. C. and Melnick, J. L. (eds.). Karger, Basel, pp. 122–130.

DiRienzo, J. M. et al. (1991). Use of randomly cloned DNA fragments for the identification of oral spirochetes. Oral Microbiol. Immunol. *6*:88–96.

Dix, K. et al. (1990). Species-specific oligodeoxynucleotide probes for the identification of periodontal bacteria. J. Clin. Microbiol. *28*:319–323.

Docke, W. D. et al. (1991). Cytomegalovirus infection and common variable immunodeficiency. Lancet *338*:1597.

Doebbeling, B. N. et al. (1988). Prospective evaluation of the Gen-Probe assay for detection of Legionellae in respiratory secretions. Eur. J. Clin. Microbiol. Infect. Dis. *7*:748–752.

Dolter, J. et al. (1990). Evaluation of five rapid systems for the identification of *Neisseria gonorrhoeae*. Diagn. Microbiol. Infect. Dis. *13*:265–267.

Downey, G. O. et al. (1988). Condylomatous carcinoma of the vulva with special reference to human papillomavirus DNA. Obstet. Gynecol. *72*:68–73.

Downie, J. C. et al. (1989). HIV-1 antibody testing strategy: evaluation of ELISA screening and Western blot profiles in a mixed low risk/high risk patient population. J. Virol. Methods *26*:291–304.

Duggan, M. A. et al. (1989). Human papillomavirus DNA determination of anal condylomata, dysplasias, and squamous carcinomas with *in situ* hybridization. Am. J. Clin. Pathol. *92*:16–21.

Duggan, M. A. et al. (1990a). A comparison of cervical scrapes for HPV typing by dot-blot hybridization obtained by wood and plastic spatulas. J. Virol. Methods *29*:267–278.

Duggan, M. A. et al. (1990b). Nonisotopic human papillomavirus DNA typing of cervical smears obtained at the initial colposcopic examination. Cancer *66*:745–751.

Dular, R. (1990). Gene probe detection of human and cell culture mycoplasmas. In: Gene Probes for Bacteria. Macario, A. J. L. and Conway de Macario, E. (eds.). Academic Press, San Diego, pp. 417–457.

Dupouy-Camet, J. et al. (1990). Preventing congenital toxoplasmosis [letter]. Lancet *336*:1018.

Durst, M. et al. (1985). The physical state of human papillomavirus type 16 DNA in benign and malignant genital tumors. J. Gen. Virol. *66*:1515–1522.

Durst, M. et al. (1987). Papillomavirus sequences integrate near cellular oncogenes in some cervical carcinomas. Proc. Natl. Acad. Sci. U.S.A. *84*:1070–1074.

Dutilh, B. et al. (1988). Detection of *Chlamydia trachomatis* by *in situ* hybridization with sulphonated total DNA. Ann. Inst. Pasteur/Microbiol. *139*:115–128.

Dutilh, B. et al. (1990). Detection of *Chlamydia trachomatis* with DNA probes. In: Gene Probes for Bacteria. Macario, A. J. L. and Conway de Macario, E. (eds.). Academic Press, San Diego, pp. 45–68.

Easton, A. J. and Eglin, R. P. (1988). The detection of coxsackievirus RNA in cardiac tissue by *in situ* hybridization. J. Gen. Virol. *69*:285–291.

Ebeling, F. et al. (1990). Recombinant immunoblot assay for hepatitis C virus antibody as predictor of infectivity [letter]. Lancet *335*:982–983.

Ebeling F. et al. (1991). Second generation RIBA to confirm diagnosis of HCV infection [letter]. Lancet *337*:912–913.

Echeverria, P. et al. (1990). Detection of diarrheogenic *Escherichia coli* using nucleotide probes. In: Gene Probes for Bacteria. Macario, A. J. L. and Conway de Macario, E. (eds.). Academic Press, San Diego, pp. 96–141.

Edelstein, P. H. (1986). Evaluation of the Gen-Probe DNA probe for the detection of legionellae in culture. J. Clin. Microbiol. *23*:481–484.

Edelstein, P. H. et al. (1987). Retrospective study of Gen-Probe rapid diagnostic system for detection of Legionellae in frozen clinical respiratory tract samples. J. Clin. Microbiol. *25*:1022–1026.

Eggerding, F. A. et al. (1991). Detection of rubella virus gene sequences by enzymatic amplification and direct sequencing of amplified DNA. J. Clin. Microbiol. *29*:945–952.

Ehrlich, G. D. et al. (1989). Prevalence of human T-cell leukemia/lymphoma virus (HTLV) type II infection among high-risk individuals: type-specific identification of HTLVs by polymerase chain reaction. Blood *74*:1658–1664.

Einsele, H. et al. (1988). *In-situ* hybridization for detection of cytomegalovirus in liver from patients after marrow grafting. Lancet *ii*:634–635.

Einsele, H. et al. (1991). Polymerase chain reaction to evaluate antiviral therapy for cytomegalovirus disease. Lancet *338*:1170–1172.

Eisenbach, K. et al. (1990). Polymerase chain reaction amplification of a repetitive DNA sequence specific for *Mycobacterium tuberculosis*. J. Infect. Dis. *161*:977–981.

Eisenstein, B. I. (1990). New molecular techniques for microbial epidemiology and the diagnosis of infectious diseases. J. Infect. Dis. *161*:595–602.

Elfassi, E. et al. (1984). Evidence of extrachromosomal forms of hepatitis B viral DNA in a bone marrow culture obtained from a patient recently infected with hepatitis B virus. Proc. Natl. Acad. Sci. U.S.A. *81*:3526–3528.

Ellner, P. et al. (1988). Rapid detection and identification of pathogenic mycobacteria by combining radiometric and nucleic acid probe methods. J. Clin. Microbiol. *26*:1349–1352.

Emmons, R. W. (1981). Colorado tick fever. In: Viral Zoonoses (Section B), Vol. I, CRC Handbook Series in Zoonoses. Beran, G. W. and Steele, J. H. (eds.). CRC Press, Boca Raton, FL, pp. 113–124.

Engel, J. N. et al. (1990). Cloning and characterization of RNA polymerase core subunits of *Chlamydia trachomatis* by using the polymerase chain reaction. J. Bacteriol. *172*:5732–5741.

Enomoto, N. et al. (1990). There are two major types of hepatitic C virus in Japan. Biochem. Biophys. Res. Commun. *170*:1021–1025.

Erice, A. et al. (1987). Ganciclovir treatment of cytomegalovirus disease in transplant recipients and other immunocompromised hosts. J. Am. Med. Assn. *257*:3082–3087.

Escaich, S. et al. (1991). Comparison of HIV detection by virus isolation in lymphocyte cultures and molecular amplification of HIV DNA and RNA by PCR in offspring of seropositive mothers. J. AIDS *4*:130–135.

Esteban, J. I. et al. (1989). Hepatitis C virus antibodies among risk groups in Spain. Lancet *2*:294–297.

Esteban, J. I. et al. (1990). Evaluation of antibodies to hepatitis C virus in a study of transfusion-associated hepatitis. N. Engl. J. Med. *323*:1107–1112.

Estes, M. K. and Tanaka, T. (1989). Nucleic acid probes for rotavirus detection and characterization. In: DNA Probes for Infectious Diseases. Tenover, F. C. (ed.). CRC Press, Boca Raton, FL, pp. 79–100.

Eversole, L. R. et al. (1987). Demonstration of human papillomavirus DNA in oral condyloma acuminatum. J. Oral Pathol. *16*:266–272.

Ezaki, T. et al. (1988). Simple genetic method to identify viridans group streptococci by colorimetric dot hybridization and fluorometric hybridization in microdilution wells. J. Clin. Microbiol. *26*:1708–1713.

Ezaki, T. et al. (1990). Evaluation of the microplate hybridization method for rapid identification of Legionella species. Eur. J. Clin. Microbiol. Inf. Dis. *9*:213–217.

Farber, J. M. et al. (1991). Neonatal listeriosis due to cross-infection confirmed by isoenzyme typing and DNA fingerprinting. J. Infect. Dis. *163*:927.

Farci, P. et al. (1988). Acute and chronic hepatitis delta virus infection: direct or indirect effect on hepatitis B virus replication? J. Med. Virol. *26*:279–288.

Fay, O. et al. (1987). Anti-delta antibody in various HBsAg positive Argentine populations. J. Med. Virol. *22*:257–262.

Feinman, S. V. et al. (1988). Serum HBV-DNA (hepatitis B virus DNA) in acute and chronic hepatitis B infection. Clin. Invest. Med. *11*:286–291.

Feinstone, S. M. et al. (1975). Transfusion-associated hepatitis not due to viral hepatitis type A or B. N. Engl. J. Med. *292*:767–770.

Felber, B. K. et al. (1990). Bioassays for the detection of HIV-1 and practical applications. In: Monogr. Virol. Human Immunodeficiency Virus: Innovative Techniques, Vol. 18. Khan, N. C. and Melnick, J. L. (eds.). Karger, Basel, pp. 91–104.

Fenoll, A. et al. (1990). Identification of a typical strains of *Streptococcus pneumoniae* by a specific DNA probe. Eur. J. Clin. Microbiol. Infect. Dis. *9*:396–401.

Ferenci, P. et al. (1990). One year treatment of chronic non-A, non-B hepatitis with interferon α-2b. J. Hepatol. *11*:S50–S53.

Fermand, J. P. et al. (1990). Detection of Epstein-Barr virus in epidermal skin lesions of an immunocompromised patient. Ann. Intern. Med. *112*:511–515.

Fife, K. H. et al. (1987). Symptomatic and asymptomatic cervical infections with human papillomavirus during pregnancy. J. Infect. Dis. *156*:904–911.

Filipenko, D. et al. (1988). *In situ* detection of rubella RNA and antigens in cultured cells. J. Virol. Methods *22*:109–118.

Finlayson, J. S. and Tankersley, D. L. (1990). Anti-HCV screening and plasma fractionation: the case against. Lancet *335*:1275–1275.

Fiore, L. et al. (1990). Characterization by T1-oligonucleotide fingerprinting of three strains of human hepatitis A virus isolated in Italy. Eur. J. Epidemiol. *6*:29–33.

Fluit, A. C. et al. (1990). A probe for the detection of methicillin-resistant *Staphylococcus aureus*. Eur. J. Clin. Microbiol. Infect. Dis. *9*:605–608.

Follett, E. A. C. (1991). HCV confirmatory testing of blood donors. Lancet *338*:1024.

Forsey, T. et al. (1990). Differentiation of vaccine and wild mumps viruses using the polymerase chain reaction and dideoxynucleotide sequencing. J. Gen. Virol. *71*:987–990.

Fournier, J. G. et al. (1985). Detection of measles virus RNA in lymphocytes from peripheral blood and brain perivascular infiltrates of patients with subacute sclerosing panencephalitis. N. Engl. J. Med. *313*:910–915.

Fox, G. M. et al. (1991). Detection of herpesvirus DNA in vitreous and aqueous specimens by the polymerase chain reaction. Arch. Opthalmol. *109*:266–271.

Fox, J. D. et al. (1990). Human herpesvirus 6 in salivary glands. Lancet *336*:590–593.

Franciosi, R. A. and Tattersall, P. (1988). Fetal infection with human parvovirus B19. Hum. Pathol. *19*:489–491.

Franzen, L. et al. (1984). Analysis of clinical specimens by hybridization with probe containing repetitive DNA from *Plasmodium falciparum*. A novel approach to malaria diagnosis. Lancet *i*:525–528.

Fries, J. W. U. et al. (1990). Genus- and species-specific DNA probes to identify mycobacteria using the polymerase chain reaction. Mol. Cell. Probes *4*:87–105.

Fries, J. W. et al. (1991). Detection of untreated mycobacteria by using polymerase chain reaction and specific DNA probes. J. Clin. Microbiol. *29*:1744–1747.

Frost, E. H. et al. (1991). Typing *Chlamydia trachomatis* by detection of restriction fragment length polymorphism in the gene encoding the major outer membrane protein. J. Infect. Dis. *163*:1103–1107.

Frydenberg, J. et al. (1987). Cloning of *Mycoplasma pneumoniae* DNA and expression of P1-epitopes in *Escherichia coli*. Isr. J. Med. Sci. *23*:759–762.

Fuchs, P. G. et al. (1989). Human papillomavirus 16 DNA in cervical cancers and in lymph nodes of cervical cancer patients: a diagnostic marker for early metastases? Int. J. Cancer *43*:41–44.

Fujita, K. et al. (1990). Detection of human papillomavirus 16 and 18 DNA in cervical dysplasia and cervical carcinoma by *in situ* hybridization. Acta Histochem. Cytochem. *23*:113–126.

Fukushima, M. et al. (1990). The physical state of human papillomavirus 16 DNA in cevical carcinoma and cervical intraepithelial neoplasia. Cancer *66*:2155–2161.

Gabuzda, D. H. and Hirsch, M. S. (1987). Neurologic manifestations of infection with human immunodeficiency virus: clinical features and pathogenesis. Ann. Intern. Med. *107*:383–391.

Gama, R. E. et al. (1988a). Polymerase chain reaction amplification of rhinovirus nucleic acids from clinical material. Nucleic Acids Res. *16*:9346.

Gama, R. E. et al. (1988b). Amplification of rhinovirus specific nucleic acids from clinical samples using the polymerase chain reaction. J. Med. Virol. *28*:73–77.

Garfinkel, L. I. et al. (1989). DNA probes specific for *Entamoeba histolytica* possessing pathogenic and nonpathogenic zymodemes. Infect. Immunol. *57*:926–931.

Garson, J. A. et al. (1990a). Detection of hepatitis C viral sequences in blood donations by "nested" polymerase chain reaction and prediction of infectivity. Lancet *335*:1419–1422.

Garson, J. A. et al. (1990b). Detection by PCR of hepatitis C virus in factor VIII concentrates. Lancet *335*:1473.

Garson, J. A. et al. (1990c). Enhanced detection by PCR of hepatitis C virus RNA. Lancet *336*:878–879.

Garson, J. A. et al. (1991). Improvement of HCV genome detection with "short" PCR products. Lancet *338*:1466–1467.

Garuti, G. et al. (1989). Detection and typing of human papillomavirus in histologic specimens by *in situ* hybridization with biotinylated DNA probes. Am. J. Clin. Pathol. *92*:604–612.

Gebhart, C. J. et al. (1989). Species-specific cloned DNA probes for the identification of *Campylobacter hyointestinalis*. J. Clin. Microbiol. *27*:2717–23.

Genesca, J. et al. (1990). Clinical correlation and genetic polymorphism of the immunodeficiency virus proviral DNA obtained after polymerase chain reaction amplification. J. Infect. Dis. *162*:1025–1030.

Gerber-Huber, S. N. et al. (1988). Sensitive detection and typing of human papillomavirus DNA in gynecological cell scrapings by slot-blot hybridization. Virchow's Arch. B: Cell Pathol. *56*:119–125.

Gerding, D. N. et al. (1986). *Clostridium difficile*-associated diarrhea and colitis in adults. Arch. Intern. Med. *146*:95–100.

Gerritzen, A. and Scholt, B. (1990). A nonradioactive riboprobe assay for the detection of hepatitis B virus DNA in human sera. J. Virol. Methods *30*:311–318.

Giacchino, R. et al. (1987). Hepatitis B virus (HBV)-DNA-positive hepatocellular carcinoma following hepatitis B virus infection in a child. J. Med. Virol. *23*:151–155.

Gicquelais, K. G. et al. (1990). Practical and economical method for using biotinylated DNA probes with bacterial colony blots to identify diarrhea-causing *Escherichia coli*. J. Clin. Microbiol. *28*:2485–2490.

Gilden, D. H. et al. (1987). Varicella-zoster virus infection of human mononuclear cells. Virus Res. *7*:117–129.

Gilden, D. H. et al. (1988). Persistence of varicella-zoster virus DNA in blood mononuclear cells by patients with varicella or zoster. Virus Genes *2*:299–305.

Gillespie, D. et al. (1989). Probes for quantitating subpicogram amounts of HIV-1 RNA by molecular hybridization. Mol. Cell. Probes *3*:73–86.

Giovanninni, M. et al. (1990). Maternal-infant transmission of hepatitis C virus and HIV infections: a possible interaction. Lancet *335*:1166.

Giri, I. and Yaniv, M. (1988). Study of the E2 gene product of the cottontail rabbit papillomavirus reveals a common mechanism of transactivation among papillomaviruses. J. Virol. *62*:1573–1581.

Gius, D. and Laimins, L. A. (1989). Activation of human papillomavirus type 18 gene expression by herpes simplex virus type 1 viral transactivators and a phorbol ester. J. Virol. *63*:555–563.

Gleaves, C. A. et al. (1989). Direct detection of cytomegalovirus from bronchoalveolar lavage samples by using a rapid *in situ* DNA hybridization assay. J. Clin. Microbiol. *27*:2429–432.

Glew, S. S. et al. (1992). HLA antigens and cervical carcinoma (correspondence). Nature (London) *356*:22.

Gloss, B. et al. (1987). The upstream regulatory region of the human papillomavirus 16 contains an E2-protein-independent enhancer which is specific for cervical carcinoma cells and regulated by glucocorticoid hormones. EMBO J. *6*:3735–3743.

Gobel, U. et al. (1987). Synthetic oligonucleotide probes complementary to rRNA for group- and species-specific detection of mycoplasmas. Isr. J. Med. Sci. *23*:742–746.

Godfrey-Faussett, P. et al. (1991). Tuberculosis pericarditis confirmed by DNA amplification. Lancet *337*:176–177.

Goeser, T. et al. (1990). Washing with 8 mol/l urea to correct false-positive anti-HCV results. Lancet *336*:878.

Goldstein, S. C. et al. (1987). Human cytomegalovirus (HCMV) enhances bovine papilloma virus (BPV) transformation *in vitro*. J. Med. Virol. *223*:157–164.

Goltz, S. P. et al. (1990). The use of nonradioactive DNA probes for rapid diagnosis of sexually transmitted bacterial infections. In: Gene Probes for Bacteria. Macario, A. J. L. and Conway de Macario, E. (eds.). Academic Press, San Diego, pp. 1–44.

Gomes, S. A. et al. (1985). *In situ* hybridization with biotinylated DNA probes: a rapid diagnostic test for adenovirus upper respiratory infections. J. Virol. Methods *12*:105–110.

Gomes, T. A. T. et al. (1987). DNA probes for identification of enteroinvasive *Escherichia coli*. J. Clin. Microbiol. *25*:2025–2027.

Gomez-Rubio, M. et al. (1990). Prolonged treatment (18 months) of chronic hepatitis C with recombinant α-interferon in comparison with a control group. J. Hepatol. *11*:S63-S67.

Gonzalez, A. et al. (1984). Minichromosomal repetitive DNA in *Trypanosoma cruzi*: its use in a high-sensitivity parasite detection assay. Proc. Natl. Acad. Sci. U.S.A. *81*:3356–3360.

Goodenow, M. et al. (1989). HIV-1 isolates are rapidly evolving quasispecies: evidence for viral mixtures and preferred nucleotide substitutions. J. AIDS *2*:344–352.

Gopal, M. R. et al. (1990). Detection by PCR of HHV-6 and EBV DNA in blood and oropharynx of healthy adults and HIV-seropositives. Lancet *335*:1598–1599.

Gottstein, B. and Mowatt, M. R. (1991). Sequencing and characterization of an *Echinococcus multilocularis* DNA probe and its use in the polymerase chain reaction. Mol. Biochem. Parasitol. *44*:183–194.

Goudsmit, J. et al. (1991). Genomic diversity and antigenic variation of HIV-1: links between pathogenesis, epidemiology and vaccine development. FASEB J. *5*:2427–2436.

Gouvea, V. et al. (1990). Serotypes and electropherotyes of human rotavirus in the USA: 1987–1989. J. Infect. Dis. *162*:362–367.

Govindarajan, S. and Valinluck, B. (1985). Serum hepatitis B virus-DNA in chronic hepatitis B and delta infection. Arch. Pathol. Lab. Med. *109*:398–399.

Govindarajan, S. et al. (1985). Delta agent superinfection. Rapidly progressive liver disease in a hepatitis B virus carrier. Arch. Pathol. Lab. Med. *109*:395–397.

Govindarajan, S. et al. (1986). Serum hepatitis B virus DNA in acute hepatitis B. Am. J. Clin. Pathol. *86*:352–354.

Gowans, E. J. et al. (1981). Detection of hepatitis B virus DNA sequences in infected hepatocytes by *in situ* hybridization. J. Med. Virol. *8*:67–78.

Gowans, E. J. et al. (1983). Patterns in single- and double-stranded hepatitis B virus DNA and viral antigen accumulation in infected liver cells. J. Gen. Virol. *64*:1229–1239.

Gowans, E. J. et al. (1985). Cytoplasmic (but not nuclear) hepatitis B virus (HBV) core antigen reflects HBV DNA synthesis at the level of the infected hepatocyte. Intervirology *24*:220–225.

Granato, P. A. and Franz, M. R. (1989). Evaluation of a prototype DNA probe test for the noncultural diagnosis of gonorrhea. J. Clin. Microbiol. *27*:632–635.

Granato, P. A. and Franz, M. R. (1990). Use of the Gen-Probe PACE system for the detection of *Neisseria gonorrhoeae* in urogenital samples. Diagn. Microbiol. Infect. Dis. *13*:217–221.

Granum, P. E. et al. (1984). Enterotoxin formation by *Clostridium perfringens* during sporulation and vegetative growth. Int. J. Food Microbiol. *1*:43–49.

Gratton, C. A. et al. (1990). Comparison of a DNA probe with culture for detecting *Chlamydia trachomatis* directly from genital specimens. Mol. Cell. Probes *4*:25–31.

Gravitt, P. et al. (1991). A direct comparison of methods proposed for use in widespread screening of human papillomavirus infections. Mol. Cell. Probes *5*:65–72.

Gray, J. J. et al. (1990). Differentiation between specific and non-specific hepatitis C antibodies in chronic liver disease. Lancet *335*:609–610.

Green, E. P. et al. (1989). Sequence and characteristics of IS900, an insertion element identified in a human Crohn's disease isolate of *Mycobacterium paratuberculosis*. Nucleic Acids Res. *17*:9063–9073.

Green, K. Y. et al. (1988). Prediction of human rotavirus serotypes by nucleotide sequence analysis of the VP7 protein gene. J. Virol. *62*:1819–1823.

Greene, W. C. (1991). The molecular biology of human immunodeficiency virus type 1 infection. N. Engl. J. Med. *324*:308–317.

Gregoire, L. et al. (1989). Amplification of human papillomavirus DNA sequences by using conserved primers. J. Clin. Microbiol. *27*:2660–2665.

Griffin, N. R. et al. (1990). Demonstration of multiple HPV types in normal cervix and in cervical squamous cell carcinoma using the polymerase chain reaction on paraffin wax embedded material. J. Clin. Pathol. *43*:52–56.

Griffith, B. P. et al. (1990). Cellular localization of cytomegalovirus nucleic acids in guinea pig salivary glands by *in situ* hybridization. J. Virol. Methods *27*:145–158.

Griffiths, P. D. and Grundy, J. E. (1987). Molecular biology and immunology of cytomegalovirus. Biochem. J. *241*:313–324.

Grillner, L. (1987). Screening of blood donors for cytomegalovirus (CMV) antibodies: an evaluation of different tests. J. Virol. Methods *17*:135–139.

Grimont, F. and Grimont, P. A. D. (1986). Ribosomal ribonucleic acid gene restriction patterns as potential taxonomic tools. Ann. Inst. Pasteur/Microbiol. *137*:165–175.

Grimont, P. A. D. et al. (1985). DNA probe specific for *Legionella pneumophila*. J. Clin. Microbiol. *21*:431–437.

Grimpel, E. et al. (1991). Use of polymerase chain reaction and rabbit infectivity testing to detect *Treponema pallidum* in amniotic fluid, fetal and neonatal sera, and cerebrospinal fluid. J. Clin. Microbiol. *29*:1711–1718.

Grody, W. W. et al. (1987). *In situ* viral DNA hybridization in diagnostic surgical pathology. Hum. Pathol. *18*:535–543.

Grody, W. W. et al. (1988). Detection of cytomegalovirus DNA in classic and epidemic Kaposi's sarcoma by *in situ* hybridization. Hum. Pathol. *19*:524–528.

Grose, C. and Weiner, C. P. (1990). Prenatal diagnosis of congenital cytomegalovirus infection: two decades later. Am. J. Obstet. Gynecol. *163*:447–450.

Grossman, S. R. and Laimins, L. A. (1989). E6 protein of human papillomavirus type 18 binds zinc. Oncogene *4*:1089–1093.

Grossman, S. R. et al. (1989). Intracellular localization and DNA-binding properties of human papillomavirus type 18 E6 protein expressed with a baculovirus vector. J. Virol. *63*:366–374.

Grover, C. M. et al. (1990). Rapid prenatal diagnosis of congenital *Toxoplasma* infection by using polymerase chain reaction and amniotic fluid. Clin. Microbiol. *28*:2297–2330.

Grussendorf Conen, E. I. and Cremer, S. (1990). The demonstration of human papillomavirus 16 genomes in the nuclei of genital cancers using two different methods of *in situ* hybridization. Cancer *65*:238–241.

Grussendorf-Conen, E. I. et al. (1987). Detection of human papillomavirus-6 in primary carcinoma of the urethra in men. Cancer *60*:1832–1835.

Guerin-Reverchon, I. et al. (1989). A comparison of methods for the detection of human papillomavirus DNA by *in situ* hybridization with biotinylated probes on human carcinoma cell lines. Application to wart sections. J. Immunol. Methods *123*:167–176.

Guerin-Reverchon, I. et al. (1990). Study of stringency conditions for human papillomavirus DNA detection on cell lines, frozen and paraffin-embedded tissue sections by *in situ* hybridization with biotinylated probes. Histochemistry *93*:637–643.

Guerrant, R. L. (1986). Amebiasis: introduction, current status, and research questions. Rev. Infect. Dis. *8*:218–227.

Gupta, S. et al. (1989). Detection of hepatitis delta virus in serum and liver tissue by molecular hybridization. Validation of a rapid spot-hybridization technique. Am. J. Clin. Pathol. *92*:218–221.

Haase, A. et al. (1984). Detection of viral nucleic acids by *in situ* hybridization. Meth. Virol. *7*:189–226.

Habibi, B. and Garretta, M. (1990). Screening for hepatitis C virus antibody in plasma for fractionation. Lancet *335*:855–856.

Hackel, C. et al. (1990). Specific identification of *Mycobacterium leprae* by the polymerase chain reaction. Mol. Cell. Probes *4*:205–210.

Hada, H. et al. (1986). State of hepatitis B viral DNA in the liver of patients with hepatocellular carcinoma and chronic liver disease. Liver *6*:189–198.

Hadchouel, M. et al. (1988). Detection of mononuclear cells expressing hepatitis B virus in peripheral blood from HBsAg positive and negative patients by *in situ* hybridization. J. Med. Virol. *24*:27–32.

Haikala, O. J. et al. (1983). Rapid detection of rotavirus in stool by latex agglutination: comparison with radioimmunoassay and electron microscopy and clinical evaluation of the test. J. Med. Virol. *11*:91–97.

Halbert, C. L. et al. (1992). The E6 and E7 genes of human papillomavirus type 6 have weak immortalizing activity in human epithelial cells. J. Virol. *66*:2125–2134.

Hallam, N. et al. (1989). Detection and typing of human papillomavirus infection of the uterine cervix by dot blot hybridization: comparison of scrapes and biopsies. J. Med. Virol. *27*:317–321.

Halpern, M. S. et al. (1983). Viral nucleic acid synthesis and antigen accumulation in pancreas and kidney of Peking ducks infected with duck hepatitis B virus. Proc. Natl. Acad. Sci. U.S.A. *80*:4865–4869.

Halpern, M. S. et al. (1986). Viral antigen expression in the pancreas of DHBV-infected embryos and young ducks. Virology *150*:276–282.

Halstead, S. B. (1988). Pathogenesis of dengue: challenges to molecular biology. Science *239*:476–481.

Hamburger, J. et al. (1991). Highly repeated short DNA sequences in the genome of *Schistosoma mansoni* recognized by a species-specific probe. Mol. Biochem. Parasitol. *44*:73–80.

Hammond, G. W. et al. (1982). Comparison of direct and indirect enzyme immunoassays with direct ultracentrifugation before electron microscopy for detection of rotaviruses. J. Clin. Microbiol. *16*:53–59.

Hammond, G. et al. (1987). DNA hybridization for diagnosis of enteric adenovirus infection from directly spotted human fecal specimens. J. Clin. Microbiol. *25*:1881–1885.

Hampson, S. J. et al. (1989). DNA probes demonstrate a single highly conserved strain of *Mycobacterium avium* infecting AIDS patients. Lancet *335*:65–68.

Han, R. et al. (1992). Linkage of regression and malignant conversion of rabbit viral papillomas to MHC class II genes. Nature (London) *356*:66–68.

Hance, A. J. et al. (1989). Detection and identification of mycobacteria by amplification of mycobacterial DNA. Mol. Microbiol. *3*:843–849.

Hara, Y. et al. (1990). Human papillomavirus infection of the uterine cervix analyzed by nonisotopic *in situ* hybridization. J. Med. Virol. *31*:120–128.

Harju, L. et al. (1990). Affinity-based collection of amplified viral DNA application to the detection of human immunodeficiency virus type 1, human cytometalovirus and papillomavirus type 16. Mol. Cell. Probes *4*:223–235.

Harper, M. E. et al. (1986). Detection of lymphocytes expressing human T-lymphotropic virus type III in lymph nodes and peripheral blood from infected individuals by *in situ* hybridization. Proc. Natl. Acad. Sci. U.S.A. *83*:772–776.

Harrison, T. J. (1990). Hepatitis B virus DNA in peripheral blood leukocytes: a brief review. J. Med. Virol. *31*:33–35.

Hart, C. et al. (1988). Direct detection of HIV RNA expression in seropositive subjects. Lancet *ii*:596–599.

Hart, G. (1986). Syphilis tests in diagnostic and therapeutic decision making. Ann. Intern. Med. *104*:368–376.

Hartford, T. and Sneath, P. H. A. (1990). Experimental error in DNA-DNA pairing: a survey of the literature. J. Appl. Bacteriol. *68*:527–542.

Hartskeerl, R. A. et al. (1989). Polymerase chain reaction for the detection of *Mycobacterium leprae*. J. Gen. Microbiol. *135*:2357–2364.

Hasan, F. et al. (1990). Hepatitis C associated hepatocellular carcinoma. Hepatology *12*:589–591.

Haseltine, W. A. (1991). Molecular biology of the human imunodeficiency virus type 1. FASEB J. *5*:2349–2360.

Hawkins, R. E. et al. (1992). Association of mycoplasma and human immunodeficiency virus infection: detection of amplified *Mycoplasma fermentans* DNA in blood. J. Inf. Dis. *165*:581–585.

Hayashi, Y. et al. (1990). A novel diagnostic method of *Pneumocystis carinii*. *In situ* hybridization of ribosomal ribonucleic acid with biotinylated oligonucleotide probes. Lab. Invest. *63*:576–580.

Heijtink, R. A. et al. (1987). Detection of HBV-DNA in liver biopsy and serum: its significance in the selection of hepatitis B patients for antiviral therapy. Antiviral Res. *7*:329–340.

Heimer, R. et al. (1992). Detection by polymerase chain reaction of human immunodeficiency virus type 1 proviral DNA sequences in needles of injecting drug users. J. Inf. Dis. *165*:781–782.

Heino, P. et al. (1989). Detection of human papillomavirus (HPV) DNA in genital biopsy specimens by *in situ* hybridization with digoxigenin-labeled probes. J. Virol. Methods *26*:331–338.

Helland, A. et al. (1992). HLA antigens and cervical carcinoma (correspondence). Nature (London) *356*:23.

Hemmings, L. and McManus, D. P. (1989). The isolation by differential antibody screening of *Echinococcus multilocularis* antigen gene clones with potential for immunodiagnosis. Mol. Biochem. Parasitol. *33*:171–182.

Hemmings, L. and McManus, D. P. (1991). The diagnostic value and molecular characterization of an *Echinococcus multilocularis* antigen gene clone. Mol. Biochem. Parasitol. *44*:53–62.

Henchal, E. A. et al. (1987). Detection of dengue virus RNA using nucleic acid hybridization. J. Virol. Methods *15*:187–200.

Henderson, B. R. et al. (1987). Detection of specific types of human papillomavirus in cervical scrapes, anal scrapes, and anogenital biopsies by DNA hybridization. J. Med. Virol. *21*:381–393.

Hendrix, M. G. R. et al. (1989). The presence of CMV nucleic acids in arterial walls of atherosclerotic and non-atherosclerotic patients. Am. J. Pathol. *134*:1151–1157.

Hendrix, M. G. R. et al. (1990). High prevalence of latently present cytomegalovirus in arterial walls of patients suffering from grade III atherosclerosis. Am. J. Pathol. *136*:23–28.

Hendrix, M. G. R. et al. (1991). Cytomegalovirus nucleic acid distribution within the human vascular tree. Am. J. Pathol. *138*:563–567.

Hermans, P. W. M. et al. (1990). Specific detection of *Mycobacterium tuberculosis* complex strains by polymerase chain reaction. J. Clin. Microbiol. *28*:1204–1213.

Herrero, R. et al. (1990). Sexual behavior, venereal diseases, hygiene practices, and invasive cervical cancer in a high-risk population. Cancer *65*:380–386.

Herring, A. J. et al. (1982). Rapid diagnosis of rotavirus infection by direct detection of viral nucleic acid in silver-stained polyacrylamide gels. J. Clin. Microbiol. *16*:473–477.

Herrmann, G. and Hubner, K. (1987). *In situ* hybridization with HBV cDNA as a sensitive method for the diagnosis of hepatitis B infection in persistent acute hepatitis. Hepato-gastroenterology *34*:148–151.

Hewlett, I. K. et al. (1989). Co-amplification of multiple regions of the HIV-1 genome by the polymerase chain reaction: potential use in multiple diagnosis. Oncogene *4*:1149–1151.

Hewlett, I. K. et al. (1990). Assessment by glue amplification and serological markers of transmission of HIV-1 from hemophiliacs to their sexual partners and secondarily to their children. J. AIDS *3*:714–720.

Hide, G. and Tait, A. (1991). The molecular epidemiology of parasites. Experientia *47*:128–142.

Higgs, T. W. et al. (1990). Type-specific human papillomavirus detection in formalin-fixed paraffin-embedded tissue sections using nonradioactive deoxyribonucleic acid probes. Lab. Invest. *63*:557–567.

Higgins, G. D. et al. (1991a). Presence and distribution of human papillomavirus sense and antisense RNA transcripts in genital cancers. J. Gen. Virol. *72*:885–895.

Higgins, G. D. et al. (1991b). Increased age and mortality associated with cervical carcinomas negative for human papillomavirus RNA. Lancet *338*:910–913.

Hilborne, L. H. et al. (1987). Direct *in situ* hybridization for rapid detection of cytomegalovirus in bronchoalveolar lavage. Am. J. Clin. Pathol. *87*:766–769.

Hill, W. E. et al. (1983). Detection and enumeration of virulent *Yersinia enterocolitica* in foods by DNA colony hybridization. Appl. Environ. Microbiol. *46*:636–641.

Hillyer, C. D. et al. (1990). The risk of cytomegalovirus infection in solid organ and bone marrow transplant recipients: transfusion of blood products. Transfusion *30*:659–666.

Hino, O. et al. (1986). Hepatitis B virus integration site in hepatocellular carcinoma at chromosome 17;18 translocation. Proc. Natl. Acad. Sci. U.S.A. *83*:8338–8342.

Hirochika, H. et al. (1987). Enhancers and trans-acting E2 transcriptional factors of papillomaviruses. J. Virol. *61*:2599–2606.

Hjelm, L. N. et al. (1990). Detection of restriction fragment length polymorphisms in clinical isolates and serially passaged *Pseudomonas aeruginosa* strains. J. Clin. Microbiol. *28*:2178–2182.

Hjerpe, A. et al. (1990). Use of cervical Cytobrush samples for dot-blot detection and Southern blot typing of human papillomaviruses using subgenomic probes. Anal. Quant. Cytol. Histol. *12*:299–305.

Ho, D. D. et al. (1985). Isolation of HTLV-III from cerebrospinal fluid and neural tissues of patients with neurologic syndromes related to the acquired immunodeficiency syndrome. N. Engl. J. Med. *313*:1493–1497.

Ho, D. D. et al. (1989). Quantitation of human immunodeficiency virus type 1 in the blood of infected persons. N. Engl. J. Med. *321*:1621–1625.

Ho, L. et al. (1991). Sequence variants of human papillomavirus type 16 in clinical samples permit verification and extension of epidemiological studies and construction of a phylogenetic tree. J. Clin. Microbiol. *29*:1765–1772.

Ho, M. et al. (1985). Epstein-Barr virus infections and DNA hybridization studies in posttransplantation lymphoma and lymphoproliferative lesions: the role of primary infection. J. Infect. Dis. *152*:876–886.

Ho-Terry, L. et al. (1988). Diagnosis of fetal rubella infection by nucleic acid hybridization. J. Med. Virol. *24*:175–182.

Ho-Terry, L. et al. (1990). Diagnosis of fetal rubella virus infection by polymerase chain reaction. J. Gen. Virol. *71*:1607–1611.

Hoffman, S. L. et al. (1984). Duodenal string-capsule culture compared with bone-marrow blood and rectal swab cultures for diagnosing typhoid and paratyphoid fever. J. Infect. Dis. *149*:157–161.

Hoffman, S. L. et al. (1986). Bone marrow aspirate culture superior to streptokinase clot culture and 8 mL 1:10 blood-to-broth culture for diagnosis of typhoid fever. Am. J. Trop. Med. Hyg. *35*:836–839.

Hofmann, H. et al. (1988). Infectivity of medical staff for hepatitis B. Infection *16*:171–174.

Hohenadl, C. et al. (1991). Strand-specific detection of enteroviral RNA in myocardial tissue by *in situ* hybridization. Mol. Cell. Probes *5*:11–20.

Hollander, H. and Stringari, S. (1987). Human immunodeficiency virus-associated meningitis: clinical course and correlations. Am. J. Med. *83*:813–816.

Holliman, R. E. et al. (1990). Diagnosis of cerebral toxoplasmosis in association with AIDS using the polymerase chain reaction. Scand. J. Infect. Dis. *22*:243–244.

Holodniy, M. et al. (1991). Detection and quantitation of human immunodeficiency virus RNA in patient serum by use of the polymerase chain reaction. J. Infect. Dis. *163*:862–866.

Honigberg, B. M. (ed.) (1989). Trichomonads Parasitic in Humans. Springer-Verlag, New York.

Hooton, T. M. et al. (1988). Prevalence of *Mycoplasma genitalium* determined by DNA probes in men with urethritis. Lancet *i*:266–268.

Hopfl, R. et al. (1991). Skin test for HPV type 16 proteins in cervical intraepithelial neoplasia. Lancet *337*:373–374.

Horsburgh, C. R., Jr. et al. (1990). Concordance of polymerase chain reaction with human immunodeficiency virus antibody detection. J. Infect. Dis. *162*:542–545.

Hosein, B. et al. (1991). Improved serodiagnosis of hepatitis C virus infection with synthetic peptide antigen from capsid protein. Proc. Natl. Acad. Sci. U.S.A. *88*:3647–3651.

Houard, S. et al. (1989). Specific identification of *Bordetella pertussis* by the polymerase chain reaction. Res. Microbiol. *140*:477–487.

Houghton, M. et al. (1991). Molecular biology of the hepatitis C viruses: implication for diagnosis, development and control of viral disease. Hepatology *14*:381–387.

Huang, C. and Deibel, R. (1988). Nucleic acid hybridization for detection of cell culture-amplified adenovirus. J. Clin. Microbiol. *26*:2652–2656.

Huang, Y. Q. et al. (1992). HPV-16 related DNA sequences in Kaposi's sarcoma. Lancet *339*:515–518.

Hudson, J. B. et al. (1990). Immortalization and altered differentiation of human keratinocytes *in vitro* by the E6 and E7 open reading frames of human papillomavirus type 18. J. Virol. *64*:519–526.

Hufert, F. T. et al. (1989). Detection of HIV-1 DNA in different subsets of human peripheral blood mononuclear cells using the polymerase chain reaction. Arch. Virol. *106*:341–345.

Hughes, J. H. et al. (1984). Latex immunoassay for rapid detection of rotavirus. J. Clin. Microbiol. *20*:441–447.

Hyde, F. W. et al. (1989). Detection of antigens in urine of mice and humans infected with *Borrelia burgdorferi*, etiologic agent of Lyme disease. J. Clin. Microbiol. *27*:58–61.

Hyman, H. C. et al. (1987). DNA probes for detection and identification of *Mycoplasma pneumoniae* and *Mycoplasma genitalium*. J. Clin. Microbiol. *25*:726–728.

Hyypia, T. (1985). Detection of adenovirus in nasopharyngeal specimens by radioactive and nonradioactive DNA probes. J. Clin. Microbiol. *21*:730–733.

Hyypia, T. (1989). Identification of human picornaviruses by nucleic acid probes. Mol. Cell. Probes *3*:329–343.

Ichimura, H. et al. (1988). Influence of hepatitis delta virus superinfection on the clearance of hepatitis B virus (HBV) markers in HBV carriers in Japan. J. Med. Virol. *26*:49–55.

Ideo, G. et al. (1990). Intrafamilial transmission of hepatitis C virus. Lancet *335*:353.

Ikeda, Y. et al. (1990). Antibody to superoxide dismutase, autoimmune hepatitis, and antibody tests for hepatitis C virus [letter]. Lancet *335*:1345–1346.

Ikenberg, H. et al. (1990). Human papillomavirus DNA in invasive carcinoma of the vulva. Obstet. Gynecol. *76*:432–438.

Imagawa, D. T. et al. (1989). Human immunodeficiency virus type 1 infection in homosexual men who remain seronegative for prolonged periods. N. Engl. J. Med. *320*:1458–1462.

Imai, H. et al. (1992). Detection of HIV-1 RNA in heparinized plasma of HIV-1 seropositive individuals. J. Virol. Methods *36*:181–184.

Imazeki, F. et al. (1986). Integration of hepatitis B virus DNA in hepatocellular carcinoma. Cancer *58*:1055–1060.

Imes, G. D., Jr. and Neafie, R. C. (1976). Babesiosis. In: Pathology of Tropical and Extraordinary Diseases. Binford, C. H. and Connor, D. H. (eds.). AFIP, Washington, D. C., pp. 301–302.

Impraim, C. C. et al. (1987). Analysis of DNA extracted from formalin-fixed paraffin-embedded tissues by enzymatic amplification and hybridization with sequence-specific oligonucleotides. Biochem. Biophys. Res. Commun. *142*:710–716.

Innis, M. A. et al. (eds.) (1990). PCR Protocols. A Guide to Methods and Applications. Academic Press, San Diego.

Itakura, S. et al. (1990). Analysis with restriction endonucleases recognizing 4- or 5-base-pair sequences of human adenovirus type 3 isolated from ocular diseases in Sapporo, Japan. J. Clin. Microbiol. *28*:2365–2369.

Iwasaka, T. et al. (1988). Combined herpes simplex virus type 2 and human papillomavirus type 16 or 18 deoxyribonucleic acid leads to oncogenic transformation. Am. J. Obstet. Gynecol. *159*:1251–1255.

Jackson, A. C. et al. (1989). Detection of rabies virus RNA in the central nervous system of experimentally infected mice using *in situ* hybridization with RNA probes. J. Virol. Methods *25*:1–12.

Jackson, D. P. et al. (1989). Extraction and amplification of DNA from archival hematoxylin and eosin sections and cervical cytology Papanicolaou smears. Nucleic Acids Res. *17*:10134.

Jackson, J. B. et al. (1987). Failure to detect human cytomegalovirus DNA in IgM-seropositive blood donors by spot hybridization. J. Infect. Dis. *156*:1013–1016.

Jackson, J. B. et al. (1988). Rapid and sensitive viral culture method for human immunodeficiency virus type 1. J. Clin. Microbiol. *26*:1416–1418.

Jackson, J. B. et al. (1990). Absence of HIV infection in blood donors with indeterminate Western blot tests for antibody to HIV-1. N. Engl. J. Med. *322*:217–222.

Jagow, J. A. and Hill, W. E. (1986). Enumeration by DNA colony hybridization of virulent *Yersinia enterocolitica* colonies in artificially contaminated food. Appl. Environ. Microbiol. *51*:441–443.

Jagow, J. A. and Hill, W. E. (1988). Enumeration of virulent *Yersinia enterocolitica* colonies by DNA colony hybridization using alkaline treatment and paper filters. Mol. Cell. Probes *2*:189–195.

Jamieson, G. A. et al. (1991). Latent herpes simplex virus type 1 in normal and Alzheimer's disease brains. J. Med. Virol. *33*:224–227.

Janda, J. M. and Duffey, P. S. (1988). Mesophilic aeromonads in human disease: current taxonomy, laboratory identification, and infectious disease spectrum. Rev. Infect. Dis. *10*:980–997.

Jansen, R. W. et al. (1990). Molecular epidemiology of human hepatitis A virus defined by an antigen-capture polymerase chain reaction method. Proc. Natl. Acad. Sci. U.S.A. *87*:2867–2871.

Janssen, H. P. et al. (1987). Comparison of *in situ* DNA hybridization and immunological staining with conventional virus isolation for the detection of human cytomegalovirus infection in cell cultures. J. Virol. Methods *17*:311–318.

Japour, A. J. et al. (1991). Detection of human immunodeficiency virus type 1 clinical isolates with reduced sensitivity to zidovudine and dideoxyinosine by RNA-RNA hybridization. Proc. Natl. Acad. Sci. U.S.A. *88*:3092–3096.

Jarrett, R. F. et al. (1990). Detection of human herpesvirus-6 DNA in peripheral blood and saliva. J. Med. Virol. *32*:73–76.

Jasmer, D. P. et al. (1990). DNA probes distinguish geographic isolates and identify a novel DNA molecule of *Babesia bovis*. J. Parasitol. *76*:834–841.

Jenison, S. A. et al. (1989). Human antibodies react with an epitope of the human papillomavirus type 6b L1 open reading frame which is distinct from the type-common epitope. J. Virol. *63*:809–818.

Jensen, J. S. et al. (1989). Detection of *Mycoplasma pneumoniae* in simulated clinical samples by polymerase chain reaction. APMIS 97:1046–1048.

Jerse, A. E. et al. (1990). Oligonucleotide probe for detection of the enteropathogenic *Escherichia coli* (EPEC) adherence factor of localized adherent EPEC. J. Clin. Microbiol. *28*:2842–2844.

Jewers, R. J. et al. (1992). Regions of human papillomavirus type 16 E7 oncoprotein required for immortalization of human keratinocytes. J. Virol. *66*:1329–1335.

Jiang, X. et al. (1987). Detection of hepatitis A virus by hybridization with single stranded RNA probes. Appl. Environ. Microbiol. *53*:2487–2495.

Jiang, X. et al. (1989). *In situ* hybridization for quantitative assay of infectious hepatitis A virus. J. Clin. Microbiol. *27*:874–879.

Jilg, W. et al. (1985). What's new in hepatitis delta virus? Pathol. Res. Pract. *180*:431–436.

Jilg, W. et al. (1990). Identification of type A and B isolates of Epstein-Barr virus by polymerase chain reaction. J. Virol. Methods *30*:319–322.

Jin, O. et al. (1990). Detection of enterovirus RNA in myocardial biopsies from patients with myocarditis and cardiomyopathy using gene amplification by polymerase chain reaction. Circulation *82*:8–16.

Johnson, G. K. and Robinson, W. S. (1991). Human immunodeficiency virus-1 (HIV-1) in the vapors of surgical power instruments. J. Med. Virol. *33*:47–50.

Johnson, J. D. et al. (1990). Detection of *Toxoplasma gondii* using the polymerase chain reaction. Biochem. Soc. Trans. *18*:665.

Johnson, J. E. et al. (1991). Typing and molecular characterization of human papillomaviruses in genital warts from South African women. J. Med. Virol. *33*:39–42.

Jones, D. M. et al. (1990). DNA probes for typing *Neisseria meningitidis*. Lancet *336*:53–54.

Jones, J. F. et al. (1988). T-cell lymphomas containing Epstein-Barr viral DNA in patients with chronic Epstein-Barr virus infections. N. Engl. J. Med. *318*:733–741.

Josephs, S. F. et al. (1991). HHV-6 reactivation in chronic fatigue syndrome. Lancet *337*:1346–1347.

Judd, R. et al. (1991). Human papillomavirus type 6 detected by polymerase chain reaction in invasive sinonasal papillary squamous cell carcinoma. Arch. Pathol. Lab. Med. *115*:1150–1153.

Kadival, G. V. et al. (1986). Sensitivity and specificity of enzyme linked immunosorbent assay in the detection of antigen in tuberculous meningitis cerebrospinal fluids. J. Clin. Microbiol. *23*:901–904.

Kahlon, J. et al. (1987). Detection of antibodies to herpes simplex virus in the cerebrospinal fluid of patients with herpes simplex encephalitis. J. Infect. Dis. *155*:38–44.

Kaklamani, E. et al. (1991). Hepatitis B and C viruses and their interaction in the origin of hepatocellular carcinoma. JAMA *265*:1974–1976.

Kaltenboeck, B. et al. (1991). Detection and strain differentiation of *Chlamydia psittaci* mediated by a two-step polymerase chain reaction. J. Clin. Microbiol. *29*:1969–1975.

Kamath, K. R. and Murugasu, R. (1974). A comparative study of four methods for detecting *Giardia lamblia* in children with diarrheal disease and malabsorption. Gastroenterology *66*:16–21.

Kanda, T. et al. (1988a). Human papillomavirus type 16 open reading frame E7 encodes a transforming gene for rat 3Y1 cells. J. Virol. *62*:610–13.

Kanda, T. et al. (1988b). Immortalization of primary rat cells by human papillomavirus type 16 subgenomic DNA fragments controlled by the SV40 promoter. Virology *165*:321–325.

Kandolf, R. (1988). The impact of recombinant DNA technology on the study of enterovirus heart disease. In: Coxsackievirus: A General Update. Benclinelli, M. and Friedman, H. (eds.). Plenum Press, New York, pp. 303–305.

Kandolf, R. et al. (1987). *In situ* detection of enteroviral genomes in myocardial cells by nucleic acid hybridization: an approach to the diagnosis of viral heart disease. Proc. Natl. Acad. Sci. U.S.A. *84*:6272–6276.

Kaneko, K. et al. (1990). Rapid diagnosis of tuberculous meningitis by polymerase chain reaction (PCR). Neurology *40*:1617–1618.

Kaneko, S. and Miller, R. H. (1990). Characterization of primers for optimal amplification of hepatitis B virus DNA in the polymerase chain reaction assay. J. Virol. Methods *29*:225–230.

Kaneko, S. et al. (1989a). Rapid and sensitive method for the detection of serum hepatitis B virus DNA using the polymerase chain reaction technique. J. Clin. Microbiol. *27*:1930–1933.

Kaneko, S. et al. (1989b). Detection of serum hepatitis B virus DNA in patients with chromic hepatitis using the polymerase chain reaction assay. Proc. Natl. Acad. Sci. U.S.A. *86*:312–316.

Kaneko, S. et al. (1990a). Detection of hepatitis B virus DNA using the polymerase chain reaction technique. J. Clin. Lab. Anal. *4*:479–482.

Kaneko, S. et al. (1990b). Detection of serum hepatitis C virus RNA. Lancet *336*:976.

Karlsson, M. (1990). Aspects of the diagnosis of Lyme borreliosis. Ph.D. thesis, Karolinska Hospital, Stockholm, Sweden.

Karlsson, I. M. et al. (1988). Characterization of antibody response in patients with Borrelia meningitis. Serodiagn. Immunother. Intern. Dis. *2*:375–386.

Karlsson, M. et al. (1989). Comparison of Western blot and enzyme-linked immunosorbent assay for diagnosis of Lyme borreliosis. Eur. J. Clin. Microbiol. Infect. Dis. *8*:871–877.

Karlsson, M. et al. (1990). Cultivation and characterization of spirochetes from cerebrospinal fluid of patients with Lyme borreliosis. J. Clin. Microbiol. *28*:473–479.

Karmali, M. A. (1989). Infection by verocytotoxin-producing *Escherichia coli*. Clin. Microbiol. Rev. 2:15–38.

Kasher, M. S. and Roman, A. (1988). Alterations in the regulatory region of the human papillomavirus type 6 genome are generated during propagation in *Escherichia coli*. J. Virol. *62*:3295–3300.

Katakura, K. et al. (1991). Tunicamycin-resistant variants from five species of *Leishmania* contain amplified DNA in extrachromosomal circles of different sizes with a transcriptionally active homologous region. Mol. Biochem. Parasitol. *44*:233–244.

Kato, N. et al. (1991). Identification of toxigenic *Clostridium difficile* by the polymerase chain reaction. J. Clin. Microbiol. *29*:33–37.

Katz, B. Z. et al. (1989). Latent and replicating forms of Epstein-Barr virus DNA in lymphomas and lymphoproliferative diseases. J. Infect. Dis. *160*:589–598.

Katzenstein, A. L. A. and Peiper, S. C. (1990). Detection of Epstein-Barr virus genomes in lymphomatoid granulomatosis: analysis of 29 cases by the polymerase chain reaction technique. Mod. Pathol. *3*:435–441.

Kaufman, L. (1992). Laboratory methods for the diagnosis and confirmation of systemic mycoses. Clin. Infect. Dis. *14* (Suppl. 1):S23–S29.

Kaufman, R. H. et al. (1988). Human papillomavirus and herpes simplex virus in vulvar squamous cell carcinoma *in situ*. Am. J. Obstet. Gynecol. *158*:862–869.

Keath, E. J. et al. (1989). DNA probe for the identification of *Histoplasma capsulatum*. J. Clin. Microbiol. *27*:2369–2372.

Keay, S. et al. (1987). Ganciclovir treatment of cytomegalovirus infections in iatrogenically immunocompromised patients. J. Infect. Dis. *156*:1016–1021.

Keh, W. C. and Gerber, M. A. (1988). *In situ* hybridization for cytomegalovirus DNA in AIDS patients. Am. J. Pathol. *131*:490–496.

Keller, G. H. and Manak, M. M. (1990). Detection of HIV-1 nucleic acid in clinical samples using target DNA amplification and nonisotopic probes. In: Monogr. Virol. Human Immunodeficiency Virus: Innovative Techniques, Vol. 18. Khan, N. C. and Melnick, J. L. (eds.). Karger, Basel, pp. 28–60.

Keller, G. H. et al. (1990). Detection of hepatitis B virus DNA in serum by polymerase chain reaction amplification and microtiter sandwich hybridization. J. Clin. Microbiol. *28*:1411–1416.

Kellogg, D. E. and Kwok, S. (1990). Detection of human immunodeficiency virus. In: PCR Protocols: A Guide to Methods and Applications. Innis, M. A. et al. (eds.). Academic Press, San Diego, pp. 337–347.

Kellogg, D. E. et al. (1990). Quantitation of HIV-1 proviral DNA relative to cellular DNA by the polymerase chain reaction. Anal. Biochem. *189*:202–208.

Kemnitz, J. et al. (1989). Rapid identification of viral infection in liver, heart, and kidney allograft biopsies by *in situ* hybridization [letter]. Am. J. Surg. Pathol. *13*:80–82.

Kennedy, D. H. and Fallon, R. J. (1979). Tuberculosis meningitis. J. Am. Med. Assn. *241*:264–268.

Kennedy, L. et al. (1988). Human papillomavirus — a study of male sexual partners. Med. J. Aust. *149*:309–311.

Kenton, J. H. et al. (1991). Rapid electrochemiluminescence assays of polymerase chain reaction products. Clin. Chem. *37*:1626–1632.

Khan, A. M. and Wright, P. J. (1987). Detection of flavivirus RNA in infected cells using photobiotin-labeled hybridization probes. J. Virol. Methods *15*:121–130.

Kidd, A. H. et al. (1985). Specific detection and typing of adenovirus types 40 and 41 in stool specimens by dot-blot hybridization. J. Clin. Microbiol. *22*:934–939.

Kido, S. et al. (1991). Detection of varicella-zoster virus (VZV) DNA in clinical samples from patients with VZV by the polymerase chain reaction. J. Clin. Microbiol. *29*:76–79.

Kiehlbauch, J. A. et al. (1991). Restriction fragment length polymorphisms in the ribosomal genes for species identification and subtyping of aerotolerant *Campylobacter* species. J. Clin. Microbiol. *29*:1670–1676.

Kikuta, H. et al. (1991). Frequent detection of human herpesvirus 6 DNA in peripheral blood mononuclear cells from kidney transplant patients. J. Infect. Dis. *163*:925.

Kilvington, S. et al. (1991). Differentiation of *Acanthamoeba* strains from infected corneas and the environment by using restriction endonuclease digestion of whole-cell DNA. J. Clin. Microbiol. *29*:310–314.

Kimpton, C. P. et al. (1988). Detection of cytomegalovirus by dot blot DNA hybridization using probes labeled with ^{32}P by nick translation or random hexanucleotide priming. Mol. Cell. Probes 2:181–188.

Kimura, H. et al. (1990). Detection and direct typing of herpes simplex virus by polymerase chain reaction. Med. Microbiol. Immunol. *179*:177–184.

Kiyabu, M. T. et al. (1989). Detection of human papillomavirus in formalin-fixed invasive squamous carcinomas using the polymerase chain reaction. Am. J. Surg. Pathol. *13*:221–224.

Kiyosawa, K. et al. (1991). Intrafamilial transmission of hepatitis C virus in Japan. J. Med. Virol. *33*:114–116.

Klein, G. et al. (1974). Direct evidence for the presence of Epstein-Barr virus DNA and nuclear antigen in malignant epithelial cells from patients with poorly differentiated carcinoma of the nasopharynx. Proc. Natl. Acad. Sci. U.S.A. *71*:4737–4741.

Klein, R. S. et al. (1991). Occupational risk for hepatitis C virus infection among New York dentists. Lancet *338*:1539–1542.

Klenk, H. D. and Rott, R. (1988). The molecular biology of influenza virus pathogenicity. Adv. Virus Res. *34*:247–281.

Knecht, H. et al. (1990). Detection of Epstein-Barr virus DNA by polymerase chain reaction in lymph node biopsies from patients with angioimmunoblastic lymphadenopathy. Br. J. Haematol. *75*:610–614.

Knight, A. I. et al. (1990). Identification of a UK outbreak strain of *Neisseria meningitidis* with a DNA probe. Lancet *335*:1182–1184.

Koch, W. C. and Adler, S. P. (1990). Detection of human parvovirus B19 DNA by using the polymerase chain reaction. J. Clin. Microbiol. *28*:65–69.

Kohno, H. (1986). *Clostridium difficile* enterotoxin and immunological assay in fecal specimens. Microecol. Ther. *16*:169–180.

Kolberg, J. A. et al. (1989). The specificity of pilin DNA sequences for the detection of pathogenic *Neisseria*. Mol. Cell. Probes *3*:59–72.

Koropchak, C. M. et al. (1989). Investigation of varicella-zoster infection of lymphocytes by *in situ* hybridization. J. Virol. *63*:2392–2395.

Koulos, J. et al. (1991). Human papillomavirus detection in adenocarcinoma of the anus. Mod. Pathol. *4*:58–61.

Koutsky, L. (1991). Role of epidemiology in defining events that influence transmission and natural history of anogenital papillomavirus infections. J. Natl. Cancer Inst. *83*:978–979.

Koutsky, L. A. et al. (1988). Epidemiology of genital human papillomavirus infection. Epidemiol. Rev. *10*:122–163.

Krambovitis, E. et al. (1984). Rapid diagnosis of tuberculous meningitis by latex particle agglutination. Lancet *ii*:1229–1231.

Krebs, H. B. (1989). Management of human papillomavirus-associated genital lesions in men. Obstet. Gynecol. *73*:312–316.

Krebs, H. B. and Schneider, V. (1987). Human papillomavirus-associated lesions of the penis: colposcopy, cytology, and histology. Obstet. Gynecol. *70*:299–304.

Krebs, H. B. and Helmkamp, B. F. (1990). Does the treatment of genital condylomata in men decrease the treatment failure rate of cervical dysplasia in the female sexual partner? Obstet. Gynecol. *765*:660–663.

Kristiansen, B. E. et al. (1991). Rapid diagnosis of meningococcal meningitis by polymerase chain reaction. Lancet *337*:1568–1569.

Kubo, Y. et al. (1989). A cDNA fragment of hepatitis C virus isolated from an implicated donor of post-transfusion non-A, non-B hepatitis in Japan. Nucleic Acids Res. *17*:10367–10372.

Kuhns, M. C. et al. (1988). A new assay for the quantitative detection of hepatitis B viral DNA in human serum. In: Viral Hepatitis and Liver Disease. Alan R. Liss, New York, pp. 258–262.

Kuhns, M. C. et al. (1989). Quantitation of hepatitis B viral DNA by solution hybridization: comparison with DNA polymerase and hepatitis Be antigen during antiviral therapy. J. Med. Virol. *27*:274–281.

Kuijper, E. J. et al. (1987). Application of whole-cell DNA restriction endonuclease profiles to the epidemiology of *Clostridium difficile*-induced diarrhea. J. Clin. Microbiol. *25*:751–753.

Kukla, B. A. et al. (1987). Use of species-specific DNA probes for detection and identification of trypanosome infection in tsetse flies. Parasitology *95*:1–16.

Kulkosky, J. and Skalka, A. M. (1990). HIV DNA integration: observations and inferences. J. AIDS *3*:839–851.

Kulski, J. K. et al. (1987). DNA sequences of human papillomavirus types 11, 16 or 18 in invasive cervical carcinoma of Western Australian women. Immunol. Cell. Biol. *65*:77–84.

Kumar, A. et al. (1988). Non-radioactive oligonucleotide probe for detection of clinical enterotoxigenic *Escherichia coli* isolates of bovine origin. Ann. Inst. Pasteur/Microbiol. *139*:315–323.

Kuo, G. et al. (1989). An assay for circulating antibodies to a major etiologic virus of human non-A, non-B hepatitis. Science *244*:362–364.

Kuo, M. Y. P. et al. (1988). Molecular cloning of hepatitis delta virus RNA from an infected woodchuck liver: sequence, structure, and applications. J. Virol. *62*:1855–1861.

Kurman, R. J. et al. (1988). Analysis of individual human papillomavirus types in cervical neoplasia: a possible role for type 18 in rapid progression. Am. J. Obstet. Gynecol. *159*:293–296.

Kurtzman, G. T. et al. (1987). Chronic bone marrow failure due to persistent B19 parvovirus infection. N. Engl. J. Med. *317*:287–294.

Kusunoki, S. et al. (1991). Application of colorimetric microdilution plate hybridization for rapid genetic identification of 22 *Mycobacterium* species. J. Clin. Microbiol. *29*:1596–1603.

Kwok, S. (1990). Procedures to minimize PCR-product carry-over. In: PCR Protocols: A Guide to Methods and Applications. Innis, M. A. et al. (eds.). Academic Press, San Diego, pp. 142–145.

Kwok, S. et al. (1987). Identification of human immunodeficiency virus sequences by using *in-vitro* enzymatic amplification and oligomer cleavage detection. J. Virol. *61*:1690–1694.

Kwok, S. et al. (1990). Effects of primer-template mismatches on the polymerase chain reaction: human immunodeficiency virus type 1 model studies. Nucleic Acids Res. *18*:999–1005.

Lacey, C. J. N. et al. (1991). Human papillomavirus type 16 infection of the cervix: a comparison of differing DNA detection modes and the use of monoclonal antibodies against the major capsid protein. Genitourin. Med. *67*:87–91.

Lai, M. E. et al. (1989). Hepatitis B virus DNA in the serum of Sardinian blood donors negative for the hepatitis B surface antigen. Blood *73*:17–19.

Lai, M. Y. et al. (1988a). Reactivation of hepatitis B virus in anti-HBe-positive chronic active type B hepatitis: molecular and immunohistochemical studies. Hepato-gastroenterology *35*:17–21.

Lai, M. Y. et al. (1988b). Status of hepatitis B DNA in hepatocellular carcinoma: a study based on paired tumor and nontumor liver tissues. J. Med. Virol. *25*:249–258.

Lai-Goldman, M. et al. (1988). Detection of human immunodeficiency virus (HIV) infection in formalin-fixed, paraffin-embedded tissues by DNA amplification. Nucleic Acids Res. *16*:8191.

Lakeman, F. D. et al. (1987). Detection of antigen to herpes simplex virus in cerebrospinal fluid from patients with herpes simplex encephalitis. J. Infect. Dis. *155*:1172–1178.

Lambert, P. F. et al. (1987). A transcriptional repressor encoded by BPV-1 shares a common carboxyterminal domain with the E2 transactivator. Cell (Cambridge, Mass) *50*:69–78.

Lampertico, P. et al. (1990). Detection of hepatitis B virus DNA in formalin-fixed, paraffin-embedded liver tissue by the polymerase chain reaction. Am. J. Pathol. *137*:253–258.

Lancaster, W. D. et al. (1986). Papillomaviruses in anogenital neoplasms. In: The Human Oncogenic Viruses. Luderer, A. A. and Weetall, H. H. (eds.). Humana Press, pp. 153–183.

Langenberg, A. et al. (1988). Detection of herpes simplex virus DNA from genital lesions by *in situ* hybridization. J. Clin. Microbiol. *26*:933–937.

Larzul, D. et al. (1987). Non-radioactive hepatitis B virus DNA probe for detection of HBV DNA in serum. J. Hepatol. *5*:199–204.

Larzul, D. et al. (1988). Detection of hepatitis B virus sequences in serum by using *in vitro* enzymatic amplification. J. Virol. Methods *20*:227–237.

Larzul, D. et al. (1989). A highly sensitive detection method for hepatitis B viral DNA sequences: monitoring of antiviral therapy. In: Molecular Probes: Technology and Medical Applications. Albertini, A. (ed.). Raven Press, New York, pp. 97–104.

Larzul, D. et al. (1990). An automatic modified polymerase chain reaction procedure for hepatitis B virus DNA detection. J. Virol. Methods *27*:49–60.

Laskay, T. et al. (1991). Generation of species-specific DNA probes for *Leishmania aethiopica*. Mol. Biochem. Parasitol. *44*:279–286.

Laure, F. et al. (1985). Hepatitis B virus DNA sequences in lymphoid cells from patients with AIDS and AIDS-related complex. Science *229*:561–563.

Laure, F. et al. (1988). Detection of HIV-1 DNA in infants and children by means of the polymerase chain reaction. Lancet *ii*:538–540.

Le Bourgeois, P. et al. (1989). Genome comparison of *Lactococcus* strains by pulsed-field gel electrophoresis. FEMS Microbiol. Lett. *59*:65–70.

LeBar, W. et al. (1989). Comparison of DNA probe, monoclonal antibody enzyme immunoassay and cell culture for the detection of *Chlamydia trachomatis*. J. Clin. Microbiol. *27*:826–828.

Lebech, A. M. et al. (1991). Comparison of *in vitro* culture and polymerase chain reaction for detection of *Borrelia burgdorferi* in tissue from experimentally infected animals. J. Clin. Microbiol. *29*:731–737.

Lee, S. D. et al. (1986). Prevention of maternal-infant hepatitis B virus transmission by immunization: the role of serum hepatitis B virus DNA. Hepatology *6*:369–373.

LeFebvre, R. B. et al. (1989). Characterization of *Borrelia burgdorferi* isolates by restriction endonuclease analysis and DNA hybridization. J. Clin. Microbiol. *27*:636–639.

Lefrere, J. J. et al. (1990a). Polymerase chain reaction testing of HIV-1 seronegative at-risk individuals. Lancet *335*:1400–1401.

Lefrere, J. J. et al. (1990b). PCR testing in HIV-1 seronegative hemophilia. Lancet *335*:1386.

Lehn, H. et al. (1985). Papillomavirus genomes in human cervical tumors: analysis of their transcriptional activity. Proc. Natl. Acad. Sci. U.S.A. *82*:5540–5544.

Lehn, H. et al. (1988). Physical state and biological activity of human papillomavirus genomes in precancerous lesions of the female genital tract. J. Gen. Virol. *69*:187–196.

Lehtomaki, K. et al. (1986). Rapid diagnosis of respiratory adenovirus infections in young adult men. J. Clin. Microbiol. *24*:108–111.

Leigh, T. R. et al. (1989). Immunofluorescence staining improves the diagnostic sensitivity of sputum induction in *Pneumocystis carinii* pneumonia. Thorax *44*:891.

Lentz, E. B. et al. (1988). Detection of antibody to cytomegalovirus-induced early antigens and comparison with four serological assays and presence of viruria in blood donors. J. Clin. Microbiol. *26*:133–135.

Lenzi, M. et al. (1990). Type 2 autoimmune hepatitis and hepatitis C virus infection. Lancet *335*:258–259.

Lenzi, M. et al. (1991). Antibodies to hepatitis C virus in autoimmune liver disease: evidence for geographic heterogeneity. Lancet *338*:277–280.

Leon, A. et al. (1991). Second generation RIBA to confirm diagnosis of HCV infection [letter to Editor]. Lancet *337*:912.

Lettau, L. A. et al. (1987). Outbreak of severe hepatitis due to delta and hepatitis B viruses in parenteral drug abusers and their contacts. N. Engl. J. Med. *317*:1256–1262.

Levine, M. M. (1987). *Escherichia coli* that cause diarrhea: enterotoxigenic, enteropathogenic, enteroinvasive, enterohemorrhagic, and enteroadherent. J. Infect. Dis. *155*:377–389.

Lewandowski, G. et al. (1990). The use of *in situ* hybridization to show human papillomavirus deoxyribonucleic acid in metastatic cancer cells within lymph nodes. Am. J. Obstet. Gynecol. *163*:1333–1337.

Lewis, D. J. M. et al. (1990). Total genomic DNA probe to detect *Giardia lamblia*. Lancet *336*:257.

Ley, C. et al. (1991). Determinants of genital human papillomavirus infection in young women. J. Natl. Cancer Inst. *83*:997–1003.

Li, J. J. et al. (1990). HIV-1 DNA proviral sequences in fresh urine pellets from HIV-1 seropositive persons. Lancet *335*:1590–1591.

Liaw, Y. F. et al. (1987). Changes of serum hepatitis B virus DNA in two types of clinical events preceding spontaneous hepatitis Be antigen seroconversion in chronic type B hepatitis. Hepatology *7*:1–3.

Liaw, Y. F. et al. (1988). Changes of serum HBV-DNA in relation to serum transaminase level during acute exacerbation in patients with chronic type B hepatitis. Liver *8*:231–235.

Lie-Injo, L. E. et al. (1983). Hepatitis B virus DNA in liver and white blood cells of patients with hepatoma. DNA *2*:301–308.

Lie-Injo, L. E. et al. (1985). Hepatitis B virus (HBV) DNA in leucocytes in acquired immune deficiency syndrome. Cytobios *44*:119–128.

Lieberman, H. M. et al. (1983). Detection of hepatitis B virus DNA directly in human serum by a simplified molecular hybridization test: comparison to the HBeAg/anti-HBe status in HBsAg carriers. Hepatology *3*:285–291.

Lifson, A. R. et al. (1990). Detection of human immunodeficiency virus DNA using the polymerase chain reaction in a well characterized group of homosexual and bisexual men. J. Infect. Dis. *161*:436–439.

Lin, H. J. et al. (1990). Evidence for intrafamilial transmission of hepatitis B virus from sequence analysis of mutant HBV DNAs in two Chinese families. Lancet *336*:208–212.

Lin, J. H. et al. (1987). An oligonucleotide probe for the detection of hepatitis B virus DNA in serum. J. Virol. Methods *15*:139–149.

Lin, M. et al. (1987). cDNA probes of individual genes of human rotavirus distinguish viral subgroups and serotypes. J. Virol. Methods *15*:285–289.

Linder, E. et al. (1987). Detection of *P. carinii* in lung derived samples using monoclonal antibodies to an 82 kDa parasite component. J. Immunol. Methods *98*:57–62.

Littler, E. et al. (1991). Diagnosis of nasopharyngeal carcinoma by means of recombinant Epstein-Barr virus proteins. Lancet *337*:685–689.

Litwin, C. M. et al. (1991). Molecular epidemiology of *Shigella* infections: plasmid profiles, serotype correlation, and restriction endonuclease analysis. J. Clin. Microbiol. *29*:104–108.

Lizardi, P. M. et al. (1988). Exponential amplification of recombinant RNA hybridization probes. BioTechnology *6*:1197–1202.

Lobbiani, A. et al. (1990). Hepatitis B virus transcripts and surface antigen in human peripheral blood lymphocytes. J. Med. Virol. *31*:190–194.

Loche, M. and Mach, B. (1988). Identification of HIV-infected seronegative individuals by a direct diagnostic test based on hybridization to amplified viral DNA. Lancet *ii*:418–421.

Lockett, S. J. et al. (1991). Automated fluorescent image cytometry. DNA quantitation and detection of chlamydial infections. Anal. Quant. Cytol. Histol. *13*:27–44.

Lohr, J. M. and Oldstone, M. B. A. (1990). Detection of cytomegalovirus nucleic acid sequences in pancreas in type 2 diabetes. Lancet *336*:644–648.

Lombardo, J. M. et al. (1989). Evaluation of an atypical HIV type 1 antibody. Arch. Pathol. Lab. Med. *113*:1245–1249.

Lomeli, H. et al. (1989). Quantitative assays based on the use of replicatable hybridization probes. Clin. Chem. *35*:1826–1831.

Lorincz, A. T. et al. (1987). Oncogenic association of specific human papillomavirus types with cervical neoplasia. J. Natl. Cancer Inst. *79*:671–677.

Lou, Y. K. et al. (1991). Papillomavirus DNA and colposcopy instruments. Lancet *338*:1601–1602.

Low, S. H. et al. (1990). Prevalence of human papillomavirus types 16 and 18 in cervical carcinomas: a study by dot and Southern blot hybridization and the polymerase chain reaction. Jpn. J. Cancer Res. *81*:118–123.

Lucotte, G. and Reveilleau, J. (1989). Identification of HIV-1 infected seropositive subjects by a direct diagnostic test involving hybridization of amplified viral DNA. Mol. Cell. Probes *3*:299–306.

Ludlam, C. A. et al. (1989). Antibodies to hepatitis C virus in hemophilia. Lancet *ii*:560–561.

Lunkenheimer, K. et al. (1990). Detection of lassa virus RNA in specimens from patients with lassa fever by using the polymerase chain reaction. J. Clin. Microbiol. *28*:2689–2692.

Lurain, N. S. et al. (1986). Rapid detection of cytomegalovirus in clinical specimens by using biotinylated DNA probes and analysis of cross-reactivity with herpes simplex virus. J. Clin. Microbiol. *24*:724–730.

Macario, A. J. L. and Conway de Macario, E. (eds.) (1990). Gene Probes for Bacteria. Academic Press, San-Diego.

Macnab, J. C. M. et al. (1986). Human papillomavirus in clinically and histologically normal tissue of patients with genital cancer. N. Engl. J. Med. *315*:1052–1058.

Madejon, A. et al. (1990). Detection of HDV-RNA by PCR in serum of patients with chronic infection. J. Hepatol. *11*:381–384.

Magrath, I. (1989). Infectious mononucleosis and malignant neoplasia. In: Infectious Mononucleosis. Schlossberg, D. (ed.). Springer-Verlag, New York, pp. 142–171.

Mahalingam, R. et al. (1990). Latent varicella-zoster viral DNA in human trigeminal and thoracic ganglia. N. Engl. J. Med. *323*:627–631.

Mahbubani, M. H. et al. (1990). Detection of *Legionella* with polymerase chain reaction and gene probe methods. Mol. Cell. Probes *4*:175–187.

Maki, H. et al. (1991). Use of universal and type-specific primers in the polymerase chain reaction for the detection and typing of genital papillomaviruses. Jpn. J. Cancer Res. *82*:411–419.

Makris, A. et al. (1991). Measurement of hepatitis B viral DNA in serum by solution hybridization and comparison with the dot-blot hybridization technique. Hepato-gastroenterol. *38*:53–55.

Makris, M. et al. (1990). Hepatitis C antibody and chronic liver disease in hemophilia. Lancet *335*:1117–1119.

Malloy, D. C. et al. (1990). Detection of *Borrelia burgdorferi* using the polymerase chain reaction. J. Clin. Microbiol. *28*:1089–1093.

Manos, M. et al. (1990). Looking for human papillomavirus type 16 by PCR. Lancet *i*:734.

Mariotti, M. et al. (1990a). Failure to detect evidence of human immunodeficiency virus type 1 (HIV-1) infection by polymerase chain reaction assay in blood donors with isolated core antibodies (anti p24 or p17) to HIV-1. Transfusion *30*:704–706.

Mariotti, M. et al. (1990b). DNA amplification of HIV-1 seropositive individuals and in seronegative at risk individuals. AIDS *4*:633–637.

Martin, P. (1990). Hepatitis C: from laboratory to bedside. Mayo Clin. Proc. *65*:1372–1376.

Martinez, A. J. (1985). Free-Living Amebas: Natural History, Prevention, Diagnosis, Pathology and Treatment of Disease. CRC Press, Boca Raton, FL.

Mason, P. R. (1979). Serodiagnosis of *Trichomonas vaginalis* infection by the indirect fluorescent antibody method. J. Clin. Pathol. *32*:1211–1215.

Matlashewski, G. (1989). The cell biology of human papillomavirus transformed cells. Anticancer Res. *9*:1447–1456.

Matlashewski, G. et al. (1987). Human papillomavirus type 16 DNA cooperates with activated *ras* in transforming primary cells. EMBO J. *6*:1741–1746.

Matsukura, T. et al. (1986). Cloning of monomeric human papillomavirus type 16 DNA integrated within cell DNA from a cervical carcinoma. J. Virol. *58*:979–982.

Matsukura, T. et al. (1989). Both episomal and integrated forms of human papillomavirus type 16 are involved in invasive cervical cancers. Virology *172*:63–72.

Matsumoto, H. et al. (1988). Analysis of integrated hepatitis B virus DNA and cellular flanking sequences clones from a hepatocellular carcinoma. Int. J. Cancer *42*:1–6.

Matsuyama, Y. et al. (1985). Discordance of hepatitis Be antigen/antibody and hepatitis B virus deoxyribonucleic acid in serum. Gastroenterology *89*:1104–1108.

Mattsson, J. G. et al. (1991). Detection of *Mycoplasma bovis* and *Mycoplasma agalactiae* by oligonucleotide probes complementary to 16S rRNA. Mol. Cell. Probes *5*:27–35.

Mazurek, G. H. et al. (1991). Chromosomal DNA fingerprint patterns produced with IS6110 as strain-specific markers for epidemiologic study of tuberculosis. J. Clin. Microbiol. *29*:2030–2033.

McClelland, M. et al. (1987). Restriction endonucleases for pulsed field mapping of bacterial genomes. Nucleic Acids Res. *15*:5985–6005.

McClintock, J. T. et al. (1989). Comparison of *in situ* hybridization and monoclonal antibodies for early detection of cytomegalovirus in cell culture. J. Clin. Microbiol. *27*:1554–1559.

McDonnell, P. J. et al. (1987). Detection of human papillomavirus type 6/11 DNA in conjunctival papillomas by *in situ* hybridization with radioactive probes. Hum. Pathol. *18*:115–119.

McDougall, J. K. (1988). *In situ* hybridization for viral gene detection. ISI Atlas of Science: Biochemistry, pp. 6–10.

McElrath, M. J. et al. (1991). Latent HIV-1 infection in enriched populations of blood monocytes and T cells from seropositive patients. J. Clin. Invest. *87*:27–30.

McFadden, J. J. et al. (1987). Crohn's disease — isolated mycobacteria are identical to *Mycobacterium paratuberculosis*, as determined by DNA probes that distinguish between mycobacterial species. J. Clin. Microbiol. *25*:796–801.

McFarlane, I. G. et al. (1990). Hepatitis C virus antibodies in chronic active hepatitis: pathogenetic factor or false-positive result? Lancet *335*:754–757.

McHugh, T. M. et al. (1988a). Simultaneous detection of antibodies to cytomegalovirus and herpes simplex virus by using flow cytometry and a microsphere-based fluorescence immunoassay. J. Clin. Microbiol. *26*:1957–1961.

McHugh, T. M. et al. (1988b). Relation of circulating levels of human immunodeficiency virus (HIV) antigen, antibody to p24 and HIV-containing immune complexes in HIV-infected patients. J. Infect. Dis. *158*:1088–1091.

McLaughlin, G. L. et al. (1986). Detection of *Babesia bovis* using DNA hybridization. J. Protozool. *33*:125–128.

McLaughlin, G. L. et al. (1991). Amplification of repetitive DNA for the specific detection of *Naegleria fowleri*. J. Clin. Microbiol. *29*:227–230.

McNearney, T. et al. (1990). Limited sequence heterogeneity among biologically distinct human immunodeficiency virus type 1 isolates from individuals involved in a clustered infectious outbreak. Proc. Natl. Acad. Sci. U.S.A. *87*:1917–1921.

McNichol, P. J. and Dodd, J. G. (1990). Detection of human papillomavirus DNA in prostate gland tissue by using the polymerase chain reaction amplification assay. J. Clin. Microbiol. *28*:409–412.

McNichol, P. J. et al. (1989). Comparison of filter *in situ* deoxyribonucleic acid hybridization with cytologic, colposcipic and histopathologic examination for detection of human papillomavirus infection in women with cervical intraepithelial neoplasia. Am. J. Obstet. Gynecol. *160*:265–270.

McReynolds, L. A. et al. (1991). Filarial DNA probes in Jakarta. Parasitol. Today *7*:65–67.

Meanwell, C. A. (1988). The epidemiology of human papillomavirus infection in relation to cervical cancer. Cancer Surv. *7*:481–497.

Meisels, A. and Fortin, R. (1976). Condylomatous lesions of the cervix and vagina. I. Cytological patterns. Acta Cytol. *20*:505–509.

Melchers, W. J. G. et al. (1988). Prevalence of genital HPV infections in a regularly screened population in the Netherlands in relation to cervical cytology. J. Med. Virol. *25*:11–16.

Melchers, W. et al. (1989a). Increased detection rate of human papillomavirus in cervical scrapes by the polymerase chain reaction as compared to modified FISH and Southern blot analysis. J. Med. Virol. *27*:329–335.

Melchers, W. J. G. et al. (1989b). Optimization of human papillomavirus genotype detection in cervical scrapes by a modified filter *in situ* hybridization test. J. Clin. Microbiol. *27*:106–110.

Melki, R. et al. (1988). Nucleic acid spot hybridization with nonradioactive labeled probes in screening for human papillomavirus DNA sequences. J. Med. Virol. *26*:137–143.

Melnick, J. L. (1984). Enteroviruses. In: Viral infections of humans. Evans, A. S. (ed.). Plenum Press, New York, pp. 187–251.

Melnick, J. L. (1990). Enteroviruses: polioviruses, coxsackiviruses, echoviruses, and newer enteroviruses. In: Virology. Fields, B. N. and Knipe, D. M. (eds.). Raven Press, New York, pp. 549–605.

Melnick, J. L. et al. (1983). Cytomegalovirus antigen within human arterial smooth muscle cells. Lancet *ii*:644–647.

Merrill, J. E. and Chen, I. S. Y. (1991). HIV-1, macrophages, glial cells, and cytokines in AIDS nervous system disease. FASEB J. *5*:2391–2397.

Meyer, G. and Enders, G. (1988). Correlation of cytomegalovirus (CMV) detection in urine by tissue culture virus isolation, early antigen immunofluorescence test and nucleic acid hybridization. Infection *16*:153–157.

Michitaka, K. et al. (1988). Change of hepatitis B virus DNA distribution associated with the progression of chronic hepatitis. Liver *8*:247–253.

Mifflin, T. E. et al. (1987). Comparison of radioactive (^{32}P and ^{35}S) and biotinylated probes for detection of cytomegalovirus DNA. Clin. Biochem. *20*:231–235.

Miliotis, M. D. et al. (1989). Development and testing of a synthetic oligonucleotide probe for the detection of pathogenic *Yersinia strains*. J. Clin. Microbiol. *27*:1667–1670.

Miller, C. A. et al. (1988). Detection of bacteria by hybridization of rRNA with DNA-latex and immunodetection of hybrids. J. Clin. Microbiol. *26*:1271–1276.

Miller, G. (1991). Molecular approaches to epidemiologic evaluation of viruses as risk factors for patients who have chronic fatigue syndrome. Rev. Infect. Dis. *13* (Suppl. 1):S119–122.

Miller, G. et al. (1987). Some recent developments in the molecular epidemiology of Epstein-Barr virus infections. Yale J. Biol. Med. *60*:307–316.

Minor, P. et al. (1990). Antibody to hepatitis C virus in plasma pools. Lancet *336*:188.

Miotti, P. G. et al. (1985). Comparative efficiency of commercial immunoassays for the diagnosis of rotavirus gastroenteritis during the course of infection. J. Clin. Microbiol. *22*:693–698.

Mirelman, D. et al. (1990). Repetitive DNA elements characteristic of pathogenic *Entamoeba histolytica* strains can also be detected after polymerase chain reaction in a cloned nonpathogenic strain. Infect. Immunol. *58*:1660–1663.

Mitrani-Rosenbaum, S. et al. (1988). Papillomaviruses in lesions of the lower genital tract in Israeli patients. Eur. J. Cancer Clin. Oncol. *24*:725–731.

Mitsuda, T. et al. (1989). Demonstration of mother-to-infant transmission of hepatitis B virus by means of polymerase chain reaction. Lancet *ii*:886–888.

Miyamura, T. et al. (1990). Detection of antibody against antigen expressed by molecularly cloned hepatitis C virus cDNA: application to diagnosis and blood screening for posttransfusion hepatitis. Proc. Natl. Acad. Sci. U.S.A. *87*:983–987.

Moestrup, T. et al. (1985). Hepatitis B virus DNA in the serum of patients followed up longitudinally with acute and chronic hepatitis B. J. Med. Virol. *17*:337–344.

Moller, B. et al. (1987). Serological assessment of HBeAg and HBV DNA: its prognostic relevance in acute hepatitis B. Liver *7*:298–305.

Moncla, B. J. et al. (1990). Use of synthetic oligonucleotide DNA probes for the identification of *Bacteroides gingivalis*. J. Clin. Microbiol. *28*:324–327.

Montone, K. T. et al. (1990). Detection of Epstein-Barr virus genomes by *in situ* DNA hybridization with a terminally biotin-labeled synthetic oligonucleotide probe from the EBV *Not*I and *Pst*I tandem repeat regions. Mod. Pathol. *3*:89–96.

Morace, G. et al. (1985). Detection of hepatitis B virus DNA in serum by a rapid filtration hybridization assay. J. Virol. Methods *12*:235–242.

Mori, J. et al. (1989). Dot blot hybridization assay of B19 virus DNA in clinical specimens. J. Clin. Microbiol. *27*:459–464.

Morotomi, M. et al. (1989). Oligonucleotide probe for detection and identification of *Campylobacter pylori*. J. Clin. Microbiol. *27*:2652–2655.

Morris, B. J. et al. (1990). Automated polymerase chain reaction for papillomavirus screening of cervicovaginal lavages: comparison with slot blot hybridization in a sexually transmitted disesase clinic population. J. Med. Virol. *32*:22–30.

Morrissey, D. V. et al. (1989). Nucleic acid hybridization assays employing dA-tailed capture probes. I. Multiple capture methods. Anal. Biochem. *81*:345–359.

Mortimer, P. P. et al. (1983). A human parvovirus-like virus inhibits hematopoietic colony formation *in vitro*. Nature (London) *302*:426–429.

Mortimer, P. P. et al. (1989). Hepatitis C virus antibody [letter]. Lancet *ii*:798.

Moser, D. R. et al. (1989). Detection of *Trypanosoma cruzi* by DNA amplification using the polymerase chain reaction. J. Clin. Microbiol. *27*:1477–1482.

Moureau, P. et al. (1989). *Campylobacter* species identification based on polymorphism of DNA encoding rRNA. J. Clin. Microbiol. *27*:1514–1517.

Mucenski, C. M. et al. (1986). Evaluation of a synthetic oligonucleotide probe for diagnosis of *Plasmodium falciparum* infections. Am. J. Trop. Med. Hyg. *35*:912–920.

Munoz, N. et al. (1988). Does human papillomavirus cause cervical cancer? The state of the epidemiological evidence. Br. J. Cancer *57*:1–5.

Murray, P. A. et al. (1991). DNA probe detection of periodontal pathogens in HIV-associated periodontal lesions. Oral Microbiol. Immunol. *6*:34–40.

Musial, C. E. et al. (1988). Identification of mycobacteria from culture using the Gen-Probe rapid diagnostic system for *Mycobacterium avium* complex and *Mycobacterium tuberculosis* complex. J. Clin. Microbiol. *26*:2120–2123.

Musiani, M. et al. (1990a). Rapid detection of cytomegalovirus DNA in urine samples with a dot blot hybridization immunoenzymatic assay. J. Clin. Microbiol. *28*:2101–2103.

Musiani, M. et al. (1990b). *In situ* detection of cytomegalovirus DNA in biopsies of AIDS patients using a hybrido-immunocytochemical assay. Histochemistry *94*:21–25.

Nago, R. et al. (1988). Detection of herpes simplex virus type 1 in herpetic ocular diseases by DNA-DNA hybridization using a biotinylated DNA probe. J. Med. Virol. *25*:259–270.

Naher, H. et al. (1991). Evidence for genetic HIV variants from detection of HIV-DNA. Lancet *338*:519–520.

Nakao, T. et al. (1991). Typing of hepatitis C virus genomes by restriction fragment length polymorphism. J. Gen. Virol. *72*:2105–2112.

Nair, P. V. et al. (1986). A pilot study on the effects of prednisone withdrawal on serum hepatitis B virus DNA and HBeAg in chronic active hepatitis B. Hepatology *6*:1319–1324.

Nakagomi, O. et al. (1989). Identification of rotavirus genogroups by RNA-RNA hybridization. Mol. Cell. Probes *3*:251–261.

Nakhleh, R. E. et al. (1991). *In situ* hybridization for the detection of Epstein-Barr virus in central nervous system lymphomas. Cancer *67*:444–448.

Nalesnik, M. A. et al. (1988). The pathology of posttransplant lymphoproliferative disorders occurring in the setting of cyclosporine A-prednisone immunosuppression. Am. J. Pathol. *133*:173–192.

Nantulya, V. M. (1991). Molecular diagnosis of parasites. Experientia *47*:142–145.

Naoumov, N. V. et al. (1988). Rapid diagnosis of cytomegalovirus infection by *in-situ* hybridization in liver grafts. Lancet *i*:1361–1363.

Naoya Kato et al. (1990). Detection of hepatitis C virus ribonucleic acid in the serum by amplification with polymerase chain reaction. J. Clin. Invest. *86*:1764–1767.

Narayanan, R. P. et al. (1983). Evaluation of ELISA as a diagnostic test in pulmonary tuberculosis. Indian J. Tuberc. *30*:29–34.

Nassau, E. et al. (1976). The detection of antibodies to *Mycobacterium tuberculosis* by microplate ELISA. Tubercle *57*:67–71.

Nasseri, M. and Wettstein, F. O. (1984). Differences exist between viral transcripts in cottontail rabbit papillomavirus-induced benign and malignant tumors as well as nonvirus-producing and virus-producing tumors. J. Virol. *51*:706–712.

Nastasi, A. et al. (1990). rRNA probing of chromosomal DNA of epidemic and sporadic isolates of *Salmonella enterica* subsp. *Enterica* serovar Kottbus from Northern and Southern Italy. Eur. J. Epidemiol. *6*:407–411.

Naylor, P. H. et al. (1990). Diagnostic assays for human immunodeficiency virus infection and for clinical progression by use of synthetic p17 peptide epitopes. Monographs in Virol. Human Immunodeficiency Virus: Innovative Techniques, Vol. 18. Khan, N. C. and Melnick, J. L. (eds.). Karger, Basel, pp. 74–90.

Negrini, B. P. et al. (1990). Oral contraceptive use, human papillomavirus infection, and risk of early cytological abnormalities of the cervix. Cancer Res. *50*:4670–4675.

Negro, F. et al. (1984). Hepatitis B virus DNA (HBV-DNA) in anti-HBe positive sera. Liver *4*:177–183.

Negro, F. et al. (1988). Chronic HDV (hepatitis delta virus) hepatitis. Intrahepatic expression of delta antigen, histologic activity and outcome of liver disease. J. Hepatol. *6*:8–14.

Negro, F. et al. (1989). Intrahepatic markers of hepatitis delta virus infection: a study by *in situ* hybridization. Hepatology *10*:916–920.

Neumann, R. et al. (1986). Use of biotinylated DNA probes in screening cells obtained from cervical swabs for human papillomavirus DNA sequences. Acta Cytol. *30*:603–607.

Neumann, R. et al. (1990). Comparison of the detection of cervical human papillomavirus infection by filter DNA hybridization of cytologic specimens and by *in situ* DNA hybridization of tissues. Acta Cytol. *34*:115–118.

Newland, J. W. and Neill, R. J. (1988). DNA probes for Shiga-like toxins I and II and for toxin-converting bacteriophages. J. Clin. Microbiol. 26:1292–1297.

Nicholson, W. J. et al. (1992). Comparison of hepatitis B virus subtyping of d/y determinants by radioimmunoprecipitation assay and the polymerase chain reaction. J. Med. Virol. *36*:21–27.

Nicotra, B. et al. (1986). *Branhamella catarrhalis* as a lower respiratory tract pathogen in patients with chronic lung disease. Arch. Int. Med. *146*:890–893.

Niedobitek, G. et al. (1989). Identification of Epstein-Barr virus-infected cells in tonsils of acute infectious mononucleosis by *in situ* hybridization. Hum Pathol. *20*:796–799.

Niel, C. et al. (1986). Direct detection and differentiation of fastidious and nonfastidious adenoviruses in stools by using a specific nonradioactive probe. J. Clin. Microbiol. *24*:785–789.

Nielsen, A. L. (1990). Human papillomavirus type 16/18 in uterine cervical adenocarcinoma *in situ* and adenocarcinoma. A study by *in situ* hybridization with biotinylated DNA probes. Cancer *65*:2588–2593.

Nielsen, S. L. et al. (1990). Detection of *Borrelia burgdorferi* DNA by the polymerase chain reaction. Mol. Cell. Probes *4*:73–79.

Nishibuchi, M. et al. (1988). Evaluation of a nonisotopically labeled oligonucleotide probe to detect the heat-stable enterotoxin gene of *Escherichia coli* by the DNA colony hybridization test. J. Clin. Microbiol. *26*:784–786.

Nobuyuki Kato et al. (1990a). Molecular cloning of the human hepatitis C virus genome from Japanese patients with non-A, non-B hepatitis. Proc. Natl. Acad. Sci. U.S.A. *87*:9524–9528.

Nobuyuki Kato et al. (1990b). Sequence diversity of hepatitis C viral genomes. Mol. Biol. Med. *7*:495–501.

Nocera, D. et al. (1990). Characterization by DNA restriction endonuclease analysis of *Listeria monocytogenes* strains related to the Swiss epidemic of listeriosis. J. Clin. Microbiol. *28*:2259–2263.

Nonogaki, H. et al. (1990). Estrogen receptor localization in normal and neoplastic epithelium of the uterine cervix. Cancer *66*:2620–2627.

Noordhoek, G. T. et al. (1990). Polymerase chain reaction and synthetic DNA probes: a means of distinguishing the causative agents of syphilis and yaws? Infect. Immunol. *58*:2011–2013.

Noordhoek, G. T. et al. (1991). Detection by polymerase chain reaction of *Treponema pallidum* DNA in cerebrospinal fluid from neurosyphilis patients before and after antibiotic treatment. J. Clin. Microbiol. *29*:1976–1984.

Norder, H. et al. (1990). Typing of hepatitis B virus genomes by a simplified polymerase chain reaction. J. Med. Virol. *31*:215–221.

Novick, D. M. et al. (1988). Hepatitis D virus and human immunodeficiency virus antibodies in parenteral drug abusers who are hepatitis B surface antigen positive. J. Infect. Dis. *158*:795–803.

Nuovo, G. T. (1990). Human papillomavirus DNA in genital tract lesions histologically negative for condylomata. Analysis by *in situ* Southern blot hybridization and the polymerase chain reaction. Am. J. Surg. Pathol. *14*:643–651.

Nuovo, G. T. (1991). Comparison of Bouin solution and buffered formalin fixation on the detection rate by *in situ* hybridization of human papillomavirus DNA in genital tract lesions. J. Histotechnol. *14*:13–17.

Nuovo, G. T. and Richart, R. M. (1989). A comparison of slot blot, Southern blot, and *in situ* hybridization analyses for human papillomavirus DNA in genital tract lesions. Obstet. Gynecol. *74*:673–678.

Nuovo, G. T. et al. (1988). Histologic correlates of papillomavirus infection of the vagina. Obstet. Gynecol. *72*:770–774.

Nuovo, G. T. et al. (1989). Correlation of histology and human papillomavirus DNA detection in condyloma acuminatum and condyloma-like vulvar lesions. Am. J. Surg. Pathol. *13*:700–706.

Nuovo, G. T. et al. (1990a). Detection of human papillomavirus DNA in penile lesions histologically negative for condylomata. Analysis by *in situ* hybridization and the polymerase chain reaction. Am. J. Surg. Pathol. *14*:829–836.

Nuovo, G. T. et al. (1990b). Human papillomavirus detection in cervical lesions nondiagnostic for cervical intraepithelial neoplasia: correlation with Papanicolaou smear, colposcopy, and occurrence of cervical intraepithelial neoplasia. Obstet. Gynecol. *75*:1006–1011.

Nuovo, G. T. et al. (1991). Occurrence of multiple types of human papillomavirus in genital tract lesions. Anaylsis by *in situ* hybridization and the polymerase chain reaction. Am. J. Pathol. *138*:53–58.

Nuovo, G. T. et al. (1992). Histological distribution of polymerase chain reaction — amplified human papillomavirus 6 and 11 DNA in penile lesions. Am. J. Surg. Pathol. *16*:269–275.

O'Shea, S. et al. (1990). HIV excretion patterns and specific antibody responses in body fluids. J. Med. Virol. *31*:291–296.

Ogle, J. W. et al. (1988). Development of resistance in *Pseudomonas aeruginosa* to imipenem, norfloxacin, and ciprofloxacin during therapy: proof provided by typing with a DNA probe. J. Infect. Dis. *157*:743–748.

Ohshima, A. et al. (1990). Molecular characterization by RNA-RNA hybridization of a serotype 8 human rotavirus with "super-short" RNA electropherotype. J. Med. Virol. *30*:107–112.

Ohshima, K. et al. (1990). Analysis of Epstein-Barr viral genomes in lymphoid malignancy using Southern blotting, polymerase chain reaction and *in situ* hybridization. Virchow's Arch. B: Cell. Pathol. *59*:383–390.

Oka, S. et al. (1990). Quantitative analysis of human immunodeficiency virus type 1 DNA in asymptomatic carriers using the polymerase chain reaction. Biochem. Biophys. Res. Commun. *167*:1–8.

Okamoto, H. et al. (1991). Nucleotide sequence of the genomic RNA of hepatitis C virus isolated from a human carrier: comparison with reported isolates for conserved and divergent regions. J. Gen. Virol. *72*:2697–2704.

Okamoto, H. et al. (1992). Typing hepatitis C virus by polymerase chain reaction with type-specific primers: application to clinical surveys and tracing infectious sources. J. Gen. Virol. *73*:673–679.

Okamoto, Y. et al. (1991). Genomic sequences of respiratory syncytial virus in otitis media with effusion. Lancet *338*:1025–1026.

Okano, M. et al. (1988). Epstein-Barr virus and human diseases: recent advances in diagnosis. Clin. Microbiol. Rev. *1*:300–312.

Old, L. J. et al. (1966). Precipitating antibodies in human serum to an antigen present in cultured Burkitt's lymphoma cells. Proc. Natl. Acad. Sci. U.S.A. *56*:1699–1704.

Olive, D. M. and Sethi, S. K. (1989). Detection of human rotavirus by using an alkaline phosphatase-conjugated synthetic DNA probe in comparison with enzyme-linked immunoassay and polyacrylamide gel analysis. J. Clin. Microbiol. *27*:53–57.

Olive, D. M. et al. (1989a). Direct detection of human cytomegalovirus in urine specimens from renal transplant patients following polymerase chain reaction amplification. J. Med. Virol. *29*:232–237.

Olive, D. M. et al. (1989b). Polymerase chain reaction assay for detection of human cytomegalovirus. J. Clin. Microbiol. *27*:1238–1242.

Olive, D. M. et al. (1990). Use of an alkaline phosphatase-labeled synthetic oligonucleotide probe for detection of *Campylobacter jejuni* and *Campylobacter coli*. J. Clin. Microbiol. *28*:1565–1569.

Olson, K. E. et al. (1990). Detection of dengue viral RNA in mosquito vectors by mixed phase and solution hybridization. Mol. Cell. Probes *4*:307–320.

Omata, M. et al. (1991). Resolution of acute hepatitis C after therapy with natural beta interferon. Lancet *338*:914–915.

Oren, I. and Sobel, J. D. (1992). Human herpes virus type 6. Clin. Infect. Dis. *14*:741–746.

Orth, G. (1986). Epidermodysplasia verruciformis, a model for understanding the oncogenicity of human papillomaviruses. Ciba Found. Symp. *120*:157–174.

Ostergaard, L. et al. (1990). Use of polymerase chain reaction for detection of *Chlamydia trachomatis*. J. Clin. Microbiol. *28*:1254–1260.

Ostrow, R. S. and Faras, A. J. 1987). The molecular biology of human papillomaviruses and the pathogenesis of genital paillomas and neoplasms. Cancer Metast. Rev. *6*:383–395.

Ostrow, R. S. et al. (1986). Detection of papillomavirus DNA in human semen. Science *231*:731–735.

Ott, M. et al. (1991). Pulsed field electrophoresis of genomic restriction fragments for the detection of nosocomial *Legionella pneumophila* in hospital water supplies. J. Clin. Microbiol. *29*:813–815.

Ou, C. Y. et al. (1988). DNA amplification for direct detection of HIV-1 in DNA of peripheral blood mononuclear cells. Science *239*:295–297.

Owen, R. J. et al. (1992). Ribosomal RNA gene patterns of *Helicobacter pylori* from surgical patients with healed and recurrent peptic ulcers. Epidemiol. Infect. *108*:39–50.

Ozawa, K. et al. (1986). Replication of the B 19 parvovirus in human bone marrow cell cultures. Science *233*:883–886.

Pak, C. Y. et al. (1988). Association of cytomegalovirus infection wtih autoimmune type 1 diabetes. Lancet *ii*:1–4.

Palefsky, J. M. et al. (1991). Detection of human papillomavirus DNA in anal intraepithelial neoplasia and anal cancer. Cancer Res. *51*:1014–1019.

Pallesen, G. et al. (1991). Expression of Epstein-Barr virus latent gene products in tumor cells of Hodgkin's disease. Lancet *337*:320–322.

Pang, S. et al. (1990). High levels of unintegrated HIV-1 DNA in brain tissue of AIDS dementia patients. Nature (London) *343*:85–89.

Panke, E. S. et al. (1991). Comparison of Gen-Probe DNA probe test and culture for the detection of *Neisseria gonorrhoeae* in endocervical specimens. J. Clin. Microbiol. *29*:883–888.

Pao, C. C. et al. (1990). Detection and identification of *Mycobacterium tuberculosis* by DNA amplification. J. Clin. Microbiol. *28*:1877–1880.

Papillomaviruses. (1986). Ciba Found. Symp. 120. John Wiley & Sons, New York.

Parkkinen, S. et al. (1989). Sandwich hybridization in solution: a rapid method to screen HPV16 DNA in cervical scrapes. Mol. Cell. Probes *3*:1–11.

Parsons, L. M. et al. (1989). DNA probes for the identification of *Haemophilus ducreyi.* J. Clin. Microbiol. *27*:1441–1445.

Parvaz, P. et al. (1987). Prevalence and significance of hepatitis B virus antigens: expression in peripheral blood mononuclear cells in chronic active hepatitis. Clin. Immunol. Immunopathol. *43*:1–8.

Pasculle, A. W. et al. (1989). Laboratory and clinical evaluation of a commercial DNA probe for the detection of *Legionella* spp. J. Clin. Microbiol. *27*:2350–2358.

Patel, R. J. et al. (1990). Sequence analysis and amplification by polymerase chain reaction of a cloned DNA fragment for identification of *Mycobacterium tuberculosis*. J. Clin. Microbiol. *28*:513–518.

Pater, M. M. and Pater, A. (1985). Human papillomavirus types 16 and 18 sequences in carcinoma cell lines of the cervix. Virology *145*:313–318.

Pau, C. P. et al. (1991). Misidentification of HIV-2 proteins by Western blots. Lancet *337*:616.

Paya, C. V. et al. (1990). Early diagnosis of cytomegalovirus hepatitis in liver transplant recipients: role of immunostaining, DNA hybridization and culture of hepatic tissue. Hepatology *12*:119–126.

Peiper, S. C. et al. (1990). Detection of Epstein-Barr virus genomes in archival tissues by polymerase chain reaction. Arch. Pathol. Lab. Med. *114*:711–714.

Pellegrino, M. G. et al. (1987). A sensitive solution hybridization technique for detecting RNA in cells: application to HIV in blood cells. BioTechniques *5*:452–459.

Peng, H. W. et al. (1988) Assessment of HBV persistent infection in an adult population in Taiwan. J. Med. Virol. *24*:405–412.

Penneys, N. S. et al. (1991). Herpes simplex virus DNA in occult lesions: demonstration by the polymerase chain raction. J. Am. Acad. Dermatol. *24*:689–692.

Perl, D. P. (1975). The pathology of rabies in the central nervous system. In: The Natural History of Rabies, Vol 1. Baer, G. M. (ed.). Academic Press, New York, pp. 235–272.

Permeen, A. M. Y. et al. (1990). Detection of Epstein-Barr virus DNA in nasopharyngeal carcinoma using a non-radioactive digoxigenin-labeled probe. J. Virol. Methods *27*:261–268.

Persing, D. H. et al. (1990). Detection of *Borrelia burgdorferi* infection in *Ixodes dammini* ticks with the polymerase chain reaction. J. Clin. Microbiol. *28*:566–572.

Peterkin, P. I. et al. (1991). Detection of *Listeria monocytogenes* by direct colony hybridization on hydrophobic grid-membrane filters by using a chromogen-labeled DNA probe. Appl. Environ. Microbiol. *57*:586–591.

Peterson, E. M. et al. (1986). Typing of herpes simplex virus with synthetic DNA probes. J. Infect. Dis. *153*:757–762.

Peterson, E. M. et al. (1989). Direct identification of *Mycobacterium tuberculosis, Mycobacterium avium*, and *Mycobacterium intracellulare* from amplified primary cultures in BACTEC media using DNA probes. J. Clin. Microbiol. *27*:1543–1547.

Peto, R. and zur Hausen, H. (eds.) (1986). Viral Etiology of Cervical Cancer. Banbury Report 21. Cold Spring Harbor Laboratory, Cold Spring Harbor, New York.

Petrie, B. L. et al. (1987). Nucleic acid sequence of cytomegalovirus in cells cultured from human arterial tissue. J. Infect. Dis. *155*:158–159.

Pfaller, M. A. et al. (1990). The use of biotyping and DNA fingerprinting in typing *Candida albicans* from hospitalized patients. Diagn. Microbiol. Infect. Dis. *13*:481–489.

Pfister, H. (1990). Papillomaviruses and Human Cancer. CRC Press, Boca Raton, FL.

Phelps, W. C. et al. (1988). The human papillomavirus type 16 E7 gene encodes transactivation and transformation functions similar to those of adenovirus E1A. Cell (Cambridge, Mass) *53*:539–547.

Phelps, W. C. et al. (1992). Structure function analysis of the human papillomavirus type 16 E7 oncoprotein. J. Virol. 66:2418–2427.

Philip, A. et al. (1987). An agar culture technique to quantitate *Trichomonas vaginalis* from women. J. Infect Dis. *155*:304–308.

Phillips, R. E. et al. (1991). Human immunodeficiency virus genetic variation that can escape cytotoxic T cell recognition. Nature (London) *354*:453–459.

Picken, R. N. et al. (1988). DNA probes for mycobacteria. I. Isolation of DNA probes for the identification of *Mycobacterium tuberculosis* complex and for mycobacteria other than tuberculosis (MOTT). Mol. Cell. Probes *2*:111–124.

Pierre, C. et al. (1991). Use of a reamplification protocol improves sensitivity of detection of *Mycobacterium tuberculosis* in clinical samples by amplification of DNA. J. Clin. Microbiol. *29*:712–717.

Pilotti, S. et al. (1990). Vulvar carcinomas: search for sequences homologous to human papillomavirus and herpes simplex virus DNA. Mod. Pathol. *3*:442–448.

Pipittajan, P. et al. (1991). Molecular epidemiology of rotaviruses associated with pediatric diarrhea in Bangkok, Thailand. J. Clin. Microbiol. *29*:617–624.

Pistello, M. et al. (1991). Hepatitis C virus seroprevalence in Italian hemophiliacs injected with virus-inactivated concentrates: five year follow-up and correlation with antibodies to other viruses. J. Med. Virol. *33*:43–46.

Pitcher, D. et al. (1990). An investigation of nosocomial infection with *Corynebacterium jeikeium* in surgical patients using a ribosomal RNA gene probe. Eur. J. Clin. Microbiol. Infect. Dis. *9*:P643–P648.

Plikaytis, B. B. et al. (1990). Rapid and sensitive detection of *Mycobacterium leprae* using a nested-primer gene amplification assay. J. Clin. Microbiol. *28*:1913–1917.

Podbielski, A. et al. (1990). Identification of group A type 1 streptococcal M protein gene by a non-radioactive oligonucleotide detection method. Med. Microbiol. Immunol. *179*:255–262.

Pohl, C. et al. (1987). A human monoclonal antibody that recognizes viral polypeptides and *in vitro* translation products of the genome of the hepatitis D virus. J. Infect. Dis. *156*:622–629.

Pohl-Koppe, A. et al. (1992). The diagnostic significance of the polymerase chain reaction and isoelectric focusing in herpes simplex virus encephalitis. J. Med. Virol. *36*:147–154.

Pollack, Y. et al. (1985). Detection of *Plasmodium falciparum* in blood using DNA hybridization. Am. J. Trop. Med. Hyg. *34*:663–667.

Pollard, D. R. et al. (1989). A polymerase chain reaction (PCR) protocol for the specific detection of *Chlamydia* spp. Mol. Cell. Probes *3*:383–389.

Pollard, D. R. et al. (1990a). Detection of the aerolysin gene in *Aeromonas hydrophila* by the polymerase chain reaction. J. Clin. Microbiol. *28*:2477–2481.

Pollard, D. R. et al. (1990b). Rapid and specific detection of verotoxin genes in *Escherichia coli* by the polymerase chain reaction. J. Clin. Microbiol. 28:540–545.

Pontisso, P. et al. (1984). Detection of hepatitis B virus DNA in mononuclear blood cells. Br. Med. J. *288*:1563–1566.

Ponzetto, A. et al. (1985). Epidemiology of hepatitis delta virus (HDV) infection. Eur. J. Epidemiol. *1*:257–263.

Ponzetto, A. et al. (1987). Titration of the infectivity of hepatitis D virus in chimpanzees. J. Infect. Dis. *155*:72–78.

Popescu, N. C. et al. (1987a). Human papillomavirus type 18 DNA is integrated at a single chromosome site in cervical carcinoma cell line SW 756. J. Virol. *51*:1682–1685.

Popescu, N. C. et al. (1987b). Integration sites of human papillomavirus 18 DNA sequences on HeLa cell chromosomes. Cytogenet. Cell. Genet. *44*:58–62.

Porter-Jordan, K. et al. (1990). Nested polymerase chain reaction assay for the detection of cytomegalovirus overcomes false positives caused by contamination with fragmented DNA. J. Med. Virol. *30*:85–91.

Portnoy, D. A. et al. (1981). Characterization of plasmids and plasmid associated determinants of *Yersinia enterocolitica* pathogenesis. Infect. Immunol. *31*:775–782.

Posnett, E. S. and Ambrosio, R. E. (1989). Repetitive DNA probes for the detection of *Babesia equi*. Mol. Biochem. Parasitol. *34*:75–78.

Powell, K. F. et al. (1990). Non-invasive diagnosis of herpes simplex encephalitis. Lancet *335*:352–358.

Pozzi, G. et al. (1989). DNA probe for identification of *Streptococcus pneumoniae*. J. Clin. Microbiol. *27*:370–372.

Psallidopoulos, M. D. et al. (1989). Integrated proviral human immunodeficiency virus type 1 is present in CD4+ peripheral blood lymphocytes in healthy seropositive individuals. J. Virol. *63*:4626–4631.

Puchhammer-Stockl, E. et al. (1990). Establishment of PCR for the early diagnosis of herpes simplex encephalitis. J. Med. Virol. *32*:77–82.

Purcell, R. H. et al. (1984). Hepatitis delta virus infection of the liver. Semin. Liver Dis. *4*:340–346.

Qavi, H. B. et al. (1989). Demonstration of HIV-1 and HHV-6 in AIDS-associated retinitis. Curr. Eye Res. *8*:379–387.

Quibriac, M. et al. (1987). Non radioactive spot hybridization test for detection of hepatitis B virus DNA in serum. Ann. Inst. Pasteur/Virol. *138*:377–384.

Quibriac, M. et al. (1989). Comparison of a non-radioactive hybridization assay for detection of hepatitis B virus DNA with the radioactive method. Mol. Cell. Probes *3*:209–212.

Quinti, I. et al. (1990). Hepatitis C virus antibodies in gammaglobulin. Lancet *336*:1377.

Rahav, G. et al. (1990). Development of sensitive methods for the detection of mycobacteria by DNA probes. FEMS Microbiol. Lett. *72*:29–34.

Rakusan, T. A. et al. (1991). Limitations in the laboratory diagnosis of vertically acquired HIV infection. J. AIDS *4*:116–121.

Randazzo, D. N. and Michalski, F. J. (1988). Comparison of antibodies for rapid detection of cytomegalovirus. J. Clin. Microbiol. *26*:369–370.

Ranki, A. et al. (1987). Long latency precedes overt seroconversion in sexually transmitted human-immunodeficiency-virus infection. Lancet *2*:589–593.

Ranki, M. et al. (1990). Use of AffiProbe HPV test kit for detection of human papillomavirus DNA in genital scrapes. J. Clin. Microbiol. *28*:2076–2081.

Rappuoli, R. et al. (1988). Molecular epidemiology of the 1984–1986 outbreak of diphtheria in Sweden. N. Engl. J. Med. *318*:12–14.

Rasing, L. A. J. et al. (1990). The value of immunohistochemistry and *in situ* hybridization in detecting cytomegalovirus in bone marrow transplant recipients. APMS *98*:479–488.

Rasshofer, R. et al. (1988). Demonstration of hepatitis D virus RNA in patients with chronic hepatitis. J. Infect. Dis. *157*:191–195.

Rayfield, M. et al. (1988). Mixed human immunodeficiency virus (HIV) infection in an individual: demonstration of both HIV type 1 and type 2 proviral sequences by using polymerase chain reaction. J. Infect. Dis. *158*:1170–1176.

Razin, S. (ed.) (1981). Mycoplasma infections of man. Isr. J. Med. Sci. *17*:509–686.

Razin, S. and Freundt, E. A. (1984). The mycoplasmas. In: Bergey's Manual of Systematic Bacteriology, Vol 1. Krieg, N. R. and Holt, J. G. (eds.). Williams & Wilkins, Baltimore, pp. 740–741.

Razin, S. et al. (1987). DNA probes for detection and identification of mycoplasmas (Mollicutes). Isr. J. Med. Sci. *23*:735–741.

Re, M. C. et al. (1989). Rapid detection of HIV-1 in clinical samples by co-culture with heat-shocked cells. J. Virol. Methods *26*:313–318.

Reagan, D. R. et al. (1990). Characterization of the sequence of colonization and nosocomial candidemia using DNA fingerprinting and a DNA probe. J. Clin. Microbiol. *28*:2733–2738.

Redfield, D. C. et al. (1983). Detection of herpes simplex virus in clinical specimens by DNA hybridization. Diagn. Microbiol. Infect. Dis. *1*:117–128.

Reesink, H. W. et al. (1990). Mother-to-infant transmission and hepatitis C virus. Lancet *335*:1216–1217.

Reeves, W. C. et al. (1989a). Human papillomavirus infection and cervical cancer in Latin America. N. Eng. J. Med. *320*:1437–1441.

Reeves, W. C. et al. (1989b). Epidemiology of genital papillomaviruses and cervical cancer. Rev. Infect. Dis. *11*:426–439.

Reid, D. M. et al. (1985). Human parvovirus-associated arthritis: a clinical and laboratory description. Lancet *i*:422–425.

Reid, R. (1987). Human papillomaviruses. Obstetrics and Gynecology Clinics of North America, Vol. 14. W. B. Saunders, Philadelphia.

Reid, R. and Campion, M. J. (1988). The biology and significance of human papillomavirus infections in the genital tract. Yale J. Biol. Med. *61*:307–325.

Reid, R. et al. (1987). Sexually transmitted papillomaviral infections. The anatomic distribution and pathologic grade of neoplastic lesions associated with different viral types. Am. J. Obstet. Gynecol. *156*:212–222.

Reizenstein, E. et al. (1990). DNA hybridization for diagnosis of pertussis. Mol. Cell. Probes *4*:299–306.

Resnick, R. M. et al. (1990). Detection and typing of human papillomavirus in archival cervical cancer specimens by DNA amplification with consensus primers. J. Natl. Cancer Inst. *82*:1477–1484.

Richman, D. D. et al. (1987). Detecting human immunodeficiency virus RNA in peripheral blood mononuclear cells by nucleic acid hybridization. J. Infect. Dis. *156*:823–827.

Rijntjes, P. J. M. et al. (1985). Hepatitis B virus DNA detected in formalin-fixed liver specimens and its relation to serologic markers and histopathologic features in chronic liver disease. Am. J. Pathol. *120*:411–418.

Riley, L. K. and Caffrey, C. J. (1990). Identification of enterotoxigenic *Escherichia coli* by colony hybridization with non-radioactive digoxigenin-labled DNA probes. J. Clin. Microbiol. *28*:1465–1468.

Riou, G. et al. (1990). Association between poor prognosis in early-stage invasive cervical carcinomas and non-detection of HPV DNA. Lancet *335*:1171–1174.

Rizzetto, M. et al. (1977). Immunofluorescence detection of a new antigen/antibody system (delta/anti-delta) associated with hepatitis B virus in liver and serum of HBsAg carriers. Gut *18*:997–1003.

Rizzetto, M. et al. (1980a). Epidemiology of HBV-associated delta agent: geographical distribution of anti-delta and prevalence in polytransfused HBsAg carriers. Lancet *i*:1215–1218.

Rizzetto, M. et al. (1980b). The hepatitis B virus-associated delta antigen: isolation from liver, development of solid-phase radioimmunoassay for delta antigen, and antidelta and partial characterization of delta antigen. J. Immunol. *125*:318–324.

Rizzetto, M. et al. (1984). Hepatitis delta virus infection. In: Viral Hepatitis and Liver Disease. Vyas, G. N. (ed.). Grune & Stratton, New York, pp. 371–379.

Roberts, M. C. et al. (1987a). DNA probes for the detection of mycoplasmas in genital specimens. Br. J. Med. Sci. *23*:618–620.

Roberts, M. C. et al. (1987b). Whole chromosomal DNA probes for rapid identification of *Mycobacterium tuberculosis* and *Mycobacterium avium* complex. J. Clin. Microbiol. *25*:1239–1243.

Roberts, W. H. et al. (1988). *In situ* DNA hybridization for cytomegalovirus in colonoscopic biopsies. Arch. Pathol. Lab. Med. *112*:1106–1009.

Robey, S. S. et al. (1988). Comparison of immunoperoxidase and DNA *in situ* hybridization techniques in the diagnosis of cytomegalovirus colitis. Am. J. Clin. Pathol. *89*:666–671.

Robins-Browne, R. M. et al. (1989). Evaluation of DNA colony hybridization and other techniques for detection of virulence in *Yersinia* species. J. Clin. Microbiol. *27*:644–650.

Robinson, W. S. et al. (1990). Hepadnaviruses in cirrhotic liver and hepatocellular carcinoma. J. Med. Virol. *31*:18–32.

Rodger, S. M. et al. (1981). Molecular epidemiology of human rotavirus in Melbourne, Australia, from 1973 to 1979, as determined by electrophoresis of genome ribonucleic acid. J. Clin. Microbiol. *13*:272–278.

Rogers, B. B. et al. (1990). Analysis of DNA in fresh and fixed tissue by the polymerase chain reaction. Am. J. Pathol. *135*:541–548.

Rogers, M. F. et al. and the New York City Collaborative Study of Maternal HIV Transmission and Montefiore Medical Center HIV Perinatal Transmission Study Group (1989). Use of the polymerase chain reaction for early detection of the proviral sequences of human immunodeficiency virus in infants born to seropositive mothers. N. Engl. J. Med. *320*:1649–1654

Roggendorf, M. et al. (1987). Characterization of proteins associated with hepatitis delta virus. J. Gen. Virol. *68*:2953–2959.

Roggendorf, M. et al. (1989). Antibodies to hepatitis C virus [letter]. Lancet *ii*:324–325.

Rogler, C. E. et al. (1985). Deletion in chromosome 11p associated with a hepatitis B integration site in hepatocellular carcinoma. Science *230*:319–322.

Roingeard, Ph. et al. (1992). Infection due to hepatitis delta virus in Africa: report from Senegal and review. Clin. Infect. Dis. *14*:510–514.

Rolfs, A. (1990). Early findings in the cerebrospinal fluid of patients with HIV-1 infection of the central nervous system. N. Engl. J. Med. *323*:418–419.

Roman, A. and Fife, K. H. (1989). Human papillomaviruses: are we ready to type? Clin. Microbiol. Rev. *2*:166–190.

Romanczuk, H. et al. (1991). The viral transcriptional regulatory region upstream of the E6 and E7 genes is a major determinant of the differential immortalization activities of human papillomavirus types 16 and 18. J. Virol. *65*:2739–2744.

Romet-Lemmone, J. L. et al. (1983). Hepatitis virus B infection in cultured human lymphoblastoid cells. Science *221*:667–669.

Rosa, P. A. and Schwan, T. G. (1989). A specific and sensitive assay for the Lyme disease spirochete *Borrelia burgdorferi* using the polymerase chain reaction. J. Infect. Dis. *160*:1018–1029.

Rose, B. R. et al. (1991). Papillomavirus DNA and prognosis in cervical cancer. Lancet *337*:489.

Rosenberg, Z. F. and Fauci, A. S. (1991). Immunopathogenesis of HIV infection. FASEB J. *5*:2382–2390.

Rossau, R. et al. (1989). Specific *Neisseria gonorrhoeae* DNA probes derived from ribosomal RNA. J. Gen. Microbiol. *135*:1735–1745.

Rossi, P. and Moschese, V. (1991). Mother-to-child transmission of human immunodeficiency virus. FASEB J. *5*:2419–2426.

Rossier, E. et al. (1987). Sensitivity and specificity of enzyme immunofiltration and DNA hybridization for the detection of HCMV-infected cells. J. Virol. Methods *15*:109–120.

Rossner, M. T. (1992). Review: hepatitis B virus X-gene product: a promiscuous transcriptional activator. J. Med. Virol. *36*:101–117.

Rotbart, H. A. (1990a). Enzymatic RNA amplification of the enteroviruses. J. Clin. Microbiol. *28*:438–442.

Rotbart, H. A. (1990b). PCR amplification of enteroviruses. In: PCR Protocols. A Guide to Methods and Application. Innis, M. A. et al. (eds.). Academic Press, San Diego, pp. 372–377.

Rotbart, H. A. (1991). Nucleic acid detection systems for enteroviruses. Clin. Microbiol. Rev. *4*:156–168.

Rotbart, H. A. and Levin, M. J. (1989). Progress toward the development of a panenteroviral nucleic acid probe. In: DNA Probes for Infectious Diseases. Tenover, F. C. (ed.). CRC Press, Boca Raton, FL, pp. 193–209.

Rotbart, H. A. et al. (1988). Nonisotopic oligomeric probes for the human enteroviruses. J. Clin. Microbiol. *26*:2669–2671.

Rowley, A. H. and Wolinsky, S. M. (1989). Direct detection of herpesvirus DNA sequences in clinical samples by *in vitro* enzymatic amplification. Pediatr. Res. *25*:189A.

Rowley, A. H. et al. (1990). Rapid detection of herpes simplex virus DNA in cerebrospinal fluid of patients with herpes simplex encephalitis. Lancet *335*:440–441.

Rubin, F. A. (1990). Nucleic acid probes for the identification of *Salmonella*. In: Gene Probes for Bacteria. Macario, A. J. L. and Conway de Macario, E. (eds.). Academic Press, San Diego, pp. 323–351.

Rubin, F. A. et al. (1985). Development of a DNA probe to detect *Salmonella typhi*. J. Clin. Microbiol. *22*:600–605.

Rubin, F. A. et al. (1988). Evaluation of a DNA probe for identifying *Salmonella typhi* in Peruvian and Indonesian bacterial isolates. J. Infect. Dis. *157*:1051–1053.

Rubin, F. A. et al. (1989). Use of a DNA probe to detect *Salmonella typhi* in the blood of patients with typhoid fever. J. Clin. Microbiol. *27*:1112–1114.

Rubino, S. et al. (1991). Molecular probe for identification of *Trichomonas vaginalis* DNA. J. Clin. Microbiol. *29*:702–706.

Russell, R. R. B. (1991). Genetic analysis and genetic probes for oral bacteria. In: Aspects of oral molecular biology. Ferguson, D. B. (ed.). Front. Oral Physiol. *8*:57–76.

Ryu, W. S. et al. (1992). Assembly of hepatitis delta virus particles. J. Virol. *66*:2310–2315.

Saag, M. S. et al. (1988). Extensive variation of human immunodeficiency virus type 1 *in vivo*. Nature (London) *334*:440–444.

Sadeghi, S. B. et al. (1989). Human papillomavirus infection. Frequency and association with cervical neoplasia in a young population. Acta Cytol. *33*:319–323.

Saez-Llorens, X. et al. (1989). Simultaneous detection of *Escherichia coli* heat-stable and heat-labile enterotoxin genes with a single RNA probe. J. Clin. Microbiol. *27*:1684–1688.

Saito, H. et al. (1989). Identification and partial characterization of *Mycobacterium avium* and *Mycobacterium intracellulare* by using DNA probes. J. Clin. Microbiol. *27*:994–997.

Saito, H. et al. (1990). Identification of various serovar strains of *Mycobacterium avium* complex by using DNA probes specific for *Mycobacterium avium* and *Mycobacterium intracellulare*. J. Clin. Microbiol. *28*:1694–1697.

Saito, I. et al. (1989). Detection of Epstein-Barr virus DNA by polymerase chain reaction in blood and tissue biopsies from patients with Sjogren's syndrome. J. Exp. Med. *169*:2191–2198.

Saito, I. et al. (1990). Hepatitis C virus infection is associated with the development of hepatocellular carcinoma. Proc. Natl. Acad. Sci. U.S.A. *87*:6547–6549.

Sakuma, S. et al. (1988). Human papillomavirus DNA in condylomata acuminata from Japanese males. Diagn. Microb. Infect. Dis. *10*:23–29.

Saldanha, J. A. et al. (1987). Use of biotinylated probes in serum hepatitis B virus DNA detection. J. Virol. Methods *16*:339–342.

Saldanha, J. et al. (1989). Detection of hepatitis delta virus RNA in chronic liver disease. J. Hepatol. *9*:23–28.

Salimans, M. M. M. et al. (1989a). Detection of parvovirus B19 DNA in fetal tissues by *in situ* hybridization and polymerase chain reaction. J. Clin. Pathol. *42*:525–530.

Salimans, M. M. M. et al. (1989b). Rapid detection of human parvovirus B19 DNA by dot-hybridization and the polymerase chain reaction. J. Virol. Methods *23*:19–28.

Saltzman, R. L. et al. (1990). Quantitation of cytomegalovirus DNA by blot hybridization in blood leukocytes of viremic patients. J. Virol. Methods *30*:67–78.

Sanchez-Quijano, A. et al. (1990). Hepatitis C virus infection in sexually promiscuous groups. Eur. J. Clin. Microbiol. Infect. Dis. *9*:610–612.

Saracco, G. et al. (1987). Rapidly progressive HBsAg-positive hepatitis in Italy. The role of hepatitis delta virus infection. J. Hepatol *5*:274–281.

Saracco, G. et al. (1990). A randomized controlled trial of interferon α-2b as therapy for chronic non-A, non-B hepatitis. J. Hepatol. *11*:S43-S49.

Sarafian, S. K. et al. (1991). Molecular characterization of *Haemophilus ducreyi* by ribosomal DNA fingerprinting. J. Clin. Microbiol. *29*:1949–1954.

Sato, H. et al. (1989). Expression of human papillomavirus type 16 E7 gene induces DNA synthesis of rat 3Y1 cells. Virology *168*:195–199.

Savitt, E. D. et al. (1990). DNA probes in the diagnosis of periodontal microorganisms. Arch. Oral Biol. (Suppl.) 35:153S-159S.

Savva, D. and Holliman, R. E. (1990a). Diagnosis of toxoplasmosis using DNA probes [letter]. J. Clin. Pathol. *43*:260–261.

Savva, D. and Holliman, R. E. (1990b). PCR to detect toxoplasma. Lancet *336*:1325.

Savva, D. et al. (1990). Polymerase chain reaction for detection of *Toxoplasma gondii*. J. Med. Microbiol. *32*:25–31.

Saxon, P. et al. (1986). Introduction of human chromosome 11 via microcell transfer controls tumorigenic expression of HeLa cells. EMBO J. *5*:3461–3466.

Schalla, W. D. et al. (1986). Use of dot-immunobinding and immunofluorescence assays to investigate clinically suspected cases of chancroid. J. Infect. Dis. *153*:879–897.

Schantz, P. M. and Gottstein, G. (1986). *Echinococcus* (hydatidosis). In: Immunodiagnosis of Parasitic Diseases. Walls, K. F. and Schantz, P. M. (eds.). Academic Press, New York, pp. 69–107.

Scheffner, M. et al. (1990). The E6 oncoprotein encoded by human papillomavirus types 16 and 18 promotes the degradation of p53. Cell *63*:1129–1136.

Scherf, A. et al. (1991). Multiple infections and unusual distribution of block 2 of the *MSA* 1 gene of *Plasmodium falciparum* detected in West African clinical isolates by polymerase chain reaction analysis. Mol. Biochem. Parasitol. *44*:297–300.

Schirm, J. et al. (1987). Rapid detection of infectious cytomegalovirus in blood with the aid of monoclonal antibodies. J. Med. Virol. *23*:31–40.

Schmidhuber, S. et al. (1987). A taxonomic study of *Streptococcus mitis, Streptococcus oralis,* and *Streptococcus sanguis.* Syst. Appl. Microbiol. *10*:74–77.

Schneider, A. et al. (1985). Papillomavirus infection of the lower genital tract: detection of viral DNA in gynecological swabs. Int. J. Cancer *35*:443–448.

Schneider, A. et al. (1987a). Human papillomaviruses in women with a history of abnormal Papanicolaou smears and in their male partner. Obstet. Gynecol. *69*:554–562.

Schneider, A. et al. (1987b). Interferon treatment of human genital papillomavirus infection: importance of viral type. Int. J. Cancer *40*:610–614.

Schneider-Gaedicke, A. and Schwarz, E. (1986). Different human cervical carcinoma cell lines show similar transcription patterns of human papillomavirus type 18 early genes. EMBO J. *5*:2285–2292.

Schneider-Maunoury, S. et al. (1987). Integration of human papillomavirus type 16 DNA sequences: a possible early event in the progression of genital tumors. J. Virol. *61*:3295–3298.

Schneweis, K. E. et al. (1989). Comparison of different methods for detecting human immunodeficiency virus in human immunodeficiency virus-seropositive hemophiliacs. J. Med. Virol. *29*:94–101.

Schnittman, S. M. et al. (1989). The reservoir for HIV-1 in human peripheral blood is a T cell that maintains expression of CD4. Science *245*:305–308.

Scholl, D. R. et al. (1990). Clinical application of novel sample processing technology for the identification of salmonellae by using DNA probes. J. Clin. Microbiol. *28*:237–241.

Schon, H. J. et al. (1991). Papanicoloaou test and enzyme-linked *in-situ* hybridization. A combined diagnostic system for papillomavirus infections with high prognostic value. Eur. J. Clin. Chem. Clin. Biochem. *29*:131–138.

Schrier, R. D. et al. (1985). Detection of human cytomegalovirus in peripheral blood lymphocytes in a natural selection. Science *230*:1048–1051.

Schuster, V. et al. (1986a). Nucleic acid hybridization for detection of herpes viruses in clinical specimens. J. Med. Virol. *19*:277–286.

Schuster, V. et al. (1986b). Detection of herpes simplex virus and adenovirus DNA by dot blot hybridization using *in vitro* synthesized RNA transcripts. J. Virol. Methods *13*:291–299.

Schwartz, E. et al. (1985). Structure and transcription of human papillomavirus 16 and 18 sequences in cervical carcinoma cells. Nature (London) *314*:111–114.

Schwartz, D. A. and Caldwell, E. (1991). Herpes simplex virus infection of the placenta. The role of molecular pathology in the diagnosis of viral infection of placental-associated tissues. Arch. Pathol. Lab. Med. *115*:1141–1144.

Scott, A. A. et al. (1988). Detection of cytomegalovrius in shell vial cultures by using a DNA probe and early nuclear antigen monoclonal antibody. J. Clin. Microbiol. *26*:1895–1897.

Scotto, J. et al. (1983). Detection of hepatitis B virus DNA in serum by a simple spot hybridization technique: comparison with results for other viral markers. Hepatology *3*:279–284.

Seal, L. A. et al. (1991). Comparison of standard culture methods, a shell vial assay, and a DNA probe for the detection of herpes simplex virus. J. Clin. Microbiol. *29*:650–652.

Selvey, L. et al. (1989). Male partners of women with genital human papillomavirus infection. Med. J. Aust. *150*:479–482.

Sethabutr, O. et al. (1990). Serotyping of human group A rotavirus with oligonucleotide probes. J. Infect. Dis. *162*:368–372.

Shafritz, D. A. (1991). Variants of hepatitis B virus associated with fulminant liver disease. N. Engl. J. Med. 324:1737–1739.

Shah, K. V. (1990). Papillomavirus infections of the respiratory tract, the conjunctiva, and the oral cavity. In: Papillomaviruses and Human Cancer. Pfister, H. (ed.). CRC Press, Boca Raton, FL, pp. 74–90.

Shankar, P. et al. (1989). Identification of *Mycobacterium tuberculosis* by polymerase chain reaction. Lancet *335*:423.

Shankar, P. et al. (1991). Rapid diagnosis of tuberculous meningitis by polymerase chain reaction. Lancet *337*:5–7.

Shapshak, P. et al. (1990). The detection of HIV by *in situ* hybridization. Mod. Pathol. *3*:146–153.

Shaul, Y. et al. (1984). Cloning and analysis of integrated hepatitis virus sequences from a human hepatoma cell line. J. Virol. *51*:776–787.

Shaul, Y. et al. (1986). Integration of hepatitis B virus DNA in chromosome-specific satellite sequences. J. Virol. *59*:731–734.

Shaw, G. M. et al. (1985). HTLV III infection in brains of children and adults with AIDS encephalopathy. Science *227*:177–182.

Shibata, D. and Klatt, E. C. (1989). Analysis of human immunodeficiency virus and cytomegalovirus infection by polymerase chain reaction in the acquired immunodeficiency syndrome. Arch. Pathol. Lab. Med. *113*:1239–1241.

Shibata, D. et al. (1988). Detection of cytomegalovirus DNA in peripheral blood of patients infected with human immunodeficiency virus. J. Infect. Dis. *158*:1185–1192.

Shibata, D. et al. (1989). Human immunodeficiency viral DNA is readily found in lymph node biopsies from seropositive individuals. Analysis of fixed tissue using the polymerase chain reaction. Am. J. Pathol. *135*:697–702.

Shibata, M. et al. (1990). Human cytomegalovirus infection during childhood: detection of viral DNA in peripheral blood by means of polymerase chain reaction. Med. Microbiol. Immunol. *179*:245–253.

Shih, J. W. K. et al. (1991). Strain analysis of hepatitis B virus on the basis of restriction endonuclease analysis of polymerase chain reaction products. J. Clin. Microbiol. *29*:1640–1644.

Shih, L. N. et al. (1990). Detection of hepatitis B viral DNA by polymerase chain reaction in patients with hepatitis B surface antigen. J. Med. Virol. *30*:159–162.

Shirasawa, H. et al. (1988). Transcriptional differences of the human papillomavirus type 16 genome between precancerous lesions and invasive carcinomas. J. Virol. *62*:1022–1027.

Shirasawa, H. et al. (1989). Structure and expression of an integrated human papillomavirus type 16 genome amplified in a cervical carcinoma cell line. J. Gen. Virol. *70*:1913–1919.

Shoebridge, G. I. et al. (1990). Polymerase chain reaction testing of HIV-1 seronegative at risk individuals. Lancet *336*:180–181.

Simmonds, P. et al. (1990). Hepatitis C quantification and sequencing in blood products, hemophiliacs, and drug users. Lancet *336*:1469–1472.

Singer, R. H. et al. (1989). Detection of HIV-1-infected cells from patients using non-isotopic *in situ* hybridization. Blood *74*:2295–2301.

Sjobring, U. et al. (1990). Polymerase chain reaction of detection of *Mycobacterium tuberculosis*. J. Clin. Microbiol. *28*:2200–2204.

Skidmore, S. (1990). Recombinant immunoblot assay for hepatitis C antibody. Lancet *335*:1346.

Skidmore, S. J. et al. (1990). Serological evidence that dry heating of clotting factor concentrates prevents transmission of non-A, non-B hepatitis. J. Med. Virol. *30*:50–52.

Skrabanek, P. (1988a). Cervical cancer in nuns and prostitutes: a plea for scientific continence. J. Clin. Epidemiol. *41*:577–582.

Skrabanek, P. (1988b). Cervical cancer screening: the time for reappraisal. Can. J. Publ. Health *79*:86–89.

Skrabanek, P. and Jamieson, M. (1985). Eaten by worms: a comment on cervical screening. N. Zealand Med. J. August:654.

Smedile, A. et al. (1986). Type D hepatitis: the clinical significance of hepatitis D virus RNA in serum as detected by a hybridization-based assay. Hepatology *6*:1297–1302.

Smedile, A. et al. (1990). Riboprobe assay for HDV RNA: a sensitive method for the detection of the HDV genome in clinical serum samples. J. Med. Virol. *30*:20–24.

Smith, R. F. (1986). Detection of *Trichomonas vaginalis* in vaginal specimens by direct immunfluorescence assay. J. Clin. Microbiol. *24*:1107–1108.

Smith, Th. F. and Shelley, C. D. (1988). Detection of IgM antibody to cytomegalovirus and rapid diagnosis of this virus infection by the shell vial assay. J. Virol. Methods *21*:87–96.

Smotkin, D. and Wettstein, F. O. (1986). Transcription of human papillomavirus type 16 early genes in a cervical cancer and a cancer-derived cell line and identification of the E7 protein. Proc. Natl. Acad. Sci. U.S.A. *83*:4680–4684.

Snijders, P. J. F. et al. (1990). The use of general primers in the polymerase chain reaction permits detection of a broad spectrum of human papilloma genotypes. J. Gen. Virol. *71*:173–181.

Somers, K. D. et al. (1990). Human papillomavirus DNA in squamous cell carcinoma of the larynx and tongue. Proc. Am. Assn. Cancer Res. *31*:326A.

Sousa, R. et al. (1990). Control of papillomavirus gene expression. Biochim. Biophys. Acta *1032*:19–37.

Spadoro, J. P. et al. (1990). Single copies of HIV proviral DNA detected by fluorescent *in situ* hybridization. BioTechniques *9*:186–195.

Spalholz, B. A. and Howley, P. M. (1989). Papillomavirus-host cell interactions. Adv. Viral Oncol. *8*:27–53.

Spector, S. A. and Spector, D. H. (1985). The use of DNA probes in studies of human cytomegalovirus. Clin. Chem. *31*:1514–1520.

Spence, M. R. et al. (1980). The clinical and laboratory diagnosis of *Trichomonas vaginalis* infection. Sex. Transm. Dis. *7*:168–171.

Spencer, E. et al. (1983). Characteristics and analysis of electropherotypes of human rotavirus isolated in Chile. J. Infect. Dis. *148*:41–50.

Spitzer, M. et al. (1989). The multicentric nature of disease related to human papillomavirus infection of the female lower genital tract. Obstet. Gynecol. *73*:303–307.

Staal, S. P. et al. (1989). A survey of Epstein-Barr virus DNA in lymphoid tissue. Frequent detection in Hodgkin's disease. Am. J. Clin. Pathol. *91*:1–5.

Stalhandske, P. and Pettersson, U. (1982). Identification of DNA viruses by membrane filter hybridization. J. Clin. Microbiol. *15*:744–747.

Stalhandske, P. et al. (1983). The use of molecular hybridization for demonstration of adenoviruses in human stools. Curr. Top. Microbiol. Immunol. *104*:299–306.

Stalhandske, P. et al. (1985). Detection of adenoviruses in stool specimens by nucleic acid spot hybridization. J. Med. Virol. *16*:213–218.

Stamm, W. E. and Holmes, K. K. (1984). *Chlamydia trachomatis* infections of the adult. In: Sexually Transmitted Diseases. Holmes, K. K. et al. (eds.). McGraw-Hill, New York, pp. 258–270.

Stein, D. S. and Libertin, C. R. (1989). Molecular analysis of viridans and nutritionally deficient (variant) streptococci causing sequential episodes of endocarditis in a patient. Am. J. Clin. Pathol. *91*:620–624.

Stevens, D. A. et al. (1990). Application of DNA typing methods to *Candida albicans* epidemiology and correlations with phenotype. Rev. Infect. Dis. *12*:258–266.

Stirk, P. R. and Griffiths, P. D. (1987). Use of monoclonal antibodies for the diagnosis of cytomegalovirus infection by the detection of early antigen fluorescent foci (DEAFF) in cell culture. J. Med. Virol. *21*:329–337.

Stockl, E. et al. (1988a). Application of a dot blot hybridization assay, for the diagnosis of CMV infection or reactivation. Zbl. Bakt. Hyg. A *270*:288–294.

Stockl, E. et al. (1988b). Potential of *in situ* hybridization for early diagnosis of productive cytomegalovirus infection. J. Clin. Microbiol. *26*:2536–2540.

Stoeckl, E. et al. (1989). Efficiency of the polymerase chain reaction for the detection of human immunodeficiency virus type 1 (HIV-1) DNA in the lymphocytes of infected persons: comparison to antigen-enzyme-linked immunosorbent assay and virus isolation. J. Med. Virol. *29*:249–255.

Stoker, N. G. (1990). The polymerase chain reaction and infectious diseases: hopes and realities. Trans. R. Soc. Trop. Med. Hyg. *84*:755–756.

Stoler, M. H. and Broker, T. R. (1986). *In situ* hybridization detection of human papillomavirus DNAs and messenger RNAs in genital condylomas and a cervical carcinoma. Hum. Pathol. *17*:1250–1258.

Stoler, M. H. et al. (1989). Differentiation-linked human papillomavirus types 6 and 11 transcription in genital condylomata revealed by *in situ* hybridization with message-specific RNA probes. Virology *172*:331–340.

Storey, A. et al. (1988). Comparison of the *in vitro* transforming activities of human papillomavirus types. EMBO J. *7*:1815–1820.

Street, D. A. et al. (1982). Evaluation of an enzyme-linked immunosorbent assay for the detection of antibody to *Trichomonas vaginalis* in sera and vaginal secretion. Br. J. Vener. Dis. *58*:330–333.

Strickler, J. G. et al. (1990). Comparison of *in situ* hybridization and immunocytochemistry for detection of cytomegalovirus and herpes simplex virus. Hum. Pathol. *21*:443–448.

Strike, D. G. et al. (1989). Expression in *Escherichia coli* of seven DNA fragments comprising the complete L1 and L2 open reading frames of human papillomavirus type 6b and localization of the "common antigen" region. J. Gen. Virol. *70*:543–555.

Stull, T. L. et al. (1988). A broad spectrum probe for molecular epidemiology of bacteria: ribosomal RNA. J. Infect. Dis. *157*:280–286.

Sullender, W. M. et al. (1990). Differentiation of respiratory syncytial virus subgroups with cDNA probes in a nucleic acid hybridization assay. J. Clin. Microbiol. *28*:1683–1687.

Sumazaki, R. et al. (1989). Detection of hepatitis B virus in serum using amplification of viral DNA by means of the polymerase chain reaction. J. Med. Virol. *27*:304–308.

Sun, N. C. J. et al. (1989). Bone marrow examination in patients with AIDS and AIDS-related complex (ARC). Morphologic and *in situ* hybridization studies. Am. J. Clin. Pathol. *92*:589–594.

Sweeney, K. G. et al. (1985). *In vitro* susceptibility of isolates from patients with *Branhamella catarrhalis* pneumonia compared with those of colonizing strains. Antimicrob. Agents Chemother. *27*:499–502.

Swierkosz, E. M. et al. (1987). Improved DNA hybridization method for detection of acyclovir-resistant herpes simplex virus. Antimicrob. Agents Chemother. *31*:1465–1469.

Syrjanen, K. J. (1987). Biology of human papillomavirus (HPV) infections and their role in squamous cell carcinogenesis. Med. Biol. *65*:21–39.

Syrjanen, K. J. (1989). Epidemiology of human papillomavirus (HPV) infections and their associations with genital squamous cell cancer. APMIS *97*:957–970.

Syrjanen, K. et al. (eds.) (1987). Papillomaviruses and human disease. Springer-Verlag, Berlin.

Syrjanen, K. et al. (1988). Factors associated with progression of cervical human papillomavirus (HPV) infections into carcinoma *in situ* during a long term prospective follow-up. Br. J. Obstet. Gynecol. *95*:1096–1102.

Syrjanen, S. et al. (1990). Colposcopy, punch biopsy, *in situ* DNA hybridization, and the polymerase chain reaction in searching for genital human papillomavirus (HPV) infections in women with normal PAP smears. J. Med. Virol. *31*:259–266.

Syrjanen, S. M. (1987). Human papillomavirus infections in the oral cavity. In: Papillomaviruses and Human Disease. Syrjanen, K. et al. (eds.). Springer-Verlag, Berlin, pp. 104–137.

Syrjanen, S. M. (1990). Basic concepts and practical applications of recombinant DNA techniques in detection of human papillomavirus (HPV) infections. APMIS *98*:95–110.

Syrjanen, S. M. et al. (1987). Detection of human papillomavirus DNA in anogenital condylomata in man using in situ DNA hybridization applied to paraffin sections. Genitourin. Med. *63*:32–39.

Syrjanen, S. M. et al. (1988). Human papillomavirus (HPV) DNA sequences in oral precancerous lesions and squamous cell carcinoma demonstrated by *in situ* hybridization. J. Oral Pathol. *17*:273–278.

Tagawa, M. et al. (1987). Duck hepatitis B virus replicates in the yolk sac of developing embryos. J. Virol. *61*:2273–2279.

Takada, N. et al. (1992). HCV genotypes in different countries. Lancet *339*:808.

Takamatsu, K. et al. (1990). Hepatitis C virus RNA in saliva. Lancet *336*:1515.

Takeuchi, K. et al. (1990). The putative nucleocapsid and envelope protein genes of hepatitis C virus determined by comparison of the nucleotide sequences of two isolates derived from an experimentally infected chimpanzee and healthy human carriers. J. Gen. Virol. *71*:3027–3033.

Takiff, H. E. et al. (1985). Detection of enteric adenoviruses by dot-blot hybridization using a molecularly cloned viral DNA probe. J. Med. Virol. *16*:107–118.

Tamura, I. et al. (1987). Prevalence of antibody to delta antigen among HBV carriers in Japan. J. Med. Virol. *22*:217–221.

Tanaka, Y. et al. (1988). Frequent integration of hepatitis B virus DNA in noncancerous liver tissue from hepatocellular carcinoma patients. J. Med. Virol. *26*:7–14.

Tannich, E. and Burchard, G. D. (1991). Differentiation of pathogenic from nonpathogenic *Entamoeba histolytica* by restriction fragment analysis of a single gene amplified *in vitro*. J. Clin. Microbiol. *29*:250–255.

Tanskanen, E. I. et al. (1990). Pulsed-field gel electrophoresis of *Sma*I digests of lactococcal genomic DNA, a novel method of strain identification. Appl. Environ. Microbiol. *56*:3105–3111.

Taylor, D. E. and Hiratsuka, K. (1990). Use of non-radioactive DNA probes for detection of *Campylobacter jejuni* and *Campylobacter coli* in stool specimens. Mol. Cell. Probes *4*:261–271.

Taylor-Wiedeman, J. et al. (1991). Monocytes are a major site of persistence in human cytomegalovirus in peripheral blood mononuclear cells. J. Gen. Virol. *72*:2059–2064.

Taylor, D. N. et al. (1988). Clinical and microbiologic features of *Shigella* and enteroinvasive *Escherichia coli* infections detected by DNA hybridization. J. Clin. Microbiol. *26*:1362–1366.

Telenti, A. et al. (1990). Detection of Epstein-Barr virus by polymerase chain reaction. J. Clin. Microbiol. *28*:2187–2190.

Tenover, F. C. et al. (1990). DNA probe culture confirmation assay for identification of thermophilic *Campylobacter* species. J. Clin. Microbiol. *28*:1284–1287.

Terpstra, W. J. et al. (1990). Detection of *Leptospira, Haemophilus* and *Campylobacter* using DNA probes. In: Gene Probes for Bacteria. Macario, A. J. L. and Conway de Macario, E. (eds.). Academic Press, San Diego, pp. 296–329.

Terry, R. M. et al. (1987). Demonstration of human papillomavirus types 6 and 11 in juvenile laryngeal papillomatosis by *in-situ* DNA hybridization. J. Pathol. *153*:245–248.

Thaler, M. M. et al. (1991). Vertical transmission of hepatitis C virus. Lancet *338*:17–18.

Thell, K. W. et al. (1981). Rapid, simple methods of preparing rotaviral double-stranded ribonucleic acid for analysis of polyacrylamide gel electrophoresis. J. Clin. Microbiol. *14*:273–280.

Thierry, D. et al. (1990). Characterization of a *Mycobacterium tuberculosis* insertion sequence, IS6110, and its application to diagnosis. J. Clin. Microbiol. *28*:2668–2673.

Thomas, D. P. (1990). Immunoglobulins and hepatitis C virus. Lancet 335:1531.

Thompson, C. H. and Roman, A. (1987). Expression of human papillomavirus type 6 E1, E2, L1 and L2 open reading frames in *Escherichia coli*. Gene 56:289–295.

Thompson, C. H. et al. (1992). Detection of HPV DNA in archival specimens of cervical cancer using in situ hybridization and the polymerase chain reaction. J. Med. Virol. *36*:54–59.

Tidy, J. and Farrell, P. J. (1989). Retraction: human papillomavirus subtype 16b. Lancet *ii*:1535.

Tidy, J. A. et al. (1989a). High rate of human papillomavirus type 16 infection in cytologically normal cervices. Lancet *i*:434.

Tidy, J. A. et al. (1989b). Relation between infection with a subtype of HPV 16 and cervical neoplasia. Lancet *i*:1225–1227.

Tilton, R. C. et al. (1988). DNA probe versus culture for detection of *Mycoplasma pneumoniae* in clinical specimens. Diagn. Microbiol. Infect. Dis. *10*:109–112.

Ting, Y. and Manos, M. (1990). Detection and typing of genital human papillomavirus. In: PCR Protocols: A Guide to Methods and Application. Innis, M. A. et al. eds. Academic Press, San Diego, pp. 356–367.

Tirasophon, W. et al. (1991). A novel detection of a simple *Plasmodium falciparum* in infected blood. Biochem. Biophys. Res. Commun. *175*:179–184.

Todd, J. A. et al. (1989). A rapid DNA probe test for detecting human papillomavirus types 6/11 and 16 in biopsy specimens. Mol. Cell. Probes *3*:273–288.

Tomita, Y. et al. (1987). Expression of human papillomavirus types 6b and 16 L1 open reading frames in *Escherichia coli*: detection of a 56,000 dalton polypeptide containing genus-specific (common) antigens. J. Virol. *61*:2389–2394.

Tompkins, L. S. (1989). Nucleic acid probes in infectious diseases. In: Current Clinical Topics in Infectious Diseases. Remington, J. S. and Swartz, M. N. (eds.). Blackwell Scientific, pp. 174–193.

Tompkins, L. S. et al. (1987). Molecular epidemiology of *Legionella* species by restriction endonuclease and alloenzyme analysis. J. Clin. Microbiol. *25*:1875–1880.

Tong, M. J. et al. (1987). Persistence of serum hepatitis B virus deoxyribonucleic acid in hepatitis B surface antigen-positive patients with chronic persistent hepatitis treated with prednisone. Gastroenterology *92*:862–866.

Tonjum, T. et al. (1989). Differentiation of some species of Neisseriaceae and other bacterial groups by DNA-DNA hybridization. APMIS *97*:395–405.

Torelli, G. et al. (1990). Presence of antibodies against HHV-6 virus and of sequences of HHV-6 genome in patients with non-AIDS related Hodgkin and non-Hodgkin lymphomas. Proc. Am. Assn. Cancer Res. *31*:329A.

Torok, T. J. et al. (1992). Prenatal diagnosis of intrauterine infection with parvovirus B19 by the polymerase chain reaction technique. Clin. Infect. Dis. *14*:149–155.

Torres, M. J. et al. (1991). Evaluation of a DNA probe of plasmid origin for the detection of *Neisseria gonorrhoea* in cultures and clinical specimens. Mol. Cell. Probes *5*:49–54.

Towner, K. J. (1990). Detection of TEM β-lactamase genes using DNA probes. In: Gene Probes for Bacteria. Macario, A. J. L. and Conway de Macario, E. (eds.). Academic Press, San Diego, pp. 459–483.

Tozuka, S. et al. (1989). State of hepatitis B virus DNA in hepatocytes of patients with noncarcinomatous liver disease. Its special relationship with necroinflammatory activity and the stage of disease. Arch. Pathol. Lab. Med. *113*:20–25.

Tramont, E. C. (1990). *Treponema pallidum* (syphilis). In: Principles and Practice of Infectious Diseases, 3rd Ed. Mandell, G. L. et al. (eds.). John Wiley & Sons, New York, pp. 1794–1808.

Tully, J. G. (1985). Newly discovered Mollicutes. In: The Mycoplasmas, Vol. IV (Mycoplasma pathogenicity). Razin, S. and Barile, M. F. (eds.). Academic Press, New York, pp. 1–26.

Tydall, R. L. et al. (1989). Effect of thermal additions on the density and distribution of thermophilic amoebae and pathogenic *Naegleria fowleri* in a newly created cooling lake. Appl. Environ. Microbiol. *55*:722–732.

Tzianabos, T. et al. (1989). Detection of *Rickettsia rickettsii* DNA in clinical specimens by using polymerase chain reaction technology. J. Clin. Microbiol. *27*:2866–2868.

Uccini, S. et al. (1989). Interaction of HIV and EBV at lymphoid tissue level: immunohistochemistry and *in situ* hybridization. APMIS (Suppl.) *8*:28–32.

Ulrich, P. P. et al. (1988). Assessment of human immunodeficiency virus in cocultures of peripheral blood mononuclear cells from healthy seropositive subjects. J. Med. Virol. *25*:1–10.

Ulrich, P. P. et al. (1990). Detection, semiquantitation, and genetic variation in hepatitis C virus sequences amplified from the plasma of blood donors with elevated alanine aminotransferase. J. Clin. Invest. *86*:1609–1614.

Ulrich, W. et al. (1986a). Detection of CMV-infected cells in biopsies of human renal allografts by *in situ* hybridization. Transpl. Proc. *18*:1377–1378.

Ulrich, W. et al. (1986b). The histopathologic identification of CMV infected cells in biopsies of human renal allografts. An evaluation of 100 transplant biopsies by *in situ* hybridization. Pathol. Res. Pract. *181*:739–745.

Unger, E. R. et al. (1991). Comparison of ^{35}S and biotin as labels for *in situ* hybridization: use of an HPV model system. J. Histochem. Cytochem. *39*:145–150.

Urdea, M. S. et al. (1989). Application of a rapid non-isotopic nucleic acid analysis system to the detection of sexually transmitted diseases causing organisms and their associated antimicrobial resistance. Clin. Chem. *35*:1571–1575.

Vaglia, A. et al. (1990). Needlestick hepatitis C virus seroconversion in a surgeon. Lancet *336*:1315–1316.

Valentine, J. L. et al. (1991). Detection of *Helicobacter pylori* by using the polymerase chain reaction. J. Clin. Microbiol. *29*:689–695.

Vallejos, H. et al. (1987). Characterization of human papillomavirus types in condylomata acuminata in children by *in situ* hybridization. Lab. Invest. *56*:611–15.

Van Damme-Jongsten, M. et al. (1989). Cloning and sequencing of the *Clostridium perfringens* enterotoxin gene. Antonie van Leeuwenhoek J. Microbiol. *56*:181–190.

Van Damme-Jongsten, M. et al. (1990). Synthetic DNA probes for detection of enterotoxigenic *Clostridium perfringens* strains isolated from outbreaks of food poisoning. J. Clin. Microbiol. *28*:131–133.

van den Brule, A. J. C. et al. (1989). Use of anti-contamination primers in the polymerase chain reaction for the detection of human papillomavirus genotypes in cervical scrapes and biopsies. J. Med. Virol. *29*:20–27.

van den Brule, A. J. C. et al. (1990a). General primer-mediated polymerase chain reaction permits the detection of sequenced and still unsequenced human papillomavirus genotypes in cervical scrapes and carcinomas. Int. J. Cancer *45*:644–649.

van den Brule, A. J. C. et al. (1990b). Rapid detection of human papillomavirus in cervical scrapes by combined general primer-mediated and type-specific polymerase chain reaction. J. Clin. Microbiol. *28*:2739–2743.

van der Poel, C. L. et al. (1989). Antihepatitis C antibodies and non-A, non-B post-transfusion hepatitis in the Netherlands. Lancet *ii*:297–298.

van der Poel, C. L. et al. (1990). Anti-HCV and transaminase testing of blood donors [letter]. Lancet *336*:187–188.

van der Poel, C. L. et al. (1991). Confirmation of hepatitis C virus infection by new four-antigen recombinant immunoblot assay. Lancet *337*:317–319.

van Eys, G. J. J. M. et al. (1989). Identification of "Old World" *Leishmania* by DNA recombinant probes. Mol. Biochem. Parasitol. *34*:53–62.

van Eys, G. J. J. M. et al. (1991). Characterization of serovars of the genus *Leptospira* by DNA hybridization with Hardjobovis and Ichterohaemorrhagiae recombinant probes with special attention to serogroup Sejroe. J. Clin. Microbiol. *29*:1042–1048.

van Sickle, M. et al. (1990). Detection of human papillomavirus DNA before and after development of invasive vulvar cancer. Obstet. Gynecol. *76*:540–542.

van Steenbergen, T. J. M. et al. (1991). Microflora and bacterial DNA restriction enzyme analysis in young adults with periodontitis. J. Periodontol. *62*:235–241.

Vandenberg, F. M. et al. (1989). Detection of *Campylobacter pylori* in stomach tissue by DNA *in situ* hybridization. J. Clin. Pathol. *42*:995–1000.

Vandenvelde, C. et al. (1990). Fast multiplex polymerase chain reaction on boiled clinical samples for rapid viral diagnosis. J. Virol. Methods *30*:215–228.

Vary, P. H. et al. (1990). Use of highly specific DNA probes and the polymerase chain reaction to detect *Mycobacterium paratuberculosis* in Johne's disease. J. Clin. Microbiol. *28*:933–937.

Venkatesan, M. et al. (1988). Development and testing of invasion-associated DNA probes for detection of *Shigella* spp. and enteroinvasive *Escherichia coli*. J. Clin. Microbiol. *26*:261–66.

Vento, S. et al. (1989). Hazards of interferon therapy for HBV-seronegative chronic hepatitis [letter]. Lancet *ii*:926.

Vermund, S. H. et al. (1989). Molecular diagnosis of genital human papillomavirus infection: comparison of two methods used to collect exfoliated cervical cells. Am. J. Obstet. Gynecol. *160*:304–308.

Victor, T. et al. (1991). Improved method for the routine identification of toxigenic *Escherichia coli* by DNA amplification of a conserved region of the heat-labile toxin A subunit. J. Clin. Microbiol. *29*:158–161.

Vilgalys, R. and Hester, M. (1990). Rapid genetic identification and mapping of enzymatically amplified ribosomal DNA from several *Cryptococcus* species. J. Bacteriol. *172*:4238–4246.

Virtanen, M. et al. (1983). Novel test for rapid viral diagnosis: detection of adenovirus in nasopharyngeal mucus apirates by means of nucleic acid sandwich hybridization. Lancet *i*:381–383.

Volsky, D. J. et al. (1990). Titration of human immunodeficiency virus type 1 (HIV-1) and quantitative analysis of virus expression *in vitro* using liquid RNA-RNA hybridization. J. Virol. Methods *28*:257–272.

von Knebel Doeberitz, M. et al. (1991). Influence of chromosomal integration on glucocorticoid-regulated transcription of growth-stimulating papillomavirus genes E6 and E7 in cervical carcinoma cells. Proc. Natl. Acad. Sci. U.S.A. *88*:1411–1415.

von Krogh, G. et al. (1987). Advantage of human papillomavirus typing in clinical evaluation of genitoanal warts. J. Am. Acad. Dermatol. *18*:495–503.

von Krogh, G. et al. (1988). Advantage of human papillomavirus typing in the clinical evaluation of genitoanal warts. Experience with the *in situ* deoxyribonucleic acid hybridization technique applied on paraffin sections. J. Am. Acad. Dermatol. *18*:495–503.

Vousden, K. H. et al. (1988). The E7 open reading frame of human papillomavirus type 16 encodes a transforming gene. Oncogene Res. *3*:167–175.

Wagatsuma, M. et al. (1990). Analysis of integrated human papillomavirus type 16 DNA in cervical cancers: amplification of viral sequences together with cellular flanking sequences. J. Virol. *64*:813–821.

Wages, J. M., Jr. et al. (1991). Clinical performance of a polymerase chain reaction testing algorithm for diagnosis of HIV-1 infection in peripheral blood mononuclear cells. J. Med. Virol. *33*:58–63.

Wakefield, A. E. et al. (1990). Detection of *Pneumocystis carinii* with DNA amplification. Lancet *336*:451–453.

Walboomers, J. M. M. et al. (1988). Sensitivity of *in situ* detection with biotinylated probes of human papillomavirus type 16 DNA in frozen tissue sections of squamous cell carcinoma of the cervix. Am. J. Pathol. *131*:587–594.

Walker, J. et al. (1989). Human papillomavirus genotype as a prognostic indicator in carcinoma of the uterine cervix. Obstet. Gynecol. *74*:781–785.

Walker, R. C. et al. (1986). Comparison of culture, cytotoxicity assays, enzyme-linked immunosorbent assay for toxin A and toxin B in the diagnosis of *Clostridium difficile*-related enteric disease. Diagn. Microbiol. Infect. Dis. *5*:61–69.

Wang, J. T. et al. (1991). Hepatitis C virus RNA in saliva of patients with posttransfusion hepatitis C infection. Lancet *337*:48.

Wang, K. S. et al. (1986). Structure sequence and expression of the hepatitis delta (δ) viral genome. Nature (London) *323*:508–514.

Wank, R. and Thomssen, C. (1991). High risk of squamous cell carcinoma of the cervix for women HLA-DQw3. Nature (London) *352*:723–725.

Ward, V. K. et al. (1990). Detection of an arbovirus in an invertebrate and a vertebrate host using the polymerase chain reaction. J. Virol. Methods *30*:291–300.

Watanabe, S. et al. (1989). Human papillomavirus type 16 transformation of primary human embryonic fibroblasts requires expression of open reading frames E6 and E7. J. Virol. *63*:965–969.

Watt, R. M. et al. (1986). Rapid assay for immunological detection of *Trichomonas vaginalis*. J. Clin. Microbiol. *24*:551–555.

Webb, D. H. et al. (1987). A one-step method for detecting and typing human papillomavirus DNA in cervical scrape specimens from women with cervical dysplasia. J. Infect. Dis. *156*:912–919.

Webb, L. et al. (1990). Detection of murine typhus infection in fleas by using the polymerase chain reaction. J. Clin. Microbiol. *28*:530–534.

Weiland, O. et al. (1990). Therapy of chronic post-transfusion non-A, non-B hepatitis with interferon α-2b: Swedish experience. J. Hepatol. *11*:S57–S62.

Weiner, A. J. et al. (1987). Hepatitis delta (δ) cDNA clones: undetectable hybridization to nucleic acids from infectious non-A, non-B hepatitis materials and hepatitis B DNA. J. Med. Virol. *21*:239–247.

Weiner, A. J. et al. (1990a). Detection of hepatitis C viral sequences in non-A, non-B hepatitis. Lancet *335*:1–3.

Weiner, A. J. et al. (1990b). HCV testing in low-risk population. Lancet *336*:695.

Weiss, L. M. and Movahed, L. A. (1989). *In situ* demonstration of Epstein-Barr viral genomes in viral-associated B-cell lymphoproliferations. Am. J. Pathol. *134*:651–659.

Weiss, L. M. et al. (1987). Epstein-Barr viral DNA in tissues of Hodgkin's disease. Am. J. Pathol. *129*:86–91.

Weiss, L. M. et al. (1989a). Detection of Epstein-Barr viral genomes in Reed-Steinberg cells of Hodgkin's disease. N. Engl. J. Med. *320*:502–506.

Weiss, L. M. et al. (1989b). Analysis of lymphoepithelioma and lymphoepithelioma-like carcinoma for Epstein-Barr viral genomes by *in situ* hybridization. Am. J. Surg. Pathol. *13*:625–631.

Weiss, L. M. et al. (1990). *In situ* hybridization studies for viral nucleic acids in heart and lung allograft biopsies. Am. J. Clin. Pathol. *93*:675–679.

Welch, D. et al. (1990). Detection of plasmid DNA from all *Chlamydia trachomatis* serovars with a two-step polymerase chain reaction. Appl. Environ. Microb. *56*:2494–2498.

Weller, I. V. D. et al. (1982). The detection of HBV-DNA in serum by molecular hybridization: a more sensitive method for the detection of complete HBV particles. J. Med. Virol. *9*:273–280.

Wernars, K. and Notermans, S. (1990). Gene probes for detection of food-borne pathogens. In: Gene Probes for Bacteria. Macario, A. J. L. and Conway de Macario, E. (eds.). Academic Press, San Diego, pp. 353–388.

Wesley, I. V. et al. (1991). Oligodeoxynucleotide probes for *Campylobacter fetus* and *Campylobacter hyointestinalis* based on 16S rRNA sequences. J. Clin. Microbiol. *29*:1812–1817.

Wetherall, B. L. and Johnson, A. M. (1990). Nucleic acid probes for *Campylobacter* species. In: Gene Probes for Bacteria. Macario, A. J. L. and Conway de Macario, E. (eds.). Academic Press, San Diego, pp. 256–293.

Wetherall, B. L. et al. (1988). Detection of *Campylobacter pylori* DNA by hybridization with non-radioactive probes in comparison with a ^{32}P-labeled probe. J. Med. Microbiol. *26*:257–263.

Whipple, D. et al. (1990). Identification of restriction fragment length polymorphism in DNA from *Mycobacterium paratuberculosis*. J. Clin. Microbiol. *28*:2561–2564.

White, D. G. et al. (1985). Human parvovirus arthropathy. Lancet *i*:419–421.

Whitley, R. J. (1988). Herpes simplex virus infections of the central nervous system. Am. J. Med. *85*:61–67.

Wickenden, C. et al. (1987). Prevalence of HPV DNA and viral copy numbers in cervical scrapes from women with normal and abnormal cervices. J. Pathol. *153*:127–135.

Wilbur, D. C. et al. (1988). Detection of infection by human papillomavirus in genital condylomata. A comparison study using immunocytochemistry and *in situ* nucleic acid hybridization. Am. J. Clin. Pathol. *89*:505–510.

Wilczynski, S. P. et al. (1988a). Human papillomaviruses and cervical cancer: analysis of histopathologic features associated with different viral types. Hum. Pathol. *19*:697–704.

Wilczynski, S. P. et al. (1988b). Identification of HPV16 early genes retained in cervical carcinomas. Virology *166*:624–627.

Wilde, J. et al. (1990). Removal of inhibitory substances from human fecal specimens for detection of group A rotaviruses by reverse transcriptase and polymerase chain reactions. J. Clin. Microbiol. *28*:1300–1307.

Wilde, J. et al. (1991). Improved detection of rotavirues shedding by polymerase chain reaction. Lancet *337*:323–326.

Wiley, C. A. et al. (1986). Localization of cytomegalovirus proteins and genome during fulminant central nervous system infection in an AIDS patient. J. Neuropathol. Exp. Neurol. *45*:127–139.

Wilkins, E. G. L. and Ivanyi, J. (1990). Potential value of serology for diagnosis of extrapulmonary tuberculosis. Lancet *336*:641–644.

Wilkinson, A. H. et al. (1989). Increased frequency of posttransplant lymphomas in patients treated with cyclsporine, azathioprine, and predinsone. Transplantation *47*:293–296.

Wilkinson, H. W. et al. (1986). Evaluation of commercial gene probe for identification of *Legionella* cultures. J. Clin. Microbiol. *23*:217–220.

Williams, P. H. (1990). DNA probes for *Escherichia coli* isolates from human extraintestinal infections. In: Gene Probes for Bacteria. Macario, A. J. L. and Conway de Macario, E. (eds.). Academic Press, San Diego, pp. 143–165.

Williamson, A. L. et al. (1991). The detection of human papillomarvirus in esophageal lesions. Anticancer Res. *11*:263–266.

Wilson, K. H. et al. (1988). Species-specific oligonucleotide probes for rRNA of *Clostridium difficile* and related species. J. Clin. Microbiol. *26*:2484–2488.

Wilson, R. W. et al. (1990). Longitudinal study of human papillomavirus infection of the female urogenital tract by *in situ* hybridization. Arch. Pathol. Lab. Med. *114*:155–159.

Wolber, R. A. and Lloyd, R. V. (1988). Cytomegalovirus detection by nonisotopic *in situ* DNA hybridization and viral antigen immunostaining using a two-color technique. Hum. Pathol. *19*:736–741.

Wolber, R. A. et al. (1988). Ultrastructural localization of viral nucleic acids by *in situ* hybridization. Lab. Invest. *59*:144–151.

Wolber, R. et al. (1990). Anal cloacogenic and squamous carcinomas. Comparative histologic analysis using *in situ* hybridization for human papillomavirus DNA. Am. J. Surg. Pathol. *14*:176–182.

Wolf, H. et al. (1973). Epstein-Barr viral genomes in epithelial nasopharyngeal carcinoma cells. Nature New Biol. *244*:245–247.

Wolinsky, S. M. et al. (1992). Selective transmission of human immunodeficiency virus type-1 variants from mothers to infants. Science *255*:1134–1137.

Wong, D. C. et al. (1990). Non-specificity of anti-HCV test for seroepidemiological analysis. Lancet *336*:750–751.

Wood, J. R. et al. (1987). Hepatitis B virus deoxyribonucleic acid in serum during hepatitis Be antigen clearance in corticosteroid-treated severe chronic active hepatitis B. Gastroenterology *93*:1225–1230.

Woods, G. L. and Yam, P. (1989). Detection of herpes simplex virus in clinical specimens using a DNA probe after centrifugal inoculation of A549 cells. J. Virol. Methods *23*:339–343.

Woods, G. L. et al. (1990). Evaluation of a nonisotopic probe for detection of *Chlamydia trachomatis* in endocervical specimens. J. Clin. Microbiol. *28*:370–372.

Wren, B. W. and Tabaqchali, S. (1990). Detection of pathogenic *Yersinia enterocolitica* by the polymerase chain reaction. Lancet *336*:693.

Wren, B. et al. (1990). Rapid identification of toxigenic *Clostridium difficile* by polymerase chain reaction. Lancet *i*:423.

Wu, S. W. P. et al. (1991). Lack of mycobacterial DNA in Crohn's disease tissue. Lancet *337*:174–175.

Wu, T. C. and Fung, J. C. (1983). Evaluation of the usefulness of counterimmunoelectrophoresis for diagnosis of *Clostridium difficile*-associated colitis in clinical specimens. J. Clin. Microbiol. *17*:610–613.

Wu, T. C. and Mounts, P. (1988). Transcriptional regulatory elements in the noncoding region of human papillomavirus type 6. J. Virol. *62*:4722–4729.

Wu, T. C. et al. (1990). Detection of EBV gene expression in Reed-Sternberg cells of Hodgkin's disease. Int. J. Cancer *46*:801–804.

Xu, L. et al. (1990). The application of polymerase chain reaction to the detection of rotaviruses in faeces. J. Virol. Methods *27*:29–37.

Yaginuma, K. et al. (1985). Hepatitis B virus integration in hepatocellular carcinoma DNA: duplication of cellular flanking sequences at the integration site. Proc. Natl. Acad. Sci. U.S.A. *82*:4458–4462.

Yaginuma, K. et al. (1987). Multiple integration site of hepatitis B virus DNA in hepatocellular carcinoma and chronic active hepatitis tissues from children. J. Virol. *61*:1808–1813.

Yam, W. C. et al. (1991). Restriction fragment length polymorphism analysis of *Vibrio cholerae* strains associated with a cholera outbreak in Hong Kong. J. Clin. Microbiol. *29*:1058–1059.

Yamakawa, K. et al. (1989). Identification of rotaviruses by dot-blot hybridization using an alkaline phosphatase-conjugated synthetic oligonucleotide probe. Mol. Cell. Probes *3*:397–401.

Yamamoto, L. J. et al. (1991). Herpes simplex virus type 1 DNA in cerebrospinal fluid of a patient with Mollaret's meningitis. N. Engl. J. Med. *325*:1082–1085.

Yamashiroga, H. M. et al. (1988). Herpes viridae in the coronary arteries and aorta of young trauma victims. Am. J. Pathol. *130*:71–79.

Yan, W. et al. (1991). Pulsed-field gel electrophoresis of *Campylobacter jejuni* and *Campylobacter coli* genomic DNA and its epidemiologic application. J. Infect. Dis. *163*:1068–1072.

Yanez, M. A. et al. (1986). Determination of mycobacterial antigens in sputum by enzyme immunoassay. J. Clin. Microbiol. *23*:822–825.

Yee, C. et al. (1985). Presence and expression of human papillomavirus sequences in human cervical carcinoma cell lines. Am. J. Pathol. *119*:361–366.

Yehle, C. O. et al. (1987). A solution hybridization assay for ribosomal RNA from bacteria using biotinylated DNA probes and enzyme-labeled antibody to DNA:RNA. Mol. Cell. Probes *1*:177–193.

Yoffe, B. et al. (1986). Hepatitis B virus DNA in mononuclear cells and analysis of cell subsets for the presence of replicative intermediates of viral DNA. J. Infect. Dis. *153*:471–477.

Yoffe, B. et al. (1990). Extrahepatic hepatitis B virus DNA sequences in patients with acute hepatitis B infection. Hepatology *12*:187–192.

Yokosuka, O. et al. (1986). Hepatitis B virus RNA transcripts and DNA in chronic liver disease. N. Engl. J. Med. *315*:1187–1192.

Yokota, H. et al. (1986). A quantitative assay for the detection of hepatitis D virus DNA employing a biotin-labeled DNA probe and the avidin-β-galactosidase complex. Biochem. Biophys. Acta *868*:45–50.

Yolken, R. H. (1988). Nucleic acids or immunoglobulins: which are the molecular probes of the future? Mol. Cell. Probes *2*:87–96.

Yoneda, N. et al. (1990). Detection of Epstein-Barr virus genome in benign polyclonal proliferative T cells of a young male patient. Blood *76*:172–177.

Yoshimura, H. H. et al. (1987). Investigation of association of mycobacteria with inflammatory bowel disease by nucleic acid hybridization. J. Clin. Microbiol. *25*:45–51.

Young, L. S. et al. (1989). The polymerase chain reaction: a new epidemiological tool for investigating cervical human papillomavirus infection. Br. Med. J. *298*:14–18.

Young, N. S. et al. (1984a). Characterization of a virus that causes transient aplastic crisis. J. Clin. Invest. *73*:224–230.

Young, N. et al. (1984b). Direct demonstration of the human parvovirus in erythroid progenitor cells infected *in vitro*. J. Clin. Invest. *74*:2024–2032.

Zaia, J. A. et al. (1990). Comparative analysis of human cytomegalovirus a —sequence in multiple clinical isolates by using polymerase chain reaction and restriction fragment length polymorphism assays. J. Clin. Microbiol. *28*:2602–2607.

Zanetti, A. R. et al. (1990). Hepatitis C virus RNA in symptomless donors implicated in post-transfusion non-A, non-B hepatitis. Lancet *336*:448.

Zappa, U. et al. (1990). Comparison of serological and DNA probe analyses for detection of suspected periodontal pathogens in subgingival plaque samples. Arch. Oral Biol. *35* (Suppl.):161S–164S.

Zarski, J. P. et al. (1989). Comparison of a quantitative standardized HBV-DNA assay and a classic spot hybridization test in chronic active hepatitis B patients undergoing antiviral therapy. Res. Virol. *140*:203–291.

Zazzi, M. et al. (1992). Low human immunodeficiency virus titer and polymerase chain reaction false-negatives. J. Infect. Dis. *165*:779–780.

Zerbini, M. et al. (1987). Dot immunoperoxidase assay using monoclonal antibody for direct detection of cytomegalovirus in urine samples. J. Clin. Microbiol. *25*:2197–2199.

Zerbini, M. et al. (1990). Rapid screening for B19 parvovirus DNA in clinical specimens with a digoxigenin-labeled DNA hybridization probe. J. Clin. Microbiol. *28*:2496–2499.

Zhang, Q. Y. et al. (1990). Genetic diversity of penicillin-binding protein 2 genes of penicillin resistant strains of *Neisseria meningitidis* revealed by fingerprinting of amplified DNA. Antimicrob. Agents Chemother. *34*:1523–1528.

Zignego, A. L. et al. (1990). Amplification of hepatitis delta virus RNA sequences by polymerase chain reaction: a tool for viral detection and cloning. Mol. Cell. Probes *4*:43–51.

Zolg, J. W. et al. (1990). Detection of pyrimethamine resistance in *Plasmodium falciparum* by mutation-specific polymerase chain reaction. Mol. Biochem. Parasitol. *39*:257–265.

zur Hausen, H. (1976). Condylomata acuminata and human genital cancer. Cancer Res. *36*:530.

zur Hausen, H. (1977). Human papillomaviruses and their possible role in squamous cell carcinomas. Curr. Top. Microbiol. Immunol. *78*:1–30.

zur Hausen, H. (1987). Intracellular surveillance of persisting viral infections: human genital cancer results from deficient cellular control of papillomavirus gene expression. Lancet *ii*:489–491.

zur Hausen, H. (1989a). Papillomaviruses as carcinomaviruses. In: Advances in Viral Oncology, Vol. 8. Klein, G. (ed.). Raven Press, New York, pp. 1–26.

zur Hausen, H. (1989b). Papillomaviruses in anogenital cancer as a model to understand the role of viruses in human cancers. Can. Res. *49*:4677–4681.

zur Hausen, H. et al. (1970). EBV DNA in biopsies of Burkitt's tumors and anaplastic carcinoma of the nasopharynx. Nature (London) *288*:1056–1058.

Zyzik, E. et al. (1986). Assay of hepatitis B virus genome titers in sera of infected subjects. Eur. J. Clin. Microbiol. *5*:330–335.

Part 7
Selected Aspects of Gene Level Evaluation in Hematology

Chapter 7.1
Hemoglobinopathies

7.1.1 SICKLE CELL DISEASE

Molecular genetic aspects of various hemoglobinopathies is probably the most thoroughly investigated area of hematology. **Sickle cell anemia (SS)** has long been considered a paradigm of a "molecular disease" (for a review, see Powars et al., 1990). It is apparent that the **variable expression** of SS disease manifested in the degree of vascular involvement and ensuing tissue alterations in multiple organs is genetically determined. Although the major genetic alteration accounting for the production of **sickle hemoglobin (Hb S)** is the substitution of valine for glutamic acid in the β^S chain, the flanking DNA regions of the sickle gene involving the promoter or suppressor areas influence the **polymerization** kinetics of Hb S (Powars et al., 1990).

Restriction analysis has demonstrated **polymorphisms in the β^S-gene cluster** related to the observed variability in the manifestation of the disease. In particular, the presence or absence of **α-thalassemia-2** appears to change the polymerization kinetics of Hb S; the **interaction of other hemoglobins (Hb F** or **Hb A$_2$)** with Hb S, and the alteration of the **interaction of red cells** carrying Hb S with the **endothelium**, are all genetically determined and influence the clinical manifestations of the disease (Powars et al., 1990). Certain **haplotype combinations**, such as the **Senegal chromosome**, are associated with milder forms of SS disease, lower overall morbidity, and a lower relative risk of acute and recurrent clinical events leading eventually to major organ failure (Powars et al., 1990). On the opposite, in the **SC** individuals the presence of the Central African Republic **(CAR) polymorphism** of the β^S chromosome carries the risk of greater morbidity, whereas the **Benin haplotype** leads to the condition of SS and SC patients that is intermediate between that with the **Senegalese** and the CAR haplotypes.

The variability of clinical manifestation of SS disease and the beneficial effects of the earliest therapeutic intervention on the health of the patient underscore the need for a molecular genetic evaluation of the underlying globin gene abnormalities. While **postnatal diagnosis** of SS anemia has been traditionally determined by **Hb electrophoresis**, at the present time molecular diagnostics can be performed on amniocytes, chorionic villi, or fetal blood cells obtained percutaneously from the umbilical cord early in pregnancy. Initially, **indirect linkage analysis** was used (Kan and Dozy, 1978), whereas later **direct identification** of mutations was developed using restriction analysis and **oligonucleotide probes** (e.g., Boehm et al., 1983; Chang and Kan, 1972; Chehab et al., 1987; Conner et al., 1983; Orkin et al., 1982), and, most recently, diagnosis has become possible by **polymerase chain reaction (PCR)** (Chehab and Kan, 1990; Posey et al., 1989; Saiki et al., 1985; Skogerboe et al., 1991).

The **original amplification procedure** developed by Saiki and coworkers (1985) was combined with the analysis of the β-globin amplified product by **solution hybridization** with specific oligonucleotide probes and **subsequent digestion** with a restriction enzyme to determine the genotype. This procedure, in fact, became the prototype for the evaluation of other genetic and infectious diseases by the PCR. Subsequently, a **PCR-allele-specific oligonucleotide (ASO)** protocol using **nonradioactive probes** labeled with **horseradish peroxidase** has been described for the detection of **sickle** and **β-thalassemia mutations** (Saiki et al., 1988). An additional advantage of the PCR-ASO method is in the **ease** of identifying the Hb A allele in the presence of high amounts of Hb F.

Furthermore, **combining PCR-ASO with Hb electrophoresis** can provide an **accurate early diagnosis** of S or C traits and β^0-thalassemia without the need to electrophorese the parental samples (Skogerboe et al., 1991). With the present set of primers, Hb electrophoresis was able to identify more hemoglobin abnormalities than the PCR-ASO method described. However, considering the above-described features and the possibility to enhance the PCR assay with other primers, the amplification protocol provides a means to analyze and definitively diagnose abnormal electrophoretic variants. Moreover, PCR methodology allows simultaneous screening for several genetic abnormalities in one blood sample (Skogerboe et al., 1991).

With adequate consideration given to the **potential drawback**s of REA for the SS mutation — **incomplete digestion** of the sample, or, as in other applications, **plasmid contamination** — this assay system appears to be specific for SS anemia (Posey et al., 1989). *Mst*II and *Cvn*I produce larger fragments of DNA than the previously used *Dde*I, which, in addition, required up to 20 times more sample DNA (1 µg vs. up to 20 µg DNA per assay).

An ingenious use of **two different fluorochromes** to label oligonucleotide primers for a PCR assay allows **direct color detection** of a mutant globin sequence (Chehab and Kan, 1990). This procedure, which can be automated, is also applicable to screening for the sickle gene in **dried blood spot** samples (Rubin et al., 1989). In fact, the *A, S,* and *C* **alleles** of the β-globin gene in dried blood collected on **Guthrie cards** for **phenylketonuria** screening have been detected by PCR (Skogerboe et al., 1991). The amplification step was followed by hybridization of the amplified product with **antisense ASOs** because they produced a **sevenfold greater signal** than sense ASO probes. Only one quarter of a Guthrie card blood spot was utilized per assay. Virtually complete agreement of the PCR-ASO dot blot with Hb electrophoresis was observed (80 out of 81) (Skogerboe et al., 1991).

7.1.2 THALASSEMIAS

The current status of molecular biology of the α-and β-globin gene clusters has been widely reviewed (e.g., Bunn and Forget, 1986; Higgs et al., 1989; Kazazian, 1990; Steinberg, 1988; Weatherall et al., 1989). For a lucid, excellent discussion of the molecular basis of the thalassemia syndromes and the effects of particular genetic alterations in the α- and β-globin genes on the results of **prenatal testing** at the gene level, a review by Kazazian (1990) is highly recommended.

7.1.2.1 α-THALASSEMIA

α-Thalassemia is caused by defective **α-globin synthesis**, and the **deletions** of either one or both α-globin genes on **chromosome 16** account for over 95% of α-thalassemia cases. Although a large number of the deletions affecting both α-globin genes are known, the predominance of one type of deletion is characteristic of a particular geographic region (Kazazian, 1990). Prenatal diagnosis of α-thalassemia at the molecular genetic level is indicated in pregnancies where both parents are carriers of a double α-gene deletion chromosome. This is accomplished mostly by **Southern analysis** using a **ζ-globin probe**, because this gene is part of the α-globin gene cluster, and such a probe detects **breakpoint fragments** that span the deletions (Kazazian, 1990). In fact, immunological detection of embryonic ζ-globin chains in adults can be used to identify α-thalassemia-1 haplotype due to a large (>17.5 kb) deletion (Chui et al., 1986).

Understandably, an α-globin probe shows no complementarity to the DNA of an α^0-**thalassemia (Bart's hydrops fetalis)** due to the absence of the target α-globin DNA in that gene. **PCR assays** for α-thalassemia use primers for both the β-globin and α-globin sequences simultaneously (Cai et al., 1989; Mullis and Faloona, 1987). Whereas failure of the assay to produce amplification products of the α-globin gene is interpreted as indicative of α-thalassemia, the situation is more complicated in **α-thalassemia trait** cases, in which only a **reduced number** of, rather than complete absence of, α-globin genes is characteristically present (Kazazian, 1990). Because **even a single** α-globin gene per diploid genome will support PCR amplification with complementary primers, the diagnosis of α-thalassemia trait would **require quantitation** of the amplified product.

Another confusing situation arises in the **large Filipino deletion** $(-,-/\alpha^0$-thal-1) that **eliminates the entire α-globin gene complex** (Chang and Kan, 1984), because Southern analysis in such persons may produce patterns indistinguishable from normal (Kazazian, 1990). A protocol combining PCR amplification with **dual restriction enzyme analysis** was developed to address this diagnostic challenge exemplifying the finesse of REA (Lebo et al., 1990). The logic of the analysis was as follows. First, primers specific for an **identical sequence** in both the α_1- and α_2-globin genes, and primers for **the unique sequences** in each α-globin gene region, were used. These allowed the distinction of fetuses with no α-globin genes from the fetuses and parents with only α_1- but not α_2-globin genes as well as from other situations in which both α_2- and α_1-globin genes are present. Because the α^0-thal-1 haplotypes lack both α_1- and α_2-globin sequences, the amplified DNA from a hydrops fetalis (Bart's) fetus would show no α_2- or α_1-globin gene fragments. On the other hand, α^+-thal-1 haplotypes, compatible with life, retain the $3'\alpha_1$-globin gene sequences, but lack the $3'\alpha_2$-globin gene sequences, this difference allowing the differential diagnosis to be made (Lebo et al., 1990).

The preliminary Hb electrophoresis on parents allows the detection of **HbH**, that is, the **presence of a single functional α-globin gene** in one of the parents. Using two different restriction enzymes (*Bgl*II and *Asp*718, an isoschizomer of *Kpn*I), and testing two different ζ-globin probes, the ambiguities in assigning the observed haplotypes were resolved (Lebo et al., 1990). Differentiation between the large deletion haplotype and a normal haplotype by PCR-dual restriction analysis amounted to the identification of the normally present 12-kb 3′ ζ-globin fragment absent in at-risk Filipino fetuses (Lebo et al., 1990).

By using only PCR amplification (Kropp et al., 1989) with allele-specific α-globin primers, a distinction between the homozygous α^+-thal-2 $(-,\alpha/-,\alpha)$ and heterozygous α^0-thal-1 $(-,-/\alpha,\alpha)$ haplotypes can be established. The double-digest restriction enzyme analysis enables unambiguous differentiation between the most common α-thalassemia haplotypes (Lebo et al., 1990). For prenatal diagnosis of α-thalassemia see Section 1.3.4.2.2.

7.1.2.2 β-THALASSEMIA

β-thalassemia syndromes are caused by defective production of **β-globin chains**. Reflecting the relatively simpler complement of the two β-globin genes per diploid genome, compared to the four α-globin genes, essentially two conditions, **trait** and **disease**, are recognized in β-thalassemia, compared to the four α-thalassemia states (Kazazian, 1990; Steinberg, 1988). The presence of one nonfunctional β-globin gene causes mild hematological abnormalities (elevated Hb A_2, reduced MCH and MCV) compatible with good health and a normal life span. In contrast, **β-thalassemia major**, or **Cooley's anemia**, manifested by a spectrum of symptoms of varying clinical severity, occurs when both β-globin genes are defective. These conditions range from mild, accidentally discovered, to severe anemia requiring frequent blood transfusions (Kazazian, 1990; Schwartz et al., 1988).

Abnormalities of **β-globin gene expression** have been largely due to **deletions** or **substitutions** of either single nucleotides or short or long nucleotide fragments within or near the β-globin gene (Kazazian, 1990; Schwartz et al., 1988; Steinberg, 1988). The well-established particular **geographic distribution** of β-thalassemia has been linked to preferential genetic selection for the trait favored by the lower morbidity of affected persons in malaria infections (see Kazazian, 1990). Determining the molecular biology of the β-globin gene cluster has long been a field of intensive research that has contributed significantly to our understanding of the regulation of eukaryotic gene expression in general (Choi and Engel, 1988; Kazazian, 1990; Orkin, 1990; Steinberg, 1988).

Haplotype analysis in various ethnic groups indicated that about 10 haplotypes are commonly seen on normal β-globin gene clusters and that most of these are represented in β-thalassemia chromosomes (Kazazian, 1990). Over **90 point mutations** resulting in β-thalassemia have been recognized so far. Only a few ethnic group-specific alleles are known to account for over 90% of β-thalassemia genes (Kazazian, 1990). **Variations in the clinical phenotype** of β-thalassemia have been traced to mutations in **regulatory sequences**, outside the β-globin gene, that control the **transcriptional efficiency** of the β-globin gene. Other types of mutations in β-thalassemia affect **RNA modification** at the **cap (7-methylguanosine) site**, **RNA cleavage**, **polyadenylation** and **splicing**, and **translation** of the mRNA into globin (reviewed in Kazazian, 1990).

Abnormalities in the γ-globin gene regulation accounting for cases of **hereditary persistence of fetal hemoglobin (HPFH)** have been defined by PCR-ASO analysis to arise from base changes in the A_γ promoter (Gottardi et al., 1990). **Point mutations** in the γ promoter itself affect both γ- and β-globin gene expression. Mutations in the **regulatory sequences** of upstream promoters of the γ-globin gene have been implicated in the HPFH phenotypes (Gilman, 1988).

Prenatal diagnosis of β-thalassemia can be made by PCR amplification of the β-globin gene, followed with ASO hybridization (Kazazian and Boehm, 1988). Due to significant amplification of target sequences, hybridization with ASO can be shortened, with the entire procedure completed within 2 h (Kazazian, 1990). Alternatively, **REA** of the amplified product can detect the presence or absence of a restriction site produced by a mutation. This approach is informative in five of seven common Mediterranean alleles, but only in a few of the β-thalassemia alleles of other groups (Kazazian, 1990). See also Section 1.3.4.2.1.

Sometimes **PCR with direct sequencing** is used to define a mutation, when an initial screening for the predominant type of mutations in a given ethnic group fails to provide the answer. Sequencing of the DNA can establish whether the same mutation that is detected in a parent is present in the fetus. At Johns Hopkins University, the protocol then calls for haplotyping of the parents of the individual carrying the unknown mutation (Kazazian, 1990).

When PCR is combined with **dot-blot** hybridization and/or **REA** of the amplified product, the result can be obtained within a week of fetal sampling. Amplifications using **nonradioactive probes** have been developed (Cai et al., 1989; Saiki et al., 1985, 1986, 1988). This approach with or without direct sequencing has been successfully applied to the characterization of mutations and prenatal diagnosis of β-thalassemia all over the world (e.g., Alfarano et al., 1990; Amselem et al., 1988; Diaz-Chico et al., 1988; Kaplan et al., 1990; Kulozik et al., 1988; Rund et al., 1991; Sutton et al., 1989; Thein et al., 1990).

Other procedures such as **restriction primer extension** of oligonucleotide probes (Gao et al., 1988) and the **amplification refractory mutation system (ARMS)** (Old et al., 1990) have been described in application to β-thalassemia diagnosis. The latter approach, developed for the **identification of multiple mutations**, is based on **specific priming** of PCR as opposed to PCR followed by ASO hybridization (Newton et al., 1989a,b). To diagnose a specific mutation, two oligonucleotide primers are used that are identical in sequence except for the **terminal 3′ nucleotides**. In the absence of a perfect match, that particular nucleotide will not support amplification. The ARMS method using 7 primers was capable of detecting the mutation in all 73 at-risk cases, proving to be as reliable as haplotype analysis (Old et al., 1990). When unknown mutations occur, **RFLP analysis** or **direct DNA sequencing** must be used. Restriction analysis with synthetic oligonucleotide probes complementary to tandem repeats of different HVRs (see Part 1) has been used also to follow engraftment in bone marrow transplantation in homozygous β-thalassemia patients by demonstrating donor-specific fragments in the recipients (Ugozzoli et al., 1989).

By applying sequential PCR amplification with two pairs of primers to testing individual unfertilized human **oocytes** and the first **polar bodies** isolated from them, the genetic defect of the β-globin gene responsible for sickle cell disease and β-thalassemia can be diagnosed (Holding and Monk, 1989; Monk and Holding, 1990). This approach may offer an option of selecting eggs without this defect for fertilization. Some of the ethical issues of dealing with human embryos in preimplantation diagnosis are avoided (Monk and Holding, 1990) by testing the first polar body of an oocyte as opposed to the amplification and analysis of DNA in single blastomeres from cleavage stage embryos (Holding and Monk, 1989). Testing for thalassemic phenotypes caused by **Hb E** mutations common in Southeast Asia has been thoroughly discussed (Anderson and Ranney, 1990).

Chapter 7.2

Coagulopathies

7.2.1 HEMOPHILIA A

Hemophilia A (HA) is caused by defects in **Factor VIII (FVIII)**, accounting for about 85% of inherited coagulation disorders. The FVIII gene on **chromosome Xq28** has been cloned and over 500 hemophilic FVIII genes have been studied so far (for a review, see Furie and Furie, 1990, 1992). The molecular basis of FVIII abnormalities has been traced to **gross deletions insertional mutations**, or **point mutations** in the FVIII gene, resulting either in a **stop codon** (frequently due to the mutation of C to T of the **CpG dimers**), **nonsense mutations**, or in substitutions leading to **missense mutations**. A precise description of the molecular defect in the more than 500 hemophilic FVIII genes analyzed has been gained only in about 70% of these (Furie and Furie, 1990, 1992). The most frequently used approach is **restriction analysis** using *Taq*I. In addition to the gross gene deletions, insertions, and point mutations, one can expect to find, when other techniques are used, **small gene alterations** leading to **alterations in FVIII function**, **stability**, and regulation of its **activity** through interaction with other coagulation factors.

The major emphasis in the practical application of gene level analysis of FVIII dysfunction lies in assessment of **carrier status** and **prenatal diagnosis** (Sadler, 1990; Sommer and Sobell, 1987). **Restriction fragment length polymorphism (RFLP) analysis** using probes **within the FVIII gene** (*Bcl*I, *Xba*I, and *Bgl*I) or the **extragenic probes** (*Bgl*II/DX13 and *Taq*I/St14) has been helpful in evaluating the carrier state. The *Bcl*I, *Xba*I, and *Taq*I/St14 RFLPs allow **100% carrier detection** in some HA families (e.g., Brocker-Vriends et al., 1987; Suehiro et al., 1988). Despite similar phenotypic manifestations, the molecular basis of FVIII deficiency is very heterogeneous. By using **polymerase chain reaction (PCR) with direct sequencing**, the molecular basis of **individual cases** of HA can be established (e.g., Youssoufian et al., 1988), whereas the **PCR-allele-specific oligonucleotide (ASO)** approach is helpful in defining structural alterations in the FVIII gene in large hemophilia **pedigrees** (e.g., Levinson et al., 1990).

New RFLPs at the **DXS115 locus** allowing improved characterization of the carrier status of HA have been studied in a Caucasian population and used for more accurate carrier detection in an HA pedigree (Jedlicka et al., 1990). The **large size** and relative **complexity** of the FVIII gene somewhat limit the diagnostic utility of the PCR-direct sequencing approach, denaturing gel electrophoresis, and mismatch analysis by chemical cleavage in detecting FVIII gene mutations, leaving RFLP analysis as the **method of choice** at the present time for prenatal diagnosis and detection of heterozygous carriers of HA (Jedlicka et al., 1990). A new family of **polymorphic markers** has been described in the FVIII gene (Lalloz et al., 1991). These are the $(CA)_n$ **repeats** in intron 13 of the FVIII gene, which are closely linked to the established RFLPs and show X-linked mendelian inheritance in HA pedigrees. The intron 13 $(CA)_n$ repeat is considered to be the **most informative marker** so far available for FVIII gene analysis that can be performed within a day using **PCR** amplification (Lalloz et al., 1991). Other **dinucleotide repeats** have been noticed during screening of the FVIII gene clones and their applicability for HA testing is under study (Lalloz et al., 1991).

Although convenient and informative, restriction analysis for the detection of FVIII gene mutations following PCR amplification may be misleading because **restriction of short fragments** of the amplified DNA is **less reproducible** and **predictable** than that of total genomic DNA and the restriction is **frequently incomplete** (Sampietro et al., 1990). To overcome this problem, the incorporation of an **internal control** of restriction based on coamplification of a segment of DNA containing a **nonpolymorphic restriction site** is proposed (Sampietro et al., 1990). Following restriction of the PCR-amplified product, the disappearance of the band from the control segment is taken as a confirmation of the completeness of restriction of the sample. A fragment generated by *Bcl*I restriction of the β-globin gene cluster is suggested as a **nonpolymorphic internal control**. The convenience of the proposed internal control is determined by the differential mobility of the digests. The fragments resulting from the polymorphic diagnostic system using *Bcl*I migrate in the gel above or below those of the nonpolymorphic control and the interference of this control is avoided (Sampietro et al., 1990). For prenatal diagnosis of HA see Section 1.3.5.1.

7.2.2 HEMOPHILIA B

Hemophilia B (HB) is caused by a **diminished level** or **dysfunction** of **factor IX (FIX)**, and in at least one third of cases a specific **point mutation** accounts for the defective FIX circulating in the blood. The **FIX gene** on **chromosome Xq26-27.3** has been cloned and various probes have been produced (for a review, see Furie and Furie, 1990, 1992). Whereas a **missense mutation** does not impair either synthesis or plasma-clearing kinetics of FIX, it accounts for the **altered function** of the FIX antigen. **Gross deletions** within the FIX gene lead to a diminished level or even absence of FIX from the circulation, and even **small deletions** are known to produce severe HB phenotypes. Nondeletion mutations in the protease domain of FIX that alter the active site conformation have been described (Ludwig et al., 1992).

Similar to HA, **insertion mutations** are recognized in HB, although the source of the inserted DNA is not known, whereas in HA the inserted DNA is known to be derived from the L1 repetitive elements scattered throughout the genome (Furie and Furie, 1990). Again, the **CpG islands** appear to be **mutational hot spots**, similar to what occurs in the FVIII gene (and many other genes; see Part 1), although no specific regions of instability can be identified in the FIX gene. Because partial and complete gene deletions account for the majority of HB mutations, **restriction analysis** has been helpful in defining these. Newer methods for the identification of point mutations, such as **PCR** amplification with **direct sequencing**, allow fast and precise characterization of FIX gene alterations. This approach was used, in one case, for example, to define the **substitution** of **glutamine for Arg-333**, which led to a dysfunctional FIX with altered substrate or cofactor binding (Tsang et al., 1988).

Intragenic FIX probes are sometimes uninformative (as in Japanese HB patients), but it is hoped that the need for coagulation studies on fetal blood (Yoshioka et al., 1988) will be reduced when responsible HB mutations are defined at the FIX gene level. The direct sequencing of PCR-amplified genomic DNA fragments has allowed the identification of independently occurring mutations in four unrelated Chinese patients with HB; all were found in exon 8 of the FIX gene (Wang et al., 1990). For prenatal diagnosis of HB see Section 1.3.5.2.

7.2.3 VON WILLEBRAND DISEASE

von Willebrand disease (vWD) is the most common inherited bleeding disorder of humans, with prevalence estimates ranging from 0.5 per million (references in Sadler and Davie, 1987) to 1% of the general population (Rodeghiero et al., 1987). More than **20 distinct** clinical and laboratory **subtypes** of vWD are now recognized, presenting essentially in **two forms — a**

quantitative (type II) or a **qualitative (type I)** abnormality of plasma **von Willebrand factor (vWF)** (Ginsburg, 1990). **Type III** is a recessive form of vWD with variable presentation.

Diagnosis by conventional laboratory methods suffers from **low sensitivity** and **specificity** and **high variability**, which ensures the high practical utility of gene-level diagnostic techniques when these become available (Ginsburg, 1990; Triplett, 1990). The vWF gene was cloned in 1985 by several groups (Ginsburg et al., 1985; Lynch et al., 1985; Sadler et al., 1985; Verweij et al., 1985). Its structure has been defined (Mancuso et al., 1989). The molecular biology of vWF and the genetics of vWD have been reviewed in great detail (e.g., Ginsburg, 1990; Lollar, 1990; Meyer et al., 1990; Ruggeri, 1990).

Southern analysis in the majority of vWD patients is normal, although in some cases **large deletions** of the entire *vWF* gene have been noted (Ginsburg, 1990). Although numerous **vWF gene RFLPs** have been defined, the existence of a **pseudogene** may complicate the interpretation in linkage studies. The pseudogene has about 3% divergence in nucleotide sequence from the **authentic gene**; based on the difference between the **lengths of simple repeats** in the gene and pseudogene, the differentiation of these loci by PCR has been developed (Mancuso et al., 1991).

In most vWD patients studied to date, the **molecular defect** appears to reside **within the gene itself**, although extragenic sites may be defined in the future (Ginsburg, 1990). **Point mutations** are known that affect subtle conformational characteristics of vWF that interfere in its interaction with other components of the coagulation system (e.g., Ware et al., 1991). Only a few vWD pedigrees have been studied by RFLP analysis so far. **Defects in the assembly** of large multimeric forms of vWF in the IIA subtype may account for 10 to 15% of cases of vWD (Ginsburg, 1990). With the introduction of PCR-based assays for the detection of point mutations affecting distinct vWF functional domains, marked improvement in the sensitivity and specificity of diagnosis of vWD is expected (Ginsburg, 1990).

A PCR-based method distinguishing **mRNA expression** from the two *vWF* alleles using DNA sequence polymorphisms within exons of the *vWF* gene has been developed for the analysis of type I and type III vWD patients (Nichols et al., 1991). Using peripheral blood platelet RNA, this **comparative DNA** and **RNA PCR-RFLP** approach can be applied not only to the evaluation of vWD, but can also be used to analyze defects at the level of gene expression in other genetic disorders (Nichols et al., 1991). A **pedigree study** using this method was able to identify family members suffering from severe vWD carrying two abnormal alleles as well as asymptomatic individuals with only one abnormal allele. So far the molecular defect accounting for the most common variant of vWD (type I), characterized by a quantitative deficiency of vWF, still remains unclear. For prenatal diagnosis of vWD see Section 1.3.5.3.

7.2.4 OTHER COAGULATION FACTORS

The molecular biology of coagulation factors is being developed at a fast pace and although gene level defects have been recognized in a number of coagulopathies (for a discussion, see e.g., Burk et al., 1991; Gaillard-Sanchez et al., 1990; Langner et al., 1990; Nowak, 1987; Vane et al., 1990; H. Yamada et al., 1990), gene level diagnostics for the majority of these have not yet been developed to the point of introduction into routine clinical laboratories.

Lymphoid and Myeloid Malignancies

Molecular genetic aspects of hematological malignancies have been continuously covered over the past several years in a vast number of reviews, symposia, and monographs (e.g., Cossman, 1990; Lindsten and Pettersson, 1991; Orlic, 1989; Peschle, 1987; Sawyers et al., 1991; Sorg, 1990), and no attempt is made here to give adequate credit to the enormous body of literature on the subject. Only **selected methodological issues** in the application of molecular biological tools to the diagnosis and monitoring of lymphoid and myeloid malignancies will be briefly discussed. This chapter is intended to emphasize the continuity, benefits, and limitations of the molecular biological approach to hematological conditions within the general framework of disease analysis at the gene level addressed in this book.

7.3.1 DNA CONTENT AND CELL CYCLE ANALYSIS

The evaluation of DNA and RNA content either by **flow cytometry (FCM)** or **image analysis (ICM)** complements immunophenotyping and traditional histopathological evaluation of **lymphoma**, monitoring of **minimal residual disease (MRD)** in lymphoma and **acute lymphoblastic leukemia (ALL)**, in some cases offering a superior alternative (e.g., Andreeff 1990a,b; Barlogie et al., 1987; Burke et al., 1990; Hiddemann et al., 1990; Jaffe, 1990; Look et al., 1987; Naeim et al., 1990). Although FCM demonstrates the **heterogeneity** of **acute myeloid leukemia (AML)**, which further helps to define more accurate prognostic characteristics, the measurement of DNA content is subject to significant error due to **variability of sample composition** (Andreeff, 1990a).

Refinements in FCM methods for estimating the **DNA, RNA**, as well as **double-stranded RNA (dsRNA) content** in **leukemias** have better defined the cell cycle kinetics in AML and ALL, but the **practical utility** of these measurements is not always apparent (Andreeff, 1990a; Jackson et al., 1990). Some of the **limitations of DNA FCM** include its inherent **insensitivity to balanced translocations; in addition, nonuniform dye binding** by chromatin DNA due to **stereohindrance effects** accounts in part for the apparent lack of prognostic significance of DNA aneuploidy in AML (Andreeff, 1990a). In contrast, **cytogenetic evaluation** shows at least 10 different types of reproducible chromosomal abnormalities (Naeim et al., 1990). DNA aneuploidy shows **no relation** to **FAB subtype, white blood cell count, S-phase index**, and the **amount of blast forms** in the bone marrow in AML and ALL patients, although cases with aneuploid DNA stemlines tend to have longer remissions (Andreeff, 1990a; Hiddemann et al., 1990). **RNA indices**, in contrast, show marked differences between AML and ALL. The significance of DNA aneuploidy as a prognostic factor in ALL remains debatable (Andreeff, 1990a).

Restriction fragment length polymorphism (RFLP) analysis and *in situ* hybridization **(ISH)** with **chromosome-specific probes** have convincingly demonstrated the clonal origin of leukemic cells defined by monoclonal antibodies or morphology (Andreeff, 1990a). The reproducible, close correlation between **DNA hyperploidy** and expression of the **common ALL antigen (CALLA)** emphasizes the validity of **immunological phenotyping** of ALL for patient management. The complexity of acute leukemias, which frequently display the presence of **multiple lineages**, necessitates the combined use of DNA and RNA FCM analysis together with immunophenotyping (Andreeff, 1990a). Diagnosis of **chronic myeloid leukemia (CML)** and **chronic lymphoblastic leukemia (CLL)** does not present any particular challenge using traditional morphological and immunophenotyping methods. DNA aneuploidy is not detectable in CLL although RNA content is typically low (Andreeff, 1990a).

CML is often characterized by multiple lineages and cytogenetically is identified by the **t(9,22) translocation** in about 90 to 95% of adult cases carrying a favorable prognosis (Naeim et al., 1990). FCM of CML cells shows a characteristic pattern, and their **myeloblastic** transformation is accompanied by a **high RNA index**, whereas **lymphoblastic** transformation is characterized by a **low RNA index**, emphasizing the practical value of FCM analysis in the early detection of transformation events (Andreeff, 1990a).

Immunophenotyping of **non-Hodgkin's lymphomas (NHLs)** by FCM based on **stage-specific** and **lineage-specific** antigenic determinants is an established diagnostic modality that utilizes a vast number of **monoclonal markers** (for a review see, e.g., Andreeff, 1990b). DNA FCM of lymphoma, similar to the situation in leukemia, detects DNA aneuploidy reflecting **numerical cytogenetic abnormalities**. A serious limitation of FCM is that translocations characteristic of specific morphological types of lymphoma, such as **t(8;14)**, **t(8;22)**, and **t(2;8)** in **diffuse small noncleaved cell**, **t(14;18)** in **follicular** lymphomas, **deletion 6(q21)** of **diffuse large lymphoma**, as well as **trisomy 12** and **t(11;14)** in **diffuse small lymphocytic (well-differentiated) lymphomas**, cannot be detected (Andreeff, 1990b). Nevertheless, DNA/RNA FCM in lymphomas, when combined with immunophenotyping, can be used to screen for submicroscopic bone marrow involvement with lymphoma cells.

The determination of **cell cycle kinetics** in lymphoma by DNA/RNA FCM contributes to the prognostic evaluation of patients (see also Cowan et al., 1989). The detection of MRD in lymphoma based on **B cell clonal excess** reveals 1 to 5% of lymphoma cells in BM specimens, comparable in sensitivity to Southern analysis but certainly inferior to PCR (Andreeff, 1990b; Naeim et al., 1990). Only limited data are available on the use of FCM in lymphomas for the evaluation of oncogene expression demonstrating cell cycle variation (c-*myc*, *p21*, *ras*, *bcl*-2, and c-*myb*) (Andreeff, 1990b).

7.3.2 MYELODYSPLASTIC SYNDROME AND MINIMAL RESIDUAL DISEASE

The significant cellular heterogeneity of **myelodysplastic syndrome (MDS)** usually involving two or more hematopoietic cell lineages has been supported by FCM studies (Andreeff, 1990a; List et al., 1990). The fundamental biological characteristic of MDS lies in the **impairment of differentiation capacity** at the level of the multi- or pluripotent stem cells, which accounts for the diverse hematological manifestations (List et al., 1990). **Multiple chromosomal abnormalities** consistently involving **chromosomes 5, 7, and 8** have been correlated with a particular clinical course and overall survival, **DNA hypoploidy** being associated with the poorer survival (see also Neuman et al., 1992). It appears, however, that chromosomal aberrations per se do not initiate MDS but should be viewed as markers of a broader **genomic instability** (List et al., 1990). The consistent cytogenetic abnormalities in MDS appear to correlate with disturbances in the regulation of **hematopoietic growth factors**; however, the utility of assessment of growth factors for patient management remains to be demonstrated in a large series (Yunis, 1987). In patients with therapy-related myelodysplasia or overt acute nonlymphocytic leukemia characteristic chromosomal aberrations were noted on chromosomes 7, 5, 17p, 21 q, and rearrangements of 11q23 (Pedersen-Bjergaard et al., 1990).

None of these could stand as an independent prognostic factor being overshadowed by other aspects of the disease. In patients with AML and MDS the most common chromosomal aberrations appear to be chromosomal loss and deletions involving chromosomes 5 and 7 suggesting the importance of tumor suppressor genes on these chromosomes in the development of malignant myeloid disorders (Neuman et al., 1992).

Minimal residual disease (MRD) evaluation constitutes part of monitoring of leukemia, traditionally performed by DNA ploidy determinations, or by immunological assessment of the persistence of the abnormal phenotype (Andreeff, 1990a). Among other techniques for MRD monitoring are the detection of chromosomal abnormalities caused by **premature chromosome condensation** and the identification of the specific translocations t(9,22), t(8,14), and t(4,11) by the **PCR**.

7.3.2.1 PCR IN MONITORING MRD

PCR amplification with primers specific for the *bcr/abl* rearranged sequences can be used to detect the presence of cells positive for the **Philadelphia (Ph) chromosome** in CML patients following bone marrow transplantation (BMT). Amplification is targeted for the transcripts of the *bcr/abl* **hybrid sequence** as the most sensitive measure of persistent presence, and, therefore, expression of malignant cells. In patients with no expression of the *bcr/abl* rearrangement detectable by **reverse transcription** of total RNA followed **with PCR amplification**, the complete remission could be observed 5 to 7 years following BMT (Morgan et al., 1989). It is suggested that multiple PCR analyses should be performed in individual patients following BMT, because, despite an initial negative result, some CML patients have been shown to become PCR positive and eventually demonstrate clinical relapse (Bartram et al., 1990). The presence of *bcr-abl* **mRNA**, identifiable by PCR in Ph^1-ALL patients, predicts hematological relapse apparently due to the high proliferative potential of the Ph^1 clone (Maurer et al., 1990; Miyamura et al., 1990).

The discrepancy emerging from the PCR assessment of MRD following BMT in CML patients may reflect differences in therapeutic protocols, sampling variability, or other technical aspects of the evaluation (Bartram et al., 1989; Gabert et al., 1989; Lange et al., 1989; Morgan et al., 1989; Roth et al., 1989; Sawyers et al., 1990). PCR was able to detect more consistently, however, the presence of residual leukemic cells in CML patients treated with interferon-α (Bartram et al., 1990).

PCR analysis for the *bcr* gene can also be used in **ALL**. The majority of ALL cases show **T cell receptor (*TCR*)-δ gene rearrangements**. Furthermore, **specific immunogenotypes** can be recognized in the area of the **V-J junction**, **imprecise VDJ joining** and extensive **insertion of N-region nucleotides** (Bartram et al., 1990; Yokota et al., 1991). One approach uses, therefore, a limited set of oligomer primers and obviates the need for cloning and sequence analysis of the amplified products. Children in complete clinical, hematological, as well as Southern blot-documented remission were tested by PCR analysis using **patient-specific *TCR*-δ probes**, and 9 out of 11 patients were found to have residual leukemic cells (Bartram et al., 1990). Quantitative estimates of the positive cells showed the presence of 10^{-3} to 10^{-5} neoplastic cells in the bone marrow in PCR-positive children. In contrast, peripheral blood specimens were positive in only 2 of the 11 cases. It should be noted, however, that some of the residual leukemic cells, although testing positive in the PCR assay, may in fact, have lost their capacity to proliferate and cause clinical relapse (Bartram et al., 1990).

Furthermore, possible clonal variations of the immunogenotype that occur in ALL patients may escape detection by PCR using **clonospecific probes** (Bartram et al., 1990; Yokota et al., 1991). Detection of MRD by PCR during complete clinical remission precedes clinical relapse in some cases, as shown in a series of 55 bone marrow and peripheral blood samples from 27 ALL patients (Yokota et al., 1991; see also Campana et al., 1990). Residual leukemic cells can be detected 6 to 18 months following successful induction of remission, emphasizing the need for maintenance therapy (M. Yamada et al., 1990). The **limited value of a single PCR evaluation** that may fail to detect MRD due to the focal nature of the disease and absence of leukemic cells in a single BM sample should be fully recognized (M. Yamada et al., 1990).

Caution is advocated in extrapolating the (so far) limited experience in PCR monitoring of MRD to the development of patient management strategies until large-scale studies firmly establish the clinical significance of the detection of residual leukemic cells at the level of PCR amplification (Yokota et al., 1991). It is suggested that this modality be used in conjunction with clinical, morphological, and immunological evaluation of patients. In fact, a comparative study by **double color immunofluorescence analysis** using **cCD3/TdT markers** and PCR amplification of rearranged TCR-δ genes reaffirmed the absence of false-positive results in predicting relapse by this technique. It may fail, however, in 25 to 30% of cases. Similar data for PCR assay are still being gathered (Campana et al., 1990). The importance of using different techniques for monitoring MRD cannot be overemphasized.

A different approach to PCR detection of residual leukemic cells is based on the identification of the **complementarity-determining region III (CDRIII)** sequences (M. Yamada et al., 1990). In this assay, the primers are constructed homologous to consensus sequences in the **variable (V_H)** and **joining (J_H)** segments that flank the **intervening diversity (D)** segment in the rearranged heavy-chain immunoglobulin locus (M. Yamada et al., 1990). Amplified CDRIII segments specific for leukemic cells are sequenced and diagnostic oligonucleotide probes are constructed that do not cross-hybridize with the CDRIII sequences of normal B lymphocytes.

So far, **leukemia-specific chromosomal aberrations** have been characterized at the molecular level only in a limited number of patients, and the development of these **novel B lineage probes** complements the PCR assays developed earlier for lymphoblastic leukemias (Bartram et al., 1990; D'Auriol et al., 1989; Hansen-Hagge et al., 1989; M. Yamada et al., 1989; Yokota et al., 1991). The sensitivity of these PCR protocols is comparable to that using T cell-specific rearrangements, being about 10^{-5} cells (M. Yamada et al., 1990). **Quantification** of leukemic cells in BM was done by multiplying the number of PCR-positive cells by the ratio of lymphocytes estimated in the specific BM sample by morphological analysis. The assumption was made that all the enumerated lymphoid cells were of the B lineage, which probably overestimated the amount of residual disease (M. Yamada et al., 1990).

The **CDRIII amplification method** with **phage library quantitation** of the positive sequences detects 1 leukemic cell among 100,000 normal cells, representing 2 to 3 orders of magnitude improvement over **TdT assay**, **cytofluorimetric** analysis with lineage-specific monoclonal antibodies, and **Southern analysis** of heavy-chain immunoglobulin rearrangements (M. Yamada et al., 1990). The failure of the amplified CDRIII sequences to hybridize with the leukemia-specific probes in some cases was considered to reflect the presence of CDRIII sequences in normal B cells, or the existence of different or mutated sequences derived from multiple heavy-chain immunoglobulin rearrangements within the leukemic population. It appears that quantitation of leukemic cells by PCR allows the prediction of a clinical relapse and justifies its use in managing leukemic patients (M. Yamada et al., 1990).

It has been suggested that the presence of **more than two Ig heavy-chain rearrangements** detectable also by Southern analysis may have led to an underestimation of the true amount of MRD by the CDRIII-PCR technique (Beishuizen et al., 1991). However, because the V_H-V_H **gene replacement** appears to be the most common event leading to **multiple rearrangements** at the Ig heavy-chain locus, and accounts for the majority of clonal evolution effects, **no false-negative** results are produced when using the CDRIII-specific diagnostic probes (Rovera et al., 1991).

The other **potential limitation** of the CDRIII-PCR protocol is related to the specific **primer sequence homology** to a V_H region and may lead to **amplification of normal B cell sequences** (Hakim et al., 1991). It does not appear to present a problem because the CDRIII-PCR assay showed a high success rate (93%) using the V_H consensus primers, the more so that in a series of over 100 CDRIII sequences no two identical sequences could be detected (Rovera et al., 1991). The specificity of PCR amplification of the leukemia-specific sequences apparently can be improved by using primers completely homologous to the **subgroup 5 variable-region** segments (Hakim et al., 1991).

In conclusion, monitoring of MRD by PCR amplification offers a highly sensitive technique capable of detecting and **predicting the clinical behavior** of residual leukemic cells. Fine aspects of primer sequence selection still require experimental and clinical **verification in larger series** of patients. Used in conjunction with other evaluation methods, PCR quantitation of MRD in leukemia will become a routine monitoring tool significantly improving patient management.

7.3.3 CYTOGENETIC ANALYSIS

7.3.3.1 MYELOID MALIGNANCIES

The well-documented **t(9;22) (q34.1;q11.21) translocation** found in 90 to 95% of adult CML (Ph-positive) cases is associated with rearrangement of c-*abl* and formation of an ***abl-bcr* fusion gene** that gives rise to the aberrant **8.5-kb chimeric *bcr-abl* mRNA** and the **210-kDa Bcr-Abl fusion protein**, rather than **145-kDa Abl product** (Bernstein, 1988; Dube et al., 1990; Naeim et al., 1990). The presence of t(9;22) and the *bcr* rearrangement is associated with reportedly **more favorable prognosis** of CML patients. Numerous contradictory reports, however, fail to demonstrate a significantly different clinical course, as well as question the diagnosis of CML and ALL in such patients (e.g., Bernstein, 1988; Botti and Verma, 1990; Dreazen et al., 1988; Dube et al., 1990; Epner and Koeffler, 1990; Gorska-Flipot et al., 1990; Leibowitz and Young, 1989; Tien et al., 1990).

The translocation of *abl* into *bcr* occurs in different introns in CML and ALL, so that a 7.0-kb chimeric mRNA and a 190-kDa fusion protein is produced in Ph^1-positive patients with ALL (see Dreazen et al., 1988; Epner and Koeffler, 1990). The exact position of the *abl* insertion may vary (e.g., Dreazen et al., 1988). Although 100% of Ph-positive patients with CML have a *bcr/abl* rearrangement, **karyotyping** alone used for the diagnosis of CML and in monitoring treated patients appears to be **inferior to Southern analysis** combined with **dilution studies** and **densitometric scanning of autoradiographs**. The latter approach allows the **quantitation** in an individual patient of the dynamics in the relative number of cells affected by the *bcr/abl* recombination events (Ayscue et al., 1990).

Further improvement can be achieved using **PCR** (see below). In addition to the t(9;22) translocation, other chromosomal aberrations such as **duplication of the Ph chromosome**, **trisomy 8, 19, and 21, isochromosome (17q), t(15,17), loss of the Y chromosome**, and many other chromosomal aberrations can be detected in myeloid malignancies by cytogenetic techniques (Bernstein, 1988; Naeim et al., 1990). An updated list of chromosomal abnormalities in different forms of myeloid malignancies has been published (Dube et al., 1990; Naeim et al., 1990). In the course of progression of MDS, **sequential cytogenetic studies** revealed a karyotypic evolution paralleling the clinical course of the patients, which supports the notion that **unstable karyotype** can be associated with a poor prognosis (Suciu et al., 1990).

7.3.3.2 LYMPHOID MALIGNANCIES

Chromosomal translocations involving **antigen receptor complexes** of **T cells (14q11)** or **B cells (14q32)** are seen in two thirds of **NHL** and one third of **lymphocytic leukemia** cases (Ersboll and Schultz, 1989; Naeim et al., 1990; Waldmann, 1987). A combination of **IgH** and ***TCR* rearrangements** found in some NHL and lymphocytic leukemia cases reflects their origin at the early (pre-B and pre-T) stage of lymphoid differentiation. A characteristic cytogenetic aberration in **Burkitt's lymphoma, t(8,14),** involves the juxtaposition of c-*myc* from **8q24** to the **joining** or **switch region** of the **immunoglobin heavy chain**, resulting in dysregulation of c-*myc* gene expression (e.g., Butturini and Gale, 1988a,b; Lenoir and Bornkamm, 1987; Nowell and Croce, 1990; Schwartz and Witte, 1988). Likewise, c-*myc* is rearranged with *TCR-α* in T cell lymphomas.

Interestingly, whereas the presence of the translocation **t(1;19)** in **ALL** patients with the **pre-B immunophenotype** treated from 1979 to 1984 was definitely associated with an adverse prognosis, in a more recent and extensive study **neither this nor other translocations were associated with a worse outcome** (Raimondi et al., 1990). Although no conclusions regarding long-term disease control can be drawn from the limited number of pre-B cases and relatively short observation period (an average of 3 years), the adverse prognostic implications of the translocation t(1;19) and other chromosomal translocations appear to be offset by more effective chemotherapy. A larger series of pre-B cases of ALL emphasized the worse prognosis for **non-T, non-B cell childhood ALL** with chromosomal translocations, the t(1;19) translocation being largely responsible for the poor prognosis of the pre-B subgroup (Crist et al., 1990).

In contrast, **T cell childhood ALL** appears to have a nonrandom occurrence, frequency, and degree of immunophenotype association with **t(1;14) (p34;q11)** (Carroll et al., 1990). It is possible that more sensitive techniques such as the PCR can detect residual leukemic cells, offer better monitoring, and give more definitive predictive tools than even high-resolution cytogenetic analysis. In patients with **B cell CLL**, chromosomal analysis offers prognostic indicators of poorer survival when **chromosome 14q** rather than **chromosome 13q** is present as well as when **trisomy 12** and a higher percentage of cells in metaphase with chromosomal abnormalities are detected (Juliusson et al., 1990). The clinical course of patients who were found to acquire aberrations of the short arm of chromosome 17 was rapidly progressive with a short survival (Rodriquez et al., 1991). Allelic loss of the p53 gene may be related to this chromosomal abnormality in NHL patients and evaluation for the p53 gene may define patients with a poor prognosis.

The most common **follicular-type lymphoma** constitutes up to 40% of NHLs and the majority of patients have a **t(14,18) (q32;q21)** that is accompanied by a rearrangement of the *bcl*-2 **oncogene** from **18q21** to the IgH chain location on **chromosome 14q32** (Naeim et al., 1990; Nowell and Croce, 1990). A number of chromosomal aberrations recognized by cytogenetic techniques have been described in lymphomas (for a listing, see Naeim et al., 1990), and the best laboratory approach is to combine cytogenetic analysis with other modalities such as morphological, immunological, and molecular biological evaluation of specific gene rearrangements.

7.3.3.3 HODGKIN'S DISEASE

Although cytogenetic studies in Hodgkin's disease (HD) are few, the emerging impression is that **chromosomes 14q, 11q,** and **6q** are each involved in slightly over 30% of cases, and **8q** in 18% (Cabanillas, 1988). The **most frequent breakpoint** is at **14q32** — the site of the IgH chain genes. Analysis of 18 cases of HD with immunoglobulin and *TCR* gene probes revealed only germ lines (no rearrangement) with the immunoglobulin gene probes and in the majority of cases (15 of 18) with the *TCR* gene probes (Roth et al., 1988). Although in one case a rearrangement of the *TCR*-β gene was observed, the **weight of the data does not support the frequent occurrence of monoclonal rearrangements of immunoglobulin or *TCR* genes in HD**. Similarly, in another study, only 2 of 23 HD patients had immunoglobulin rearrangements and none showed *TCR*-β gene rearrangements (Bernard et al., 1990). Furthermore, no correlation could be noted between the presence of rearranged bands and the number of **Reed-Sternberg (RS)** cells. The cytogenetic abnormality that sets HD apart from other lymphomas appears to be that involving **chromosome 11q23** (Cabanillas, 1988).

7.3.4 *IN SITU* HYBRIDIZATION

The diagnosis of conditions in which the characteristic pathognomonic alteration is traceable to only a few cells among apparently unaffected cell populations, such as RS cells in HD can be accomplished by ISH analysis, which is capable of demonstrating **individual cell involvement** with a pathological process. In fact, ISH allows the detection of Epstein-Barr virus (EBV) in lymphocytes of infectious mononucleosis and Burkitt's lymphoma patients, and of RS cells in HD and hairy cell leukemia (Peiper et al., 1990; Strickler and Copenhaver, 1990; Weiss et al., 1987a,b, 1989; Wolf et al., 1989).

The long-recognized capacity of EBV to transform and/or immortalize T cells (e.g., Ishihara et al., 1989; Jones et al., 1988; Stevenson et al., 1986) and B cells (e.g., Klein and Raab-Traub, 1987; Wakasugi et al., 1987) has been ascribed to the activation of **EBV latent membrane protein (LMP)** (Cohen et al., 1989; Dawson et al., 1990; Gregory et al., 1991). Attempts have been made to develop more sensitive and specific immunological methods, surpassing the conventional immunofluorescence technique, for the diagnosis of EBV in clinical laboratories (Nonoyama et al., 1988; Pearson, 1988), however, inherent **antigenic cross-reactivities** between EBV, HSV1 and HSV2, and cytomegalovirus (CMV) plague even the better monoclonal antibody methods (Balachandran et al., 1987). The more direct gene level detection of EBV in hematological diseases is, therefore, highly desirable.

ISH for EBV DNA has repeatedly demonstrated the viral nucleic acids in RS cells in HD (Staal et al., 1989; Weiss et al., 1987a, 1989; T. C. Wu et al., 1990), lymphoproliferative conditions associated with heart-lung transplants (Randhawa et al., 1989), hairy leukoplakia (De Souza et al., 1990), B and T cell lymphomas, as well as benign lymphadenitis (Ohshima et al., 1990) and HIV-associated persistent generalized lymphadenopathy (Uccini et al., 1989). ISH EBV probes constructed with terminally biotin-labeled synthetic oligonucleotides were able to demonstrate the viral genomes in lymphocytes of a patient with acute, fatal, disseminated EBV infection (Montone et al., 1990). EBV gene expression can be visualized in interphase nuclei of a latently infected human lymphoma cell line (Lawrence et al., 1989).

ISH combined with confocal microscopy is capable of defining a **spatial topography** of specific chromosomal regions and genes in **hematopoietic cells** (van Dekken et al., 1990a,b). FCM analysis of **filter ISH** stained hemopoietic cells can be used for evaluation of **β globin expression** (Bayer and Bauman, 1990). Specific molecular probes derived from particular types of tumors can be used in ISH for diagnostic purposes and differentiation of characteristic gene expression (e.g., c-*myc*) in morphologically similar disorders (Seibel and Kirsch, 1989).

ISH has also been helpful in delineating fine events in **megakaryocyte gene expression** (Gewirtz, 1990). Detecting numerical chromosomal abnormalities (**monosomy 9** or **trisomy 9**) in **hematological neoplasias** by interphase cytogenetic analysis with ISH in BM or peripheral blood cells appears to be a clinically useful method (Anastasi et al., 1990). ISH was able to demonstrate chromosomal aberrations and characterize specific gene rearrangements in HD-derived T cells, using DNA probes for TCRA, TCRB, MET oncogene and rRNA (Fonatsch et al., 1990). With the help of specific probes for the *bcr/abl* fusion gene, the ISH technique and Southern analysis help to elucidate the **variant translocations** in CML (Verma et al., 1989).

Although the application of ISH to diagnostic pathology is growing, the utility of some specific probes in clinical diagnostic laboratories is not quite clear. An example of a useful probe is one for the **myeloperoxidase gene**, which identifies the characteristic **17q translocation breakpoint** in **acute promyelocytic leukemia** (van Tuinen et al., 1987). Later studies pointed to the specificity of translocation t(15;17) associated with **retinoic acid receptor α gene truncation**, the latter apparently playing a crucial role in the leukemogenesis of acute promyelocytic leukemia (Alcalay et al., 1991; de The et al., 1990). This gene can be specifically targeted for treatment modalities.

ISH using **antisense RNA/mRNA hybridization** is clinically useful in detecting low levels of expression of the c-*abl* oncogene in NHL (Greil et al., 1989). The optimized method allows a semiquantitative estimate of v-*abl*, bcr/*abl*, and c-*abl* gene expression in fixed tissue, correlation with morphology, and the detection of weak expression in only a fraction of cells in the tumor tissue associated with the predicted relapse. It is suggested that evaluations by anti-oncoprotein antibodies be carried out in parallel with ISH (Greil et al., 1989).

EBV can be easily identified by **PCR** under the same conditions in which ISH is informative as well as when ISH fails to detect the virus. Some of the disorders already tested for EBV by PCR include infectious mononucleosis (spleen, lung, and lymph nodes), immunoproliferative disorders of immunosuppressed patients, as well as nasopharyngeal carcinoma (Gopal et al., 1990; Peiper et al., 1990; Telenti et al., 1990), lymphomatoid granulomatosis (Katzenstein and Peiper, 1990), angioimmunoblastic lymphadenopathy (Knecht et al., 1990), and lymphoepithelial carcinoma of the stomach (Burke et al., 1990).

7.3.5 GENE REARRANGEMENTS

Molecular biological characterization of antigen receptor gene rearrangements in the ontogeny of T and B cells provided a foundation for the use of defined gene probes in clinical evaluation of lymphomas and leukemias. This field has evolved over the last 6 to 7 years into an established branch of practical molecular diagnostics that finds increasing application in routine practice. Only major methodological aspects of the evaluation of gene rearrangements will be outlined here.

The traditional approach to the analysis of gene rearrangements has been **Southern blotting**, which is being progressively complemented, and in some cases may even be replaced, with **PCR**-based techniques (see below). Briefly, the molecular events in the maturation of T and B lymphocytes, but not cells of nonhematopoietic lineages, lead to rearrangements of immuno-globulin genes (B cells) and T cell receptor (*TCR*) genes that differ from the nonrearranged, germ-line configuration of these genes found in stem cells (e.g., Burns, 1989; Cossman, 1990; Harrington, 1990; Lindsten and Pettersson, 1991; Orlic, 1989; Sklar, 1990; Sorg, 1990). The predominant proliferation of an individual cell carrying a **unique rearrangement** of its antigen receptor genes gives rise to a clone that can be identified and distinguished from cells with nonrearranged configuration. This recognition is based on **analysis of restriction patterns** of DNA from respective cell populations digested with the **same restriction enzyme**. The presence of fragments (bands) differing from the predominant **germ-line band** pattern identifies clones of cells with rearrangements.

In Southern analysis clonal population can be identified only if they constitute **no less than 1%** of the total cells (Burns, 1989; Sklar, 1990). Because neoplastic hematological disorders are usually accompanied by clonal proliferation of cells, Southern analysis of gene rearrangements is used essentially in three situations: to assess the **clonality** (monoclonality vs. oligo- or polyclonality), to determine **cell lineage** (B vs. T vs. nonhematopoietic differentiation), and to evaluate the existence of the **multiclonal** or **multilineage origin** of tumor cells in some hematopoietic disorders (Sklar, 1990). Analysis of gene rearrangements by Southern blotting can be performed on a small, even poorly preserved tissue sample, without diagnostic histology (0.1 to 1.0 mm^3 of wet tissue). Rearrangements can be detected in the **absence** of the target **gene expression** and, certainly, in the absence of **clonal antigenic markers** detectable by immunophenotyping (Sklar, 1990).

Judicial use of rearrangement analysis can be a helpful adjunct in diagnostically challenging cases, provided **inherent limitations** of Southern blot analysis are recognized. In evaluating cases with possible **multilineage origins** specific immunoglobulin or *TCR* probes may suggest the tumor cell origin at an early stage, **prior to gene rearrangement** when malignant transformation took place (Sklar, 1990). Furthermore, **multiclonal derivation** can be estab-lished on the basis of the specific pattern of rearranged bands assignable to different clones of lymphocytes. The possibility of **somatic mutations** in **follicular B cells** may complicate the interpretation of rearrangements by restriction patterns due to the appearance or elimination of expected restriction sites characteristic of **nonmutated cells**. Because *TCR* **genes are not known to undergo somatic mutation** at a frequency affecting interpretation, the evidence of multiclonality based on *TCR* gene rearrangements is **less prone to misinterpretation**, than immunoglobulin gene rearrangements in B cells (Sklar, 1990).

Evidence of **monoclonality per se does not imply malignancy**, because a number of clinically **benign conditions** — lymphomatoid papulosis, Sjögren's disease, pseudolymphoma of the orbit and skin, systemic Castleman's disease, pityriasis lichenoides et varioliformis acuta (PLEVA, or Mucha-Habermann disease), angioimmunoblastic lymphadenopathy — may have clonal populations of T and/or B cells, and **precise distinction between benignity and malignancy in these conditions is not straightforward** (reviewed in Burns, 1989; Sklar, 1990).

DNA analysis has been found to help in resolving uncertainties arising in the diagnosis of **peripheral T cell lymphomas (PTCLs)** (Winberg et al., 1988). Particular difficulties were encountered in PTCL with an unremarkable immunophenotype, those lacking expression of the usual pan-T cell-associated antigen, and in differentiation between PTCL and pseudo-PTCL. Again, DNA hybridization helped in differentiating tumors with polymorphous large-cell proliferations, including PTCL, SIg-negative large-cell lymphoma of the B cell type, and lymphocyte depletion-type HD (Winberg et al., 1988).

The major **interpretive problems** in Southern analysis of gene rearrangements have been grouped essentially into two categories (Burns, 1989). Situations may arise in which the bands present in restriction patterns do not reflect rearrangements that occur when **genetic polymorphisms** of immunoglobulin or *TCR* genes exist, which can be distinguished by **digesting nonlymphoid DNA** and **comparing the patterns**. Other sources of irrelevant bands such as cross-hybridization with pseudogenes, bacterial plasmid contamination, and incomplete digestion should be excluded when analyzing Southern blots by incorporation of **appropriate controls** (Burns, 1989).

The **absence of expected rearranged bands** may be due to either **incorrect diagnosis**, the presence of **target cells below 1%** of the total cell population, or **comigration** of the rearranged band with the germ-line band. Again, it is also possible that in some cases, at least of T cell lineage, despite the overtly malignant characteristics the cells **fail to reveal clonal gene rearrangements by Southern analysis** (Sklar, 1990). Furthermore, assignment of lineage only on the basis of gene rearrangement evidence may be ambiguous because immunoglobulin gene rearrangement can sometimes occur in **T cell-derived tissues** and even in **myeloid leukemias** (Sklar, 1990).

A large, multiinstitutional study of the use of gene rearrangements in the diagnosis of lymphomas and leukemias demonstrated a high correlation with conventional immunotyping and morphological diagnosis (Cossman et al., 1991). To ensure that reproducible, useful information is obtained by gene rearrangement analysis, a **set of guidelines** has been proposed based on that study. Among these is the recommendation that bands on Southern blots in order to be considered rearranged must **equal the intensity of the 3% positive sensitivity control** (Cossman et al., 1991).

Next, the use of **three restriction enzymes** (*Eco*RI, *Hin*dIII, and *Bam*HI) is proposed in every case to **exclude partial digestion** and **cross-hybridization**. It is recommended that a rearrangement be established with at least two of the three enzymes or that two rearrangements be observed with one single enzyme (two rearranged alleles). Furthermore, any J_k **rearrangement** was regarded as **evidence of B cell lineage** (it is rarely seen in T cell neoplasms), whereas J_H and C_T β **rearrangements** can be observed in both common (pre-B) and T cell acute lymphoblastic leukemia and lymphoblastic lymphoma (Cossman et al., 1991).

To achieve **better coordination** between conventional morphological and immunohisto-chemical studies and DNA hybridization, sections prepared as **fresh-frozen** preparations can be used to **extract DNA** for subsequent **Southern analysis** (A. M. Wu et al., 1990a). The quality and quantity of DNA recorded from such sections were affected by sample size and artifacts produced by ice crystals. In addition to ensuring **greater sharpness of bands**, the alkaline transfer method saved time compared to the standard Southern high salt transfer method. Furthermore, hybridization could be performed on **dried gels** instead of on filters, the main benefit being the elimination of two overnight transfers and overnight prehybridization (A. M. Wu et al., 1990a).

The **yields** of high molecular weight DNA suitable for Southern analysis of antigen receptor gene rearrangements from frozen and ethanol-fixed, paraffin-embedded tissues were comparable (A. M. Wu et al., 1990b). The DNA extracted from **formalin-fixed, paraffin-embedded tissue** was more **degraded** and produced variable results in Southern analysis, with less than half of the results being interpretable. Also, DNA extracted from **ethanol-fixed tissue** stored for 2 years gave yields and results on restriction analysis similar to those of freshly fixed tissue (A. M. Wu et al., 1990b).

The **length of formalin fixation** apparently is the contributing factor to progressive DNA degradation, because under optimal fixing conditions Southern blots were essentially similar in both fixed and unfixed tissue, except for somewhat reduced electrophoretic mobility of DNA from formalin-fixed tissue (Dubeau et al., 1988). Suboptimally fixed tissue apparently led to autolysis, resulting in degraded DNA. Other investigators have also demonstrated the possibility of studying gene rearrangements in formalin-fixed tissue (Goelz et al., 1985; Warford et al., 1988).

In addition to the possibility of using routine fresh-frozen and ethanol- or formalin-fixed tissue for the evaluation of **antigen receptor gene rearrangements** in clinical laboratories, the demonstration of acceptability of **biotinylated DNA probes** for these studies enables a routine immunohistochemistry laboratory to perform gene rearrangement studies in (Samoszuk et al., 1990; Walts et al., 1990).

7.3.6 ANALYSIS OF GENE REARRANGEMENTS AND CHROMOSOME TRANSLOCATIONS BY PCR

The sensitivity of PCR amplification far exceeds (by four orders of magnitude) that of Southern analysis, the latter typically showing a 1% level of sensitivity. At the junction of rearranged gene segments **unique sequences** are produced, which represent clonal markers **more specific than positions of bands on Southern blots** (Sklar, 1990). Oligonucleotide primers can be constructed to complement sequences spanning the junction region, and those containing nonspecific DNA sequences flanking the junctional region. Using such primers, a specific amplification of the clonal *TCR-γ* gene rearrangement can be obtained in bone marrow cells with a sensitivity of one neoplastic cell out of 10^5 to 10^6 cells (Sklar, 1990). Although by far exceeding Southern analysis, PCR may also be prone to **sampling artifacts** and the practicality of its use in monitoring MRD will have to be established in clinical studies. By combining PCR amplification with the discriminatory analysis of the amplified product by **denaturing gradient gel electrophoresis**, the clonality of lymphoid cells encountered at low levels in skin biopsies from early **mycosis fungoides** can be demonstrated by a rapid, nonradioactive approach (Sklar, 1990; Bourguin et al., 1990).

PCR has been applied to the detection of **residual *bcr/abl* translocation** in CML (Dobrovic et al., 1988; Hughes and Goldman, 1990; Kawasaki et al., 1988; Maurer et al., 1990), CML following BMT (Gabert et al., 1989; Roth et al., 1992), and in Ph-positive ALL (Hooberman et al., 1989; Maurer et al., 1991; Saglio et al., 1991). The **translocation t(14;18)** at both the *mbr* (Cotter et al., 1990a,b; Crescenzi et al., 1988; Lee et al., 1987) and *mcr* (Cotter et al., 1990a; Ngan et al., 1988) loci can be detected in lymphoma tissue and even in the peripheral blood (Cotter, 1990). The **precise size of the amplified product** can be used as a clonality marker in addition to establishing the unique clonal nature of the "N", **random DNA insertion segment** in the translocation (Cotter, 1990).

The *bcr-abl* **fusion sequences** can be rapidly and reliably detected by the **hybridization protection assay (HPA)** combined with PCR (Dhingra et al., 1991). HPA uses **acridinium ester-labeled oligonucleotides** (Gen-Probe, Inc., San Diego, CA) complementary to the *bcr-abl* junction sequences. One set of probes detects amplified transcripts encoding **p210**$^{bcr-abl}$, whereas the other set detects **p190**$^{bcr-abl}$. The resulting hybrids are detected by **chemoluminescence** following appropriate incubation and washing steps. The **differential hydrolysis** maximizes the chemoluminescence from hybridized probe, while reducing the signal from unhybridized probe. HPA is used as an **alternative to Southern blotting** for the characterization of the amplified product. HPA is performed in solution, uses nonradioactive reagents, and requires no physical separation of the free probe from the hybridized probe (Dhingra et al., 1991). In addition to markedly reducing the background signal, this method displays a sensitivity equal to that obtained with radiolabeled probes. Because the presence of just one cell in a million cells is sufficient to give a positive result, Ph-positive cells can be rapidly (within 1 day) detected when monitoring MRD.

A PCR-based identification of **rare neoplastic B lymphocytes** in a population of normal cells has been described using **consensus primers** of the **chain-determining region 3 (*CDR3*)** of the variable and joining regions of the immunoglobulin heavy-chain gene (Brisco et al., 1990). A novel approach to identifying **leukemia-associated translocations**, such as t(6;11), uses PCR amplification of specific target sequences from small numbers of **microdissected chromosome fragments** (Cotter et al., 1991). These translocations are mapped in relation to the positions of known genes such as *CD3D, THY1,* and *ETS1*. The exquisite sensitivity of PCR allowed the detection of the **lymphoma-specific t(14;18) translocation** in occult cases of follicular small cleaved cell lymphoma with a relapse in the central nervous system (Shibata et al., 1990).

7.3.7 EVALUATION OF ONCOGENES IN HEMATOPOIETIC NEOPLASIA

Involvement of oncogenes in hematopoietic neoplasias has long been established although the exact roles given to individual oncogenes in various clinically and morphologically definable disorders have not been fully identified yet. Nevertheless, it is apparent that **multiple** oncogenes and growth factors cooperate in the course of malignant transformation of hematopoietic cells, and the **relative prominence** of a given set of oncogenes and growth factors can be related to the clinically relevant behavior of a malignancy (e.g., Saglio et al., 1986; Schwartz and Witte, 1988). Most of the evidence of oncogene involvement in hematological malignancies has been gathered in experimental systems, and extrapolations to clinical situations in many cases await further confirmation. The evaluation of some oncogenes (e.g., *bcr/abl*) was discussed above. Detailed discussions of oncogene involvement in hematological neoplasias have been presented (e.g., Arlinghaus and Kloetzer, 1987; Boehm et al., 1987; Butturini and Gale, 1988a,b; Cossman, 1990; De Klein, 1987; Dooley et al., 1988; Ihle and Askew, 1989; Jucker et al., 1990; Ngan and Berinstein, 1990; Olsen et al., 1988; Orlic, 1989; Perlmutter et al., 1988; Peschle, 1987; Pierce, 1989; Rosson and Reddy, 1988; Saglio et al., 1986; Schwartz and Witte, 1988; Sheridan and Reis, 1990; Sorg, 1990; Westbrook, 1987). A summary of major findings follows.

7.3.7.1 c-*abl*

Oncogene activation is most frequently associated with chromosomal translocations leading to the **juxtaposition** of a modified protooncogene and one of the antigen receptor genes (Ngan and Berinstein, 1990). The prototypical c-*abl* oncogene forming a **fused** *abl/bcr* gene in

t(9;22) translocation is diagnosed and monitored in Philadelphia chromosome-associated malignancies (Rosson and Reddy, 1988; also see above). The potential of v-*abl* alone to induce a process resembling a myeloproliferative disease in humans originating from lymphoid-myeloid pluripotent hematopoietic stem cells has been convincingly demonstrated, although the involvement of additional events in this process is also suggested (Chung et al., 1991).

Measuring the *abl* **mRNA** by **Northern analysis** and the respective protein by **anti-Abl antibody** is helpful in CLL, because the majority of *abl*-positive cases display a more aggressive course of disease (Arlinghaus and Kloetzer, 1987; Greil et al., 1988; Westbrook, 1987).

7.3.7.2 bcl-2

Translocation **t(14;18)** leads to activation of the *bcl*-2 (B cell leukemia-lymphoma) protooncogene, the product of which is present in transformed cells harboring this translocation (Ngan et al., 1988; Tsujimoto et al., 1985, 1987; Weiss et al., 1987a,b). **Chromosome 18 DNA probes** identify the *bcl*-2 gene in virtually all follicular neoplasms and in about a third of diffuse large-cell lymphomas (Aisenberg et al., 1988; Ngan et al., 1988; Weiss et al., 1987b). The Bcl-2 protein can be easily detected in frozen sections of lymphoid tissues and tumors using monoclonal antibodies in routine immunohistochemical procedures (Ngan et al., 1988).

In lymphomas other than the follicular type (e.g., diffuse lymphomas of small B lymphocytes, B-CLL, and T cell neoplasms), Southern analysis identifies only germ-line configurations of *bcl*-2 (Aisenberg et al., 1988). It can also reveal sequential *bcl*-2 and c-*myc* activation in FNA specimens during follicular lymphoma progression, particularly when tumors lack significant sclerosis or cell degeneration (Lee et al., 1989; Williams et al., 1990). Rearrangements of the c-*myc* oncogene accumulate along the transition of initial NHL to a high-grade malignancy (Lee et al., 1989; also see below). Although a role for the *bcl*-2 gene products in signal transduction had been proposed, Bcl-2 could be separated from G proteins, indicating that Bcl-2 proteins do not belong to the G class of signal proteins (Monica et al., 1990). On the other hand, it appears that the expression of *bcl*-2 leads to coupling of the growth factor receptors triggering inositol phosphate production, which is one of the major pathways of signal transduction (Haldar et al., 1990). ***bcl*-2 is not strictly specific for the t(14;18) translocation** — it is also expressed in mucosa-associated lymphoid tissue (Isaacson et al., 1991) as well as in follicular lymphomas without the t(14;18) translocation (Pezzella et al., 1990).

It appears that the expression of *bcl*-2 in myeloid cells and their precursors is regulated in a differentiation-related manner (Delia et al., 1992). It is manifested in a progressive reduction of the *bcl*-2 expression with the induction of differentiation toward the monocytic and granulocytic lineages. However, high expression of *bcl*-2 can be observed in myeloma so the issue of the *bcl*-2 expression vis-à-vis the state of differentiation remains to be resolved (Pettersson et al., 1992).

7.3.7.3 bcl-1

Although *bcl*-1 is associated with the **t(11;14)** translocation located at 11q13 (Koduru et al., 1989) and is usually found in diffuse small- and large-cell lymphomas, CLL, and multiple myeloma (Ngan and Berinstein, 1990), the *bcl*-1 locus seems to be only rarely (ca. 4%) rearranged in B cell CLLs and NHLs (Athan et al., 1991). When it is rearranged, it predominantly occurs in low-grade B cell neoplasms, and **is not associated with clinical aggressiveness**, advanced clinical stage, or large cell transformation (Richter's syndrome). These observations do **not support the clinical usefulness** of evaluating *bcl*-1 rearrangements (Athan et al., 1991).

7.3.7.4 c-myc

Activation of c-*myc* resulting from a chromosomal translocation is another classic example of involvement of oncogenes in tumor development (Klein and Klein, 1985). The **t(8;22)** translocation in Burkitt's lymphoma results in alteration of c-*myc* and leads to its activation (Cesarman et al., 1987; Szajnert et al., 1987; Zajac-Kaye et al., 1988). Emergence of a follicular lymphoma and a clonally related pre-B cell lymphoblastic lymphoma out of a follicular lymphoma associated with translocation t(14;18) led to the activation of c-*myc* (de Jong et al., 1988). In a leukemia T cell line with the **t(8;14) (q24;q11)** translocation, the *myc* **regulatory regions** do not appear to be involved in *myc* deregulation, and it is the **juxtaposition** of the *TCR*-α locus to a germ-line *myc* that results in *myc* deregulation (Finver et al., 1988).

Experimental transformation of Burkitt's lymphoma cells into a B-lymphoblastoid cell line can be achieved using carcinogen stimulation (Numoto, 1988). The **augmented expression** of c-*myc* following this transformation occurs at the transcriptional level, apparently through activation of the normal c-*myc* alleles, giving some insight into the pattern of regulation of c-*myc* in lymphomatoid malignancies (Eick and Bornkamm, 1989; Numoto, 1988). The **transcriptional deregulation** of c-*myc* in Burkitt's lymphoma may result from the position and number of mutations produced in and around the *myc* locus (Morse et al., 1989). The level of expression of the c-Myc protein in lymphomas has been studied by **immunohistochemical techniques** (Mitani et al., 1988) and **FCM** (Holte et al., 1989). It was found to correlate with the overall DNA synthetic and ploidy characteristics. More clinical studies are needed to establish whether the c-Myc protein level should be used as a practical tool in managing patients with lymphoid malignancies. At the gene level, the pattern of c-*myc* rearrangements in primary gastrointestinal lymphomas is different from those usually seen in node-based follicle center-cell lymphomas and is characteristic of aggressive lymphomas (van Krieken et al., 1990). In acute myeloid leukemia, dysregulation of c-*myc* leads to enhanced stability of c-*myc* mRNA (as well as c-*myb* mRNA) resulting from transcriptional aberration nuclear transport abnormalities and changes in posttranscriptional regulation (Baer et al., 1992).

Interestingly, studies on c-*myc* amplification reveal a **sequential transition** of these amplified sequences from submicroscopic circular extrachromosomal DNA (**episomal form**), to **double minutes**, and later to a **specific chromosomal site** (von Hoff et al., 1990). This shift of the c-*myc* sequences is associated with a more rapid proliferation of the cells carrying the amplified sequences. It is possible that prevention of integration of the oncogene sequences into chromosomal sites may influence the progression of tumors (von Hoff et al., 1990).

7.3.7.5 *ras*

Point mutations in N-*ras* or Ki-*ras* genes are found in one third of leukemia cells of patients with AML, some cases of ALL, and occasionally in Burkitt cell lines (reviewed in Butturini and Gale, 1988a,b; Sheridan and Reis, 1990). Ha-*ras* rearrangements associated with activation of the oncogene are present in T cell tumors (Ihle et al., 1989), and in some cases of myeloid leukemia (AML and CML) (Mise et al., 1987; Toksoz et al., 1987). N-*ras* mutations in **codon 12** have been reported in lymphoid leukemias more frequently than other mutations (Nishida et al., 1987), whereas the mutations at **codon 13** are rare in acute and chronic leukemias, but occur in MDS (Collins et al., 1989; Nishida et al., 1987; Wodnar-Filipowicz et al., 1987).

In general, *ras* mutations are rare events that occur in myeloid blast crisis at a late stage in CML (Le Maistre et al., 1989). N- or Ki-*ras* mutations in codons 12 or 13 appear to be frequent in chronic myelomonocytic leukemia, and may be helpful in prognosis if confirmed in large studies (Hirsch-Ginsberg et al., 1990). The detection of point mutations of N- and Ki-*ras* oncogenes, in particular, by **oligonucleotide hybridization**, or newer **PCR**-based assays may be useful in following the progression of the leukemic clone during and after **therapy** of acute nonlymphocytic leukemia (Pedersen-Bjergaard et al., 1988; Senn et al., 1988a) and in acute myelomonocytic leukemia (Senn et al., 1988b).

Conflicting reports assert, however, that comparison of patients with and without N-*ras* mutations does not support the notion that the presence of N-*ras* point mutations clearly identifies a unique clinical or biological subset of AML patients (Radich et al., 1990). *ras* mutations, particularly N-*ras,* are a well-recognized feature of **myelodysplastic syndrome** (e.g., Bartram, 1988; Carter et al., 1988; Hirai et al., 1987, 1988; Janssen et al., 1987; Mano et al., 1989; Padua et al., 1988). They have been found to appear in some cases only later in the course of disease, whereas in other patients the presence of the mutation is detected well before the overt manifestation of MDS (van Kamp et al., 1992).

In aggressive, diffuse, CALLA-positive lymphomas, a **coexpression** of N-ras^{p21} and c-*erb*B-2 (*neu*) oncogene products has been detected (Imamura et al., 1990). A **synergistic action** of c-*myc* and Ha-*ras* oncogenes has been demonstrated in EBV-immortalized human B lymphocytes and the transformed phenotype correlated with the level of Ras oncoprotein expression (Nasi et al., 1990).

7.3.7.6 OTHER ONCOGENES

c-*raf*-1 protooncogene has been studied in NHL and ALL patients and tumor-derived cell lines (Chenevix-Trench et al., 1989a; Sariban et al., 1989). c-*myb* has been implicated in the development of T lymphocyte proliferation and expression of the c-*myb* gene is altered in some leukemias and NHL (Barletta et al., 1987; Ohyashiki et al., 1988; Venturelli et al., 1990), and in erythropoietin-induced erythroid differentiation (Todokoro et al., 1988). c-*myb* antisense oligonucleotides inhibit normal hematopoiesis *in vitro* (Gewirtz and Calabreta, 1988). The enhanced expression of c-*myb* and **B-*myb*** (but not **A-*myb***) is correlated with the proliferative activity of T and B lymphocytes, but not monocytes and granulocytes (Golay et al., 1991), despite the suppression of c-*myb* in myeloid cell lines treated with tumor necrosis factor α (Schachner et al., 1988).

c-*fms* is associated with monocytic lineage differentiation induced in cultured cells (e.g., Imaizumi and Breitman, 1988; Radzun et al., 1988; Sherr, 1988; Sherr et al., 1988). No c-*fms* expression is found in acute lymphoid leukemias whether of T or B origin, but it appears to be a **specific marker of leukemogenesis in myeloid cells** (Rambaldi et al., 1988). In AML, c-*fms* reaches high levels of expression at the M5 stage (Dubreuil et al., 1988).

c-*fes* mRNA can be demonstrated in AML and CML, and at much lower levels in ALL (Smithgall et al., 1990). It is believed that an **RNA protection assay** for c-*fes* mRNA may be **used clinically to diagnose myeloid leukemia**, or to predict eventual **conversion** of ALL cases to AML (Smithgall et al., 1990). The **pim** protooncogene has been reported in human myeloid leukemias (Nagarajan et al., 1986), whereas **c-*ets*-1** is rearranged in some cases of ALL (Goyns et al., 1987). **c-*fos*** is highly expressed in acute leukemias with a monocytic phenotype (FAB M4/M5), in some B-lymphoid leukemias, with low levels found in AML, the majority of B- and all T-lymphoid leukemias, as well as in erythroleukemia (Pinto et al., 1987). c-*fos* can be used as an additional marker of myelomonocytic forms of leukemia.

A number of protooncogenes are expressed in Hodgkin's disease (**c-*myc*, *p53*, c-*jun*, pim-1, *lck*, c-*syn*, c-*raf*, N-*ras*,** and **c-*met***), the combinations of which are not found to be expressed in untransformed hematopoietic cells (Jucker et al., 1990). Another gene, ***blk***, has been described that codes for **B lymphoid kinase** and is specifically expressed in the B cell lineage (Dymecki et al., 1990). In T cell lymphomas/leukemias, the protooncogenes ***Tcl*-1, *Tcl*-2, *Tcl*-3,** and ***Tcl*-5** have been found is association with specific chromosomal aberrations (for a review, see Ngan and Berinstein, 1990). In adult T cell leukemias, *p53* mutations appear to be associated with an acute phase of the disease and they may play a role in the transition from the chronic phase to acute leukemia (Fenaux et al., 1992; Sakashita et al., 1992).

The expression of protooncogenes in hematological malignancies not only involves **more than one oncogene**, but the interaction of hematopoietic cells with the **bone marrow stroma** and growth factors released from it apparently affects the **level of protooncogene expression** (Wetzler et al., 1990). When assessing the levels of expression of individual protooncogenes in BM and peripheral blood cells, the intrinsic differences in their expression in these two compartments should be considered for clinically relevant estimates to be made (Preisler et al., 1987).

Attempts have been made to elucidate whether any significant associations exist between a **particular allele** of protooncogenes **L-*myc*** or **c-*myb*** revealed by RFLP analysis and the disease manifestations in leukemias and lymphomas (Chenevix-Trench et al., 1989b,c). Evidence of some variation at the L-*myc* locus has been obtained and correlated with survival to old age and susceptibility to NHL and ALL, whereas no such correlation could be found for c-*myb* or **c-*mos*** (Chenevix-Trench et al., 1989b,c). Deletions of the *RB* gene leading to the absence of the RB110 product have been observed in primary leukemias (Chen et al., 1990).

In addition to the above-mentioned oncogenes, a host of other oncogenes and growth factors are known to be involved in regulation of the hematopoietic system. These have been thoroughly reviewed in the literature cited above; however, very little, if any, testing of these regulatory factors at the gene level has been performed in the clinical laboratory setting so far. The availability of molecular probes for a number of growth factors (Dooley et al., 1988; Perlmutter et al., 1988; Pierce, 1989) and the use of some of these in the therapy of hematological disorders (e.g., Brandt et al., 1988; Laver and Moore, 1989) will undoubtedly lead to evaluation of the growth factor genes and many other protooncogenes at the gene level as part of patient management.

7.3.8 MULTIPLE MYELOMA

Approximately one third of aspirated or biopsied bone marrow cells in multiple myeloma (MM) are aneuploid and their significantly higher RNA content has been observed by FCM (Tafuri et al., 1991). Although DNA aneuploidy appears to be correlated with adverse prognosis, RNA content cannot reliably predict response to chemotherapy (Tafuri et al., 1991).

The most common chromosomal abnormalities observed in patients with malignant monoclonal gammopathies were in **chromosomes 1, 14**, and **12**; these abnormalities, however, had little correlation with the survival of patients (Lisse et al., 1988). Interestingly, **FCM** studies have detected the expression of the pre-B cell antigen CALLA by myeloma tumor cells as well as the presence of megakaryocytic, myelomonocytic, and even erythroid cell surface markers, supporting the view that there exists a "common primary neoplastic lesion for all hematologic cancers" (Epstein et al., 1990).

Analysis of established human myeloma cell lines revealed no alterations of the ***met, raf, abl, erb*B, HER-2 (*neu*), *fos*, *myb*-7, *fms*, L-*myc*, *sis*,** or ***myb*-1 genes** (Fourney et al., 1990). The expression of **c-*myc*** mRNA, however, was notably enhanced but not due to amplification of the gene. When the level of expression of some growth-regulated protooncogenes (c-*myc*, **c-*myb***, and *p53*) is related to the level of expression of histone H3 (which is suppressed in myeloma cells compared to normal bone marrow cells) it becomes apparent that their level of expression is markedly increased in myelomatous plasma cells (Palumbo et al., 1989). Comparison of the expression of the Ha-*ras* p21 product in **monoclonal gammopathies of undetermined significance (MGUS)** with that in MM cells revealed a significantly higher level of expression in MM cells as determined by FCM (Danova et al., 1990). It appears at this time that the oncogene expression or DNA ploidy offers an additional tool for predicting the clinical course of MM patients, complementing a combination of poor performance status, infections before diagnosis, renal impairment, serum calcium, severe anemia, Bence-Jones proteinuria, and other traditional parameters (San Miguel et al., 1989).

7.3.9 CONCLUDING REMARKS

Analysis of disease-specific associations at the gene level in hematological malignancies has given ample evidence that certain consistent genotypic alterations can be used for clinical diagnosis and patient management. Addressing the utility of molecular genetic analysis in clinical practice in a study of 175 hematolymphoid lesions, Davis and coworkers (1991) acknowledge the high sensitivity and specificity of genotyping, **although, in their view, little additional help is gained from it in unequivocally malignant cases** recognized by conventional analysis. Its use should be restricted, according to these authors, to cases in which diagnosis is uncertain after conventional analysis. It seems, however, that the role of genotyping, and the identification of malignant cells present at low levels, as in monitoring MRD, using simplified, nonradioactive formats will be growing.

Consonant with this position is the recommendation to use molecular biological techniques as an adjunct to cytogenetic evaluation in Philadelphia-positive leukemia. However, the role of PCR in the diagnosis and management of this condition is expected to expand particularly in ALL, since Ph[1]-positive patients are at a particularly high risk (Morgan and Wiedemann, 1992).

Another modality very likely to assume prominence in the evaluation of hematological malignancies is **DNA fingerprinting**, the more so that chemotherapy does not appear to interfere with the specific patterns obtained with the Jeffrey's probes (Helminen, 1992). Other probe/enzyme combinations will definitely be used as well.

References

Aisenberg, A. C. et al. (1988). The *bcl*-2 gene is rearranged in many diffuse B-cell lymphomas. Blood *71*:969–972.

Alcalay, M. et al. (1991). Translocation breakpoint of acute promyelocytic leukemia lies within the retinoic acid receptor α locus. Proc. Natl. Acad. Sci. U.S.A. *88*:1977–1981.

Alfarano, A. et al. (1990). Screening of β-thalassemia mutations by PCR and ASO analysis in an Italian population of mixed geographic origin. Hematologica *75*:506–509.

Amselem, S. et al. (1988). Determination of the spectrum of β-thalassemia genes in Spain by use of dot-blot analysis of amplified β-globin DNA. Am. J. Hum. Genet. *43*:95–100.

Anastasi, J. et al. (1990). Detection of numerical chromosomal abnormalities in neoplastic hematopoietic cells by *in situ* hybridization with a chromosome-specific probe. Am. J. Pathol. *136*:131–139.

Anderson, H. M. and Ranney, H. M. (1990). Southeast Asian immigrants: the new thalassemias in Americans. Semin. Hematol. *27*:239–246.

Andreeff, M. (1990a). Flow cytometry of leukemia. In: Flow Cytometry and Sorting, 2nd Ed. Wiley-Liss, New York, pp. 697–724.

Andreeff, M. (1990b). Flow cytometry of lymphoma. In: Flow Cytometry and Sorting, 2nd Ed. Wiley-Liss, New York, pp. 725–743.

Arlinghaus, R. B. and Kloetzer, W. S. (1987). Oncogenes and their involvement in chronic myelogenous leukemia. J. Clin. Lab. Anal. *1*:229–237.

Athan, E. et al. (1991). *Bcl*-1 rearrangement. Frequency and clinical significance among B-cell chronic lymphocytic leukemias and non-Hodgkin's lymphomas. Am. J. Pathol. *138*:591–599.

Ayscue, L. H. et al. (1990). *bcr/abl* recombinant DNA analysis versus karyotype in the diagnosis and therapeutic monitoring of chronic myeloid leukemia. Am. J. Clin. Pathol. *94*:404–409.

Baer, M. R. et al. (1992). Defective *c-myc* and *c-myb* mRNA turnover in acute myeloid leukemia cells. Blood *79*:1319–1326.

Balachandran, N. et al. (1987). Antigenic cross-reactions among herpes simplex virus types 1 and 2 Epstein-Barr virus, and cytomegalovirus. J. Virol. *61*:1125–1135.

Barletta, C. et al. (1987). Relationship between the c-*myb* locus and the 6q– chromosomal aberration in leukemias and lymphomas. Science *235*:1064–1067.

Barlogie, B. et al. (1987). Characterization of hematologic malignancies by flow cytometry. Analyt. Quantit. Cytol. Histol. *9*:147–155.

Bartram, C. R. (1988). Mutations in *ras* genes in myelocytic leukemias and myelodysplastic syndromes. Blood Cells *14*:533–538.

Bartram, C. R. et al. (1989). Minimial residual leukaemia in chronic myeloid leukemia patients after T-cell depleted bone marrow transplantation. Lancet *i*:1260.

Bartram, C. R. et al. (1990). Detection of minimal residual leukemia by polymerase chain reactions. Bone Marrow Transplant *6* (Suppl. 1):4–8.

Bayer, J. A. and Bauman, J. G. J. (1990). Flow cytometric detection of β globin mRNA in murine hemopoietic tissues using fluorescent *in situ* hybridization. Cytometry *11*:132–143.

Beishuizen, A. et al. (1991). Detection of mimimal residual disease in childhood leukemia with the polymerase chain reaction [letter to Editor]. N. Engl. J. Med. *324*:772–773.

Bernard, D. J. et al. (1990). Genotypic analyses of Hodgkin's disease. Mol. Biol. Med. *7*:503–509.

Bernstein, R. (1988). Cytogenetics of chronic myelogenous leukemia. Semin. Hematol. *25*:20–34.

Boehm, C. D. et al. (1983). Prenatal diagnosis using DNA polymorphisms. Report on 95 pregnancies at risk for sickle-cell disease or β-thalassemia. N. Engl. J. Med. *308*:1054–1058.

Boehm, T. L. J. et al. (1987). Oncogene amplifications, rearrangements, and restriction fragment length polymorphisms in human leukemia. Eur. J. Cancer. Clin. Oncol. *23*:623–629.

Botti, A. C. and Verma, R. S. (1990). The molecular biology of acute lymphoblastic leukemia [review]. Anticancer Res. *10*:519–26.

Bourguin, A. et al. (1990). Rapid, nonradioactive detection of clonal T-cell receptor gene rearrangements in lymphoid neoplasms. Proc. Natl. Acad. Sci. U.S.A. *87*:8536–540.

Brandt, S. J. et al. (1988). Effect of recombinant human granulocyte-macrophage colony-stimulating factor on hematopoietic reconstitution after high-dose chemotherapy and autologous bone marrow transplatation. N. Engl. J. Med. *318*:869–876.

Brisco, M. J. et al. (1990). Development of a highly sensitive assay, based on the polymerase chain reaction, for rare B-lymphocyte clones in a polyclonal population. Br. J. Hematol. *75*:163–167.

Brocker-Vriends, A. H. J. T. et al. (1987). Genotype assignment of hemophilia A by use of intragenic and extragenic restriction fragment length polymorphisms. Thromb. Hemostasis *57*:131–136.

Bunn, H. F. and Forget, B. G. (1986). Hemoglobin: molecular, genetic and clinical aspects. W. B. Saunders, Philadelphia.

Burk, C. D. et al. (1991). A deletion in the gene for glycoprotein IIb associated with Glanzmann's thrombosthenia. J. Clin. Invest. *87*:270–276.

Burke, A. P. et al. (1990). Lymphoepithelial carcinoma of the stomach with Epstein-Barr virus demonstrated by polymerase chain reaction. Mod. Pathol. *3*:377–380.

Burns, B. F. (1989). Molecular genetic markers in lymphoproliferative disorders. Clin. Biochem. *22*:33–39.

Butturini, A. and Gale, R. P. (1988a). Oncogenes and human leukemias. Int. J. Cell Cloning *6*:2–24.

Butturini, A. and Gale, R. P. (1988b). Oncogenes in chronic lymphocytic leukemia. Leukemia Res. *12*:89–92.

Cabanillas, F. (1988). A review and interpretation of cytogenetic abnormalities identified in Hodgkin's disease. Hematol. Oncol. *6*:271–274.

Cai, S. P. et al. (1989). Rapid prenatal diagnosis of β-thalassemia using DNA amplification and non-radioactive probes. Blood *73*:372–374.

Campana, D. et al. (1990). The detection of residual acute lymphoblastic leukemia cells with immunologic methods and polymerase chain reaction: a comparative study. Leukemia *4*:609–614.

Carroll, A. J. et al. (1990). The t(1;14)(p34;q11) is nonrandom and restricted to T cell acute lymphoblastic leukemia: a pediatric oncology group study. Blood *76*:1220–1224.

Carter, G. et al. (1988). The Ha-*ras* polymorphism in myelodysplasia and acute myeloid leukemia. Leuk. Res. *12*:385–391.

Cesarman, E. et al. (1987). Mutations in the first exon are associated with altered transcription of c-*myc* in Burkitt lymphoma. Science *238*:1272–1275.

Chang, J. C. and Kan, Y. W. (1972). A sensitive new prenatal test for sickle cell anemia. N. Engl. J. Med. *307*:P30–32.

Chang, J. C. and Kan, Y. W. (1984). Deletion of the entire human α-globin gene cluster. Clin. Res. *32*:549A.

Chehab, F. F. and Kan, Y. W. (1990). Detection of sickle cell anemia mutation by color DNA amplification. Lancet *335*:15–17.

Chehab, F. F. et al. (1987). Detection of sickle cell anemia and thalassemias. Nature (London) *329*:293–294.

Chen, Y. C. et al. (1990). Deletion of the human retinoblastoma gene in primary leukemias. Blood *76*:2060–2064.

Chenevix-Trench, G. et al. (1989a). Allelic variation of the c-*raf*-1 protooncogene in human lymphoma and leukemia. Oncogene *4*:507–510.

Chenevix-Trench, G. et al. (1989b). Restriction fragment length prolymorphisms of L-*myc* and *myb* in human leukemia and lymphoma in relation to age-selected controls. Br. J. Cancer *60*:872–874.

Chenevix-Trench, G. et al. (1989c). The *Eco* RI RFLP of c-*mos* in patients with non-Hodgin's lymphoma and acute lymphoblastic leukemia, compared to geriatric and non-geriatric controls. Int. J. Cancer *43*:1034–1036.

Choi, O. R. and Engel, J. D. (1988). Developmental regulation of β globin switching. Cell (Cambridge, Mass) *55*:17–26.

Chui, D. H. K. et al. (1986). Embryonic ζ-globin chains in adults: a marker for α-thalassemia-1 haplotype due to a >17.5 k6 deletion. N. Engl. J. Med. *314*:76–79.

Chung, S. W. et al. (1991). Leukemia initiated by hemopoietic stem cells expressing the v-*abl* oncogene. Proc. Natl. Acad. Sci. U.S.A. *88*:1585–1589.

Cohen, J. I. et al. (1989). Epstein-Barr virus nuclear protein 2 is a key determinant of lymphocyte transformation. Proc. Natl. Acad. Sci. U.S.A. *86*:9558–9562.

Collins, S. J. et al. (1989). Rare occurrence of N-*ras* point mutations in Philadelphia chromosome positive chronic myeloid leukemia. Blood *73*:1028–1032.

Conner, B. J. et al. (1983). Detection of sickle cell β^S-globin allele by hybridization with synthetic oligonucleotides. Proc. Natl. Acad. Sci. U.S.A. *80*:278–282.

Cossman, J. (ed.) (1990). Molecular Genetics in Cancer Diagnosis. Elsevier, New York.

Cossman, J. et al. (1991). Gene rearrangements in the diagnosis of lymphoma/leukemia. Guidelines for use based on a multinstitutional study. Am. J. Clin. Pathol. *95*:347–354.

Cotter, F. E. (1990). The role of the *bcl*-2 gene in lymphoma. Br. J. Hematol. *75*:449–453.

Cotter, F. E. et al. (1990a). Direct sequence analysis of the 14qt and 18q-chromosome junctions in follicular lymphoma. Blood *76*:131–135.

Cotter F. E. et al. (1990b). Minimal residual disease in leukemia and lymphoma. Ann. Oncol. *1*:167–170.

Cotter, F. E. et al. (1991). Gene mapping by microdissection and enzymatic amplification: heterogeneity in leukemia associated breakpoints on chromosome 11. Genes Chromosomes Cancer *3*:8–15.

Cowan, J. S. (1990). The histopathologic classification of non-Hodgkin's lymphomas: ambiguities in the working formulation and two newly reported categories. Semin. Oncol. *17*:3–10.

Cowan, R. A. et al. (1989). DNA content in high and intermediate grade non-Hodgkin's lymphoma — prognostic significance and clinicopathological correlations. Br. J. Cancer *60*:904–910.

Crescenzi, M. et al. (1988). Thermostable DNA polymerase chain amplification of t(14;18) chromosome breakpoints and detection of minimal residual disease. Proc. Natl. Acad. Sci. U.S.A. *85*:4869–4873.

Crist, W. M. et al. (1990). Poor prognosis of children with pre-B acute lymphoblastic leukemia is associated with the t(1;19)(q23;p13): a pediatric oncology group study. Blood *76*:117–122.

D'Auriol, L. et al. (1989). *In vitro* amplification of T cell γ gene rearrangements: a new tool for the assessment of minimal residual disease in acute lymphoblastic leukemias. Leukemia *3*:155–158.

Danova, M. et al. (1990). *Ras* oncogene expression and DNA content in plasma cell dyscrasias: a flow cytofluorimetric study. Br. J. Cancer *62*:781–785

Davis, R. E. et al. (1991). Utility of molecular genetic analysis for the diagnosis of neoplasia in morphologically and immunophenotypically equivocal hematolymphoid lesions. Cancer *67*:2890–2899.

Dawson, C. W. et al. (1990). Epstein-Barr virus latent membrane protein inhibits human epithelial cell differentiation. Nature (London) *344*:77–80.

Delia, D. et al. (1992). *bcl*-2 proto-oncogene expression in normal and neoplastic human myeloid cells. Blood *79*:1291–1298.

de Jong, D. et al. (1988). Activation of the c-*myc* oncogene in a precursor-B-cell blast crisis of follicular lymphoma, presenting as composite lymphoma. N. Engl. J. Med. *318*:1373–1378.

De Klein, A. (1987). Oncogene activation by chromosomal rearrangement in chronic leukemia myelocytic. Mutat. Res. *186*:161–172.

De Souza, Y. G. et al. (1990). Diagnosis of Epstein-Barr virus infection in hairy leukoplakia by using nucleic acid hybridization and noninvasive techniques. J. Clin. Microbiol. *28*:2775–2778.

de The, H. et al. (1990). The t(15;17) translocation of acute promyelocytic leukemia fuses the retinoic acid receptor α gene to a novel transcribed locus. Nature (London) *347*:558–561.

Dhingra, K. et al. (1991). Hybridization protection assay: a rapid, sensitive, and specific method for detection of Philadelphia chromosome-positive leukemias. Blood *77*:238–242.

Diaz-Chico, J. C. et al. (1988). Mild and severe β-thalassemia among homozygotes from Turkey: identification of the types by hybridization of amplified DNA with synthetic probes. Blood *71*:248–251.

Dobrovic, A. et al. (1988). Detection of the molecular abnormality in chronic myeloid leukemia by use of the polymerase chain reaction. Blood *72*:2063–2065.

Dooley, D. C. et al. (1988). Granulocyte-monocyte progenitor cells from human peripheral blood: modulation of growth *in vitro* by T lymphocytes and monocytes. Int. J. Cell Cloning *6*:45–59.

Dreazen, O. et al. (1988). Molecular biology of chronic myelogenous leukemia. Semin. Hematol. *25*:35–49.

Dube, I. D. et al. (1990). Chromosome abnormalities in chronic myeloid leukemia. A model for acquired chromosome changes in hematological malignancy. Tumor Biol. *11* (Suppl. 1):3–24.

Dubeau, L. et al. (1988). Studies on immunoglobulin gene rearrangement in formalin-fixed, paraffin-embedded pathology specimens. Am. J. Pathol. *130*:588–594.

Dubreuil, P. et al. (1988). c-*fms* expression is a molecular marker of human acute mycloid leukemias. Blood *72*:1081–1085.

Dymecki, S. M. et al. (1990). Specific expression of a tyrosine kinase gene, *blk*, in B lymphoid cells. Science *247*:332–336.

Eick, D. and Bornkamm, G. W. (1989). Expression of normal and translocated c-*myc* alleles in Burkitt's lymphoma cells: evidence for different regulation. EMBO J. *8*:1965–1972.

Epner, D. E. and Koeffler, H. P. (1990). Molecular genetic advances in chronic myelogenous leukemia. Ann. Intern. Med. *113*:3–6.

Epstein, J. et al. (1990). Markers of multiple hematopoietic-cell lineages in multiple myeloma. N. Engl. J. Med. *322*:664–668.

Ersboll, J. and Schultz, H. B. (1989). Non-Hodgkin's lymphomas: recent concepts in classification and treatment. Eur. J. Hematol. *42* (Suppl. 48):15–29.

Fenaux, P. et al. (1992). Mutation of the *p53* gene in acute myeloid leukemia. Br. J. Hematol. *80*:178–183.

Finver, S. N. et al. (1988). Sequence analysis of the *MYC* oncogene involved in the t(8;14)(q24;q11) chromosome translocation in a human leukemia T-cell line indicates that putative regulatory regions are not altered. Proc. Natl. Acad. Sci. U.S.A. *85*:3052–3056.

Fonatsch, C. et al. (1990). Chromosomal *in situ* hybridization of a Hodgkin's disease-derived cell line (L540) using DNA probes for TCRA, TCRB, MET, and rRNA. Hum. Genet. *84*:427–434.

Fourney, R. et al. (1990). Elevated c-*myc* messenger RNA in multiple myeloma cells lines. Dis. Markers *8*:117–124.

Furie, B. and Furie, B. C. (1990). Molecular basis of hemophilia. Semin. Hematol. *27*:270–285.

Furie, B. and Furie, B. C. (1992). Molecular and cellular biology of blood coagulation. N. Engl. J. Med. *326*:800–806.

Gabert, J. et al. (1989). Detection of residual *bcr/abl* translocation by polymerase chain reaction in chronic myeloid leukemia patients after bone-marrow transplantation. Lancet *ii*:1125–1128.

Gaillard-Sanchez, I. et al. (1990). Successful detection by *in situ* cDNA hybridization of three members of the serpin family: angiotensinogen, α₁ protease inhibitor, and antithrombin III in human hepatocytes. Mod. Pathol. *3*:216–222.

Gao, Q. S. et al. (1988). Restriction primer extension method of labeling oligonucleotide probes and its application to the detection of Hb E genes. Hemoglobin *12*:691–697.

Gewirtz, A. M. (1990). Use of *in situ* hybridization to study human megakaryocyte gene expression. In: Differentiation of Megakaryocytes. Wiley-Liss, New York, pp. 105–117.

Gewirtz, A. M. and Calabreta, B. (1988). A c-*myb* antisense oligodeoxynucleotide inhibits normal human hematopoiesis *in vitro*. Science *242*:1303–1306.

Gilman, J. G. (1988). Expression of $^Gγ$ and $^Aγ$ globin genes in human adults. Hemoglobin *12*:707–716.

Ginsburg, D. (1990). The von Willebrand factor gene and genetics of von Willebrand's disease. Mayo Clin. Proc. *66*:506–515.

Ginsburg, D. et al. (1985). Human von Willebrand factor (vWF): isolation of complementary DNA (cDNA) clones and chromosomal localization. Science *228*:1401–1406.

Goelz, S. E. et al. (1985). Purification of DNA from formaldehyde fixed and paraffin embedded human tissue. Biochem. Biophys. Res. Commun. *130*:118–126.

Golay, J. et al. (1991). Expression of c-*myb* and B-*myb*, but not A-*myb*, correlates with proliferation in human hematopoietic cells. Blood *77*:P149–158.

Gopal, M. R. et al. (1990). Detection by PCR of HHV-6 and EBV DNA in blood and oropharynx of healthy adults and HIV-seropositives. Lancet *335*:1598–1599.

Gorska-Flipot, I. et al. (1990). Molecular pathology of chronic myelogenous leukemia. Tumor Biol. *11* (Suppl. 1):25–43.

Gottardi, E. et al. (1990). Molecular diagnosis of $^Aγ$ hereditary persistence of fetal hemoglobin using polymerase chain reaction and oligonucleotide analysis. Hematologica *75*:17–20.

Goyns, M. H. et al. (1987). The c-*ets*-1 proto-oncogene is rearranged in some cases of acute lymphoblastic leukemia. Br. J. Cancer *56*:611–613.

Gregory, C. D. et al. (1991). Activiation of Epstein-Barr virus latent genes protects human B cells from death by apoptosis. Nature (London) *349*:612–614.

Greil, R. et al. (1988). *abl* oncogene expression in non-Hodgkin's lymphomas: correlation to histological differentiation and clinical status. Int. J. Cancer *42*:529–538.

Greil, R. et al. (1989). *In situ* hybridization for the detection of low copy numbers of c-*abl* oncogene nRNA in lymphoma cells: technical approach and comparison with results with anti-oncoprotein antibodies. Lab. Invest. *60*:574–582.

Hakim, I. et al. (1991). Detection of minimal residual disease in childhood leukemia with the polymerase chain reaction [letter]. N. Engl. J. Med. *324*:773–774.

Haldar, S. et al. (1990). Role of *bcl*-2 in growth factor triggered signal transduction. Cancer Res. *50*:7399–7401.

Hansen-Hagge, T. E. et al. (1989). Detection of minimal residual disease in acute lymphoblastic leukemia by *in vitro* amplification of rearranged T-cell receptor δ chain sequences. Blood *74*:1762–1767.

Harrington, D. S. (1990). Molecular gene rearrangement analysis in hematopathology. Am. J. Clin. Pathol. *93* (Suppl. 1):S38-S43.

Helminen, P. (1992). Does chemotherapy of hematological malignancies affect DNA fingerprint pattern? Leukemia Res. *16*:133–138.

Hiddemann, W. et al. (1990). Analysis of the cellular DNA and RNA content in acute leukemias by flow cytometry. J. Cancer Res. Clin. Oncol. *116*:507–512.

Higgs, D. R. et al. (1989). A review of the molecular genetics of the human α-globin gene cluster. Blood *73*:1081–1104.

Hirai, H. et al. (1987). A point mutation at codon 13 of the N-*ras* oncogene in myelodysplastic syndrome. Nature (London) *327*:430–432.

Hirai, H. et al. (1988). Relationship between an activated N-*ras* oncogene and chromosomal abnormality during leukemic progression from myelodysplastic syndrome. Blood *71*:256–258.

Hirsch-Ginsberg, C. et al. (1990). *RAS* mutations are rare events in Philadelphia chromosome-negative/*bcr* gene rearrangement-negative chronic myelogenous leukemia, but are prevalent in chronic myelomonocytic leukemia. Blood *76*:1214–1219.

Holding, C. and Monk, M. (1989). Diagnosis of β-thalassemia by DNA amplification in single blastomeres from mouse preinplantation embryos. Lancet *ii*:532–535.

Holte, H. et al. (1989). Levels of Myc protein, as analyzed by flow cytometry, correlate with cell growth potential in malignant B-cell lymphomas. Int. J. Cancer *43*:164–170.

Hooberman, A. L. et al. (1989). Unexpected heterogeneity of *BCR-ABL* fusion mRNA detected by polymerase chain reaction in Philadelphia chromosome-positive acute lymphoblastic leukemia. Proc. Natl. Acad. Sci. U.S.A. *86*:4259–4263.

Hughes, T. and Goldman, J. M. (1990). Improved results with PCR for chronic myeloid leukemia. Lancet *ii*:812.

Ihle, J. N. and Askew, D. (1989). Origins and properties of hematopoietic growth factor-dependent cell lines. Int. J. Cell Cloning *7*:68–91.

Ihle, J. N. et al. (1989). Activation of the c-H-*ras* protooncogene by retrovirus insertion and chromosomal rearrangement in a Moloney leukemia virus-induced T-cell leukemia. J. Virol. *63*:2959–2966.

Imaizumi, M. and Breitman, T. R. (1988). Changes in c-*myc*, c-*fms*, and N-*ras* proto-oncogene expression associated with retinoic acid-induced, monocytic differentiation of human leukemia HL60/MRI cells. Cancer Res. *48*:6733–6738.

Imamura, N. et al. (1990). Co-expression of N-*ras* p21 and c-*erb*B-2 (*neu*) oncogene products by common ALL antigen-positive aggressive diffuse lymphoma. Lancet *336*:825–826.

Isaacson, P. G. et al. (1991). *bcl*-2 expression in lymphomas. Lancet *337*:175–176.

Ishihara, S. et al. (1989). Clonal T-cell lymphoproliferation containing Epstein-Barr (EB) virus DNA in a patient with chronic active EB virus infection. Jpn. J. Cancer Res. *80*:99–101.

Jackson, J. F. et al. (1990). Favorable prognosis associated with hyperdiploidy in children with acute lymphocytic leukemia correlates with extra chromosome 6. A pediatric oncology group study. Cancer *66*:1183–1189.

Jaffe, E. S. (1990). The role of immunophenotypic markers in the classification of non-Hodgkin's lymphomas. Semin. Oncol. *17*:11–19.

Janssen, J. W. G. et al. (1987). *RAS* gene mutations in acute and chronic myelocytic leukemia, chronic myeloproliferative disorders, and myelodysplastic syndromes. Proc. Natl. Acad. Sci. U.S.A. *84*:9228–9232.

Jedlicka, P. et al. (1990). Improved carrier detection of hemophilia A, using novel RFLPs at the DXS115 (767) locus. Hum. Genet. *85*:315–318.

Jones, J. F. et al. (1988). T-cell lymphomas containing Epstein-Barr viral DNA in patients with chronic Epstein-Barr virus infections. N. Engl. J. Med. *318*:723–741.

Jucker, M. et al. (1990). Heterogenous expression of proto-oncogenes in Hodgkin's disease derived cell lines. Hematol. Oncol. *8*:191–204.

Juliusson, G. et al. (1990). Prognostic subgroups in B-cell chronic lymphocytic leukemia defined by specific chromosomal abnormalities. N. Engl. J. Med. *323*:720–724.

Kan, Y. W. and Dozy, A. M. (1978). Antenatal diagnosis of sickle-cell anemia by DNA analysis of amniotic-fluid cells. Lancet *ii*:910–912.

Kaplan, F. et al. (1990). β-Thalassemia genes in French-Canadians; haplotypes and mutation analysis of Portneuf chromosomes. Am. J. Hum. Genet. *46*:126–132.

Katzenstein, A. L. A. and Peiper, S. C. (1990). Detection of Epstein-Barr virus genomes in lymphomatoid granulomatosis: analysis of 29 cases by the polymerase chain reaction technique. Mod. Pathol. *3*:435–441.

Kawasaki, E. S. et al. (1988). Diagnosis of chronic myeloid and acute lymphocytic leukemias by detection of leukemia-specific mRNA sequences amplified *in vitro*. Proc. Natl. Acad. Sci. U.S.A. *85*:5698–5702.

Kazazian, H. H., Jr. (1990). The thalassemia syndromes: molecular basis and prenatal diagnosis in 1990. Semin. Hematol. *27*:209–228.

Kazazian, H. H., Jr. and Boehm, C. D. (1988). Molecular basis and prenatal diagnosis of β-thalassemia. Blood *72*:1107–1116.

Klein, C. and Raab-Traub, N. (1987). Human neonatal lymphocytes immortalized after microinjection of Epstein-Barr virus DNA. J. Virol. *61*:1552–1589.

Klein, G. and Klein, E. (1985). Evolution of tumors and the impact of molecular oncology. Nature (London) *315*:190–195.

Knecht, H. et al. (1990). Detection of Epstein-Barr virus DNA by polymerase chain reaction in lymph node biopsies from patients with angioimmunoblastic lymphadenopathy. Br. J. Hematol. *75*:610–614.

Koduru, P. R. K. et al. (1989). Molecular analysis of breaks in *BCL*-1 proto-oncogene in B-cell lymphomas with abnormalities of 11q13. Oncogene *4*:929–934.

Kropp, G. L. et al. (1989). Selective enzymatic amplification of α2·globin DNA for detection of the hemoglobin Constant Spring mutation. Blood *73*:1987–1992.

Kulozik, A. E. et al. (1988). Rapid and non-radioactive prenatal diagnosis of β-thalassemia and sickle cell disease: application of the polymerase chain reaction (PCR). Br. J. Hematol. *70*:455–458.

Lalloz, M. R. A. et al. (1991). Hemophilia A diagnosis by analysis of a hypervariable dinucleotide repeat within the factor VIII gene. Lancet *338*:207–211.

Lange, W. et al. (1989). Detection by enzymatic amplification of *bcr-abl* mRNA in peripheral blood and bone marrow cells of patients with chronic myelogenous leukemia. Blood *73*:1735–1741.

Langner, K. D. et al. (1990). Molecular biology of proteins involved in blood coagulation. Behring Inst. Mitt. *86*:146–169.

Laver, J. and Moore, M. A. S. (1989). Clinical use of recombinant human hematopoietic growth factors. J. Natl. Cancer Inst. 81:1370–1382.

Lawrence, J. B. et al. (1989). Highly localized tracks of specific transcripts within interphase nuclei visualized by *in situ* hybridization. Cell (Cambridge, Mass) *57*:493–502.

Le Maistre, A. et al. (1989). *RAS* oncogene mutations are rare late stage events in chronic myelogenous leukemia. Blood *73*:889–891.

Lebo, R. V. et al. (1990). Prenatal diagnosis of α-thalassemia by polymerase chain reaction and dual restriction enzyme analysis. Hum. Genet. *85*:293–299.

Lee, J. T. et al. (1989). Sequential *bcl*-2 and c-*myc* oncogene rearrangements associated with the clinical transformation of non-Hodgkin's lymphoma. J. Clin. Invest. *84*:1454–1459.

Lee, M. S. et al. (1987). Detection of minimal residual cells carrying the t(14;18) by DNA sequence amplification. Science *237*:175–178.

Leibowitz, D. and Young, K. S. (1989). The molecular biology of CML: a review. Cancer Invest. *7*:195–203.

Lenoir, G. M. and Bornkamm, G. W. (1987). Burkitt's lymphoma, a human cancer model for the study of the multistep development of cancer: proposal for a new scenario. Adv. Viral Oncol. *7*:173–206.

Levinson, B. et al. (1990). Molecular analysis of hemophilia A mutations in the Finnish population. Am. J. Hum. Genet. *46*:53–62.

Lindsten, J. and Pettersson, U. (eds.) (1991). Etiology of Human Disease at the DNA Level. Raven Press, New York.

Lisse, I. M. et al. (1988). Occurrence and type of chromosomal abnormalities in consecutive malignant monoclonal gammapathies: correlation with survival. Cancer Genet. Cytogenet. *35*:27–36.

List, A. F. et al. (1990). The myelodysplastic syndromes: biology and implications for management. J. Clin. Oncol. *8*:1424–1441.

Lollar, P. (1990). The association of factor VIII with von Willebrand factor. Mayo Clin. Proc. *66*:524–534.

Look, A. T. et al. (1987). Prognostic value of cellular DNA content in acute lymphoblastic leukemia of childhood. N. Engl. J. Med. *317*:1666.

Ludwig, M. et al. (1992). Hemophilia B caused by five different nondeletion mutations in the protease domain of factor IX. Blood *79*:1225–1232.

Lynch, D. C. et al. (1985). Molecular cloning of cDNA for human von Willebrand factor: authentication by a new method. Cell (Cambridge, Mass) *41*:49–56.

Mancuso, D. J. et al. (1989). Structure of the gene for human von Willebrand factor. J. Biol. Chem. *264*:19514–19527.

Mancuso, D. J. et al. (1991). Human von Willebrand factor gene and pseudogene: structural analysis and differentiation by polymerase chain reaction. Biochemistry *30*:253–269.

Mano, H. et al. (1989). Mutations of N-*ras* oncogene in myelodysplastic syndromes and leukemias detected by polymerase chain reaction. Jpn. J. Cancer Res. *80*:102–106.

Maurer, J. et al. (1990). Molecular diagnosis of the Philadelphia chromosome in chronic myelogenous and acute lymphoblastic leukemias by PCR. Dis. Markers *8*:211–218.

Maurer, J. et al. (1991). Detection of chimeric BCR-ABL genes in acute lymphoblastic leukemia by the polymerase chain reaction. Lancet *337*:1055–1058.

Meyer, D. et al. (1990). Von Willebrand factor: structure and function. Mayo Clin. Proc. *66*:516–523.

Mise, K. et al. (1987). Localization of c-Ha-*ras*-1 oncogene in the t(7p–;11p+) abnormality of two cases with myeloid leukemia. Cancer Genet. Cytogenet. *29*:191–199.

Mitani, S. et al. (1988). Expression of c-*myc* oncogene product and *ras* family oncogene products in various human malignant lymphomas defined by immunohistochemical techniques. Cancer *62*:2085–2093.

Miyamura, K. et al. (1990). Prediction of clinical relapse after bone-marrow transplantation by PCR for Philadelphia-positive acute lymphoblastic leukemia. Lancet *336*:890.

Monica, K. et al. (1990). Small G proteins are expressed ubiquitously in lymphoid cells and do not correspond to Bcl-2. Nature (London) *346*:189–191.

Monk, M. and Holding, C. (1990). Amplification of a β-hemoglobin sequence in individual human oocytes and polar bodies. Lancet *335*:985–988.

Montone, K. T. et al. (1990). Detection of Epstein-Barr virus genomes by *in situ* DNA hybridization with a terminally biotin-labeled synthetic oligonucleotide probe from the EBV *Not*I and *Pst*I tandem repeat regions. Mod. Pathol. *3*:89–96.

Morgan, G. J. et al. (1989). Polymerase chain reaction for detection of residual leukemia. Lancet *i*:928–929.

Morgan, G. J. and Widemann, L. M. (1992). The clinical application of molecular techniques in Philadelphia-positive leukemia. Br. J. Hematol. *80*:1–5.

Morse, B. et al. (1989). Somatic mutation and transcriptional deregulation of *myc* in endemic Burkitt's lymphoma disease: heptamer-nonamer recognition mistakes? Mol. Cell. Biol. *9*:74–82.

Mullis, K. B. and Faloona, F. (1987). Specific synthesis of DNA *in vitro* via a polymerase-catalyzed chain reaction. Methods Enzymol. *155*:335–350.

Naeim, F. et al. (1990). Recent advances in diagnosis and classification of leukemias and lymphomas. Dis. Markers *8*:231–264.

Nagarajan, L. et al. (1986). Localization of the human *pim* oncogene (PIM) to a region of chromosome 6 involved in translocations in acute leukemia. Proc. Natl. Acad. Sci. U.S.A. *83*:2556–2560.

Nasi, S. et al. (1990). Induction of the neoplastic phenotype of EBV-established B lymphocytes by the human Ha-*ras* oncogene. Oncogene *5*:117–122.

Neuman, W. L. et al. (1992). Chromosomal loss and deletion are the most common mechanisms for loss of heterozygosity from chromosomes 5 and 7 in malignant myeloid disorders. Blood *79*:1501–1510.

Newton, C. R. et al. (1989a). Analysis of any point mutation in DNA, the amplification refractory mutation system (ARMS). Nucleic Acids Res. *17*:2503–2516.

Newton, C. R. et al. (1989b). Amplification refractory mutation system for prenatal diagnosis and carrier assessment in cystic fibrosis. Lancet *ii*:1481–1483.

Ngan, B. Y. and Berinstein, N. (1990). Oncogene involvement in lymphoid malignancy. Tumol Biol. *11* (Suppl. 1):78–93.

Ngan, B. Y. et al. (1988). Expression in non-Hodgkin's lymphoma of the *bcl*-2 protein associated with the t(14;18) chromosomal translocation. N. Engl. J. Med. *318*:1638–1644.

Nichols, W. C. et al. (1991). Severe von Willebrand disease due to a defect at the level of von Willebrand factor mRNA expression: detection by exonic PCR-restriction fragment length polymorphism analysis. Proc. Natl. Acad. Sci. U.S.A. *88*:3857–3861.

Nishida, J. et al. (1987). Activation mechanism of the N-*ras* oncogene in human leukemias detected by synthetic oligonucleotide probes. Biochem. Biophys. Res. Commun. *147*:870–875.

Nonoyama, M. et al. (1988). A novel EBNA-1 titration method and putative anti-EBNA-1 protein. J. Virol. Methods *21*:161–170.

Nowak, J. S. (1987). Genetic variability of complement receptor on human erythrocytes. J. Genet. *66*:133–138.

Nowell, P. C. and Croce, C. M. (1990). Chromosome translocations and oncogenes in human lymphoid tumors. Am. J. Clin. Pathol. *94*:229–237.

Numoto, M. (1988). The same external signal differentially induced the c-*myc* expression in Burkitt lymphoma and B-lymphoblastoid cell lines. Eur. J. Cancer Clin. Oncol. *24*:1727–1735.

Ohshima, K. et al. (1990). Analysis of Epstein-Barr viral genomes in lymphoid malignancy using Southern blotting, polymerase chain reaction and *in situ* hybridization. Virchow's Arch. B: Cell. Pathol. *59*:383–390.

Ohyashiki, K. et al. (1988). *myb* oncogene in human hematopoietic neoplasia with 6q-anomaly. Cancer Genet. Cytogenet. *33*:83–92.

Old, J. M. et al. (1990). Rapid detection and prenatal diagnosis of β-thalassemia: studies in Indian and Cypriot populations in the UK. Lancet *336*:834–837.

Olsen, L. C. et al. (1988). Hematopoiesis, myeloid leukemia and growth factors. Int. J. Biochem. *20*:883–888.

Orkin, S. H. (1990). Globin gene regulation and switching: circa 1990. Cell (Cambridge, Mass) *63*:665–672.

Orkin, S. H. et al. (1982). Improved detection of the sickle mutation by DNA analysis and its application to prenatal diagnosis. N. Engl. J. Med. *307*:33–36.

Orlic, D. (ed.) (1989). Molecular and cellular controls of hematopoiesis. Ann. N.Y. Acad. Sci. *554*:250.

Padua, P. A. et al. (1988). *ras* mutations in myelodysplasia detected by amplification, oligonucleotide hybridization, and transformation. Leukemia *2*:503–510.

Palumbo, A. P. et al. (1989). Altered expression of growth-regulated protooncogenes in human malignant plasma cells. Cancer Res. *49*:4701–4704.

Pearson, G. R. (1988). ELISA tests and monoclonal antibodies for EBV. J. Virol. Methods *21*:97–104.

Pedersen-Bjergaard, J. et al. (1988). Point mutation of the *ras* protooncogenes and chromosome aberrations in acute nonlymphocytic leukemia and preleukemia related to therapy with alkylating agents. Cancer Res. *48*:1812–1817.

Pedersen-Bjergaard, J. et al. (1990). Chromosome aberrations and prognostic factors in therapy-related myelodysplasia and acute nonlymphocytic leukemia. Blood *76*:1083–1091.

Peiper, S. C. et al. (1990). Detection of Epstein-Barr virus genomes in archival tissues by polymerase chain reaction. Arch. Pathol. Lab. Med. *114*:711–714.

Perlmutter, R. M. et al. (1988). Specialized protein tyrosine kinase proto-oncogenes in hematopoietic cells. Biochim. Biophys. Acta *948*:245–262.

Peschle, C. (ed.) (1987). Normal and neoplastic blood cells: from genes to therapy. Ann. N.Y. Acad. Sci. *511*.

Pettersson, M. et al. (1992). Expression of the *bcl*-2 gene in human multiple myeloma cell lines and normal plasma cells. Blood *79*:495–502.

Pezzella, F. et al. (1990). Expression of the *bcl*-2 oncogene protein is not specific for the 14;18 chromosomal translocation. Am. J. Pathol. *137*:225–232.

Pierce, J. H. (1989). Oncogenes, growth factors and hematopoietic cell transformation. Biochim. Biophys. Acta *989*:179–208.

Pinto, A. et al. (1987). c-*fos* oncogene expression in human hematopoietic malignancies is restricted to acute leukemias with monocytic phenotype and to subsets of B cell leukemias. Blood *70*:1450–1457.

Posey, Y. F. et al. (1989). Prenatal diagnosis of sickle cell anemia. Hemoglobin electrophoresis versus DNA analysis. Am. J. Clin. Pathol. *92*:347–351.

Powars, D. et al. (1990). The variable expression of scikle cell disease is genetically determined. Semin. Hematol. *27*:360–376.

Preisler, H. D. et al. (1987). Differing patterns of human proto-oncogene expression in peripheral blood and bone marrow acute leukemia cells. Cancer Res. *47*:3747–3751.

Radich, J. P. et al. (1990). N-*RAS* mutations in adult *de novo* acute myelogenous leukemia: prevalence and clinical significance. Blood *76*:801–807.

Radzun, H. J. et al. (1988). Modulation of c-*fms* proto-oncogene expression in human blood monocytes and macrophages. J. Leukocyte Biol. *44*:198–204.

Raimondi, S. C. et al. (1990). Cytogenetics of pre-B cell acute lymphoblastic leukemia with emphasis on prognostic implications of the t(1;L19). J. Clin. Oncol. *8*:1380–1388.

Rambaldi, A. et al. (1988). Expression of the macrophage colony stimulating factor and c-*fms* genes in human acute myeloblastic leukemia cells. J. Clin. Invest. *81*:1030–1035.

Randhawa, P. S. et al. (1989). The clinical spectrum, pathology, and clonal analysis of Epstein-Barr virus-associated lymphoproliferative disorders in heart-lung transplant recipients. Am. J. Clin. Pathol. *92*:177–185.

Rodeghiero, F. et al. (1987). Epidemiological investigation of the prevalence of von Willebrand's disease. Blood *69*:454–459.

Rodriguez, M. A. et al. (1991). Chromosome 17p and *p53* changes in lymphoma. Br. J. Hematol. *79*:575–582.

Rosson, D. and Reddy, E. P. (1988). Activation of the *abl* oncogene and its involvement in chromosomal translocations in human leukemia. Mutat. Res. *195*:231–243.

Roth, M. S. et al. (1988). Rearrangement of immounoglobulin and T-cell receptor genes in Hodgkin's disease. Am. J. Pathol. *131*:331–338.

Roth, M. S. et al. (1989). Detection of Philadelphia chromosome positive cells by the polymerase chain reaction following bone marrow transplant for chronic myelogenous leukemia. Blood *74*:882–885.

Roth, M. S. et al. (1992). Prognostic significance of Philadelphia chromosome-positive cells detected by the polymerase chain reaction after allogeneic bone marrow transplant for chronic myelogenous leukemia. Blood *79*:276–282.

Rovera, G. et al. (1991). Detection of minimal residual disease in childhood leukemia with the polymerase chain reaction [letter]. N. Engl. J. Med. *324*:774–775.

Rubin, E. M. et al. (1989). Newborn screening by DNA amplification of dried blood spots. Hum. Genet. *82*:134–136.

Ruggeri, Z. M. (1990). Structure and function of von Willebrand factor: relationship to von Willebrand's disease. Mayo Clin. Proc. *66*:847–861.

Rund, D. et al. (1991). Evolution of a genetic disease in an ethnic isolate: β-thalassemia in the Jews of Kurdistan. Proc. Natl. Acad. Sci. U.S.A. *88*:310–314.

Sadler, J. E. (1990). Recombinant DNA methods in hemophilia A: carrier detection and prenatal diagnosis. Semin. Thromb. Hemost. *16*:341–347.

Sadler, J. E. and Davie, E. W. (1987). Hemophilia A, hemophilia B, and von Willebrand's disease. In: The Molecular Basis of Blood Diseases. Stamatoyannopoulos, G. et al. (eds.). W. B. Saunders, Philadelphia, pp. 575–630.

Sadler, J. E. et al. (1985). Cloning and characterization of two cDNAs coding for human von Willebrand factor. Proc. Natl. Acad. Sci. U.S.A. *82*:6394–6398.

Saglio, G. et al. (1986). Proto-oncogene activation in human hematologic malignancies. Hematologica *71*:497–509.

Saglio, G. et al. (1991). Detection of Ph[1]-positive acute lymphoblastic leukemia by PCR. Lancet *338*:958.

Saiki, R. K. et al. (1985). Enzymatic amplification of β-globin genomic sequences and restriction site analysis for diagnosis of sickle cell anemia. Science *230*:1350–1354.

Saiki, R. K. et al. (1986). Analysis of enzymatically amplified β-globin and HLA-DQ-α DNA with allele-specific oligonucleotide probes. Nature (London) *324*:163–166.

Saiki, R. K. et al. (1988). Diagnosis of sickle cell anemia and β-thalassemia with enzymatically amplified DNA and nonradioactive allele-specific oligonucleotide probes. N. Engl. J. Med. *319*:537–541.

Sakashita, A. et al. (1992). Mutations of the *p53* gene in adult T cell leukemia. Blood *79*:477–480.

Samoszuk, M. K. et al. (1990). Evaluation of biotinylated DNA probes for the detection of gene rearrangements in clinical samples. Am. J. Clin. Pathol. *94*:729–733.

Sampietro, M. et al. (1990). Restriction of polymerase chain reaction products for carrier detection and prenatal diagnosis of hemophilia A: description of an internal control. Thromb. Hemost. *63*:527–528.

San Miguel, J. F. et al. (1989). Prognostic factors and classification in multiple myeloma. Br. J. Cancer *59*:113–118.

Sariban, E. et al. (1987). Expression of the c-*raf* protooncogene in human hematopoietic cells and cell lines. Blood *69*:1437–1440.

Sawyers, C. L. et al. (1990). Molecular relapse in chronic myelogenous leukemia patients after bone marrow transplantation detected by polymerase chain reaction. Proc. Natl. Acad. Sci. U.S.A. *87*:563–567.

Sawyers, C. L. et al. (1991). Leukemia and the disruption of normal hematopoiesis. Cell (Cambridge, Mass.) *64*:337–350.

Schachner, J. et al. (1988). Suppression of c-*myc* and c-*myb* expression in myeloid cell lines treated with recombinant tumor necrosis factor α. Leukemia *2*:749–753.

Schwartz, E. et al. (1988). Overview of the β-thalassemias: genetic and clinical aspects. Hemoglobin *12*:551–564.

Schwartz, R. C. and Witte, O. N. (1988). The role of multiple oncogenes in hematopoietic neoplasia. Mutat. Res. *195*:245–253.

Seibel, N. L. and Kirsch, I. R. (1989). Tumor detection through the use of immunoglobulin gene rearrangements combined with tissue *in situ* hybridization. Blood *74*:1791–1795.

Senn, H. P. et al. (1988a). Mutation analysis of the N-*ras* proto-oncogenes in active and remission phase of human acute leukemias. Int. J. Cancer *41*:59–64.

Senn, H. P. et al. (1988b). Relapse cell population differs from acute onset clone as shown by absence of the initially activiated N-*ras* oncogene in a patient with acute myclomonocytic leukemia. Blood *72*:931–935.

Sheridan, B. L. and Reis, M. D. (1990). Oncogene involvement in myelodysplasia and acute myeloid leukemia. Tumor Biol. *11* (Suppl. 1):44–58.

Sherr, C. J. (1988). The role of the CSF-1 receptor gene (c-*fms*) in cell transformation. Leukemia 2:132S-42S.

Sherr, C. J. et al. (1988). Colony-stimulating factor-1 receptor (c-*fms*). J. Cell Biochem. *38*:179–187.

Shibata, D. et al. (1990). Detection of occult CNS involvement of follicular small cleaved lymphoma by the polymerase chain reaction. Mod. Pathol. *3*:71–75.

Sklar, J. (1990). What can DNA rearrangements tell us about solid hematolymphoid neoplasms? Am. J. Surg. Pathol. *14* (Suppl. 1):16–25.

Skogerboe, K. J. et al. (1991). Genetic screening of newborns for sickle cell disease: correlation of DNA analysis with hemoglobin electrophoresis. Clin. Chem. *37*:454–458.

Smithgall, T. E. et al. (1990). Detection of c-*fes* proto-oncogene mRNA in human myeloid leukemias by RNase protection assay. Proc. Am. Assn. Cancer Res. *31*:314A.

Sommer, S. S. and Sobell, J. L. (1987). Application of DNA-based diagnosis to patient case: the example of hemophilia A. Mayo Clin. Proc. *62*:387–404.

Sorg, C. (ed.) (1990). Molecular Biology of B Cell Developments. S. Karger, Basel, p. 174.

Staal, S. P. et al. (1989). A survey of Epstein-Barr virus DNA in lymphoid tissue. Frequent detection in Hodgkin's disease. Am. J. Clin. Pathol. *91*:1–5.

Steinberg, M. H. (1988). Review: thalassemia: molecular pathology and management. Am. J. Med. Sci. *296*:308–321.

Stevenson, M. et al. (1986). Immortalization of human T lymphocytes after transfection of Epstein-Barr virus DNA. Science *233*:980–984.

Strickler, J. G. and Copenhaver, C. M. (1990). *In situ* hybridization in hematopathology. Am. J. Clin. Pathol. *93* (Suppl. 1):S44-S48.

Suciu, S. et al. (1990). Results of chromosome studies and their relation to morphology, course, and prognosis in 120 patients with *de novo* myelodysplastic syndrome. Cancer Genet. Cytogenet. *44*:15–26.

Suehiro, K. et al. (1988). Carrier detection in Japanese hemophilia A by use of three intragenic and two extragenic factor VIII DNA probes: a study of 24 kindreds. J. Lab. Clin. Med. *112*:314–318.

Sutton, M. et al. (1989). Polymerase chain reaction amplification applied to the determinatoin of β-like globin gene cluster haplotypes. Am. J. Hematol. *32*:66–69.

Szajnert, M. F. et al. (1987). Clustered somatic mutations in and around first exon of non-rearranged c-*myc* in Burkitt lymphoma with t(8;22) translocation. Nucleic Acids Res. *15*:4553–4565.

Tafuri, A. et al. (1991). DNA and RNA flow cytometric study in multiple myeloma. Clinical correlations. Cancer *67*:449–454.

Telenti, A. et al. (1990). Detection of Epstein-Barr virus by polymerase chain reaction. J. Clin. Microbiol. *28*:2187–2190.

Thein, S. L. et al. (1990). The molecular basis of β-thalassemia in Thailand: application to prenatal diagnosis. Am. J. Hum. Genet. *47*:369–375.

Tien, H. F. et al. (1990). Chromosome and *bcr* rearrangement in chronic myelogenous leukemia and their correlation with clinical states and prognosis of the disease. Br. J. Hematol. *75*:469–475.

Todokoro, K. et al. (1988). Down-regulation of c-*myb* gene expression is a prerequisite for erythropoietin-induced erythroid differentiation. Proc. Natl. Acad. Sci. U.S.A. *85*:8900–8904.

Toksoz, D. et al. (1987). *ras* gene activation in a minor proportion of the blast population in acute myeloid leukemia. Oncogene *1*:409–413.

Triplett, D. A. (1990). Laboratory diagnosis of von Willebrand's disease. Mayo Clin. Proc. *66*:832–840.

Tsang, T. C. et al. (1988). A factor IX mutation, verified by direct genomic sequencing, causes hemophilia B by a novel mechanism. EMBO J. *7*:3009–3015.

Tsujimoto, Y. et al. (1985). Involvement of the *bcl*-2 gene in human follicular lymphoma. Science *228*:1440–1443.

Tsujimoto, Y. et al. (1987). Characterization of the protein product of *bcl*-2, the gene involved in human follicular lymphoma. Oncogene *2*:3–7.

Uccini, S. et al. (1989). Interaction of HIV and EBV at lymphoid tissue level: immunohistochemistry and *in situ* hybridization. APMIS Suppl. *8*:28–32.

Ugozzoli, L. et al. (1989). Genotypic analysis of engraftment in thalassemia following bone marrow transplantation using synthetic oligonucleotides. Bone Marrow Transplant. *4*:173–180.

van Dekken, H. et al. (1990a). Spatial topography of a pericentromeric region (1Q12) in hemopoietic cells studied by *in situ* hybridization and confocal microscopy. Cytometry *11*:570–578.

van Dekken, H. et al. (1990b). Three-dimensional reconstruction of pericentromeric (1Q12) DNA and ribosomal RNA sequences in HL60 cells after double-target *in situ* hybridization and confocal microscopy. Cytometry *11*:579–585.

van Kamp, H. et al. (1992). Longitudinal analysis of point mutations of the N-*ras* proto-oncogene in patients with myelodysplasia using archived blood smears. Blood *79*:1266–1270.

van Krieken, J. H. J. M. et al. (1990). Molecular genetics of gastrointestinal non-Hodgkin's lymphomas: unusual prevalence and pattern of c-*myc* rearrangements in aggressive lymphomas. Blood *76*:797–800.

van Tuinen, P. et al. (1987). Localization of myeloperoxidase to the long arm of human chromosome 17: relationship to the 15;17 translocation of acute promyelocytic leukemia. Oncogene *1*:319–322.

Vane, J. R. et al. (1990). Regulatory functions of the vascular endothelium. N. Engl. J. Med. *323*:27–36.

Venturelli, D. et al. (1990). Down-regulated c-*myb* expression inhibits DNA synthesis of T-leukemia cells in most patients. Cancer Res. *50*:7371–7375.

Verma, R. S. et al. (1989). Molecular characterization of variant translocations in chronic myelogenous leukemia. Oncogene *4*:1145–1148.

Verweij, C. L. et al. (1985). Construction of cDNA coding for human von Willebrand factor using antibody probes for colony-screening and mapping of the chromosomal gene. Nucleic Acids Res. *13*:4699–4717.

von Hoff, D. D. et al. (1990). Double minutes arise from circular extrachromosomal DNA intermediates which integrate into chromosomal sites in human HL-60 leukemia cells. J. Clin. Invest. *85*:1887–1895.

Wakasugi, H. et al. (1987). Epstein-Barr virus-containing B-cell line produces an interleukin 1 that it uses as a growth factor. Proc. Natl. Acad. Sci. U.S.A. *84*:804–808.

Waldmann, T. A. (1987). The arrangement of immunoglobulin and T cell receptor genes in human lymphoproliferative disorders. Adv. Immunol. *40*:247–321.

Walts, A. E. et al. (1990). Diagnosis of malignant lymphoma in effusions from patients with AIDS by gene rearrangements. Am. J. Clin. Pathol. *94*:170–175.

Wang, N. S. et al. (1990). Point mutations in four hemophilia B patients from China. Thromb. Hemost. *64*:302–306.

Ware, J. et al. (1991). Identidication of a point mutation in type IIB von Willebrand disease illustrating the regulation of von Willebrand factor affinity for the platelet membrane glycoprotein IB-IX. Proc. Natl. Acad. Sci. U.S.A. *88*:2946–2950.

Warford, A. et al. (1988). Southern blot analysis of DNA extracted from formol-saline fixed and paraffin wax embedded tissue. J. Pathol. *154*:313–320.

Weatherall, D. J. et al. (1989). The hemoglobinopathies. In: The Metabolic Basis of Inherited Disease. Scriver, C. R. et al. (eds.). McGraw-Hill, New York, pp. 2281–2339.

Weiss, L. M. et al. (1987a). Epstein-Barr viral DNA in tissues of Hodgkin's disease. Am. J. Pathol. *129*:86–91.

Weiss, L. M. et al. (1987b). Molecular analysis of the t(14;18) chromosomal translocation in malignant lymphomas. N. Engl. J. Med. *317*:1185–1189.

Weiss, L. M. et al. (1989). Detection of Epstein-Barr viral genomes in Reed-Sternberg cells of Hodgkin's disease. N. Engl. J. Med. *320*:502–506.

Westbrook, C. A. (1987). The *ABL* oncogene in human leukemias. Blood Rev. *2*:1–8.

Wetzler, M. et al. (1990). Constitutive and induced expression of growth factors in normal and chronic phase chronic myelogenous leukemia Ph[1] bone marrow stroma. Cancer Res. *50*:5801–5805.

Williams, M. E. et al. (1990). Fine-needle aspiration of non-Hodgkin's lymphoma. Southern blot analysis for antigen receptor, *bcl*-2 and c-*myc* gene rearrangements. Am. J. Clin. Pathol. *93*:754–759.

Winberg, C. D. et al. (1988). T-cell-rich lymphoproliferative disorders: morphologic and immunologic differential diagnoses. Cancer *62*:1539–1555.

Wodnar-Filipowicz, A. et al. (1987). Glycine-cysteine substitution at codon 13 of the N-*ras* proto-oncogene in a human T cell non-Hodgkin's lymphoma. Oncogene *1*:457–461.

Wolf, B. C. et al. (1989). The detection of Epstein-Barr virus in hairy cell leukemia cells by filter and *in-situ* hybridization. Mod. Pathol. *2*:106A.

Wu, A. M. et al. (1990a). Genotype and phenotype: a practical approach to the immunogenetic analysis of lymphoproliferative disorders. Hum. Pathol. *21*:1132–1141.

Wu, A. M. et al. (1990b). Analysis of antigen receptor gene rearrangements in ethanol and formaldehyde-fixed, paraffin-embedded specimens. Lab. Invest. *63*:107–114.

Wu, T. C. et al. (1990). Detection of EBV gene expression in Reed-Sternberg cells in Hodgkin's disease. Int. J. Cancer *46*:801–804.

Yamada, H. et al. (1990). Lack of gene deletion for complement C4A deficiency in Japanese patients with systemic lupus erythematosis. J. Rheumatol. *17*:1054–1057.

Yamada, M. et al. (1989). Detection of minimal disease in hematopoietic malignancies of the B-cell lineage by using third-complementarity-determining region (CDRIII) specific probes. Proc. Natl. Acad. Sci. U.S.A. *86*:5123–5127.

Yamada, M. et al. (1990). Minimal residual disease in childhood B-lineage lymphoblastic leukemia. Persistence of leukemia cells during the first 18 months of treatment. N. Engl. J. Med. *323*:448–455.

Yokota, S. et al. (1991). Use of polymerase chain reactions to monitor minimal residual disease in acute lymphoblastic leukemia patients. Blood *77*:331–339.

Yoshioka, A. et al. (1988). Prenatal diagnosis of hemophilia BM. Jpn. J. Hum. Genet. *33*:395–400.

Youssoufian, H. et al. (1988). Moderately severe hemophilia A resulting from GLU-GLY substitution in exon 7 of the factor VIII gene. Am. J. Hum. Genet. *42*:867–871.

Yunis, J. J. (1987). High-resolution chromosme analysis in the prognosis and clincial evolution of "*de novo*" myelodysplasia. Hematologica *76* (Suppl. 1):34–37.

Zajac-Kaye, M. et al. (1988). A point mutation in the c-*myc* locus of a Burkitt lymphoma abolishes binding of a nuclear protein. Science *240*:1776–1780.

Index

INDEX

A

B

hypermethylation and, gastrointestinal disease
    and, 199
non-small cell lung cancer and, 231
pancreatic carcinoma and, 212
reverse cholesterol transport and, 178–179
thyroid neoplasms and, 360
uterine cancer and, 291, 294
Chromosome 11p
bladder cancer and, 321, 322
breast cancer and, 286
hepatocellular carcinoma and, 205
loss of heterozygosity at, 203, 205
    adrenal adenoma and, 365–366
    head and neck tumors and, 236
    pancreatic carcinoma and, 212
renal cell carcinoma and, 314
Chromosome 11p13-15
breast cancer and, 275
gastrointestinal disease and, 189
Chromosome 11q
non-small cell lung cancer and, 231
pancreatic carcinoma and, 212
renal cell carcinoma and, 315
Chromosome 11q13
breast cancer and, 282
hepatocellular carcinoma and, 207
MEN 1 and, 368
Chromosome 12
renal cell carcinoma and, 315
thyroid neoplasms and, 360
Chromosome 12q, gastrointestinal disease and, 188
Chromosome 12q13, breast cancer and, 276
Chromosome 13
breast cancer and, 275
c-*myc* and, 192
Chromosome 13p, non-small cell lung cancer and,
    231
Chromosome 13q, hepatocellular carcinoma and, 205
Chromosome 13q14, retinoblastoma and, 414
Chromosome 14
ataxia-telangiectasia and, 107
renal cell carcinoma and, 316, 317
Chromosome 14p, non-small cell lung cancer and,
    231
Chromosome 15
Marfan syndrome and, 397–398
prostate cancer and, 307
Chromosome 15p, non-small cell lung cancer and,
    231
Chromosome 16
adult polycystic kidney disease and, 84
α-thalassemia and, 606
uterine smooth muscle tumors and, 294
Chromosome 16p, hepatocellular carcinoma and, 205
Chromosome 16p13, polycystic kidney disease and,
    318
Chromosome 16q
hepatocellular carcinoma and, 204–205
prostate cancer and, 307
Chromosome 17
NF1 and, 412
non-small cell lung cancer and, 231

osteosarcoma and, 398
renal cell carcinoma and, 315
topoisomerases and, drug resistance and, 113
uterine smooth muscle tumors and, 294
Chromosome 17p
bladder cancer and, 322
breast cancer and, 275, 284
gastrointestinal disease and, 188
hepatocellular carcinoma and, 205
lung carcinoma and
    non-small cell, 231
    small cell, 227
ovarian cancer and, 300
renal cell carcinoma and, 315
Chromosome 17q
breast cancer and, 275–276
hepatocellular carcinoma and, 205
Chromosome 18
breast cancer and, 285
renal cell carcinoma and, 315
uterine smooth muscle tumors and, 294
Chromosome 18q, gastrointestinal disease and, 188
DCC gene and, 194
Chromosome 19
DNA repair assessment and, 103
familial hypercholesterolemia and, 174
Chromosome 19p
non-small cell lung cancer and, 231
renal cell carcinoma and, 315
Chromosome 21
Alzheimer's disease and, 430
Down syndrome and, 429–430
uterine smooth muscle tumors and, 294
Chromosome 22
gliomas and, 423
meningiomas and, 424
uterine smooth muscle tumors and, 294
Chromosome 22q, gastrointestinal disease
    and, 188
Chromosome 22q11, DiGeorge syndrome and, 86
Chromosome analysis, 34–45
automated karyotyping in, 36
genetic linkage analysis in, 34–36
in situ hybridization in, 37–45, see also *In situ*
    hybridization
Chromosome assignment, 7
Chromosome dissection, 34–35
Chromosome fingerprinting, see also Fingerprinting
    analysis
in epidemiological analysis, 458
Chromosome instability syndromes, see also specific
    syndromes
prenatal diagnosis of, 83
Chromosome jumping, 34
CF gene and, 216
Chromosome-mediated DNA transfer, 7
"Chromosome painting", 36, 43
Chromosome paints, 36
Chromosome primers, specific, 36
Chromosome-specific libraries, ISH and, 41–42
Chromosome-specific "painting", single-copy ISH
    and, 42

F

variable, see Minisatellite variable repeats
(MVRs); Variable repeats
Repeat sequences, see also Repetitive sequences
"core" consensus, multiple loci probes and, 48
11-nucleotide, direct, 206, 207
tandem, see Tandem repeat sequences
telomeric, 51
reduction in length of, 189
Repetitive DNA, 8–9
Repetitive DNA-ISH, 41–42
Repetitive sequence fingerprinting, 7
Repetitive sequences, see also Repeat sequences
hybridization to, competitive suppression of, 41
interspersed, PCR and, see Interspersed repetitive
sequences-polymerase chain reaction (IRS-
PCR)
major, 50
Replication
HBV infection and, 206
inhibition of, antibiotics and, 98
slippage in, hypervariable loci and, 46
topoisomerases and, 101
Replicational banding, 41
Replication code, 9
Replication initiation, radioresistant, ataxia-
telangiectasia and, 107
Reporter molecules, 38–39
alkaline phosphatase as, 457
radioactive, 60
Reproducibility, fetal sexing and, 74
Respiratory pathogens, 485–508, see also specific
type
Respiratory syncytial virus, 485
Respiratory viruses, 485–499, see also specific virus
"Restricted *Alu*-PCR", 50
Restriction analysis
B19 parvovirus and, 550
Chlamydia species and, 511
EBV and, 497
Entamoeba histolytica and, 484
HBV and, 464
hemophilia B and, 612
hereditary amyloidosis and, 86
Legionella species and, 500
sickle cell disease and, 605
Restriction digestion, PCR coupled to, see Poly-
merase chain reaction coupled to restriction
digestion (PCRD)
Restriction endonuclease analysis
in oral diseases, 553
quality control in, 64–65
Restriction endonuclease digestion, 20, 456, see also
Southern blotting
Restriction endonuclease mapping, *myc* and, small
cell lung cancer and, 228
Restriction enzyme allele recognition, 7
Restriction enzyme digestion, reproducibility of,
DNA fingerprinting analysis and, 53–54
Restriction enzyme inhibitors, 64
Restriction enzyme isoform genotyping, *apo* E and,
169

Restriction enzymes, probes paired with,
hypervariable loci and, 46
Restriction fragment length polymorphism (RFLP),
6, 7, 47
of *apo* B gene, 171
in astrocytomas, 423–424
in cystic fibrosis diagnosis, prenatal, 221
in familial hypercholesterolemia diagnosis, 175
in hemophilia A, 611
in lymphoid malignancies, 615–616
M. tuberculosis and, 502–503
in MEN 1, 368
multiple-locus pattern of, 48
in muscular dystrophies, 393
in ovarian tumors, 301
PCR amplification of target sequences combined
with, 47
in polycystic kidney disease, 318
in *RB* gene abnormalities, 417
in renal cell carcinoma, 317
rRNA in, 457–458
Southern blotting and, 20
Restriction fragment length polymorphism (RFLP)
markers, alumorphs versus, 51
Restriction isotyping, *apo* E and, 169
Restriction pattern(s)
Cryptococcus species and, 501
in renal oncocytomas, 318
Restriction site, artificial, 31
Restriction site polymorphisms, in hemophilia A,
79–80
Retention mechanismxs, chemotherapy and, 114
Retinitis pigmentosa, 164
Retinoblastoma, 11, 15, 104, 412–418, see also *RB*
gene
chromosome aberrations in, 414
general features of, 413–414
RB protein and, 415–416
Retinoic acid receptor, hepatocellular carcinoma and,
207
Retinoic acid receptor (RAR α), 10
Retransformation, 12
Reverse allele-specific oligonucleotide (ASO)
probes, 21
Reverse cholesterol transport, genetic defects in,
178–181
Reverse dot-blot technique, 49
Reverse transcriptase activity, of HBV, 207
Reverse transcription (RT), PCR combined with, 48
endocrine disorders and, 371
HCV and, 475
MRD and, 617
respiratory syncytial virus and, 485
rubella virus and, 550
Reversible binding, acridines and, 98
Revertants, "flat" phenotype in, 12
Reye's syndrome, AS-PCR in, 22
R5452 antibody, 191
R553 Stop mutation, 220
RFLP, see Restriction fragment length polymor-
phism